Y0-BUD-709

Magnetism and Magnetic Materials—1976

(Joint MMM-Intermag Conference, Pittsburgh)

AIP Conference Proceedings
Series Editor: Hugh C. Wolfe
Number 34

Magnetism and Magnetic Materials—1976

(Joint MMM-Intermag Conference, Pittsburgh)

Editors

J. J. Becker
General Electric Co.

G. H. Lander
Argonne National Laboratory

American Institute of Physics
New York 1976

JOINT MMM-INTERMAG CONFERENCE ORGANIZATION

GENERAL CHAIRMAN – E.W. Pugh

PROGRAM COMMITTEE – R. Alben and C.D. Graham, Jr. (co-chairmen), J. Axe, R.C. Barker, J.J. Becker, J. Bonner, A.I. Braginski, J.I. Budnick, G.S. Cargill, III., R.B. Clover, J.F. Dillon, Jr., B. Gustard, F.B. Hagedorn, A.A. Halacsy, A.B. Harris, G.H. Lander, F.E. Luborsky, J.C. Mallison, A. Malozemoff, R. Potter, R. Silberglitt, G.R. Slemon, C.M. Varma.

PUBLICATION CHAIRMEN – J.J. Becker, G.H. Lander, F.E. Luborsky.

PUBLICITY CHAIRMAN – R.M. Josephs

FINANCE CHAIRMAN – D.I. Gordon

SECRETARY – R.E. Watson

LOCAL COMMITTEE – F.E. Werner, Chairman; L. Berger, S.K. Bhate, W.J. Carr, S. Charap, T.H. Gray, A. Goldman, K. Narasimhan, R.W. Patterson, D.V. Ratnam, F.C. Schwerer, J.W. Shilling, W.V. Bratkowski, Winifred Carr.

CONFERENCE ON MAGNETISM AND MAGNETIC MATERIALS (MMM) ORGANIZATION

STEERING COMMITTEE – E.W. Pugh, Chairman; R.E. Watson, R.L. White, C.D. Graham, F.E. Luborsky, C.E. Patton.

ADVISORY COMMITTEE – R.L. White, Chairman; M.K. Wilkinson, Secretary.

Term Expires 1976:
S.H. Charap, P.L. Donoho, F.B. Hagedorn, I.S. Jacobs, A.V. Pohm, E.W. Pugh, M.P. Sarachik, B.F. Stein, E.J. Torok, F.E. Werner.

Term Expires 1977:
J.L. Archer, G. Bate, C.D. Graham, K. Lee, C.E. Patton, J.J. Rhyne, P.W. Shumate, Jr., R.E. Watson, M.K. Wilkinson, R. Wolfe.

Term Expires 1978:
R.S. Alben, J.C. Bonner, R.B. Clover, D.I. Gordon, G.H. Lander, F.E. Luborsky, J.C. Suits, G.P. Vella-Coleiro, R.L. White, R.M. White.

SPONSORING SOCIETY REPRESENTATIVES – H.C. Wolfe (AIP), J.M. Lommel (IEEE).

IEEE MAGNETICS SOCIETY ADMINISTRATIVE COMMITTEE

OFFICERS – F.E. Luborsky, President; F.J. Friedlaender, Vice President; D.I. Gordon, Secretary-Treasurer.

Term Expires 1976:
B.F. DeSavage, W.D. Doyle, G.A. Fedde, C.E. Johnson, Jr., W. Kayser, R.J. Parker, A.V. Pohm, G.R. Slemon.

Term Expires 1977:
R.C. Byloff, R.F. Elfant, D.I. Gordon, C.D. Graham, Jr., F.E. Luborsky, C.E. Patton, G.F. Pittman, Jr., P.W. Shumate.

Term Expires 1978:
G. Almasi, G. Bate, P.O. Biringer, F.J. Friedlaender, R.O. McCary, E.W. Pugh, J.W. Shilling, L.J. Varnerin, Jr.

EX-OFFICIO – H.F. Storm, J.J. Suozzi, E.J. Torok.

IEEE HEADQUARTERS REPRESENTATIVE – R.M. Emberson.

CONFERENCE EXECUTIVE COMMITTEE CHAIRMAN – J.M. Lommel.

COOPERATING SOCIETY REPRESENTATIVES – J.W. Shilling (Met. Soc. AIME), J.O. Dimmock (ONR), D.H. Jones (ASTM).

IUPAP REPRESENTATIVE – M.B. Stearns

L.C. Catalog Card No. 76- 47106
ISBN 0-88318-133-9
ERDA CONF–760619

JOINT MMM-INTERMAG CONFERENCE

June 15-18 1976 Pittsburgh, Pennsylvania

Sponsored by

The American Institute of Physics

The Magnetics Society of the Institute of Electrical and Electronics Engineers

In Cooperation with

The Office of Naval Research; The American Physical Society; The Metallurgical Society of the American Institute of Mining, Metallurgical, and Petroleum Engineers; The American Society for Testing and Materials

The Conference is especially grateful to

THE OFFICE OF NAVAL RESEARCH

for its support of the expenses of foreign and interdisciplinary speakers under Contract N00014-76-G0006

Contributions to the Conference from the following firms are gratefully acknowledged:

Allegheny Ludlum Steel Corporation
Allied Chemical Corporation
Ampex Corporation
Applied Magnetics Corporation
The Arnold Engineering Company
Eastman Kodak Company
Eriez Magnetics, Inc.
Ford Motor Company
General Electric Company
G.T.E. Automatic Electric, Inc.
General Magnetic Company
Hewlett Packard
Indiana General Division of Electronic Memories & Magnetics Corporation
International Business Machines Corporation
Magnetic Metals Company
Magnetics Division of Spang Industries, Inc.
National Micronetics, Inc.
The Permanent Magnet Company, Inc.
Pfizer Minerals, Pigments & Metals Division Pfizer, Inc.
Raytheon Company
Rockwell International, Electronics Research Division
Sperry Univac, Inc.
Texas Instruments, Inc.
United States Steel Corporation
Walker Scientific, Inc.
Westinghouse Electric Corporation
Xerox Corporation

The following firms had exhibits at the Conference:

FW Bell Inc.
Columbus, Ohio
Burroughs Corp.
Detroit, Michigan
West Lake Village, California
Ceramic Magnetics, Inc.
Fairfield, New Jersey
Intermagnetics General
Guilderland, New York
LDJ Electronics, Inc.
Troy, Michigan
Magnetic Conference Book Exhibit
Maple Glen, Pennsylvania
Oxford Instruments, Inc.
Annapolis, Indiana
RFL Industries, Inc.
Boonton, New Jersey
Thomas & Skinner, Inc.
Indianapolis, Indiana

All papers in this volume, and in previous Proceedings of the Conference on Magnetism and Magnetic Materials published in this series, have been reviewed for technical content. The selection of referees, review guidelines, and all other editorial procedures are in accordance with standards prescribed by the American Institute of Physics.

INTRODUCTION TO THE PROCEEDINGS OF THE FIRST

JOINT MMM-INTERMAG CONFERENCE

The first Joint MMM-Intermag Conference was held in Pittsburgh, Pennsylvania on June 15-18, 1976. It combined, for the first time, the Conference on Magnetism and Magnetic Materials (MMM) with the International Magnetics Conference (Intermag) and thus is a dream come true for many persons in the field of magnetism. This first joint conference also represents the culmination of years of planning and hard work by many individuals in the Magnetics Society of the IEEE and the MMM Conference Advisory Committee.

The MMM Conference dates back to June 1955 when it was first convened in Pittsburgh under the co-Chairmanship of R. M. Bozorth and J. E. Goldman. The Intermag Conference adopted its present name for its conference held in Washington D.C. in April 1963; however, it traces its origins back to the Conference on Magnetic Amplifiers which was convened for the first time in Syracuse, New York, in April 1956, under the Chairmanship of H. W. Lord. This joint conference therefore represents the 22nd annual MMM Conference and the 21st annual Intermag Conference. These two conferences will again function independently in 1977 and 1978 and then be combined into a joint conference for the second time in 1979. The possibility of further joint meetings after 1979 is now under study by the organizing committees.

The proceedings of this first Joint MMM-Intermag Conference is being published in two volumes: one in the IEEE Transactions on Magnetics and the other in the AIP Conference Proceedings. Both volumes are sent to all Conference attendees; however, subscribers to one or the other of these publications will have to purchase the companion volume in order to have a complete proceedings. Both volumes do have the complete Table of Contents.

The success of this first Joint MMM-Intermag Conference can be judged only partially by the published proceedings or by statistics which show that 44 invited papers and 280 contributed papers were presented by individuals from 20 different countries and that 747 persons (including 667 full-fee registrants) attended the meeting. The real success lies in the opportunity afforded scientists, technologists, and engineers from all over the world to discuss topics of mutual interest, to learn to understand each other better, and thus to return to their own laboratories invigorated and better prepared to contribute to this rapidly evolving field of modern technology.

Emerson W. Pugh
General Chairman
Joint MMM-Intermag Conference

Note: The companion volume to AIP Conference Proceedings No. 34, containing the other half of the proceedings, is IEEE TRANSACTIONS ON MAGNETICS, Volume MAG-12, No. 6, November 1976 (Product No. JH3 479-3).

Table of Contents of AIP Conference Proceedings No. 34

Section 5 ASPECTS OF SUPERCONDUCTIVITY

Section 6 SURFACES AND FINE PARTICLES

Section 7 DOMAIN WALLS

Section 8 BUBBLE PHYSICS

Section 12 RARE EARTH SYSTEMS; RESONANCE

Section 13 MICROWAVE DEVICES

Section 14 SYMPOSIUM ON AMORPHOUS MAGNETISM

Section 15 AMORPHOUS MAGNETIC ALLOYS: FUNDAMENTAL STUDIES

MAGNETICS

NOVEMBER 1976 VOLUME MAG-12 NUMBER 6

A PUBLICATION OF THE IEEE MAGNETICS SOCIETY

JOINT MMM-INTERMAG CONFERENCE

PAPERS

Memory Devices

Bubble Device and System Design

Bubble Processing and Circuit Design

Bubble Device Characterization and Testing

Note: The number appearing to the left of the title refers to the Conference program paper number. Sessions have been regrouped according to topic.
* Abstract only.

Recording Heads

Digital Recording

Recording Media

Devices, Sensors and Measurements

Power Devices

Silicon Iron

Transportation

Magnetic Separation

Applied Superconductivity

Amorphous Transition Metal-Metalloid Alloys

Hard Magnetic Materials

Anisotropy and Hard Magnetic Materials

High Fields

Field Calculations

Field Calculations, Eddy Currents

RELAXATION IN ANTIFERROMAGNETIC RESONANCE†

Sergio M. Rezende*
Departamento de Fisica, Universidade Federal de Pernambuco, Recife, Brazil
and
Department of Physics, University of California, Santa Barbara, California 93106

ABSTRACT

The recent progress made in the understanding of the mechanisms that govern the relaxation of the antiferromagnetic resonance (AFMR) is reviewed. At temperatures small compared with T_N the sources of broadening can be pit/imperfection two-magnon scattering and radiation damping. The former has been clearly identified in MnF_2. The latter increases rapidly with frequency and can result in very large widths, particularly in materials where the AFMR falls in the infrared region. At higher temperatures the relaxation is dominated by four- and six-magnon scattering processes. Calculations based on spin wave theory agree quantitatively with linewidth data in several antiferromagnets at temperatures up to 0.8 T_N.

INTRODUCTION

The mechanisms that govern the relaxation of the ferromagnetic resonance (FMR) have been understood for more than a decade[1]. Despite the fact that the origins of linewidth in antiferromagnetic resonance (AFMR) should be essentially the same as in FMR, little contact between theory and experiment was made in this case until relatively recently. In the 1972 Magnetism Conference, Kotthaus and Jaccarino[2] (KJ) reviewed what was known about AFMR linewidths at the time and reported their own work in MnF_2. They found that by applying a large magnetic field, thus reducing the spin wave gap, the linewidths at relatively low microwave frequencies were orders of magnitude smaller than those previously measured[3,4] with zero or small fields at millimeter frequencies. With reduced linewidths and increased resolution, those authors were able to identify quantitatively the conditions under which two-magnon scattering becomes a dominant temperature independent mechanism and to measure the temperature dependence of the linewidth in MnF_2 with great accuracy. That work stimulated further theoretical investigations of questions which remained unexplained, such as the sources of the large linewidths observed at higher frequencies and the origins of the temperature dependence of the relaxation. It has since been found that magnetic-dipole radiation can account for the former, whereas magnon-magnon scattering explains the latter effect.

It is the purpose of the present paper to review the recent advances made in the understanding of the sources of line broadening in antiferromagnetic insulators. We shall restrict the discussion to two-magnon scattering, radiation damping, magnon-magnon and magnon-phonon scattering, for which detailed comparison with experimental data on a large variety of antiferromagnets has been made.

TWO-MAGNON IMPERFECTION SCATTERING

Due to the dipole-dipole interaction, the spin wave energy in a magnetic material depends on the direction of the wavevector. This gives rise to an energy band[5] in which many small-wavenumber magnons can be degenerate with the k=0 (uniform) mode or with the magnetostatic modes. The presence of surface pits or imperfections in the crystal can create non k-conserving potentials which allow the uniform or the magnetostatic modes to scatter into the degenerate manifold[5]. This two-magnon scattering can be caracterized by a perturbing Hamiltonian

$$\mathcal{H}^{(2)} = \sum_k F(\vec{k})\,(\alpha_0\alpha_k^\dagger + \alpha_0^\dagger\alpha_k) \tag{1}$$

where $\alpha_k^\dagger$ and α_k are the creation and annihilation magnon operators and $F(\vec{k})$ represents the k dependence of the scattering potential. The linewidth ΔH of the 0-mode due to (1) is obtained from perturbation theory,

$$\Delta H_{pit} = \frac{2\pi}{\gamma\hbar^2} \sum_k |F(k)|^2 \delta(\omega_0 - \omega_k) \tag{2}$$

and is clearly a function only of the scattering potential and of the number of degenerate magnon modes. The two-magnon scattering from surface pits in the sample had been identified as an important source of line broadening in FMR of single crystals[1] and was predicted as a possible relaxation mechanism in AFMR[5] many years ago. However, until the KJ work in MnF_2, this prediction had not been verified in antiferromagnets. One of the reasons for the long time elapsed between the theoretical prediction and the observation is the presence of other mechanisms that produce disproportionately large widths at low temperatures. Another reason is that the two-magnon relaxation rate in AFMR itself is smaller than in FMR, due to the fact that the magnon band is reduced by a factor H_A/H_C (H_A = anisotropy field, H_C = spin-flop field). KJ were able to identify the contribution of the two-magnon dipolar induced pit scattering to the linewidth in samples prepared with controlled surface polishing. Fig. 1 presents some of their experimental results. The insert shows a typical AFMR spectrum as observed in a MnF_2 disk under an applied field close to the spin-flop value (93 kOe), at 23.1 GHz, T = 4.2 K. The larger width of the uniform mode is due to radiation damping, which will be discussed in the next section. The several magnetostatic modes are not radiation broadened and provide very convenient means to verify the contribution of pit scattering to the linewidth. This is so because each mode is displaced from the bottom of the energy band differently. Therefore, each one has a different number of degenerate magnons into which to decay and should have a correspondingly different linewidth. The comparison between the actual

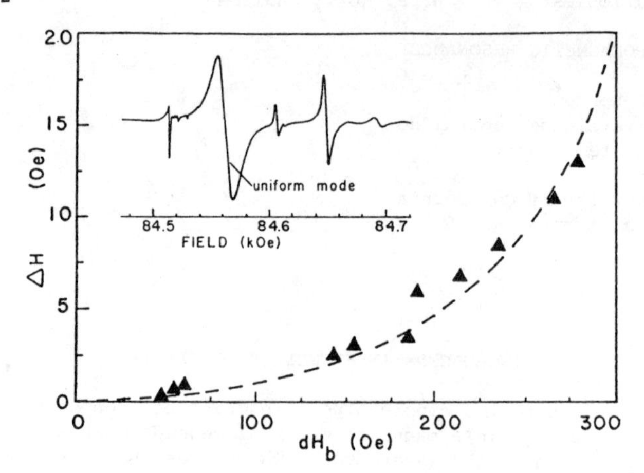

Fig. 1. Linewidth ΔH of magnetostatic modes in an MnF_2 disk as a function of their displacement dH_b from the bottom of the spin-wave band. The dashed line represents the result of two-magnon pit scattering theory. Insert shows typical AFMR spectrum at 23.1 GHz, (After Kotthaus and Jaccarino[2]).

linewidths of the various modes (triangles), as a function of the displacement dH_b from the bottom of the band, with the results based on the Loudon-Pincus two-magnon[5] theory shows excellent agreement. These results confirmed the validity of the two-magnon pit scattering theory in antiferromagnets. They also confirmed the prediction that this mechanism could not account for the large linewidths measured in most AFMR experiments, such as in MnF_2 at high frequencies[3,4] in FeF_2[6], NiF_2[7], CoF_2[7,8], and many others[7,8].

Small scale imperfections, such as caused by impurities in substitutionally doped crystals, can also be a source of two-magnon scattering due to a local change in the exchange constant or in the anisotropy. MnF_2 and other transition metal fluorides of the rutile type also provide ideal systems to conduct studies of this mechanism since they can be produced with impurities that are magnetically different than the host metal ion, though chemically very similar. Kotthaus and Jaccarino[2] made linewidth studies in MnF_2 with Fe^{2+}, Co^{2+}, Ni^{2+} and Zn^{2+} impurities. They found general qualitative agreement with the calculations of two-magnon imperfection scattering, but the quantitative results were not very conclusive. It seems clear that further investigations of this mechanism are desired, in which care is taken to separate out the contribution from radiation damping, which had not been identified at the time of the KJ work.

RADIATION DAMPING

Recent linewidth studies in antiferromagnets[9,10] have shown that (classical) magnetic dipole radiation can explain a number of intriguing observations made in AFMR experiments for many years. The most puzzling is, perhaps, the surprisingly large widths of homogeneously broadened lines in materials with very high resonance frequencies[3,6]. Others include the dependence of the linewidth on sample size[4], and the large difference between the linewidths of the uniform mode (k=0) and of the magnetostatic modes[2]. The radiative damping of a spin system was discussed by several authors[11] well before 1960. However, since its contribution to the resonance linewidth in EPR and FMR at low frequencies is usually much smaller than the contributions due to intrinsic relaxation processes, this mechanism remained either unnoticed or was deliberately separated out when present in experiments. The fact that the radiation increases with frequency makes its contribution to the linewidth large in antiferromagnets with high anisotropy, since the AFMR frequencies are $\omega_0/\gamma = H_C \pm H_0$, where $H_C^2 = 2H_EH_A + H_A^2$, H_E being the exchange field and H_0 the applied field (parallel to the symmetry axis in a two-sublattice system). This radiation broadening occurs predominantly in the uniform mode resonance, because in the magnetostatic modes the varying phase of the spin precession throughout the sample greatly reduces the radiation[9].

If we assume that the dimensions of the sample in an AFMR experiment are much smaller than the wavelength of the radiation, so that the spins precess in phase and radiate like a point dipole, the radiated power is easily seen to be proportional to ω_0^4 and to the square of the transverse oscillating moment. Since the stored energy (Zeeman, exchange and anisotropy) is also proportional to the square of the moment and to ω_0, the radiative relaxation time $T = U/P_{rad}$ can be expressed in terms of a linewidth, which in free space conditions is

$$(\Delta H)_0^{rad} = \frac{4}{3} M_s V \frac{H_A}{H_C} \left(\frac{\omega_0}{c}\right)^3 \tag{3}$$

Where M_s is the sublattice magnetization and V is the volume of the sample. When the sample is inside a shorted waveguide with dimensions comparable to the wavelength, the radiative linewidth becomes dependent on the geometry of the experimental conditions. In particular, it is proportional to the factor $\cos^2(2\pi\ell/\lambda_g)$ where ℓ is the distance of the sample to the reflecting short and λ_g is the guide wavelength. The original Kotthaus-Jaccarino experiments in MnF_2 were made with the MnF_2 samples mounted near the shorted end of the guide (ℓ=0), therefore resulting in a full radiative linewidth. Since even at relatively low-frequencies (23 GHz) the radiative broadening of the uniform mode is much larger than the intrinsic width in MnF_2, it completely dominates the relaxation of the AFMR. These conclusions were confirmed in later experiments by Sanders, et al.[10]

The resonance frequency of MnF_2 and several other antiferromagnets with a small applied field falls in the millimeter or far-infrared range of the spectrum, where the condition sample dimension << wavelength is not satisfied. In this case, the phases of both the driving and the radiated fields vary over the sample resulting in relatively smaller linewidths. The calculation of the radiation damping in this situation can be done similarly to that of the superradiance by an extended collection of atomic sources, developed by Rehler and Eberly[12]. The radiative linewidth can be written as

$$(\Delta H)^{rad} = \mu (\Delta H)_0^{rad} \tag{4}$$

where μ is a shape factor, which depends on the frequency the sample geometry and the electromagnetic configuration of the excitation of the resonance. In the usual geometry of a thin disk with thickness h and radius a, under a dc field perpendicular to its plane and excited by an rf field propagating parallel to the symmetry axis, the shape factor in the small wavelength limit (λ < a) becomes approximately[12]

$$\mu_d \simeq \frac{3}{2(ka)^2} \left[1 + \frac{\sin^2(\sqrt{\varepsilon}\, kh)}{(\sqrt{\varepsilon}\, kh)^2}\right] \tag{5}$$

where ε is the dielectric constant of the sample and $k = \omega_0/C$. Note that in the limit where $(\sqrt{\varepsilon}\, kh) \gg 1$, (3) - (5) give $(\Delta H)^{rad} \simeq 2\pi\omega_0 M_s H_A h/(C\, H_C)$, which shows that the radiation broadened linewidth is proportional to the thickness of the disk and to ω_0, rather than ω_0^3 as in the large wavelength case, Eq. (3).

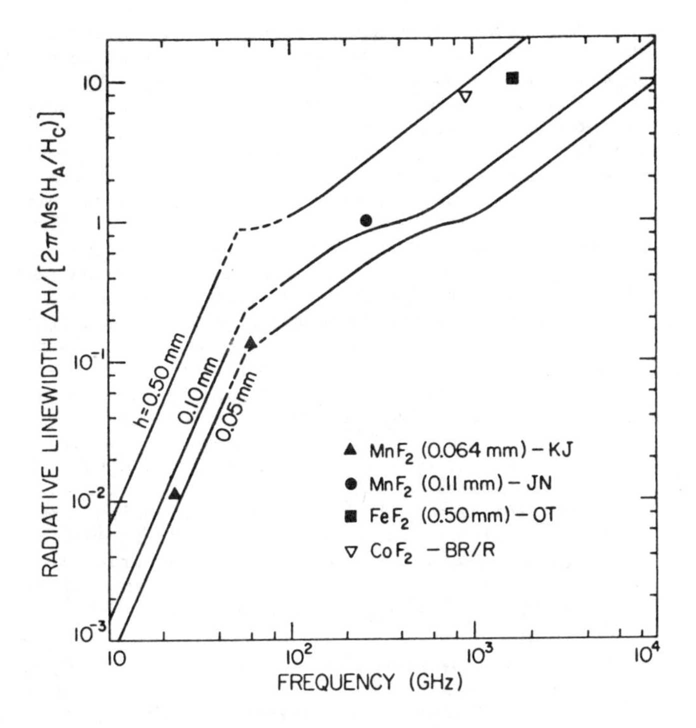

Fig. 2. Comparison of the low- and high-frequency approximation calculations of the radiation damping linewidth with low temperature linewidth data for: MnF_2 (Kotthaus and Jaccarino[2] and Johnson-Nethercot[3]), FeF_2 (Ohlman-Tinkham[6]) and CoF_2 (Brunner-Renk[14] and Richards[7,8]).

Fig. 2 shows plots of a normalized radiative linewidth as a function of frequency calculated both in the large (Eq. (3) with a = 1.5 mm) and in the small wavelength (Eq. (5)) limits. The latter is plotted for $\varepsilon = 6$, a typical dielectric constant of the fluorides. Also shown in the figure are the low temperature linewidth data for several antiferromagnets. We note that the experimental data are taken with samples inside waveguides, but we expect that the radiation is essentially the same as in free space because at high frequencies the guides are usually oversized. The simplest case study is MnF_2, which has been extensively investigated by several authors over a wide frequency range. As can be seen in the figure, the data of Kotthaus and Jaccarino[2] at different microwave bands and the one of Johnson-Nethercot[3] (JN) at millimeter frequencies can be entirely explained by the radiation damping calculation.

FeF_2 is another simple antiferromagnet, which was studied in detail by Ohlman and Tinkham[6] (OT). Its resonance falls in the far-infrared range (52.3 cm^{-1}) due to the large anisotropy field (H_A = 200 kOe, H_C = 520 kOe). Though the ratio H_A/H_C is larger than in MnF_2, the two-magnon pit scattering linewidth is about 100 - 500 Oe for typical surface polishing. This is too small to account for the measured linewidths, which are in excess of 14 kOe (1.5 cm^{-1}). We also note that the original OT data show unequivocably a size-dependent linewidth which is difficult to explain by intrinsic mechanisms, even when propagation effects are considered. The comparison of the OT data with the calculation in Fig. 2 demonstrates that radiation damping can also be the dominant relaxation mechanism of the AFMR in FeF_2. The k=0 magnon linewidth in this material has also been studied by light scattering[13]. In these experiments the wavevector of the excitation is comparable to that of light, so that negligible radiation damping would be expected. Unfortunately, the experimental resolution is not sufficient to confirm this assertion in the case of FeF_2.

A more difficult case to analyze is that of CoF_2, another widely studied antiferromagnet[7,8,14,15]. The 36.6 cm^{-1} line (at 4.2 K; H=0) has been studied in detail with far-infrared laser spectroscopy techniques[14] but the experiments cannot distinguish whether the line is homogeneous or inhomogeneously broadened. Most likely both mechanisms are present. The latter could be caused by static inhomogeneities, which result in large fluctuations of the effective anisotropy field due to the strong admixture of phonon excitation in the magnetic modes. Though the sample thickness is not given in the original references[7,8,14], we plot the experimental point in Fig. 2. The comparison with the theoretical results shows that radiation damping can be the source of homogeneous broadening in CoF_2.

There exists a number of other antiferromagnets with large AFMR frequencies, in which radiation damping may dominate the relaxation at small temperatures. Since the experimental configuration, and in particular, the sample shape, have to be known in detail for any conclusive study, the calculations presented here have not been applied to other materials. It seems, however, that in light of the results presented, further studies made with varying frequency and sample shape should reveal that radiation is the major source of the AFMR linewidth in other antiferromagnets.

MAGNON-MAGNON AND MAGNON-PHONON RELAXATION

In usual AFMR experiments, the linewidth is observed to increase with temperature. This is a result of the thermally induced relaxation dominating over the temperature independent mechanisms. The form and the magnitude of the temperature dependence varies from material to material. In MnF_2, for example, from 0.1 T_N to 0.8 T_N, this variation is over five orders of magnitude whereas in FeF_2, which has the same magnetic structure, it is less than three orders of magnitude. The most obvious source of this temperature dependent relaxation is the scattering among various elementary excitations, for which the Bose factors increase rapidly with temperature. Since processes involving three magnons (arising from the dipolar interaction) or two-magnons and one phonon cannot conserve both energy and momentum to relax the k=0 mode, the lowest order processes available are four-magnon, three magnon-one phonon and four magnon-one phonon.

The magnon-magnon interaction is a result of the nonlinearity that appears in the representation of the spin operators in terms of the magnon operators. If the dipole-dipole interaction is neglected, the exchange and the anisotropy energies give rise to an interaction Hamiltonian that contains terms with even numbers of magnons

$$\mathcal{H}_{mm} = \mathcal{H}^{(4)} + \mathcal{H}^{(6)} + \ldots \tag{6}$$

A general n-magnon term gives rise to the scattering process illustrated in Fig. 3a, where the modes can be from the same or from different magnon branches.

The magnon-phonon scattering arises from the magnetoelastic interaction, which can be represented by

$$\mathcal{H}_{me} = \sum_i b_{\alpha\beta\gamma\delta} S_i^\alpha S_i^\beta \frac{\partial R_i}{\partial x_\delta} \tag{7}$$

where S_i^α is the α-component of the spin operator at lattice site i, R_i is the γ-component of the displacement operator and $b_{\alpha\beta\gamma\delta}$ are magnetoelastic constants. By expanding S_i and R_i in terms of magnon and phonon operators, one obtains several N magnon-1 phonon terms. Fig. 3b illustrates a 4 magnon-1 phonon process that

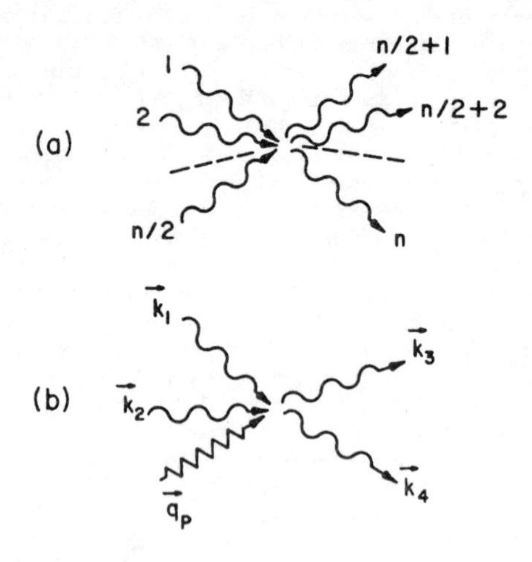

Fig. 3. Scattering processes that can relax the k=0 AFMR magnon: (a) n-magnon process; (b) four magnon-one phonon process.

can relax a k=0 magnon.

The relaxation rate of the k_1 mode due to a process in which n-particles are annihilated and m-particles are emitted is given by first order perturbation theory[16,17],

$$\eta_1^{n-m} = \frac{2\pi}{\hbar^2}\frac{1}{\bar{n}_1} \sum_{k_2 k_3} |c^{nm}|^2 e^{\beta\hbar(\omega_2+\omega_3+\ldots\omega_n)} \times$$

$$\times \; \bar{n}_2\bar{n}_3\ldots\bar{n}_{n+m}\delta(\omega)\Delta(k) \tag{8}$$

where $\beta = 1/k_B T$, $\bar{n}_k$ is the thermal number of the k-mode with frequency ω_k, $\delta(\omega)$ and $\Delta(k)$ are the energy and momentum conservation delta functions and C^{nm} is the interaction coefficient which takes into account all possible n-m processes.

The lowest order process in (6), which involves four magnons, was studied by several authors for many years[18-24]. The various results showed considerable disagreement as to the temperature dependence of the relaxation rate of the k=0 mode, and none could convincingly explain the existing experimental data. This was partly due to the analytical approximations used in the calculations at higher temperatures and partly because the experimental dependence at low temperatures was obscured by the large temperature independent component. With the availability of the accurate data on MnF_2, White, Freedman and Woolsey[25] showed that the intrinsic temperature dependence could be very well described by a four-magnon scattering up to a temperature of the order of $T_N/4$. More recently White, Rezende and Miranda[16,17] showed that above this temperature six-magnon processes dominate the relaxation and can explain the data in MnF_2 at a higher temperature range. These authors also showed that with convenient approximations, the relaxation rate for the AFMR with frequency $\hbar\omega_0 \ll k_B T$, due to an n-magnon process in the high temperature range $k_B T \gg \hbar\omega_k$ (the energy of the significant magnons in the integrations), can be expressed analytically by

$$\eta_1^{(n)}\Big|_{\text{high-T}} = \alpha^{(n)} \left(\frac{T}{T_N}\right)^{n-2} \tag{9}$$

where $\alpha^{(n)}$ is an appropriate factor which depends on the frequency and on the material parameters. This expression shows that higher order magnon scattering becomes increasingly important as the temperature approaches T_N, as expected. Eventually when eight-magnon dominates over six-magnon, and ten over eight, and so on, the spin wave picture loses meaning and should not be attempted to explain the results. However, this does not seem to occur up to temperatures as large as 0.8 T_N. Interestingly enough, below this temperature, there may be a large range in which six-magnon processes give a larger relaxation rate than four-magnon processes. This is mainly due to the fact that the six-magnon coupling coefficient is comparatively larger than the four-magnon one, as a result of cancellation effects among the several terms that exist in the latter. It has also been verified that the analytical approximations made in the relaxation calculations are very sensitive to temperature and material parameters. Therefore, numerical calculations are preferable to compare with experimental data. The numerical results for MnF_2, shown in Fig. 4, are in excellent agreement with the Kotthaus-Jaccarino data. In this calculation there is no adjustable parameter. The magnon energy renormalization with temperature, which is known for MnF_2, was introduced in the calculation but its effects are small below 35 K.

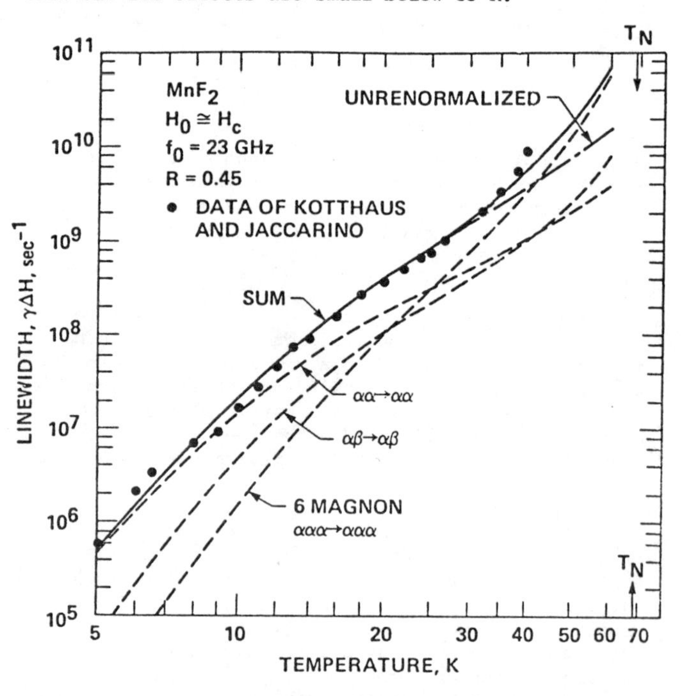

Fig. 4. Comparison of the four- and six-magnon calculations with the experimental data[2] for MnF_2.

Fig. 5 shows the comparison between theoretical and experimental results for $GdAlO_3$, an orthorombic antiferromagnet with T_N = 3.87 K but which behaves essentially as an uniaxial system under the conditions of the experiment[26]. In this calculation the energy renormalization of the zone boundary magnons is used as an adjustable parameter. The calculations[17] show that in this case the six-magnon decay rate does not overcome the four-magnon rate because the thermal population of the large-k magnons is small and higher order processes are not favored.

Since the Fe^{2+} ion in FeF_2 has non-zero orbital angular momentum, the spin-orbit and crystal field couplings give rise to a strong magnetoelastic interaction, which is accurately known. As a consequence, magnon-phonon processes are expected to be more important than in crystals with s-state magnetic ions,

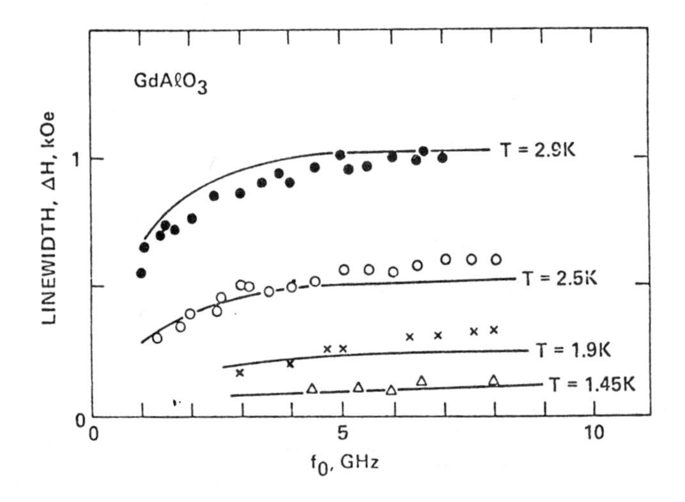

Fig. 5. Comparison of the theoretical (lines) and experimental (points from Ref. 26) frequency dependence of the linewidth in $GdAlO_3$.

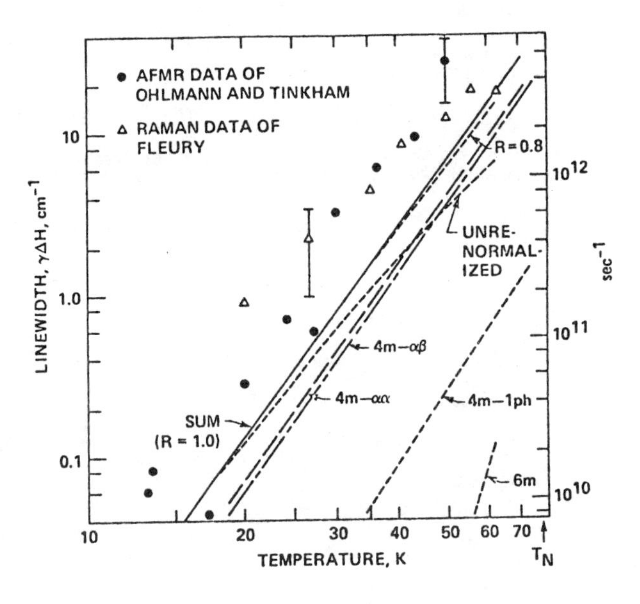

Fig. 6. Comparison of various relaxation processes with experimental data for FeF_2 (Refs. 6 and 13).

such as MnF_2 and $GdAlO_3$. However, the theoretical results[17,27] shown in Fig. 6 demonstrate that even in this material the magnon-magnon scattering prevails over magnon-phonon processes. Note also that due to the very large gap in FeF_2 the magnon population is small and the relaxation is entirely dominated by four-magnon processes.

Calculations have also been performed[17,28] for Rb_2MnF_4 and K_2MnF_4, which behave as almost ideal two-dimensional antiferromagnets. The interesting feature of these sytems is that small k states dominate the integrations, which now involves $\int kdk$ rather than $\int k^2dk$ as in three dimensions. Also the energy shows no dispersion for k perpendicular to the magnetic planes. As a result of these two facts the thermal number of the relevant magnons is large. This, combined with cancellation effects in the four-magnon coupling coefficient, make the six-magnon scattering the dominant relaxation mechanism, which can, as shown in Fig. 7, explain the experimental data[29].

CONCLUSION

A considerable understanding of the mechanisms that govern the relaxation of the AFMR has been achieved. At low temperatures the sources of homogeneous line boradening can be pit or imperfection two-magnon scattering and radiation damping. The latter is now reasonably well understood in both cases where the dimensions of the sample are either smaller or larger than the wavelength of the radiation. This mechanism completely dominates the low-temperature linewidth at high frequencies and was responsible for obscuring the temperature dependence of the experimental data in several materials for many years. The role of impurities on the AFMR linewidth is understood only qualitatively and needs further investigation. The temperature dependence of the linewidth is due primarily to four- and six-magnon scattering processes. Magnon-phonon and magnon-exciton[17] processes give usually negligible contributions. The theoretical predictions of spin wave theory agree quantitatively with linewidth data in MnF_2, FeF_2, $GdAlO_3$ and the two-dimensional antiferromagnets Rb_2MnF_4 and K_2MnF_4 up to $0.8\ T_N$.

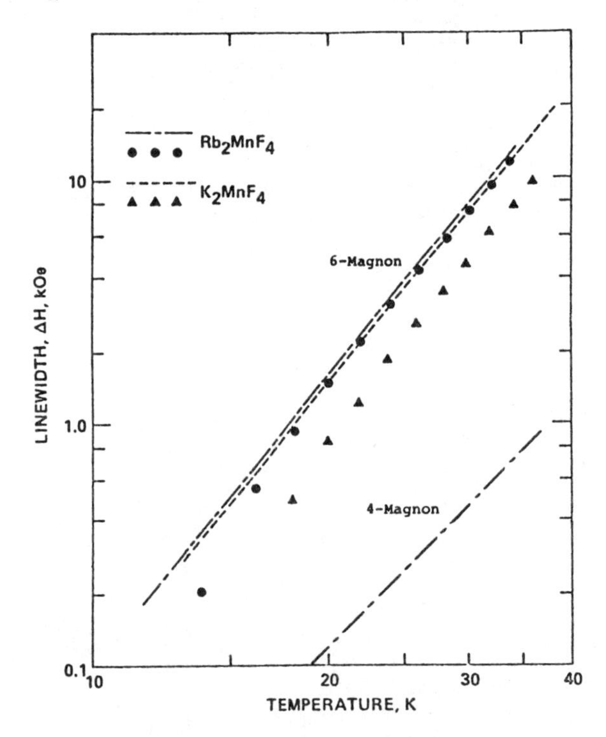

Fig. 7. Comparison of the four- and six-magnon calculation for the two-dimensional antiferromagnets, Rb_2MnF_4 and K_2MnF_4, with the experimental results[29].

ACKNOWLEDGMENTS

I would like to express my gratitude to Dr. R.M. White and Dr. V. Jaccarino who contributed decisively to my involvement and understanding of the subject discussed in this paper.

REFERENCES

†Supported in part by Conselho Nacional de Desenvolvimento Cientifico e Tecnologico, Banco Nacional do Desenvolvimento Economico and CAPES (Brazilian Government) and by the National Science Foundation, Grants GF 42494 and DMR 75-03847.
*J.S. Guggenheim Memorial Foundation Fellow.

1. See, for example, M. Sparks, Ferromagnetic Relaxation Theory (McGraw-Hill, New York, 1974) and C.W. Haas and H.B. Callen in Magnetism I, ed. by G.T. Rado and H. Suhl (Academic Press, New York, 1973).
2. J.P. Kotthaus and V. Jaccarino, Magnetism and Magnetic Materials - 1972, AIP Conf. Proc. 10, 57 (1973); Phys. Rev. Lett. 28, 1649 (1972).
3. F.M. Johnson and A.H. Nethercot, Jr., Phys. Rev. 114, 705 (1959).
4. K.C. O'Brien, J. Appl. Phys. 41, 3713 (1970).
5. R. Loudon and P. Pincus, Phys. Rev. 132, 673 (1963).
6. R.C. Ohlman and M. Tinkham, Phys. Rev. 123, 425 (1961)
7. P.L. Richards, J. Appl. Phys. 34, 1237 (1963).
8. P.L. Richards, J. Appl. Phys. 35, 850 (1964).
9. S.M. Rezende, E.A. Soares and V. Jaccarino, Magnetism and Magnetic Materials - 1973, AIP Conf. Proc. 18, 1083 (1974).
10. R.W. Sanders, D. Paquette, V. Jaccarino, and S.M. Rezende, Phys. Rev. B10, 132 (1974).
11. Among others, V.L. Ginzburg, Zh. eksp. i. teoret. fiz. 13, 33 (1943); N. Bloembergen and R.V. Pound, Phys. Rev. 95, 8 (1954); S. Bloom, J. Appl. Phys. 27, 785 (1956); 28, 800 (1957).
12. N.E. Rehler and J.H. Eberly, Phys. Rev. A3, 1735 (1971).
13. P.A. Fleury, Proc. 2nd. Int. Conf. Light Scattering (ed. M. Balkanski, Flamarion, Paris 1971). p. 151.
14. H. Brunner and K.F. Renk, J. Appl. Phys. 41, 2250 (1970).
15. S.J. Allen and H.J. Guggenheim, Phys. Rev. B4, 950 (1971).
16. R.M. White, S.M. Rezende, and L.C.M. Miranda, Magnetism and Magnetic Materials - 1974, AIP Conf. Proc. 24, 172 (1975).
17. S.M. Rezende and R.M. White, Phys. Rev. B, to be published.
18. V.N. Genkin and V.M. Fain, Zh. eskp. teor. fiz. 41, 1522 (1961) [Sov. Phys. - JETP 14, 1086 (1962)].
19. K. Tani, Prog. Theor. Phys. (Kyoto) 30, 580 (1963); 31, 335 (1964).
20. P. Pincus, J. Phys. Radium 23, 536 (1962).
21. J. Solyom, Zh. eksp. teor. fiz. 55, 2355 (1968) [Soviet Phys. - JETP 28, 1251 (1969)].
22. M.I. Kaganov, V.M. Tsukernik, and I.Y. Chupis, Fiz. Metal. Metaloved. 10, 797 (1960) [Phys. Metal. Mettalog. 10, 154 (1960)].
23. M.G. Cottam and R.B. Stinchcombe, J. Phys. C 3, 2326 (1970).
24. A.B. Harris, D. Kumar, B.I. Halperin, and P.C. Hohenberg, Phys. Rev. B3, 961 (1971).
25. R.M. White, R. Freedman, and R.B. Woolsey, Phys. Rev. B10, 1039 (1974).
26. H. Rohrer, Magnetism and Magnetic Materials - 1974 AIP Conf. Proc. 24, 268 (1975).
27. E.F. Sarmento, B. Zeks and S.M. Rezende, Magnetism and Magnetic Materials - 1975, AIP Conf. Proc. 29, 414 (1976).
28. S.M. Rezende and B. Zeks, Phys. Letters 54A, 135 (1975).
29. H.W. de Wijn, L.R. Walker, S. Geschwind, and H.J. Guggenheim, Phys. Rev. B8, 299 (1973).

CRYSTAL FIELD EFFECTS IN RARE EARTH COMPOUNDS*

B. Lüthi
Physics Dept., Rutgers University, New Brunswick, NJ 08903

ABSTRACT

We review the effect of the crystal field on thermal properties in intermetallic rare earth compounds. In substances where single ion effects are predominant, we give experimental results for the specific heat, thermal expansion, magnetic susceptibility and elastic constants. All these thermodynamic derivatives show clear cut crystal field effects (TmSb). Coupling constants deduced from these measurements are discussed. In substances, where the ions are interacting, we concentrate on elastic constants. We show how quadrupole-quadrupole coupling constants g' can be determined from the temperature dependence of certain symmetry elastic constants (rare earth pnictides, TmCd, TmZn, $PrCu_2$). We argue that g' is in large part due to conduction electron aspherical Coulomb charge scattering. A valence fluctuation system (TmSe) will be mentioned. Finally magnetic field effects on elastic constants will be given, with particular emphasis on rotational strain effects.

I. INTRODUCTION

In rare earth compounds the unfilled 4f shell gives rise to interesting magnetic properties[1]. The ground state of the 4f electrons is given by Hund's rule and because of strong spin orbit interaction, the lowest J-multiplet is separated from the next higher by 10^3 cm^{-1} or more. The splitting of the J multiplets by crystal electric fields (CEF) is much smaller, about 10^2 cm^{-1}. One expects therefore CEF effects on various physical properties. Such effects can be divided into 3 categories: a) resonance effects b) thermal effects c) transport properties.

a) Resonance effects: The most important technique in detecting directly transitions between CEF levels is inelastic neutron scattering. In this experiment magnetic dipole transitions are measured by varying incident neutron energy and temperature[2]. In intermetallic compounds this technique is very useful because optical experiments are difficult. In fact we are unaware of any (far infrared) optical experiment in rare earth intermetallic compounds. One should also mention EPR experiments[3], which is the only tool to measure the small CEF splitting for S-state ions.

b) Thermal Effects: Because of the CEF splitting of the order of 100-200K, effects on various thermodynamic quantities are possible. There are the well known Schottky anomaly in specific heat and Van Vleck magnetic susceptibility. In the last few years similar effects on thermal expansion, magnetostriction and elasticity were found. It is the purpose of this paper to review these recent developments and to show what can be learned from such experiments.

c) With electrical resistivity, thermal conductivity, thermoelectric power and magnetoresistance in intermetallic rare earth compounds, one can study various scattering processes (e.g. exchange scattering, aspherical Coulomb charge scattering). These effects are just going to be explored in detail now[4]. Space does not allow to review these experiments here, however we shall make contact with these electron scattering processes later (IIIb).

In the next section we present an outline of the theory. In the results section we give representative examples of various thermal effects. We give applications to various phenomena such as cooperative Jahn-Teller effect, valence fluctuations, and magnetic field effects. We give a general discussion of the various coupling constants determined from these experiments.

II. THEORY

We give formulas for the specific heat, thermal expansion, magnetic susceptibility and elastic constants. They will be written in such a way as to express common features.

Consider first the case of noninteracting rare earth ions, whose ground state J multiplet is split by the CEF of the ligands. We denote the energy levels $E_i = E(\Gamma_i)$ with Γ_i the representation index. The Helmholtz free energy per unit volume of such an assembly of ions can be expressed as

$$F = -kTN \ln \sum_i \exp(-E_i/kT) \qquad (1)$$

From this we can get all the quantities of interest as derivatives of F. The specific heat due to the crystal field split ions is:

$$C = -T\frac{\partial^2 F}{\partial T^2} = \frac{N}{kT^2}\left[\langle E^2\rangle - \langle E\rangle^2\right] \qquad (2)$$

where the statistical average $\langle X\rangle$ is defined as $\langle X\rangle = \frac{\sum_i X_i \exp(-E_i/kT)}{\sum_i \exp(-E_i/kT)}$. Likewise the thermal expansion due to the 4f ions is[5]:

$$\beta = \frac{1}{V}\left(\frac{\partial V}{\partial T}\right)_p = -\kappa\frac{V\partial^2 F}{\partial V\partial T} = \frac{\kappa N}{kT^2}\left[\langle E^2\gamma\rangle - \langle E\rangle\langle E\gamma\rangle\right] \qquad (3)$$

Here κ is the compressibility and $\gamma_i = -\frac{\partial \ln E_i}{\partial \ln V}$ is the Grüneisen parameter for the CEF level E_i. Note the similar structure of equations (2) and (3). β is proportional to C only if all the γ_i which contribute to β are the same. For the magnetic susceptibility we get:

$$\chi_m = -\frac{\partial^2 F}{\partial H^2} = -N\left\{\langle\frac{\partial^2 E}{\partial H^2}\rangle - \frac{1}{kT}\langle\left(\frac{\partial E}{\partial H}\right)^2\rangle + \frac{1}{kT}\langle\frac{\partial E}{\partial H}\rangle^2\right\} \qquad (4)$$

Here the first term is called the Van Vleck term and the last two are the Curie-Weiss terms. Similarly to the magnetic susceptibility one can calculate the elastic constants as strain susceptibilities for a given symmetry strain ε_Γ[6]:

$$c_\Gamma = \frac{\partial^2 F}{\partial \varepsilon_\Gamma^2} = -c_o g_\Gamma^2 \chi_s = N\left\{\langle\frac{\partial^2 E}{\partial \varepsilon_\Gamma^2}\rangle - \frac{1}{kT}\left[\langle\left(\frac{\partial E}{\partial \varepsilon_\Gamma}\right)^2\rangle - \left(\langle\frac{\partial E}{\partial \varepsilon_\Gamma}\rangle\right)^2\right]\right\} \qquad (5)$$

Again note the similar structure of equations (4) and (5). The expressions (2)-(5) can only be calculated if one knows the CEF levels E_i and their dependence on volume V, applied field H and strain ε_Γ. The energy levels E_i

can be determined from the CEF Hamiltonian, which for cubic symmetry is[7]

$$H_{CEF} = B_4O_4 + B_6O_6 \qquad (6)$$

O_4, O_6 are the Stevens Operators. The CEF parameters B_4, B_6 have been determined by inelastic neutron scattering for many intermetallic rare earth compounds. Therefore E_i is very often known experimentally. One can in principle determine the E_i also from the thermal effects listed above, but this is only reliable if the number of levels is small. A nice example is SmSb where one has for cubic symmetry for $J=5/2$ only 2 levels Γ_7, Γ_8. From specific heat and elastic constant measurements[8] one obtains a splitting of 65K. We shall not persue this any further here[9]. With the Zeeman term $H_Z = -g\mu_B J^Z H$ one can determine $E_i(H)$. Analogously one can determine the volume and strain dependence of the energy levels by constructing the magnetoelastic Hamiltonian. We shall do this for cubic symmetry, where one has 3 types of symmetry strains: The volume strain $\varepsilon_V=\varepsilon_{xx}+\varepsilon_{yy}+\varepsilon_{zz}$ which forms the basis of a Γ_1 representation and which is used for the Grüneisenparameters γ_i. The corresponding symmetry elastic constant is the bulk modulus $c_B=1/3\,(c_{11}+2c_{12})=1/\kappa$. The $c_{11}-c_{12}$ elastic constant is based on the 2 Γ_3 strains $\varepsilon_2=1/\sqrt{2}\,(\varepsilon_{xx}-\varepsilon_{yy})$, $\varepsilon_3=1/\sqrt{6}\,(2\varepsilon_{zz}-\varepsilon_{xx}-\varepsilon_{yy})$. The c_{44} elastic constant belongs to $\varepsilon_{xy}, \varepsilon_{xz}, \varepsilon_{yz}$ (Γ_5). For these three symmetries the magnetoelastic interaction is given by[10,6]:

$$\begin{aligned} H_{me}(c_B) &= -g_1\sqrt{\frac{c_B}{N}}\,N\varepsilon_V \\ H_{me}(c_{11}-c_{12}) &= -g_2\sqrt{\frac{c_{11}-c_{12}}{N}}\sum_i(\varepsilon_2O_2^2+\varepsilon_3O_2^0)_i \\ H_{me}(c_{44}) &= -g_3\sqrt{\frac{c_{44}}{N}}\sum_i(J_xJ_y+J_yJ_x)_i\,\varepsilon_{xy}+\cdots \end{aligned} \qquad (7)$$

The strain dependence of the energy levels can be determined analogously to the case of the Zeeman term by solving the secular equation: $|E(\Gamma_i)-E-\langle\Gamma_j|H_{me}(c_\Gamma)|\Gamma_i\rangle|=0$. Since the strains are small, one can use perturbation theory to get $E_i(\varepsilon_\Gamma)$. A quantitative interpretation of thermal expansion and elastic constants gives the corresponding magnetoelastic coupling constants g_Γ.

Next we consider the case of interacting ions, since many rare earth compounds show magnetic and structural transitions. Although one can have complicated forms of interactions, especially in intermetallic compounds[11], we consider only exchange and quadrupole-quadrupole interactions:

$$H=\sum_{ij}(J_{ij}\underline{J}_i\,\underline{J}_j+G_{ij}O_{\Gamma i}O_{\Gamma j}) \qquad (8)$$

where J_i and $O_{\Gamma i}$ are the spin angular and quadrupole operators for the rare earth ion i. We do not specify here the nature of the J_{ij}, G_{ij} except to remark that in intermetallic compounds they can both be mediated by conduction electrons[12].

Within the framework of molecular field theory, equation (8) can describe magnetic and quadrupolar phase transitions, the important features depending on the relative strength of J_{ij} and G_{ij}[13]. This interaction can also have an effect on elastic constants even in the high temperature phase. One obtains[14]

$$\frac{c_\Gamma}{c_o}=\frac{1-(g_\Gamma^2+g')\chi_s}{1-g'\chi_s} \qquad (9)$$

Here χ_s is the strain susceptibility defined in equation (5), g_Γ the magnetoelastic coupling constants (Eqn. (7) and g' is the k=0 component of the quadrupole interaction $g'=\sum_i G_{ij}$, with possible self energy corrections included (see below).

III. RESULTS AND DISCUSSION

We give typical results of the various effects mentioned in the introduction. We divide them into a) single ion effects b) structural phase transition c) magnetic field effects.

a) Single ion Effects

specific heat and thermal expansion: TmSb is a compound where interactions between the rare earth ions can be neglected. This was shown in the system $Tm_xY_{1-x}Sb$ by susceptibility measurements[15] and in TmSb by neutron scattering[16]. In figure 1 we show specific heat and thermal expansion measurements for TmSb. These results are taken from the literature[5,8]. In both cases one sees clear evidence of CEF effects. The full lines are calculated curves with eqn. (2), (3) and with the level scheme $\Gamma_1-\Gamma_4(25K)-\Gamma_5^2(56K)-\Gamma_1(115K)-\cdots$. The agreement between experiment and theory is very good. One interesting feature is the value and sign of the Grüneisenconstants $(\Gamma_1-\Gamma_4)\gamma_1=-13$, $(\Gamma_1-\Gamma_5)\gamma_2=+1.05$ compared to point charge model predictions of $\gamma=5/3$. The CEF parameters in the rare earth monopnictides do not deviate much from an effective point charge model[17]. Elastic constants: As an example of CEF effects on elasticity we give results for TmSb in figure 2. These results were interpretated[8] using eqn. (5) for the $c_{11}-c_{12}$ and c_{44} mode. From the fit one gets magnetoelastic coupling constants g_2, g_3. A proper choice for the background elastic constant c_o is important. In the case of TmSb we used corresponding results for LaSb with an appropriate scaling[8]. Note that

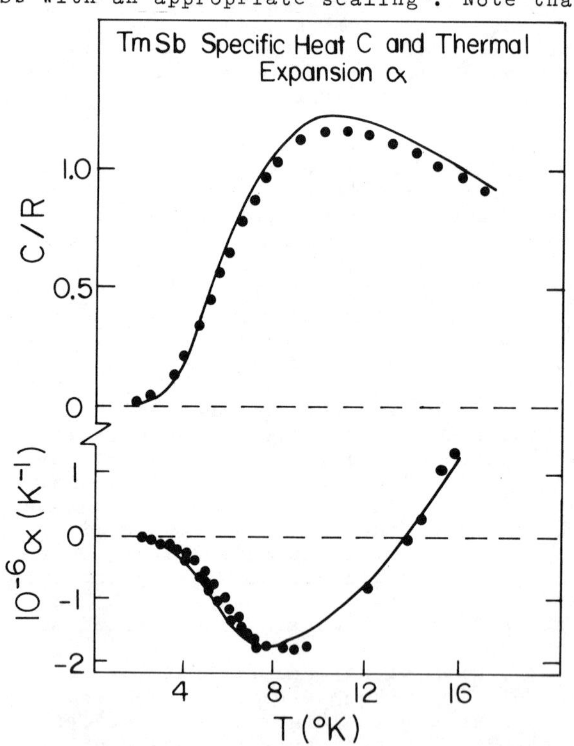

Fig. 1. Specific heat and thermal expansion for TmSb. Full lines are calculated curves based on eqs. (2) and (3). (Taken from refs. 5 and 8.

the CEF effects explain the results very well, in particular the pronounced minimum of the c_{44} mode at 9K and the more shallow minimum for the $c_{11}-c_{12}$ mode. The linear magnetoelastic theory (eqn. (7)) seems to be sufficient to explain $c_{ij}(T)$ quantitatively for many different cases (PrSb, SmSb, TmSb, TmTe, CeBi, $PrSn_3$). The only major disagreement we have encountered so far is for the c_{44} mode of PrSb where we measured a minimum at 60K while the calculated χ_s gives a minimum at 26K[6]. Higher order coupling effects do not account for this discrepancy. Another kind of discrepancy we found in $Pr_{.05}La_{.95}Al_2$ for the c_{44} mode where an elastic constant minimum, accompanied by an attenuation maximum, points to a crystal defect anelastic relaxation mechanism[18].

Next we compare the magnetoelastic coupling constants with predictions from point charge calculations. This we do in table I for g_2 for NaCl-and CsCl-structure materials. Here we listed materials which show also magnetic and structural transitions to be discussed below. Point charge expressions for g_2 for coordination 6 and 8 read:

$$\text{coord. 6 (NaCl)} \quad g_2=9\sqrt{\frac{N}{6c_o}}\,\alpha_J\,\langle r^2\rangle\,\frac{Ze^2}{R^3}\,(1-\sigma_2)$$
$$\text{coord. 8 (CsCl)} \quad g_2=-8\sqrt{\frac{N}{6c_o}}\,\alpha_J\,\langle r^2\rangle\,\frac{Ze^2}{R^3}\,(1-\sigma_2) \qquad (10)$$

α_J is the Stevens factor[7], $\langle r^2\rangle$ the 4f integral, Z the effective ligand charge, R the ion-ligand distance and σ_2 the shielding factor. For the LnSb series we took corresponding values used for the B_4 estimate[17]: Ze=-1.3, $\langle r^2\rangle$ based on relativistic calculations and σ_2 interpolated Sternheimer shielding factors. For TmCd, TmZn we took σ_2=0, Ze=-1. For the NaCl structure materials one notices better agreement than given previously[8]. However we stress that the elastic constant results give only g_2^2, the sign of g_2 could still be opposite as in the case of the Grüneisenparameter γ_1 for TmSb above. For CsCl-structure materials the point charge estimate does not agree at all with measured ones[19,20], similar to B_4[21].

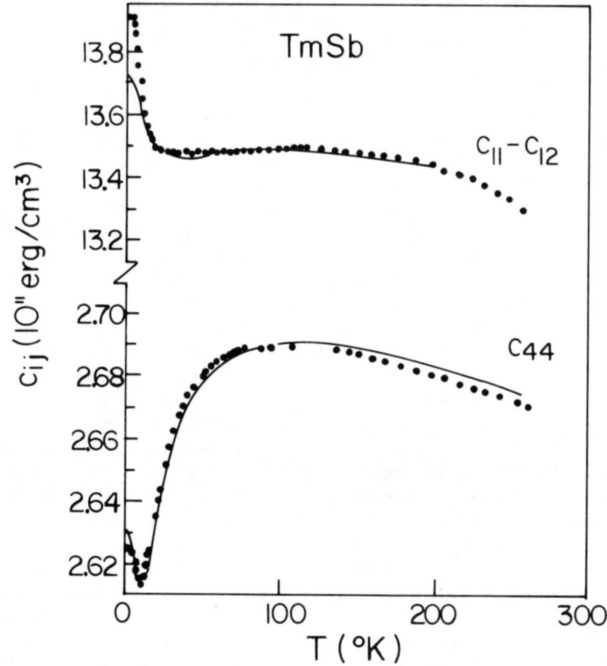

Fig. 2. Elastic constants with a calculated fit based on eq. (5) for TmSb (from ref. 8).

TABLE I

	β_J	T_a(K)		g_2^2(mK)		g'(mK)
		exp.	calc.	exp.	calc.	exp.
PrSb	-	-	-	4.9	8	
SmSb	+	2.11	-	38	22.8	
DySb	-	9.5	-	1.3	0.43	-0.4
HoSb	-	5.25	4.5	0.11	0.06	0.6
ErSb	-	3.53	-	<0.2	0.08	
TmSb	+	-	-	1.2	1.2	
TbP	+	7.08	-			13
TmTe	+	-	-	2.3		
TmCd	+	3.16	2.3	0.34	11.9	1.34
TmZn	+	10	-	8.9	11.4	-9.6
ErZn	-	20	3.7	0.8	0.67	2.1

CEF effects on elastic constants are not limited to intermetallic rare earth compounds. An effect of this nature has been observed for the c_{66} mode in $TmVO_4$.[22] In addition, such effects have been identified in the paramagnetic phase of the transition metal compound $FeCl_2$,[23] where spin orbit and trigonal crystal field give rise to a splitting of the lowest orbital triplett of the order of 400 cm^{-1}. A somewhat related effect is also the effect electrons have on elastic properties of a multivalley semiconductor.[24]

b) Cooperative Jahn-Teller effect

A nice application of CEF effects on the elasticity of a solid is the structural transition induced by a magnetic ion-lattice interaction. In cases where the ground state level has a non Kramers degeneracy ($\Gamma_3, \Gamma_4, \Gamma_5, \Gamma_8$ for cubic point symmetry), the strain susceptibility χ_s (eqn. (5)) has a divergence, so c_Γ=0 for $g_\Gamma^2\chi_\Gamma$(Ta)=1. The atoms no longer have a restoring force under the strain ε_Γ, they shift to new equilibrium positions, i.e. a structural transition occurs for Ta. Taking the interaction between ions into account (eqns. 8,9) the condition for a structural transition is modified to:

$$1 = (g_\Gamma^2 + g')\chi_s(Ta) \qquad (11)$$

Calculated T_a based on eqn. (11) are given in table I.

Excellent reviews on the cooperative J-T effect, especially for rare earth vanadates and transition metal compounds, have been given before[13,25]. We concentrate therefore on intermetallic compounds. In these materials pure structural transitions are relatively rare. We are only aware of 2 cases[20,26]: TmCd and $PrCu_2$. In many cases there is a structural transition coinciding with the magnetic transition[8,27,28]. The temperature dependence of the symmetry mode c_Γ can also in such cases be well described by eqn. (9) as shown before[8] and discussed now. In figure 3 we show $(c_{11}-c_{12})$(T) for HoSb for 3 different samples. HoSb has an antiferromagnetic transition at T_N=5.25K. One notices for all 3 samples a strong softening of c.40% from high T down to 5.6K where the ultrasonic echo is strongly damped. Although all 3 samples show qualitatively the same c_Γ(T), the quantitative behavior is somewhat sample dependent at least for the interval 7-20K. This one should keep in mind while determining g_2^2 and g'. For our results in fig. 3 we get 2 sets of parameters (g^2=0.13mK, g'=0.75mK for 2 samples; g_2^2=0.10mK, g'=0.82mK for the 3rd sample.) This analysis gives an indication of systematic errors encountered in intermetallic rare earth compouds. Note that transition temperatures reported in the literature for such compounds vary appreciably. It is believed that rare earth impurities, nonstochiometry, methods of growing and preparing single crystals are important factors, determining the physical constants. Additional uncertainties, influencing the parameter fit are: background choice c^0_Γ (discussed before[8,19]), uncertainty of CEF levels, short range order or critical effects. All these factors contribute to the uncertainty of g_Γ^2, g' and g' having a somewhat larger one because it enters eqn. (9) both in numerator and denominator, c_Γ being less sensitive to g' then to g_Γ^2 changes.

The values of g' listed in table I show large variations. The following physical mechanisms can contribute to g': Self energy terms, direct quadrupole-quadrupole terms, optical (k=0) phonon coupling, indirect electron coupling. The self energy terms are difficult to estimate. For a one dimensional phonon mode with one J-T ion per unit cell, a Debye approximation gives[13] $g'=-\frac{1}{3}g_\Gamma^2$. In the case of doubly degenerate ($c_{11}-c_{12}$) or triply degenerate (c_{44}) modes the self energy terms are difficult to estimate. A pseudo spin Hamiltonian approximation has been made for DySb[29]. Direct quadrupole-quadrupole interactions should not contribute in the case of rare earths with localized 4f wave-functions. For ions with inversion symmetry, such as NaCl-, CsCl-structure, k=0 optic phonons cannot give a coupling to g'[14]. The conduction electron aspherical Coulomb charge scattering seems to be an important contribution to g' for rare earth intermetallics (for insulators g' is either 0 as for $NiCr_2O_4$, or can be explained by self energy correction terms alone[13,25]). In an oversimplified picture one can write the indirect exchange and aspherical Coulomb charge scattering as

$$H=\sum_{kk'} j_{ex}\underline{sJ} + \sum_{kk'\ell\ell'} V_Q a^+_{k'\ell'} a_{k\ell} \qquad (12)$$

where in the second term ℓ,ℓ'=s,p,d. Interactions of this kind have been derived[4,11] and applied for magnetic and structural transition[12] and to transport effects[4,12]. Eqn.(12) is just the microscopic analogue of eqn. (8). By transforming away the conduction electron coordinates, one arrives at an ion-ion interaction of the form of eqn. (8), which in its simplest form exhibits RKKY behavior, but which in detail is a complicated function of kk'.[30]

There exist simple arguments[31,32] to predict easy axis of magnetisation and the type of distortion to expect for cubic materials. Let us consider the 4th order CEF term B_4O_4 of Eqn. (6). If one takes the angular momentum operator as a classical vector, this term is like an anisotropy energy $-B_4(J_x^2J_y^2+J_y^2J_z^2+J_z^2J_x^2)$. For $B_4<0$ one has a (100) easy axis, a tetragonal distortion and a $c_{11}-c_{12}$ soft mode. For $B_4>0$ one has a (111) easy axis, a trigonal distortion and a c_{44} soft mode. For B_4 similar point charge expressions as for g_2 give[7]:

$$\text{coord. 6 (NaCl)}\ B_4=\frac{7}{16}\frac{Ze^2}{R^5}\langle r^4\rangle\,\beta$$
$$\text{coord. 8 (CsCl)}\ B_4=-\frac{7}{18}\frac{Ze^2}{R^5}\langle r^4\rangle\,\beta \qquad (13)$$

The sign of B_4 depends on the sign of β, Ze and on the coordination. This simple argument explains easy axis, main distortion and soft mode for all the compounds listed in table I which exhibit a phase transition. Using the β-values in table I, Ze<0 for the pnictides[17] we get a trigonal distortion and a soft c_{44}-mode for TbP as observed[28,33], but a tetragonal distortion and a soft $c_{11}-c_{12}$-mode for DySb and HoSb as observed.[8,27,33] For CsCl-structure materials we have[20,21] Ze<0 and for both TmCd, TmZn this gives a (100) easy axis, a tetragonal distortion and a soft $c_{11}-c_{12}$ mode as observed[19,20,34]. Similar conclusions for easy axis and structural distortions have been put forth for some rare earth monochalcogenides[35]. The only exceptions to these simple arguments are CeSb[32] and ErS[35].

Finally we should mention an interesting theoretical conjecture, based on ε-expansion, which predicts for some classes of antiferromagnets a first order phase transition[36]. In the pnictide series, to which this conjecture also applies, it predicts a first order transition for the following materials listed in table I: TbP and ErSb. While the case of ErSb is not yet settled[36], we have clear cut evidence for a first order transition in TbP, based on specific heat (latent heat) and elastic constant measurements[28].

c) Valence Fluctuation Effects

A great deal of interest has been focused recently on valence fluctuations[37-39]. We discuss briefly the case of temporal fluctuations between 2 configurations such as $Tm^{2+}(4f^{13}, J=7/2)$ and $Tm^{3+}(4f^{12}, J=6)$. Possible candidates, based on X-ray photoemission experiments,[38] are TmSe and TmTe. In figures 4 and 5 we show results of elastic constant measurements for TmTe and TmSe. We investigated 3 single crystals for each compound, all giving the same T-dependence for c_{ij}. No thermal expansion corrections were applied to the results. The lattice constants 6.346Å(TmTe) and 5.71Å(TmSe) indicate good stochiometry[40]. The main results of our experiments shown in figure 4.5 are the noticeable CEF effect for TmTe and the striking absence of such effects for TmSe. The results of figure 4 we can interpret for T<100K with a CEF effect for Tm^{2+}. A Tm^{3+} configuration would give CEF effects on elastic constants as shown for TmSb in figure 2, not compatible with our results for TmTe. The full line in figure 4 is a fit to a χ_s with a Tm^{2+} level scheme as indicated in the figure caption. The shallow minima in c_{44} and c_L at 140K are not explained yet. As in the case of $Pr_{.5}La_{.5}Al_2$ mentioned before they could be due to an anelastic relaxation mechanism[18]. There is no sign of CEF effects for TmSe (Fig. 5) but only an effect due to the onset of magnetic order at $T_N \simeq 3K$.

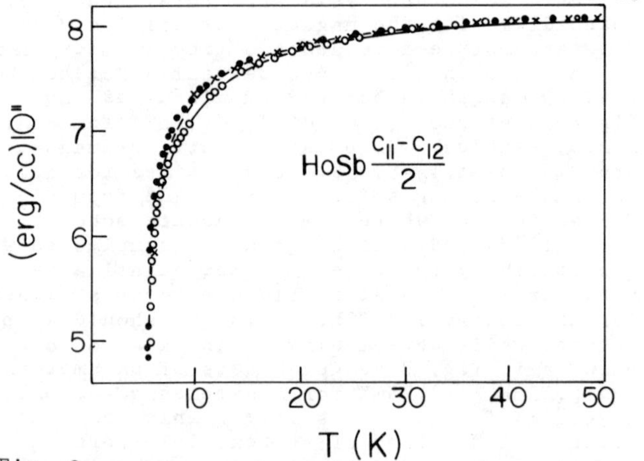

Fig. 3. $c_{11}-c_{12}$ mode for HoSb for 3 different samples (o, •, x) and calculated fit based on eq. (9).

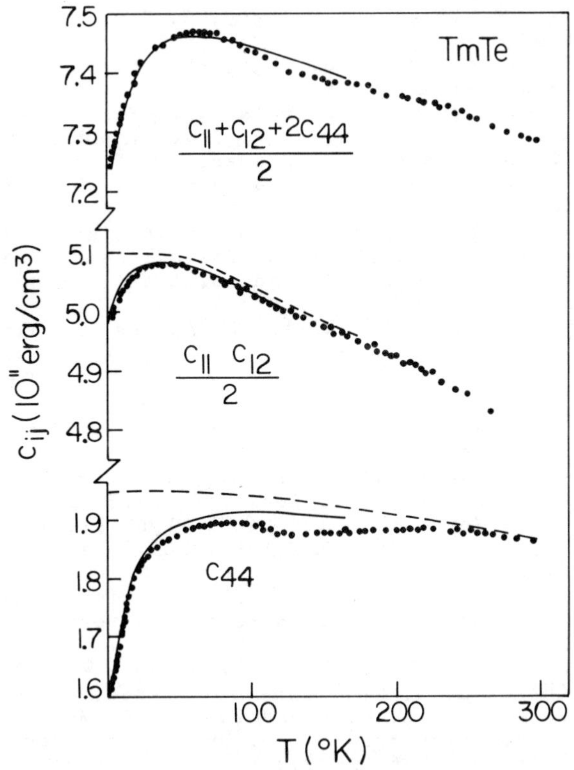

Fig. 4. Elastic constants for TmTe. Full line calculations based for Tm^{2+} ion on CEF levels Γ_6-Γ_8(10K) - Γ_7 (17K), g_2^2= 2.3mK, g_3^2= 37.3mK.

Our results clearly show a well defined valency of TmTe(Tm^{2+}), in agreement with recent other experiments.[41] On the other hand TmSe might well show interconfigurational fluctuations. In such a case a low frequency soundwave ($10^7 s^{-1}$) cannot sample the fast 2^+-3^+ fluctuations hence showing no CEF effects. Our experiment could equally well be explained by broadened CEF level due to formation of a virtual bound state.

d) Magnetic Field Effects on Elastic Constants

PrSb and TmSb, are ideally suited to test higher order magnetoelastic interactions because of the absence of magnetic or structural transitions, the cubic symmetry and the knowledge of relevant parameters g_2, g_3, B_4, B_6. Recently a theory for magnetoelastic interactions in the presence of a magnetic field, taking into account effects due to the asymmetric part of the strain tensor, has been developed for cubic rare earth CEF systems[42-44]. This followed a previous theory for magnetically ordered phases[45]. Terms, involving rotational strain interactions can be easily derived by expanding the CEF potential in terms of a general distortion. With x,y,z denoting the position of the magnetic ion in cubic octaheder symmetry we get, keeping only ε_{xz}, ω_{xz} strains:

$$\delta V=\frac{3Ze^2}{R^3}(xz+zx)\varepsilon_{xz}-\frac{21Ze^2}{4R^3}(x^2-z^2)\varepsilon_{xz}\omega_{xz}$$

$$+\frac{3Ze^2}{8R^3}(3y^2-r^2)(\varepsilon_{xz}^2+\omega_{xz}^2) \qquad (14)$$

The first term in (14) is the linear magnetoelastic from eq. (7), whereas the other involve both the symmetric (ε_{xz}) and antisymmetric (ω_{xz}) part of the strain tensor. Note that for tetragonal symmetry already linear terms in ω_{xz} appear in eq. (14). Eq. (14) does not show

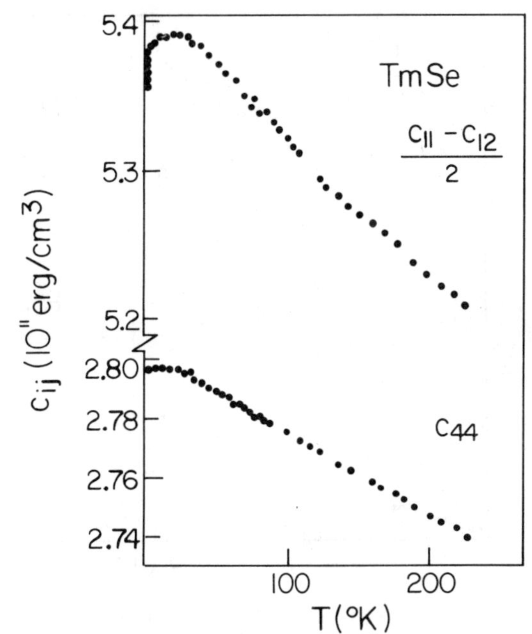

Fig. 5. Elastic constants for TmSe. No CEF-effect noticeable.

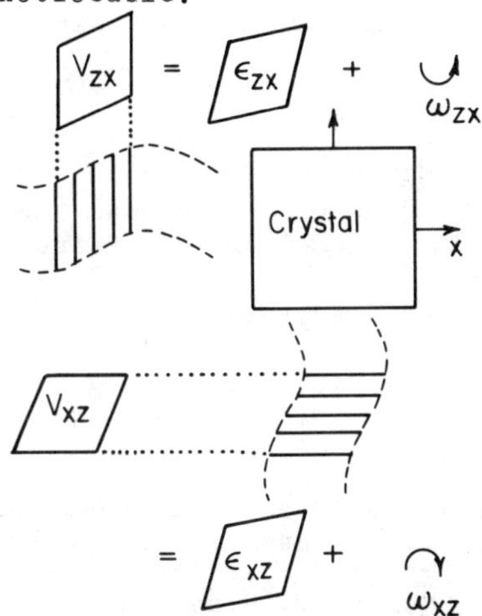

Fig. 6. c_{44}-mode propagating along x and z axis. Figure indicates that (k_z, R_x) has opposite rotational deformation than (k_x, R_z)-mode (from ref. 42).

separation into pure rotational and (finite) strain contributions[42].

One consequence of such rotational and finite strain interactions are the different sound velocities a c_{44} mode experiences in a magnetic field, depending on the geometry of propagation direction and polarization vector with respect to $\underline{H}$[42,45] (k_zR_x, k_xR_z). This is a consequence of the ω_{xz} strains contributing differently in the two geometries, as illustrated in fig. 6. Preliminary experimental results[46] for TmSb are shown in fig. 7 for the 2 modes of fig. 6. One clearly sees the difference of the 2 sound velocities, which would be identical in linear magnetoelastic theory. Also shown are calculations based on the theory of rotational interactions[42], with no adjustable parameters. Clearly the salient features (H^2-dependence for low fields, correct order of magnitude for both modes k_zR_x, k_xR_z) are well reproduced. For further experimental tests of these theories (for higher H and also c_{11}-c_{12} modes) see Ref. 46.

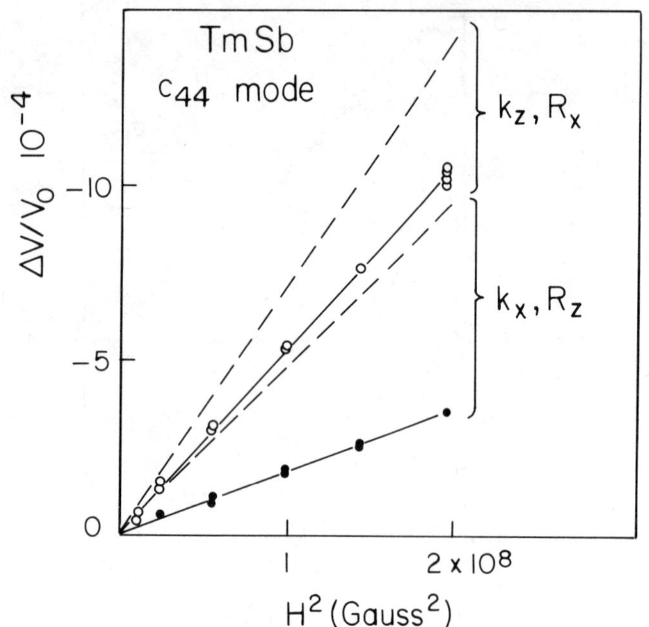

Fig. 7. Rotational effect for c_{44}-mode in TmSb at T=1.9K and as a function of H^2. Dashed curve theoretical calculations based on theory of ref. 42.

ACKNOWLEDGEMENT

Many friends and colleagues contributed to this work, amongst them K. Andres, E. Bucher, M. E. Mullen, H. R. Ott and P. S. Wang.

REFERENCES

* Supported by National Science Foundation.
1. Magnetic Properties of Rare Earth Metals, R. J. Elliott,editor Plenum London & New York 1972.
2. Proceedings of first conference on CEF effects in Metals and Alloys, Montreal 1974.
3. Proceedings of Conference of Haute-Nendaz 1973, B. Cohen, B. Giovannini, editors; R. H. Taylor, Adv. in Phys. 24, 681 (1975).
4. P. Fulde and I. Peschel, Adv. in Phys. 21, 1 (1972); A Fert and A. Friederich, Magnetism and Magnetic Materials 1974, AIP Conf. Proc. 24 p. 466.
5. H. R. Ott and B. Lüthi, Phys. Rev. Lett. 36, 600 (1976).
6. B. Lüthi, M. E. Mullen and E. Bucher, Phys. Rev. Lett. 31, 95 (1973).
7. K. R. Lea, M. J. M. Leask and W. P. Wolf, J. Phys. Chem. Solids 23, 1381 (1972); M. T. Hutchings, Solid State Phys. 16, 227 (1964), F. Seitz, D. Turnbull, editors, Acad. Press.
8. M. E. Mullen, B. Lüthi, P. S. Wang, E. Bucher, L. D. Longinotti, J. P. Maita and H. R. Ott Phys. Rev. B10, 186 (1974).
9. See for example W. E. Wallace, Rare Earth Intermetallics, Acad. Press (1973).
10. E. R. Callen and H. B. Callen, Phys. Rev. 129, 578 (1963).
11. L. L. Hirst, Z. Phys. 244, 230 (1971); ref. 4.
12. P. Fulde, Z. Phys. B20, 89 (1975); M.J. Sablik, H. H. Teitelbaum and P. M. Levy, Magnetism and Magnetic Materials 1972, AIP Conf. Proc. 10 p. 548.
13. See G. A. Gehring and K. A. Gehring, Rep. Prog. Phys. 38,1 (1975) and references contained therein.
14. Ref. 8, 13 and P. M. Levy, J. Phys. C6, 3545 (1973).
15. B. R. Cooper and O. Vogt, Phys. Rev. B1, 1218 (1970).
16. R. J. Birgeneau, E. Bucher, L. Passell and K. C. Turberfield, Phys. Rev. B4, 718 (1971).
17. R. J. Birgeneau, E. Bucher, J. P. Maita, L. Passell and K. C. Turberfield, Phys. Rev. B8, 5345 (1973).
18. A. S. Nowick and B. S. Berry, Anelastic Relaxation in Crystalline Solids, Acad. Press
19. P. Morin, A. Waintal and B. Lüthi, to be published.
20. B. Lüthi, M. E. Mullen, K. Andres, E. Bucher and J. P. Maita, Phys. Rev. B8, 2639 (1973).
21. P. Morin, thèse Université Grenoble, CNRS A. O. 9323.
22. R. L. Melcher, E. Pytte and B. A. Scott, Phys. Rev. Lett. 31, 307 (1973).
23. G. Gorodetsky, A. Shaulov, V. Volterra and J. Makovsky, Phys. Rev. B13, 1205 (1976).
24. R. W. Keyes in Solid State Phys. 11, 149 (1960) F. Seitz, D. Turnbull, editors, Acad. Press.
25. R. L. Melcher in Physical Acoustics Vol. 12, W. P. Mason, R. N. Thurston, editors, Acad. Press, NY.
26. K. Andres, P. S. Wang, Y. H. Wong and B. Lüthi, H. R. Ott, this conference.
27. E. Bucher, R. J. Birgeneau, J. P. Maita, G. P. Felcher and T. O. Brun, Phys. Rev. Lett. 28, 746 (1972).
28. E. Bucher, J. P. Maita, G. W. Hull Jr., L. D. Longinotti, B. Lüthi and P. S. Wang to be published.
29. D. K. Ray and A. P. Young, J. Phys. C6, 3353 (1973).
30. P. A. Lindgård, B. N. Harmon and A.J. Freeman, Phys. Rev. Lett. 35, 383 (1975).
31. G. T. Trammell, Phys. Rev. 131, 932 (1963).
32. K. W. H. Stevens and E. Pytte, Solid State Commun. 13, 101 (1973).
33. F. Levy Phys. Kondens. Mater. 10, 85 (1969).
34. H. R. Ott and K. Andres, Solid State Commun. 15, 1341 (1974).
35. L. J. Tao, J. B. Torrance and F. Holtzberg Solid State Commun. 15, 1025 (1974).
36. P. Bak, S. Krinsky and D. Mukamel, Phys. Rev. Lett. 36, 52 (1976).
37. L. L. Hirst, Magnetism and Magnetic Materials 1974 Conf. Proc. 24 p. 11.
38. M. Campagna, E. Bucher, G. K. Wertheim, D. N. E. Buchanan and L. D. Longinotti, Phys. Rev. Lett. 32, 885 (1974).
39. D. Wohlleben and B. R. Coles in Magnetism V G. T. Rado, H. Suhl, editors, Acad. Press 1974.
40. E. Bucher, K. Andres, F. J. deSalvo, J. P. Maita, A. C. Gossard, A. S. Cooper and G. W. Hull Jr., Phys. Rev. B11, 500 (1975).
41. R. Suryanarayanan, G. Güntherodt, J. L. Freeouf and F. Holtzberg, Phys. Rev. B12, 4215 (1975).
42. V. Dohm and P. Fulde, Z. Phys. B21, 369 (1975).
43. P. Thalmeier and P. Fulde, Z. Phys. B22, 359 (1975).
44. V. Dohm, Z. Phys. B23, 153 (1976).
45. R. L. Melcher in: Proc. Internat. School of Physics "E. Fermi", Varenna Course LII London 1972, E. Burstein, editor.
46. P. S. Wang and B. Lüthi, to be published.

MAGNETIC POLYMERS+

M. F. Thorpe, Becton Center, Yale University, New Haven, CT., 06520

Recent experiments[1] on poly(metal phosphinates) suggest that amorphous magnetic polymers are interesting new materials. In this talk we discuss some experiments that might yield useful information about the magnetic and conformational properties of magnetic polymers. Static measurements, like specific heat and susceptibility, can give information about the magnetic ordering. However, neutron measurements of the wave vector dependent susceptibility $X(\vec{k})$ and the scattering law $S(\vec{k},\omega)$ can give more detailed information about the magnetic properties and more importantly, about the conformations of the polymer chains (in particular, the chain length and end-to-end distance).

We set up some simple models in order to investigate the kind of behavior to be expected.[2] We show that the form of $X(\vec{k})$ is modified when the correlation length associated with the magnetic ordering becomes comparable with the chain length. The neutron scatterin law $S(\vec{k},)$ measures essentially the spin wave density of states at low temperatures although there is some $\vec{k}$ dependence which reflects the amount of conformational disorder. A full account of this work is given in reference 2.

+Work supported in part by the N.S.F.

[1]J.C. Scott et al., Phys. Rev. B12, 356 (1975).
[2]M.F. Thorpe, Phys. Rev. B13, 2186 (1976)

HIGHLY MAGNETOSTRICTIVE RARE EARTH ALLOYS

A. E. Clark
U. S. Naval Surface Weapons Center
White Oak, Silver Spring, MD 20910

Within the last few years, huge magnetostrictions have been achieved at room temperature in the Laves phase RFe_2 alloys (R = rare earth). Strains $> 2000 \times 10^{-6}$, over ten times those of previously known polycrystals, have been observed in $TbFe_2$ and $SmFe_2$. The rare earth-Fe_2 compounds, while exhibiting the largest known room temperature magnetostrictions, possess a wide range of cubic magnetic anisotropies, ranging from the largest positive value of 2×10^7 erg/cm^3(for $DyFe_2$) to the largest negative value of -7.6×10^7 erg/cm^3 (for $TbFe_2$). Room temperature values of magnetization, magnetic anistropy, and magnetrostriction are listed in the table below:[1,2,3]

	M (emu/cm^3)	K ($\times 10^7$ergs/cm^3)	λ_p ($\times 10^{-6}$)	λ_{111} ($\times 10^{-6}$)
$SmFe_2$	∿400	--	-2340	-2100
$TbFe_2$	800	-7.6	2600	2460
$DyFe_2$	800	2.1	640	1260
$ErFe_2$	300	-0.3	- 340	- 300

Here λ_p $(=\lambda_{\parallel} - \lambda_{\perp})$ denotes the fractional change in length of a polycrystal as a magnetic field is rotated from perpendicular to parallel to the measurement direction.

An unprecedented magnetostriction anistropy, $\lambda_{111} >> \lambda_{100}$, characterizes these compounds. For $DyFe_2$,[3] $\lambda_{100} = 0 \pm 4 \times 10^{-6}$. The huge rare earth magnetoelastic interaction is effectively shorted out in the cubic Laves phase RFe_2 compounds when M∥[100]. This $\lambda_{111} >> \lambda_{100}$ anistropy dictates a magnetostrictive strain that depends almost exclusively on domain wall motion rather than on magnetization rotation. An atomic model for the anisotropic magnetostriction, based upon the symmetry at the rare earth site, has been proposed.[3]

The ternary alloy, $Tb_{1-x}Dy_xFe_2$($x \simeq 0.7$), possess a high magnetostriction/anisotropy ratio. Magnetomechanical coupling measurements on this alloy reveal coupling factors, k's ≃ 0.6, making these materials attractive for magnetostrictive transducer applications. Prototype transducers have been designed at both government and industrial laboratories.[4] In addition, the elastic moduli of this alloy are found to be strongly field dependent. The fractional change in Young's modulus, $\Delta E/E$ = 150% at 10 kHz. The sound velocity increase upon magnetization, associated with this modulus change, is 57%.[5]

Many applications of rare earth magnetostrictive materials are currently emerging. These include acoustic delay lines, variable frequency resonators, active millisecond valves, and micropositioning devices.

1. A. E. Clark, Conf. on Mag. and Mag. Mat'ls, Nov. 1973; AIP Conf. Proc. 18, 1015 (1974).
2. A. E. Clark, J. R. Cullen, and K. Sato, Conf. on Mag. and Mag. Mat'ls, Nov. 1974; AIP Conf. Proc. 24, 670 (1975).
3. A. E. Clark, J. R. Cullen, O. D. McMasters, and E. R. Callen, Conf. on Mag. and Mag. Mat'ls, Dec. 1975, AIP Conf. Proc. 29, 192 (1976).
4. R. W. Timme, Workshop on Magnetostrictive Transducers, USRD/NRL, Orlando, Florida, Nov. 1975.
5. H. T. Savage, A. E. Clark, J. M. Powers, IEEE Trans. on Magnetics, Mag.-11, 1355 (1975).

COMMENTS ON SPIN- AND CHARGE-DENSITY OSCILLATIONS IN METALS

Berend Kolk*
Serin Physics Laboratory
Rutgers University
Piscataway, N.J. 08854

ABSTRACT

The exchange interaction of conduction electrons with local moments gives rise to a repopulation of states at the Fermi level and also to a perturbation of the conduction-electron wave functions. In the RKKY theory this perturbation is not taken into account for electrons in the repopulated states. It is found that the perturbation of these electrons leads to a net charge oscillation as well as to an anomalous behavior of spin-density oscillation around the local moment.

I INTRODUCTION

It is generally assumed that the exchange interaction between conduction electrons and the electrons which contribute to a local moment only gives rise to spin-density oscillations.[1] It is shown in Sec.II, however, that the exchange interaction also leads to variations in the net charge density as a result of the perturbation of the wave functions of electrons in the repopulated states at the Fermi level. The perturbation of these electrons is neglected in the RKKY theory. This perturbation affects also the spin density oscillations. The shapes of the charge- and spin-density oscillations is found to depend on the net concentration of parallel local moments. Analysis of the available isomershift and magnetic hyperfine shift data [2-5] of FeSi indicates the existence of charge oscillations in iron (see Sec.IV). The anomalous shape of the spindensity oscillation observed in this system agrees fairly well with the modified RKKY theory given in Sec.II. In fitting the data the charge- and spindensity oscillations arising from the shielding of the Si impurities by conduction electrons (Sec.III) are taken into account.

II CHARGE OSCILLATIONS DUE TO EXCHANGE INTERACTIONS

The diagonal terms, $\vec{k}=\vec{k}'$, in the exchange integral, $J(\vec{k},\vec{k}')$, lower the energies of the conduction electrons with spin parallel (+) to the local moment so that the antiparallel (-) spin electrons near the Fermi surface flip their spin to occupy the lower energy (+) spin states (Zener repopulation).[1] This leads to an excess of spin (+) electrons. The off-diagonal terms, $\vec{k}\neq\vec{k}'$, in $J(\vec{k},\vec{k}')$ do not affect the energies but yield a density redistribution which in first order approximation is given by

$$\rho_{\pm}(\vec{k},\vec{r}) = \rho_o(\vec{k},\vec{r}) \mp \sum_n \sum_{\vec{k}'>\vec{k}}^{\infty} \frac{J_n(\vec{k},\vec{k}')S_n^z/N}{E_{\vec{k}} - E_{\vec{k}'}} \times \left[\psi^*(\vec{k}',\vec{r})\psi(\vec{k},\vec{r})+\psi^*(\vec{k},\vec{r})\psi(\vec{k}',\vec{r})\right], \quad (1)$$

where $\rho_o(\vec{k},\vec{r}) = |\psi(\vec{k},\vec{r})|^2$, $\psi(\vec{k},\vec{r})$ and $E_{\vec{k}}$ are the unperturbed conduction-electron wave functions and energies, N the number of lattice sites, and S_n^z the z component of the spin operator of the local moment at the nth site.

The conduction-electron polarization

$$p(r) \equiv \{\rho_+(r) - \rho_-(r)\}/2\rho_o(r) \quad (2)$$

owing to the exchange interaction may be written using Eq.(1) as

$$p_{ex}(r) = \left\{ 2\sum_{o}^{k_F} \Delta\rho(\vec{k},\vec{r}) + \sum_{k_F^-}^{k_F^+} \rho_o(\vec{k},\vec{r}) + \sum_{k_F^-}^{k_F^+} \Delta\rho(\vec{k},\vec{r}) \right\} /2\rho_o(r) \quad (3)$$

where $\Delta\rho(\vec{k},\vec{r}) = \rho_+(\vec{k},\vec{r}) - \rho_o(\vec{k},\vec{r})$, k_F is the average value of $k_F^{\pm}$, i.e. by definition $k_F^{\pm} = k_F (1 \pm \varepsilon)$. The first term in Eq.(3) represents the perturbation of the charge densities in the (+) and (-) band up to $k = k_F$, the second term accounts for the Zener repopulation and the third one for the perturbation of these repopulated states.

Besides these spin-density variations the exchange interaction also generates a relative difference in the total charge density,

$$q(r) \equiv \{\rho_+(r) + \rho_-(r) -2\rho_o(r)\}/2\rho_o(r) \quad (4)$$

given by

$$q_{ex}(r) = \sum_{k_F^-}^{k_F^+} \Delta\rho(\vec{k},\vec{r})/2\rho_o(r) \quad (5)$$

In the original RKKY theory [6-9] Eq.(1) is evaluated for a free-electron metal and for $J_n(k,k') = J_o e^{i(\vec{k}-\vec{k}').\vec{R}_n}$ which results in

$$\rho_{\pm}(\vec{r}) = \rho_o(r) \mp (2m/\hbar^2) J_o V \sum_n S_n^z/N \times (z_n^{\pm} \cos z_n^{\pm} - \sin z_n^{\pm})/32\pi^3 r^4, \quad (6)$$

where $z_n^{\pm} = 2k_F^{\pm}|\vec{r}-\vec{R}_n|$, $\vec{R}_n$ is the lattice vector of the nth site, V the crystal volume and m the effective mass of the conduction electron. The well known RKKY formula for the conduction electron polarization is obtained from Eqs.(2) and (6) by using the approximation $k_F^+ = k_F^- = k_F$;

$$p_{RKKY}(r) = -P \sum_n S_n^z\{z_n\cos z_n - \sin z_n\}/z_n^4 \quad (7)$$

where $P = 9\pi n\, J_o/E_F$ and $n = k_F^3 V/6\pi^2 N$ is the number of conduction electrons per atom. Expression (6) includes the states due to the Zener repopulation as shown by Yosida[8]. In Eq.(7), however, the contribution arising from the perturbation of these repopulated states (the third term in Eq.(3)) is neglected as a result of the approximation $k_F^+ = k_F^- = k_F$. It is obvious from Eq.(5) that in this approximation $q_{RKKY}(r) = 0$. The perturbation of the repopulated states can easily be taken into account by inserting $k_F^{\pm} = k_F(1 \pm \varepsilon)$ in Eq.(6). This yields for the conduction-electron polarization

$$p_{ex}(r)=-P \sum_n S_n^z \cos\varepsilon z_n \{z_n \cos z_n - \sin z_n\}/z_n^4 + O(\varepsilon^2) \qquad (8)$$

and for the charge-density redistribution

$$q_{ex}(r)=P \sum_n S_n^z \sin\varepsilon z_n \{z_n \sin z_n + \cos z_n\}/z_n^4 + O(\varepsilon^2) \qquad (9)$$

where

$$\varepsilon = J_o/2E_F \sum_n S_n^z/N \qquad (10)$$

The difference in the phase factors $z_n^{\pm}$ in Eq.(6) for the (+) and (-) charge densities gives rise to interference effects as shown in Fig. 1, where the spin and charge oscillations generated by a single local moment are displayed for typical values $J_o = 1.2eV$, $E_F = 6.5eV$, $a = 2.7$ Å (lattice constant) and assuming parallel local moments with $S_n^z = 1$ at each lattice site. Hence, the modified expression, Eq.(8), gives a quite anomalous behavior of the spin density oscillations (Fig.1b) compared with the original RKKY oscillations (Fig.1a). The charge variations shown in Fig. 1c can lead to measurable isomershifts in Mössbauer spectra (see Sec.IV).

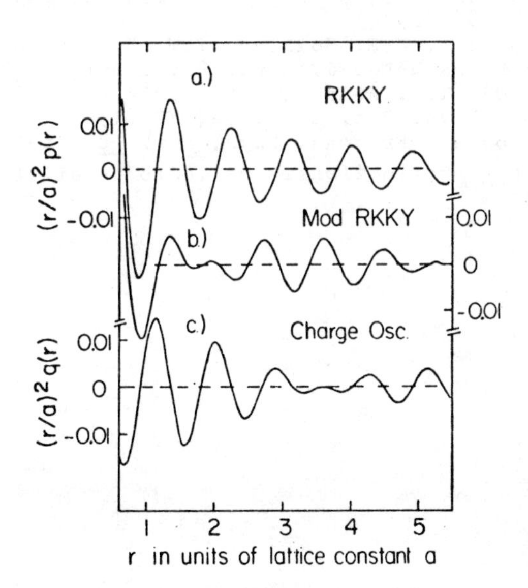

Fig. 1. The conduction-electron polarization $p(r)$ and charge distribution $q(r)$ arising from a single local moment calculated for values of J_o, E_F and a given in the text. a) Eq. 7, b) Eq. 8, and c) Eq. 9

It is interesting to note that the spin- and charge-density variations depend on the fraction of local moments with parallel spin since $\varepsilon \sim \sum S_n^z/N$. In the case of only one local moment $\varepsilon \sim S^z/N \approx 0$ which inserted in Eqs.(8) and (9) yields the original RKKY formula, Eq.(7), and $q_{ex}(r) = 0$. Hence, the spin density oscillation generated by one moment will vary between that represented in Fig.1a when the total number of local moments is one, and that in Fig.1b when each lattice site is occupied by an atom with a local moment.

III SHIELDING EFFECTS

In the free-electron approximation the charge redistribution of the conduction electrons due to shielding of an impurity is for $k_F r \gg 1$ given by Friedel oscillations[1,10]

$$q_{sh}(r) = Q\cos(z+\theta)/z^3 \qquad (11)$$

where Q and θ are related to the difference in valence, or charge, between the impurity and the host atoms. In a ferromagnetic host the charge oscillations cannot be described adequately by these Friedel oscillations as has been attempted[3], because of similar arguments as given in Sec.II. When the difference in number of occupied states in the (+) and (-) band is taken into account the charge variations are given by

$$q_{sh} = Q\cos\varepsilon z\, \cos(z+\theta)/z^3 + O(\Delta Q, \Delta\theta) \qquad (12)$$

where $\Delta Q = Q^+ - Q^-$ and $\Delta\theta = \theta^+ - \theta^-$. Furthermore these charge variations generate a conduction-electron polarization

$$p_{sh} = -Q\sin\varepsilon z\, \sin(z+\theta)/z^3 + O(\Delta Q, \Delta\theta) \qquad (13)$$

For $\varepsilon = 0$ the original Friedel oscillations and $p_{sh} = 0$ are obtained. The charge- and spin-density oscillations given by expressions (12) and (13) have a behavior similar to that shown by curve b) and c) in Fig. 1.

Expression (13) describes the spin density outside the impurity. The spin density at the impurity site due to shielding has been studied by Daniel and Friedel[11] and by Campbell[12].

IV SPIN AND CHARGE OSCILLATIONS IN IRON

FeSi has been used to study the spin- and charge-density oscillations in iron, because the Si solutes do not carry a local moment and do not disturb the host moments[13]. The Si solute is therefore considered to act as a magnetic hole, which gives rise to spin and charge oscillations similarly to an Fe atom, but with opposite sign[2]. Spin-density oscillations of the s conduction electrons give rise to changes in the magnetic hyperfine fields at neighboring Fe nuclei of a Si atom via the Fermi-contact interaction which can be measured with the aid of the Mössbauer effect or NMR. Typical results of magnetic hyperfine shifts are shown in Fig. 2.

It is interesting to note that the data cannot be fitted with the aid of the RKKY theory, as has been attempted[2,3]. Stearns[2] at-

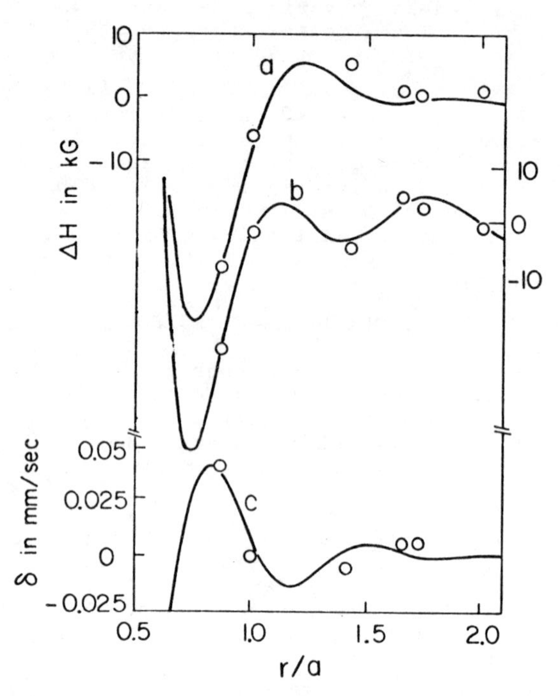

Fig. 2. Least-squares fit of FeSi hyperfine shift data from a) Ref.2 Set A, b) Ref.2 Set B. Isomershift data (c) from Ref.s 3 and 4.

tributes this to the hybridization of the 3d levels with the Fe conduction band (interband mixing)[1,14]. However, there are alternative explanations possible. In Sec.II we showed that the RKKY theory has to be modified which yields an "anomalous" behavior for the spin-density oscillations as shown in Fig.1b. Furthermore we saw in Sec.III that shielding of an impurity also may contribute to the spin density. Therefore the available experimental hyperfine shift data were fit with $p(r) = p_{ex}(r) + p_{sh}(r)$, using Eq.s 8 and 13. In Eq. 13 $\theta = 0.51$ radians was used. This value was derived in Ref.10 from resistivety measurements of Si impurities in copper. Fair fits were obtained (see Fig. 2) with reasonable values of J_o, k_F and $2\rho_o(0)$, assuming[15] $\Sigma\, S_n^{\ z}/N = 1$. The fit of Set A of Stearns (curve a) yielded $J_o = 1.5$ eV, $k_F = 1.3$ Å^{-1} $2\rho_o(o) = 2.6\ a_o^{-3}$ and $Q/P = 0$. Set B of Stearns (curve b) yielded $J_o = 2.1$ eV, $k_F = 1.3$Å^{-1}, $2\rho_o(0) = 3.0\ a_o^{-3}$ and $Q/P = 0.1$. Fits of the data given in Ref.s 4 and 5 gave roughly the same results as those of Set B of Stearns.

The fact that $Q/P \ll 1$ seems to support Stearns statement that the effects arising from the shielding of the Si impurities are small. In that case the observed isomer shifts[3,4] must arise mainly from the charge oscillations generated by the exchange interaction. Curve c, for instance, is calculated with $J_o = 3.2$ eV, $k_F = 1.6$ Å^{-1}, $2\rho_o(0) = 4.8 a_o^{-3}$ and $Q/P = 0.2$, using Eq.s 9 and 12.

Hence, the hyperfine shift and isomer-shift data of FeSi may be explained by taking into account the perturbation of the wave functions of electrons in the repopulated states which in the RKKY and other theories have been neglected.

A more extensive study of spin and charge variations in metals is in progress, in which a better approximation for the exchange integral $J_n(\vec{k},\vec{k}')$ and more realistic wave functions for the conduction electrons are used.

ACKNOWLEDGEMENTS

I am much indebted to Gary Collins for his assistence in fitting the FeSi data and to Noémie Koller for her support and encouragement.

REFERENCES

* Work supported by NSF Grant DMR 76-02053
1 For a review on spin- and charge-density oscillations in metal we refer to the paper of R.E. Watson in Hyperfine Interactions, Ed.s A.J. Freeman and R.B. Frankel, Academic Press 1967, p.413
2 M.B. Stearns, Phys. Rev. B 4, 4081 (1971) and Phys. Rev. B 13, 1183 (1976)
3 G. Gruner, I. Vincze and L. Cser, Sol. State Comm. 10, 347 (1972)
4 T.E. Cranshaw, C.E. Johnson, M.S. Ridout and G.A. Murray, Phys. Lett. 21, 481 (1966)
5 E.F. Mendis and L.W. Anderson, Phys. Status Sol. 41, 375 (1971)
6 M.A. Ruderman and C. Kittel, Phys. Rev. 96, 99(1954)
7 Y. Kasuya, Progr. Theoret. Phys. (Kyoto) 16, 45(1956)
8 K. Yosida, Phys. Rev. 106, 893(1957)
9 A.W. Overhauser, J. Appl. Phys. 34, 1019S (1963); T.A. Kaplan, Phys. Rev. Lett. 14, 499(1965)
10 W.Kohn and S.M. Vosko, Phys. Rev. 119, 912 (1960)
11 E. Daniel and J. Friedel, J. Phys. Chem. Solids 24, 1601 (1963)
12 J.A. Campbell, J. Phys.C (Solid St. Phys.) 2, 1338(1969)
13 A. Arrott, M.F. Collins, T.M. Holden, G.G. Low and R. Nathans, J. Appl. Phys. Suppl. 37, 1194 (1966); T.M. Holden, J.B. Comly and G.G. Low, Proc. Phys. Soc. Lond. 92, 726(1967)
14 Watson and Freeman, Phys. Rev. 186, 625 (1969)
15 This approximation is valid for a dilute FeSi system

USE OF OPTICAL DATA TO DETERMINE INDIVIDUAL CONTRIBUTIONS TO THE MAGNETIC SUSCEPTIBILITY OF METALS

C. L. Foiles*
Michigan State University, East Lansing, Mich. 48824

ABSTRACT

Magnetic studies of a set of ternary alloys, binary alloy hosts containing Mn impurities, indicate a better understanding of conduction electron behavior in the hosts is necessary. Residual scattering eliminates powerful Fermiology techniques as a means of gaining this understanding and necessitates more indirect methods. The use of magnetic susceptibility data which is separated into individual contributions by means of complementary optical data appears to be a promising method. Application of this method to copper and silver is used to clarify its limitations and reliability.

INTRODUCTION

The magnetic properties of Mn impurities in dilute alloys exhibit many diverse effects and it is generally accepted that the inter-impurity interactions play a major role in producing these effects. Reliable separation of the relative importances of the impurity distribution and the form of these interactions is difficult but magnetic studies of alloy films in amorphous and crystalline states appear to provide a crude separation.[1,2] Systematic changes in the magnetic susceptibility of the impurity are associated with the state of the alloy. Of potentially greater significance, these changes are consistent with predictions based upon a free electron model of impurity coupling.[3] This form of coupling is also consistent with neutron scattering done on bulk samples.[4]

Interesting parallels to the amorphous-crystalline studies occur when binary alloys which undergo atomic order-disorder transitions are used as the alloy host.[5] However, the free electron predictions for these alloys fail. Since this failure is even qualitative, any general empirical guidance suggested by the apparent separation of influences above is challenged by these alloy data unless it can be shown (i) that the Mn impurities preferentially occupy certain sites and produce a long range order which is consistent with a free electron model of coupling or (ii) that the conduction electrons in the binary hosts deviate from free electron behavior more severely than the noble metals do. X-ray studies reported at this conference[6] provide quantitative measurements of the long range order parameter, S. These measurements of S as a function of Mn concentration indicate no preferential occupancy and eliminate item (i).

The present paper considers item (ii) and explores the possible use of magnetic susceptibility data complemented by optical property data as a basis for judging conduction electron behavior. The choice of this basis is not as strange as it initially appears. The set of binary alloys can be prepared with a high degree of order ($S \geq 0.8$) but they still retain a large residual scattering. This scattering is sufficient to render the standard Fermiology methods inoperable. Comparisons of transport properties can be misleading due to the nature of the scattering processes. Previously we demonstrated that a sign change in the measured thermoelectric power for Cu_3Au does not indicate any change in the sign of the diffusion contributions.[7] Similar arguments based upon high field and low field effects can be advanced for the observed behavior of the Hall effect. An equilibrium property becomes the logical choice. For Cu_3Au, an archetypal alloy, the electronic specific heat is only slightly altered by ordering ($\lesssim 3\%$)[8] but the magnetic susceptibility shows a significant change ($\sim 20\%$).[9] Thus, magnetic susceptibility becomes the chosen property.

RESULTS AND DISCUSSION

The free electron predictions and experimental results for 5 alloys plus the comparison systems Cu and Ag are listed in table I. The magnetic susceptibility is a sum of contributions and predictions were made using a common, approximate separation of terms

$$\chi_t = \chi_c + \chi_P + \chi_L + \chi_{\nu\nu} \qquad (1)$$

where χ_c is the diamagnetic contribution from the core electrons, χ_P is the spin contribution from the conduction electrons, χ_L is the diamagnetic contribution from the conduction electrons and $\chi_{\nu\nu}$ is a Van Vleck term associated with the band structure. The calculations of Hurd and Coodin[10] were averaged with the empirical values of Klemm[11] to determine χ_c. A free electron model was used for χ_P and χ_L with the former being multiplied by (m^*/m_o) and the latter being multiplied by (m_o/m^*) if the free conduction electron mass m_o was replaced by an effective mass m^*. $\chi_{\nu\nu}$

Table I. Magnetic Susceptibility. The references denote sources for experimental data: (a) references listed in reference 5, (b) present investigation, and (c) reference 13. m^*/m_o is unity unless otherwise noted. OS and DOS denote atomically ordered and disordered states, respectively.

Material	Magnetic Susceptibility (10^{-6} emu/mole) Eqn. (1)	Exp.	Ref.
Cu_3Au	-15	-12.5 (DOS) -15 (OS)	(a,b)
Au_3Cu	-25.2	-23 (DOS) -21.6 (OS)	(a)
Ag_3Mg	-13.8	-13.9 (DOS)	(b)
Cu_3Pt	-18	-4.6 (DOS) -8 (OS)	(b)
$Cu_{0.83}Pt_{0.17}$	-13	-6.6 (DOS) -13.1 (OS)	(b)
Cu	-8.6		
Cu (m^*/m_o=1.5)	-4.3	-4.2	(c)
Ag	-19.1	-18.9	(c)

was set equal to zero. The Hurd and Coodin calculations used self-consistent solutions of the Hartree-Fock-Slater equations; however, since these results are combined with free electron estimates for the other contributions and since a concise term is needed to contrast this set of values with another set, the term free electron values will be adopted for the predicted results listed in table I.

Two features of table I deserve explicit attention. First, the observed and predicted values are in remarkably good agreement for every system except Cu_3Pt. Second, the significance of this agreement is questionable without some verification of individual terms. This latter point is clearly illustrated by alternate sets of contributions for the noble metals.[12,13] These alternate sets predict similar final results by using larger $|\chi_c|$ values, by increasing χ_p by a factor near 2 and by reducing χ_L to nearly zero (see tables II and III).

Optical data can provide some information about each contribution except $\chi_{\nu\nu}$. Cohen[14] has noted that in the absence of interband contributions the real part of the dielectric constant can be written as

$$\varepsilon_1 = 1 + 4\pi N\alpha_1 - \frac{\omega_p^2}{\omega^2} \qquad (2)$$

where N is the number of atoms per unit volume, α_1 is the electric polarizability, ω_p is the plasma frequency and ω is the frequency of light. Cohen further noted that α_1 is the sum of a core electron contribution α_o and a conduction electron contribution. An old argument by Kirkwood[15], which has been revived by Dorfman[16], produces the relation

$$\chi_c = -3.11(10^6)\sqrt{k\,\alpha_o} \quad \text{(emu/mole)} \qquad (3)$$

with k being the total number of core electrons. If α_1 is equated to α_o, then an upper limit on $|\chi_c|$ can be determined from optical data.

The many sets of optical data available for Ag and Cu permit this test to be applied. Ag has an initial onset of interband transitions near 4 eV and thus is a particularly useful example. Two interpretational features of these optical data should be clarified before examining the results listed in tables II and III. One, although the real and imaginary parts of the dielectric constant are related via a Kramers-Kronig relation, independent determination of these parts is essential. The key issue in this problem lies in the extrapolation scheme used to extend observed values to the ω=0 and ω=∞ limits. Jan and Vishnubhatla[17] discuss the problem for some binary alloys. Dr. Jan generously provided their original data; fitting the CuZn data to eqn. (2) we found the observed α_1 could vary by a factor of 5 depending upon the extrapolation scheme. Thus independent determination of ε_1 is essential. Two, freedom from surface contamination is essential. This point is demonstrated by the study of Mathewson, etal[18] on Ag and by the study of Roberts[19] on Cu. In the former study a vacuum environment gave data which fit eqn. (2) and enabled α_1 to be determined. As the samples were subjected to increasing amounts of the general laboratory atmosphere, the values of ε_1 changed systematically. Equation (2) was still obeyed but the α_1 values increased. A factor of 2 or so was common. Mathewson, etal showed that this change for Ag was consistent with the formation of a thin sulfide layer (~20 Å). Roberts reported related effects in Cu and suggested oxidation occurred.

Using the vacuum environment data of Mathewson, etal for Ag and that of Roberts for Cu produce the values of χ_c in tables II and III respectively. In each case these values support the values used with the free electron model.

Optical data also provide an indirect test for χ_p values. The parameter ω_p^2 in eqn. (2) is proportional to $1/m_{opt}$. If a free electron model were truly valid, then m_{opt} would equal m* as determined from the electronic specific heat and hence be an independent measure of the density of states. χ_p is directly proportional to the density of states. The preceding "if" is necessitated by three well known considerations; Fermi surface anisotropy, electron-phonon interactions (EPI) that lead to an enhancement λ in the electronic specific heat which is not present in χ_p nor in m_{opt}, and electron-electron interactions which alter the relation between χ_p and m*. Let us consider each effect in turn.

In a generalized model which allows for Fermi surface anisotropy, m_{opt} can be related to an integral of electron velocity over the Fermi surface while m* can be related an integral of the reciprocal of this same velocity over the Fermi surface.[20] At an earlier stage in the study of metals, the ratio m^*/m_{opt} was used as a crude measure of Fermi surface anisotropy. That measure may prove useful for the binary alloys of interest here and we note that the ratios for Ag and Cu equal unity within experimental uncertainty.

Table II. Contributions to the magnetic susceptibility of silver. The term "8 cone" denotes references 12 and 13; other terms and sources are identified in the text. All values are in 10^{-6} emu/mole.

Property or Model	χ_c	χ_p	χ_L	χ_t	χ_{exp}
"8 cone"	-42	20.5	-0.3	-21.8	
"8-cone"	-38.5	19.2	0	-19.3	-18.9±0.1
free electron	-25±1	8.8	-2.9	-19.1±1	
$4\pi N\alpha_1$	-24.9				
$m^*_{(\lambda=0)}=0.95\ m_o$		8.4			
CESR		6.6±0.6			
PRFE(χ_{LP})			(-3.8±0.4)		

Table III. Contributions to the magnetic susceptibility of copper. The term "8 cone" denotes references 12 and 13; other terms and sources are identified in the text. All values are in 10^{-6} emu/mole.

Property or Model	χ_c	χ_p	χ_L	χ_t	χ_{exp}
"8 cone"	-19.7	15.6	-0.4	-4.5	
"8 cone"	-18.1	14.6	-0.4	-3.9	-4.2±0.1
free electron					
(m^*/m_o=1.0)	-13.2±1	6.9	-2.3	-8.6±1	
(m^*/m_o=1.5)	-13.2±1	10.4	-1.5	-4.3±1	
$4\pi N\alpha_1$	-14.9				
$m^*=1.38m_o$ (λ=0)		9.5			
dHvA		10.0			
CESR		10.1±1			
		6.9±0.6			

The possibility of λ being included in m* but not in m_{opt} complicates the preceding anisotropy measure. For the noble metals a direct measure of λ is not yet possible. However, a number of indirect tests indicate that λ is probably less than 0.15 for these metals.[21,22] A value of this size causes minor decreases in the m^*/m_{opt} ratios and we conclude that these ratios are essentially unity, the free electron value.

The effects of electron-electron interactions are too complex to be discussed here. Fortunately, the importance of these interactions for present considerations can be put into context by considering two independent measurements of χ_P, conduction electron spin resonance (CESR) and integrals evaluated from detailed de Haas van Alphen (dHvA) data. Randles, using the difference between g factors measured by CESR and de Haas van Alphen studies to estimate electron-electron interactions, integrated dHvA data to estimate χ_P for Cu.[23] CESR data alone can measure χ_P if an independent susceptibility exists to provide a calibration. The results from two different calibrations[24,25] are included in tables II and III.

Although significant scatter occurs for the χ_P values, a concensus emerges. Use of m^*/m_{opt} can provide a crude measure of anisotropy and an upper limit for χ_P values. This limit is consistent with various experimental values and with the free electron estimates but is inconsistent with values used in the alternate sets.

Finally, polar reflection Faraday effect (PRFE) data can be used in conjunction with optical constant data to determine χ_{LP}, the Landau Peierls portion of χ_L. In an isotropic Fermi surface model χ_L and χ_{LP} are identical. Space restrictions do not permit an adequate discussion of the relation between these data and we simply note several features. Once again, a separation of intraband and interband effects is necessary. The intraband portion of the PRFE is needed and this requires data to be taken at energies sufficiently below the onset of interband transitions to produce a saturation in the PRFE data. A more thorough discussion of this topic has been given for silver[26] and the end result is listed in table II.

Before concluding, the possible complication of $\chi_{\nu\nu}$ must be considered. This term is always paramagnetic and is generally important when occupancy of a band is just beginning or just concluding. However, Walstedt and Yafet[27] have considered the s-d hybridization in Cu and they estimate $\chi_{\nu\nu}$ in this metal may be as large as $4(10^{-6})$emu/mole. They indicate a smaller value occurs in Ag. Terms of this size alter some details in final values but do not alter the basic patterns in tables II and III nor the considerations discussed above.

CONCLUSIONS

Inspection of the trends in tables II and III indicate optical data generally support the susceptibility contributions predicted from a free electron model. The agreement for χ_c values is striking. The χ_P values are more scattered but m^*/m_{opt} ratios suggest upper limits for these values; the limits are consistent with independent measurements and favor free electron values. No simple conclusion about χ_L can be made from the single χ_{LP} entry.

The general support for free electron values in Ag and Cu increases the promise of crude electronic structure characterization for the binary alloys listed in table I. Reliable optical data for these alloys is needed to complete this characterization.

REFERENCES

* Work supported by National Science Foundation.

1. Dietrich Korn, Z. für Physik 214, 136 (1968).
2. D. Korn, Z. für Physik 187, 463 (1965).
3. W.C. Kok and P.W. Anderson, Phil. Mag. 24, 1141 (1971).
4. N. Ahmed and T.J. Hicks, J. Phys. F 5, 2168 (1975).
5. T.W. McDaniel and C.L. Foiles, Solid State Comm. 14, 835 (1974).
6. paper 6C-5.
7. C.L. Foiles in Thermal Conductivity 14, edited by P.G. Klemens and T.K. Chu (Plenum Press, New York and London, 1976).
8. Dougles L. Martin, Can. J. Phys. 46, 923 (1968).
9. see table I. Previous work listed in reference 5.
10. C.M. Hurd and P. Coodin, J. Phys. Chem. Solids 28, 523 (1967).
11. Tabulation given by E. König in Landolt-Börnstein Tables, Vol. II pt 2. section 1.6.
12. E. Borchi and S. De Gennaro, Phys. Rev. B5, 4761 (1972).
13. R. Dupree and C.J. Ford, Phys. Rev. B8, 1780 (1973).
14. M.H. Cohen, Phil. Mag. 3, 762 (1958).
15. J.G. Kirkwood, Phys. Z. 33, 57 (1931).
16. Ya. G. Dorfman, Diamagnetism and the Chemical Bond (American Elsevier Publishing Co., Inc., New York, 1965).
17. J-P Jan and S.S. Vishnubhatla, Can. J. Phys. 45, 2505 (1967) and J-P Jan, private communication.
18. A.G. Mathewson, etal, J. Phys. F 2, L39 (1972).
19. S. Roberts, Phys. Rev. 118, 1509 (1960).
20. Frederick Wooten, Optical Properties of Solids (Academic Press, New York, 1972).
21. N.E. Christensen, Phys. Stat. Sol. B54, 551 (1972).
22. G. Grimvall, Phys. Kond. Mat. 11, 279 (1970).
23. D.L. Randles, Proc. R. Soc. London A 331, 85 (1972).
24. P. Monod and S. Schultz, Phys. Rev. 173, 645 (1968).
25. S. Schultz, M.R. Shanabarger and P.M. Platzman, Phys. Rev. L. 19, 749 (1967) and M. Shanabarger, 1970 Thesis (unpublished).
26. C.L. Foiles, Phys. Rev. (in press).
27. R.E. Walstedt and Y. Yafet, Solid State Comm. 15, 1855 (1974).

NUCLEAR ORIENTATION OF VERY DILUTE ^{54}Mn IN PALLADIUM-HYDROGEN ALLOYS

J. R. Thompson and J. O. Thomson
Department of Physics, University of Tennessee, Knoxville, Tennessee 37916

ABSTRACT

The behavior of very dilute ^{54}Mn in PdH_x has been studied by nuclear orientation methods. The alloy PdH_x is expected to become a pseudo-noble metal when the Fermi level is above the Pd "d" bands. Reported specific heat and susceptibility measurements suggest that this is the case in the β phase for $x \gtrsim 0.6$, where PdH_x is diamagnetic. In this work, $x \cong 0.66$. We measured the anisotropy $\langle W(T,H_a)\rangle$ of the 0.835 MeV γ ray as a function of temperature ($T \gtrsim 0.012$K) and applied magnetic field, $H_a \leq 18.8$kG. From measurements in large applied fields, we find that the hyperfine coupling, expressed as a hyperfine field is (-430 ± 20) kG. Within experimental error, this is the same as that in pure Pd^{54}Mn. This is similar to the ^{57}Fe Mössbauer results in Pd Fe where the Fe hyperfine field is unchanged by hydrogenation.

We find that it is much more difficult to orient Mn in PdH_x than in pure Pd. As in our previous study of the unhydrogenated alloy, we find that it is necessary to assume a distribution in direction for the local field on the Mn ions. To model this behavior, we let $\vec{H}_{local} = \vec{H}_a + \vec{H}_2$ where $\vec{H}_2$ is randomly oriented magnetic field, with a Lorentzian distribution in magnitude. The data are well described by the model with 700-900 Gauss for the half width of the distribution, a factor of ≈ 4 larger than that found in the unhydrogenated alloy.

INTRODUCTION

In this work we use Nuclear Orientation (N.O.) methods to investigate the magnetic behavior of dilute Mn in Palladium Hydride, PdH_x. Recently there has been substantial interest in hydrogen-loaded Pd which becomes a "pseudo-noble metal" and finally a superconductor when the hydrogen concentration x is quite large. As is well known, pure Pd has a large "d" density of states at the Fermi surface, resulting in an exchange enhanced susceptibility and a large electronic specific heat, linear in T. PdH_x, however, for $x \geq 0.6$ which corresponds to the fcc β phase is diamagnetic[1] and the electronic specific heat is substantially smaller.[2] Furthermore, the electronic mean free path is substantially reduced and the impurities Cr and Fe in $PdH_{0.7}$ show a resistance behavior characteristic of the classical Kondo effect.[3] In $Pd_{1-c}Mn_c$ where c = 4-10% the addition of $\simeq$70% H changes the magnetic character of the alloy from a spin glass to an antiferromagnet.[4]

The Mn impurity may be described by the Hamiltonian

$$H = g\mu_B\vec{S}\cdot\vec{H} + A\vec{I}\cdot\vec{S} - g_N\mu_N\vec{I}\cdot\vec{H}$$

where $\vec{S}$ and $\vec{I}$ are the electronic and nuclear spins, respectively and A is the hyperfine coupling constant. At sufficiently low temperature, the nuclear I_z substates are occupied unequally, which, upon radioactive decay of the ^{54}Mn nucleus, produces an anisotropic γ ray radiation pattern. The distribution of radiation is given by $W(\theta,T) = 1 + A_2U_2F_2Q_2P_2(\cos\theta) + A_4U_4F_4Q_4P_4(\cos\theta)$, where θ is the angle between the γ propagation vector and the local quantization axis, i.e., the local magnetic field. P_2 and P_4 are Legendre polynomials. The quantities A_2U_2 and A_4U_4 are factors characteristic of the nuclear decay, Q_2 and Q_4 depend upon the counting geometry and F_2 and F_4 are temperature dependent factors related to the probability of occupation of various I_z substates. For $T \to \infty$, W=1 for all angles θ, which is the isotropic case. As $T \to 0$, $W(\theta=0) \to 0$.

The experimentally measured quantity is the ratio of γ count rates at temperature T to that at infinite temperature (which in practice is ∿0.5K). This ratio, denoted by $\langle W\rangle$, is an ensemble average of $W(\theta)$ and is measured at zero angle with respect to the applied magnetic field. Since the Mn nucleus is oriented predominantly by the hyperfine interaction with the Mn ion, N.O. provides a quite local probe of the impurity in the Pd matrix.

EXPERIMENTAL

The experiments were performed on two separately prepared samples. Sample I contained approximately 9 ppm of Fe; Sample II, approximately 1 ppm of Fe. No Mn was apparent under x-ray fluorescence, indicating a concentration of less than ∿1 ppm; the ^{54}Mn content was far smaller than this. The samples were hydrogenated under pressure in a furnace which was slowly cooled from 500°C to 25°C. This normally provided an atomic ratio $x \equiv [H]/[Pd]$ of 0.65-0.7 as determined by the weight change.

The experiments were performed in a ^{3}He-^{4}He dilution refrigerator, wherein the primary thermometry was by Nuclear Orientation of ^{60}Co in a single crystal of hcp cobalt. The relative count rate $\langle W(T,H_a)\rangle$ was monitored using either a 3"x3" or a 5"x7" NaI detector located below the cryostat on the axis of symmetry of a superconducting magnet.

RESULTS AND DISCUSSION

In figure 1, we show, for sample I, the measured γ ray anisotropy $\langle W\rangle$ plotted versus inverse temperature T^{-1} for various applied fields H_a. The behavior of sample II is very similar qualitatively. Several features should be noted:

(1) For sufficiently large H_a, the anisotropy $\langle W\rangle$ is nearly independent of H_a;

(2) For lower applied fields, e.g., H_a=430G, $\langle W\rangle$ is much closer to unity than one might expect, in view of the fact that the magnetic moment of the Mn ion should be well saturated in this temperature range;

(3) In the lower fields, $\langle W\rangle$ at first decreases fairly rapidly, but for $T^{-1} \gtrsim 50K^{-1}$ it changes rather slowly. This is the temperature region where the P_4 term in W should become important.

These features can be explained via a model in which we introduce an internal effective magnetic field $\vec{H}_2$ which is random in magnitude and direction. We assume here a Lorentzian distribution for the probability $P(H_2)$ of observing an internal field of magnitude $H_2 > 0$. Thus $P(H_2) = (H_1/2\pi)[(H_2)^2+(H_1)^2]$ where H_1 is the halfwidth of the distribution. The Mn spin $\vec{S}$ then is subject to $(\vec{H}_a + \vec{H}_2)$ which produces a local quantization axis not parallel to the lab axis defined by $\vec{H}_a$. This leads to an angular ensemble averaging of W. We denote this angularly averaged quantity as $\langle W\rangle$. Since the P_4 term varies more rapidly with angle, it is more severely attentuated than the P_2 term which accounts for point (3) above. When, however, $|\vec{H}_a| >> |\vec{H}_2|$, the quantization axis is essentially parallel to $\vec{H}_a$, so the full anisotropy W is observed. This qualitatively accounts for the basic behavior of the data.

For large applied fields where $\vec{H}_a$ dominates the

internal field, the nucleus experiences the sum of $\vec{H}_a$ and $\vec{H}_{sat}$ where $|\vec{H}_{sat}| = A \cdot S/g_N\mu_N$. For both samples we find $H_{sat} = (-430 \pm 20)$kG where the sign of the hyperfine field is obtained by observing for very large H_a, $|\vec{H}_a + \vec{H}_{sat}|$ decreases as $\vec{H}_a$ increases. The negative sign is expected since Mn should have little or no positive orbital contribution to H_{sat}. This value of H_{sat} is the same within experimental error as that observed in a non-hydrogenated PdMn alloy[5]. This suggests that either H_{sat} has an insignificant contribution from local polarization of the conduction electron spins or, more likely, that the hydrogenation does not change significantly the magnitude of this contribution. This is similar to the Mössbauer Effect results for (PdFe)H with very low Fe concentrations in which it was found that the magnitude of the saturation hyperfine field for ^{57}Fe is unaffected by hydrogenation.[6]

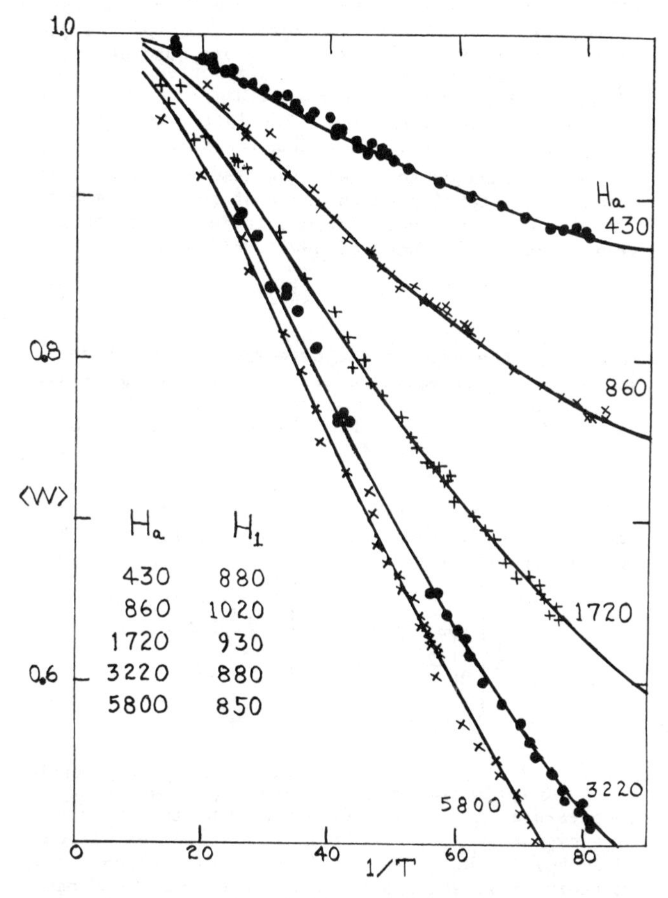

Figure 1. A plot of the ensemble averaged γ ray anisotropy <W> versus inverse temperature 1/T in K^{-1} for various applied fields H_a (in Gauss) as indicated. The points are experimental values. The solid lines are calculated assuming an angular distribution of internal fields, where the half-width of the distribution, H_1, is shown in Gauss for each case.

The addition of hydrogen, however, increases substantially the difficulty in orienting the Mn with an applied magnetic field. In our model, the magnitude of H_1, the characteristic magnitude of the internal field is substantially larger than in the pure Pd host. We have made numerical calculations of $\langle W(H_a,T)\rangle$ using the hyperfine coupling strength found above. We assume that the field H_2 is random in direction with a Lorentzian distribution in magnitude. The calculated anisotropies are shown as solid curves in figure 1, where the best value of H_1 is chosen for each curve. The values of H_1 are reasonably constant provided the applied field is not too small. For sample I (∿9 ppm of Fe) one has $H_1 = (900 \pm 10\%)$G; for sample II (∿1 ppm of Fe), $H_1 = (750 \pm 10\%)$G. The field in I may receive some contribution from the Fe content but this is not conclusive. It is clear, however, that these values are a factor of 3-4 larger than those in pure Pd.[5]

In very low applied fields, ($H_a \lesssim 215$G) the value of <W> is close to unity indicating that the Mn nuclei sense a nearly isotropic distribution of fields.

The origin of this internal disaligning field may be an anisotropic RKKY-like interaction. Then H_1 is a measure of the strength of interaction and should vary with impurity concentration although the detailed dependence may be complicated by mean free path effects. Localized spin fluctuations may also contribute significantly. The exictance of a distribution of internal parameters is corroborated in the specific heat measurements[7] on PdMn which are, of course, at much higher impurity concentrations. In the specific heat work, which provided the important experimental result that S≛2.3 for the Mn spin, the experimental behavior is interpreted in terms of a distribution of effective g values. The Nuclear Orientation measurements give evidence for a spatial distribution of spins in hydrogenated PdMn as well as in the nonhydrogenated alloy. One may speculate, therefore, that low impurity alloys may display a "spin glass" behavior at sufficiently low temperature.

To conclude, we find that although the saturation hyperfine field in PdMn is essentially unchanged by the addition of hydrogen, the disaligning mechanism is substantially stronger in the alloyed Pd.

ACKNOWLEDGMENTS

We wish to express our appreciation to L. Hulett for analysis of the samples.

REFERENCES

1. H. C. Jamieson and F. D. Manchester, J. Phys. F: Metal Phys. 2, 323-336 (1972).
2. C. A. Macliet and A. I. Schindler, Phys. Rev. 146, 463-467 (1966).
3. J. Mydosh, Phys. Rev. Lett. 33, 1562-1566 (1974).
4. J. P. Burger and D. S. McLachlan, Solid State Comm. 13, 1563-1566 (1973).
5. J. O. Thomson and J. R. Thompson, A.I.P. Conf. Proc. of 21st Magnetism and Magnetic Materials Conf. 29, 342-3 (1975).
6. J. S. Carlow and R. E. Meads, J. Phys. F: Metal Phys. 2, 982-994 (1972).
7. G. J. Nieuwenhuys, Adv. in Phys. 24, 515-591 (1975).

ANISOTROPY OF CONDUCTION ELECTRON g-FACTOR IN Au USING THE DE HAAS-VAN ALPHEN EFFECT*

G. W. Crabtree and L. R. Windmiller
Argonne National Laboratory, Argonne, Illinois 60439
and
J. B. Ketterson
Northwestern University, Evanston, Illinois 60201, and
Argonne National Laboratory, Argonne, Illinois 60439

ABSTRACT

Conduction electron g-factors in Au were measured by an absolute amplitude technique. The gain of the field modulation detection system was calibrated and absolute amplitudes were deduced from the size of the measured signal. g-factors were derived using measurements of the frequency, mass, and Dingle temperature and values of the curvature factor from both KKR phase shift and Fourier series parameterizations of the surface. The measured g factors show a large anisotropy, from a low of ∿ 1.2 on the neck at <111> to a high of ∿ 2.4 for belly and dogsbone orbits.

INTRODUCTION

Recently there has been increasing interest in the conduction electron g factor of metals (as opposed to semiconductors and semimetals where a considerable literature exists). Conduction electron spin resonance (CESR) has shown that g factors averaged over the Fermi surface differ from the free electron value (g = 2) in many metals,[1-4] but this technique cannot measure the anisotropy of the g factor for individual cyclotron orbits. Early work with the de Haas-van Alphen (dHvA) effect showed large anisotropies in extremal area g factors for certain transition metals.[5,6] This work relied on the occurrence of spin splitting zeros and therefore gave results for only a few isolated orbits. Later, the harmonic content of dHvA oscillations was exploited to measure g factors for orbits where magnetic interaction is not too severe.[7-9]

In an earlier paper,[9] we discussed an absolute amplitude technique which is applicable to all extremal orbits for which the dHvA effect can be observed, and compared results for the neck orbits in Au with harmonic content results. The application of this technique to the remaining orbits in Au is the subject of this paper. Finally, the very recent observation of "Larmor waves"[10] where phase coherence is maintained by the Larmor precession of electron spins drifting along the field direction may allow measurements of g factors for cyclotron orbits whose drift velocity (rather than cross sectional area) is extremal.

EXPERIMENTAL

Our experiments were performed on Au single crystals 1 mm diameter by 1 mm long with a resistance ratio of ∿ 2000. Oscillations were observed in fields up to 7.2 T and temperatures between 1-2 K by conventional field modulation techniques.[11,12] Fourier transforms were available through the use of an on-line PDP-11 mini-computer system for digitizing and processing data.[13,14] All data were taken in the (110) plane.

A modulation frequency of 18 Hz was used to insure complete penetration of the sample by the modulation field. Complete penetration was verified by three techniques: First, plots of dHvA signal amplitude versus modulation frequency at constant field and temperature were constructed. At ∿400 Hz the plots deviate from linearity indicating that part of the sample is shielded from the full modulation field at these frequencies. Further tests for distributions in the amplitude and phase of the modulation field over the sample caused by eddy current shielding were made by observing the Bessel function zeros and the quadrature signals respectively. An amplitude distribution produces an imperfect Bessel function zero (i.e., there is no value of applied modulation amplitude which forces $J_n(\delta\phi)$ in Eq. 1 below to be zero over the whole sample simultaneously) while a phase distribution produces a finite signal in the quadrature channel (i.e., there is no choice of quadrature channel phase which produces zero signal for the whole sample simultaneously). No observable quadrature signal or degration of the Bessel function zero was observed below ∿100 Hz for the <110> or <111> directions where tests were made. The latter two tests appear to be more sensitive than the first.

Our technique for measuring absolute amplitudes depends on establishing the gain of our field modulation detection system. The magnitude of the observed signal for the fundamental dHvA oscillation is[9,11]

$$|V| = G\, \vec{m}\cdot\hat{u}\; J_n(\delta\phi) A_1 \tag{1}$$

where G is the gain of the detection system, $\vec{m}$ is a vector in the direction of the oscillating magnetization, $\hat{u}$ is the direction of the pick up coil, J_n is the nth order Bessel function corresponding to the detection harmonic (the second in this experiment) $\delta\phi = \frac{2\pi F}{H^2}h$ where F is defined below and h is the amplitude of the modulation field, and A_1 is the absolute amplitude of the fundamental dHvA oscillation. This may be written as[9,15]

$$A_1 = \left\{ \frac{\lambda T F}{H^{1/2}} \frac{\exp(-\frac{\alpha m X}{H})}{\sinh\frac{\alpha m T}{H}} \right\} \frac{\cos(\pi\frac{gm}{2})}{\left(\frac{\partial^2 A}{\partial k_H^2}\right)^{1/2}} \tag{2}$$

where F is the frequency of the oscillation in field, m is the cyclotron effective mass, X is the Dingle temperature, A is the area encolsed by the extremal area cyclotron orbit defined by the field direction, $(\partial^2 A/\partial k_H^2)$ is the curvature factor (which depends only on geormetry), and α and λ are products of fundamental constants. All of the quantities in curly brackets above can be measured by conventional means: F from the oscillation frequency, m from the temperature dependence of the amplitude, and X from the field dependence of the amplitude. The curvature factor must be generated from a model Fermi surface. Two such models based on Fourier series[16] and KKR phase shift[17] parameterizations, respectively, were used in this work.

The calibration technique was fully described earlier[9] but will be summarized here. If at some orientation the g factor is known, then F, m, and X can be measured conventionally, the curvature factor can be generated from a model, and the value of A_1 for any H and T can be computed from Eq. (2). In Eq. (1), $J_n(\delta\phi)$ and $\vec{m}\cdot\hat{u}$ can be determined from experimental parameters and the geometry of the surface ($\delta\phi$ depends on F, $\vec{m}\cdot\hat{u}$ on $1/F(\partial F/\partial\theta)$) so that G can be found from the ratio of the observed signal to the computed value of $\vec{m}\cdot\hat{u}\; J_n(\delta\phi)A_1$. This procedure is independent of magnetic interaction so long as H and T are adjusted to keep A_1 small enough to prevent second order corrections to A_1 from being significant.[9] This condition is most severe for belly orbits but can easily be met with sufficient care.

On the neck orbits in Au, there are five angles where g is reliably known: two spin splitting zeros of A_1, two zeros of A_2 (the amplitude of the 2nd dHvA harmonic), and the region near <111> where extensive harmonic ratio measurements have been done. Using the zeros of A_2 (at the zeros of A_1, the signal from A_2 is too weak for reliable amplitude measurement) and the region near <111>, we have determined values for G which agree to within 5%.

In locating the zeros of A_2 on the neck, the phase information contained in the Fourier transform was very useful. Magnetic interaction produces a finite second harmonic signal even at a zero of A_2; however, the relative phase of the observed second harmonic and the fundamental goes through 180° when A_2 goes through zero.[9,14] This feature allows the zero of A_2 to be easily located within 0.25°. For measuring amplitudes, Fourier transforms were not required except near <100> where the rosette and belly orbits were simultaneously present. At other orientations, signal amplitudes were measured directly from chart recordings of the oscillations. In these cases, modulation levels were set to zero the second harmonic; Fourier transforms showed that higher harmonics were weak enough to be ignored. When using Fourier transforms to measure amplitudes, the effect of variation of the signal amplitude across the data window vanishes to first order[14] and was therefore ignored.

RESULTS

With the gain of the system determined, the same procedure may be used to solve for $\cos(\pi gm/2)/(\partial^2A/\partial k_H^2)^{1/2}$ at an arbitrary orientation. Figure 1 shows the results for the rosette, dogsbone, and belly orbits and for neck orbits near <111>. The error bars reflect statistical errors but do not include any possible systematic errors. Near <111>,

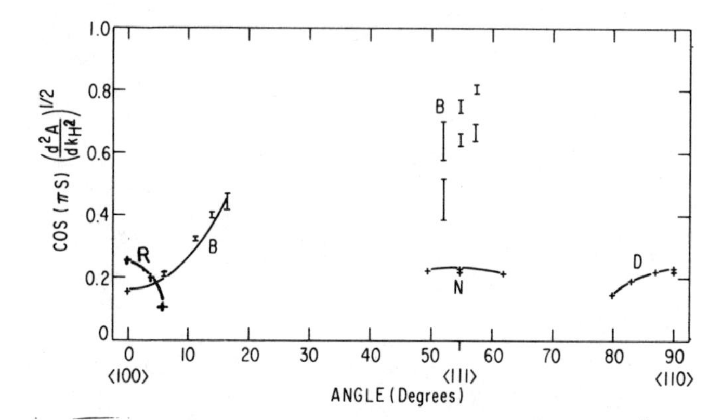

Fig. 1 Experimental values of $\cos(\pi S)/(\partial^2A/\partial k_H^2)^{1/2}$ derived from absolute amplitude measurements for rosette (R), belly (B), dogsbone (DB) and neck (N) orbits. $S \equiv gm/2$.

the belly signal is strongly modulated by the neck through magnetic interaction. The resulting peaks and valleys in the belly amplitude were analyzed separately and both are shown.

Figure 2 shows curvature factors as generated by Fourier series[16] and KKR phase shift[17] schemes. Surprisingly, there is significant disagreement for many angles, even though both schemes fit the symmetry point and turning point areas to better than 0.1%. The most serious disagreements occur far from symmetry

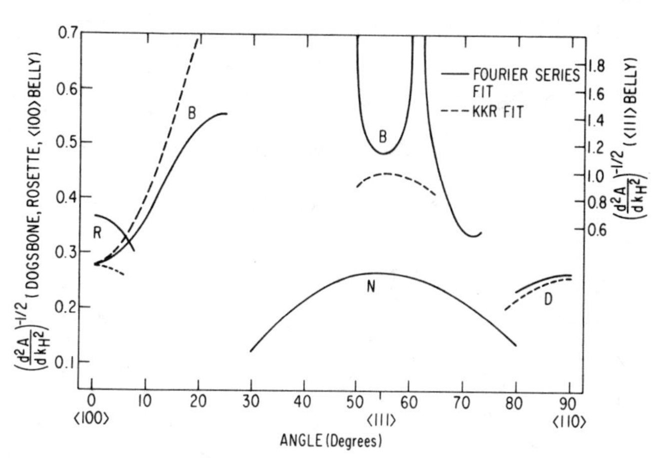

Fig. 2 Curvature factors generated from Fourier series[16] (solid line) and KKR phase shift[17] (dotted line) parameterizations of the Au Fermi surface for rosette (R), belly (B), dogsbone (DB), and neck (N) orbits. For neck orbits, the two schemes give the same results. The right hand scale refers only to the belly orbits near <111>.

points where we do not report experimental data. For belly orbits, we argue as follows that the KKR curvature factors are more reliable. At angles where central and non-central orbits merge $\partial^2A/\partial k_H^2$ must go through zero. Joseph, et al.,[5] observe a non-central belly orbit merging with the central orbit at 23° from <100>. They do not observe any non-central orbits in the region near <111>. The KKR results shown in Fig. 2 are consistent with these observations: an apparent singularity near 23° (corresponding to a zero of $\partial^2A/\partial k_H^2$) and finite behavior near <111>. The Fourier series results do not show this behavior.

In spite of the differences shown in Fig. 2, the data of Fig. 1 may be inverted to give a fairly accurate picture of the g-factor anisotropy. On the neck and dogsbone orbits, the difference in g-factor resulting from the use of the KKR or Fourier series curvature factor is negligibly small; on the belly the difference is less than 10%. The rosette orbit shows the greatest sensitivity. The 25% difference between KKR and Fourier series curvature factors at <100> is the largest symmetry point difference for any of the orbits in Au. As a result, we do not report a g-factor for the rosette at this time.

Figure 3 shows the g-factors resulting from the KKR curvature factors and the data of Fig. 1. The dotted and solid lines shown for the belly and dogsbone orbits result from ambiguities in inverting $\cos(\pi gm/2)$. In principle, the experiment cannot choose between these solutions; however, all published results from CESR show a positive g-shift for Au,[2,4] indicating the solid lines are probably correct. Also shown are earlier harmonic ratio results for neck orbits and a spin splitting zero result for the belly at 71.6° from <100>.[9] (Alternate choices of g for neck orbits are greater than 13 and are not shown.) The belly g-factors near <111> result from a simple average of the peak and valley amplitudes of the belly signal as modulated by the neck orbit. This produces a large uncertainty (perhaps as much as 10%) in the g-factor. Since the g-factor is rather flat in this region, the errors may mask a broad extremum.

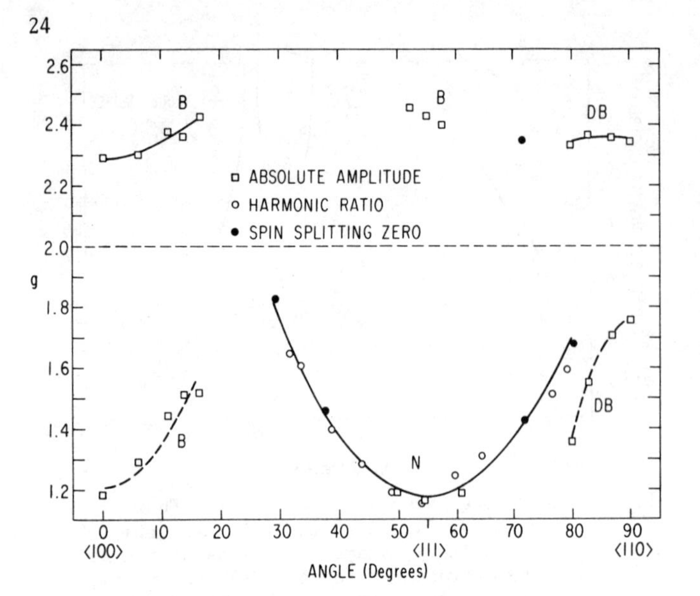

Fig. 3 Derived g-factors for belly (B), dogsbone (DB), and neck (N) orbits. The solid and dotted lines are smooth curves drawn through the data.

CONCLUSION

The large anisotropy of the g factor in Au is thought to be due to spin orbit effects rather than electron-electron interactions.[18] Therefore the anisotropy can be calculated using a relativistic band structure theory.[19] Such a program is currently under way at Argonne.

The present experiments in gold demonstrate the effectiveness of this absolute amplitude method for measuring g-factors. We hope the comparison of this data with calculations will lead to more detailed understanding of the paramagnetic response of Au. The method could be applied to other transition metals where electron-electron interactions are important.

REFERENCES

*Work performed under the auspices of the U. S. Energy Research and Development Administration.

1. S. Schultz, M. Shanabarger, and P. M. Platzman, Phys. Rev. Letters 19, 749 (1967).
2. R. Dupree, C. T. Forwood, and M. J. A. Smith, phys. stat. sol. 24, 525 (1967).
3. R. Monot, A. Chatelain, and J. P. Borel, Phys. Letters 34, 57 (1971).
4. P. Monod and A. Janossy, prviate communication and to be published.
5. A. S. Joseph, A. C. Thorsen, and F. A. Blum, Phys. Rev. 140A, 2046 (1965).
6. L. R. Windmiller and J. B. Ketterson, Phys. Rev. Letters 21, 1076 (1968).
7. D. L. Randles, Proc. Roy. Soc. A331, 85 (1972).
8. H. Alles, R. J. Higgins, and D. H. Lowndes, Phys. Rev. 12, 1304 (1975).
9. G. W. Crabtree, L. R. Windmiller, and J. B. Ketterson, J. Low Temp. Phys. 20, 655 (1975).
10. A. Janossy and P. Monod, private communication and to be published.
11. R. W. Stark and L. R. Windmiller, Cryogenics 8, 272 (1968).
12. L. R. Windmiller and J. B. Ketterson, Rev. Sci. Instr. 39, 1672 (1968).
13. L. R. Windmiller, J. B. Ketterson, and J. C. Shaw, Report No. ANL-7907 (National Technical Information Service, U. S. Dept. of Commerce, Washington, D. C., 1972).
14. G. W. Crabtree, Ph.D. Thesis, University of Illinois at Chicago Circle, 1974 (unpublished).
15. I. M. Lifshitz and A. M. Kosevich, Zh. Eksp. Teor. Fiz. 29, 730 (1955) [Soviet Phys.-JETP 2, 636 (1956)].
16. B. Bosacchi, J. B. Ketterson, and L. R. Windmiller, Phys. Rev. B4, 1197 (1971).
17. Jerry C. Shaw, J. B. Ketterson, and L. R. Windmiller, Phys. Rev. B5, 3894 (1972).
18. J. W. Garland, private communication.
19. D. D. Koelling, unpublished.

MAGNETIC AND ELECTRICAL PROPERTIES OF INTERNALLY OXIDIZED CuFe ALLOYS*+

F. R. Fickett
National Bureau of Standards, Boulder, CO. 80302

ABSTRACT

A series of CuFe alloys containing 1-100 at. ppm Fe has been internally oxidized. In nearly all cases the specimens become ferromagnetic. We have measured the remanent moment and low field susceptibility at room temperature, and the electrical resistivity at 273 K and 4 K. These data combined with Curie temperature measurements and scanning electron microscopy indicate the presence of very small single domain grains of copper ferrite within the copper matrix. The annealing process appears to be capable of sweeping most of the iron impurities from the matrix in spite of their very low concentrations.

INTRODUCTION

It has been known for some time that, by annealing very pure copper under the proper conditions of temperature and pressure, one can produce very large decreases in the, already low, impurity resistivity.[1] This effect was thought to result from oxygen diffusing into the copper and tying up the natural iron impurities in some manner, probably as oxide molecules. The relatively large, neutral molecules would give only very little scattering of the conduction electrons. By this technique, polycrystalline copper can be produced with a residual resistance ratio (RRR ≡ R at room temperature/R at liquid helium temperature) of 20,000 and we have made single crystals with ratios as high as 64,000. The low resistivity at cryogenic temperatures is desirable for copper used to stabilize superconducting materials. These superconducting composites have potential applications in a large number of devices ranging from small medical magnets to power transmission cables and controlled thermonuclear reactors.[2] In this paper we present the results of a study to determine the actual mechanism of internal oxidation of copper and to investigate the properties of copper which has been thus treated.

SAMPLE PREPARATION

A series of thirteen alloys with 1-100 at. ppm Fe was prepared by successive dilutions of a master alloy with pure (∿0.6 ppm Fe, RRR = 1900) copper. The master contained 690 ppm Fe, determined by atomic absorption measurements. It was prepared by melting copper rod and iron powder in graphite under a mixture of argon and hydrogen (5%). The master boule was cut into 6 mm diameter rods which were then swaged to 3 mm. After a stress relief anneal in vacuum, the 4 K electrical resistivity of segments of the rods was measured to evaluate homogeneity. The iron distribution was found to be homogeneous to ∿1%. Several of the alloys prepared from the master were analyzed chemically and gave results consistent with those calculated. Furthermore, the linear behavior observed for the 4 K electrical resistivity and magnetic susceptibility with calculated iron content indicates a proper relative dilution. The homogeneity of each alloy was determined to be ∿1 % as described above.

Complete internal oxidation of the 3 mm alloy specimens was accomplished by a 4 day anneal at 1000° C at a pressure of 7×10^{-7} atm maintained by an air leak into the vacuum furnace. The time was chosen to be far in excess of that required for complete oxidation of this size specimen predicted by earlier experiments[1]--about 4 hours. At 1000° C Fe oxidizes to FeO at $\sim 10^{-15}$ atm of oxygen and FeO oxidizes to Fe_3O_4 at $\sim 10^{-13}$ atm in the free state[3]. Further oxidation of the iron to Fe_2O_3 does not occur in copper because the formation of Cu_2O commences at a lower pressure.

ELECTRICAL RESISTIVITY

The electrical resistivity at 4 K is a sensitive indicator of total impurity content. Our measurements were made by a conventional four-probe dc technique. Pressure contacts to fine copper wires were used for the voltage leads and to copper foil for the current leads. This was necessary because even minute amounts of solder would have made our later magnetic susceptibility measurements meaningless. A lack of current dependence of the specimen resistance showed that no appreciable heating occurred at the pressure contact. The results of the resistivity measurements are shown in Fig. 1. The unoxidized alloys show a linear behavior below 90 ppm with a very high slope which, we feel, is typical of single impurity scattering. The slope decreases as iron content increases and is lower at higher temperatures (76 K, 273 K). The only truly surprising thing is that the deviation from linearity occurs at such a low iron content. The dramatic effect of internal oxidation is obvious here. In every case the resistivity is reduced to an insignificant value on this scale, 0.1 nΩ-cm at $n_2 = 0$. The data still show a slight dependence on iron concentration, 4.8×10^{-3} nΩ-cm/ppm, probably a measure of oxide particle scattering.

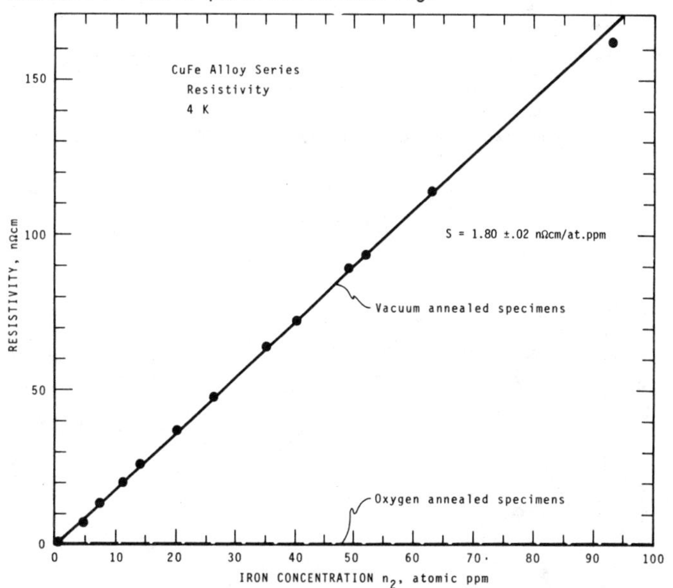

Fig. 1. Electrical resistivity of CuFe alloys at 4 K.

MAGNETIC MEASUREMENTS

Our magnetic measurements were made with a very sensitive apparatus consisting of a superconducting flux transformer and an rf biased SQUID (superconducting quantum interference device) detector.[4] It is a moment measuring device with a sensitivity of 2×10^{-8} emu. A reentrant dewar tube allows measurements to be made with the sample at any temperature and a small solenoid provides the external field for susceptibility measurements.

When the alloy specimens are removed from the furnace they show a fairly large remanence ∿0.4 emu/g for $n_2 = 63$ ppm. Unoxidized specimens have no remanence This is apparently a chemical remanent magnetization (CRM) phenomenon.[5] In spite of this remanence a linear moment-field curve is observed. The susceptibility and its variation with iron content is shown in Fig. 2. The pure copper has a susceptibility of 8.3×10^{-8} emu/g at room temperature. Following the susceptibility measurement, the specimens were magnetized at high field, removed from the field and their magnetization measured. The results are shown in Fig. 3 in which the left hand axis plots the moment per gram of iron and the right hand axis the moment per gram of specimen. The saturated remanence of Fe_3O_4 is ~128 emu/g of Fe.

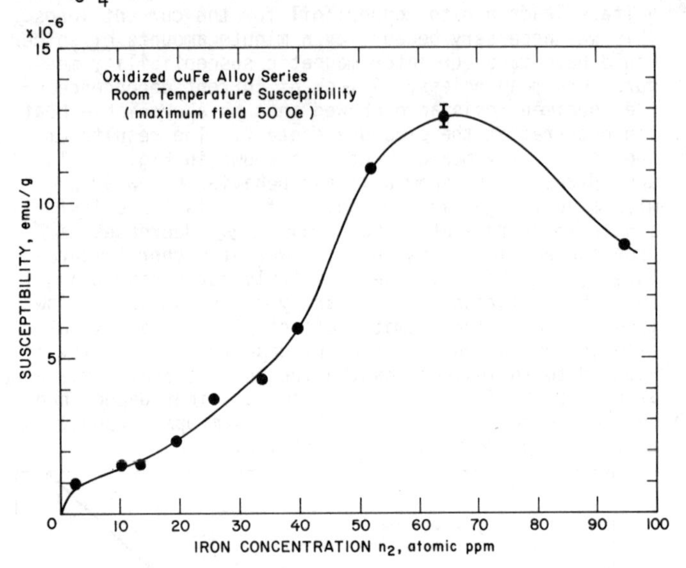

Fig. 2. Susceptibility of internally oxidized CuFe alloys.

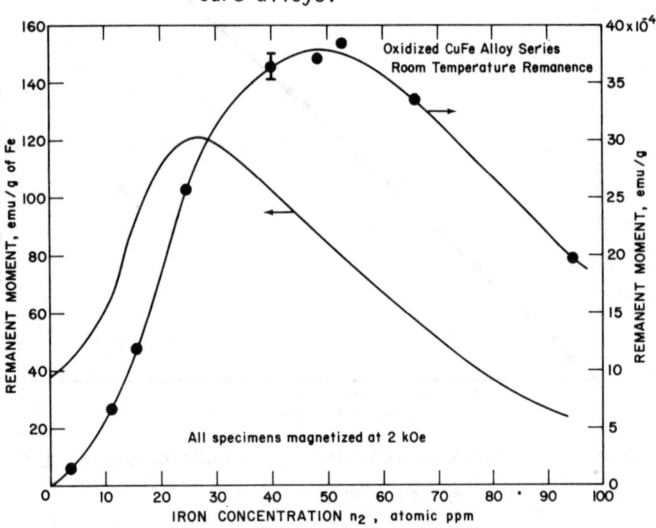

Fig. 3. Remanent moment of internally oxidized CuFe alloys.

OTHER MEASUREMENTS[6]

The Curie temperature of two specimens with 14 and 93 ppm Fe was measured in a controlled atmosphere microbalance apparatus to be 530 and 450°C respectively. A brief high temperature deoxidation of the 93 ppm specimen with hydrogen increased its Curie temperature to that of Fe_3O_4, 580°C, and reoxidation returned it to its previous value. This lower temperature is consistent with that observed for copper ferrite.[7] An attempt was made to look for the ferrite particles in the high iron content specimen by SEM (scanning electron microscopy). Large (~ 5 μm) Cu_2O crystals were seen, as were some smaller (~0.1μm) inclusions, probably octahedral in shape, which we have tentatively identified as the ferrite.

CONCLUSIONS

Internal oxidation effectively purifies copper for low temperature electrical applications. The internal oxidation occurs in two steps. In the first, FeO is formed throughout the specimen. Nucleation of Fe_3O_4 crystals then occurs and the FeO (and/or unoxidized Fe) migrates to the site from a large radius surrounding the nucleus. The magnetic properties indicate that this crystal growth occurs for all iron concentrations from 4-100 ppm Fe. The particles are apparently of single domain size and the stability of the remanence argues against any superparamagnetism in these alloys. Above about 50 ppm Fe a decreasing remanance and a decreasing Curie temperature suggest that the crystals are forming as copper ferrite. Note that this production of "ferromagnetic copper" can occur under conditions which are frequently described in the literature as "annealing in a good vacuum."

REFERENCES

+Work supported by the International Copper Research Association (INCRA)

1. F. R. Fickett, Materials Science and Engineering 14, 199 (1974).
2. S. Foner and B. B. Schwartz, eds., Superconducting Machines and Devices--Large Systems Applications (Plenum Press, New York, 1974).
3. F. D. Richardson and J.H.E. Jeffes, J. Iron and Steel Inst. 160, 261 (1948).
4. F. R. Fickett and D. B. Sullivan, J. Phys. F: Metal Physics 4, 900 (1974).
5. F. D. Stacey and S. K. Banerjee, The Physical Principles of Rock Magnetism (Elsevier Scientific Publishing Co., New York, 1974) Ch. 9.
6. I am most grateful for the assistance of E. Larson of the University of Colorado (CIRES) and D. Watson of the U.S. Geological Survey in making the measurements described here.
7. T. Nagata, Rock Magnetism (Maruzen Company Ltd., Tokyo, 1961) p. 116.

FERMI SURFACE OF FERROMAGNETIC NICKEL*

R. Prasad and S. K. Joshi
University of Roorkee, Roorkee, India

S. Auluck
Punjab Agricultural University, Ludhiana
141004 (Pb.) India

The fermi surface of ferromagnetic nickel is reexamined in the light of the recent data of Stark pertaining to the large Γ centered sp↓ and sp↑ fermi surface sheets using a modified interpolation scheme. The original model of Hodges et al. is unable to give a satisfactory geometrical representation for these sheets although the fit to the X-centered d↓ pocket and the neck is excellent. In order to fit the larger sheets we require a shift in E_F of about 0.020 Ryd. Zornberg has also calculated the fermi surface using Mueller's interpolation scheme and here we require a shift in E_F of 0.007 Ryd to fit the large sheets. For these the experimental uncertainty is less than 1% which is 0.001 Ryd. In view of this, we have adjusted the interpolation scheme parameters to fit the large as well as the small fermi surface sheets. The goodness of the fit is demonstrated by the fact that the shift in E_F of 0.001 Ryd is sufficient to fit the data. We note, in passing, that the corresponding value for Wang and Callaway's work is 0.018 Ryd. We conclude that our fit to the data is excellent and within the experimental uncertainty. We await more data to judge the validity of our model.

*Work supported by the CSIR (India) under the Silver Jubilee Scheme.

TUNNELING BETWEEN FERROMAGNETS

J. C. Slonczewski
IBM T. J. Watson Research Center
Yorktown Heights, NY 10598

ABSTRACT

Recently, M. Julliere [Physics Letters 54A, 225 (1975)] has reported that the low-temperature tunneling current between electrodes of Fe and Co separated by Ge depends on whether the moments are parallel or antiparallel. We apply the transfer-matrix, WKB, and more general models to the problem of tunneling between ferromagnetic electrodes separated by an insulating barrier. We find in certain limits a valve effect described by the effective conductance per unit area $g_0(1 + P_1P_2\cos\theta)$, where P_1 and P_2 are fractional spin polarization coefficients for the electrodes and θ is the angle between their moments. Electron tunneling transmits an exchange torque per unit area between the ferromagnets which is given by $P_1P_2g_0 F \hbar f/2\pi e^2$, where F is the Fermi energy, e is the electron charge and f is a dimensionless constant of order unity which depends on band structures. For a uniform barrier this torque is ordinarily small. However, if tunneling is concentrated in very small regions of the barrier as is sometimes suggested, the torque twists the micro-magnetic structure diminishing the strength of the valve action. In some models of tunneling the polarization is a bulk property; however, in the WKB model it is attributed only to the degree in which the exchange field penetrates into the classically forbidden region of the barrier.

MÖSSBAUER AND NMR STUDIES OF SITE SUBSTITUTION AND MAGNETIC STRUCTURE OF $Fe_{3-x}Mn_xSi$ ALLOYS+

K. Raj*
Yale University, New Haven and The Univ. of Conn., Storrs, Conn. 06268

V. Niculescu, T. J. Burch, and J. I. Budnick
The University of Connecticut, Storrs, Conn. 06268

R. B. Frankel
Francis Bitter National Magnet Lab., *MIT, Cambridge, Mass. 02139

ABSTRACT

The NMR measurements in $Fe_{3-x}Mn_xSi$ alloys show increasingly complex spectra for $x > 0.75$ due to overlapping signals from Fe, Si and Mn nuclei in various sites. Mössbauer experiments were undertaken to identify the Fe^{57} resonances. The combined NMR and Mössbauer results have been used to elucidate the site substitution and magnetic structure of these alloys for $(0 < x < 1.75)$.

INTRODUCTION

The Do_3-type crystal structure of Fe_3Si (Fig. 1) has A,B,C and D sites; the A and C sites are equivalent and, together with the B site, are occupied by Fe, whereas the D site is occupied by Si. The near neighbor configurations distinguishing these sites are shown in Table I. It has been observed that transition metal atoms enter the Fe_3Si lattice selectively[1,2,3] and in particular, Mn enters the B sites. Neutron diffraction

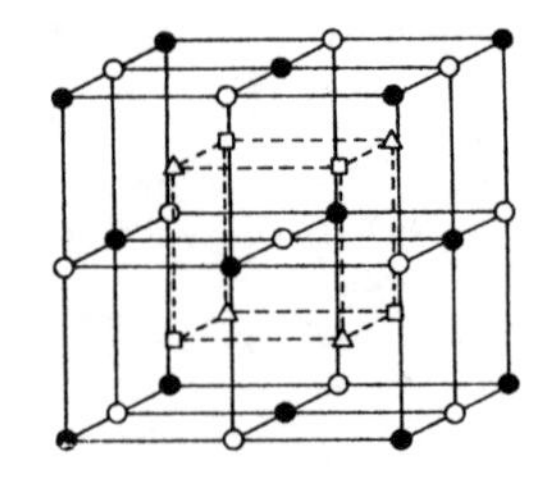

●A □B ○C △D

Fig. 1. Crystal structure of Fe_3Si. Site occupation A,C:Fe, B:Fe, D:Si

TABLE I

Fe_3Si Neighbor Configurations

Site	1nn	2nn	3nn
A,C	4B,4D	6A,C	12A,C
B	8A,C	6D	12B
D	8A,C	6B	12D

studies made by Booth et al.[4] and Babanova et al.[5] on $Fe_{3-x}Mn_xSi$ alloys show that Mn also enters the A,C sites for $x > 0.75$. For $x < 0.75$, the alloys are ferromagnetic; but for $0.75 < x < 1.75$ the alloys show ferromagnetic behavior above about 72 K and complex spin structure below this temperature. The structural and magnetic properties of some $Fe_{3-x}Mn_xSi$ alloys are summarized in Table II.

An NMR investigation of $Fe_{3-x}Mn_xSi$ alloys in the composition range $0 < x < 1.6$ was carried out by Niculescu et al.[6] and their results were found to be consistent with the previous description of site moment[4] and site occupation.[4,5] The spectra observed in the low frequency region were found to become increasingly complex for $x > 0.75$ due to overlapping intensities from Fe(B), Fe(A,C), Si and possibly Mn(A,C) resonances. Thus the analysis of moment behavior and site occupation could not be carried out precisely at higher x. Mössbauer experiments were undertaken so that the Fe sites could be identified unambiguously and together with NMR results, an understanding of the correlations between the internal fields and site magnetic moments could be extended.

TABLE II

A summary of structural and magnetic properties of some $Fe_{3-x}Mn_xSi$ alloys.

Alloy	Crystal Structure	Magnetic Structure	Ordering Temp. (K)	Magnetic Moment (μ_B)
Fe_3Si (9)	Do_3	Ferro-	805	Fe(B):2.20 Fe(A,C):1.35
Fe_2MnSi (4)	$L2_1$	Ferro-complex	205 73	Mn,Fe(B):1.8 Mn,Fe(A,C):0.4
Mn_3Si (10,11)	Do_3	Triangular Helical	30 27	Mn(B):1.72 Mn(A,C):0.19

EXPERIMENTAL RESULTS

The Mössbauer data were acquired on several $Fe_{3-x}Mn_xSi$ alloys up to $x = 1.75$ at room temperature, 77 K and 4.2 K. In Fig. 2a Mössbauer spectra are shown corresponding to $x = 0.15$, 0.50 and 0.75 at 77 K. At these compositions the alloys are ferromagnetic at all temperatures studied. The presence of various Fe sites with different populations is clearly seen. The spectra at 4.2 K were virtually identical except for the $x = 0.75$ sample which showed broadened lines. This may be an indication of a spin transition occurring in the sample below 77 K.

Fig. 2b shows Mössbauer spectra for alloys with $x = 1.0$, 1.5 and 1.75 in the complex spin structure region at 4.2 K. In general, decreasing splittings are observed in the spectra with higher Mn concentrations. At 77 K smaller splittings, and at room temperature single lines, are observed. It should be noted that the spectra for all compositions are not strictly symmetric implying that either different Fe sites have different isomer shifts or that small quadrupolar effects exist in these alloys.

Mössbauer spectra obtained in external magnetic fields up to 60 kG on these alloys allowed the identification of low field sites and indicate that the signs of the Fe hyperfine fields at all the sites are negative. The values of Fe internal fields obtained from the spectra along with the site identification are summarized in Table III. The identification of sites is made by extending the systematics of Fe and Mn internal fields obtained previously.[6] It should be noted that, in the composition range $(x < 0.75)$ where NMR Fe lines can be identified clearly, the agreement between the Mössbauer and NMR internal field[6] values is good.

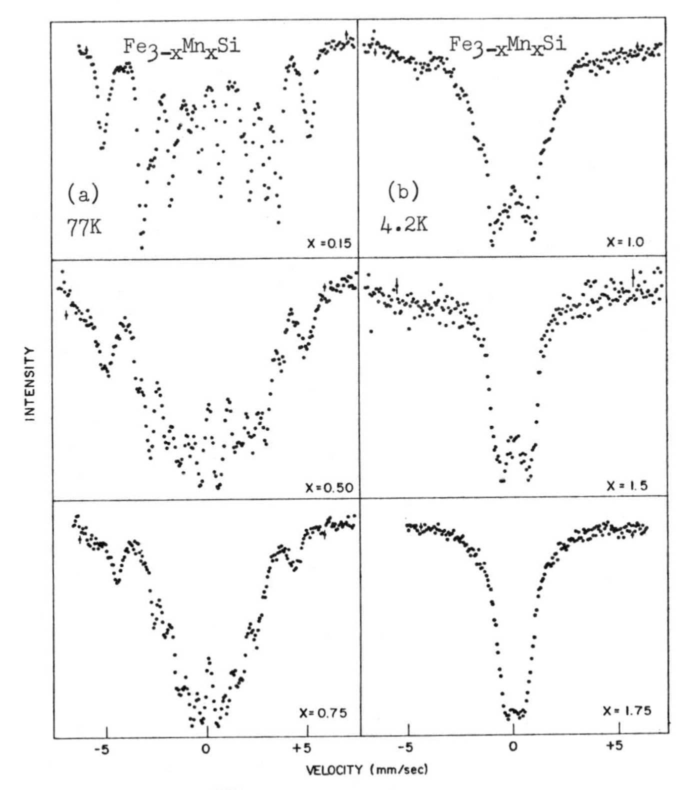

Fig. 2. Fe^{57} Mossbauer spectra in $Fe_{3-x}Mn_xSi$ alloys (a) 77K (b) 4.2K

TABLE III

A summary of Fe internal fields determined from Mössbauer spectroscopy at various sites in $Fe_{3-x}Mn_xSi$ alloys as a function of composition (x).

Alloy composition (x)	Site	Internal Field (kG) 4.2K	Internal Field (kG) 77K
0.15	Fe(B) Fe(A,C)	324 213,180	320 210
0.50	Fe(B) Fe(A,C)	300,280 206,165,125	295,275 200,160,120
0.75	Fe(B) Fe(A,C)	282 161,117	275 158,120,90
1.00	Fe(B) Fe(A,C)	255(weak) 102,63	220 70,35
1.50	Fe(A,C)	53	25
1.75	Fe(A,C)	27	single line

Since the inherent resolution of the NMR data is better, we have used NMR Fe internal fields where available. The identification of the Fe internal fields in the higher Mn concentration alloys permits the separation of the Mn resonances in the NMR spectra.[6] The maximum in the NMR field distribution is associated with the Mn(A,C) resonance (see Fig. 3). This maximum becomes increasingly intense with increasing x, entirely consistent with the increase in the fractional occupation of Mn in the (A,C) sites for $x > 0.75$. The extremely low intensity Si resonance is also believed to be present in the NMR distribution shown in Fig. 3.

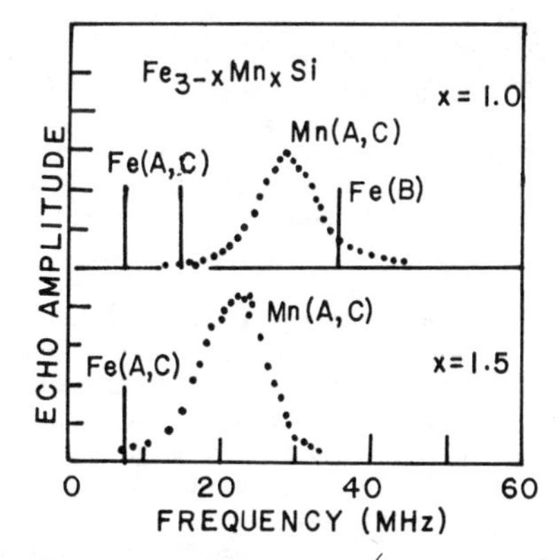

Fig. 3. Low frequency NMR spectra[6] at two compositions, x = 1.0 and 1.5, in $Fe_{3-x}Mn_xSi$ alloys. The Fe resonances revealed by Mössbauer measurements are shown by vertical lines and do not represent the actual intensities. The assignment of these resonances to various sites is discussed in the text.

DISCUSSION

The internal fields measured in the $Fe_{3-x}Mn_xSi$ alloys can be related to the spin structure and magnetic moments at various sites. Both (A,C) and B sites will be discussed separately and the NMR data will be used to supplement the discussion.

The (A,C) Site: Due to the selective substitution of the Mn atoms, the Fe(A,C) atoms will have Fe in their 1nn shell replaced by Mn. Neutron diffraction[4] has shown a linear decrease in the average Fe(A,C) moment as a function of concentration from $1.35\mu_B$ at x = 0 to $0.4(1)\mu_B$ at x = 0.75. The most probable 1nn configuration for Fe(A,C) at x = 0.75 is 3 Mn, 1 Fe and 4 Si. If this shift is due principally to 1nn, then 3 Mn neighbors decrease the Fe(A,C) moment by $1.35 - 0.4 = 0.95(1)\mu_B$, or the decrease per Mn atom is $0.32\mu_B$.

If we now assume that the resolved lines in the Fe(A,C) spectrum are from (A,C) sites with different numbers of Mn 1nn, (contributions to the internal fields from Mn substituted in more distant shells, 4nn and higher, are thought to be small) the data of Table III and also NMR results show that the average shift due to 1 Mn 1nn is about 40 kG. Thus we find that the field shift of 40 kG corresponds to a magnetic moment change of $0.32\mu_B$ on the Fe(A,C) atoms. These changes are consistent with the value of 125 kG/μ_B predicted by Watson and Freeman[7] for Fe.

For $x > 0.75$, the Mn atoms begin to enter the 2nn and 3nn shells and the effect of this substitution on the Fe(A,C) moment is expected to be smaller. Table III indicates that the average Fe(A,C) internal field decreases smoothly between x = 1.0 and 1.75. The same trend is observed for the Mn(A,C) internal field as measured by NMR.[6] The Mn(A,C) field decreases from 28 to 21 kG between x = 1.0 and 1.6. A more definite relationship between the internal fields and magnetic moments in this concentration range requires the knowledge, not yet available, of the atomic moments and the complex spin structure.

The B Site: Some conclusions can also be drawn regarding the magnetic moment behavior of the Fe(B) site. For $x > 1$, the Fe(B) field distribution could not be detected because of the very small number of Fe(B) atoms in the alloys. In the low composition range the satellites observed for both Fe(B) and Mn(B) are predominantly associated with the perturbations in the neighbor[8] shells, unlike the Fe(A,C) structure which arises from the changes in the Fe(A,C) moment. This conclusion is consistent with the neutron diffraction data which indicates a constant moment of $2.3(3)\mu_B$ for both Fe(B) and Mn(B) between $x = 0$ and 0.75. In this composition range both Fe(B) and Mn(B) moments are deduced to be parallel. For $x > 0.75$, the magnetic moment on the remaining fewer Fe(B) atoms would be sensitive to the occupation of Mn(A,C) in the 1nn shell to Fe(B). Between $x = 0.75$ and $x = 1.0$ the Fe(B) internal field shift is too large to be explained by conduction electron polarization alone. Therefore a small decrease in the Fe(B) moment is probably produced by replacing its 1nn Fe(A,C) by Mn. The Mn(B) internal field decreases from 230 kG at $x = 0.75$ to 203 kG at $x = 1.6$. If this entire change, 27 kG, is attributed to a change in the magnetic moment, Mn(B) will have a magnetic moment of about $1.9\mu_B$ at the highest concentration. On the other hand the bulk magnetization[4] at 4.2 K decreases rapidly with concentration. Some antiferromagnetic alignment, or canting, of the near constant Mn(B) moments is thus required to explain this decrease in the bulk magnetization to be consistent with the Mn(B) moment change deduced from the shift in the hyperfine field. According to reference (11) the moment of Mn(B) in Mn_3Si is $1.7\mu_B$. This fact leads us to believe that the spin structure in the unexplored region $1.6 < x < 3$ continues to be complex. An investigation of this region is underway.

REFERENCES

+ Supported in part by the University of Connecticut Research Foundation.
* Supported by NSF.

1. T. J. Burch, T. Litrenta and J. I. Budnick, Phys. Letters 33, 421 (1974).
2. S. Pickart, T. Litrenta, T. J. Burch and J. I. Budnick, Phys. Letters 53A, 321 (1975).
3. A. Switendick, Bull. Am. Phys. Soc. 21, 441 (1976).
4. J. G. Booth, J. E. Clark, J. D. Ellis, P. J. Webster and S. Yoon, Proc. Int. Conf. Mag. Moscow (1973), Vol. IV, p.577; S. Yoon and J. G. Booth, Phys. Letters 48A, 381 (1974).
5. Ye. N. Babanova, F. A. Sidorenko, P. V. Gel'd and N. I. Basov, Fiz. Metal. Metalloved 38, 586 (1974).
6. V. Niculescu, K. Raj, T. J. Burch and J. I. Budnick, Phys. Rev. B13, 3167 (1976).
7. R. E. Watson and A. J. Freeman, Hyperfine Interaction, Ed. A. J. Freeman and R. B. Frankel (Academic Press, New York, 1967), p. 53.
8. K. Raj, V. Niculescu, J. I. Budnick and S. Skalski, AIP Conf. Proc. 29, 348 (1975).
9. A. Paoletti and L. Passari, Nuovo Cimento 32, 1449 (1964).
10. J. I. Budnick, V. Niculescu, W. A. Hines, A. H. Menotti, K. Raj and T. J. Burch, AIP Conf. Proc. 29, 437 (1975).
11. S. Tomiyoshi and H. Watanabe, J. Phys. Soc., Japan 39, 295 (1975). These authors report a helical spin structure and the moment values given in Table II.

MAGNETIC PROPERTIES AND ORDER IN CO-GA ALLOYS WITH 3d SUBSTITUTION FOR GA

R Cywinski and J G Booth
Department of Pure and Applied Physics,
University of Salford, Salford M5 4WT, UK.

ABSTRACT

The equiatomic alloy CoGa is almost completely ordered in the B2 (CsCl) structure and is paramagnetic. The partial substitution of a 3d transition metal TM for Ga results in series having the general formulae $Co_2Ga_{2-x}TM_x$ and the magnetic properties and crystallographic states of order of these series have been determined using magnetization, X-ray, neutron diffraction and resistivity techniques. In some of the series the appearance of a (111) line in the powder diffraction patterns signifies the establishment of double ordering of the $L2_1$ type. In the case of TM = Ti this order appears to be established for small values of x. In all the series ferromagnetism is observed beyond a critical concentration and this appears to be related primarily to the electron concentration rather than to any change in ordering which may occur.

INTRODUCTION

3d transition metal substitution for Ga in the equiatomic B2 CoGa alloy results in a series of alloys which may be characterised by the general formula $Co_2Ga_{2-x}TM_x$ for $0 \leqslant x \leqslant 1$. Previous studies for which TM = Ti[1], and V and Cr[2] indicate that a (111) Bragg peak in the neutron diffraction spectrum occurs above a particular value of x, suggesting the partial establishment of double ordering of the $L2_1$ type. Thus some site selectivity appears to be exercised by the substitutional transition metal on entering Ga sublattice.

The onset of ferromagnetism is observed beyond a critical value of x for each series and this value appears to be unrelated to any observed change in crystallographic ordering. The magnetic properties of the alloys were therefore assumed to be primarily related to the electron concentration, and the substitution TM = Mn and Fe as described here provides a test of this conclusion in cases where the substitutional transition metal usually has a substantial magnetic moment.

EXPERIMENTAL TECHNIQUES

The appropriate proportions of spectroscopically pure constituent metals were melted in an argon arc furnace. Weight losses were typically of the order of 0.3%, the actual alloy compositions thus being close to the nominal ones. The resulting ingots were powdered and annealed under vacuum in quartz ampoules at 830°C for 24 hours and subsequently water quenched.

X-ray powder diffraction spectra of all specimens were obtained and showed all the alloys to be single phase. The relative X-ray scattering lengths of the constituents however are not sufficiently different to allow determinations of crystallographic order between transition metals. The X-ray density, ρ_x, was compared with the bulk density, ρ_a, of the powdered metals, measured by the standard displacement technique and the vacancy concentration estimated.

The site occupation was determined by neutron diffraction experiments carried out at AERE Harwell using the Curran diffractometer (neutron wavelength = 1.37 Å). Bragg patterns were obtained at 4.2 K, and the nuclear and magnetic scattering were determined by a 'field on - field off' technique (H = 10 KOe). Magnetization and susceptibility measurements were made in the range 4.2 - 300 K using a vibrating sample magnetometer, and above room temperature by a Sucksmith ring balance.

RESULTS AND DISCUSSION

The structures have been determined in terms of a unit cell of four interpenetrating f.c.c. sublattices A, B, C, D with origins at the points (0,0,0), $(\frac{1}{4},\frac{1}{4},\frac{1}{4})$, $(\frac{1}{2},\frac{1}{2},\frac{1}{2})$ and $(\frac{3}{4},\frac{3}{4},\frac{3}{4})$ as shown in Fig.1. If A=C, and B=D the B2 (CsCl) structure obtains and this can be indexed on a unit cell half the size of that illustrated. However the lattice parameters have usually been referred to the double unit cell in order to discuss the series as a whole. For the Fe and Mn series the variation in lattice parameter was small as indicated in Fig.2. The establishment of any double ordering, eg of the type $A=C\neq B\neq D$ results in a non-zero structure factor for the (111) Bragg peak.

From the neutron diffraction spectra the atomic order was determined to be the singly ordered B2 structure in the ranges of concentration up to x = 0.6 for Mn and x = 0.8 for Fe. A small (111) peak appears at these values of x, suggesting the establishment of some $L2_1$ order. The intensities of the (200) line

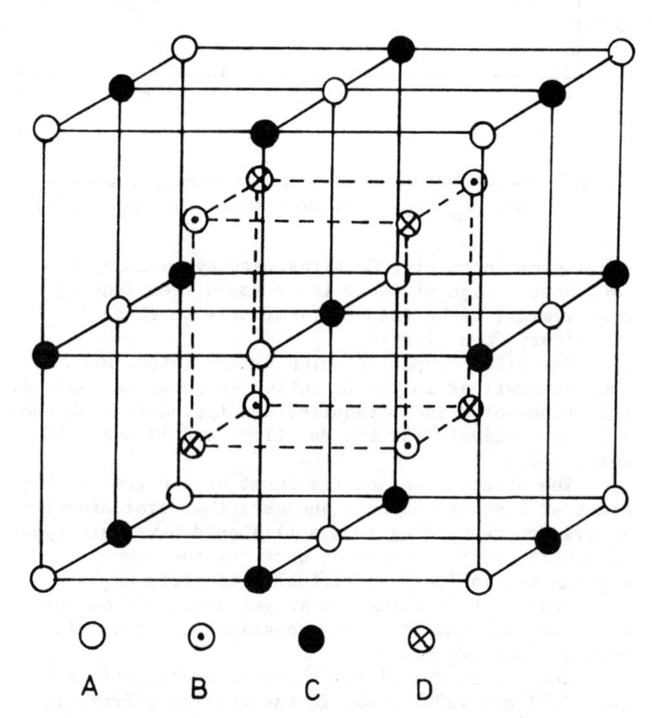

Fig 1 Unit cell formed by four inter-penetrating face centred cubic sublattices.
B2 structure: $A = C \neq B = D$,
$L2_1$ structure: $A = C \neq B \neq D$

as evaluated from the diffraction patterns (and normalized to the (220) line) were compared with values calculated theoretically assuming a perfectly ordered B2 structure in which the Mn and Fe atoms enter solely Ga sites. Discrepancies between the observed and calculated values were small and may be explained by including in this simple model a disorder parameter ε which reflects a small percentage of transition

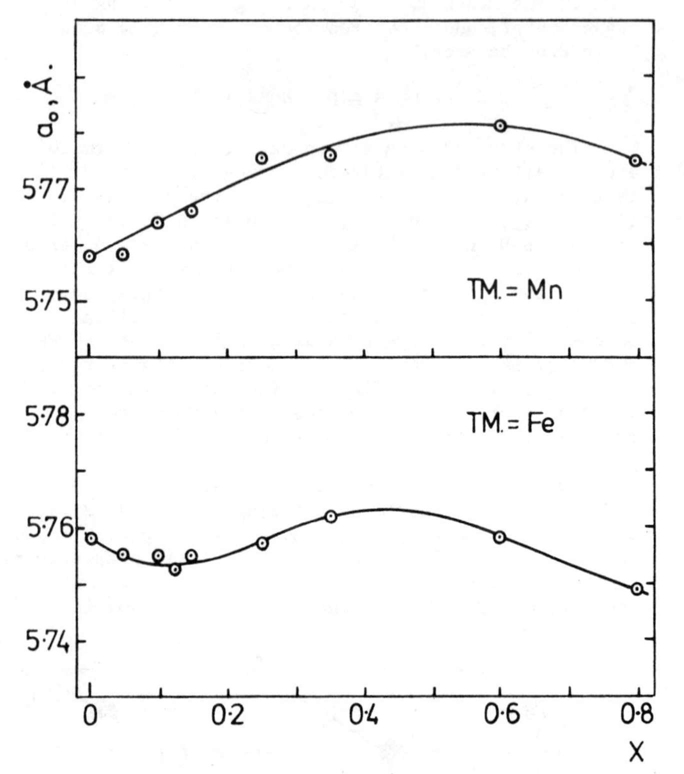

Fig 2 Variation of lattice parameter, a_o, with x for $Co_2Ga_{2-x}TM_x$ alloys where TM = Mn and Fe.

metal atoms occupying Co sites, and by assuming the small percentage of vacancies γ found from density measurements to be disposed similarly to those in the binary CoGa alloys[3].

The site occupation which is consistent with all the observations in the B2 region is shown in Table 1. The values of ε and γ required are typically 1.5% and 2.5% respectively for the Mn alloys and 1% and 3.5% for the Fe.

The observations of the onset of a degree of $L2_1$ order at high values of x suggests that stoichiometic alloys (corresponding to x = 1) should have this type of double order. Previous work for the compound Co_2GaMn has indicated a virtually complete $L2_1$ ordering[4]. No information is yet available on the alloy Co_2GaFe but it seems probable that it should contain some $L2_1$ order.

The magnitudes of the magnetic contributions to the (200) and (220) peaks in the neutron diffraction spectra indicate that magnetic moments exist on Co, Mn and Fe sites. The absence of a (111) line precludes any reasonable estimate of the respective sizes of these moments. Even for those alloys for which a (111) line is observed an accurate estimate is prevented by the absence of the (200) line. Webster[4] has reported magnetic moments of 0.5 μ_B and 3.0 μ_B on Co and Mn atoms respectively in the stoichiometic alloy Co_2GaMn.

TABLE I

Proposed model of site occupancy of the B2 phase of $Co_2Ga_{2-x}Mn_x$ and $Co_2Ga_{2-x}Fe_x$ alloy series (□ = vacancy).

Sites A,C	Sites B,D
$Co_{(1-2\gamma-\varepsilon)}$	$Ga_{(1-\frac{x}{2})(1-\gamma)}$
$□_{2\gamma}$	$Co_{\gamma+\varepsilon}$
TM_{ε}	$TM_{\frac{x}{2}(1-\gamma)-\varepsilon}$

Magnetic measurements (Fig.3) show that the ferromagnetism becomes established beyond x = ∿0.075 for Mn and x = ∿0.06 for Fe. The Curie points illustrated in Fig.2 were determined from Arrott plots. Thus the occurrence of ferromagnetism beyond a critical concentration in these alloys does not appear to be related to any obvious feature of the crystallographic ordering. The variation of the Curie temperatures for the systems in which TM = Ti[1], V and Cr[2], and also for Mn and Fe clearly suggests that the onset of ferromagnetism is linked to an electron concentration parameter, and this view is further strengthened by the Rhodes Wohlfarth plot (Fig.4) in which the ratio of the paramagnetic to ferromagnetic moment is plotted as a function of Curie temperature. The results follow closely upon the solid line representing an itinerant model.

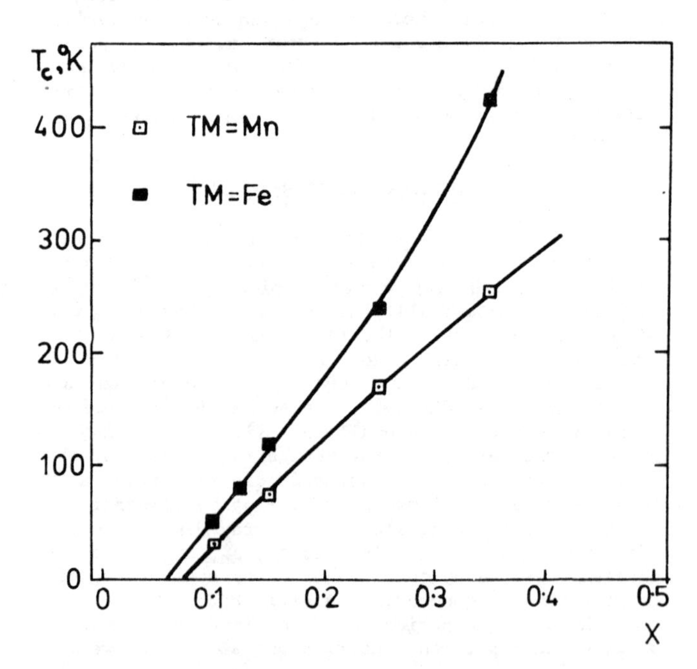

Fig 3 Variation of Curie temperature, T_c, with x for $Co_2Ga_{2-x}TM_x$ alloys where TM = Mn and Fe.

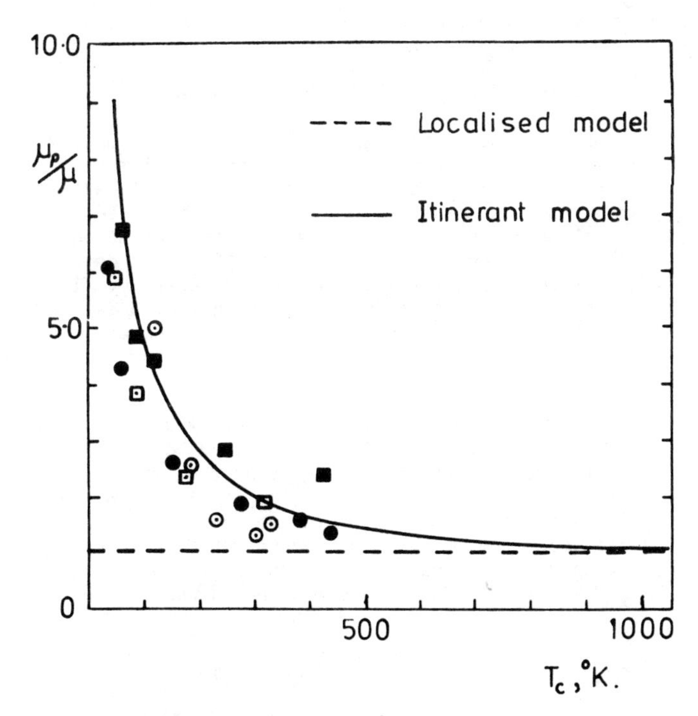

Fig 4 Rhodes-Wohlfarth plot (μ_p/μ vs. T_c) for $Co_2Ga_{2-x}TM_x$ alloys where TM = V(⊙), Cr(●), Mn(⊡) and Fe(■).

ACKNOWLEDGEMENTS

The neutron diffraction aspects of this work have been carried out at AERE Harwell under a grant from the SRC Neutron Beam Research Unit. The authors are grateful to the University Support Section at Harwell for their invaluable assistance.

REFERENCES

1. J G Booth and R G Pritchard, J.Phys.F.(Metal Physics) 5, 347, (1975).
2. R Cywinski and J G Booth, J.Phys.F.(Metal Physics) 6, L75, (1976).
3. E Wachtel, V Linse and V Gerold, J.Phys.Chem. Solids, 34, 1461, (1973).
4. P J Webster, J.Phys.Chem.Solids, 32, 1221, (1971).

FERRO-NONMAGNETIC TRANSFORMATION OF Mn ALLOYS WITH CaC_2 TYPE STRUCTURE[†)]

K. Adachi, T. Ido*, H. Watarai**, T. Shimizu and K. Sato***
Faculty of Engineering, Nagoya University, Nagoya 464, Japan

ABSTRACT

The magnetization, magnetic susceptibility, electrical resistivity and lattice parameter for the substituted $(Au_{1-x}Pd_x)_2Mn$ and $Au_2(Mn_{1-x}Cr_x)$ systems were measured.

In the case of $(Au_{1-x}Pd_x)_2Mn$, x=0.02 and 0.05, H_{th} and c/a increase with increasing of Pd, while χ_i and a decrease.

On the other hand for $Au_2(Mn_{1-x}Cr_x)$, x=0.05 and 0.10, the H_{th} decreases with increasing Cr and for x=0.10 the H_{th} tends to zero at the T*=240 K. However a and c/a do not change with Cr. The helical angle which is estimated from our experimental results increases with increasing Pd, but decreases with increasing Cr.

INTRODUCTION

Magnetization curve of Au_2Mn with CaC_2 type crystal structure shown in Fig. 1 is known as metamagnetism,[1,2] showing the linear dependence of magnetization in a weak magnetic field, the abrupt increase above a threshold field and the saturation in a high field. Also, it is reported[3] that the compound has a helical spin structure in the c-plane with a propagation vector along the c-axis. The helical angle, due to competition of the exchange interactions,[4] may possibly change by the substitution of magnetic or nonmagnetic elements.

The purpose of this paper is to present our data on the lattice constant, magnetization and electrical resistivity of substituted systems of $(Au_{1-x}Pd_x)_2Mn$ and $Au_2(Mn_{1-x}Cr_x)$.

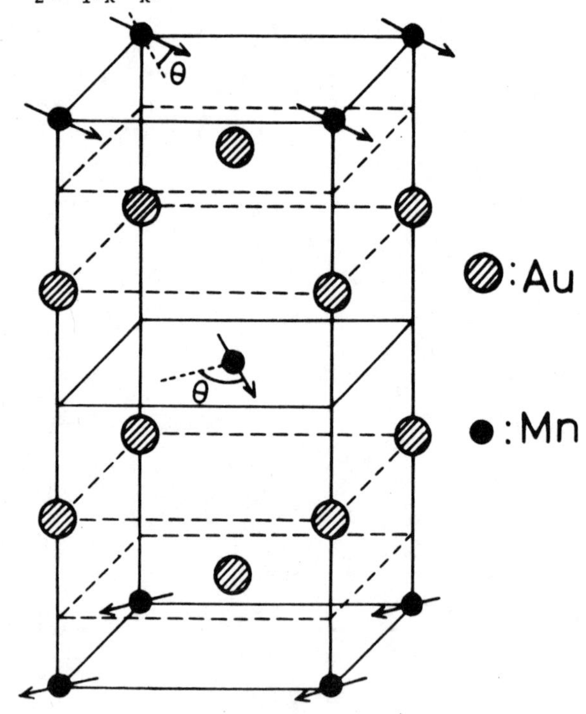

Fig. 1 Crystal structure of Au_2Mn, CaC_2 type (D_{4h}^{17} - I_4/mmm)
● Mn: $(0, 0, 0)_I$, $(1/2, 1/2, 1/2)_{II}$
o Au: (0, 0, z), (1/2, 1/2, 1/2-z)
(0, 0, $\bar{z}$), (1/2, 1/2, 1/2-$\bar{z}$)
θ is the helical angle.

SAMPLE PREPARATION AND X-RAY ANALYSIS

The compounds $(Au_{1-x}Pd_x)_2Mn$ and $Au_2(Mn_{1-x}Cr_x)$ were prepared from mixtures of the high purity elements (>4N) by electric and plasma jet furnaces. To attain composition homogeneity, the button was remelted several times on alternate sides and then annealed at 650°C (below decomposition point 730°C) for 5 days. The samples for X-ray analysis were prepared by filing from the bulk and then annealed for another 3 days at 500∿550°C. For both systems X-ray and chemical analysis indicate the pure CaC_2 type structure with nominal composition.

The CaC_2 type samples having more than 5 at.% of Pd and 10 at.% of Cr could not be prepared.

The concentration dependence of the lattice parameter is shown in Fig. 2. For $(Au_{1-x}Pd_x)_2Mn$ the a-axis decreases with increasing paradium content x, while the c-axis does not change; consequently the ratio c/a increases monotonously with x. However, for the $Au_2(Mn_{1-x}Cr_x)$, both a and c/a do not depend on the substitution.

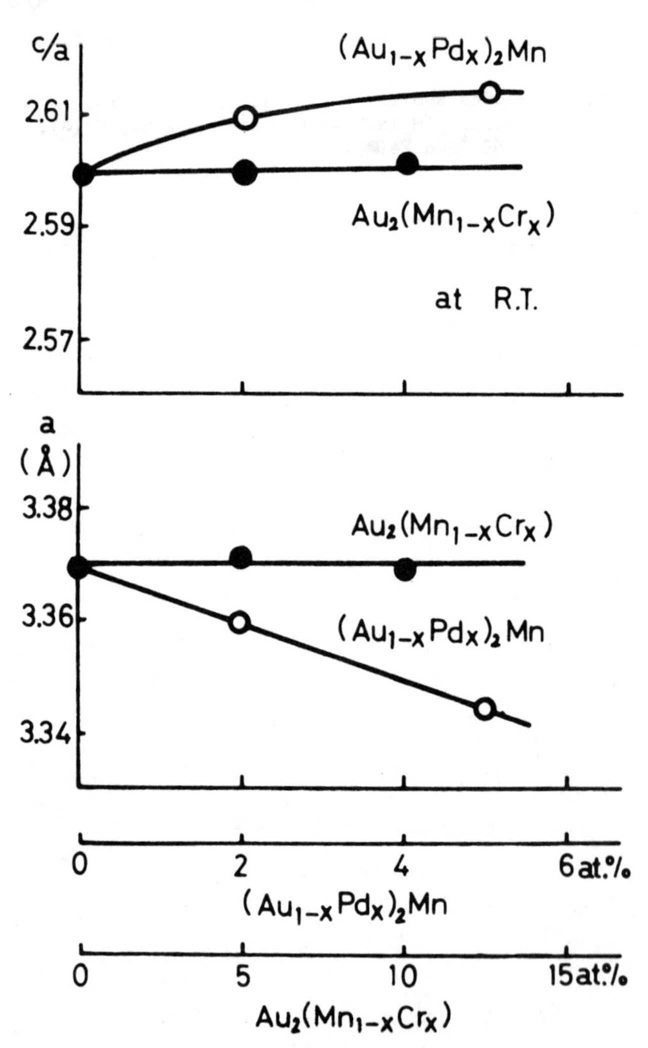

Fig. 2 Lattice constant of $(Au_{1-x}Pd_x)_2Mn$ and $Au_2(Mn_{1-x}Cr_x)$ at room temperature
o for $(Au_{1-x}Pd_x)_2Mn$
● for $Au_2(Mn_{1-x}Cr_x)$

MAGNETIZATION AND ELECTRICAL RESISTIVITY

The magnetization and susceptibility were measured by means of magnetoinductive and magnetic balance methods.

Fig. 3(a) and (b) show the magnetization curves of $Au_2(Mn_{1-x}Cr_x)$ for x=0.10 at various temperatures. As shown in Fig. 3(a), typical metamagnetic behavior is seen in a low temperature region, i.e. the magnetization in low field region shows the linear change with field strength, but when the field approaches some critical value, the magnetization increases abruptly

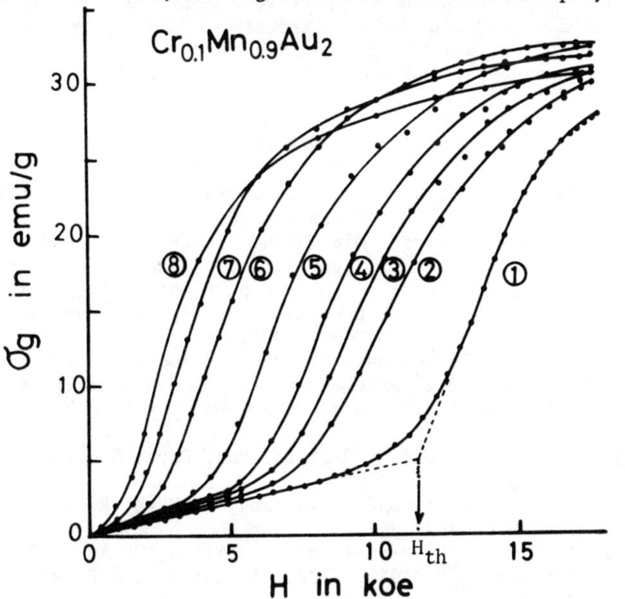

Fig. 3(a)

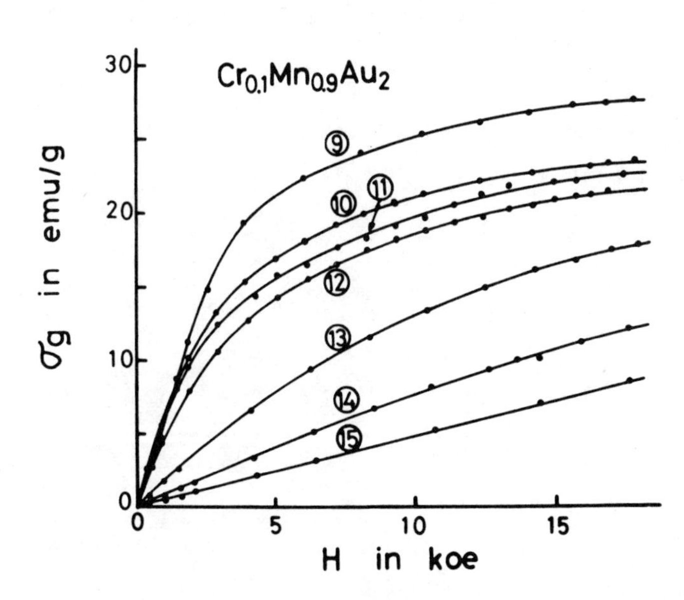

Fig. 3 (b)

Fig. 3(a), (b) Magnetization curves of $Au_2(Mn_{0.9}Cr_{0.1})$ at various temperatures: ① 4.2K, ② 77K, ③ 93K, ④ 110K, ⑤ 136K, ⑥ 174K, ⑦ 190K, ⑧ 211K, ⑨ 248K, ⑩ 276K, ⑪ 292K, ⑫ 298K, ⑬ 309K, ⑭ 335K, ⑮ 356K

and this is similar to the behavior of ferromagnetic material as seen in Fig. 3(b). Defining a threshold field denoted by H_{th} as a cross point of the extrapolation of the magnetization in low and high magnetic fields as shown by the dotted line in Fig. 3(a), H_{th} is given as a function of x and T. The result is given in Fig. 4. H_{th} decreases with increasing x and T, and for x=0.10 it tends to zero at T*=230±2 K. In T>T*,

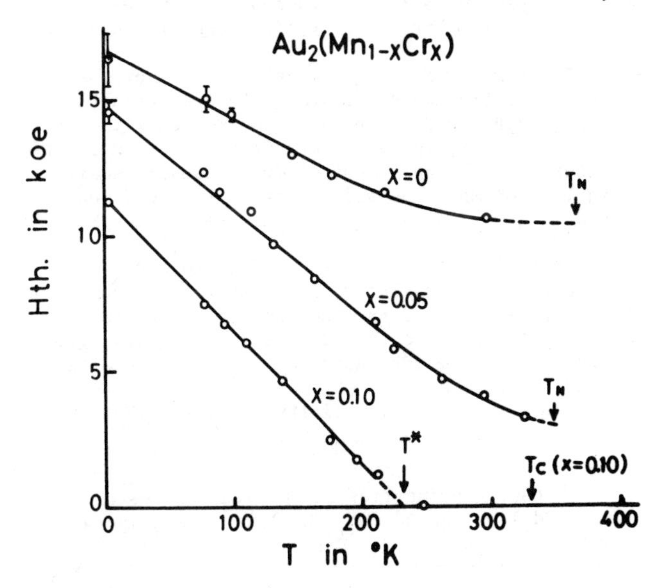

Fig. 4 Threshold field vs. temperature for $Au_2(Mn_{1-x}Cr_x)$. The helical magnetism changes to the ferromagnetism at T=T*.

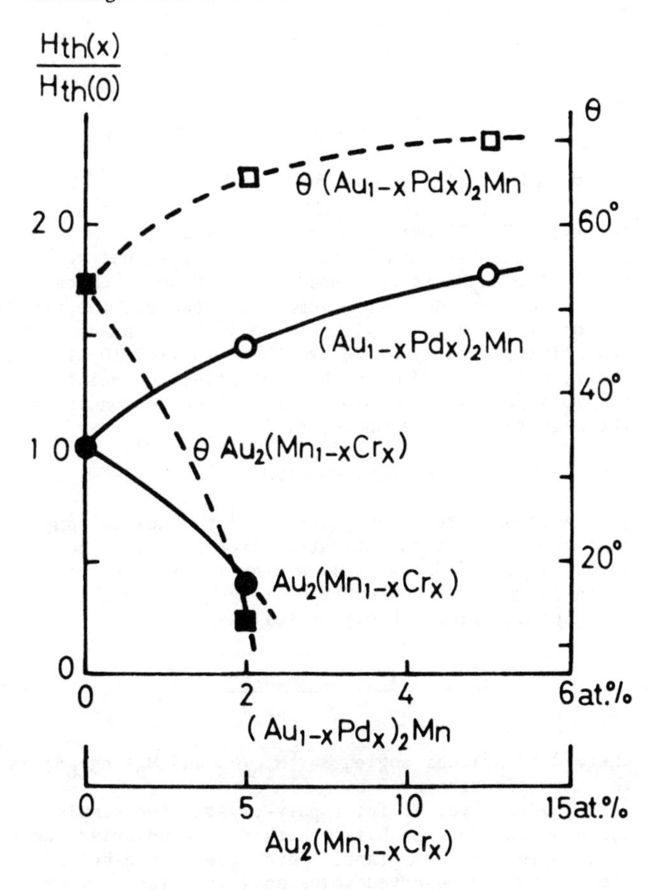

Fig. 5 Threshold field and helical angle vs. composition for $(Au_{1-x}Pd_x)_2Mn$ and $Au_2(Mn_{1-x}Cr_x)$ at room temperature.

$Au_2(Mn_{0.90}Cr_{0.10})$ shows ferromagnetic behavior and the Curie point was ascertained to be 330 K by H/σ-σ^2 plot. However the initial magnetic susceptibility (χ_i), which is defined by a slope of magnetization in low field region, is obtained as a function of x and T. The χ_i of the substituted systems increases with the increasing of T. For x=0.10 $\chi_i^{-1}(T)$ tends to zero at T*=240±10 K.

The paramagnetic as well as the antiferromagnetic susceptibilities were measured in a weak magnetic field. The paramagnetic Curie point θ_p decreases with the increasing of x, while the effective moment p_{eff} increases, though the moment of Cr is usually smaller than that of Mn. The Néel point (T_N) depends on the field, and dT_N/dH is estimated to be -3.2 deg/kOe for Au_2Mn.

On the other hand, the H_{th} for $(Au_{1-x}Pd_x)_2Mn$ system at room temperature increases monotonously with the increasing of Pd as shown in Fig. 5 and this is contrary to the case of Cr. Also, χ_i, θ_p and p_{eff} all decrease with the increasing of x. These results are listed in Table I.

The electrical resistivity (ρ) of $Au_2(Mn_{1-x}Cr_x)$ for x=0 and 0.10 were measured from 4.2 to 1050 K by means of the 4-terminals method. The residual resistivity for Au_2Mn is obtained to be 0.85±0.02 μΩ-cm indicating exact stoichiometry. However, the residual one for x=0.10 is given to be 12.5±0.1 μΩ-cm. The resistivity anomaly at T_N (x=0) and T_c (x=0.10) is not clear, but the $\rho(T)$ curve has an inflection point close to 370 K for x=0, and 350 K for 0.10 corresponding to the magnetic transition point.

TABLE I

Magnetic properties and helical angles of $(Au_{1-x}Pd_x)_2Mn$ and $Au_2(Mn_{1-x}Cr_x)$ at room temperature.

	$(Au_{1-x}Pd_x)_2Mn$			$Au_2(Mn_{1-x}Cr_x)$
x	0	0.02	0.05	0.05
$\chi_i\times10^{-4}$ [emu/g]	4.32	3.99	3.5	
$H_{th}(x)/H_{th}(0)$	1.0	1.5	1.8	0.4
θ_p [K]	383±10	370±10	362±10	380±10
p_{eff} [μ_B]	4.4±0.1	4.3±0.1	4.2±0.1	4.5±0.1
θ [deg]	53	66	70	13

DISCUSSION

According to Nagamiya et al.[5)] the helical angle characterizes the magnetization process, i.e. the threshold field H_{th} as well as initial susceptibility χ_i in the c-plane. From them it is given on the basis of molecular field theory as follows,

$$\cos\theta = \frac{1}{2}\left(\sqrt{\frac{1-2a-a^2}{2a+a^2}} - 1\right), \qquad (1)$$

where θ is helical angle, $a=\chi_i^c H_{th}/M_0$ and $M_0(=ng\mu_B S)$ is the magnetization at 0 K.

The observed χ_i for a poly-crystalline sample contains the susceptibility with a perpendicular component to the helical plane. Then, the χ_i can be expressed by the observed value as $\chi_i^c=p\cdot\chi_i(obs)$, where a factor p should be taken as 2/3<p<1. The p value is chosen so as to fit χ_i for Au_2Mn at 0 K, in which the angle was determined by N. D. already as $\theta_0=45°$.

Making use of the Brillouin function $(M(T)=M_0B_J(T))$ and observed value χ_i as well as H_{th}, the θ can be estimated at room temperature to be 66° for $(Au_{0.98}Pd_{0.02})_2Mn$ and 70° for $(Au_{0.95}Pd_{0.05})_2Mn$ respectively. This fact indicates that the magnetic structure of the $(Au_{1-x}Pd_x)_2Mn$ changes from the metamagnetism of helical spin arrangement to the antiferromagnetism of a co-linear one with the substitution of Pd.

On the other hand, for the Cr system the angle decreases with an increasing of x and T and then for x=0.10 the angle tends to zero, i.e. there is a transfer from metamagnetic to ferromagnetic spin structure. Such composition and temperature dependence of θ may be caused by sign and by the strength of the superexchange interaction, which depend on a change of conduction electron number and atomic distance by substitution on the basis of a double exchange mechanism.

REFERENCES

* Present address: Tokyo Shibaura Electric Co., Ltd., Kawasaki 210, Japan.

** Present address: Mitsubishi Electric Co., Ltd., Sagamihara 229, Japan.

*** Present address: Physics Department, College of Liberal Arts, Toyama University, Toyama 930, Japan.

†) The details and analysis of experimental data will be reported elsewhere, and the theoretical explanation will also be given by one of the authors (K. Adachi).

1) J. H. Smith and R. Street, Proc. Phys. Soc. 70B (1957) 1089.
2) A. J. P. Meyer and P. Taglang, J. Phys. Radium 17 (1956) 457.
3) A. Herpin, P. Mériel and J. Villain, CR Acad. Sci. (France) 239 (1959) 1334; A. Herpin and P. Mériel J. Phys. Radium 22 (1961) 337.
4) A. Yoshimori, J. Phys. Soc. Japan 14 (1959) 807.
5) T. Nagamiya, K. Nagata and Y. Kitano, J. Phys. Soc. Japan 17S (1961) 10; Progr. theor. Phys. 27 (1962) 1253; Y. Kitano, T. Nagamiya, Progr. theor. Phys. 31 (1964) 1.

ANOMALOUS FIELD DEPENDENCE OF MAGNETIC ANISOTROPY CONSTANT OF Mn ALLOY

K. Sato*, T. Ido**, O. Ishihara***, C. Tokura and K. Adachi
Faculty of Engineering, Nagoya University, Nagoya 464, Japan

ABSTRACT

A single crystal of the ordered phase of Au_4Mn was prepared by the Bridgeman and annealing methods.

The magnetic anisotropy of ordered Au_4Mn was measured by the torque method utilizing this crystal below the Curie temperature. The magnetic easy direction was determined to be the tetragonal <001> axis.

The four-fold torque amplitude first increases and then decreases with increasing field strength. This behavior can be understood as an effect of the twin structured model of the ordered Au_4Mn. The uniaxial anisotropic energy at absolute zero was determined to be 3.0×10^6 erg/cc.

INTRODUCTION

According to previous reports [1-4] the ordered-disordered transformation temperature of the intermetallic compound Au_4Mn is 680 K. Below this point the crystal has the Ni_4Mo type of superstructure based on f.c.t. as shown in figure 1(a), while above this point it is disordered f.c.c..[5] As for the magnetic property of this compound, both the ordered and the disordered phases manifest the ferromagnetic state, of which the Curie temperature for the ordered ferromagnetic state is 370 K, while for the disordered state it is 120 K.

The spin direction of the ordered Au_4Mn has been suggested recently by a neutron diffraction experiment with a powdered sample,[6] in which the easy axis of the ferromagnetic moment is parallel to the tetragonal C axis.

The purpose of this paper is to present our data on the magnetic torque measurements for a single crystal of the ordered phase Au_4Mn and we will propose the twin structured model to explain the observed torque curve of this crystal.

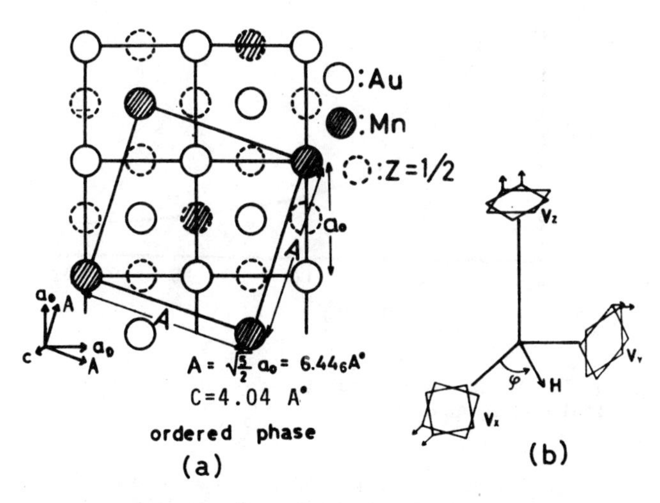

Fig. 1 Crystal structure of ordered phase Au_4Mn,
(a): Ni_4Mo type (C^5_{4h} - I4/m)
• Mn: (0, 0, 0), (1/2, 1/2, 1/2)
o Au: (1/5, 2/5, 0), (4/5, 3/5, 0), (3/5, 1/5, 0), (2/5, 4/5, 0), (7/10, 9/10, 1/2), (3/10, 1/10, 1/2), (1/10, 7/10, 1/2), (9/10, 3/10, 1/2)
(b): Three dimentional twin domain, V_ν: Volume of domain with C-axis, ν=X, Y, Z.

CRYSTAL PREPARATION AND X-RAY ANALYSIS

A mixture of gold (99.9%) and manganese (99.9%) was sealed in an evacuated quartz tube and melted at 1373 K (1100°C); then, the homogenized button crystal was moved at a velocity of 2.5∿3.0 cm/hr in a furnace which was heated to 1423 K (1150°C). The temperature gradient was 40 deg/cm. Such was the preparation of a single crystal of the disordered phase, i.e. the Bridgeman method. Then the disordered single crystal which was ground into sphere (dia.=4.000±0.002 mmø) was quickly heated up to 773 K (500°C) by an electric furnace with a temperature gradient of 30 deg/cm, and was kept two days at this temperature. The sample was then removed from the furnace until it cooled to 573 K (300°C). The sample was annealed for ten days at this temperature, and then quenched to room temperature.

Since the curve of the temperature vs. time of the electrical resistivity of the sample approaches saturation, and the residual resistivity of the ordered state is sufficiently small as compared with the disordered state, the crystal may be considered to be well-ordered. X-ray Laue patterns at various locations of the sample confirmed the sample as a single crystal of the ordered phase. However, since the ordered phase of Au_4Mn has tetragonal symmetry shown in figure 1(a), each principal axis has two kinds of crystal domains.

Therefore, the total number of crystal domains in the single crystal is six, as shown in figure 1(b) and each domain is a single crystal of f.c.t.. This also was confirmed by splitting of the Laue spots for our sample. Further, from the powder method, the lattice constant of Au_4Mn was estimated to be A=6.446±0.004 Å, C=4.04±0.004 Å, and C/A=0.62; these were in very good agreement with other results[2,5]

TORQUE MEASUREMENTS AND DISCUSSION

The magnetic torque curves of the ordered phase in the (001) plane were measured in the temperature range 4.2∿350 K. Figure 2 shows the magnetic field dependence of the magnetic torque curve for the (001) plane at 4.2 K. A two-fold torque amplitude which is separated numerically from observed torque curves was negligibly small. This indicates that the volumes of the twin domains are almost equal, as is also assumed in the latter paragraphs. In the figure, the torque amplitude increases gradually with the increasing magnetic field, while when the field is stronger than 8.1 kOe, the amplitude decreases gradually with an increasing field as a function of 1/H as shown in figure 3, and the shape of the torque curve approaches to a sinusoidal wave.

The easy direction of magnetization was determined to be the direction of the original cubic axis <100> and <010> corresponding to the C axis for each domain of the ordered state (See Fig. 1(b)).

As mentioned in INTRODUCTION, since the ordered structure of Au_4Mn is a large distorted f.c.t. with C/A=0.62, the magnetic anisotropy due to the tetragonal symmetry may be uniaxial. Such is the case when the observed torque curve is separated into two of uniaxial anisotropy, and the anisotropy energy in the (001) plane can be expressed as follows,

$$E(\theta_1, \theta_2) = V_1\{-K_u\cos^2\theta_1 - M_sH\cos(\phi-\theta_1)\} + V_2\{K_u\cos^2\theta_2 - M_sH\cos(\theta_2-\phi)\} \quad (1)$$

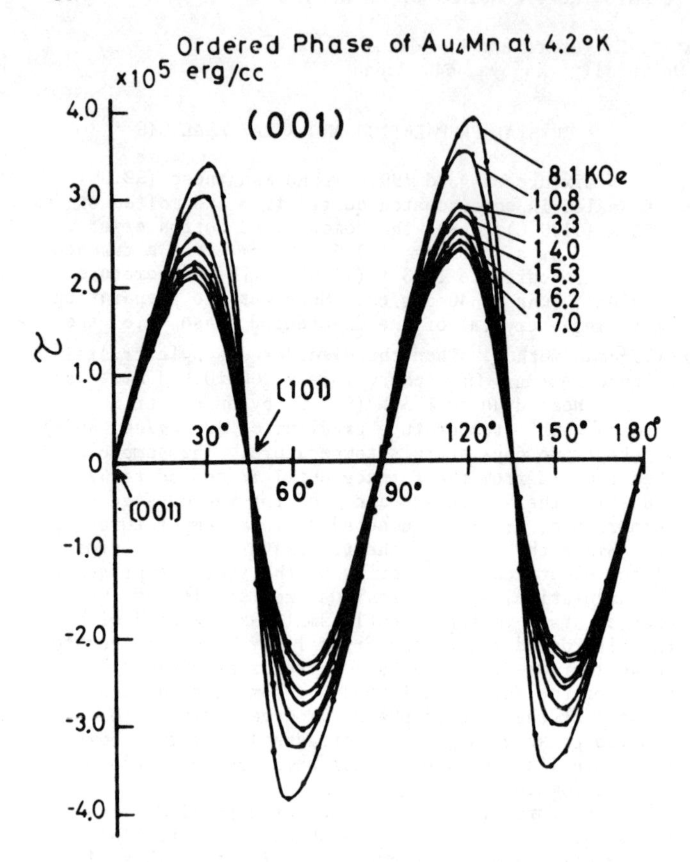

Fig. 2 Magnetic field dependence of magnetic torque curve at 4.2 K.

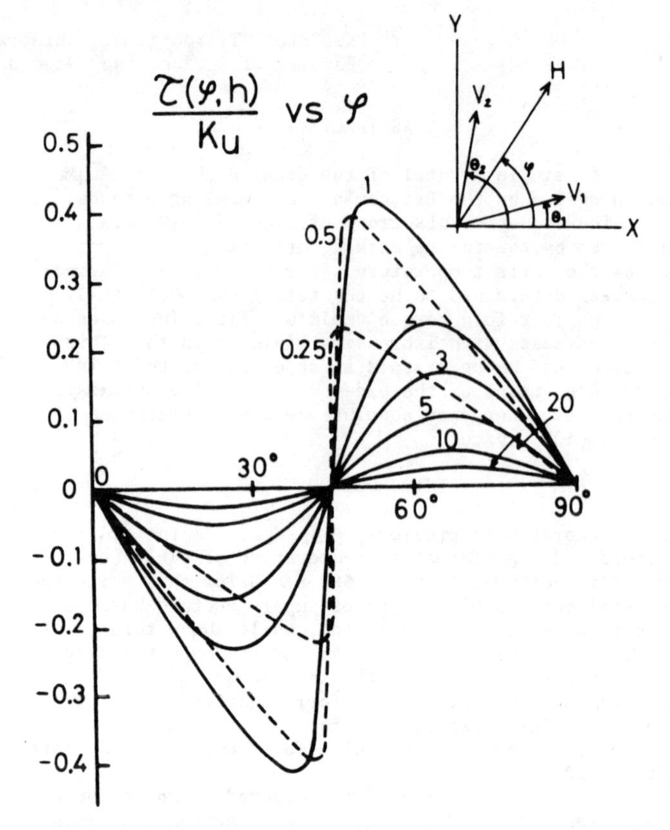

Fig. 4 Calculated torque curve for twin structure, dotted lines: $h=MH/K_u=0.25$, 0.5, solid lines: h=1, 2, 3, 5, 10, 20.

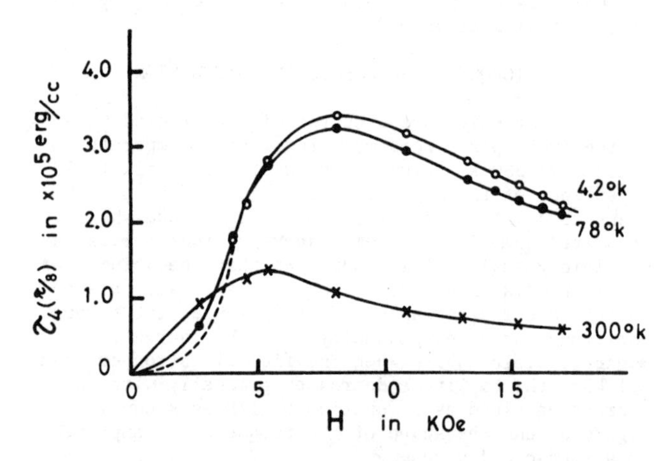

Fig. 3 Magnetic field dependence of 4-fold torque amplitude at various temperatures.

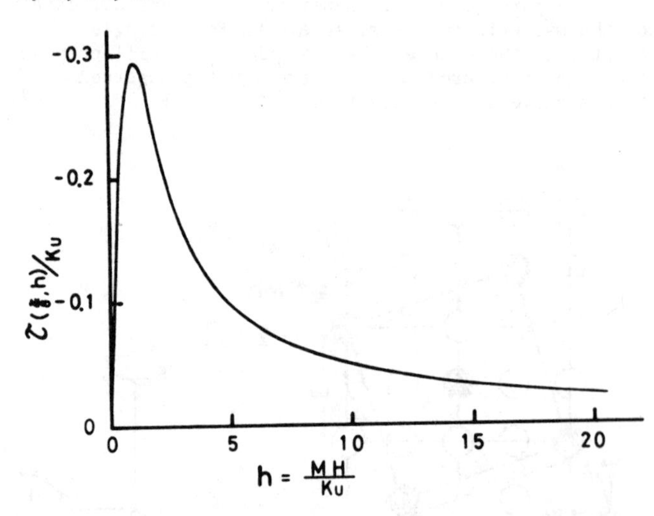

Fig. 5 Field dependence of 4-fold amplitude of calculated torque curve.

where θ_1 and θ_2 are angles between the magnetization and X-axis in domains 1 and 2 respectively, ϕ is the field direction to the same axis as shown in figure 1, and K_u is the uniaxial anisotropy constant.

Since the moment is parallel to the C-axis[6] as mentioned in INTRODUCTION, the domains mentioned above can be reduced from six to three.

The stable direction (θ_1, θ_2) of the magnetization in the magnetic field is determined by the equilibrium conditions for eq. (1).

Then the torque curve $\tau(\phi, h)$ is given to be

$$\tau(\phi, h) = -\partial E/\partial \phi. \tag{2}$$

Assuming V_1, V_2 and V_3 as volumes of the domains, in which the axis is parallel to X, Y and Z respectively, as mentioned before in this section, we can assume that $V_1=V_2=V_3=V/3$. The calculated torque curves are given in figure 4, where the ordinate is $\tau(\phi, h)/K_u$ and the abssissa is ϕ. The numbers in the figure give the values of $h=MH/K_u$. The torque amplitude (τ/K_u) for $h<1$ shown by the dotted lines increases with increasing h value, while in the case of $h>1$ it decreases. Also the torque curve approaches a sinusoidal wave for a large value of h.

Making use of the expansion of the expression (1) in the high magnetic field limit ($\phi \simeq 0$), an approximate formula of the amplitude is obtained as follows,

$$\tau(\pi/8,\ h) = (2/3)VK_u[(1/4)h^{-1} - (5/64)h^{-3} + O(h^{-5})] \qquad (3)$$

and it shows that the amplitude of the torque curve decreases with an increasing of the field as a function of 1/h as shown in figure 5. This behavior is in agreement with the experiments (Figure 3 and Calculated curve). Also the uniaxial anisotropy constant at any temperature is determined from the tangent of τ-1/h relation in the high field limit and $K_u(0)$ is estimated to be 3.0×10^6 erg/cc.

ACKNOWLEDGEMENT

The authors wish to thank Dr. N. Kunitomi, Professor of Osaka University for his communication on the unpublished data of N.D., and Mr. J. P. Heald of M. A., American University for his editorial assistance. The authors also wish to thank Mr. T. Shimizu of Nagoya University for his technical assistance.

REFERENCES

* Present address: Physics Department, College of Liberal Art, Toyama University, Toyama 930, Japan.

** Present address: Tokyo Shibaura Electric Co., Ltd., Kawasaki 210, Japan.

*** Present address: Nippon Denso Co., Ltd., Kariya, Aichi 448, Japan.

1) B. Raub, U. Zwicker and H. Bzur, Z. Metallkde. 44 (1953) 312.
2) A. Kussmann and E. Raub, Z. Metallkde. 47 (1956) 9.
3) A. J. P. Meyer, J. Phys. Radium 20 (1959) 43.
4) T. Ido, O. Ishihara, T. Kasai, K. Sato and K. Adachi, Rep. Toyoda Phys. Chem. Res. Inst. 23 (1970) 54.
5) D. Watanabe, J. Phys. Soc. Japan 15 (1960) 1251.
6) T. Azuma and N. Kunitomi, Private communication.

MAGNETIC INVESTIGATIONS ON THE $MnNiSi_{1-x}Ge_x$ COMPOUNDS CRYSTALLIZING IN THE Co_2P TYPE STRUCTURE

K. S. V. L. Narasimhan
Department of Chemistry, University of Pittsburgh
Pittsburgh, Pa. 15260

ABSTRACT

MnNiSi crystallizes in the orthorhombic Co_2P type structure with a Curie temperature of 616 K and saturation magnetic moment of 2.51 μ_B. Ge was substituted for silicon to increase the Mn-Mn distance, the magnetic moment, and Curie temperatures. It was possible to obtain single phase materials of $MnNiSi_{1-x}Ge_x$ up to x = 0.8. As x is increased the magnetic moment increases from 2.51 μ_B to a maximum of 2.90 but the Curie temperature decreases continuously as x increases. Magnetization versus field measurements at 4.2 K and room temperature show a normal ferromagnetic behavior for all the compounds investigated except for $MnNiSi_{.2}Ge_{.8}$. This compound has a metamagnetic behavior at room temperature and magnetization versus temperature at 6 kOe indicates a sharp peak in the magnetization at 370 K. This anomalous behavior is absent in $MnNiSi_{.3}Ge_{.7}$.

INTRODUCTION

MnNiSi crystallizes in the orthorhombic Co_2P type structure with the atoms occupying three inequivalent positions,[1] $Mn_{.9}Ni_{.9}Ge$ exists as hexagonal Ni_2In type structure[2] and both of these compounds are ferromagnetic. Recently several manganese-containing compounds with rare earths silicon and germanium were found to possess large magneto-crystalline anisotropy[3] and magnetic properties sensitive to the Mn-Mn distance.[4,5] Substitution of Ge for Si in MnNiSi was attempted to vary the Mn-Mn distance and possibly alter the magnetic moment and magnetic ordering.

EXPERIMENTAL

All the compounds were prepared under an argon atmosphere in an induction furnace using Al_2O_3 crucibles and Ta susceptors. The ingots formed were induction melted again several times in a water cooled copper boat under a purified argon atmosphere. losses of manganese (usually less than 0.5%) were compensated by adding the required excess. The ingots obtained were annealed for 4 days at 750°C and quenched in water.

X-ray measurements on the powders were carried out using a Picker diffractometer using Cu K_α radiation. For x = 0 to 0.8 all the lines observed could be indexed to an orthrhombic Co_2P type structure. The lattice constants obtained for MnNiSi are in good agreement with those of Johnson and Frederick.[6] For x = 0.9 a mixture of phases were present with the primary phase having the orthorhombic Co_2P type structure. Attempts to form MnNiGe ended in a failure however, at a composition of $Mn_{.9}Ni_{.9}Ge$ single phase material crystallizing in the Ni_2In type structure was obtained.

Magnetic measurements were carried out by the Faraday method using two sets of apparatus. The low temperature apparatus covered the range of 2 K to 300 K and applied fields can be varied up to 21.6 kOe; the high temperature apparatus was used to cover the range from 300 K to 1200 K in applied fields of 1 to 6 kOe. Measurements were carried out on loose powders sealed in a pyrex tubing under a helium atmosphere. Curie temperatures (T_c) were obtained from the square of magnetization (σ) versus temperature plots extrapolated to zero magnetization.

RESULTS AND DISCUSSION

Table I shows the lattice constants and magnetic data for the various ternaries investigated. The increase in the lattice constants as Ge is substituted is indicative of the larger size of Ge. The increase is linear in all the lattice parameters up to x = .7 but show a sudden jump at a composition of x = .8 and for compositions beyond x = .8 two phases were obtained. The magnetic moment per formula unit[6] increases ($\sim$.35 μ_B) on substituting Ge showing a broad maximum extending from x = 0.2 to 0.6, but the Curie temperatures (T_c) decrease linearly. (see Fig. 1 and 2).

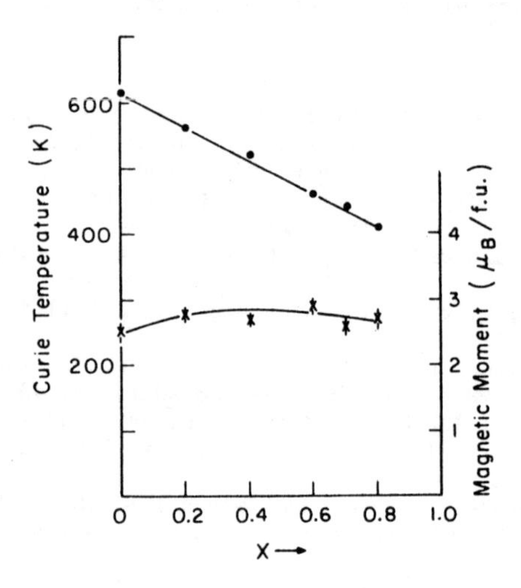

Fig. 1 The Variation of Curie Temperature and Magnetic Moment (4.2 k, 20 kOe) as a function of x for $MnNiSi_{1-x}Ge_x$.

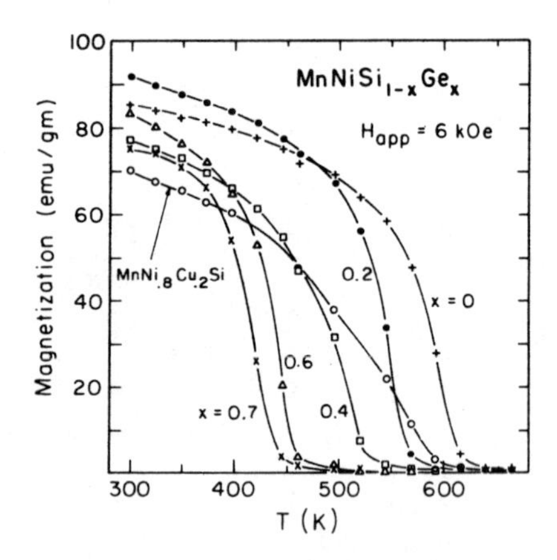

Fig. 2 Magnetization versus temperature for $MnNiSi_{1-x}Ge_x$ and $MnNi_{.8}Cu_{.2}Si$ compounds.

Castelliz[7] analysed the T_c of Mn containing compounds and found that a plot of T_c versus the ratio of lattice spacing to the 3d metal radius shows a maximum at a critical 3d-3d metal distance of 2.8Å. $MnNiSi_{1-x}Ge_x$ compounds would fall on the left of the maximum of Castelliz's curve and one would expect T_c to increase with increasing lattice spacing i.e. with the Ge substitution, which is contradictory to what is actually observed. These differences may arise due to the differences in bonding of Ge and Si to the metal atom.

The variation of σ versus field at 4.2 K and room temperature and also the shape of the σ versus temperature curve for the composition x = 0 to .7 indicate a simple ferromagnetic behavior. The Curie temperature of 616 K observed for the MnNiSi agrees well with that reported by Johnson and Frederick[8] (622 K). The σ versus T curve for the $MnNiSi_{.2}Ge_{.8}$ under an applied field of 6 kOe shows transitions at 158 K and 370 K (see Fig. 3 & 4). The transition

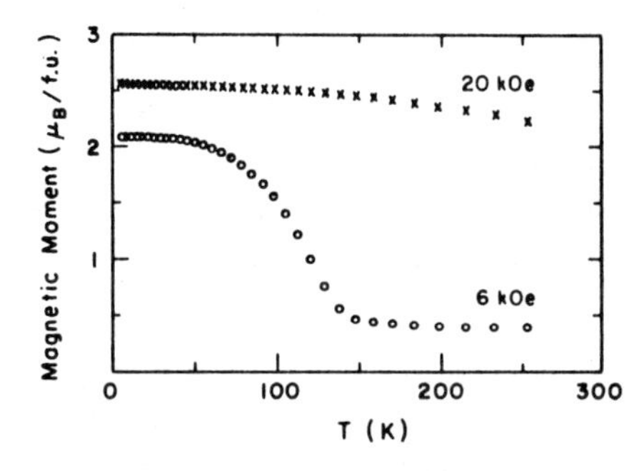

Fig. 3 Magnetization versus temperature at an applied field of 6 and 20 kOe for $MnNiSi_{.2}Ge_{.8}$.

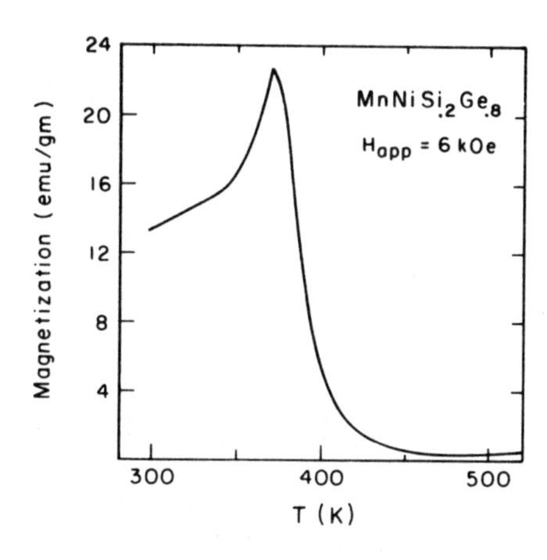

Fig. 4 Magnetization versus temperature for the $MnNiSi_{.2}Ge_{.8}$ in an applied field of 6 kOe.

observed at 158 K disappears in an applied field of 20 kOe. The σ versus field curve (Fig. 5) for this composition shows a ferromagnetic behavior at 4.2 K and a metamagnetic behavior at room temperature. Based on the above results we can conclude that in an applied field of 6 kOe the compound $MnNiSi_{.2}Ge_{.8}$ is ferromagnetic below 158 K and antiferromagnetic between 158 K and 370 K. At a higher applied field (∿20 kOe) the antiferromagnetic structure is destroyed and the compound behaves as a simple ferromagnet. This unusual behavior may be related to the

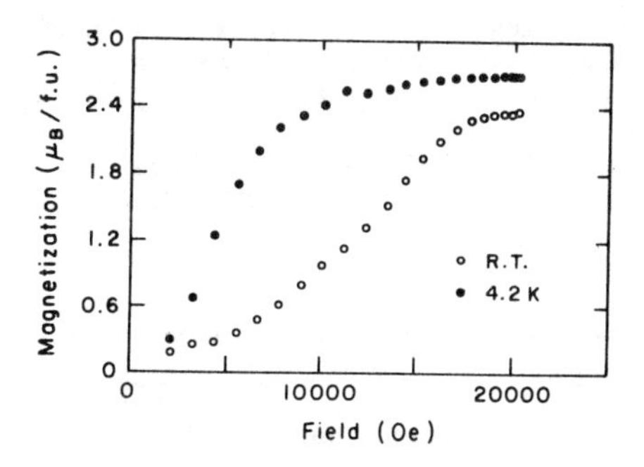

Fig. 5 Magnetization versus applied field for $MnNiSi_{.2}Ge_{.8}$ at room temperature and 4.2 K.

lattice constant jump observed at this composition. The anomalous transition observed for $MnNiSi_{.2}Ge_{.8}$ is similar to that previously reported for MnAs and other NiAs type structures, wherein a ferro to antiferromagnetic transition occurs either with an increase or decrease in temperature. Kittel[9] proposed a mechanism by which at a critical value for the lattice parameter exchange constant goes linearly through zero, and this exchange inversion results in a magnetic ordering change. The lattice constants as a function of temperature are currently being studied for $MnNiSi_{.2}Ge_{.8}$ near the transition.

Acknowledgements: The author is thankful to Dr. W. E. Wallace for many stimulating discussions.

REFERENCES

* This work was assisted by a grant from the Army Research Office-Durham.

1. V. Johnson and W. Jeitschko, "Ternary Equiatomic Transition Metal Silicides and Germanides," J. Solid State Chem. 4, 123-130 (1972).
2. A. E. Austin, "Magnetic Properties of Fe_5Ge_3-Mn_5Ge_3 Solid Solutions," J. App. Phys. 40, 1381-2 (1969).
3. K.S.V.L. Narasimhan (unpublished).
4. K.S.V.L. Narasimhan, V.U.S. Rao, R. L. Bergner and W. E. Wallace, "Magnetic Properties of RMn_2Ge_2 Compounds (R = La, Ce, Pr, Nd, Gd, Tb, Dy, Ho, Er and Th)," J. App. Phys. 46, 4957-60 (1975).
5. K.S.V.L. Narasimhan, V.U.S. Rao, W. E. Wallace and I. Pop, "Magnetic Properties of RMn_2X_2 Compounds (R = Rare Earth, Y or Th and X = Ge,Si)." AIP Conf. Proc. No. 29, 594 1975.
6. The moment observed may arise from both the Mn atoms and the Ni atoms. Attempts to prepare $MnNi_{1-x}Cu_xSi$ ternaries to isolate the Mn moment were unsuccessful due to the limited solubility of Cu(x=.2). The results obtained for the $MnNi_{.8}Cu_{.2}Si$ are shown in Table I and Fig. 1.
7. L. Castelliz, "Beitrag Zum Ferromagnetismus von Legierungen der Ubergangsmetalle mit Elementen der B-Gruppe" Z. Metallkde. 46, 198-203 (1955).

8. V. Johnson and C. G. Frederick, "Magnetic and Crystallographic Properties of Ternary Manganese Silicides with Ordered Co_2P Structure," Phys. Stat. Sol. (a) 20, 331-5 (1973).
9. C. Kittel, "Model of Exchange-Inversion Magnetization," Phys. Rev. 120, 335-42 (1960).

Table I. Lattice Constants and Magnetic Data for $MnNiSi_{1-x}Ge_x$ Compounds and $MnNi_{.8}Cu_{.2}Si$

Composition	Lattice constants(Å) ± .005 a	b	c	Curie temperature (K)	Magnetic moment μ_B/formula unit 4.2 K, 21.6 kOe	300 K, 20 kOe
MnNiSi	5.907	3.618	6.916	616	2.51	2.32
$MnNiSi_{.8}Ge_{.2}$	5.927	3.629	6.928	560	2.80	2.63
$MnNiSi_{.6}Ge_{.4}$	5.962	3.651	6.963	520	2.71	2.51
$MnNiSi_{.4}Ge_{.6}$	5.999	3.685	7.004	460	2.90	2.76
$MnNiSi_{.3}Ge_{.7}$	6.014	3.687	7.023	440	2.58	2.31
$MnNiSi_{.2}Ge_{.8}$	6.049	3.715	7.063	410	2.68	2.36
$Mn_{.9}Ni_{.9}Ge$*	4.078		5.420	273	0.44	.24
$MnNi_{.8}Cu_{.2}Si$	5.873	3.616	6.899	585	2.02	-

* Crystallizes in hexagonal Ni_2In type structure.

EPITAXIAL GROWTH OF CrO_2 FILMS

S. Ishibashi, T. Namikawa and M. Satou
The Graduate School at Nagatsuta, Tokyo Institute of Technology, 2-12-1,
O-okayama, Meguro-ku, Tokyo, Japan

ABSTRACT

Thin films of single crystal CrO_2 were prepared in order to examine their crystallographic and magnetic properties as bit oriented memory materials.

The samples were grown epitaxially on {100}, {110}, {111} and {001} surfaces of rutile single crystals by decomposition of CrO_3.

The crystallographic orientation, surface morphology and magnetic properties of these films were investigated.

The results obtained by reflection electron diffraction showed that these films were all in the form of single crystals and the crystal axes of CrO_2 were parallel to that of the substrates.

By electron microscopy, it was indicated that the surface morphologies of the films grown on {001} and {111} rutile substrates were quite different from those of the films grown on the {100} and {110}.

Magnetic measurements showed the easy axis of magnetization lay almost parallel to the c-axis of CrO_2.

INTRODUCTION

Chromium dioxide, with the rutile-type crystal structure, is a ferromagnetic transition metal oxide, and it is well known that CrO_2 fine particles are useful for the magnetic recording material.

CrO_2 particles [1-4] have been prepared by the thermal decomposition of CrO_3 under a high pressure of oxygen gas, or by the vapor deposition of CrO_2Cl_2.

Epitaxially grown thin layers of CrO_2 were prepared by DeVries[5], in 1966. He obtained the layers by a kind of modified liquid phase epitaxy.

Stoffel[9] examined the magneto-optic properties these films,and suggested that magneto-optic read out of information stored on CrO_2 was feasible. Afterward, a few papers on CrO_2 thin films[6-8] were presented.

The purpose of this study is to obtain epitaxially grown CrO_2 single crystals by deposition from vaporized CrO_3, and to examine the surface morphology of CrO_2, which has the possibility to apply to optical bit oriented memory materials.

EXPERIMENTAL

The apparatus for the growth of the thin films is shown in Fig. 1. The starting material, CrO_3, was loaded in a platinum crucible placed in a 3ml pressure bomb, made of Hastelloy C. The bomb was heated at the rate of 60, 120, 240 and 360°C/hr and was kept at 400°C for 2 hours. The total pressure of reaction was controlled by both the temperature and the amount of loaded CrO_3. The pressure of oxygen gas liberated from the reaction $CrO_3 \rightarrow CrO_2 + 1/2\ O_2$ was 300-350 kg/cm^2. Rutile single crystals grown by Verneuil's method with {100}, {110},{111} and {001} surface were used as substrates, and were purchased from the Nakazumi Crystal Co. As it was considered that the condition of the surface of the substrates was the most important factor for epitaxial growth, the substrate crystals were polished carefully, cleaned and annealed for 5 hours at 1000°C in air. Then they were supported inside the crucible as illustrated in Fig. 1.

The surface morphologies of the CrO_2 thin films were observed with a scanning electron microscope(HFS-2, HITACHI).

The orientations of the over grown films relative to the rutile substrate were confirmed by reflection electron diffraction.

Magnetic properties, such as saturation magnetization at room temperature, coercive force in various crystallographic directions, and Curie temperature, were measured using a Foner type vibrating sample magnetometer in a 5000 Oe magnetic field.

RESULTS AND DISCUSSION

The prepared CrO_2 film thickness ranged from 0.5 μm to 1.5 μm. The film thickness increased as heating rate decreased. This indicates that the film thickness can be controlled by the total amount of CrO_3 diposited on the substrate, which is controlled by heating rate.

The adhesion of CrO_2 films to the substrate was fairly strong.

A. CrO_2 thin films grown on the rutile crystal

1) CrO_2 films on a {100} rutile substrate

Fig. 2(a) shows a scanning electron micrograph of the prepared CrO_2 film surface, and it is obvious that there exists a distinctly oriented growth. The structure could also be easily observed by optical microscope.

The reflection electron diffraction pattern of this film is shown in Fig. 2(b). The pattern indicates that the grown CrO_2 film is in the form of a single crystal, and the {100} of the grown film is parallel to the {100} of the rutile substrate, and the axes of the grown crystal are parallel to the axes of the rutile crystal.

2) CrO_2 films on {110} rutile substrates

The surface morphology of the films is shown in Fig. 3(a). A unidirectional elongated prismatic structure is observed. The structure also could be easily observed by optical microscope.

The reflection electron diffraction pattern obtained from this film, shown in Fig. 3(b), indicates that the grown film is in the form of a single crystal, and the {110} of the grown film is parallel to the {110} of the rutile substrate, and the axes of the grown crystal are parallel to the axes of the rutile crystal. The direction of the elongated structure of CrO_2 single crystal films was determined using reflection electron diffraction patterns by setting the direction of elongated structure to the fixed direction. The result obtained shows that the direction is parallel to $\langle 001 \rangle$ of CrO_2. This direction coincides with the long axis of a fine particle [1].

3) CrO_2 films on {111} rutile substrates

Fig. 4(a) shows the surface morphology of the prepared films. This reveals the presense of the oriented elongated structure on the surface. This microstructure could not be clearly observed by optical microscope.

The reflection electron diffraction pattern of this film indicates that the film is in the form of a single crystal, and the {111} of the grown film is parallel to the {111} of the rutile substrate, and the axes of the grown crystal are parallel to the axes of the rutile crystal(Fig. 4(b)).

4) CrO_2 films on {001} rutile substrates

The surface morphology of the grown film is shown in Fig. 5(a). The morphology of this film was very different from that of {100} and {110} films. This film has a moderately smooth surface, and shows a brownish-black metallic luster in reflection.

The reflection electron diffraction pattern from such surface, as shown in Fig. 5(b), indicates that this film is also in the form of a single crystal, and the {001} of the grown film is parallel to the {001} of the rutile substrate, and the axes of the grown crystal are parallel to the axes of the rutile crystal.

B. CrO_2 crystallites grown on the rutile crystal

Fig. 6 shows the micrograph of the substrate surface near the point indicated as A in Fig. 1, where a very little quantity of vaporized CrO_3 were deposited. These micrographs show many epitaxially grown crystallites exist on the substrate. It is considered that the continuously grown thin films described previously are formed by the coalescing these epitaxially grown crystallites; therefore, it is of interest to investigate this stage of growth.

As shown in Fig. 6(a), many tetragonal pyramidal crystallites and coalesced crystallites of CrO_2 standing perpendicular to the substrate can be observed on the {001} rutile substrate. In this micrograph, the growth of specific planes is observed. From electron diffraction photos and electron micrographs taken from the <001> direction of the same film, as shown in Fig. 6(c), these specific planes were found to be {100} and {110}. The {100} planes tend to develop more rapidly than the {110} planes on the {001} rutile substrate. This means that the growth rate of <110> is faster than that of <100> on the rutile.

When <001> of the rutile crystal lay in the plane, many epitaxially grown crystallites were grown in the form of an elongated tetragonal structure with <001> as dominating growth direction (Fig. 6(d)).

In the case that <001> of rutile crystal lay at intermediate angles between perpendicular and parallel to the plane of the substrate, crystallites, which grew inclined to the surface of the substrate, can be observed (Fig. 6(b)). The inclining angle of the crystallites corresponded to the angle between substrate plane and <001> of rutile.

From these results, it is considered that the differences in the surface morphology between these films are mainly based on the angle between <001> of the rutile crystal and the plane of substrate.

C. Magnetic properties of the films

Fig. 7 shows magnetization curves of a {100} film grown on the rutile substrate. These magnetization curves varied according to the crystallographic direction of CrO_2. In the <001> direction of CrO_2, a rectangular hysteresis loop was obtained. From these data, it seems that the easy axis of magnetization lies almost parallel to the <001> of CrO_2.

The Curie temperature of these CrO_2 thin films was determined to be about 120 °C. This value is nearly equal to the value of CrO_2 particles.

CONCLUSIONS

Thin films of single crystal CrO_2 were grown epitaxially on {100}, {110}, {111} and {001} surfaces of rutile single crystals by decomposition of CrO_3. The direction of the crystal axes of CrO_2 was parallel to the same indices of the rutile substrates.

The CrO_2 single crystal grown on the rutile exhibited the elongated prismatic structure, and the elongated direction was <001>.

Observation of the crystal habit of CrO_2 films grown on the {001} rutile surface indicated the growth rate of <110> was faster than that of <100>.

The difference between these films in surface morphology was considered to be based on the crystal habit of the CrO_2 on the substrates.

The data on the magnetization curves showed the easy axis of magnetization of CrO_2 lay almost parallel to the c-axis.

ACKNOWLEDGMENT

The authors are grateful to Dr. S. Tochihara of TMP Co., Ltd. for kind advice in the preparation of this paper.

REFERENCES

[1] T.J.Swoboda, Paul Arthur, Jr., N.L.Cox, J.N.Ingraham, A.L.Oppegard and M.S.Sadler, "Synthesis and Properties of Ferromagnetic Chromium Oxide", J.Appl.Phys., 32(3,Suppl),374S-75S,1961.

[2] B.Kubota,"Decomposition of Higher Oxides of Chromium Under Various Pressure of Oxygen", J.Am.Ceram.Soc.,44(5),239-48,1961.

[3] F.J.Darnell and W.H.Cloud,"Magnetic Properties of Chromium Dioxide",Bull.Soc.Chim.France,1164-66,1965.

[4] R.D.Shannon,B.L.Chamberland and Carol G.Frederick, "Effect of Foreign Ions on the Magnetic Properties of Chromium Dioxide",J.Phys.Soc.Japan,31(6)1650-56, 1971.

[5] R.C.DeVries,"Epitaxial Growth of CrO_2", Mat.Res.Bull.,1,83-93,1966.

[6] A.Funke und P.Roseman,"Verfahren zur Herstellung von Chromdioxid-Schichten fur Speicherelemente", D.D.R.Pat.98273.

[7] G.Elbinger und A.Lerm,"Verfahren zur Herstellung von ferromagnetischen Chromdioxidschichten fur Speichersysteme",D.D.R.Pat.100638.

[8] L.Ben-Dor and Y.Shimony,"Low Pressure Growth and Properties of CrO_2 on Al_2O_3",J.cryst.Growth, 24/25, 175-178,1974.

[9] A.M.Stoffel,"Magnetic and Magneto-optic Properties of FeRh and CrO_2",J.Appl.Phys.,40(3),1238-39,1969.

FIGURE CAPTIONS

Fig. 1 Apparatus for epitaxial growth of CrO_2 thin films and enlargement of the substrate holder

Fig. 2 (a) Scanning electron micrograph of CrO_2 thin film grown epitaxially on {100} rutile crystal (b) Reflection electron diffraction pattern corresponding to (a) (100KV)

Fig. 3 (a) Scanning electron micrograph of CrO_2 thin film grown epitaxially on {110} rutile crystal (b) Reflection electron diffraction pattern corresponding to (a) (50KV)

Fig. 4 (a) Scanning electron micrograph of CrO_2 thin films grown epitaxially on {111} rutile crystal (b) Reflection electron diffraction pattern corresponding to (a) (100KV)

Fig. 5 (a) Scanning electron micrograph of CrO_2 thin film grown epitaxially on {001} rutile crystal (b) Reflection electron diffraction pattern corresponding to (a) (50KV)

Fig. 6 Scanning electron micrographs of CrO_2 crystallites grown on the rutile crystals (a){001} surface (b){111} surface (c){001} surface, observation from <001> (d){110} surface

Fig. 7 Magnetization curves of the CrO_2 thin film epitaxially grown on {100} rutile substrate film thickness: 1.0 μm

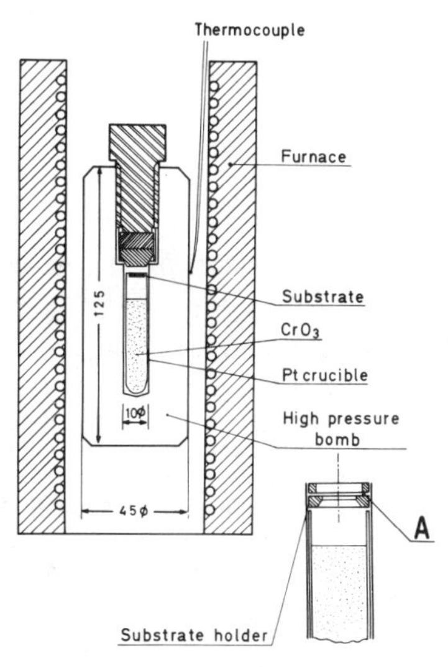

Fig. 1 Apparatus for epitaxial growth of CrO_2 thin films and enlargement of the substrate holder

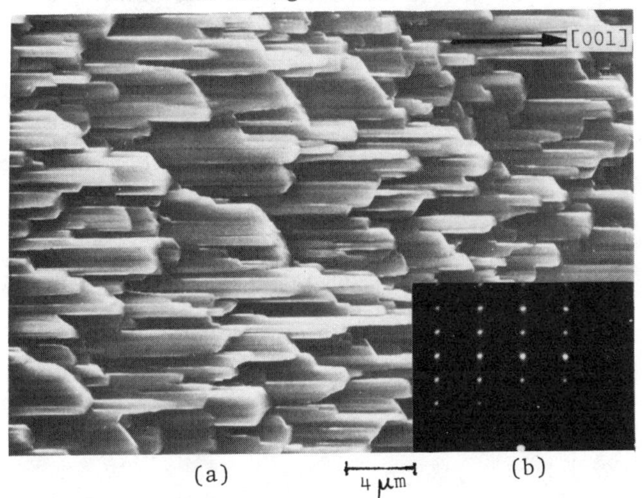

Fig. 2 (a) Scanning electron micrograph of CrO_2 thin film grown epitaxially on {100} rutile crystal
(b) Reflection electron diffraction pattern corresponding to (a) (100KV)

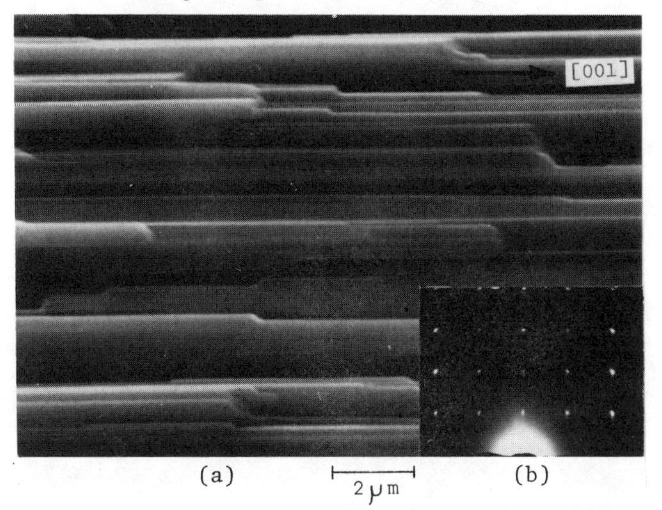

Fig. 3 (a) Scanning electron micrograph of CrO_2 thin film grown epitaxially on {110} rutile crystal
(b) Reflection electron diffraction pattern corresponding to (a) (50KV)

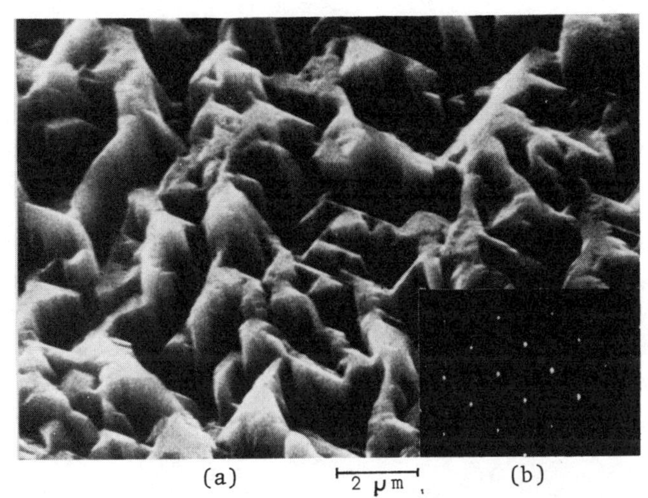

Fig. 4 (a) Scanning electron micrograph of CrO_2 thin films grown epitaxially on {111} rutile crystal
(b) Reflection electron diffraction pattern corresponding to (a) (100KV)

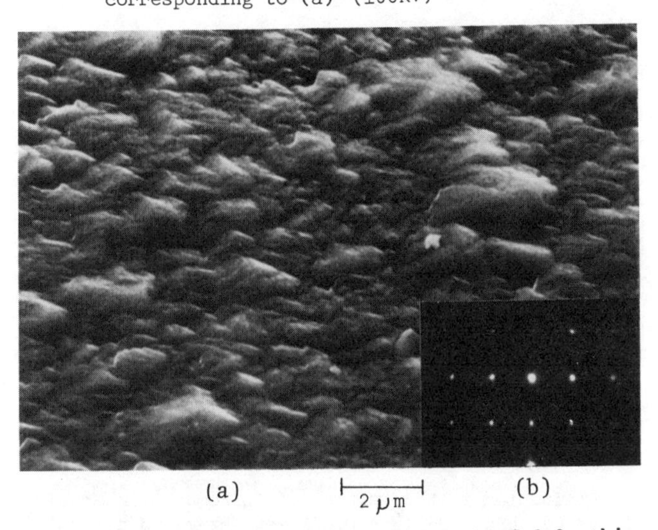

Fig. 5 (a) Scanning electron micrograph of CrO_2 thin film grown epitaxially on {001} rutile crystal
(b) Reflection electron diffraction pattern corresponding to (a) (50KV)

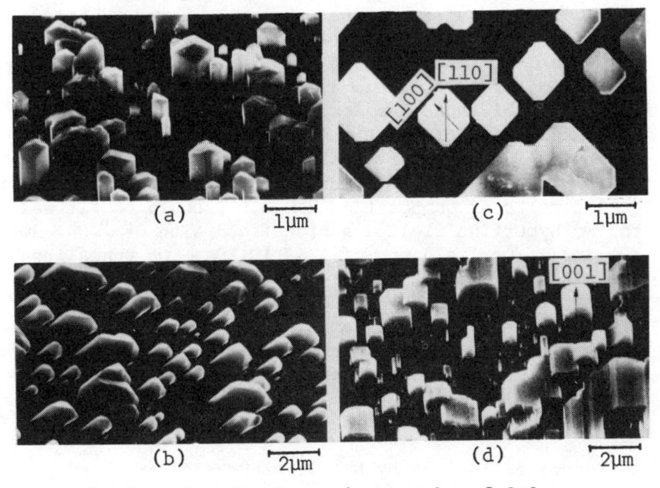

Fig. 6 Scanning electron micrographs of CrO_2 crystallites grown on the rutile crystals
(a){001} surface (b){111} surface
(c){001} surface, observation from <001>
(d){110} surface

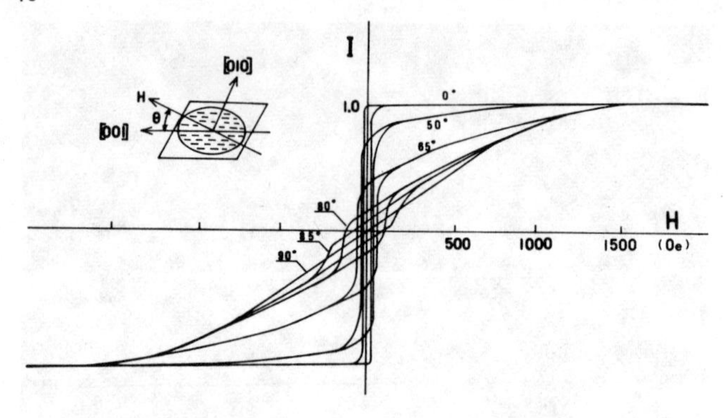

Fig. 7 Magnetization curves of the CrO_2 thin film epitaxially grown on {100} rutile substrate. Film thickness: 1.0μm.

^{119}Sn MAGNETIC HYPERFINE STRUCTURE IN Pd_2MnSb

M. Tenhover, P. Boolchand, C.S. Kim and S. Jha
Physics Dept., University of Cincinnati†
Cincinnati, Ohio 45221

ABSTRACT

Induction melted samples of ferromagnetic Heusler alloys $Pd_2MnSb_{1-x}Sn_x$ ($x \leq 0.02$) were examined using ^{119}Sn Mössbauer spectroscopy as a function of heat treatment and composition 'x'. Air cooled melts of said alloys examined at room temperature generally revealed narrow single lines ($\Gamma \sim 0.9$ mm/s) having an Isomer-shift (IS) of +1.45(4) mm/s wrt $BaSnO_3$. Below T_c however, broad internal field distributions were observed in the air cooled melts, results which are attributed to crystallographic disorder[1] in this system. A sample of composition x = 0.01, annealed for 3 days at 700°C, to promote crystallization of the $L2_1$ phase, showed a clear magnetic hyperfine structure yielding an internal field of |220 ± 10|kOe at Sn in Pd_2MnSb at 4.2°K. Spectra of air cooled melts of composition x = 0.02, could be fit well to two hyperfine fields, a high field site of |204 ± 20| kOe and a low field one of |94 ± 10|kOe. On annealing the x = 0.02 sample, the high field component became better defined and the low field one nearly vanished precipating pockets of elemental Sn. The low field site observed in these experiments seems to represent Sn in a defect site formed on air cooling the Pd_2MnSb melt.

REFERENCES

† Supported in part by NSF Grant-DMR-74-24061 and NASA Grant NSG 3091.

1. P.J. Webster and R.S. Tebble, Phil. Mag 16, 347 (1967).

EFFECTS OF HYDROSTATIC PRESSURE ON THE MAGNETIC PROPERTIES OF DISORDERED MONOSILICIDE $Fe_xCo_{1-x}Si$ ALLOYS

J. Beille and D. Bloch
Laboratoire de Magnétisme, C.N.R.S.,
166X, 38042-Grenoble-Cedex, France
V. Jaccarino
Physics Dept., University of California,
Santa Barbara, Calif. 93106, U.S.A.
J.H. Wernick
Bell Laboratories, Murray Hill, N. Jersey 07974, U.S.A.

ABSTRACT

Disordered $Fe_xCo_{1-x}Si$ alloys are known to be ferromagnetic in an iron concentration range starting at x = 0.2. We have measured their magnetization in magnetic fields up to 40 kOe at 4.2 K and under hydrostatic pressures up to 8 kbar, as well as their zero pressure susceptibility at 4.2 K in high magnetic fields of 150 kOe. We have also determined the Curie temperatures of the ferromagnetic samples under hydrostatic pressure up to 8 kbar from the temperature variation of the initial susceptibility. We determined a critical iron concentration of x ≃ 0.95 for ferromagnetism in FeSi rich alloys. From some of our results, the ferromagnetism in the CoSi rich alloys seems to appear in a rather homogeneous way, whereas the FeSi rich alloys in contrast exhibit local effects.

NUCLEAR MAGNETIC RESONANCE STUDY OF FERROMAGNETIC $Mn_{1-x}Cu_xSb$

T. Rajasekharan* and K. V. S. Rama Rao
Department of Physics
Indian Institute of Technology, Madras 600 036, India

ABSTRACT

The effect of introduction of Cu atoms in small quantities into a matrix of MnSb has been studied by the zero-field Nuclear Magnetic Resonance (NMR) technique. The compounds studied are $Mn_{0.98}Cu_{0.02}Sb$, $Mn_{0.97}Cu_{0.03}Sb$ and $Mn_{0.96}Cu_{0.04}Sb$. At 300 K, all materials give a group of five lines which are the quadrupolar split ^{55}Mn resonance lines. The frequencies of these lines are about 1-1.5MHz below the corresponding ^{55}Mn resonance lines in MnSb. The introduction of Cu atoms which are assumed to occupy the Mn sites in the NiAs type MnSb lattice causes a decrease in the exchange bonds between the Mn atoms and consequently causes a reduction in the hyperfine field experienced by the ^{55}Mn nuclei. A broad line with a peak around 231 MHz in $Mn_{0.98}Cu_{0.02}Sb$ and at 233.5 MHz in $Mn_{0.96}Cu_{0.04}Sb$ is assigned to unresolved satellite lines caused by the Mn atoms with Cu atoms in the neighbourhood.

INTRODUCTION

In continuation of the study of the effect of introduction of different elements into the intermetallic compound MnSb by the zero-field NMR technique in this laboratory[1], we report in this paper the study of the $Mn_{1-x}Cu_xSb$ system.

MnSb has a NiAs type crystal structure[2] shown in Fig. 1. It is ferromagnetic with a Curie temperature of 587 K[3]. The observation of NMR in MnSb was reported for the first time by Tsujimura et.al[4]. They reported ^{55}Mn, ^{123}Sb and ^{121}Sb resonances at 273 K. We have reported previously the effect of the substitution of Sb atoms by Sn atoms in MnSb[1]. NMR investigation of $Mn_{1+x}Sb_{1-y}Sn_y$ has been carried out by Bouma[5].

EXPERIMENTAL PROCEDURES

The specimens of $Mn_{1-x}Cu_xSb$ (x = 0.02, 0.03 and 0.04) were prepared using Mn, Sb

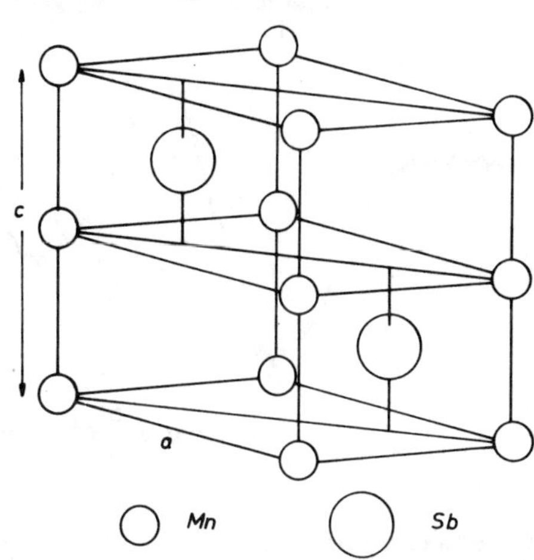

Fig. 1. Crystal structure of Mn Sb

and Cu of 99.99% purity as starting materials. The components in the powder form were mixed in stoichiometric proportions and heat treated in evacuated quartz tubes at 620 C for 24 hrs, at 860 C for 75 hrs and then cooled slowly to room temperature. The ingot was powdered and again heated at 860 C for 75 hrs. The product was powdered and well annealed at 400 C. X-ray photograph showed the presence of a single phase, and no lines from free Mn, Sb or Cu elements were detected.

An externally quenched superregenerative spectrometer was employed for studying the zero-field NMR. The samples were used in the powder form. The first derivatives of the signals were recorded using a PAR Lock-in-Amplifier and a strip chart recorder. The frequencies were measured using a Marconi R.F.Signal Generator in conjunction with a Hewlett-Packard frequency counter.

EXPERIMENTAL RESULTS

A recording of the quadrupolar split ^{55}Mn resonance signals in MnSb is shown in Fig. 2. In the figure, each domain wall line has a domain line at approximately 0.5 MHz towards the higher frequency side[6]. The domain wall lines have a dispersion shape and the domain lines have a mixture of absorption

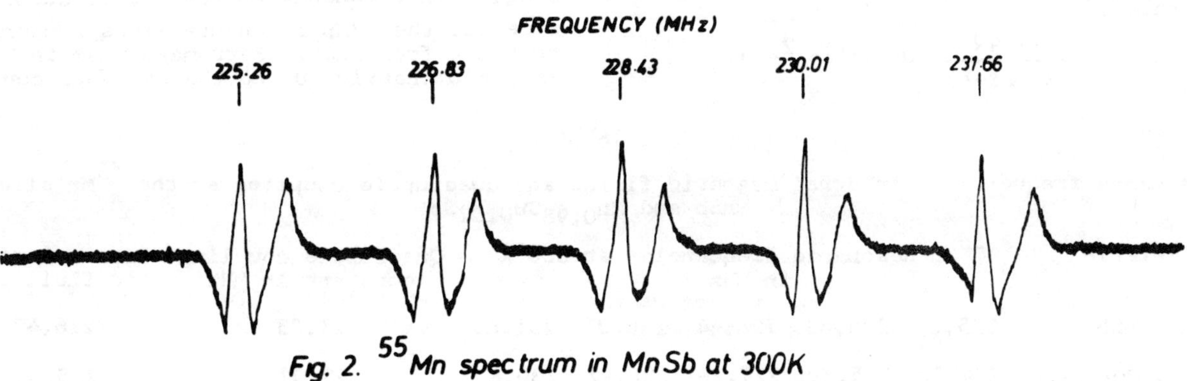

Fig. 2. ^{55}Mn spectrum in MnSb at 300K

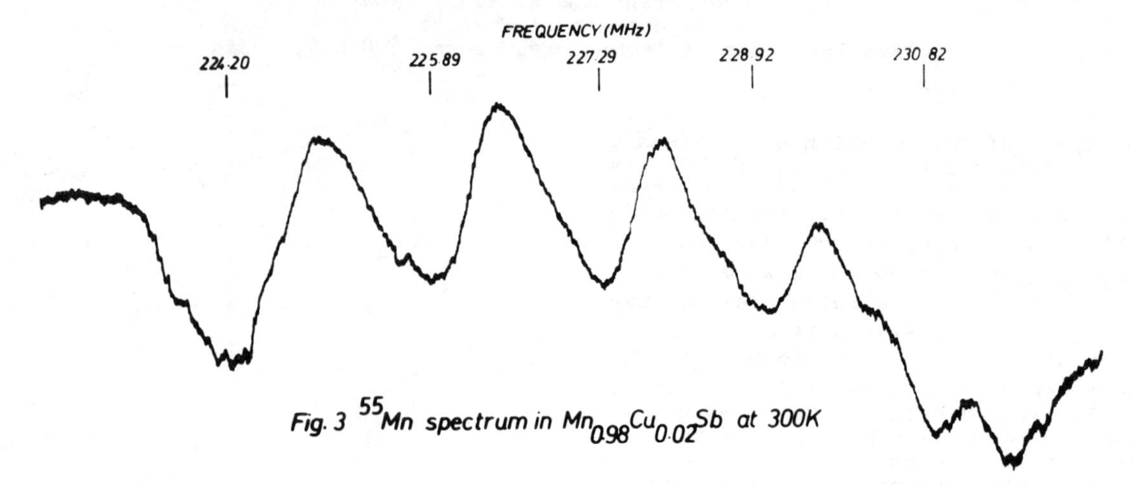

Fig. 3 ^{55}*Mn spectrum in* $Mn_{0.98}Cu_{0.02}Sb$ *at 300K*

and dispersion shape[7].

The NMR spectrum obtained from the material $Mn_{0.98}Cu_{0.02}Sb$ is shown in Fig. 3. The spectrum was recorded with a much larger amount of the sample compared to MnSb as the signals in this material are much weaker. There are five lines at frequencies 224.20, 225.89, 227.29, 228.92 and 230.82 MHz. These frequencies are listed in Table I along with the frequencies of the resonances from ^{55}Mn nuclei in domain walls in MnSb. A broad line whose frequency is around 231 MHz is also observed.

From the frequencies of the five lines observed in $Mn_{0.98}Cu_{0.02}Sb$, it is inferred (on comparison with the frequencies of the ^{55}Mn lines from MnSb) that these are due to ^{55}Mn nuclei. It is also inferred that they are from nuclei in the domain walls because of their line shapes and also from the fact that signals from the nuclei in domains are generally two orders of magnitude weaker than those from nuclei in domain walls[8]. It is observed that the lines from the material $Mn_{0.98}Cu_{0.02}Sb$ occur at frequencies about 1 MHz lower than those from MnSb.

Table I also lists the quadrupole coupling constants for the ^{55}Mn nuclei in MnSb and $Mn_{0.98}Cu_{0.02}Sb$ and the hyperfine fields at their sites. The quadrupole coupling constants are obtained from the separation of the outer most satellites using the formula[9],

$$\Delta\nu = \frac{3e^2qQ}{2I(2I-1)}\,(m-\tfrac{1}{2})(3\mu^2-1).$$

$\Delta\nu$ is the separation between the satellites corresponding to the transitions m to (m-1) and -m to -(m-1), where m is the magnetic quantum number. $\mu = \cos\theta$, where θ is the angle between the c-axis and the direction of magnetization. The formula is applicable in cases of axially symmetric field gradients. Since the magnetization lies in the basal plane for annealed MnSb below 520 K[10], θ is taken as 90° in the calculations. The quadrupole coupling constant for the ^{55}Mn nuclei in $Mn_{0.98}Cu_{0.02}Sb$ is slightly larger than that of ^{55}Mn nuclei in MnSb.

A spectrum obtained in $Mn_{0.96}Cu_{0.04}Sb$ is shown in Fig. 4. It is observed that the ^{55}Mn lines are shifted by around 1.4 MHz to the low frequency side with respect to the lines from MnSb. It is also observed that the ^{55}Mn lines are decreased in intensity and that the broad line increases in intensity and shifts to a higher frequency of 233.5 MHz compared to $Mn_{0.98}Cu_{0.02}Sb$. The results in $Mn_{0.97}Cu_{0.03}Sb$ are intermediate between those in $Mn_{0.98}Cu_{0.02}Sb$ and $Mn_{0.96}Cu_{0.04}Sb$. The ^{123}Sb spectrum could not be observed in the range 180-210 MHz, probably because they were too weak to be observed.

INTERPRETATION OF THE RESULTS

In the materials $Mn_{1-x}Cu_xSb$, there is a shift in frequency to the low frequency side for the ^{55}Mn resonance lines compared to those from MnSb. Ferromagnetism in MnSb is predominantly due to the exchange coupling

TABLE I

Resonance frequencies, internal magnetic fields and quadrupole coupling at the ^{55}Mn sites in MnSb and $Mn_{0.98}Cu_{0.02}Sb$

Material	^{55}Mn Resonance frequencies at 300 K in MHz					Quadrupole coupling constant in MHz	Internal magnetic field in kOe
MnSb	225.26	226.83	228.44	230.01	231.66	21.33	216.47
$Mn_{0.98}Cu_{0.02}Sb$	224.20	225.89	227.29	228.92	230.82	22.07	215.38

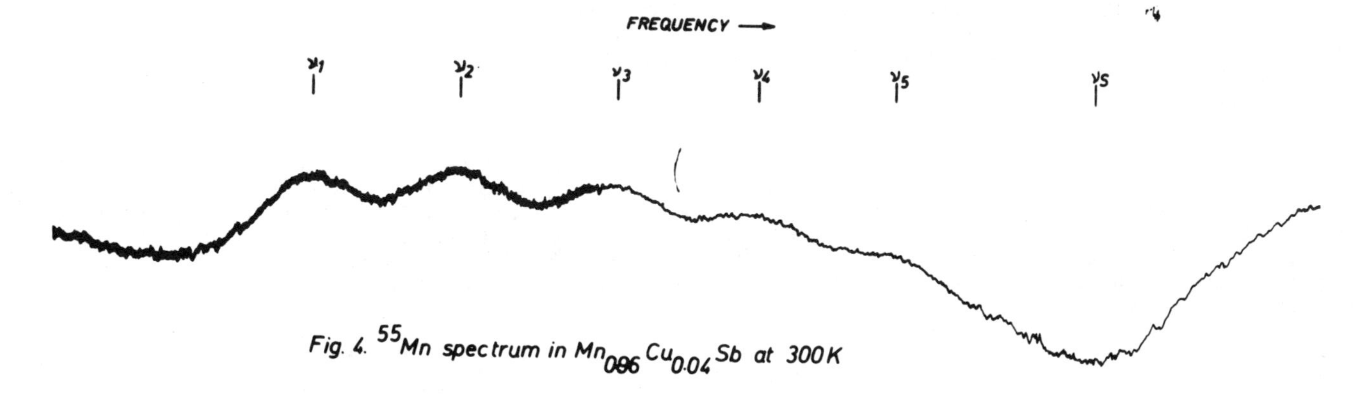

Fig. 4. ^{55}Mn spectrum in $Mn_{0.96}Cu_{0.04}Sb$ at 300K

between the Mn atoms in the neighbouring c-planes because the Mn-Mn distance is the shortest in the c-direction (2.89 Å)[5]. The introduction of copper atoms, which are assumed to occupy the Mn sites in the NiAs type MnSb lattice, causes a decrease in the exchange bonds between the Mn atoms and causes fluctuations in tthe average spin value of the Mn ions[11]. Consequently, there is a change in the hyperfine field experienced by ^{55}Mn nuclei.

If the introduction of Cu strains the lattice, we would expect random variations in the direction of the electric field gradient axis. This would cause line broadening. In the first order, the satellites would be affected while the $-\frac{1}{2} \rightarrow \frac{1}{2}$ transition remains unperturbed[12]. As the strains increase, the satellites would be affected severely. It is observed that the central line as well as the satellites are equally broadened in the ^{55}Mn spectra of $Mn_{1-x}Cu_xSb$. This suggests that the lower signal to noise ratio of the resonance lines from $Mn_{0.98}Cu_{0.02}Sb$ compared to those from MnSb is not due to the straining of the lattice. Hence we infer that the reduction in intensity of the lines is due to the fact that the Mn atoms with Cu atoms in the neighbourhood have a different hyperfine field at their nuclei. As there are many satellite lines possible for the different quadrupolar split lines, these satellite lines are not resolved and they form the broad line at the high frequency end of the five line spectrum. This is corroborated by the observation that the intensity of the broad line increases with an increase of copper content. Though we are unable to make any quantitative estimate of the hyperfine fields at the Mn sites with Cu neighbours, we can conclude that the introduction of Cu into the neighbouring sites of Mn enhances the hyperfine field at such Mn nuclei. In Mn_2Sb also, introduction of Cu into the Mn_{II} sub-lattice[11] causes a satellite line to appear on the higher frequency side of the Mn_{II} resonance line.

ACKNOWLEDGEMENTS

The authors are grateful to Prof. C. Ramasastry (Indian Institute of Technology, Madras) and Prof. Alarich Weiss (Technische Hochschule, Darmstadt) for their interest in this work. One of the authors (KVS) is grateful to Prof. R. D. Spence (Michigan State University, East Lansing) for introducing him to the line of zero-field NMR.

REFERENCES

* National Science Talent Scholar

1. T. Rajasekharan and K. V. S. Rama Rao, Proc.Nucl.Physics and Solid State Physics Symposium, Calcutta, December,(1975)
2. A.J. Cornish, Acta Metallurgica 6, 371 (1958)
3. B.T.M. Willis and H.P. Rooksby, Proc.Phy. Soc. B, 67, 290 (1954)
4. A. Tsujimura, T. Hihara and Y. Koi, J.Phy. Soc.Japan 17 1078 (1962)
5. J. Bouma, Thesis, University of Groningen (1972)
6. V. Nagarajan and R. Vijayaraghavan, J.Phy.Soc.Japan 33 88 (1972)
7. R.L.Streever and L.H.Bennet, Phy.Rev. 131 2000 (1963)
8. A. M. Portis and R. H. Lindquist, Magnetism, Ed.G.T.Rado and H.Suhl II A 362 (Academic Press, 1965)
9. A. Abragam, The Principles of Nuclear Magnetism 234 (Oxford at the Clarendon Press, 1961)
10. G. A. Murray and W. Marshall, Proc. Phy. Soc. 86 315 (1965)
11. N.M.Kovtun, V.M.Khmara, E.M.Morozov and G.A.Troitskii, Soviet Phys. Solid State 13 536 (1971)
12. M. H. Cohen and F. Reif, Solid State Physics, Ed. F. Seitz and D. Turnbull 5 363-400 (Academic Press, 1957)

THE MAGNETIC RESISTANCE ANISOTROPY IN NICKEL AND IRON BASED ALLOYS

J.W.F. Dorleijn and A.R. Miedema

Philips Research Laboratories
Eindhoven, The Netherlands

ABSTRACT

Ferromagnetic metals display the phenomenon of the resistivity anisotropy: the value of the resistivity depends on the way in which a particular metal is magnetized, i.e. parallel or perpendicular to the current.

In nickel as well as iron alloys a simple description of this resistivity anisotropy, $(\rho_{\parallel} - \rho_{\perp})/\rho_{\parallel}$, can be obtained in terms of the two-current model. The conduction electrons are classified according to their spin direction, leading to two independent parallel currents. An impurity metal, dissolved in Ni or Fe, is characterized by two specific residual resistivities ($\rho^{\uparrow}$ and $\rho^{\downarrow}$), one for either group of current carriers.

From experiments at low temperature we derived that the two currents produce anisotropy effects of opposite signs. Hence large effects only occur when one of the two currents dominates ($\rho^{\uparrow} \ll \rho^{\downarrow}$).

Experimental data for $\rho^{\uparrow}$, $\rho^{\downarrow}$ and $(\rho_{\parallel} - \rho_{\perp})/\rho_{\parallel}$ are reported for alloys based on Ni with Al, Au, Co, Cr, Cu, Fe, Mn, Pt, Re, Rh, Ru, Si, Sn, Ti, V and Zn and for alloys based on Fe with Al, Be, Co, Cr, Ga, Ge, Ir, Mn, Mo, Ni, Os, Pt, Re, Rh, Ru, Si, Ti, V, W and Zn. The total impurity concentration is always smaller than 3 at %.

The consequences of the present low temperature results for applications at room temperature are discussed.

THE TWO-CURRENT MODEL

It was suggested by Mott[1] that the unusual temperature dependence of the resistivity of ferromagnetic metals, e.g. the difference between Ni and Pd, can be explained in terms of a so-called two-current model. More recently this model was used by a number of investigators in order to describe data on a variety of transport properties in ferromagnetic metals, thus providing strong support for the validity of the model. See for instance a review by Fert and Campbell[2]. Unambiguous evidence for the validity of the two current model can be found in experiments on the resistivity of ternary ferromagnetic alloys: extremely large deviations from Matthiessen's rule do occur at low temperatures (see below). Ternary nickel alloys were studied by Fert and Campbell[2,3], Leonard et al[4] and on a large scale by the present authors[5]; a few experiments for ternary cobalt base and iron base alloys were reported by Durand, Loegel and Gautier[6,7], and Campbell et al[8], respectively. The two-current model is also essential in a description of the thermoelectric power (see ref. 9 and further references given there), the Hall effect and the magnetoresistance[2,10]. Parameters that characterize an impurity in the two current model, i.e. the two residual resistivities, were determined by studies of the temperature dependence of binary alloys[2,3,6,8,11] as well as from data on ternary alloys at low temperatures[4,5].

In the two current model the conduction electrons are divided into two groups. One group is associated with electrons having their magnetic moment parallel to the net magnetization $4\pi M_s$ (indicated by $\uparrow$) while for the other group the magnetic moment is antiparallel to the magnetization and labelled by $\downarrow$. One can safely assume that at least at low temperatures $\uparrow$ electrons can only be scattered to $\uparrow$ states since the angular momentum will be conserved. This results in a picture of two independent currents in parallel, as indicated in fig. 1. Depending on the source of resistivity (e.g. a particular solute element or phonons) the specific resistivities can be quite different for the two bands; a difference by a factor of 10 is quite realistic.

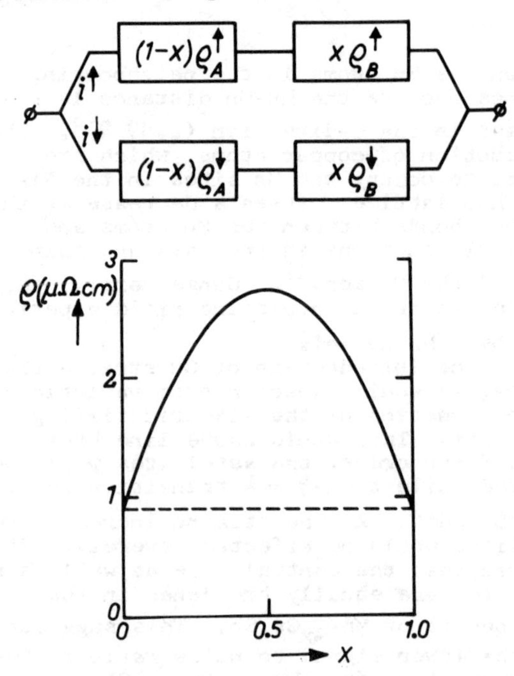

Fig. 1. Top: model of two current conduction in a ternary alloy $M_{99}A_{1-x}B_x$. Bottom: a numerical example of the resistivity in a ternary alloy $M_{99}A_{1-x}B_x$ as a function of x; for values of parameters see text.

In fig. 1 we illustrate the implications of the two-current model for the residual resistivity of a set of ternary alloys of the type $M_{99}A_{1-x}B_x$, where M stands for Ni or Fe, A or B are different solute elements and x varies between 0 and 1. For non-interacting impurities the residual resistivities within the up or down band can be added as is sketched in fig. 1:

$$\rho^{\uparrow} = (1-x)\rho_A^{\uparrow} + x\rho_B^{\uparrow}$$
$$\rho^{\downarrow} = (1-x)\rho_A^{\downarrow} + x\rho_B^{\downarrow} \quad (1)$$

The total resistivity is found from

$$\rho = \rho^{\uparrow}\rho^{\downarrow} / (\rho^{\uparrow} + \rho^{\downarrow}) \quad (2)$$

which differs from

$$\rho = (1-x)\rho_A + x\rho_B \quad (3)$$

from which one may see that Matthiessen's rule is not valid.

As an illustration fig. 1 gives the example of a ternary alloy for which we assume: $\rho_A^{\uparrow} = 1$, $\rho_A^{\downarrow} = 10$ (so $\rho_A = 10/11$) for impurity A; the reverse: $\rho_B^{\uparrow} = 10$, $\rho_B^{\downarrow} = 1$ (so $\rho_B = 10/11$) for impurity B, with all numbers in $\mu\Omega$cm/at%.

One immediately sees from eq. (1) and (2) that the resistivity of a ternary alloy with 0.5 at% A and 0.5 at% B is much larger than 10/11 $\mu\Omega$ cm since $\rho^{\uparrow} = 5.5$, $\rho^{\downarrow} = 5.5$, so $\rho = 2.75\ \mu\Omega$ cm.

Albeit that the example of fig. 1 may seem rather exaggerated, in practice effects of a comparable order of magnitude do occur, as we illustrate for both nickel and iron base alloys in fig. 2. For both $Ni_{99}Co_{1-x}Rh_x$ and $Fe_{99}Co_{1-x}Mn_x$ the deviations from Matthiessen's rule (the dashed line in fig. 2) are about a factor of 2. In other systems the deviations can be even larger (e.g. in $Fe_{99}Al_{1-x}Mo_x$ and in $Fe_{99}Al_{1-x}V_x$) but they can also be very small (e.g. in $Fe_{99}Ru_{1-x}V_x$, $Ni_{99}Au_{1-x}Co_x$ and $Ni_{99}Rh_{1-x}Ru_x$).

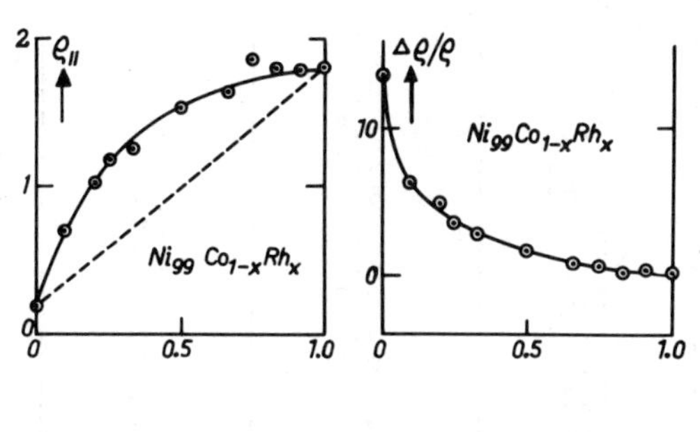

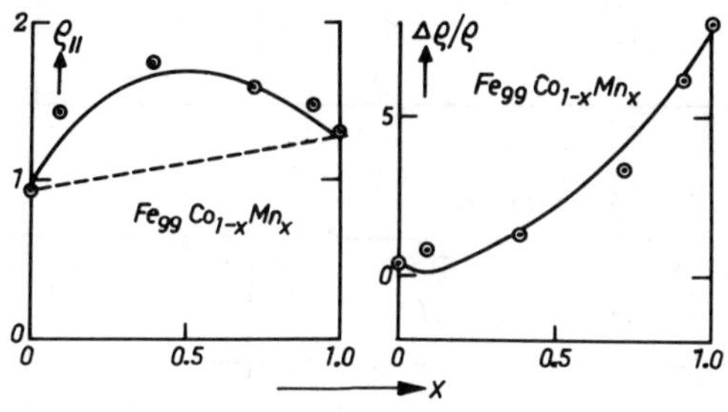

Fig. 2. The resistivity (in $\mu\Omega$ cm) at 4.2 K and the resistivity anisotropy (in %) in $Ni_{99}Co_{1-x}Rh_x$ and in $Fe_{99}Co_{1-x}Mn_x$ as a function of x.

Most of the actual measurements were done on alloys with a total solute concentration of 3 at %. For various alloys (with constant x) $\rho_{//}$ was found to vary linearly with the total solute concentration. The solid lines are calculated in the two current model, i.e. for the FeCoMn system using the values for the parameters $\rho^{\uparrow}_{Co}$, $\rho^{\downarrow}_{Co}$, $\rho^{\uparrow}_{Mn}$ and $\rho^{\downarrow}_{Mn}$ obtained in a best fit procedure of 250 different Fe-based alloys. The dashed lines would be expected according to Matthiessen's rule.

In fig. 1 we have calculated the residual resistivities for a set of ternary alloys from the sub-band resistivities of two impurities. It goes without saying that one can reverse this procedure: having observed deviations from Matthiessen's rule in a system $M_{99}A_{1-x}B_x$ for a number of compositions x, one can derive $\rho^{\uparrow}_A$, $\rho^{\downarrow}_A$, $\rho^{\uparrow}_B$ and $\rho^{\downarrow}_B$. If the two current model is a good approximation the results for $\rho^{\uparrow}_A$ and $\rho^{\downarrow}_A$ derived from the systems $M_{99}A_{1-x}B_x$ and $M_{99}A_{1-x}C_x$ should not differ significantly, which is indeed the case.

We have obtained experimental information on the residual resistivity of about 250 binary and ternary nickel-based alloys and about 150 iron-based alloys. The sub-band resistivities for a given solute element have been obtained by making a computer-fit of all data in terms of the two current relations (1) and (2). The results are given in table I. Since the resistivity of ferromagnetic alloys depends on the angle between the current and the magnetization, the data all apply to the same geometry, i.e. the current is parallel to the total induction B but extrapolated to B = 0. From the large variations in the ratio $\alpha = \rho^{\downarrow}/\rho^{\uparrow}$ for different solute elements one may conclude that there is little or no relation between the residual resistivities and the band structure of pure nickel and iron. For instance in binary NiRu only 7% of the current is carried by ↑ electrons whereas in NiFe or in NiCo alloys more than 90% is. For iron based alloys too large differences in the ratio $\beta = i^{\uparrow}/i_{total}$ occur between impurities: it equals 90% for FeV and about 10% for FeAl binary alloys.

In a certain sense at elevated temperatures binary alloys can be considered as ternary alloys in which one of the solute elements is replaced by phonons. As a consequence large deviations from Matthiessen's rule also occur as a function of temperature. One may likewise characterize the thermal resistivities of pure iron and nickel by values of $\beta = i^{\uparrow}/i_{total}$. For pure iron this value is about 0.3. At room temperature large deviations from Matthiessen's rule do occur for FeV and FeMo while no deviations are found for FeSi and FeCo, from which we derive β is about 0.3 for pure iron.

Similarly, for nickel, one can derive $\beta \approx 0.7$ for phonons. In table II we included for iron base alloys also some solute elements which have only been studied as binary alloys. For these cases $\rho^{\uparrow}$ and $\rho^{\downarrow}$ have been estimated from deviations from Matthiessen's rule at room temperature.

The value of β for phonons indicates how to define which are the ↑ and which are the ↓ carriers. One may expect that for iron and nickel at room temperature the phonon resistivity is the higher for that spin current that corresponds to the higher density of states. Information for the density of states can be obtained from theoretical band structure calculations [13,14]; in both nickel and iron the density of sates is smaller for the majority spin carriers. Our convention agrees with this theoretical expectation. However, we note that an interchange of ↑ and ↓ is a matter of notation only and has no consequences for our analysis.

THE RESISTANCE ANISOTROPY

Ferromagnetic metals display the phenomenon of the resisitivity anisotropy, i.e., the resistivity depends on the angle between the spontaneous magnetization and the current. This phenomenon is of technical interest for recording applications [15]. The phenomenon of the resistivity anisotropy is extensively investigated in Ni-based alloys. It is well known that effects of practical interest (a relative change in resistivity of a few percents) can be realized in NiFe or in NiCo alloys [16]. With the exception of FeV alloys [17] there is no information about the effect in iron-based alloys. We investigated extensively the phenomenon of the resistivity anisotropy in the nickel and iron alloys mentioned above, with the aim to establish a correlation (if any) in terms of the two-current model between the sub-band resistivities and the resistivity anisotropy effect of a given solute element.

The experimental procedure is clear from fig. 3. The resistivity in the two extreme orientations of the external field with respect to the current is recorded as a function of the field in the region of complete magnetic saturation (no Weiss domains). At liquid helium temperatures the dependence of the resistivity on the field originates from the Lorentz force on the conduction electrons. In order to separate this Kohler resistivity from the anisotropy effect one has to extrapolate the experimental data to B = 0. In

this respect the analysis at low temperatures is different from the one at elevated temperatures. At room temperature Lorentz forces are of minor importance since they depend on B/ρ and ρ is of the order of $10\ \mu\Omega$ cm. As a result the magnetoresistance is negative since an external field suppresses the magnetic excitations (spin waves) which contribute to the room temperature resistivity. At zero temperature there are no magnetic excitations so that the Lorentz force effects dominate.
Hence we feel that the definition of the anisotropy effect at relatively high temperature should be at $H = 0$ rather than at $B = 0$. Since the range of extrapolation, which equals $4\pi M_s$, is much larger for iron than for nickel-based alloys, the experimental uncertainties involved are correspondingly larger.

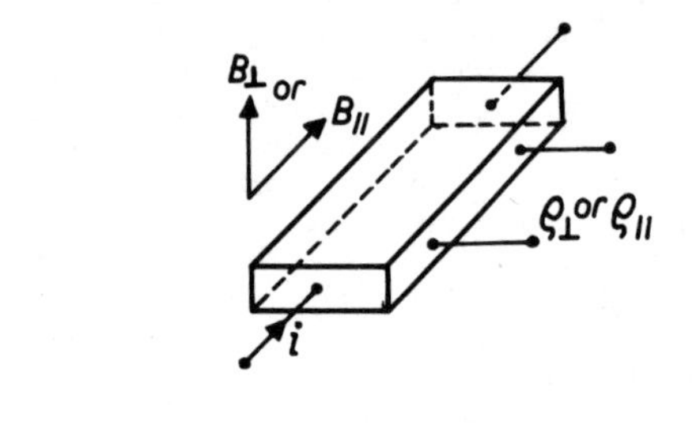

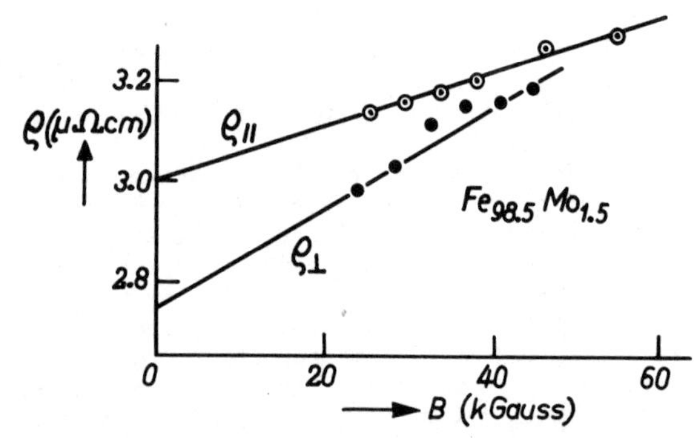

Fig. 3. The resistivity anisotropy in $Fe_{98.5}Mo_{1.5}$. Two extreme cases do occur: $B // i$ and $B \perp i$; $B = H_{ext} - H_{dem} + 4\pi M_s$. For a discussion of the extrapolation see [12].

As said above the residual resistivities of table I are obtained from experiments at 4.2 K in the configuration $\vec{B} // \vec{i}$, extrapolated to $B = 0$. Clearly one can repeat this analysis for the configuration $\vec{B} \perp \vec{i}$. However, it is more straightforward to do the analysis in terms of $(\Delta\rho/\rho)$, $(\Delta\rho/\rho)^\uparrow$ and $(\Delta\rho/\rho)^\downarrow$, i.e. decomposing the observed anisotropy effect into the contributions from the two spin currents. We define the anisotropy effect as

$$(\Delta\rho/\rho) = (\rho_{//} - \rho_\perp)/\rho_{//} \tag{4}$$

We prefer relation (4) because the parallel orientation clearly is the reference direction for which there is no coupling of the two currents through normal or anomalous Hall effects.

In a two-current model the two currents contribute to the total effect according to the corresponding current fraction:

$$\frac{\Delta\rho}{\rho} = \frac{i^\uparrow}{i_{total}}\frac{\Delta\rho^\uparrow}{\rho^\uparrow} + \frac{i^\downarrow}{i_{total}}\frac{\Delta\rho^\downarrow}{\rho^\downarrow} \tag{5}$$

where in case of a ternary alloy

Table I. The specific sub-band resistivities at 4.2 K calculated in the two-current model for different solute elements in Ni or in Fe and the resistivity anisotropy effect in binary alloys. For asterisked iron-based alloys values of $\rho^\uparrow$ and $\rho^\downarrow$ were estimated, see text; $i^\uparrow/i_{total} = \rho^\downarrow/(\rho^\uparrow + \rho^\downarrow)$.

Alloys based on nickel.

	$\rho^\uparrow$ in $\mu\Omega$cm/at %	$\rho^\downarrow$ in $\mu\Omega$cm/at %	$i^\uparrow/i_{total}$	$(\Delta\rho/\rho)$ in %
Al	3.4	5.8	.63	+3.8
Au	.44	2.6	.86	+7.5
Co	.20	2.6	.93	+13.5
Cr	29.	6.1	.17	-0.35
Cu	1.3	3.8	.74	+6.8
Fe	.44	4.8	.92	+12.5
Mn	.83	5.2	.86	+7.8
Pt	3.6	.85	.19	+0.40
Re	24.	7.5	.24	-0.50
Rh	8.0	2.3	.23	+0.05
Ru	72.	5.4	.07	-0.6
Si	5.0	6.4	.56	+2.1
Sn	4.4	7.2	.62	+2.9
Ti	7.6	7.2	.49	+0.55
V	14.	6.4	.31	+0.15
Zn	1.3	2.9	.70	+4.6

Alloys based on iron.

	$\rho^\uparrow$ in $\mu\Omega$cm/at %	$\rho^\downarrow$ in $\mu\Omega$cm/at %	$i^\uparrow/i_{total}$	$(\Delta\rho/\rho)$ in %
Al	48.	5.6	0.11	-0.30
Be^x	29.	4.7	0.14	+0.45
Co	4.5	1.2	0.21	+0.40
Cr^x	2.6	7.0	0.73	+5.6
Ga^x	44.	5.4	0.11	+0.15
Ge^x	49.	7.9	0.14	0.0
Ir^x	20.	2.2	0.10	+0.30
Mn	1.5	8.5	0.85	+8.0
Mo	2.3	11.	0.83	+8.4
Ni	17.	2.4	0.13	+1.6
Os	4.3	13.	0.75	+6.8
Pt^x	12	1.5	0.11	+0.60
Re	2.7	8.7	0.77	+8.3
Rh	6.4	1.1	0.15	+1.0
Ru	2.8	7.3	0.72	+4.7
Si	36.	6.4	0.15	-0.20
Ti^x	4.4	6.6	0.60	+3.0
V	1.0	7.5	0.88	+11.0
W^x	1.8	7.5	0.81	+8.0
Zn^x			0.35	+3.6

$$\Delta\rho^\uparrow = (1-x)\,\Delta\rho_A^\uparrow + x\cdot\Delta\rho_B^\uparrow$$
$$\Delta\rho^\downarrow = (1-x)\,\Delta\rho_A^\downarrow + x\,\Delta\rho_B^\downarrow \tag{6}$$

We have determined values of $(\Delta\rho/\rho)$ for different solute elements per sub-band by analysing sets of ternary alloys as in fig. 2 in terms of eq. (5) and (6), using the sub-band resistivities from table I.

For solutes in nickel the results given in table II exhibit a clear pattern. For all elements in nickel the anisotropy is positive for the ↑ sub-band

Table II. The resistivity anisotropy effect within each sub-band for different solute elements in Ni, arranged in the sequene of decreasing $i^{\uparrow}/i_{total} = \rho^{\downarrow}/(\rho^{\uparrow}+\rho^{\downarrow})$.
Note: These values and the ones in table I for nickel alloys differ slightly from those given in ref. 18 because the experimental material has been extended.

	$i^{\uparrow}/i_{total}$	$(\Delta\rho/\rho)^{\uparrow}$ in %	$(\Delta\rho/\rho)^{\downarrow}$ in %
Co	.93	+14.	-1.5
Fe	.92	+14.	-2.2
Mn	.86	+9.6	-3.3
Au	.86	+8.1	-2.2
Cu	.74	+9.7	-2.1
Zn	.70	+7.8	-2.9
Al	.63	+7.3	-2.3
Sn	.62	+6.3	-2.7
Si	.56	+6.2	-3.2
Ti	.49	+4.2	-3.2
V	.31	+7.9	-3.1
Re	.24	+6.5	-2.4
Rh	.23	+7.8	-2.3
Pt	.19	+5.8	-0.8
Cr	.17	+5.8	-1.7
Ru	.07	+7.6	-1.1

and negative for the ↓ sub-band. In approximate terms $(\Delta\rho/\rho)^{\uparrow}$ lies in the region of +10% while the mean value of $(\Delta\rho/\rho)^{\downarrow}$ is about -2%. The approximately constant values of $(\Delta\rho/\rho)$ for each sub-band for different elements are particularly interesting since there are large differences in the actual resistivity values (see table I).
As to any rule the general rule of unique sign for $\Delta\rho/\rho$ for each current too has got an exception. Jaoul et al[20] report the negative sign for $\Delta\rho/\rho$ for Ir; we checked this for a series of NiAuIr alloys.

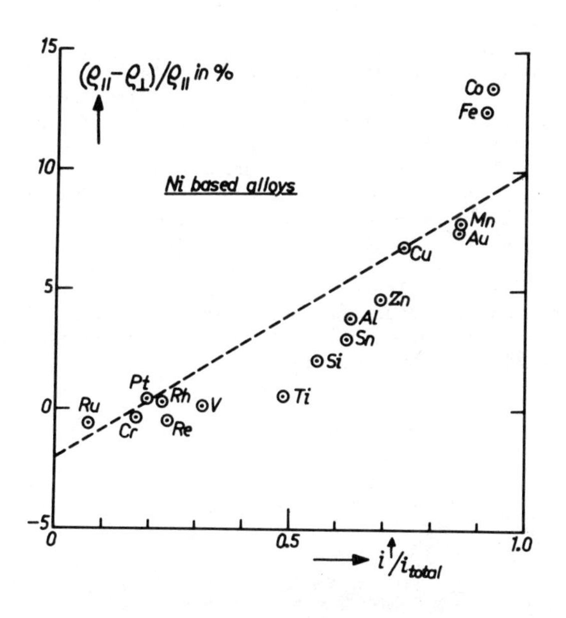

Fig. 4. The resistivity anisotropy of binary Ni alloys vs. $i^{\uparrow}/i_{total}$. The straight line corresponds to values of $(\Delta\rho/\rho)^{\uparrow}$ and $(\Delta\rho/\rho)^{\downarrow}$ independent of the impurity metal.

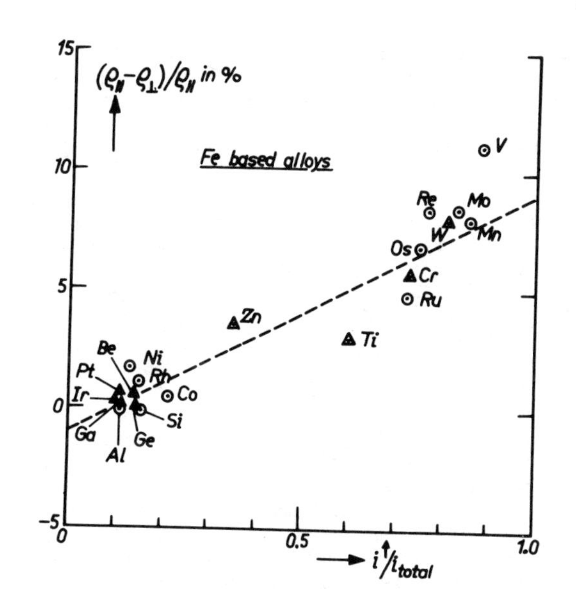

Fig. 5. Same as Fig. 4, but for Fe-based alloys. O..Solute elements for which $i^{\uparrow}/i_{total}$ has been determined in experiments on ternary alloys; Δ Solute elements for which $i^{\uparrow}/i_{total}$ has been determined from deviations from Matthiessen's rule at room temperature.

If $(\Delta\rho/\rho)^{\uparrow}$ and $(\Delta\rho/\rho)^{\downarrow}$ were true constants there would be a linear correlation between the measured resistance anisotropy of a binary system and the ratio $\beta = i^{\uparrow}/i_{total}$. This linear dependence is indeed followed for the majority of elements dissolved in nickel, as may be clear from fig. 4. The fairly large deviation from the average behaviour for NiFe and NiCo was already fully clear from table II. A plot of $(\Delta\rho/\rho)$ vs. β, but now for iron-based alloys, is shown in fig. 5. The straight line corresponds to $(\Delta\rho/\rho)^{\uparrow} = +9\%$ and $(\Delta\rho/\rho)^{\downarrow} = -1\%$. The detailed analysis of data on $(\Delta\rho/\rho)$ for ternary alloys, which results in a table for iron--based alloys analogous to table II, indeed gives values for $(\Delta\rho/\rho)^{\uparrow}$ and $(\Delta\rho/\rho)^{\downarrow}$ varying around +9% and -1%, respectively for all solute elements included in table I.

A simple result as we have obtained here for the resistivity anisotropy of different solute elements in nickel and iron has not been predicted theoretically. Our result can be simply visualized as follows: for conduction electrons in the majority spin band of both iron and nickel, impurities look like oblate ellipsoids with their short axes parallel to the magnetization direction. However as seen by the conduction electrons of the minority spun band, impurities look like prolate ellipsoids[21].

EXTENSION TO ROOM TEMPERATURES

From the parameters derived in the foregoing section for solute elements in nickel and iron one can in principle predict the anisotropy effect for any alloy composition, but only at low temperatures. At room temperature one has to take account of phonons and magnons. Both phonons and magnons will introduce an additional resistivity which has a different value for the two spin currents. In addition magnon and electron-electron scattering processes may cause a mixing of the two spin currents through momentum--conserving spin flip scattering processes. In the analysis of experimental data it is diffcult to separate these two contributions to the temperature dependent resistivities unambiguously. Values for $\rho^{\uparrow\downarrow}$ in combination with a corresponding value for $\alpha = \rho^{\downarrow}_{phonon} / \rho^{\uparrow}_{phonon}$ lead to equivalent descriptions of the temperature dependent resistivity of pure nickel or iron, deviations from Matthiessen's rule for alloys and temperature dependent anisotropy effects. For the moment being one may prefer at room temperature the simple solution of no spin flip scattering and $\alpha_{phonon} = 2$, $\rho(300) = 7\ \mu\Omega cm$ for phonons in nickel and $\alpha_{phonon} = 0.5$, $\rho(300) = 10\ \mu\Omega cm$ for phonons in iron.

It will be clear that in order to have a large resistivity anisotropy effect near room temperature the resistivity must be as small as possible compared to that of the impurity selected. This involves alloys with impurity concentrations that can no longer be treated as dilute alloys with isolated scattering centres. The interaction between impurity centres becomes important; as an example in FeV alloys the effect is considerably reduced at V concentrations above 10 at % V (Being only $\Delta\rho/\rho = 2\%$ for $Fe_{80}V_{20}$). Hence FeV alloys are not attractive. On the other hand the anisotropy effect increases above the dilute-alloy values of about 13% [18] in concentrated NiCo and NiFe alloys. Jaoul and Campbell [19] report the effect to increase (at low temperatures) to about 24 and 17% for concentrations near 30 at % Co or Fe. Hence it is not surprising that at room temperature the strongest effects are met in NiFe and in NiCo alloys [16].

CONCLUSION

We have demonstrated that the resistivity anisotropy in ferromagnetic iron and nickel finds a simple description in the model of independent conduction by two spin currents. The description of the anomalous Hall effect, too, becomes of a comparable simplicity, as we shall report elsewhere [10,12]. This proves the usefulness of the two-current model, i.e. among the large number of groups of charge carriers which undoubtedly will be present in complicated metals like nickel and iron (s- and d-like carriers, electrons and holes, spin $\uparrow$ and spin $\downarrow$ carriers), the division according to spin direction is the most important one.

As far as the search for materials of practical interest is concerned we conclude that NiCo and NiFe present accidental favourable cases. In situations where additional requirements are to be satisfied (e.g. low magnetostriction, low or high saturation magnetization etc.) other choices of solute elements may be more appropriate.

ACKNOWLEDGEMENT

The authors wish to thank Mrs. P. Hokkeling, A. Kahle and A.F.J. Vlaminckx for preparing the alloys studied and Mr. T. Derijck for rolling the samples.

REFERENCES

1. N.F. Mott, Proc. Roy..Soc. 156, 368 (1936); Adv.Phys. 13, 325 (1964).
2. A. Fert and I.A. Campbell, J. de Phys. 32, Colloque C1, 46 (1971).
3. A. Fert and I.A. Campbell, Phys.Rev.Lett. 21, 1190 (1968).
4. P. Leonard, M.C. Cadeville, J. Durand and F. Gautier J.Phys.Chem.Solids 30, 2169 (1969).
5. J.W.F. Dorleijn and A.R. Miedema, J.Phys.F Metal Phys. 5, 487 (1975).
6. J. Durand and F. Gautier, J.Phys.Chem.Solids 31, 2773 (1970).
7. B. Loegel and F. Gautier, J.Phys.Chem.Solids 32, 2723 (1971).
8. I.A. Campbell, A. Fert and A.R. Pomeroy, Phil.Mag. 15, 977 (1967).
9. M.C. Cadeville and J. Roussel, J.Phys.F Metal Phys. 1. 686 (1971).
10. J.W.F. Dorleijn and A.R. Miedema, Proc. IUPAP Magn. Conf. Amsterdam 1976, to be published as supplement to Physica.
11. F.C. Schwerer and J.W. Conroy, J.Phys.F Metal Phys. 1, 877 (1971).
12. J.W.F. Dorleijn and A.R. Miedema, to be published in Philips Research Reports.
13. R.A. Tawil and J. Callaway, Phys.Rev. B 7, 4242 (1973).
14. K.J. Duff and T.P. Das, Phys.Rev. B 3, 192 (1971).
15. D.A. Thompson, L.T. Romankiw and A.F. Mayadas, IEEE Trans.Mag. 11, 1039 (1975).
16. T.R. McGuire and R.I. Potter, IEEE Trans.Mag. 11, 1018 (1975).
17. N. Sueda and H. Fujiwara, J.Sci. Hiroshima Univ. A 35, 59 (1971).
18. J.W.F. Dorleijn and A.R. Miedema, J.Phys.F Metal Phys. 5, 1543 (1975).
19. O. Jaoul and I.A. Campbell, J.Phys.F Metal Phys. 5, L 69 (1975).
20. O. Jaoul, I.A. Campbell and A. Fert, preprint.
21. Here we use the expression ellipsoids for bodies with the symmetry of the different d wave functions. An interpretation where the cross-section of an impurity center becomes of importance may be realistic since for the present alloys the Born approximation of weak scattering is not acceptable. The mean free path for conduction electrons in alloys that contain 1% solute is of the order of 100 atomic distances or less. When we do no longer consider a solute element as a small perturbation we obtain a picture in which the cross-section of an impurity and hence its shape become relevant.

GALVANOMAGNETIC EFFECTS IN MnBi FILMS

K. Okamoto*, M. Tanaka, S. Matsushita, and Y. Sakurai,
Department of Control Engineering, Faculty of Engineering Science
Osaka University, Toyonaka, Osaka 560 JAPAN
S. Honda, and T. Kusuda
Department of Electronics, Faculty of Engineering, Hiroshima University
Hiroshima 730 JAPAN

ABSTRACT

Transverse Hall effect, planar Hall effect, transverse magnetoresistance effect, and also resistivity were measured on polycrystalline MnBi thin films which have the uniaxial magnetic anisotropy normal to their film planes(i.e. perpendicular anisotropy). It was found that these MnBi films exhibit a remarkable extraordinary Hall effect at room temperature and that the hysteresis loop of the extraordinary Hall effect (V_H-H loop) agrees well with the magnetization curve obtained by the Faraday effect. The Hall resistivity ρ_H(= $V_H d/I$) of the film decreases and the V_H-H loop change in shape from rectangular to linear as the temperature decreases.

Our experimental data on the transverse Hall effect in MnBi films are considerably different from those reported earlier.

INTRODUCTION

The galvanomagnetic effects[1,2] in ferromagnetic thin films have been studied extensively by many investigators during the past ten years[3,4,5]. A few works have, however, been done on films which have the uniaxial magnetic anisotropy perpendicular to the film planes(we designate such films as the perpendicular anisotropy films). Chen et al.[6] measured the Hall effect on vapor-deposited MnBi polycrystalline films. But they did not describe the hysteresis characteristics of the Hall effect in detail. Recently it was found that amorphous Gd-Co films with perpendicular anisotropy exhibit a remarkable extraordinary Hall effect[7]. It was also reported that the existence of the magnetic domain walls account for the transverse magnetoresistance effect in the Gd-Co films[8].

In the present work we investigate the galvanomagnetic effects in polycrystalline MnBi thin films with perpendicular anisotropy with special emphasis on their hysteresis loops.

SAMPLE PREPARATION

MnBi thin films were prepared by successive vacuum-depositions[9] of Bi and Mn onto a clean glass substrate using a mask and subsequent annealing process(270°C, 6 hours). The deposition and the annealing were performed in the vacuum of 10^{-8}-10^{-7} torr and 10^{-9}-10^{-8} torr, respectively[10]. The evaporated atomic ratio of Mn to Bi, R(Mn/Bi), was 1.5 and the film thickness d of the produced MnBi films were approximately 1000 Å. After the annealing process, two pairs of electrodes (one for the control current I, and the other for the measurement of the Hall voltage V_H) were made on the each film by successive vacuum-depositions of Cr and Cu using another mask. The Cr-layer acts as an adhesive agent between the Cu-layer and the glass substrate. And then SiO layer(1000 Å in thickness) was overcoated on the surface of the MnBi film in order to protect from corrosion. Finally each film was cut out with glass substrate piece and then four lead wires were soldered on the electrodes. Each pair of wires was twisted together in order to avoid picking up induction. Fig.1 illustrates a sample fabricated in such a manner.

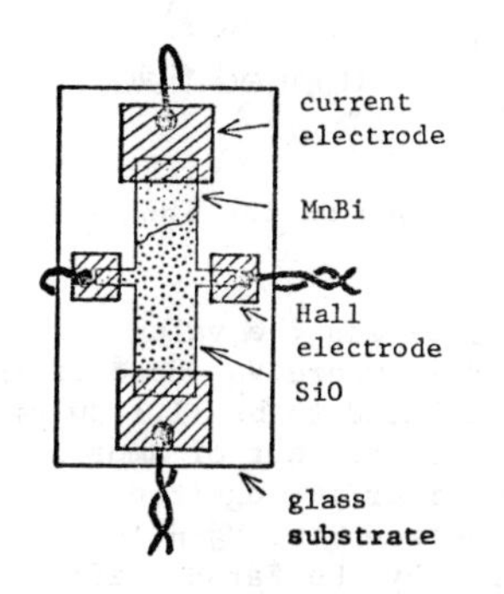

Fig.1. Sample for measurement(the effective area is 2 x 6 mm).

MEASUREMENT

The transverse Hall effect(THE), the planar Hall effect(PHE), and their temperature dependences were measured on a sample. The Hall effect and the Faraday effect were observed simultaneously and their hysteresis loops were compared. The temperature dependence of the electrical resistivity ρ was also measured. Finally the transverse magnetoresistance effect(TME) was investigated and its hysteresis loop was compared with that of the THE. Every hysteresis loop of THE, PHE and TME was recorded by an X-Y recorder. Fig.2 illustrates the schematic diagram for the measurement of these three galvanomagnetic effects.

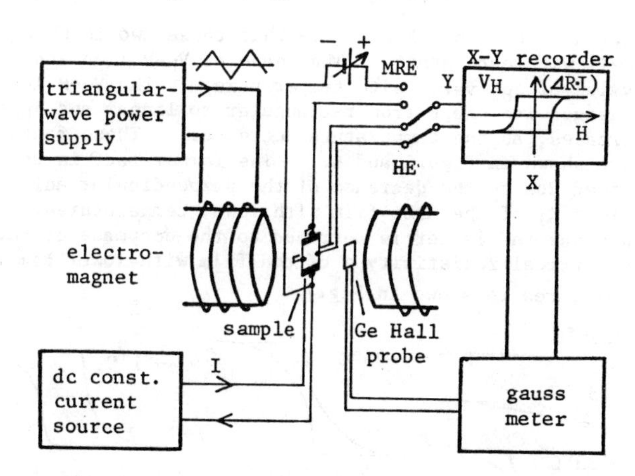

Fig.2. Schematic diagram for measurement of the galvanomagnetic effects.

In the case of PHE measurement, the magnetic field was applied in the plane of the film at 45° to the current I in order to obtain the maximum planar Hall voltage[2]. The experimental procedure was essentially the same as reported earlier[7,8].

RESULTS AND DISCUSSION

Fig.3 shows a typical hysteresis loop of THE(designate as the V_H-H loop). It is evident from this figure that the extraordinary Hall effect plays a dominant role and that the ordinary Hall effect is negligibly small because V_H is almost constant in the high magnetic field (10-20 kOe)[1]. The sign of the extraordinary Hall

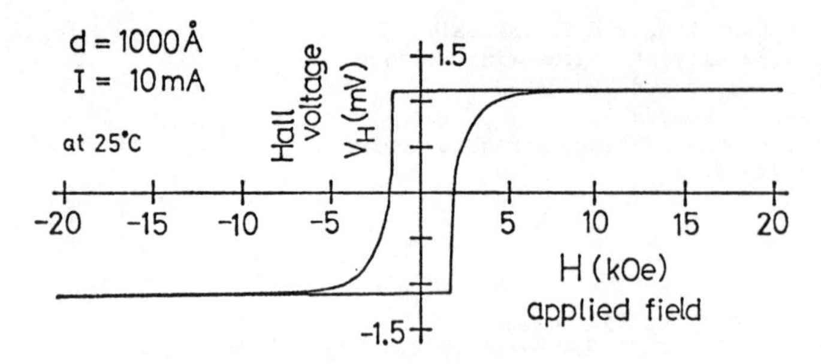

Fig.3. A typical hysteresis loop of the THE(V_H-H loop) in a MnBi film.

coefficient R_1 is positive ($R_1 > 0$) and the value of the Hall resistivity ρ_H(= $V_{Hs} \cdot d/I$, where V_{Hs} denotes the saturation value of V_H) is calculated to be +1.1 μΩ·cm at 25°C. This value is a fraction of that of amorphous Gd-Co films but is in the same order in magnitude.

Fig.4 shows a comparison between the V_H-H loop and the magnetization curve obtained by the Faraday effect.

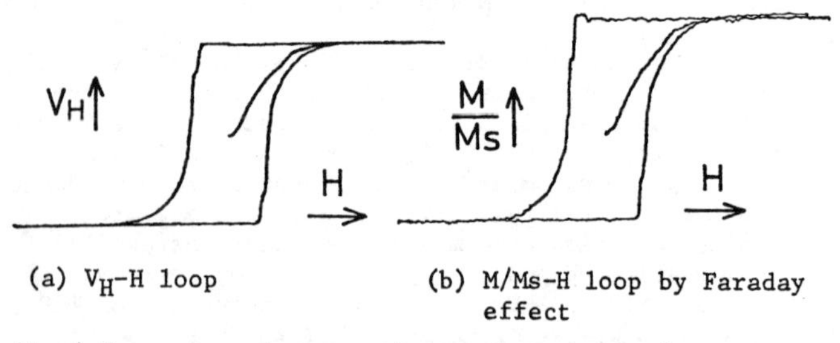

(a) V_H-H loop (b) M/Ms-H loop by Faraday effect

Fig.4.Comparison between the two hysteresis loops obtained by the Hall effect and the Faraday effect.

It is clear from this figure that these two loops agree well with each other. The shape of V_H-H loop and the value of ρ_H vary with temperature. The V_H-H loop changes its shape from rectangular to linear and ρ_H decreases, as the temperature decreases. These facts are shown in Figs.5 and 6. The former fact is understood due to the decrease of the perpendicular anisotropy K_1 of the MnBi film with lower temperatures. Whereas the latter is ascribed to the decrease of the electrical resistivity ρ of the film with lower temperatures as shown in Fig.7.

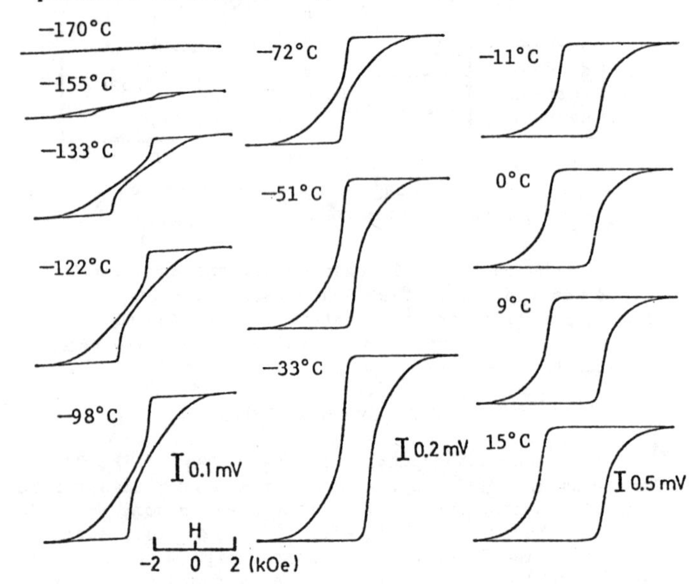

Fig.5. Temperature dependence of the V_H-H loop in a MnBi thin film. (film thickness: 700 Å, I= 10mA)

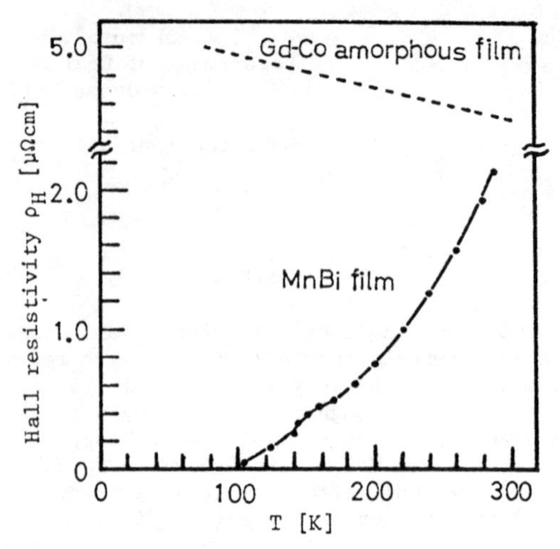

Fig.6. Temperature dependence of ρ_H.

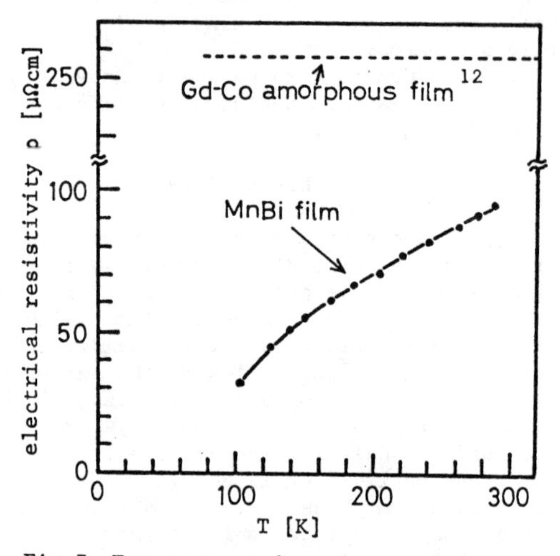

Fig.7. Temperature dependence of ρ.

In order to examine the relationship between ρ_H and ρ, both measured values were plotted on a double-logarithmic scale. The result is indicated in Fig.8. It is seen from this figure that there is a relationship of $\rho_H \propto \rho^{3.3}$. This relationship, however, is not always specific to MnBi film. Because the film investigated is not regarded as a perfect polycrystalline MnBi film, but it is supposed to form a columnar structure described in ref.10.

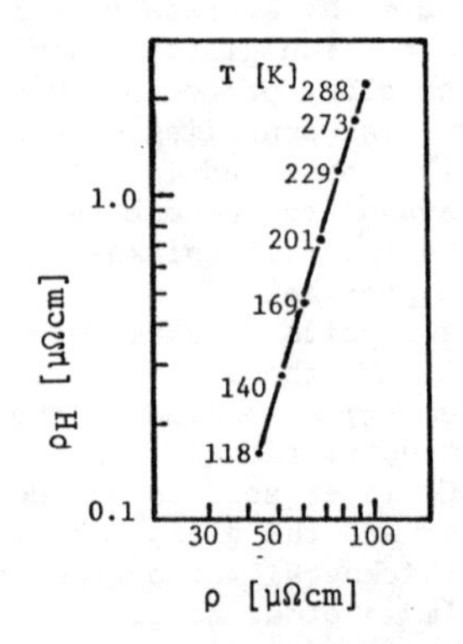

Fig.8. Relation of ρ_H and ρ.

It has been known that the magnetic anisotropy K_1 of MnBi film reduces with the temperature[11]. Therefore, it is conceivable that the magnetization vector rotates more easily toward the film plane by a field applied parallel to the film plane. This suggests that the PHE should increase as the temperature decreases. Our data on the measurement of PHE shown in

Fig.9 supports the above supposition. Because the shape of V_{PH}-H loop (where V_{PH} denotes the planar Hall voltage) becomes steeper with temperature drop.

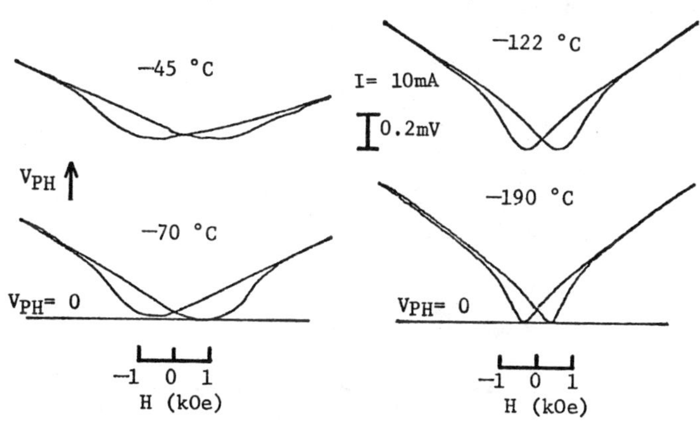

Fig.9. Temperature dependence of the planar Hall effect. (V_{PH}: planar Hall voltage, $\angle(I,H)= 45°$)

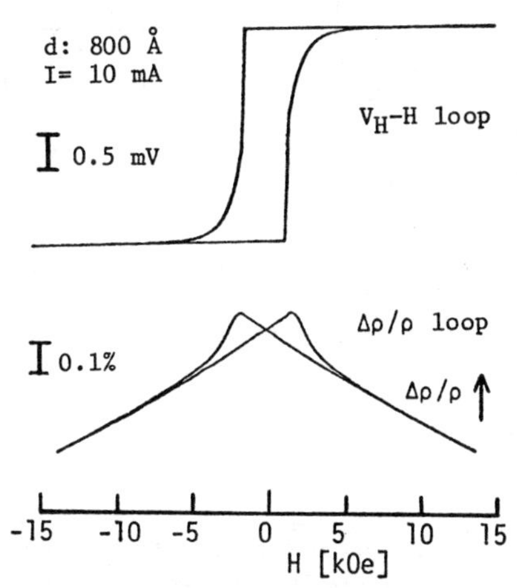

Fig.10. Comparison between the V_H-H loop and the $\Delta\rho/\rho$-H loop(TME loop).

Fig.10 shows the hysteresis loop of the TME, or the $\Delta\rho/\rho$-H loop($\Delta\rho/\rho$ is the fractional change in ρ), in comparison with the V_H-H loop at room temperature. This data shows a similar tendency to that of amorphous Gd-Co perpendicular anisotropy film[8] reported previously except that $\Delta\rho/\rho$ does not saturate at the field where V_H saturates. This exceptional result is supposed to be due to either of the following facts. That is, the direction of H_K (anisotropy field) tilts to the film plane, or there exists the dispersion of magnetic anisotropy in the film. In any case, the cause is ascribed to the imperfect perpendicular anisotropy of the sample film. It is expected that MnBi film with perfect perpendicular anisotropy will exhibit the TME similar to that of Gd-Co sputtered films.

The butterfly-like loop in Fig.10 is considered as the domain wall effect described in the previous paper[8].

CONCLUSION

The experimental results of the galvanomagnetic effects in MnBi thin films have been described. In this experiment we found the following facts: 1) the extraordinary Hall effect plays a dominant role and the ordinary Hall effect is negligibly small in the transverse Hall effect, 2) the Hall resistivity ρ_H is comparatively large at room temperature($\rho_H \simeq +10^{-6}\Omega cm$), 3) the hysteresis loop of the extraordinary Hall effect (V_H-H loop) agrees well with that of Faraday effect (or the magnetization curve), 4) ρ_H and the electrical resistivity ρ reduce keeping a certain relationship when the temperature decreases, 5) the V_H-H loop changes in shape from rectangular to linear as the temperature drops, 6) the planar Hall effect becomes remarkable as the temperature decreases, and 7) the transverse magnetoresistance effect is mainly due to the existence of magnetic domain walls below saturation. Among these experimental results, 1), 2), 3) and 7) are qualitatively similar to those obtained for amorphous Gd-Co films with perpendicular anisotropy.

The measurement of the THE enables us to determine the magnetization characteristics and that of the PHE will enable us to aquire some informations about the magnetic anisotropy in the perpendicular anisotropy films. These galvanomagnetic methods are advantageous because of their high S/N ratio(negligibly low noise level), sufficiently high output level, easily adaptable for measurement in a wide temperature range and simple setup of apparatus.

From the practical point of view, the extraordinary Hall effect of MnBi thin film is attractive. Because it can be applied to the non-destructive readout of information bit written in the film by the use of the "remanence Hall voltage"(the Hall voltage produced by the residual magnetization M_r).

REFERENCES

* Present adress: Faculty of Education, Kagawa University, Takamatsu 760, Kagawa JAPAN

1. J.P. Jan, "Solid State Physics", F. Seitz and D. Turnbull, Academic Press, New York, 1957, vol.5, pp.1-96.
2. S. Legvold, "Magnetic Properties of Rare Earth Metals", R.J. Elliott, Plenum Press, London, 1972, p.335.
3. T.K. Wu and E.F. Kuritsyna, Sov. Phys.-Dokl., vol.10, pp.51-53, Jul. 1965.
4. G. Kneer, IEEE Trans. on Mag., vol MAG-2, 747, Dec. 1966.
5. A.A. Hirsch and Joseph Kleefeld, IEEE Trans. on Mag., MAG-7 No.3, 733, 1971.
6. D. Chen, Y. Gondo, and M.D. Blue, J.Appl.Phys., vol. 36, 1261, Mar. 1965.
7. K. Okamoto, T. Shirakawa, S. Matsushita, and Y. Sakurai, IEEE Trans. on Mag., MAG-10 No.3, 799. 1974.
8. K. Okamoto, T. Shirakawa, S. Matsushita, and Y. Sakurai, AIP Conf. Proc. No.24, 113, 1974.
9. D. Chen, J. Appl. Phys. 42, 3625, 1971.
10. S. Honda and Tetsuzo Kusuda, J. Appl. Phys., vol.45, No.6, 2689, June 1974.
11. C. Guillaud, J. Phys. Radium, 12, 143, 1951.
12. T. Shirakawa et al., in this issue.

TRANSVERSE MAGNETORESISTANCE EFFECT IN C-PLANE OF MnBi FILM

M.Masuda, S.Yoshino and H.Tomita
Mie University, Tsu, Mie, Japan 514
and
S.Uchiyama
Nagoya University, Nagoya, Japan 464

ABSTRACT

The electrical resistance in the c-plane of MnBi films is found to be decreased by the magnetic field applied parallel to the c-axis. This negative magnetoresistance effect is explained by considering the contribution of domain wall resistance. Since the magnetization direction in the wall is deflected from the easy direction, the electrical resistivity across the wall differs from that in c-plane of the domain. The application of the magnetic field parallel to the c-axis of the film results in the decrease of the number of domain walls, and thus the electrical resistance is decreased. The theoretical result agrees well quantitatively with the experiment.

INTRODUCTION

It was reported in our recent work[1] that oriented MnBi films with easy axis normal to the film plane have the stripe magnetic domains. Furthermore, the period of the stripe domain or the number of magnetic walls per unit length as well as the resultant magnetization curve accompanied by the hysteresis phenomena was well interpreted theoretically by introducing a term of the wall coercivity into Kooy and Enz's stripe domain theory.[2] This modified stripe domain theory is also expected to explain the hysteresis phenomenon in other magnetic properties of MnBi films which is related to the magnetization process of wall displacement.

In this work, therefore, the Galvanomagnetic effect, particularly the anormalous transverse magnetoresistance effect in MnBi films, was examined in relation to the wall displacement process.

Generally, the anormalous magnetoresistance in magnetic substances is known to be related to the magnetization process, that is, to come from (1) the growth of magnetic domains due to the wall displacement, (2) the rotation of magnetization—orientation effect—in a lower field range where the magnetization does not reach saturation yet, and (3) the increase in the spontaneous magnetization—forced effect—in a higher applied field range.[3]

In the present experiment, as the external field smaller than $4\pi I_s$ is applied normal to the film plane, only the wall displacement process occurs and the magnetic spin in the magnetic domains only reverses its direction parallel to the c-axis, the fact which does not contribute to the change in resistance.

Consequently, the magnetoresistance effect observed here must be interpreted in connection with the domain wall; the interpretation will be shown in the text.

EXPERIMENTAL PROCEDURE

MnBi films were prepared with Chen's method,[4] where Bi and then Mn layers were deposited successively in vacuum onto cleaned glass substrates, and then annealed. The films prepared were polycrystalline, but their c-axes were well aligned to the film normal.

The samples were rectangular with the dimension of 8mmx2mm and 600Å- 1μm in thickness, and were provided with the arrangement of electrodes as shown in Fig.1, so that both the transverse magnetoresistance and the transverse Hall voltage could be measured. The electrodes were connected to conducting wires with Dotite conducting paste. These two measurements were made at room temperature, and the schematical arrangement of the electrical circuits is shown in Fig.2(a) and (b).

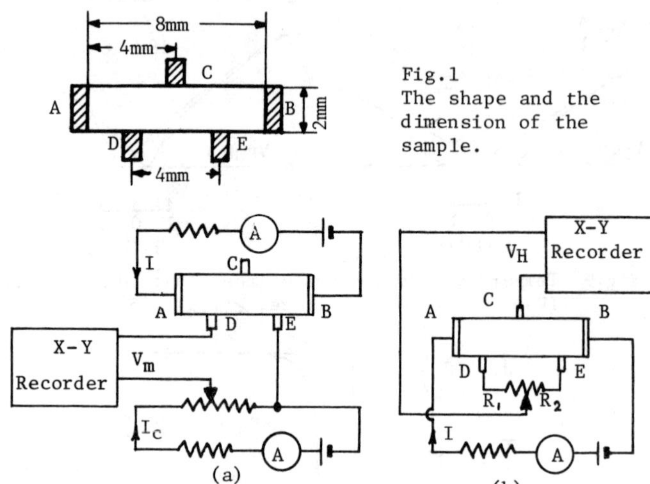

Fig.1 The shape and the dimension of the sample.

Fig.2 The schematical arrangement of electrical circuit for the measurement of Galvanomagnetic effect in MnBi films.
(a) transverse magnetoresistance effect
(b) transverse Hall effect

The electrical d.c. current with the density of J_x was sent between the electrodes A and B, and the magnetic field H_z was applied normal to the film plane, namely parallel to the c-axis of MnBi crystallites. The voltage drop V_m due to the magnetoresistance was detected between the electrodes D and E, where the ohmic voltage drop was compensated by the compensating ohmic voltage I_cR_c. The Hall voltage V_H was detected in y-direction. where the offset voltage due to the ohmic voltage drop was compensated by the use of balancing resistors R_1 and R_2 between the electrodes D and E. Both V_m and V_H were directly fed to the Y terminal of the X-Y recorder, since its input resistance was over 1MΩ and its sensitivity was 100 μV/cm. The magnetic field was detected by the InSb Hall generator, and the output of the generator was fed to the X terminal of the recorder.

EXPERIMENTAL RESULT

Fig.3(a) and(b) show respectively the voltage drop V_m due to the magnetoresistance and the Hall voltage V_H as a function of the applied field H_z for a sample of 3000Å MnBi film, where H_z is normalized by $4\pi I_s$ which is equal to 7700 Gauss.[5] Both of them have hysteresis phenomena, and the hysteresis curve of V_m (A→B→C→D→E→F→A) coresponds to the hysteresis curve of the transverse Hall voltage V_H (A′→B′→C′→D′→E′→F′→A′). Since I=30mA, I_c=5.4mA, and R_c=20.8Ω were chosen as compensating parameters for Fig.3(a), 0.1mV of the voltage V_m corresponds to 3.33×10^{-3}Ω, and the ohmic resistance of the sample is 3.75Ω which gives 7.0×10^{-5}Ω-cm as specific resistivity.

The magnitude of the magnetic field H_z was changed within ±5400 Oe (h=0.7) just above saturating field so that the magnetic walls of stripe domain might make a continuous displacement, because V_H-h curve showed larger hysteresis with discontinuous change similar to Barkhausen jump in the demagnetization region B′→C′ or E′→F′ if maximum value of H_z was so large as to saturate the magnetization completely.

A comparison of V_m-h curve with V_H-h curve leads to

the remarkable points: (1) In the range of $H_z>0$, for example, Vm is larger for the ascendant curve F´→A´ and smaller for the descendant curve B´→C´ of V_H curve. (2) Vm does not change in the saturation range A´→B´ and D´→E´ of V_H curve.

This result suggests that the magnetoresistance effect in the MnBi film is related closely to the number of magnetic walls. which is larger in the ascendant magnetization F´→A´ than in the descendant magnetization B´→C´ as reported earlier by us.[1]

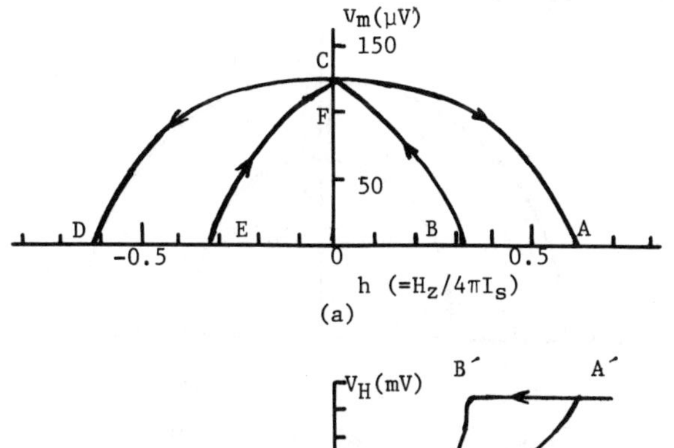

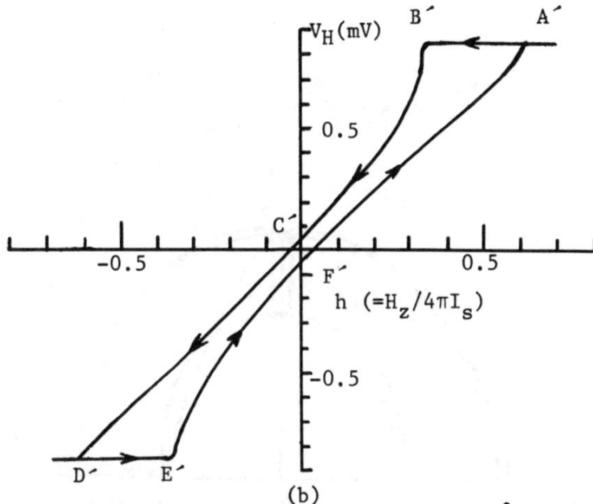

Fig.3 The Galvanomagnetic effect of 3000Å MnBi film.
(a) transverse magnetoresistance effect
(b) transverse Hall effect

DISCUSSION

As is well known, the resistivity of the single crystal ferromagnet with hexagonal symmetry depends both on magnetization direction and on current direction, as given in the following expession;[6]

$$\rho(\alpha_i,\beta_i) = \rho_0 + \rho_1\beta_3{}^2 + \rho_2(1-\alpha_3{}^2) + \rho_3\beta_3(1-\alpha_3{}^2) + \rho_4\alpha_3\beta_3(\alpha_1\beta_1+\alpha_2\beta_2) + \rho_5(\alpha_1\beta_1+\alpha_2\beta_2)^2 + \ldots, \quad (1)$$

where α_i and β_i are respectively the direction cosines of the magnetization I_s and the current density J with respect to the crystallographic axes [01̄10], [21̄1̄0] and [0001] which are parallel to x,y and z directions respectively.

Since in the present experiment, the current flows in c-plane of every crystallite, that is, $\beta_3=0$,

$$\rho(\alpha_i,\beta_i) = \rho_0 + \rho_2(1-\alpha_3{}^2) + \rho_5(\alpha_1\beta_1+\alpha_2\beta_2)^2 + \ldots \quad (2)$$

First, let us consider the resistivity in the magnetic domains, where rhe magnetization is parallel or antiparallel to the c-axis, namely $\alpha_3=\pm1$, $\alpha_1=\alpha_2=0$. Then we have

$$\rho = \rho_0 \quad (3)$$

Eq.(3) shows that resistivity is the same in the magnetic domains where the magnetization is parallel or antiparallel to the c-axis, and that the change in volume of those domains makes no contribution to the change in the transverse magnetoresistance.

Next, let us estimate the resistance across a magnetic wall of Bloch type, where we let x-axis and z-axis respectively normal to the wall plane and the film plane, and y-axis be parallel to the wall and the film plane. Generally, α_is are not zero in the wall, because the magnetization changes its direction gradually. If we neglect the terms of higher order than the second term in Eq.(2) for simplicity, then we have for the resistivity in the wall

$$\rho = \rho_0 + \rho_2(1-\alpha_3{}^2). \quad (4)$$

Then, the resistance R_w across the magnetic wall is given as

$$R_w = (1/S)\int_{-\delta/2}^{+\delta/2}\rho\, dx = (1/S)\rho_0\delta + (1/S)\int_{-\delta/2}^{+\delta/2}\rho_2(1-\alpha_3{}^2)\, dx = (1/S)\rho_0\delta + (1/S)\int_{-\delta/2}^{+\delta/2}\rho_2\cos^2\theta\, dx, \quad (5)$$

where S is the area of the magnetic wall under speculation, $\delta(=\pi\sqrt{A/K_u})$ the width of Bloch wall, and θ the angle of the spin rotation about the x-axis. Since the way of the spin rotation is given in Bloch wall as follows[7]

$$dx = \sqrt{A/K_u}\, d\theta/\cos\theta, \quad (6)$$

and the boundary condition is that $\theta=+\pi/2$ at $x=\delta/2$, and $\theta=-\pi/2$ at $x=-\delta/2$,

$$R_w = \frac{\rho_0\delta}{S} + \frac{2}{\pi}\frac{\delta}{S}\rho_2. \quad (7)$$

This Equation implies that the average resistivity in the wall is higher by $2\rho_2/\pi$ than in the domain.

Next we consider a sample of MnBi film which is D in thickness, b in width, and ℓ in length, and assume for simplicity that all the stripe domain walls are aligned parallel to one another with a period $2w=w_1+w_2$ and their planes are normal to the current density J as shown in Fig.4(a), where w_1 and w_2 are widths of the domains whose magnetization is parallel and antiparallel to the z-axis respectively. Taking into account that the total number of the wall is given by $2\ell/(w_1+w_2)=\ell/w$, the resistance R across the whole sample is

$$R = \frac{\rho_0\ell}{S} + \frac{2}{\pi}\frac{\delta}{S}\frac{\ell}{w}\rho_2. \quad (8)$$

In the above equation, the first term is the usual electrical resistance, and the second term which depends on w contributes to the magnetoresistance, because w is a function of the applied field. Therefore we have for the magnetoresistance of MnBi film,

$$R_m(H_z) = \frac{2}{\pi}\frac{\ell\delta}{bD}\frac{\rho_2}{w(H_z)}. \quad (9)$$

If the current density J is not normal to, but makes an angle ϕ with the magnetic wall normal, as shown in Fig.4(b), we can consider as follows.
The total number and the width of the walls in the direction of current density J are $(\ell/w)(1/\sec\phi)$ and

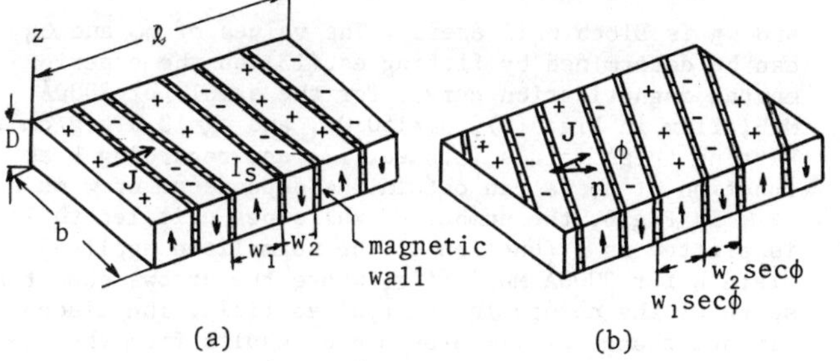

Fig.4 The stripe domain configurations.

$\delta\sec\phi$ respectively. Therefore, the magnetoresistance for this case is also given by eq.(9). Moreover, this result shows that eq.(9) is valid for the magnetoresistance of the actual magnetic domain configuration where the angle ϕ is different from place to place in the film.

Since no report was ever made on the value of ρ_2 for MnBi, we have determined it experimentally in this work. First the film is saturated by the field which is applied normal to the film plane, and then the field direction is rotated by an angle ψ from the film normal. The resistivity for this situation is derived from eq.(4) as

$$\rho = \rho_0 + \rho_2\sin^2\theta, \tag{10}$$

where θ is the angle between the magnetization I_s and the film normal, and is given by the equation

$$(H_k - 4\pi I_s)\sin 2\theta = 2H\sin(\psi - \theta). \tag{11}$$

In the above equation, the anisotropy field $H_k=2K_u/I_s$ $=3.3\times10^4$Oe, and $4\pi I_s$=7700 Oe for MnBi film.

Fig.5 shows an example of the dependence of the resistivity ρ on the field direction ψ, where H =9.3 KOe. This experimental curve is well explained by the combination of eqs.(10) and (11) where $\rho_2=1.7\times10^{-6}$ Ω-cm.

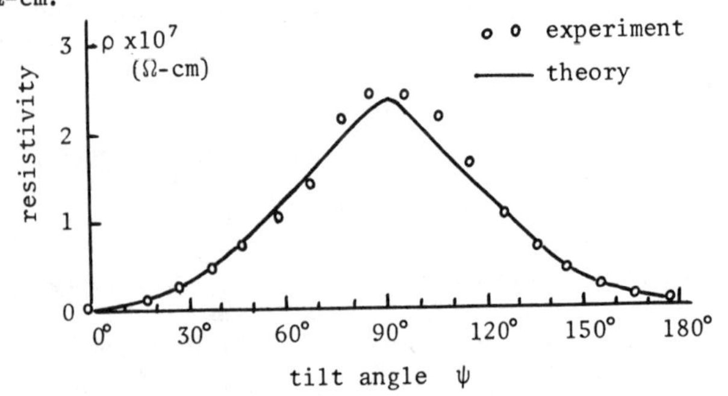

Fig.5 The resistivity ρ as a function of the tilt angle ψ of the applied field in 3000Å MnBi film.

Last, a discussion will be made on $w(H_z)$ in eq.(9). In our earlier work,[1] by introducing the term of the wall coercivity H_w into the Kooy and Enz' stripe domain theory, we succeeded in interpreting the hysteresis phenomena in $w(H_z)$ curve as well as the magnetization curve. The result is given by the following equations;

$$w = \frac{w_0}{\sqrt{f(K)}}\{\sqrt{1+a^2/f(K)} \mp a/\sqrt{f(K)}\} \tag{12}$$

- for increasing the magnitude of applied field
+ for decreasing the magnitude of applied field

$$h = k + \frac{8wf'(k)}{\pi^3 D(1+\sqrt{\mu})} \pm h_w \tag{13}$$

+ for increasing the magnitude of applied field
- for decreasing the magnitude of applied field

$$f(k) = \sum_{n=1}^{\infty}(1/n^3)\sin^2\{n\pi(k+1)/2\}, \tag{14}$$

where $h=H_z/4\pi I_s$, $k=I/I_s$, $w_0=(\pi/4I_s)\sqrt{\sigma_B D(1+\sqrt{\mu})/2}$, $a=(\pi H_w/4)\sqrt{(D/\sigma_B)(1+\sqrt{\mu})/2}$, $h_w=H_w/4\pi I_s$, $\sqrt{\mu}=\sqrt{1+(2\pi I_s^2/K_u)}$,

and σ_B is Bloch wall energy. The values of h_w and σ_B can be determined by fitting eq.(13) on the experimental magnetization curve. For the sample of 3000Å MnBi film in Fig.3(b), H_w=150 Oe, and σ_B=13.0 erg/cm^2. Putting these values into eq.(12) and regarding k as a function of h, we can obtain the dependence of w on h.

In Fig.6, the number of walls per unit length 1/w is plotted as a function of the normalized applied field h for 3000Å MnBi film, where the arrows show the sense of the change in the applied field. The discontinuous change of 1/w around h=0 results from the assumption made in this theory that the normalized coercivity is constantly $+h_w$ and $-h_w$ respectively for the plus and minus values of k.

In Fig.7, an example of the experimental magnetoresistance curve (broken curve) for 3000Å MnBi film is compared with that calculated form eq.(9) (solid curve), where the following parameters are used; ℓ=0.4cm, b=0.2 cm, D=3x10^{-5}cm, $\rho_2=1.7\times10^{-6}\Omega$-cm, and δ=11.2x10^{-7}cm. The value of δ was obtained by putting $K_u=9.1\times10^6$erg/cm^3 in the equation $\delta=\pi\sigma_B/4K_u$, which is derived from the equations $\sigma_B=4\sqrt{AK_u}$ and $\delta=\pi\sqrt{A/K_u}$ for Bloch wall. It is seen that the present theory interprets satisfactorily the hysteresis curve of the magnetoresistance in the MnBi film except around h=0.

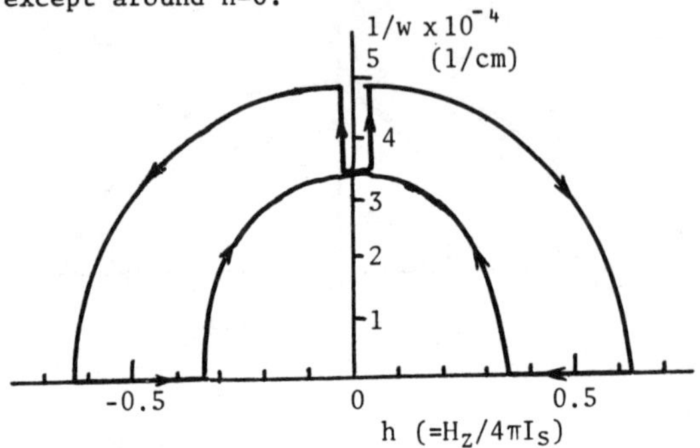

Fig.6 The number of walls 1/w per unit length as a function of the applied field in 3000Å MnBi film.

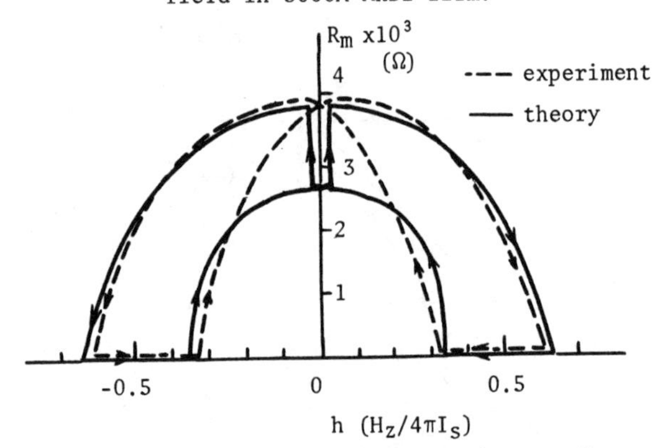

Fig.7 The transverse magnetoresistance R_m as a function of the applied field in 3000Å MnBi film.

CONCLUSION

In this work the transverse magnetoresistance in MnBi film was investigated in relation to the wall displacement. The experimental results showed that the transverse magnetoresistance varied with the applied magnetic field, accompanied by the hysteresis. The hysteresis phenomenon was interpreted theoretically by taking into account that the resistivity in the magnetic wall is higher than that in the magnetic domain, and that the dependence of the number of walls on the applied field is accompanied by the hysteresis.

REFERENCES

1. M. Masuda, S. Yoshino, H. Tomita and S. Uchiyama, Japan. J. appl. Phys. 15, 283 (1976).
2. C. Kooy and U. Enz, Philips Res. Repts. 15, 7 (1960)
3. E. Englert, Ann. Physik 14, 589 (1932).
4. D. Chen, J. appl. Phys. 42, 3625 (1971)
5. D. Chen, J. F. Ready and E. Bernal, J. appl. Phys. 39, 3916 (1968).
6. R. B. Birss, "Symmetry and magnetism", North-Holland Amsterdam (1966) Ch.5, 233.
7. S. Chikasumi, "Physics of magnetism", John Wiley, New York (1964) Ch.9, 190.

MAGNETIC PROPERTIES AND TEXTURE OF UNALLOYED NON-ORIENTED ELECTRICAL STEEL

P.K. Rastogi, Inland Steel Research Laboratories
3001 East Columbus Drive, East Chicago, Indiana 46312

ABSTRACT

A study was made to compare the magnetic properties and texture of a rephosphorized 0.1 Wt% carbon steel with those of a rephosphorized 0.06 Wt% carbon steel. The results indicate that the 0.1 Wt% carbon steel is superior to the 0.06 Wt% carbon steel with respect to 15 and 17 kG permeabilities and 15 kG core loss in the cold rolled and decarburized condition. It is also noted that the 0.1 Wt% carbon steel exhibits a greater planar magnetic anisotropy than 0.06 Wt% carbon steel. Besides, anomalous loss for both steels is found to be about 20% of the total loss, which is generally considered high for non-oriented steels. The significant improvement in permeability at 15 and 17 kG is quantitatively explained in terms of a texture model. The reduction in the 15 kG core loss is attributed to the decrease in the hysteresis loss through the development of a favorable texture.

INTRODUCTION

It is apparent from the work of Stefán et al[1] that the magnetic properties of unalloyed cold rolled electrical steel containing initial carbon content up to 0.10 Wt% could be considerably improved by promoting a significant degree of decarburization in the hot rolled condition through a special annealing process. In addition, 0.1 Wt% carbon content permits the use of standard ferromanganese alloy in steelmaking, which is cheaper than the medium carbon ferromanganese alloy previously used in 0.05/0.06 Wt% carbon steels. In view of these advantages, a study was made to compare the magnetic properties and texture of a rephosphorized 0.1 Wt% carbon steel with those of a rephosphorized 0.06 Wt% carbon steel.

MATERIAL PREPARATION AND EXPERIMENTAL PROCEDURE

Two 17.3 kg vacuum melted ingots of rephosphorized steels A and B containing 0.09/0.10 and 0.05/0.06 Wt% carbon were used in this investigation. The chemical compositions of the two ingots are presented in Table I. The ingots were hot rolled to about 2.54 mm thick strips using a finishing temperature of approximately 870°C and a simulated coiling temperature of 710°C. The hot rolled samples originating from Steel A were annealed for 8-hours at 790°C to promote decarburization through existing surface oxide, after which the carbon level was found to be in the range of 0.035 to 0.046 Wt%. Following pickling, the hot rolled samples of both steels were cold rolled to approximately 0.64 mm thick sheets.

Longitudinal and transverse Epstein samples, 210.0 mm x 22.5 mm, were prepared from each cold rolled sheet steel and subsequently decarburized for 2-1/2 hours at 790°C in an atmosphere of 10% H_2 + 90% Ar with a dew point in excess of +12°C to achieve a carbon level $\leq$0.004 Wt%. Magnetic measurements were made on decarburized longitudinal (L), transverse (T) and mixed (L+T) Epstein packs containing equal number of longitudinal and transverse strips. The A.C. permeability at 15 and 17 kG was measured on a 3/4 size standard Epstein frame,[2] while the total core loss and D.C. hysteresis loss were measured only at 15 kG. In addition, the electrical resistivity was measured to compute the classical eddy current loss.

The grain size was characterized in terms of the average number of intercepts/mm which is inversely proportional to the average grain size. Texture was monitored by the inverse pole figure technique in terms of the pole density, $(I/I_R)_{hkl}$, of eight {hkl} planes measured at the mid-plane of the decarburized samples.

TABLE I

Chemical Compositions (Wt%) of Low Carbon Steels

Steel	C	Mn	P	S	Si
A	0.092	0.37	0.14	0.025	0.04
B	0.054	0.38	0.15	0.024	0.03

RESULTS AND DISCUSSION

Data in Table II show that Steel A is superior to Steel B with respect to the permeability at 15 and 17 kG and the 15 kG core loss in the decarburized condition. It is noted that for longitudinal and mixed Epstein packs the 15 kG permeability of Steel A is about 22% higher than that for Steel B, whereas the improvement in this property for the transverse Epstein pack is 9%. At 17 kG, the improvement in permeability of Steel A relative to Steel B for longitudinal, transverse and mixed Epstein packs is about 74, 15 and 39%, respectively. In addition, it is seen from Table II that at 17 kG the permeability ratio of Steel A is particularly much higher than that of Steel B, indicating that the former is magnetically more anisotropic than the latter. In contrast to the large improvement in permeability, the improvement in the 15 kG core loss ranges only between 4 to 6%, as can be seen from the data presented in Table II.

It is known that the 17 kG permeability of non-oriented steel is primarily dependent upon texture, while at 15 kG it is a function of both grain size and texture. Since the grain size is essentially the same for both steels, the improvement in permeability at 15 and 17 kG is attributed to a significant increase in texture parameter, (T_p), as shown in Table III. The above findings are qualitatively consistent with the results of previous investigations.[3,4] However, it should be pointed out that the higher value of T_p for Steel A does not indicate that it is a strongly textured steel, as shown from the data in Table III. The reduction in the 15 kG core loss is caused by the decrease in the hysteresis loss through the development of a favorable texture. This is also consistent with the results of earlier investigations.[3,5]

According to the classical analysis of A.C. total core loss,[6] the anomalous loss (P_A) is defined as follows:

$$P_A = P_T - (P_H + P_E)$$

where P_T, P_H and P_E refer to the total core loss, hysteresis loss and classical eddy current loss, respectively. Data in Table II suggest that for both steels P_H and P_E together constitute approximately 80% of the total loss. However, the computed anomalous loss (P_A) is about 20% of the total loss, which is generally considered high for non-oriented low carbon steels. This loss appears to be primarily related to the local eddy current losses at moving domain boundaries.[7] In addition, an attempt has been made to compare the computed value of anomaly factor ($\eta = PA{+}PE/P_E$) of these steels, which is usually a function of the ratio of domain size to the sheet thickness.[7] Since the anomaly factor is found to be the same, it is inferred that there is essentially no difference

TABLE II

Magnetic Properties, Grain Size and Resistivity of Decarburized Steels

Steel	Direction	A.C. Permeability (G/Oe) 15 kG	17 kG	P_T*	P_H*	P_E*	P_A*	Average Grain Size (Int/mm)	Resistivity μ-ohm-cm
				(watts/lb) at 15 kG)					
A	L	3740	900	4.35	-	-	-	35.1	15.39
	T	2800	470	4.59	-	-	-		
	L+T	3340	613	4.47	1.62	2.00	0.85		
	(L/T)**	(1.33)	(1.91)						
B	L	3090	516	4.65	-	-	-	36.9	15.34
	T	2570	409	4.78	-	-	-		
	L+T	2740	442	4.67	1.81	2.02	0.84		
	(L/T)**	(1.20)	(1.26)						

*P_T, P_H, P_E and P_A refer to total core loss, hysteresis loss, classical eddy current loss and anomalous loss respectively. ** Values in parenthesis denote permeability ratios.

TABLE III

Mid-Plane Pole Densities $(I/I_R)_{hkl}$ of Cold Rolled and Decarburized Steels

Steel	200	211	220	310	222	321	420	332	Tp*
A	0.56	1.07	1.47	0.61	2.49	0.94	0.96	1.23	0.62
B	0.62	1.61	0.11	0.33	8.98	0.53	0.21	2.52	0.09

(Column headers 200–332 are under "Plane".)

*T_p is defined as,

$$T_p = \frac{(I/I_R)_{200} + (I/I_R)_{220} + (I/I_R)_{310} + (I/I_R)_{420}}{(I/I_R)_{211} + (I/I_R)_{222} + (I/I_R)_{321} + (I/I_R)_{332}}$$

in the average domain size of these steels for the same thickness.

CONCLUSIONS

The following conclusions are derived from the results of this investigation: 1) In cold rolled and decarburized condition, the 0.092 Wt% carbon steel is superior to the 0.054 Wt% carbon steel with respect to permeability at 15 and 17 kG, 15 kG core loss and hysteresis loss. The improvement in these properties is related to the formation of a superior texture. 2) The 0.092 Wt% carbon steel exhibits, in general, a greater magnetic anisotropy in terms of permeability ratio than the 0.054 Wt% carbon steel. 3) The anomalous loss at 15 kG constitutes about 20% of the total loss, which is generally considered high for non-oriented low carbon steels.

REFERENCES

1. M. Stefán, Z. Hegedüs, F. Balázs and G. Juhász, U. S. Patent No. 3,870,574, March 11, 1975.
2. D. E. Jonquet, Unpublished Information, Inland Steel Company, 1975.
3. P. K. Rastogi, Unpublished Information, Inland Steel Company, 1975.
4. P. K. Rastogi, AIP Conference Proceeding No.29 1976. (Accepted for Publication)
5. P. K. Rastogi, AIP Conference Proceedings No. 24,724 (1975).
6. ASTM Special Technical Publications, No. 371,32 (1969).
7. F. Brailsford, Physical Principles of Magnetism, F. Van Nostrand, 238-240, (1966).

THE INFLUENCE OF GRAIN STRUCTURE AND NONMAGNETIC INCLUSIONS ON THE MAGNETIC PROPERTIES OF HIGH-PERMEABILITY FE-NI-ALLOYS

W. Kunz[+] and F. Pfeifer[+], Vacuumschmelze GmbH, D-6450 Hanau, FRG

ABSTRACT

For permalloys, the coercive force H_c was investigated as a function of grain size and nonmagnetic inclusions. H_c can be described as a linear function of 3 contributions: $H_c = H_{cd} + H_{cn} + H_{co}$. H_{cd}, the contribution of grain boundaries, is found to be proportional to the inverse grain diameter $1/d_K$. H_{cn} is proportional to the number of inclusions, and H_{co}, the remaining term, is probably due to magnetostrictive forces. Depending on melting and deoxidizing conditions, H_{cn} is found to depend on micron- or submicron- size inclusions. As particles with a diameter s of about δ (Bloch wall thickness) should give a maximum contribution to H_c, one has to suppose that in the second case, the submicron-size particles give rise to stress fields of about δ. The different behaviour of the particles is probably due to differences in the thermal expansion coefficients of matrix and inclusions.

INTRODUCTION

The magnetic properties of soft magnetic alloys are determined by the anisotropy constants (K_1-magnetocrystalline and λ-magnetostrictive anisotropy) and by defects in the crystalline structure, due to internal stresses, grain boundaries and nonmagnetic inclusions[1]. For permalloys, the anisotropy constants are nearly zero. This is achieved by an appropriate compositional variation and annealing treatment[2]. The present objective was to find a quantitative relation between the magnetic properties - e. g. the coercive force H_c- and grain size d_K and the number n of inclusions per cm^2 in the ferromagnetic matrix. It extends previous experiments on 50 % NiFe[3].

In this report it shall be shown that for permalloys, H_c can be described as a linear superposition of the perturbing factors mentioned above.

$$H_c = H_{cd} + H_{cn} + H_{co} \quad (1)$$

H_{cd} and H_{cn} indicate the influences of grain boundaries and nonmagnetic inclusions respectively, while H_{co} is probably due to magnetostrictive forces. According to Mager[4], H_{cd} is a linear function of the inverse grain diameter $1/d_K$.

$$H_{cd} = H_{co} + m \cdot 1/d_K \quad (2)$$

The intercept with the H_c - axes is given by H_{co}. The slope m can be expressed in terms of the wall energy γ_w and the saturation induction B_s .

$$m \approx \frac{9\pi}{8} \cdot \frac{\gamma_w}{B_s} \quad (3)$$

A rough estimate for K_1 and the Bloch wall thickness may be obtained from (3), as

$$\gamma_w \approx \sqrt{kT_c \cdot \frac{K_1}{a}} \quad (4)$$

(k... Boltzmann constant, T_c... Curie temperature, and a...lattice parameter)

$$\text{and } \delta \approx \sqrt{\frac{k \cdot T_c}{a \cdot K_1}} \quad (5)$$

For permalloys, δ is about 1 µm.

As to the influence of nonmagnetic inclusions, there is a complicated relationship between H_c and number, size, chemical nature, and morphology of the particles[5,6,7]. According to Dijkstra and Wert[5], H_{cn}, the contribution of inclusions to H_c, should be maximum for a particle size s in the range of the Bloch wall thickness δ . In this report, $H_{cn} \sim n$ should apply as a rough approximation. Introducing a scale factor F, H_{cn} can be written explicitly

$$H_{cn} = F \cdot n \quad (6)$$

where n means the number of "effective" particles per cm^2 with $s \sim \delta$. F expresses the contribution of a single "effective" particle to the coercive force H_{cn}. Eq. (1) can be now written explicitly:

$$H_c = H_{co} + m \cdot 1/d_K + F \cdot n \quad (7)$$

H_c, d_K, and n can be measured, while H_{co}, m, and F are calculated as the coefficients of a linear regression of the form $z = a_o + a_1x_1 + a_2x_2$.

EXPERIMENTAL PROCEDURE

The experiments reported here, were carried out on permalloys (77 % Ni, 4 % Mo, 4,5 % Cu[8], balance Fe) using standard melting techniques (open, vacuum, electro-slag, vacuum arc remelting). The material was deoxidized with C, Mn, Si and with additions of Mg and Ca in some cases. It was processed to 1 mm strip thickness by an appropriate hot and cold rolling treatment and subsequently annealed in a pure dry hydrogen atmosphere (4 hrs). The magnetic and metallographic data were varied by the variation of the final anneal between 750 and 1200° C. H_c was measured by a dc method. Grain boundaries and etch-pits (see fig. 2) were made visible by electrochemical processing. (Struers V2A electrolyte, current I = 3 A at room temperature.) The average grain diameter was determined by counting the intercepts with a linear pattern. The number of etch-pits in an area of 1 mm^2 was counted. The number of submicron-size inclusions (0,02 µm - 0,5 µm), counted in an electron microscope, was found to agree rather well with the number of etch-pits. Micron-size particles were made visible by mechanical polishing.

RESULTS AND DISCUSSION

The variations of the melting and deoxidation conditions gave rise to striking differences in magnetic and metallographic results. Two types of behaviour were found, characterized as "type 1"

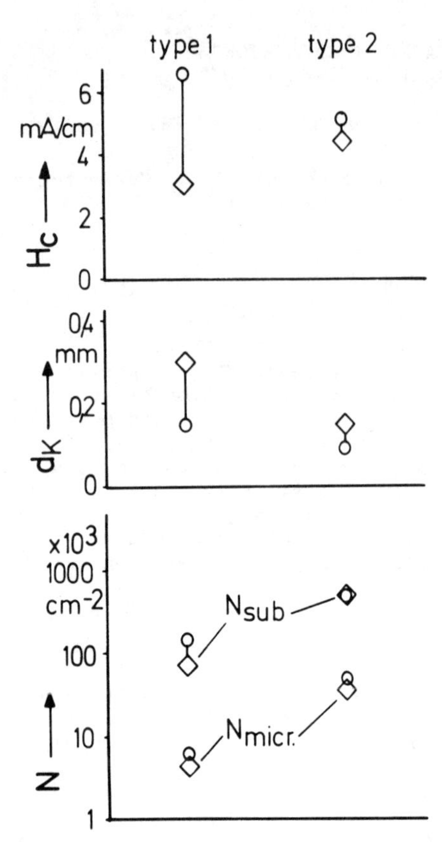

Fig. 1. Coercive force H_c, average grain diameter d_K, and density of inclusions in the sub-micron (N_{sub}) and the micron ($N_{micr.}$) range for type 1- and type 2- material in 1 mm strip thickness. (○ 1100° C, ◇ 1200° C)

and "type 2" (see fig. 1 and fig. 2). Raising the annealing temperature from 1100 to 1200° C, the magnetic and metallographic data are significantly improved for type 1 material, whereas for type 2 materials, these values remain nearly unaffected. After a 1100° C anneal, H_c is lower in the case of type 2 material in spite of its finer grain size and higher number of micron- and submicron-size inclusions respectively. The optical micrographs of type 1- and type 2-material are given in fig. 2. Fine dispersed etch-pits are observed in the first case. In the other material, the slag particles are bigger and often aligned. This may be due to the different properties of different deoxidizers: Inclusions, containing Mn, tend to dissociate and coagulate respectively at higher annealing temperatures. Particles, containing Mg, give rise to rather stable inclusions, often ordered in lamellar form, which are nearly unaffected by a heat treatment.

The simple theoretical model proposed is now to be applied to the experimental results reported here. In fig. 3, H_c and the metallographic data vs. annealing temperature are compiled for a sample of type 1. Several specimens show a characteristic behaviour, as illustrated here. Raising the annealing temperature from 950 to 1050° C, H_c does not decrease as would be expected. Surprisingly, the submicron-size inclusions show a corresponding behaviour, whereas the grain grows regularly and the micron-size inclusions decrease! The anomaly of H_c gives rise to a deviation from linearity in the $H_c(1/d_K)$-plot (fig. 4, broken line). As shown in fig. 3, the particle contribution to H_c can be described only by submicron-size inclusions. $H_{cn} = F \cdot n$ is given by the arrows in fig. 4. They mark the contribution of the inclusions to every measured value of H_c. The unbroken line gives a significantly better fit than the broken line. Considering micron-size inclusions, one gets no better linear correlation in the H_c $(1/d_K)$-plot. Surprisingly, submicron particles seem to play an important role for type 1-permalloys. This may be due to stress fields, evoked by the submicron inclusions. For type 2 - material, the inclusions are nearly unaffected by the annealing temperature. Therefore, the particle correction gives rise to a parallel shift in the H_c $(1/d_K)$-plot. As the magnetostrictive contribution to H_c is nearly independent on the melting and deoxidizing conditions one may assume the same H_{co} for both types of permalloys. Considering submicron inclusions as the "effective" particles - as for type 1 material- the particle-correction H_{cn} would lead to values, an order of magnitude too large. Considering micron-size inclusions, one gets the same F as in case 1 for submicron inclusions and reasonable values of H_{cn}. The different behaviour of permalloys, investigated here, may be interpreted in terms of the different chemical structure of the nonmagnetic inclusions. For 50 % NiFe[3], such differences were not observed. The stress fields, assumed in the first case, are obviously due to different thermal expansion coefficients of inclusions[9] and NiFe-matrix[10]. In the second case, these coefficients[9] agree rather well and do not give rise to additional stress fields.

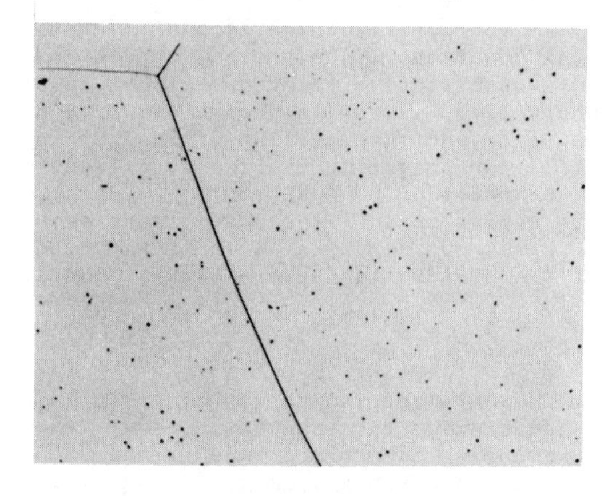

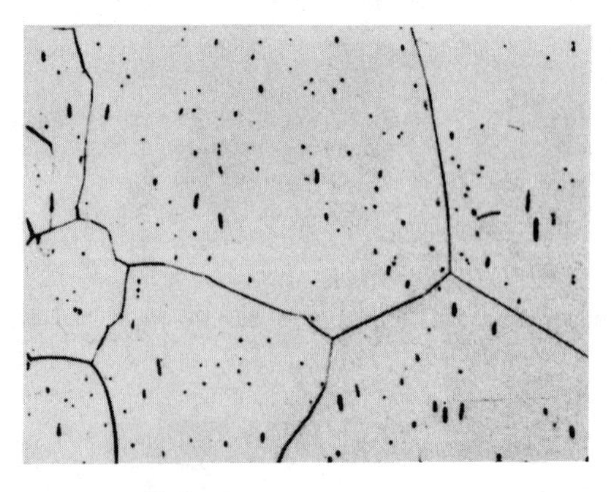

Fig. 2. Optical micrographs of etched submicron inclusions for type 1- and type 2- material after 1200° C annealing.

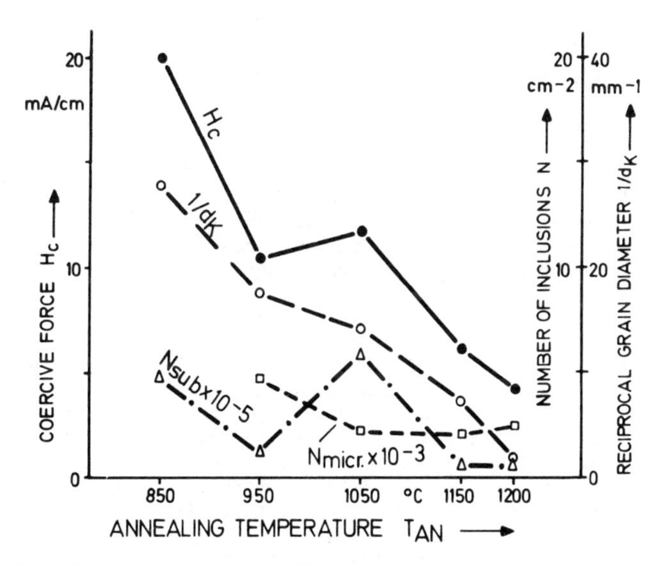

Fig. 3. Coercive force H_c, reciprocal grain diameter $1/d_K$, and number of particles in the submicron (N_{sub}) and in the micron ($N_{micr.}$) range as a function of the annealing temperature T_{An} for type 1 - material.

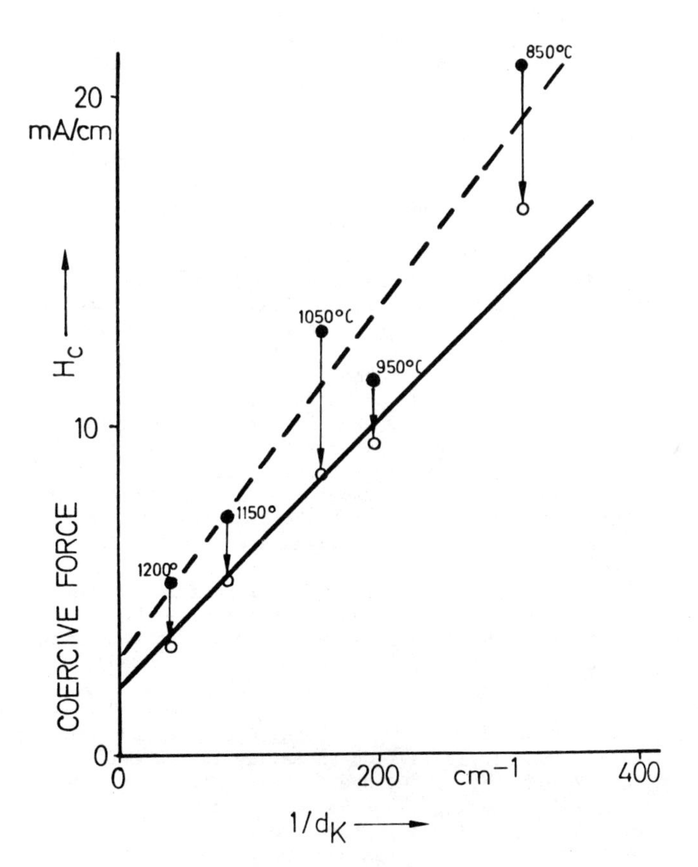

Fig. 4. Coercive force H_c as a function of the reciprocal grain diameter $1/d_K$ without considering (●) and with considering (○) inclusions in the submicron range for type 1 - material.

ACKNOWLEDGEMENT

The authors are indebted to Mrs. Pfeiffer, Mrs. Wilke-Dörfurt, and Mr. Baer.

REFERENCES

+ work partially supported by the Bundesministerium für Forschung und Technologie of the Federal Republic of Germany. The authors alone are responsible for the contents.

1. E. Kneller: Ferromagnetismus, Springer Verlag Berlin-Göttingen-Heidelberg, 1962
2. F. Pfeifer: Z.Metallkde 57 (1966) 240-249
3. E. Adler, H. Pfeiffer: IEEE Trans. Magnetics Vol MAG 10 Nr. 2 (1974) 172-174
4. A. Mager: Ann. Phys. Leipzig 11 (1952) 15
5. L.J.Dijkstra and C.Wert: Phys.Rev.79(1950)979
6. L. Néel: Ann. Univ. Grenoble 2 , 299(1945/46)
7. A. Mager: Z.f.Angew. Phys. 14,4 (1962)230-237
8. F. Assmus and F. Pfeifer: Metall 5/6, 7. Jahrg. 1953 S 189-191
9. J.D'Ans and E. Lax: Taschenbuch f. Chemiker u. Physiker. Springer-Verlag, Berlin-Göttingen-Heidelberg, 1949
10. R.M.Bozorth: Ferromagnetism, D. Van Nostrand Company, Inc Toronto-New York-London, 1951

MICROSTRUCTURE AND MAGNETIC PROPERTIES OF SPINEL FERRITES

Raja K. Mishra and G. Thomas
Department of Materials Science and Engineering and
Materials and Molecular Research Division
Lawrence Berkeley Laboratory
University of California, Berkeley, California 94720

ABSTRACT

Achieving suitable magnetic properties in ceramic ferrites through thermomechanical treatments rather than through varying the processing and fabrication parameters alone are investigated. The high temperature phase transformation in lithium ferrite ($LiFe_5O_8$) spinel and the defects in $LiFe_5O_8$ and $NiFe_2O_4$ are studied using high voltage transmission electron microscopy in an effort to characterize the microstructures. Results show that a dispersion of paramagnetic $LiFeO_2$ particles in $LiFe_5O_8$ matrix gives rise to increased squareness of the hysteresis curve and increased coercivity. Annealing treatments of sintered ferrites remove undesirable intra-granular {100}¼<110> cation stacking faults and improve hysteresis loop parameters.

INTRODUCTION

The microstructure-sensitive magnetic properties of commercially used spinel ferrites are conventionally controlled by varying processing parameters[1] so as to change porosity, grain size, grain distribution etc. On the other hand, great success has been made in designing of metallic alloys for mechanical and physical applications through thermomechanical treatments to obtain sutiable microstructures. This has been possible solely due to our understanding of the microstructural features, (such as defects and phases) and their effects on the material properties[2]. An analogous situation does not exist for ceramic materials, specifically because of the lack of detailed microstructural information on many ceramic systems. Recent developments of experimental tools such as high voltage transmission electron microscopy and ion-bombardment technique[3] for thinning non-metallic specimens have opened up the field of microstructural charactaerization in ceramic systems. In the present work, a high voltage electron microscopy study of microstructures in lithium ferrite and nickel ferrite has been carried out. The effect of these microstructures on the magnetic hysteresis curve has been studied. The significance and importance of these preliminary observations are discussed.

EXPERIMENT

High temperature phase transformations in lithium ferrite are studied by heating thin discs of $LiFe_5O_8$ single crystal in air, vacuum (10^{-5} torr) and oxygen (760 torr) at 1200°C for different lengths of time. Thin foil specimens for examination in the microscope are prepared from the center of these discs by mechanical polishing followed by ion bombardment. Defects in single crystal lithium ferrite and polycrystalline nickel ferrite are studied by preparing thin foils from as-received as well as annealed materials. The electron microscope observations are made in a Hitachi Hu-650 high voltage electron microscope operating at 650 kV. The dynamic hystersis loops are taken at 60Hz using torroidal specimens. The size of the specimens used is 1 cm O. D., 0.6 cm I.D. and 0.2 cm thick. Twenty turns of copper magnetic wire are used in both primary and secondary windings.

RESULTS AND INTERPRETATION

A. Phase transformation and magnetic properties: The phase transformation in air is described below in detail since air is the most economical atmosphere to maintain during any thermomechanical treatment. The features of the transformation are similar in other atmospheres. However, the transformation kinetics are quite different. The reaction proceeds faster in vacuum and slower in oxygen than in air.

On heating a 2mm thick disc of $LiFe_5O_8$ in air at 1200°C for about 25 minutes, one sees small octahedral shaped particles of a second phase homogenously dispersed as shown in Fig. 1a. The particles are of $LiFeO_2$ phase with a lattice parameter roughly half of that of spinel. Particles with an average size of 2500Å or less remain coherent with the matrix. Dislocation networks form at the interface to relieve the strain as the praticles grow larger.

Figure 1b shows the interfacial dislocations of Burgers vector ½<110> at the $LiFeO_2$-$LiFe_5O_8$ interface. Measurement shows that approximeately 10% of the spinel transforms to $LiFeO_2$. This $LiFeO_2$ later transforms to a lithium deficient spinel structure leaving behind incoherent grain-boundaries as in Fig. 1c.

The hysteresis loops corresponding to microstructures in Fig. 1a-c are given in Fig. 2a-c and Fig. 2d is the hysteresis curve for the as-received single phase material. The coercivity of the two phase microstructures in Fig. 2a or 2b is higher than that of single phase $LiFe_5O_8$. Also coercivity of polygranular single phase spinel as in Fig. 1c is higher than that of single phase crystal. $LiFeO_2$ phase is paramagnetic[4] and thus it is not surprising that its dispersion in a ferrimagnetic phase increases H_c by acting as a barrier to the domain wall motion[5]. Quantitative investigation of the dependence of H_c on particle size, volume fraction and coherency strain are in progress to verify Haasen's results.[5] A new and significant observation in Fig. 2 is that the squareness of the hysteresis loop, defined as $Br/4\pi M_s$ is higher for the two phase microstructures of $LiFeO_2$-$LiFe_5O_8$. Reduction of the value of M_s in Fig. 2c is due to reduction of Fe^{+3} to Fe^{+2}.

B. Defects and magnetic hysteresis:

Figure 3 shows the cation stacking faults in a specimen prepared from the vicinity of the surface of flux-grown $LiFe_5O_8$ single crystal. The faults are on {110} planes with ¼<110> as the displacement vector[6]. The defect density decreases rapidly with increasing distance from the surface. The hysteresis loops of faulted and unfaulted materials are shown in Fig. 4. Presence of faults increases the coercivity. In the absence of any data on the magnetic domain configurations, a discussion on the interaction of the cation faults with the domains is not possible at this stage. It may be noticed that the M_s value of faulted and unfaulted materials are the same. Figure 5 shows a faulted grain in a polycrystalline $NiFe_2O_4$. More than 30% of the grains are seen to be faulted in the as-received crystals. On annealing the material in air for 12 hours at 850°C, the fault density decreases considerably, leaving about 95% of the grains fault-free with no other microstructural modifications. There is a change in the hysteresis curve as in Fig. 6.

DISCUSSION

Applications of ferrites in computer memory cores and in microwave device components such as latching. devices[7] require good squareness of the B-H loop. A necessary condition for good squareness is the dominance of the anisotropy energy over the magnetostrictive energy and this dictates the choice of materials. Proper processing is also used to enhance this effect[8]. The present results suggest yet another way of achieving it. In the absence of data on other magnetic properties of the two-phase systems, evaluation of the usefulness of these materials compared to the currently used ones is not possible, but it seems to be a step in the right direction in materials technology.

The result that H_c increases in a two-phase microstructure is not new[9], but the approach is new for ceramic magnets. It may be possible to use a similar approach to achieve better properties in hard magnetic ferrites than presently exist.

The polycrystalline nickel ferrites in the as-fabricated state are highly defective and this may be the reason for the poor performance of this material. (The samples examined here were from a rejected batch of commercial $NiFe_2O_4$ manufactured by Countis Industries, San Luis Obispo, California). All the magnetic properties of the annealed material have not been studied yet, but the data presented here suggests that a proper post-processing thermomechanical treatment may provide ways of improving the properties; thus reducing waste of material.

In conclusion, this preliminary study of the effect of microstructure on magnetic hysteresis shows that proper thermomechanical treatment of ceramic ferrites may be the step beyond the presently existing processes[1] which will bring about cheaper, better and more useful ceramic ferrites.

REFERENCES

* This report was done with support from the United States Energy Research and Development Administration. Any conclusions or opinions expressed in this report represent solely those of the author(s) and not necessarily those of The Regents of the University of California, the Lawrence Berkeley Laboratory or the United States Energy Research and Development Administration.

1. M. Sugimoto, "Recent Advancement in the Field of High Frequency Ferrites" in AIP Conference Proc. 10 (2), Ed. by C. D. Graham Jr. and J. J. Rhyne, American Institute of Physics, New York, 1973, pp 1335-1349.
2. G. Thomas, "Utilization and Limitations of Phase Transformations and Microstructures in Alloy Design for Strength and Toughness", in Proc. of Battelle Colloquium on Fundamental Aspects of Structural Alloy Design, Ed. by R. I. Jaffee, Seattle, Washington, 1975, In Press. Lawrence Berkeley Laboratory Report #4175.
3. D. J. Barber, Thin Foils of Non-metals Made for Electron Microscopy by Sputter-Etching, J. Mat. Sc. 5, pp 1-8, 1970.
4. J. C. Anderson et al., The Magnetic Susceptibilities of $LiFeO_2$, J. Phys. Chem. Sol. 26, pp 1555-1560, 1965.
5. P. Haasen, "Mechanical, Magnetic and Superconductor Hardening by Precipitates", Mat. Sci. Eng., 9, pp 191-196, 1972.
6. O. Van der Biest and G. Thomas, "Cation Stacking Faults in Lithium Ferrite Spinel", Phys. Stat. Sol. 24(a), pp 65-77, 1974.
7. G. M. Argentina and P. D. Baba, "Microwave Lithium Ferrite - An Overview", IEEE Trans. Microwave Theory and Tech., MTT-22, pp 652-658, 1974.
8. P. D. Baba et al. "Fabrication and Properties of Microwave Lithium Ferrites", IEEE Trans. Magnetics, MAG-8, pp 83-94, 1972.
9. H. G. Brion et al. "Pinning of Bloch Walls by Non-Magnetic Particles", Int. J. Magnetism, 5, pp 109-110, 1973.

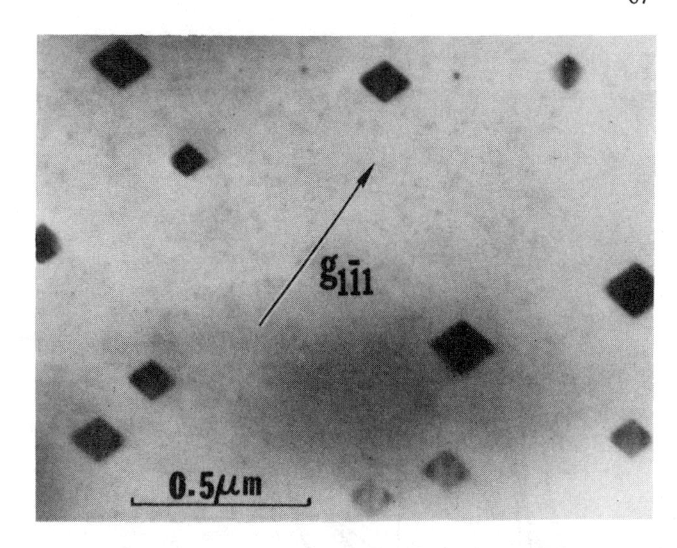

Fig. 1(a) Octahedral coherent precipitates of $LiFeO_2$ dispersed in $LiFeO_8$.

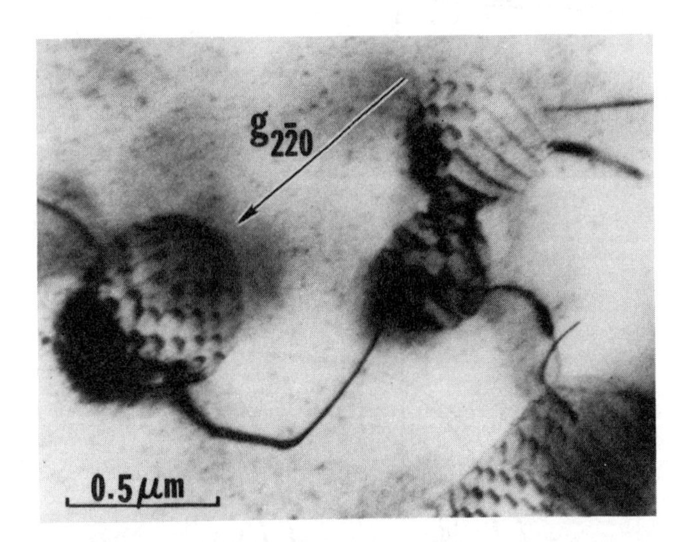

Fig. 1(b) Semicoherent $LiFeO_2$ particles with interfacial dislocations of Burgers vector ½<110>.

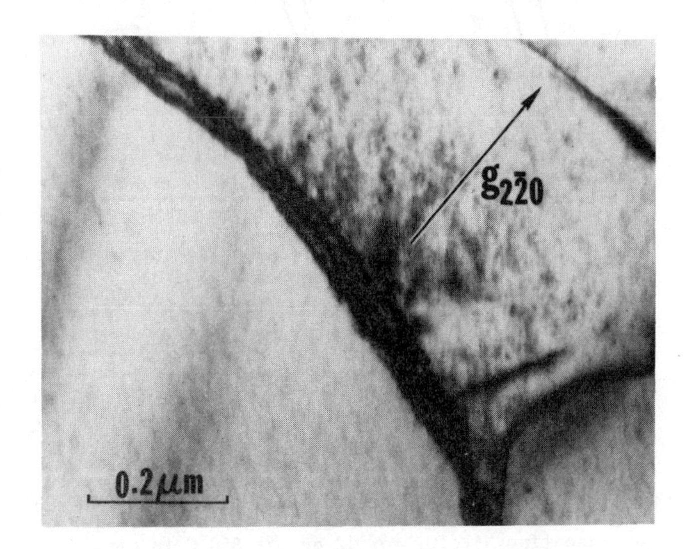

Fig. 1(c) Incoherent grain boundaries in the transformed spinel.

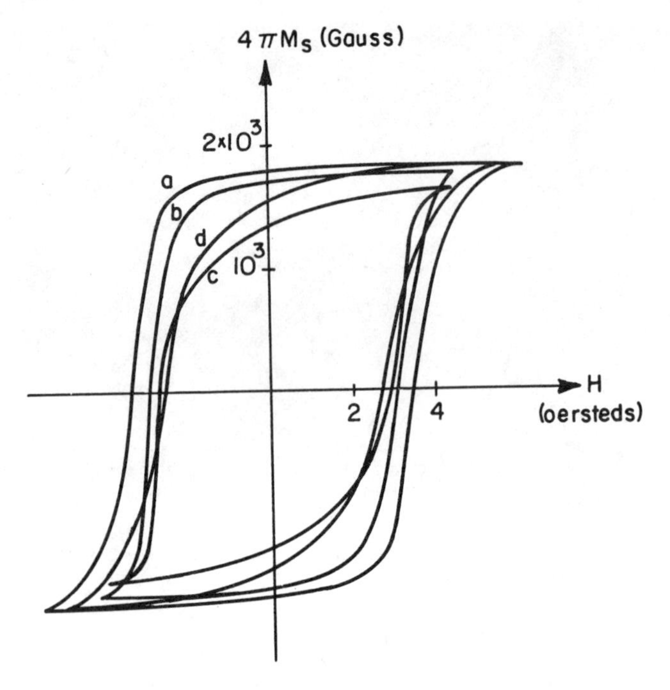

Fig. 2 (a-c) Hysteresis loops corresponding to the microstructures in Fig. 1(a-c). (d) Hysteresis loop of the single phase $LiFe_5O_8$ single crystal. All experiments on discs with normal along <110>.

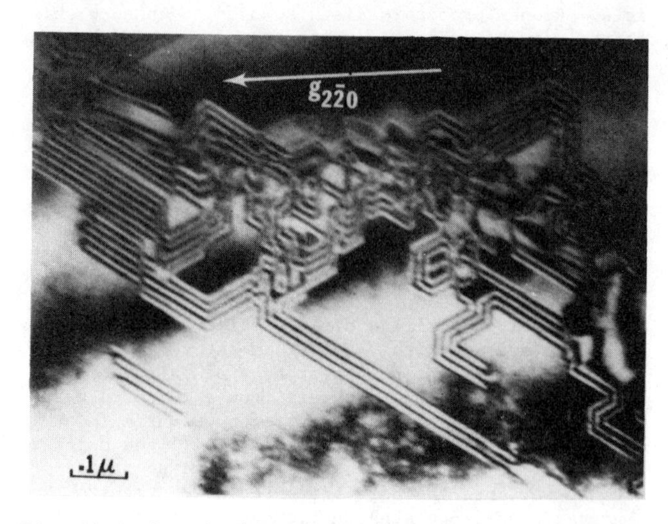

Fig. 3 Cation stacking faults in $LiFe_5O_8$. Faults lie on {110} planes with ¼<110> displacement vectors.

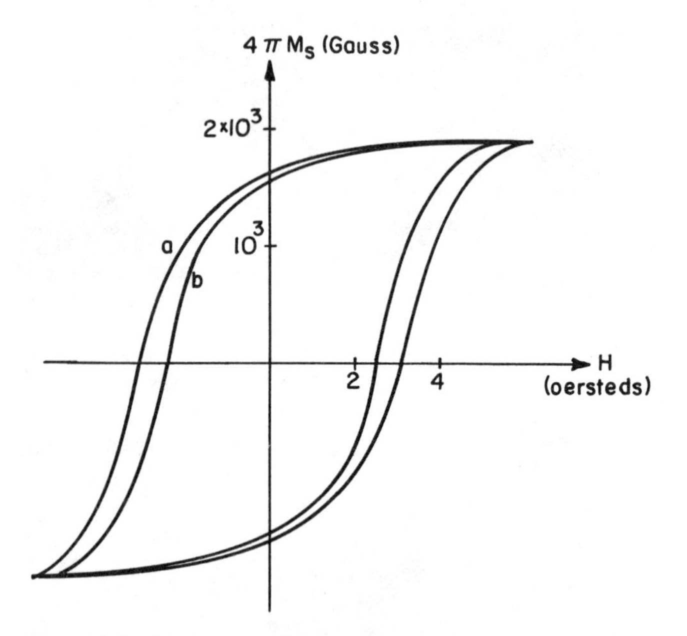

Fig. 4(a) Hysteresis loop corresponding to the microstructure in Fig. 3. (b) Same as Fig. 2(d).

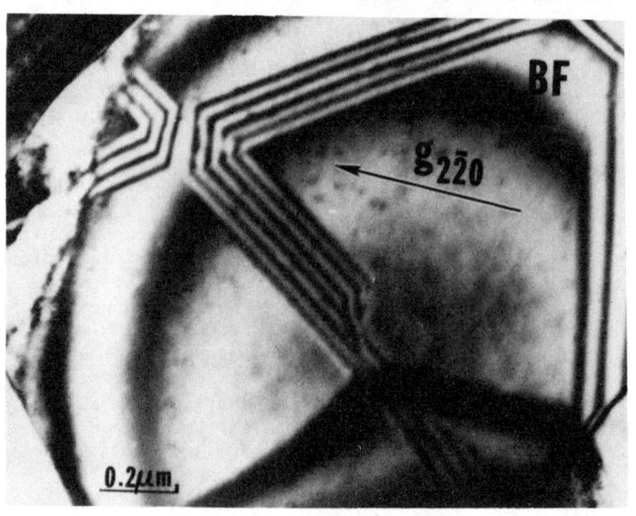

Fig. 5 {110}¼<110> cation fault inside a grain in polycrystalline $NiFe_2O_4$. Grain size is approximately 1.5 microns.

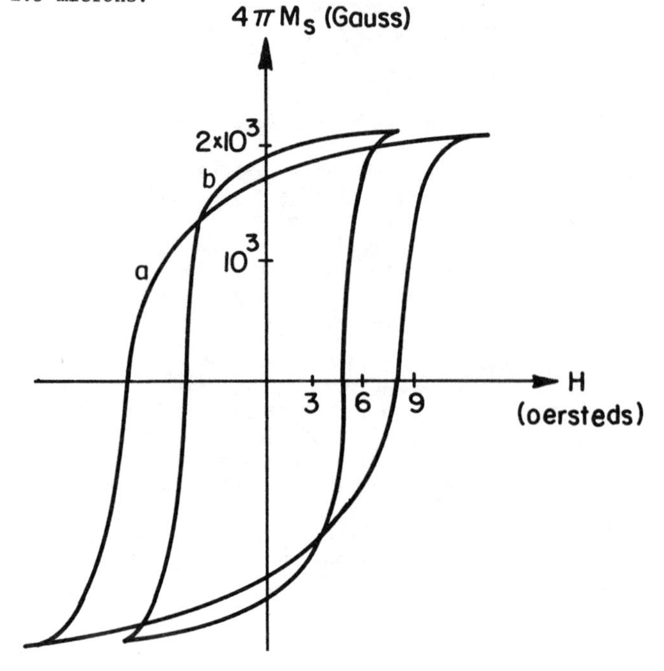

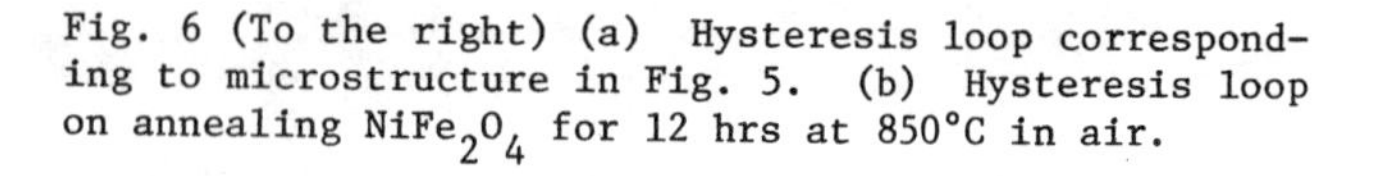

Fig. 6 (To the right) (a) Hysteresis loop corresponding to microstructure in Fig. 5. (b) Hysteresis loop on annealing $NiFe_2O_4$ for 12 hrs at 850°C in air.

TEMPERATURE DEPENDENCE OF THE COERCIVITY OF PEARLITE

F. C. Schwerer, C. E. Spangler, Jr., and J. F. Kelly
U. S. Steel Research, Monroeville, PA 15146

Coercivities have been measured at temperatures from -196°C to 770°C for four carbon steels (0.08 to 1.6 weight percent carbon) that had been processed so that cementite, Fe_3C, occurred primarily as lamellar pearlite. For medium- and high-carbon steels, coercivities are strongly temperature dependent, exhibiting maxima at temperatures near 210°C, the Curie temperature of Fe_3C. Reversals of remanent magnetization near 210°C have been reported for similar steels. Energy-minimization calculations were performed for a model that takes into account magnetocrystalline and magnetostatic energy for two known crystallographic orientations of cementite and ferrite (bcc-Fe with C in solution) in pearlite. These calculations predict temperature-dependent phenomena that are in good qualitative agreement with experimental observations.

ANALYSIS OF INFORMATION SIGNAL OF PLATED WIRES

J. Przyluski and R. Liwak
Warsaw Technical University, Warsaw, Poland

Analysis of the information signal of plated wires in the tunnel memory has been described. Output signal of the film is of importance as a basic, utility parameter. NiFe plated wires ($\alpha_{90} < 2°$, wire diameter 100 μm, thickness of film 0.9 μm) have been studied. Saturation flux ϕ_s and anisotropy field H_k have been obtained from typical B-H loop traces. Moreover, the ratio H_w/H_k in the films (H_w = amplitude of word field in tunnel matrix) and the ratio ϕ/ϕ_s, depending on the write parameters in the matrix, have been measured by using the pulse tester. An equation has been derived from the rotational model of switching. The influence of the write parameter, the adjacent cells and the real form of the word pulse on flux ϕ in the easy axis has been taken into consideration. The results of measurements are in good accord with the theoretical curves over the range up to $H_w = 0.75H_k$.

CROSSTIE DYNAMIC NUCLEATION THRESHOLDS AND BLOCH LINE MOBILITY MEASUREMENTS IN THIN PERMALLOY FILMS

G. Cosimini and J. H. Judy
University of Minnesota, Minneapolis, MN 55455

The crosstie memory stores information as crosstie-Bloch line pairs along Néel walls in thin Permalloy films. The basis of this memory is a crosstie shift register which propagates data by a sequential nucleation and annihilation of crosstie-Bloch line pairs. The data rate is limited by the crosstie dynamic thresholds and the Bloch line mobility. The purpose of this paper is to present measurements of these limitations using Bitter colloid and magneto-optical Faraday observation techniques. For a field-induced Néel wall in a 350 Å thick Permalloy film having an anisotropy field of 4.7 oe., nucleation fields of 0.7,1.3,3.2 oe. (Faraday) and 2.7,3.7,4.7 oe. (Bitter) were measured with pulse widths of 100,10,1 ns respectively. The difference between these measurements is attributed to the interaction between the Bloch lines and the Bitter particles. Bloch line mobilities as high as 500 m/s-oe were measured.

SUPERCONDUCTIVITY IN COMPOUNDS WITH ANTI-Th_3P_4 STRUCTURE

F. Hulliger* and H.R. Ott*
Laboratorium für Festkörperphysik
Eidgenössische Technische Hochschule Zürich
8093 Zürich, Switzerland

ABSTRACT

The majority of the compounds with Th_3P_4 structure exhibit nonmetallic and/or ferro- or antiferromagnetic properties. Superconductivity in the metallic chalcogenides La_3X_4 (X=S, Se, Te) which contain one excess valence electron per formula unit was discovered recently and fairly high values for the critical temperature T_c were reported. The only nonmagnetic pnictides (thorium pnictides) are valence compounds and at least the phosphide and the arsenide are semiconductors. A large number of rare-earth pnictides of composition Ln_4Pn_3 adopt the anti-Th_3P_4 structure. Again most of these compounds order magnetically. We have now detected superconductivity in the La-compounds below 1 K. Changes of T_c by substituting part of the cations or the anions are investigated and discussed.

RESULTS AND DISCUSSION

The normal, body-centered cubic Th_3P_4 structure occurs among chalcogenides and pnictides of the kind

$M^{2+}Ln_2^{3+}X_4$ (M=Ca,Sr,Ba,Eu; Ln=La,... Gd; X=S,Se)

M_3X_4 (M=La ... Sm, U,Np,...; X=S,Se,Te)

$An_3^{4+}Pn_4$ (An=Th,U,Np,..; Pn=P,As,Sb,Bi(pnigogen))

The representatives of the first group are nonmetallic whereas the binary chalcogenides are all metallic. With the exception of the La- and Eu^{3+} compounds, these phases undergo ferro- or antiferromagnetic ordering at low temperatures. The ternary La chalcogenides are diamagnetic valence compounds while the La_3X_4 compounds contain one excess valence electron per formula unit. Superconductivity in the latter was discovered recently and rather high values for the critical temperature T_c have been reported[1] (see below). The only nonmagnetic pnictides, those with thorium, are again normal valence compounds and at least the phosphide and the arsenide are semiconductors[2].

A large number of rare-earth pnictides of composition Ln_4Pn_3 adopt the anti-Th_3P_4 structure[3] where the "cation" is now found in roughly octahedral coordination while the "anion" has a coordination number of 8. Again most of these compounds order magnetically. In the La compounds we have now detected superconductivity. For the arsenide (which we have synthesized for the first time), the antimonide and the bismuthide rather sharp superconducting transitions have been observed in the range between 0.1 K and 0.7 K (see table I). As with the Ln_3X_4 compounds the higher T_c is obtained with the lighter anions while the contrary results with the lighter cations: If part of the lanthanum is replaced by yttrium (La_3YSb_3) then the transition temperature drops. For pure Y_4Sb_3 we therefore expect a very low critical temperature.

As turned out during our preparative work these phases display an appreciable range of homogeneity which might be as wide as that observed in the normal Th_3P_4-type compounds. There the binary chalcogenides are homogeneous within the concentrations of La_3X_4 to La_2X_3. In order to emphasize the structural features we prefer referring the formula to the primitive cell: $La_{6-z}\square_zX_8$ with $z<2/3$. The corresponding homogeneity range in our phases $La_8Pn_{6-z}\square_z$ greatly hampered our experimental work. As we did not succeed in getting monophase ternary alloys the influence of substitutions is less significant since a change in excess-electron concentration is caused by both the anion substitution and the deviation from the ideal 4:3 stoichiometry. Nevertheless, a drastic increase of T_c is observed on substituting for part of the anions: La_4Bi_2Pb has a T_c of one order of magnitude higher than pure La_4Bi_3 (see also table I). This substitution reduces (in the case of ideal 4:3 stoichiometry) the number of excess valence electrons per primitive cell from 6 to 4 whereas a pnigogen deficiency analogous to that in the case of the La chalcogenides (i.e. up to the limit of La_3Bi_2) would increase this number from 6 to 8.

COMPOUND	T_c (mK)
La_4As_3	650 ± 30
La_4Sb_3	250 ± 5
La_4Bi_3	155 ± 3
La_4Bi_3 [1)]	127 ± 2
La_4Bi_3 [2)]	170 ± 5
$La_4(Sb_{.5}Bi_{.5})_3$	158 ± 2
La_3YSb_3	130 ± 3
La_4Bi_2Pb	2300 ± 100
La_4BiPb_2	2400 ± 300

Table I: Superconducting transition temperatures for various anti-Th_3P_4-type compounds.
[1)] Bi-rich, [2)] La-rich

Forgetting about the preparation problems the outstanding feature of this group of compounds is just the possibility of varying the excess-electron concentration within a wide range. This concentration could formally be reduced to zero since La_4Ge_3, La_4Sn_3 and La_4Pb_3 also exist with this anti-Th_3P_4 structure. We expect them, however, to be semimetals. Partial substitution of the cations is the alternative way for varying the d-electron concentration responsible for the occurrence of superconductivity. Thus, mixtures of La_4Pn_3 with hypothetical Sr_3LaPn_3, Ba_3LaPn_3 or $NaSrLa_2Pn_3$ may serve to reduce the excess-electron concentration while partial substitution of La by Th will increase it.

Finally we point out that in the normal Th_3P_4-type La chalcogenides the high transition temperatures (see table II) are caused by two excess-electrons per primitive cell. In that case a reduction of the electron concentration is possible by simple cation subtraction up to the normal valence compound La_2X_3 or by mixing with the semiconducting Th_3P_4-type compounds La_3PnX_3[4)], as well as by partial substitution of La by Sr or Ba. We wonder whether an increase of the electron concentration, say by forming $La_{3-z}Th_zX_4$ mixed crystals would

COMPOUND	T_c (K)
La_3S_4	8.06
La_3Se_4	7.80
La_3Te_4	5.30
$La_{2.4}Y_{.6}S_4$	4.77
$La_{2.4}Y_{.6}Se_4$	3.92

Table II: Superconducting transition temperatures for some compounds with Th_3P_4 structure. The results are taken from ref. 1.

shift the critical temperature T_c to still higher values. Th_3Te_4, a hypothetical compound analogous to the magnetic Th_3P_4-type U_3Te_4, would contain as much as 8 excess-electrons per primitive cell.

In conclusion we may say that the anti-Th_3P_4-type La pnictides exhibit similar features with respect to the occurrence of superconductivity as the already known La chalcogenides with the Th_3P_4 structure, except that their critical transition temperature T_c is, unfortunately, at least one order of magnitude lower.

REFERENCES

* Work supported in part by the Schweizerischer Nationalfonds zur Förderung der wissenschaftlichen Forschung

1. E. Bucher, K. Andres, F.J. di Salvo, J.P. Maita, A.C. Gossard, A.S. Cooper and G.W. Hull, Phys.Rev. B11, 500 (1975)
2. I.H. Warren, J. Electrochem. Soc. 112, 510 (1965)
3. R.J. Gambino, J. Less-Common Met. 12, 344 (1967)
4. F. Hulliger, unpublished.

SUPERCONDUCTIVITY - A PROBE OF THE MAGNETIC STATE OF LOCAL MOMENTS IN METALS*

M. B. Maple**
Institute for Pure and Applied Physical Sciences
University of California, San Diego
La Jolla, California 92093

ABSTRACT

The superconducting properties of exemplary matrix-impurity systems in the three distinct regimes of magnetic character of the impurity which have been identified are reviewed. The three regimes can be distinguished by the detailed behavior of the depressions of (1) the superconducting transition temperature T_c as a function of impurity concentration n and (2) the specific heat jump ΔC at T_c as a function of T_c. These systematics of superconductivity in the presence of local moments appear to be sufficiently well established that it is possible to (1) ascertain whether the solute spin is long-lived (magnetic) or short-lived (nonmagnetic) compared to thermal fluctuation lifetimes at superconducting temperatures, (2) determine the sign and magnitude of the conduction electron-impurity spin exchange interaction parameter $\mathcal{J}$ and the temperature dependence of the exchange scattering of conduction electrons by long-lived solute spins, (3) derive, in favorable cases, information pertaining to the energy level structure of rare earth ions in the crystalline electric field of their superconducting metallic host, and (4) observe magnetic-nonmagnetic transitions of an impurity induced by the application of an external pressure or variation of the composition of a binary alloy matrix.

---

*A thorough discussion of the conference presentation may be found in APPLIED PHYSICS 9, 179 (1976).

**Supported by the U.S. Energy Research and Development Administration under Contract No. ERDA E(04-3)-34PA227.

SUPERCONDUCTIVITY OF THE HYDRIDES AND DEUTERIDES OF HfV_2*

P. Duffer, D. M. Gualtieri and V. U. S. Rao
Department of Chemistry, University of Pittsburgh
Pittsburgh, PA 15260

ABSTRACT

Current interest in the superconductivity of metal hydrides and deuterides has encouraged an investigation of the influence of hydrogen and deuterium absorption on the crystallographic and superconducting properties of the C15 ($MgCu_2$-type) intermetallic compound, HfV_2. Room temperature absorption of hydrogen (H) and deuterium (D) by HfV_2 at 2.5 x 10^6 Pa. was found to produce alloys of the compositions $HfV_2H_{4.53} \pm 0.02$ and $HfV_2D_{4.55} \pm 0.02$ which were stable upon removal to air. These fully hydrided and deuterated alloys were brought to equilibrium in various mixtures with HfV_2 to produce a range of alloy concentrations. Within the limits of measurement, the lattice parameters of the hydrides and deuterides of the same concentration were found to be identical, and the lattice expansion arising from absorption was found to vary linearly with hydrogen or deuterium concentration. These specimens, which were powdered by the absorption process, were tested for superconductivity to the compositions $HfV_2H_{1.59}$ and $HfV_2D_{1.00}$ by an inductance bridge method. Alloying with hydrogen was found to decrease T_c by 4.7 K per mole of hydrogen per mole of HfV_2. The deuterides exhibit a more rapid decrease of T_c of 6.8 K per mole deuterium. Upper critical field measurements on $HfV_2D_{0.5}$ indicate a depression of H_{c2} in comparison to HfV_2. These characteristics of the hydrides and deuterides are discussed on the basis of their lattice and electronic properties.

INTRODUCTION

In recent years there has been considerable interest in the superconducting properties of metallic hydrides owing to the possibility of finding high T_c in these materials via an attractive pairing interaction arising from the high frequency lattice modes of the proton lattice. So far, detailed studies have been carried out on the hydrides and deuterides of thorium[1] and palladium.[2,3,4]

It is interesting to note that although Pd itself is not superconducting, Pd-H and Pd-D exhibit T_c values as high as 9 and 11 K, respectively. It has been shown[5] that the appearance of superconductivity in Pd-H(D) is not just a result of the quenching of spin fluctuations. It has its origin in the enhanced electron-electron interaction arising from the high frequency optic phonon modes associated with the proton lattice.

Another interesting feature is the discovery that $Th_4H(D)_{15}$ exhibits no detectable isotope effect[1] while Pd-H(D) exhibits a pronounced opposite isotope effect.[3] Different models[5,6] have been proposed to explain the opposite isotope effect in Pd-H(D).

In view of these interesting features, it is necessary to perform experiments on a variety of metallic hydrides, in order to understand the mechanism of superconductivity in these materials. The hydrides of the cubic Laves (C15) phase alloys $Zf_{1-x}Hf_xV_2$ are well suited for a detailed investigation. It had been found earlier[7] that ZrV_2 absorbs large amounts of hydrogen even at room temperature and moderate pressures. In our laboratory $Zr_{1-x}Hf_xV_2H_y$ compositions have been prepared with y ranging from y = 5.7 for x = 0 to y = 4.5 for x = 1. The samples retain the cubic C15-structure but there is considerable increase in lattice parameter (up to 8% increase in the fully hydrided samples).

In this paper, the superconducting characteristics of the hydrides and deuterides of HfV_2 are presented. HfV_2 itself[8] has T_c = 9.9 K and H_{c2} = 200 kOe at 2.4 K and it is therefore of interest to study its superconducting properties after hydrogenation.

EXPERIMENTAL

Sample Preparation.

Quantities of HfV_2 were prepared by induction melting of the elements in a water-cooled copper boat under a titanium-gettered argon atmosphere. The purity of the starting materials was 99.9%. These samples were induction remelted several times to ensure homogeneity and were placed in evacuated quartz tubing for heat treatment of 800°C for 100-400 hours. Samples were alloyed with high-purity hydrogen and deuterium at room temperature and 2.5 x 10^6 Pa. in a fixed volume system, and the quantity of absorbed gas was noted by the pressure change in the fixed system volume. Absorption at these temperature and pressure conditions resulted in alloys of the composition $HfV_2H_{4.53} \pm 0.02$ and $HfV_2D_{4.55} \pm 0.02$. Within the limits of errors, the amounts of H_2 and D_2 absorbed are identical.

A range of hydrogen and deuterium concentrations was produced by equilibrating various ratios of the fully hydrided or deuterated alloys with virgin HfV_2 in quartz tubes at 450°C for 48-72 hours. The tubes were slightly evacuated before sealing, to prevent

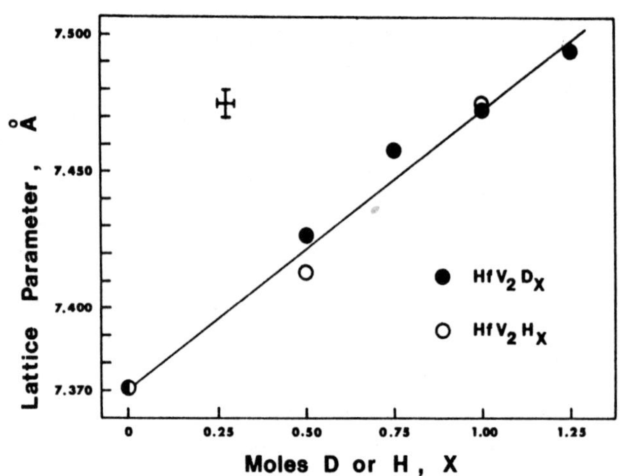

Fig. 1. Lattice parameters of HfV_2H_x and HfV_2D_x.

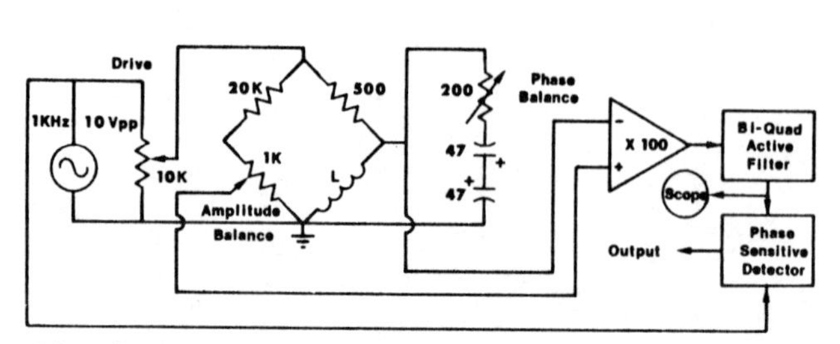

Fig. 2. Schematic of the inductance bridge system for measuring T_c.

bursting at the treatment temperature, but not fully evacuated to preclude the possibility of hydrogen or deuterium desorption. After equilibration, portions of the alloy samples were resealed in a helium atmosphere for measurement and the rest reserved for crystallographic study. X-ray diffraction revealed a C15 ($MgCu_2$-type) structure for all alloys. The lattice parameters of the fully hydrided and deuterated alloys were found to be identical, with a = 7.830 ± 0.005 Å as compared to 7.370 ± 0.005 Å for HfV_2. The lattice parameters of the hydrides and deuterides appear to vary linearly with concentration of H or D (Fig. 1). The straight line fit to the data of Fig. 1 was drawn with the assumption of linear expansion to a = 7.830 Å for hydrogen or deuterium concentration of 4.54. Within the limits of error, the lattice parameters of the hydrides and deuterides are identical.

Superconducting Measurements.

All hydrided and deuterated samples were powdered by the absorption process. Conventional resistivity measurements were therefore not possible. Instead, an inductance bridge circuit was designed[9] (Fig. 2) which produces a voltage signal proportional to the imbalance of a bridge having an inductance coil in one leg. The sample to be tested for superconductivity is placed in this coil, and the bridge balanced for both resistive and reactive components to give a zero voltage output, as measured on an oscilloscope. The superconducting transition causes an imbalance voltage across the bridge, which is amplified, filtered, and detected. The output voltage is recorded on a chart recorder as a function of temperature, as illustrated in Fig. 3. The center of the superconducting transition was noted as the temperature of greatest dV/dT; that is, the temperature where d^2V/dT^2 is zero. As shown in Fig. 3, this method yields a value of T_c for HfV_2 which is identical to that obtained from a resistivity measurement of the same alloy melt. The temperature of transition from the superconducting to normal state can be measured to within 0.1 K.

Critical temperature data for various hydrides and deuterides is presented in Fig. 4. The effect of both hydrogen and deuterium absorption is to decrease T_c. Hydrogen absorption reduces T_c by 4.7 K per mole, whereas deuterium absorption reduces T_c by 6.8 K per mole. The upper critical field, H_{c2} was measured for one deuteride of composition $HfV_2D_{0.5}$ in an Intermagnetics General 120 kOe superconducting solenoid and compared with the H_{c2} data of a virgin HfV_2 specimen. As noted in Fig. 4, T_c of $HfV_2D_{0.5}$ is 6 K as compared to 9.9 K for HfV_2. There is a corresponding decrease in H_{c2} in comparison to the virgin HfV_2 (Fig. 5).

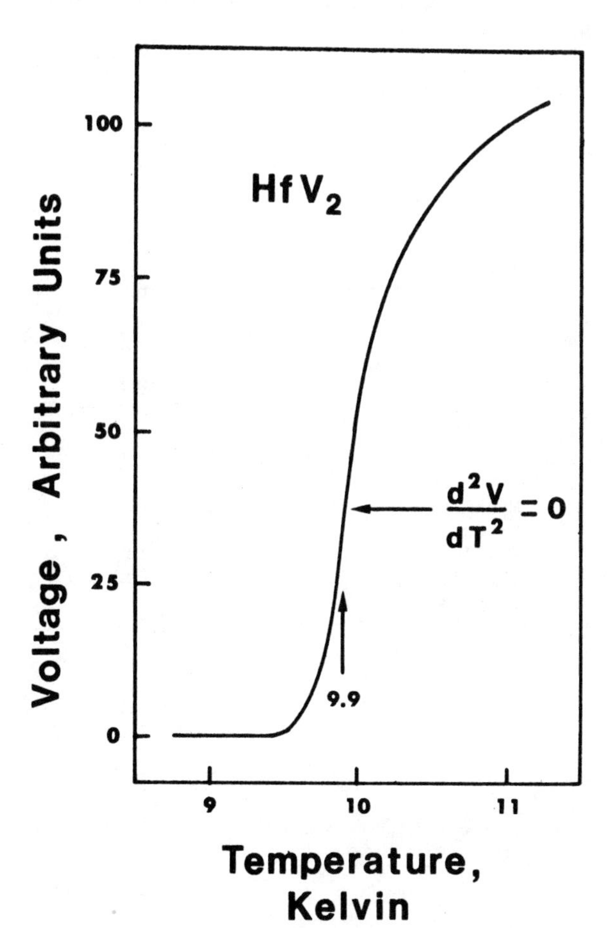

Fig. 3. Superconducting transition of a HfV_2 specimen as measured by the inductance bridge method. The center of the transition was noted as the temperature at which d^2V/dT^2 was zero.

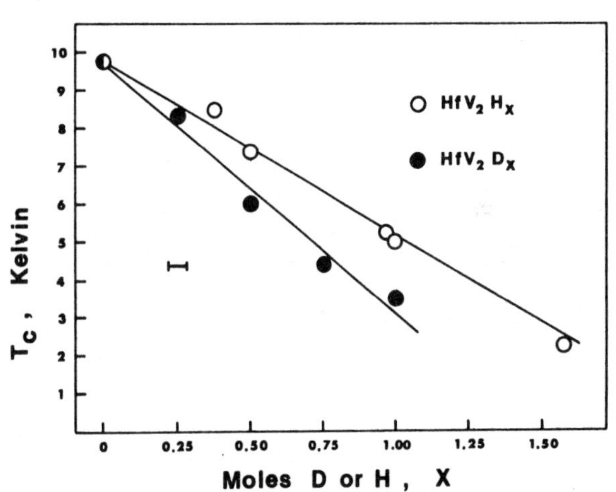

Fig. 4. T_c of HfV_2H_x and HfV_2D_x as a function of composition.

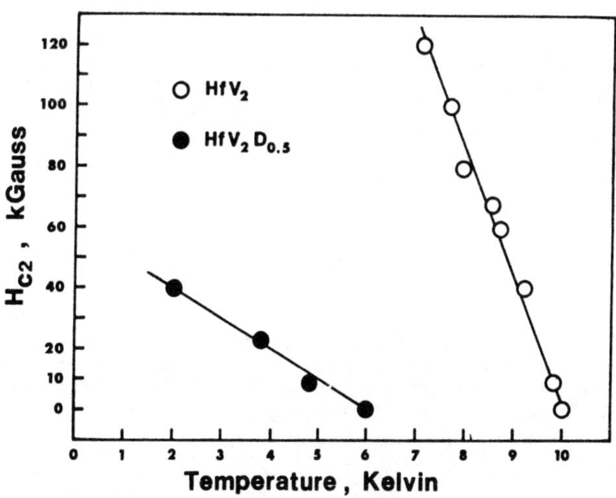

Fig. 5. H_{c2} of HfV_2 and $HfV_2D_{0.5}$ as a function of temperature.

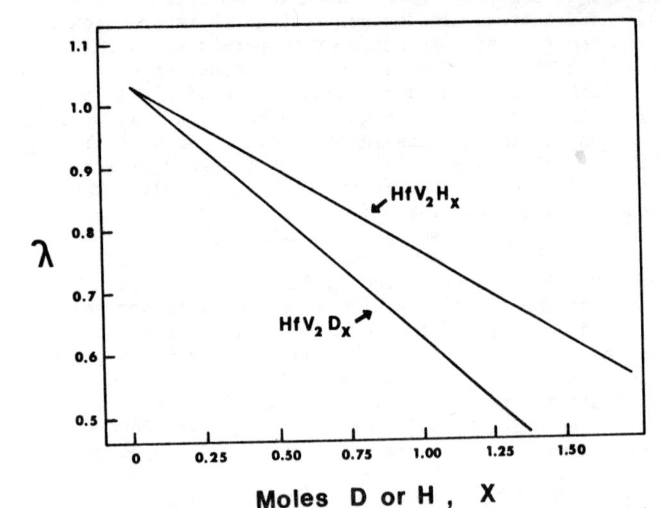

Fig. 6. The electron-phonon coupling constant λ, as a function of composition in HfV_2H_x and HfV_2D_x.

DISCUSSION

Two important features of the results are (a) the rapid decrease of T_c upon the introduction of hydrogen and deuterium into HfV_2 and (b) a pronounced isotope effect, with the hydrides showing a higher T_c than the deuterides of same composition (Fig. 4). This is opposite to what is observed[3] in the PdH(D) system.

It is interesting to note that, if one denotes by $\Delta T_c(x)$, the reduction in T_c from the parent HfV_2 upon the addition of x moles of H or D, one observes that

$$\Delta T_c(x) \propto M^{0.5} \quad (1)$$

where M is the mass of H or D, as the case may be. A detailed interpretation of this result may have to await experimental information on the electronic structure and phonon spectra of HfV_2 and its solutions with H and D. Nevertheless, since the lattice parameters of HfV_2H_x and HfV_2D_x are very nearly the same (Fig. 1), one may surmise that the acoustic phonon characteristics associated with the metal atom vibrations should be nearly the same for the two cases and therefore may not be responsible for the pronounced difference in T_c between HfV_2H_x and HfV_2D_x. Hence the difference must arise from the high frequency local modes associated with the H or D atoms.

McMillan's expression[10] for strong-coupling superconductors may be written

$$T_c = \frac{\theta_D}{1.45} \exp\left[\frac{1.04(1+\lambda)}{\lambda - \mu^*(1+0.62\lambda)}\right] \quad (2)$$

where θ_D is the Debye temperature, μ^* is an electron-electron interaction parameter and λ is the mass enhancement factor arising from electron-phonon interaction. It is of interest to obtain λ as a function of H or D concentration in HfV_2H_x and HfV_2D_x.

$\mu^* \sim 0.13$ for transition metals.[10,11] Low temperature specific heat measurements[12] on HfV_2 yield $\theta_D = 190$ K. It may be argued that θ_D for the hydrides and deuterides of HfV_2 might be expected to be different from that of HfV_2. However, in the Pd-H system, very little change in θ_D was observed upon hydrogenation, from low temperature specific heat measurements.[13] For the present, we shall therefore set $\theta_D \sim 190$ K for all the compositions HfV_2H_x and HfV_2D_x presently under discussion. With the above assumptions, the parameter λ has been calculated from the measured T_c for the various compositions and is shown in Fig. 6. It should be pointed out that the λ's obtained in this manner are averages obtained for the metal and H(D) sites.

For pure metals one may follow McMillan[10] in writing

$$\lambda = \frac{n(E_F) \langle I^2 \rangle}{M \langle \omega^2 \rangle} , \quad (3)$$

where $n(E_F)$ is the density of states at the Fermi level, $\langle I^2 \rangle$ is an average over the Fermi surface of the square of the electron-phonon matrix element, M is the atomic mass, and $\langle \omega^2 \rangle$ is the second moment of the phonon frequencies as defined by McMillan.[10] For alloys containing H, D or other light atoms, it has been suggested[14,15] that λ can be separated into contributions arising from the light and heavy atoms. However, a detailed interpretation of the measured λ values for HfV_2H_x and HfV_2D_x must await heat capacity and other measurements on these systems.

REFERENCES

* Supported by U.S. Energy Research and Development Administration.

1. C. B. Satterthwaite and I. L. Toepkė, Phys. Rev. Lett. 25, 741 (1970).
2. T. Skoskiewicz, Phys. Stat. Sol. (a) 11, K123 (1972).
3. B. Stritzker and W. Buckel, Z. Physik 257, 1 (1972).
4. B. Stritzker, Z. Physik 268, 261 (1974).
5. B. N. Ganguly, Z. Physik 265, 433 (1973).
6. R. J. Miller and C. B. Satterthwaite, Phys. Rev. Lett. 34, 144 (1975).
7. A. Pebler and E. A. Gulbransen, Trans. Met. Soc. AIME 239, 1593 (1967).
8. K. Inoue, K. Tachikawa and Y. Iwasa, Appl. Phys. Lett. 18, 235 (1971); V. Sadagopan, E. Pollard and H. C. Gatos, Solid State Commun. 3, 97 (1965).
9. D. M. Gualtieri (to be published).
10. W. L. McMillan, Phys. Rev. 167, 331 (1968).
11. I. R. Gomersall and B. L. Gyorffy, Phys. Rev. Lett. 33, 1286 (1974).
12. O. Rapp and L. J. Vieland, Phys. Lett. 36A, 369 (1971).
13. C. A. Mackliet and A. I. Schindler, Phys. Rev. 146, 463 (1966).
14. J. C. Phillips, in "Superconductivity in d- and f-Band Metals," AIP Conf. Proc. No. 4 (edited by D. H. Douglass, 1972), pp 339-357.
15. B. M. Klein and D. A. Papaconstantopoulos, Phys. Rev. Lett. 32, 1193 (1974).

ANALYSIS OF CALORIMETRICALLY OBSERVED SUPERCONDUCTING TRANSITION-TEMPERATURE ENHANCEMENT IN Ti-Mo(5 at.%)-BASED ALLOYS

J. J. White and E. W. Collings*
BATTELLE, Columbus Laboratories, Columbus, Ohio 43201

ABSTRACT

The low temperature specific heats of four alloys based on Ti-Mo(5 at.%) have been measured and analyzed. The four samples were (1) as-quenched with a $\beta + \omega$ structure, $T_c < 1.5$ K, (2) compressively deformed with an α'' martensite structure, $T_c = 3.2$ K, (3) solution strengthened by 1% Al addition with an α' martensite structure, $T_c = 3.1$ K, and (4) solution strengthened by 3% Al addition with an α' martensite structure, $T_c = 2.8$ K. The analysis employed a nonlinear fit for the usual parameters plus an asymmetric distribution of transition temperatures to account for the transition rounding. The procedure to be described yields accurate values for T_c, and an equivalent (or homogeneous) specific-heat-jump height, ΔC, which is an indicator of the fraction of material participating in the superconductivity. These results combined with earlier observations indicate that the martensite phase, whether produced by deformation or otherwise, is the seat of an enhanced superconducting transition-temperature.

INTRODUCTION

The structures retained by Ti-Mo after quenching into iced brine from an anneal in the single-phase-bcc (β) region at 1300°C are indicated in Fig. 1(a). For Mo concentrations less than about 4-1/2 at.%, quenching yields an hcp-based martensitic structure (α') formed by spontaneous shear transformations from the parent bcc lattice. Above 15-20 at.% Mo, the β structure is essentially retained on quenching; while between 5 and at least 10 at.% Mo, the bcc lattice supports submicroscopic ω-phase precipitation (hexagonal in structure, and of particle size within the range of about 70-300 Å) the volume fraction of which decreases continuously as the Mo concentration increases. The superconducting transition temperatures of Ti-Mo alloys, as well as those of Ti-T_2 (another transition element) alloys in general, depend strongly on solute-concentration-dependent [or electron/atom-ratio ($\mathfrak{z}$)-dependent] microstructure. As depicted in Fig. 1(b), T_c has its maximal value within the bcc regime, decreasing as $\mathfrak{z}$ tends towards 5 on one hand and towards 4.1 on the other; the latter effect being the result of an increasing abundance of the ω-phase precipitate.

The martensitic regime is characterized by an enhanced T_c, as indicated by the separate branch in Fig. 1(b). In this context two forms of martensite are to be considered: the so-called "thermal martensite", α', referred to above, which occurs spontaneously as the sample temperature is lowered through a critical transus; and "deformation martensite", α'', which results when an otherwise (ω+β)-phase alloy is heavily deformed. The structure of α'', whose occurrence in Ti-Mo has been discussed by several authors and most recently in References [1-3], although reported by some to be bcc or bct [2], is generally considered to be hcp. In any event both α' and α'' are characterized by high levels of atomic disorder.

This study represents an analysis and comparison of the calorimetrically observed superconducting transitions in martensitically transformed alloys of, and based on, Ti-Mo(5 at.%), henceforth TM-5. On one hand α'' was produced simply by compressive deformation of the as-quenched (β+ω) TM-5; while on the other, the presence of small amounts of Al in solid solution led on quenching to large volume fractions of α'.

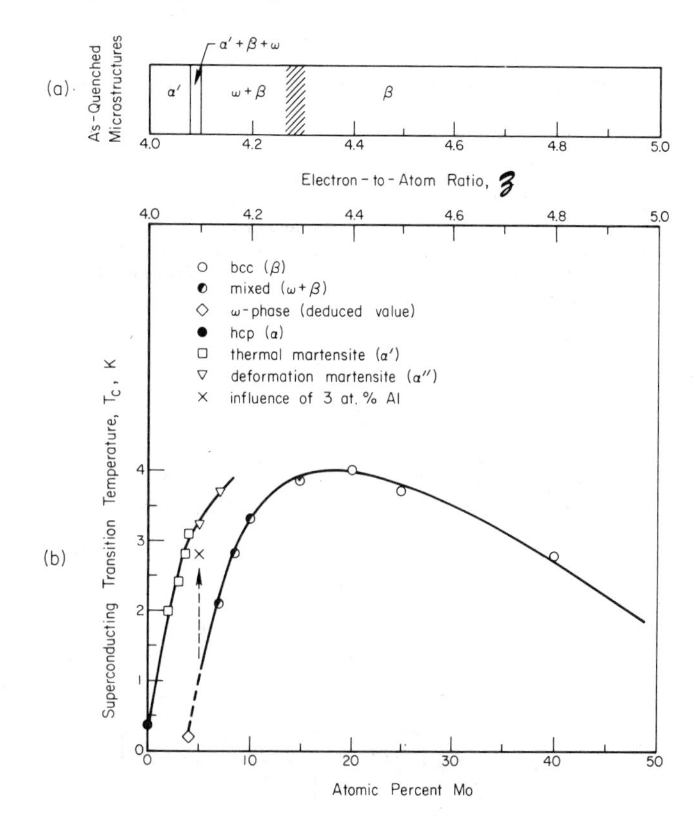

Fig. 1. Concentration-dependent [or electron/atom-ratio ($\mathfrak{z}$) dependent] microstructures (a), and superconducting transtion temperature (b), in quenched Ti-Mo alloys.

EXPERIMENTAL PROCEDURES

Samples of TM-5, $(TM\text{-}5)_{99}Al_1$, and $(TM\text{-}5)_{97}Al_3$ were prepared by the multiple arc melting of high-purity ingredients. Prior to measurement all samples were subjected to solution heat treatment and quenched into iced brine. TM-5 itself was subsequently deformed and re-measured. Low-temperature specific heat was measured on 30g samples using a conventional adiabatic calorimeter. The experimental results when plotted in the format C/T versus T^2 yield a linear region and a "specific heat jump" corresponding to the superconducting transition. If the transition is sharp and the jump abrupt its relative height, $\Delta C/\gamma T$, yields immediately the mass-fraction of material participating in the superconducting transition. On the other hand, if the transition is appreciably rounded, as it is in compositionally or structurally inhomogenous samples, the experimental data elude simple analysis for either T_c or fraction transformed, and sophisticated curve-fitting procedures must be invoked.

* Supported by the Air Force Materials Laboratory, Wright-Patterson Air Force Base, under Contract AF33(615)69-C-1594; and the Air Force Office of Scientific Research (AFSC), under Grants 71-2084 and 75-2786.

EXPERIMENTAL RESULTS

As discussed elsewhere [4] Ti-Mo (7 to 40 at.%) alloys all yield sharp superconducting transitions. Although not observable in our equipment, the T_c of (β+ω) TM-5 is estimated (Fig. 1) to be about 1K. After deformation of the sample (to α'') the transition, albeit broad, rises to the vicinity of 3K [5], a value which is consistent with those of the quenched (α') (TM-5)-Al alloys. The specific heat results for the as-quenched and quenched-plus-deformed TM-5 are given in Fig. 2. Based on the experimental data alone it would be difficult to estimate the degree of completeness of the superconducting transition. The specific-heat anomalies in the Al-containing alloys are considerably sharper, nevertheless curve-fitting is again an essential first step towards a proper interpretation of the influence of the martensitic structure on the superconducting transition.

METHOD OF ANALYSIS

The specific heat data was analyzed in terms of two models, one for the unrounded data and one for the rounded data near T_c. In the regions unaffected by the rounding, we assume [6]

$$C_- = Ae^{-B/T} + \beta T^3 \quad T<T_c \tag{1}$$

and,

$$C_+ = \gamma T + \beta T^3 \quad . \quad T>T_c \tag{2}$$

The rounding is attributed to a distribution of transition temperatures [7,8]. Thus we assume

$$C = gC_- + (1-g)C_+, \tag{3}$$

where

$$g = \int_T^\infty F(T_c)dT_c \tag{4}$$

and

$$F(T_c) = \frac{1}{\Gamma}\sqrt{\frac{\ell n2}{\pi}} \exp\left[-\frac{\ell n2}{\Gamma^2}(T_c-T_{co})^2\right] \tag{5}$$

The Gaussian distribution, Eq. (5), is made asymmetric about T_{co} by changing the Γ in the exponential term to

$$\Gamma_+ = \Gamma(1+f) \qquad T_c>T_{co} \tag{6}$$

and,

$$\Gamma_- = \Gamma(1-f), \qquad T_c<T_{co} \tag{7}$$

where f, the asymmetry factor, may assume values from -1 to +1.

The seven parameters, A, B, γ, β, T_{co}, and f, may be optionally determined or held fixed by a non-linear least squares fit computer program called NONLIN4 [9]. The random error in the data is assumed to be 1% of the measured specific heat. A major option in using NONLIN4 is the choice of a one- or two-part fit. In a one-part fit all seven parameters are analyzed together. In a two-part fit the parameter sets A, B, γ, β and T_{co}, Γ, f are treated separately. The authors prefer the two-part approach; however, the quantitative differences in the resulting parameter estimates are seldom significant. Future plans for NONLIN4 include the incorporation of a $\gamma'T$ and an exact BCS term [10] in Eq. (1); and possibly a deconvolution [11] for $F(T_c)$ instead of the asymmetric Gaussian model.

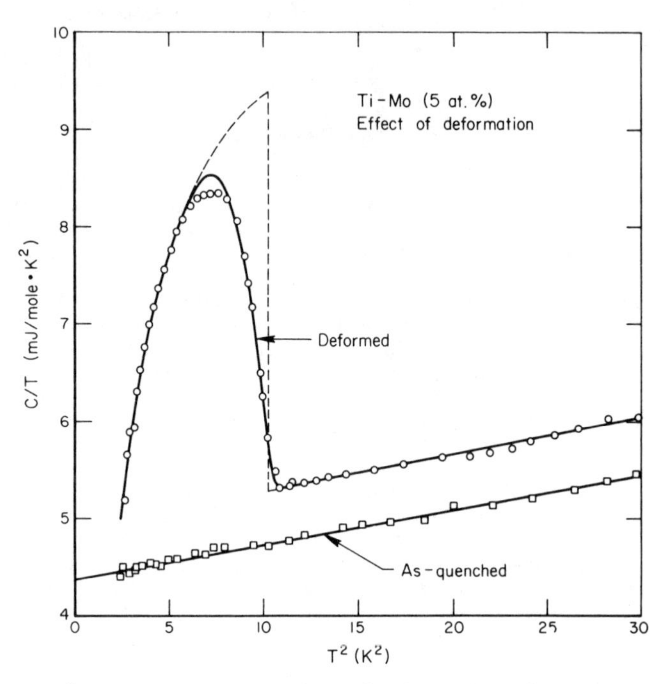

Fig. 2. Low temperature specific heat data for the superconducting alloy Ti-Mo(5 at.%) before and after deformation. The solid curve through the as-quenched data is the usual two-parameter least squares fit. The solid curve through the deformed data is the two-part, seven-parameter least squares fit, which yielded f = -0.71 ±0.11. The dashed curve indicates the homogeneous specific heat.

RESULTS AND DISCUSSION

The results of the analysis of the Ti-Mo(5 at.%) are given in Figs. 2-4 and Table I. Case 1 is a fit to high temperature data on the as-quenched TM-5. Cases 2 and 3 refer to the quenched plus deformed basic binary for a symmetric distribution (f = 0.0) and a minimizing asymmetric distribution (f = -0.71). Likewise, Cases 4-7 correspond to fits of the data on the aluminum-alloyed TM-5.

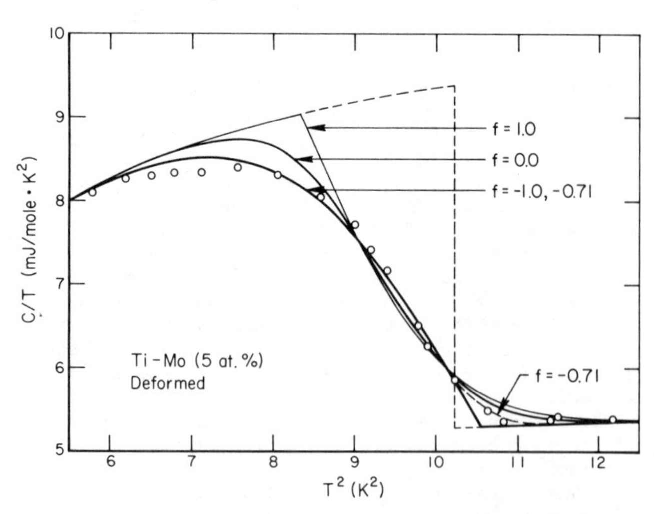

Fig. 3. Low temperature specific heat data for the superconducting alloy Ti-Mo(5at.%) after deformation. The solid curves give the best two-part, seven-parameter fits with f = -1.0, 0.0, and 1.0, respectively. The dashed curve indicates the optimal fit, yielding f = -0.71 ±0.10.

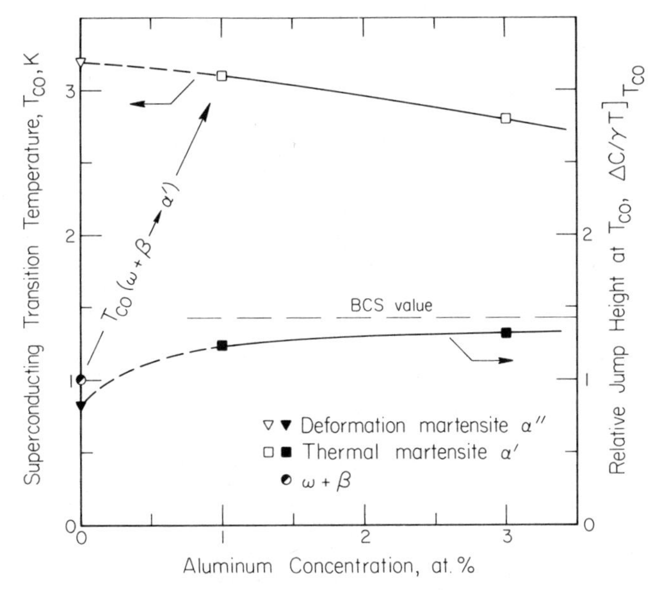

Fig. 4. Upper curve -- influence of Al additions on the superconducting transition temperature in martensitic Ti-Mo(5 at.%)-Al alloys. Lower curve -- relative height of the "specific heat jump" compared to the BCS value of 1.43 [10].

For each of the three samples exhibiting a superconducting transition, it was found that the choice of a large negative value of the asymmetry parameter f gave a substantial reduction in χ^2 of approximately 50%. The effect of f on $\Delta C/\gamma T_{co}$ is quite small, whereas the influence on T_{co} is large in comparison with the width of the rounded region.

Figure 3 gives an expanded version of the rounded region of Fig. 2. The best fits of the data for f = -1.0, 0.0, and 1.0 indicate the shapes of rounded regions characteristic of various distributions $F(T_c)$. Introduction of a BCS term into Eq. (1) may reduce the asymmetry somewhat because the BCS linear region [10] near T_c will predict a greater homogeneous specific heat than the $Ae^{-B/T}$ term. Data with the appearance of the f = 1.0 fit have been observed [5] in a deformed sample of TM-7.

Figure 4 compares the fitted ("asymmetrical") T_{co} values for the quenched-plus-deformed TM-5, and the as-quenched material. The martensite (either α'' or α') exhibits a clearly enhanced transition temperature. It is interesting to note that T_{co} for α'' (TM-5, deformed) is quite consistent with that noted for the Al-alloyed α' material, and lies on a back-extrapolation of that data. The T_{co} values plotted represent modal values of $F(T_c)$ -- the fraction of material actually transforming within a given temperature range centered on T_{co} is of course contained in our analysis but has not yet been explicitly expressed. There is no question, however, but that the enhanced T_{co} is a property of the martensitic structure, and is undoubtedly associated with the high density of lattice defects inherent in it. A possible mechanism, that of localized soft-phonon enhancement, has been suggested elsewhere [5]. The slight decrease of T_{co} with addition of Al may be due to either an electronic dilution of the lattice through Al substitution (density-of-states effect) or to a reduction of the soft-phonon effect through a solution strengthening by the Al.

The lower curve in Fig. 4 compares the relative height of the specific heat jump at T_{co} with the BCS-predicted value [10]. That surprisingly close agreement is obtained suggests that practically all of the evidence for superconductivity in each as-quenched specimen is contained in the data already at hand, and that proceeding to a lower temperature range would be unlikely to reveal any further transitions. A lower-lying transition is, on the other hand, expected in the deformed TM-5.

We conclude that the bulk of the material is responsible for the observed broadened transitions and suggest that just as the T_c enhancement arises through the existence of lattice defects, the broadening is a result of a wide variation of defect environments.

ACKNOWLEDGMENTS

The authors acknowledge Dr. J. C. Ho for making the calorimetric data available.

REFERENCES

[1] P. Gaunt and J. W. Christian, Acta Met. 7, 534 (1959).

[2] M. J. Blackburn and J. C. Williams, TMS-AIME 242, 2461 (1968).

[3] M. K. Koul and J. F. Breedis, Acta Met. 18, 579 (1970).

[4] J. C. Ho and E. W. Collings, in Titanium Science and Technology, ed. by R. I. Jaffee and H. M. Burte (Plenum, N.Y., 1973), p. 815.

[5] J. C. Ho and E. W. Collings, J. Appl. Phys. 42, 5144 (1971).

[6] N. E. Phillips, Phys. Rev. 134, A385 (1964).

[7] J. J. White, J. Phys. C 7, L317 (1974).

[8] S. D. Bader, N. E. Phillips, and E. S. Fisher, Phys. Rev. B12, 4929 (1975).

[9] J. J. White, Appl. Phys. 5, 57 (1974).

[10] B. Mühlschlegel, Z. Physik 155, 313 (1959).

[11] E. Bucher, F. Heiniger, and J. Muller, in Proceedings of the Ninth International Conference of Low-Temperature Physics, Columbus, Ohio, 1965, ed. by J. G. Daunt, et al. (Plenum, N.Y., 1965), p. 482.

TABLE I. RESULTS FOR A TWO-PART SEVEN-PARAMETER FIT OF LOW TEMPERATURE SPECIFIC HEAT DATA ON THE SUPERCONDUCTING ALLOY Ti-Mo(5 at.%)

Case	Sample	Asymmetry Factor, f	T_{co} (K)	Γ (K)	A (mJ/mole.K)	B (K)	γ (mJ/mole.K^2)	β (mJ/mole.K^4)	$\frac{\Delta C}{\gamma T_{co}}$	χ^2
1	As-quenched	--	1.0	--	--	--	4.38 ±0.01	(3.56 ±0.08) x 10^{-2}	--	14.6[a]
2	Deformed	0.0	3.040 ±0.006	0.179 ±0.010	100.2 ±2.4	3.99 ±0.05	4.91 ±0.05	(3.73 ±0.27) x 10^{-2}	0.81	137.2[b]
3	Deformed	-0.71 ±0.10	3.197 ±0.023	0.191 ±0.008	100.2 ±2.4	3.99 ±0.05	4.91 ±0.05	(3.73 ±0.27) x 10^{-2}	0.84	63.4[b]
4	1% Al	0.0	3.068 ±0.001	0.042 ±0.002	176.4 ±2.1	4.99 ±0.02	5.07 ±0.03	(4.32 ±0.12) x 10^{-2}	1.23	123.9[c]
5	1% Al	-0.61 ±0.09	3.100 ±0.005	0.044 ±0.001	176.4 ±2.1	4.99 ±0.02	5.07 ±0.03	(4.32 ±0.12) x 10^{-2}	1.24	63.5[c]
6	3% Al	0.0	2.700 ±0.002	0.062 ±0.003	167.4 ±2.5	4.72 ±0.03	4.81 ±0.03	(5.18 ±0.14) x 10^{-2}	1.29	74.3[d]
7	3% Al	-0.90 ±0.37	2.832 ±0.021	0.057 ±0.005	167.4 ±2.5	4.72 ±0.03	4.81 ±0.03	(5.18 ±0.14) x 10^{-2}	1.32	33.9[d]

a) 29 data points; (b) 46 data points; (c) 55 data points; (d) 47 data points.

CRITICAL CURRENT DENSITY AND FLUX PINNING IN Nb_3Ge

A. I. Braginski,* Michael R. Daniel,† and G. W. Roland*
Westinghouse Research Laboratories
Pittsburgh, Pennsylvania 15235

ABSTRACT

High self-field critical current densities, J_c, of the order of 10^6 A/cm^2 between 4 and 14 K have been reported for chemical vapor deposited (CVD) films of Nb_3Ge. Our results of J_c measurements in a wide field range and film composition analyses suggest that high values of J_c are due primarily to flux pinning on dispersed tetragonal (σ) phase. Single phase (A15) Nb_3Ge samples exhibited lower J_c of the order of 10^4 to 10^5 A/cm^2. The ac loss measurements corroborate the proposed interpretation.

INTRODUCTION

High self-field and low-field critical current densities, J_c, of the order of 10^6 A/cm^2 between 4 and 14 K, have been observed in A15 Nb_3Ge films having high critical temperatures, T_c = 20 to 23 K. Such films have been so far synthesized by sputtering,[1,2] evaporation,[3] or chemical vapor deposition (CVD).[4,5] It is generally assumed that in the brittle A15 compounds the effective fluxoid pinning, leading to high J_c, occurs predominantly on grain boundaries. The purpose of this work was to determine the origin of high J_c's in CVD-grown, Nb_3Ge thick films. These films have coarse grains, of the order of 1 μm and above, as opposed to fine-grained sputtered Nb_3Ge films, and Nb_3Sn layers formed by the diffusion process. One could suspect, therefore, that grain boundaries alone may not insure a concentration of flux pinning centers sufficient for attaining the high J_c values.

SAMPLES AND THEIR CHARACTERIZATION

The Nb-Ge superconducting layers were deposited on 50 cm long, 1.27 cm wide, and 50 μm thick Hastelloy B tape sections.[6] The standard deposition temperature was T_d = 900°C, but deposition experiments in the range 750 to 1000°C have also been performed to obtain layers differing in microstructure. At lower T_d smaller grain sizes could be anticipated. The lower temperature limit of 750°C was imposed by the chemistry of the deposition process and the upper limit by equipment limitations. The deposit thickness was in the range of 5 to 20 μm. The chemical composition was determined by electron microprobe. For T_d = 900°C the average deposit composition varied along the tape such that the molar ratio s = Nb/Ge in the solid monotonically increased from one end of the tape to the other. The x-ray phase analysis was performed on powdered samples removed from the substrates by dissolving Hastelloy in HNO_3. The unit cell edge of the A15 phase, a_o, was determined from Debye-Scherrer photographs by taking the average of the cell sizes determined from the 622, 600, 610, and 611/532 reflections. The chemical vapor deposition conditions at T_d = 900°C were usually set such that distinct phase regions appeared in sequence along the tape. Starting from the Ge-richer end at the position coordinate x = 0, there first appeared a two-phase region (A15 + σ), where σ denotes the tetragonal Nb_5Ge_3 modification. In this region the cell edge of the A15 phase was constant. The concentration of σ depended upon s set in the vapor phase. This concentration, estimated from the x-ray intensity of characteristic lines, often fluctuated but generally decreased with increasing x.

Beyond a certain point, x = x_b, only the A15 phase was present, which extended to the Nb-richer end, x = x_{max}. In this single phase region the cell edge monotonically increased with x thus reflecting an increasing Ge-deficiency in the A15 structure. Hence, at $x \simeq x_b$, the deposit composition was closest to Nb_3Ge.

The a_o (x) dependence typical of the above situation is shown in Fig. 1. While the CVD deposition may occur far from thermodynamic phase equilibrium, the a_o (x) dependence for the A15 implies that traversing the tape from x = 0 to x_{max} is equivalent to traversing a corresponding section of the Nb-Ge binary phase diagram, starting in the A15 + σ phase field, and at the knee of the a_o (x) curve crossing the boundary into the A15 solid solution field.

The microstructure of deposits was examined by optical microscopy, scanning electron microscopy (SEM) and transmission electron microscopy (TEM). The deposits exhibited the [100] columnar texture. The column diameter, estimated by SEM, was assumed to be the grain size, d. This d value represented the upper limit of the single crystallite size determined by TEM. In the two-phase region d was generally larger than in the single phase region. Typical data are shown in Table I. The presence of the second phase (σ) could not be detected by TEM and SEM, suggesting a high degree of dispersion in the A15 matrix.

CRITICAL CURRENT DENSITY DETERMINATION

Most of the critical current density data were obtained from magnetization measurements performed using a Foner-type magnetometer (P.A.R. Model 155). Measurements were made on disc samples, 3 mm in diameter, punched out from the tape. The magnetic field, H, was applied normal to the disc face. Previous calculations have shown the demagnetizing field of these discs to be less than 1 kG for film thicknesses h $\leq$ 10 μm.[7] Thus, we were able to calculate low field J_c from the magnetization after Fietz and Webb.[8] Experiments have confirmed that excellent agreement exists with direct, transport current, measurements at H $\geq$ 5 kG. The magnetometer measurements were performed in the field range up to 70 kilogauss. Self-field J_c measurements and measurements at H > 70 kilogauss** were perfomed by the four-point transport method on short tape sections. The disc sample temperatures were measured with an accuracy of ± 0.3 K between 4.2 K and the critical temperature, T_c. The four-point transport measurements were performed in liquid helium, i.e., at 4.2 K, and in the pumped liquid hydrogen range 13.8 to 20.3 K where the accuracy was ± 0.05 K.

In addition to J_c determinations the ac losses, p, were determined using the electronic wattmeter*† method at f = 60 Hz, 4.2 K, and in fields up to a peak value of H_p = 2 kilogauss. The tape sample section length was typically 2.5 to 4.0 cm. The short sample arrangement used was similar to that devised by Bussiere.[9] For samples which exhibited the p (H) dependence obeying Bean's model of the critical state, p (μWcm^{-2}) = $(0.5f/12\pi^2)$ H_p^3 (G^3)/J_c (Acm^{-2}), and thus the value of low-field J_c could also be estimated by ignoring its field dependence.

RESULTS AND DISCUSSION

To correlate the critical current density with the microstructure and grain size of nearly single phase layers having an approximately stoichiometric composition Nb_3Ge, J_c was measured for samples synthesized in the temperature range 750 to 1000°C. The results are shown in Table I. In this range of synthesis temperatures the grain size varies by less than one order of magnitude and maximum J_c's at 4.2 K and 5 kilogauss are uniformly low, of the order of 10^5 A/cm^2. In contrast, much higher J_c values, of the order of 10^6 A/cm^2, are

TABLE I. Grain Size (from SEM), d, Critical Temperature, T_c, and Maximum Critical Current Density, J_c, Determined for Nb_3Ge Layers Deposited at Various Temperatures, T_d

T_d °C	Single Phase A15 Deposits: d μm	T_c (midpoint) K	J_c @ 4.2 K, 5 kG 10^6 A/cm^2	Two Phase, A15 + σ, Deposits: d μm	T_c (midpoint) K	J_c @ 4.2 K, 5 kG 10^6 A/cm^2
750	< 1	16.0 to 17.0	0.25	0.4 to 1	14 to 19	3.0
800	0.2 to 2	19.0 to 21.0	0.25	0.4 to 2	19 to 20	1.8
900	0.5 to 2	21.0 to 21.5	0.4	0.5 to 3	19 to 21	2.4
950	0.5 to 2	21.0 to 21.7	0.1	0.5 to 3	19 to 20	0.40
1000	1 to 3	19.0 to 20.5	0.06	2 to 5	16 to 20	0.34

characteristic of samples containing the σ ohase except for T_d = 1000°C. On the other hand, Table I shows that the σ phase is not inhibiting the A15 grain growth so that it should not enhance the concentration of flux pinning centers through the grain size control. The data of Table I suggest, therefore, that flux pinning may occur on nonsuperconducting particles of dispersed σ phase in analogy to other known cases of superconductor hardening by precipitation.[10] Flux pinning on grain boundaries may still represent a secondary contribution. However, lower J_c values for T_d = 950 and 1000°C may represent the effect of analytically verified significant diffusion of nickel from the substrate rather than that of the large grain size.

The correlation of J_c with the concentration and dispersion of σ phase was difficult to establish. All attempts failed to visualize the dispersed σ phase by the etching or anodizing of micrographic sections. Electron microscopy (TEM and SEM) was of no help either. This lack of success could be only tentatively attributed to a high degree of dispersion. The average concentration of the σ phase has been estimated from the powder x-ray diffraction data. However, electron microprobe traces of tape sections indicated that an upper layer of these σ-phase-containing deposits, (∿ 10% of the total thickness) was generally richer in germanium and consisted mostly of σ phase. This stratification resulted from the deposition shut-off procedure. Hence, the average concentrations measured represent only an upper bound to the amount of σ phase dispersed in the A15 matrix. For σ-phase-containing samples of Fig. 1, and high-J_c samples in general, such upper bound was 10 ± 5 vol. %.

An indirect but meaningful correlation could be established, however, due to the fact that the average Nb/Ge ratio in deposits grown at T_d = 900°C increased monotonically with x, as already discussed. Figure 1 shows a typical dependence of a_o, T_c, the upper critical field, H_{c2}, and J_c upon x. One can assume that the dramatic decrease of J_c with increasing x in the two-phase region reflects a correlation with the amount of σ phase dispersed such that effective pinning of the fluxoid lattice becomes possible either through direct perturbation of the superconducting parameters or through the associated internal stresses. The region of pure, nearly stoichiometric Nb_3Ge at $x \simeq x_b$ is at the knee of the a_o (x) curve. Accordingly, H_{c2} and T_c reach a maximum there. These H_{c2} values (at 4.2 K) were extrapolated from data collected at the pumped hydrogen temperatures by using the Werthamer expression for a dirty Type II superconductor with no paramagnetic limiting.[11] The variations of H_{c2} and T_c vs. x are too slight to significantly affect J_c. They are inconsistent with the observed sharp drop of J_c (x), but may be responsible for the local maximum of J_c at $x \simeq x_b$ (Fig. 1), a feature only occasionally observed.

Since both the precipitate concentration and dispersion (pinning center size and spacing) affect J_c, the determination of an "optimum" average σ concentration for a highest J_c was not attempted. However, as could be expected, at high σ concentrations J_c is degraded. This is caused not only by the reduction of the net content of the superconducting medium (which would imply a linear J_c decrease) but also by the deterioration of superconducting parameters due, e.g., to the internal stress. The degradation is seen in Figure 2, which shows the variation of J_c and T_c over a

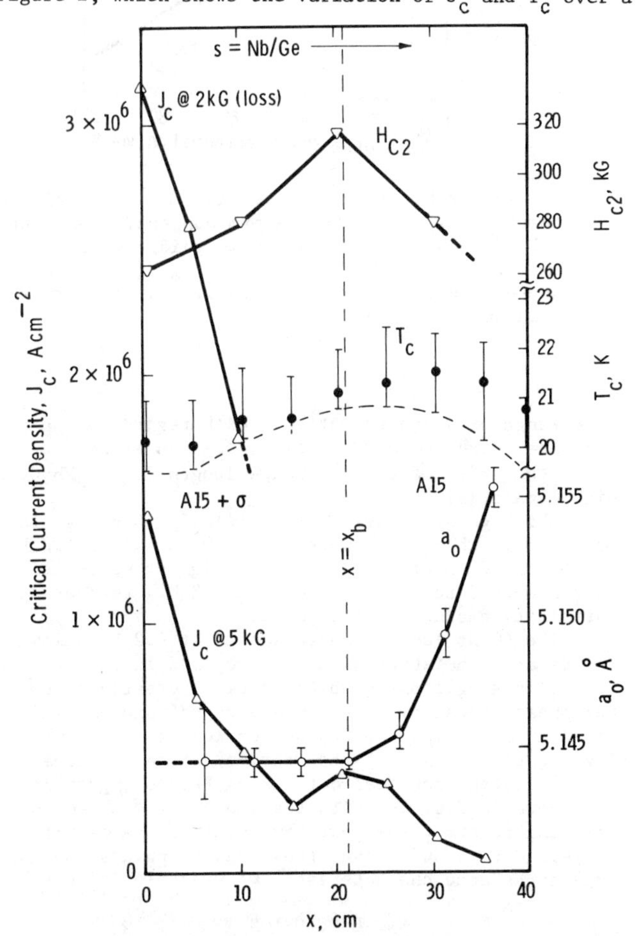

Fig. 1. The A15 phase cell edge, a_o, midpoint critical temperature, T_c, upper critical field, H_{c2}, and critical current density, J_c, vs. the Nb-Ge tape position coordinate, x. The T_c bars represent the superconducting transition temperature range, from onset to tail. Sample series 201, T_d = 900°C.

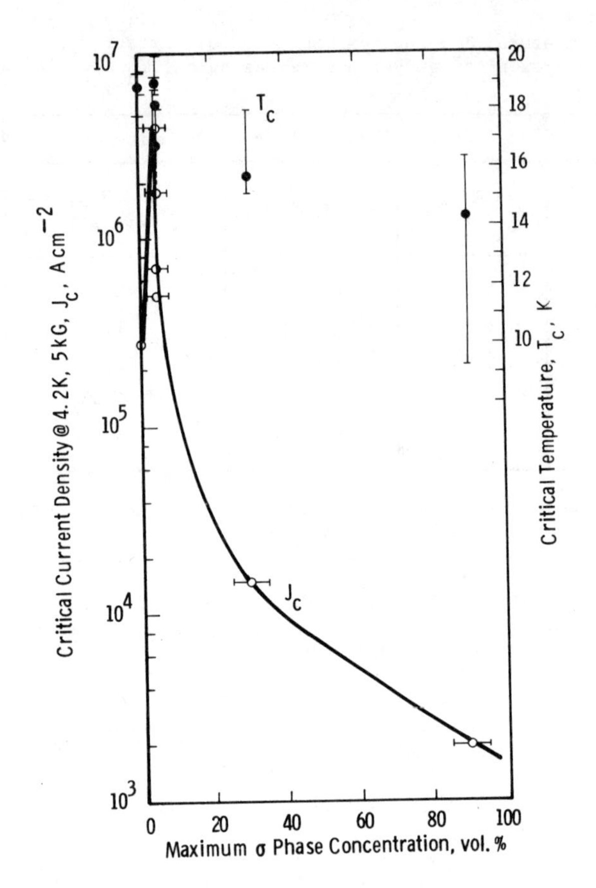

Fig. 2. The critical current density, J_c, and critical temperature, T_c, vs. the maximum concentration (upper bound) of σ-phase. Line drawn to guide the eye. The T_c bars represent the superconducting transition temperature range, from onset to tail. Sample series 279 and 86, T_d = 750°C.

wide range of σ concentration. With regard to the "optimum" σ phase particle size, one can expect it to be of the order of the coherence length ξ. In Nb_3Ge a high-κ material, ξ ∿ 30 Å.

The highest values of low field J_c (5 kG) so far observed in this study were ∿ 3 x 10^6 A/cm^2 at 4.2 K and ∿ 1 x 10^6 A/cm^2 at 14 K. The highest self-field values were 5 to 8 x 10^6 A/cm^2 at 4.2 K. Further J_c optimization can be anticipated.

The 60 Hz loss characteristics at 4.2 K in low fields are consistent with the proposed interpretation. Losses in single phase Nb_3Ge exceeded greatly those of two-phase deposits. The σ-phase doped samples exhibited a cubic field dependence of losses thus conforming to Bean's model of the critical state. Values of J_c calculated from loss data using Bean's expression are shown in Fig. 1. With the amount of σ phase decreasing to zero, the power exponent of the peak field increased to 4 and above, thus indicating a strong J_c (H_p) dependence characteristic of weak flux pinning.

CONCLUSION

In coarse-grained Nb_3Ge deposited by chemical vapor deposition high critical current densities are characteristic of samples containing several volume percent of the precipitated tetragonal Nb_5Ge_3 (σ) phase. While the quantitative correlation of J_c with the second phase dispersion and concentration is yet to be determined, the present results strongly suggest that flux pinning occurs primarily on dispersed, non-superconducting σ particles. The presence of this σ phase does not inhibit the grain growth of Nb_3Ge, and flux pinning on grain boundaries plays a secondary role, perhaps more significant at lower deposition temperatures. Further enhancement of J_c through the control of precipitate dispersion, concentration, and also of A15 grain size, can be anticipated.

ACKNOWLEDGMENT

We acknowledge the contribution of M. A. Janocko and A. T. Santhanam in the H_{c2} measurement and electron microscopy, respectively. M. T. Miller, P. A. Piotrowski, H. C. Pohl, R. P. Storrick, D. Sunseri, and A. L. Foley offered valuable technical assistance.

REFERENCES

* Supported by ERDA under Contract No. E(11-1)-2522.
† Supported in part by AFOSR under Contract No. F44620-74-C-0042.
** At the National Magnet Laboratory.
*† Low power factor wattmeter, Model IH744, constructed by the Brookhaven National Laboratory.
1. J. R. Gavaler, Appl. Phys. Lett. 23, 480 (1973).
2. L. R. Testardi, J. H. Wernick, and W. A. Royer, Solid State Comm. 15, 1 (1974).
3. Y. Tarutani, M. Kudo, and S. Taguchi, Proc. 5th Int. Cryo. Eng. Conf., ICEC5, IPC Sci. & Tech. Press, London (1974), p. 477.
4. L. R. Newkirk, F. A. Valencia, A. L. Giorgi, E. G. Szklarz, and T. C. Wallace, IEEE Trans. MAG-11, 221 (1975).
5. A. I. Braginski and G. W. Roland, Appl. Phys. Lett. 25, 762 (1974).
6. G. W. Roland and A. I. Braginski, in "Advances in Cryo. Eng.," Vol. 22 (in press).
7. Michael R. Daniel and M. Ashkin (Westinghouse Research Laboratories) - submitted to Cryogenics.
8. W. A. Fietz and W. W. Webb, Phys. Rev. 178, 657 (1969).
9. O. Horigami, J. F. Bussiere, and Y. Tanaka, Cryogenics 15, 660 (1975).
10. A. M. Campbell and J. E. Evetts, Adv. Phys. 21, 199 (1972).
11. N. R. Werthamer, E. Helfand, and P. Hohenberg, Phys. Rev. 147, 295 (1966).

EFFECT OF STRAIN ON THE CRITICAL CURRENT OF Nb_3Sn AND NbTi MULTIFILAMENTARY COMPOSITE WIRES*

J. W. Ekin and A. F. Clark
Cryogenics Division, Institute for Basic Standards
National Bureau of Standards, Boulder, Colorado 80302

ABSTRACT

The critical currents of flexible Nb_3Sn and NbTi composite wires have been observed to decrease as a function of strain. Characteristic samples of the data are presented along with a brief summary and intercomparison of the results for each wire type. In the NbTi wires the decrease commenced at strains of about 0.5%, but did not become appreciable (i.e. greater than ~5%) until strains exceeded about 1.5%. The effect is almost totally reversible and the magnitude of the decrease is not strongly dependent on sample configuration or stabilization material. In the Nb_3Sn composites the degradation in critical current is relatively much larger, becoming significant at strains ranging from 0.1% to 0.3% depending on the reinforcement technique used in the wires' construction. At high strains, the effect in Nb_3Sn is only partially reversible. Also in the Nb_3Sn composites, stress-induced resistivities as high as 10^{-10} Ωcm were observed at currents well below I_c. This leads to significant Joule heating of the superconducting wire and ambiguities in the operational definition of "critical current".

INTRODUCTION

Recently it has been shown that the critical current of multifilamentary superconducting wires can be significantly affected by the stress and strain the wire experiences at cryogenic temperatures.[1,2] In this report a brief summary of these and other data is presented along with an intercomparison of the results for three general types of multifilamentary wires: copper-stabilized NbTi, aluminum-stabilized NbTi, and experimental multifilamentary Nb_3Sn.

The apparatus in which these experiments were performed consisted of a simple solenoidal magnet with an integral load train for applying stress to 2 meter lengths of wire. Critical-current measurements were performed at magnetic fields ranging to 8 x 10^6A/m (100 kOe) with the sample bent into a hairpin geometry over two 2.54 cm diameter rollers in the magnet bore. Voltage was detected along a 3 cm section of the wire perpendicular to the applied field at the center of the magnet. The wire ends were led out of the magnet bore, wrapped around a 3.8 cm diameter friction grip, anchored, and soldered to two 600 A current leads. Stress was applied to the wire using dead-weight loading. Strains were measured with a clip gauge over a 1.2 cm gauge length outside the magnet. Overall accuracy of the critical measurements was about 0.5%.

RESULTS ON NbTi

A typical set of voltage-current characteristics for the NbTi conductors is shown in Fig. 1. This particular wire consisted of 180 filaments of NbTi embedded in an OFHC copper matrix. Other characteristics are given in Table I. Also shown in Fig. 1 are three dashed reference lines corresponding to several critical current criteria: an electric field along the wire of 3 x 10^{-6}V/cm, and two resistivity levels of 10^{-10}Ωcm and 10^{-11}Ωcm (normalized to the cross-sectional area of NbTi in the composite wire). As seen from Fig. 1, the voltage rise is sufficiently sharp for NbTi that there are only small differences between the various methods of defining critical current. As will be discussed, however, this is not true for Nb_3Sn, and so for consistency the electric-field criterion has been used throughout the following discussion.

From Fig. 1, it can be seen that the critical current for this particular NbTi wire was found to be about 230 A at low stress. However as stress on the wire was increased, the critical current, I_c, systematically degraded until eventually a critical current of less than 190 A was reached just prior to fracture. These results have been plotted as a function of strain in Fig. 2 and are indicated by the curve labeled NbTi. Note that the decrease in I_c commences at about 0.5% strain (stresses of about 3 x 10^8Pa), but does not become appreciable (i.e. greater than 5%) until strains exceeded about 1-1/2% (stresses of about 6 x 10^8Pa). Other copper-stabilized NbTi wires generally exhibited slightly greater I_c degradation, but in all cases, the results were within about 10% of those shown in Fig. 2.

When the load applied to the wire was removed, a substantial recovery took place. As seen by the dashed curve in Fig. 2, the recovery in critical current upon load removal was to within about 4% of the initial zero-stress value over the entire range of strain. That is, the critical current degradation in NbTi is nearly reversible, even at strains approaching the fracture point of the wire.

Similar results were obtained in aluminum-stabilized NbTi wires. In an effort to compare the effects of aluminum stabilization vs. copper stabilization directly, two wires were tested which were alike in every way, except one was stabilized by high purity aluminum (RRR≅5000), the other by OFHC copper (RRR≅70). Both consisted of NbTi tubes embedded in a 5056 aluminum alloy matrix with the stabilization material on the inside of the tubes. (Further details are given in Table I.) Results for the two wires showed their degradation characteristics to be identical to within 1%, and differed from the curve labeled NbTi in Fig. 2 by less than 3%.

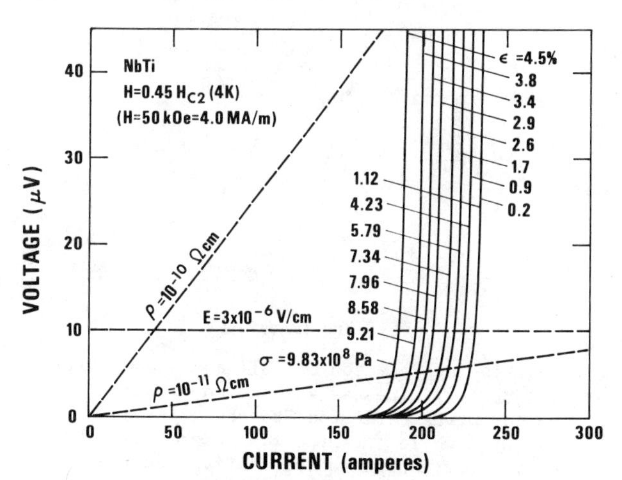

Fig. 1 Voltage-current characteristics for a typical NbTi multifilamentary wire as a function of stress and strain. Also shown are 3 dashed lines corresponding to several critical-current criteria: an electric field along the wire of 3 μV/cm, and resistivities of 10^{-10} and 10^{-11}Ωcm normalized to the cross sectional area of NbTi in the conductor.

TABLE I

SAMPLE CHARACTERISTICS

Sample	Size	No. Fil.	Fil. Diam.	Fil. Twist Length	Composition
Copper-stab. NbTi	0.53 x 0.68 mm	180	30 μm	1.3 cm	36% Nb-55 Ti 64% OFHC Copper
Aluminum-stab. NbTi, 5056 Al matrix	0.56 x 0.66 mm	54 tubes	69 μm O.D. 39 μm I.D.	1.3 cm	40% Nb-55 Ti 17% Al 43% 5056
Copper-stab. NbTi, 5056 Al Matrix	0.56 x 0.64 mm	54 tubes	69 μm O.D. 39 μm I.D.	1.7 cm	46% Nb-55 Ti 21% Cu 33% 5056
Nb_3Sn	0.33 x 0.66 mm	3553	3.6 μm	1.1 cm	12% Nb_3Sn 5% Nb 11% Cu 65% Bronze 7% Ta
Nb_3Sn (Reinforced)	7 strands 0.15 mm diam. 1-304 Stainless 6-multifil. Nb_3Sn	6 x 240	6 μm	1.3 cm cable twist Length= 0.18 cm	20% Nb_3Sn 7% Nb 73% Bronze

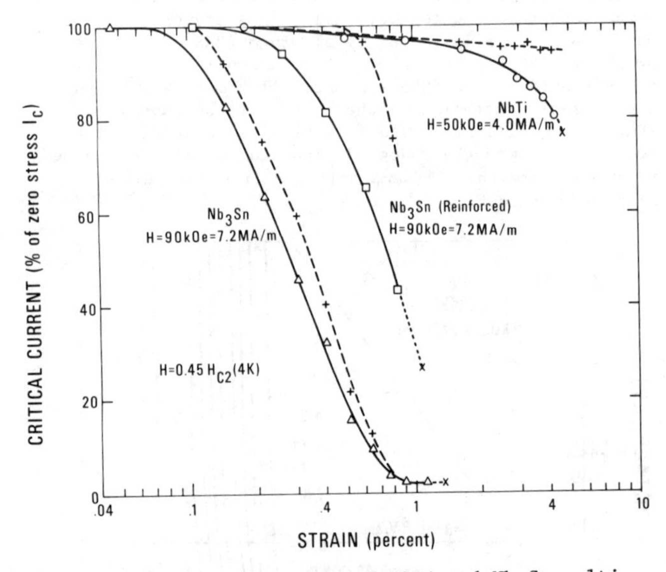

Fig. 2 Relative degradation of NbTi and Nb_3Sn multifilamentary wires as a function of strain. Solid curves correspond to the critical current measured while the wire was under load, dashed curves correspond to the critical current measured after removing load from the wire. Point of fracture is indicated by an X. Critical-current criteria used was an electric field along the wire of 3 μV/cm.

RESULTS ON Nb_3Sn

Results for two Nb_3Sn wires are also presented in Fig. 2, one typical of the data obtained on unreinforced wires, the other typical of the results on reinforced cables. Sample characteristics of each are presented in Table I.

Three points are noted. First, as seen in Fig. 2, the threshold for significant degradation in Nb_3Sn occurs at lower strains than for NbTi. For the unreinforced wire, the critical current starts to decrease at strains in the range 0.1 to 0.2% (stresses of 1 to 2 x 10^8Pa). For the reinforced wire, on the other hand, degradation onset was delayed to stresses and loads about twice as large, i.e., strains of about 0.3% (stresses of about 3 x 10^8Pa). This is mainly a result of the reinforced-cabled construction used in this second type of wire. Most of the load is supported by a central stainless steel reinforcing strand, and actual strain in the Nb_3Sn is reduced by helical coiling of the Nb_3Sn strands around the reinforcing strand.

Second, note that at high strains the critical-current recovery upon load removal is relatively smaller in Nb_3Sn than in NbTi. This may be seen by comparing the dashed recovery curves in Fig. 2. However, note also that at low strain the recovery in the Nb_3Sn wires was nearly complete, indicating behavior very similar to that observed in the ductile NbTi wires in the low-strain range (i.e. where I_c degradation is limited to less than ~15%).

The last point concerns the behavior of Nb_3Sn cables at low current. A significant amount of strain-induced resistivity was observed in the Nb_3Sn wires at currents well below the critical current. This may be seen in a typical set of voltage-current characteristics for Nb_3Sn, shown in Fig. 3. When the applied load is small, the curves show very little voltage at low currents, exhibiting the usual sharp up-turn as the critical current is approached. However, as load on the wire is increased, the curves tend toward a more linear shape, indicating substantial resistivity at very low currents. This has two secondary effects. First, at high strain, the resistivity becomes large enough that vastly different critical-current values are obtained depending on how "critical-current" is defined. This may be seen for the several different critical-current criteria shown as dashed lines in Fig. 3. Second, and most important, the Joule heating accompanying this resistivity can be substantial and may significantly affect magnet refrigeration requirements.

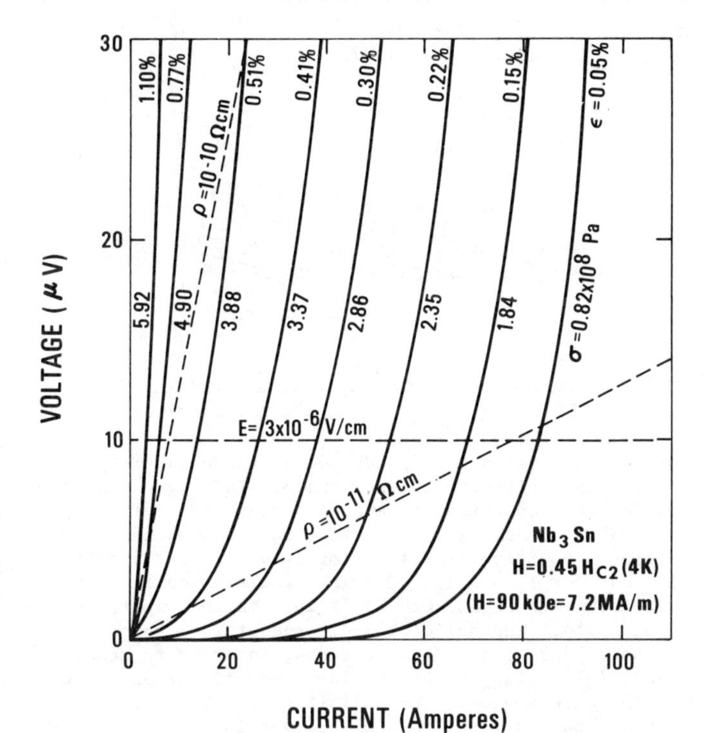

Fig. 3 Voltage-current characteristics for a typical Nb_3Sn multifilamentary wire as a function of stress and strain. Also shown are 3 dashed lines corresponding to several critical-current criteria: an electric field along the wire of 3 μV/cm, and resistivities of 10^{-10} and 10^{-11}Ωcm normalized to the cross sectional area of Nb_3Sn in the conductor.

SUMMARY

The principal results on each type of wire may be summarized as follows:

1) Preliminary data on NbTi wires indicate that the critical-current of the wire decreases when subjected to strains greater than ∿0.5% (stresses of 3×10^8 Pa). The effect is almost totally reversible and does not become large until strains of 1-1/2% or more are obtained (stresses of ∿6×10^8 Pa).

2) Aluminum-stabilized NbTi conductors have critical-current degradation properties similar to those of copper-stabilized NbTi conductors. The decrease in I_c is almost totally reversible and amounts to less than 5% at strains less than about 1-1/2% (stresses of ∿6×10^8 Pa).

3) Data on experimental Nb_3Sn wires show the critical-current degradation to start at strains as low as 0.1% (stresses of 1×10^8 Pa). At higher strains the degradation becomes large, irreversible, and is characterized by wire resistivities as high as 10^{-10} Ωcm. In these short sample tests, however, reinforced-cabling techniques have been effective in delaying the onset of degradation to strains above 0.3% (stresses above 3×10^8 Pa).

ACKNOWLEDGMENTS

The authors are grateful to M. J. Superczynski and F. R. Fickett for useful discussions relating to these results; and to the personnel of Airco, Alcoa, IGC, and Supercon for supplying the wires studied.

REFERENCES

*Work supported by the Naval Ship Research and Development Center.

1. J. W. Ekin, F. R. Fickett, and A. F. Clark, "Effect of stress on the critical current of NbTi multifilamentary composite wire", Intl. Cryogenic Material Conf., Kingston, Ont., Aug. 1975.
2. J. W. Ekin, "Effect of stress on the critical current of Nb_3Sn multifilamentary composite wire", Applied Phys. Letters 28 (1976) (to be published).

DESIGN OF DOUBLE HELIX CONDUCTORS FOR SUPERCONDUCTING AC POWER TRANSMISSION†

M. Garber, J. F. Bussiere and G. H. Morgan
Brookhaven National Laboratory*, Upton, N. Y. 11973

ABSTRACT

Coaxial cable conductors in the form of helical tape windings have been proposed in order to make Nb_3Sn cables which have flexibility and the ability to take up thermal contraction. For ac power transmission the axial magnetic fields which occur in a simple helical construction produce a number of undesirable consequences. It has been shown that these problems can be avoided by using double layer windings of opposite helicity, with 45° as the optimum helix angle. However, smaller values than this are desirable for mechanical reasons, and this paper extends the theory to include pitch angles $< 45^{\circ}$. Measurements on short cable models are shown to be in reasonable agreement with calculation. The effect of current flow around the superconductor tape edges, which occurs in helical windings, is analyzed and it is shown that appreciable ac loss can arise if laminated tape with non-superconductive edges is used indiscriminately.

INTRODUCTION

The original Brookhaven National Laboratory conceptual design of a Nb_3Sn superconducting coaxial cable for ac power transmission envisioned segmented tape conductors wound in simple helical form.[1] The aim was to provide flexibility and the ability to take up thermal contraction. However, helical current flow produces axial magnetic fields which lead to several undesirable consequences. It has been shown, for example, that non-zero axial flux can generate large eddy current losses in metal enclosures (such as cryostat walls) external to the cable proper.[2] In addition, with simple helical conductors this flux would produce large voltage drops along the length of the outer conductor, necessitating high voltage insulation on the outside of the cable.[3] Finally, axial magnetic fields in the inner core region of the cable can produce significant eddy current heating in normal metals which are present for mechanical support and current stabilization.[3]

The authors referred to have pointed out that it is not possible to eliminate all of these problems simultaneously by a simple adjustment of the inner and outer helix pitch angles. Double windings of opposite helicity have been proposed in place of the simple helices. Although it was initially suggested that the tapes of a double helix be transposed periodically to ensure balanced currents,[2] this would be difficult, if not impractical, for comparatively fragile Nb_3Sn tapes. It has been shown that a simple overlay of the helices will work as well if the winding pitch angle is approximately 45°.[3] In this paper we calculate the behavior for smaller pitch angles. This is of some importance since smaller angles are desired in order that the radial contractions of the conductors and dielectric are properly matched upon cooldown. We show that for pitch angles in the range 20° to 45° the problems discussed above can be reduced to acceptable levels.

In addition, we discuss the current flow perpendicular to the tape axes which occurs in the individual segments of a double helix. The losses which arise because of this are analyzed as a function of pitch angle and tape construction.

DESCRIPTION OF CURRENTS AND FIELDS

The cable design considered is shown in Fig. 1. The conductors are numbered 1 to 4 from inner to outer.

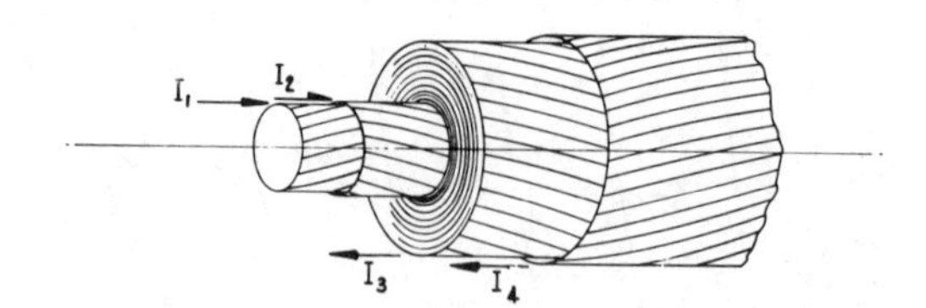

Fig. 1. Double helix coaxial cable schematic. $I_1 + I_2 = I_3 + I_4 = I$, the single phase line current.

Dielectric insulation is located in the annular space of the coax, i.e., between 2 and 3. The region inside conductor 1 contains mechanical support for the tapes and refrigerant. It is referred to below as the core region. Positive direction for I_1 and I_2 is opposite to that for I_3 and I_4. Current flow between individual tape segments is assumed negligible in the superconductive state. The helicities of conductors 1 and 2 are opposite and likewise for 3 and 4, in order to reduce the axial fields in the annular space of the coax and in the core region. The helicities of conductors 2 and 3 are shown the same but this convention is not essential. If we denote the cable current by I, then:

$$I_1 + I_2 = I \tag{1}$$

$$I_3 + I_4 = I \tag{2}$$

In addition, each helix pair is connected at its ends, so that the voltage drops per meter of conductor satisfy

$$V_1 = V_2 \tag{3}$$

$$V_3 = V_4 \tag{4}$$

These voltages can be expressed as functions of the currents and the various pitches, P_i, and diameters, D_i, with the aid of Faraday's law. In mks units:

$$V_1 = (j\omega\mu_o\pi/4)D_1^2P_1^{-1}(I_1P_1^{-1} - I_2P_2^{-1} + I_3P_3^{-1} - I_4P_4^{-1}) + (j\omega\mu_o/2\pi)\, I_1 \ln D_2/D_1 \tag{5}$$

$$V_2 = (-j\omega\mu_o\pi/4)D_2^2P_2^{-1}[(D_1^2/D_2^2)I_1P_1^{-1} - I_2P_2^{-1} + I_3P_3^{-1} - I_4P_4^{-1}] \tag{6}$$

$$V_3 = (-j\omega\mu_o\pi/4)P_3^{-1}(I_1P_1^{-1}D_1^2 - I_2P_2^{-1}D_2^2 + I_3P_3^{-1}D_3^2 - I_4P_4^{-1}D_3^2) + (j\omega\mu_o/2\pi)I_4 \ln D_4/D_3 \tag{7}$$

$$V_4 = (j\omega\mu_o\pi/4)P_4^{-1}(I_1P_1^{-1}D_1^2 - I_2P_2^{-1}D_2^2 + I_3P_3^{-1}D_3^2 - I_4P_4^{-1}D_4^2) \tag{8}$$

The P_i are taken as positive quantities in these expressions; ω is the angular frequency and $j = \sqrt{-1}$. In Eqs. (5) and (6), voltages due to flux linkages outside conductor 2 are not written down since they cancel in Eq. (3). Flux linkages and resistance in the superconducting surfaces are neglected in the calculation. In reference 3 these voltages were written in terms of the coefficients of inductance of the conductors. Eqs. (1) to (4), with the voltages given by Eqs. (5) to (8), completely determine I_1, I_2, I_3 and I_4 as functions of the pitches, diameters, and I.

Referring to Fig. 2, the magnetic field at the surface of the inner conductor (outside 2) is entirely azimuthal, and $= I/\pi D_2$. We shall call this reference field H_θ. As can be shown by numerical calculation the individual helix currents are within 10% of I/2 for the range of practical cases considered below. Thus in the annulus between 1 and 2 the azimuthal field component is $\approx \frac{1}{2}H_\theta$ and there is an axial component $Hz \approx \frac{1}{2}H_\theta \tan\varphi$, where φ is the pitch angle, i.e., the angle between the tape axis and the coax axis. There is a small axial field in the core region, H_o, which cannot be calculated accurately under the equal helix current approximation. It will be calculated exactly in the next section. The fields for the outer conductor are similar and will not be considered further.

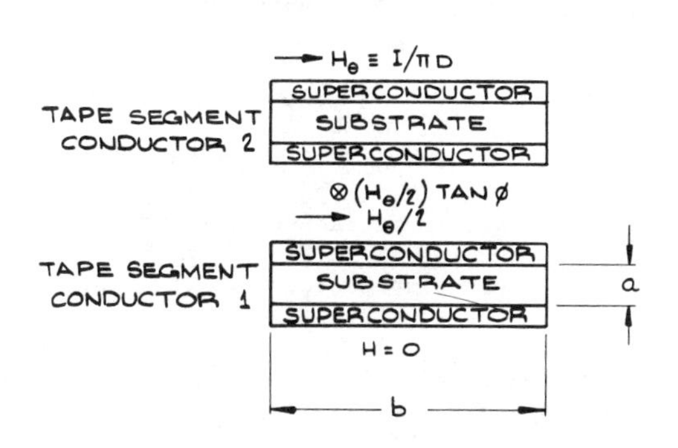

Fig. 2. Magnetic fields on inner conductor tape segments.

VOLTAGE IN OUTER CONDUCTOR; CORE FIELD

The purpose of the double helix is to reduce to insignificant levels the quantities H_o, the axial magnetic field in the core region and Φ, the total axial flux across the cable cross section. These are given by:

$$H_o = I_1P_1^{-1} - I_2P_2^{-1} + I_3P_3^{-1} - I_4P_4^{-1} \quad (9)$$

and

$$\Phi = (\pi\mu_o/4)(I_1P_1^{-1}D_1^2 - I_2P_2^{-1}D_2^2 + I_3P_3^{-1}D_3^2 - I_4P_4^{-1}D_4^2) \quad (10)$$

Equations (1) through (8) can be manipulated to give H_o and Φ without explicitly solving for the I's. The result is:

$$H_o = - I_1(P_1D_2^2+P_2D_1^2)^{-1}[(2/\pi^2)P_1P_2 \ln D_2/D_1 - (D_2^2-D_1^2)] \quad (11)$$

$$\Phi = (\pi\mu_o/4)I_4(P_3+P_4)^{-1}[(2/\pi^2)P_3P_4 \ln D_4/D_3 - (D_4^2-D_3^2)] \quad (12)$$

The conditions for the pitches if $H_o = 0$ and $\Phi = 0$, are obtained by setting the bracketed quantities equal to zero. As was pointed out in reference 3, they imply a pitch angle $\varphi = \tan^{-1}\pi D/P \approx 45^o$. In practice it will be necessary to employ smaller pitch angles in order that a tight coaxial structure results upon cooldown. Therefore, we are interested in the magnitude of H_o and Φ for other pitch angles. In practice Nb_3Sn tapes are thin (less than 0.2 mm, typically), and we may treat $D_2-D_1 = \Delta D_1$ and $D_4-D_3 = \Delta D_4$ as small quantities. Also, to simplify the discussion let $P_1=P_2=\alpha_1\pi D_1$ and $P_4=P_3=\alpha_4\pi D_4$. Deviations from this assumption would lead to terms in ΔP in the expressions which follow. As these are unimportant in practice, they have not been included. For H_o we get:

$$H_o \approx (I_1/\alpha_1\pi D_1)(\Delta D_1/D_1)(\alpha_1^2 - 1) \quad (13)$$

As $I/\alpha_1\pi D_1$ is the axial field for single helix conductors, this field is reduced by $(\Delta D_1/2D_1)(\alpha_1^2 - 1)$ for the double helix. Taking reasonable limits for $\Delta D_1/D_1 \lesssim 10^{-2}$ and $20^o \le \varphi \le 45^o$ $(1 \le \alpha_1 \le 2.7)$ we find that H_o is less than 3% of the single helix value. Eddy current losses in support metals in the core region[3] are thus reduced 10^3, i.e., to an insignificant level.

The quantity Φ determines the longitudinal voltage in the outer conductor, $V_4 = j\omega\,\Phi/P_4$. With the same assumptions as before we obtain

$$V_4 \approx (\mu_o\omega/4\pi)I_4(\Delta D_4/D_4)(1 - \alpha_4^{-2}) \quad (14)$$

We see that for all pitch angles less than 45° and $\Delta D_4/D_4 \le 10^{-2}$, $I \le 10^4$, and $\omega = 377$, V_4 is less than 2×10^{-3} V/m, which is entirely acceptable in practice.

MEASUREMENTS OF FIELDS IN A MODEL COAXIAL CABLE

Ac magnetic fields were measured in a coaxial model in which each conductor was a double helix. The overlying double helix layers were wound in direct contact, using uninsulated tapes. The Nb_3Sn surfaces were clad with a thin (0.03 mm) copper layer for stabilization; this copper served in effect to separate the superconducting layers. In addition to being easy to make, this method of construction is important since it leads to low radial heat resistance in the conductors of the double helix. The model was 800 mm long and was shorted at one end through a copper ring. The Nb_3Sn tapes were 3.2 mm wide and 0.13 mm thick. The average gap between adjacent tape segments was about 0.1 mm.

Results are given in Table I. Measured quantities were proportional to current (1 kA to 5 kA) and were independent of frequency (except for V_4)(40 to 150 Hz). In addition to H_o, Φ, and V_4 which have been defined previously, the quantities H_A, the axial field in the annular space between 2 and 3 (Fig. 1) and Φ_i, the axial flux across the cross section of the inner double helix were also measured. Formulas for H_o, Φ, and V_4 have been given. Subject to the previously mentioned approximations, expressions for Φ_i and H_A are as follows:

$$\Phi_i \approx (\mu_o/4)I_1\Delta D_1(\alpha_1 + \alpha_1^{-1}) \quad (15)$$

$$H_A \approx [\pi(D_3^2 - D_2^2)]^{-1}[I_1\Delta D_1(\alpha_1 + \alpha_1^{-1}) + I_4\Delta D_4(\alpha_4 + \alpha_4^{-1})] \quad (16)$$

The currents calculated numerically from Eqs. (1) to (8) are $I_1 = .94(I/2)$ and $I_4 = .92(I/2)$. Pitch and diameter data are given in the Table. It is difficult to determine the quantities ΔD_1 and ΔD_4 since the tape thickness is comparable to these quantities and since the double helix layers are not in uniform contact. Values for ΔD_1 and ΔD_4 were picked which give the best agreement between all five measured and calculated quantities. The values thus chosen for ΔD_1 and ΔD_4 are physically reasonable, but the difference between them seems somewhat large. It may be due to experimental error, to variations in pitch, and to the assumption in the calculations that the tape thickness is zero.

TABLE I

Measured and calculated fields in a double helix model coax. $D_1 = 19.2$ mm, $D_4 = 28.9$ mm, $P_1 = 112$ mm$(\varphi_1 = 28^o)$ $P_4 = 242$ mm $(\varphi_4 = 21^o)$. Values of $\Delta D_1 = .17$ mm and $\Delta D_4 = .35$ mm were picked to give best fit. I = 1000 A.

	Measured	Calculated
$\Phi(Wb/m^2)$	10.5×10^{-8}	11.5×10^{-8}
$\Phi_i(Wb/m^2)$	5.9×10^{-8}	6.0×10^{-8}
H_o(A/m)	92 (1.2 Oe)	91
H_A(A/m)	590(7.4 Oe)	500
V_4(V/m)	0.22×10^{-3}	0.18×10^{-3}

LOSSES IN CRYOGENIC ENCLOSURES

It is likely that the coaxial cable of each phase of a superconducting transmission line will be enclosed in a pipe to contain the supercritical helium refrigerant. For pitch angles other than 45° the axial flux, Φ, will lead to eddy current losses in the pipe. The calculation of these losses is complicated by the fact that the induced currents may be large enough to alter significantly the value of Φ calculated previously; i.e., the voltages V_3 and V_4 now include a term proportional to the current in the enclosure. Solving the modified equations results in the following:

$$P/\ell \approx \frac{(\mu_o \omega/4\pi)(2\delta_x^2 D_x/t^*)(\alpha_4 - \alpha_4^{-1})^2 (\Delta D_4)^2 I_4^2}{(D_x^2 - D_4^2)^2 + (2\delta_x^2 D_x/t^*)^2} \qquad (17)$$

where P/ℓ is the loss per unit length of coax; D_x, δ_x are the diameter and skin depth of the enclosure; $t^* = t$, the enclosure wall thickness when $t \lesssim \delta_x$ and $t^* = \delta_x$ when $t \gtrsim \delta_x$. This expression is greatest when the terms in the denominator are equal. The maximum acceptable value of P/ℓ is of order 10^{-2} W/m for typical superconducting cable designs. Substituting for numerical evaluation $\Delta D_4 \lesssim 2 \times 10^{-4}$ m, $(\alpha_4 - \alpha_4^{-1})^2 \leq 6$ (i.e., $45^o \geq \varphi \geq 20^o$), $I \leq 10^4$ A, $D_x \geq 0.1$ m, and $P/\ell \leq 10^{-2}$ W/m we find $\delta_x^2/t^* \geq 0.1$ m. For pure metals, $t^* \approx \delta \ll 10$ mm, and it would be necessary to keep close to $\alpha = 1$ ($\varphi = 45^o$). However, for virtually all alloy constructional materials this condition, which is conservative in view of the approximations made, is easy to satisfy. For example, assuming a stainless steel enclosure with $D_x = 0.1$ m, $t = 2$ mm, $\delta = 30$ mm ($\rho = 20$ μΩ cm), $I = 10^4$ A, $\varphi = 20^o$ and $\Delta D_4 = 0.2$ mm we find $P/\ell \leq 2.4 \times 10^{-3}$ W/m. In this example the first term in the denominator of Eq. (17) is negligible; Eq. (17) reduces to the expression obtained when the contribution of the enclosure currents to Φ is negligible.

LOSSES DUE TO EDGE CURRENTS

In contrast to the behavior in an ohmic conductor, current flow in superconducting tapes is confined to a much narrower surface layer (except for currents close to the critical current). The surface current density is perpendicular to the direction of the surface magnetic field and is equal to it in magnitude (in mks units). For conductor 1 (Fig. 2) there is little or no field on the inner surface, $H_o \approx 0$. Hence, the current, I_1, flows entirely on the outer tape surfaces, parallel to the tape axes. The magnetic field in the space between 1 and 2 is therefore perpendicular to the tape axes of 1. On the other hand for conductor 2 there is a component of field parallel to the tape axes of magnitude $H_\theta \sin\varphi$ on each surface, and therefore, a circulation of current around the tape segments. Similar remarks may be made for conductors 3 and 4 and it is not necessary to consider them explicitly. If conductor 2 consists of a solid superconductor the circulating currents cause only a slight increase in the hysteretic loss. However, as indicated in Fig. 2, there is often in practice a core or substrate of different material which extends to the tape edge. This results, for example, when wide sheets of Nb_3Sn are slit into narrower tapes. The substrate usually consists of unreacted superconductive Nb or Nb-1% Zr or a resistive alloy such as Hastelloy. If the substrate is resistive the axial field, $H_\theta \sin\varphi$, will cause a fraction of the circulating current to cross through the substrate and the rest to flow along the inner superconducting surfaces (i.e., at the superconductor - substrate boundary). If, on the other hand, the substrate is itself superconducting the circulating current will flow entirely through the edge region of the substrate. In either case significant additional losses can result, as we now show.

<u>Normal metal substrates</u>. This problem can be solved by application of standard eddy current formulas.[5] The superconductive surfaces impose the boundary condition $E \approx 0$ at the interfaces; that is, there is little or no current flow parallel to the interface in the normal metal. The current flow in the resistive layer is therefore identical to that in a slice of width a taken from an infinite slab of thickness b with a field $H_\theta \sin\varphi$ applied along both surfaces of the slab. The eddy current losses in the substrates in a length ℓ of helix 2, divided by the conductor surface area, $\pi D_2 \ell$, are:

$$P/A = (2\mu_o \rho\omega)^{\frac{1}{2}} (a/b) H^2 \quad \text{for } b/2 \gg \delta \qquad (18)$$

$$P/A = (\mu_o^2 \omega^2/12\rho)\, a b^2 H^2 \quad \text{for } b/2 \ll \delta \qquad (19)$$

where ρ is the substrate resistivity and H is the rms value of the parallel field ($= H_\theta \sin\varphi$). For the component of field perpendicular to the tape axes, currents cross the substrates only at the ends of the cable, and $P/A \approx 0$.

For normal metals of high conductivity the above formulas give unacceptably large losses. For substrates of poor conductivity the skin depth δ, will be $\gtrsim 10$ mm and therefore comparable to the tape width; hence, the second of the above expressions applies. Compared to the usual a^3 loss dependence for a normal slab with either one or no superconducting face[5] the loss per unit area now varies as $ab^2 \sin^2\varphi$; that is, it is enhanced by the factor $(b^2/a^2)\sin^2\varphi$. For numerical illustration let $a = 12$ μm (1/2 mil), $b = 6$ mm (1/4") $H_\theta = 5 \times 10^4$ A/m, $\varphi = 30^o$, and $\rho = 2 \times 10^{-8}$ Ωm (approximate value for Nb-1%Zr in the normal state). Then $P/A = 25$ μW/cm². This is too large compared with hysteretic losses in Nb_3Sn of 10 μW/cm²,[4] but it is apparent from Eq. (19) that this number can be brought down to an acceptable value by reducing a, b, φ, and increasing ρ. An alternative solution is to employ tape segments which are fully enclosed by a superconductor layer.

Fig. 2 does not show the normal metal cladding usually present for superconductor stabilization. This can produce significant losses if the metal is too thick. However, these losses are not greatly affected by the helix angles and accordingly this problem is not discussed here.

<u>Superconductive substrates</u>. Losses for the case of a superconductive substrate (say Nb or Nb-1% Zr below 9 K) can similarly be calculated provided one again makes the simplifying assumption that no currents flow parallel to the superconductor - substrate interfaces. In this case the current pattern in the substrate is given by the critical state model for a section of width a taken from an infinite slab of thickness b. The loss is hysteretic. Provided flux does not penetrate to the center of the tape during a cycle, the loss per unit area of the coaxial cable at 60 Hz is (in mks units)

$$P/A = 2.85 \times 10^{-4}\, (a/b)\, (H^3/J_c) \qquad (20)$$

where J_c is the critical current density of the substrate, and $H = H_\theta \sin\varphi$ is the rms value of the axial field. For a given peak field amplitude, (i.e., transport current) the maximum loss for the substrate will occur when flux penetrates to the center of the tape. If $H_\theta \sin\varphi = 25$ A/mm rms and $b = 6$ mm, full penetration will occur for $J_c = 1.2 \times 10^7$ A/m². The temperature at which this occurs can be estimated assuming that the critical current density, J_c, decreases linearly with temperature:

$$J_c(T) = J_o\,(1 - T/T_{cH}) \qquad (21)$$

where T_{cH} is the critical temperature of the substrate in the presence of the axial field. T_{cH} is obtained by assuming a linear decrease of H_{c2} with temperature. Assuming $T_c = 9$ K and $H_{c2} = 0.4$T at 4.2 K one obtains $T_{cH} = 8.47$ K. Substituting this value and $J_o = 5\times10^9 A/m^2$ in Eq. (21), the temperature for maximum substrate loss is 8.45 K. At this temperature the contribution to the cable loss will be $\sim 74\ \mu W/cm^2$, i.e., higher than in the normal state (see above). However, this loss is reduced to an acceptable level of $< 5\ \mu W/cm^2$ by lowering the temperature a small amount, to $\lesssim 8.15$ K. If the maximum temperature of the cable does not exceed this value, edge losses would therefore be acceptable for a Nb substrate provided other effects such as flux jumping, etc., associated with flux entering the edge do not degrade the overall performance of the Nb_3Sn.

As before this effect can be eliminated if the tape segments are fully enclosed, i.e., the edges are Nb_3Sn. Attempts to produce such tapes are being made both commercially and in our laboratory.

CONCLUSIONS

Model calculations and experiments have been carried out for double helix, tape wrapped, superconducting cables. The cables were made by a simple overlay of opposing helices. Over a wide range of pitch angles, $20^o \le \varphi \le 45^o$, the effects of axial flux in producing ac losses in normal metals, which are present both outside and in the core region of the cable, are reduced to insignificant amounts. Voltage drop along the outer conductor is likewise reduced to an acceptably small value.

Finally, the flow of current perpendicular to the axes of the tape segments of a double helix is discussed. This can lead to significant losses in cables made with presently available commercial tapes. There are several possible options for eliminating or reducing this problem: the tapes may be fully enclosed by the Nb_3Sn layer; the substrate may be made highly resistive; finally if the substrate is made of superconductive material other than Nb_3Sn, the cable must be operated at temperatures not too close to the transition temperature of the substrate.

ACKNOWLEDGEMENTS

The authors thank E.B. Forsyth and D.H. Gurinsky for stimulating this work and T.D. Barber and S.W. Pollack for experimental help.

REFERENCES

1. Forsyth, E.B. et al, IEEE Trans. PAS-92,494(1973).
2. Sutton, J., Cryogenics 15, 541 (1975).
3. Morgan, G.H., and Forsyth, E.B., Considerations of Voltage Drop and Losses in the Design of Helically-Wound Superconducting AC Power Transmission Cables, Cryo. Eng. Conf., Kingston, Ontario, July 1975.
4. Bussiere, J.F., Garber, M., and Suenaga, M., IEEE Trans., Mag.-11, 324 (1975)
5. Lammeraner, J., and Stafl, M., Iliffe Books, Ltd., London, (1966).

† Work supported by the Energy Research and Development Administration, the Electric Power Research Institute and the National Science Foundation.

* Operated by Associated Universities, Inc., under contract to the U.S. Energy Research and Development Administration.

MÖSSBAUER SPECTROSCOPY OF PdH Fe ALLOYS

M. Weber and C.R. Abeledo
Observ. Nac. de Fisica Cosmica, Buenos Aires, Argentina

R.B. Frankel and B.B. Schwartz
Francis Bitter National Magnet Laboratory,* MIT
Cambridge, Mass. 02139

ABSTRACT

Large changes in the electron susceptibility of PdH alloys have been observed with increasing H concentration, with important consequences for the magnetic and superconducting properties of these alloys.[1] To observe the effect of the changing spin susceptibility, we have studied PdH Fe alloys with 2, 1 and <0.1 at.% Fe by Mössbauer spectroscopy in external magnetic fields. Mydosh[2] has reported spin glass behavior in the high Fe concentration alloys and Kondo phenomena in the dilute Fe alloys. For hydrogen-metal ratios <0.5, α and β phases, with low and high hydrogen content respectively, coexist. For temperatures below T_c of the α phase but above T_c of the β phase, we observe a superposition of a six line spectrum and a single line corresponding to the α and β phases respectively.[3] In an applied field, the β phase magnetizes and both phases have hyperfine fields which extrapolate to the same saturation value ($H^s_{hf} = -308$ kOe for 1% Fe). For H/metal ratios >0.5, only the β phase is present and slightly lower H^s_{hf} values are obtained. These results indicate that hydrogenation of FePd alloys has little effect on the iron moment but has a marked effect on the polarization of the Pd spins and consequently on the coupling between spins. The low Fe concentration samples were studied as Co^{57} in Pd sources. For a 10 mCi source (~0.1% Fe in Pd) with H/Pd > 0.5, a Brillouin function with parameters J = 2.5, g = 2 and $H^s_{hf} = -300$ kOe gave the best fit to the data. The hyperfine field is the same obtained in the nonhydrogenated case. For a 1 mCi source (~0.01% Fe in Pd) with H/Pd, lower moments and hyperfine fields are observed suggesting Kondo quenching of the moment. These results suggest competition between Kondo and exchange effects in determining the Fe moment in dilute alloys.

REFERENCES

* Supported by the National Science Foundation.

1. R.J. Meller and C.B. Satterthwaite, Phys. Rev. Letters 34, 144 (1975).
2. J.A. Mydosh, Phys. Rev. Letters 33, 1562 (1974).
3. J.S. Carlow and R.E. Meads, J. Phys. F: Metal Phys. 2, 982 (1972).

MAGNETIC RESONANCE OBSERVATIONS OF QUANTUM SIZE EFFECTS IN SMALL PARTICLES

W. D. Knight*
Physics Department, University of California, Berkeley, CA., 94720

ABSTRACT

Information on the electronic structure of small particles is obtained from complementary NMR and CESR experiments. At low temperatures where the average level spacing is larger than kT the particles which contain an odd number of electron spins are paramagnetic and provide CESR signals with enhanced relaxation times; NMR signals are strongly shifted and broadened. For even particles the electron spins are imperfectly paired in the ground state, where the depairing parameter $\hbar/\tau\Delta$ depends primarily on the spin orbit coupling and scattering at the particle boundaries. In the small particle regime the CESR lines are broadened by exchange with external paramagnetic impurities near the particle surfaces, and by nuclear hyperfine interactions when the number of atoms per particle is small. The NMR linewidths are dominated by the effects of exchange broadening from the impurities via the RKKY interaction. The residual spin paramagnetism measured by NMR in the even particles provides a measure of the electron spin relaxation time.

INTRODUCTION

Since the original calculations of Kubo[1] outlined the interesting properties which would be exhibited by systems of small particles, a number of theoretical[2-5] and experimental[6-11] works have elaborated on the idea. This paper is an attempt to summarize the main features of the related published work up to the present time as it relates to small particles of metallic elements. The quantum size effects are manifest under conditions when the average energy level spacing Δ in the particles satisfies the condition $\Delta \geqslant kT$. Under these conditions it is possible to distinguish between those particles which contain even and odd numbers of conduction electrons by their respective magnetic properties: at low temperatures the spin pairing in even particles tends to reduce their magnetic susceptibility, as may be observed in nuclear magnetic resonance (NMR) experiments. The odd particles of monovalent metals contain a single unpaired spin which is paramagnetic and which may be observed by NMR, by conduction electron spin resonance (CESR), or by conventional magnetic susceptibility experiments. The electron spin relaxation time is governed by spin orbit coupling[12] at the surfaces of the particles and thus by particle size. The spin orbit coupling parameter also governs the mixing of electronic states so that the ground state of even particles is not purely singlet which results in a residual Pauli paramagnetism[5,11] at the lowest temperatures. It is well known that the relative surface/volume ratio in small particles gives rise to important effects: in the magnetic resonance experiments these are principally caused by paramagnetic impurities in the embedding matrix which couple to that part of the conduction electron wave function which leaks out of the metal surface boundary. This exchange interaction broadens the CESR, and also couples the nuclei to the impurities so that the NMR is broadened by the resulting spin density (RKKY) oscillations,[13] which are proportional to the impurity magnetization. The nuclear spins are also relaxed by the coupling to the dynamical fluctuations of the impurity spins[14] in addition to the usual hyperfine coupling.[15] In the low temperature range where the conduction electron spin flips are inhibited by quantum level effects, the nuclear spins may still be relaxed by the impurity spins. The nuclei are uniquely sensitive to electronic phenomena under conditions when the electron spins are not directly observable, i.e., the residual NMR shift in even particles is a measure of the conduction electron spin relaxation time in a range where the corresponding CESR line is too broad to observe experimentally.

It will be seen in what follows that the NMR and CESR experiments are in many respects complementary, because of the hyperfine interactions. Thus, the electron and nuclear spins mutually see average local magnetization fields and fluctuating fields causing relaxation. The average electron spin polarization is measured by the NMR shift in even particles, while the itinerant electrons sample effective fields of the nuclear spins and can show a hyperfine broadened line.

THE CONDUCTION ELECTRON SPIN RESONANCE

The CESR occurs at the frequency

$$\omega_z = g\mu_B H/\hbar \tag{1}$$

in the applied field H, where the observed g-factor is shifted from the free electron value by $\Delta g = g - g_f$ resulting in a corresponding resonance shift $\Delta g/g$ which identifies the metal in question. The g-shift is related to the spin-orbit coupling[12] constant λ in the metal and an appropriate energy band gap in the metal,

$$\Delta g \sim \lambda/\Delta E, \tag{2}$$

and is generally small compared to unity. If Δg is anisotropic or varies with position within a metal, the CESR line may be broadened in proportion to this variation $\langle \Delta g \rangle$, unless line narrowing results from a short relaxation time.[16] The principal spin relaxation process is based on the spin orbit coupling and may be estimated from the relation of Elliott

$$1/\tau \simeq v_F \Delta g^2/d, \tag{3}$$

where v_F is the Fermi velocity and d the particle diameter. This spin relaxation time provides an important contribution to the linewidth until the particle size is reduced below the limit where the parameter ρ is of the order unity,

$$\rho \simeq \hbar/\tau\Delta. \tag{4}$$

In this quantum size effect regime τ no longer retains the same meaning as a relaxation time,[2] but rather becomes a parameter reflecting the size of the spin-orbit coupling. The average level spacing for the small particle is

$$\Delta = 4E_F/3N, \qquad (5)$$

with N equal the number of conduction electrons in the small particle and E_F the Fermi energy. Introducing atomic volume, particle diameter, and metallic cell radius r_s we have

$$(4\pi/3)(d/2)^3 = (4\pi/3)Nr_s^3 = N\Omega. \qquad (6)$$

Combining Eqs. (3)-(6),

$$\rho = 3N^{2/3}\Delta g^2/4r_s k_F, \qquad (7)$$

or in terms of the particle diameter

$$\rho = (3/16)(\Delta g^2/r_s k_F)(d/r_s)^2. \qquad (8)$$

A principal result of the calculation of Kawabata is the residual CESR linewidth $\delta\omega_e$ and shift of resonance $\Delta\omega_e$ at low temperatures for $\hbar/\tau\Delta << 1$ and $\hbar\omega_z/\Delta << 1$,

$$\delta\omega_e/\omega_z \sim \hbar/\tau\Delta; \; \Delta\omega_e/\omega_z \sim -\hbar/\tau\Delta \qquad (9)$$

which according to Eqs. (7) or (8) vanishes for small N or small d and varies as d^2 in contrast to the result Eq. (3) for larger particles. See Fig. (1). For pure metals with small spin orbit coupling and small g-shift one then expects the CESR linewidth to be extremely small in the regime of small ρ. In fact one hoped that by employing samples of small particles one could overcome the problem of excessive linewidth in bulk materials presumed to arise from large spin orbit couplings. Up to now, however, this hope of detecting heretofore unobserved CESR in other metals has not been realized, and we are faced with the problem of deciding whether the foregoing theoretical framework is adequate.

In particular, experiments[8] suggest that paramagnetic impurities and electron-nuclear magnetic hyperfine interactions may have pronounced effects on the observed resonances. We may assume that the effects of impurities within the volume of the small particles are negligible, because the known impurity concentration in the samples is much smaller than would average to one impurity per particle. However, it is well known that most samples, especially those which have been irradiated, contain paramagnetic centers, the important ones for our purposes being those which are in the layer approximately one lattice constant a_o thick in the matrix surrounding the metal particle. Thus the effective concentration, c, of impurities with respect to the volume of the particle will be the concentration in the matrix, c_m, multiplied by the surface/volume ratio, $c = 6c_m a_o/d$. The CESR linewidth from this interaction, including a reduced value of the exchange constant appropriate to an overlap integral which involves the leakage of conduction electron wavefunctions at the particle boundary, is the result of an inhomogeneous broadening appropriate to some average exchange constant J, and the proper expression is,[8] assuming the same g-value for impurity and conduction electron

$$\delta\omega_i/\omega_z = 4S(S+1)c_m J(a_o/d)/3kT$$

$$= 2S(S+1)c_m J(a_o/r_s)/3kTN^{1/3}. \qquad (10)$$

Alternative expressions are given by Gordon[8] which apply to the case of strong exchange and the coupled mode resonance, which will apply in some cases.

Although one might still hope to observe Kawabata line-narrowing at low temperatures in pure samples free of paramagnetic impurities where electron-phonon interactions are frozen out, a contribution from the electron-nuclear hyperfine interaction will remain. One expects this to produce an inhomogeneously broadened line[8] consisting of spin packets, each packet possessing an enhanced relaxation time appropriate to a Kawabata "width" as in Eq. (9). Thus we have[17]

$$\delta\omega_h = (a/2\hbar)/N^{1/2} \qquad (11)$$

where a is the s-electron hyperfine coupling constant.

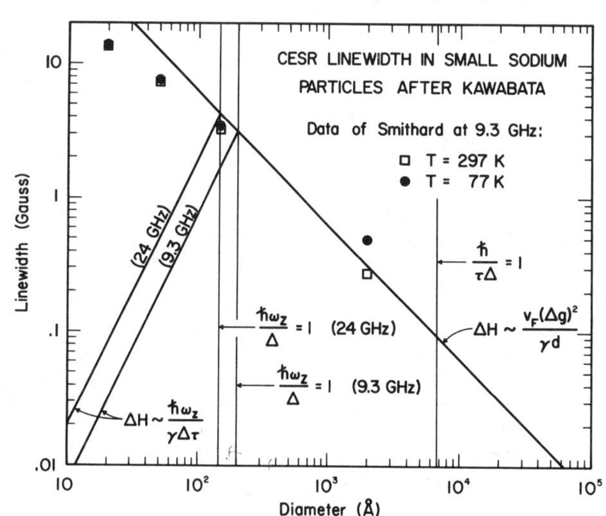

Fig. 1 CESR linewidth vs particle size in the region of quantum size effects. For diameters larger than the condition $\hbar\omega_z/\Delta = 1$ the indicated linewidth arises from the Elliott mechanism and surface scattering. In the case of silver the linewidths and $\hbar/\tau\Delta$ are larger than sodium by a factor of approximately 500. The experimental data do not show the Kawabata narrowing below 100 Å in sodium.

THE NUCLEAR MAGNETIC RESONANCE

The NMR occurs at the frequency

$$\omega = \mu H/I\hbar \qquad (12)$$

with μ and I being the nuclear magnetic moment and spin, respectively. The resonance is shifted in metals by the electron-nuclear hyperfine interaction

$$K = \delta\omega/\omega = a\chi_p\Omega I/g\mu_B\mu \qquad (13)$$

where a is the s-electron hyperfine coupling constant in the metal. The Pauli susceptibility, χ_p, in the non-interacting electron approximation is

$$\chi_p = (3/2)\mu_B^2 N/E_F = 2\mu_B^2/\Delta, \qquad (14)$$

where Δ is the average level separation given by Eq. (5).

Among the significant properties of the small particles is the contrasting behavior of those containing even vs odd numbers of conduction electrons. In the even case the ground state of the electronic system is a singlet, and the susceptibility indicated by Eq. (14) for normal bulk material vanishes because of the spin pairing. For the odd particles, the ground state is appropriate to the single unpaired spin[3] ($\chi/\chi_p = \Delta/2kT$). See Fig. 2. For small spin orbit coupling, the symmetries of the statistical distribution of energy levels in the particles results in a distribution of near neighbor levels varying directly with neighbor spacing. This gives a low temperature behavior[3] for even particles (see Fig. 2),

$$\chi(T)/\chi_p \simeq 3.8\ kT/\Delta, \tag{15}$$

in which the levels available for excitation are in proportion to the temperature. This applies to the so-called orthogonal statistical ensemble.

In the case of intermediate spin-orbit coupling the symmetries are typical of the symplectic distribution in which neighbor levels are probable as the fourth power of neighbor spacing in which case[3] the leading low temperature term is proportional to T^4,

$$\chi(T)/\chi_p \simeq 2.02 \times 10^3\ (kT/\Delta)^4, \tag{16}$$

at low temperatures. For the equal level spacing case we would have an exponential behavior at low temperatures.[2]

The larger spin-orbit coupling has a further effect which was not calculated by Denton et al.: in the even particles the mixing of the singlet ground state wavefunctions with excited triplet states results in a non-zero susceptibility at T = 0. This corresponds to imperfect spin pairing and results also in a non-zero NMR shift. The depairing parameter is, within a constant factor, equal to ρ as in Eq. (4). Thus for $\rho << 1$ we have the following approximate low temperature behavior for K and χ

$$\chi(T)/\chi_p \simeq 1.4\hbar/\tau\Delta + 3.8kT/\Delta. \tag{17}$$

The first term represents the residual susceptibility,[5] and the term linear in T corresponds to the orthogonal case.[3] The numerical coefficients in Eq. (17) are of order unity but the precise values remain to be determined. Shiba's result for $\chi(0)/\chi_p$ must be approximately correct on physical grounds and by analogy the similar case of the superconductor,[11,18] but his temperature dependence (Fig. 3) is probably incorrect owing to the fact that he considered only the case of equal level spacing, and used the grand canonical ensemble which does not conserve electron number and permits of excitations not allowed in the analyses of Kubo and Denton. On the other hand, although Denton et al. use more realistic energy level distributions and consequently arrive at better forms for the temperature variation of χ, they omit consideration of the effects of the electron dynamics and the resulting residual susceptibility $\chi(0)$. Equation (17) then should be useful as an approximate interpolation formula until a unified theoretical treatment of all the important effects is available.

The NMR linewidth can have several origins (for example, nuclear dipole-dipole coupling and nuclear quadrupole coupling with crystalline field gradients). For small particles in matrices at low temperatures, it is found that the important broadening can be described by a term arising from indirect coupling of the nuclear magnetic moments to impurity spins in the matrix via the RKKY spin density oscillations in the conduction electron system. The width may be estimated following the arguments in the development of Eq. (10) for the CESR impurity width, with the addition of the K factor (from Eq. (13)) accounting for the hyperfine interaction and a factor arising from an average over the RKKY range function,[13]

$$\delta\omega/\omega = c_m KJ(a_o/d)/3kT. \tag{18}$$

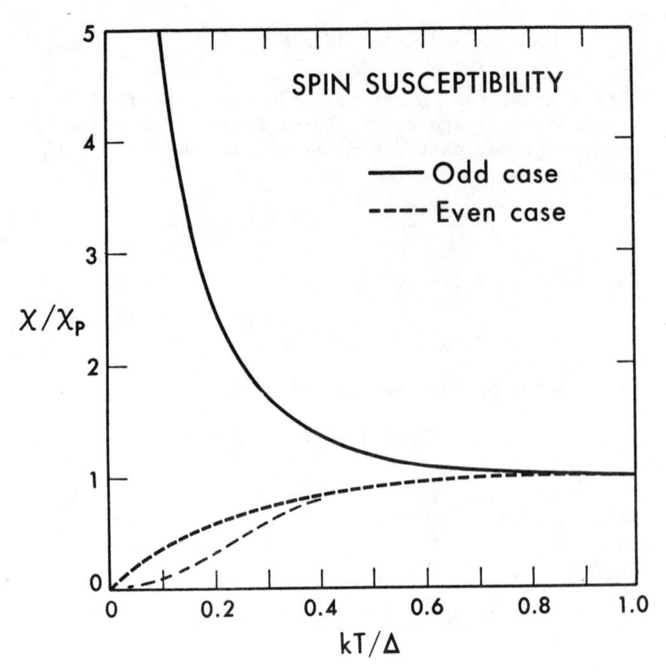

Fig. 2 Conduction electron spin susceptibility after Denton et al. Spin pairing in the ground state causes the even particle susceptibility to vanish at T = 0. The low temperature dependence for the even case is slower for the symplectic ensemble in which level repulsion reduces the available excitations at low temperatures as shown in the lowest curve. The orthogonal case gives $\chi(T) \propto T$ at low temperatures for small spin orbit coupling. The odd particles exhibit a Curie-like paramagnetism.

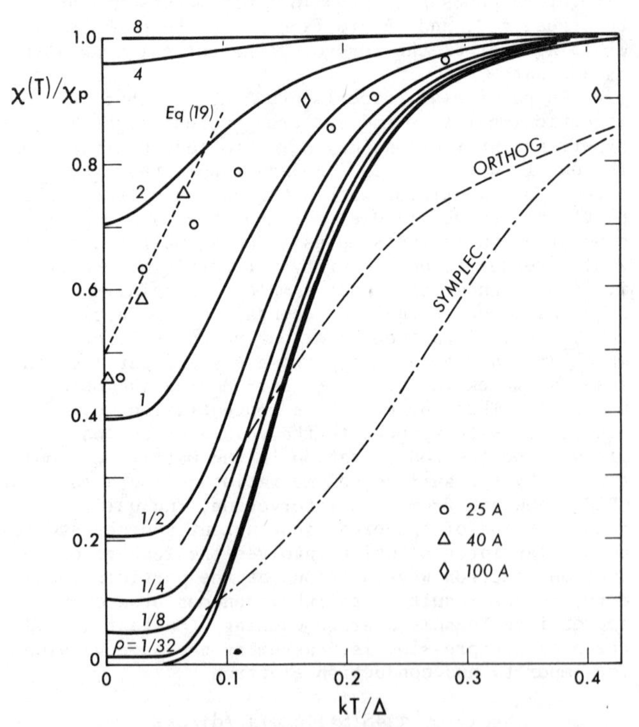

Fig. 3 Spin susceptibility after Shiba. The solid curves are taken directly from Shiba's paper and indicate the residual paramagnetism $\chi(0)$ as a function of the parameter $\rho = \pi\hbar/\tau\Delta$.[11] The experimental points are given for three particle sizes. The dotted line is a plot of Eq. (19) with $\hbar/\tau\Delta = 1/3$. The dashed curves are taken from Fig. 2 for comparison. The results for NMR in aluminum are expected to follow the orthogonal curve.

The broadening, which arises from the external surface layer of impurity spins, is inversely proportional to the diameter of the particle, provided that c_m is large enough to provide an average of one or more impurity spins per particle, and follows the susceptibility temperature dependence of the impurity spins.

According to the simplest picture we should expect the nuclear spin relaxation rate[19] to diminish at low temperatures in small particles as the level spacing increases and the conduction electron spin relaxation rate is reduced, consistent with Eq. (9). However, paramagnetic impurities may couple indirectly to the nuclear spins, and provide relaxation for them.[14] The behavior of superconducting small particles has been investigated,[20,21] but will not be discussed here.

EXPERIMENTAL RESULTS

In the following we refer to selected experimental results without attempting to be exhaustive. The first report[6] of a quantum size effect indicated by CESR the presence of paramagnetism in the odd particles in a sample of lithium, and vanishing Pauli paramagnetism in the even particles. Similar results have since been obtained on small particle samples of aluminum[9b,10a] above the superconducting transition temperature. In these cases, Eqs. (9) and (15) are expected to hold, since the spin orbit couplings are small. Certainly the condition $\chi(0) = 0$ seems to hold for even particles. The CESR result for aluminum[9b] appears to be consistent with Eq. (9) with appropriate size distributions and assuming g-shifts and linewidths of similar magnitudes. In view of the fact that Kawabata uses the Δg value appropriate to the bulk material, the positive value of Δg reported for small particles of aluminum opens the question of altered g values within the volume of a small particle.

An extensive study of silver particles[9a] supported in a solid inert gas matrix indicates the expected odd particle paramagnetism and narrowed CESR lines with position and width consistent with Eq. (9). In these experiments the single atom resonances with hyperfine structure were observed, and after annealing small particle resonances occurred with asymmetric lines between g = 2.0075 and 2.01. These are interpreted as arising from particles in the size distribution with diameters below 20 Å with an inferred $\Delta g = +.032$ for bulk silver. Experiments at 450 MHz, 9300 MHz, and 35 GHz confirm the ω_z dependence in Eq. (9). The fact that the linewidth of around 30 gauss is relatively independent of sample is attributed to the sharp cutoff of the quantum size effect narrowing above 20 Å. A long low-field tail on the lines which was easily saturated at low microwave powers and varied from sample to sample, appears to be associated with the smallest particles. No NMR results are available.

The experiments on sodium were performed in irradiated-annealed samples of NaN_3 and showed no direct indication of Kawabata line narrowing.[7,8] The size distribution of particles ranged from around 10 Å to 200 Å or more. The effects of size and temperature are related to the impurity and phonon broadening, the former corresponding[8] to Eq. (10) with $c_m J = 4 \times 10^{-8}$ e.v. in the 100 Å range. See Fig. 4. That no evidence was found for Kawabata line narrowing is perhaps not surprising in view of the role played by matrix impurities. A broad (30 gauss) line appeared as a background under the sharper signal at the same g-value and was easily saturated at low microwave powers at the lowest temperatures. This was attributed to the smallest particles not affected by impurities, with long relaxation times and inhomogeneous hyperfine broadening, according to Eq. (11), which is equivalent to $175/N^{1/2}$ gauss for sodium. See Fig. 5. This is consistent with the observed width for particles containing around 100 atoms (d = 25 Å).

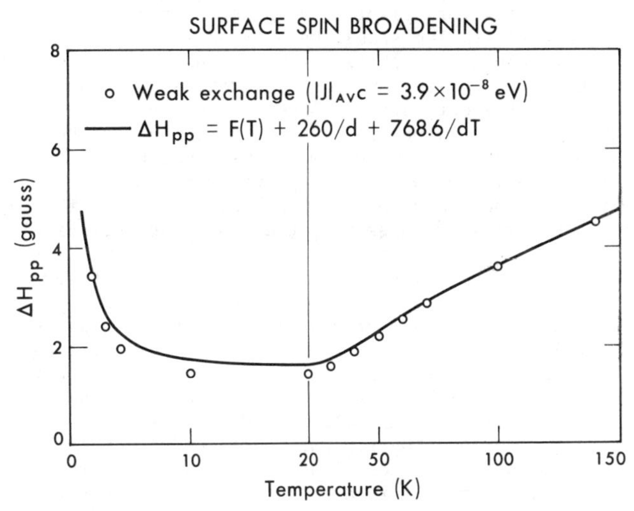

Fig. 4 CESR linewidth as a function of temperature, after Gordon. Above 20 K the linewidth increases according to the electron-phonon coupling for a distribution of particles of diameter less than 200 Å. Below 10 K the broadening is consistent with Eq. (10) and is inversely proportional to diameter and temperature.

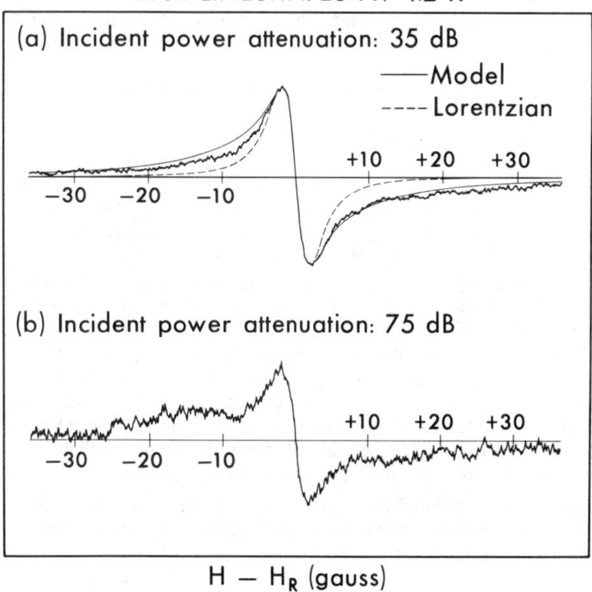

Fig. 5 CESR linewidth at low microwave power, after Gordon. The narrow central portion of the line falls on the curve of Fig. 4 which represents measurements at relatively high microwave powers, where the broad line is saturated. The broad line is attributable to hyperfine interactions in the smallest particles which are not affected by matrix impurities.

It is unfortunate that the experiments on silver and sodium dealt with wide particle size distributions and that the exact nature and distribution of the paramagnetic impurities was not known in sodium. It is also puzzling that hyperfine broadening, which should be observed according to Eq. (11), seemed to show up in sodium but not in silver. Of course the saturable low field tail in the silver experiments was nearly 100 gauss wide, and this might very well be hyperfine broadening, as Eq. (11) gives $400/N^{\frac{1}{2}}$ gauss for silver, and we recall moreover that the hyperfine line should be a broad envelope of easily saturable packets which is consistent with the observations. It should be realized, also, that the experimental data are not extensive enough, nor is the theoretical situation clear with respect to the appropriate correlation time to use in the estimate[16] of motional narrowing. The position of the silver line does not correspond to the value found in the bulk material,[22] suggesting that in Ag as well as Al particles the g-shift may vary within the particles.

The NMR results are perhaps somewhat less perplexing. As we have already indicated, the metals lithium and aluminum provide results in clear agreement with the condition $\chi(0) = 0$ for even particles.[10] For copper the existence of the low field tail on the line gives ample evidence for the onset of paramagnetism[10b] in the odd particles and for the result $\chi(0) \neq 0$ for even particles. For a narrow distribution of particle sizes[11] the result indicates more precisely that $\chi(0)$ is a function of particle size as implied by the first term of Eq. (17), and the low temperature dependence appears to favor the orthogonal ensemble according to the second term of Eq. (17). See Fig. 3. It should be noted particularly that the value of ρ so indicated for d = 40 Å corresponds to a spin relaxation time $\tau = 5 \times 10^{-13}$, which is equivalent to an average spin flip per two hundred surface encounters of the conduction electrons. This would correspond to a CESR line considerably too broad to observe, emphasizing the fact that the NMR results provide information on electron spin relaxation rates in a range which is inaccessible to direct CESR measurements.

As we mentioned earlier, the NMR linewidth and relaxation time of the small particle depend primarily on the matrix impurities. The linewidth can be derived by combining Eqs. (17) and (18), and it appears[11] that the spin density oscillations in even particles are less effective in producing NMR linewidth as the spin pairing becomes more complete. The experimental results[10b,11] for the nuclear spin relaxation rate indicate no reduction in small particles corresponding to quantum size reduction in the electron spin flip rate. As Alloul and Bernier have pointed out, the NMR nuclear spin relaxation rate in a metallic system containing paramagnetic impurities proceeds via the RKKY spin density oscillations coupled to the spin flips of the impurities. In the small particle system at low temperatures, this appears to be the primary remaining nuclear spin relaxation mechanism. This will disappear only when the impurity concentration and spin flip rate are greatly reduced.

CONCLUSIONS

The experimental evidence favors the expected quantum size effects as to: difference in behavior of even and odd particles; reduction in Pauli susceptibility for even particles; Curie susceptibility for odd particles; role of spin orbit coupling in both static and dynamic characteristics of the conduction electron spin system; complementary aspects of NMR and CESR in elucidating the electron spin relaxation rates; role of impurity broadening and RKKY spin density oscillations; the contributions of the hyperfine interactions. Unfortunately, although the interactions with paramagnetic impurities in the matrix are interesting problems, they are also somewhat of a hindrance in measuring and understanding the properties of isolated particles, for which one would wish to understand more completely the metal/vacuum boundary. Furthermore, the ideal experiment should involve a series of controlled narrow size distributions. With a better knowledge of the behavior of the pure particles, one could then hope to address the problem of interactions between particles and the embedding matrix, and the interesting question of leakage of the conduction electron wavefunctions out of the metal and into the matrix. This paper has omitted discussion of a number of important features of the magnetic resonance experiments, and several related questions remain. For example, the importance of exchange in the conduction electron system and contributions of orbital shifts in the NMR have been emphasized by Walstedt and Walker.[13] The importance of many body effects and motional narrowing have emerged in some of the related TESR experiments of Schultz et al.[16] In these experiments both line narrowings and large g-shifts were observed under conditions which raise the question of the nature of many body effects in small particle systems. The question of the sign of Δg as observed by TESR and in small particles of aluminum and silver is unresolved at the present time, and if the sign is uncertain the appropriate magnitude is yet to be decided.

Finally, it is worth reemphasizing the fact that the measured properties of small particles can not be independent of the embedding matrix, even if the matrix is free of paramagnetic impurities.

REFERENCES

*Supported in part by the National Science Foundation.

1. R. Kubo, J. Phys. Soc Japan 17, 975 (1962).
2. Arisato Kawabata, J. Phys. Soc. Japan 29, 902 (1970).
3. R. Denton, B. Mühlschlegel, and D.J. Scalapino, Phys. Rev. B7, 3589 (1973).
4. B.W. Holland, Colloque Ampere XIV, (North Holland, Amsterdam, 1967), p. 468.
5. H. Shiba, J. Low Temp. Phys. 22, 105 (1976).
6. C. Taupin, J. Phys. Chem. Solids 28, 41 (1967); J. Charvolin, J.P. Cohen-Addad, and C. Froidevaux, Solid State Comm. 5, 357 (1967).
7. M.A. Smithard, Solid State Commun. 14, 411 (1974).
8. D.A. Gordon, to be published in Phys. Rev., May, 1976; Thesis, University of California, Berkeley (1975).
9. R. Monot. (a) Thesis, Ecole Polytechnique Federale, Lausanne, Switzerland; (b) Colloque Ampere XVIII, (North Holland, Amsterdam, 1974), p. 319; (c) R. Monot, C. Narbel, and J.-P. Borel, Il Nuovo Cimento, Series II 19B, 253 (1974).
10. S. Kobayashi, T. Takahashi, and W. Sasaki, J. Phys. Soc. Japan, (a) 31, 1442 (1971); (b) ibid. 32, 1234 (1972).
11. P. Yee, and W.D. Knight, Phys. Rev. B11, 3261 (1975); P. Yee, Thesis, University of California, Berkeley (1973).
12. R.J. Elliott, Phys. Rev. 96, 266 (1954); Y. Yafet, Solid State Phys. 14, 1 (1963).
13. R.E. Walstedt and L.R. Walker, Phys. Rev. 11B, 3280 (1975).
14. H. Alloul and P. Bernier, to be published.
15. J. Korringa, Physica 16, 601 (1950).
16. D. Lubzens, M.R. Shanabarger, and S. Schultz, Phys. Rev. Lett. 29, 1387 (1972).
17. C. Kittel, Introduction to Solid State Physics, 4th edition, (John Wiley, New York, 1971), Ch. 17.
18. A.A. Abrikosov and L.P. Gor'kov, Zh. Eksp. Teor. Fiz. 42, 1088 (1962); JETP 15, 752 (1962).
19. A. Overhauser, Phys. Rev. 89, 689 (1953).
20. B. Mühlschlegel, D.J. Scalapino, and R. Denton, Phys. Rev. B6, 1767 (1976).
21. S. Kobayashi, T. Takahashi, and W. Sasaki, J. Phys. Soc. Japan 36, 714 (1974).
22. S. Schultz, M.R. Shanabarger, and P.M. Platzmann, Phys. Rev. Lett. 19, 749 (1967).

THE BEHAVIOR OF SPINS NEAR THE SURFACE OF A HEISENBERG MAGNET

D. L. Mills
Department of Physics, University of California, Irvine, Ca. 92717

ABSTRACT

This paper reviews the theoretical description of the surface region of Heisenberg magnets, both at low temperatures in the spin wave regime, and near the ordering temperature. The results of new studies of instabilities in the spin configuration at the surface are presented, and the experimental data on magnetic properties of the surface are reviewed briefly.

INTRODUCTION

By this time, one may say that the properties of Heisenberg magnets are quite well understood from both the theoretical and experimental point of view. Of course, many questions remain unanswered particularly in highly anisotropic materials, but nonetheless in the last two decades, we have seen a most successful effort aimed at elucidating spin behavior in insulators described by a Heisenberg model Hamiltonian.

This paper is concerned with the behavior of spins in and near the surface of such magnets. A considerable theoretical literature exists on this topic, and the literature contains rather intriguing predictions. While the last decade has been characterized by an explosive expansion of reliable and reproducible experimental studies of the outermost atomic layers of crystals, rather little attention has been devoted to the possibility of probing magnetic properties near the surface. In the view of the present author, this is unfortunate, particularly since presently available experimental methods should enable a number of important questions to be investigated. Nonetheless, rather intriguing data does exist at this time, and we hope that the next decade will be one in which activity in this area increases.

The outline of this paper is as follows. We begin with a review of the theoretical picture of the behavior of spins near the surface of the Heisenberg ferromagnet, for the case where the surface spins are also ferromagnetically aligned. Then we turn to a description of some new work by Demangeat and the present author which shows that the ferromagnetic arrangement of the surface spins may not be stable in some circumstances, and outlines some properties of the resulting surface spin arrangement. The emphasis in these portions of the paper is on the Heisenberg ferromagnet, although brief mention is made of some earlier work on the Heisenberg antiferromagnet. We also confine our attention to exchange-coupled spin arrays, and we ignore the dipolar interactions crucial to long wavelength excitations in the spin system. We conclude with a discussion of the data presently available that provides information about the behavior of surface spins.

We remind the reader of the behavior of spins in the infinitely extended ferromagnet before we enter the description of the surface region.

At temperatures T low compared to the bulk Curie temperature T_c, there are well defined elementary excitations that obey Bose-Einstein statistics: the spin waves. When a spin wave is excited, the transverse spin component $S_+(\vec{\ell},t)$ varies in a wave-like manner from site to site,

$$S_+(\vec{\ell},t) = \mathcal{S}\exp[\,i\,k\cdot\ell - i\Omega(k)t\,] \quad , \qquad (1)$$

where in the long wavelength limit (in zero magnetic field) $\Omega(k) = Dk^2$, with D the spin wave exchange stiffness. Thermal excitation of spin waves causes $\langle S_z\rangle$ to fall below + S, to give the well known result[1]

$$S - \langle S_z\rangle = \Delta_\infty(T) = \frac{\zeta\left(\frac{3}{2}\right)}{8\pi^{3/2}n}\left(\frac{k_BT}{\hbar D}\right)^{\frac{3}{2}} \quad , \qquad (2)$$

with n the number of spins per unit volume.

As the transition temperature is approached from below, the magnetization vanishes in a power law fashion, with

$$\langle S_z\rangle \sim (T_c - T)^{\beta} \quad , \qquad (3)$$

where mean field theory yields 1/2 for the critical exponent β, and improved theories provide the value .33.[2] Since the onset of magnetic order is a second order phase transition, near T_c either above or below, large amplitude fluctuations in the magnetization occur. These fluctuations may be studied by the method of neutron scattering.[3] Since one finds that the cross section for magnetic diffuse scattering integrated over energy transfer diverges as T_c is approahced from either side,[3] it is important to note one need not analyze the energy spectrum of the scattered neutrons to obtain information about the critical scattering, although such data is of great interest.

II. MAGNETIC PROPERTIES OF THE SURFACE REGION; THE CASE WHERE FERROMAGNETIC ORDER IS STABLE IN THE SURFACE

For a semi-infinite sample at temperatures low compared to T_c, spin waves remain the elementary excitations of the system. However, they are more complex in nature than in an infinitely extended material, where periodic boundary conditions are appropriate. To discuss the modes in the presence of a surface, we follow an earlier paper here[4] and consider a film of thickness L. We have in mind the limit $L\to\infty$, with one surface held fixed (in the x y plane) as the limit is taken, so we have a semi-infinite spin array in the end.

There are two distinct classes of spin waves for finite L. The first are standing spin waves in the film. In simple cases, these waves are superpositions of solutions like that in Eq. (4), where the plane waves have $\vec{k} = \vec{k}_\parallel \pm \hat{z}\,k_z$. The allowed values of $\vec{k}_\parallel$ are provided by periodic boundary conditions applied in the two directions parallel to the surface, and the allowed values of k_z are found by the requirement that the equations of motion for surface spins be satisfied.[4] For fixed $\vec{k}_\parallel$, one finds a sequence of allowed values $k_z(1)$, $k_z(2)$, ..., spaced by roughly $2\pi/L$. The minimum and maximum frequency for a given $\vec{k}_\parallel$ will be denoted by $\Omega_m(\vec{k}_\parallel)$ and $\Omega_M(\vec{k}_\parallel)$ respectively.

Under a variety of circumstances, one finds surface spin waves in the semi-infinite sample. These are modes which, for a given $\vec{k}_\parallel$, have spin deviation localized near the surface, with frequency either below $\Omega_m(\vec{k}_\parallel)$ or above $\Omega_M(\vec{k}_\parallel)$.[5] The spin deviation in these modes has the form

$$S_+(\vec{\ell},t) = \mathcal{S}\exp\left[i\vec{k}_\parallel\cdot\vec{\ell}_\parallel - \alpha(\vec{k}_\parallel)\ell_z - i\Omega_s(\vec{k}_\parallel)t\right] \quad (4)$$

where for a given site, we write $\ell = \ell_\parallel + \hat{z}\,\ell_z$.

For the long wavelength, low frequency surface spin waves of interest in the study of the thermodynamics of the material at low temperatures, there are two features one finds[6]:

(i) To lowest order in $k_\parallel^2$, $\Omega_s(\vec{k}_\parallel)$ and $\Omega_m(\vec{k}_\parallel)$ are equal. The "binding energy" of the surface spin wave (the amount by which $\hbar\Omega_s(k)$ lies below $\hbar\Omega_m(k)$) makes its appearance first in the terms of order $k_\parallel^4$.

(ii) For small $k_\parallel$, $\alpha(k_\parallel)$ is the order of $a_o k_\parallel^2$, where a_o is the lattice constant.

These properties are in striking contrast to other well known surface waves, such as the Rayleigh waves of elasticity theory. For Rayleigh waves, the leading term in the dispersion relation (that proportional to $k_\parallel$) lies below that of any bulk excitation with the same $k_\parallel$, and the attenuation constant is $\sim k_\parallel$ rather than $a_o k_\parallel^2$. Thus, by comparison, the surface spin waves are very weakly bound surface modes.[8]

To calculate the thermodynamic properties of the surface region, one recognizes the surface spin waves are boson-like excitations of the system. In addition, it is crucial to recognize that the standing spin waves are redistributed in frequency by the surfaces (i.e. the values of $k_z(n)$ are shifted away from $2\pi n/L$ allowed by periodic boundary conditions applied normal to the film). This produces a change in the density of states of the volume waves that leads to corrections to the thermodynamic properties of the semi-infinite sample proportional to the surface area; there are also singular contributions at $\Omega_m(\vec{k}_\parallel)$ and $\Omega_M(\vec{k}_\parallel)$.[4]

A dramatic example of the role of the surface induced shift in volume spin wave density of states is provided by the calculation of the mean spin deviation $\Delta_{\ell_z}(T)$ near the surface[9] ($\Delta_{\ell_z}(T) = S - \langle S_{\ell_z}\rangle$). From remark (ii) above, one expects the contribution from the surface waves to the deviation from $\Delta_\infty(T)$ to extend a distance λ_T^2/a_o into the bulk from the surface, where λ_T is the wavelength of a spin wave with energy $k_B T$. The calculation shows instead that $\Delta_{\ell_z}(T)$ falls from the value $2\Delta_\infty(T)$ at the surface to $\Delta_\infty(T)$ in the much smaller distance λ_T; the anomalously long range contribution from the weakly bound surface waves is precisely cancelled by a surface-induced anti-resonance in the bulk spin wave density of states near $\Omega_m(\vec{k}_\parallel)$. The precise formula that obtains may be written[9]

$$\Delta_{\ell_z}(T) - \Delta_\infty(T) = \frac{1}{4\pi n\alpha}\left\{1 + 2\sum_{m=1}^{\infty}\exp\left[-n\left(\frac{m\pi}{\alpha}\right)^{\frac{1}{2}}\right] \times \cos\left[n\left(\frac{m\pi}{\alpha}\right)^{\frac{1}{2}}\right]\right\} \quad (5)$$

where $n = 2\ell_z - 1$, and $\alpha = D/a_o^2 k_B T$.

The physical picture that emerges from this analysis is that in the presence of the surface, there is increased thermal disorder in the spin system at the surface; the mean spin deviation rises from the bulk value $\Delta_\infty(T)$ to $2\Delta_\infty(T)$ as one approaches the surface, and there is in addition an increase in this magnon contribution to the specific heat.[4,9]

The spin wave analysis is valid only for temperatures well below T_c. For T near T_c, the behavior of the surface region has been described theoretically both within the framework of mean field theory, and also by more rigorous methods. The physical picture that emerges is consistent with the one just described. That is to say, since the spins in the surface are coupled to fewer neighbors than are those in the bulk, there is a greater degree of thermal disorder in and near the surface than in the bulk.

For example, in mean field theory,[10] unless the exchange interactions in the surface are considerably stronger than those in the bulk,[11] at the surface $\langle S_z\rangle$ vanishes linearly with (T_c-T), in contrast to the more gradual $(T_c-T)^{\frac{1}{2}}$ behavior in the bulk. The transition from the surface behavior to that characteristic of the bulk occurs in a distance the order of the coherence length $\xi = a_o|1-T/T_c|^{-\frac{1}{2}}$. Furthermore, while neutron studies show strong critical scattering from the bulk spins near T_c, as discussed earlier, the theory shows that a particle which probes only the outermost layer of the semi-infinite spin array (a low energy electron perhaps) would not experience critical scattering in the near vicinity of T_c. The particle must penetrate a distance the order of the coherence length to probe the critical fluctuations. The picture that emerges from the mean field theory (for both the ferromagnet and the antiferromagnet) is one in which the surface spins are pulled into order at T_c only by virtue of their coupling to the bulk spins; the large amplitude critical fluctuations present in the bulk are subdued in the surface.

Several very pretty papers have appeared which present treatments of the surface region near T_c by rigorous methods, for both Ising and Heisenberg ferromagnets.[12,13] We do not have space to review the results in detail here, although the physical picture that emerges from the analysis is consistent with that provided by mean field theory. Of course as one knows well, the quantitative predictions of mean field theory require modification near T_c.

We do wish to call attention the the remarkable numerical calculations displayed the second paper by Binder and Hohenberg.[13] These authors consider a finite Heisenberg magnetic cube, and calculate the magnetization as a function of temperature and distance from the surface, for a wide range of temperatures. The results are in excellent quantitative accord with the spin wave theory,[9] up to temperatures the order of 0.6 T_c.

This section has focused attention on the spin wave regime, and also on the static properties near T_c. A study of spin dynamics near the surface, has recently been presented by Kumar and Maki.[14] These authors construct a time dependent Landau-Ginzburg equation for the semi-infinite magnet, and study its solutions near T_c.

We should mention that in the spin wave regime, the surface of the Heisenberg antiferromagnet has been studied theoretically.[15] There are two qualitative features of the surface spin behavior here that are new. There is a gap in the bulk spin wave spectrum, with the minimum bulk spin wave excitation energy (in zero external magnetic field) given by $\Omega_o = \gamma(H_A^2 + 2\ H_A H_E)^{\frac{1}{2}}$, with H_A and H_E the anisotropy and exchange fields respectively. There may be surface spin waves in the gap between zero frequency and Ω_o. At $k_{\parallel} = o$, these surface waves may range in frequency from $\Omega_o/\sqrt{2}$ to very close to Ω_o, depending on the surface geometry. These waves may be observable in infrared absorption studies. If a magnetic field is applied to the material, then an antiferromagnetic surface spin wave may "go soft" at a magnetic field below that where the bulk spin wave frequency touches zero at $\vec{k} = o$. In such a case, one may have a "surface spin flop" transition at fields where the bulk remains antiferromagnetic.[16]

III. INSTABILITIES IN THE SURFACE SPIN ARRANGEMENT: THE PHENOMENON OF MAGNETIC SURFACE RECONSTRUCTION

Quite clearly, if we consider a semi-infinite Heisenberg ferromagnet with all exchange interactions of ferromagnetic sign, then below T_c the spins in and near the surface will order ferromagnetically, and the ground state is that with spins fully aligned.

However, if the exchange interactions in the surface differ in sign from those in the bulk,[17] or in the more commonly encountered situation where exchange interactions with both ferromagnetic and antiferromagnetic are present simultaneously, then it is not clear that the surface spin configuration need be the same as that in the bulk. In the latter case, Blandin[18] has pointed out that in a ferromagnet, the ferromagnetic spin arrangement may be unstable in the surface, if the surface spins experience a greater fraction of antiferromagnetic interactions than spins in the bulk. Blandin illustrated this point through study of a semi-infinite plane of spins.

Recently, Demangeat and the present author have completed a detailed study of surface spin instabilities in a semi-infinite three dimensional model crystal.[19] In this section, we present a summary of this work.

We have considered a semi-infinite FCC crystal, with nearest and next nearest neighbor exchange J_1 and J_2 respectively. The surface is a (100) surface, and $J_1 > o$, $J_2 < o$. This crystal structure and choice of exchange interactions is appropriate to the europium chalcogenides. If $J_1 > |J_2|$, the bulk of the crystal orders ferromagnetically.

For a bulk spin, the ratio of antiferromagnetic to ferromagnetic exchange couplings experienced by a spin is 0.5, while in the surface the ratio is 0.625. Thus, if we increase the ratio $|J_2|/J_1$, we expect the surface spins will become unstable with respect to ferromagnetism before the bulk spins.

The stability of the ground state may be explored through study of the surface spin wave dispersion relations;[17,18] one requires the excitation energy $\hbar\Omega_s(\vec{k}_{\parallel})$ be positive for all $\vec{k}_{\parallel}$ for the ferromagnetism to be stable in the surface. We find this criterion requires $|J_2| < 0.86\ J_1$ if the exchange interactions in the surface are identical to those in the bulk. For $0.86\ J_1 < |J_2| < J_1$, ferromagnetism is stable in the bulk, but not in the surface. The critical value of $|J_2|$ required for the surface instability may be lowered substantially if the exchange interactions in the surface differ from the bulk.

In the region where the ferromagnetic arrangement is unstable in the surface, the surface spins cant into an arrangement reminescent of the spin flop transition in a ferromagnet. There are two distinct sublattices in the surface. In one the surface spins are canted at an angle $+\theta_o$ relative to the bulk magnetization, and in the other the angle is $-\theta_o$. Thus, we have here a magnetic analogue of the well known phenomenon of surface reconstruction;[21] the unit cell that characterizes the configuration of spins in the surface layer is larger than that in the bulk.

Within the framework of mean field theory, we have studied the nature of the reconstructed state, as a function of temperature and magnetic field. We find a surface transition temperature $T_s < T_c$ above which $\theta_o(T)$ vanishes. That is to say, $\theta_o(o)$ is finite, but decreases to zero as $T \rightarrow T_s$, to remain zero for all $T > T_s$. At all temperatures, the reconstructed spin arrangement is confined to the outermost atomic layers of the crystal; the "scissors angle" $\theta_{\ell_z}(T)$ falls to zero within four or five layers of spins. Furthermore, application of a magnetic field suppresses the transition temperature T_s, as well as the value of $\theta_o(o)$. These results are described in detail in a forthcoming publication.[19]

We have underway currently a study of the spin dynamics of a model semi-infinite ferromagnet, with surface reconstruction present.[2] The model we have under study is the one employed by Trullinger and Mills.[17] This model has the simplifying feature that all spins in the ground state are aligned ferromagnetically, save for those in the outermost layer. We have derived the dispersion relation for surface spin waves in this structure, and a study of spin correlations at low temperatures is underway.

The surface spin wave dispersion relation in the reconstructed state differs in a striking and qualitative manner from that described in section II. We suppose an external magnetic field of strength H_o is present, so a "Zeeman gap" $g\mu_B H_o$ exists in the bulk spin wave dispersion relation. Then the surface spin wave spectrum consists of two branches. The frequency of the upper branch approaches $g\mu_B H_o$ as $k_\parallel \to o$, but the lower branch has a frequency which vanishes linearly with $k_\parallel$, as $k_\parallel \to o$. The spin deviation associated with spin waves on the lower branch is confined entirely to the surface layer, for small $k_\parallel$. At $k_\parallel = o$, the mode consists of a rigid precession of both surface sublattices about the bulk magnetization. Symmetry considerations require this mode to have zero frequency in the absence of special anisotropy fields in the surface; we call this a surface Goldstone mode. This surface Goldstone mode has no analogue in the theory of crystallographic surface reconstruction, and since symmetry arguments dictate its existence in the magnetic reconstruction problem, we expect it is a general feature and not a consequence of the particular model we use.

The existence of the lower branch with frequency that vanishes as $k_\parallel \to o$ suggests that large amplitude spin fluctuations occur in the surface, in the presence of magnetic surface reconstruction. We have underway an investigation of the nature of these fluctuations with a spin wave theory description that includes surface anisotropy fields.

IV. EXPERIMENTAL STUDIES OF MAGNETIC SURFACES

Wigen and his colleagues have reported[23] the direct observation of a surface spin wave, under conditions where the exchange energy plays a dominant role. These authors study microwave resonance absorption in a thin YIG film deposited on a non-magnetic substrate. When the external magnetic is aligned parallel to the film surface, they find an absorption line on the high field side of the main ferromagnetic resonance line. They argue that spins at the YIG/substrate boundary experience a surface pinning field antiparallel to the external field, for this field direction. This produces a surface spin wave localized near the interface. This identification has been checked through study of the dependence of the line intensities on film thickness: the intensity of the main line decreases linearly with film thickness as expected, with that of the presumed surface mode is independent of film thickness, as long as the film is greater than 500 Å thick. So far as we know, this work constitutes the only clear observation of a surface spin wave, under conditions where exchange interactions play a dominant role. Anomalous absorption lines have been reported in powdered samples of antiferromagnetic MnF_2 by Tennant et al.,[24] but as these authors point out, it is not yet clear that surface magnons are responsible for the lines.

Low energy electron beams are a powerful probe of the outermost layers of a crystal. Such beams may be used to explore magnetic properties of the surface. It is important to note that spin polarized beams are not required for this purpose, although they would prove most valuable. For example, the onset of antiferromagnet order in the surface (or a reconstructed spin configuration) would lead to additional spots in a LEED pattern of magnetic origin separated from the crystallographic Bragg peaks. Unfortunately, with exception of Palmberg's work cited below, little attention has been devoted to magnetic studies by the surface physics community.

Several years ago, Palmberg observed the onset of antiferromagnet order in the surface of NiO.[25] From the data on the half order Bragg peaks of magnetic origin, one may make inferences about the temperature variation of the sublattice magnetization near the surface,[10,26] after certain assumptions about the nature of the electron-surface spin interaction are made. It is a pity that further studies of this type have not been carried out. In the particular case of NiO, it would be most intriguing to study the diffuse scattering near T_c as a function of beam energy; at high beam energies, the electrons should penetrate deeply enough to allow bulk critical fluctuations to be sampled, while only the outermost surface layer or two is explored at low energies. Such a study should be quite feasible, particularly since one may observe the half order Bragg peaks readily.

Another intriguing set of studies are the spin polarized photoemission experiments on the europium chalcogenides carried out by the Zurich group.[27] The data is interpreted to suggest that below the bulk ordering temperature, the outermost layers behave rather like a Heisenberg paramagnet. Unfortunately, it is not possible to obtain detailed information about the surface spin structure from the data. It would be quite fascinating to explore these surfaces with low energy electron beams, as Palmberg did for NiO. In the ferromagnetic materials, the use of spin polarized beams would be particularly useful to discriminate magnetic from non-magnetic scattering near the crystallographic Bragg peaks. The photoemission data has provided an impetus for the theoretical work on surface spin reconstruction,[17-19] although at this time the theory is far from explicit contact with the photoemission data.

We feel that there is one other experimental method which may yield useful information about surface magnetism, at low temperatures. In low temperature physics, a method of cooling He^3 relies on coupling between the He^3 nucleus and electron spins in a cooling salt.[28,29] Theoretical analyses show that the outermost layer or two of spins play the dominant role in providing the coupling, in the usual circumstances.[29] Thus, we have here a method of probing surface spins experimentally, if the phenomenon is exploited for this purpose rather than simply as a means of cooling liquid He^3. It has been proposed that this coupling should play a role in limiting the nuclear relaxation times in liquid He^3 in contact with a magnetic substrate.[29] This should offer a more direct method of studying the coupling when compared with thermal transport studies, since phonons do not influence spin relaxation in the liquid He^3. Recently Saito[30] has studied nuclear relaxation of liquid He^3 in contact with a magnetic salt which undergoes a phase transition at 0.43 K. Saito finds the phase transition has a dramatic effect on nuclear relaxation in the liquid, although the detailed interpretation is not yet clear. This is a question we shall investigate in the near future.

Thus, to summarize this section, the existing experimental data on magnetic properties of the outermost layers of crystals is quite sparse at the moment. The information to date offers an intriguing glimpse at surface spin behavior, and we hope the area will attract increasing attention from experimentalists in the future.

REFERENCES

† Supported by Grant No. AFOSR 76-2887 of the Air Force Office of Scientific Research, Office of Aerospace Research, U.S.A.F.

1. See Chapter 4 of C. Kittel, Quantum Theory of Solids, (John Wiley and Sons, New York, 1963).
2. H. Eugene Stanley, Introduction to Phase Transitions and Critical Phenomenon (Clarendon Press, Oxford, 1971).
3. See the article by R. J. Elliott and W. Marshall, Rev. Mod. Physics 30, 75 (1958).
4. D. L. Mills, Phys. Rev. B1, 264 (1970).
5. See Figure (2) of D. L. Mills, Journal de Physique, P. C1 in Supplement No. 4 to Volume 31 (1970).
6. These points are discussed in somewhat greater length in D. L. Mills, Comments on Solid State Physics IV, 28 (1972) and IV, 95 (1972).
7. See P. 105 of L. Landau and E. M. Lifshitz, Theory of Elasticity, (Pergamon Press, Oxford, 1959).
8. These characteristics of surface spin waves may be altered by the presence of surface anisotropy fields.
9. A. A. Maradudin and D. L. Mills, J. Phys. Chem. Solids 28, 1855 (1967).
10. D. L. Mills, Phys. B3, 3887 (1971)
11. D. L. Mills, Phys. Rev. B8, 4424 (1973) and R. A. Weiner, Phys. Rev. B8, 4427 (1973).
12. K. Binder and P. Hohenberg, Phys. Rev. B6, 3461. (1972), M. N. Barber, Phys. Rev. B8, 407 (1973), T. C. Lubensky and M. H. Rubin, Phys. Rev. Letters 31, 1469 (1973), Phys. Rev. B11, 4533 (1975).
13. K. Binder and P. Hohenberg, Phys. Rev. B9, 2194 (1974).
14. P. Kumar and K. Maki, Phys. Rev. B13, 2011 (1976).
15. W. Saslow and D. L. Mills, Phys. Rev. 171, 488 (1968), T. Wolfram and R. de Wames, Phys. Rev. 185, 762 (1969).
16. D. L. Mills, Physical Review Letters 20, 18 (1968). This calculation has been corrected in one important respect by F. Keffer and H. Chow, Phys. Rev. Letters 31, 1061 (1974). Keffer and Chow suggest an intimate connection between the surface spin flop and bulk spin flop transition, based on their improved calculation.
17. S. E. Trullinger and D. L. Mills, Solid State Communications 12, 819 (1973)
18. A. Blandin, Solid State Communications 13, 1537 (1973).
19. C. Demangeat and D. L. Mills, preprint entitled "Surface Spin Instabilities in the Heisenberg Ferromagnet".
20. See Figure 2(a) of Reference (17).
21. See P. 40 of G. A. Somorjai, Principles of Surface Chemistry (Prentice Hall, Englewood Cliffs, N.J., 1972).
22. C. Demangeat and D. L. Mills (to be published).
23. J. T. Yu, R. A. Turk and P. E. Wigen, Phys. Rev. B11, 420 (1975).
24. W. E. Tennant et al., Proc. of 20th Conference of Magnetism and Magnetic Materials (San Francisco, 1974).
25. P. W. Palmberg et al., Physical Review Letters, 21, 682 (1968).
26. T. Wolfram et al., Surface Science 28, 45 (1971).
27. K. Sattler and H. C. Siegmann, Phys. Rev. Letters 29, 1565 (1972), M. Campagna et al., AIP Conf. Proc. 18/2, 1388 (1974) and F. Meier et al., Solid State Communications 16, 401 (1975).
28. A. J. Leggett and M. Vuorio, J. Low Temp. Phys. 3, 359 (1970).
29. D. L. Mills and M. T. Beal-Monod, Phys. Rev. A10, 343 (1974) and Phys. Rev. A10, 2473 (1974).
30. S. Saito, Phys. Rev. Letters 36, 975 (1976).

ELECTRONIC STRUCTURES AND MAGNETIC PROPERTIES OF TRANSITION- AND NOBLE-METAL CLUSTERS

K. H. Johnson
Massachusetts Institute of Technology
Cambridge, Massachusetts 02139

ABSTRACT

The electronic structures and magnetic properties of small Cu, Ni, Pd, Pt, and Fe clusters are described on the basis of spin-polarized molecular-orbital calculations carried out by the self-consistent-field-Xα-scattered-wave (SCF-Xα-SW) method. The results for the clusters are compared with the band structures, densities of states, and magnetic properties of the corresponding bulk crystalline metals and are used as a basis for explaining the properties of the active centers of supported metal catalysts. The effects of alloying on cluster electronic structure and magnetism are described for Cu-Ni and Cu-Fe systems, respectively. The possible implications of these results on the local bulk and surface electronic structures of Cu-Ni alloys, the catalytic activity and selectivity of supported bimetallic clusters, and the Kondo problem in dilute Cu-Fe alloys are discussed.

EXPERIMENTS ON MAGNETIZATION OF CHARGED FERROMAGNETIC METAL SURFACES

G. Bayreuther, Horst Hoffmann, and J. Reffle
Universität Regensburg, 84 Regensburg, W-Germany

ABSTRACT

Ferromagnetism in transition metals is very often discussed by means of the itinerant electron model and the related Slater-Pauling-curve. In this model the density of states of 4s and 3d electrons at the Fermi level with spin up and spin down is responsible for the magnetic moment of the sample. In the present inverstigations we tried to change the Fermi level by electric charging of the sample, following former experiments of Steenbeck[1]. The ferromagnetic sample forms one of the electrodes of a capacitor. Its surface can be charged with electrons thus changing the position of the Fermi level in a very thin (≈ 0.5 Å) surface layer. Changes of the magnetic moment therefore are to be expected only for the first atomic layer at the surface of the ferromagnetic sample. To observe the related moment of the whole sample requires experiments with very thin films.

Two sets of experiments were carried out:

i) Mössbauer experiments to observe a shift in the hyperfine field splitting due to charging of the surface
ii) high sensitivity magnetic balance measurements to detect a change of the magnetic moment of the sample during charging.

The Mössbauer experiment needed an increase of sensitivity by about two orders of magnitude compared to the method usually applied in Mössbauer investigations. We used the ferromagnetic film as absorber and registered the conversion and Auger electrons by a 2π-proportional counter. The sensitivity then was large enough to detect 1/100 of a paramagnetic monolayer of iron.

A magnetic balance was constructed to be used in a zero methode. By charging a capacitor with one ferromagnetic electrode the change of the magnetic moment of the sample could by investigated by lock-in technique.

The absolute sensitivity of the balance was 10^{-8}g, allowing to observe changes of the magnetic moment of our arrangement of $\Delta m = 10^{-8}$Gauß cm^3.

If we assume all electrons and holes, which can be added to the surface, to contribute one Bohr's magneton to the magnetic moment of the sample then the sensitivity of both arrangements (Mössbauer experiment and magnetic balance) is large enough to detect one tenth of the expected effect.

At present none of both measurements show a change of the magnetic moment due to the charging.

That means either the electrons are bound to surface states and do not enter the d- and s-bands or the contribution of the additional electrons to the magnetic moment is less than ten percent of the value assumed by the simple model, in contrast to the result of Steenbeck. Details of the experiments will be published elsewhere.

1 K. Steenbeck. J.de Physique 32, C1-1096 (1971)

RKKY OSCILLATIONS IN THE SPIN DENSITY AT THE SURFACE OF A FERROMAGNETIC METAL

R. L. Kautz
Francis Bitter National Magnet Laboratory,* Massachusetts Institute of Technology, Cambridge, Massachusetts 02139 and Magnetic Theory Group, Department of Physics, Northwestern University, Evanston, Illinois 60201

Brian B. Schwartz
Francis Bitter National Laboratory,* Massachusetts Institute of Technology, Cambridge, Massachusetts 02139

ABSTRACT

In recent years great progress has been made in the understanding of the surface behavior of simple metals and semiconductors.[1] The experimental techniques developed for surface studies have recently been used to examine the surfaces of magnetic materials. In the case of magnetic transition metals, some experiments yield results which do not agree with simple models for transition metals. We have developed a model which considers the surface of a spin-polarized electron gas, in which the conduction electrons are polarized either by an external uniform magnetic field or by an internal molecular field.[2] The two models yield results which can be applied to the surface susceptibility of alkali metals and the conduction ferromagnetic gadolinium. The results obtained are accurate self-consistent solutions based on the density-functional formalism of Hohenberg and Kohn[3] and Kohn and Sham[4] as extended to the spin polarized case by von Barth and Hedin[5] and Rajagopal and Callaway.[6] Our treatment neglects the periodic potential and omits band structure effects, however, in principle the density functional formalism can be extended to include the periodic potential. We obtain the detailed behavior of the separate up- and down-spin electrons at the surface including the Friedel oscillations as a function of the bulk magnetization. The work function, surface magnetization and surface energy difference as a function of the magnetization is obtained. Although our model omits complex band structure effects of transition metal surfaces, it can be used to describe ferromagnetic gadolinium where the electron polarization is due to an exchange interaction with the 4f moment.

REFERENCES

* Supported by the National Science Foundation.

1. N.D. Lang and W. Kohn, Phys. Rev. B 3, 4555; J.A. Applebaum and D.R. Hamann, Phys. Rev. B 6, 2166 (1972).
2. K.L. Kautz and B.B. Schwartz, a complete paper on this subject will appear in Phys. Rev. B (1976).
3. P. Hohenberg and W. Kohn, Phys. Rev. 136, B864 (1964).
4. W. Kohn and L.J. Sham, Phys. Rev. 140, A1133 (1965).
5. U. von Barth and L. Hedin, J. Phys. C 1629 (1972).
6. A.K. Rajagopal and J. Callaway, Phys. Rev. B7, 1912 (1973).

THE MICROMAGNETICS OF IDEALLY SOFT FERROMAGNETIC CYLINDERS AND SPHERES

A. Arrott, B. Heinrich, M. Press and T.L. Templeton
Department of Physics, Simon Fraser University
Burnaby, British Columbia, Canada V5A 1S6

ABSTRACT

The forms of the divergence-free unit vector fields of spheres and cylinders are discussed. The symmetries of the minimum energy states of ferromagnetic bodies without anisotropy are described. A speculative picture of the magnetization process is put forth.

INTRODUCTION

The terms in the micromagnetic expression for the energy of ferromagnetic bodies are well formulated mathematically. Yet there does not exist, as far as the authors are aware, a description of a complete magnetization reversal process in a finite geometry with dimensions greater than the critical dimensions for single domain behavior. The best descriptions invoke domain walls without describing their nucleation. The problem lies in the difficulty of handling a problem in any detail when a three dimensional grid is needed to solve the differential equations even with the fastest of modern computers. In order to stay within the limits of computing techniques we have examined problems for finite geometry with cylindrical symmetry[1].

Inasmuch as the essential problem is the competition between exchange energy trying to maintain uniform alignment and dipole-dipole torques trying to force div $\vec{M} = 0$, we have considered problems in which the anisotropy and magnetostriction are neglected. The prototype of such problems is Néel's treatment of the energy of a demagnetized sphere. He calculated the energy for $\vec{M} = M_S\hat{\phi}$. This satisfies div $\vec{M} = 0$ and $\vec{M}\cdot\hat{n} = 0$, where $\hat{n}$ is the surface normal. Néel put a hole of radius a down the axis, where a is of the order of an atomic distance. This was to avoid a logarithmic infinity in the exchange energy

$$E_{ex} = A\int(\text{curl } M)^2 dV \tag{1}$$

Néel's solution is an equilibrium solution of the torque equations of micromagnetics, but it is neither a minimum energy nor a stable state of the sphere. Yet it does show that torque-free divergence-free solutions do exist.

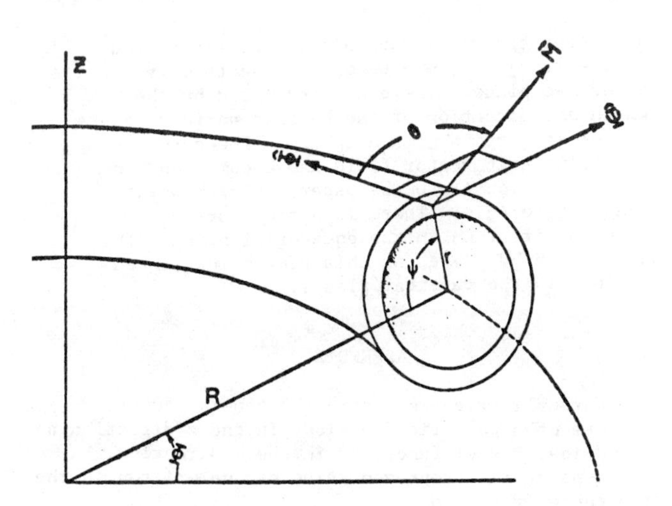

Fig. 1 The Coordinate system for the toroidal shell.

TOROIDAL SHELL

Arrott, Heinrich and Bloomberg[2] have discussed two variations on Neel's sphere with a hole down its axis. One of these is the cylinder with a hole down its axis and the other is the toroidal shell which has two holes, one along the axis of cylindrical symmetry and the other in the cross section of the toroid. The coordinate system for the toroidal shell is shown in fig. 1. In the absence of an applied field the magnetization is in the $\pm\hat{\Phi}$ direction as long as r is small compared to R. The constraints div $\vec{M} = 0$ and $\vec{M}\cdot\hat{n} = 0$ have a unique solution:

$$\vec{M} = M_s(\hat{\psi}\sin\theta + \hat{\Phi}\cos\theta) \tag{2}$$

where

$$\sin\theta(\psi) = \sin\theta(0)\,\frac{R - r}{R - r\cos\psi} \tag{3}$$

The exchange energy is a function of the single variable $\theta(0)$, the angle the magnetization makes with the $\hat{\Phi}$ axis for $\psi = 0$. For large R/r the energy is minimized for $\theta(0) = 0$ and the magnetization simply curls about the z axis. But for $R/r < 1.54$ the energy is lower if $\theta(0) = \pm\pi/2$. The magnetization is up (or down) at $\psi = 0$, but not down (or up) at $\psi = \pi$ as $\sin\theta(\pi) = \pm(R - r)/(R + r)$. Hysteresis in the toroidal shell occurs because to reverse the magnetization one must produce magnetic charge to get $\theta(\pi)$ to go beyond the angle given above.

CONCENTRIC TOROIDAL SHELLS

Next visualize a space occupied by concentric toroidal shells all with the same R. At $\psi = 0$ the magnetization would be in the $\pm\hat{\Phi}$ direction. At the z axis the magnetization would be in the $\pm\hat{z}$ direction. If in each shell Eq. 3 applied, this would be a divergence-free unit vector field. This solution is over-constrained in that $\vec{M}\cdot\hat{n} = 0$ need only apply on the outer surface. Nevertheless this does illustrate the basic symmetry of the solutions we seek. With this constraint relaxed, the divergence-free solutions are no longer unique. One is looking for divergence-free solutions that minimize the exchange energy. The Neel solution for the sphere or cylinder gives a logarithmic singularity all along the axis. The solutions we seek will have the magnetization along the axis at the axis and the only singularities will be at the north and south poles. (During the irreversible part of a magnetization process these singularities will likely propagate down the axis.)

DIVERGENCE-FREE FIELDS

There is of course no difficulty in writing down divergence-free fields until one restricts these to being unit vector fields. To create a unit vector field that is divergence-free one proceeds as follows. Find the set of divergence-free fields that satisfy the boundary conditions, have cylindrical symmetry and lie in the R-z plane. Normalize the field so that its maximum magnitude does not exceed unity. Then add a component in the $\hat{\phi}$ direction that makes it a unit vector. The component in the $\hat{\phi}$ direction is automatically divergence-free because it does not depend on the ϕ coordinate in a system with cylindrical symmetry.

The vector fields that do this for spherical geometry are the divergence-free vector spherical harmonics of the first kind with index m = 0. These are

$$\vec{A}^{\circ}_{\ell}(\vec{qr}) = \left(\frac{\ell}{2\ell+1}\right)^{\frac{1}{2}} j_{\ell+1}(qr)\, \vec{Y}^{\circ}_{\ell,\ell+1} - \left(\frac{\ell}{2\ell+1}\right)^{\frac{1}{2}} j_{\ell-1}(qr)\, \vec{Y}^{\circ}_{\ell,\ell-1}$$

where the j_ℓ are spherical Bessel functions and the $\vec{Y}^{\circ}_{\ell,\ell+1}$ and $\vec{Y}^{\circ}_{\ell,\ell-1}$ are vector spherical harmonics. The properties of these functions are discussed in detail by Ford and Werner in their study of helicon oscillations in a sphere[3]. We have programmed the computer to produce pictures of these vector fields. The spherical harmonics with $\ell = 0$ vanish, but this is just Néel's solution if one proceeds to complete the unit vector field by adding the component in the $\hat{\phi}$ direction. For a particular ℓ one needs the boundary conditions to determine the q's. For $\vec{M}\cdot\hat{n} = 0$ on a spherical surface, it is required that the q's produce zeros of $j_\ell(qr)$ at the radius r. The members of the set of q's for each ℓ are labelled by the index n_ℓ. The divergence-free unit vector field generated from the vector spherical harmonic with indices $\ell = 1$ and $n_\ell = 1$ has the symmetry of the toroidal shells as shown in fig. 2. This field has a lower energy than Néel's solution in that it diminishes the coefficient of the logarithmic divergence of the energy on the axis. For at the origin of the sphere the magnetization points straight along the axis. To wipe out the singularity all along the axis requires an infinite set of vector spherical harmonics. These produce a unit vector all along the axis and the singularities occur only at the poles. Such sets are not unique and it is still necessary to minimize the exchange energy, Eq. 1.

RELAXATION METHODS

Our present approach is to use the symmetry we gather from this analysis to generate starting fields for relaxation calculations. These use the torque equations generated by minimizing the exchange energy with the constraint div $\vec{M} = 0$. In particular we have been concentrating on the behavior of the cylinder in that the coordinate system is then more suitable for computer analysis. Some notes concerning cylinders are given in the appendix. Though we have had success previously using relaxation methods for solving liquid crystal elasticity problems[4] with two simultaneous partial differential equations, so far we have not successfully handled the three simultaneous equations employed here. The problem seems to lie in the nature of the Lagrangian multiplier used in handling the constraint. Some of us remain optimistic.

SPECULATION

Even without the detailed solutions one can visualize how a cylinder or sphere might magnetize. Starting from the demagnetized state in which the magnetization is up along the axis, the sphere would magnetize in a field in the up direction by a general rotation of the vectors. The line of pure circulation about the $\hat{z}$ axis would move in the process to the outside surface. But if the field were applied in the opposite direction the line of pure circulation would shrink towards the axis and the energy would increase indefinitely. Yet the singular points at the poles have finite energies which simply double if they are moved inward from the surfaces. At a critical field the cylinder (or sphere) becomes unstable with respect to keeping the singularities at the surface. Then the singularities would propagate down the axis, reversing the magnetization and repelling the line of pure circulation toward the outside surface. We are not optimistic about how long it will take to find out if this picture is valid.

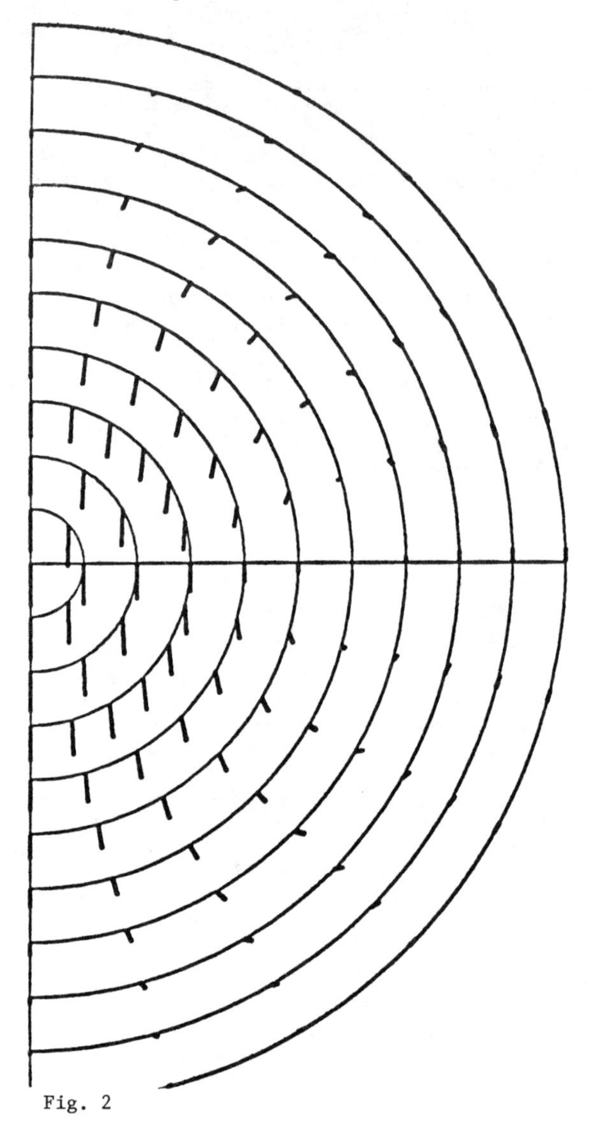

Fig. 2

The vector spherical harmonic of the first kind for $\ell = 1$, $n_\ell = 1$, m = 0 has components only in the $\hat{\rho}$-$\hat{z}$ plane. These are indicated by the length and direction of the bars at various points in a cross section of the sphere. The corresponding unit vector field has a component in the $\hat{\phi}$ direction (into the paper) at each point except the origin. There is a ring purely in the $\hat{\phi}$ direction in the equatorial plane. The pattern can be viewed as this ring being linked by rings in the vertical planes.

APPENDIX

For cylindrical symmetry the general functions are products of hyperbolic functions in the z direction and cylindrical Bessel functions in the ρ direction. A simple polynomial representation can be written in the divergence-free form

$$\vec{A} = \sum\sum \alpha_{h\ell}\left[(h+1)\rho^{h-1} z^{\ell}\hat{z} - \ell\, \rho^{h} z^{\ell-1}\,\hat{\rho}\right]$$

where ρ and z are cylindrical coordinates. The unit

vector is again formed by normalizing A to unity at its maximum length and adding the vector $(1 - A^2)^{1/2}\,\hat{\phi}$.

The simplest unit vector field that is divergence-free and satisfies $\vec{M}\cdot\hat{n} = 0$ on the surface of a cylinder is

$$n_z = \left[1 - \left(\frac{2z}{z_o}\right)^2\right]\left[1 - \frac{3\rho}{2\rho_o}\right]$$

$$n_\rho = \frac{\rho_o}{z_o}\cdot\frac{2z}{z_o}\cdot\frac{\rho}{\rho_o}\left(1 - \frac{\rho}{\rho_o}\right)$$

$$n_\varphi = \left(1 - n_z^2 - n_\rho^2\right)^{1/2}$$

This has the symmetry of the concentric toroidal shells. Though it has lower energy than Néel's solution it is not at all torque free.

The singularity at the poles is the same for a cylinder or a sphere. If the origin of coordinates is taken at the pole, the field in the region of the pole can be written as a superposition of terms like

$$n_z^{h\ell} = \left(\frac{1}{h\rho}\tanh \ell\rho + \frac{\ell}{h}\,\mathrm{sech}^2\ell\rho\right)\tanh h\rho$$

$$n_\rho^{h\ell} = -\tanh \ell\rho\ \ \mathrm{sech}^2 hz$$

In the limit as ℓ and h get large, the singularity along the z = 0 axis reduces to a point and has finite energy.

REFERENCES

1. B. Heinrich and A.S. Arrott, AIP Conference No. 24, Magnetism and Magnetic Materials - 1974, p. 702 (1975).

2. A.S. Arrott, B. Heinrich, D.S. Bloomberg, IEEE MAG-10, p. 950 (1974).

3. G.W. Ford and S.A. Werner, Phys. Rev. 8, 3704 (1975).

4. M.J. Press and A.S. Arrott, J. de Phys. 36, C1-177; J. de Phys. 37, April (1975) in press.

EFFECT OF THE SECOND ANISOTROPY CONSTANT K_2 ON THE ORIENTATION DEPENDENCE OF DOMAIN WALL ENERGY*

Robert S. Williams, Juan C. Figueroa, and C. D. Graham, Jr.,
Dept. of Metallurgy and Materials Science and Laboratory for Research on the Structure of Matter, University of Pennsylvania, Philadelphia, Pa. 19174

ABSTRACT

In a well-known paper, Lilley[1] worked out the dependence of the domain wall energy and thickness on the crystallographic orientation of the wall, using the usual continuum approximation and considering only a single anisotropy constant K_1. Using computer methods, we have extended Lilley's work by including the effects of the second anisotropy constant K_2, and report here the results for 180° walls. The inclusion of the K_2 term leads to the following conclusions: For ⟨100⟩ easy directions, there is a range of values of K_2/K_1 for which {110} rather than the usual {100} is the minimum-energy plane of a 180° domain wall. For ⟨111⟩ easy directions, {110} is always the minimum-energy plane. For ⟨110⟩ easy directions, which do not occur if $K_2=0$, {111} is the minimum -energy plane except when $|K_2/K_1|>13$, when {100} becomes minimum.

INTRODUCTION

In the standard continuum approximation the energy per unit area of a magnetic domain wall in an unstrained crystal is given by [1]:

$$\gamma = 2\sqrt{A}\ \sin\theta' \int_{\phi_1}^{\phi_2} f_a^{\frac{1}{2}}\ d\phi \tag{1}$$

where A is the exchange constant, θ' is the angle between the magnetization vector $\vec{M}$ and the wall normal, and ϕ_1, ϕ_2 are the angles made by the projection of $\vec{M}$ initial and $\vec{M}$ final on some chosen reference direction in the plane of the wall. Physically, ϕ_1 and ϕ_2 correspond to $\vec{M}$ lying parallel to easy directions. f_a is the magneto-crystalline anisotropy energy, which can be represented in a cubic crystal by [2]

$$f_a = K_1(\alpha_1^2\alpha_2^2 + \alpha_2^2\alpha_3^2 + \alpha_3^2\alpha_1^2) + K_2(\alpha_1^2\alpha_2^2\alpha_3^2) + \ldots \tag{2}$$

where α_1, α_2, α_3 are the direction cosines of $\vec{M}$ with respect to the cubic axes.

Lilley[1] thoroughly examined the dependence of wall energy and thickness on the crystallographic orientation of the wall for all the physically possible cases with $K_2=0$. We have extended Lilley's work to cases where K_2 is not zero, and in this paper we report the results for 180° walls. Because of the large magnetostatic energy associated with free poles, we make the usual assumption that θ' is constant throughout the wall. For 180° walls, $\theta'=90°$. The expression for γ can then be written as:

$$\gamma = 2\sqrt{AK_1} \int_0^{180}\Big[(\alpha_1^2\alpha_2^2 + \alpha_2^2\alpha_3^2 + \alpha_3^2\alpha_1^2) + K_2/K_1(\alpha_1^2\alpha_2^2\alpha_3^2)\Big]^{\frac{1}{2}} d\phi \tag{3}$$

We can thus examine the effect of the ratio K_2/K_1 on the wall energy, for any possible crystallographic orientation of the wall.

PROCEDURE

The easy direction [PQR], the wall normal, [HKL], and a vector [ABC] equal to the cross product of [HKL] and [PQR], each specified with respect to the cubic crystallographic axes, define a set of cartesian coordinate axes, X', Y', and Z' respectively (see fig. 1). $\vec{M}$ then rotates from X' to -X' in the X'-Y' plane, and ϕ is the angle between $\vec{M}$ and the X' axis. The direction cosines of $\vec{M}$, which are just the coordinates of a unit vector in the $\vec{M}$ direction, are then $\cos\phi$, $\sin\phi$, and 0. The X', Y', Z' coordinate axis can be rotated into the cubic crystallographic axes by a unitary transformation represented by the matrix:

$$T = \begin{pmatrix} P & A & H \\ Q & B & K \\ R & C & L \end{pmatrix} \tag{4}$$

Applying this transformation to the vector ($\cos\phi$, $\sin\phi$, 0) yields the direction cosines with respect to the cubic crystallographic axes:

$$T \begin{pmatrix} \cos\phi \\ \sin\phi \\ 0 \end{pmatrix} = \begin{pmatrix} \alpha_1 \\ \alpha_2 \\ \alpha_3 \end{pmatrix} \tag{5}$$

A PDP-8 computer was programed to numerically integrate eq.(3) for various values of K_2/K_1, and for a variety of domain wall orientations, by computing T and solving for the α's for small ($\pi/250$) increments in ϕ. The zero of energy was taken to be the state with all moments parallel to an easy axis.

There are three possible easy axes consistent with cubic symmetry: <100>, <111>, <110>. A 180° domain wall can lie in any crystallographic plane containing the easy axis. A {110} plane is a possible wall for each of the three easy axes; accordingly we designate the wall orientation in each case by its angle of rotation (Θ) away from {110} about the easy axis. The domain wall energy is given as γ/γ_0, where $\gamma_0 = 2\sqrt{AK_T}$, with $K_T = \sqrt{K_1^2 + K_2^2}$, and the ratio K_2/K_1 may be expressed as $\Psi = \tan^{-1}(K_2/K_1)$.

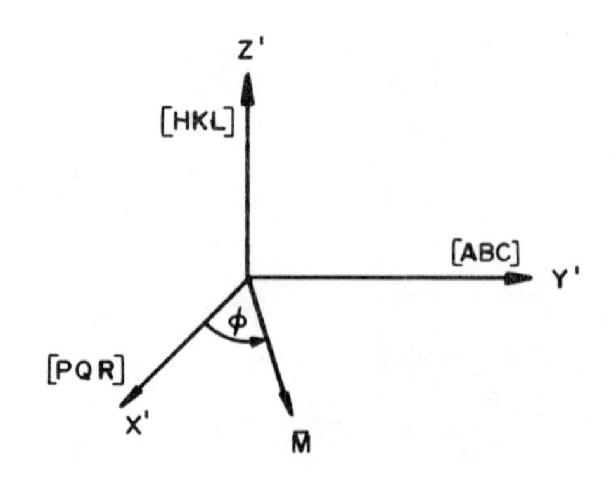

Fig. 1. Axes defining $\vec{M}$ in the domain wall whose normal is Z'.

RESULTS

Figs. 2-4 are plots of γ/γ_0 vs Θ, for various values of Ψ. The minimum in γ/γ_0 is the lowest energy wall orientation, and is the orientation taken by a 180° wall in an infinite strain-free crystal.

Fig. 5 is a polar diagram of the reduced wall energy γ/γ_0 for certain low-index planes, plotted as the radius r, and $\Psi = \tan^{-1}(K_2/K_1)$. A unit circle on this diagram represents all possible values of K_2/K_1, with the normalization condition $\sqrt{K_1^2 + K_2^2} = 1$.

Briefly the features of these calculations are:

(a) <100> easy direction (Fig. 2)
$K_1 \geq 0$, $-9 < K_2/K_1 < +\infty$

For $K_2=0$, the {100} plane has minimum energy. This remains true except when $K_2/K_1<-6.6$, where {110} becomes the plane of minimum energy. At $K_2/K_1 \approx -6.6$, the wall energy varies by no more than 4% for any possible orientation Θ.

(b) <111> easy direction (Fig. 3)
$K_1 \leq 0$, $K_2/K_1 > -9/4$
or $K_1 \geq 0$, $K_2/K_1 < -9$

The minimum energy plane is always {110} and the variation of energy with orientation is always small.

(c) <110> easy direction (Fig. 4)
$K_1 \leq 0$, $K_2/K_1 < -2.5$

This case arises only when $K_2 \neq 0$, and hence was not considered by Lilley. The {111} plane has minimum energy, except when $K_2/K_1<-13$, where {100} becomes minimum. The variation of energy with wall orientation is particularly large for this case (<110> easy direction).

For any given values of K_1 and K_2 the minimum energy 180° wall can be determined from Fig. 5. The absolute energy is just the energy shown times $2\sqrt{AK_T}$.

We have not considered any magnetostriction contribution to the energy and the calculation assumes an isolated wall in an infinite crystal. In a real situation, magnetostatic energy and total wall area minimization would have to be included[3,4].

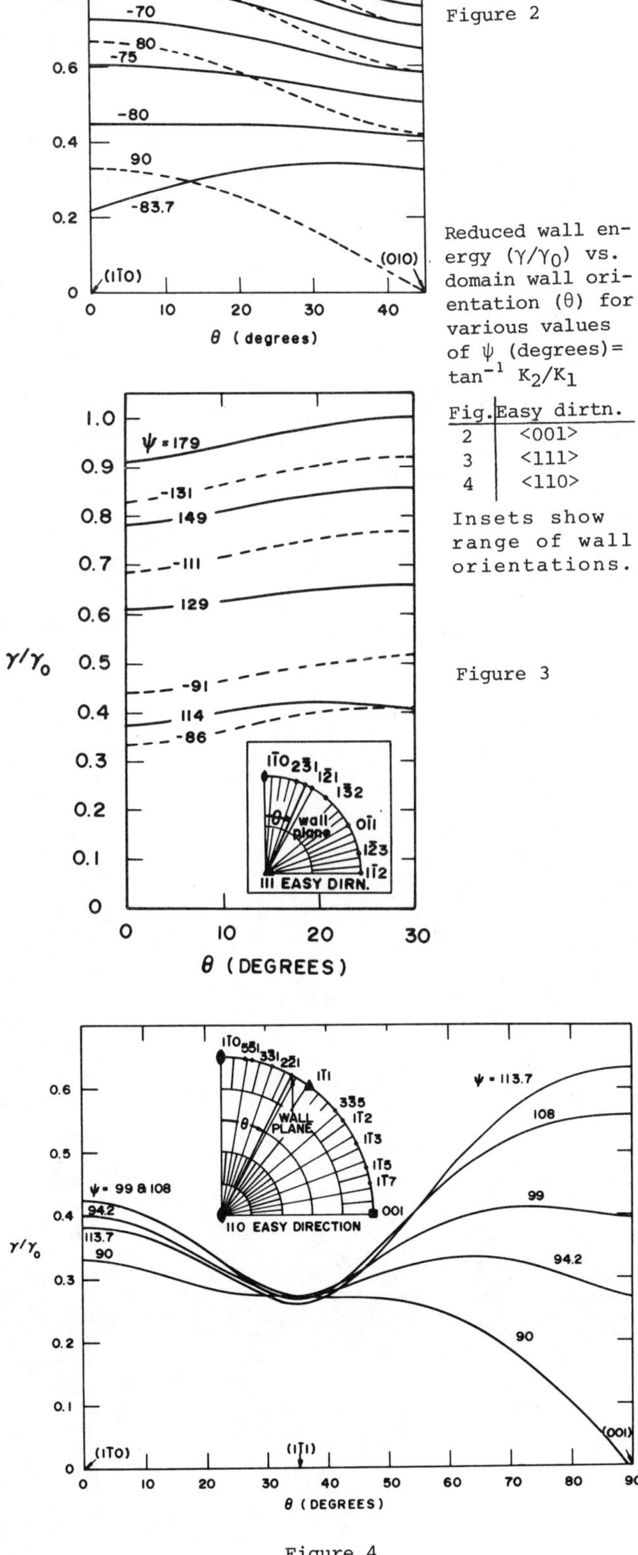

Figure 2

Reduced wall energy (γ/γ_0) vs. domain wall orientation (θ) for various values of ψ (degrees) = $\tan^{-1} K_2/K_1$

Fig.	Easy dirtn.
2	<001>
3	<111>
4	<110>

Insets show range of wall orientations.

Figure 3

Figure 4

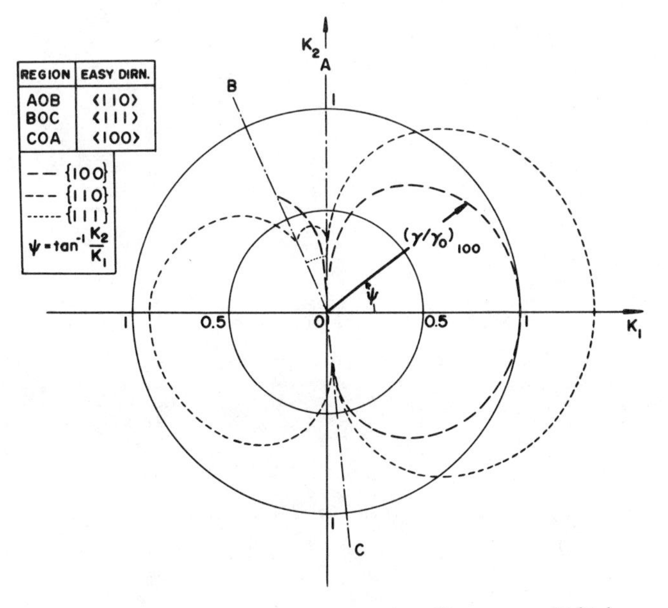

Fig. 5. Polar diagram of reduced wall energy (γ/γ_0) versus ψ

APPLICATION TO REAL MATERIALS

The cases of interest, in the sense that they lead to qualitatively different behavior from the usual $K_2=0$ situation, are $K_1>0$, $K_2>0$, $6.6<|K_2/K_1|<9$; and $K_1<0$, $K_2>0$, $2.5<K_2/K_1<\infty$. Since K_2 is generally small compared to K_1 in cubic materials, these cases do not arise frequently. The first of the two cases does occur near 500°K in pure nickel, when K_1 changes sign from negative to positive while K_2 is negative[5]. A similar situation arises below room temperature for several compositions in the systems $MnTi_xFe_{2-x}O_4$[6] and $Fe^{2+}_{1+x}Ti_xFe^{3+}_{2-2x}O_4$.[7] We are not aware of any examples of the second of the two cases, although we have not made an extensive search of the literature.

REFERENCES

* Work supported by National Science Foundation.

1. B.A. Lilley, Phil. Mag. 42, (1950), p. 792.
2. B.D. Cullity, Introduction to Magnetic Materials, Addison-Wesley Publishing Co., Reading, Mass., 1972, p. 211.
3. C.D. Graham, Jr., P.W. Neurath, J. Appl. Phys., 28, 888-891, (1957).
4. John E.L. Bishop, IEEE Trans. on Magnetics, Vol. Mag.-12, No. 3, May (1976).
5. G. Aubert, J. Appl. Phys. 39, 504, (1968).
6. J. Smit, F.K. Lotgeving, R.P. van Stapele, J. Phys. Soc. Japan, 17, Suppl.B-1,268 (1962).
7. Y. Syono, Japan. J.Geophys. 4, 71 (1965).

A MODEL FOR THE CALCULATION OF THE INDUCTION ABOVE THE KNEE OF THE MAGNETIZATION CURVE FROM POLE DENSITY TEXTURE MEASUREMENTS

Jack M. Shapiro, Inland Steel Research Laboratories, East Chicago, Indiana 46312

ABSTRACT

Nominally non-oriented electrical steels can have a range of permeability at high magnetic inductions. It is shown that such variation is due to the actual non-randomness of the crystal texture. The theoretical relationship between the knee of the B-H curve and the orientation of a single crystal is adapted to the case where the texture is measured by the X-ray pole density method at a series of depths in sheet samples. The results are given by:

$$\frac{B_k}{B_s} = \sum_{j=1}^{m} A_j N_j I_j / I_{rj} \Bigg/ \sum_{j=1}^{m} N_j I_j / I_{rj} \quad ,$$

where B_k = theoretical value of the knee of the magnetization curve; B_s = saturation induction; m = total number of diffracting planes, each of index j; N_j = multiplicity of occurrence of plane j; I_j, I_{rj} = intensity of diffraction from all grains of orientation j in the unknown sample and random samples, respectively; A_j = averaged reciprocal of the sum of direction cosines between the applied field and the cube axes for orientation j. Experimentally, B_k is obtained by extrapolating the induction from fields up to 100 Oe. The theory is confirmed with texture data from a relatively fine-grained 1% Si steel.

INTRODUCTION

It is known that nominally non-oriented electrical steels can have a wide range of permeability at high magnetic inductions. Such variations are due to the non-randomness of the distribution of orientations of the individual grains in the material. In the following, an expression relating the texture measured by the X-ray pole density technique to the knee of the d-c magnetization curve of strip steel samples is derived and tested.

The theoretical relationship between the magnetization curve and the orientation of single crystals or polycrystalline samples with strong textures has been confirmed experimentally. The magnetic behavior of a single crystal rod at low inductions can be understood by assuming that the volumes of domains with magnetization in each of the six easy directions are so proportioned that the sum of the $\vec{B}$ vectors is equal to the net magnetization. At a sufficiently high induction, the magnetization is distributed only among those three directions closest to the applied field, as shown in Figure 1, so that

$$\vec{B} = B\alpha_1\vec{X}_1 + B\alpha_2\vec{X}_2 + B\alpha_3\vec{X}_3 \tag{1}$$

where $\vec{B}$ is the net magnetization of the crystal, and α_i are the direction cosines of $\vec{B}$ with the 3 easy directions: X_1, X_2, X_3. Each component is also equal to f_iB_s, where f_i is the volume fraction of domains with magnetization in the direction $\vec{X}_i$ and B_s is the saturation induction.

$$f_iB_s = B\alpha_i. \tag{2}$$

$$\text{Since} \quad \sum_{i=1}^{3} f_i = 1,$$

$$\text{therefore} \quad B = \frac{B_s}{\alpha_1 + \alpha_2 + \alpha_3} \tag{3}$$

Equation (3) can be considered to define the "knee" of the magnetization curve. This relation was derived by Gorter,[1] and experimentally noticed by Kaya.[2] The calculation for single crystals was also extended to the entire magnetization curve beyond the knee.[3] The derivation was shown to apply to polycrystalline material by Stewart[4] for relatively moderate inductions above the knee and to cube-on-face textured silicon-iron sheet by Foster and Kramer.[5]

THEORY

In the examples given above, the orientation and size, if necessary, of every grain could be determined because the grain diameter was greater than the thickness of the sheet material. In finer grained, non-oriented electrical steels, such detailed orientation measurements cannot be made; however, an approximate measure of the volume fraction of several orientations can be obtained using the pole density technique. Here a sheet sample is rotated such that the sheet surface is always in the reflecting plane of X-rays. The intensity from a given reflecting plane {hkl} is compared with that of a random sample. Since the depth of penetration of MoKα X-rays in iron is <0.001", then to determine the texture throughout the sample, the measurements must be made at several depths by grinding and polishing. Aside from the limitations due to the "short-sightedness" of X-rays, there is another approximation involved. A particular grain may be so oriented that none of its planes will diffract. Thus, the orientation of this grain will not be considered. Equation (3) will be adapted to a polycrystalline sample in which the orientation of all grains can be determined and then to the experimental configuration described above with its specific approximations.

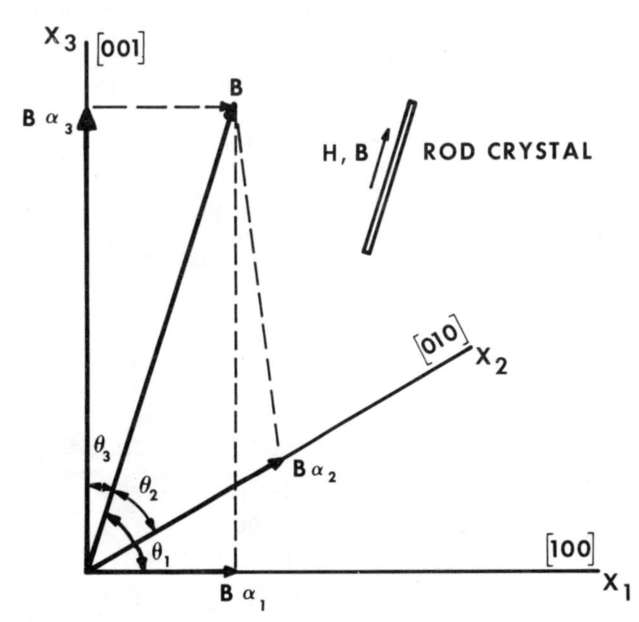

FIGURE 1 RESOLUTION OF THE MAGNETIC INDUCTION VECTOR ALONG THE <100> DIRECTIONS FOR A LONG ROD CRYSTAL

Let each grain of a sample of n grains be labeled with subscript k. For each grain, we can define

$$A_k = \frac{1}{\alpha_1 + \alpha_2 + \alpha_3}$$

and a volume, V_k. Hence for all grains we define the average value of B/B_s at the knee as:

$$J = \sum_{k=1}^{n} V_k A_k \Big/ \sum_{k=1}^{n} V_k \qquad (4)$$

For the pole density measurement at the depth of interest, using m diffracting planes, the volume fraction of material with plane {hkl} or j parallel to the sheet surface is approximated by:

$$\frac{N_j I_j}{I_{rj}} \Big/ \sum_{j=1}^{m} N_j I_j / I_{rj}$$

where I_j = intensity of diffraction from all grains with orientation j in the sample, I_{rj} = corresponding intensity of diffraction from the random sample, and N_j = multiplicity of occurrence of plane j.[6] That is, we approximate our sample by a finite number of orientations, M, each of volume fraction, V_j.

The orientation of each of the m regions is not fixed by the pole density measurement. Only the normal has been determined. Thus, we do not know how to weight the direction <h'k'l'> in the plane {hkl} in order to redefine A_k into A_j. If we assume that all directions are distributed randomly in that plane, then we should average $\frac{1}{\alpha_1 + \alpha_2 + \alpha_3}$ for all directions that are possible in the plane. Each plane will have a particular variation of direction around the normal and hence a different average, A_j. For example, the average

$$A\{100\} = \frac{1}{2\pi} \int_0^{2\pi} \frac{d\theta}{|\sin\theta| + |\cos\theta|} = 0.794.$$

For any other plane, this calculation becomes more difficult, in part because $\alpha_1, \alpha_2, \alpha_3$ must be positive; i.e., there are discontinuities in the slope of the α's at certain angles θ, in the plane. We have chosen to approximate the average $A_j(\theta)$ in all directions in a plane with the mean of highest and lowest $A_j(\theta)$. These values correspond to the easiest and hardest directions in which to magnetize a grain with the given {hkl} plane parallel to the sheet surface. Table I lists the easy and hard directions for the planes of interest, the values of A_j and the means. It is seen that $A_{\{100\}}$ has a different value from that calculated above. It is also seen that {211}, {222} and {321} have relatively low values of A, while {220}, {310} and {420} have higher values of A. One can thus assign the simple description "good" or "bad" to any of these planes with reference to the effect of its presence on the knee of the magnetization curve.

The average value of B/B_s at the knee of the B-H curve can now be expressed in terms of data generated from pole density measurements as:

$$J = \sum_{j=1}^{m} A_j \frac{N_j I_j}{I_{rj}} \Big/ \sum_{j=1}^{m} \frac{N_j I_j}{I_{rj}} \qquad (5)$$

If for every plane $I_j = I_{rj}$, the sample appears random and the value of B/B_s for a random sample, J_r, is 0.725.

EXPERIMENTAL RESULTS AND DISCUSSION

X-ray and magnetic data suitable for analysis were obtained on hot and cold rolled (and annealed) 1% Si steels. These materials showed considerable variation both in magnetic properties with direction and in pole density with depth.[7] Samples ground to four depths between the surface and mid-plane were used for the X-ray measurements. Values of $B_k = JB_s$ were calculated using Equation (5) and are listed in Table II.

TABLE II

Predicted and Measured Values of B, in Tesla

Predicted B_k Equation 5	Predicted B(H) Equation 7	Measured $\bar{B}$
1.534	1.765	1.769
1.499	1.744	1.747
1.494	1.741	1.737
1.524	1.759	1.757
1.529	1.762	1.752
1.533	1.764	1.768
1.511	1.751	1.757
1.493	1.740	1.749

TABLE I

Values of A for Various Planes

Diffracting Plane j	Easy Direction	Hard Direction	Max. Aj	Min. Aj	Mean Aj	Nj	NjAj
(200)	[100]	[110]	1.000	0.707	0.854	6	5.12
(211)	[120]	[111]	0.745	0.578	0.661	24	15.86
(220)	[100]	[111]	1.000	0.578	0.789	12	9.47
(222)	[110]	[112]	0.707	0.614	0.660	8	5.28
(310)	[100]	[265]	1.000	0.621	0.810	24	19.44
(321)	[103]	[111]	0.746	0.578	0.662	48	31.78
(420)	[100]	[365]	1.000	0.597	0.799	24	19.18
(411)	[140]	[122]	0.825	0.600	0.713	24	17.11

Strip samples were cut parallel, transverse and at 45° to the rolling direction for d-c magnetic testing at 8000 ampere/metre (100 Oe) with a Fahy permeameter and an electronic hysteresigraph. 8000 A/m was chosen as the field strength in order to exclude the effects of grain size, carbides and other inclusions on the induction measurement. Average values were calculated as $\bar{B} = \frac{1}{4}(B_0 + 2B_{45} + B_{90})$, where the subscript refers to the orientation of the test strip. $\bar{B}$ is shown in Table II.

The induction obtained was consistently higher than that predicted for the knee, since the magnetizing force needed to overcome the impediments to domain wall motion was also sufficient to cause some domain rotation. In order to estimate the induction at 8000 A/m from the predicted value at the knee, the following empirical equation was used:

$$\frac{B_s - B(H)}{B_s - B'_k} = 10^{-cH} \qquad (6)$$

Figure 2 shows that Equation (6) is a good fit to the data above 4800 A/m. For the steel under study B_s is estimated as 2.11 tesla, and c was found to be 2.79 x 10^{-5} for a number of samples. Thus at H = 8000 A/m

$$B(H) = 0.40\ B_s + 0.60\ B'_k \qquad (7)$$

The values of B_k given in Table II were then substituted for B'_k to estimate B(H), also shown in Table II. Figure 3 shows the predicted values of B(H) as a function of the average measured values $\bar{B}$. A least squares fit of B(H) to $\bar{B}$ yields:

$$B(H) = 1.74 + 0.900\ \bar{B} \qquad (8)$$

which is also drawn in Figure 3.

Figure 3 shows that B(H) is comparable to the measured $\bar{B}$ in both absolute magnitude and slope. The scatter is probably derived from the two basic approximations associated with the X-ray technique: the finite number of orientations considered, and averaging the easy and hard directions within each orientation, in lieu of information on the directional anisotropy of each orientation. B'_k is probably a good approximation for B_k; otherwise B(H) would differ from $\bar{B}$ by a constant. It may, therefore, be concluded that the model presented here, in conjunction with the simple pole density technique, provides a means of estimating the induction at or above the knee of a directionally averaged magnetization curve.

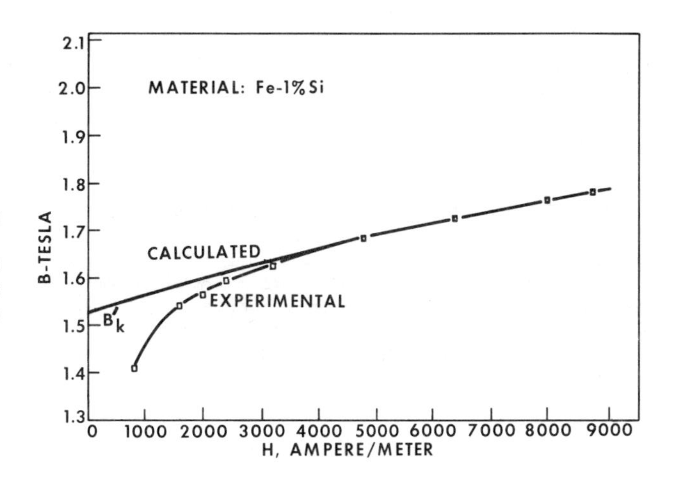

FIGURE 2 RELATIONSHIP BETWEEN EXPERIMENTAL CURVE AND THAT CALCULATED FROM

$$\frac{B_s - B(H)}{B_s - B'_k} = 10^{-2.79\times10^{-5}H}$$

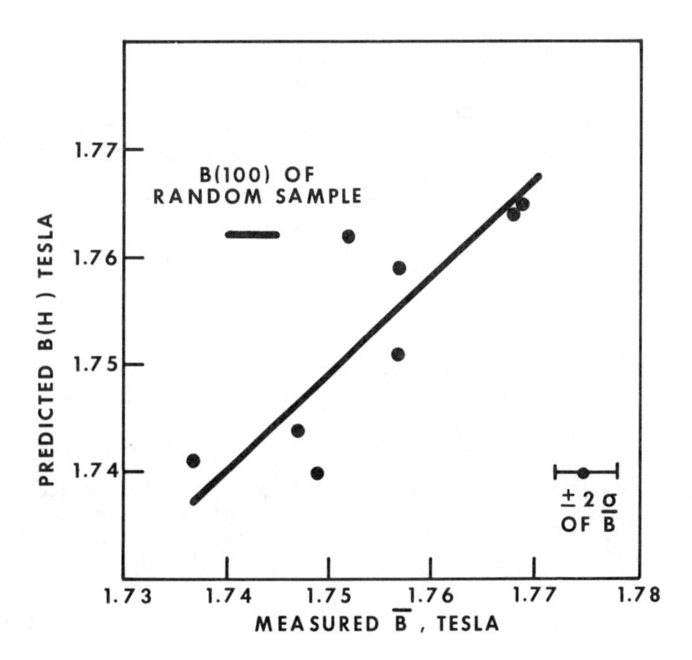

FIGURE 3 PREDICTED VALUES OF B(H) AS A FUNCTION OF MEASURED $\bar{B}$

REFERENCES

1. C. J. Gorter, Nature, 132, 517 (1933).
2. Referred to by K. H. Stewart, Ferromagnetic Domains, Cambridge, 31, (1954).
3. H. Lawton and K. H. Stewart, Proc. Phys. Soc. A., 63, 848 (1950).
4. K. H. Stewart, Ferromagnetic Domains, Cambridge, 33 (1954).
5. K. Foster and J. J. Kramer, J. of Applied Physics, Suppl. 31, 2335 (1960).
6. R. M. Horta, W. T. Roberts and D. V. Wilson, Trans. AIME 245, 2525 (1969).
7. I. F. Hughes and J. M. Shapiro, Met. Trans. 2, 3347 (1971).
8. R. M. Bozorth, Ferromagnetism, Van Nostrand, Princeton, 77 (1951).

MAGNETIC DOMAIN WALL MASS IN METALLIC CRYSTALS

W. J. Carr, Jr.
Westinghouse Research Laboratories
Pittsburgh, Pennsylvania 15235

ABSTRACT

It is shown that the pressure exerted on a domain wall resulting from the magnetic field of eddy currents, in a metal, leads to wall inertia in addition to damping. In an iron crystal at room temperature the domain wall mass introduced by this effect is larger than the Döring mass for sheet thickness greater than about 10^{-3} cm. An estimate is made of the time required to accelerate a domain wall from rest to its final velocity.

RESULTS

The equation of motion for each point on a slowly moving simple domain wall, in a uniform perfect crystal, with no difference in anisotropy energy or magnetostriction on either side of the wall is given by[1]

$$m\dot{v} = C\varepsilon - \frac{1}{2}(\underset{\sim}{H}^{+} + \underset{\sim}{H}^{-}) \cdot (\underset{\sim}{M}^{+} - \underset{\sim}{M}^{-}) - \beta v. \quad (1)$$

In this equation m is a "mass" per unit area, v is the velocity normal to the wall, $\dot{v}$ the acceleration, C the wall curvature, ε the wall energy per unit area, $\underset{\sim}{H}^{\pm}$ and $\underset{\sim}{M}^{\pm}$ are respectively the magnetic field and magnetization on the positive and negative sides of the wall, and β is the damping constant for all losses except the eddy current loss. The field $\underset{\sim}{H}$ can be written as the sum of an applied field $\underset{\sim}{H}_A$, a "demagnetizing" field $\underset{\sim}{H}_D$ and an eddy current field $\underset{\sim}{H}_e$. In a metal the eddy currents play an important role in damping the wall motion. Since there is no explicit eddy current drag written into the equation of motion (the force $-\beta v$ applies only to the non-eddy current part) it follows that the former must result from the term $-\underset{\sim}{H}_e \cdot (\underset{\sim}{M}^{+} - \underset{\sim}{M}^{-})$. The reason eddy current damping cannot, in general, be treated explicitly comes from the fact that eddy current dissipation is not localized within the wall but occurs mainly on the outside.

The purpose of the present calculation is to point out that in a metal the force density $-\underset{\sim}{H}_e \cdot (\underset{\sim}{M}^{+} - \underset{\sim}{M}^{-})$ not only contributes to wall damping but also produces an inertial term which defines an additional wall mass. The mass on the left-hand side of (1) is defined[1] by

$$m = \frac{1}{v}\frac{\partial \varepsilon}{\partial v}, \quad (2)$$

which for low velocity is just the Döring[2,3] mass, since in this case the velocity dependence of ε is given by the velocity dependence of the demagnetizing energy within the wall. In a physical sense this mass arises because in a moving wall the spins have a component normal to the wall, which increases with velocity, producing an internal demagnetizing field. Therefore an acceleration of the wall requires a buildup of energy in the demagnetizing field, leading to inertia. By analogy the same should occur with the eddy current field. To illustrate the eddy current effect we assume, for simplicity, a long plane 180° wall extending through the thickness of an infinite sheet, and consider only an average value of H_e. The plane wall approximation is valid for small applied fields where a small amount of curvature allows $C\varepsilon$ to balance the forces tending to bend the wall. The magnetization on either side of the wall is in the plane of the sheet, and equation of motion (1) reduces to

$$m\dot{v} = -(\underset{\sim}{H}_A + \bar{\underset{\sim}{H}}_e) \cdot (\underset{\sim}{M}^{+} - \underset{\sim}{M}^{-}) - \beta v, \quad (3)$$

where $\bar{\underset{\sim}{H}}_e$ indicates the average of $\underset{\sim}{H}_e$ over the wall area.

The eddy current field is defined by the equations

$$\text{curl } \underset{\sim}{H}_e = 4\pi \underset{\sim}{j} \quad (4)$$

and

$$\text{div } \underset{\sim}{H}_e = 0 \quad (5)$$

where j is the current density, and displacement current has been neglected. In the Maxwell equation

$$\text{curl } E = -\dot{\underset{\sim}{B}} \quad (6)$$

let B be written as $\underset{\sim}{B}_0 + \underset{\sim}{B}_1$ where $\underset{\sim}{B}_1$ results from wall motion and $\underset{\sim}{B}_0$ results from a changing applied field. The part $\underset{\sim}{B}_1$ is non-vanishing only in the neighborhood of the wall, whereas $\underset{\sim}{B}_0$ exists over the whole crystal and outside space. Let $\underset{\sim}{E}_1$ satisfy

$$\text{curl } \underset{\sim}{E}_1 = -\dot{\underset{\sim}{B}}_1, \quad (7)$$

and inside the crystal let j_1 be defined by $\underset{\sim}{j}_1 = \sigma \underset{\sim}{E}_1$, where σ is the conductivity. The current density in turn defines a magnetic field H_1 given by

$$\text{curl } \underset{\sim}{H}_1 = 4\pi j_1 \quad (8)$$

$$\text{div } \underset{\sim}{H}_1 = 0. \quad (9)$$

By comparison with (4) and (5), $\underset{\sim}{H}_1$ is observed to be the eddy current field due to wall motion.

By forming the scalar product of both sides of (8) with $\underset{\sim}{E}_1$, and the scalar product of both sides of (7) with $\underset{\sim}{H}_1$, one obtains by subtracting these two equations and using a vector transformation

$$\frac{1}{4\pi}\text{div } \underset{\sim}{H}_1 \times \underset{\sim}{E}_1 = \underset{\sim}{E}_1 \cdot \underset{\sim}{j}_1 + \frac{1}{4\pi}\underset{\sim}{H}_1 \cdot \dot{\underset{\sim}{B}}_1 \quad (10)$$

and integration over all space gives

$$0 = \int \frac{j_1^2}{\sigma} dV + \frac{1}{4\pi}\int \underset{\sim}{H}_1 \cdot \dot{\underset{\sim}{B}}_1 \, dV. \quad (11)$$

The vector $\dot{\underset{\sim}{B}}_1$ may be replaced with $\dot{\underset{\sim}{H}}_1 + 4\pi \dot{\underset{\sim}{M}}_1$ where $\dot{\underset{\sim}{M}}_1$ is assumed to equal $\dot{M}$, which vanishes except inside the wall. (For simplicity $\underset{\sim}{M}$ is assumed constant outside the wall which is the case for weak applied fields and large anisotropy.)

In general

$$\dot{\underset{\sim}{M}} = \frac{\partial \underset{\sim}{M}}{\partial t} = \frac{D\underset{\sim}{M}}{dt} - \underset{\sim}{v} \cdot \nabla \underset{\sim}{M} \quad (12)$$

where DM/dt is the time derivative following a point in the wall. In a uniform perfect crystal, the latter is non-vanishing only because M in the wall can depend upon velocity, and

$$\dot{\underset{\sim}{M}} = \dot{\underset{\sim}{v}} \cdot \frac{\partial \underset{\sim}{M}}{\partial \underset{\sim}{v}} - \underset{\sim}{v} \cdot \nabla \underset{\sim}{M} = \dot{v}\frac{\partial \underset{\sim}{M}}{\partial v} - v\frac{\partial}{\partial x}\underset{\sim}{M} \quad (13)$$

where the x axis is normal to the wall.

At low velocity only the normal component of $\underset{\sim}{M}$ changes with velocity,[2,3] in first order approximation, and since the eddy current field at the wall will not have a normal component $\underset{\sim}{H}_1 \cdot \partial\underset{\sim}{M}/\partial v$ vanishes. Then (11) becomes

$$0 = \int \frac{j_1{}^2}{\sigma}\, dV + \frac{1}{8\pi}\frac{\partial}{\partial t}\int H_1{}^2\, dV - v\int \underset{\sim}{H}_1 \cdot \frac{\partial \underset{\sim}{M}}{\partial x}\, dV \qquad (14)$$

For a thin wall the eddy current field is substantially constant through the wall thickness and in the last integral $\underset{\sim}{H}_1$ can be taken outside the integral over x leading to $- v\, \bar{\underset{\sim}{H}}_1 \cdot (\underset{\sim}{M}^+ - \underset{\sim}{M}^-)\, S$ where S is the wall area. Then (14) can be rearranged to give

$$\bar{\underset{\sim}{H}}_1 \cdot (\underset{\sim}{M}^+ - \underset{\sim}{M}^-) = \frac{1}{vS}\int \frac{j_1{}^2}{\sigma}\, dV + \frac{1}{8\pi vS}\frac{\partial}{\partial t}\int H_1{}^2\, dV. \qquad (15)$$

The first term on the right-hand side of (15) gives the dissipation and the second term determines the inertia. When the second term is small, it may be approximated by the use of the value of H_1 calculated for constant velocity. The eddy current loss at constant velocity has been considered by Williams, Shockley and Kittel,[4] who obtain

$$\frac{1}{vS}\int \frac{j_1{}^2}{\sigma}\, dV = \beta' v \qquad (16)$$

where

$$\beta' = \frac{256}{\pi}\sigma M_s{}^2\, d \sum_{n\ \mathrm{odd}} \frac{1}{n^3}$$

in units of c.g.s. e.m.u., where M_s is the saturation magnetization and d the sheet thickness. The eddy current field for this case is directed along the long dimension of the wall (z axis) and given by[5]

$$H_1 = -\, 64\, \sigma M_s v d \sum_{n\ \mathrm{odd}} \frac{1}{n^2} \sin\frac{n\pi}{2} \cos\frac{n\pi y}{d}\, e^{-\frac{n\pi}{d}|x|} \qquad (18)$$

where the origin for y and x is the center of the wall. For the average over the wall (18) gives

$$\bar{H}_1 = -\,\frac{128}{\pi}\sigma M_s v d \sum_{n\ \mathrm{odd}} \frac{1}{n^3}\,, \qquad (19)$$

and $- \bar{H}_1\, 2M_s$ is equal to $\beta' v$ as Eq. (15) demands for an unaccelerated wall. If (18) is used as the zeroth order approximation for an accelerated wall, then

$$\frac{1}{8\pi vS}\frac{\partial}{\partial t}\int H_1{}^2\, dV = m'\dot{v} \qquad (20)$$

where

$$m' = \frac{512}{\pi^2}\sigma^2 M_s{}^2 d^3 \sum_{n\ \mathrm{odd}} \frac{1}{n^5} \qquad (21)$$

and in first approximation

$$\bar{\underset{\sim}{H}}_1 \cdot (\underset{\sim}{M}^+ - \underset{\sim}{M}^-) = \beta' v + m'\dot{v}. \qquad (22)$$

The total eddy current field is $\underset{\sim}{H}_e = \underset{\sim}{H}_{eo} + \underset{\sim}{H}_1$, where $\underset{\sim}{H}_{eo}$, which comes from the currents induced by the changing applied field, acts simply to reduce the applied field. If an effective applied field is defined by

$$\underset{\sim}{H}'_A = \underset{\sim}{H}_A + \bar{\underset{\sim}{H}}_{eo} \qquad (23)$$

the equation of motion (3) becomes approximately

$$(m + m')\,\dot{v} = -\, \underset{\sim}{H}'_A \cdot (\underset{\sim}{M}^+ - \underset{\sim}{M}^-) - (\beta + \beta')\, v. \qquad (24)$$

For an imperfect or non-uniform crystal the equation may be generalized by adding additional terms to the right-hand side.[1] For an iron crystal m is roughly[6] 10^{-10} gm/cm^2 while m' at room temperature is 1.5 d^3 gm/cm^2. It follows that m' becomes comparable with m in a sheet of thickness 10^{-3} cm, and for thicker sheets m' can be much larger than m. The upper limit on thickness for which the expression (21) for m' (which is based upon a plane wall approximation) can be used depends upon the magnitude of the applied field. The maximum pressure tending to bend the wall is given by the applied pressure $H_A 2M_s$. Setting this equal to $C\varepsilon$ where $C \lesssim 1/d$ (since for larger curvature the wall is no longer approximately plane) gives $d \lesssim \varepsilon/(2M_s H_A)$. The smallest applied field which can be employed is given by the dynamic coercive force of the crystal, which typically puts an upper limit on d of several millimeters[4] to satisfy the plane wall approximation.

For an oscillating wall it is of interest to examine the angular frequency ω where $m'\dot{v}$ becomes comparable with $\beta' v$. This occurs for $m'\omega = \beta'$, which can be written approximately as $\pi\delta = d$ where $\delta = (2\pi\omega\sigma)^{-\frac{1}{2}}$ is the classical skin depth calculated for permeability equal to unity. A relaxation phenomena might be difficult to observe, however, because near this frequency the wall is shielded from the applied field by the eddy current field H_{eo} and H'_A becomes strongly dependent on frequency. A similar question of some interest is the time required for a constant field to accelerate a propagating wall to a steady velocity, after the wall encounters a barrier in the crystal which slows it down. This time T in order of magnitude is given by calculating $2\pi/\omega$ in the above equation, i.e., $T \sim 2\pi m'/\beta' \simeq \pi\sigma d^2$. This acceleration time, on the time scale of interest, is quite short, being roughly $10^{-4}\, d^2$ sec. for iron at room temperature.

REFERENCES

1. W. J. Carr, Jr., A. I. P. Conf. Proc. 24, 747 Magnetism and Magnetic Materials, San Francisco, 1974.
2. W. Döring, Z. Naturforsch 3a, 373 (1948).
3. R. Becker, J. Phys. et Rad. 12, 332 (1951).
4. H. J. Williams, W. Shockley and C. Kittel, Phys. Rev. 80, 1090 (1950).
5. W. J. Carr, Jr., J. Appl. Phys. 30, 90S (1959).
6. C. Kittel and J. K. Galt, Solid State Physics 3, Seitz and Turnbull, Academic Press, N. Y. (1956).

OBSERVATION OF DYNAMIC DOMAIN STRUCTURE IN 4%Si-Fe SINGLE CRYSTALS WITH THE (110)[001] AND (100)[001] ORIENTATIONS UNDER AC FIELD EXCITATION

K. Narita and M. Imamura
Faculty of Engineering, Kyushu University, Fukuoka 812, Japan

ABSTRACT

The stroboscopic observation of dynamic domain structure in two types of 4%Si-Fe single crystal with the (110)[001] and (100)[001] orientations has been made under ac field excitation, using the Kerr magneto-optic effect. The changes of the magnetization at the surface and of the flux in the volume were measured simultaneously using a photomultiplier and a search coil wound around the specimen.

The variation of the number of 180° domain walls was investigated over the wide frequency range 10-400 Hz at various flux densities. By plotting the degree of magnetic saturation at the surface against the degree of saturation in the volume, the amount of domain wall bowing was estimated in each specimen.

The experimental results indicated that the wall bowing occurs even below 10 Hz and can be a main cause of the nonlinear relation of the iron loss per cycle to the frequency in the low frequency range. It was also shown that the dynamic wall profile is strongly affected by the direction of the crystallographic axis to the surface.

INTRODUCTION

In grain-oriented silicon-iron sheet materials, the nonlinearity of loss per cycle versus frequency curves in the low frequency range has been of practical interest for many years.

Haller and Kramer[1] reported the dynamic domain width reduction for both induction and frequency increases, making the observation of dynamic domain size variation in a Si-Fe single crystal with the (001)[100] orientation. Bishop[2] discussed the concept of the domain wall bowing at low frequencies. He showed that the eddy current loss calculated by the wall bowing model is smaller than that obtained by the plane wall model. Both the domain width reduction and the wall bowing imply the nonlinearity of the loss per cycle versus frequency curves.

The observation of dynamic domain structure at lower frequencies under larger inductions, where the curvature is prominent, will be indispensable to make clear the dominant cause of this phenomena. In this paper the variation of the number of walls with increasing frequency at various flux densities, in two types of Si-Fe single crystal with the (110)[001] and (100)[001] orientations, is shown. From the results of the measurements of alternating domain wall movements, the influence of the directional difference of the crystallographic axis to the surface upon the dynamic wall profile was discussed. The results of loss measurements for those specimens will be reported in a separate paper.

EXPERIMENTAL PROCEDURES

The specimens in the form of rectangular plates were cut from single crystals grown by the Bridgman method. They were carefully finished by mechanical polishing and finally annealed at 1100 °C in vaccum. Those surfaces were coated with $(1/4)\lambda$ thick ZnS layer. The dimensions of the (100) and (110) specimens were 28x5.85x0.55 mm and 26x5.7x 0.56 mm, respectively. The Si concentration of each specimen was about 4 wt% and the misorientation between the crystal face and the sheet surface was within 0.5°.

The optical system consisting of a Kerr effect microscope with a stroboscopic disc-shutter driven by motor was used to observe and photograph the surface domain appearance during magnetization reversal. The observation over the frequency range 10-400 Hz was realized by this system. The photomultiplier was used to measure the change of the magnetization at the surface with the variation of the flux in the volume.

The long magnetizing yokes were connected to both sides of the specimens which were magnetized by the solenoid coil.

RESULTS OF OBSERVATIONS

The observations of dynamic domain structure were made under ac field excitation at flux densities of 2, 5,10 and 15 kGs, increasing the frequency from 10 to 400 Hz after ac demagnetization at 50 Hz.

In Fig.1, the variations of the number of walls in the (110) and (100) specimens are shown. At flux densities of 2 and 5 kGs the number of walls in both specimens is independent of frequency. At flux densities of 10 and 15 kGs, in the (110) specimen it decreases once from the beginning values and increases to the saturation value, and in the (100) specimen it decreases with increasing frequency. These two specimens show the opposite variation of the wall number against frequency at higher flux densities.

Photographs of Fig.2 were taken at the wall position of maximum magnetization at the surface during magnetization reversal. It is clearly found that the wall displacement at the surface becomes larger with increasing frequency and at higher frequencies walls disappears from the surface. These results indicate that the wall bowing occurs continuously from lower

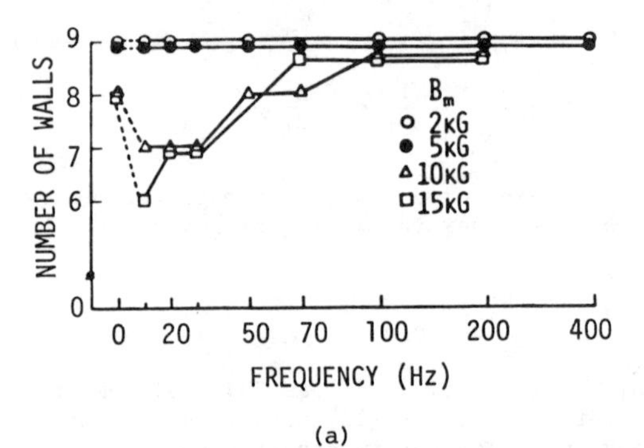

(a)

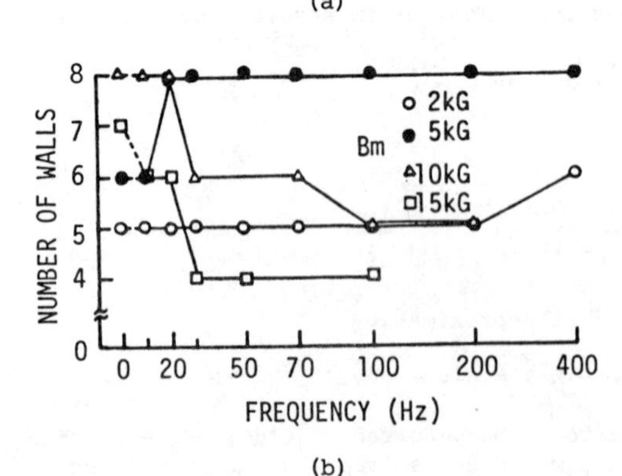

(b)

Fig. 1. Variation of the number of walls with frequency at various flux densities. (a) In the (110)[001] specimen, (b) In the (100)[001] specimen.

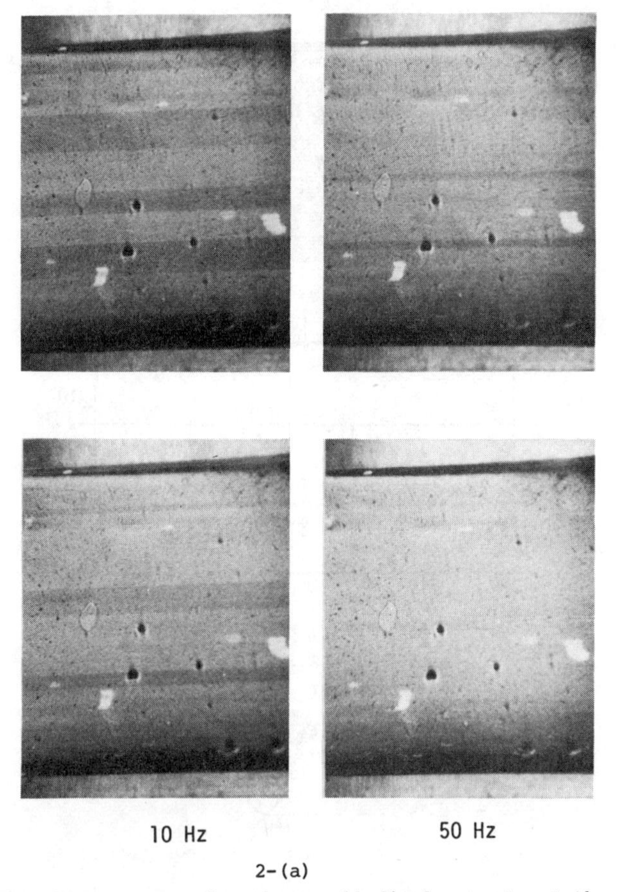

2-(a)

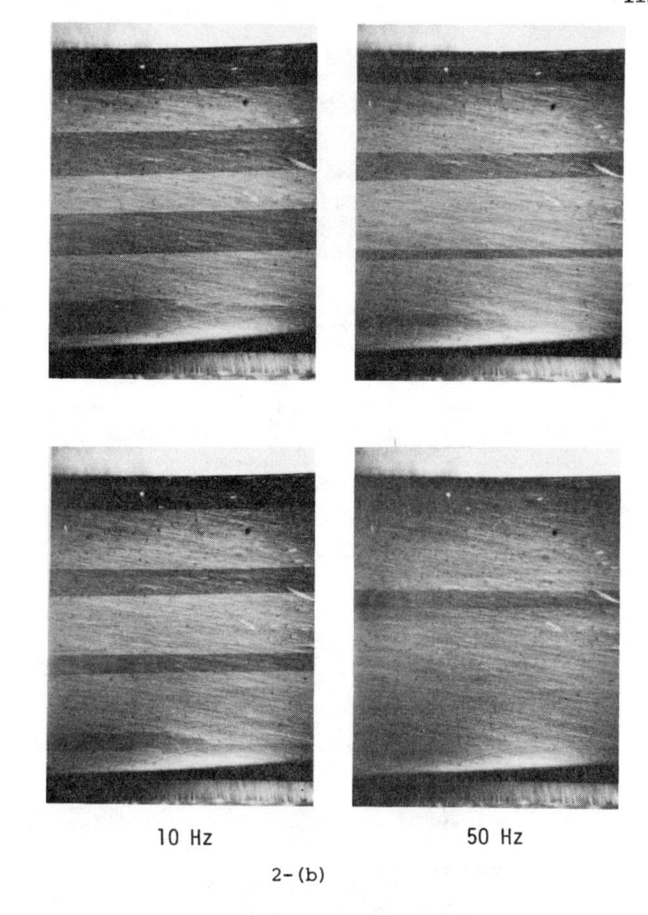

2-(b)

Fig. 2. Increase of maximum wall displacements at the surface with increasing frequency under the peak induction constant at 10 kGs. (a) The (110)[001] specimen, (b) The (100)[001] specimen.

frequencies and the merging of the neighbouring walls occurs when the wall bowing becomes severe. In Fig.3, the degree of the wall bowing for these specimens is shown, where $\hat{B}$sur denotes the maximum flux density at the surface given by Overshott and Thompson[3] and Bs the saturation flux density. The point where $\hat{B}$sur/Bs = 1 corresponds to the disappearance of walls from the surface. These figures show that the wall bowing in the (110) plane occurs easily in comparison with that in the (100) plane.

The average movements of walls were studied using the photomultiplier. Those results are shown in Figs. 4(A), (B) and (C). In Fig.(a) of each of these figures, the sinusoidal wave advanced in phase is the current of excitation and the one delayed is the flux. The wave of the small amplitude shows a signal from the photomultiplier, which is equivalent to the flux density at the surface (Bsur). Figures (b) and (c) have been obtained by plotting Bsur against each phase of the flux in the volume. In Fig.(b), d is the thickness of the specimen and δ is the average domain width. The domain wall profiles have been drawn by the approximation of biquadratic curve, considering the relation between the amplitude of the flux density and the wall position at the surface. By comparing Fig.4(A) with Figs.(B) and (C), the difference of the wall motion in each specimen under ac field excitation can be known.

DISCUSSION

As the frequency was increased, the number of walls remained constant at low flux densities and its variation at higher flux densities was smaller than that expected from previous papers.[1,4] In the specimens studied here, the wall bowing will play a dominant role

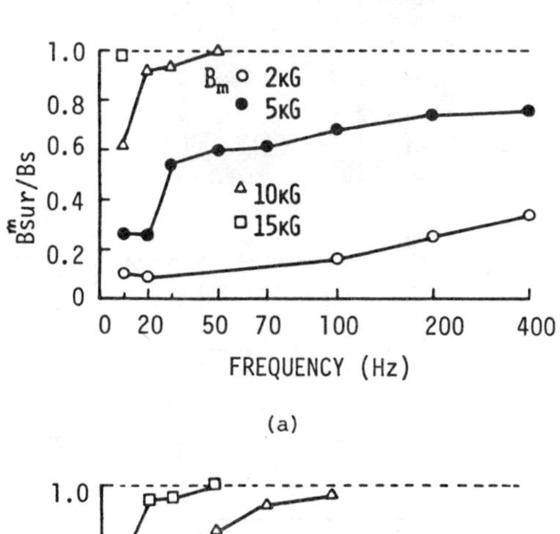

(a)

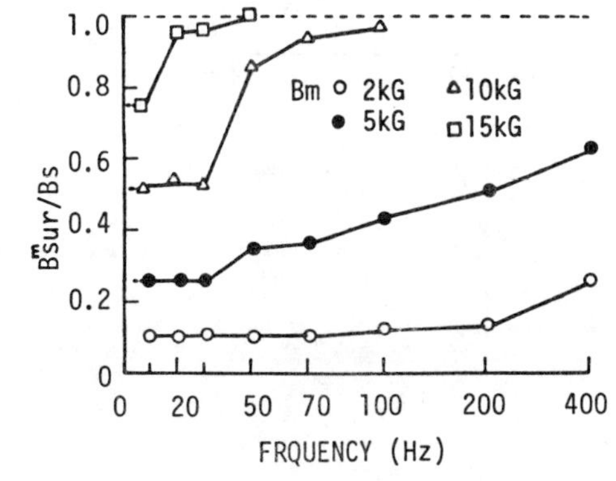

(b)

Fig. 3. Frequency dependence of ($\hat{B}$sur/Bs) at various flux densities. (a) In the (110)[001] specimen, (b) In the (100)[001] specimen.

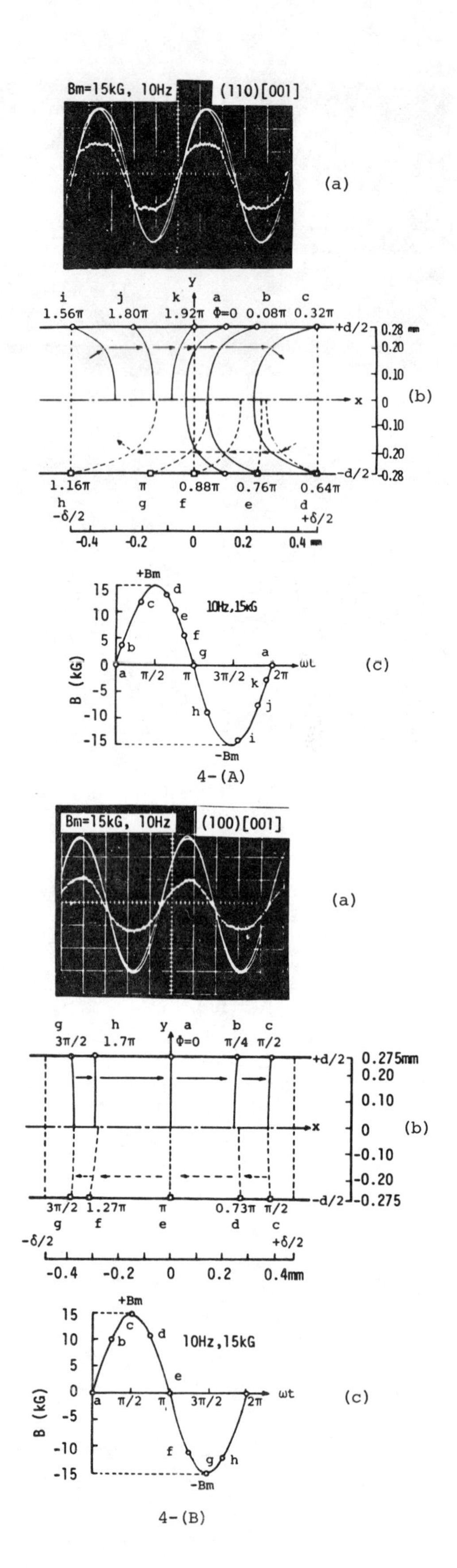

to reduce the eddy current loss.

Graham[5] noted that the product of the wall energy γ and the wall area σ is nearly constant as the wall orientation changes from (110) to (100) in the (110) specimen and it increases monotonously in the (100) specimen as the wall inclines from (100). The differ-

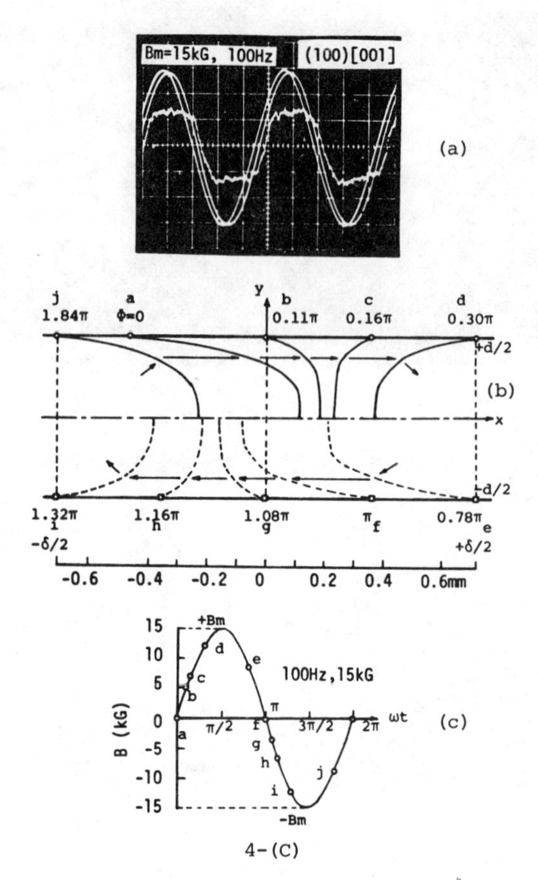

Fig. 4. Wall movements at the surface measured by the photomultiplier. (A) The (110)[001] specimen, Bm = 15 kGs, f = 10Hz, (B) The (100)[001] specimen, Bm = 15kGs, f = 10Hz, (C) The (100)[001] specimen, Bm = 15kGs, f = 100Hz. In each of Figs. 4-(A), 4-(B) and 4-(C), (a) Relation between the current of excitation, the flux in the volume and the magnetization at the surface, (b) Mean wall positions at the surface against each phase angle of the flux in the volume, (c) Amplitudes of the flux density corresponding to each wall position showed in Fig. (b).

ences between the wall profiles of the two specimens suggest that the variation of $\gamma \times \sigma$ with the change of the wall orientation from the crystal face must be considered to decide the dynamic wall profile in the frequency range investigated here. It can be considered from our experimental results that the 180° domain walls in the (100) specimen are rigid in comparison with those in the (110) specimen.

CONCLUSION

It has been shown by the direct observation of dynamic domain structure in two types of Si-Fe single crystal that in the (110) specimen the wall bowing occurs even below 10 Hz and in the (100) specimen the wall remains almost straight at low frequencies, but bends markedly at high frequencies.

REFERENCES

1. T.R. Haller and J.J. Kramer, J.Appl. Phys., 41, 1034 (1970).
2. J.E.L. Bishop, J.Phys. D: Appl. Phys., 6, 97 (1973).
3. K.J. Overshott and J.E. Thompson, Proc.IEE, 115, 1840 (1968).
4. E.W. Lee, Proc.IEE, 107C, 257 (1960).
5. C.D. Graham, Jr. and P.W. Neurath, J.Appl. Phys., 28, 888 (1957).

MEASUREMENTS ON THE μ^*-TENSOR METHOD OF WILLIAMS, BOZORTH, SHOCKLEY. TEMPERATURE (5.5-500 K) AND SURFACE ORIENTATION (0.01°-5.41°) DEPENDENCE

U.Keyser, O.Schärpf, Ch.Schwink, Institut A für Physik,
Technische Universität, 33 Braunschweig, FR Germany

ABSTRACT

We describe the application of an electron optical shadow method to measure the magnetic normal field component at the surface of stripe domains in Ni-single crystals, and the method to obtain the required precise orientation of the surface. The results of all our measurements can be represented by one formula. The equations of the μ^*-tensor method applied to the ⟨110⟩-Ni crystal are given. Comparison shows that the μ^*-tensor method can give "semiquantitative agreement with all observed features" (as was reported by Shockley[1]) only near room temperature. The temperature and surface orientation dependence of the normal field component predicted by this method shows great inconsistency with the measured values, particularly at low temperatures.

INTRODUCTION

In studies on ferromagnetic domains, the problem of surface charges always remains and of the resulting stray field energy. Williams, Bozorth and Shockley[2] have presented an expression for this energy if one considers the fact that the internal field produced by free poles on the surface rotates the domain magnetization out of the easy directions of magnetization. This was achieved by solving a potential equation for the case of a medium with anisotropic permeability, given by a tensor μ^*. This appears to be the only anisotropic potential theoretical boundary value problem to date, for which a solution has been attempted. Whether this solution really describes the physics of the problem was hitherto not quite clear, because only very indirect methods were available to test it. Only semiquantitative agreement with all observed features at room temperature could be demonstrated by Shockley[1].

METHOD OF MEASUREMENT

The boundary condition of Shockleys solution (see equ.(12)) yields a certain expression for the resulting field strength at the surface outside the crystal. This field strength outside the crystal at the surface of single domains is accessible to measurement when the electron optical shadow method by L.Marton et al.[3] is applied which has been developed further[4,5] by using a two-stage electrostatic electron microscope and a simpler way of quantitative evaluation of the magnetic surface charges[6]. A very thin wire placed behind the focus of the objective lens of the electron microscope is used as a shadow object. Without any stray field this shadow wire should be projected as a magnified straight rod on the screen. Its magnification V_s' depends on its distance from the electron projection centre. When this wire is normal to the direction of the investigated magnetically charged surface, its shadow is displaced with a displacement proportional to the normal field component on this surface. This is again proportional to the magnetic charge density on the plane surface provided there is no volume pole density. If the field is inhomogeneous, the displacement is different for different parts of the shadow and this results in a distortion of the shadow, as shown in fig.1. If the displacement of the shadow is known (Δx in

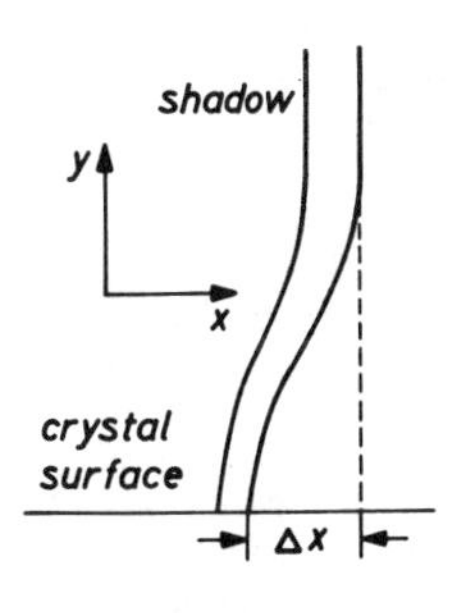

Fig.1. Distortion of the shadow by a field, decreasing with distance from the surface.

fig.1) the quantitative evaluation of H_y (= the normal field component) can be achieved by the eqation

$$H_y \cdot L = \frac{\Delta x \sqrt{2mU}}{(V_s - V_{pr}) \cdot f \cdot \sqrt{e}} \qquad (1)$$

with L=crystal width, U=beam voltage, m,e = mass and charge of the electron, f=focal length of the objective lens, V_{pr}= magnification of the projective lens, $V_s = V_s' V_{pr}$ (i.e.total shadow magnific.).

OBJECT OF MEASUREMENT

Nickel is a material in which the μ^*-tensor μ_{ik} is extremely sensitive to a temperature variation (μ^*=3.12 for 5.5 K and μ^*=123.3 for 400 K, see equ.(8)) and therefore seems very suitable as a means of exploring the μ^*-tensor method. According to a theory by Néel (for iron crystals)[7] one can obtain a regularly arranged domain configuration with large disc-shaped single domains all parallel to each other and perpendicular to the crystal axis over the whole cross section. Applying this theory to nickel[8], many investigations show that one obtains the same disc-shaped structure for cylindrical and prismatic Ni-crystals with ⟨110⟩ as axis. Our measurements using the shadow method show that with this kind of structure one can obtain surfaces without magnetic stray fields but magnetically saturated, if the effective easy directions of magnetization lie exactly in the (1$\bar{1}$0)-surface of

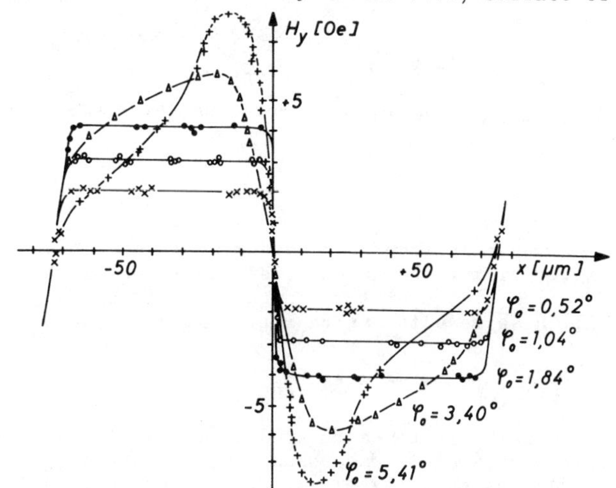

Fig.2. Normal stray field at the crystal surface for different angles φ_o. At angles $\varphi_o > 2°$ deviations from uniform pole density above the domains can be seen. Domain boundaries at x = 0, ±75 μm.

the Ni-crystals with the ⟨110⟩ axis. If the surface normal deviates slightly from ⟨1$\bar{1}$0⟩ by a small angle φ_0 of 0.01°-5.41°, (tilting axis = crystal axis) our measurements (fig.2) also show that for angles $\varphi_0 < 2°$ strips of uniform pole density of equal width and alternating sign are actually produced. This is a very suitable configuration for measurement using the electron optical shadow method and it was precisely for this configuration that the μ^*-tensor method was developed (but for 180°-walls, whereas we have 71°-walls).

To achieve the precise orientation of the crystal surfaces we used a set-up shown in fig.3 for grinding and polishing the crystals of up to 10 cm length and 4 mm width. It consists of a crystal holder (=a cone rotatable in a brass body by a micrometer gauge with an accuracy of 0.001° for correction of φ_0). The brass body is tiltable by means of two micrometer gauges, for correction of the surface position relative to the grinding jaws. This allows a correction of the surface orientation in the direction of the crystal axis by grinding. The feed for grinding can be adjusted by simultaneously moving both micrometer gauges. The grinding jaws are made of stainless steel. To control the grinding operation from time to time the whole set-up can be mounted on an adapted commercial X-ray counter goniometer, developed for orientation control of quartz crystal cuts by Bragg reflection, with an accuracy of 0.005°. Before annealing the crystal the surface is electropolished. This is to carry out with extreme care to maintain the precise orientation of the surface.

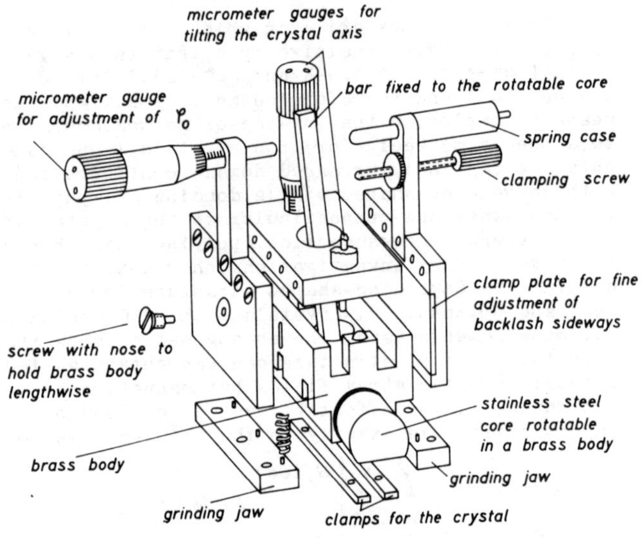

Fig.3 Crystal holder for grinding large crystals with well defined orientation.

RESULTS

The results of our measurements are shown in fig.4, giving the normal field component as a function of temperature for different angles φ_0. All these results can be represented by one formula

$$H_y(T, \varphi_0) = 56.6\cdot(1-\frac{T}{685})(\sin\varphi_0)^{0.6} \qquad (2)$$

allowing H_y to be calculated in the whole investigated temperature range and range of angle φ_0, within the accuracy of our measurements; this latter is better than can be indicated by experimental uncertainties in the figure (better than 5%). (Error of the numbers in (2): 56.6±0.1; 585±5; 0.600±0.004)

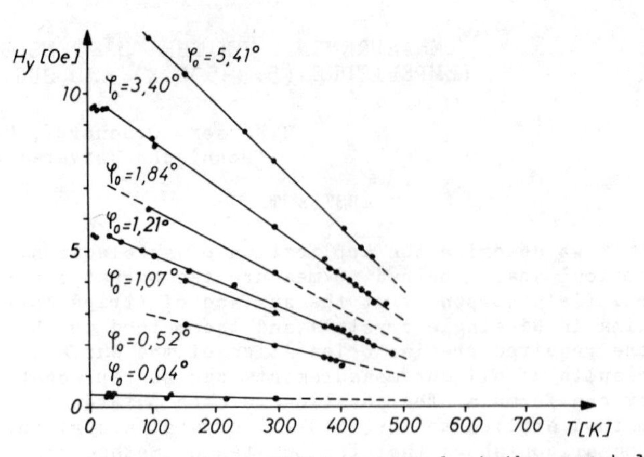

Fig.4. Normal stray field measured at the crystal surface (plateaus resp. maxima of curves of fig.2) as a function of temperature for different angles φ_0

DISCUSSION

To apply the μ^*-tensor method for the ⟨110⟩-Ni-single crystal we first have to determine χ^* [8]. To obtain the crystal anisotropy energy in a system of the suitable symmetry we transform the (x',y',z') system of the cubic axis of the crystal thus that the $\bar{z}$-axis of the new system ($\bar{x},\bar{y},\bar{z}$) coincides with an easy direction {111} of magnetization i.e. with a cubic body diagonal. In this system the anisotropy energy is

$$E_K = K_1(-\frac{2}{3}\sin^2\theta + \frac{7}{12}\sin^4\theta - \frac{1}{3}\sqrt{2}\sin^3\theta\cos\theta\cos 3\Phi); \qquad (3)$$

with θ, Φ = spherical coordinates with θ=0 at the $\bar{z}$-axis. The magnetostatic energy is

$$E_s = -J_s(H_{\bar{x}}\cos\Phi\sin\theta + H_{\bar{y}}\sin\Phi\sin\theta) - H_{\bar{z}}J_s\cos\theta \qquad (4)$$

For $\theta \ll 1$ omitting terms of order >3 in θ the sum of these two energies is

$$E \approx -\frac{2}{3}K_1\theta^2 - J_s(H_{\bar{x}}\cos\Phi + H_{\bar{y}}\sin\Phi)\theta - H_{\bar{z}}J_s \qquad (5)$$

with the minimum at

$$\theta = -\frac{3}{4}\frac{J_s}{K_1}(H_{\bar{x}}\cos\Phi + H_{\bar{y}}\sin\Phi) \qquad (6)$$

At Φ = 0 we obtain

$$J_x = J_s\theta = \frac{3}{4}\frac{J_s^2}{|K_1|}H_{\bar{x}} = \chi^* H_{\bar{x}} \qquad (7)$$

$$\chi^* = 3J_s^2/(4|K_1|), \quad \mu^* = 1 + 4\pi\chi^*, \qquad (8)$$
$$\chi_{ik} = (\mu_{ik} - \delta_{ik})/(4\pi)$$

Now one can express the problem by the following equations. The μ^*-tensor in a system where the easy direction coincides with the 1st-axis has to be transformed in the (x,y,z) system with x in the direction of the crystal axis and y normal to the crystal surface. This is achieved by

$$\mu_{ik} = T\begin{pmatrix}1 & 0 & 0\\ 0 & \mu^* & 0\\ 0 & 0 & \mu^*\end{pmatrix}T^{-1} \qquad (9)$$

with

$$T = \begin{pmatrix}\cos\vartheta & 0 & -\sin\vartheta\\ \sin\vartheta\sin\varphi_0 & \cos\varphi_0 & \cos\vartheta\sin\varphi_0\\ \sin\vartheta\cos\varphi_0 & -\sin\varphi_0 & \cos\vartheta\cos\varphi_0\end{pmatrix} \qquad (10)$$

and with ϑ = angle of the easy direction of magnetization with the crystal axis, i.e. $\cos^2\vartheta = 2/3$, $\sin^2\vartheta = 1/3$ for the ⟨110⟩ Ni single crystal.

The resulting potential equation is:

$$\mathrm{div}(\underline{\mu}^* \mathrm{grad}\,\varphi) = \mu_{11}\,\partial^2\varphi/\partial x^2 + \mu_{12}\,\partial^2\varphi/\partial x\partial y + \mu_{22}\,\partial^2\varphi/\partial y^2 = 0 \qquad (11)$$

the boundary condition at the surface of the crystal can be written as:

$$-\left.\frac{\partial\varphi}{\partial y}\right|_{y=+0} + \mu_{12}\left.\frac{\partial\varphi}{\partial x}\right|_{y=-0} + \mu_{22}\left.\frac{\partial\varphi}{\partial y}\right|_{y=-0} = 4\pi\sigma(x) \qquad (12)$$

The solution giving stripe domains of uniform pole density is:

$$\varphi = \frac{4\sigma Ad}{\pi}\sum_{n=0}^{\infty}\frac{1}{(2n+1)^2}\sin\left(\frac{2\pi(2n+1)\eta}{d}\right)e^{-\frac{2\pi(2n+1)|\zeta|}{d}} \qquad (13)$$

with $\sigma = J_s \sin\vartheta \sin\varphi_0$ and $d/2$ = domain width. For the potential outside the crystal one has to set $\eta = x$, $\zeta = y > 0$; for the potential inside the crystal $y < 0$, $\eta = x+ay$, $\zeta = by$;

$$\mu_{11} = 1+4\pi\chi^*/3, \qquad \mu_{12} = (-4\pi\sqrt{2}\chi^*\sin\varphi_0)/3,$$
$$\mu_{22} = 1+4\pi\chi^*(1-\tfrac{1}{3}\sin^2\varphi_0), \qquad a = -\mu_{12}/\mu_{22},$$
$$b = \sqrt{\mu_{11}\mu_{22}-\mu_{12}^2}\Big/\mu_{22}, \qquad A = 2/(1+\mu_{22}b) \qquad (14)$$

The temperature dependence of A,b is given in fig.5 for $\varphi_0 = 1^\circ$, that of a for $\varphi_0 = 0.5^\circ$ and 1°. The dependence of A,b on φ_0 is only very weak. The field strength at the surface outside the crystal in the centre of the domain ($\eta = d/4$, $\zeta = 0$) is $H^*_{y=+0} = 2\pi A\sigma$ and is given in fig.6 for different angles φ_0 together with two measured curves (dotted). This field strength shows great inconsistency with

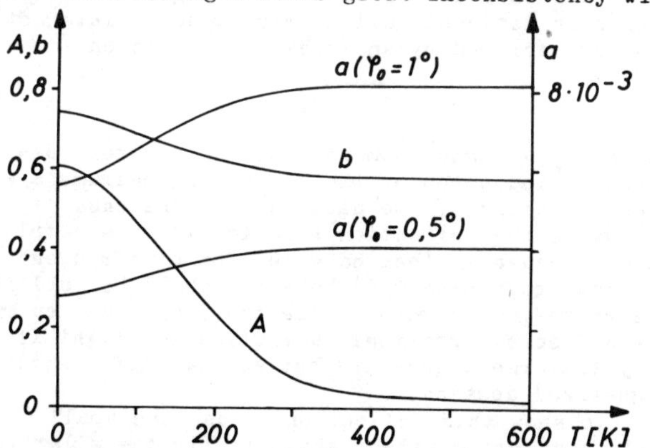

Fig.5. Temperature dependence of A,b for $\varphi_0 = 1^\circ$ and of a for $\varphi_0 = 0.5^\circ$ and 1°.

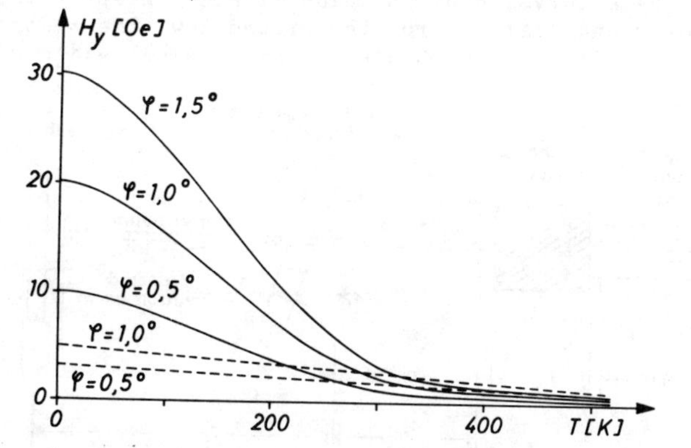

Fig.6. Field strength of the normal field outside the crystal at the crystal surface as a function of T for different angles φ_0 calculated with the μ^*-tensor method together with two measured curves (dotted).

the measured values, particularly at low temperatures. A systematic error in the measurements seems impossible. Probably the μ^*-tensor method is not complete. To get more hints for the reason of this discrepancy we now measure with iron single crystals. The solution will be of relevance for other problems of magnetic properties of surfaces (cf.[10]), too.

REFERENCES

1. W.Shockley,Phys.Rev. 73(1948)1246.
2. H.J.Williams, R.M.Bozorth and W.Shockley, Phys.Rev.75(1949)155.
3. L.Marton and S.H.Lachenbruch,J.appl.Phys.20 (1949)1171-82
4. Ch.Schwink,Optik 12(1955)481-496.
5. O.Schärpf,Ch.Schwink und W.Fellner,Z.angew.Phys. 28(1969)158-165.
6. O.Schärpf,Diss.Universität München,1967.
7. M.L.Néel,J.Phys.Radium 5(1944)241-251 and 265-76.
8. Ch.Schwink and O.Grüter,phys.stat.sol.19(1967)217 H.Spreen,phys.stat.sol.24(1967)413-429.
9. H.D.Dietze,Techn.Mitt.Krupp 15(1957)169-177.
10. P.C.Hohenberg,K.Binder,AIP Conf.Proc.24(1975)300

SUPPLEMENTARY REMARKS TO THE μ^*-TENSOR METHOD

O.Schärpf and U.Keyser

The μ^*-tensor method seems to be not complete for the problem of surfaces of ferromagnets with spontaneous magnetization,as it is equivalent to the image method in the case of isotropic dielectric media without spontaneous polarization. It describes the behavior if there is a fixed charge on the surface. In our case the charge on the surface itself is not fixed because the spontaneous magnetization itself is rotated and therewith the effective surface charge is diminished.We try the following approximation: We assume that the effective surface charge density adjusts itself due to two counteracting reasons: 1. $\sigma_s = (J_s/\sqrt{3})\sin\varphi_0$ (primary surface charge), 2. stray field proportional to the final effective surface charge σ_{eff} which tries to rotate the direction of spontaneous magnetization back into a position parallel to the surface.These two effects must be in equlibrium, giving

$$\sigma_{eff} = (J_s/\sqrt{3})\sin\varphi_0 - \sigma_{eff}2\pi Ab\chi_{22}C$$

With this effective surface charge (setting $C=4\pi$) one obtains the dependence of the normal component of the field on temperature given in the figure

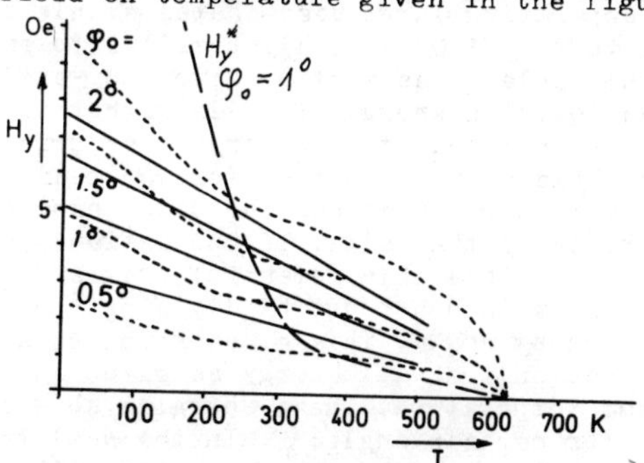

by the dotted curves for some angles.The lines drawn are the measured curves. The hatched curve (————) is the normal component behavior calculated with the μ^*-tensor method for $\varphi_0 = 1^\circ$.

ZIGZAG STRUCTURES OF BLOCH WALLS IN BULK FeSi(2.5%) SINGLE CRYSTALS MEASURED BY NEUTRON SMALL ANGLE SCATTERING

O.Schärpf and K.Brandt
Institut A für Physik, Technische Universität,33 Braunschweig, FR Germany

ABSTRACT

Bloch walls are refracting planes for neutrons. Using this property one can study the zigzag structure of the walls in the interior of bulk crystals, which have hitherto not been investigated in any other way. We use a small angle scattering set-up similar to that described by W.Schmatz et al.[1].Our measuring object is an iron single crystal with only one type of wall: 90° Bloch walls normal to ⟨110⟩ and all parallel to each other.This type of wall has to be a zigzag wall[3] whose zigzag structure is also visible at the surface. We describe the method to find the direction of the two refracting planes of the zigzag. We observe the variation of the zigzag angle by an applied field. Some observed features can only be understood by a distribution of zigzag angles in a range of $\pm 3^{\circ}$.

INTRODUCTION

In investigating domain configurations one sometimes encounters zigzag structures of Bloch walls on the surface of iron single crystals[2] (fig. 1). The wall energies depend on the orientation of the walls as can be shown by calculations[3]. In FeSi(2.5%) single crystals of the Néel type[4,2] with ⟨110⟩ as axis one obtains regularly arranged 90°

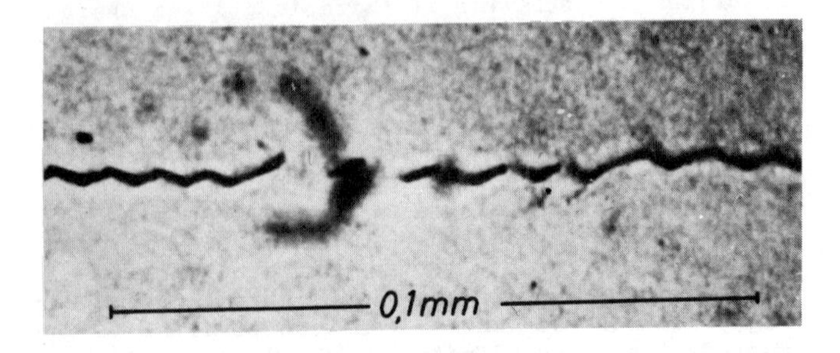

Fig.1.Zigzag structure of a FeSi(2.5%) 90°Bloch wall.Bitter colloid pattern on the (1$\bar{1}$0) surface

walls separating large disc-shaped single domains normal to the ⟨110⟩ axis all parallel to each other over the whole cross section of our crystals of 15 mm diameter. The energy of walls with this orientation is higher than that of any other orientations nearby. This energy can therefore be reduced by zigzag structures which on the other hand enlarges the surface of the walls. If one defines the angles as in fig.2, this enlargement of the surface can be taken into account by dividing the energy expression by cos ψ. This gives an orientation dependence on the wall energy as given in fig.3 for room temperature. These energies show flat minima for certain angles ψ. In the Néel type crystal the wall has to be normal to the ⟨110⟩ direction as also can be observed.But the observed zigzag structure on the surface shows that it consists of pieces which are preferently oriented in the directions where the energy has a minimum. Hitherto this is only observed on the surface.As it is caused by orientation dependence of the wall energy it ought to be found also in the interior of bulk crystals.

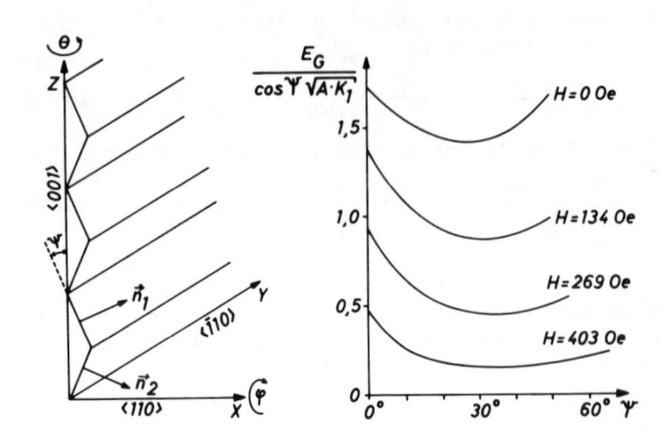

Fig.2.Zigzag structure: definition of angles and directions

Fig.3.Wall energy of 90°-walls as a function of the angle ψ for different applied fields at room temperat.

OBSERVATION OF REFRACTION AS BASIS OF THE METHOD OF MEASUREMENT

Neutrons are an almost perfect tool in investigating such structures as their magnetic interaction is dominant. Often this magnetic interaction can be described by an index of refraction

$$n = \sqrt{1 \pm \mu_n B/E} \tag{1}$$

where μ_n = magnetic moment of the neutron, B = magnetic induction of the refracting medium, E = kinetic energy of the neutrons. In the case of 90° walls the refraction behavior is more complicated because we then have neutron birefringence. The laws governing this behavior shall be published elsewhere[5]. Here we only use the fact that neutrons are deflected from their direction of flight if they traverse a boundary between two differently magnetized domains.

To show this deflection we use the small angle scattering set-up shown in fig.4 similar to that described in[1]. Thermal neutrons from a 1 MW reactor (5×10^{12} n/cm²s at the core edge) are filtered by a curved neutron guide[6] thereby keeping the γ-ray and fast neutron background low. The neutrons are collimated to obtain a beam of small width of

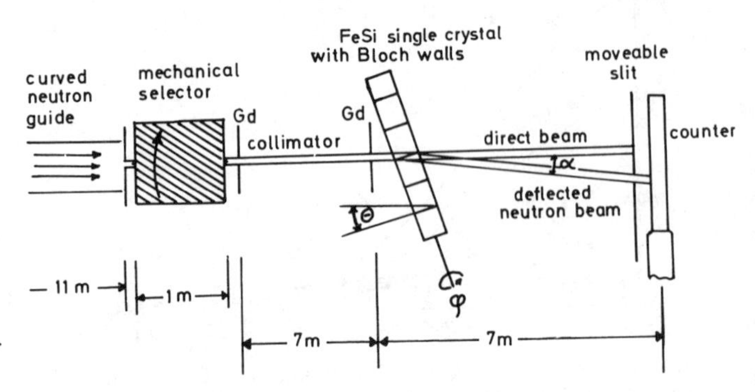

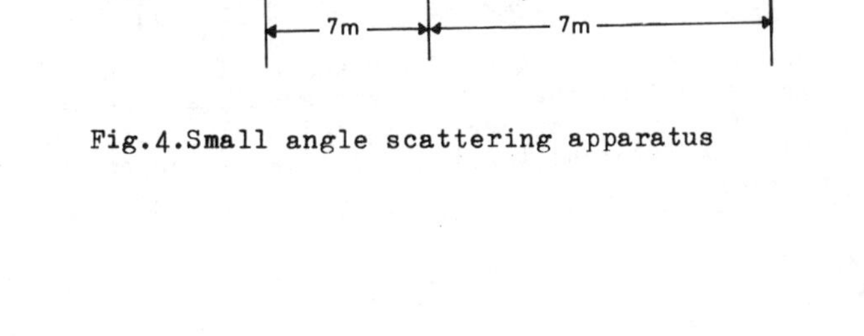

Fig.4.Small angle scattering apparatus

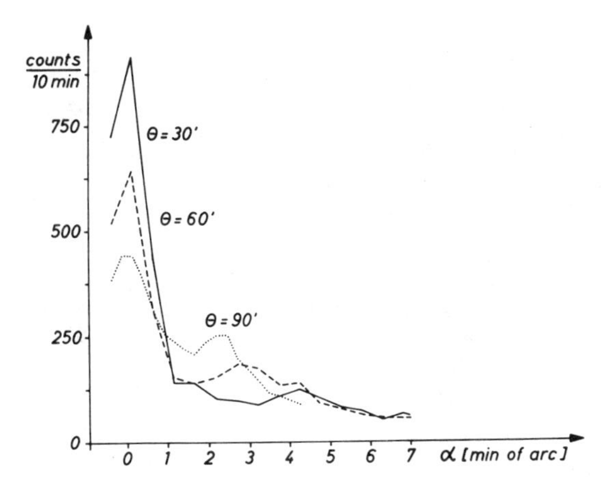

Fig.5.Direct and refracted beam: behavior on variation of θ showing different deflections α; error= $\sqrt{\text{counts}}$, H=130 Oerstedt, λ = 4.04 Å

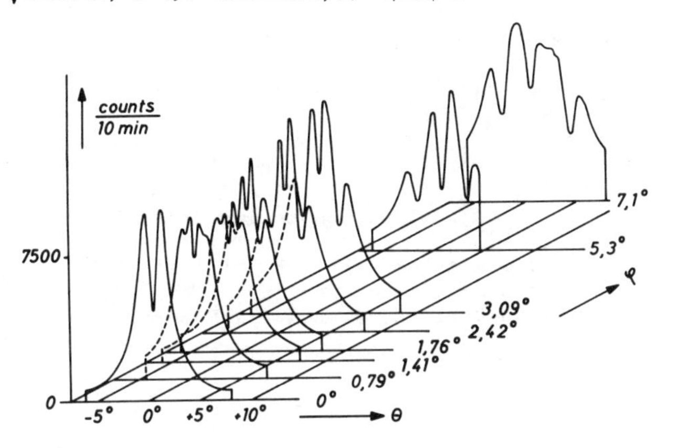

Fig.6.Intensity refracted into a fixed slit at α =2' as a function of θ for different angles φ ;error = $\sqrt{\text{counts}}$,H=130 Oerstedt,selector removed

1' (minute of arc). This is accomplished by two slits which are 1mm wide of Gd-foils 0.025 mm thick and 7 m apart in a vacuum arrangement. Thin foils allow to avoid edge scattering. The collimated beam of neutrons falls on the Bloch walls in our test object, a crystal with Bloch walls of well-defined orientation described above. By refraction on the Bloch walls as boundaries between two media of different index of refraction part of the beam is deflected. To measure this deflection we moved a slit before a counter in steps of 0.5' and counted at each step for 10 minutes.As the deflection depends on the wavelength of the neutrons we used a mechanical selector allowing to select the wavelength in the range 1.5 Å < λ < 10 Å with a resolution of 10% [7]. The results are shown in fig.5 for some grazing angles θ. Alteration of the grazing angle of incidence θ changed the angle of deflection α which can be seen as displacement of the side maximum in fig.5.

METHOD TO FIND THE DIRECTION OF THE REFRACTING PLANE

We chose a fixed angle of deflection α of 2' by proper positioning of the moveable slit. As θ is varied the deflected beam can be seen by the counter with a certain angle θ through the fixed slit.As the neutrons are unpolarized two symmetric angle positions θ of the wall give the same angle of deflection for the spin up respectively spin down part of the neutron beam.Therefore we obtain a pattern with two maxima by this process (fig.6 curve for φ = 0°). These maxima appear if the beam is refracted exactly into the fixed slit.The minimum between these two deflected beams appears at the grazing angle zero where there are not any neutrons hitting the wall.Thus one can find the direction of the refracting plane. It is not necessary to use monochromatic neutrons and the mechanical selector can be removed to obtain more intensity.

MEASUREMENT OF THE ZIGZAG ANGLE

The curve of fig.6 for φ = 0° with two maxima appears only if the angle of rotation φ (fig.4) is selected in such a way that the zigzag structure is parallel to the neutron beam, i.e. if the neutron beam incidence coincides with the <1$\bar{1}$0> direction of the crystal. This corresponds to a grazing angle zero for both pieces of the zigzag structure (φ=0°). We controlled this also by an X-ray counter goniometer. If we give the directions of the normals $\vec{n}_1$ and $\vec{n}_2$ of the both pieces of the zigzag structure in the crystal in stereographic projection (fig.7a) one immediately can see what happens, if we rotate by an angle φ about the axis (the angles and directions are indicated in fig.7): Then there are two angles θ symmetrically to the position θ=0° and φ=0 where a part of the zigzag plane is parallel to the incident neutron beam (= has grazing angle zero). This angle is after rotation by φ=90° exactly the angle 2ψ between the two zigzag directions. The results in this case (φ=90°) are given in fig.8 for two different field strengths (measured by a Hall probe as tangential field at the surface). In the range $0° \leq \varphi \leq 90°$ the following relation holds

$$\tan \theta = \pm \tan \psi \sin \varphi \qquad (2)$$

For some angles φ the results of such measurements are shown in fig.6. There is a difference in that what appears in fig.6 and in fig.8: In fig.6 we find minima where the wall pieces are parallel to the incident beam whereas in fig.8 maxima appear in this case. This can be well understood if one assumes, that the zigzag angle caused by the flat minimum of the energy in fig.3 is not very sharply defined. The orientation of the wall pieces is therefore given as a line in the stereographic projection as in fig.7b instead of a point for the normal of the wall pieces as in fig.7a.

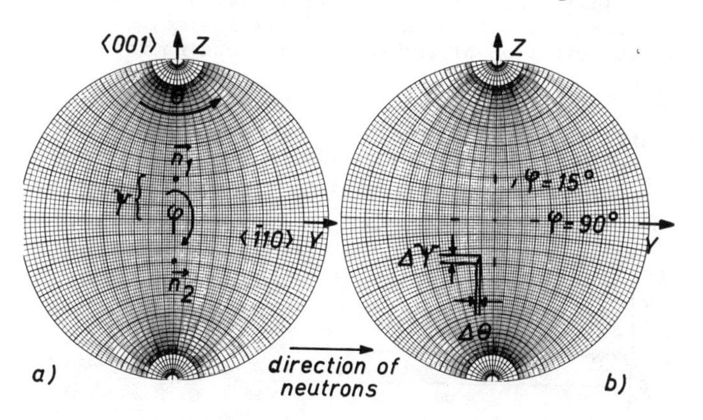

Fig.7.Zigzag structure in stereographic projection a)ideal zigzag structure with definitions of angles and directions. b)Lines indicating the range of zigzag directions showing the effect of this distribution on the θ-dependence of the refraction into a fixed slit.

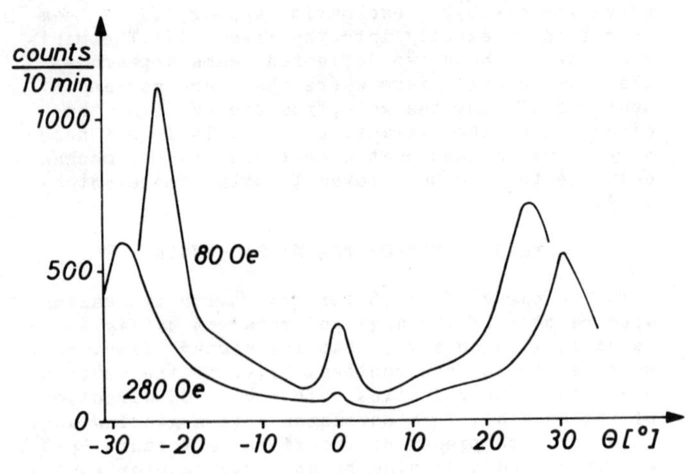

Fig.8. Intensity refracted into a fixed slit at $\alpha = 3'$ for $\varphi = 90^\circ$ and for two different applied fields, selector removed, error = $\sqrt{\text{counts}}$

This shall indicate that not only one direction of this normals is possible but a certain range of directions. On the other hand the wall orientation in the ⟨1$\bar{1}$0⟩ direction is much more sharply defined because by a deviation from that direction the wall would become magnetically charged giving a high contribution to stray field energy because of the high magnetization of iron. By the distribution of the zigzag angles the minima and maxima at large φ are superimposed such that only broader maxima result in fig.8. This behavior starts at angles $\varphi = 16^\circ$. Stereographic projection immediately shows that then the range of $\Delta\psi$ should be $\pm 3^\circ$ as $\varphi = 16^\circ$ and $\Delta\psi = \pm 3^\circ$ gives a range of θ of $\Delta\theta = \pm 1^\circ$, (= the width of the minimum that is smeared out by the distribution of ψ).

Fig.9 shows the behavior of the angles θ for the minima or maxima, respectively, giving grazing angle zero for different φ together with curves calculated by equ.(2) for different zigzag angles ψ. The best fit is here achieved for $\psi = 28^\circ$. Fig.8 shows that the zigzag angle depends on an applied field, which was predicted by A.Hubert[3] by means of the minima of the energy shown in fig.3. Fig.8 shows that at 80 Oerstedt $\psi = 25^\circ$, at 280 Oerstedt $\psi = 29^\circ$. These angles are not precisely those of the minima of fig.3. This stems probably from the flatness of the minima, giving a distribution of angles as discussed above.

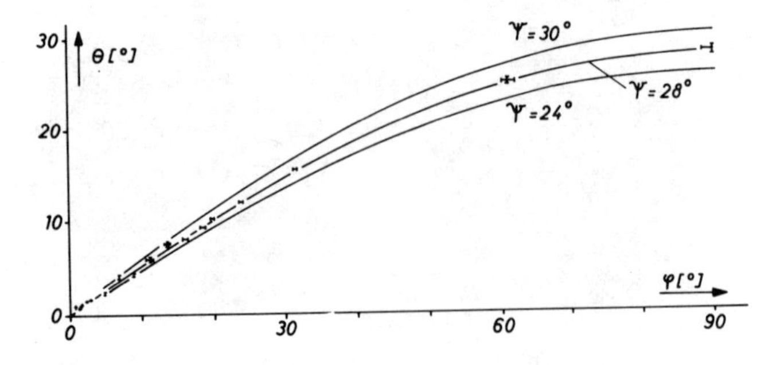

Fig.9. θ-positions of the minima (for $\varphi < 16^\circ$) respectively maxima (for $\varphi > 16^\circ$) as a function of the angle of rotation φ, together with calculated curves, to determine the zigzag angle ψ; error = $\sqrt{\text{counts}}$, H= 130 Oerstedt, selector removed

In these measurements we only use the fact that neutrons are refracted by 90° walls. The laws of this behavior describing neutron birefringence giving 4 beams in traversing one wall and also considering the fact that with flat incidence the neutrons are also influenced by traversing a wall as a helical magnetic structure are published elsewhere[5]. We expect that this behavior can give further information on the microscopic wall structure.

The authors wish to thank Prof.Dr.Ch.Schwink. The experiments are performed at the FMRB Braunschweig. We also wish to thank the FMRB reactor group, especially Prof.Dr.W.Heintz, Director of the FMRB.

REFERENCES

1. W.Schmatz, T.Springer, J.Schelten and K.Ibel, J.appl. Cryst.7,96(1974)
2. G.Dedié, J.Niemeyer, Ch.Schwink, phys.stat.sol.(b) 43,163(1971)
3. C.D.Graham, J.Appl.Phys.29,1451(1958)
L.Spaček, Ann.d.Physik(7),5,217(1960)
A.Hubert, Theorie der Domänenwände in geordneten Medien, Springer Verlag Berlin 1974
4. M.L.Néel, J.Phys.Radium 5,241 and 265(1944)
5. Conference on Neutron Scattering in Gatlinburg June 1976
6. O.Schärpf, D.Eichler, J.Phys.E, Sci.Instr.6,774(1973)
7. O.Schärpf, R.Seifert, Proceedings of the Neutron Diffraction Conference, RCN-234 Report, Petten 1975, p.90

INTERACTIONS OF MOBILE IMPURITIES WITH A SINGLE DOMAIN WALL IN IRON WHISKERS

B. Heinrich and A.S. Arrott
Department of Physics, Simon Fraser University
Burnaby, British Columbia, Canada V5A 1S6

ABSTRACT

The magnetic after effect is studied by measuring the magnetic response of an iron whisker with a single 180 degree wall in the presence of mobile interstitial atoms. The frequency dependence of both the in phase and out phase component of the small amplitude a.c. susceptibility are fitted to an exponential memory function with a single time constant at each temperature. The interactions of the wall with impurities is used also to investigate the properties of the wall.

INTRODUCTION

Many texts on magnetism devote a chapter to the subject of the magnetic after effect. Kneller[1], for example, lists 160 references to theory and experiment going all the way back to Ewing. The general problem has required a good deal of ingenuity because many factors must be considered to explain the observations on typical commercial materials, where the effect can be important. Twenty years ago Schreiber[2] applied Néel's[3] development of the subject to the quite tractable problem of the single crystal picture frame with one 180 degree wall. Yet Schreiber's analysis seems to have been overlooked by workers in the field. The experiment has not been carried out. Maringer[4] a decade later pointed out that the use of picture frames should be ideal for studying the magnetic after effect.

Our interest in this problem arose from our studies of the behavior of iron whiskers at the Curie temperature. Samples are held near 1040 K for the several week duration of the experiment. Upon return to low temperatures the magnetic susceptibility exhibits quite dramatic evidences of interactions of the one 180 degree domain wall with some mobile impurities.

THEORY

When mobile impurities such as carbon are present in iron, Néel's theory predicts and experiments[5] confirm a preference of the interstitial atom for certain of the octahedral sites depending on the local direction of the magnetization. If a domain wall is held stationary for sufficient time, the pattern of site occupation creates a stabilization field such that if the wall is slightly displaced it will experience a restoring force that will decrease with time.

To drive the magnetization at frequency ω and amplitude m an addition field H_s is necessary. This field can be described in terms of a memory function $\phi(t)$ that relates its value at time t to the rate of change of magnetization at all previous times,

$$H_s(t) = \int_{-\infty}^{t} \phi(t-t') \frac{dm(t')}{dt'} dt' \qquad (1)$$

If the process is controlled by a single relaxation time

$$\phi(t) = \frac{1}{\chi_s} \exp(-t/\tau_s) \qquad (2)$$

and the stabilization field is

$$H_s(t) = \left[\frac{\omega^2\tau_s^2}{1+\omega^2\tau_s^2} + i\,\frac{\omega\tau_s}{1+\omega^2\tau_s^2}\right] \frac{1}{\chi_s}\, m\, e^{i\omega t} \qquad (3)$$

which we abbreviate by the notation

$$H_s(t) = (\alpha_s + i\omega\beta_s)\, m\, e^{i\omega t} \qquad (4)$$

inasmuch as the stiffness coefficient α and the viscosity coefficient β have a very simple interpretation in a spring model with damping.

The response of an iron whisker with a single 180 degree domain wall can be accurately described using a spring model. In the absense of the impurity effect the applied field is balanced by the demagnetizing field plus the eddy current field as expressed by

$$(\alpha_D + i\omega\beta_e)\, m\, e^{i\omega t} = h\, e^{i\omega t} \qquad (5)$$

where $\alpha_D = 1/4\pi D$ is calculable from the magnetostatics of a medium with infinite susceptibility and β_e is calculable in the range of frequencies where the shape of the wall does not distort from the uniform quadratic bow it takes at low frequencies. These things have been well established[6].

The total response of the iron whisker with mobile impurities is given by a balance of these fields including the stabilization field. The response is given by Eq. (5) but with β_e replaced by $\beta_e + \beta_s$ and α_D replaced by $\alpha_D + \alpha_s$.

EXPERIMENTS

The sample investigated in some detail in this work is the etched whisker described in an accompanying paper on the behavior at the critical temperature[7]. As the present work considers only the motion of the domain wall when it is in the center of the whisker, the fact that the corners have been rounded is of consequence only in that it results in some small uncertainties in the absolute values of D and of the conversion factor from voltage to wall displacement. The whisker is used as the core of a transformer with concentric windings. The current in the primary is sufficiently low that the wall amplitude is generally between 5 and 10 nm. The measuring frequency is varied from 50 Hz to 20 kHz. The parameters $1/4\pi D$, β_e, χ_s and τ_s are found from fitting the data up to 8 kHz. Above 10 kHz β_e is slightly frequency dependent. As can be judged from fig. 1 and fig. 2, the single relaxation time model provides a quite good account of the data. The temperature dependence of τ_s and χ_s are shown in fig. 3. It is clear from this example that the oscillation of a single 180 degree wall is a suitable tool for investigating the mobility of interstitial impurities.

On the other hand, the interaction of the wall with mobile interstitial impurities presents an opportunity to investigate some of the properties of the domain wall[8]. If we let the impurities reach equilibrium with a stationary wall and then measure the restoring force in a time very short compared to the relaxation time τ_s, we can obtain directly a detailed comparison with the application of Néel's theory to a single 180 degree wall. One of the methods we have used is to measure the a.c. susceptibility during the application of a pulsed field superimposed upon the a.c. field. The temperature is low enough and the frequency high enough that β_s is negligible while the frequency is low enough that the contribution from β_e is also negligible except at the instability points. As we are controlling fields and not magnetization, an instability sets in when the rate of decrease of the restoring field with wall displacement becomes greater than the rate of increase of the demagnetizing field with wall displacement. It so happens that our demagnetizing field is sufficiently small to produce a large jump in magnetization at the instability points.

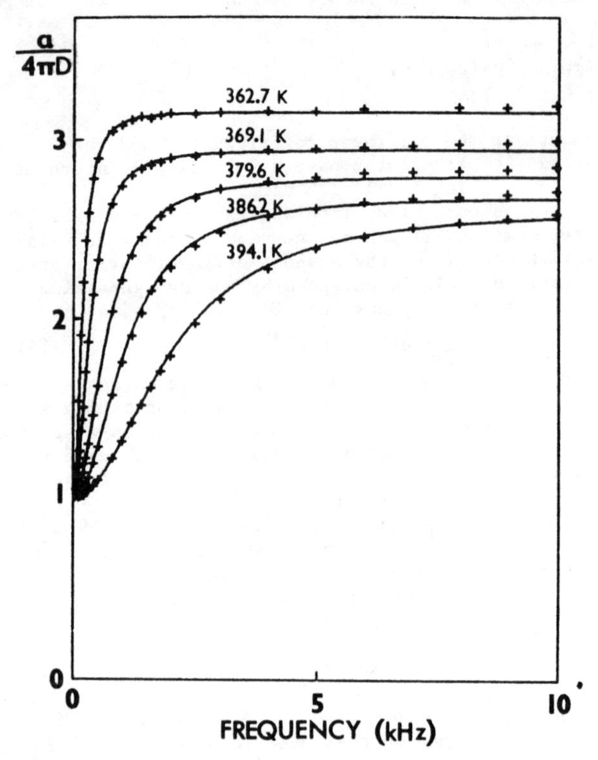

Fig. 1

The magnetic stiffness parameter α from the small amplitude a.c. magnetic response at several temperatures. The solid lines are fits to $\alpha = \alpha_D + \alpha_s = 1/4\pi D + \chi_s \cdot \omega^2\tau_s{}^2/(1+\omega^2\tau_s{}^2)$.

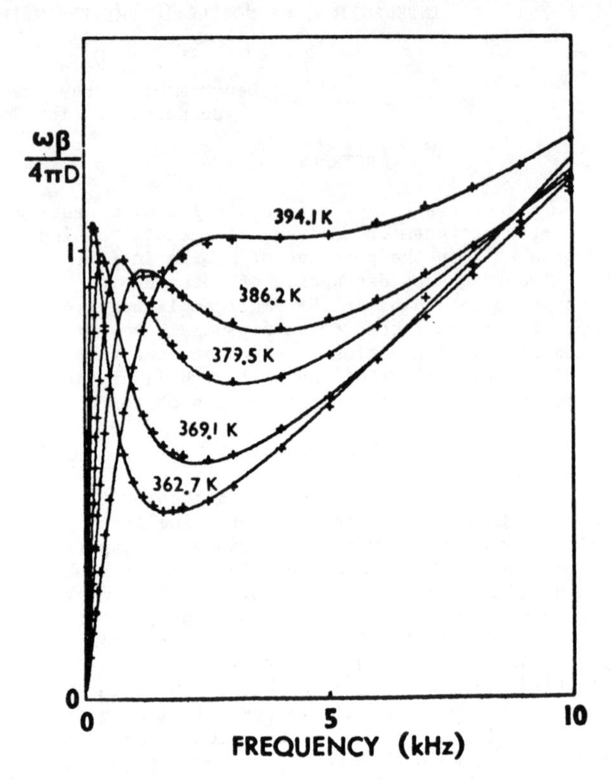

Fig. 2

The magnetic viscosity ωβ from the small amplitude a.c. magnetic response at the several temperature. The solid lines are fits to $\omega\beta = \omega\beta_e + \chi_s\ \omega\tau_s/(1+\omega^2\tau_s{}^2)$ using the same χ_s's and τ_s's as in fig. 1.

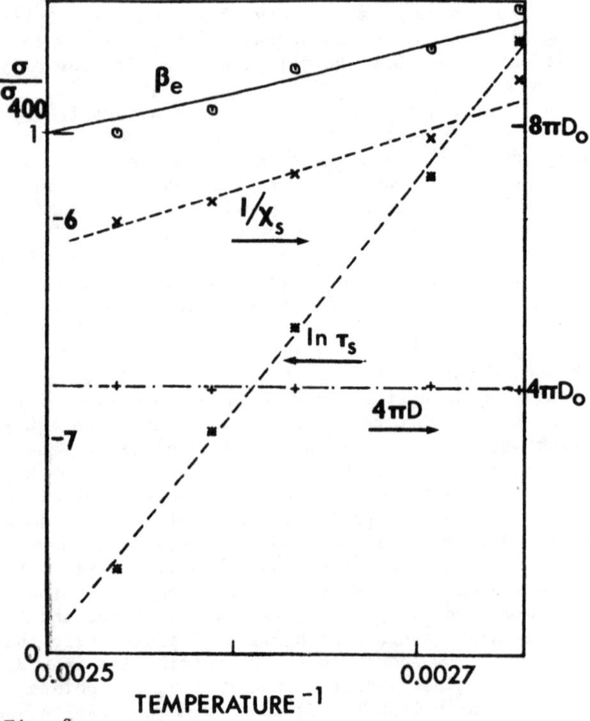

Fig. 3

Schematic illustration of the stabilization field H_s, the sum of H_o and H_D, and sum of $H_s + H_o + H_D$ as a function of the wall displacement (which is proportional to the change in magnetization). H_1 and H_2 are the values of an additional applied field at which instabilities occur. A similar curve is shown in ref. 5.

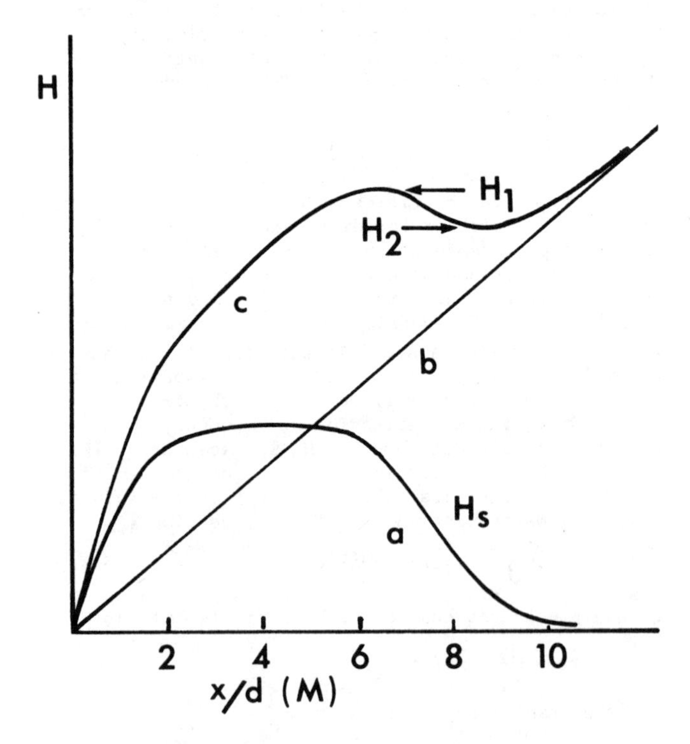

Fig. 4

The dependence of the parameters D, β_e, χ_s and $\ell n\ \tau_s$ upon the inverse temperature. The solid line compares β_e to the resistivity.

The effective field opposing the changing applied field is the sum of $H_s(0)$, the stabilization field for short times, H_o the applied field at which the wall was stabilized and H_D the demagnetizing field. These fields are shown schematically in fig. 4. If the changing applied field reaches H_1 the wall will move rapidly to a new equilibrium position, while if the field is then decreased to H_2 it will jump back. If the a.c. field includes the fields H_1 and H_2 there will be two spikes in the response during each cycle. The spike is independent of frequency for low frequencies where the eddy currents completely determine the speed of the wall in the instability region. Such an instability provides a sensitive caliper for measuring the effective width of a domain wall. As the change in magnetization is directly proportional to the wall displacement one can use fig. 4 to predict the response to an applied field that varies linearly in time on a scale fast compared to the the relaxation time. A typical result is shown in fig. 5.

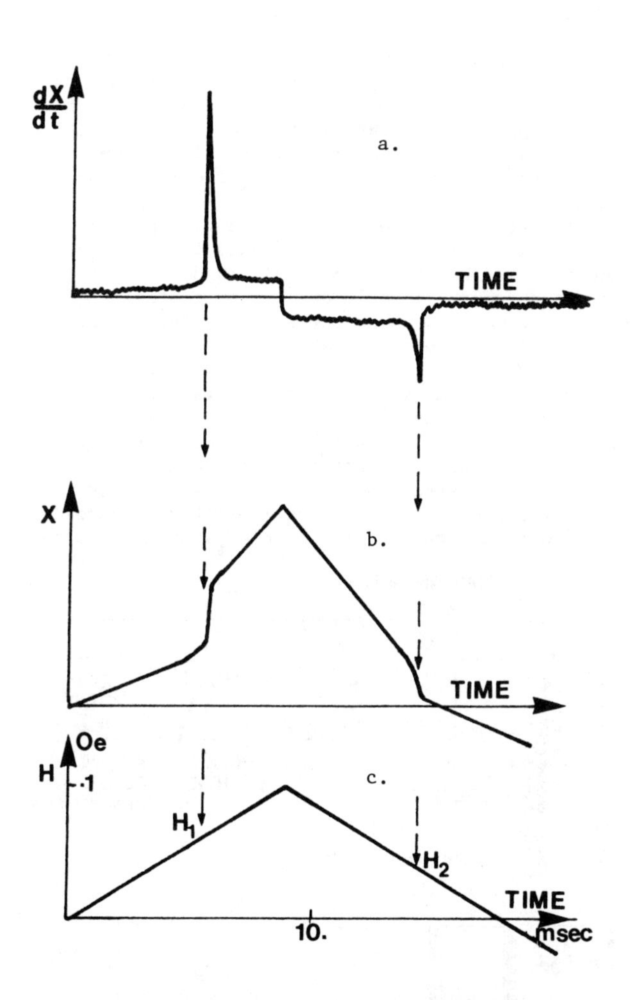

Fig. 5

The response of the 180 deg wall to a triangular pulse in the externally applied field. The voltage in the secondary coil is proportional to the rate of change of the magnetization which is a function of the position of the wall. The function is approximately linear. a. The time dependence of the voltage in the secondary during the first cycle of the triangular pulse after the wall was stabilized; b. The integral of the voltage; c. The applied field.

There is however a problem in that the wall is bowed. As soon as the force is non-linear in the wall displacement it is not as simple to apply the theory as it would be if the wall were straight. For example the measured hysteresis (the difference between the H_1 and H_2) is apparently larger than we would have predicted using the theory for a straight wall. This is currently being investigated.

ACKNOWLEDGEMENT

The authors are indebted to Professor F. J. Friedlaender for encouraging them to pursuing this investigation.

REFERENCES

1. E. Kneller, Ferromagnetismus, Springer-Verlag Berlin 1962.

2. F. Schreiber, Zeit. angew. Phys. 9 ,203 (1957).

3. L. Néel, J. de Phys. 13, 249 (1952).

4. R.E. Maringer, J. Appl. Phys. 35, 2375 (1964).

5. G.W. Rathenau in Magnetic Properties of Metals and Alloys, Am. Soc. for Metals, Cleveland, 1959, pp. 168.

6. A. Arrott, B. Heinrich and D.S. Bloomberg, AIP Conference Proc. 10, 941 (1973); D.S. Bloomberg and A.S. Arrott, Can. J. Phys. 53, 1454 (1975).

7. B. Heinrich and A.S. Arrott, this conference.

CREEPING MAGNETIZATION REVERSAL IN SEMI-HARD FERROMAGNETIC WIRES

Ivan J. Garshelis
Research Associates, Inc., Linden, NJ 07036

ABSTRACT

Creeping magnetization reversal in small second quadrant easy axis bias fields, by hard axis field pulses, is observed in several semi-hard Fe Co X alloy wires of high squareness ratios. The creeping reversal is activated by the circumferential fields associated with longitudinally conducted alternating current pulses. Creep rate depends strongly on field combinations and extends over a range exceeding 10^6. Threshold conditions are not definite. Under certain combinations of hard and easy axis fields, a precipitous rise in creep rate occurs after a large number of current pulses. Unipolar pulses are markedly less efficacious and show diminishing relative effect in decreasing bias fields. The strong dependence of creep rate on field combinations, together with the radial gradient of the hard axis field, establish the progression of reversal from the surface towards the center of the specimens, an action apparently aided by the formation of circumferential domains.

INTRODUCTION

Creeping reversal of magnetization has been the subject of many previous studies. Most of these have been concerned with this occurence in thin (< 150 nm) films[1] and plated wires,[2] each of which approximates a two dimensional magnetic structure. Primary interest has been directed towards magnetic coatings with low magnetoelastic and magnetocrystalline energies, endowed by processing with low dispersion uniaxial anisotropies typically due to directional order. In plated wires, the easy axis direction of most frequent interest has been circumferential. In both of these forms the creeping reversal is activated by the periodic application of hard axis fields while the previously magnetized material is under a constant second quadrant easy axis field. The field combinations at which creep occurs are substantially below those which define the static threshold curves.

Creeping reversal from longitudinal remanence has also been observed in solid iron wires[3] under constant, small, second quadrant fields upon the passage of longitudinally conducted sinusoidal alternating currents.

We present observations of creeping reversal under somewhat similar conditions in some semi-hard ferromagnetic alloy wires. These materials, characterized in Table I, all possess a longitudinal easy axis deriving from crystallographic texture due to cold drawing and the second phase particles precipitated during subsequent heat treatment. Thus, although these materials have appreciable crystal anisotropies with cubic symmetry, in wire form they are effectively uniaxial. Since H_K far exceeds H_C, it is assumed that reversal normally proceeds by 180^0 wall motion.

EXPERIMENTAL

Specimens described in Table I were each prepared as the central conductor in a coaxial arrangement as shown in Fig. 1(a). The specimen and its outer sleeve were connected together at the bottom end to provide a continuous current path having no significant external magnetic field. The assembly containing the specimen to be tested was placed within the small "B" search coil centrally located within a long solenoid as shown in Fig. 1(b). An identical, proximately placed "H" coil was connected in series opposition so that their combined output signal was due only to axial magnetization changes within the specimen.

The specimen was first magnetized to technical

TABLE I

Specimen	1	2	3	4
Material	Nibcolloy 85Co12Fe 3Nb	Vacozet 200 85Co12Fe 3Nb	Vacozet 655 55Co29Fe 12Ni3Ti1Al	Remendur 48Co48Fe 4V
Treatment	700 C	650 C	600 C	580 C
Dia/L	.6/150 mm	.6/150	.6/150	1.27/190
M_S	1.75 T	1.75	1.65	2.1
M_R	1.5 T	1.5	1.45	1.78
H_C	1680 A/m	1670	5270	3730
λ_S	$+5 \times 10^{-6}$	+5	+48	+65
K_1 (j/m³)	-10,000	-10,000		-15,000
Ref.	(4)	(5)	(5)	(6)

saturation along its easy axis by a momentary 64,000 A/m field. A constant direct current was then passed through the solenoid to establish an appropriate bias field antiparallel to the remanent magnetization.

Current pulses of either alternating or unipolarity were then conducted through the specimen assembly. Within the specimen, the circumferential fields from these longitudinally conducted current pulses provided the hard axis field cycling necessary to activate the creeping reversal of the easy axis magnetization. The range of observation included circumferential fields having peak magnitudes at the specimen surface of nearly 16,000 A/m, requiring peak current densities which sometimes exceeded 100 A/mm². Sinusoidal current waves were thus precluded since crest factors from 15-25 were necessary to prevent more than a few °C temperature rise in the specimen. Suitable current pulses were obtained from the transient flow of charge in a series RLC circuit operated synchronously at power line (60 Hz) frequency by an electronic switch. Unipolar pulses were selected by diodes which shunted unwanted

RMS AMMETER
PEAK VOLTMETER
PULSE COUNTER
AC-DC PULSE GENERATOR
150-200 mm
~3 mm
SPECIMEN
INSULATION
BRAIDED SLEEVE
(a)

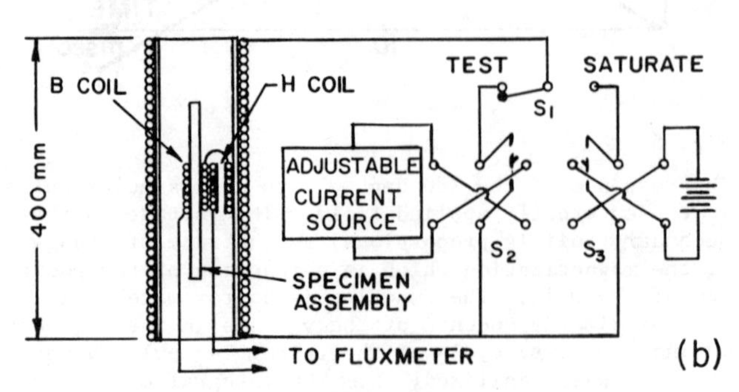

Fig. 1. Specimen assembly, apparatus and schematic experimental arrangement.

alternate pulses through an equivalent resistor.

Typical pulses had a smoothly declining rate of rise from a maximum of ≈ 3 A/μsec to a relatively flat peak (within 95% of maximum for ≈ 10 μsec), followed by a fall from 90% to 10% in ≈ 50 μsec. Although the peak was generally of insufficient duration to obtain uniform current density over the entire cross section, adequacy of the uniformity was indicated by both the rapidly diminishing $d\phi/dt$ during the flat portion of the current pulses and the eventual reversal of the magnetization over nearly all of the area under appropriate field combinations.

Variations in pulse shape noticeably affected the pulse count required for particular degrees of reversal, markedly so in very small bias fields. Loading of the search coils during pulsing caused similar effects.

OBSERVATIONS

The most obvious immediate observation is that changing field combinations varies the creep rate i.e., the magnetization change per pulse, over an enormous range. The specific combinations are different for each material, but the range extends from nearly complete reversal by just one pulse to an easily measurable creep still proceeding after 10^6 pulses.

It was not found possible to determine precise creeping thresholds since the current pulses caused some loss in magnetization even in moderate first quadrant fields. Although this loss was found to stabilize after at most a few hundred pulses, its magnitude and the pulse count required for stabilization both slowly but continuously increase with initial increases in second quadrant fields. With further field increases, there usually follows a rapid increase in the acceleration of the creep rate, ultimately reaching a value different in each material but more or less constant and somewhat independent of the peak pulse magnitude.

Thus, a bona fide creep range may be convincingly defined as the region in which the average creep rate, taken over some arbitrary number of pulses e.g., 10,000, increases linearly with bias field. Extrapolating this line back to zero creep rate establishes a threshold, however artificial. An alternative closely equivalent method is to arbitrarily choose some fixed fraction e.g., ½, of the peak acceleration as the indication of the onset of creep. Fig. 2 shows creep thresholds determined in this manner. Each point on these curves

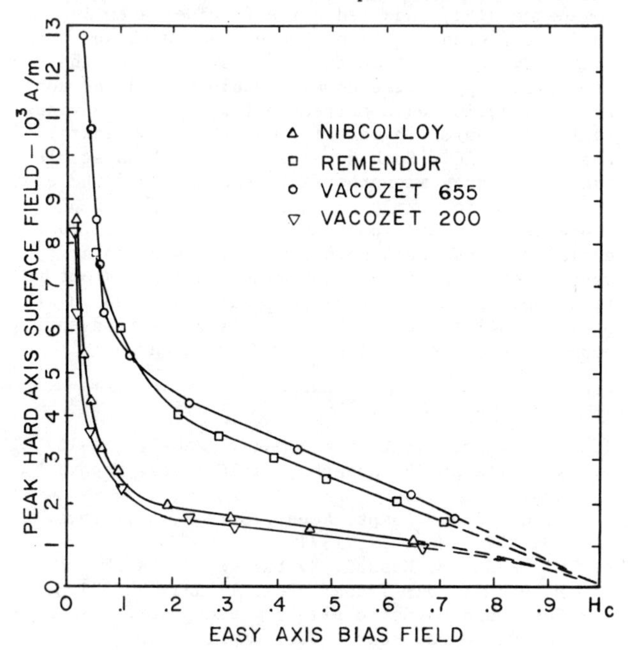

Fig. 2. Threshold curves above which creeping occurs.

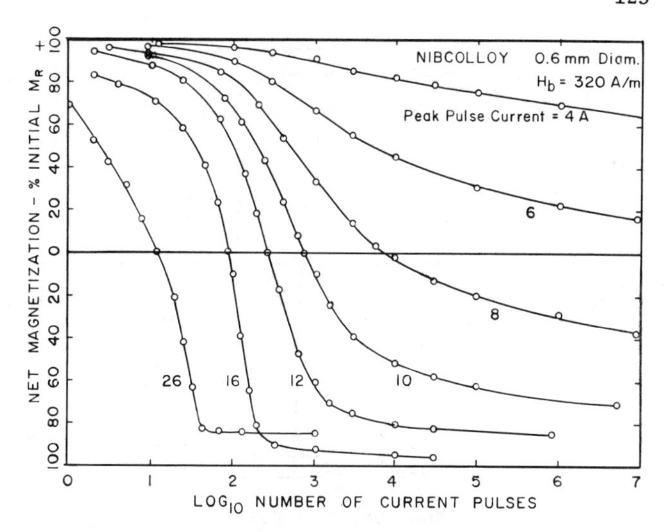

Fig. 3. Remanent magnetization following N bipolar pulses of peak value indicated.

would be shifted slightly to the left with increasingly patient observers, since the average creep rate curves make their sharp upward turn at lower bias fields when the average is taken over larger numbers of pulses.

Further manifesting the lack of clear thresholds is the failure of the magnetization to attain limiting values in all combinations of fields. This is seen in Fig. 3 by the diminishing but still finite creep rates under some field combinations even after 10^7 current pulses. Also seen in this figure is the permanent loss of some easy axis magnetization by large current pulses with circular fields comparable to H_K over part of the cross section.

The radial gradient of the circular field and the variation in creep rate with field combinations suggests that the magnetization reversal takes place progressively from the surface towards the center. This is supported by the data shown in Fig. 4.

In two specimens of Nibcolloy (prepared in like manner to specimen #1), the reversal was interrupted after 10 and 15,000 pulses under the conditions respectively indicated. The specimens were then etched in acid to successively smaller diameters with periodic interruptions for measurement of the remanent flux. The rise and subsequent fall of the remanent flux take place approximately along the lines with slopes of (-)

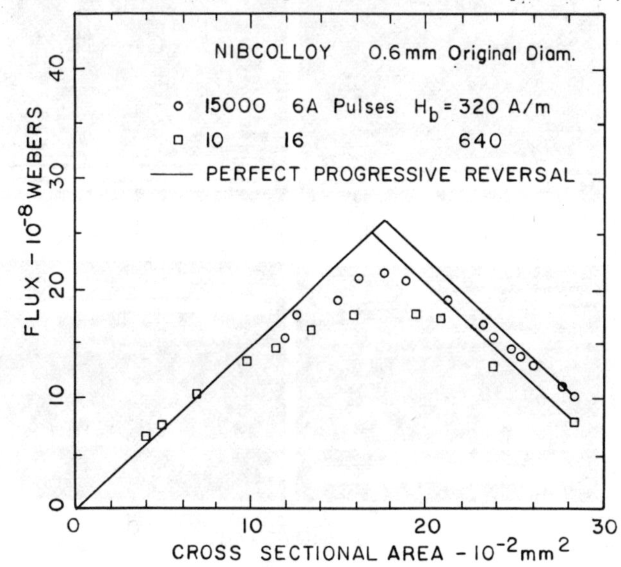

Fig. 4. Remanent flux after etching to cross sectional areas indicated.

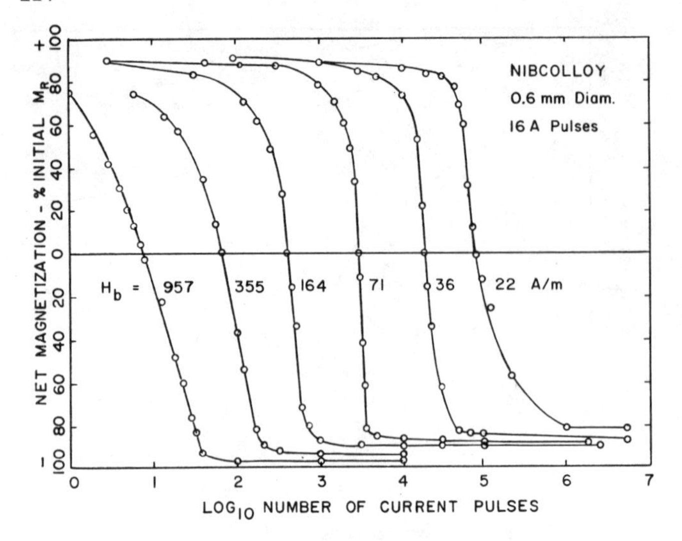

Fig. 5. Remanent magnetization following N 16A peak bipolar pulses in bias fields shown. Peak $H_t = 8488$ A/m at surface.

and (+) M_R except for the transition zone in which the reversal was actively taking place.

The non-monotonic variations in creep rate with pulse count shown in Fig. 3, (e.g. with 8 - 16A pulses) were found only in the 85% Co alloys. The creep rate in the other materials follow an approximate logarithmic decrement throughout most of the reversal in all field combinations thus far studied. This is the expected variation for radially progressive reversal because of the gradient in the circular field and the diminishing unreversed area.

The unusual nature of this creep rate variation and the range of conditions in which it occurs are illustrated in Fig. 5. The creep rate at the highest bias field shown follows the normal logarithmic decrement. In somewhat smaller bias fields e.g., 355 A/m, the reversal proceeds at an approximately constant rate. In still smaller fields the rate is very low for substantial numbers of pulses after which it rises precipitously and completes most of the remaining reversal at a high rate. This reversal route occurs in a limited range of easy axis fields and only at relatively high pulse currents. This may be seen by comparing the curves shown in Figs. 3 and 5.

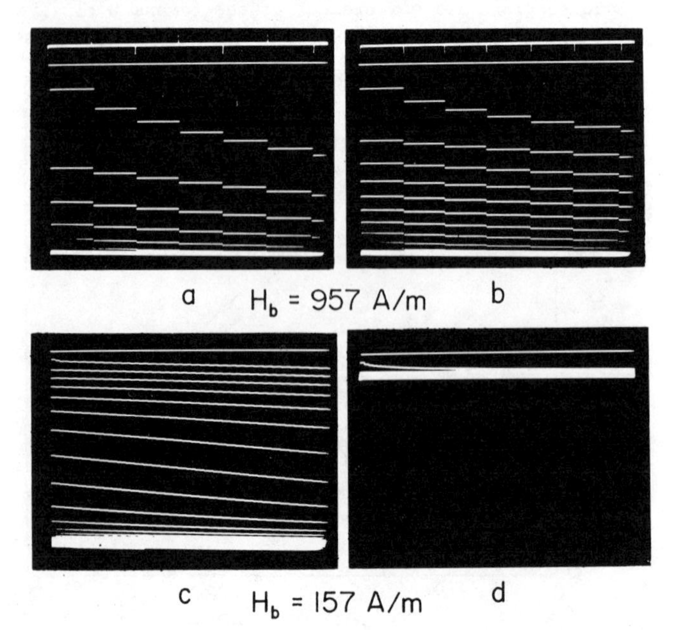

Fig. 6. Comparison of magnetization changes caused by 16A peak bipolar (left) and unipolar (right) current pulses in easy axis bias fields indicated. Peak $H_t = 8488$ A/m. Specimen #1.

The ascension of the limiting values of the curves in Fig. 5, indicate an unreversed central core whose radius apparently increases with decreasing bias fields. From this data, another creep threshold criteria may be established. Thresholds derived in this manner are considerably smaller than those shown in Fig. 2. The large pulse counts required before the onset of substantial creep (>200,000 with $H_b \simeq .01\ H_C$) emphasize the need for patient observation in determining thresholds. At these low levels, the bias field magnitude loses definite meaning due to the spatial and time variations of the specimen's demagnetizing field.

The creeping rate for unipolar pulses has no unusual variations. It was found to follow a logarithmic decrement in all materials under all field combinations. Under conditions for which the creep rate for bipolar pulses also followed a logarithmic decrement, the exponent of the unipolar decrement is always larger. This is clearly shown in the oscillograms of Fig. 6. The top lines in (a) and (b) show the sequence of the current pulses. The next line down shows the starting position of the magnetization. Obviously, the magnetization change caused by the first pulse is the same regardless of subsequent pulse polarity. Although the creep rate decrement is obviously greater with the unipolar pulses (b), the magnetization reversal is eventually brought to approximately the same degree of completion as with the bipolar pulses (a). Under the smaller bias field conditions shown in (c) and (d), the reversal with unipolar pulses reaches barely 14% of the completion achieved with bipolar pulses. (c) shows the unusual growth in creep rate, exceeding in this case three times the approximately constant rate of the early stages of reversal.

DISCUSSION

The indefinite thresholds and the variations in creep rate over the range of materials, field combinations, pulse count and polarity indicates that creep occurs by more than a single mechanism. Locally and initially near the surface, small circumferential domains may form in response to the field of the enclosed current. This is suggested by the loss of some axial remanence with conducted current pulses even in first quadrant fields. Once nucleated, and with an interim stability derived from the cubic symmetry of the crystal anisotropy, these domains would expand and contract radially during each current pulse. Inward advance would be favored in material magnetized antiparallel to the bias field. Such domains would be less stable in the more highly magnetostrictive alloys and therefore less likely to form a cooperative network capable of concerted advance. Under unipolar pulses, circumferential domains would have always the same polarity and consequently their wall movements during any one pulse rise would mostly retrace the paths taken during the previous pulse decay, thus providing fewer adventitious opportunities for irreversible inward advance.

REFERENCES

1. W. Kayser, IEEE Trans. Magn., MAG-3, 141-167, 1967
2. W. Doyle and R. Josephs, IEEE Trans. Magn., MAG-8 306-309, 1972
3. D. Cozmita, Compt. Rend., Ser. B, Sci. Phys. 273, 904-907, Nov. 22, 1971
4. M. Okada, M. Kassai, Y. Suzuki, T. Sasaki and Z. Henmi, JIM Trans., Vol. 13 No. 6, 391-395 1972
5. Semi-hard Magnetic Alloys, Vacuumschmelze GMBH Hanau, Data Sheet G 006, April 1975
6. Product of Carpenter Technology. A Higher V content than Remendur as described in M.R. Pinnel and J.E. Bennett, Met. Trans., Vol. 5, 1273-1283 1974

INTRA SITE CATION ORDERING AND CLUSTERING IN NATURAL Mn-Zn FERRITES

R. H. Vogel and B. J. Evans
Department of Chemistry
The University of Michigan, Ann Arbor, MI 48109

and

L. J. Swartzendruber
National Bureau of Standards, Gaithersberg, Maryland 20760

ABSTRACT

Cooling rates of 10^{-3} K/yr. permit naturally occurring Mn-Zn ferrites (franklinites) to be in thermodynamic equilibrium at ambient temperatures with respect to both electrostatic and elastic energies. The elastic strain energy resulting from the occupancy of the A sites by ions of differing sizes is minimized by clustering of Mn and Zn in different regions of a crystal. Using ^{57}Fe NGR, these regions have been observed directly in franklinite with low Mn^{3+} contents and their structures elucidated. The Mn rich region has a local crystal chemistry very similar to that of $MnFe_2O_4$ and the Zn rich region has a local structure similar to that of $ZnFe_2O_4$. The two regions are crystallographically coherent. Laboratory heat treatments destroy the Mn-Zn ordering and alter the magnetic properties. The Fe spin configurations are also observed to be collinear. The discovery of the intrasite ordering in these materials proves that Goodenough's criteria for square B-H loops in ferrites can be realized. For samples containing Mn^{3+}, the effects of the Jahn-Teller distortions do not permit a direct observation of Mn-Zn intrasite ordering.

INTRODUCTION

Direct evidence for the proposal that the square B-H loops in ferrites can result from chemical inhomogeneities in grains without crystallographic discontinuities[1] has been lacking. Recent reviews[2] of materials for ferrite memory cores indicate that this proposal is still without a firm experimental basis despite the considerable volume of indirect evidence[3,4]. In addition to grain boundaries, Goodenough has shown that chemical inhomogeneities can also serve as nucleation sites for reverse domains.[1] The chemical inhomogeneity believed to be responsible for the square loops in actual memory core ferrites is clusters of Jahn-Teller ions such as Cu^{2+} and Mn^{3+}. Above some critical concentration, the local Jahn-Teller distortions of Mn^{3+} and Cu^{2+} will interact cooperatively to distort the spinel structure from cubic to tetragonal symmetry. Below this critical concentration, the distortions are uncorrelated and the elastic strain energy is rather large. The clustering of the Mn^{3+} and Cu^{2+} ions results in a decrease in the elastic strain; the Jahn-Teller distortions play no critical role in the magnetization processes. The requirement that the cluster regions have the same cubic symmetry as the surrounding spinel matrix seems, however, to contradict the very basis on which the clustering is predicated.

Chemical inhomogeneities resulting from clusters of S-state ions would provide a better fit to the theoretical model. Clustering of S-state ions is driven primarily by differences in size, charge state or both, but the enthalpy change associated with such clustering or ordering is rather small for 2-3 spinels such as the Mn-Zn ferrites and is likely to be smaller in magnitude than the TΔS entropy term at temperatures where ion diffusion is sufficient to permit clustering to occur. Clustering has been observed to occur in Sb^{5+} substituted $LiFe_5O_8$[5] but in this material the oxidation numbers of the ions differ by 4.

For sufficiently slow cooling rates over a broad temperature range, clustering is expected to occur even in 2-3 spinels. The lowest cooling rates are observed in natural metamorphic settings and evidence for the existence of clustering in natural Mn-Zn ferrites are presented in this paper. Mn-Zn ferrites occur in nature as the mineral franklinite. The occurrence of franklinites as fissure-and inclusion free octahedra with edges as much as 5 cm in length and having very narrow composition ranges is evidence of their having being slowly cooled over several thousands years, at least. Further evidence for the thermal equilibrium of franklinites is provided by the macroscopic geologic parameters of the deposit in which they occur.[6]

EXPERIMENTAL

^{57}Fe nuclear gamma-ray resonance(NGR) measurements, x-ray powder and single crystal diffraction measurements and wet chemical analysis were performed on five specimens of franklinites. Specimens were chosen primarily on the basis of crystal perfection, with composition being of secondary importance. A typical octahedral crystal weighed approximately 150 gm and was about 3 cm long on each edge. Thin sections were cut parallel to selected crystallographic planes using a diamond section machine and a Laue back-reflection camera for alignment. The thin sections were polished on diamond impregnated laps. Laue back-reflection photographs of the crystals and polished thin section exhibited sharp diffraction spots and no indications of the presence of more than one phase. Portions of the polished thin sections were powdered and used in the lattice constant determinations, wet chemical analysis and NGR measurements. Lattice constants were determined using a powder diffractometer and Mn filtered, Fe Kα radiation and a silicon internal standard.

The ^{57}Fe NGR measurements were performed on a constant acceleration spectrometer with an electromechanical velocity transducer and a 512 channel multichannel analyser operated in the multi-scaling mode. A 25 mCi Co^{57}/Rh source was employed, and Fe metal and sodium nitroprusside were used to calibrate the spectrometer. The Néel temperatures were determined using the thermal scanning NGR technique.[7] The Néel temperatures have an estimated error of ±5 K. The NGR applied field spectra were obtained in a superconducting solenoid in which the source and absorber temperature were separately controlled with a temperature stability of 1 K. Platinum resistor thermometer and an iron-constantan thermocouple were used for temperature control in the Néel temperature determinations and semiconductor diodes were used for temperature control during the applied field measurements. The NGR absorbers were in the form of thin discs prepared from acetone/Duco cement slurries of the powdered thin sections and were either mounted on mylar films or held between rigid boron nitride wafers.

RESULTS

The lattice constants and chemical compositions of the samples are listed in Table I. The ^{57}Fe NGR spectra at 298 K and at 50 K in an applied magnetic field of 50 kG are shown in Figs. 1, 2 and 3. The solid lines are the results of fitting the data points by a least-mean-squares technique to a number of different Lorentzian lines. The strong structure in the residual of the fit of two Lorentzian lines to the spectrum in Fig. 1. shows this fit to be unsatisfactory; the presence of at least one other quadrupole doublet is indicated. Several fits of two quadrupole doublets were attempted and the resulting parameters were observed to approximate those of $ZnFe_2O_4$[8] and $MnFeO_4$[9] at 298 K. The spectra of the low Mn^{3+} samples were therefore fitted to two quadrupole doublets with quadrupole splittings (ΔE_Q) and isomer shifts (δ) constrained to those of pure $ZnFe_2O_4$ and $MnFe_2O_4$, e.g., $\Delta E_Q = 0.58$ and $\delta = 0.31$ for $MnFe_2O_4$ and $\Delta E_Q = 0.35$ and $\delta = 0.37$ for $ZnFe_2O_4$. The intensities were free variables. The Chi-square of this fit was as small as or smaller than that of any other fit that corresponded to physically reasonable models of possible crystal chemistries. Especially noteworthy in Table I is the larger asymmetries in the areas of the two apparent lines for low Mn^{3+} samples compared to the high Mn^{3+} samples.

If the sample used to obtain the spectrum in Fig. 1 is annealed under vacuum at 600°C for 12 hours, the spectrum shown in Fig. 2 is obtained. It is obvious from the residual, the asymmetry in the line is much reduced; and the area ratio is observed to increase from 0.808 to 0.864.

At 77 K, the spectrum of all of the samples are qualitatively similar consisting of a broad apparent six line pattern. The average magnetic hyperfine field (H_{eff}) for sample 3, $Mn_{.487}Zn_{.699}Fe_{1.814}O_4$ in the absence of an applied field is 497 kG; in a 50 kG field at 50 K, the spectrum shown in Fig. 3 is obtained and two patterns are observed. One pattern is rather broad and has the smaller splitting corresponding to H_{eff} of 343 kG and the more widely split, sharper pattern has a H_{eff} of 467 kG. Both H_{eff} values are reduced below the zero-field values indicating that all Fe ions are on the B site and that the sample is ferrimagnetic. Fig. 3 is typical of the spectrum obtained for all samples but the relative widths and intensities of the lines of the two subspectrum varied considerably from sample to sample. For the low Mn^{3+} samples in the 50 kG applied field, the area ratios of the subspectrum with the large splitting to that with the smaller splitting were observed to be quite similar to the ratio of the areas of $MnFe_2O_4$ pattern to $ZnFe_2O_4$ pattern in the 298 K paramagnetic spectra; for example $A_{MnFe_2O_4}/A_{ZnFe_2O_4} = 0.32$ at 298°K and $A_{(large\ H_{eff})}/A_{(small\ H_{eff})} = 0.59 \pm 0.1$ at 50 K in a 50 kG field. This was not true of the high Mn^{3+} samples.

DISCUSSION

The presence of two patterns in the 298 K spectra of the low Mn^{3+} franklinites with ^{57}Fe NGR parameters appropriate to those of pure $ZnFe_2O_4$ and $MnFe_2O_4$ is interpreted as indicating the existence of two different regions in these samples having the corresponding compositions and local structures. The area ratio of the two doublets, $A_{MnFe_2O_4}/A_{ZnFe_2O_4}$ of 0.32 for $Mn_{.347}Zn_{.730}Fe_{1.923}O_4$ is very close to the Mn-Zn A site occupancy ratio of 0.37, indicating a high degree of ordering. That some kind of ordering is indeed present is confirmed by the change in the spectrum upon annealing the sample in the laboratory. Further direct evidence is provided by the low temperature applied field spectra in which the area ratios and relative magnitudes of the hyperfine fields are those expected on the basis of the 298 K spectra. Indirect evidence is provided by the value of the Néel temperature which is considerably lower than that of synthetic ferrites with similar compositions[10], and by the fact that upon annealing, the Néel temperature increases from 178 K to 233 K. A low Néel temperature and an increase in T_N upon annealing are in accord with the preposed structural and chemical ordering.

Although the Mn^{3+} content of these samples never reach the critical value necessary to cause cooperative Jahn-Teller distortions of the lattice, there are sufficient uncertainties in the magnitude and direction of the individual Jahn-Teller distortions such that samples with $Mn^{3+} > 0.08$ do not exhibit local structure comparable to those with the pure materials $ZnFe_2O_4$ and $MnFe_2O_4$. Thus, samples with the higher Mn^{3+} content are not susceptible to this type of analysis.

CONCLUSION

Unfortunately, naturally occurring ferrites similar to the samples studied in this paper would have little industrial importance, particularly with Néel temperatures of approximately 178 K. However, they served to demonstrate the presence of intrasite cation ordering and clustering in the less favorable case of the A sites, which share no polyhedral elements and are completely separated and therefore make plausible the existence of clustering and intrasite ordering on the B sites where electrostatic and structural factors provide stronger driving forces for ordering and clustering. Local ordering and clustering on the B sites are, however, difficult to demonstrate directly. Indeed, the present study provides direct and unambiguous evidence only for A site ordering and clustering but the A site order is so high that there must surely be some B site ordering. This ordering observed in these ferrites provide excellent opportunities for studies such as the effects of chemical inhomogeneities on reverse domain nucleation, B-H loop squareness and other magnetic properties. These studies also show that naturally occurring materials can be uniquely useful in magneto-crystal chemical studies of ferrites.

	COMPOSITION	LATTICE a_o (nm)	δ^a mm s^{-1}	ΔE_Q mm s^{-1}	DOUBLET AREA RATIO
1	$(Zn_{.730}Mn_{.270})\ [Mn_{.076}Fe_{1.924}]\ O_4$	0.8451	0.369	0.435	0.808
2	$(Zn_{.721}Mn_{.279})\ [Mn_{.040}Fe_{1.960}]\ O_4$	0.8457	0.365	0.420	0.915
3	$(Zn_{.699}Mn_{.301})\ [Mn_{.186}Fe_{1.814}]\ O_4$	0.8447	0.364	0.459	0.929
4	$(Zn_{.678}Mn_{.322})\ [Mn_{.570}Fe_{1.430}]\ O_4$	0.8476	0.367	0.471	0.903
5	$(Zn_{.665}Mn_{.335})\ [Mn_{.187}Fe_{1.813}]\ O_4$	0.8469	0.368	0.435	0.914

[a]Relative to an Fe metal absorber

Table I Composition, lattice constants, and single quadrupole doublet ^{57}Fe NGR parameters of Zn-Mn-Ferrites at 298 K.

REFERENCES

1. Goodenough, J.B., J. Appl. Phys. 36, 2342 (1965).
2. Greifer, A.P., IEEE Trans. Magnetics MAG-5, 774 (1969).
3. Rogers, D.B., Germann, R.W. and Arnott, R.J., J. Appl. Phys. 36, 2338 (1965).
4. Cervinka, L., Hosemann, R. and Vogel, W., Acta Crmpt. A26, 277 (1970).
5. Evans, B.J. and Swartzendruber, L.J., J. Appl. Phys. 42, 1628 (1971).
6. Metsger, R.W., Tennant, C.B., and Rodda, J.L. Bull. Geol. Soc. Am. 69, 775 (1958).
7. Preston, R.S., Hanna, S.S., Heberle, J. Phys. Rev. 128, 2207 (1962).
8. Evans, B.J., Hafner, S.S. and Weber, H.-P., J. Chem. Phys. 55, 5282 (1971).
9. Konig, V., Solid State Comm. 9, 425 (1971).
10. Vogler, G., Phys. Stat. Sol. (b) 43, K161 (1971).

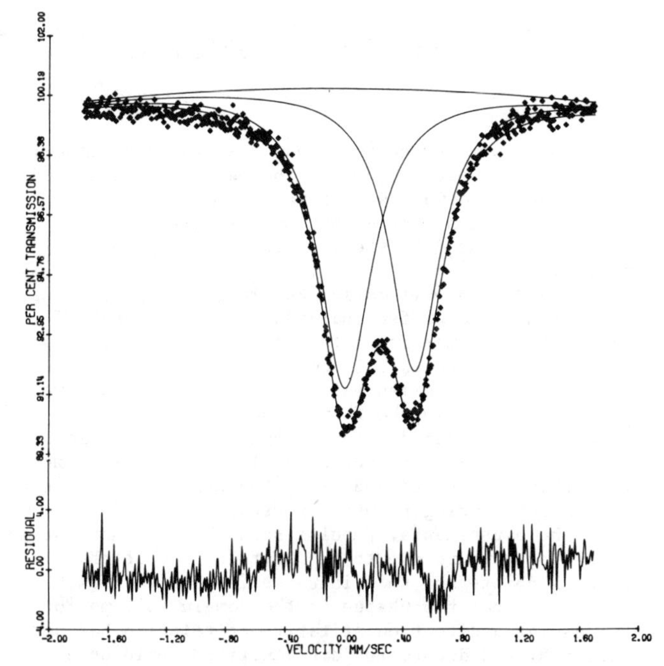

Fig. 2 ^{57}Fe NGR spectrum of sample in Fig. 1 after annealing at 600 C, under vacuum, for 12 hours. The solid line is the result of fitting two Lorentzian shaped lines to the spectrum. Note the much weaker structure in the residual.

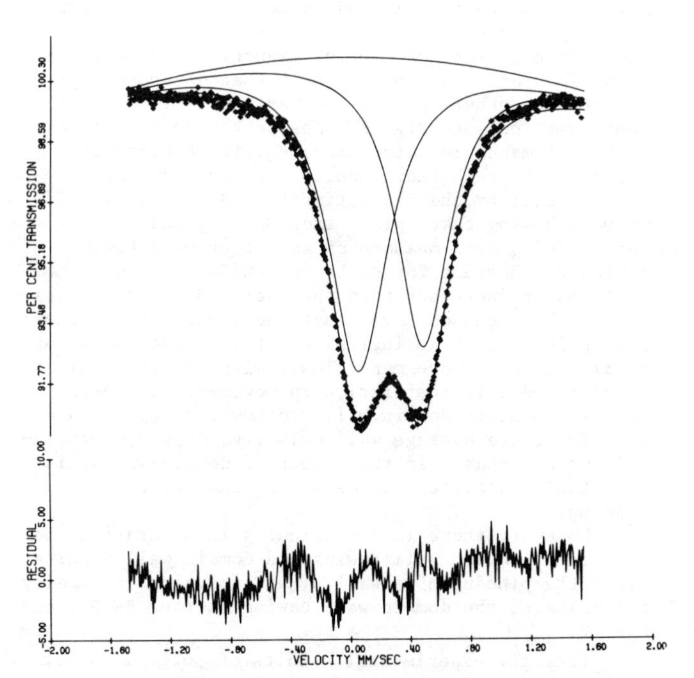

Fig. 1 ^{57}Fe NGR spectrum of $Mn_{.346}Zn_{.730}Fe_{1.924}O_4$ at 298 K. Solid line is a result of fitting two Lorentzian-shaped lines to the spectrum. Note the strong structure in the residual.

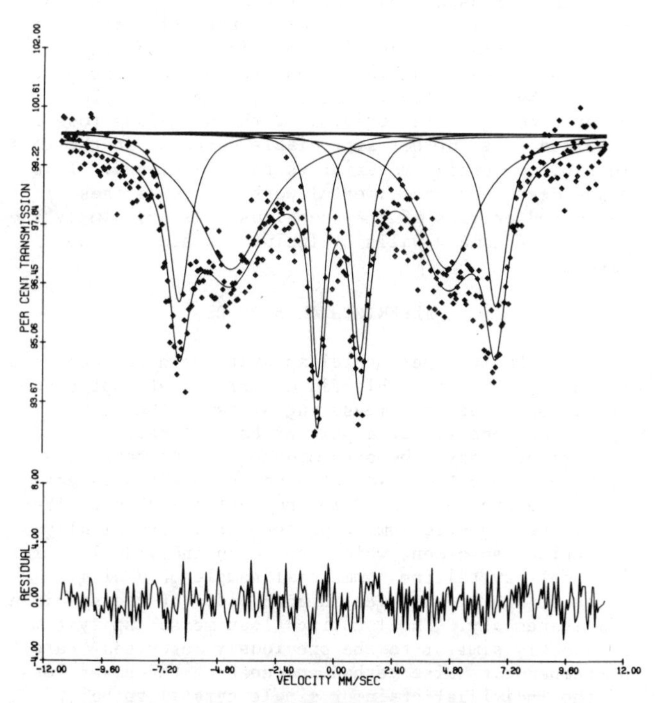

Fig 3 ^{57}Fe NGR spectrum of $Mn_{.319}Zn_{.721}Fe_{1.960}O_4$ at 50 K in a 50 kG applied field. The solid line is the result of fitting two 4-line patterns to the spectrum. Note the absence of the $\Delta M=0$ lines and the absence of any structure in the residual.

THE DEPENDENCE OF POWER LOSS ON DOMAIN WALL MOTION

Mrs. S. Hill and K.J. Overshott
Wolfson Centre for Magnetics Technology, University College, Cardiff CF2 3AD, UK

ABSTRACT

The cause of the anomalous loss in soft magnetic materials has often been attributed in the published literature to domain wall pinning. An experimental program has been carried out to assess the change in power loss caused by the introduction of pinning points.

The domain wall motion and localised power loss have been measured for individual grains in polycrystalline commercial material and for single crystals of 3% grain-oriented silicon-iron. The orientation of both types of specimen was (110) [001]. Domain wall pinning points, in the form of scratches or holes, were then introduced into the specimen. The domain wall motion and localized power loss of the specimens were then remeasured and hence the effect of the introduction of pinning points assessed.

The experimental results show that an increase in power loss occurs due to the introduction of the pinning points and this increase is qualitatively dependent upon the change in the domain wall motion; but as yet a direct quantitative correlation has not been achieved due to the difficulty in defining a satisfactory criterion of wall pinning. The experimental results to be presented also show marked differences between the behaviour of commercial material and single crystals when subjected to the introduction of pinning points.

INTRODUCTION

An increased knowledge of the shape of the loss per cycle against frequency characteristic has shown that the anomalous loss is, in modern commercial grain-oriented 3% silicon-iron, responsible for over 50% of the total loss. Therefore, a more complete understanding of the origins of the anomalous loss is required if a further appreciable decrease in the total loss of commercial material is to be achieved. This paper describes experimental work on one of these suggested origins of the anomalous loss[1,2], namely the pinning of domain walls by inclusions etc. in the material.

EXPERIMENTAL METHODS

Two basic experimental apparatuses have been used in this study: a double-sided Kerr magneto-optic bench and a localized loss measuring system. The former apparatus consists of a pair of basic longitudinal Kerr magneto-optic benches mounted in the horizontal plane one on either side of a specimen which is part of a yoke system mounted in the vertical plane. Therefore, the magnetic domain pattern of a sinusoidally magnetized specimen, which can be an individual grain in a polycrystalline commercial sample or a single crystal, can be observed on both sides of the specimen simultaneously. The localized loss measuring system is exactly similar to the previously described[3] rate-of-temperature rise method and enables the power loss of the individual grain or single cyrstal to be measured under the same conditions of frequency and flux density as the domain wall motions are observed. Therefore, it is possible to make comparisons between the domain wall motion and power loss of an individual grain or a single crystal.

The peak position of the domain walls have been measured for flux densities up to 1.0T over a range of frequency from 20 to 150 Hz. It has been suggested that the fact that the walls disappear from the surface of the material before complete saturation in the volume of the material is reached is due to the phenomenon of wall bowing[4], see Fig. 1. The peak degree of

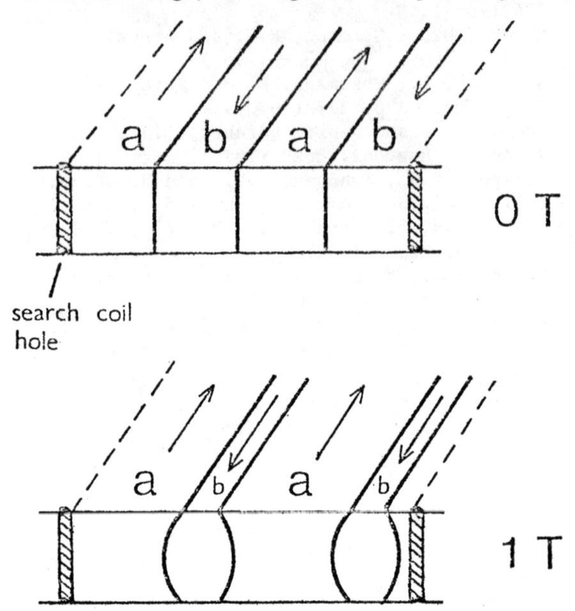

Fig 1 Domain wall bowing. The walls move faster near the surface, and disappear from the surface before the material is saturated

magnetic saturation, as indicated by the domain surface, B_s, can be written as (a-b)/(a+b) where a is the combined width of domains magnetized one way between two search coil holes and b that of those magnetized the other way, at the maximum wall displacement position, see Fig. 1. The peak true or volume degree of magnetic saturation, B_b, is obtained by dividing the peak flux density, measured by the search coil, by the saturation flux density, 2T. If no wall bowing takes place then $B_b = B_s$ and the factor B_b/B_s is a measure of amount of wall bowing taking place since for no bowing $B_b/B_s = 1$ and as the wall bowing increases then the factor B_b/B_s decreases.

It has been observed that the occurrence of pinning points, such as inclusions etc., causes adjacent domain walls to move more slowly with applied flux density and this restriction in movement in general causes the other domain walls to move at faster rates. Therefore, the average wall velocity, cm/sec, normalized for any change in the number of domain walls, is a suitable parameter for assessing the effect of pinning.

However, there is a difficulty in separating the effects of domain wall bowing and domain wall pinning since the pinning of a wall by an inclusion can modify the value of the domain wall bowing factor, B_b/B_s, see Fig. 2.

From the experimental results of power loss, it is possible to define two definitive parameters, namely the loss per cycle and the extrapolated static hysteresis loss which is obtained by extrapolating the loss per cycle against frequency characteristic to zero frequency[5].

It has earlier been shown by Sharp and Overshott[5] that this is not a particularly exact method of ascertaining the static hysteresis loss, therefore the term extrapolated static hysteresis loss is used.

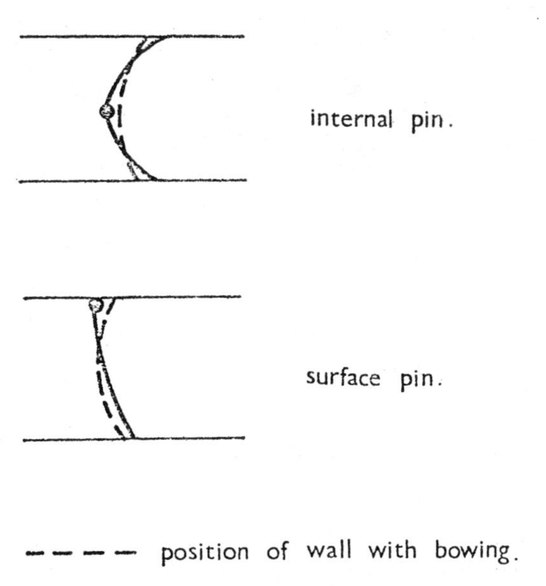

Fig. 2 Effects of pinning and bowing on positon of wall

Therefore, from the experimental measurements it is possible to define four parameters which correlate the domain wall motion and the power loss, namely the domain wall bowing factor B_b/B_s, average wall velocity, total loss per cycle and the extrapolated static hysteresis loss.

The average wall velocity, $\bar{\upsilon}$, has been chosen since changes in this parameter due to the introduction of pinning points is easy to discern. It is hoped that the correlation between the increase in eddy current loss and the more pertinent parameter, $\overline{\upsilon^2}$, can be investigated later.

EXPERIMENTAL RESULTS

The domain wall motion and power loss of individual grains in polycrystalline commercial material and single crystals of 3% grain-oriented silicon-iron have been measured and then artificial pinning points have been introduced into these specimens. The domain wall motion and power loss of the specimens have been remeasured after the introduction of the artificial pinning points and therefore an assessment of their effect on the wall motion and power loss can be made.

The pinning points have taken two forms; namely a surface scratch made in the rolling direction or alternatively a hole drilled completely through the specimen. The specimens were not strain relief annealed after the introduction of the pinning points. These two types of artificial pinning point have a different effect on the domain wall motion and hence on the power loss. The surface scratch tends to retard the motion of the walls in proximity to the scratch but only on one side of the specimen. The opposite end of the wall which is on the other surface tends to move at a faster rate to compensate for the retardation in motion of the pinned end. The introduction of a hole tends to completely pin the motion of the wall or walls, close to the hole, throughout the thickness of the specimen.

The variation of the domain wall bowing factor, B_b/B_s, with frequency is shown in Fig. 3 for a sample in the initial condition and after the introduction of a pinning point. The percentage change in average wall velocity caused by the introduction of pinning points can be obtained from the measurements of domain wall motion.

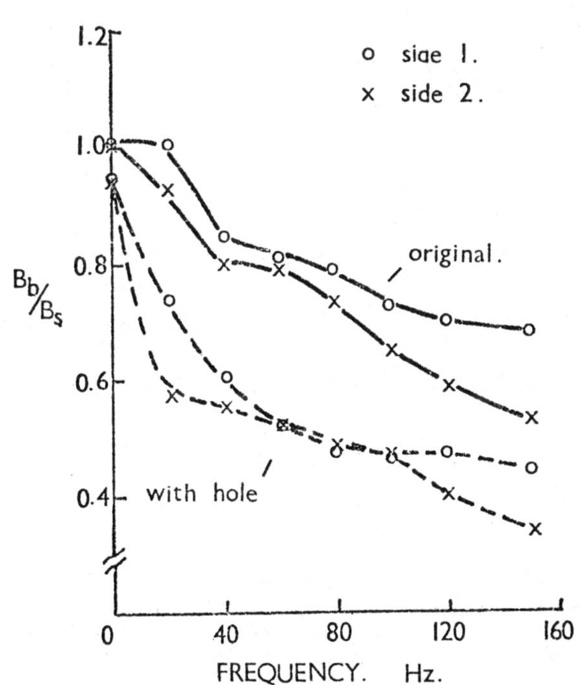

Fig. 3 Increase in wall bowing effect due to introduction of hole in single crystal sample

The change in power loss caused by the introduction of pinning can be obtained by considering the changes in the loss per cycle against frequency characteristic, see Fig. 4. The general changes caused by the pinning points are summarized for four typical samples in Table I.

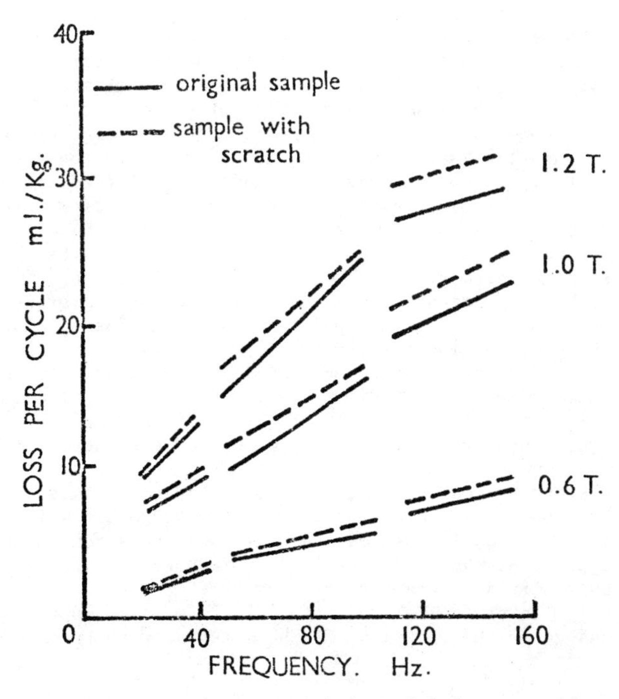

Fig. 4 Effect of introduction of hole on power loss per cycle against frequency characteristic for single crystal sample

Type of Sample	Type of Pinning Point	Conditions	Change in Wall Bowing Factor	Change in Normalised Average Wall Velocity		Change in Total Loss	Change in Extrapolated Static Hysteresis Loss
				Pinned Walls	Unpinned Walls		
Single crystal	Scratch	All	-20%	-11%	+25%	+10%	No change
Single crystal	Hole	All	-34%	-57%	+45%	+20%	Increase (50%)
Commercial	Scratch	> 100 Hz	0%	n.c.	n.m.d.	0%	Decrease (100%)
		< 100 Hz	-15%	n.c.	n.m.d.	-20%	
Commercial	Hole	> 80 Hz	-8%	n.c.	n.m.d.	+170%	Large increase (270%)
		< 80 Hz	+12%	n.c.	n.m.d.	+180%	

n.c. = no consistent behaviour. n.m.d. = no consistent and measurable difference.

TABLE I Change in parameters caused by introduction of pinning points

DISCUSSION

In the single crystals the pinned walls are slowed down while the unpinned walls increase in velocity but in the commercial materials the walls close to the pinning points behave inconsistently since some walls are retarded and others increase their mobility. In this latter material, the velocity of the unpinned walls is not substantially changed which implies that the motion of the domain walls in commercial material is controlled by effects exterior to the individual grain being considered, that is the behaviour of the rest of the material. In addition, the domain wall motion in commercial material is less uniform and more varied than in single crystals.

In general, the amount of wall bowing increases with the introduction of pinning points, that is the wall bowing factor B_b/B_s decreases, which is to be expected when Fig. 2 is considered. If domain wall bowing is caused by lack of flux penetration then these changes should be purely due to the wall pinning.

In the single crystal samples the effect of pinning points is to increase the total power loss and a hole causes a change in the extrapolated static hysteresis loss. It should be remembered that the method of obtaining the static hysteresis loss by extrapolation is not accurate[3] and that a large change in this loss only makes a small variation in the value of the total power loss. In addition, the traditional method of separating losses at power frequencies into static hysteresis loss and eddy current loss has recently been criticised[1,2]. The results for the single crystals, shown in Table I, are consistent in so much as the greatest changes in wall bowing, average wall velocity and extrapolated hysteresis loss produces the greatest increase in total power loss.

The results for individual grains in polycrystalline commercial material show that the principal mechanism in this material is the change in extrapolated static hysteresis loss which occurs on the introduction of the holes or scratches. The value of the hysteresis loss controls the change in total loss and this result is not surprising since the domain wall motion, that is the eddy current loss, does not change appreciably and is controlled by factors exterior to the individual grain being considered.

In the commercial material there is an appreciable change in the behaviour of the material with frequency that is the effect of pinning points is frequency dependent. The same effect is observed for the single crystal specimens but is not appreciable or important. The behaviour of the specimens is to a small extent dependent upon an inconsistent manner on flux density.

CONCLUSIONS

The principal results of this preliminary investigation of the effect of the introduction of artificial pinning points into single crystal specimens and individual grains in polycrystalline commercial samples of 3% grain-oriented silicon-iron can be summarized as follows:

(a) the pinning points increase the amount of domain wall bowing.

(b) the pinning points in the single crystal specimen increase the total power loss since some walls are retarded but the majority move at greatly increased average wall velocity.

(c) the pinning points do not cause a consistent change in the domain wall motion in commercial material since the behaviour of the individual grains being considered is controlled by factors exterior to these grains.

(d) the introduction of pinning points into commercial materials modifies the value of the extrapolated static hysteresis loss and hence the total power loss.

Pinning points in 3% grain-oriented silicon-iron modify the behaviour of the domain walls and vary the power loss but it is not possible, as yet, from this study to state categorically that the anomalous loss is principally caused by the pinning of the domain wall motion.

REFERENCES

1. J.W. Shilling and G.L. Houze, Jr., IEEE Trans. Mag., 10, 195-223, 1974.
2. K.J. Overshott, IEEE Trans. Mag., Proc. of Intermag/MMM, 1976.
3. M.R.G. Sharp, R. Phillips and K.J. Overshott, Proc. IEE, 120, 822-4, 1973.
4. K.J. Overshott, I. Preece and J.E. Thompson, Proc. IEE, 115, 1840-5, 1968.
5. M.R.G. Sharp and K.J. Overshott, Proc. IEE, 12, 1451-3, 1973.

GRADIENTLESS PROPULSION AND STATE SWITCHING OF BUBBLE DOMAINS

B. E. Argyle, S. Maekawa*, P. Dekker**, and J. C. Slonczewski

IBM Thomas J. Watson Research Center, Yorktown Heights, New York 10598

ABSTRACT

We investigate a class of phenomena which we call bubble automotion whereby a bubble or a bubble lattice is propelled by a time-modulated field which is homogeneous rather than by locally applied field gradients. Linear translation of certain kinds of bubbles occurs when bias modulation $H_b+h_z(t)$ having certain amplitudes and large $|dh_z/dt|$ is superimposed on a steady in-plane field H_{ip} which is less than H_b. Pulsed h_z excitations produce characteristic automotion vectors that can uniquely distinguish for the first time among certain wall states (S,L,P) having a common winding number S but different arrangements of L Bloch lines and P Bloch points. A pair of S=1 bubbles we call σ_+ and σ_-, translate typically one micron per pulse in opposite directions orthogonal to H_{ip}, i.e., parallel to $\pm\vec{H}_b \times \vec{H}_{ip}$. A model consistent with our measurements has the static configurations $(1,2,0)_+$ and $(1,2,0)_-$ where the ± refers both to the direction of automotion and the two arrangements (N^+,S^-) and (N^-,S^+) of Bloch line polarity and sense of twist. The automotion arises because the translational coercive force couples to motions of the wall radius and the Bloch lines. Unichiral bubbles $\chi_\pm=(100)_\pm$ of either handedness (±) do not automote.

Excitations outside the range for automotion can cause changes in wall structure. We show how to switch at will among σ_-, σ_+, χ_-, and χ_+ states.

INTRODUCTION

In the past, dynamic propulsion of bubble domains has required gradients of magnetic field with a component parallel to the direction of motion. In bubble devices, the necessity for gradients implies the use of structures[1] (current lines, T and I bars of permalloy, or ion-implanted regions) whose dimensions are on a scale comparable to the bubble diameter or distance between bubbles. An avenue for simplifying devices through diminished reliance on such structures would ensue from a technique of propagating domains by means of homogeneous fields.

Works of Boxall[2] and Hubbell[3] showed how bubbles could propagate parallel to the strip lines carrying currents in the form of pulse trains or sinusoidal a.c. Although the gradient component parallel to the motion direction clearly vanished, the results were interpreted qualitatively in terms of possible skew deflection[4,5] of bubbles in the gradient components normal to the current lines.

A direct attack on gradientless propagation of domains was made by Malozemoff and Slonczewski[6] who showed how a homogeneous in-plane field pulse acting on a 5μm diameter individual domain should theoretically cause some tenths of a micron displacement. Although displacements of this order of magnitude were observed, some specifics, such as direction and reversibility of the predicted motion were not in accord with experiment.

Some very recent work[7] shows how a finite lattice of bubbles may rotate continuously when excited by a continuous train of pulses or sinusoidal a.c. In this case, use of a large pancake coil made contributions of gradients seem less significant. However, the only models proposed again assumed irreversible deflections in the gradient component, this time radial with respect to the coil axis. Still other work[8] shows how homogeneous bias-field pulses produce a limited number of bubble jumps following gradient propagation at high drive.

In Section I of this report our experiments[9a] show that superposition of two homogeneous fields - constant in-plane field plus a time-varying bias field - induces continuous linear motion of certain types of bubbles. Since the bubbles pass singly across the symmetry axis of the coil, gradients are clearly excluded. We call this gradientless propulsion effect "automotion". By this means we find a "spectrum" of propagation modes with characteristic mean velocity vectors. These vectors differentiate between certain bubbles having different winding numbers as well as between certain bubble types having the same winding number, thus serving to identify new wall-structural characteristics.

In Section II we give a quantitative theory for automotion of two bubble states (designated σ_+ and σ_-), each having but one pair of unwinding vertical Bloch lines (BLs). We show the essential roles played by coercivity and by asymmetric wall motions caused by the presence of BLs. These $\sigma_\pm$-states and two unichiral states ($\chi_\pm$) comprise a quartet having winding number unity (S=1). According to our interpretation of the data $\sigma_\pm$ automote in opposite directions and $\chi_\pm$ are immobile under the same conditions.

Once the automotion effect is understood it can serve as a tool for investigating structural transitions between states having different automotion characteristics. In Section III we describe such a study for our S=1 quartet, using a single bias-field pulse to achieve each transition. Broad operating margins for each switch are established in the space of pulse amplitude versus rise time. Our interpretation of these switches in terms of nucleation, propagation and annihilation of BLs lends further support to our state assignments. Our interpretation of the $\chi^+ \leftrightarrow \chi^-$ switch[9b] in particular, confirms the Bloch-line model of velocity-saturated bubble collapse.

Other state switching work is reported at this conference; in contrast to our work, it involves state switching and discrimination of bubbles with different winding number in ion-implanted films.[10] In addition, two of us (S.M. and P.D.) report effects of transient Bloch-line motion on gradient deflection.[11]

In application to digital storage, automotion at modest field values offers potential means for propagating bubbles either individually or within a lattice while reducing the amount of deposited structures needed for propagation. When the structures replaced are current conductors as in the bubble lattice file device,[12] an automotion drive would reduce the on-chip heating by the use of externally heat-sinked coils. Sensing information encoded in the wall configuration is also afforded by the automotion effect.

I. EXPERIMENT AND RESULTS

Our experiment (Fig. 1) is capable of supplying steady uniform fields $H_b\hat{z}$ normal to the film and $H_{ip}\hat{y}$ in the plane of the film while a flat spiral coil and pulse generator provide a bias modulation field $h_z(t)\hat{z}$ which is gradientless near the center.[13] A pair of strip lines with dimensions shown, though not necessary for the automotion effect, was used to determine velocity and winding number of the various bubble states by their angle of skew propagation.[4,5] Bias pulses $h_z(t)$ with amplitudes $0<h_a<70$ Oe, durations $50ns<T<100\mu s$, and rise and fall times usually 15ns (but also variable) were available by means of a suitable pulse generator. Gradient pulses having amplitudes ∇H_z up to 1 Oe/μm and durations $50ns<T'<5\mu s$ were provided by gold conductor lines deposited on glass.

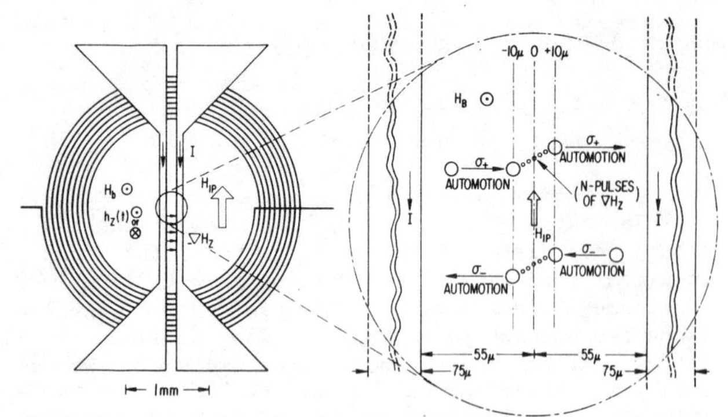

Fig. 1. Experiment: gradientless bubble propulsion (automotion) and measurement of wall winding number.

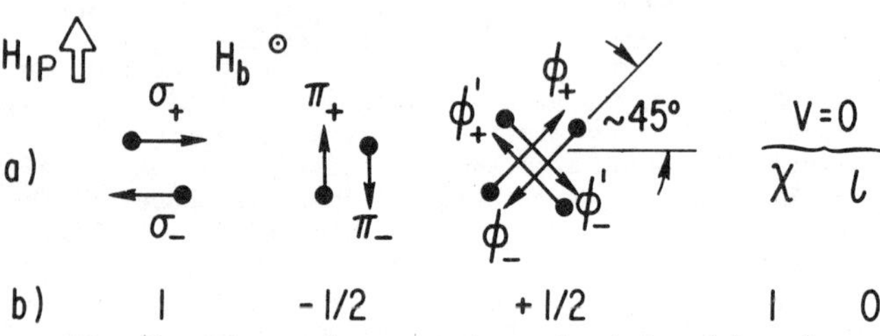

Fig. 2. Observed automotion velocities (a) and winding numbers (b).

Numerous sample films were investigated, but one film for which the most detailed observations were made has the following static properties: composition $Gd_1Y_1Tm_1Fe_{4.2}Ga_{0.8}O_{12}$, thickness 3.84μm, $4\pi M$=188G, anisotropy constant K_u=10,400 ergs/cm^3, wall-width parameter Δ=0.072μm, material length ℓ=0.91μm, $Q=H_k/4\pi M_s \simeq 8$, and crystallographic tilt $\simeq$ 0.4 deg. The bubble diameter was nominally d=6μm with radial compliance $\partial r/\partial H_z$ = 0.175μm/Oe. Measured dynamic parameters include gyromagnetic ratio $\gamma=1.27\times10^7$/s Oe, mobility μ=1750 cm/s Oe, saturation velocity V_s=1100 cm/s Oe, and Gilbert damping parameter α=0.045.

To investigate automotion, a few bubbles were nucleated in the presence of in-plane fields, typically H_{ip}= 12 Oe, by saturating the film and subsequently reducing H_b to the range of bubble stability while applying negative bias pulses $h_z(t)$ typically 10 to 20 Oe amplitude and 10μs duration. Most of these bubbles were caused to move (automote) visibly across the microscope field by keeping H_{ip}=12 Oe and applying bias-field pulse trains of typical amplitude h_a=-6 Oe, duration 1μs, and repetition rate 100 sec^{-1}. Most bubbles moved orthogonal to the in-plane fields. We call these σ for "senkrecht". Among these approximately equal numbers $(\sigma_\pm)$ moved in opposite directions, i.e., along $\pm \vec{H}_{ip} \times \vec{H}_b$. Other bubbles occasionally observed were either inert or exhibited other directionality of automotion.

The occurrence of discrete vectors of velocity in various automoting bubbles suggested naming these states according to each one's directionality as indicated in Fig. 2a. In addition to σ-states automoting transversely to $\vec{H}_{ip}$, we found $\Pi_\pm$-states (Π for "parallel") that automote parallel to $\vec{H}_{ip}$. A third set designated $\phi_\pm$ translated at approximately +45° and +225° to $\vec{H}_{ip} \times \vec{H}_b$, and $\phi'_\pm$ at -45° and -225° to $\vec{H}_{ip} \times \vec{H}_b$. The "inert" bubbles ι are those that jiggled randomly over distances less than 1μm. The automotion direction reverses on reversing the sign of $\vec{H}_{ip}$, but doesn't depend on the sign of the pulse, $\pm h_a$. The Π and φ states are relatively unstable against transitions (during excitation) such as $\Pi_+ \rightleftarrows \Pi_-$ and $\phi_+ \rightleftarrows \phi_-$ as well as Π (or φ) → ι, and occasionally Π → σ.

The experimental sequence of automotion, then gradient propagation, again followed by automotion, was conducted on each bubble as indicated in Fig. 1. This procedure served to determine the winding number S of each automotive state that survived the gradient test. The nominal S values are given in Fig. 2b.

Half-integer values belonging to Π and φ can arise from the presence of Bloch points,[14] and their presence is consistent with the unstable behavior of these states during automotion. Inert bubbles usually have S=1 (χ-state) or S=0 (state ι), but occasionally $S\geq2$.

Because σ bubbles are more stable, detailed studies of their gradient deflections were made under different conditions of $\vec{H}_{ip}$ and $\nabla\vec{H}_z$. The magnitudes $|\nabla H_z|$ were in every case made sufficient to overcome coercivity (∿ 0.4 Oe) but insufficient to induce velocities in excess of saturation (∿ 1100 cm/sec) where Bloch curve generation can lead to dynamic conversion of state. When H_{ip}=0, the data show S∿1 with considerable scatter. When $H_{ip} \simeq$ 10 Oe, we found $S \lesssim 1$ with scatter or S=1.0 with very small scatter (e.g., with std. dev.=0.15) depending on whether the gradient force $-\nabla\vec{H}_z$ is opposite to - or along the direction of automotion. Two of us (S.M. and P.D.) have further explored and explained the mentioned deviations from S=1 as well as a turnaround anomaly that is seen in σ bubbles but not in nonautomoting χ bubbles.[11]

The velocity of automotion is given by data of displacement δX per bias pulse for σ-bubbles in Fig. 3. These δX are the same for both σ_+ and σ_-, decrease with H_{ip}, increase with $-h_a$, and are essentially independent of pulse duration T for $T \gtrsim$ 1μs. Automotion disappears when pulse rise and fall times are beyond about 120ns. Otherwise, the margins for automotion are rather broad, i.e., $5<H_{ip}<50$ Oe, $2.5 \lesssim -h_a \lesssim 10$ Oe, and 100ns<T<∞. Optimum conditions produced δX=1μm. Pulse trains having pulse separations equal to T=0.34μs have produced σ velocities as large as 150 cm/sec consistent with optimum single-pulse data (1μm/pulse). The theoretical curves in Fig. 3 are based on the theory given in Section II.

Changes in bubble state occurring as also indicated in Fig. 3 appear erratic and uncontrollable while pulses $h_z(t)$ are continued. However, controlled state changes by single - or at most two - pulses of special shape are produced and described in Section III.

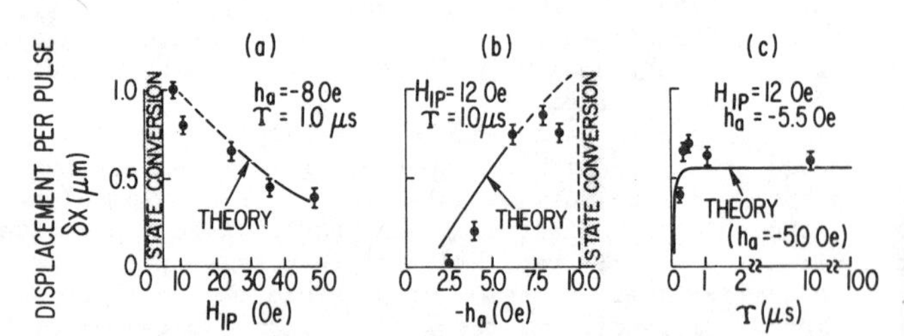

Fig. 3. Automotion velocity of σ-bubbles versus (a) in-plane field H_{ip}, (b) bias-pulse amplitude h_a, and (c) pulse duration T. Curves from theory in Section II.

Similar automotions of isolated bubbles were observed in numerous compositions, e.g., $(Y_{1.92}Sm_{0.1}Ca_{0.98})(Fe_{4.02}Ge_{0.98})O_{12}$, $(Y_{2.35}Eu_{0.65})(Fe_{3.8}Ga_{1.2})O_{12}$ and in an ion-implanted region of the latter composition dosed with $10^{16}H^+/cm^2$ at 25 keV. Others have reported that a capping layer can suppress one of the two σ-states, which one depending on the top or bottom positioning of the layer relative to the bias-field sign.[15]

We have also produced automotion of bubble lattices after finding suitable films (usually thick films and presumably having therefore layered gradients) and/or finding other means suitable for nucleating σ-bubbles all of one sign. Thus, we could study the dependence of σ-velocity on lattice density as well as on H_b, H_{ip}

and h_a. Generally, we found that the optimum H_{ip} is the same as for isolated bubbles, but that lattices require larger pulse amplitudes, e.g., h_a = -25 Oe in our principal film when bubble lattice spacing was D ≃ 8μm. The optimum h_a decreased with increasing D. At fixed h_a the automotion velocity increased with D from a few tenths of a micron per pulse at D ≃ 8μm to the expected one micron per pulse at D equal several bubble diameters.

Preliminary work shows that sinusoidal bias modulation can also automote both lattices and isolated bubbles. A σ-type of (isolated) motion occurs in our principal sample in the frequency range 2-10MHz with a peak position that increases with increasing H_{ip}. The optimum values of H_{ip}, H_b and amplitude are nominally the same as for pulsed bias modulation.

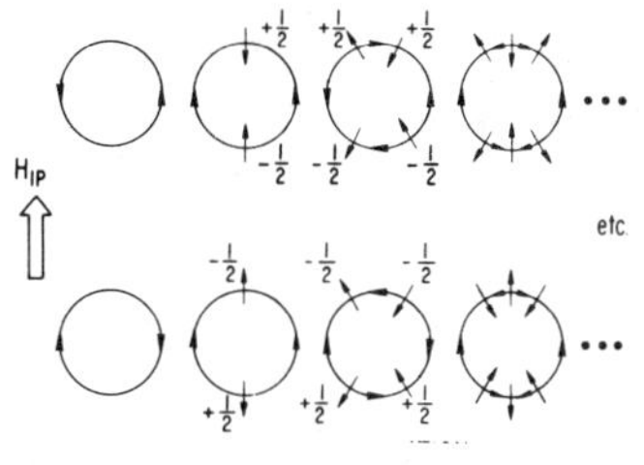

Fig. 4. Configurations satisfying unity winding $S=1+(1/2)(n_+-n_-)=1$ as candidates for bidirectional pairs of σ bubbles.

Wall Configuration for σ-Bubbles. Fig. 4 shows a class of possible wall structures available to satisfy unity winding (S=1) as in σ-bubbles. That they appear in conjugated pairs satisfies the observed duality of bidirectional motions. In each case the normal wall regions contribute unity winding while the vertical BLs, each numbering $n_+=n_-$, contribute ±1/2. They are arranged in clusters paired on opposite sides of the bubble which would unwind spontaneously if brought together. Consequently, these structures satisfy $S=1+1/2(n_+-n_-)=1$. The following phenomenological arguments coupled with another type of experiment involving reversing the in-plane field eliminates all structures but those for which $n_\pm$ is odd.

Configurations with even $n_\pm$ have oppositely oriented normal-wall segments. This means that one BL pair flanks and must sustain the wall segment that is oriented unfavorably with respect to the in-plane field H_{ip}. We show in Appendix A that this configuration has a low threshold H_{ip} ≃ 10 Oe for an instability with respect to the automotion environment whereby the members of the "flanking" pairs of BLs can separate from their attractive BL neighbors, meet each other, and annihilate. This process, as illustrated in Fig. 5, reverses the "unfavorable" part of the bubble wall so that now both sides of the bubble are favorably oriented.

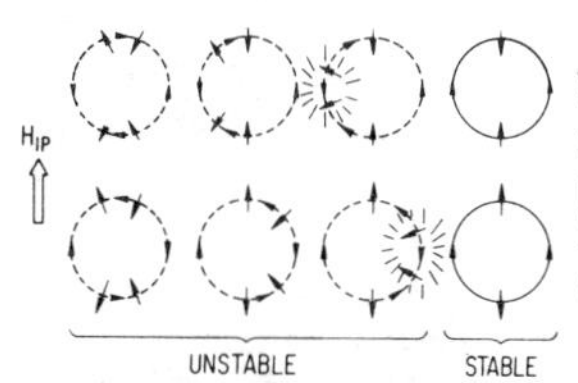

Fig. 5. Instability and decay products for cases $n_+=n_-=2$ pairs of BLs. Progression left-to-right is for increasing in-plane field.

Thus far, the unichiral structures and the odd-pair arrangements remain as candidates for the σ-bubbles. We eliminate unichiral states by an experiment performed after halting automotion by turning off the bias pulses. The experiment consists of varying H_{ip} in two steps. In the first step, H_{ip} is reduced to zero and the bubble position does not change. In the second step, H_{ip} is restored to its former magnitude with either (1) the same or (2) reversed sign. Following the second step, a change in position ("jump") of the σ bubble is observed if and only if the sign of H_{ip} is reversed. The jump is always in the direction of previous automotion and amounts to about one micrometer. This result logically rules out the unichiral possibility for a σ state because its axial symmetry under the condition H_{ip}=0 makes the unichiral structure necessarily indifferent to the sign of the second step in H_{ip}.

Finally then, only bubbles having odd-numbered pairs of BLs remain to account for σ-automotions. Numerical results of a theory in Section II based on the simple case of one unwinding BL pair ($n_+=n_-=1$) confirm most of the observed behavior of automoting σ-bubbles. The theory for the jump due to in-plane field reversal is also presented in Section II. Though the mean observed jump is consistent with the two-BL model, the predictions of larger jump displacements if more pairs are present are supported by an occasional observation of an extra large jump, e.g., 3μm, associated with H_{ip}-reversal.

II. DYNAMICAL THEORY OF σ AUTOMOTION

Figure 6a illustrates the mid-plane cross-section for the two configurations chosen to represent σ-bubbles. The magnetization $\vec{M}(\vec{x})$ (of constant magnitude) is parallel to $\vec{H}_b$ outside the domain and antiparallel inside the domain. Within the wall, $\vec{M}$ rotates continuously as a function of $\vec{x}$ (not shown). Dipole-dipole interactions orient $\vec{M}_{xy}$, the projection of M onto the film plane, tangentially with respect to most of the circular boundary, as indicated by the arrows in Fig.8a. Bloch lines, lying parallel to the z-axis and having N and S magnetic polarities, separate wall regions of opposing $\vec{M}_{xy}$ orientation. The field H_{ip} stabilizes their positions at the ±y extremities of the cylinder.

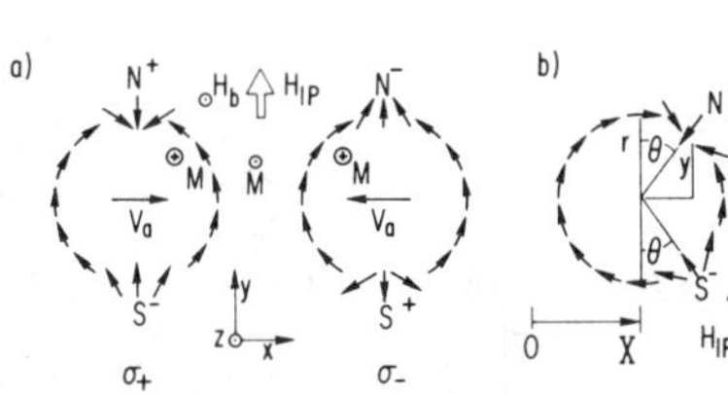

Fig. 6. (a) Static moment distributions for σ_+ and σ_- bubbles in applied field H_{ip}. (b) Dynamical variables for σ_+.

The winding or revolution number S is the number of rotations $\vec{M}_{xy}(\vec{x})$ makes when $\vec{x}$ circulates once around the domain.[4] The states σ_+ and σ_-, shown in Fig. 6a, satisfy S=1. Local winding numbers ±1/2 and ∓1/2 are associated with N and S, respectively, for the states $\sigma_\pm$.

Application of the bias pulse $h_z(t)$ tends to induce Larmor precession of the spins (gyrotropic effect) in the domain wall. In the normal portion of the wall, this precession is inhibited by the local demagnetizing effect of the wall which tends to orient the wall moment tangentially to the wall surface. It follows that at sufficiently low drive amplitudes and long times, gyrotropic effects within the normal wall regions are small and their dynamical reactions are understandable in terms of the conventional mobility and coercivity. However, within the BL region, spins form many angles with respect to the wall-surface and thus can precess more easily. Because of an inherent rigidity of the BL region, the spin precessions there are interpretable as motion of the BL center. The velocity V_{BL} produces a dynamic reaction force[16,17] (neglecting BL damping)

$$\vec{F}_\pm = \pm\, 2\pi M\gamma^{-1}\, \hat{z} \times \vec{V}_{BL} \tag{1}$$

per unit length of BL. Here $\hat{z}$ is the unit vector normal to the film, γ is the gyromagnetic ratio, and M is the saturation magnetization. This force must be balanced by the constraining force requiring the BL to remain within the wall surface.

Because the sign of $\vec{F}$ depends on the sense of BL winding, the BLs circulate in opposite directions, traversing the wall towards each other. Upon leaving their positions of equilibrium with respect to the in-plane field, an additional force due to H_{ip} acts on the BLs which, combined with the force of constraint along the wall normal, gives a net force applied to the BLs. The gyrotropic reaction of each BL against the wall produces y-components that cancel and x-components that add according to the signs of Eq. (1).

The net dynamic reaction of both BLs is the x-component (per unit film thickness)

$$F_x = \mp 4\pi M\gamma^{-1}\dot{y} - 8\alpha M\gamma^{-1}Q^{-1/2}\dot{x} \quad (2)$$

where the $\mp$ signs account for the opposite configurations in $\sigma_\pm$ bubbles. The last term accounts for BL viscous damping[16,17] with the Gilbert damping parameter α and the parameter $Q \equiv K/2\pi M^2$. (x,y) are BL coordinates.

Additional forces acting during the motion include dynamic gyrotropic contributions of the normal wall regions,[16,17] viscous damping and coercive forces. The wall gyrotropic effect destroys the symmetry about y=0 implicit in Eq. (2). Fortunately, however, to neglect it, seems reasonable whenever the deflection during gradient propagation is small. (In the sample under study it is ≤ 25°.) Reactions of the BLs against the wall are more significant because their instantaneous velocities are an order of magnitude larger. Resultant symmetry about y=0 allows a natural choice of dynamic variables r, Θ, and X (Fig. 6b) specifying cylinder radius, BL positions and bubble position, respectively.

We now account for the effects of coercive force, H_c. The net coercive pressure which opposes wall motion (Fig. 7) is expressible as

$$P_c = -2MH_c \,\mathrm{sgn}\, V_n \quad (3)$$

where the coercive field H_c (>0) is a constant and $V_n = \dot{r}+\dot{X}\cos\beta$ is the normal wall velocity at any point (β) on the wall. By integrating over the cylinder surface we find the total x-component of coercive reaction force[17]

$$F_c = -8MH_c r(\mathrm{sgn}\,\dot{X})(1-(\dot{r}/\dot{X})^2)^{1/2'} \quad (4)$$

where $\cdot = d/dt$ and the function $u^{1/2'}$ extends the usual meaning $u^{1/2}$ for u>0 to the region u<0 where we want $u^{1/2'} \equiv 0$. An interesting property of F_c is that its value depends strongly on the ratio of two kinds of motion, $\dot{X}$ and $\dot{r}$. The effects of this dependence are exemplified in Fig. 7 and in numerical computations described below and in Fig. 8 for all phases of the automotion cycle.

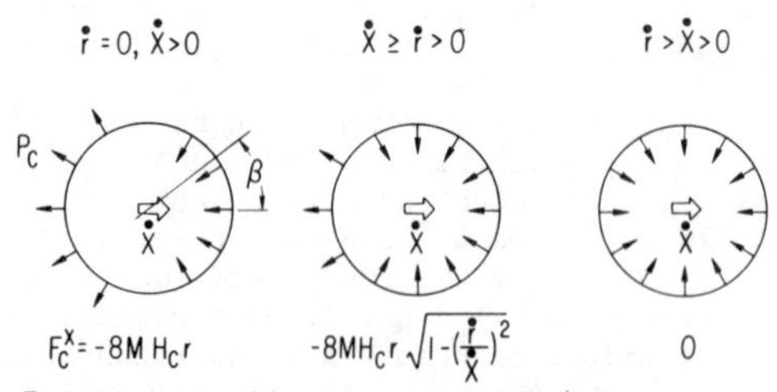

Fig. 7. Local wall coercivity (arrows) depending on relative wall radial and translational velocities; F_c^x is net translational coercive force.

As no net external force acts on the bubble, the reaction F_x in Eq. (2) and the coercive force F_c in Eq. (4) are balanced against viscous damping of the bubble. $2\pi M\alpha r\dot{X}/\Delta\gamma$. (The effect of in-plane field on Δ has been neglected because of the small H_{ip} being considered.) Collecting and integrating these force terms gives an expression for displacement δX induced by a single bias pulse

$$\delta X = \mp \frac{2\Delta}{\alpha}\int_0^\infty \frac{\dot{y}}{r}\,dt - \frac{4\Delta}{\pi Q^{1/2}}\int_0^\infty \frac{\dot{x}}{r}\,dt - \frac{4\Delta\gamma H_c}{\pi\alpha}\int_0^\infty \left[1-\left(\frac{\dot{r}}{\dot{X}}\right)^2\right]^{1/2'} \mathrm{sgn}\,\dot{X}\,dt \quad (5)$$

where x and y are BL coordinates. The gyrotropic reaction of BLs in the first term does not contribute to δX unless the BL motion correlates with the radial motion, because the initial and final BL positions are the same and therefore $\int_0^\infty \dot{y}\,dt = 0$. The BL damping reaction in the second term gives at most,

$$|\delta X| < (4\Delta|\delta r|/\pi Q^{1/2}r)/(1+4\Delta/\pi Q^{1/2}r) \quad (6)$$

where δr is an appropriate value of the radius change during an application of the bias pulse since $\dot{x} = \dot{X} + d(r\sin\Theta)/dt$. However, this value is negligibly small because Q>>1 and Δ<0.1μm is usual in bubble materials. Thus, we conclude that δX is not appreciable unless (1) a strong nonlinear coupling between the BL motion and the radial motion is taken into account or (2) the effect of coercivity is explicitly taken into account in the bubble motion. Numerical integration shows that the coercivity must be present to give an appreciable δX, as described below.

We consider now the nonlinear bubble motion in detail by writing equations of motion for the variables Θ, r, and X. For this purpose a formalism of bubble motion with a small number of BLs is useful.[17] Accordingly, balancing the components of conservative force derived from the energy W against the dynamic reaction forces derived by superposing BL reaction forces (with BL damping neglected) onto normal wall forces gives

$$\partial W/\partial r = -\gamma^{-1}4\pi M(\alpha\Delta^{-1}r\dot{r}) \pm \gamma^{-1}4\pi M(\dot{X}\cos\Theta + r\dot{\Theta}) - 8MH_c r\,\arcsin'(\dot{r}/|\dot{X}|) \quad (7)$$

$$\partial W/\partial\Theta = \mp \gamma^{-1}4\pi Mr(\dot{X}\sin\Theta + \dot{r}) \quad (8)$$

$$\partial W/\partial X = \mp \gamma^{-1}4\pi M(\dot{r}\cos\Theta - r\sin\Theta\,\dot{\Theta}) - 4\pi M\alpha r(2\Delta\gamma)^{-1}\dot{X} + F_c(\dot{r},\dot{X}), \quad (9)$$

where the function arcsin' u is defined as 1/2·πsgn u for |u| ≥ 1. Eq. (7) shows that the BL motion influences the radial motion. Eq. (8) originates from the fact that the BL motion is tangential to the wall. Equation (9) with $\partial W/\partial X=0$ is equivalent to Eq. (5) before integration. The energy of the bubble is (per unit film thickness)

$$W = 2\pi r^2 Mh_z(t) - 4\pi rM\Delta H_{ip}\cos\Theta + 1/2\cdot C_0(r-r_0)^2 \quad (10)$$

where r_0 is the equilibrium radius with $h_z(t)=0$ and $H_{ip}=0$. The usual potential energy of the bubble with a constant bias H_b is assumed to be a parabolic form with a constant $C_0/2$. The equation tells us that the equilibrium radius depends on H_{ip}, and is expressed as

$$r = r_0 + 4\pi MH_{ip}\Delta C_0^{-1}. \quad (11)$$

We have confirmed this in-plane field dependence of the static radius experimentally. The static forces are obtained by differentiating the energy (10); and their substitution into Eqs. (7)-(9) gives three simultaneous differential equations (not shown) governing Θ(t), r(t), and X(t).

To weigh the relative importance of (1) correlation of r with $\dot{y}$ in the first two terms of Eq. (5), with that of (2) coercivity in the last term, we first set H_c=0 and integrate numerically. The results gave negligible values of δX for all reasonable parameter values, confirming thus, that coercivity is essential.

Additionally, the computations with $H_c=0$ revealed oscillations in each dynamical variable, implying therefore special significance for the dynamical coercive force. Oscillations in bubble radius and position having asymmetry in the ratio $|\dot{r}/\dot{X}|$ with respect to forward and backward motions imply that the dynamic coercive force is asymmetric. We therefore solved the three simultaneous equations governing r, Θ, and X using the Runge-Kutta method[18] while taking the velocities ($\dot{r}$ and $\dot{X}$) in the coercive terms self-consistently. We also utilized a linearized approximation to the bubble restoring force so that the net effective field acting on the wall is

$$h_{eff} = h_z(t) + \kappa[r(t) - r_1] \quad (12)$$

where $\kappa \equiv C_0/4\pi M r_1$ and r_1 is the equilibrium radius when $h_z(t)=0$ given by Eq. (11).

Typical numerical results for motions within a σ_+-bubble are shown in Fig. 8, where BL damping was included for sake of completeness. The characteristic factor of coercive reaction $f_c \equiv -\,\mathrm{sgn}\,\dot{X}[1-(\dot{r}/\dot{X})^2]^{1/2}$, acting on the bubble center is also shown as it resulted from the self-consistent calculations.

Discussion about the motions of r, Θ and X can be focused by considering each of four different time phases 1-4 in Fig. 8. Due to gyrotropic reaction of the BLs, the bubble center moves forward in phase 1. The radial motion is generally faster than $\dot{X}$ because the in-plane field restrains the BL motion. Therefore, the coercive force is relatively ineffective on $\dot{X}$. During phase 2, the radius approaches the new equilibrium value r_2 and the effective field h_{eff} is small. Thus the influence of H_{ip} on the BLs overcomes the effect of h_{eff}, and the BLs are forced back toward their equilibrium position, Θ=0. Responding to this return motion of the BLs, the bubble moves backward. Because $\dot{r}$ is small at this phase, the coercive force has more effect on $\dot{X}$. Similar events, though not simply sign changes, take place during phases 3 and 4 after the end of the pulse.

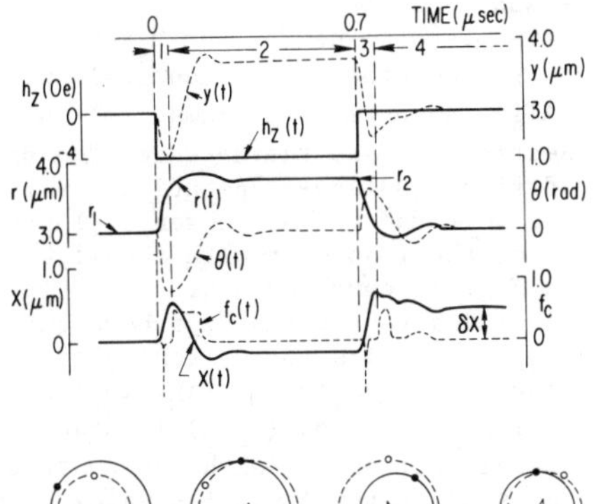

Fig. 8. (a) Typical results of theory. Note $f_c \equiv F_c/8MH_c$. (b) Behavior of bubble and BLs during four time-phases in (a). Arrows indicate bubble-center motion between beginning (dashed) and end (solid) of phase.

The solid curves of Fig. 3 represent numerical results for δX. The instantaneous wall velocity, except for transients of less than 10ns duration, is less than the threshold

$$V = V_s + (\pi/2)\Delta\gamma H_{ip} \quad (13)$$

at which the structure of a wall may change due to nucleation of horizontal BLs. (See Appendix B.) Here V_s is the observed saturation velocity[19,20] in the absence of H_{ip}. The dashed part of the curves in Fig. 4 represent regions where there is a possibility that the theory may break down as the computed velocity exceeds the above threshold during transients of more than 10ns.

It is straightforward to show that the σ_+ and σ_- bubbles obey the symmetry relations $\Theta(\sigma_+) = -\,\Theta(\sigma_-)$ and $\dot{X}(\sigma_+) = -\,\dot{X}(\sigma_-)$ as observed. We also find an equivalence between $+h_a$ and $-h_a$ bias pulses, i.e., both yield similar automotion velocities. However, the velocities do depend on the pulse rise- and fall-times τ. If τ is too great (e.g., τ ≳ 300 ns depending on H_{ip}), automotion is destroyed because gyrotropic BL velocity is small due to a smaller effective drive, h_{eff}. The observed "bubble jump" response to sudden reversal of $\vec{H}_{ip}$ is also understandable, e.g., by Eq. (5), and further verifies our identification of σ. The N and S BLs simply traverse opposite sides of the bubble without a state change, causing the bubble to jump once in the previous automotion direction. Equation (5) gives an estimate $[(\alpha/4\Delta)+|\gamma H_c/\pi\dot{X}|]^{-1} \approx 1.9\mu m$ assuming (1) r and $\dot{X}$ are constant during the rapid BL interchange, (2) BL damping is negligible, and (3) $\dot{X}$ has an average 500 cm/s estimated as half the saturation velocity. If clusters of n BLs are involved, the "jump" would be n times as large. Usually n ≈ 1 is observed; occasionally n ≈ 2,3.

In summary, we find that automotion is an asymmetric coercive reaction to oscillation of the bubble center caused by bias-driven oscillations of the coupled bubble radius and BL positions. It's not surprising therefore that, as shown in Fig. (3), (a) larger h_a produce larger δX through larger induced amplitudes of r and Θ motions, and (b) that δX is essentially independent of pulse duration T beyond the time (∿200 ns) of relaxation of the initial response that proceeds from the flank of the pulse.[21] Large BL motions can also produce state changes when Θ reaches π/2 where the BLs can unwind, a subject for Section III.

III. CONTROLLED DYNAMIC CHANGES IN S=1 WALL STATES

Erratic changes in automotion state that occur at excitations outside the range of automotion (Fig. 3) do not usually involve changes in S; they seem to be due to collisions of BLs. Collisions can occur when amplitudes h_a are sufficient to make the BLs traverse a quarter of the bubble rim where they meet and unwind. We call this event a state change although total winding number S isn't affected. The erratic behavior could be associated either with changes in the automotion directionality ($\sigma_+ \rightleftarrows \sigma_-$) or with loss and/or reappearance of automotion ($\sigma \rightleftarrows \chi$).

Another observation is that when H_{ip} is removed and h_z- pulsing is continued, automotion doesn't always reoccur by reapplication of H_{ip}. This suggests that σ-bubbles lose their BLs and become, for example, unichiral χ-states, again retaining S=1.

A third observation is that if the bias pulses are subsequently increased to a sharply defined amplitude while $H_{ip}>0$, all χ bubbles jump and immediately become active as σ-automotion states. Thus χ-to-σ conversions are demonstrated.

These conversions initially found during studies of automotion inspired theoretical and experimental investigation regarding controllability and margins for state conversions. A systematic study presented elsewhere[9b] and briefly below includes a phenomenological guide to expected combinations of magnitude and shape of $h_z(t)$ and H_{ip} where specific transitions may occur. The measurements support the ideas and reveal margins for state changes in the principal sample described in Section I.

Figure 9 shows the static moment configurations for the four S=1 states of interest. The notation (S,L) signifies the number L of Bloch lines present in a configuration having winding number S. The $\sigma_\pm$ and $\chi_\pm$ states are thus denoted by $(1,2)_\pm$ and $(1,0)_\pm$, respectively.

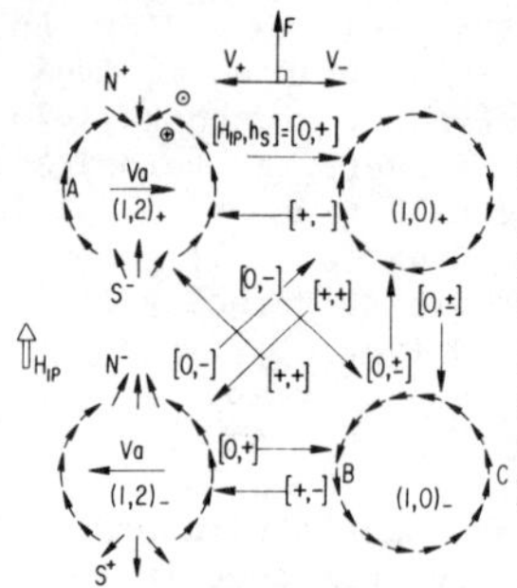

Fig. 9. σ and χ states $(1,2)_\pm$ and $(1,0)_\pm$ and bracketed combinations of H_{ip} (0 or -) and h_s(+ or -) required for switching.
Inset, top: force-velocity relationship for ± vertical BLs (Eq. 14).

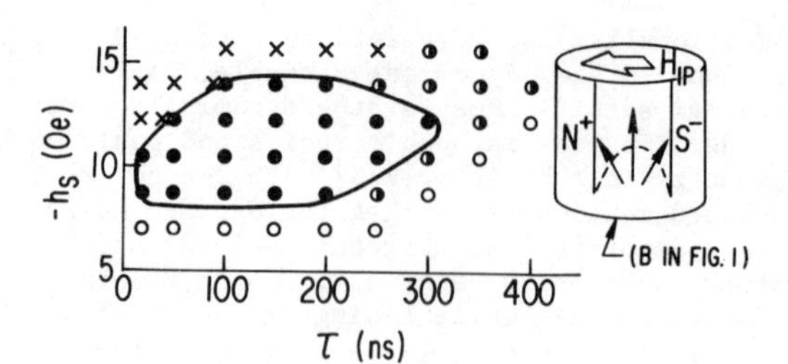

Fig. 10. Operating margins for $\chi_+ \to \sigma_+$ (full dots) at H_{ip}=34 Oe. Open circles: no switch. Half-open circles: mixture of no switch and $\chi_+ \to \sigma_+$. Crosses: resulting state either σ_+ or σ_- or S=0.

Switching Mechanisms. Ideally, switching between any two states (except $(1,2)_+ \leftrightarrow (1,2)_-$) is done most simply by a single step $h_z(t)$ of amplitude h_s and a fast linear rise time τ_r followed by plateau-and fall times great enough (e.g. >0.5μs) to return the bubble radius adiabatically to its initial value without further state changes. The fast leading edge produces pressure on the wall of the bubble tending to expand (h_s<0) or contract it (h_s>0). It also produces forces on existing BLs and may nucleate new BLs. Any force $\vec{F}$ per unit length on a BL (which may be curved) with a twist angle $\pm\pi$ imparts the velocity[16,17]

$$\vec{V}_\pm = \pm(\gamma/2\pi M)\vec{e} \times \vec{F} , \tag{14}$$

where $\vec{e}$ is the unit vector tangent to the BL and the sign depends on the direction of the gyrovector within the BL.[16] Figure 9 indicates the geometric relationships between $\vec{F}$ and $\vec{V}_\pm$ for vertical BLs ($\vec{e} = \hat{e}_z$).

State changes may thus be induced by $h_z(t)$. The sign of h_s is chosen according to the state change desired. In addition, we set H_{ip}=0 if the final state is to be $(1,0)_\pm$ and set H_{ip}>0 if it is to be $(1,2)_\pm$. The direction of H_{ip} is immaterial. The combinations of H_{ip} (0 or +) and h_s (+ or -) required for the 10 distinct state conversions are indicated within brackets in Fig. 9.

Considering $\sigma \to \chi$, it is apparent from Fig. 9 that the switch $(1,2)_+ \to (1,0)_-$ occurs if the two BLs move leftward until they collide and annihilate at A. This sequence requires H_{ip}=0 to release the BLs and h_s<0 to exert the radially outward forces that produce tangential motions leftward following Eq. (14). The other three $\sigma \to \chi$ transitions $(1,2)_\pm \to (1,0)_\pm$ require the signs indicated.

Considering $\chi \to \sigma$, the sequence of events for $(1,0)_- \to (1,2)_+$ resembles that for $(1,2)_+ \to (1,0)_-$ in reverse order. The field H_{ip}>0 produces an essential assymmetry. Wherever it opposes $\vec{M}$ (Fig. 9, point B), it tends to destabilize the wall structure. For example, when H_{ip} is above a threshold $4\sqrt{2\pi A}/h$ (Appendix B), it induces a horizontal BL having 2π twist;[22] at arbitrary H_{ip} it decreases the critical velocity for BL nucleation. In a straight wall the induced 2πBL would be horizontal; in a bubble it is horizontal only at the wall segment where H_{ip} is tangent and is curved elsewhere. At the part of the wall where $\vec{M}$ is aided by $\vec{H}_{ip}$ (point C) the critical velocity is increased. Thus, the bulge of a curved BL, nucleated at B in the critical circle near one surface of the film can be propagated toward the other surface (inset, Fig. 10) by an appropriate combination of amplitude h_s and rise-time τ_r. There, the Bloch curve may be severed into a pair of vertical unwinding BLs by the mechanism of "punch through."[24,23,20] This pair with signatures N^+ and S^-, constitutes the σ_+-state $(1,2)_+$. The field H_{ip} returns these BLs to the equilibrium position shown in Fig. 9.

Direct chiral reversal $\chi_+ \leftrightarrow \chi_-$ (demonstrated for the first time by two of us, P. D. and J. C. S.) happens when H_{ip}=0 and the pressure $2Mh_z(t)$ rises rapidly enough to nucleate a circular BL along a critical circle[22] near one surface which subsequently moves towards the other surface. After punch-through the sense of chirality has changed. The process may repeat itself several times during one bias pulse depending on h_s and τ_r.

Results. Using the garnet film studied in Section I we fixed the bias H_b=52 Oe, obtaining a radius r_o=3μm. We distinguished among σ_+, σ_-, and χ (non-automoting) initially and after each attempted switch by testing for automotion. In addition, gradient propagation confirmed the S=1 character of χ. Values of H_{ip} and h_s were adjusted until a state test indicated a switch had occurred, thus confirming the scheme of Fig. 9.

While absence of automotion cannot by itself demonstrate that the χ state is a doublet, its use in switching combinations can. For example, the fact that the two transitions in Fig. 9 $\sigma_+ \to \chi \to \sigma_-$ and $\sigma_- \to \chi \to \sigma_+$ have different outcomes induced by the same switch conditions [+,-] acting on the χ shows the intermediate χs must be different and justifies attaching different subscripts, - and + respectively, to χ.

The effects of changing H_{ip} and h_s on switches involving χ_+ and χ_- show they have identical switching thresholds. Also, absence of a "jump" following an H_{ip} reversal[9a] (Section II) and the absence of a sideways displacement during gradient-propagation reversal,[11] both prove that vertical BLs are absent. Therefore identification of the χ-doublet with the $(1,0)_\pm$ structures shown in Fig. 1 seems certain.

After confirming the switching scheme qualitatively, we measured the margins for switching in the space h_s vs τ_r as shown for example, in Fig. 10 for the case $\chi_+ \to \sigma_+$. We utilized expansion (h_s<0) in order to avoid inconvenient bubble collapses, although contraction (h_s>0) produced similar results. The value H_{ip}=34 Oe was used because it produced wide (h_s,τ_r) margins. With H_{ip}<22 Oe no combination of h_s and τ_r produces the intended transition. With H_{ip}>50 Oe, S=1→S=0 state changes narrow the h_s range considerably. Margins for $\sigma_+ \to \chi_-$ and $\chi_+ \leftrightarrow \chi_-$ with H_{ip}=0 are different. The reader is referred to Ref. 9b for detailed descriptions.

Thresholds and margins for $\chi_- \to \chi_+$ reversals involving 1, 2 and 3 horizontal BL nucleations[9b] are noteworthy. Analysis in terms of the horizontal Bloch-line mechanism gives critical times τ_n for the nth chiral switch in terms of the mean or saturation velocity of a wall containing one horizontal BL. The observed h_s-τ_r thresholds are consistent with this mechanism and therefore confirm an earlier BL-model of velocity-saturated bubble collapse.[19]

ACKNOWLEDGEMENT

We are indebted to E. A. Giess for sample preparation, and to A. P. Malozemoff, J. C. DeLuca and D. Cronemeyer for helpful discussions and access to unpublished results.

APPENDIX

Appendix A. According to Hubert[24] and Slonczewski,[17] two BLs of equal winding sense in a flat wall have an equilibrium separation s_o specified by a balance of forces: repulsive exchange F_{ex} versus attractive magnetostatic F_m. An in-plane field H_{ip} opposing the wall magnetization outside the BL-region exerts an additional force $F_H=2\pi\Delta M\ H_{ip}$ causing s_o and thus the area with favorable magnetization to increase. Thus, a threshold field H_{ip}* destabilizing the BL bound-state occurs when $F_{ex}(s_o)+F_m(s_o)+F_H(H_{ip}*)=0$ is satisfied for any s_o. Accordingly, we find $H_{ip}*=9.1$ Oe using film parameters of Section I.

Appendix B. The critical wall velocity in an in-plane field H_{ip} is obtained by considering the relation[19] $V=\gamma(2M)^{-1}dW/d\bar{\psi}$ between velocity V, energy W, and momentum $\gamma^{-1}2M\bar{\psi}$ of a straight wall. When a horizontal BL moves a distance dz the energy change per unit wall area due to H_{ip} is $dW=\pm 2\pi\Delta MH_{ip}\sin\phi dz/h$ in notations of Ref. 17. The (+) and (-) signs refer to the cases of wall moments oriented parallel and antiparallel to H_{ip}, respectively. $\bar{\psi}$ also changes with dz as $d\bar{\psi}=2h^{-1}\phi dz$. Combining these expressions and including the stored energy[19] in the BL we find (taking $\sin\phi/\phi\simeq 1$) $V=V_s\pm 1/2\cdot\pi\Delta\gamma H_{ip}$.

Appendix C. An H_{ip} opposing the magnetization within a flat wall, can nucleate[22] two horizontal BLs, each at the critical lines near the surfaces.[19] Once nucleated these BLs may shift toward the midline of the wall where they join and become a 2πBL, a configuration stable with respect to further increases in H_{ip}. Balancing the BL force due to H_{ip} against that due to BL energy stored in the local stray field (Eqs. 4.3, 4.6, and 4.7 of Ref. 19) provides the necessary H_{ip} to equilibrate the 2πBL. From this force equation $2\pi\Delta MH_{ip}=8\pi AM(2\pi/K)^{1/2}/h$ we estimate the critical field $H_{ipc}=4\sqrt{2\pi A}/h$ for the appearance of a horizontal 2πBL.

REFERENCES

*Permanent address: Tohoku Univ., Sendai 980, Japan
**Permanent address: Delft Univ. of Tech., Delft, The Netherlands

1. A. H. Bobeck and E. Della Torre, Magnetic Bubbles, North Holland Publishing Co., Amsterdam, 1975. Also, Hsu Chang, Magnetic Bubble Technology, IEEE Press, New York, 1975.
2. B. A. Boxall, IEEE Trans. Magn. MAG-10, 648 (1974).
3. W. C. Hubbell, AIP Conf. Proc. 24, 552 (1974).
4. J. C. Slonczewski, A. P. Malozemoff and O. Voegeli, AIP Conf. Proc. 10, 458 (1972).
5. W. J. Tabor, A. H. Bobeck, G. P. Vella-Coleiro and A. Rosencwaig, AIP Conf. Proc. 10, 442 (1973).
6. A. P. Malozemoff and J. C. Slonczewski, AIP Conf. Proc. 24, 603 (1974).
7. B. E. Argyle, J. C. Slonczewski and O. Voegeli, IBM J. Res. & Dev. 20, 101 (1976).
8. A. P. Malozemoff and S. Maekawa, J. Appl. Phys., to be published, July, 1976.
9a. B. E. Argyle, J. C. Slonczewski, P. Dekker and S. Maekawa, to be published.
9b. P. Dekker and J. C. Slonczewski, to be published.
10. T. J. Beaulieu, et al., this Conf., paper 7A-2.
11. S. Maekawa and P. Dekker, this Conf., paper 7A-5.
12. B. A. Calhoun, J. S. Eggenberger, L. L. Rosier and L. F. Shew, IBM J. of Res. & Dev. 20, 368 (1976).
13. It is important to adjust the film-to-coil spacing so that there is no strong ripple in the field profile $h_z(r)$ near the edge of the pancake coil. (See Ref. 7). Otherwise gradients in $h_z(r)$ propel extraneous bubbles towards the center region and disturb the automotion experiment.
14. J. C. Slonczewski, AIP Conf. Proc. 24, 613 (1974).
15. J. Gietscher and G. R. Henry, private communication.
16. A. A. Thiele, J. Appl. Phys. 45, 377 (1974).
17. J. C. Slonczewski, J. Appl. Phys. 45, 2705 (1974).
18. H. Margenau and G. M. Murphy, Mathematics of Physics and Chemistry, D. van Nostrand Company, Inc., New YOrk 1952.
19. J. C. Slonczewski, J. Appl. Phys. 44, 1759 (1973).
20. A. P. Malozemoff, J. C. Slonczewski and J. C. DeLuca, AIP Conf. Proc. 29, 58 (1975).
21. Note: The resonance frequency ω_r and decay time τ derivable from a small amplitude approximation ($\theta<<1$) are $\omega_r\simeq\gamma\sqrt{\Delta\kappa H_{ip}}$ and $\tau=(\gamma\alpha H_{ip}/2)^{-1}$.
22. A. Hubert, J. Appl. Phys. 46, 2276 (1975).
23. F. Hagedorn, J. Appl. Phys. 45, 3129 (1974).
24. A. Hubert, AIP Conf. Proc. 18, 178 (1973).

WALL STATES IN ION-IMPLANTED GARNET FILMS

T. J. Beaulieu, B. R. Brown, B. A. Calhoun, T. Hsu
IBM Corporation, San Jose, California 95193
and
A. P. Malozemoff
IBM Corporation, Yorktown Heights, New York 10598

ABSTRACT

A model is presented which explains the wall states, and transitions between states, observed for bubbles propagating under the influence of an in-plane field in ion-implanted garnet films. We have identified nine transitions in which the deflection angle changes, six of which are accounted for by switching of the capping layer and injection of a Bloch point (BP). It was necessary to invoke three different wall structures for S=+1 bubbles depending upon the relative orientation of bubble velocity and the in-plane field. Two different S=1/2 bubbles containing one BP are identified which differ in location of the Bloch point and the associated capping structure. There is one state containing two vertical Bloch lines and two BP's, one on each line; and there is a single S=0 state.

I. INTRODUCTION

The winding number S of a bubble is related to the angle χ between the field gradient and the direction of motion by the equation[1]

$$\sin\chi = \frac{8SV}{\gamma d \Delta H} \qquad (1)$$

where γ is the gyromagnetic ratio, V is the bubble velocity, and ΔH the field difference across the bubble diameter d. In garnet films which have been ion-implanted to suppress hard bubbles,[2] the number of different wall structures which can exist is quite limited. At zero in-plane field, only bubbles with S=+1 are stable; and at sufficiently large values of in-plane field, H_p, only S=0 bubbles are stable.[3] At intermediate values of H_p, additional states corresponding to fractional S values have been observed.[4,5] Some observations of the changes in deflection angle χ, as an in-plane field perpendicular to the field gradient was increased, have been reported.[6] State changes for an in-plane field parallel to the drive gradient have also been reported in as-grown EuYIG.[7]

We have measured the deflection angles for both increasing and decreasing in-plane fields for several different orientations of the in-plane field and have observed nine discrete, irreversible changes in deflection angle. Following a brief description of our experimental techniques, we present these results for the in-plane field parallel and perpendicular to the drive field gradient. Section IV discusses the different mechanisms responsible for these state changes, and Section V covers the detailed wall structures of S=+1 bubbles moving in an in-plane field.

II. EXPERIMENTAL PROCEDURE

Two different methods for measuring deflection angles were used. The overlay technique[8] requires no additional device fabrication and is, therefore, convenient for surveying large numbers of samples. For small displacements, the drive gradient is nearly constant, and mobility changes are easy to observe. The deflectometer[9] provides much higher angular resolution at the expense of drive-field non-uniformities and the necessity for device fabrication. Most of the data in this work were taken using the deflectometer at ambient temperature. Bubbles were sequentially propagated back and forth by means of pulsing pairs of current conductors which were deposited over a 1 μm thick spacer layer on the garnet surface. Conductors were 1 μm thick, 8.5 μm wide, spaced on 15 μm centers. Nominal propagation pulse width was 4 μs with 1 μs rise and fall time. State transitions can be recognized by discrete changes in deflection angle. Deflection angle χ and in-plane field orientation ϕ are measured clockwise from ∇H_z when looking along the direction of H_B.

Except as noted, all measurements were made on ion-implanted chips of YSmCaGe garnet with properties as given in Table I.

TABLE I

Properties of $Y_{1.92}Sm_{0.1}Ca_{0.98}Fe_{4.02}Ge_{0.98}O_{12}$ Chip

Implantation Dosage	2×10^{14} Ne/cm^2
Implantation Energy	80 Kev
Thickness h (by SEM)	5.95 μm
Stripe-Width W_s	5.53 μm
Characteristic Length ℓ	0.598 μm
Collapse Field H_o	95.3 Oe *
$4\pi M_s$	176 G
Q	4.54
K_u	5650 ergs/cm^3 **
K_1	-2050 ergs/cm^3 **
g	2.042 **
α	0.062 **

* As-grown wafer
** FMR

III. OBSERVATIONS OF DEFLECTION ANGLE

The observed deflection angles as a function of in-plane field H_p for $H_p || \nabla H_z$ are given in Fig. 1 for a drive current of 35 mA ($\Delta H \approx 3.4$ Oe nominal drive across the 6 μm diameter bubble). As H_p increases, the S=+1 deflection angle decreases slightly until, at H_p=110 Oe, transition A1 occurs in which S=+1→S=0. Reduction of H_p causes no change in deflection angle of S=0 until $H_p \sim 40$ Oe where transition B1 occurs corresponding to (0,2)→(1/2,2,1). The three digit notation (S,ℓ,p) signifies a revolution number S, ℓ vertical Bloch lines, and p Bloch points. When no ambiguity can arise, we use abbreviated notation in which ℓ and/or p are not specified. The 1/2 state shows a gradual increase in deflection angle as H_p is reduced. At H_p near zero, 1/2 converts to (1,0); and at large H_p, 1/2 converts to (0,2), transition C1, at the same field as for S=+1 decay, A1. The feature in Fig. 1 we wish

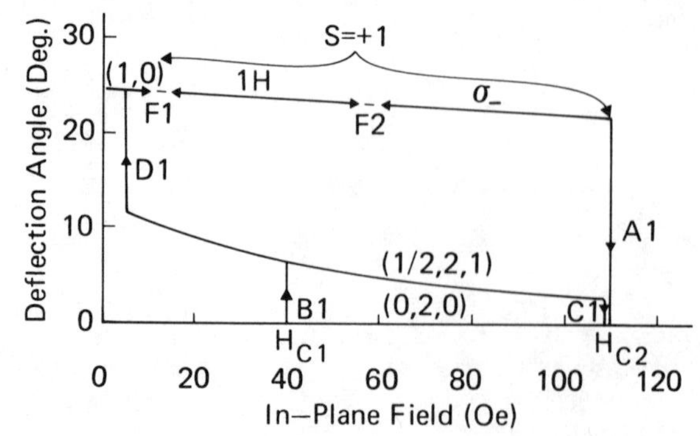

Fig. 1. Bubble deflection angles versus in-plane field for $H_p || \nabla H_z$ in an ion-implanted 5.95 μm thick chip of $Y_{1.92}Sm_{0.1}Ca_{0.98}Fe_{4.02}Ge_{0.98}O_{12}$ with a drive current of 35 mA at a bias field of 88 Oe.

to emphasize is that for $H_p||\nabla H_z$, transition A1 always results in $+1\rightarrow(0,2)$. The fine structure of S=+1, transitions F1 and F2, will be discussed in Section V.

Rotating H_p to give $H_p\perp\nabla H_z$ gives the results in Fig. 2, where all other parameters are identical to those of Fig. 1. The S=+1 deflection angle increases monotonically with H_p until A2 converts S=+1 into a state we label (1/2,2,1)* or 1/2* for short. Although the deflection angles of 1/2 and 1/2* are identical, the observation that the two states are formed and destroyed by different transitions shows that they are different states. Increasing H_p to 140 Oe converts 1/2* to (0,2). Transitions B1 and C1 are similar to Fig. 1. If, before transition E, the value of H_p is reduced, 1/2* will convert by transition B2 to (1,2,2). As will be discussed, this state consists of two vertical Bloch lines (VBL's) with one Bloch point (BP) on each line. The (1,2,2) state is stable between transitions D2 and C2 and can only be obtained from 1/2* by means of transition B2. Even though 1/2* and (1,2,2) could not be created for $H_p||\nabla H_z$, these states are stable for this orientation of H_p and can be obtained by generating with $H_p\perp\nabla H_z$ and then rotating H_p. We have found that the behavior of a given state does not depend on the conditions of its creation. We emphasize that for $H_p\perp\nabla H_z$, the decay of S=+1 always leads to 1/2*, not (0,2) as for $H_p||\nabla H_z$.

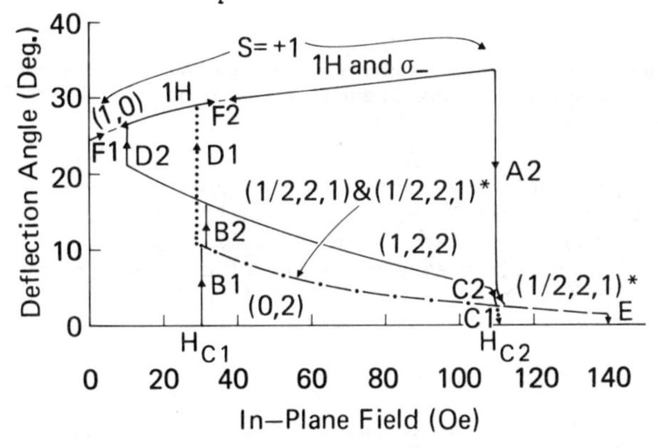

Fig.2. Bubble deflection angles for $H_p\perp\nabla H_z$ for the same chip as in Fig.1. All other conditions are identical to those in Fig.1.

The state S=+1 in Figs. 1 and 2 has a deflection angle of 24 degrees at H_p=0. Equation (1) gives a χ of 22.7 degrees for S=+1 if we use the measured wall mobility of 1040 cm/sec-Oe. The wall mobility calculated from the FMR value of α=0.062 is 1912 cm/sec-Oe.

Before treating the detailed dependence of deflection angle on H_p for bubbles having fractional S-state, we must account for those effects which are independent of S. Since χ for S=+1 at H_p=0 decreases only slightly as the drive current is increased from 35 to 50 mA, we will ignore coercivity in what follows. Putting $V=(\mu/2)\Delta H\cos\chi$, where μ is the wall mobility, we can rewrite Eq. (1) in the more convenient form

$$\tan\chi = \frac{4\mu S}{\gamma d} \tag{2}$$

The collapse field was observed to be a function of the magnitude and orientation of the in-plane field. The resulting variation of bubble diameter at constant bias gives rise to changes in deflection angle. We assume the change in diameter is independent of S. In addition, Eq. (2) is only valid for circular bubbles. The in-plane field causes elliptical deformation of the bubble[9] which produces changes in deflection angle so that the observed $\chi_T=\chi_g+\chi_\varepsilon$, where χ_g is given by Eq. (2) and χ_ε is proportional to the deformation of the bubble. For H_p either parallel or perpendicular to ∇H_z, it can be shown by an extension of the analysis in reference 9 that

$$\tan\chi_T = (1 + k\varepsilon)\tan\chi_g \tag{3}$$

where ε is the ellipticity and k is a proportionality constant which is independent of S. Two other results of the analysis can be used to evaluate ε. The deflection angle of S=0 bubbles is given by

$$\chi(0) = \frac{3}{4}\varepsilon\sin 2\phi \tag{4}$$

where ϕ is the orientation of H_p. Measurements of $\chi(0)$ for the chip of Figs. 1 and 2 as a function of ϕ for H_p=80 Oe gave ε=0.21. The difference in deflection angle for +1 bubbles with H_p perpendicular or parallel to ∇H_z is given by

$$\chi_\perp - \chi_{||} = \frac{3}{4}\varepsilon\sin 2\chi_0 \tag{5}$$

where χ_0 is the deflection angle for H_p=0. Measurements of +1 bubbles gave ε=0.24±0.05 at H_p=80 Oe. In contrast to Josephs et al,[6] these results indicate that elliptical deformation of the bubble accounts for most of the variation of the +1 deflection angle with H_p. Using Eq. (3) and the fact that χ_g is given by Eq. (2), we can correct the measured χ for a bubble of effective S-value, S_e, using the equation

$$\tan\chi(S_e) = S_e \cdot \tan\chi(+1) \tag{6}$$

where we measure $\tan\chi(+1)$ for each value of H_p for which we wish to calculate $\chi(S_e)$. Hasegawa[4] has calculated S_e for bubbles containing one and two BP's. For 1/2 bubbles, $S_e=1/2-|z|/h$ where z is the displacement of the BP from the midplane of the film. For (1,2,2) bubbles containing two BP's, $S_e=1-2|z|/h$. The stable location of a BP occurs where H_p and the radial stray field of the bubble sum to zero.

Figure 3 compares the measured deflection angles (points) and the calculated curves using Eq. (6) and the S_e values as described above. The agreement unambiguously identifies 1/2 and 1/2* as containing one BP, and (1,2,2) as containing two BP's.

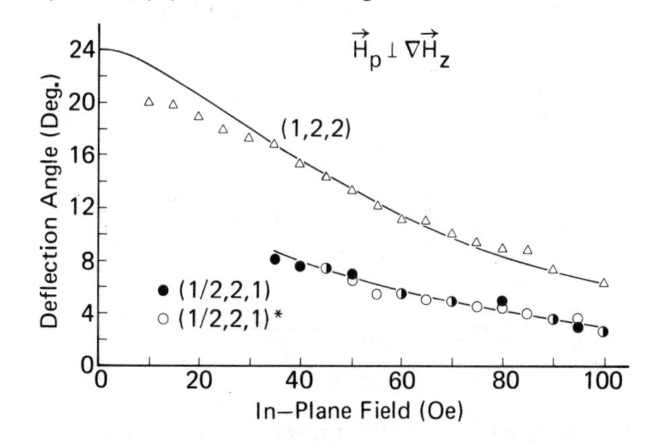

Fig.3. Comparison of measured deflection angles (points) and calculated curves versus in-plane field for the implanted YSmCaGe film of Fig.1. Drive current 35 mA, bias field 88 Oe.

IV. TRANSITION MECHANISMS

A. Cap-Switching Processes

From Fig. 4, we observe that transitions A1, A2, C1, and C2 occur at approximately the same value of H_p, roughly independent of orientation ϕ. Similar behavior is observed for transitions B1 and B2. A common process is suggested for these two groups of transitions, and because of the lack of ϕ-dependence, it is clear that the gyrotropic forces play little role in this process. We identify four transitions A1, A2, C1, and C2 as a cap-switch process in which the ion-implanted capping layer of the film becomes saturated. Such a process will occur when H_p dominates the stray field in the capping layer. The low field processes B1 and B2 are the reverse cap-switch process in which the capping layer switches from a saturated configuration to a formation having a closure domain which is compatible with the stray field of the bubble.[10] We introduce the nomenclature "capped Bloch line" to indicate a Bloch line which terminates under the portion of the capping layer which switches; i.e., under the closure domain. For cap-switch processes, we postulate injection of a BP onto a capped BL whenever the cap switches. The required energy presumably arises from the change in magnetostatic energy accompanying the cap-switch itself. Considering the exchange interaction between the capping layer and the underlying VBL's, it is clear on topological grounds that a BP singularity must exist any time the cap magnetization opposes the magnetization in the VBL.[3]

Schematic representations of two pairs of states which undergo cap-switches are shown in Fig. 5. The effect of the in-plane field is to stabilize two vertical Bloch lines on the extremities of the bubble as determined by the in-plane field direction. Transitions B1 and B2 involve injection of a BP onto the capped VBL as the cap switches from the saturated to the umbrella configuration which has the closure domain. The high field processes C1 and C2 cause the injection of one additional BP onto a VBL which already contains a BP. The two BP's on the same VBL annihilate each other, thus completing processes C1 and C2. Also in Fig. 5, we see that 1/2 and 1/2* have structures which differ in cap configuration and the location of the BP's on different lines.

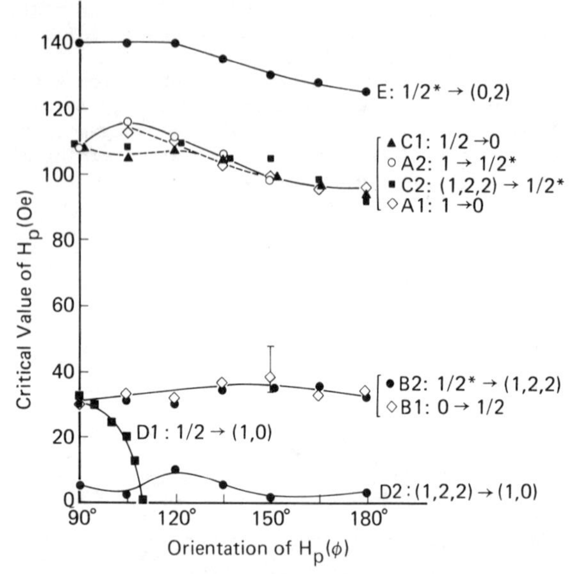

Fig.4. Dependence of the in-plane field at which state switching occurs on the azimuthal orientation ϕ of the in-plane field. Ion-implanted YSmCaGe chip of Fig.1, drive current 35mA, bias field 88 Oe. Angles are measured clockwise from ∇H_z, looking along the direction of H_B.

We expect cap-switch processes to be affected by the level of ion-implantation. Chips of a 3.88 μm film of $Y_{1.52}Eu_{0.3}Tm_{0.3}Ca_{0.88}Ge_{0.88}Fe_{4.12}O_{12}$ ($4\pi M_s$=240 G, ℓ=0.62 μm, H_o=103 Oe) were implanted with 2×10^{14} Ne+ ions/cm^2 using energies of 25, 50, and 80 Kev. The upper cap-switch field H_{C2} decreased slightly from 115 Oe for 25 Kev to 109 Oe for 80 Kev. The lower cap-switch field H_{C1} increased from 31 Oe to 56 Oe for 25 and 80 Kev, respectively. Transition D1, which is not a cap-switch, showed no variation with implantation level.[11]

We have ignored dynamic effects in our discussion of the cap-switching process. In general, the difference between the upper and lower cap-switch fields decreases

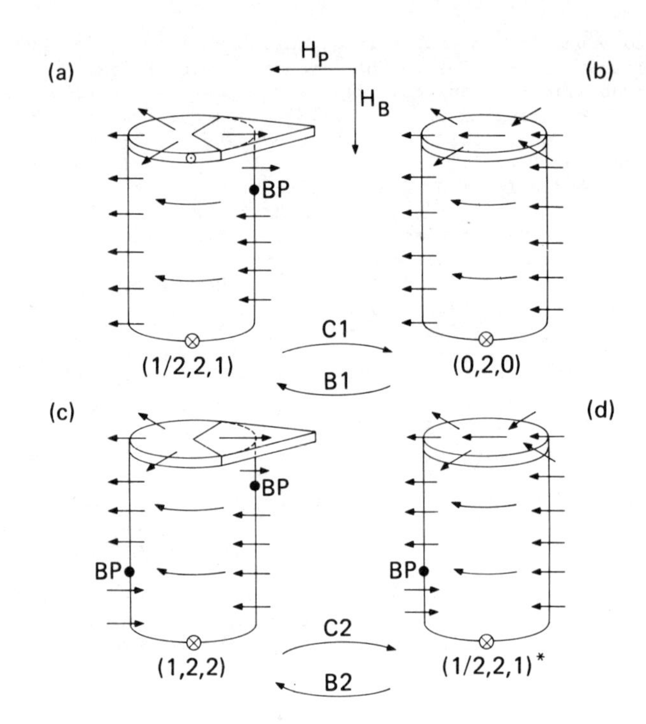

Fig.5. Schematic representations of four states derived by cap-switching processes. Transitions C1 and C2 occur at $H_p = H_{C2}$; and transitions B1 and B2 occur at $H_p = H_{C1}$.

as bubble velocity increases, but there are significant differences in this behavior for different garnet compositions. For example, in our YSmCaGe garnet, the high field transition decreases with velocity much more rapidly than the low field transition increases. In YEuTmCaGe garnets, the converse is true.

B. Bloch Point Annihilation

With the structure of the 1/2* bubble shown in Fig. 5d, it is natural to identify the transition 1/2*→0 (E in Fig. 2) as the annihilation of the BP as it approaches the lower surface (no capping layer) of the garnet film. In a stationary bubble for the YSmCaGe chip studied, E occurs at H_p=143 Oe and, as shown in Fig. 4, is only slightly dependent on ϕ when the bubble is in motion. Thus, the gyrotropic forces play little role in this transition.

From the field at which transition E occurs statically and the condition that the BP is located where H_p equals the radial stray field from the bubble, we calculate that the BP is 0.23 μm from the surface when E occurs. Slonczewski[12] has pointed out that a Bloch line is perturbed by the presence of a BP for a distance of about one linewidth. Thus, the perturbed wall magnetization reaches the surface when the BP is half a linewidth from the surface. For our sample, ℓ=0.598 μm and Q=4.54 so that $\pi\Lambda/2=\pi\ell/Q^{1/2}$=0.22 μm. This agreement is probably fortuitous, but it does support the proposed mechanism for transition E.

C. Bloch Line Annihilation

From Figs. 1, 2, and 4, we notice that transition D1 displays very strong ϕ-dependence which is shown in more detail in Fig. 6. In all cases, 1/2 was observed to decay to (1,0) when two conditions were met: (1) H_p was reduced below a critical value, and (2) the bubble velocity had a positive component along the $\vec{H}_p\times\vec{H}_B$ direction. From the inset of Fig. 6, we notice that such a velocity places the mobile VBL near the unstable g-force node. The gyrotropic force acting on a VBL has been calculated by Slonczewski[13]

$$F_G = \pm \frac{2\pi M}{|\gamma|} \hat{z} \times \vec{V} \quad (7)$$

where $\hat{z}$ is along the bias field direction and the sign is (+) if the sense of the VBL is the same as that of the adjacent wall magnetization. Equating this gyrotropic force on the VBL to the in-plane field restoring force $2M\pi\Delta H_p$, we find to lowest order the critical wall velocity for the 1/2→(1,0) transition

$$V_W = \Delta \cdot \gamma \cdot H_p \quad (8)$$

where Δ is the wall width parameter and γ is the gyromagnetic ratio. Using the fact that the linear wall mobility $\mu=\gamma\Delta/\alpha$, we can rewrite Eq. (8) in terms of the drive field ΔH acting across the bubble diameter

$$\Delta H = 2\alpha H_p \quad (9)$$

where α is the Gilbert damping parameter. For $\vec{V}=\vec{H}_p\times\vec{H}_B$, once H_p falls below the value given by Eq. (9), the 1/2 bubble will decay by line annihilation to (1,0). The detailed dependence on in-plane field orientation is given in Fig. 6. The asymmetry about ϕ=90° is attributed to the radial velocity $\dot{r}$>0 present in our deflectometer. The resulting gyrotropic force $F_{G,\dot{r}}$ either assists (ϕ<90°) or opposes (ϕ>90°) the g-force $F_{G,V}$ due to the translational velocity. Hence, for ϕ<90°, we expect less stability for 1/2 than for ϕ>90°. For ϕ>90°, 1/2 becomes stable down to $H_p \approx 0$ for ϕ between 110° and 160°. Figure 7 plots the critical H_p versus drive current and field for H_p aligned at ϕ=100°, perpendicular to the 1/2 trajectory. The nominal drive

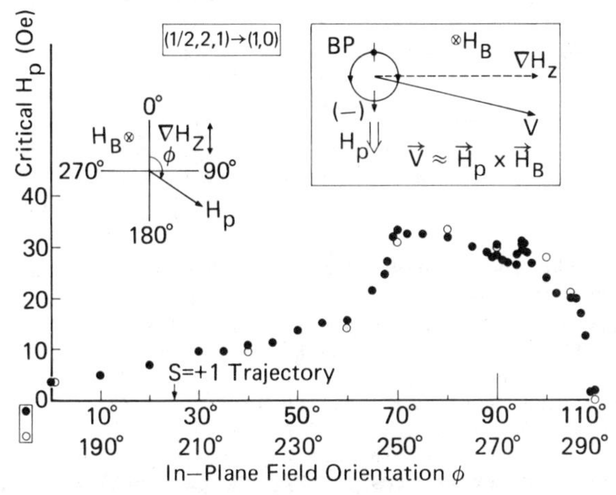

Fig.6. Variation of the critical in-plane field for the 1/2→(1,0) transition with azimuthal orientation of H_p. Bubble velocity was always as shown in the inset when decay was observed.

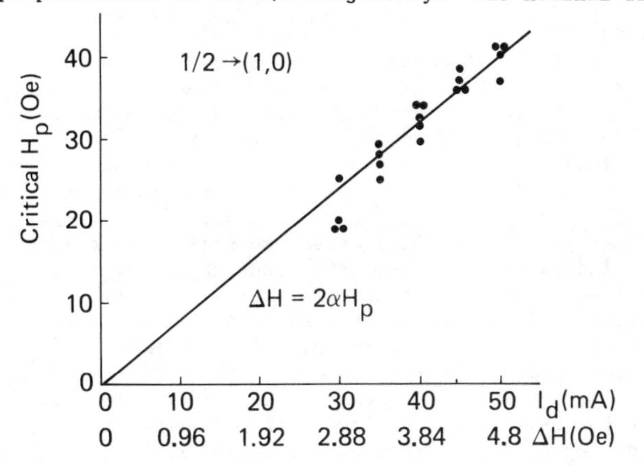

Fig.7. Dependence of the critical field of Fig.6 on drive current I for $H_p \perp V$ (ϕ=100°).

ΔH on the bubble is calculated midway between the powered conductors and ignores $\dot{r}$. The curve is a plot of Eq. (9) with slope corresponding to α=0.06 (from FMR) which is in reasonable agreement with α=0.11 from the mobility measurement.

Transition D2, in which (1,2,2)→S=1, displays only slight ϕ-dependence, again suggesting a quasi-static process. As H_p→0, the BP's on the two VBL's of the (1,2,2) bubble approach the mid-plane and the lines become gyrotropically inactive. These lines are, however, magnetostatically attractive and would be expected to unwind as they collide.

V. DEPENDENCE OF S=+1 WALL STRUCTURE ON $\vec{H}_p$ AND $\vec{V}$

In this section, we discuss the influence of an in-plane field on the wall structure of the S=+1 bubble.[14] Depending on the direction of propagation, and the actual drive conditions, two structures other than the simple unichiral (1,0,0) structure are identified. Both structures are necessary in order to explain the experimental observations.

Referring to transitions A1 and A2 of Figs. 1 and 2, we observe that the nominal S=+1 bubble decays to either 1/2* or (0,2) as H_p is increased through the upper cap-switch field H_{C2}. This dynamic behavior provides a clue to the detailed Bloch line structure of the +1 bubble in the presence of an in-plane field. It is known that an in-plane field directed antiparallel to the magnetization in a planar domain wall can quasi-statically twist the spins near both surfaces to form Bloch loops at a threshold field given by[14]

$$H_p = 4\sqrt{2\pi A}\, h^{-1} \quad (10)$$

which is 8 Oe in the YSmCaGe sample. Once formed, these loops give rise of a 2π horizontal Bloch line (HBL) which is subject to the usual gyrotropic forces.[15]

For a cylindrical domain, this 2π HBL formation leads quite naturally to the wall structure shown in Fig. 8a, which we label 1H. For the velocity shown, the gyrotropic force on the 2π HBL is upward. The opposite initial chirality would result in the 2π HBL located on the back of the bubble with the gyrotropic force still directed upward. The capping layer, due to a large exchange repulsion, resists punch-thru at the top surface for drive currents used in this study. As the cap

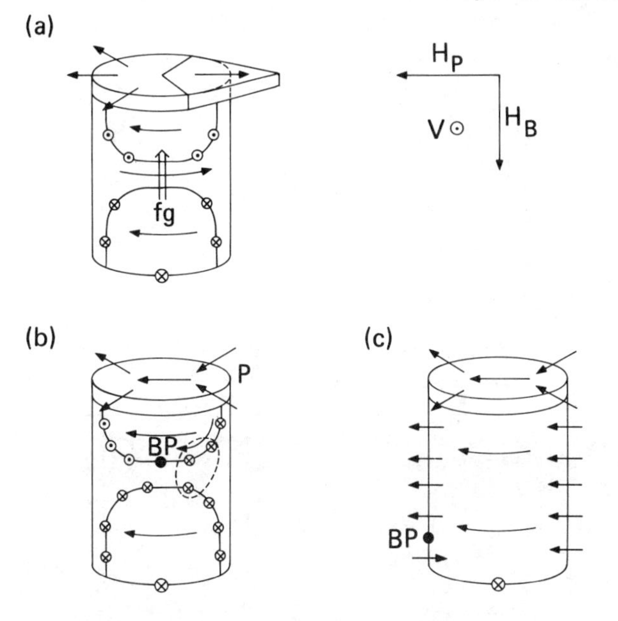

Fig.8. (a) The 1H bubble showing a 2π horizontal Bloch line. (b) transient state which exists just after 1H undergoes a cap-switch at P. The dotted circle denotes the region where unraveling of the upper and lower loops is expected as the BP advances. (c) The final state 1/2*.

of the 1H bubble switches at $H_p=H_{C2}$, a BP is introduced at point P, as shown in Fig. 8b. This BP travels down the upper Bloch loop as shown, reversing the sign of the line as it progresses. The two Bloch loops unravel behind the moving BP and the resulting state, 1/2*, shown in Fig. 8c, has the BP on the front VBL (with respect to the direction of H_p) even though the cap-switch occurred at the back of the bubble. This BP position is stable with respect to the bubble stray field; whereas, no position on the back or right-hand VBL is stable.

The 1H structure and its high-field cap-switch provide a natural explanation of transition A2 of Fig. 2 in which +1→1/2* for $\vec{V}=\vec{H}_p\times\vec{H}_B$. The question arises as to how to explain the cap-switch A1 of Fig. 1 in which +1→0 for $H_p||\nabla H_z$. Dynamic results cannot distinguish between g-forces on 1H during the cap-switch process and the possibility of a different structure for the parent S=+1 bubble. In order to sort out the detailed wall structure as a function of bubble velocity and in-plane field orientation, we make use of fact that even quasi-static switching of the ion-implanted capping layer introduces a Bloch point onto the underlying Bloch line. From Fig. 4, it is clear that the center of the operating margin for the YSmCaGe chip is $H_p \sim 70$ Oe. Thus, if we stop an S=+1 bubble operating under a certain H_p, and raise H_p to 135 Oe, the static cap-switch field for this chip, we can then reset H_p=70 Oe and propagate the bubble gently in order to determine its final state by means of its deflection angle. This technique greatly simplifies the identification of states because in the absence of g-forces, the 2π HBL relaxes to the film mid-plane; i.e., it does not disappear as in the case of the loops present in the dynamic conversion process.[16] Thus, any time we observe a final state 1/2* after a static cap-switch, we deduce that the parent S=+1 bubble was 1H. When the static cap-switch process was applied to bubbles initially propagating with $H_p||\nabla H_z$, we observed the following: For I=35 mA and $H_p \lesssim 55$ Oe, transition F2 of Fig. 1, final states 1/2* were observed for S=+1 initially propagating either parallel or anti-parallel to H_p. Larger values of H_p lead to (0,2) for either direction of velocity. Increasing the drive to 50 mA caused F2 to be reduced to ~35 Oe. For an explanation of these results, we refer to Fig. 9 which shows the

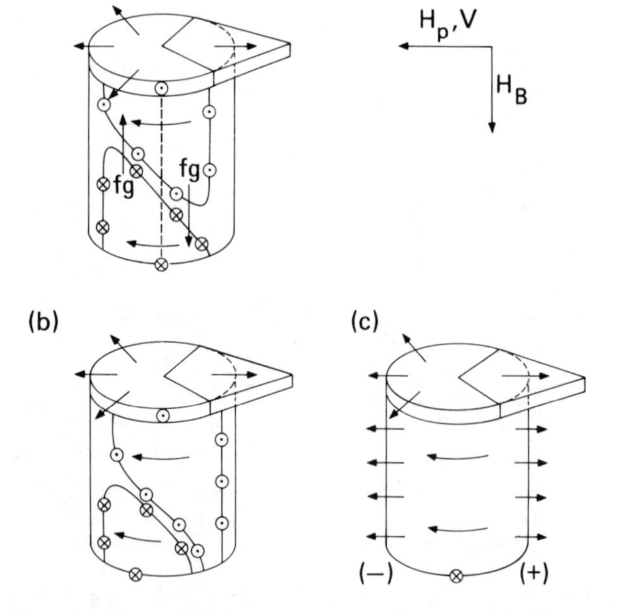

Fig.9. (a) Sheared Bloch loop structure which occurs for 1H propagating with V ǁH_p. (b) Punch-thru if drive or H_p is sufficiently large. (c) Final state σ− which exists after punch-thru and collapse of the lower loop in (b).

1H bubble propagating with V parallel to H_p. In the vicinity of H_{C2}, H_p is large, so to first order, the loops of the 1H bubble are held on the side of the bubble as shown. Because of the nodes in the g-force which exist on the sides of the bubble where V_n=0, we see that the loops tend to be sheared for V parallel to H_p. For V anti-parallel to H_p, the g-forces reverse direction. Thus, half of the loop is being pushed down, favoring punch-thru, for either direction of velocity. Once punch-thru at the lower surface has occurred, as in Fig. 9b, and the velocity is again reduced to zero, the lower Bloch loop collapses leaving a bubble as shown in Fig. 9c, which is labeled (1,2,0)− or σ−. The complementary state, labeled σ+, in which the two VBL's point inward, is not compatible with the cap magnetization for the direction of bias field shown.[17] Figure 10 shows the result of a static cap-switch of the σ− bubble. The BP is introduced

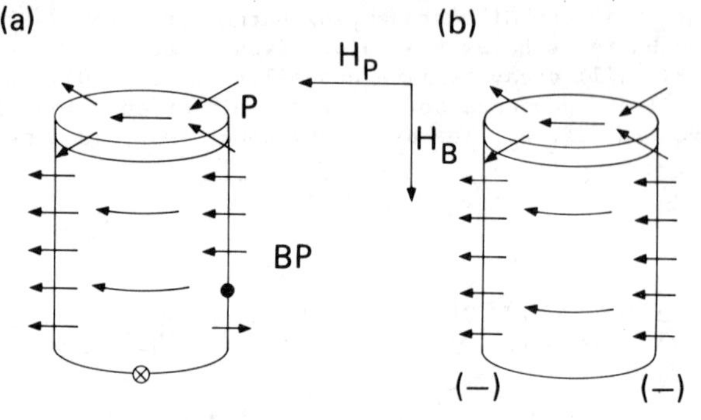

Fig.10. (a) The σ− bubble just after undergoing a cap-switch at P. (b) The resulting final state (0,2).

below point P as the cap switches, and it travels down the VBL and is lost at the bottom surface since there is no statically stable location for the BP on that line. The resulting state is seen to be (0,2). Thus, whenever we observe a final state (0,2), we deduce that the parent S=+1 bubble must have been σ−. Returning to the earlier results for $H_p||\nabla H_z$, we conclude that for I=35 mA and $H_p \lesssim 55$ Oe, the parent S=+1 bubble is 1H for bubble velocity either parallel or anti-parallel to H_p. For larger values of H_p, the parent S=+1 bubble must have been σ−, based on the results of the static cap-switch. Raising I to 50 mA resulted in punch-thru of 1H→σ− whenever H_p exceeded 35 Oe.

Thus, not only does increased drive favor punch-thru, but so too does increased H_p. The physical rationale for increased H_p favoring punch-thru is unclear but may be associated with the fact that H_p compresses the 2π HBL thereby enabling it to more closely approach the film surface before the repulsive potential is felt.

Returning to the case of $H_p \perp \nabla H_z$, there is the possibility of punch-thru of the 2π HBL at the lower surface of the 1H bubble if the velocity in Fig. 8a is reversed to $\vec{V}=-(\vec{H}_p\times\vec{H}_B)$. Observations of the static cap-switch process for $H_p \perp \nabla H_z$ yielded the following: For I=35 mA and $H_p \lesssim 35$ Oe, transition F2 of Fig. 2, the final state observed after a static cap-switch was 1/2* for $\vec{V}=\pm(\vec{H}_p\times\vec{H}_B)$, indicating that the parent state was 1H. For H_p>35 Oe, $\vec{V}=\vec{H}_p\times\vec{H}_B$ yielded 1/2*; whereas, $\vec{V}=-(\vec{H}_p\times\vec{H}_B)$ yielded (0,2) after the static cap-switch. This indicates that $\vec{V}=-(\vec{H}_p\times\vec{H}_B)$ causes punch-thru at the lower film surface in which 1H→σ−. If the drive current is increased to I=50 mA, the corresponding critical H_p for 1H→σ− conversion for $\vec{V}=-(\vec{H}_p\times\vec{H}_B)$ is reduced to about 20 Oe.

These results are remarkable in that they indicate that the S=+1 bubble has two different structures depending upon the direction of motion for $H_p \perp \nabla H_z$ (and sufficiently large H_p). For $\vec{V}=\vec{H}_p\times\vec{H}_B$, S=+1 is actually the 1H bubble; whereas, for $\vec{V}=-(\vec{H}_p\times\vec{H}_B)$, it is the σ− bubble. Even one step is sufficient to cause conversion between 1H and σ-, depending upon the direction of motion. The process of 1H→σ- is punch-thru, as already discussed. A proposed mechanism for the reverse process σ-→1H is shown schematically in Fig. 11. The VBL labeled (+) in Fig. 11a is gyrotropically unstable for $\vec{V}=\vec{H}_p\times\vec{H}_B$. In addition, we note that if this (+) line lies on the right-hand side of the bubble, its magnetization is unfavorably oriented with respect to H_p. We expect this magnetostatic misalignment energy to be reduced as the (+) VBL rotates toward the front of the bubble, since $\dot{r}>0$. We also expect nucleation of a horizontal Bloch loop to occur behind the rotating (+) line. At a later instant, t_2, shown in Fig. 11b, the lower portions of the two VBL's collide; and since they are unwinding, we expect formation of the 1H structure shown in Fig. 11c.

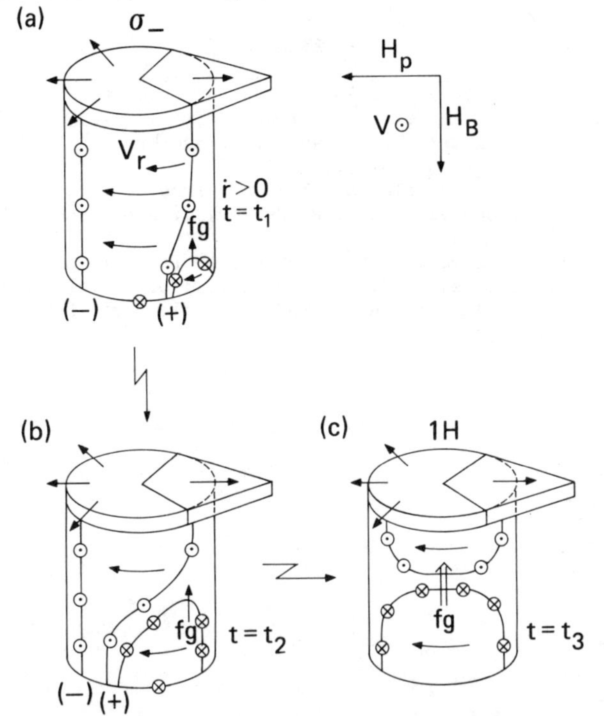

Fig.11. Proposed conversion of $\sigma_- \to$ 1H. The details are discussed in the text.

For large values of H_p, we do not find the σ- bubble stable even for one step if $\vec{V}=\vec{H}_p\times\vec{H}_B$. This appears to be in conflict with the related transition 1/2→(1,0) discussed in Figs. 6 and 7, for which 1/2 is stablized for $\vec{V}=\vec{H}_p\times\vec{H}_B$ if H_p is sufficiently large. At the present time, we do not have a satisfactory explanation for this difference.

For conditions under which we expect to have 1H or σ- depending upon the direction of propagation of the S=+1 bubble, we have observed two important differences confirming the existence of two distinct states. The mobility of 1H is about 20 percent less that that of σ- if the measurement is made using propagation pulse durations comparable to the expected transit time of the 2π HBL (0.2-0.5 μs). Stopping the respective bubbles, reducing H_p to zero, and subjecting them to a square collapsing bias field pulse, we observed that 1H did not jump, but σ- did jump in the direction of previous motion. The theory of bias jumps will be discussed elsewhere.[18]

Further confirmation of the existence of the 1H bubble was obtained by subjecting a unichiral (1,0,0) bubble to the static cap-switch process. Quasi-statically increasing H_p to H_{C2} always resulted in a final state 1/2* as confirmed by propagation of the final state, and by subsequent observation of transition B2 as H_p was reduced to H_{C1}.

VI. SUMMARY

We have found nine different transitions in ion-implanted garnet films, six of which are accounted for by cap-switching and injection of a BP. We found it necessary to invoke three different wall structures for S=+1 bubbles depending upon the direction of motion in an in-plane field. We have determined the static and dynamic conditions under which these different wall structures are stable. There are two 1/2 states containing one BP which differ in the location of the BP and in the associated capping structure. There is one state containing two BP's, one on each line; and there is a single S=0 state.

These results are applicable at sufficiently low velocities such that, for wall magnetization approximately parallel to the in-plane field, there is no generation of Bloch loops. At higher velocities, there will be additional complications.

ACKNOWLEDGEMENTS

The authors are grateful for many helpful discussions with P. Dekker, J. C. Slonczewski, B. E. Argyle, and S. Maekawa, and with R. L. White of Stanford University. We are also grateful to J. Engemann for permission to quote his unpublished results, and to D. Y. Saiki and D. Johnson to device fabrication.

REFERENCES

1. J. C. Slonczewski, A. P. Malozemoff, and O. Voegeli, AIP Conf. Proc. 10, 458 (1972).
2. R. Wolfe, J. C. North, and Y. P. Lai, Appl. Phys. Lett. 22, 683 (1973).
3. T. Hsu, AIP Conf. Proc. 24, 624 (1974).
4. R. Hasegawa, AIP Conf. Proc. 24, 615 (1974).
5. D. C. Bullock, AIP Conf. Proc. 18, 232 (1973).
6. R. M. Joseph, B. F. Stein, and W. R. Bekebrede, AIP Conf. Proc. 29, 65 (1975).
7. O. Voegeli, C. A. Jones, and J. A. Brown, Paper 5A-6 (21st Annual Conf. on Magnetism and Magnetic Materials, 1975), to be published.
8. B. R. Brown, AIP Conf. Proc. 29, 69 (1975).
9. T. J. Beaulieu and B. A. Calhoun, Appl. Phys. Lett. 28, 290 (1976).
10. R. Wolfe and J. C. North, Appl. Phys. Lett. 25, 122 (1974).
11. J. Engemann (unpublished).
12. J. C. Slonczewski, AIP Conf. Proc. 24, 613 (1974), and private communication.
13. J. C. Slonczewski, J. Appl. Phys. 45, 2705 (1974).
14. P. Dekker and J. C. Slonczewski, to be published.
15. A. A. Thiele, J. Appl. Phys. 45, 377 (1974).
16. F. B. Hagedorn, J. Appl. Phys. 45, 3129 (1974).
17. G. R. Henry and J. Gitschier, private communication.
18. A. P. Malozemoff and S. Maekawa, to be published in J. Appl. Phys.

TIME DEPENDENT EFFECTS IN MAGNETIC DOMAIN WALL DYNAMICS

G. P. Vella-Coleiro
Bell Laboratories, Murray Hill, N. J. 07974

ABSTRACT

A number of experiments[1-4] have been reported recently which indicated that when a pulsed field was applied to a domain wall in an epitaxial garnet film the response consisted of an initial motion at high velocity followed by conversion to a low velocity mode. We have studied the initial high velocity phase with the use of a mode-locked argon ion laser whose output pulse duration was ~1 ns. A pulsed magnetic field having rise and fall times of ~2 ns was applied to an isolated magnetic bubble domain in the same direction as the static bias field. The repetition rate was ~3 kHz, and the instantaneous bubble diameter was measured with a microscope and filar eyepiece as the time delay between the laser and field pulses was varied. Data obtained in a $LuGd_2Al_{0.6}Fe_{4.4}O_{12}$ film having a thickness $h = 9.4\ \mu m$, saturation magnetization $4\pi M = 189G$, and gyromagnetic ratio $\gamma = 1.4 \times 10^7\ s^{-1}\ Oe^{-1}$, showed a peak velocity of 8000 cm/s with a pulse amplitude of 13.1 Oe. Using pulse amplitudes of 24 and 44 Oe, peak velocities of 9300 and 9500 cm/s, respectively, were observed. In the latter two cases, the wall was observed to move backwards while the field pulse was on, in agreement with calculations based on the Walker model.[5] For all three pulse amplitudes, large deviations from the calculated displacements were observed for times $t > 10$ ns after the beginning of the field pulse.

Similar effects were observed in a $(YSmCa)_3(GeFe)_5O_{12}$ film having $h = 8.1\ \mu m$, $4\pi M = 160G$, $\gamma = 1.8 \times 10^7\ s^{-1}\ Oe^{-1}$. The peak velocities measured with pulse amplitudes of 12.8, 24 and 44 Oe were 5500, 9600, and 10,600 cm/s, respectively. For the highest amplitude pulse, the Walker model indicated oscillatory motion at $t \approx 6$ ns, and the data showed a large decrease in velocity at this time. After the termination of the 24 and 44 Oe pulses the wall remained practically stationary for more than 60 ns, perhaps due to a nonuniform spin precession giving rise to different parts of the wall moving out of phase with one another.

Wall motion was also studied in two films of the general composition $(YEuCa)_3(GeSiFe)_5O_{12}$, having large gyromagnetic ratios due to angular momentum compensation.[6] In one of these films with $g = 9.6$, $h = 8\ \mu m$, $4\pi M = 275G$, peak velocities of 28,000 and 44,000 cm/s were observed with pulse amplitudes of 80 and 128 Oe, respectively. These fields were insufficient to reach the Walker breakdown region, and no velocity oscillations were observed. The other film had the parameter values $g \approx 30$, $h = 9.3\ \mu m$, $4\pi M = 170G$. Using pulse amplitudes of 46 and 176 Oe, peak velocites of 25,000 and 95,000 cm/s, respectively, were observed without oscillations, since the field amplitudes were below the Walker critical field in this case also. The most striking feature of the data in these high-g films is the occurrence of well-behaved wall motion at very high velocites in high drive fields for relatively long periods of time, i.e., many tens of nanoseconds.

In all four films described above it was found necessary, in order to fit the data to the Walker model, to use velues of the Gilbert damping parameter α considerably greater than those obtained from FMR, when the wall velocity approached the Walker critical value. The increase in the value of α ranged from a factor of 5 in the LuGd film, to a factor of 2 in the YSm film.

The conclusion to be drawn from our experiments is that velocities up to the Walker critical velocity occur in epitaxial garnet films. In materials with $g \approx 2$, these high velocities persist for only ~10 ns, but during this time a wall displacement of the order of 1 μm can occur. It appears, therefore, that the fast transient should be taken into consideration in any wall motion experiment where wall displacements of a few μm or less are observed, even if the time scale of the experiment is much longer than 10 ns.

A fuller account of these results will be published elsewhere.

REFERENCES

1. F. H. deLeeuw, J. Appl. Phys. 45, 3106 (1974).
2. G. P. Vella-Coleiro, AIP Conf. Proc. 24, 595 (1975).
3. G. J. Zimmer, L. Gal and F. B. Humphrey, AIP Conf. Proc. 29, 85 (1976).
4. G. P. Vella-Coleiro, AIP Conf. Proc. 29, 64 (1976).
5. N. L. Schryer and L. R. Walker, J. Appl. Phys. 45, 5406 (1974).
6. R. C. LeCraw, S. L. Blank and G. P. Vella-Coleiro, Appl. Phys. Lett. 26, 402 (1975); G. P. Vella-Coleiro, S. L. Blank and R. C. LeCraw, Appl. Phys. Lett. 26, 722 (1975).

DOMAIN WALL DISPLACEMENT UNDER PULSED MAGNETIC FIELD

S. Konishi, K. Mizuno, F. Watanabe, and K. Narita
Faculty of Engineering, Kyushu University, Fukuoka 812, Japan

ABSTRACT

The domain wall displacement has been studied as a function of pulse width in garnet films using the bubble collapse method. The wall velocity defined from the increments of displacement distance and pulse width shows very large velocity for short pulse width and small steady-state velocity over about 100 ns. The initial large velocity appears spurious and to be caused by the wall relaxation. The Bloch line damping, the Bloch line annihilation loss, and the free precession are responsible for the steady-state velocity.

The domain wall velocity in bubble materials has been studied extensively for several years by many authors using the bubble collapse method and the transport method. These methods provide the wall or bubble displacement distance for a given pulse width. The wall velocity is defined by the ratio of the distance to pulse width. Such an average velocity is reasonable only when the wall relaxation phenomena at rising and falling edges of pulse do not affect significantly the net wall displacement. The effect is reduced by applying a fairly long pulse and making the wall transverse a fairly long distance. However, in the above-mentioned two methods, the wall displacement is limited physically and technically, and is about one bubble diameter or so in practical cases.

Recently, clear evidence of wall relaxation phenomena has been reported by Vella-Coleiro[1] and by Malozemoff.[2] Vella-Coleiro observed a remarkable increase of average velocity with decreasing pulse width using the bubble collapse technique. He attributed it to the indication of the velocity breakdown which will occur at about 10 ns after the rising edge of pulse. Malozemoff, using the bubble transport technique with a pulsed light source, observed the wall velocity relaxation of fairly long period after the drive pulse is terminated. The phenomena is attributed to the rewinding process of wound-up horizontal Bloch lines. These observations have made the many data of velocity obtained previously obscure.

This report presents the wall displacement distance per pulse as a function of pulse width in bubble collapse, from which the steady-state wall velocity is deduced reasonably. The mechanisms for the steady-state wall velocity are discussed in fairly wide range of drive field.

The samples are non-implanted and implanted $Y_{1.57}Eu_{0.78}Tm_{0.65}Ga_{1.05}Fe_{3.95}O_{12}$ garnet films. The thickness of the LPE film is about 6.2 μm, the $4\pi M$ is 200 gauss, the gyromagnetic constant γ is 1.1×10^7/sec·Oe, respectively, for the non-implanted film. Three small drive coils (5-, 10-, and 50-turn coils) are used. The 5-turn and 10-turn coils yield 50 and 122 Oe/A, respectively. The 50-turn coil yields 370 Oe/A. The inner diameter is 1 mm for the 5-turn coil, and 0.5 mm for both 10- and 50-turn coils. The rise time of pulse field for the 5-turn coil is about 1 ns, and longer for the other two coils. To make easy the precise control of bias field, a second bias coil of a small number of turns is wound around the usual bias coil.

The critical wall displacement distance for the dynamic bubble collapse is calculated using the Callen-Josephs approximation for the force function.[4] The result is,

$$\Delta r = \frac{4}{3} h \left(4\pi M - \sqrt{4\pi M H_{co}}\right)^{1/2} \cdot \left\{1 + \frac{H_{co} - H_b}{4(4\pi M - \sqrt{4\pi M H_{co}})}\right\}^{1/2} \cdot \frac{\sqrt{H_{co} - H_b}}{H_b}, \quad (1)$$

where Δr is the critical distance, h is the film thickness, H_{co} is the static bubble collapse field, and H_b is the bias field. For the practical cases, the following relation gives a useful approximation.

$$\Delta r \cong \frac{4}{3} h \left(4\pi M - \sqrt{4\pi M H_{co}}\right)^{1/2} \cdot \sqrt{H_{co} - H_b}/H_b \quad (2)$$

The measurement is performed by carefully increasing the bias field until bubbles collapse, applying successively pulse fields with a given amplitude and a pulse width. The wall displacement distance per pulse is calculated using (1) or (2) at the measured critical bias field for collapse.

The wall displacement distance for the non-implanted film is shown in Fig. 1 as a function of pulse width with pulse field amplitude as a parameter.

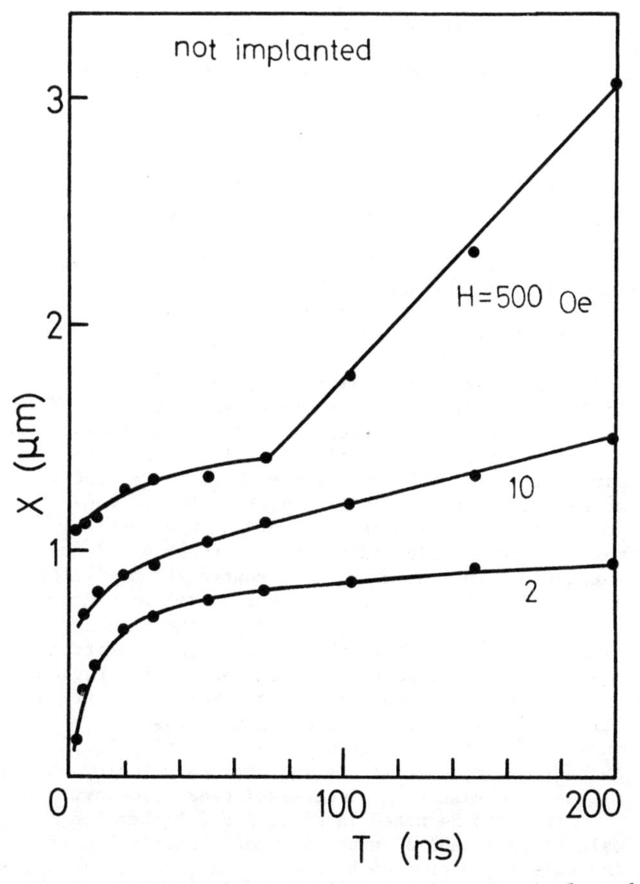

Fig. 1 Wall displacement distance in a non-implanted film as a function of pulse width with pulse field amplitude as a parameter.

As is clearly seen, the dependence of wall displacement on pulse width is fairly non-linear. It is well understood how the average velocity defined by the ratio of distance to pulse width increases with decreasing pulse width or distance.[1] However, it is more reasonable to define the velocity from the slope. For the low drive field, the distance increases nearly linearly with increasing pulse width in very short pulse width region. The measured slopes are 100×10^2 and 140×10^2 cm/sec for H=2 and 5 Oe, respectively. The clear deviations from the linear relation are initiated at 5 and 4 ns for respective drive field. For H=500 Oe, such a linear relation in short pulse width region is not observed, but the average velocity reaches 420×10^2 cm/sec for the shortest width (2.8 ns). These values far exceed the calculated Walker peak velocity[5] (V_W=3700 cm/sec) as well as the Slonczewski peak velocity[6] (V_p=760 cm/sec) for this film. It sounds quite strange, and appears to suggest that the velocity defined from a slope or average velocity under such short pulses does not give a true wall velocity. Then, let us calculate the wall displacement distance under a short pulse field assuming a one-dimensional wall. One of the basic equations for wall motion derived by Slonczewski is,

$$\frac{dx}{dt} = \frac{\gamma}{\alpha}\Delta H - \frac{\Delta}{\alpha}\frac{d\phi}{dt}\,, \tag{3}$$

where dx/dt is the wall velocity, Δ is the wall width parameter, and ϕ is the tilting angle of the magnetization in the moving wall.[6] Integrating (3) with time t, the wall displacement distance X under a pulse with width T is,

$$\begin{aligned} X &= \int_0^\infty dx/dt\ dt \\ &= \gamma\Delta HT/\alpha - \left(\int_0^{\phi_o} d\phi + \int_{\phi_o}^{o} d\phi\right)\cdot\frac{\Delta}{\alpha} \\ &= \gamma\Delta HT/\alpha\,, \end{aligned} \tag{4}$$

where ϕ_o is the maximum tilting angle, and is assumed to be less than $\pi/2$ for a short pulse. From (4), the average velocity is obtained as $\gamma\Delta H/\alpha$, which holds exactly for the field larger than the Walker critical field $2\pi M\alpha$ as well as the lower field. The condition imposed on ϕ_o implies for large field,

$$\gamma HT \cong \phi_o < \pi/2 \tag{5}$$

In other words, the maximum wall displacement is limited as $\pi\Delta/2\alpha$. The calculated value for this film is about 2 μm, which agrees with the measured knee points in Fig. 1 qualitatively. Fig. 2 shows the similar result for the implanted film. The tendency appears to be quite similar to that for the non-implanted film. The real situation in a bubble wall may be more complicated because of twisting of magnetization, but the concept presented above will be applicable, too, even for the horizontal Bloch line nucleation and shrinking process for short pulse fields. In conclusion, though these considerations, the slope or average velocity in short pulse region in Figs. 1 and 2 gives an apparent velocity. Almost all of the net wall displacement will take place during a fairly long period after the drive field is terminated, even if the instantaneous velocity reaches the Walker or Slonczewski peak velocity.

It is to be noted in Figs. 1 and 2 that the wall displacement distance increases nearly linearly for pulses longer than about 100 ns. Then the steady-state average velocity is defined from the increment of wall displacement distance between T=100 and 200 ns. The velocity thus determined is shown in Fig. 3 as a function of pulse field amplitude for both films. The closed circles show the steady-state velocity in the non-implanted film, and open circles show the velocity in the implanted film. Both plots give a fairly large mobility for H $\lesssim$ 20 Oe (30$\sim$40 cm/sec·Oe). However,

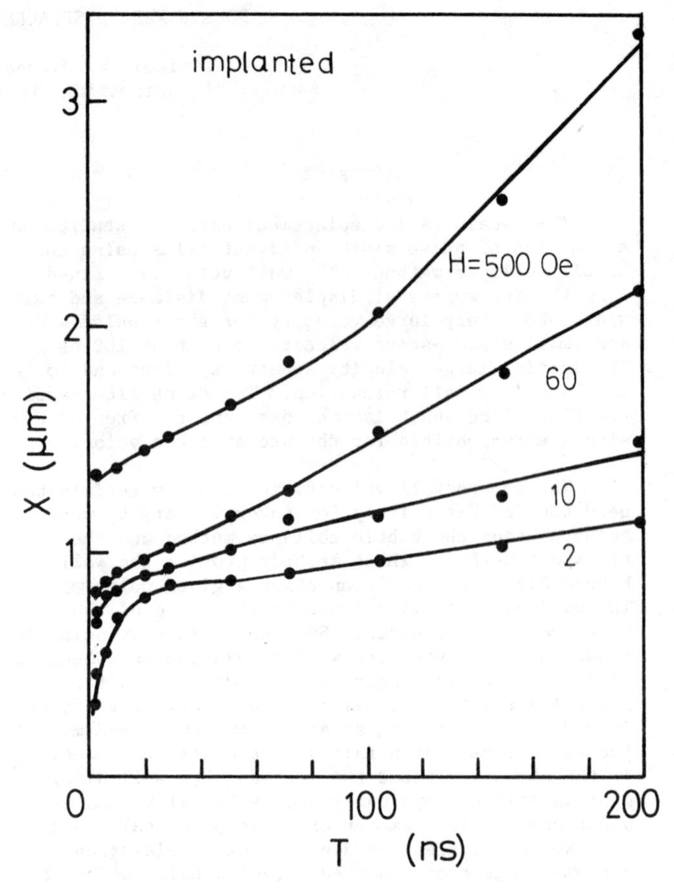

Fig. 2 Wall displacement distance in an implanted film as a function of pulse width with pulse field amplitude as a parameter.

the value is much smaller than the mobility $\gamma\Delta/\alpha$=1400 cm/sec·Oe expected for these films, and is close to the mobility damped by an oscillating horizontal Bloch line (10 cm/sec·Oe).[7] The threshold field for nucleation and instability of horizontal Bloch lines is considered to be too small, because even the Walker threshold field is 2.6 Oe. In larger drive field, the velocity appears to consist of two components, that is, a component independent of drive (660 and 500 cm/sec, for the non-implanted and implanted films, respectively) and a component increasing linearly with increasing drive field. The mobilities give 1.3 and 1.6 cm/sec·Oe for each film. The values are in good agreement with the mobility in the hard bubble wall (1.1 cm/sec·Oe, see the lower linear plots in Fig. 3). The calculated $\alpha\gamma\Delta$, which is the wall mobility for the case of free precession, is 0.96 cm/sec·Oe, and is very close to the above values.

The mechanism responsible for a constant component of velocity is not understood well, but appears to be related to the non-uniform precession of magnetization around the drive field. The surface magnetization is considered to slip and precess under such large drive at the rate different from that in the middle part. The successive annihilation of quasi-horizontal Bloch lines at the film surface will give a constant velocity as proposed by Slonczewski.[6] Over about 800 Oe, the component disappears in the non-implanted film, which appears reasonable because the moment in the moving wall is expected to precess around the drive field rather uniformly under such a larger drive field. The implanted layer will be yet effective to cause non-uniform precession and velocity independent of drive even under such a large drive, as seen in the figure.

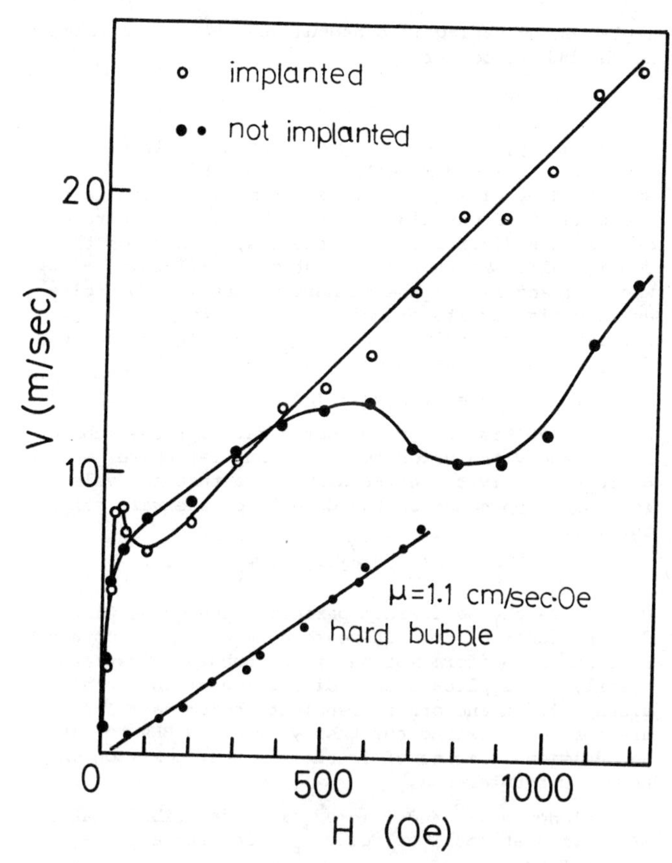

Fig. 3 Steady-state wall velocity as a function of field amplitude for both non-implanted and implanted films. The lowest linear plots show wall velocity in hard bubbles in the non-implanted film.

In summary, the wall displacement distance in bubble garnet films was measured as a function of the pulse width using the bubble collapse method. The wall velocity determined from the slope of displacement pulse width curves shows very large velocity for exceeding the Walker peak for short pulses and small velocity for longer pulses. The initial large velocity is spurious and attributable to the intrinsic wall relaxation process. The Bloch line damping, annihilation loss of Bloch lines at film surface, and the free precession appear to be responsible for the steady-state velocities.

REFERENCES

1. G. P. Vella-Coleiro, AIP Conf. Proc. 24, 595 (1975).
2. A. P. Malozemoff and J. C. DeLuca, Appl. Phys. Lett. 26, 719 (1975).
3. F. B. Humphrey, private communication.
4. H. Callen and R. M. Josephs, J. Appl. Phys. 42, 1977 (1971).
5. F. B. Hagedorn, AIP Conf. Proc. 5, 72 (1972).
6. J. C. Slonczewski, J. Appl. Phys. 44, 1759 (1973).
7. J. C. Slonczewski, J. Appl. Phys. 45, 2705 (1974).

THE "TURN AROUND EFFECT" IN BUBBLE PROPAGATION

S. Maekawa* and P. Dekker,**
IBM Thomas J. Watson Research Center
Yorktown Heights, New York 10598

ABSTRACT

The "turn around effect", i.e., the peculiar nonlinear initial motion of bubbles upon reversal of a driving gradient, is studied. A theory describes the effect local motions of Bloch lines inside the bubble wall have on the motion of the bubble center. Simple experiments with bubbles having unit winding number and either zero or two Bloch lines confirm the theory.

INTRODUCTION

The peculiar motion of bubbles upon reversal of gradient bias field, the "turn around effect" has been observed by several authors[1,2,3]. Tentatively, the phenomenon (see Fig. 1A) has been attributed to rearrangement of the positions of vertical Bloch lines (BLs) on the perimeter of the bubble[1,3]. Our present theory of gradient propagation accompanied by BL rearrangements is applied to some simple cases involving bubbles of winding number S = 1 and having two BLs. Its agreement with our experiments encourages us to semi-quantitatively predict the mechanism of the turn around effect and also to explain a phenomenon which we call "bubble creeping" (see Fig. 1B).

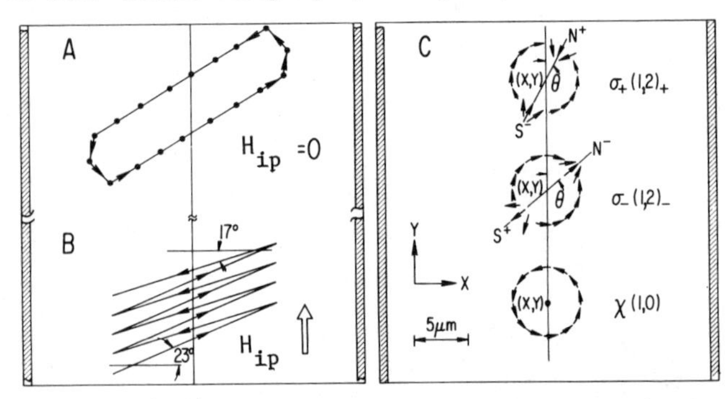

Fig. 1: A) S = 1 bubble showing two step turn around effect. B) S = 1 bubble creeping in direction of an in-plane field. Gradient and pulse direction are the same as in A. C) Configurations of bubble walls considered in this paper.

Identification of S = 1 bubbles with two BLs and no BLs has recently become possible using "automotion"[4] Upon bias field modulation, S = 1 bubbles with two BLs move at right angles to an in-plane field in either direction, depending on the combinations N^+S^- and N^-S^+. We denote the N^+S^- bubble with $(1,2)^+$ or σ_+ and the N^-S^+ bubble with $(1,2)_-$ or σ_- (see Fig. 1C). These bubbles automote in opposite directions, respectively. S = 1 bubbles having no BLs do not automote. We refer to these as (1,0) or χ bubbles. One of the features of the (1,2) bubbles is the possibility of aligning the N-S charged BL pair along the direction of an in-plane field, thus making it possible to compare the effect of different orientations on the initial bubble motion in pulsed gradient.

We find experimentally, i) "anomalous" displacement during the initial first gradient pulse following reversal, and subsequent displacements depend on the state σ_+, σ_- or χ.

ii) the turn around effect does not appear with χ bubbles and iii) a small in-plane field suppresses the turn around effect of σ_+ bubbles, but the deflection angle depends on the polarity of the gradient. Hence, a σ bubble creeps gradually in the direction of the in-plane field when gradient propagated to and fro normal to the in-plane field. These results are explained by taking into account the BL motion relative to the bubble center.

THEORY

Usually, bubble propagation as a result of an applied bias field H_z with a gradient $H' = dH_z/dz$ is specified by giving the winding number S which is independent of details of the domain wall structure such as the distribution of the BLs, if any, on the domain wall. We have the relation $S=1+(1/2)(n_+ - n_-)$ where n_+ and n_- are the numbers of BLs with positive and negative twists of magnetization, respectively. The deflection angle ρ of the bubble propagation from the gradient is expressed as [5,6].

$$\sin\rho = 2SV/\gamma r^2|H'| \quad [1]$$

where γ (>0) is the gyromagnetic ratio, r the bubble radius and V the velocity of the bubble. Here, we neglect coercivity and BL damping so that the velocity is also independent of the details of the wall structure, i.e.,

$$V = \frac{\gamma\Delta}{\alpha} r|H'| \Big/ \sqrt{1+(2\Delta S/\alpha r)^2} \quad [2]$$

where Δ is the wall width parameter and α the Gilbert damping coefficient. These results have been obtained by assuming uniform motion of the bubble. They are clearly not applicale to a discussion of the turn around effect and other transient phenomena. Thus, we are forced to extend the theory to allow BLs to have an additional degree of freedom: their position on the bubble circumference.

Slonczewski[5] and Thiele[6] first described that a BL, moving at local velocity $\vec{V}_{BL}$, imparts a gyrotropic force

$$\vec{F}_g = \pm \frac{2\pi M}{\gamma} \hat{z} \times \vec{V}_{BL} \quad [3]$$

where $\hat{z}$ is a unit vector normal to the plane of the magnetic film, M the saturation magnetization and the $\pm$ sign relates to the winding sense of the magnetization in the BL.

During steady state motion of a bubble, the BLs occupy equilibrium positions which cannot be changed by the local gyrotropic forces. However, when the BLs are present in positions which are not the dynamical equilibrium positions, the action of the local gyrotropic forces will be such as to move the BLs toward these equilibrium positions. The BLs thus will have local velocity vectors which differ from the velocity vector of the bubble center. Because the BLs are bound to move within the bubble wall, extra reaction forces will appear which may influence the resulting bubble motion to deviate from [1] and [2].

In view of the experiments we have done, we restrict the analysis to S = 1 bubbles. Fig. 1C illustrates schematically the wall structures of two S = 1 bubbles each of which includes one BL with positive, and one with negative twists ($n_+ = n_- = 1$). These bubbles are the σ_+ and σ_- bubbles respectively. Also shown is a unichiral bubble (χ). The position of the bubble center is given by (X,Y) and the position of the BLs in the σ bubbles is given by θ, which is equal for both BLs, if no radial motion is involved. The applied bias field has a gradient H'. A small in-plane field H_{ip} in the +y direction is also taken into account. Allowing BL motion relative to the bubble center, the external force is balanced by viscous drag and gyrotropic forces as [5]

$$2\pi r^2 M H' = -\frac{2\pi M}{\gamma}\alpha r\dot{X} - \frac{4\pi M}{\gamma}\dot{Y} \pm \frac{4\pi M}{\gamma} r\sin\theta\dot{\theta} , \quad [4]$$

$$0 = \frac{4\pi M}{\gamma}\dot{X} - \frac{2\pi M}{\gamma\Delta}\alpha r\dot{Y} \pm \frac{4\pi M}{\gamma} r\cos\theta\dot{\theta} , \quad [5]$$

$$2\pi r\Delta M H_{ip} \sin\theta = \mp \frac{2\pi M}{\gamma} r(\dot{Y}\cos\theta + \dot{X}\sin\theta), \quad [6]$$

Where • means d/dt, and upper and lower signs correspond to σ_+ and σ_- bubbles, respectively. [4] and [5] represent the force balances in the x and y directions, respectively. Since the in-plane field H_{ip} stabilizes the BL positions along the $\pm y$ extremities of the bubble, it appears as an additional pressure on the BLs as shown in [6]. From [4], [5] and [6] we find the following instantaneous velocities:

$$\dot{\theta} = \frac{\gamma\Delta}{2r^2\alpha}\left[\pm r^2H'(2\cos\theta + \frac{\alpha r}{\Delta}\sin\theta) - (4+\frac{\alpha^2r^2}{\Delta^2})\Delta H_{ip}\sin\theta\right], \quad [7]$$

$$\dot{X} = -\frac{\gamma\Delta}{\alpha r}\left[r^2H'\cos^2\theta \mp \Delta H_{ip}(\cos\theta - \frac{\alpha r}{\Delta}\sin\theta)\sin\theta\right], \quad [8]$$

$$\dot{Y} = \frac{\gamma\Delta}{\alpha r}\sin\theta\left[r^2H'\cos\theta \mp \Delta H_{ip}(2\sin\theta + \frac{\alpha r}{\Delta}\cos\theta)\right], \quad [9]$$

Obviously, the center velocity $(\dot{X},\dot{Y})$, and thus the bubble deflection angle, are functions of the BL position θ.

As an example, consider a σ_+ bubble which has BLs in positions $\theta = 0$ in $H_{ip} = 0$ and apply a gradient with negative polarity ($H' < 0$). Then, the BLs relax along the perimeter of the bubble with an initial angular velocity $\dot{\theta} = \pm\gamma\Delta H'/\alpha$ to their dynamical equilibrium positions $\theta_o = -\tan^{-1}(2\Delta/\alpha r)$. In the meantime, the bubble experiences anomalous motion due to BL motions. This anomalous motion is different among σ_+ and σ_- bubbles because of the difference of BL motion. The solid lines in Fig. 2 show the theoretical results of the deflection angle ρ as a function of time, obtained from [7], [8], and [9] by taking the initial condition $\theta = 0$ and $H_{ip} = 0$, where ρ is defined as

$\rho(t) = \tan^{-1}[(Y(t) - Y(0))/(X(t) - X(0))]$ and where we

used $\Delta = 0.07$ µm, $\alpha = 0.05$, $r = 3$ µm and $rH' = -0.5$Oe. In addition, the deflection angle of a unichiral bubble (χ) under the same conditions is shown as calculated from [1], which is justified because if a bubble has no BLs, [4] and [5] reduce to [1]. We find that it takes several tenths of a microsecond before σ_+ bubbles assume steady state motion. Because we expect that an in-plane field affects the BLs motion, we also calculated deflection angles under the condition $H_{ip} = 10$ Oe. The result is shown as dotted lines in Fig. 2.

Though Fig. 2 shows the reaction of σ_+ and σ_- bubbles to a negative gradient only, it can be used for positive gradient also. That is, σ_+ bubbles react upon a positive gradient as σ_- bubbles upon an equally strong negative gradient with x replaced by -x and θ by $-\theta$ in calculations.

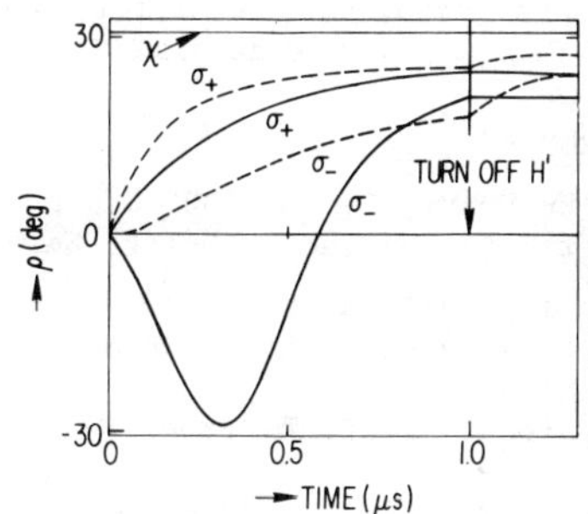

Fig. 2: Deflection angle of $\sigma_\pm$ bubbles as function of time after prealigning the BL pair normal to the gradient. Also shown is deflection of χ bubble (no BLs). For all curves: H' = -2650 Oe, Solid curves: no in-plane field. Dashed curves: H_{ip} = 10 Oe.

EXPERIMENTAL RESULTS

We used a $GdTmYFe_{4.2}Ga_{0.8}O_{12}$ film at a bias of 52 Oe to give 6 µm bubbles. Material parameters for the film are given elsewhere in these Proceedings.[4] The film is mounted face down on a glass slide provided with four parallel strip conductors, the outer ones of which were used for gradient generation. Bias compensation was accomplished by appropriately pulsing the inner strip conductors. An in-plane field H_{ip} could be applied parallel to the strip lines, i.e., normal to the z-field gradient pulses. Furthermore, a flat coil, mounted underneath the glass slide could be used for homogeneous bias field modulation with pulses of -6 Oe height, transition times less than 25 ns and a duration of 1 µs. These pulses, when applied simultaneously with H_{ip} = 10 Oe, are utilized for selecting the σ_+ or χ bubbles we want to investigate. σ_+ bubbles are easily recognized because they automote in opposite directions. χ bubbles are obtained by subjecting σ bubbles to the same z-field modulation as for automotion (typically several pulses), but in the absence of the in-plane field.

We did six experiments:

1) A σ_+ bubble is brought up to the center line between the gradient strip conductors by automotion in the +x direction. With the bubble at the center line, z-field excitation is stopped and subsequently the in-plane field is removed. The BLs in the σ_+ bubble, as a result of this procedure, are aligned along the center line, i.e., $\theta = 0$ in Fig. 1C but are free to move, once the bubble is subjected to a z-field gradient pulse. A negative gradient pulse of amplitude H' = - 2650 Oe/cm and duration τ, with no bias compensation produces its final displacement (X,Y) shown in Fig. 3A. We find that the deflection angle in this case increases with increasing τ.

2) The same experiment is repeated with a χ bubble. The deflection angle is found not to depend on pulse duration (Fig. 3A).

3) The same experiment is repeated with σ_- bubbles. We find, however, that these bubbles change state upon application of a negative gradient pulse. The resulting state is either σ_+ or χ. These state changes are prevented with the use of bias compensation. Then, we find that σ_- bubbles do not move when the gradient pulse duration is below 0.3 µs, and, as a result of τ = 0.9 µs, have considerably smaller displacements than σ_+ and χ bubbles when these are subjected to the same pulse, using bias compensation. These results are depicted in Fig. 3B.

4) A σ_+ bubble is brought, using automotion, to a position before crossing the center line. The bubble is subsequently subjected to a series of negative gradient pulses with $H_{ip} = 0$. After the bubble has travelled several steps across the center line, the gradient is reversed and the bubble travels backwards after showing the regular two-step turn around effect. The conditions for having the bubble follow the path displayed in Fig. 1A are τ = 0.25 µs, $|H'|$ = 3200 Oe/cm. In addition to the turn around effect, we find that the bubble is in the σ_+ state when moving in the +x direction, and in the σ_- state when moving in the -x direction.

5) The same experiment is repeated with a χ bubble, with $|H'|$ = 2650 Oe/cm. We find that χ bubbles under this condition do not show the turn around effect. When the intensity of gradient is increased, however, the χ bubble converts to a σ bubble and we obtain the result of experiment #4.

6) The same experiment as 4) is repeated with a σ_+ bubble and H_{ip} = 10 Oe applied perpendicular to H'. Moving in the +x direction, the σ_+ bubble deflects at 23°. Moving in the -x direction, the deflection is only 17°. Also, the bubble does not convert to σ_-, but stays σ_+. The effect, as illustrated in Fig. 1B, is that the bubble creeps gradually in the direction of

the in-plane field when propogated to and fro (bubble creeping).

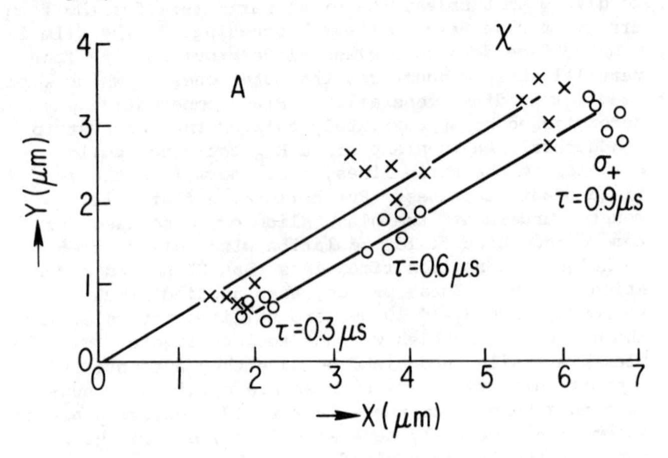

Fig. 3A: Displacements of σ_+ and χ bubbles due to H' = -2650 Oe/em as function of pulse duraction. No bias compensation.

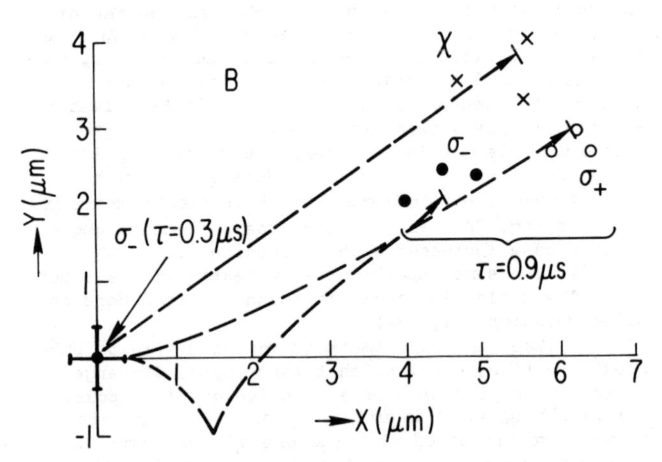

Fig. 3B: Same as Fig. 3A but with bias compensation and σ_- bubbles. Dashed curves show the theoretical trajectories of bubble propagations.

DISCUSSION

The results of the first three experiments may be summarized as follows: a σ_+ bubble, when prealigned with its BLs normal to the gradient, deflects at a smaller angle than a χ bubble. Furthermore, the deflection angle of σ_+ bubble depends on the gradient pulse duration τ, i.e., ρ (σ_+) increases with τ. whereas, ρ(χ) is constant. This result is readily understood by calculating the paths σ_+ and χ bubbles follow, using [7], [8], and [9]. In this calculation we subtracted 1000 Oe/cm from H' to compensate roughly for the coercivity which is approximately 1200 Oe/cm.[4] These paths are presented in Fig. 3B, along with the path a σ_- bubble follows. Clearly, the experiments verify the theoretically predicted positions. The experimental results presented in Fig. 3A furthermore confirm the theoretical result that the motion of a σ_+ bubble becomes almost parallel to the motion of a χ bubble under the same circumstances, after an initial deviation during the first 0.3 μs. The experiments have not resulted in a confirmation of the predicted negative deflection of a σ_- bubble during its initial 0.3 μs of motion. However, we feel that this discrepancy can be attributed to coercivity since the drive rH' is much reduced by the rotation of the BLs toward their new equilibrium positions during the initial motion.

The regular turn around effect, as exhibited by an S = 1 bubble, shown in Fig. 1A, consists of a two-step anomaly. In our experiment, gradient reversal includes a spurious bias field pulse h_z of 2.5 Oe. As shown by Malozemoff and Maekawa[7], bias field pulses cause BL pairs of unwinding nature to move in opposite sense on the perimeter of the bubble with angular velocity $\dot{\theta} \sim + \gamma h_z$. As a reaction to these local BL motions, the bubble is propelled in the direction of motion in the gradient field previous to reversal. This effect of h_z on the angular motion of the BLs is much larger than the one due to the gradient field. Fig. 4 illustrates qualitatively how the gradient drive in the -x direction competes with the propulsion due to BL motion, causing a net drive whose direction deviates considerably from the direction of the gradient, resulting in the first step. If the unwinding is not reached within the first pulse, it may be during the second so that the bubble becomes χ, or σ_+ if nucleation of a new pair of unwinding BLs has occurred. Whatever the final number of BLs, the bubble remains S = 1, irrespective of these complex processes.

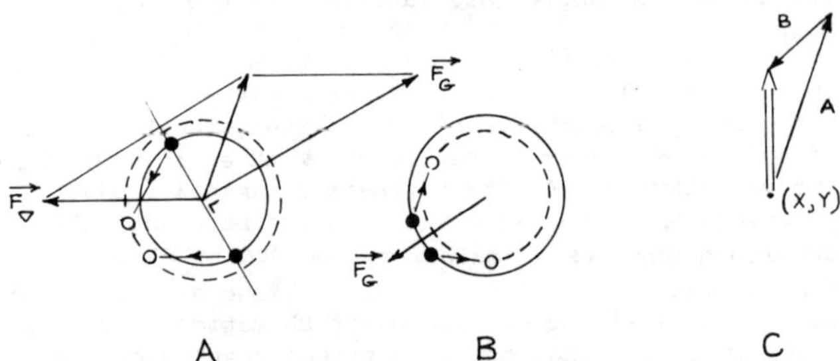

Fig. 4: Schematic diagram of competing propagating forces acting on σ bubble during turn around. A) BL motion pulls composite force away from rH' when H' is on. B) BL motion pushes bubble back at termination of H' pulse. C) Assumed displacement of bubble center due to A) and B).

The "bubble creeping" experiment, finally, may be understood as follows: the σ_+ bubble does not convert to σ_- or χ, because the in-plane field prevents BL collision and hence annihilation. When the bubble moves in the +x direction, under influence of H_{ip}=10 Oe, H' = -2650 Oe/cm, τ=0.25 μs, its deflection angle 23° corresponds to the calculated value for σ_+ at 0.25 μs (dashed curve in Fig. 2). When the σ_+ bubble moves in the -x direction, its deflection angle is the angle at the initial 0.25 μs of motion of a σ_- bubble going in the +x direction. As Fig. 2 reveals, there is a substantial difference in these values, so we find a qualitative agreement between theory and experiment.

ACKNOWLEDGMENTS

The authors are indebted to B.E. Argyle, J.C. Slonczewski and A.P. Malozemoff for discussions, J.C. DeLuca for comparison of experimental data, E.A. Giess for supplying the sample and D.C. Cronemeyer for determination of γ.

REFERENCES

* permanent address: Tohoku Univ., Sendai 980, Japan.
**permanent address: Delft Univ. of Tech., Delft, Neth.

1. A.P. Malozemoff, J. Appl. Phys. 44, 5080 (1973).
2. H. Nakanishi and C. Uemura, Jap. J. Appl. Phys. 13, 191 (1974).
3. R.M. Josephs and B.F. Stein, AIP Conf. Proc. 24, 598 (1975).
4. B.E. Argyle, J.C. Slonczewski, P. Dekker and S. Maekawa, This Conf. paper 7A-1.
5. J.C. Slonczewski, J. Appl. Phys., 45, 2705 (1974).
6. A.A. Thiele, J. Appl. Phys 45, 377 (1974).
7. A.P. Malozemoff and S. Maekawa, To be published in July 1976 issue of J. Appl. Phys.

THE EFFECTS OF IN-PLANE FIELDS ON BALLISTIC OVERSHOOT IN THE GRADIENT PROPAGATION OF MAGNETIC BUBBLE DOMAINS

J. C. De Luca and A. P. Malozemoff
IBM T. J. Watson Research Center
Yorktown Heights, New York 10598

ABSTRACT

Using high-speed photography we have studied the effects of an applied in-plane field H_p on the velocity and ballistic overshoot of S = 0 bubbles propagated by a pulsed field gradient in EuGa, GdTmGa, and SmCaGe garnet films. Peaks appear in the velocity versus drive curve. These peaks shift to higher drives with increasing in-plane field. At drives below the peaks the velocity is roughly linear with drive with very few state changes and little or no ballistic overshoot. At drives above the peaks the velocity is saturated at a value which increases with in-plane field. In this range ballistic overshoot and state changes occur quite often, indicating that Bloch lines and Bloch points can be nucleated.

INTRODUCTION

In this paper we report on the influence of in-plane fields (H_p) in the gradient propagation[1] of magnetic bubble domains in non-ion implanted garnet films. Previous investigators[2,3,4,5] have demonstrated that in-plane fields increase apparent bubble velocities. However, these earlier studies do not reveal whether the velocity increases observed are real or due in part to ballistic overshoot.[6]

Using high speed photography, we have determined the real average velocity and the size of the overshoot as a function of drive and in-plane fields. By and large we find that the in-plane field increases the true velocity and surpresses overshoot. The dependence of true velocity on drive is similar to that observed in photometric measurements on straight walls.[7,8] The velocity increases linearly with drive to a peak and then falls back to a saturation velocity. One key difference with the photometric experiments is that in one sample we observe not one but two peaks in the velocity versus drive. We can also relate some of the scatter in the data to state changes which indicate Bloch line and point nucleation.

EXPERIMENTAL PROCEDURES AND RESULTS

Our study encompassed three garnet films whose nominal compositions are $Eu_{0.65}Ga_{1.2}$, $Gd_{0.9}Tm_{1.1}$-$Ga_{0.75}$, and $Sm_{0.1}Ca_{0.98}Ge_{0.98}$ and will subsequently be referred to as EuGa, GdTmGa and SmCaGe. The material parameters of these films were previously reported in the Table of Ref. 9, where they are numbered 1, 4, and 7 respectively.

An initial S = 0 state bubble,[10,11] one whose motion is parallel to the field gradient and contains a net of two Bloch lines, was selected for all measurements. The bubble was positioned at the center of the pulse lines by a weak gradient pulse held close to the coercive limit so as not to disturb the bubble state. During both positioning and propagation an in-plane field was aligned orthogonal to the bubble velocity vector. It is known that the use of H_p increases both the ellipticity and the area of the bubble.[2,5]

We assume the gradient drive is defined as the bias field gradient times the bubble's semiaxis $r_{\parallel}$ along the direction of motion. Therefore, to keep the gradient drive constant as a function of in-plane field, we adjusted the DC bias field to keep $r_{\parallel}$ constant. This technique also reduces the static ellipticity. Furthermore high speed photography measurements indicate that proper bias compensation[1,6] preserves $r_{\parallel}$ to ±0.5μm during propagation in addition to minimizing distortions perpendicular to the motion.[12] To test for any change in the S-state of the bubble after propagation, very weak gradient pulses were used and changes in the direction of bubble motion with regard to the field gradient noted.

Results on a EuGa sample are given in Fig. 1 and the Table. The figure shows both the bubble displacement at the end of the gradient pulse (X_T) and the final position of the bubble (X_∞) as a function of drive strength at values of in-plane field $0 \leq H_p \leq 100$ Oe. The velocity in the right hand scale is defined[6] as X/T where T is the pulse width. The real average velocity is X_T/T while X_∞/T is the apparent velocity. Any difference between X_∞ and X_T is a result of bubble overshoot. For $H_p = 0$, X_T smoothly saturates with increasing drive as the overshoot increases. With $20 \leq H_p \leq 100$, however, two peaks appear as a function of gradient drive. These peaks shift to higher drives with increasing in-plane fields, finally becoming washed out for $H_p > 100$ Oe.

TABLE: Results on garnet samples, where the number refers to the Table of Ref. 9, h is film thickness, V_T is saturation velocity at no in-plane field, dV_T/dH_p is increase of V_T with in-plane field H_p for low H_p (<100 Oe), α is the Gilbert damping determined from ferromagnetic resonance linewidth, and dH/dH_p is the rate of change of the drive at the first peak with H_p.

Sample	$Eu_{0.65}$ $Ga_{1.2}$	$Gd_{0.9}$ $Tm_{1.1}$ $Ga_{0.75}$	$Sm_{0.1}$ $Ca_{0.98}$ $Ge_{0.98}$
Number	1	4	7
h(μm)	4.3	3.8	5.8
V_T(±50cm/sec-Oe)	800	1100	1750
dV_T/dH_p(cm/sec-Oe)	16	20	58
α(FMR)	0.030	0.035	0.08
dH/dH_p(±.005)	0.023	0.025	0.055

Below the lowest peak the velocity is roughly linear with the drive with little or no overshoot and few state changes. Above the two peaks ballistic overshoot and state changes from S = 0 to right and left skewing states occur more readily. Also the frequency of state conversions greatly increases as a function of increasing drive strength. The real average velocity X_T/T in this region has a saturated value increasing with in-plane field at a rate of 16cm/sec-Oe. To further examine critical and saturation velocities, bubble displacement as a function of pulse width for H_p = 40 Oe is reported with results shown in Fig. 2. The drives selected were 1.2 Oe at the first peak, 1.6 Oe between the first and second peak, 2.4 Oe at the second peak, and 3.2 Oe beyond the second peak. The data are linear except at the 1.6 Oe drive value which is between the two peaks.

Data which are typical of the results for the GdTmGa garnet film are shown in Fig. 3 for H_p = 100 Oe. At drives $\lesssim$2.5 Oe the velocity increases linearly with drive with no overshoot or state conversions recorded. Increasing the drive above 2.5 Oe results in some state conversions of the starting S = 0 bubble. This critical drive at which the onset of conversions occurs is approximately a linear function of in-plane field. Note that all bubbles which do convert remain at an apparent saturation velocity of 4000 cm/sec. Those bubbles which do not convert and remain S = 0 bubbles follow the upper portion of the velocity versus drive curve where at approximately 3.2 Oe a peak of 5500 cm/sec occurs. Beyond this peak scatter increases and bubble overshoot occurs with the velocity of the S = 0 bubbles gradually decreasing. At 5.5 Oe the velocities of both the S = 0 and the skewed bubbles merge and have the same saturation velocity of 4000 cm/sec. With drives higher than 5.5 Oe, at 6.4 Oe for example, all S = 0 bubbles convert. The probability of finding right or left skew states is about equal. For this GdTmGa sample the saturation velocity increases at a rate of 20 cm/sec with in-plane field which, as indicated in the Table, is a somewhat higher rate than that found in the EuGa.

Bubble displacement and velocity for the SmCaGe film are shown in Fig. 4 as a function of drive and $0 \leq H_p \leq 125$ Oe. A single weak peak is seen for $60 \leq H_p \leq 100$ Oe. As in the other samples, state changes and overshoot are generally suppressed below the peak. Because the peak shifts to higher drives at a greater rate with H_p than in the other samples, we find the state changes largely suppressed by $H_p \gtrsim 100$ Oe over the range of drives we have used. Up to $H_p \lesssim 80$ Oe all curves exhibit a saturation velocity which increases at a rate of approximately 58 cm/sec Oe. This is a considerably higher value than that found in the EuGa or GdTmGa samples.

DISCUSSION

High speed photography has enabled us to discern many effects in the true average velocity X_T/T that are hidden by the scatter in the apparent velocity X_∞/T. We believe that the data on the three samples is basically consistent: a) There is a low drive region with no overshoot and few state changes; such a region can be used to extract the intrinsic mobility. b) There is also a first velocity peak which tracks linearly through zero in drive versus in-plane field. Table I shows that the slope of this dependence correlates surprisingly well with the Gilbert damping constants. As will be discussed elsewhere,[13] the peak may be related to a rotational instability of the vertical Bloch lines in the S = 0 bubble, and a relationship $H = \alpha H_p$ for this instability has previously been proposed by Hsu.[14] In view of the state changes which occur beyond the peak, it may be a threshold for stable operation of devices using S = 0 bubbles.[14] (This transition has no analog in plane wall experiments.) c) In the EuGa sample, a second peak occurs. It is likely that the absence of the second peak in the other two samples is due to the prevalence of state changes which wash out the second peak. This trend can also be seen in the H_p = 80 Oe data on EuGa. The second peak may be caused by nucleation of new horizontal Bloch lines in a dynamic conversion process.[13,15,16] d) Beyond the peaks there is a saturation velocity which increases with in-plane field. We interpret this region as one of multiple Bloch line nucleation much as in the case without in-plane field.

ACKNOWLEDGEMENTS:

The authors thank M. Rozmus for experimental assistance, E. Giess and T. L. Hsu for the garnet films, R. Anderson and T. L. Hsu for the gradient pulse lines, and P. Dekker and J. Slonczewski for helpful discussions and comments.

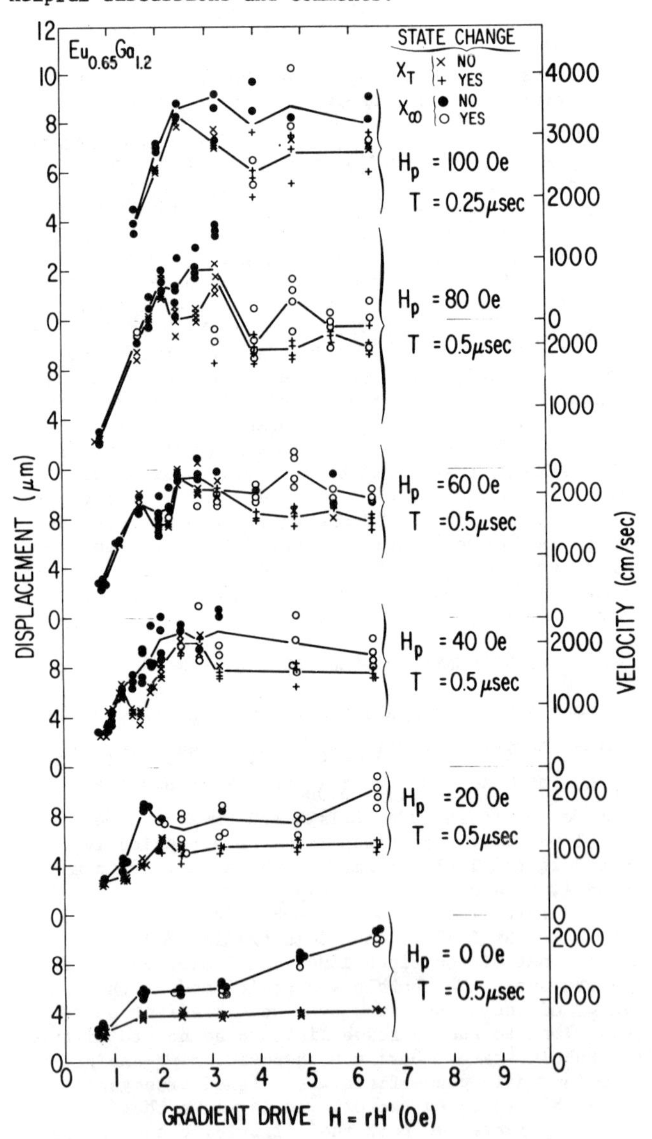

Fig. 1. Bubble displacement vs. gradient drive for $0 \leq H_p \leq 100$ Oe in the $Eu_{0.65}Ga_{1.2}$ garnet sample. Lines connect centers of gravity of X_T and X_∞ data.

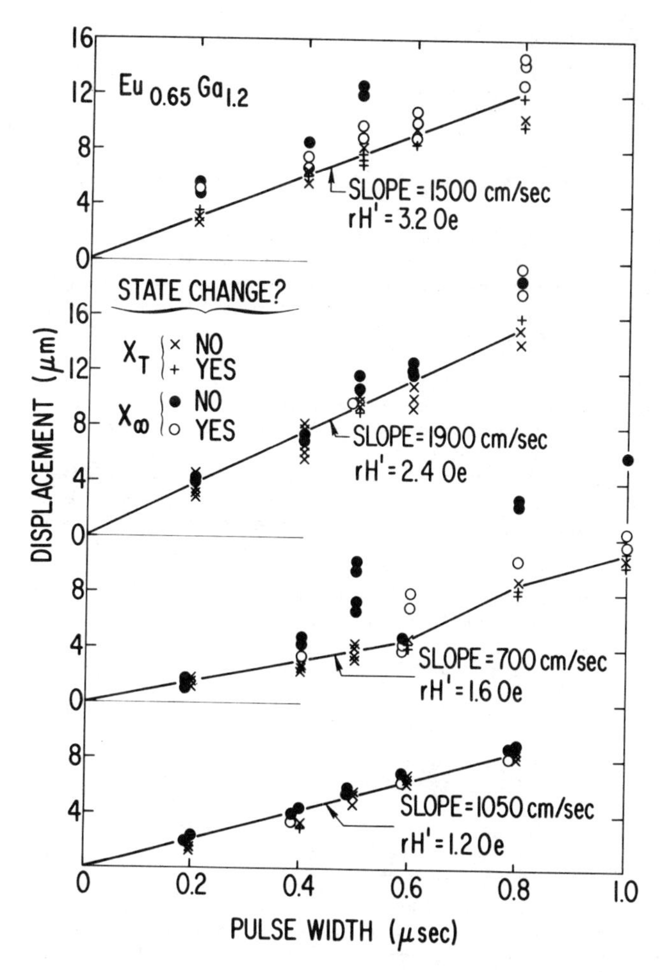

Fig. 2. Bubble displacement vs. pulse width for H_p = 40 Oe in the $Eu_{0.65}Ga_{1.2}$ garnet sample. Lines show the average slopes of X_T data except for rH' = 1.6 Oe where a noticeable nonlinearity occurs at large pulse width.

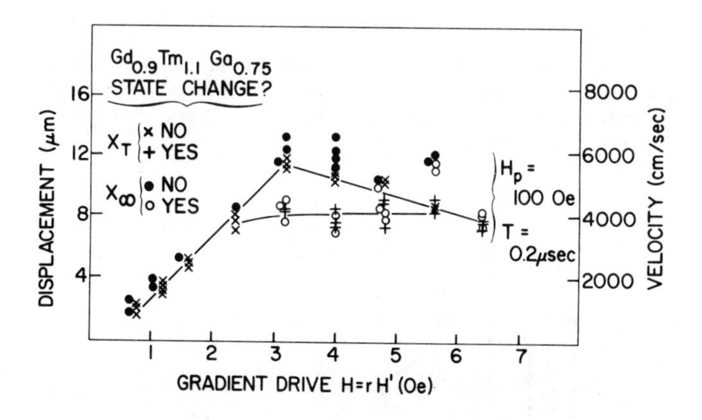

Fig. 3. Bubble displacement vs. gradient drive for H_p = 100 Oe in the $Gd_{0.9}Tm_{1.1}Ga_{0.75}$ garnet sample. Lines show X_T data for bubbles with or without state conversion.

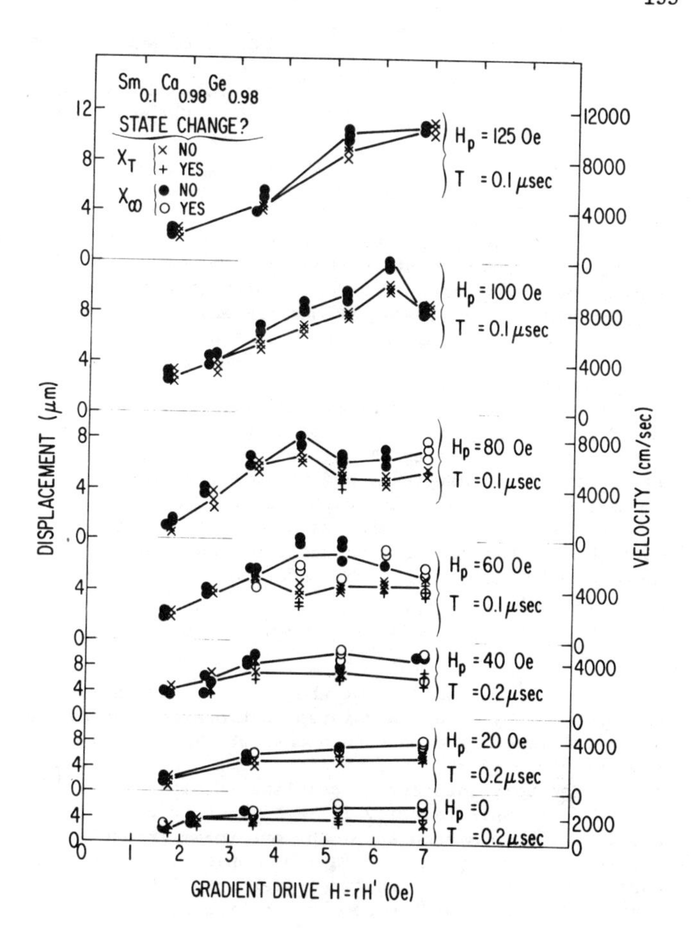

Fig. 4. Bubble displacement vs. gradient drive for $0 \le H_p \le 125$ Oe in the $Sm_{0.1}Ca_{0.98}Ge_{0.98}$ garnet sample. Lines as in Fig. 1.

REFERENCES

1. G. P. Vella-Coleiro and W. J. Tabor, Appl. Phys. Lett. 21, 7 (1972).
2. D. C. Bullock, AIP Conf. Proc. 18, 232 (1974).
3. F. H. deLeeuw, and J. M. Robertson, J. Appl. Phys. 46, 3182 (1975).
4. R. M. Josephs and B. F. Stein, AIP Conf. Proc. 18, 227 (1974).
5. G. P. Vella-Coleiro, AIP Conf. Proc. 18, 217 (1974).
6. A. M. Malozemoff and J. C. De Luca, Appl. Phys. Lett. 26, 719 (1975).
7. F. H. deLeeuw, IEEE Trans. Magn. MAG-9, 624 (1973).
8. B. E. Argyle and A. Halperin, IEEE Trans. MAG-9, 238 (1973).
9. A. P. Malozemoff, J. C. Slonczewski, and J. C. De Luca, submitted to 21st Annual Conference on Magnetism and Magnetic Materials, Philadelphia, Penn., Dec. 9-12, (1975).
10. J. C. Slonczewski, A. P. Malozemoff and O. Voegeli, AIP Conf. Proc. 10, 458 (1973).
11. A. P. Malozemoff, J. Appl. Phys. 44, 5080 (1973).
12. G. P. Vella-Coleiro, F. B. Hagedorn, and S. L. Blank, Appl. Phys. Lett. 26, 69 (1975).
13. A. P. Malozemoff and J. C. Slonczewski (to be published).
14. T. L. Hsu, Intermag Conf. London, England, April 14-17 (1975).
15. J. C. Slonczewski, J. Appl. Phys. 44, 1759 (1973).
16. F. B. Hagedorn, AIP Conf. Proc. 18, 222 (1974).

EFFECT OF ACOUSTIC WAVE ON DOMAIN WALL VELOCITY

S. Uchiyama, S. Shiomi, and T. Fujii
Nagoya University, Nagoya, Japan

ABSTRACT

Fundamental equations of motion for the spin and the lattice displacement in the magnetic domain wall are derived by taking the effect of magnetoelastic coupling into consideration. The equations are solved for the case of steady wall moving with a constant velocity under an assumption that the tilting angle of the spin in the wall is small.

The solution obtained may interprete the anomaly in the velocity-field relation appearing near the transverse acoustic velocity. The effect of lattice damping is introduced intuitively, and the correct treatment of the effect is left to the further investigation.

For small value of the quality factor q, the velocity has two possible values for a certain region of the applied field. This might infer the instability of the domain wall.

INTRODUCTION

The dynamic behavior of the magnetic domain wall has been a subject of many investigations because of the technical importance in bubble device. The velocity and the structure of the wall at high driving field may be one of the very interesting problems. Konishi et al.[1] measured the wall velocity in $YFeO_3$ up to 25000 m/sec by bubble collapse method and found out three irregular points in velocity-field relation. Konishi, Kawamoto and Wada [2] suggested that the first weak knee like velocity saturation around 4400 m/sec may be the indication of the Walker's critical velocity.[3] We thought, however, that this knee might be from the effect of acoustic wave. This seemed to be confirmed by the finding of Tsang and White [4]; namely they reported that the velocity of head-to-head walls in $YFeO_3$ saturated at a value of 4100 m/sec which agreed with the transverse acoustic velocity. In this paper, the effect of acoustic wave on the wall velocity is investigated theoretically.

FUNDAMENTAL EQUATIONS OF MOTION

Since the coupled equations of motion of spin and lattice are derived from the wall energy variation, the wall energy is considered firstly. After the work of Kittel [5], the energy density E in the wall consists of the Zeeman term E_z, the anisotropy term E_a, the exchange term E_{ex}, the demagnetizing term E_d, the magnetoelastic term E_{mel}, and the phonon term E_{ph}, namely

$$E = E_z + E_a + E_{ex} + E_d + E_{mel} + E_{ph} . \quad (1)$$

Each of these terms may be written as follows.

$$E_z = -M_s H\cos\theta , \quad (2)$$

$$E_a = K_u \sin^2\theta , \quad (3)$$

$$E_{ex} = A[(\partial\theta/\partial y)^2 + \sin^2\theta(\partial\phi/\partial y)^2] , \quad (4)$$

$$E_d = 2\pi M_s^2 \sin^2\theta \sin^2\phi , \quad (5)$$

$$E_{mel} = B_1(\partial R_y/\partial y)(\sin^2\theta\sin^2\phi - 1/3) + B_2[(\partial R_x/\partial y)\sin^2\theta\sin\phi\cos\phi + (\partial R_z/\partial y)\sin\theta\cos\theta\sin\phi] , \quad (6)$$

$$E_{ph} = (1/2)[c_{11}(\partial R_y/\partial y)^2 + c_{44}\{(\partial R_x/\partial y)^2 + (\partial R_z/\partial y)^2\}] , \quad (7)$$

where M_s is the saturation magnetization, H the applied field, K_u the uniaxial anisotropy constant, A the exchange stiffness, B_1 and B_2 the magnetoelastic coupling constants, c_{11} and c_{44} the elastic moduli and R_x, R_y, and R_z the components of lattice displacement. In deriving these equations, the wall is assumed to have a form of infinite plane parallel to the x-z plane as shown in Fig.1, and the polar coordinates θ, ϕ of the magnetization direction as well as R_x, R_y, R_z are assumed to depend only on y coordinate. Further assumptions are as follows: the drive field is applied in z-direction, the ferromagnetic material has a uniaxial anisotropy with easy axis parallel to z-axis and it has a cubic elastic and magnetoelastic symmetries for simplicity.

Fundamental equations of motion are given by the following equations.

$$\dot{\theta} + \alpha\dot{\phi}\sin\theta = -(|\gamma|/M_s \sin\theta)(\delta E/\delta\phi) , \quad (8)$$

$$\dot{\phi}\sin\theta - \alpha\dot{\theta} = -(|\gamma|/M_s)(\delta E/\delta\theta) , \quad (9)$$

$$\rho\ddot{R}_x + \beta\dot{R}_x = -\delta E/\delta R_x , \quad (10)$$

$$\rho\ddot{R}_y + \beta\dot{R}_y = -\delta E/\delta R_y , \quad (11)$$

$$\rho\ddot{R}_z + \beta\dot{R}_z = -\delta E/\delta R_z , \quad (12)$$

where dot indicates the time derivative, α is the Gilbert's damping constant, β is the lattice damping constant, γ is the gyromagnetic ratio, ρ is the density, and

$$\delta E/\delta\psi = \partial E/\partial\psi - (\partial/\partial y)[\partial E/\partial(\partial\psi/\partial y)] , \quad (13)$$

where ψ stands for θ, ϕ, R_x, R_y, or R_z.

SOLUTION FOR STEADY MOTION

Combining eqs.(1)-(7),(10) and (13), the following equation is obtained for R_x.

$$\rho\ddot{R}_x = c_{44}(\partial^2 R_x/\partial y^2) + B_2\partial(\sin^2\theta\sin\phi\cos\phi)/\partial y , \quad (14)$$

where the phonon damping term is dropped for simplicity. In case of steady wall moving in y-direction with a constant velocity v, the solution of R_x should be a function of (y-vt), and thus

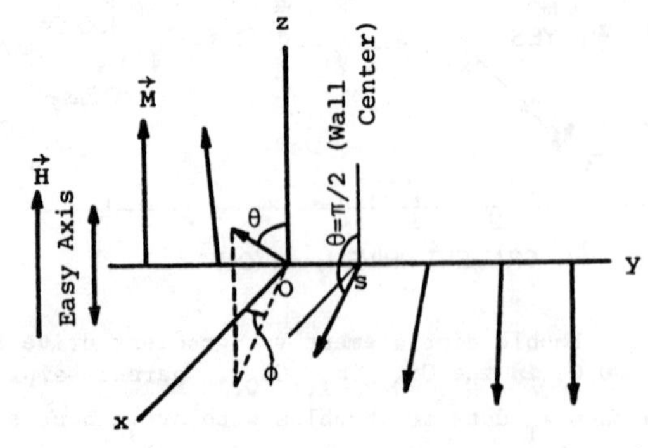

Fig.1 Coordinate system of domain wall and spins within wall

$$\ddot{R}_x = v^2(\partial^2 R_x/\partial y^2) \ . \tag{15}$$

Replacing the left hand side of eq.(14) by eq.(15),

$$\partial^2 R_x/\partial y^2 = -(4\pi M_s^2/B_2)\delta_t \partial(\sin^2\theta \sin\phi\cos\phi)/\partial y \ , \tag{16}$$

or $$\partial R_x/\partial y = -(4\pi M_s^2/B_2)\delta_t \sin^2\theta \sin\phi\cos\phi \ , \tag{17}$$

since $\partial R_x/\partial y$ should vanish for $y=\pm\infty$ or $\theta=0,\ \pi$. In eqs. (16) and (17),

$$\delta_t \equiv v_{mt}^2/(v_t^2 - v^2) \ ,$$

$$v_{mt} \equiv B_2/2M_s(\pi\rho)^{1/2},$$

and $$v_t \equiv (c_{44}/\rho)^{1/2}.$$

Here v_t is the transverse acoustic velocity. Similarly, following equations are derived from eqs.(11) and (12).

$$\partial R_y/\partial y = -(4\pi M_s^2/B_1)\delta_1 \sin^2\theta \sin^2\phi \ , \tag{18}$$

$$\partial R_z/\partial y = -(4\pi M_s^2/B_2)\delta_t \sin\theta\cos\theta\sin\phi \ , \tag{19}$$

where

$$\delta_1 \equiv v_{ml}^2/(v_1^2 - v^2) \ ,$$

$$v_{ml} \equiv B_1/2M_s(\pi\rho)^{1/2},$$

and $$v_1 \equiv (c_{11}/\rho)^{1/2}.$$

Here v_1 is the logitudinal acoustic velocity.

Substituting eqs.(17)-(19) into eqs.(8) and (9), the equations of motion for spin are derived as follows.

$$\dot{\theta} + \alpha\dot{\phi}\sin\theta = -\frac{|\gamma|}{M_s}[-2A\{\sin\theta(\frac{\partial^2\phi}{\partial y^2}) + 2\cos\theta(\frac{\partial\theta}{\partial y})(\frac{\partial\phi}{\partial y})\}$$
$$+2\pi M_s^2\{(1-\delta_t)+2(\delta_t-\delta_1)\sin^2\theta\sin^2\phi\}\sin\theta\sin 2\phi] \tag{20}$$

$$\dot{\phi}\sin\theta - \alpha\dot{\theta} = \frac{|\gamma|}{M_s}[\{K_u + A(\frac{\partial\phi}{\partial y})^2\}\sin 2\theta + M_s H\sin\theta - 2A(\frac{\partial^2\theta}{\partial y^2})$$
$$+2\pi M_s^2\{(1-\delta_t)+2(\delta_t-\delta_1)\sin^2\theta\sin^2\phi\}\sin 2\theta\sin^2\phi] \tag{21}$$

As is well known, the structure of the static 180° wall can be expressed by

$$\theta = 2\tan^{-1}[\exp\{(y-s)/\Delta_0\}] \ , \tag{22}$$

and $$\phi = 0 \ , \tag{23}$$

where s is the y-coordinate of the wall center and

$$\Delta_0 \equiv (A/K_u)^{1/2}$$

is the wall width parameter. For steady wall moving with constant velocity v, it is also known [3,6] that the form of eq.(22) holds though the wall width parameter Δ differs from Δ_0 if the magnetoelastic coupling can be neglected, namely

$$\theta = 2\tan^{-1}[\exp\{(y-s)/\Delta\}] \ . \tag{24}$$

As for ϕ, it is shown [6] to be almost constant independent of the y-coordinate. Therefore

$$\partial\phi/\partial y = 0 \ . \tag{25}$$

Let us assume that eqs.(24) and (25) hold also in the present case, then eqs.(20) and (21) become

$$\dot{\theta} = -2\pi|\gamma|M_s[(1-\delta_t)+2(\delta_t-\delta_1)\sin^2\theta\sin^2\phi]\sin\theta\sin 2\phi \tag{26}$$

$$-\alpha\dot{\theta} = 2\pi|\gamma|M_s[q\{1-(\Delta_0/\Delta)^2\}\sin 2\theta + \alpha h\sin\theta$$
$$+\{(1-\delta_t)+2(\delta_t-\delta_1)\sin^2\theta\sin^2\phi\}\sin 2\theta\sin^2\phi] \ , \tag{27}$$

where $q \equiv K_u/2\pi M_s^2$ and $h \equiv H/2\pi\alpha M_s$.

Eliminating θ from eqs.(26) and (27), we get

$$\alpha[(1-\delta_t)\sin 2\phi - h] - 2[(1-\delta_t)\sin^2\phi + q\{1-(\Delta_0/\Delta)^2\}]\cos\theta$$
$$+4\alpha(\delta_t-\delta_1)\sin^3\phi\cos\phi\sin^2\theta - 4(\delta_t-\delta_1)\sin^4\phi\sin^2\theta\cos\theta = 0 \tag{28}$$

So far as v is constant, it is not possible to satisfy this equation for all values of θ. However, if α and ϕ are very small compared to one, the third and the

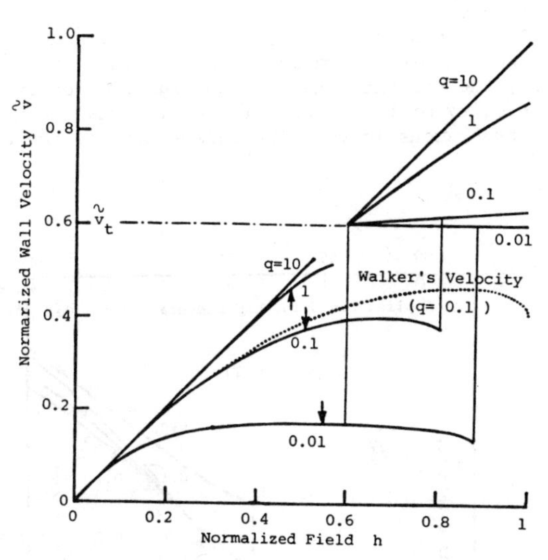

Fig.2 Dependence of wall velocity upon applied field

fourth terms in eq.(28) may be negligible. With this approximation, the next relation is obtained by equating the first term of eq.(28) to be zero.

$$\sin 2\phi = h/(1-\delta_t) \tag{29a}$$

or $$\sin^2\phi = (1/2)[1-\{1-h^2/(1-\delta_t)^2\}^{1/2}] \ . \tag{29b}$$

In the same way, the wall width parameter Δ is determined from the second term of eq.(28) as follows.

$$\Delta = \Delta_0[1+\frac{1-\delta_t}{2q}\{1-[1-h^2/(1-\delta_t)^2]^{1/2}\}]^{-1/2} \ . \tag{30}$$

Thus the wall velocity v is given by

$$v = \Delta|\gamma|H/\alpha$$

$$= \frac{\Delta_0|\gamma|H}{\alpha}[1+\frac{1}{2q}(1-\frac{v_{mt}^2}{v_t^2-v^2})\{1-[1-h^2/(1-\frac{v_{mt}^2}{v_t^2-v^2})^2]^{1/2}\}]^{-1/2}. \tag{31}$$

Since the velocity v is involved in both sides of this equation, self consistent value of v has to be looked for numerically.

RESULTS OF NUMERICAL CALCULATION

Figure 2 shows the dependence of the normarized velocity $\tilde{v}$ ($\tilde{v} = v/2\pi|\gamma|M_s\Delta_0$) on the normarized field h taking the quality factor q as a parameter. The values of other parameters used here are as follows;

$$\tilde{v}_t = v_t/2\pi|\gamma|M_s\Delta_0 = 0.60, \quad \tilde{v}_{mt} = v_{mt}/2\pi|\gamma|M_s\Delta_0 = 0.20 \ .$$

So far as q is much larger than one, the velocity is approximately proportional to the field as typically seen in this figure by q=10. Because of the magnetoelastic coupling, however, there is a region just below v_t where no real number solution for v exists. In this region, the wall may accompany the oscillatory motion. With decreasing q, the irregularity in the v-h relation becomes more appreciable. When q becomes as small as 0.1, for example, there appears a region where v has two solutions for a certain value of h. In this connection, we must make mention of the fact that the results shown in the figure are valid only for $\sin\phi \ll 1$. The arrows shown in the graph indicate the points of $\sin^2\phi = 0.1$, and only the left portion of the arrows in the lower velocity branches satisfy the condition of $\sin^2\phi \leq 0.1$. In order to make the comparison easy, the Walker's velocity ($v_{mt}=0$) for q=0.1 is shown in the figure by dotted curve.

In the case of q=0.01, the lower branch velocity shows noticeable saturation. Beyond the two-values region, the velocity jump to the higher velocity branch. The velocity of this branch is nearly equal to the transverse acoustic velocity independent of the applied field.

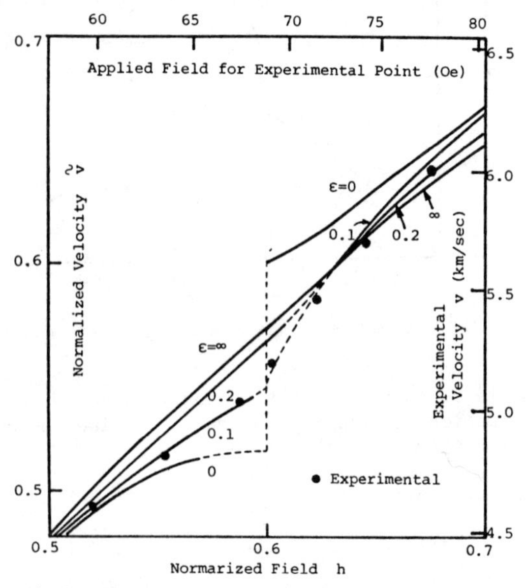

Fig.3 Effect of lattice damping on v-h relation and comparison with experiment

EFFECT OF LATTICE DAMPING

The effect of lattice damping is thought to be very important in the region where the contribution from the magnetoelastic energy becomes noticeable. Leaving the correct treatment of the effect in further investigation, the effect is introduced here intuitively, namely the parameter δ_t in eq.(30) is replaced by

$$\delta_t = Re\left(\frac{\tilde{v}_{mt}^2}{\tilde{v}_t^2 - \tilde{v}^2 + i\varepsilon\tilde{v}}\right) = \frac{\tilde{v}_{mt}^2}{\tilde{v}_t^2 - \tilde{v}^2 + \varepsilon^2\tilde{v}^2/(\tilde{v}_t^2 - \tilde{v}^2)}, \quad (32)$$

where *Re* means the real part, and ε is a parameter expressing the effect of lattice damping.

Examples of the numerical calculation are shown in Fig.3, where $\tilde{v}_t=0.6$, $\tilde{v}_{mt}=0.2$, q=1, and ε is taken as a parameter.

Closed circles shown in the same figure are the experimental results measured in a $YFeO_3$ thin plate by means of bubble collapse method in our laboratory.

DISCUSSION AND CONCLUSION

In relation to the work of Tsang and White[4], the appearance of the two velocities region in the v-h relation seems to be interesting. According to the present theory, however, the saturation velocity of the lower velocity branch is very small compared with the transverse acoustic velocity. Furthermore, the velocity of the higher velocity branch coincides with the transverse acoustic velocity. These facts contradict to the results on the head-to-head velocity[4]. In addition, the two velocities region appears only in the case of small q value, while for $YFeO_3$, q is surely larger than one.

The anomaly seen in the v-h relation near v = 4.4 (km/sec) [2] seems to be explained by the present theory, although the knee velocity is a little larger than the transverse acoustic velocity different from the theoretical prediction.

It should be noted that the longitudinal acoustic wave hardly influences the domain wall motion.

ACKNOWLEDGEMENT

The authors express our sincere thanks to Mr.A.Ikai for his experimental assistance and to Dr.S.Tsunashima and Dr.M.Takayasu for their discussion.

REFERENCES

1. S.Konishi, T.Miyama, and K.Ikeda,"Domain wall velocity in orthoferrites", Appl.Phys.Letters 27 , 258 (1975)
2. S.Konishi, T.Kawamoto, and M.Wada, "Domain wall velocity in $YFeO_3$ exceeding the Walker critical velocity", IEEE Trans.Magn. MAG-10, 642 (1974)
3. L.R.Walker (unpublished). Described by J.F.Dillon,Jr. in Treatise on Magnetism, edited by G.T.Rado and H. Suhl (Academic, New York, 1963) Vol.III, p.450
4. C.H.Tsang and R.L.White, "Observations of domain wall velocities and mobilities in $YFeO_3$", AIP Conf.Proc. 24, 749 (1974)
5. C.Kittel, "Interaction of spin waves and ultrasonic waves in ferromagnetic crystals", Phys.Rev. 110, 836 (1958)
6. N.L.Schryer and L.R.Walker, "The motion of 180°domain walls in uniform dc magnetic fields", J.Appl.Phys. 45 5406 (1974)

GROWTH REPRODUCIBILITY AND TEMPERATURE DEPENDENCIES OF THE STATIC PROPERTIES OF YSmLuCaFeGe GARNET*

G. G. Sumner and W. R. Cox, Texas Instruments Incorporated
P. O. Box 5936, M.S. 145, Dallas, Texas 75222

ABSTRACT

Over 300 garnet films of nominal composition $[Y_{1.5}Sm_{0.3}Lu_{0.3}Ca_{0.9}](Fe_{4.1}Ge_{0.9})O_{12}$ were prepared for 5 μm bubble device operation at 0.5 MHz from -25 to 75°C. Melt composition, run-to-run process control, as well as temperature and composition differentials required for reproducible growth are presented. The effects of charge imbalance and mobility changes on domain coercivity are discussed. The static bubble properties were measured from -30 to 100°C, and temperature coefficients of -0.2%/°C and -0.1%/°C were found for the bubble collapse field and material length, respectively. With domain wall coercivity as low as 0.1 Oe at 25°C and 0.4 Oe at -25°C and with q > 4 at 75°C, this material should be well suited for devices requiring moderate speed of operation over broad ranges of temperature.

INTRODUCTION

Although CaGe garnets were first reported with the advent of synthetic garnets,[1,2,3,4] they have been proposed and described as superior bubble memory materials only fairly recently.[5,6,7,8,9,10] We present here information concerning the preparation and properties of YSmLuCaFeGe garnet (YSLCaGe) with which we have gained considerable experience in the process of growing several hundred films. The objective of this work was to develop and investigate a bubble material which would satisfy the contract requirements listed in Table I.

TABLE I

Bubble Garnet Requirements

h	5.4 μm
ℓ	0.54 μm
H_o	105-130 Oe
$4\pi M_s$	200-240 G
q	> 3 (75°C)
mobility	500-600 cm/sec-Oe
$a_{film}-a_{GGG}$	0 to -0.006 Å
$\Delta H_o/\Delta T$	- 0.2%/°C
frequency	0.5 MHz

EXPERIMENTAL

The initial composition of our melt is given in Table II and is designed to give a growth velocity of about 1 μm/minute. This composition has also been found to yield good 2 μm and 1 μm bubble films with smaller additions of Ca and Ge oxides.

TABLE II

Composition for 5 μm bubble CaGe melt

Oxide	Weight (g)	Moles/1000	Mole %
Y	2.080	9.208	0.327
Sm	0.414	1.188	0.042
Lu	0.615	1.545	0.055
Ca	4.279	76.302	2.710
Fe	45.	282.	10.014
Ge	6.505	62.199	2.209
Pb	500.	2240.	79.544
B	10.	143.6	5.099

The furnace used is illustrated in Fig. 1. Two thermocouples are in contact with the crucible bottom, one for control (CT) and one for DVM read-out (ROT). Two other thermocouples sense the winding temperature for rough control (RC) and for over-temperature (OT) cut-off.

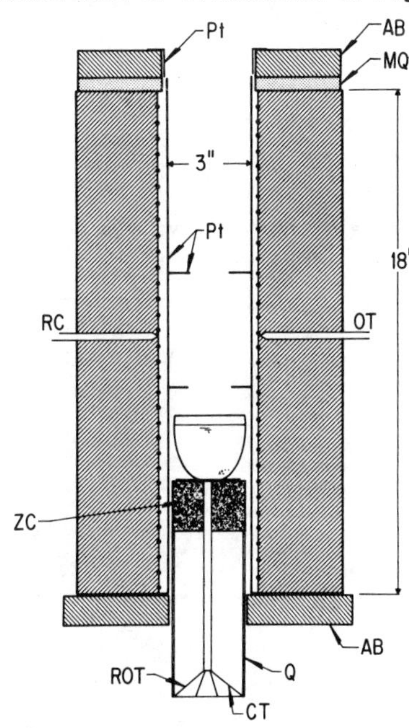

Fig. 1. LPE garnet furnace showing top and bottom asbestos boards (AB), a microquartz (MQ) top pad and the castable zirconia (Zc) crucible stand supported by a quartz (Q) tube.

It is essential for the reproducible growth of good CaGe films to control the temperature closely because of the temperature sensitivity of the distribution coefficients of the components of CaGe garnet materials.

The substrates are loaded into pure platinum holders and lowered to within a few inches above the melt surface where they remain for about 30 minutes. They are dipped into the melt while rotating at 30-100 rpm and held there to produce a layer whose properties approximate the center of our growth "target" (Fig. 2.)

PROCESS CONTROL

The two plots used for process control are ℓ vs T and growth velocity (V) vs T (Figs 3 and 4). By keeping these plots for each melt, we can follow the shifts that are occurring from run-to-run and thus predict what ℓ/V/T combinations will be suitable for the next runs.

After a melt is tuned, additions of PbO are added to compensate for continual evaporation losses. If the growth velocity is too low, it is increased with the addition of Y_2O_3, Sm_2O_3, CaO and GeO_2.

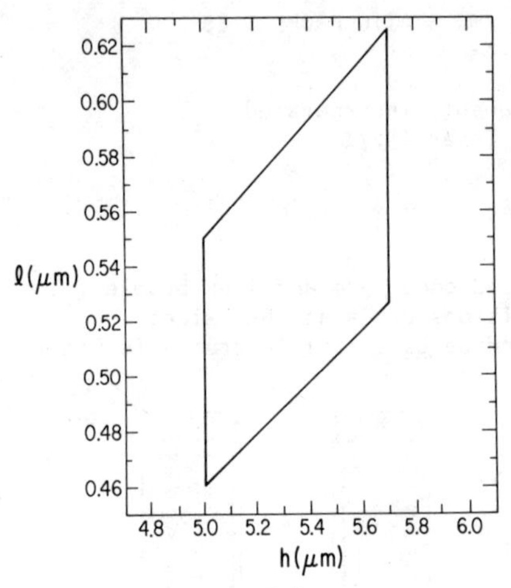

Fig. 2. Material length (ℓ) vs. thickness (h) growth target used for YSmLuCaFeGe garnet films.

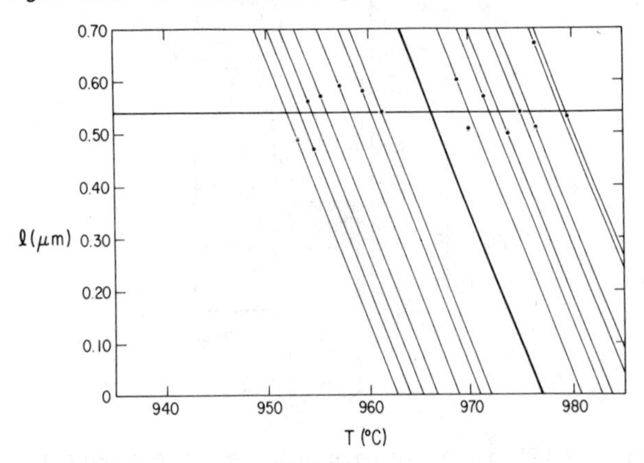

Fig. 3. Material length (ℓ) vs. growth temperature (T).

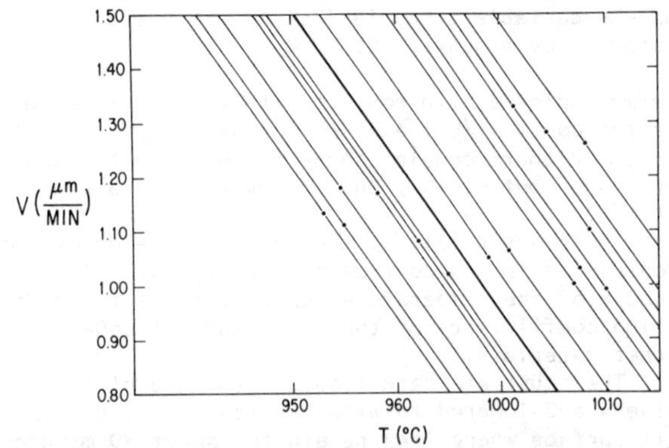

Fig. 4. Growth velocity (V) vs growth temperature (T).

The composition given in Table II will give films with a bubble mobility of about 400 cm/sec-Oe[11] and q = 6. The addition of Y_2O_3 together with CaO, GeO_2 and PbO will decrease q and increase mobility.

Table III lists the approximate composition and temperature differentials found for the melt given in Table II, although each melt will have slightly different values and these may vary with the age of the melt and the growth temperature used.

COERCIVE FORCE

We have found that one of the most difficult aspects of growing CaGe garnets may be establishing and maintaining a low coercive force, viz, $H_c < 0.2$ Oe. The coercivity appears to be primarily associated with charge imbalance between Ca^{+2} and Ge^{+4}, although a monotonic decrease in H_c is also observed as the mobility increased with increasing Y/Sm ratio. We have found that small additions of GeO_2 or CaO generally change H_c, indicating thenature and amount of further additions ofthese oxides needed to reduce the coercivity to within acceptable levels.

TABLE III

Selected oxide addition and temperature differentials

$\Delta\ell/\Delta$CaO, GeO_2	= + 100 μm/mole (each)
$\Delta V/\Delta Y_2O_3$	= + 300 μm/Min.-mole
$\Delta V/\Delta$PbO	= - 0.01 μm/Min.-g
$\Delta\ell/\Delta T$	= - 0.04 μm/deg.
$\Delta V/\Delta T$	= - 0.030 μm/Min.-deg.

STATISTICAL RESULTS

The growth statistics for hundreds of runs from several reactors are illustrated by the distribution of ℓ-values shown in Fig. 5. The average collapse field $\langle H_o \rangle$ for all device grade films was about 114 Oe with a standard deviation of 5 Oe.

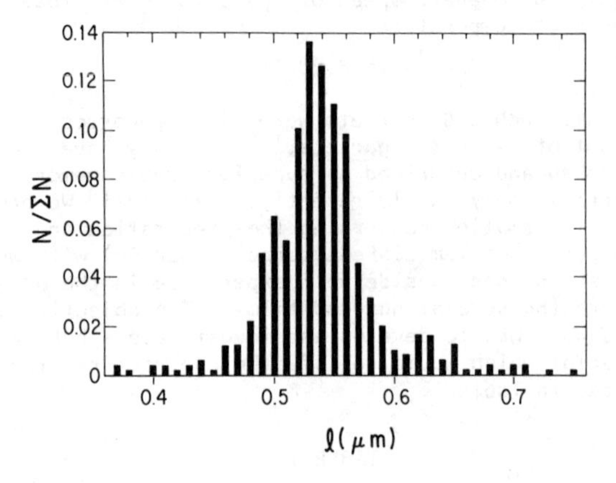

Fig. 5. Fraction of run starts (N/ΣN) vs. material length (ℓ).

STATIC FILM PROPERTIES

The temperature variations of the static bubble properties of a YSLCaGe film with a room temperature bubble mobility μ of 660 cm/sec-Oe are presented in Figs. 6-9, along with some data for a film with μ = 400 cm/sec-Oe. The domain wall (static) coercivity was measured using an a.c. bias field, HeNe laser and silicon photoconductive detector, with the samples mounted in a temperature-control chamber. The other static parameters were measured using conventional microscopic and ferromagnetic resonance techniques.[12,13] The temperature coefficients for these parameters (Fig.6-8) were calculated between 0 and 100°C and normalized to 25°C.

The temperature coefficient of the bubble collapse field H_o (Fig. 6) meets the -0.2%/°C requirement for bias magnet tracking, while the coefficient of the material length, ℓ, at -0.1%/°C, is somewhat less in magnitude than that previously reported for some Lu-free YSmCaGe films.[5,6] The magnitude of the uniaxial anisotropy field H_K (Fig. 7) dropped by roughly 20% when μ was increased from 400 to 660 cm/sec-Oe. The material q (Fig. 8) tracks H_K in its variation with μ, showing little change in temperature coefficient but a reduction in magnitude with increasing mobility. The reduction is insufficient to drop q below 3, however, even at 100°C. A two-fold reduction is seen for the domain wall coercivity H_c (Fig. 9) at 25°C as μ is increased from 400 to 660 cm/sec-Oe, but the relative increase in H_c

at low temperatures, as measured by the coercivity ratio, R_{H_c} [$\equiv H_c(-25°C)/H_c(25°C)$], does not appear to vary with μ. Since $H_c \lesssim 0.4$ Oe at -25°C for the higher mobility material, it should not be coercivity limited at the lower end of the temperature range.

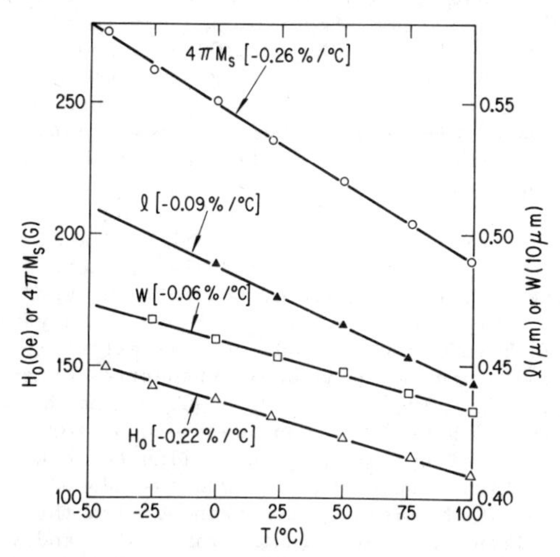

Fig. 6. Bubble collapse field (H_o), saturation magnetization ($4\pi M_s$), material length (ℓ) and strip width (W) vs. temperature (T) for an YSmLuCaFeGe garnet film with a mobility of 660 cm/sec-Oe.

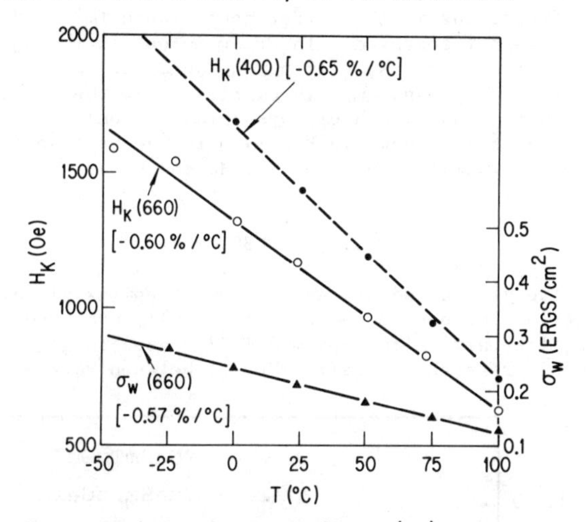

Fig. 7. Uniaxial anisotropy field (H_K) and wall energy (σ_w) vs. temperature (T) for YSmLuCaFeGe garnet films with mobilities of 660 cm/sec-Oe and 400 cm/sec-Oe.

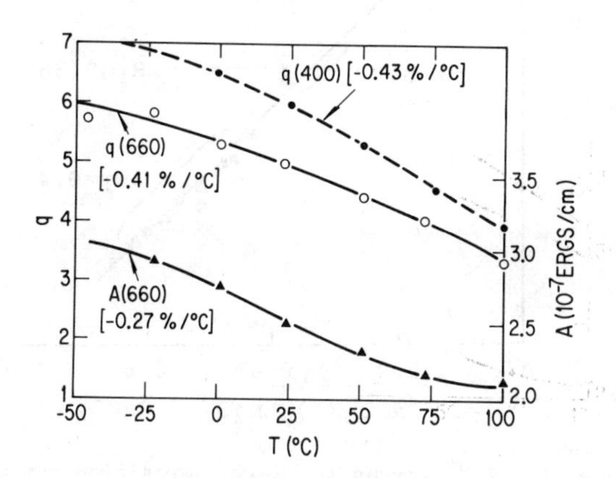

Fig. 8. Material q (q) and exchange constant (A) vs. temperature (T) for YSmLuCaFeGe garnet films with mobilities of 660 cm/sec-Oe and 400 cm/sec-Oe.

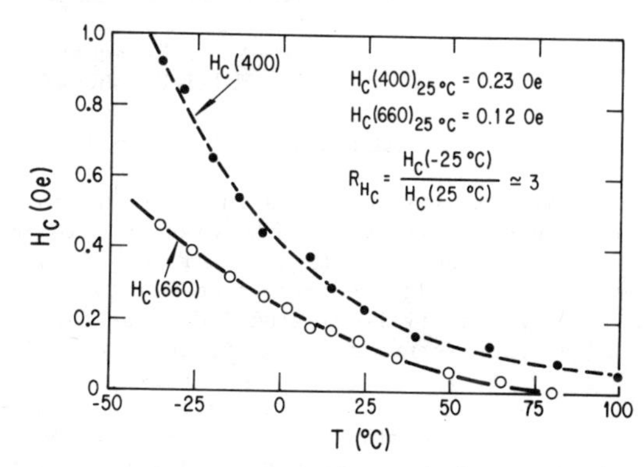

Fig. 9. Domain wall coercivity (H_c) vs. temperature (T) for YSmLuCaFeGe garnet films with mobilities of 660 cm/sec-Oe and 400 cm/sec-Oe.

The increase in μ and the decrease in H_K, q and H_c with increasing Y/Sm ratio were expected from considerations of orbital angular momentum.[5] The data presented here for the higher mobility material agrees with that shown by Ikeda, et.al.[9] for garnets of similar composition, except for their higher room temperature values of H_c (∿ 0.6 Oe). It is clear that YSLCaGe can be reproducibility grown to meet the specifications for moderate-speed device operation between -25°C and at least 75°C, but that coercivity control may be more difficult than is the case with the Ga garnet systems.

REFERENCES

* Work partially funded by Air Force Avionics Laboratory.

1. B. J. Skinner, Amer. Mineral, 41, 428 (1956).
2. A. Tauber, E. Banks and H. Kedesy, Acta Cryst. 11, 893 (1958).
3. S. Geller and C. E. Miller, Amer. Mineral, 44, 1115 (1959).
4. S. Geller, C. E. Miller and R. G. Treuting, Acta Cryst., 13, 179 (1960).
5. J. W. Nielsen, S.L. Blank, D. H. Smith, G. P. Vella-Coleiro, F. B. Hagedorn, R. L. Barns and W. A. Biolsi, J. Electronic Materials, 3, 693 (1974).
6. D. M. Heinz, R. G. Warren and M. T. Elliott, 21st Conf. Magnetism and Magnetic Materials, Philadelphia, (1975).
7. W. A. Bonner, J. E. Geusic, D. H. Smith, L. G. Van Uitert and G. P. Vella-Coleiro, Mat. Res. Bull., 8, 1223 (1973).
8. S. L. Blank, J. W. Nielsen and W. A. Biolsi, J. Electrochemical Soc. 123 (856) 1976.
9. T. Ikeda, F. Ishida, K. Ando, R. Suzuki and Y. Sugita, 7th Applied Magnetics Conf., Japan, 157 (1975) (In Japanese).
10. T. Obokata, H. Tominaga, T. Mori, and H. Inoue, 21st. Conf. Magnetism and Magnetic Materials, Philadelphia, (1975).
11. R. E. Fontana, Jr. and D. C. Bullock, Paper 3A-7, This Conference.
12. R. M. Josephs, AIP Conf. Proc. #10, Magnetism and Magnetic Materials - 1972, Am. Inst. of Phys., N.Y., 1973, p. 286.
13. R. C. LeCraw and R. D. Pierce, AIP Conf. Proc. #5, Magnetism and Magnetic Materials- 1971, Am. Inst. of Phys., N. Y. 1973, p. 200.

ACKNOWLEDGMENTS

We thank A. E. Tilton for the preparation of the substrate surfaces, R. E. Norris for the epitaxial growth of the garnet layers and J. A. Arnold for the x-ray determinations of a_o.

MAGNETIC PROPERTIES OF EPITAXIAL YCaSmTmGeIG AND SmTmIG WITH SUBMICRON BUBBLES

K. Yamaguchi, H. Inoue and K. Asama
FUJITSU LABORATORIES LTD., KAWASAKI, JAPAN

ABSTRACT

The molar composition ratios in the melt, to support submicron diameter magnetic bubbles in a YCaSmTmGeIG system, have been investigated. Films of saturation magnetization $4\pi M_s$ ranging from 511 to 919 gauss were grown on (111)-oriented GGG using conventional LPE dipping techniques. A stripe domain width S_w ranging from 1.31 to 0.71 μm and a collapse field H_o of 287 to 625 Oe for this garnet system have been obtained. Films of SmTmIG, in which $4\pi M_s$=1378 gauss, S_w = 0.56 μm, H_o = 900 Oe, were grown using similar techniques. From these experimental results growth-induced anisotropy was found to roughly approximate to the following empirical relation for a YCaSmTmGeIG system in which four ions were substituted on {24C} sites: $K_u(M_s x) = K_{uo} M_s x / M_{so} x_o$. Where K_{uo} and M_{so} are growth-induced anisotropy and saturation magnetization of SmTmIG, respectively. K_u and M_s are these notations for a YCaSmTmGeIG system. x is the variable related to the composition of Sm and Tm ions in this garnet system, while that for SmTmIG is defined as x_o. This formulation is also found to be appricable in a YCaSmLuGeIG system.

1. INTRODUCTION

Several kinds of mixed rare earth iron garnet films, which support submicron bubble domains, have been reported.[1,2] It is generally well known that Ca-Ge substituted rare earth iron garnet films are more stable thermally than those with Ga substituted. Ca-Ge substituted mixed rare earth iron garnet films, supporting submicron bubble domains, have been reported by one of the authors, Inoue et al.[3] It is the purpose of this paper to report further developments of the previous paper, concerned with melt composition and growth-induced anisotropy observed in Ca-Ge substituted mixed rare earth iron garnet systems. Rare earth ions of both SmTm and SmLu were chosen from the point of view of large growth-induced anisotropy. Growth-induced anisotropy of mixed rare earth films is considered to be attributable to rare earth ions, and it is a significant parameter in the material research of garnet films for bubble domains. However, it is very difficult to evaluate growth-induced anisotropy of garnet films containing many ions prior to film growth. Empirical formulations to evaluate growth-induced anisotropy, depending on the concentration of SmTm and SmLu in films of the $Y_u Ca_y (SmTm_{2.52})_x - Fe_{5-y} Ge_y O_{12}$(YCaSmTmGeIG) and $Y_w Ca_{0.75} (SmLu_{1.43})_x - Fe_{4.25} Ge_{0.75}$(YCaSmLuGeIG) systems, are derived. Where u=3-3.52x-y and w=2.25-2.43x. The domain wall mobility of the YCaSmTmGeIG system is discussed.

2. FILM GROWTH AND MAGNETIC PROPERTIES

Film of YCaSmTmGeIG with submicron bubble diameter have been prepared by varying the molar composition ratio of germanium oxide to iron oxide and carbonate to rare earth oxides and yttrium oxide.
In order to obtain the desired saturation magnetization and other properties, three melt compositions were provided, in which the ratio of $R_1 = GeO_2/(GeO_2+Fe_2O_3)$ as shown in Fig. 1 was different. In each case, the ratio of $R_2 = CaCO_3/(\Sigma Ln_2O_3 + CaCO_3)$ was slightly changed by adding small amounts of $CaCO_3$, where Ln_2O_3 is the sum of rare earth oxides and yttrium oxide. Saturation magnetization of films grown in various melt compositions is shown as a function of the ratios of R_1 and R_2 in Fig. 1. The melt composition to prepare films of YCaSmTmGeIG having saturation magnetization $4\pi M_s$=850 gauss, as shown in Table I, is as follows: Y_2O_3 0.9032 gms, $CaCO_3$ 0.5600 gms, Sm_2O_3 0.5314 gms, Tm_2O_3 1.470 gms, Fe_2O_3 15.310 gms, GeO_2 0.5702 gms, PbO 319.2 gms and B_2O_3 6.385 gms. Undercooling temperature is about 20°C. Rotation speeds of about 150 rpm were employed, and a growth rate of about 0.8 μm/min was obtained.

Curie temperature T_c, collapse field H_o and anisotropy field H_k were automatically measured with a photomultiplier mounted on a polarizing microscope. Typical properties of SmTmIG, YCaSmTmGeIG and TCaSmLuGeIG films are given in Table I, where ℓ is the characteristic length, h, the film thickness, σ_w, the wall energy and q, $H_k/4\pi M_s$. Among the crystals grown, an SmTmIG crystal had the smallest stripe domain width and its remarkable properties were S_w = 0.56 μm, $4\pi M_s$ = 1378 gauss, H_o = 900 Oe and H_k = 3478 Oe. This width of S_w is considered to be very close to the limit of epitaxial garnet films for bubble domains. Fig. 2 shows how the growth-induced anisotropy of YCaSmTmGeIG system changed with the concentrations x and y. In the range of x and y covered by the experiments, K_u increased linearly with x. Fig. 3 shows H_k vs.x of the garnet system.
K_u vs.x of the YCaSmLuGeIG system is almost the same as that of YCaSmTmGeIG for the same Sm concentration, and effects due to the difference between Tm^{3+} and Lu^{3+} were not observed. In order to obtain the μ_w of these garnet systems, film having an S_w of about 1.3 μm have been grown. μ_w was carried using the bubble transport method which employs parallel conductors. The results are shown in Fig. 4. In the next section, the experimental results on K_u, H_k and μ_w are discussed.

3. DISCUSSION

The growth-induced anisotropy of epitaxial garnet films for bubble domains are usually analyzed using the pair anisotropy model [4], based on growth-induced preferential pair ordering between rare earth

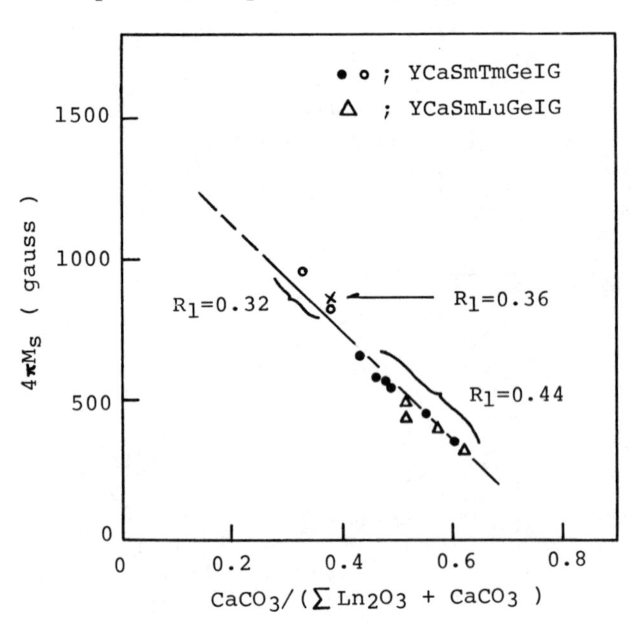

Fig. 1. $4\pi M_s$ versus the melt composition ratio of $CaCO_3/[\Sigma Ln_2O_3 + CaCO_3]$ for the different ratios of $R_1 = GeO_2/[GeO_2 + Fe_2O_3]$.

ions or yttrium and these ions. When {24C} sites consists of two kinds of ions as in $Y_{3-x}Lu_xFe_5O_{12}$ $Y_{3-x}Sm_xFe_5O_{12}$ and $Gd_{3-x}Eu_xFe_5O_{12}$, growth-induced anistropy is expressed as

$$K_u = K(1-x/3)x \quad \ldots\ldots \quad (1)$$

The experimental results of K_u, of these films, agreed with this equation [5,6]. Here we analyze the dependence of K_u on the SmTm concentration x shown in Fig. 2. In the YCaSmTmGeIG film, the {24C} sites are occupied by four kinds of ions, Y^{3+}, Ca^{2+}, Sm^{3+} and Tm^{3+}. If the two ion pair model is applied to the garnet system, ion pairs of Y^{3+}-Sm^{3+}, Sm^{3+}-Tm^{3+}, Tm^{3+}-Y^{3+} and those including Ca^{2+} must be considered. If one consider the probability of the existence of these ion pairs and the contribution of each ion pair to K_u in the garnet system, it is very difficult to discuss K_u using Eq. (1). The experimental results on $Sm_{0.85}Tm_{2.15}Fe_5O_{12}$ (SmTmIG) are therefore applied to discuss the K_u of YCaSmTmGeIG. SmTmIG has a composition of x=0.85 in the representation of $Y_{3-x}(SmTm_{2.52})_xFe_5O_{12}$(YSmTmIG) with $4\pi M_s$ = 1378 gauss and K_u = 1.91 x 10^5 erg/cm^3. The following three assumptions are introduced to analyze the experimental results.

1) Over the range of x, from 0.1 to 0.8 in the YSmTmIG system, $4\pi M_s$ = 1378 gauss.

2) The K_u of the assumed YSmTmIG system increases linearly with x,

$$K_u(x) = K_{uo}\frac{x}{x_o} \quad \ldots\ldots \quad (2)$$

3) $H_k(x)$ of YCaSmTmGeIG is essentially equal to that of YSmTmIG for the same concentration x,

$$H_k(x) = \frac{2K_{uo}}{M_{so}}\frac{x}{x_o} \quad \ldots\ldots \quad (3)$$

On the basis of the above assumptions, the $K_u(M_s x)$ of the YCaSmTmGeIG system may be expressed as follows:

$$K_u(M_s x) = \frac{1}{2} H_k(x) M_s$$

$$= K_{uo}\frac{M_s x}{M_{so} x_o}, \quad \ldots\ldots \quad (4)$$

where K_{uo} and M_{so} are the growth-induced anisotropy and the saturation magnitization of SmTmIG, respectively.

x_o = 0.85 is the SmTm concentration in this garnet system as described earlier. As shown in Fig. 2, the experimental values of $K_u(M_s x)$ of the YCaSmTmGeIG system agree with Eq. (4), therefore the three assumptions are presumed to be reasonable. A comparison between the experimental results on H_k and Eq.(3) is shown in Fig. 3. The agreements between the experimental results and Eq.(3) is good for x = 0.4 ~ 0.5. However, it is poor for x < 0.3. Although Eq.(3) is not sufficient to describe the experimental results over the wide range of SmTm concentration x, this equation is considered to be convenient in order to estimate roughly H_k of the region of large x prior to film growth. K_u of the YCaSmLuGeIG system is also compared with Eq. (4), substituting the magnetic parameters [2)]

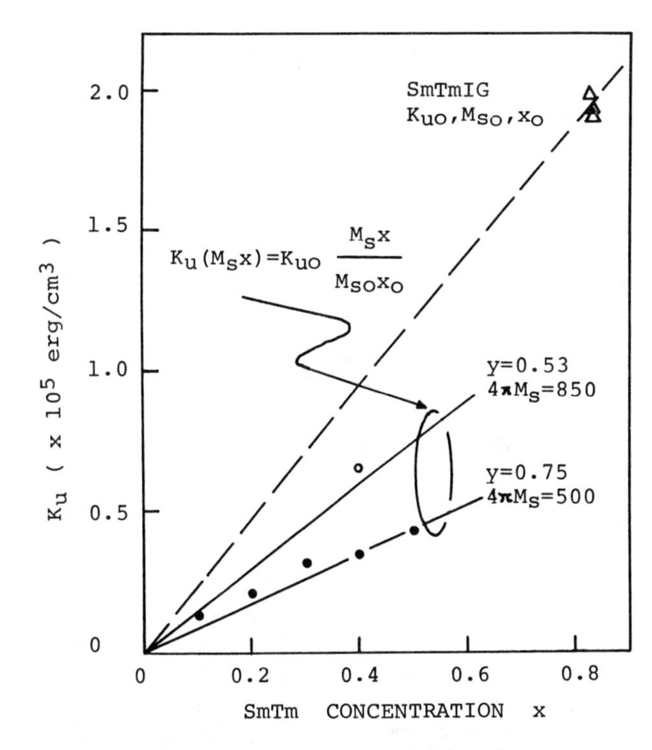

Fig. 2. K_u as a function of the SmTm concentration x in a YCaSmTmGeIG system. Two solid lines are the empirical relation, $K_u(M_s x) = K_{uo}M_s x/M_{so}x_o$, which was deduced experimentally for this garnet system. A broken line is K_u as a function of SmTm concentration x in an assumed YSmTmIG system.

Table I Magnetic properties of SmTmIG, YCaSmTmGeIG and YCaSmLuGeIG films

Material	S_w (μm)	h (μm)	H_o (Oe)	ℓ (μm)	$4\pi M_s$ (Gauss)	H_k (Oe)	g	K_u (x 10^4 erg/cm^3)	σ_w (erg/cm^2)	T_c (°C)
$Sm_{0.85}Tm_{2.15}Fe_5O_{12}$	0.56	0.87	900	0.047	1378	3478	2.52	19.1	0.71	272
	0.74	1.35	1008	0.054	1440	3630	2.52	20.4	0.89	
$Y_{1.1}Ca_{0.5}Sm_{6.4}Tm_{1.0}$-$Fe_{4.5}Ge_{0.5}O_{12}$	0.71	0.98	522	0.065	850	1956	2.31	6.58	0.37	245
	0.91	1.56	625	0.070	919	2239	2.43	8.19	0.47	
$Y_{1.2}Ca_{0.75}Sm_{0.3}$-$Tm_{0.75}Fe_{4.25}Ge_{0.75}O_{12}$	0.92	1.09	287	0.094	504	1554	3.08	3.12	0.19	224
	1.31	1.62	301	0.130	516	1896	3.67	3.89	0.28	
$Y_{1.29}Ca_{0.75}Sm_{0.4}$-$Lu_{0.56}Fe_{4.25}Ge_{0.75}$-$O_{12}$	0.81	0.63	188	0.103	434	1304	3.00	2.25	0.15	203
	1.02	1.01	242	0.118	475	1645	3.46	3.11	0.21	

of $Sm_{1.2}Lu_{1.8}Fe_5O_{12}$ (SmLuIG) required in the equation, $4\pi M_{so}$ = 1760 gauss and $K_{uo} = 3.04 \times 10^5$ erg/cm^3, and $x_o = 1.2$. Empirical relationship have also been found to be applicable in this garnet system.

μ_w is expressed as follows.

$$\mu_w = \frac{1}{\lambda/\gamma^2}\sqrt{\frac{A}{2\pi q}} \quad , \quad \ldots\ldots \quad (5)$$

where λ/γ^2 is the Landau-Lifschitz damping constant and A is an exchange constant which is assumed to be 3×10^{-7} erg/cm, considering T_c is in the vicinity of 220°C. The Landau-Lifschitz damping constant of a mixed garnet system could be approximated by the following formula.

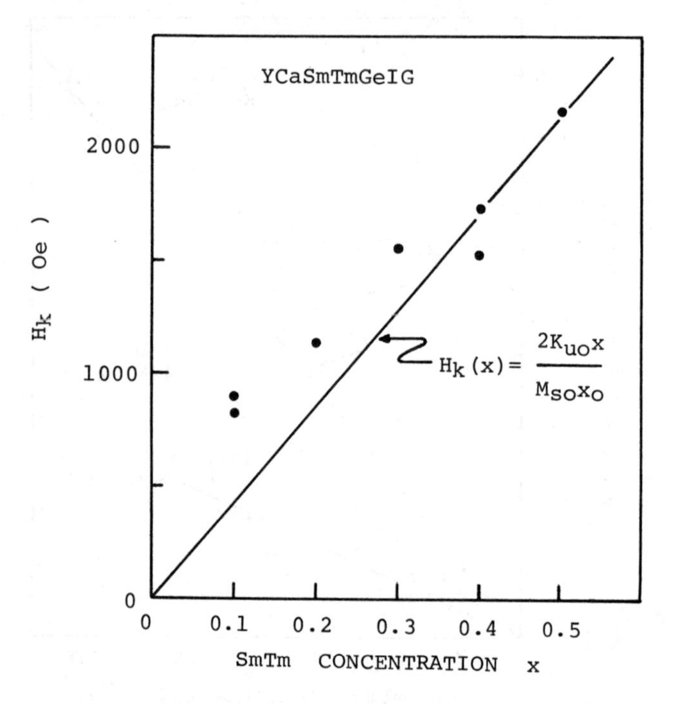

Fig. 3. H_k as a function of the SmTm concentration x in a YCaSmTmGeIG system. A solid line is the empirical relation, $H_k(x) = 2K_{uo}x/M_{so}x_o$.

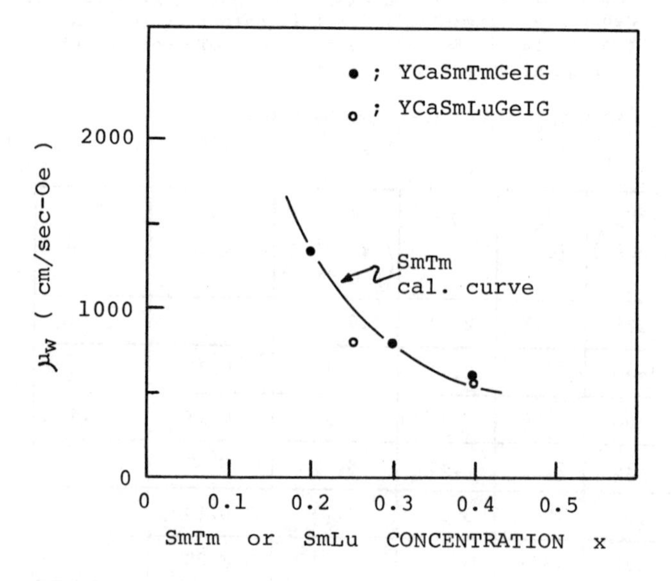

Fig. 4 Comparision between experimental results of μ_w and calculated curve based on equation (5) for a YCaSmTmGeIG system, and experimental results for a YCaSmLuGeIG system.

$$\frac{\lambda}{\gamma^2} = \frac{\sum_i C_i(\lambda/\gamma^2)_i}{\sum_i C_i} \quad , \quad \ldots\ldots \quad (6)$$

where $(\lambda/\gamma^2)_i$ are the values of pure garnet and C_i the respective compositions of the pure garnet in a mixed garnet system. μ_w was calculated using damping constant [6] of each pure garnet listed below.

YIG : 0.52 ($\times10^{-7}$ Oe^2 sec rad^{-1})
SmIG : 25
TmIG : 1.2
LuIG : 0.52

The damping constant of LuIG is assumed to be the same value as that of YIG, because Lu^{3+} is a nonmagnetic ion similar to Y^{3+} and different from Sm^{3+} and Tm^{3+} which have magnetic moment. The damping constants of YCaSmTmGeIG and YCaSmLuGeIG systems were calculated using Eq. (6) and the composition x dependence of μ_w was calculated using q obtained experimentally. The calculated curve, μ_w vs. x was smaller than the experimental ones. However, the corrected curve, obtained by multiplying the damping constants of the pure garnets by a factor of about 0.7, agree well with the experimental results, as shown in Fig. 4. The necessity for such a correction of the damping constants in the YCaSmTmGeIG and YCaSmLuGeIG systems may be attributed to the effects of Ge^{4+} or other unknown effects.

4. SUMMARY

The melt composition and the magnetic properties of YCaSmTmGeIG and YCaSmLuGeIG systems have been described. SmTmIG was grown as an end member of the YCaSmTmGeIG system and its film properties were investigated. From the experimental results, an empirical formula was introduced to calculate the growth-induced anisotropy of Ca-Ge substituted rare earth iron garnet films.

AKNOWLEDGEMENTS

The authors wish to thank T. Obokata and H. Tominaga for critical reading of the manuscript. They also thank M. Nakamura and F. Iwai for their valuable discussions and encouragement.

REFERENCES

1. E.A. Giess, C.F. Guerci, J.D. Kuptsis and H.L. Hu, Mat. Res. Bull. 8, 1061 (1973)
2. D.C. Bullock, J.T. Carlo, D.W. Mueiler and T.L. Brewer, AIP Conf. Proceedings 24, 647 (1974)
3. T. Obokata, H. Tominaga, T. Mori and H. Inoue, AIP Conf. Proceedings (to be published)
4. A. Rosencwaig, W.J. Tabor and R.D. Pierce, Phys. Rev. Letters 26, 779 (1971)
5. E.M. Gyorgy, M.D. Sturge and L.G. Van Uitert, AIP Conf. Proceedings 18, 70 (1973)
6. E.J. Heilner and W.H. Grodkiewicz, J. Appl. Phys. 44, 4218 (1973)
7. G.P. Vella-Coleiro, AIP Conf. Proceedings 10, 424 (1972)

GROWTH ATMOSPHERE EFFECTS ON Ca, Ge-SUBSTITUTED GARNET LPE FILM BUBBLE PROPERTIES*

T. Hibiya, H. Makino and Y. Hidaka
Central Research Laboratories, Nippon Electric Co., Ltd., Kawasaki 211, Japan

ABSTRACT

Growth atmosphere effects on bubble properties for $(Ca,Lu,Sm,Y)_3(Fe,Ge)_5O_{12}$, $(Ca,Eu,Lu,Y)_3(Fe,Ge)_5O_{12}$ and $(Ca,Y)_3(Fe,Ge)_5O_{12}$ garnet LPE films grown on (111) GGG substrates have been investigated. Equilibrium between the melt and gas phase was achieved, by keeping the $PbO-B_2O_3$ fluxed melt in an O_2- N_2 mixed gas atmosphere. Film lattice parameter and $4\pi M_s$ linearly increase with oxygen partial pressure, $P(O_2)$, while Curie temperature and growth-induced uniaxial anisotropy decrease with increasing $P(O_2)$ for these garnets grown at ~900°C. However, at higher growth temperature (~1000°C), the $P(O_2)$ dependence turns out to be less significant. Pt^{4+} substitution in the octahedral sites accompanied with Pb^{2+} in the dodecahedral sites accounts for these drastic changes in film properties.

INTRODUCTION

Ca, Ge-substituted garnet LPE films have been extensively investigated for use in bubble domain materials, because those garnets are known to have higher Curie temperatures which cause better temperature stability, as well as higher domain wall mobility than Ga-substituted garnet LPE films.[1,2] Employing Ca^{2+} and Ge^{4+}, whose valencies and chemical properties are quite different, makes film growth mechanism differ from that for Ga^{3+} or Al^{3+} substitution. So far as is known, bubble films have been mostly grown in air from melts containing $PbO-B_2O_3$ flux. In order to investigate growth atmosphere effects on bubble properties, we have grown Ca, Ge-substituted garnet LPE films under several oxygen partial pressures, $P(O_2)$. We have found significant $P(O_2)$ dependence of film properties, such as Curie temperature, saturation magnetization, film lattice parameter and uniaxial anisotropy. A possible mechanism, which results in such $P(O_2)$ dependence, is discussed.

EXPERIMENTAL

The film compositions studied are $(Ca,Lu,Sm,Y)_3(Fe,Ge)_5O_{12}$, $(Ca,Eu,Lu,Y)_3(Fe,Ge)_5O_{12}$ and $(Ca,Y)_3(Fe,Ge)_5O_{12}$, designated as CLSY, CELY and CY, respectively. Each film was grown on a (111) GGG substrate by the LPE dipping technique from a $PbO-B_2O_3$ fluxed melt within a vertical furnace, through which O_2-N_2 mixed gas was flowed in order to control $P(O_2)$. Total gas flow rate was 600 ℓ/hr. Before the film growths, equilibrium between the melt and gas phase was achieved by keeping the melt in a given atmosphere at temperature 30°C above the saturation temperature for at least 6 hours. The melt compositions and growth temperature ranges are listed in Table I.

Film lattice parameters, a_f, were measured by X-ray double crystal diffractometry and by the Bond method.[3] Magnetic properties, such as $4\pi M_s$, characteristic length, ℓ, and K_u were obtained by using a vibrating sample magnetometer, the Fowlis method[4] and a conventional torque method, respectively. Curie temperature, T_c, was measured by detecting an abrupt change of magneto-optical signal as temperature is slowly raised and lowered near T_c. Film composition and Pt and Pb contents were observed by electron-probe microanalysis (EPMA) and film thickness by optical interferometry.

RESULTS AND DISCUSSION

Experimental results show that melt saturation temperature does not shift with $P(O_2)$ and that concentration of garnet host constituents does not seem to be affected by $P(O_2)$. However, remarkable effects on magnetic properties due to Pt and Pb contamination are observed.

1. Pt and Pb substitution

Figure 1 shows the growth temperature, T_g, dependence of $4\pi M_s$, ℓ, T_c and a_f for CLSY-1 films, grown in various amounts of $P(O_2)$. The growth temperatures were randomly chosen so that systematic errors were reduced. In a given atmosphere, both T_c and $4\pi M_s$ decreased as T_g was lowered, thus indicating an increase in Ge^{4+} substitution with decreasing growth temperature (i.e, with increasing growth rate). On the other hand, a_f increased with lowering T_g, although increase in paired substitution of Ca^{2+} and Ge^{4+} is reported not to increase a_f.[1] It is interesting that separate curves of $4\pi M_s$, ℓ, T_c and a_f were drawn for different growth atmospheres. As is evident from Fig. 1, for films grown in every atmosphere, Ge^{4+} contents do not exceed compensation points. This is also confirmed by EPMA Ge content measurement.

Figure 2 shows the $P(O_2)$ dependence of $4\pi M_s$, T_c and a_f for CLSY-1 films grown at a constant temperature; 900°C. It is noteworthy that $4\pi M_s$ and a_f increased linearly with $P(O_2)$, while T_c decreased with increasing $P(O_2)$. The similar $P(O_2)$ dependence of film proper-

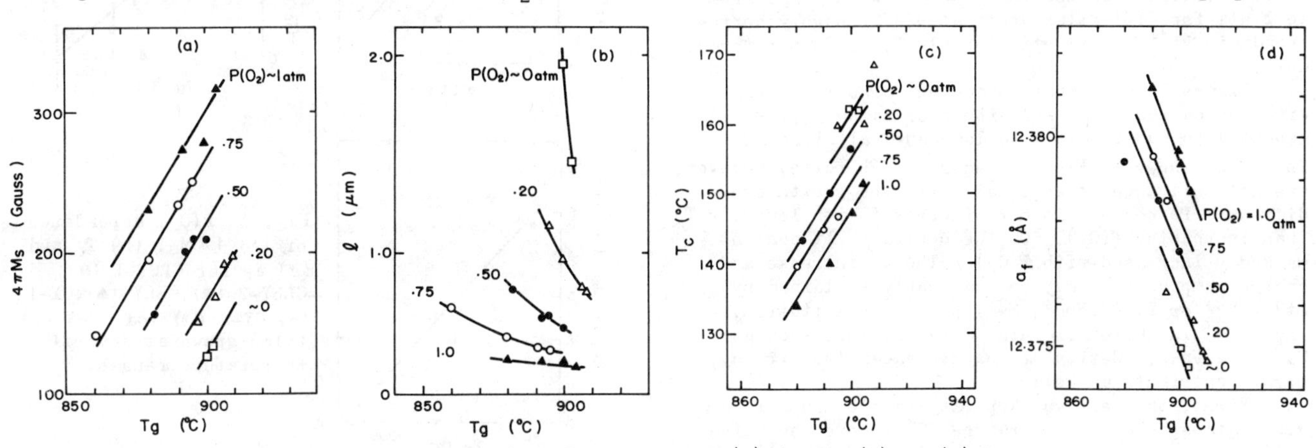

Fig. 1 Growth temperature dependence of (a) $4\pi M_s$, (b) ℓ, (c) T_c and (d) a_f for CLSY-1 films grown in various amounts of $P(O_2)$.

Table-I Melt Compositions for Garnet Films

Sample	Film composition (nominal)	Melt composition (mol %)									T_s (^oC)	Growth temperature range (^oC)
		Y_2O_3	Eu_2O_3	Sm_2O_3	Lu_2O_3	CaO	Fe_2O_3	GeO_2	PbO	B_2O_3		
CLSY-1	$Ca_1Lu_{11}Sm_2Y_7Fe_4Ge_1O_{12}$	.156	——	.043	.234	2.109	8.627	2.109	81.436	5.286	916	860 - 910
CLSY-2	$Ca_1Lu_{13}Sm_2Y_5Fe_4Ge_1O_{12}$	.111	——	.048	.293	2.515	12.649	2.660	76.743	4.981	987	965 - 980
CELY	$Ca_1Eu_4Lu_2Y_{14}Fe_4Ge_1O_{12}$	.282	.077	——	.033	2.053	7.805	2.050	82.355	5.345	929	900 - 910
CY-1	$Ca_{.8}Y_{2.2}Fe_{4.2}Ge_{.8}O_{12}$	.412	——	——	——	1.829	8.230	1.829	82.355	5.345	945	890 - 925
CY-2	$Ca_1Y_2Fe_4Ge_1O_{12}$	.489	——	——	——	2.879	9.754	2.879	78.880	5.119	1000	985 - 1000
CY-3	$Ca_{.7}Y_{2.3}Fe_{4.3}Ge_{.7}O_{12}$	.613	——	——	——	2.717	12.227	2.717	76.746	4.980	1040	1030

ties for CELY films were observed. These experiments strongly suggest that a part of octahedral Fe^{3+} is substituted for by nonmagnetic ions. The substitution should become more pronounced as $P(O_2)$ is increased and eventually, result in the lattice expansion.

Lu^{3+} can be substituted into the octahedral sites.[5,6] This is, in general, true for CLSY-1 and CELY films, since lower T_c and larger a_f values (e.g. for air-growth films) than values estimated from the nominal compositions cannot be explained, unless octahedral Lu^{3+} substitution is taken into considera- tion. However, $P(O_2)$ dependence of film properties was also observed in Lu-free system, $(Ca,Y)_3(Fe,Ge)_5O_{12}$. When growth was carried out at around 920°C, using CY-1 melt, the values of $d(4\pi M_s)/dP(O_2)$, $dT_c/dP(O_2)$ and $da_f/dP(O_2)$ were approximately equal to those of CLSY and CELY films; 180 Gauss/atm, 15°C/atm and 0.005Å/atm, respectively. Therefore, contribution of Lu^{3+} to these $P(O_2)$ dependence was excluded.

It was found that $P(O_2)$ dependence of these film properties depends on film growth temperature range, as well. At higher growth temperatures, dependence of film properties, such as $4\pi M_s$, T_c and a_f, was less significant, as shown in Fig. 3. For example, $dT_c/dP(O_2)$ was decreased to zero at 1030°C. This was also observed in CLSY-2 films. From these experimental results, the amount of substituted nonmagnetic ions in part of octahedral Fe^{3+} is concluded to be strongly dependent on not only $P(O_2)$ but also on growth temperature range.

Although Glass and Elliott[7] have reported that the Pt concentration was approximately the same in YIG films grown in wide temperature ranges, we found that Pt content evidently increased with increasing $P(O_2)$ and with decreasing T_g, as is shown in Fig. 4. The ionic radius of Pt^{4+} is 0.63Å[8] and is expected to enter octahedral sites. Since this $P(O_2)$ dependence of Pt substitution well corresponds to that of T_c and $4\pi M_s$, Pt^{4+} ions are concluded to be octahedral-site substituents. Decrease in T_c(~10°C) with $P(O_2)$ from 0 to 1 atm for CY-1 films grown at 920°C roughly corresponds to Pt^{4+} substitution of ~1wt%, 0.038 per formula unit.

Figure 5 shows that Pb content also increases with increasing $P(O_2)$ and with decreasing T_g. Pb is expected to substitute into dodecahedral sites as Pb^{2+}. As shown in Figs. 4 and 5, Pt/Pb ratio, however, was not unity, but changed all over the growth conditions. Pb content is more increased with lowering T_g than increasing $P(O_2)$. On the contrary, Pt content markedly increased with $P(O_2)$. The a_f increase and charge compensation may be reasonably explained by Pt^{4+}-Pb^{2+} or Pt^{4+}-(Pb^{2+}, M^{2+}) paired substitution, where M^{2+} is the other large divalent ion, such as Ca^{2+} and Fe^{2+}. Cation and oxygen vacancies, if any, might contribute as well.

First, the melt was kept at least 6 hours in a given atmosphere at temperature 30°C above the saturation temperature before growth in order to achieve equilibrium between melt and atmosphere. We found,

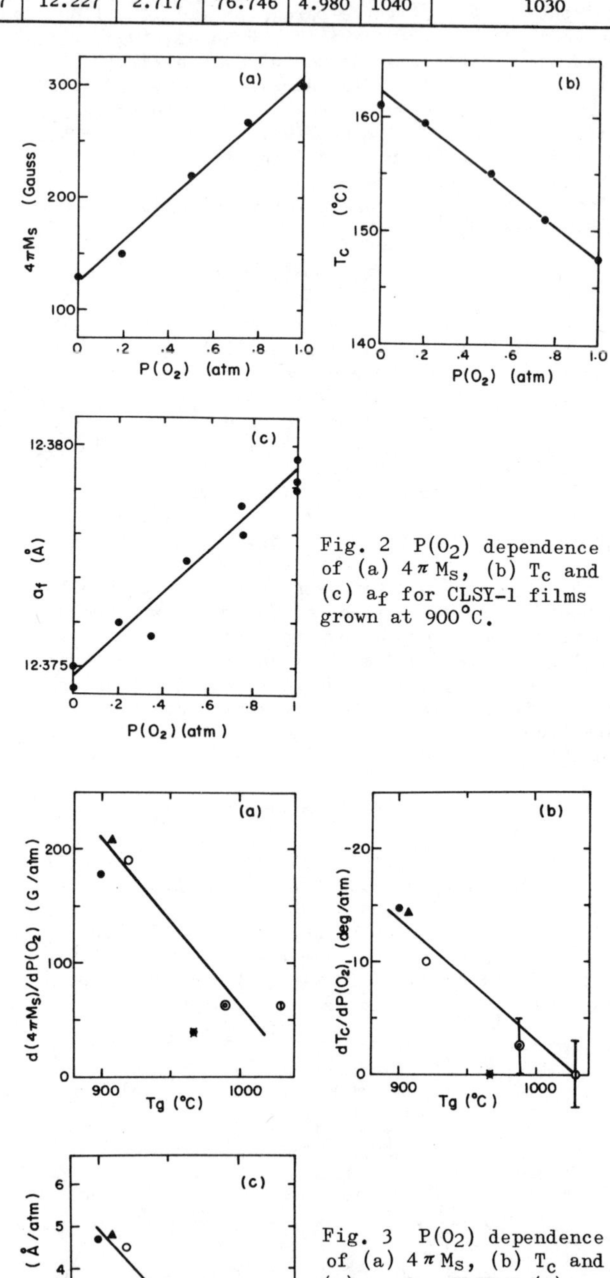

Fig. 2 $P(O_2)$ dependence of (a) $4\pi M_s$, (b) T_c and (c) a_f for CLSY-1 films grown at 900°C.

Fig. 3 $P(O_2)$ dependence of (a) $4\pi M_s$, (b) T_c and (c) a_f for CLSY-1 (●), CLSY-2 (✖), CELY (▲), CY-1 (o), CY-2 (◎) and CY-3 (⦶) films grown at several temperature ranges.

however, that only 1 hour or shorter was enough to result in growth atmosphere effects for CY-2. This implies that the melt properties and the growth mechanism may by rapidly altered according to $P(O_2)$.

The distribution coefficients of Pt^{4+} and Pb^{2+} may be sensitive to growth temperature as well as growth atmosphere. Therefore, at higher temperature, film properties $P(O_2)$ dependence is considered to become less significant.

2. $P(O_2)$dependence of growth-induced anisotropy

As shown in Fig. 6, both $P(O_2)$ and growth temperature range dependences of K_u were observed. At lower temperature, K_u decreased with $P(O_2)$. However, $P(O_2)$ dependence became less significant at higher growth temperature range (CY-3). For CY-films, K_u values changed from negative to positive with increasing growth temperature range.

Films were annealed at 1080°C for 5 hours in air. Uniaxial anisotropy constants for CLSY-1 films, which initially ranged from 18000 to 6000 erg/cm^3, depending on $P(O_2)$, decreased to 3000 to 1000 erg/cm^3, thus most uniaxial anisortopy is concluded to be growth-induced. T_c and a_f did not change beyond the experimental uncertainty. K_u was changed into constant value ~+2000 erg/cm^3 after annealing in the same condition for all CY films. This means that growth-induced in-plane anisotropy, which depends on $P(O_2)$, was annealed out so that positive stress-induced anisotropy appeared.

Elliott and Glass[9] and Plaskett et al.[10] reported that Pb^{2+} in dodecahedral sites for $Y_3Fe_5O_{12}$ and $Eu_3Fe_5O_{12}$ LPE films causes an increase in growth induced anisotropy. The exact cause of drastic $P(O_2)$ dependence of K_u observed in this work is not known at present. However, our experiment shows that Pt^{4+} -Pb^{2+} paired substitution eventually increased negative growth-induced anisotropy.

Similar $P(O_2)$ effects on T_c, ℓ, a_f and K_u were observed for $(Y,Eu,Yb)_3(Fe,Ga)_5O_{12}$ films; decreases in T_c, ℓ and K_u and an increase in a_f.[11]

Drastic changes observed in bubble properties were attributed to small but effective content change of constituents and incorporated impurities, as well as valence changes.

Molar content ratio of Ca^{2+} to Ge^{4+} in the film would shift from unity. This may be attributed to nonstoichiometry and defect structure in garnet films, such as vacancies and valence changes of Fe ion. A clue for this subject will be given through film growth experiments in various atmospheres.

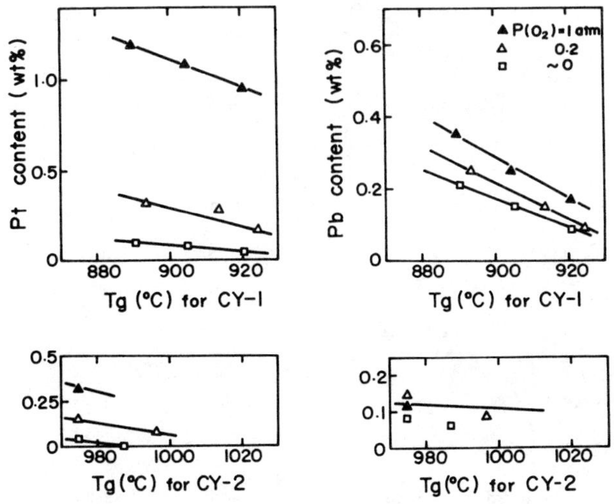

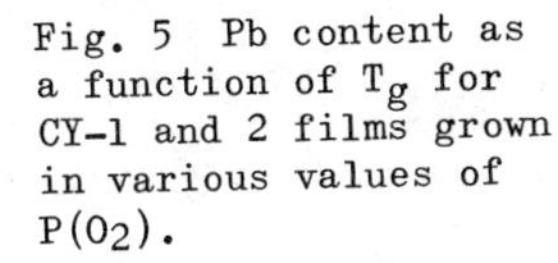

Fig. 4 Pt content as a function of T_g for CY-1 and 2 films grown in various values of $P(O_2)$.

Fig. 5 Pb content as a function of T_g for CY-1 and 2 films grown in various values of $P(O_2)$.

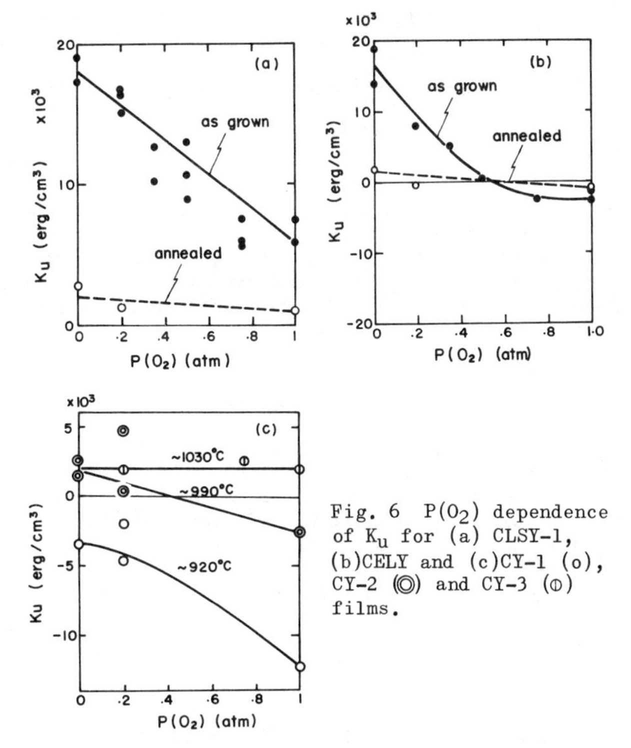

Fig. 6 $P(O_2)$ dependence of K_u for (a) CLSY-1, (b)CELY and (c)CY-1 (o), CY-2 (◎) and CY-3 (⦶) films.

ACKNOWLEDGMENT

The authors would like to thank Messrs. T. Furuoya, K. Matsumi and K. Suzuki for many valuable discussions, and Miss A. Ogura for her expert technical assistance.

REFERENCES

* Work partly supported by the Agency of Industrial Science and Technology, Ministry of International Trade and Industry.

1. W. A. Bonner, J. E. Geusic, D. H. Smith, L. G. Van Uitert and G. P. Vella-Coleiro, Mat. Res. Bull. 8, 1223 (1973).
2. J. W. Nielsen, S. L. Blank, D. H. Smith, G. P. Vella-Coleiro, F. B. Hegedorn, R. L. Barns and W. A. Biolsi, J. Electronic Mat. 3, 693 (1974).
3. H. Makino, S. Nakamura and K. Matsumi, Japan. J. Appl. Phys. 15, 415 (1976).
4. D. C. Fowlis and J. A. Copeland, AIP Conf. Proc. No. 5, 240 (1972).
5. D. M. Heinz, R. G. Warren and M. T. Elliott, 21st MMM Conf. 8A-5, Philadelphia (1975).
6. S. L. Blank, J. W. Nielsen and W. A. Biolsi, J. Electrochem. Soc. to be published.
7. H. L. Glass and M. T. Elliott, J. Cryst. Growth 27, 253 (1974).
8. R. D. Shannon and C. T. Prewitt, Acta Cryst. B25, 925 (1969).
9. M. T. Elliott and H. L. Glass, 21st MMM Conf. 6B-3, Philadelphia (1975).
10. T. S. Plaskett, E. Klokholm, D. C. Cronemeyer, P. C. Yin and S. E. Blum, Appl. Phys. Lett. 25, 357 (1974).
11. T. Hibiya, unpublished work.

$(Lu,La,Y,Ca)_3(Ge,Fe)_5O_{12}$: A TEMPERATURE COMPENSATED HIGH MOBILITY BUBBLE GARNET

R. Hiskes
Hewlett-Packard Laboratories, Palo Alto, CA 94304

ABSTRACT

The $(Lu,La,Y,Ca)_3(Ge,Fe)_5O_{12}$ garnet system is useful for practical high speed bubble devices because it contains no rare earth ions which contribute to damping; yet a combination of growth and stress induced anisotropy provides a q > 7 for 6 μm diameter bubbles. The partial substitution of Lu^{3+} on octahedral sites allows manipulation of the Néel temperature between 180°C and 200°C, with attendant H_T adjustment to -0.20%/°C over a -20°C to +120 °C temperature range to match a particular barium ferrite bias magnet. Defect free films of this composition grown by standard LPE techniques in a $PbO-B_2O_3$ solvent in the temperature range 930°C to 1000°C exhibit a coercive force < 0.4 Oe and bubble mobilities > 2600 cm/s-Oe as measured by the bubble translation technique. The magnetization can be adjusted as low as 125 Gauss for a 6 μm bubble to 190 Gauss for a 4 μm bubble, which lowers the pulse power requirements for bubble devices by a factor of four compared to that needed by competing Sm containing garnets.

INTRODUCTION

Most garnet compositions which exhibit bubble stability over a wide temperature range involve a compromise between high mobility and high enough uniaxial anisotropy to produce a q > 3. (Ca^{2+},Ge^{4+}) substituted garnets have a nearly linear temperature variation of bubble collapse field over the temperature range -20°C to +80°C.[1]

Blank et al.[2] have reported that Lu^{3+} substitution on octahedral sites can be used to adjust the temperature characteristics of these garnets to match a particular barium ferrite bias magnet, which requires $H_T = (1/H_{col})(\partial H_{col}/\partial T) = -0.18$ to -0.20%/°C. The requirements for (1) matching the bubble garnet lattice parameter to that of the $Gd_3Ga_5O_{12}$ substrate and (2) increasing the uniaxial anisotropy so that q > 3, necessitates further substitution of a cation larger than Y^{3+} into the lattice. Suitable cations which have been tried are Sm^{3+}[2] and Eu^{3+}.[3] However, Sm^{3+} and Eu^{3+} have large damping parameters which reduce bubble mobility and limit device operation to ∿500 kHz. Nevertheless, they increase the anisotropy sufficiently so that q values greater than 14 can be generated in films with 6 μm bubbles. Most of the anisotropy in these films arises from pair ordering of the large and small rare earths (i.e., Sm^{3+} and Lu^{3+}).[4] Small contributions may come from (Ca^{2+},Ge^{4+}) and (Pb^{2+},Pt^{4+}) interactions as well as from lattice mismatch.

In all instances of La^{3+} doping reported to date,[5-8] the garnet composition has been Ga^{3+} substituted rather than (Ca^{2+},Ge^{4+}) substituted, and the reported anisotropies have been very low, resulting in q < 3. However, La^{3+} has a very low damping parameter and high mobilities should be expected in these materials. We have combined La^{3+} doping with (Ca^{2+},Ge^{4+}) substitution as well as a deliberate lattice mismatch to increase the anisotropy to an acceptable value via the stress induced mechanism.

CRYSTAL GROWTH AND PHASE EQUILIBRIA

The films were grown by the LPE dipping technique (horizontal rotation) in a supersaturated, isothermal PbO-based solution. The 3.8 cm diameter $Gd_3Ga_5O_{12}$ substrates were rotated at 100 rpm during growth in ∿100 cm^3 melts contained in Pt crucibles. The $Ca^{2+}/(Ca^{2+} + Ge^{4+})$ ratio was maintained at 0.59, and the $Fe^{3+}/(Lu^{3+} + La^{3+} + Y^{3+})$ ratio was kept at 25 for all the films grown. The solute concentration was varied from 17.5 mole% to 23.0 mole% to adjust the growth temperature between 930°C and 1000°C. Growth rates of 0.10 μm/min to 1.40 μm/min were employed. La^{3+} content of the film was varied from 0.02 to 0.20 atoms per formula unit by adjusting the solution composition. A representative melt composition and set of growth conditions are shown in Tables I and II.

TABLE I. Solution Composition for the Growth of $La_{.06}Lu_{.39}Y_{1.52}Pb_{.01}Ca_{1.02}Ge_{1.02}Fe_{3.97}Pt_{.01}O_{12}$ (film composition determined by microprobe analysis).

Component	Mole Percent
PbO	72.35
B_2O_3	4.64
Fe_2O_3	16.80
GeO_2	2.27
CaO	3.26
Y_2O_3	0.17
La_2O_3	0.29
Lu_2O_3	0.22

TABLE II. Growth Conditions and Properties of La-Doped Films.

Film Composition	A*	B*
Growth Temp (°C)	956	980
Growth Rate (μm/min)	0.22	0.71
Thickness (μm)	4.1	2.8
Stripe Width (μm)	6.11	3.85
Characteristic Length (μm)	0.79	0.49
Magnetization 4πM (Gauss)	145	190
Collapse Field (Oe)	57.0	79.8
Wall Energy (ergs/cm^2)	0.12	0.14
Néel Temp (°C)	186.4	191.5
Lattice Mismatch (Å)	-0.010	-0.005
H_T (%/°C)	-0.17	-0.17
ℓ_T (%/°C)	-0.19	-0.18
M_T (%/°C)	-0.26	-0.25
H_c (Oe)	0.36	-
Mobility (cm/s-Oe)	2700	-
A (ergs/cm)	1.9×10^{-7}	2.6×10^{-7}
H_a (Oe)	1030	617
K_u (ergs/cm^3)	5942	4664
q	7.1	3.2

* A - $La_{.06}Lu_{.39}Y_{1.52}Pb_{.01}Ca_{1.02}Ge_{1.02}Fe_{3.97}Pt_{.01}O_{12}$.

B - $La_{.19}Lu_{.84}Y_{.99}Pb_{.01}Ca_{.97}Ge_{.97}Fe_{4.02}Pt_{.01}O_{12}$.

The solubility of the garnet phase increases with increasing La^{3+} content as reported by Stein et al.[8] The solubility curves for films containing 0.1 and 0.2 atoms per formula unit (Lu^{3+}) are shown in Fig. 1.

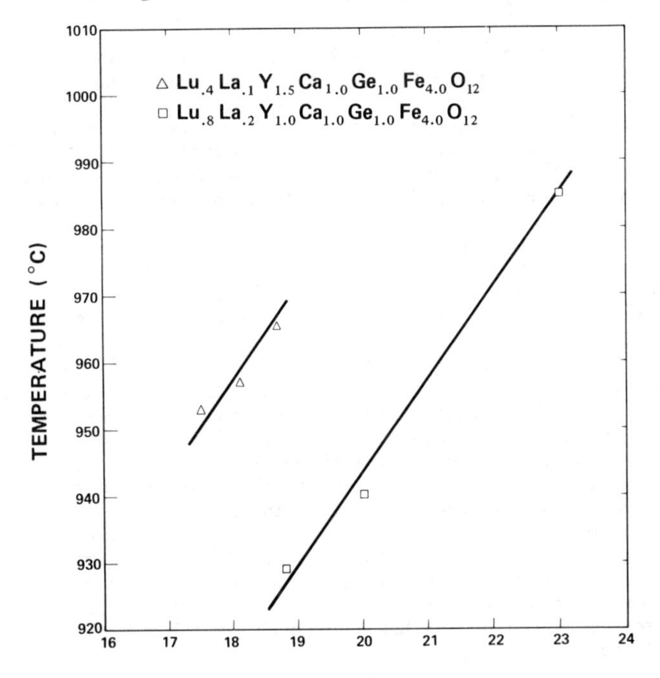

Fig. 1. Solubility of $(La,Lu,Y,Ca)_3(Ge,Fe)_5O_{12}$ in PbO-B_2O_3 solvent.

The growth rate is a linear function of supercooling (for 5-25°C supercooling), and the slope is ∿0.05 μm/min-°C at a solute concentration of 23 mole%.

The distribution coefficient for La^{3+} in this system was found by microprobe analysis to be in the range α_{La}† = 0.1 to 0.2, depending upon growth rate, La^{3+} content and temperature. α_{La} increases as growth rate, growth temperature and total La^{3+} in the solution increase.

RESULTS

The $(La,Lu,Y,Ca)_3(Ge,Fe)_5O_{12}$ garnets have been grown with properties suitable for bubble diameters 4 to 6 μm and defect densities < 5/cm^2. Representative films and their properties are shown in Table II. Double crystal x-ray measurements indicate the films are in tension with 0.005 to 0.01 Å lattice mismatch parallel to the film interface. The growth induced anisotropy accounts for a small portion of the total uniaxial anisotropy as evidenced by annealing studies in which the films were annealed in O_2 for four hours at 1250°C. K_u for film A in Table II was reduced upon annealing from 5942 ergs/cm^3 to 4223 ergs/cm^3. The lattice mismatch did not change. The anisotropy energy for film A was measured as a function of temperature before annealing and found to decrease at a rate of ∿ - 0.8%/°C in the temperature range 15 to 80°C.

Hard bubbles in these films were completely suppressed by ion implantation with 200 keV neon at a dosage of 3 x 10^{14}/cm^2. The coercive force measured by the oscillating stripe method[9] was less than 0.4 Oe for all the films measured, and bubbles exhibited mobilities > 2600 cm/s-Oe. The limiting velocity was greater than 3250 cm/s.

H_T is in the range -0.17 %/°C to -0.20 %/°C for temperatures of -20°C to +120°C, quite suitable for device applications. A 24 μm period circuit deposited on film composition A has been operated successfully and found to exhibit bubble nucleation and replication currents less than one-half that needed for the same circuit deposited on films of $(Lu,Sm,Y,Ca)_3(Ge,Fe)_5O_{12}$. The short term room temperature circuit tests at 100 kHz indicate adequate margins (∿10%) for all functions.

† α_{La} is defined as $\alpha_{La} = [La^{film}/(La + Lu + Y)^{film}]/(La^{soln}/(La + Lu + Y)^{soln}]$, where all quantities are molar.

CONCLUSIONS

$(La,Lu,Y,Ca)_3(Ge,Fe)_5O_{12}$ is a useful composition for magnetic bubble devices requiring a high data rate. Its relatively low $4\pi M$ lowers power requirements, and the room temperature q of 7 in compositions for 24 μm period circuits is adequate for bubble stability. As the characteristic length decreases, however, q decreases until it becomes marginal for successful operation of a 16 μm period device. Since the anisotropy is primarily stress induced in this composition, it is difficult to modify it without addition of another cation to increase the growth induced portion. Such an addition might include Co^{3+} or Sm^{3+} in sufficiently low concentrations so that the mobility is not significantly degraded.

ACKNOWLEDGMENTS

The author gratefully acknowledges the assistance of L. Small in the crystal growth of these films. Microprobe analysis and x-ray characterization results were supplied by T. Cass. Dynamic magnetic characterization and anisotropy measurements were performed by R. Waites and R. Lacey.

REFERENCES

1. J. W. Nielsen, S. L. Blank, D. H. Smith, G. P. Vella-Coleiro, F. B. Hagedorn, R. L. Barns and W. A. Biolsi, J. Electron. Mater. 3, 693 (1974).
2. S. L. Blank, J. W. Nielsen and W. A. Biolsi, "Preparation and properties of magnetic garnet films containing divalent and tetravalent ions," paper presented at Dallas Meeting, The Electrochemical Society, 5-9 October 1975, Dallas, Texas [J. Electrochem. Soc. 122, 263C (August 1975), Abstract No. 223].
3. W. A. Bonner, J. E. Geusic, D. H. Smith, L. G. Van Uitert and G. P. Vella-Coleiro, Mater. Res. Bull. 8, 1223 (1973).
4. A. Rosencwaig and W. J. Tabor, Sol. State Commun. 9, 1691 (1971).
5. J. Haisma, G. Bartels and W. Tolksdorf, Philips Res. Repts. 29, 493 (1974).
6. W. Tolksdorf, G. Bartels and P. Holst, J. Cryst. Growth 26, 122 (1974).
7. J. M. Robertson, W. Tolksdorf and H. D. Jonker, J. Cryst. Growth 27, 241 (1974).
8. B. F. Stein and M. Kestigian, J. Cryst. Growth 31, 366 (1975).
9. R. B. Clover, L. S. Cutler and R. S. Lacey, AIP Conf. Proc. No. 10, 388 (1973).

FACETING OF $(YEuYbCa)_3(GeFe)_5O_{12}$ ON LPE GARNET FILMS

H. Tominaga, M. Sakai, I. Hirai and K. Yamaguchi
FUJITSU LABORATORIES LTD., KAWASAKI, JAPAN

ABSTRACT

The correlations of the faceting on LPE films and the lattice mismatches between LPE films and GGG substrates were investigated, taking into consideration LPE growth conditions for $(YEuYbCa)_3(GeFe)_5O_{12}$ in a $PbO \cdot B_2O_3$ flux. The facets in this garnet system appeared, even when the mismatches were very small. In the case of a small mismatch, it was found that the tendency to faceting increased with increase in the lattice constants of LPE films and decrease in their growth rate. However, no defects such as a dislocation were observed in the faceting on LPE films using the X-ray reflection topography method. The lattice mismatches caused no harmful effects to the characteristics of bubble domains such as magnetic domain wall coercivity, domain wall energy and mobility, and it also did not effect magnetic bubble propagation.

INTRODUCTION

It has been reported[1] that lattice mismatches between the LPE films of the magnetic bubble garnets and GGG substrates cause the cracks or facets. Those defects were shown to be avoidable by controlling the lattice constants of the LPE films, so that lattice mismatches were small. The controlling method is to substitute other ions in rare earth or iron sites, such as $(ErEu)_3(GaFe)_5O_{12}$ and $(YEuYb)_3(A\ell Fe)_5O_{12}$.

However, for a $(YEuYbCa)_3(GeFe)_5O_{12}$ garnet [2], which had high wall mobility and a low temperature coefficient, the facets often appeared, even when the lattice constants of the LPE films and substrates were nearly the same. This suggests another cause of faceting in addition to lattice mismatches.

The habits of the garnet crystals grown by a flux method were found[3] to be changed by growth conditions and compositions of the melt and crystals. This suggests that faceting in LPE films is due to growth conditions. As the lattice mismatches were small in the order of ±0.005Å, the present experiment was performed to investigate the faceting on LPE garnet films of $(YEuYbCa)_3(GeFe)_5O_{12}$, from the point of view of growth conditions.

EXPERIMENTAL DETAILS AND RESULTS

Garnets of a nominal composition, $Y_{1.3}\ Eu_{0.2}\ Yb_{0.6}\ Ca_{0.9}\ Ge_{0.9}\ Fe_{4.1}\ O_{12}$, were grown on polished (111) GGG substrates under various growth conditions using the conventional LPE technique[4]. Melt compositions are listed in Table 1. The films were grown at temperatures between 860°C to 890°C and the growth rates were then changed from 1 μm/min to 3 μm/min. The occurrence of faceting and crystal perfection were observed and bubble domain characteristics measured, in selected LPE films with lattice constants in the range ±0.005Å. The surfaces of the LPE films were observed, using a differential interference microscope with x400 magnification, to determine the presence of facets. The facets are shown in Fig. 1 and Fig. 2.

The correlations between faceting and lattice mismatches, A_f-A_s, are shown in Fig. 3, 4 and 5, with regard to film thickness, dipping temperature and growth rate respectively. Where A_f and A_s are lattice constants of the films and the substrates respectively. Fig. 3 and Fig. 4 show that film thickness, dipping temperature and lattice mismatches are not directly concerned with the occurrence of facets. In Fig. 5, the growth rate and lattice mismatches are seen to be related to the occurrence of faceting. Faceting tends to occur with increase in the lattice constants of the LPE films, i.e., as the film becomes compressed, and decrease in their growth rate. The solid line in Fig. 5 represents the limits of occurrence of facets. Heare, the dipping temperature does not determine the growth rate uniquely, because many melts having various compositions were used and so the saturation temperature was changed.

The facets were observed in a thin LPE film grown for about 10 seconds. Therefore, the faceting should be discussed from the point of view of the lattice constants at the growth temperature. Since the thermal expansion coefficients of YIG and GGG are about $10.4 \times 10^{-6}/^{o}C$ and $9.2 \times 10^{-6}/^{o}C$ respectively, the lattice mismatch at growth temperature was found to be larger by 0.013Å than that at room temperature. The calculated values of YIG are shown in Fig. 5 as a broken line. Faceting is avoided by selecting the growth rate and the lattice constants of the LPE films at growth temperatures below the broken line in Fig. 5.

Distinctions between perfect and imperfect crystals, due to lattice mismatches and faceting, were not evident from the line width of the rocking curves, using the X ray double crystal method. No dislocations were observed on any LPE films using the X ray reflection topograph.

Table 1. Compositions of melts

Y_2O_3	2.00 g	Yb_2O_3	0.749 g ~ 1.749 g
Eu_2O_3	0.300 g ~ 0.401 g	$CaCO_3$	1.67 g
GeO_2	13.991 g	Fe_2O_3	23.0 g
B_2O_3	9.57 g ~ 11.0 g	PbO	383 g ~ 440 g

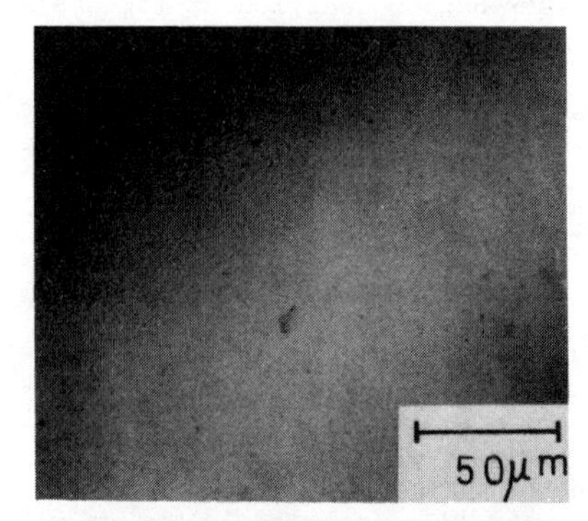

Fig. 1 Facets in the form of fine ripples.

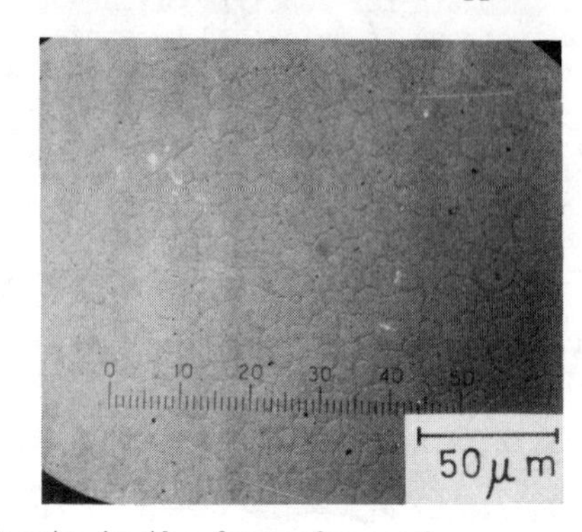

Fig. 2 Facets in the form of a mesh

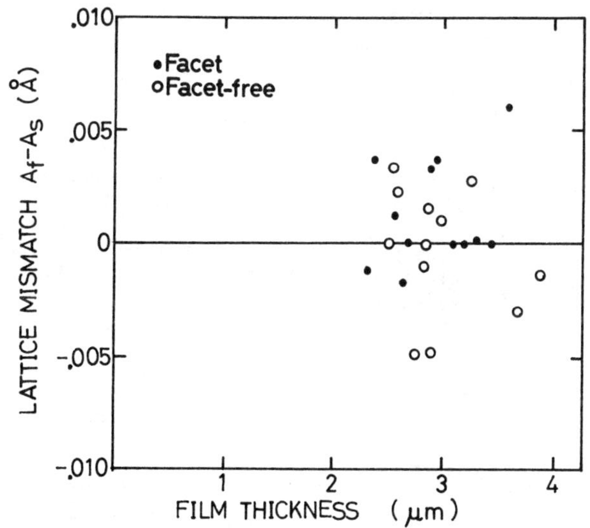

Fig. 3 Correlation of lattice mismatch, film thickness and faceting.

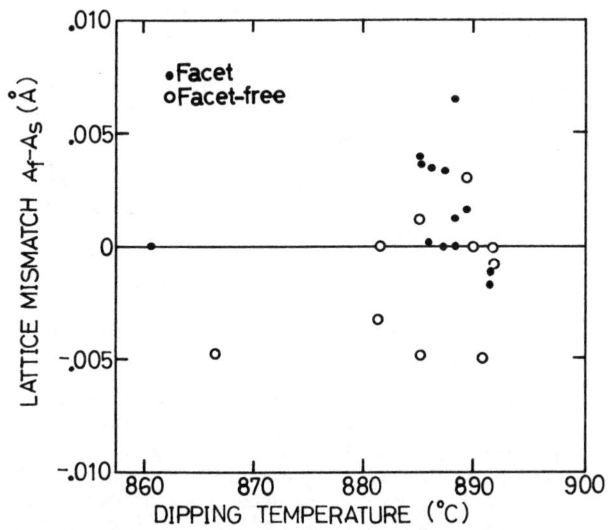

Fig. 4 Correlation of lattice mismatch, dipping temperature and film faceting.

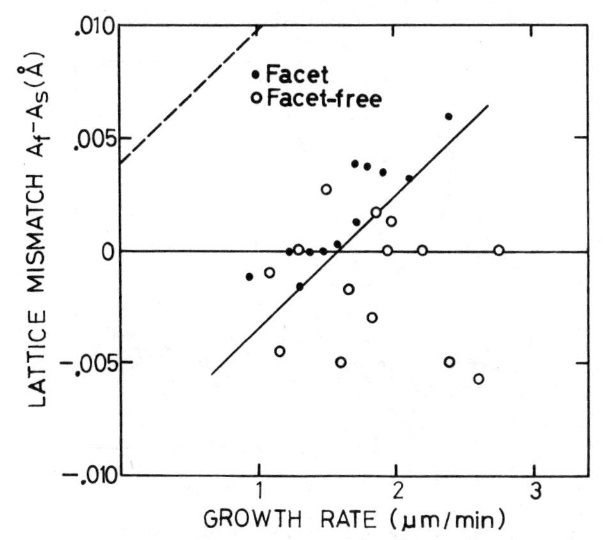

Fig. 5 Correlation of lattice mismatch, growth rate and faceting. The broken line represents lattice mismatches at a growth temperature calculated using the thermal expansion coefficient of YIG.

The effects of lattice mismatch and faceting on the characteristics of magnetic bubble domains were measured. On measuring the saturation magnetization, wall energy, domain wall mobility and coercivity, we found that they had scarcely changed. We can determine changes in the film composition from the saturation magnetization and the lattice constants. No correlation of the saturation magnetization and the faceting indicates that faceting was not influenced by crystal composition.

No change in domain wall energy, which is proportional to the square root of the magnetic anisotropy energy, indicates that the change in anisotropy energy is small by the occurrence of faceting and small lattice mismatches . Bobeck et al.[5] have found that anisotropy characteristics changed with different facets. However, magnetic anisotropy changes were not recognized in LPE films and no harmful influence to bubble propagation was observed.

The cause of faceting on LPE films is unknown. However, in previous wark[3], it was reported that the growth rate, which is proportional to super-saturation, plays an important role in determining the habits of crystals grown by the flux method; and that the crystal habits depend on the compositions of flux and crystals. It was explained that surface energy and surface diffusion rate were changed by the presence of a certain cation in the flux and changes in the compositions of crystals. In LPE growth, the occurrence of faceting may be due to the same reasons; and in epitaxial films, it may be possible to change the surface energy and the surface diffusion rate by the strain that is introduced from a lattice mismatch, and which causes faceting.

If the lattice mismatch produces dislocations in LPE films, it might be the reason for the occurrence of faceting. However, such a dislocation was not observed on the surface of the films, and crystal imperfection was not recognized. Neverthless, it is possible that lattice mismatches vary the surface mucleation energy and result in faceting.

ACKNOWLEDGMENT

The author wish to thank T. Mori for the film growth, and K. Asama, M. Nakamura, T. Namikata and F. Iwai for their valuable discussions and encouragement.

REFERENCES

1. S.L. Blank and J.W. Nielsen, J. Crystal Growth, 17, 302 (1971)
J.W. Matthews, E. Klokholm and T.S. Plaskett, AIP Conf. Proc., 10, 271 (1972)
2. T. Obokata, H. Tominaga, T. Mori and H. Inoue, AIP Conf. Proc. (to be published)
3. J.W. Nielsen and E.F. Dearborn, J. Phy. Chem. Solid, 5, 202 (1958)
A.B. Chase and J.A. Osmer, J. Crystal Growth, 5, 239 (1969)
4. E.A. Giess, J.D. Kuptsis, E.A.D. White, J. Crystal Growth, 16, 36 (1972)
5. A.H. Bobeck, D.H. Smith, E.G. Spencer, L.G. Van Uitert and E.M. Walters, Imtermag. Conf., 461 (1971)
L.G. Van Uitert, W.A. Bonner, W.H. Grodkiewicz, M.L. Pictroski and G.T. Zydzik, Mat. Res. Bull., 5, 825 (1970)

TEMPERATURE DEPENDENCE OF THE DYNAMIC PROPERTIES OF $S \approx 1$ BUBBLES IN YSmLuCaGe IRON GARNET FILMS*

R.E. Fontana, Jr. and D.C. Bullock
Texas Instruments Incorporated, P.O. Box 5936, M.S. 145, Dallas, Texas 75222

ABSTRACT

We have measured bubble velocity as a function of drive field for 5 micron diameter bubbles in Ne^+ ion implanted $(YSmLuCa)_3(FeGe)_5O_{12}$ films over a 115°C temperature range from -40°C to +75°C. From this translation data the bubble mobility, μ, and dynamic coercivity, H_c, were determined. For material with a room temperature mobility of 500 cm/sec-Oe, a mobility temperature coefficient of 1.1%/°C was calculated. Skew angle measurements were consistent with the identification of the $S \approx 1$ bubble state as the only stable wall configuration over this temperature range.

INTRODUCTION

Magnetic garnet films containing divalent and tetravalent ions have excellent temperature characteristics because of their high Curie temperatures[1]. The reported temperature variations of H_0, ℓ, $4\pi M$, and q[2,3] associated with such films suggest that bubble memory devices can operate at temperatures between -25°C and +75°C. In this work we have examined the temperature variations of bubble mobility, dynamic coercivity, and skew angle in Ne^+ ion implanted $(YSmLuCa)_3(FeGe)_5O_{12}$ films which support 5 micron diameter bubbles. This information is obviously necessary for predicting both the frequency limits and drive field requirements for field access devices which must operate over a temperature range of ±50°C about room temperature. Skew angle measurements give information on the stability of the bubble wall states over this temperature range.

EXPERIMENTAL PROCEDURE

Bubble translation experiments were performed using the technique of Vella-Coleiro and Tabor[4]. Two parallel current conductors 12.5 microns wide and separated by 100 microns were deposited on flat glass. The electrode pattern was clamped to the test sample and the entire assembly placed in a teflon temperature chamber which could be heated or cooled with N_2 gas.

Room temperature translation data obtained from a 2" diameter film are shown in Figure 1. The bubble velocity, V, and drive field, ΔH, are measured along

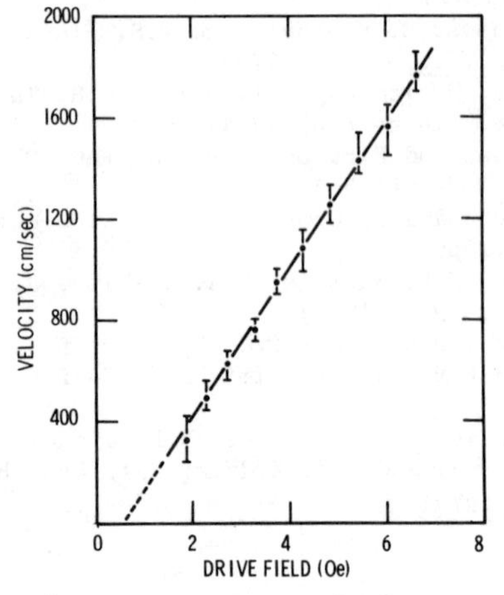

Figure 1. Room temperature velocity vs. drive field data for 5 micron diameter bubbles obtained from a 2" diameter $(YSmLuCa)_3(FeGe)_5O_{12}$ film. The mobility is 581 cm/sec-Oe. The dynamic coercivity is 0.22 Oe.

the skew direction. The drive field may be expressed as $|d\vec{\nabla} H| \cos\theta$ where θ is the skew angle, d is the bubble diameter, and $\vec{\nabla} H$ is the field gradient created by the electrode pattern. Mobility and dynamic coercivity are determined by fitting the data to the established relation[5]

$$V = \frac{\mu}{2}\left(\Delta H - \frac{8}{\pi} H_c\right) \tag{1}$$

From the data of Figure 1, a mobility of 580 cm/sec-oe and a coercivity of 0.22 Oe result. Repeatable bubble translations to within 1.5 Oe of the coercivity intercept were possible.

Velocity data points were determined from the average of four bubble translations, two to the left and two to the right, from the center line of the electrode pattern. The test bubble was always moved toward the center line in the same direction that the translation experiment would be taken. The typical spread in translation distances was ± 0.25 microns.

In order to minimize ballistic overshoot errors in our measurements, bubble translations were limited to one bubble diameter, 5 microns. Current pulse widths were kept below 0.75μs with the majority of data determined with pulse widths less than 0.5μs. The drive field was limited to constrain the bubble velocity well below the Slonczewski critical velocity[6], about 2100 cm/sec for this material.

TEMPERATURE RESULTS

The data presented in this section was taken on a $(YSmLuCa)_3(FeGe)_5O_{12}$ film with the following room temperature parameters: h = 5.1 microns, ℓ = 0.55 microns, q = 6.00, H_0 = 118 Oe, $4\pi M$ = 227 Gauss, μ = 500 cm/sec-Oe. The film was ion implanted with a Ne^+ dose of $2.5 \times 10^{14}/cm^2$ at 100 keV. Figure 2 shows the velocity vs. drive field data from -40°C to +75°C. The temperature dependence of mobility and dynamic coercivity are plotted in Figs. 3 and 4. A mobility temperature coefficient, referenced to room temperature, of 1.1%°C was calculated from the data. For 100kbit devices fabricated in our laboratory with this material, the mobility decrease by a factor

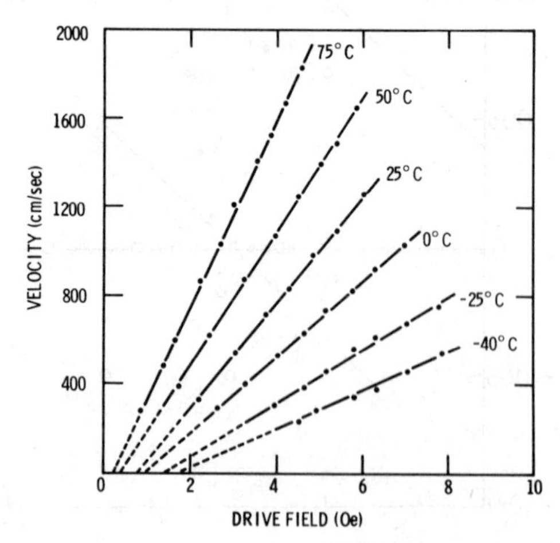

Figure 2. Velocity vs. drive field data of 5 micron diameter bubbles in a $(YSmLuCa)_3(FeGe)_5O_{12}$ film taken at 75°C, 50°C, 25°C, 0°C, -25°C, and -40°C.

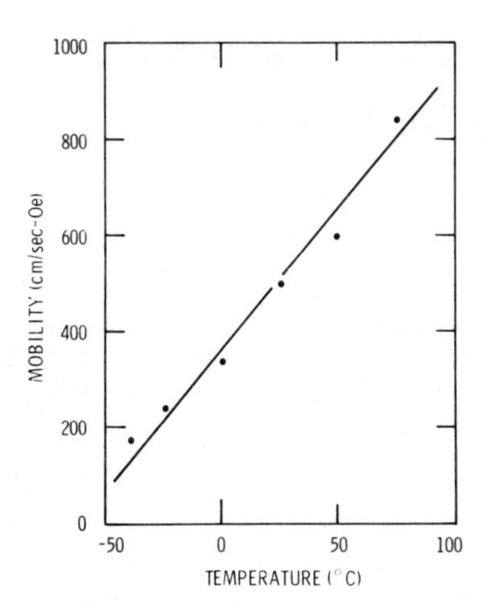

Figure 3. Temperature dependence of mobility for 5 micron diameter bubbles in a $(YSmLuCa)_3(FeGe)_5O_{12}$ film with room temperature mobility of 500 cm/sec-Oe.

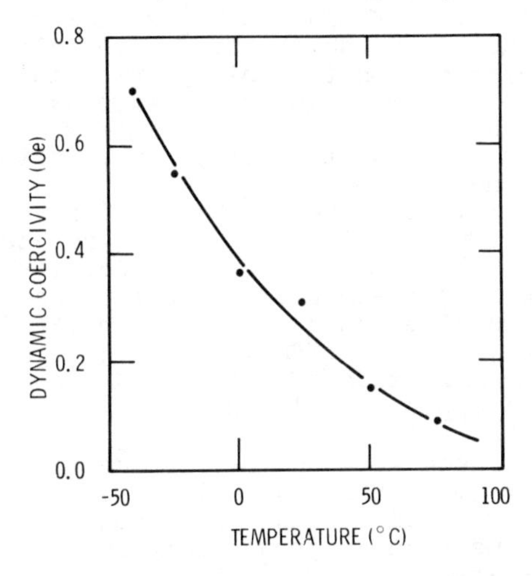

Figure 4. Temperature dependence of dynamic coercivity for 5 micron diameter bubbles in a $(YSmLuCa)_3(FeGe)_5O_{12}$ film with room temperature mobility of 500 cm/sec-Oe.

of two from +25°C to -25°C limited the maximum drive field frequency to 300 KHz. A 50 Oe in-plane drive field was required to sustain this data rate from -25°C to +75°C[7].

The bubble skew angle varied from 5° to 22° in this film. Only one deflection angle was observed at any given temperature indicating that the ion implant dosage effectively eliminated hard bubbles over this 100°C temperature range. However no S = 0 bubbles were observed. Using the translation data and Slonczewski's[8] formula for skew angle,

$$\sin \theta = \frac{8n \, |v|}{\gamma d |\vec{\nabla} H|} \qquad (2)$$

where n is the bubble wall state and γ is the gyromagnetic ratio, $1.76 \times 10^7 \ Oe^{-1} \ sec^{-1}$, the stable wall state was identified as S = 1. The data for these calculations are shown in Table 1. All measurements were made with the bubble diameter set at 4.88 microns. We consider the scatter in n to be within the accuracy of our measurements. However, if ballistic overshoot errors are present and velocities greater than actual velocities are being measured, then all the values for n would be shifted toward the n = 1.5 regime.

TABLE I
SKEW ANGLE DATA

T° (C)	θ(°)	|v| (cm/sec)	$\|d\vec{\nabla}H\|$ (Oe)	n
-25	5.6	372	4.53	1.29
0	7.1	378	3.15	1.11
25	11.6	546	3.15	1.25
50	15.0	786	3.15	1.11
75	22.5	1147	3.15	1.12

DICUSSION

Though the static parameters of garnet films containing divalent and tetravalent ions have temperature coefficients in the 0.1 - 0.25%/°C range the mobility and coercivity vary at rates 4 to 10 times greater. Film compositions must be engineered with this point in mind. To some extent, data rates can be maintained at low temperatures by increasing room temperature mobility. This is accomplished at the expense of increased skew angle and a decrease in q at higher temperatures.

Skew angle measurements indicate that only one bubble state is stabilized over this 100°C temperature range over which an almost four fold increase in mobility occurs. No wall state conversions, as evidenced by no changes in skew angle with in-plane d.c. bias field, were observed. This feature together with the bubble wall state data indicates that the wall state in $(YSmLuCa)_3(FeGe)_5O_{12}$ has a simple structure.

ACKNOWLEDGEMENTS

The authors thank G.G. Sumner for growing the epitaxial films used in this work, W.R. Cox for the static characterization of the material. The authors are grateful for their many discussions with F.G. West.

REFERENCES

* This work partially supported by Air Force Avionics Laboratory.
1. S.L. Blank, J.W. Nielsen, W.A. Biolsi, to be published in the Journal of Electro-Chemical Soc.
2. G.G. Sumner, W.R. Cox, Paper 3A-4, This Conference.
3. J.W. Nielsen, S.L. Blank, D.H. Smith, G.P. Vella-Coleiro, F.B. Hagedorn, R.L. Barnes, W.A. Biolsi, Journal of Electronic Materials, 3 693 (1974).
4. G.P. Vella-Coleiro, W.J. Tabor, Appl. Phys. Lett. 21 7 (1972).
5. A.A. Thiele, Bell System Technical Journal, 50 711 (1971).
6. J.C. Slonczewski, J. Appl. Phys., 44 1759 (1973).
7. S. Singh, Paper 8A-5, This Conference
8. J.C. Slonczewski, A.P. Malozemoff, O. Voegeli, AIP Conf. Proc. 5 458 (1972).

GROWTH-INDUCED ANISOTROPY IN BUBBLE GARNET FILMS CONTAINING CALCIUM

R. Wolfe, R. C. LeCraw, S. L. Blank and R. D. Pierce
Bell Laboratories, Murray Hill, New Jersey 07974

ABSTRACT

The growth-induced anisotropy, K_u measured on (111) LPE films of $Eu_{3-x}Ca_xSi_xFe_{5-x}O_{12}$ grown on $Sm_3Ga_5O_{12}$ substrates is large and positive (K_u = 4.0 to 8.1×10^4 ergs/cm^3 for x = 0.5). Many experiments have shown that K_u in $Eu_{3-x}Y_xFe_{5-y}Ga_yO_{12}$ films is also positive. In films of compositions $(YSmCa)_3(GeFe)_5O_{12}$, the effects of Ca and of Y on the growth-induced anisotropy are the same in sign and similar in magnitude. Since Ca^{2+} is larger in ionic radius than Eu^{3+} or Sm^{3+}, whereas Y^{3+} is smaller, the simple site selectivity model of growth-induced anisotropy would predict opposite signs of K_u for the Ca and Y additions. This contradiction shows that ionic radius is not the only relevant parameter which determines growth-induced anisotropy in garnets, although size differences have been shown to account for the observed effects within the rare earth series. The charge difference between the Ca^{2+} ions and the rare earth ions, or the pairing of Ca^{2+} and Ge^{4+} or Si^{4+} ions during crystal growth may influence the site preference of the Ca ions and the resulting anisotropy.

The growth-induced anisotropy of the $Eu_{2.5}Ca_{0.5}Si_{0.5}Fe_{4.5}O_{12}$ film, grown at 835°C at a rate of 2 μm/min, annealed out after several hours at 900°C in oxygen. This temperature is much lower than the usual annealing temperatures of 1200°C to 1300°C in O_2 which are required to reduce K_u in other bubble garnets. This is attributed to a high concentration of defects due to an imbalance in the Ca-Si concentrations in films grown at low temperatures and large growth rates.

INTRODUCTION

Growth-induced anisotropy in magnetic garnets containing two or more rare earths has been shown to arise from the nonrandom distribution of the rare earth ions on the dodecahedral sites of the garnet lattice.[1,2] During the crystal growth process, certain ions are incorporated in greater-than-average numbers on particular sites which are distinguished from the remaining sites by their configuration with respect to the growth interface. The overall cubic symmetry of the crystal is distorted by this nonrandom site occupancy and the large anisotropies of the individual rare earth ions do not average out to zero. For (111) LPE film growth, the resultant anisotropy is either uniaxial (positive anisotropy energy K_u) or planar (negative K_u).

In experiments on mixed rare earth aluminum garnets, large differences in site occupancy were observed by spin resonance measurements,[3] and it was suggested that the ionic radius was the most likely variable which determines the site selectivity. If a rare earth ion, larger in radius than the host ions, preferred certain dodecahedral sites, then ions smaller than the host ions were shown to avoid these sites. Experiments on the anisotropies of (110) facet material of a variety of mixed rare earth iron garnets[4] again were consistent with this model in which site preferences depend on the ionic radius mismatch.

The garnet compositions which have recently been most favored for bubble device applications contain Si or Ge ions as substitutes for iron rather than Ga or Aℓ.[5,6] These tetravalent ions prefer the tetrahedrally coordinated iron sites much more strongly than the trivalent Ga and Aℓ, which also populate some of the octahedral sites. As a result, less substitution is required for a given room temperature magnetic moment; the Curie temperature is higher and the temperature variations of the magnetic properties are improved. Along with the tetravalent substitutions, an approximately equal concentration of divalent ions is added to the garnet to preserve charge neutrality. The preferred ion for this purpose is Ca, a large ion which occupies the rare earth dodecahedral sites. It is therefore of interest to study the effect of this ion on growth-induced anisotropy.

Additions of Ca and Ge or Si have been made to YEu garnets[5] and to YSm garnets[6] to make useful bubble device materials. In both systems, growth-induced anisotropy was found to be positive, just as it was in $(YEu)_3(FeGa)_5O_{12}$[7,8] and $(YSm)_3(FeGa)_5O_{12}$.[9] The magnitude of positive K_u in each case was approximately the same in the CaGe materials as in the corresponding Ga-substituted garnets with the same concentration of Eu or Sm. (Exact quantitative comparisons are not feasible because K_u depends on growth conditions such as growth temperature and growth rate.)

The relevant ionic radii for ions on the dodecahedral garnet sites are as follows:[10]

Ion	Radius (Å)
Y^{3+}	1.015
Eu^{3+}	1.07
Sm^{3+}	1.09
Ca^{2+}	1.12

In spite of its large radius, Ca^{2+} appears to have the same effect on growth-induced anisotropy as Y^{3+} in these systems. To study this effect, garnet compositions with a single rare earth and Ca^{2+} rather than both Y^{3+} and Ca^{2+} were investigated.

FILM GROWTH

The anisotropy of (111) LPE films of $(Eu_{3-x}Y_x)(Ga_yFe_{5-y})O_{12}$ has been shown to be growth induced in origin and positive in sign.[7,8] For direct comparison, the system $Eu_{3-x}Ca_xSi_xFe_{5-x}O_{12}$ was studied in the present investigation. These compositions without Y have lattice constants which are too large to match the usual $Gd_3Ga_5O_{12}$ substrates. Substrates of $Sm_3Ga_5O_{12}$ were therefore used (lattice constant a_o = 12.438 Å).

Samples of nominal composition $Eu_{2.5}Ca_{0.5}Si_{0.5}Fe_{4.5}O_{12}$ were grown by the LPE method. The garnet oxide components were dissolved in a 50/1 weight ratio PbO/B_2O_3 flux. Additions of flux were made to the melt so that films could be grown at successively lower temperatures in the range 940°C to 835°C. Films were grown isothermally at 20° to 50° supercooling, with growth rates between 0.8 and 2 µm/min. The (111) samarium gallium garnet substrates were held in a vertical position by a Pt-5% Au wire and were rotated unidirectionally at 60 rpm during growth.

Electron microprobe analysis showed that the Pb content in these films was very small, even for the films grown at the lowest temperature.

RESULTS

The anisotropy field for each film was determined by ferromagnetic resonance measurements at 13.15 GHz with the applied dc magnetic fields parallel and perpendicular to the film plane.[11] This was combined with the magnetization obtained by standard bubble collapse and strip width measurements to determine the anisotropy energy K_u. The anisotropies were all positive, with room temperature values of K_u ranging from $+4.0 \times 10^4$ ergs/cm^3 for a film grown at 940°C to $+8.1 \times 10^4$ ergs/cm^3 for a film grown at 835°C.

The variation of K_u with temperature is shown in Fig. 1 for a film grown at 835°C at a rate of 2 µm/min. The anisotropy of this film as grown was 7.3×10^4 ergs/cm^3. It was reduced by a factor of three by annealing (see below) so that a reasonable temperature range could be covered before K_u became too large to measure with the available microwave spectrometer. The Curie temperature of this film was 530°K and the magnetization at room temperature was: $4\pi M_s = 410$ G.

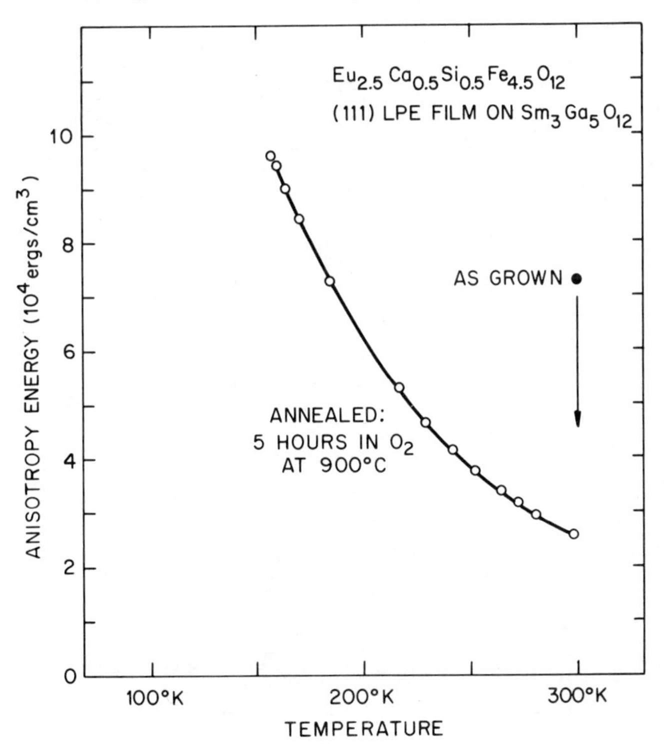

Fig. 1. Variation of anisotropy energy with temperature for a (111) LPE film of $Eu_{2.5}Ca_{0.5}Si_{0.5}Fe_{4.5}O_{12}$ on a substrate of $Sm_3Ga_5O_{12}$. The film was annealed to reduce K_u from its large as-grown value. Anisotropies were measured by ferromagnetic resonance and were not corrected for cubic and stress-induced components.

The K_u values in Fig. 1 are not corrected for the cubic anisotropy contribution, K_1 or the small stress induced component of K_u which were not measured over the temperature range. It is therefore not feasible to analyze this temperature variation in detail. However, the very rapid increase in K_u with decreasing temperature and the large magnitude of K_u indicate that the growth-induced anisotropy is associated with the Eu ions and not with the Fe ions.[12]

To verify that the observed anisotropies were growth induced, the films were annealed in an oxygen atmosphere. As shown in Fig. 2, rapid annealing took place at 900°C for the film grown at 835°C. (The residual anisotropy includes the cubic anisotropy and a small stress-induced component; the film lattice constant was well matched to the substrate both before and after annealing). This temperature is surprisingly low when compared with typical oxygen annealing temperatures of 1200°C to 1300°C previously reported for a variety of mixed garnets.[13]

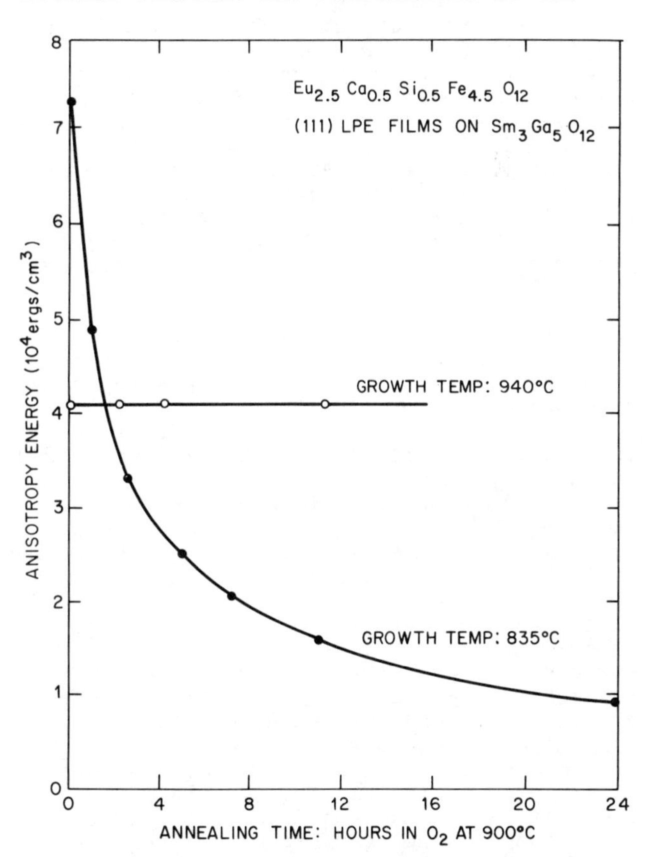

Fig. 2. Effect of annealing at 900°C on the anisotropy energy of $Eu_{2.5}Ca_{0.5}Si_{0.5}Fe_{4.5}O_{12}$ films. The film grown at 835°C at a rate of 2 µm/min. annealed rapidly at this unusually low temperature. The film grown at 940°C at a rate of 0.8 µm/min required a temperature of 1100°C for a similar annealing rate.

For films of similar composition, the annealing rate was lower for films grown at slower rates or higher temperatures. A film grown at 940°C at a rate of 0.8 μm/min showed no change in K_u after a few hours at 900°C in O_2, but large decreases in K_u were observed at 1100°C in O_2 in several hours.

DISCUSSION

The positive values of K_u found in the Ca-containing films contradicts the simplest model of growth-induced anisotropy in which ionic radius alone determines the site selectivity of the ions on the dodecahedral sites. Although this size criterion is consistent with the observed effects for mixed rare earth systems, other ionic properties such as electronic charge could be of greater importance when ions from different groups are involved. The naive assumption that the individual cations are present in the melt, and choose those sites on the growing crystal surface which are most compatible with their radius, is obviously oversimplified. The melt must contain more complex ions and molecules, the simplest of which would be a single cation combined with one or more oxygen ions. The complexes involving any of the rare earths or Y would be expected to be very similar. However, those involving divalent calcium could be quite different in shape and size, and their behavior as building blocks during crystal growth would not be related to that of the rare earth oxide complexes. The larger Ca^{2+} ions could therefore prefer the same sites which are selected by the smaller Y^{3+} ions. The pairing of Ca^{2+} ions with the charge compensating Si^{4+} or Ge^{4+} ions during crystal growth could also influence the Ca site preference. On the other hand, if Ca^{2+} and Y^{3+} do have opposite site selectivities, their effects on the local crystal fields felt by the Eu or Sm ions could be sufficiently different to result in the same sign for K_u.

The low annealing temperature observed for the $(EuCa)_3(SiFe)_5O_{12}$ films grown near 835°C is very unusual. It cannot be explained in terms of the larger lattice constant of these films compared with those grown on gadolinium gallium garnet substrates, because other compositions grown on even larger substrates such as $Nd_3Ga_5O_{12}$ show normal annealing behavior.[14] Other garnet compositions containing Ca and Si[15] required higher annealing temperatures, but these were grown at temperatures well above 900°C. The most likely explanation for the low annealing temperature is that the films grown at the lower temperatures and higher growth rates contain more defects, such as cation or oxygen vacancies, due to an imbalance in the Ca-Si concentrations. Further experiments to test this hypothesis are in progress.

ACKNOWLEDGMENT

We are grateful to F. B. Hagedorn, J. W. Nielsen and M. D. Sturge for valuable discussions, and to W. A. Biolsi for growing the LPE films.

REFERENCES

1. A. Rosencwaig and W. J. Tabor, AIP Conf. Proc. No. 5, p. 57 (1971).
2. H. Callen, Appl. Phys. Lett. 18, 311 (1971).
3. R. Wolfe, M. D. Sturge, F. R. Merritt and L. G. Van Uitert, Phys. Rev. Lett. 26, 1570 (1971).
4. E. M. Gyorgy, M. D. Sturge, L. G. Van Uitert, E. J. Heilner and W. H. Grodkiewicz, J. Appl. Phys. 44, 438 (1973).
5. W. A. Bonner, J. E. Geusic, D. H. Smith, L. G. Van Uitert and G. P. Vella-Coleiro, Mat. Res. Bull. 8, 1223 (1973).
6. J. W. Nielsen, S. L. Blank, D. H. Smith, G. P. Vella-Coleiro, F. B. Hagedorn, R. L. Barns and W. A. Biolsi, J. Electronic Materials 3, 693 (1974); S. L. Blank, J. W. Nielsen and W. A. Biolsi, to be published, J. Electrochem. Soc. June, 1976.
7. E. A. Giess, B. E. Argyle, B. A. Calhoun, D. C. Cronemeyer, E. Klokholm, T. R. McGuire and T. S. Plaskett, Mat. Res. Bull. 6, 1171 (1971).
8. F. B. Hagedorn, S. L. Blank and R. L. Barns, Appl. Phys. Lett. 22, 209 (1973).
9. F. B. Hagedorn and B. S. Hewitt, J. Appl. Phys. 45, 925 (1974).
10. R. Shannon and C. T. Prewitt, Acta. Crystallogr., Sect. B 25, 925 (1969).
11. R. C. LeCraw and R. D. Pierce, AIP Conf. Proc. No. 5, p. 200 (1971).
12. M. D. Sturge, R. C. LeCraw, R. D. Pierce, S. J. Licht and L. K. Shick, Phys. Rev. B 7, 1070 (1973).
13. A. J. Kurtzig and F. B. Hagedorn, IEEE Trans. Magnetics MAG-7, 473 (1971).
14. R. C. LeCraw, private communication.
15. R. C. LeCraw, S. L. Blank, G. P. Vella-Coleiro and R. D. Pierce, to be published in Proc. 21st Ann. Conf. on Magnetism and Magnetic Materials, Philadelphia, Dec. 1975.

MAGNETOCRYSTALLINE ANISOTROPY IN (100) LPE EUIG:Pb FILMS

E.Klokholm,* T.S.Plaskett and D.C.Cronemeyer
IBM Thomas J.Watson Research Center, Yorktown Heights
New York 10598

ABSTRACT

Previous work has shown that the incorporation of 5 wt.% Pb in (100) LPE EuIG films results in a large positive uniaxial anisotropy. Similarly, it has been observed that Pb causes large changes in the magnetocrystalline anisotropy constant, K_1. On the tensile side of the lattice mismatch, Δa, (0-1.5% Pb), the magnitude of K_1 increases with decreasing Δa until at $\Delta a = 0.0026$, K_1 attains a value of $-42,800$ ergs/cm^3. On the compressive side, $\Delta a < 0$, (Pb > 1.5%) the magnitude of K_1 decreases with increasing compressive Δa, reaching a value of -17500 ergs/cm^3 at $\Delta a = -0.013$ (5 wt% Pb). There is a discontinuity in K_1 near $\Delta a = 0$. Approaching $\Delta a = 0$ from the tensile side, K_1 extrapolates to -45000 ergs/cc, while the extrapolated value from the compressive side is -27500 ergs/cm^3. Theory predicts a linear behavior for K_1 with elastic strain; the proportionality constant being related to the magnetostriction. The observed values of K_1 with lattice mismatch are non-linear and much larger than the values predicted by the theory. It appears therefore that the variation of K_1 with Pb content in (100) LPE EuIG films is not caused by epitaxial strains, but by a structural modification of the lattice.

INTRODUCTION

Previous work[1] has shown that the incorporation of Pb in LPE EuIG (100) films causes unusual variations in the lattice mismatch and also results in large positive uniaxial anisotropies. In contrast to other LPE garnet film compostions, the lattice mismatch of EuIG as a function of Pb content is decidedly non-linear;[2] a discontinuity is observed at the Pb content for 0 lattice mismatch. For Pb contents > 1.5% in the EuIG films a large positive uniaxial anisotropy is observed; it is much larger than predicted by magnetoelastic effects or from the theory of growth anisotropy. The present work will report on the data observed for the effect of the Pb on the magnetocrystalline anisotropy constant, K_1. The results show an analogous behavior to the dependence of the uniaxial anisotropy upon the Pb content; namely, the magnetocrystalline constant, K_1, changes markedly with the incorporation of Pb during film deposition.

EXPERIMENTAL

The films were grown isothermally by the LPE tipping process from a $PbO:B_2O_3$ flux.[3] The films were deposited on (100) chemically-mechanically polished NdGG substrates. The film thicknesses were between 0.6 and 1.8 µm; the deposition rates were between 0.1 and 0.4 µm/min. The Pb content of the films was varied by changing the deposition temperature. The Pb content was measured by electron probe analysis.

The lattice mismatch, Δa, is defined as the lattice parameter of the substrate minus the undistorted lattice parameter of the film. A positive Δa indicates a film in tension and a negative Δa is for a film in compression. From the lattice parameters published for pure EuIG (12.498 A) and NdGG (12.507 A),[4] it is apparent that a pure EuIG film on NdGG is in tension ($\Delta a = 0.009$). The Δa was calculated from the differences in the lattice spacings of the film and substrate as measured by the Bragg-Brentano x-ray diffraction technique.[5] A Poisson's ratio of 0.3 was used to calculate Δa and the films were assumed to be pseudomorphic.

For this series of EuIG films, the uniaxial anisotropy constant, K_u, and K_1 were obtained from the analysis of ferromagnetic resonance, FMR, data. The samples, small (100) disks about 3 mm diameter, were placed in a cavity such that the specimens could be rotated in the plane of the film. In this manner the resonance field was measured about every five degrees as the orientation was changed from the [100] to the [110] direction. The resonance field was also measured for the magnetic field in the direction normal to the film plane, i.e., the [001] direction. For cubic ferromagnetic crystals of this orientation, in which there has been induced a uniaxial anisotropy, and also contains a crystalline contribution to the anisotropy, the resonance conditions are:

$$\omega/\gamma = H_1 + H_k + H_u - 4\pi M \tag{1}$$

for the field applied normal to the specimen plane, and

$$(\omega/\gamma)^2 = [H_2 - H_u + 4\pi M + H_k \cos 4\phi] \times [H_2 + H_k/4(3 + \cos 4\phi)] \tag{2}$$

for the field applied in the specimen plane. In Eqs. 1 and 2, H_1 is the resonance field for the perpendicular mode and H_2 is the resonance field in the film plane. The angle ϕ is measured from the [100] direction to the

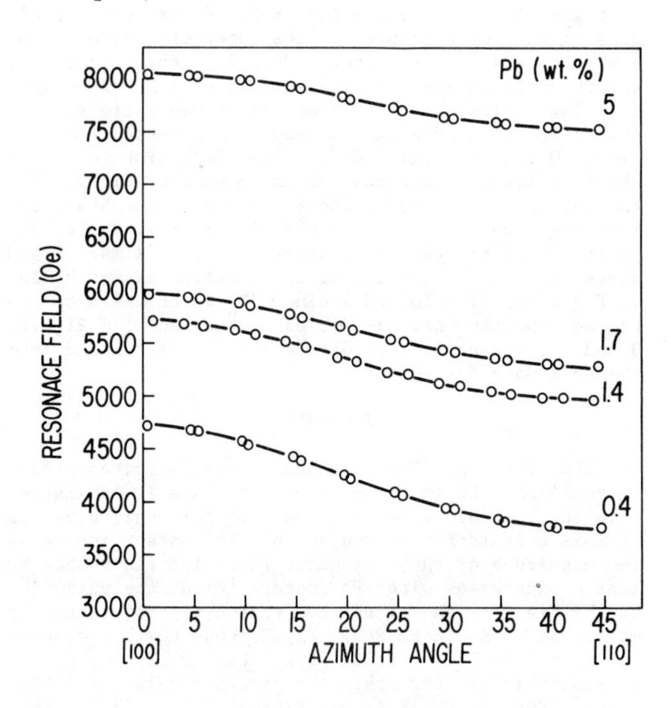

Fig. 1. In-plane resonance field as a function of the azimuth angle from the [100] to the [110] direction for specimens of varying Pb content.

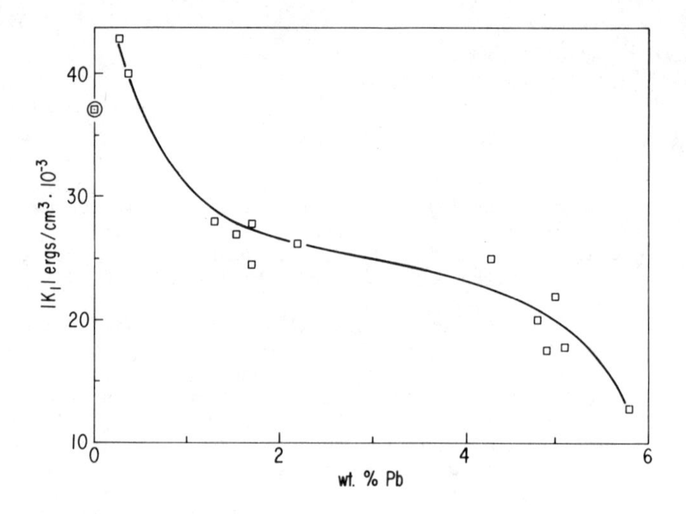

Fig. 2. Magnitude of the magnetocrystalline constant, K_1, as a function of Pb content. The encircled data point is for the value of K_1 for bulk EuIG.[4]

[110] direction. The resonance frequency, about 9 GHz, is designated by ω (rad/sec) and γ is the gyromagnetic constant. The term H_u is defined by $H_u = 2K_u/M$ and similarly $H_k = 2K_1/M$ where M is the saturation magnetization. For this work the bulk[4] value of 1172 gauss at 30°C for EuIG was used for $4\pi M$.[4] The values for K_u and K_1 were obtained from a computer program which fitted the experimental data to Eqs. 1 and 2.[3] All of the data reported for K_u and K_1 are in ergs/cm^3.

RESULTS

Figure 1 shows the in-plane FMR results for 4 specimens of varying Pb content. The uppermost curve is for 5 wt% Pb, and for the other curves in descending order, the Pb contents are 1.7, 1.4, and 0.4 wt % respectively. The ordinate is the resonance field while the abscissa is for the azimuthal angle ϕ measured from the [100], 0°, to the [110], 45°, direction. FMR data was obtained for all specimens in the manner of Fig.1. The FMR data was analyzed using Eqs.1 and 2 to obtain the results shown in Figs. 2, 3 and 4. In Fig. 2, $|K_1|$ is plotted against the Pb content of the films; Fig.3 shows $|K_1|$ as a function of the lattice mismatch, Δa. In Fig.4, $|K_1|$ is plotted against K_u which was also obtained from the analysis for each specimen. In Figs.2, 3 and 4, the encircled[4] data point is calculated from the bulk data for EuIG.[4]

DISCUSSION

The data of Fig.1 shows three important features; first, the in-plane resonance data is consistent with the fourfold symmetry required for cubic crystals of this orientation; second, as the Pb content increases the magnitude of the resonance field increases showing that K_u increases with Pb content; third, the variation of the field with ϕ indicates that K_1 persists as the magnitude of K_u increases. Fig.2 shows how $|K_1|$ appears to vary with Pb content. For about 0.3% Pb, $|K_1|$ is considerably greater than the reported value for bulk EuIG.[4] For 5% Pb $|K_1|$ has decreased to 17500. This overall change in $|K_1|$ is much greater than the changes expected for the equivalent substitution of Y on the rare earth sites or the substitution of an equivalent amount of Ga on the iron sites. It seems therefore that this decrease in $|K_1|$ with increasing Pb content cannot be caused by a simple substitution of Pb for the atoms occupying the metal sites in the garnet lattice. There is experimental and theoretical evidence that K_1 can be altered by[6] lattice strains. In particular there is published data[6] for the effect of hydrostatic pressures on K_1. The pressures used are of the same magnitude as those caused by Δa. In Fig.3 $|K_1|$ is shown as a function of Δa. These data show three distinct regions. For the films in tension, $|K_1|$ increases with decreasing Δa, that is with increasing Pb. If the curve is extrapolated to Δa = 0 a value of 45000 for $|K_1|$ is obtained. This is much greater than the reported value for bulk EuIG. At Δa ≃ 0 there is a discontinuous decrease in K_1 to about 28000. However, as the Pb content increases Δa remains ≃ 0, but K_1 decreases to about 25000. As Δa becomes compressive, $|K_1|$ decreases with increasing compressive Δa. The theory for the change in K_1, ΔK_1, with elastic strain as derived by Carr[7] gives the following relation

$$\Delta K_1 = 2\ \Delta a/a\ E/(1-v)(h_4/3 - h_3) \qquad (3)$$

for (100) films. In this equation Δa/a is the lattice strain caused by the lattice mismatch, Δa ; E is Young's modulus and v Poisson's ratio of the film. The terms in h are from the five constant description of the magnetostriction[6]; h_3 is explicitly the volume magnetostriction. Eq.3 predicts a linear dependence of ΔK_1 on Δa. From the available[6] data on hydrostatic pressure experiments,[6] h_3 is small; no data is available for h_4; however, the combination of these terms should not be larger than 10^{-7}. The largest elastic strain is about 10^{-3}. Using 3×10^{12} dyn/cm^2 for E/1-v and the above values for the strain, h_3 and h_4, the expected value for ΔK_1 is about 600. This is much less than the changes in ΔK_1 shown in Fig 3. Further, the changes in $|K_1|$ with Δa of Fig. 3 are decidedly non-linear and in fact appear to depend upon the sign of Δa. The magnitude of $|K_1|$ increases with decreasing tensile Δa and decreases with increasing compressive Δa. A comparison of the data of Fig. 3 and the predictions of Eq. 3 leads to the conclusion that the observed changes in $|K_1|$ are not due to lattice mismatch strains alone.

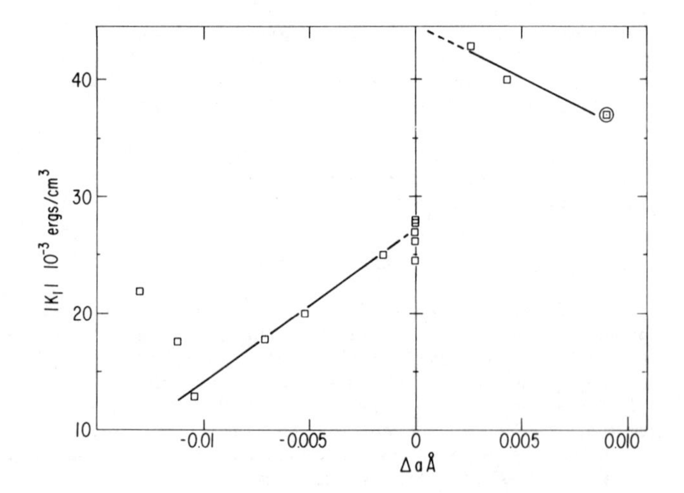

Fig. 3. Magnitude of the crystalline constant, K_1, as a function of the lattice mismatch, Δa. The encircled data point is for the bulk value of K_1 and the computed value of the lattice mismatch for 0 Pb content.[4]

Finally Fig. 4 shows $|K_1|$ plotted against K_u and the explicit dependence of K_1 and K_u upon either Pb or Δa are removed. For simple substitutional dilutions of the garnet lattice a plot of $|K_1|$ against K_u should show a smoothly varying curve. This is not the case for Fig. 4 where there is a distinct discontinuity in $|K_1|$ near $K_u = 0$; approaching $K_u = 0$ from the negative side $|K_1|$ extrapolates to 45000, while from the positive side K_1 extrapolates to 25000. This discontinuity mirrors the previously reported discontinuity in Δa vs Pb content[2] and also the data shown in Fig. 3. None of these data can be the result of a simple dilutional substitution of Pb in EuIG.

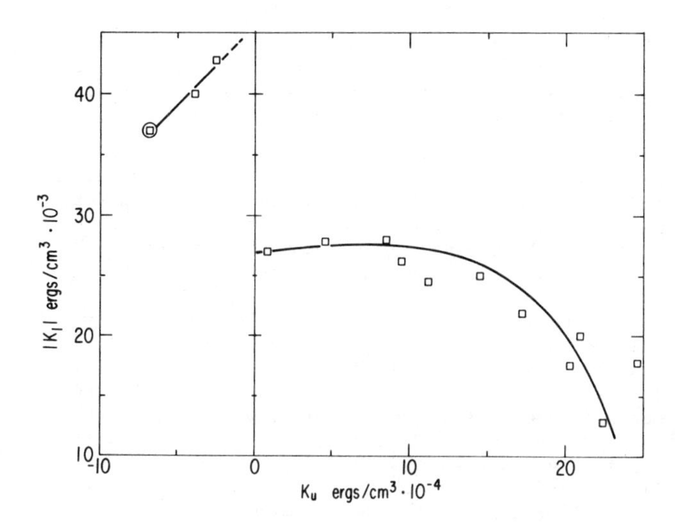

Fig. 4. Magnitude of the crystalline anisotropy constant, K_1, as a function of the induced uniaxial anisotropy constant, K_u. The encircled data point is for the published value of K_1 and the computed value of K_u using published values of the [100] magnetostriction constant and lattice mismatch for 0 Pb content.[4]

CONCLUSION

The data of this report shows:

1. The FMR data of Fig.1 indicates that the fourfold symmetry of K_1 is retained as the Pb content increases.
2. In Fig.2, the changes observed in K_1 cannot be due to a simple dilutional substitution of Pb in EuIG.
3. From the theory for the changes in K_1 with strain, the changes in K_1 with Δa of Fig 3 cannot be caused by epitaxial lattice mismatch strains.
4. The data of Fig.4 for $|K_1|$ vs K_u cannot be caused by simple dilution substitution of Pb in EuIG and/or lattice mismatch strains.

The above evidence strongly suggests that, as the Pb content increases and Δa approaches 0, EuIG undergoes a modification to a cubic structure with a lattice constant smaller than expected.

ACKNOWLEDGEMENTS

The authors are indebted to A.H.Parsons for film preparation and characterization, R.Schad, S.D. Ellman, F.Cardone and C.F. Aliotta for electron microprobe analysis and to J.M. Karasinski for x-ray diffraction analysis.

REFERENCES

* Manufacturing Res. Lab., System Products Division.

1. T.S. Plaskett, E.Klokholm, D.C. Cronemeyer, P.C.Yin, S.E.Blum, Appl.Phys.Lett. 25, 357, (1974)
2. T.S. Plaskett, E. Klokholm, D.C. Cronemyer, AIP Conf. Proc. 29, 109 (1976).
3. T.S. Plaskett, R.Ghez, AIP Conf. Proc. 24, 584 , (1975).
4. L. G. Van Uitert, E. M. Gyorgy, W. A. Bonner, W. H. Grodkiewicz, E. J. Heilner, G. J. Zydzik: Mat. Res. Bull. 6, 1185 (1971).
5. E. Klokholm, J.W. Matthews, A.F. Mayadas and J. Angilello AIP Conf. Proc. 5, 105, (1972)
6. M.I. Darby and E.D. Isaac, IEEE Trans. Mag. , Mag 10 ,259,(1974)
7. W. Carr, Handbuch der Physik Vol. XVIII/2, pp 288-292,Springer Verlag, Berlin (1966)

BUBBLE DOMAIN GARNETS - A MATURING TECHNOLOGY

J. W. Nielsen, Bell Laboratories, Murray Hill, New Jersey 07974

Bubble domain memories are now in an advanced stage of development. It appears certain that the first saleable bubble memories will contain magnetic garnet memory elements. The way in which garnet materials development has kept pace with the increasing sophistication and more stringent specifications imposed on bubble domain memories will be explored. Emphasis will be placed on the unique flexibility of the garnet system, and how, with the discoveries of growth induced anisotropy, and liquid phase epitaxy from super-cooled $PbO-B_2O_3$ solutions, the preparation of a wide variety of garnet memory materials satisfying many different memory requirements has become routine. Specific examples of the "fine tuning" of garnet films to meet device requirements are the precise control of the bubble collapse field, and its temperature dependence, through careful control of the composition of garnet-flux solutions and the kinetics of film growth.

Recent results on the scale-up of the film growth process, both in the area of garnet film per wafer and the number of wafers per run will also be discussed.

MICROWAVE MEASUREMENT OF MAGNETIC BUBBLE COLLAPSE AND STRIPOUT FIELD

Millard G. Mier, Hilmer W. S. Swenson
Electronic Research Branch
Electronic Technology Division
Air Force Avionics Laboratory
Wright-Patterson AFB, Ohio 45433

and

P. E. Wigen
Physics Department
Ohio State University
Columbus, Ohio 43210

ABSTRACT

The first use of magnetic resonance apparatus for measurement of the bubble collapse field, H_{col}, and the bubble stripout field, H_{so}, is reported. A low field peak with a distinctive line shape is observed at applied fields corresponding to H_{col} (increasing field sweep) or H_{so} (decreasing field sweep). A standard EPR apparatus with a high-Q cavity was used at 9.52 GHz. H_{col} was determined over the temperature range -50°C to 125°C and as a function of the field angle from perpendicular. These data were corroborated by vibrating sample magnetometer measurements. The ease of temperature control and the applicability to bubbles which are difficult to observe optically due to small size are attractive features of the microwave technique.

INTRODUCTION

Measurement of magnetic thin film parameters is a critical step, not only in research and development of new magnetic bubble materials[1] but also for quality control and for parameter adjustment by such techniques as annealing[2] and ion implantation.[3] A prime consideration of an evaluation process is the ease of measurement over a range of environmental parameters. Especially important is the measurement of bubble collapse field which is commonly used as a criterion for matching wafers to be used in modules.[4] In fact, the free bubble collapse field, taken as a function of temperature, determines the upper bias margin of magnetic bubble devices. Ideally, such a measurement technique will have ready application to small bubbles which are difficult to measure optically. In this report, it is shown that a microwave resonance apparatus has the desired qualities of sensitivity, speed, and temperature control. It is also readily applicable to small bubbles.

During a systematic investigation of the ferromagnetic resonance absorption characteristics of magnetic bubble materials, a low-field change in microwave susceptibility was observed as shown in Fig. 1. The shift was found to be repeatable but to depend on whether the magnetic field was decreasing or increasing. The shifts in the susceptibility were found to be correlated with the bubble collapse field or stripe collapse field for increasing field sweep and the bubble strip-out field for decreasing field sweep.

EXPERIMENTAL

A standard X-band microwave spectrometer operating at 9.52 GHz was used for these measurements. The high-field ferromagnetic resonanc spectrum was observed for [111] oriented films in one of two conditions. The internal field, H_{int}, determined when the magnetization is normal to the film surface (out-of-plane resonance) is composed of the induced anisotropy field, H_u, the cubic anisotropy field, H_K, and the demagnetization field, $H_d = -4\pi M_s$, i.e. $H_{int} = H_u + 2/3H_K - 4\pi M_s$. If ω/γ, where γ is the gyromagnetic ratio, is greater than H_{int}, one line was observed in the out-of-plane resonance orientation and another was observed for the in-plane resonance orientation. In those films for which ω/γ was less than H_{int}, an out-of-plane resonance was not observed while two modes were observed in the in-plane resonance orientation.

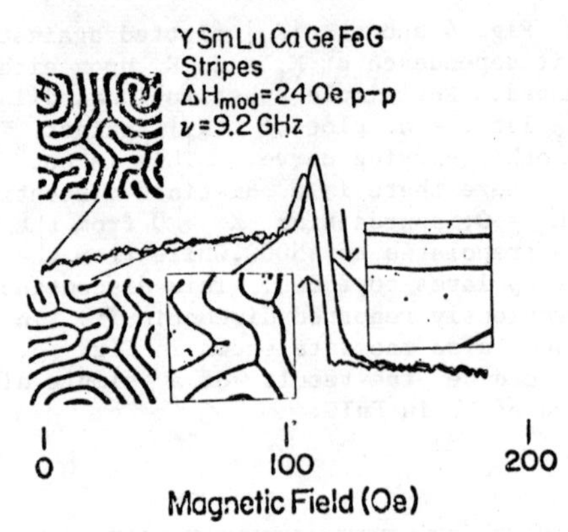

Fig. 1. The microwave signal observed in a magnetic bubble material when the starting configuration is that of stripes in the sample. The absorption is strongly dependent on the modulation amplitude.

In addition to the high field ferromagnetic resonance lines, a change in microwave susceptibility was also observed at low fields for the magnetic field in the out-of-plane orientation. Operating the spectrometer in the derivative presentation mode, this change in microwave susceptibility appeared as a peak with a distinctive nonresonance shape for up-field and down-field sweeps. The intensity of these signals is two to three orders of magnitude weaker than the high field ferromagnetic resonance signals. Figure 1 is the spectra obtained for a sample having stripe domains as the initial state at zero field. The presence of the peak at the high field end of the sweep is very dependent on the magnetic field modulation amplitude. The position of the high field cut-off is observed to increase with increased modulation amplitude and at large amplitude modulation it also tends to wash out the characteristic differences in the stripe collapse and stripout field values.

Figure 2 is the spectra obtained for a sample having a bubble lattice as the initial state at zero field with the bubble diameter decreasing with increasing field. This condition was created by applying a large (5000 Oe) field in the plane of the film and reducing it to zero. In bubble samples, the enhanced peak characteristic of stripe collapse is not present and the disappearance of the absorption is observed to occur at higher fields. As the magnetic field is reduced, the characteristic signal associated with stripe formation is observed.

Figure 3 is the spectra obtained from a sample having a bubble lattice as the initial state but with the magnetic field applied in such a direction as to increase the bubble diameter. The signal does not have the characteristic peak of the stripe collapse spectra but the collapse field is the same as that observed for the stripe collapse field.

Using these characteristic features, it is possible to quantitatively evaluate the stripe collapse, bubble collapse and strip-out fields for magnetic bubble materials at the out-of-plane orientation, as a function of the angle of the applied field and as a function of the temperature.

The angular dependence of these data are shown in Figure 4. Figure 5 shows the temperature dependence of the stripe collapse field at the out-of-plane magnetic

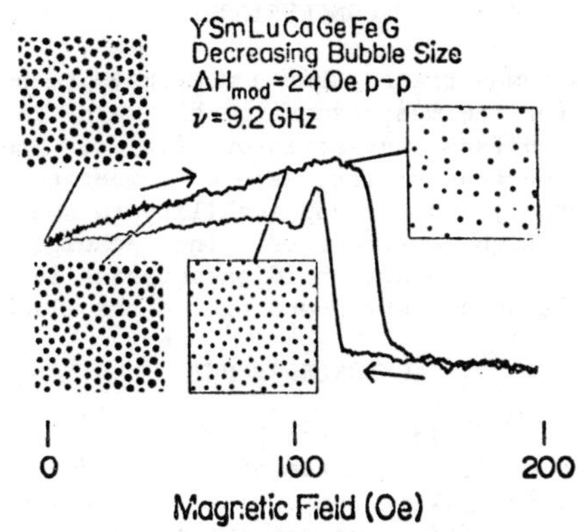

Fig. 2. The microwave signal observed in a magnetic bubble material when the starting magnetic configuration is a bubble lattice. The bubbles are decreasing in size with increasing field.

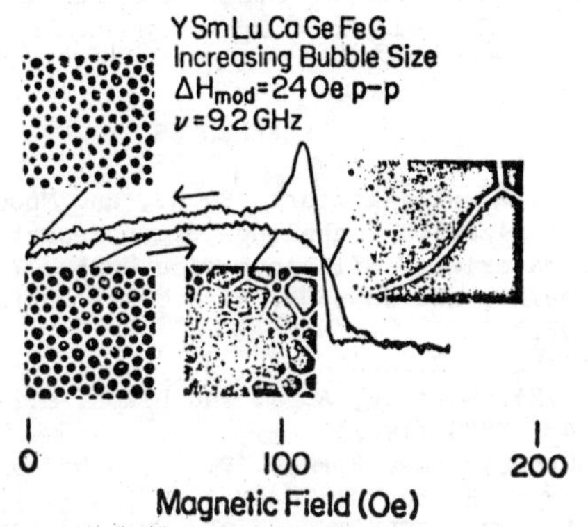

Fig. 3. Same as Fig. 2 except the bubbles are increasing in size with increasing field.

field orientation over the temperature range -60°C to +130°C on a sample of YSmLuCaFeGe garnet. Taking these data at 10°C increments required less than an hour.

These data were independently compared with results from a vibrating sample magnetometer and standard microscopic magneto-optical collapse field measurements. Figure 4 compares the angle dependence of the VSM measurements with the microwave results. Table I compares magneto-optical, VSM, and perpendicular low-field peak measurements on a number of bubble material samples

To evaluate the frequency dependence of this effect, a search was made at 24 GHz but no low field absorption was observed in any of the samples studied. However, at 15 MHz, the low field change in susceptibility is present having a different shape but occurring at similar field values as observed in Figure 6. In addition to the changes near 120 Oe in Figure 6, a near zero field resonance is also observed at these frequencies. This signal has a strong angle dependences and gets much weaker at 20 MHz.

ORIGIN OF ABSORPTION

The data indicate that the origin of the low field change in the microwave susceptibility is clearly associated with the presence of domain walls in the sample. Beginning in zero applied field, the absorption, as shown in the derivative spectra in Figures 1-3 is observed until the applied field reaches a value where stripe domain walls or bubble domain walls begin to collapse. The absorption continues to decrease with increasing magnetic field until the field value is reached where the domain have collapsed and the film is saturated. As the film is rotated in the magnetic

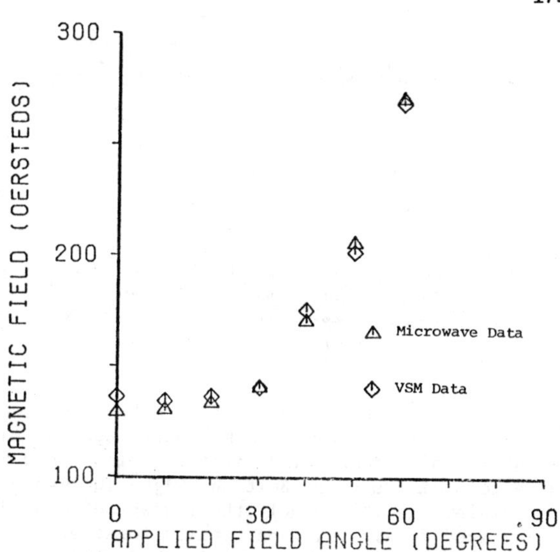

Fig. 4. Angle dependence of the collapse field as observed by vibrating sample magnetometer (VSM) and by low field microwave susceptibility changes.

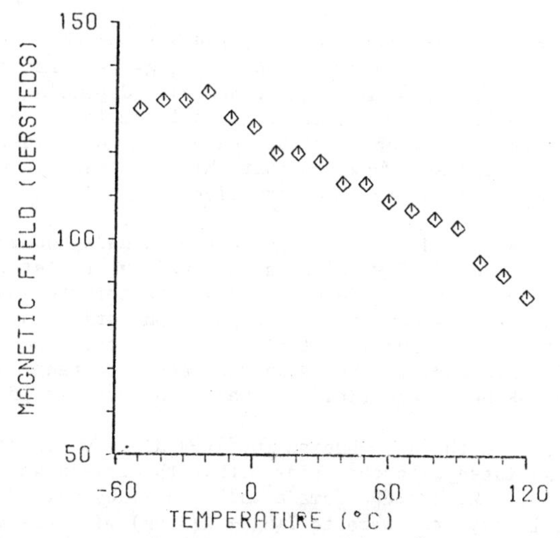

Fig. 5. Microwave measurement of the stripe collapse field over the temperature range indicated on a sample of YSmLuCaGeFe garnet.

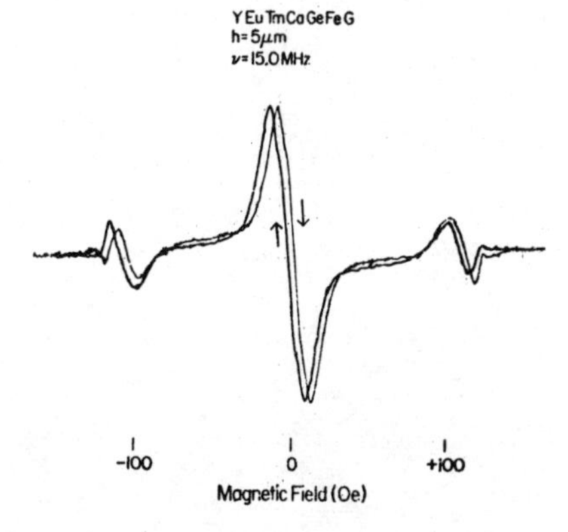

Fig. 6. The signal observed in a magnetic bubble material at 15 MHz. The resonance signal occuring at near zero field is thought to be domain wall motion resonance.

field, the same general features are observed and the saturation field value is found to be in excellent agreement with the saturation field obtained from VSM measurements as shown in Figure 4.

Table I

Bubble film characteristics and microwave collapse field measurements.

Film	Components	$4\pi M$ in Gauss	$H_\perp$ in Oe	$H_\parallel$ in Oe	H_{coll} in Oe Optical	Microwave	VSM
5421	YEuCaSiGeFe	250	500	1800	133	150	-
5444	YEuCaSiGeFe	280	600	2910	147	170	-
5696	YSm Ca Fe	200	1870	4220	112	98	112
H62	YSm Lu Ga Fe	530	-	3000 5500	350	310	310
H99	YSm Lu Ga Fe	650	-	2400 5650	400	420	440
5692	YSmLuCaGeFe	240	1800	4125	125	110	134
5702	YSmLuCaGeFe	250	1700	4100	122	115	
5804	YSmLuCaGeFe	210	2025	3925	132	115	125
330	YSmLuCaGeFe	215	1400	4200	112	100	110
Y209	YSmLuCaGeFe	~400	2500	3700	225	180	185
Y229	YSmLuCaGeFe	330	2400	3800	175	-	195

Three possible mechanisms have been considered as the origin of the low field microwave susceptibility in the bubble films. The first is a low field microwave loss discussed by Smit and Wijn[5] in unsaturated materials. In this model, the regions of oppositely magnetized domains will be coupled together by the demagnetization field. For [111] oriented garnet film the resonance frequency will occur over the range

$$\frac{\omega}{\gamma} = H_a + H_u + \frac{2}{3} H_K \pm 4\pi M \qquad (1)$$

Using the data from Table I, these resonances should occur within a narrow frequency range centered at some value between 4 to 8 GHz frequencies depending on the sample being investigated. In all samples, the absorption was observed at the 9.5 GHz used in this experiment and suggests that this mechanism is not applicable for these observations.

A second mechanism is that of domain boundary resonance of 180° walls as reviewed by Brailsford.[6] The domain boundaries will oscillate through small displacements about their mean positions driven by the rf fields. For bubble materials, these resonance frequencies are expected to be in the megahertz range and may be the near zero field resonance observed at 15 MHz.

The third mechanism involves a k = 0 crossover associated with the spins within the domain wall itself.[7] Within the domain wall, the internal fields will vary from zero to $2(H_u + 2/3\ H_K)$ with the upper value being above the 9.5 GHz frequencies used in this experiment but below the 24 GHz frequency at which no absorption was observed. The absorption at 15 MHz is also consistent with this mechanism.

CONCLUSIONS

It is demonstrated that a microwave technique is available for the measurement of the magnetic bubble and stripe collapse and strip-out fields. The technique has the sensitivity, ease of temperature dependence, accuracy, and the applicability to small bubbles desired for such measurements. The technique has the added virtue that operator idiosyncrasies in estimating bubble collapse and strip-out fields are eliminated.

ACKNOWLEDGEMENTS

The authors wish to thank Drs. W. R. Cox and J. T. Carlo of Texas Instruments, Inc., for furnishing the samples and making the optical measurements reported here. In addition our appreciation is extended to Mr. D. Miyashiro, Ohio State University, for his assistance in the VSM data and to Mr. A. Gerhardstein, Ohio State University, for his helpful assistance in the data analysis. Special thanks are due to Dr. R. C. LeCraw of Bell Telephone Laboratories for his suggestions on data interpretation.

REFERENCES

1. Shaw, R. W., Sandfort, R. M., and Moody, J. W., "Characterization Techniques Study Report - Magnetic Bubble Materials" NTIS Accession Number 741390 (July 1972), and Josephs, R. M., AIP Conf. Proc. 10, 286 (1972)
2. Smith, D. H. and North, J. C., AIP Conf. Proc. 10, 334 (1972), Kurtzig, A. J. and Dixon, M., J. Appl. Phys. 43, 2883 (1972)
3. LeCraw, R. C., Byrnes, P. A., Johnson, W. A., Levinstein, H. J., Nielsen, J. W., Spiwak, R. R., and Wolfe, R., IEEE Trans. Mag. MAG-9, 422 (1973)
4. Warren, R. G., Mee, J. E., Stearns, F. S., and Whitcomb, E. C., AIP Conf. Proc. 18, 63 (1973)
5. Smit, J. & Wijn, N.H.P., Ferrites p. 83-4 Philips Technical Library, (The Netherlands, 1959)
6. Brailsford, F., Physical Principles of Magnetism, pp. 216-219, D. Van Nostrand Co. Ltd. (London, 1966)
7. Details of this model to be published elsewhere.

HIGH TEMPERATURE ELECTRONIC STRUCTURE OF Fe_3O_4 *

Lu-San Pan and B. J. Evans
The University of Michigan, Ann Arbor, Michigan 48109

ABSTRACT

The high temperature electronic structure and conduction mechanism in Fe_3O_4 have been investigated through correlations of resistivity and magnetoresistance data with ^{57}Fe NGR. Similar to the resistivity, the relative A and B site isomer shifts are found to decrease rapidly at temperatures well below T_N. Above T_N the A and B site patterns are unexpectedly poorly resolved and this is found to be due to a reduction in the difference in the electronic structure at the two sites. $\delta_B - \delta_A$ is only 0.22 mm s^{-1} at 855 K compared to the expected value of 0.32 mm s^{-1} assuming no change in the relative electronic structures at the two sites on passing through the magnetic order disorder transition. No significant changes in the ^{57}Fe NGR parameters are observed at the conductivity maximum near 400 K. These results require a band model for the conduction mechanism above the Verwey transition and non-negligible conduction electron-ion core interactions for the A site ions above T_N.

INTRODUCTION

Despite the many detailed studies of the Verwey transition in Fe_3O_4, [1,2] controversy still surrounds even a qualitative description of the electronic structure above or below the transition temperature, T_V. The transition has been described variously as a metal-insulator, semiconductor-semiconductor, and semimetal-semiconductor. Much of the recent attention has been devoted to phenomena at or near T_V. However, neither the transition nor the electronic structure of the low temperature phase can be adequately understood in the absence of a valid description of the high temperature phase. In addition, the electrical conductivity of Fe_3O_4 displays some remarkable variations above T_V with a local maximum at ∿400 K and a minimum at ∿50 K below the Néel temperature, T_N.[3] These features of the electrical conductivity are poorly understood and have been subjected only recently to theoretical analysis.[4]

In the present investigation, ^{57}Fe nuclear gamma-ray resonance (NGR) measurements have been made on pure Fe_3O_4 from 77 K to 900 K to determine the change (if any) in the electronic structure giving rise to the conductivity maximum at ∿400 K and the minimum just below T_N. By considering the broad systematics of the data, it was also hoped that a definitive conclustion could be reached concerning the high temperature electronic structure of Fe_3O_4.

Furthermore, because the ^{57}Fe NGR spectrum of Fe_3O_4 above T_N is itself a conundrum, consisting of an apparent single line instead of two, partially resolved lines, a determination of the temperature at which the spectrum deviates from the expected behavior and a rigorous analysis of the line shape above T_N have been made. From earlier measurements[2] it was known that the ^{57}Fe NGR spectrum could be understood up to 750 K in terms of the 300 K spectrum; and consequently, the change in electronic structure giving rise to the apparent, single line above T_N must take place within 100 K of T_N. This result suggests a possible relationship between the conductivity minimum below T_N and the unusual character of the NGR spectrum above T_N. No ^{57}Fe NGR measurements have been specifically directed at determining possible changes in electronic structure associated with the conductivity maximum at ∿400 K, and this temperature region has also received close attention in the present study.

The results of this study demonstrate that there is indeed a relationship between the unusual character of the ^{57}Fe NGR spectrum above T_N and the conductivity minimum below T_N. The change in electronic structure giving rise to both phenomena appears to be the onset of significant interactions between the A site Fe ions and the conduction electrons and the concomitant increase in spin-disorder scattering. The conductivity maximum at ∿400 K results in no noticeable anomalies in the NGR spectrum and is apparently due to higher order dynamical effects and not to fundamental changes in electronic structure, validating theoretical models.[4,5] The temperature dependence of the NGR parameters from 300 K to 750 K are consistent with band conduction and confirms recent theoretical predictions to this effect.

EXPERIMENTAL

The polycrystalline Fe_3O_4 sample has been described previously, investigated with a number of different techniques and is known to be highly stoichiometric.[2]

The NGR spectrometer consisted of an electromechanical velocity drive and a 1024 channel analyzer operated in the so-called "time mode". A 50 mCi Co^{57}/Rh source was employed and maintained at 298 K during the measurements. The spectrometer was calibrated with sodium nitroprusside and iron metal. A commercial design, vacuum furnace was used in the high temperature measurements, and the sample was contained in a boron nitride cup which was in contact with a chromel/alumel thermo-couple. The temperature was controlled with a proportional controller and was stable to within ±2 K of the desired temperature.

RESULTS

The ^{57}Fe NGR spectrum of Fe_3O_4 at 856 K is shown in Fig. 1; T_N for this sample is 855 K. The solid line is the result of a least-mean-squares fit of a single, Lorentzian shaped line to the data. The line in the lower part of Fig. 1, the "residual", is the difference between the fitted line and the experimental data; and the strong structure in the residual demonstrates the inadequacy of the single line fit. Even though the asymetry in the structure of the residual indicates the presence of unresolved structure in the NGR spectrum and not a simple deviation from Lorentzian shape as might result from instrumental and sample preparation effects, attempts to fit more than one line to the spectrum produced physically meaningless results in the absence of doubtful constraints during the fitting procedure.

The spectra below 800 K were straightforward and easily understood on the basis of the 300 K spectrum of Fe_3O_4. They were fitted to two, magnetic hyperfine + electric quadrupole patterns; the magnetic

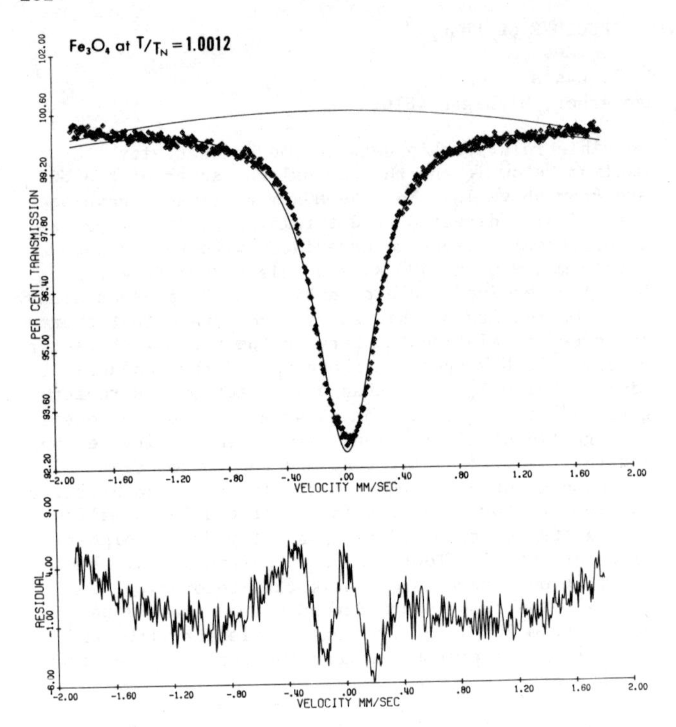

Fig. 1. ^{57}Fe NGR spectrum of Fe_3O_4 at 856 K (T_N = 855 K). The points are the experimental data and the solid line is the least squares fit of a single, Lorentzian line to the spectrum. The lower line is the difference between the fitted spectrum and the experimental points. The deviations between ∿-0.2 mm s^{-1} and ∿-0.6 mm s^{-1}, at the bottom of the absorption line, and between ∿0.1 mm s^{-1} and ∿0.3 mm s^{-1} are to be noted and compared with Fig. 5. The relative heights of the minima in the residual are also to be compared with those in Fig. 5.

hyperfine field, H(i), isomer shift, δ_i, electric quadrupole interaction, $\Delta E_Q(i)$, line-width, Γ_i, and intensities of each pattern were varied independently until a chi-square, goodness-of-fit criterion was met. The intensities and widths of lines 1, 2, and 3 in a given pattern were constrained to be equal to those lines 4, 5, and 6 of the same pattern.

None of the NGR parameters exhibited unusual behavior in the region of the conductivity maximum near ∿400 K. The difference in hyperfine fields, H(B) - H(A), increased from 30 kG at 300 K to 45 kG at 700 K in qualitative accord with the relative sublattice magnetisations. H(B) - H(A) decreased rapidly above 725 K, but the significance of this decrease is complicated by the rapid fall in magnetisation in this temperature range and will not be considered further. The ratio of the integrated intensity of the B site pattern to the A site pattern decreased from 1.90 at 300 K to 1.70 at 800 K, exhibiting no anomalies and in accord with the recoilless fractions of the two sites. Γ_B/Γ_A is nearly constant from 300 K to 750 K but shows a substantial decrease above this temperature. Surprisingly, the decrease in Γ_B/Γ_A is occasioned primarily by an increase in Γ_A. This increase in the A site line-width is almost certainly related to the isomer shift anomaly to be discussed below.

In contrast to the behavior near 400 K, δ_A exhibited an anomalous increase at 800 K as shown in Fig. 2; and both δ_A and δ_B decreased rapidly above this temperature. The difference in isomer shifts, $\delta_A - \delta_B$, demonstrates more dramatically this change in isomer shifts as shown in Fig. 3.

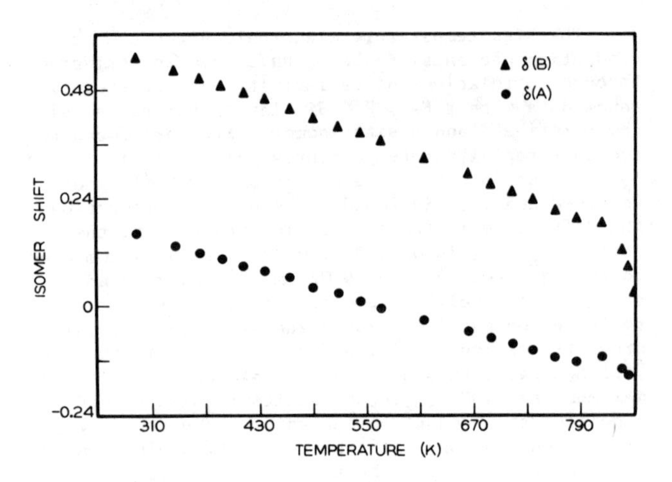

Fig. 2. The temperature dependences of the A and B site isomer shifts, δ(B) and δ(A), respectively. The isomer shifts are relative to a Co^{57}/Rh source at 298 K and are uncorrected for the second-order-Doppler shift. The increase in δ(A) at ∿800 K and the rapid decrease in δ(B) above this temperature are noteworthy. The Néel temperature is offscale on the abscissa.

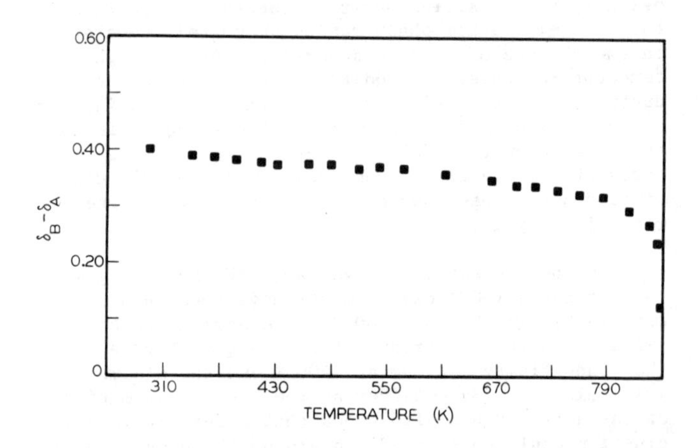

Fig. 3. The temperature dependence of the differential isomer shift at the A and B sites, $\delta_A - \delta_B$. The abrupt decrease above ∿800 K is clearly evident.

DISCUSSION

Insofar as the conductivity maximum near 400 K and the minimum approximately 50 K below T_N are concerned, only δ_i and possibly Γ_i provide insights as to their origins. None of the NGR parameters, including δ_i, shows unusual behavior in the region of the conductivity maximum and it may be safely concluded that the maximum in the conductivity is not caused by a fundamental change or cessation of such change in the electronic structure. This result provides for the theoretical suggestions that the broad maximum in conductivity is due to dynamical scattering proceese.[4,5]

The temperature dependence of δ_i near the high temperature conductivity minimum indicates that it is to be associated with a change in electronic structure. Below 750 K $d\delta_A/dT$ and $d\delta_B/dT$ have values of -5.84×10^{-4} mm s^{-1} K^{-1} and -7.34×10^{-4} mm s^{-1} K^{-1}, respectively. $\delta_A - \delta_B$ is 0.40 mm s^{-1} at 298 K and on the basis of the above slopes of δ_A and δ_B below 750 K, $\delta_A - \delta_B$ is calculated to be 0.32 mm s^{-1} at 855 K. The NGR spectrum expected on this basis is shown in Fig. 4 and is quite unlike that in Fig. 1. We can now understand the poor resolution of the NGR

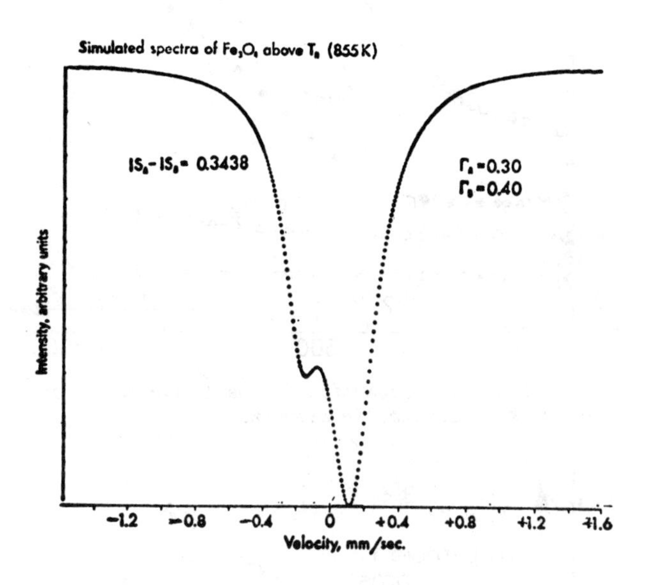

Fig. 4. Computed spectrum of Fe_3O_4 above the Néel temperature on the assumption that there are no changes in the electronic structure. The more intense line at more positive velocities is due to the B site and the weaker, partially resolved line is due to the A site. The NGR parameters used to compute this spectrum are given in the figure.

spectrum of Fe_3O_4 above T_N as being a consequence of a change in electronic structure at ~800 K as manifested in the isomer shifts shown in Figs. 2 and 3. The change in electronic structure is such that the s-like electron density at the A site increases relative to that at the B site and approaches that of the B site as T_N is approached.

Even though the spectra above T_N cannot be fitted unambiguously, spectrum simulation permits a semi-quantitative estimate to be made of the isomer shift difference. A large number of spectra were computed in which the B to A site area ratio, Γ_B and Γ_A were assigned values derived from spectra below T_N and $\delta_A - \delta_B$ was varied. These simulated spectra were then "fitted" with the ^{57}Fe NGR parameters obtained in the fit depicted in Fig. 1. The residual of such a spectrum in Fig. 5 corresponds most closely to the residual in Fig. 1 and indicates an isomer shift difference, $\delta_A - \delta_B$, of 0.22 mm s^{-1}. Thus, the change in electronic structure commencing at ~800 K results in a differential increase of 0.1 mm s^{-1} in

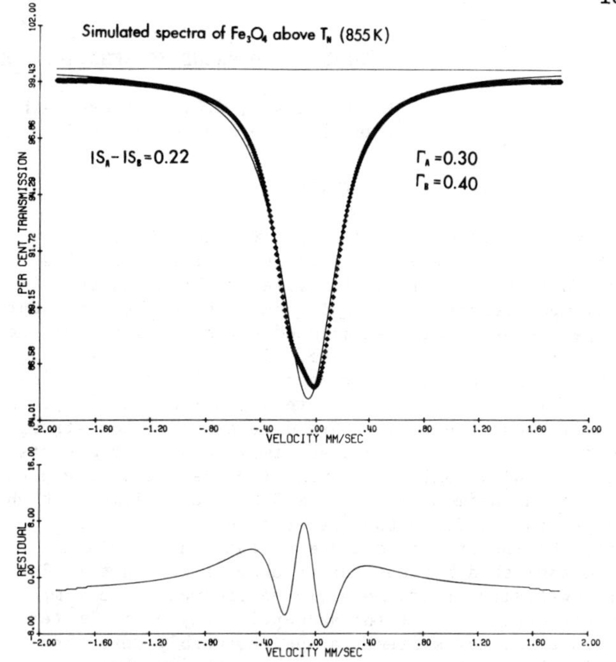

Fig. 5. Simulated spectrum of Fe_3O_4 above T_N. The points are the spectrum calculated assuming an isomer shift difference of 0.22 mm s^{-1} with all other parameters being the same as in Fig. 4. The solid line is the same as the line in Fig. 1. The similarities in the deviations of the line from the points, as indicated in Fig. 1, and the shape of the residual are remarkable.

δ_A relative to δ_B and indicates a significant conduction electron - Fe^{3+}(A) ion core interactions.

An unambiguous mechanism for this interaction is difficult to come by but there are some plausible altermatives. At temperatures as high as 800 K, there is sufficient thermal energy to excite electrons from the B site band into an A site band and the rapid drop in magnetisation lessens the restrictions imposed by the antiparallel A and B site spin directions. It is also probable that a band overlap transition occurs between two, spin-split A and B site bands. Band overlap is permitted by the symmetry of the A and B sublattices. Either of these mechanisms provides an unstrained explanation for the rapid decrease of $\delta_A - \delta_B$ as T_N is approached.

The weak temperature dependence of δ_i and Γ_i below T_N and the absence of any incipient Fe^{2+} character in the NGR spectrum above T_N makes a localized conduction mechanism for Fe_3O_4 above the Verwey transition unlikely. Thus the high temperature electronic structure of Fe_3O_4 is best described as that of a narrow-band, metallic or semimetallic, magnetic oxide. The unusual temperature dependence of the conductivity appears to be due to strong electron scattering resulting from temperature dependent magnetisations and electron phonon interactions.

REFERENCES

* Support of this investigation by the National Science Foundation is gratefully acknowledged.

1. S. Chikazumi, Tech. Rept. ISSP A737, 1975.
2. B. J. Evans, AIP Conf. Proc. 24, 73 (1974).
3. R. Parker and C. J. Tinsley, Phys. Stat. Sol. a33, 189 (1976).
4. B. Lorenz and D. Ihle, Phys. Stat. Sol. b69, 451 (1975).
5. W. Haubenreisser, Phys. Stat. Sol. b1, 619 (1961).

VALENCY AND MAGNETIC BEHAVIOUR OF YTTERBIUM IN INTERMETALLIC COMPOUNDS

J.C.P.Klaasse, W.C.M.Mattens, A.H. van Ommen, F.R. de Boer and P.F.de Châtel
Natuurkundig Laboratorium der Universiteit van Amsterdam, Amsterdam, The Netherlands

ABSTRACT

We have investigated 17 binary and some ternary intermetallic compounds of Yb in order to determine the factors controlling its valency. From susceptibility measurements between 1.4 and 1060 K the trivalent state can be firmly established by the validity of the Curie-Weiss law in 7 systems. Diamagnetism or weakly temperature dependent paramagnetism, indicative of divalent Yb, is found in 7 systems. In $YbGa_2$ and especially $YbCu_2$ the temperature dependence is too strong to classify these systems as simple Pauli paramagnets, and in $YbAl_2$, $YbAl_3$ and YbCuAl the susceptibility goes through a maximum and becomes Curie-Weiss like at high temperatures. Yb in the latter five systems is considered having an intermediate valency. A correlation is established between the valency of Yb and the electronegativity of its metallic environment. The temperature dependence of the susceptibility of the intermediate-valency systems can be reasonably described by invoking a 4f resonance on each Yb ion, but the complete lack of saturation cannot be understood within this picture. Other semi-empirical and theoretical formulae are also partially succesful, but none of them offers a complete description of the results.

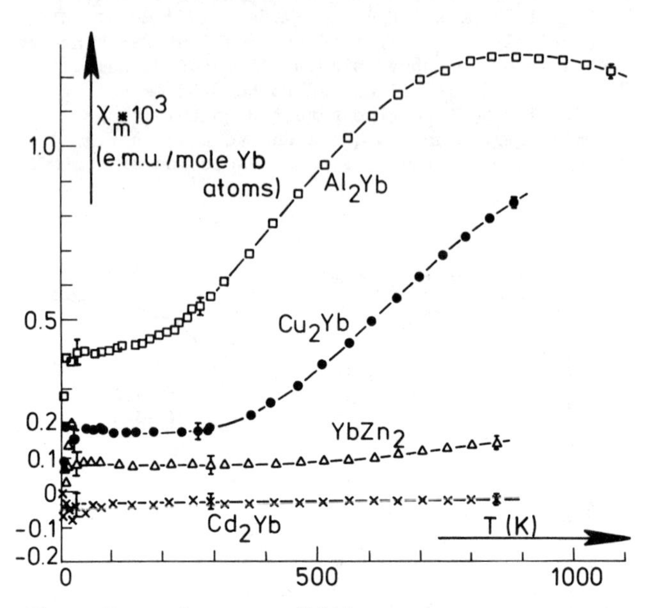

Fig.1. Magnetic susceptibility vs temperature for various Yb intermetallic compounds.

INTRODUCTION

In view of the considerable recent interest[1] in "interconfiguration-fluctuation" (ICF) systems, that is, in rare-earth systems with a non-integral occupancy of the 4f shell, it is of inportance to determine the factors controlling the valency of Ce, Sm, Eu, Tm, and Yb (the rare-earth elements occurring with two valencies and thus two different 4f occupancies) in metallic environments. Qualitatively it is clear that the most important factor determining the valency must be the electronegativity of the surrounding atoms, more electronegative partners being able to reduce the occupancy of the 4f shell and thereby stabilize the state with the higher valency. In Figs. 1 and 2 this tendency is clearly spelled out by the gradual appearance of a Curie-Weiss susceptibility, characteristic of the trivalent state ($4f^{13}$ configuration) as Yb is combined with metals of increasing electronegativity.

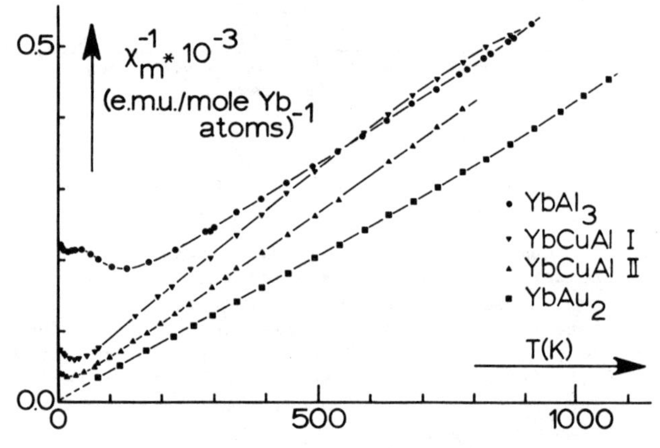

Fig.2. Inverse magnetic susceptibility vs temperature for various Yb intermetallic compounds.

ALLOY STABILITY

In an attempt to assess the role of electronegativity quantitatively, we have applied to Yb alloy systems an empirical electronegativity scale developed recently in a study of alloy stability[2]. Using the empirical formulae[2] involving electronegativity and electron density parameters (ϕ^* and $n_{WS}^{1/3}$, respectively) and atomic volumes (V_m), it is possible to determine the enthalpy of ordered Yb_xM_y alloys (no entropy term is included to account for disorder), where M is a transition or noble metal.

To determine the relative stability of intermetallic phases involving divalent and trivalent Yb, these two modifications are treated as two different metals, the parameters relevant to their alloying behaviour being determined in the following way. Since divalent Yb has the electron vonfiguration of alkaline earth metals, and since the parameters ϕ^* and $n_{WS}^{1/3}$ are related to the atomic volume, we extrapolate the ϕ^* and $n_{WS}^{1/3}$ values of Sr and Ca[2] linearly according to the volumes. The result is shown in Table I. If we ignore the 4f shell, the electronic structure of trivalent Yb is the same as that of Sc and Y. The atomic volume of hypothetical trivalent Yb can be obtained by averaging the atomic volumes of Lu and Tm and lies between those of Sc and Y. The ϕ^* and $n_{WS}^{1/3}$ values given in Table I for trivalent Yb were obtained by linear interpolation according to the volume between the Sc and Y values[2].

With the parameters given in Table I we can determine the enthalpies of $Yb_x^{2+}M_y$ compounds with respect to an x:y mixture of the pure metals Yb and M, and those of $Yb_x^{3+}M_y$ compounds with respect to an x:y mixture of pure (hypothetical) trivalent Yb and metal M. In order to determine the valency of Yb in a given Yb_xM_y compound, it is still necessary to estimate the enthalpy needed to transform normal divalent Yb into a trivalent metal. This enthalpy difference has been variously estimated to be between 5 and 13 kcal/mole[3]; in what follows we will adopt the value 8 kcal/mole.

Table I. Alloying parameters of di- and trivalent Yb.

	ϕ^* (V)	$n_{WS}^{1/3}$ $(d.u.)^{1/3}$	V_m (cm^3)
Ca	2.55	0.91	26.19
Sr	2.40	0.84	33.19
Yb^{2+}	2.575	0.92	24.87
Sc	3.25	1.27	15.06
Y	3.20	1.21	19.88
Yb^{3+}	3.22	1.235	17.95

Figure 3a shows the enthalpies determined as described above for the Yb-Ni system. In the shaded concentration range no stable compound is predicted -and none is found- outside this region trivalent Yb is expected to form stable compounds with Ni -this again is confirmed by experiment, as indicated by the vertical lines where such compounds were found[4]. The same construction for the Yb-Cu system yields Fig.3b, in which enthalpies predicted for $Yb^{2+}_xCu_y$ compounds are seen to differ by less than 1 kcal/mole from those predicted for $Yb^{3+}_xCu_y$. In view of the simplicity of

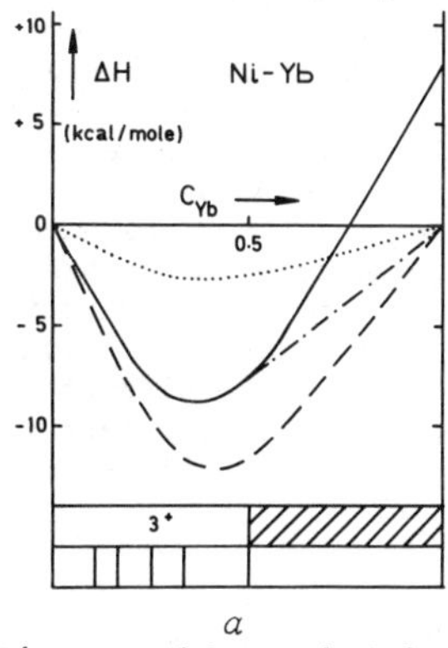

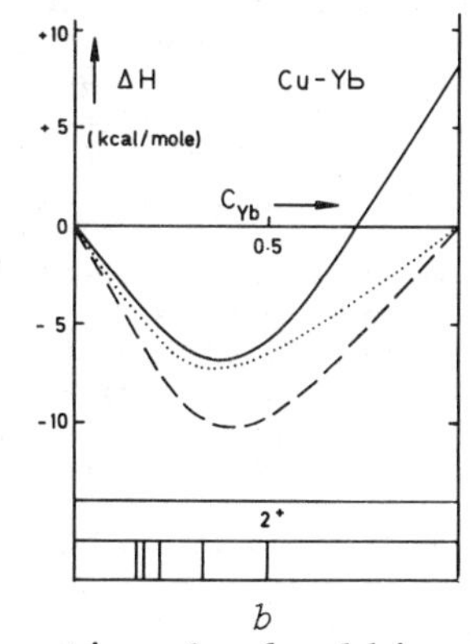

a *b*

Fig.3a and b. Enthalpies of formation of ordered binary compounds of divalent ant trivalent Yb metal with Ni and Cu. The dotted lines represent compounds of divalent Yb, the dashed lines correspond to compounds prepared from (hypothetical) trivalent Yb. The solid line once more gives the enthalpy of formation of compounds in which Yb is trivalent but now (like in reality) with divalent Yb metal as starting material.

the scheme used to calculate these enthalpies, this small difference cannot be the basis of predictions regarding the valency of Yb when combined with Cu. However, if the observation of Wohlleben and Coles[5] that a well-defined intermediate-valence region separates the regions of stability of two integral-valence states is of general validity, it is not unreasonable to expect that in some of the Yb-Cu compounds Yb will be in an intermediate-valence state. Indeed, in Fig.1 the susceptibility of $YbCu_2$ is seen to have a temperature dependence, which can be termed intermediate in a sense to be specified below.

EXPERIMENTAL RESULTS

The present experimental investigation has not been limited to the Yb-transition metal systems for which we can estimate the formation enthalpies. Table II gives a summary of the compounds we have studied. All presented susceptibility vs. temperature curves, shown in Figs.1,2,4,5 and 6, are the as-measured curves corrected as given in ref.6 for the contribution of the main impurity Yb_2O_3. All samples were prepared[6] in an induction furnace by melting together the weighed-in components in sealed-off Ta or Mo crucibles. Homogeneity was achieved by annealing just below the melting point (if known) for several days in the case of congruently melting compounds, and for a longer period (several weeks) in the case of com-

Table II. Summary of susceptibility data.

Compound	Temperature dependence of susceptibility.
YbNi	Curie-Weiss, $\mu_{eff}=(4.5\pm.1)\mu_B,\theta=(15\pm10)K$
YbPd	Curie-Weiss, $\mu_{eff}=(4.3\pm.1)\mu_B,\theta=(160\pm10)K$.
YbPt	Curie-Weiss, $\mu_{eff}=(4.0\pm.1)\mu_B,\theta=(0\pm10)K$
$YbCu_2$	slight, $\chi_m(T=0)=(.17\pm.03)x10^{-3}$emu/mole Yb at.
$YbAg_2$	slight, $\chi_m(T=0)=(.00\pm.02)x10^{-3}$emu/mole Yb at.
$YbAu_2$	Curie-Weiss, $\mu_{eff}=(4.5\pm.1)\mu_B,\theta=(5\pm10)K$.
$YbAu_3$	Curie-Weiss, $\mu_{eff}=(4.5\pm.1)\mu_B,\theta=(-10\pm10)K$.
$YbAu_4$	Curie-Weiss, $\mu_{eff}=(4.5\pm.1)\mu_B,\theta=(10\pm10)K$.
$YbZn_2$	slight, $\chi_m(T=0)=(.08\pm.02)x10^{-3}$emu/mole Yb at.
$YbCd_2$	none, $\chi_m=(-.04\pm.01)x10^{-3}$ emu/mole Yb atoms.
$YbAl_2$	slight, $\chi_m(T=0)=(.41\pm.04)x10^{-3}$emu/mole Yb at.
$YbGa_2$	slight, $\chi_m(T=0)=(-.03\pm.01)x10^{-3}$emu/mole Yb at.
$YbIn_2$	none, $\chi_m=(-.07\pm.01)x10^{-3}$ emu/mole Yb atoms.
$YbAl_3$	strong, $\chi_m(T=0)=(4.65\pm.10)x10^{-3}$emu/mole Yb at.
$YbIn_3$	slight, $\chi_m(T=0)=(.11\pm.02)x10^{-3}$ emu/mole Yb at.
$YbSn_3$	slight, just like $CaSn_3$.
$YbPb_3$	strong, just like $CaPb_3$.
YbNiGa	Curie-Weiss, $\mu_{eff}=(4.5\pm.1)\mu_B$, $\theta=(0\pm10)K$.
YbCuAl I	strong, $\chi_m(T=0)=(13.3\pm.3)x10^{-3}$ emu/mole Yb at.
YbCuAl II	strong, $\chi_m(T=0)=(25.0\pm.5)x10^{-3}$ emu/mole Yb at.

pounds forming by a diffusion reaction. All chemical analyses gave results consistent with the weighed-in amounts. X-ray diffraction patterns showed the compounds to be almost or totally single-phase, with lattice parameters in good agreement with literature values. Differences found in the magnetic properties of different melts of YbCuAl are probably to be ascribed to sample inhomogeneities, to which this nearly trivalent compound may be very sensitive.

DISCUSSION

Available lattice parameter data[7] support the view that Yb is in an intermediate valence state in those compounds where a temperature dependence unlike the Curie-Weiss law is observed. If the constant, and often rather high susceptibility (comparable or exceeding that of Pd, ($\chi(o)= +.7x10^{-3}$ e.m.u./mole) observed in all such samples below 4 K is thought of as an exchange--enhanced Pauli susceptibility, an almost completely filled 4f virtual bound state just below the Fermi energy can account for the temperature dependence. Havinga et al.[8] have shown that a Lorentzian peak in the density of states would lead to the temperature dependence of χ observed in $YbAl_2$ and $YbAl_3$. However, to explain the Curie-Weiss behaviour observed at temperatures above the susceptibility maximum in $YbAl_3$ and YbCuAl, an independent "high-temperature state",

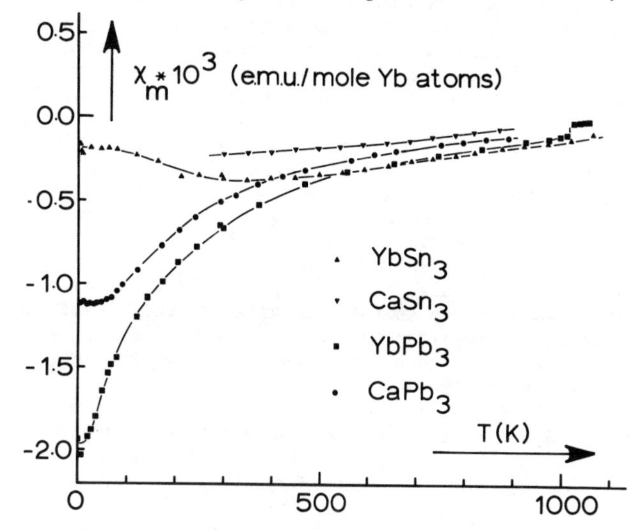

Fig.4. Magnetic susceptibility vs temperature for $YbSn_3$, $YbPb_3$ and the "dummies" $CaSn_3$ and $CaPb_3$.

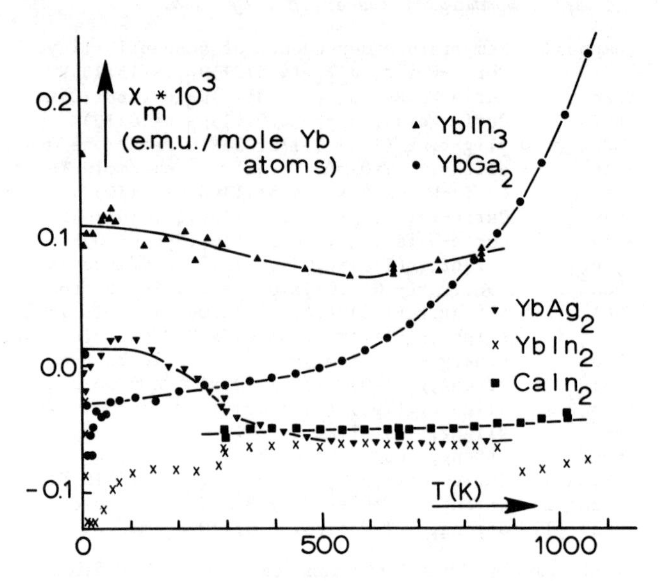

Fig.5. Magnetic susceptibility vs temperature for various Yb compounds. The curve for $CaIn_2$ is given for comparison.

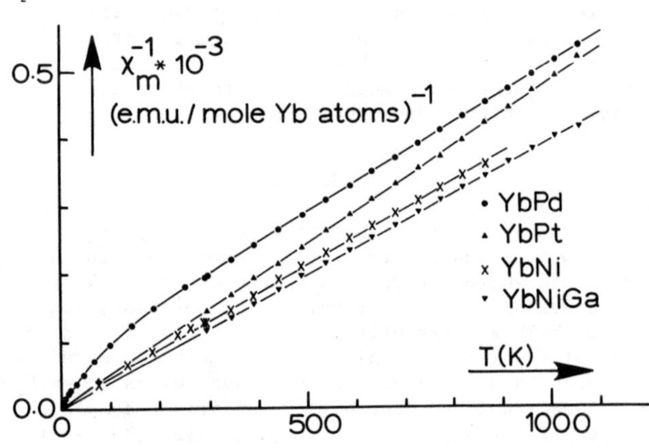

Fig.6. Magnetic susceptibility vs temperature for various trivalent Yb compounds.

involving localised moments, had to be invoked. Another difficulty with the virtual-bound-state picture arises if one wants to analyse the low-temperature magnetic isotherms. Once the Lorentzian peak in the density of states is specified, the magnetization as a function of applied field can be calculated for an exchange-enhanced Pauli paramagnet. Such calculations lead invariably to "normal" magnetization curves, i.e., a tendency to saturation. In contradiction with this prediction, the differential susceptibility of YbCuAl at 4.2 K is found to increase in fields exceeding 150 kOe (Fig.7). It is possible to resolve this discrepancy by allowing for a magnetization-dependent transfer of electrons from the virtual bound states to the conduction bands[9], but this extension of the model amounts to the introduction of a rather arbitrary new free parameter. As the same can be said about the "high-temperature state"[8] referred to above, we conclude that the metallic systems involving Yb in an intermediate valence state cannot be looked upon as simple exchange-enhanced Pauli paramagnets, and that the virtual-bound-state picture fails.

The difficulties mentioned above do not occur in the analysis of the data on less spectacular systems, such as $YbCu_2$ and $YbGa_2$, where no maximum in the susceptibility or upturn in the magnetization curves is observed. It is quite likely that the χ(T) and M(H) curves of these compounds can be consistently fitted, if a suitable density of states curve is chosen. But in view of the systematics we observe over

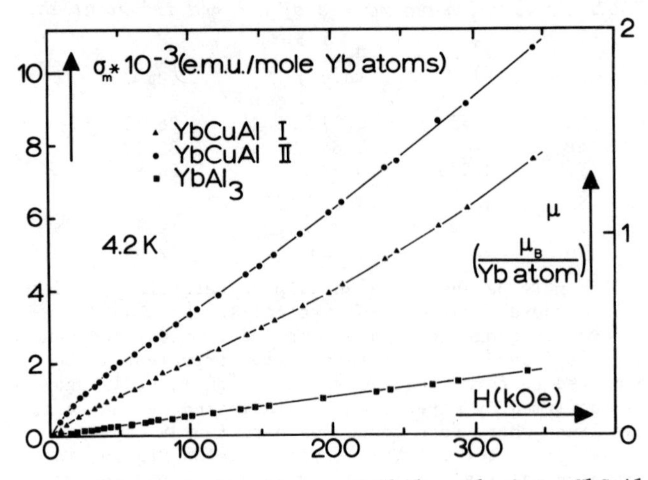

Fig.7. Gram magnetization σ of $YbAl_3$ and of two YbCuAl samples vs applied magnetic field H at 4.2 K.

a range of materials (see Figs.1 and 2), it should rather be considered as a take-off towards a maximum, which is to be followed by a Curie-Weiss-like decrease.

If this interpretation is correct, there should be a universal functional form, which, with the adjustment of one or more material-dependent parameters, accounts for all observed χ(T) curves. Two such functions have been suggested on theoretical grounds:one on the basis of a phenomenological description of interconfiguration fluctuations[10], and another resulting from Fermi-liquid theory[11]. Applying these two theoretical results to our data implies quite different physical pictures: in the former one intermediate-valence rare earth systems are considered qualitatively different from other metallic systems (e.g., transition metals), whereas the latter can be applied only if the fractional occupancy of 4f states is seen as a sign of the existence of a 4f band containing a small number of holes.

CONCLUSION

Unfortunately, it has not been possible to make a choice between these two pictures on the basis of our magnetic measurements. None of the two formulae given above is in satisfacotry agreement with the experimental χ(T) curves, and introducing new terms in either of them would increase the number of adjustable parameters and render the analysis meaningless. Further attempts to understand the observed temperature dependence will be undertaken with the use of new experimental results, such as specific heat and lattice-parameter data, which are now being collected at our laboratory.

REFERENCES

1. L.L.Hirst,AIP Conf.Proc.No.24 MMM 1974, San Francisco, p.11, M.Campagna,G.K.Wertheim and E.Bucher, Ibid,p.22.
2. A.R.Miedema,R.Boom and F.R.de Boer,J.Less-Common.Met 41 , 283 (1975).
3. K.A.Gschneidner Jr.,J.Less-Common Met.17,13 (1969).
4. K.H.J.Buschow,J.Less-Common Met.26,329(1973).
5. D.K.Wohlleben and B.R.Coles, Magnetism (Ed.G.T.Rado and H.Suhl), Vol.V,p.3, Acad.Press,NY (1966).
6. J.C.P.Klaasse,J.W.E.Sterkenburg,A.H.M.Bleyendaal and F.R.de Boer, Sol.St.Comm.12, 561 (1973).
7. K.N.R.Taylor,Adv.in Phys. 20, 551 (1971). H.Oesterreicher,J.Less-Common Met.30, 225 (1973).
8. E.E.Havinga,K.H.J.Buschow and H.J. van Daal, Sol.St.Comm. 13, 621 (1973).
9. E.E.Havinga, unpublished.
10. B.C.Sales and D.K.Wohlleben,Phys.Rev.Lett.35,1240 (1975).
11. S.Misawa,Phys.Lett.32A,153, 541 (1970).

MICROWAVE CONDUCTIVITY OF EuO AND EuSe

K. Kaski and P. Kuivalainen,
Helsinki University of Technology,
Electron Physics Laboratory,
and
T. Salo, H. Stubb and T. Stubb,
Technical Research Centre of Finland,
Semiconductor Laboratory,
SF-02150 Espoo 15, Finland

ABSTRACT

The conductivity and dielectric constant have been measured in the temperature range 1.7 to 300 K. Three samples have been studied: unannealed and annealed EuSe and unannealed EuO. At room temperature the microwave conductivity is $\sim 10^{-1}$, $8.0 \cdot 10^{-3}$ and $1.4 \cdot 10^{-3}$ $(\Omega\mathrm{cm})^{-1}$ respectively, the corresponding dc-conductivities being $3 \cdot 10^{-1}$, $\sim 10^{-5}$ and 10^{-7} $(\Omega\mathrm{cm})^{-1}$. The strong temperature dependence of the dc-conductivity is not seen in the microwave conductivity. The dielectric constants obtained from the same measurements are 24.0 and 12.1 for EuO and EuSe respectively. The strong frequency dependence of conductivity can possibly be explained by a hopping mechanism, which could depend on temperature and magnetic order.

INTRODUCTION

The Eu chalcogenides are semiconductors which order magnetically at low temperatures. EuSe is a magnetic semiconductor with a Neel temperature of 4.6 K and EuO is ferromagnetic below 69.4 K. The magnetic properties of Eu chalcogenides arise from the well localized spins of the Eu^{2+} ions. Pure and stoichiometric EuSe seems to be an insulator even at room temperature. Doping with other rare earth metals or nonstoichiometry always leads to an increase of the n-type conduction in EuSe.

Very strong influence of magnetic ordering on the electrical properties has been observed experimentally in Eu chalcogenides[1]. The most pronounced feature is the deep minimum in the electrical conductivity found near the magnetic ordering temperature. In the presence of a sufficiently large external magnetic field the minimum almost disappears indicating a very strong negative magnetoresistance. The reasons for the conductivity minimum and the negative magnetoresistance are still unknown to some extent.

Measurements of the high frequency conductivity offer a possibility to determine the dominant conduction mechanism: as known, band conduction decreases with increasing frequency whereas impurity conduction shows the opposite behaviour. Pollack[2] has pointed out that the increase with frequency in the hopping regime as contrasted with the decrease of ordinary mobility at high frequencies can be a direct test of the characteristic mode of motion. In this paper we present the results of microwave conductivity measurements in some Eu chalcogenides in order to provide information about the reasons for the conductivity minimum. We also obtained the dielectric constants of the materials at a microwave frequency.

EXPERIMENTAL PROCEDURE

The Eu chalcogenide single crystals were grown in sealed crucibles from metal-rich solutions following the method described by Reed et al.[3]. There was no intentional doping. X-ray fluorescence studies on the crystal lots used in this work revealed some 10 ppm or less of the following impurities: Si, W, Ti, Ca and Mg for EuO and Ca, K and Si for EuSe. X-ray diffraction analysis resulted in the monochalcogenide pattern plus a few unidentified peaks of low intensity. The crystals were stoichiometric within the limits of the atomic absorption measurements. Some EuSe crystals were annealed in Ar at 1600°C for 50 hours. The annealed samples contained cubic voids. Their dimensions were typically 10 μm, the total amount being a few percent by volume.

Measurements of the microwave conductivity at temperatures $1.5 \le T \le 300$ K were performed using the cavity perturbation technique[4],[5], where the conductivity is determined from the change in the quality factor of the resonant cavity caused by the sample. The measurements were performed inside a cryostat in vacuum, except for the temperature region $T \le 4.2$ K, where the cavity and the crystal were immersed in liquid helium. The microwave resonant cavity was made of copper. In order to prevent oxidation a thin gold layer was deposited using an electroless method. The resonant frequency of the empty cavity at room temperature was 23.30 GHz and the corresponding unloaded quality factor was 3200. Measurements of the dc-electrical resistivity of EuSe were reported earlier by J. Heleskivi et al.[6],[12].

EXPERIMENTAL RESULTS

The results of the microwave conductivity measurements on EuSe are shown in Fig. 1 and Fig. 2. The dc and mw conductivities were not measured on the same samples. The dc conductivity curves shown in Fig. 1 were measured on two different samples[6],[12], having roughly equal carrier concentration $4.5 \cdot 10^{18}$ cm^{-3}. The sample used for the lower curve is from the same lot as the one used in the mw measurements. The difference between the high frequency and the dc conductivity is several decades at temperatures $T < 15$ K. Unfortunately mw measurements were not possible with our system in the interval $4.2\ \mathrm{K} < T < 10$ K. Preliminary dc measurements on annealed EuSe samples resulted in a room temperature conductivity of order 10^{-5} $\Omega^{-1}\mathrm{cm}^{-1}$. The microwave conductivity of EuO was $1.4 \cdot 10^{-3}$ $\Omega^{-1}\mathrm{cm}^{-1}$ at room temperature, $3 \cdot 10^{-4}$ $\Omega^{-1}\mathrm{cm}^{-1}$ at liquid nitrogen temperature and $8 \cdot 10^{-2}$ $\Omega^{-1}\mathrm{cm}^{-1}$ at liquid helium temperature. The dc conductivity was several decades smaller than the high frequency conductivity and did not show a minimum at the transition temperature indicating the possibility that the EuO sample was of B-type according to the classification by Oliver[13].

Table 1. contains the volumes of the samples and the measured dielectric constants. The temperature dependence of the dielectric constants was small.

Sample	T/K	Volume/mm^3	Dielectric constant
EuSe (annealed)	293	0.302	12.1
EuSe (unannealed)	4.2	0.173	11.4
EuO	293	0.115	24.0

Table 1. The dielectric constants of EuSe and EuO at the frequency 23 GHz.

DISCUSSION

The dielectric constants measured in this work are approximately the same as the values 9.4 ± 0.8 and 23.9 ± 4 for EuSe and EuO respectively measured by J.D. Axe[8] at infrared frequencies. The cavity perturbation technique could not be used at higher temperatures ($T > 30$ K) on the unannealed EuSe sample because of its relatively high conductivity, which causes small skin depths. At lower temperatures where the dc conductivity is smaller the measured dielectric constant is the same as for annealed EuSe.

The measurements by Axe show that we can rule out the possibility of observing in the microwave conductivity measurements a resonance absorption due to the displacement of elastically bound ions as these resonance frequencies fall in the infrared region for EuO and EuSe. Therefore the observed high frequency conductivity must be due to charge carriers.

As mentioned above it was possible to measure by cavity perturbation technique the permittivity of unannealed EuSe only in the region $1.5\ K \leq T \leq 30\ K$, where the dc-conductivity was lower than $10^{-2}\ cm^{-1}\Omega^{-1}$. Fig. 1 shows that in the region $T \leq 15\ K$ the difference between the microwave conductivity and the dc-conductivity is several decades. Thus frequency has the same effect on conductivity as magnetic fields. The other interesting feature is the change in the microwave conductivity curvature near 15 K: For $T > 15\ K$ the microwave conductivity has approximately the same temperature dependence as the dc-conductivity. The values of the two conductivities are also rather close in this region. This indicates that the characteristic mode of motion of the charge carriers is related to band conduction at higher temperatures. In the region $T < 15\ K$ the microwave conductivity seems to be almost constant - the interval $4.2\ K < T < 10\ K$ being uncertain - whereas the dc conductivity decreases drastically. This points to the possibility that the conduction mechanism changes at low temperatures from band conduction to impurity conduction. In the hopping regime the high frequency conductivity is a sum of the dc conductivity σ_{DC} and the frequency sensitive dielectric losses, which are connected with the bound charge carriers in impurity states and proportional to the density of non-ionized impurities(2): $\sigma(\omega) = \sigma_{DC} + n_{imp}e\mu_{ac}(\omega)$. Thus the small temperature dependence of the high frequency conductivity in EuSe could be explained in the following way: although our crystals were not doped intentionally small deviations in stoichiometry could be responsible for donor type impurities. According to the bound magnetic polaron model(9) the ionization energy of these impurities depends strongly on temperature and has a maximum at T_c(10). The increase of bound charge carriers n_{imp} at low temperatures would increase the dielectric losses connected with them. This could compensate the decrease of the dc-conductivity leading to the observed small temperature dependence of the microwave conductivity. Measurements of the thermoelectric power in Eu-chalcogenides(6),(7) also indicate large changes in the carrier concentration near the magnetic ordering temperature.

The same kind of results as in this work have been obtained in the measurements of the high frequency conductivity of the magnetic semiconductor $CdCr_2Se_4$(11).

ACKNOWLEDGEMENTS

The authors are grateful to Dr. J. Heleskivi, Dr. J. Sinkkonen, Mr. M. Mäenpää and Mr. E. Lanne for helpful discussions.

REFERENCES

1. Methfessel, S. and Mattis, D.C., Magnetic Semiconductors, in Encyclopedia of Physics, Vol. 18/1, Wijn, H.P.J., Ed., Springer Verlag, Berlin, 1968, 389.
2. Pollack, M. in Proceedings of the International Conference on the Physics of Semiconductors. Exeter, July, 1962 (The Institute of Physics and the Physical Society, London, 1962), pp. 86-93.
3. Reed, T.B. and Fahey, R.E., J. Crystal Growth 8 (1971) 4, 337-340.
4. Sucher, M., Fox, J., Handbook of Microwave Measurements, Vol. II. John Wiley & Sons 1963.
5. Salo, T., Doctoral thesis, Technical Research Centre of Finland, publ. 7, 1973.
6. Heleskivi, J. and Shiosaki, T., in Proceedings of 21st Annual Conference on Magnetism and Magnetic Materials (1976).
7. Samohvalov, A.A., Afanas'ev, A.Ya., Gizhevskii, B. A., Loshsareva, N.N. and Simonova, M.I., Sov. Phys. Solid State 16, 365-366, 1974.
8. Axe, J.D., J. of Phys. and Chem. of Solids 30, 1403-1406 (1969).
9. Torrance, J.B., Shafer, M.W., McGuire, T.R., Phys. Rev. Letters 29, 1168 (1972).
10. Concalves de Silva, C.E.T., Solid State Communications 17, 677 (1975).
11. Kamata, N., Yamazaki, S., Kabashima, S., Hattand, T., Kawakuba, T., Solid State Communications 10, 905 (1972).
12. Heleskivi, J., Mäenpää, M., Stubb, T., Report 11, Semiconductor Laboratory, Technical Research Centre of Finland, September 1975.
13. Oliver, M.R., Ph.D. Thesis, MIT, June 1970.

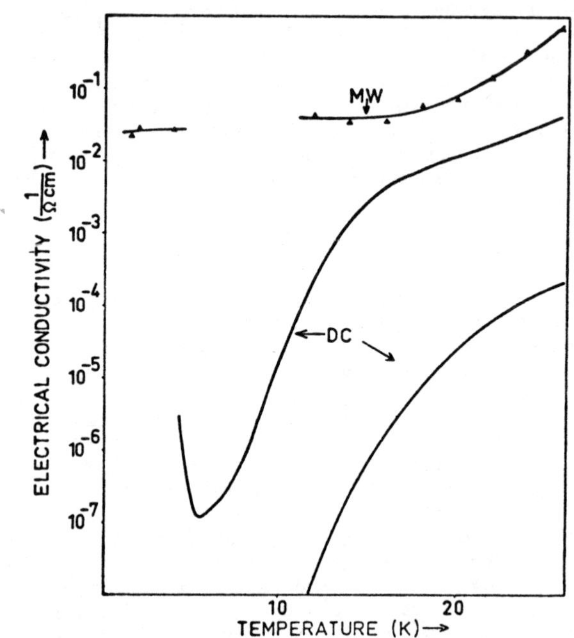

Fig 1. Dc and microwave conductivity of unannealed EuSe

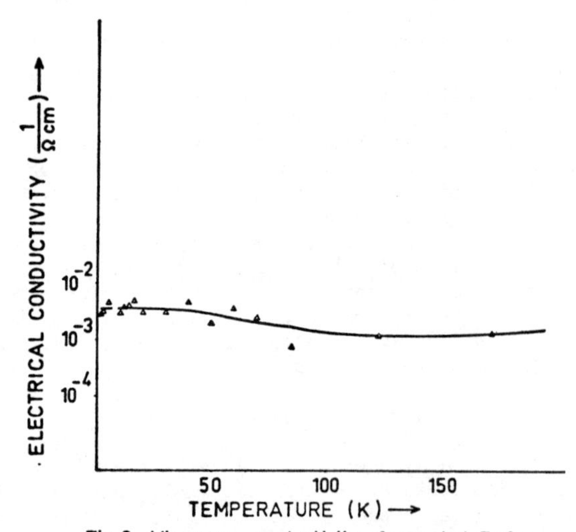

Fig. 2. Microwave conductivity of annealed EuSe

THEORY OF VARIATION FROM LOCALIZED TO ITINERANT MAGNETISM IN CUBIC INTERMETALLIC ACTINIDE COMPOUNDS

J. M. Robinson
Indiana University-Purdue University at Fort Wayne,
Fort Wayne, Indiana 46805

ABSTRACT

We derive the Fermion excitation spectrum of a cubic crystal of actinide ions with partly filled 5f shells. The crystal field and intraatomic interactions are assumed to be large compared to the one-electron interatomic hopping energy, so that Hubbard's "atomic representation" is valid. The magnetic moment, the magnetic susceptibility, and the fractional valence of the actinide ions are calculated as functions of the temperature, the external (or internal) magnetic field, and the f-f hopping energy. Assuming that the latter energy increases with decreasing lattice constant, one finds results in qualitative agreement with the variation from localized to itinerant magnetic behavior observed in intermetallic actinide compounds.

I. INTRODUCTION

Several series of cubic intermetallic actinide compounds such as NpX_2 (X=Aℓ, Os, Ir, or Ru) show a change from localized to itinerant magnetism as the actinide-actinide lattice separation d decreases.[1,2] For example, the ferromagnetic ordered moment μ_S at T=0°K decreases from $1.6\mu_B$ per Np ion in $NpA\ell_2$(d=3.37Å) to $0.4\mu_B$ in $NpOs_2$(d=3.26Å).[1] The isomer shift also decreases by 15 mm/sec, indicating increasing delocalization of the 5f electrons.[1,2] The delocalization may result from s-f hybridization or from direct overlap of the 5f wavefunctions on neighboring lattice sites. In a previous paper[3], the s-f hybridization was assumed to be dominant in those actinide intermetallics such as $UA\ell_4$ which do not order magnetically but which display "spin-fluctuation" resistivity. In the present work, which treats the magnetically ordered compounds, the f-f overlap will be considered. It is not likely that the f-f hopping energy is large enough to break down the intraatomic and crystal field level structure of the 5f ions, so that a single-particle or virtual bound state approach is not appropriate. Instead, we use the previously described standard-basis operator technique[3,4] within Hubbard's atomic representation[5], which reduces to the localized, crystal field description in the zero-bandwidth limit.

II. THE THEORETICAL MODEL

We consider a crystal of N actinide ions having partly filled 5f shells. The crystal field eigenstates of an ion at site i are each labeled by an integer p and a symbol $|ip\rangle$, where

$$|ip\rangle=|i5f^nLSJ\Gamma_\alpha\rangle \ . \quad (1)$$

In Eq. (1), the letters L, S, and J are the usual angular momentum quantum numbers, and Γ_α labels the crystal field symmetry. Following Lea, Leask, and Wolf[6], we assume that

$$|i5f^nLSJ\Gamma_\alpha\rangle = \sum_{M=-J}^{J} a_M(\Gamma_\alpha)|i5f^nLSJM\rangle \ , \quad (2)$$

where the a_M are crystal field coefficients. The effect of f-f overlap is that an ion at site j in the state $|js\rangle$ of the $5f^n$ configuration may lose an electron and change to the state $|jr\rangle$ of $5f^{n-1}$. The "emitted" electron hops to site i, changing it from $|iq\rangle$ of $5f^n$ to $|ip\rangle$ of $5f^{n+1}$, and this excitation propagates in the lattice as a Bloch wave. Note that the letters p, q, r, and s each represents a different set of quantum numbers. The energy for this "hopping" process is called $B_{ij}(pqrs)$. The Hamiltonian H of the crystal becomes

$$H=\sum_{pi}(\varepsilon_p-\zeta n_p)L^i_{pp}+\sum_{pqrs}\sum_{ij} B_{ij}(pqrs)L^i_{pq}\,L^j_{rs} \ . \quad (3)$$

In Eq. (3) the energy of the state p having n_p electrons is denoted by ε_p, ζ is the chemical potential, and L^i_{pq} is a standard-basis[4] (or atomic)[5] operator causing a transition of the ion at site i from the state q to the state p. The first term in H describes an ensemble of non-interacting ions; the second term leads to highly correlated electronic energy bands. The coefficients B_{ij} are given by the expression[5]

$$B_{ij}(pqrs)=\sum_{mm'\sigma}t_{ij}(m\,m')\ \langle ip|C^+_i(m'\sigma)|iq\rangle\langle jr|C_j(m\sigma)|js\rangle \ , \quad (4)$$

where $c_j(m\sigma)$ is the destruction operator for a single electron in an atomic f orbital of spin $\sigma=\pm1/2$ and orbital projection $\ell_z=\hbar m$. The quantity $t_{ij}(m\,m')$ is the hopping energy between two such orbitals and occurs in the tight-binding band model.[7] In practical applications the $t_{ij}(m\,m')$ are specified by a few variable parameters.[7] The matrix elements on the right-hand side of Eq. (4) may be evaluated by the methods of atomic spectroscopy,[8] and a computer program was developed to perform the evaluation for given values of the quantum numbers and crystal field coefficients of the states p, q, r, and s. Note that in the previous paper[3], the matrix elements are equal to ±1, because there the states $|ip\rangle$ are assumed to be atomic orbitals. In the present work the states $|ip\rangle$ are linear combinations of determinantal wavefunctions.

The equations of motion for the Green's functions $\langle\langle L^i_{pq};L^j_{rs}\rangle\rangle$ may now be written down and solved in the RPA approximation as previously described.[3] The poles of the Fourier-transformed Green's functions yield the energy bands, which depend upon the occupation probabilities $D_p=\langle L^i_{pp}\rangle$ of the states $|ip\rangle$. The probabilities D_p are the solutions of non-linear self-consistent equations, which are solved numerically, subject to the constraints $\Sigma D_p=1$ and $\Sigma n_pD_p=n$, where nN is the total number of f electrons. The diagonal average value of the localized magnetic moment $\vec{\mu}_i$ per atom is then given by the equation

$$\langle\vec{\mu}_i\rangle = \sum_p D_p\ \langle ip|\vec{\mu}_i|ip\rangle \ . \quad (5)$$

III. RESULTS

As an illustration of the above theory, we consider (for simplicity) a simple cubic lattice. The spin-orbit and crystal field energies are assumed to be large compared to the hopping energy so that only the states $|ip\rangle$ (p=1,2,3,4) shown in Fig. 1 need be considered. In the zero bandwidth limit, the ground state of the actinide ions is assumed to be the magnetic doublet $|i5f^3\ L=6,S=3/2,J=9/2,\Gamma_6^\pm\rangle$. We include an external (or self-consistent internal) magnetic field h along the z-axis. The field splits the energy of the doublet by $\pm\mu_0h$, where $\mu_0(=1.33\mu_B)$ is independent of the strength of the crystal field.[6] There are two excited singlet states $|i5f^2\ L=5,S=1,J=4,\Gamma_1\rangle$ and $|i5f^4\ L=6,S=2,J=4,\Gamma_1\rangle$ having energies ε_3 and ε_4, respectively. The set of states listed above is clearly the simplest choice compatible with the crystal symmetry of the intermetallic actinides. In Eq. (4), we consider only nearest neighbor interactions and make the further approximation $t_{ij}(m\ m')=\frac{t}{2}$, independent of m and m'. There are sixteen non-zero quantities $B_{ij}(pqrs)$, which are all related to the single hopping energy t by Eq. (4). The crystal field coefficients in Eq. (2) are taken from Ref. 6. The equations of motion for the Green's functions were solved and the energy spectrum and other observables were calculated by the procedure described in the last section.

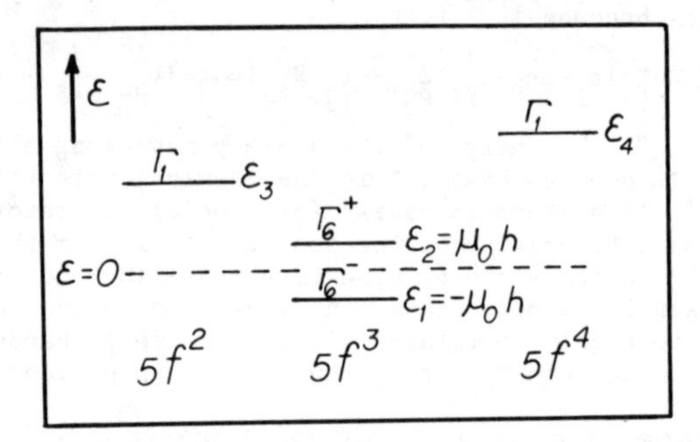

Fig. 1 The energy levels of the actinide ions in the zero-bandwidth limit. The symbols Γ_1 and $\Gamma_6^{\pm}$ denote a crystal field singlet and doublet, respectively. Here h is an internal magnetic field, ε_3 and ε_4 are parameters of the model, and $\mu_o=1.33\mu_B$.

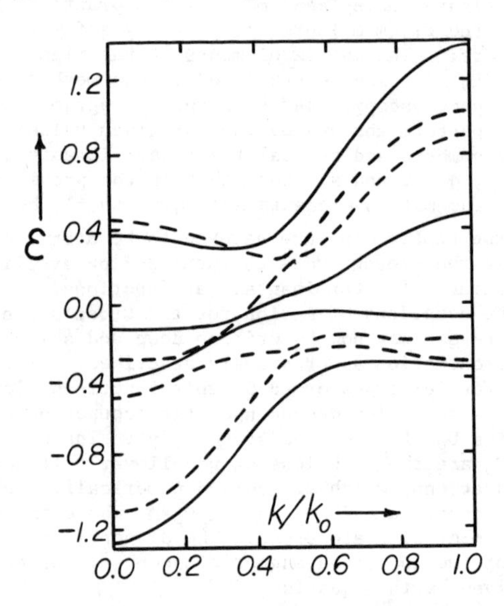

Fig. 2 The Fermion excitation spectrum (energy bands) in the [111] direction of the simple cubic Brillouin zone for the case t=4.0, $\varepsilon_3=\varepsilon_4=0.25$, and $k_BT=0.025$ (arbitrary energy units). Solid curves: $\mu_oh=0.20$. Dashed curves: $\mu_oh=0.025$. Here $k_o=\sqrt{3}\ \pi/a$.

In Fig. 2 the energy spectrum in the [111] direction of the cubic Brillouin zone is shown for the case $\varepsilon_3 = \varepsilon_4 = 0.25$ (arbitrary energy units), t=4.0, $k_BT=0.025$, and for two values of the field h. Note the wavevector-dependent magnetic splitting of the bands. There are weakly dispersed, mostly localized modes overlapped by strongly dispersed or bandlike modes.

In Fig. 3 we see the induced magnetic moment $\mu=\mu_o(D_1-D_2)$ per ion as a function of the magnetic field h in the z-direction for two values of the temperature T and for various values of the hopping energy t. As t increases, so does the field required for saturation of μ as is observed experimentally.[1] For T=0.025 and $0.25 \lesssim t \lesssim 1.5$, the slope of the $\mu(h)$ curves (i.e. the susceptibility χ) decreases as $h\to 0$. This indicates the presence of antiferromagnetic short range order which is broken up at higher fields, temperatures, and hopping energies. From curves such as those in Fig. 3 one can calculate the spontaneous ferromagnetic moment μ_s per ion as a function of T in the effective field approximation in which $h=\lambda\mu_s/\mu_o$, where λ is the effective field constant. The results for t=5.0 are shown in Fig. 4. For $\lambda=0.25$, the magnetization curve resembles a J=1/2 Brillouin curve but with a small value $(0.36\mu_B)$ of the ordered moment $\mu_s(0)$ at T=0°K. This behavior is observed, for example, in $NpOs_2$.[1] For $\lambda=0.20$, we find $\mu_s(0)=0.28\mu_B$, and the magnetization curve resembles that of a weak itinerant ferromagnet such as $U_{0.5}Pt_{0.5}$ at zero pressure.[9] Naturally, for the case t=0, $\varepsilon_3 = \varepsilon_4 = \infty$, one obtains a J=1/2 Brillouin curve with $\mu_s(T=0)=1.33\mu_B$.

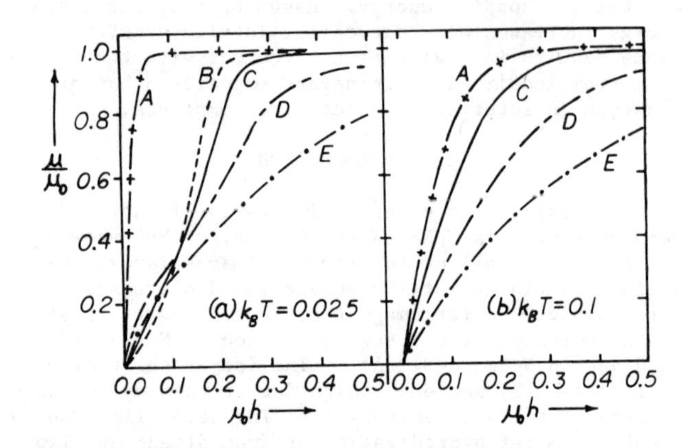

Fig. 3 The z component of the magnetic moment μ per ion as a function of the magnetic field h for (a) $k_BT=0.025$ and (b) $k_BT=0.1$ (arbitrary energy units). Curves A, B, C, D, and E correspond to the values 0.0, 1.0, 2.0, 5.0, and 10.0, respectively, of the hopping energy t. In part (b), curve B is close to curve A and is omitted for the sake of clarity.

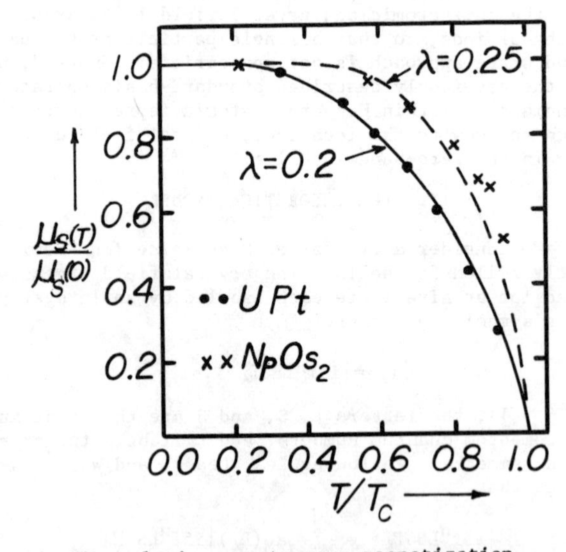

Fig. 4 The relative spontaneous magnetization $\mu_s(T)/\mu_s(0)$ as a function of T/T_c (T_c=Curie temperature). Solid curve: theory with $\varepsilon_3=\varepsilon_4=0.25$, t=5.0, and molecular field constant $\lambda=0.2$. Dashed curve: same as solid curve except $\lambda=0.25$. Dots: Experimental data on a typical weak itinerant ferromagnet, UPt, from Ref. 9. Crosses: Experimental data on $NpOs_2$, from Ref. 1.

The delocalization or fractional loss of the $5f^3$ configuration in the ground state at low temperature is given by $P=D_3+D_4$ and is shown as a function of t in Fig. 5. The gradual increase of P with increasing t correlates with the experimentally observed variation of the isomer shift with lattice constant in intermetallic actinides,[1,2] if one assumes that t increases with decreasing lattice constant. As the f electrons delocalize, they are expected[1,2] to screen less effectively the s electrons, leading to a change in the charge density at the nucleus and therefore to the observed isomer shift.

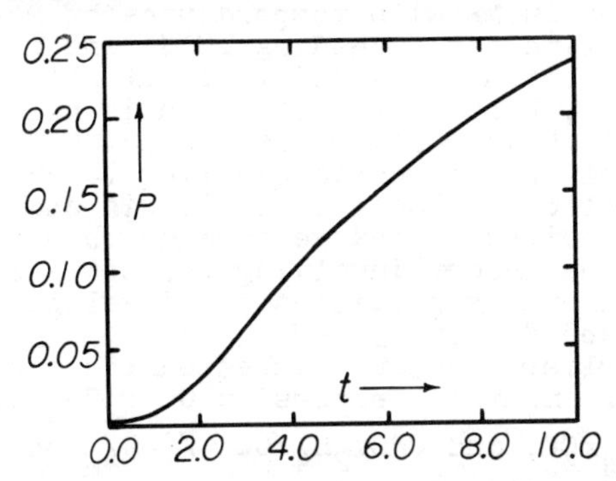

Fig. 5 The delocalization parameter P (the fraction of polar states) as a function of the hopping energy t for the case $\varepsilon_3=\varepsilon_4=0.25$, $k_BT=0.025=\mu_0 h$ (arbitrary energy units).

IV. DISCUSSION

The results described above show that the main features of the magnetically ordered intermetallic actinides can be explained by taking into account the effect of intraatomic and crystal field interactions on the band structure. In particular, the model explains how a decrease in the lattice constant leads to a slow approach to saturation of the field-induced magnetization, a small value of the ordered moment at T=0°K, and a negative isomer shift. The model also shows how the temperature dependence of the spontaneous magnetization may be either a J=1/2 Brillouin function or a function characteristic of a weak itinerant ferromagnet, depending on the relative values of the exchange and hopping energies. Thus, the experimental observation of a Brillouin function does not prove that the f electrons are fully localized. For more quantitative results, a larger number of atomic levels should be considered. A diagrammatic technique would be very useful in doing this. A more serious defect of the present model is the neglect of the s-d-f hybridization, which is expected to influence the magnetic properties in the same way as the f-f overlap. The s-d-f hybridization will be incorporated into the theory by including atomic configurations of the form $5f^n6d^m7s^p$. Also, the calculation should be carried out for the cubic Laves-phase lattice of the NpX_2 compounds, with account taken of the complicated dependence of the hopping energies t_{ij} on the quantum numbers m and m'.[7] The simplifying assumptions made in Section III are inadequate for a quantitatively correct band structure.

The present model is similar to the "minimum polarity" theory of ferromagnetic metals[10] and the Hirst model[11] of magnetic impurities in metals. A difference is that the calculation of the excitation spectrum is greatly facilitated by the standard basis operators used here.

REFERENCES

1. A. T. Aldred, B. D. Dunlap, D. J. Lam, and I. Nowik, Phys. Rev. B10, 1011(1974).
2. J. Gal, Z. Hadari, U. Atzmony, E. Bauminger, I. Nowik, and S. Ofer, Phys. Rev. B8, 1901(1973).
3. J. M. Robinson, A.I.P. Conf. Proc. (21st Annual Conf. on Magnetism and Magnetic Materials-Philadelphia 1975), to appear.
4. S. B. Haley and Paul Erdös, Phys. Rev. B5, 1106 (1972).
5. J. Hubbard, Proc. Phys. Soc. A, 277, 237(1964); 285, 542(1965).
6. K. R. Lea, M. J. Leask, and W. P. Wolf, J. Phys. Chem. Solids 23, 1381(1962).
7. J. C. Slater and G. F. Koster, Phys. Rev. 91, 1498(1954).
8. R. Stevenson, Multiplet Structure of Atoms and Molecules (W. B. Saunders Co., Philadelphia, 1965) Ch. 5, p. 84.
9. J. G. Huber, M. B. Maple, and D. Wohlleben, Journ. Magnetism and Mag. Mat. 1, 58(1975).
10. S. V. Vonsovskii, Magnetism (John Wiley and Sons, Jerusalem, 1974) vol. 2, Ch. 4, p. 661.
11. L. L. Hirst, Phys. Kondens. Materie 11, 255(1970).

CHARGE ORDER-DISORDER AND INSULATOR-METAL TRANSITION IN MAGNETITE

Tomoyasu Tanaka and Charles C. Chen,
Ohio University, Athens, Ohio 45701
and
Nino L. Bonavito,
NASA, G.S.F.C., Greenbelt, Maryland 20771
and
J. C. Foreman,
C.S.T.A., Seabrook, Maryland 20801

ABSTRACT

A Hartree-Fock self-consistent calculation of the electronic band structure of Magnetite with four sublattice assumption is carried out over a wide range of temperature including the Verwey transition point. From the crystallographic symmetry it is shown that the so called Verwey charge order is not acceptable. With the assumption of two electrons per four B-sites, three charge order parameters and six nearest neighbor exchange correlations, plus the chemical potential are introduced. When these ten parameters are consistently and numerically evaluated it is found that the allowed charge order is identical with the model proposed by Yamada*, i.e., the Δ_5-symmetry appears and the four electronic energy bands are split into two almost doubly degenerate bands separated by a large gap representing an insulating phase. As the temperature is increased each of the degenerate bands is broadened and the gap decreases, accompanied by a rapid decrease of the charge order parameters. At a temperature slightly below the point at which the charge order parameters vanish, the gap between the second and the third bands disappears, while the charge order still maintains a residual Δ_5-symmetry.
*Y. Yamada, AIP Conf. Proc. 24, 79 (1974).

MAGNETIC PROPERTIES OF $CdCr_2Se_4$

V.T.Kalinnikov, T.G.Aminov
N.S.Kurnakov Institute of General and Inorganic Chemistry of the Academy of Sciences of the U.S.S.R., Moscow, U.S.S.R.

K.P.Belov, L.I.Koroleva, M.A.Schalimova
Department of Physics, Moscow State University, Moscow, U.S.S.R.

ABSTRACT

The dependence of magnetization, σ, and resistivity, ρ, on temperature T and magnetic field H has been studies near Curie point and at somewhat higher temperatures for single crystals of $CdCr_2Se_4$ deficient in selenium or doped with indium or copper. The magnetic field has a pronounced effect on the T_c values determined by extrapolation of the steepest section of the σ(T) curves to the temperature axis. Thus, an increase of the field strength from 0.9 to 12.7 kOe leads to a 20 to 30° high temperature shift of the Curie point. The σ(H) curves measured near T_c are characterized by critical field values in the vicinity of which magnetization increases stepwise by about 80%. Copper doped crystals exhibit thermomagnetic hysteresis and those containing indium show positive magnetic resistivity in the temperature region where the critical field phenomena are observed. The experimental data are explained by the formation of ferrons, that is ferromagnetic microregions, under the action of magnetic field. The ferrons which owe their stability to the s-d exchange and Coulomb attraction are supposed to be localized at non-ionized impurity atoms. The magnetic field induced transition from a paramagnetic state to that characterized by the presence of ferrons represents the first-order phase transition.

RESULTS

Strong exchange coupling between charge carriers and localized magnetic moments (the s-d exchange) occurs in magnetic semiconductors (MS), in particular, in $CdCr_2Se_4$. As a consequence, these materials have specific magnetic and electric properties which, according to the current theory, depend on the presence of peculiar magnetic microregions characterized by higher degree of magnetic ordering called ferrons. The formation of ferrons in non-doped MS was theoretically predicted by Nagaev /I, 2/, and von Molnar and Methfessel /3/. These authors have shown that autolocalization of a charge carrier with ferromagnetic ordering of magnetic ions in its vicinity may be energetically favoured because of the s-d exchange. The probability for the formation of ferrons in ferromagnetic semiconductors is maximum near the magnetic ordering temperature and is practically zero at T=0 . Ferrons may have radii up to 50-100 Å depending on the temtemperature. Computations of probabilities for the formation of free ferrons carried out by Yanase /4/ have shown that ferrons may occur in MS with temperatures of magnetic ordering not exceeding I0°K. The formation of ferrons in doped MS eventually takes place, because the electron is hold by the Coulomb attraction near defect and the degree of ferromagnetic ordering to exist in the electron location region is stabilized by indirect exchange through this electron. The theory involving ferrons localized at impurity particles was developed by Yanase and Kasuya /5, 6/.

Earlier, we have suggested the occurrence of impurity ferrons in $CdCr_2Se_4$/Ga, $CdCr_2Se_4$/Cu, and $CdCr_2S_4$/Cu /7-9/ on the basis of indirect evidence, such as observed photoconductivity minima in the region of the Curie temperature for samples of $CdCr_2Se_4$ weakly doped with gallium /7/ and shifts of paramagnetic Curie points in $CdCrSe_4$/Cu and $CdCr_2S_4$/Cu having constant temperatures of ferromagnetic ordering /8,9/. We now report additional data supporting the occurrence of ferrons near T_c in these materials. The effects we are going to discuss are: increase of T_c with external magnetic field H accompanied by critical field phenomena observed for σ(H) curves, thermomagnetic hysteresis and positive magnetic resistance at temperatures near T_c.

Techniques for obtaining single crystals and production of Ohmic contacts likewise the data on the chemical and X-ray analysis of the crystals may be found in /7, 9-11/. The discription of eigther magnetization measurement techniques or magneto (electro) resistance determination methods is also given there.

The term "T_c increase" should be understood as follows. The Curie point is to be determined in the absence of the external magnetic field. Howe ver, the Curie point is practically determined by the extrapolation of the most rapid varing part of the curve σ(T) on the temperature axis, because of simplicity of this method, though, generally speaking, it won't be the Curie point exactly, but some characteristic temperature T_c near to the Curie point. It is in this way that we shall understand further the "T_c increase" term ΔT_c with the switching on of the external magnetic field, where the σ (T) curve is taken. The so obtained values depend markedly on the field strength in the case of single crystals of $CdCr_2Se_4$ doped with copper or indium or deficient in selenium by ca 1.5%. The σ(T) curves obtained at various H for $Cd_{0.93}Cu_{0.07}Cr_2Se_4$ and shown in Fig.1 illustrate the observed dependence. An increase of H from 0.9 to 12.7 kOe leads to an increase of T_c (ΔT_c) of about 20-30°. The data listed in the Table show that ΔT_c values vary monotonically with the degree of doping over the series of the samples studied. The Table also presents the data for single crystals of $CdCr_2Se_{3.9}$ (ΔT_c of 23°) and for stoichiometric to within detecting power of our instrument single crystals $CdCr_2Se_4$ ($\Delta T_c \sim 4°$). For comparison, similar measurements were taken with a gadolinium single crystal to obtain the ΔT_c value below 3°. Hence we conclude that the "Curie point shifts" are due to impurity or selenium deficit. All the samples studied excepting compound I feature σ(H) curves with critical H values under the conditions responsible for shifting of Curie points. The σ(H) curves for $CdCr_2Se_{3.9}$ (Fig.2) provide an example. Magnetization of $CdCr_2Se_{3.9}$ rises abruptly when approaching the critical field value (by about 80% at fields near 5 kOe) at temperatures of 143 to 164°K while at the lower temperatures, smooth variation of σ with H is observed.

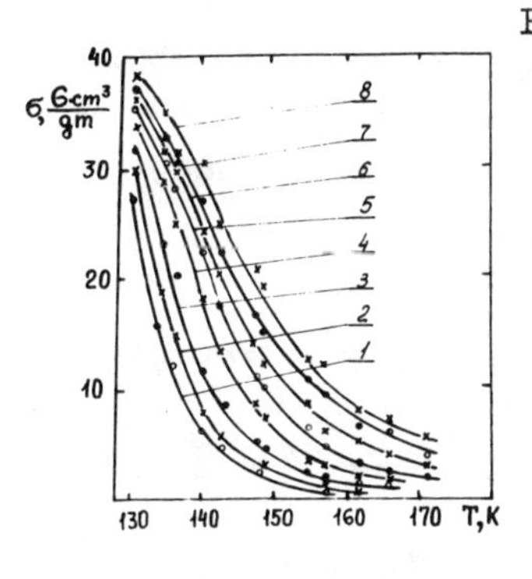

Fig.1. Magnetization, σ, of a single crystal of the composition $Cd_{0.93}Cu_{0.07}Cr_2Se_4$ as a function of temperature. The values of H (kOe) are the following: 1 - 0.9; 2 - 1.3; 3 - 2.25; 4 - 4.15; 5 - 6.2; 6 - 8; 7 - 9.6; 8 - 12.7.

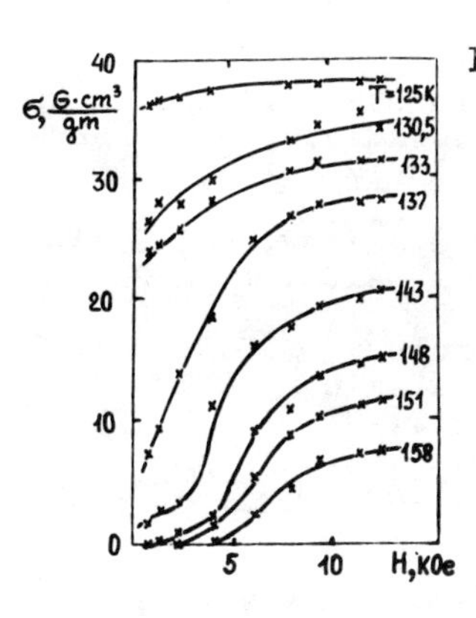

Fig.2. Magnetization, σ, of a selenium deficient single crystal of $CdCr_2Se_{3,9}$ as a function of magnetic field H in the temperature region where ΔT_c is observed.

For copper and indium doped samples together with samples having selenium deficit we investigated thermomagnetic hysteresis, as shown in Fig.3 for the composition $Cd_{0,93}Cu_{0,07}Cr_2Se_4$.

Indium doped single crystals of the composition $Cd_{1-x}In_xCr_2Se_4$ where x is 0.004 or 0.013 show positive magnetic resistance (MR) at temperatures above 140°K. Figures 4 and 5 demonstrate temperature dependence of MR for these compounds. Near the Curie point, MR is negative and goes through a minimum. As temperature rises, MR changes sign and reaches its maximum value. The external field maintains long-range magnetic ordering in the region of T_c which reduces magnetization fluctuations caused by the s-d exchange and hence diminishes scattering at such fluctuations. For that reason, negative MR values are observed near T_c characteristic to the n-type $CdCr_2Se_4$ /12, 13/. In the paramagnetic region characterized by the absence of long-range magnetic ordering, magnetic field, as shown by Nagaev /14/, may favour the magnetization fluctuations round defects thus leading to enhanced scattering and positive MR values, which is probably the case with our compounds.

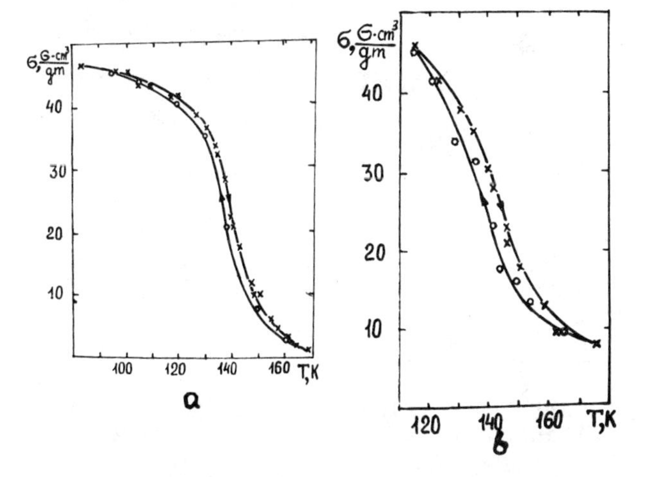

Fig.3. Single crystal of the composition $Cd_{0.93}Cu_{0.07}Cr_2Se_4$. Thermomagnetic hysteresis a) in the field of 6.2 kOe and b) in the field of 8 kOe.

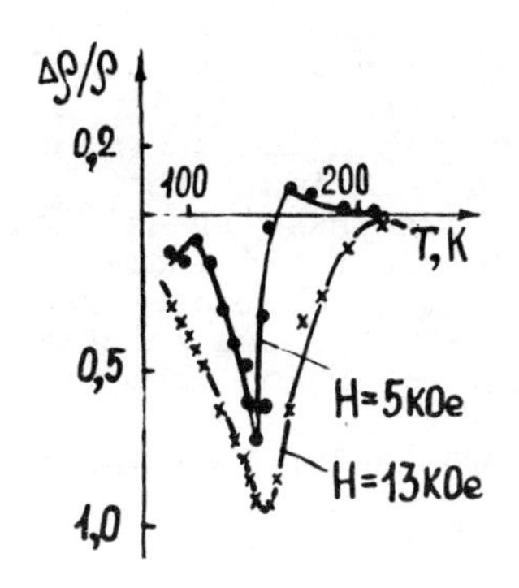

Fig.4. Temperature dependence of magnetic resistance, $\Delta\rho/\rho$, for the sample $Cd_{0.996}In_{0.004}Cr_2Se_4$.

Table

Composition	T_c, ^oK H=0.9 kOe	T_c, ^oK H=12.7 kOe	ΔT_c, ^oK ΔH=11.8 kOe
Gd	287	290	3
$CdCr_2Se_4$ [1]	140	144	4
$CdCr_2Se_{3,9}$	138	161	23
$Cd_{0.96}Cu_{0.04}Cr_2Se_4$	136.6	162	25.4
$Cd_{0.93}Cu_{0.07}Cr_2Se_4$	138.8	167	28.2
$Cd_{0.86}Cu_{0.14}Cr_2Se_4$	132	168	36
$Cd_{0.993}In_{0.007}Cr_2Se_4$	139.5	151	11.5
$Cd_{0.987}In_{0,013}Cr_2Se_4$	139,5	150,5	11
$Cd_{0.98}In_{0.02}Cr_2Se_4$	137.5	155	17.5

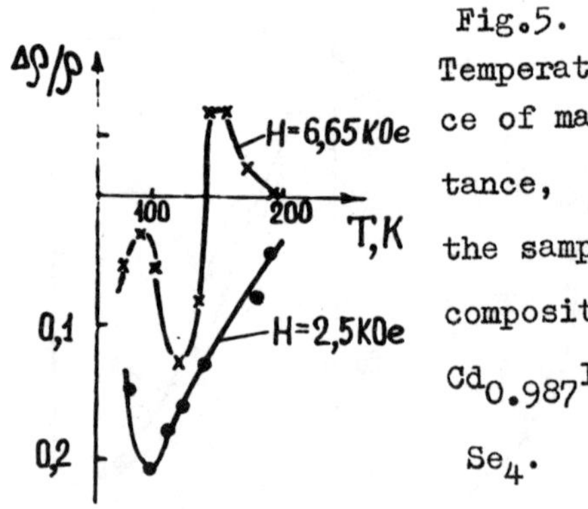

Fig.5. Temperature dependence of magnetic resistance, $\Delta\rho/\rho$, for the sample of the composition $Cd_{0.987}In_{0.013}Cr_2Se_4$.

All these findings including the critical character of σ vs. H dependence, positive magnetic resistance near T_c, and variations of T_c under the action of magnetic field, can be explained on the following hypothesis. Ferrons have maximum magnetic moment near T_c where long-range magnetic ordering undergoes destruction but short-range ordering is not jet destroied. At the higher temperatures, the ferrons, however, disappear rapidly because of thermal disordering. At temperatures slightly above T_c, magnetic field turn on enhances ordering in the vicinity of donor (acceptor) centres to a greater extent than it does on the average over the crystal, because of contributions from electrons on nonionized impurity atoms. The critical field and thermomagnetic hysteresis phenomena imply the occurrence of a first-order phase transition under the action of magnetic field from a paramagnetic state to a state characterized by the presence of impurity ferrons which, seemingly, favour long-range magnetic ordering.

We wish to thank Drs.Nagaev and Grigin for valuable discussions.

REFERENCES

1. É.L.Nagaev, ZhETF Pis. Red., 6, 484, 1967.

2. É.L.Nagaev, ZhETF 54, 228 (1968).

3. S.von Molnar, S.Methfessel, J.Appl. Phys. 38, 959, 1967.

4. A.Yanase, T.Kasuya, J.Phys.Soc. Japan 25, 1025, 1968.

5. A.Yanase, Intern. J.Magnetism 2, 99, 1972.

6. T.Kasuya, A.Yanase, Rev.Mod.Phys. 40, 684, 1968.

7. K.P.Belov, L.I.Koroleva, S.D.Batorova, M.A.Shalimova, V.T.Kalinnikov, T.G. Aminov, G.G.Shabunina, N.P.Shapsheva, ZhETF Pis.Red. 20, 191, 1974.

8. K.P.Belov, I.V.Gordeev, L.I.Koroleva, A.V.Ped'ko, Yu.D.Tret'yakov, V.A.Alferov, E.M.Smirnovskaya, Yu.G.Samsonov, Zh.ETF 63, 1321, 1972.

9. K.P.Belov, L.I.Koroleva, M.A.Shalimova, V.T.Kalinnikov, T.G.Aminov, G.G.Shabunina, Fiz.tverd.Tela 17, 3156, 1975.

10. V.T.Kalinnikov, T.G.Aminov, E.S.Vigileva, N.P.Shapsheva, PTE . 1, 227.

11. K.P.Belov, L.I.Koroleva, S.D.Batorova, V.T.Kalinnikov, T.G.Aminov, G.Yu.Shabunina. JETP Lett., 22, 140, 1975.

12. H.W.Lehmann, Phys. Rev., 163, 488, 1967.

13. P.F.Bongers, C.Haas, van A.M.Run, G.Zanmarchi, J.Appl Phys. 40, 958, 1969.

14. V.A.Kashin, É.L.Nagaev, ZhETF Pis. Red. 21. 126, 1975.

SURFACE MAGNETISM

G.S. Krinchik, V.E. Zubov and E.A. Ganshina
Department of Physics, Moscow State University,
Moscow 117234, USSR

The process of reconstruction of the magnetic structure of thin surface layers of the weak ferromagnetic insulator hematite (α - Fe_2O_3) and of the ferromagnetic semiconductors Hg Cr_2Se_4, $Cd_{1-x}Hg_xCr_2Se_4$ and $Cd_{1-x}Zn_xCr_2Se_4$ has been observed by a magneto-optical method. On the natural faces of hematite we have observed a macroscopic intermediate layer like a domain wall; the magnetic state of the surface is qualitatively different from that of the interior. The main factor in the appearance of the surface layer is the magnetodipole energy of the Fe^{3+} ions. The surface energy has been calculated for (100) and (111) faces and it was shown that for each face there are two values of the surface anisotropy energy, which depend on the type of the Fe^{3+} ions in the surface layer. Erasure of this surface magnetism by an applied field has been observed and theoretical explainations are offered. Details of this problem are discussed in reference 1.

On the natural faces of ferromagnetic semiconductors there were observed spontaneous drastic changes of magneto-optical properties that have proved the radical change of magnetic and electronic structure of the surface layer of the crystal.

In $HgCr_2Se_4$, we observed a two-times increase of the magnitude of the maximum of the equatorial Kerr effect after the sample had been kept for three or four hours at 77° K. The alternating magnetic field destroys this state and magneto-optical properties become stable. The unstable state restores after annealing of the sample at 100°C and, in our opinion, it evidences the decisive role of giant spin molecules (ferrons) present in the ferromagnetic semiconductor at $T > T_c$.

In crystals of $Cd_{1-x}Zn_xCr_2Se_4$ and $Cd_{1-x}Hg_xCr_2Se_4$, we found an essential change of the entire form of the curves $\delta(\omega)$ - a 1 ev frequency shift of the positions of singularities in these curves - that indicates a total reconstruction of the band structure of the ferromagnetic semiconductor in the surface layer.

1. G.S. Krinchik, A.P. Khrebtov, V.E. Zubov, and A.A. Askochenski. JETP Letters 17, 466, 1973;
G.S. Krinchik, V.E. Zubov. JETP 69, 707, 1975.

MAGNETIC NEUTRON SCATTERING ON THE INTERMEDIATE VALENCE SYSTEM $CePd_3$

E. Holland-Moritz M. Loewenhaupt W. Schmatz
Institut für Festkörperforschung der KFA Jülich, D-5170 Jülich, Germany
and
D. Wohlleben
II. Phys. Institut der Universität Köln, D-5000 Köln, Germany

ABSTRACT

$CePd_3$ is an intermediate valence intermetallic compound /1,2/ fluctuating between a nonmagnetic Ce^{4+} ($4f^0$, J=0) and a magnetic Ce^{3+} ($4f^1$, J=5/2)state. We report on measurements /3/ of the diffuse neutron scattering cross-section with TOF-analysis at several angles in the temperature interval 3.5 - 300 K.

We find a quasielastic magnetic line with 4f formfactor and an abnormally large, nearly temperature independent width ($\Gamma/2=18\pm2$ meV at 240 K). The lower the temperature the more the measured magnetic cross-section is reduced compared to the value calculated with the assumption that all Ce-ions are in the $4f^1$ (Ce^{3+}) state (Curie-susceptibility). However, the cross-section is consistent with the actually measured, nearly temperature independent static susceptibility /1/.

In the phenomenological interfiguration-fluctuation-model /4/ the static susceptibility of $CePd_3$ is given by

$$\chi(T) = \frac{(g\mu_B)^2 \cdot J(J+1)}{k_B(T+T_{SF})} \cdot n(T)$$

with

$$n(T) = \frac{6}{6+\exp(-E_{exc}/k_B(T+T_{SF}))}$$

Here n(T) is the average occupation of the 4f-shell (valence). If one identifies $k_BT_{SF}= \Gamma/2$ and assumes a temperature independent E_{exc}= -63 meV, the neutron scattering cross-section is also consistent with the anomalous temperature dependence of the lattice parameter /2/ according to

$$a(T) = a^{4+}(T)[1-n(T)] + a^{3+}(T)\cdot n(T)$$

/1/ W.E. Gardener, J. Penfold, T.F. Smith, I.R. Harris J. Phys. F2, 133 (1972)
/2/ I.R. Harris, M. Norman, W.E. Gardener J. Less Common Metals 29, 299 (1972)
/3/ The measurements were performed on the HFR-ILL in Grenoble and on the FRJ-2 in Jülich
/4/ B.C. Sales, D.K. Wohlleben Phys. Rev. Lett. 35, 1240 (1975)

PARAMAGNETIC ANISOTROPY MEASUREMENTS ON DEOXYMYOGLOBIN

Shunichi Kawanishi
Research Reactor Institute, Kyoto University
Yuhei Morita
Research Institute for Food Science, Kyoto University
and
Akira Tasaki
Engineering Science, Osaka University, Toyonaka, Osaka

ABSTRACT

Paramagnetic anisotropy of a single crystal of deoxymyoglobin (type A) was studied by measuring magnetic torque in the temperature range of 1.9 to 77 K. Torque curves were measured in three perpendicular planes of the crystal by the rotating magnetic field.

Result of these measurements leads to a conclusion that the temperature dependence of these three torque curves are well explained by the following spin Hamiltonian

$$\mathcal{H} = D S_z^2 + E(S_x^2 - S_y^2)$$

where D = 5.2, E = 0.4 cm^{-1} and S = 2. These obtained values of D and E are almost the same as those in deoxyhemoglobin, however, the principal axis (z axis) seemed to be parallel to the heme plane.

INDTODUCTION

Hemoprotein, such as hemoglobin, myoglobin, etc., possesses an iron ion at the center of its physiological activity and the spin state of the central iron ion is closely related to the ligand bond of these macromolecules. Deoxyhemoglobin in the venous blood, for instance, possesses high spin type Fe^{2+} ions, while oxyhemoglobin in the arterial blood had that of low spin type.

During the last decade, we have reported magnetic measurements and theoretical analysis of the electronic state of heme iron1)-4).

The profile of the ferric (Fe^{3+}) heme iron is as follows: (1) The fine structure of the lowest level ($^6A_1$) for the high spin iron ion in methemoglobin and metmyoglobin is well represented by the spin Hamiltonian DS_z when D was as chosen to 10 cm^{-1}. (2) In the cace of methemoglobin, the thermal equilibrium between high and low spin state were observed in wide temperature range.

For the ferrous compounds of hemoproteins, EPR methods are not usually applicable, since the ferrous ion has an even number of d-electrons in its outer orbit. Therefore, it can be said that the static magnetic measurements are a quite important method to study the electronic state of this ferrous iron. In the previous work[3)], we have reported paramagnetic susceptibility of deoxyhemoglobin and deoxymyoglobin. The fine structure of the lowest energy sub-level is well represented by the spin Hamiltonian, D = 5 cm^{-1}.

The magnetic field dependence of the Mossbauer effect is also well represented by the spin Hamiltonian[7)].

Since the heme plane has almost fourfold symmetry, it is widely believed that the principal axis of anisotropy coincides with the symmetry axis. We have analyzed the paramagnetic anisotropy of a single crystal of deoxyhemoglobin based on this assumption and obtained satisfactory results[4)]. Recent Mossbauer measurements made by Huynh et al[8)], however, show evidence that the principal axis lies in the plance. This result seems quite interesting, since in order to know which amino-acid can break C_{4v} symmetry, we must understand the electronic state of heme iron.

A single crystal of myoglobin (Type A) has two hemes in a unit cell, while that of hemoglobin (Form II) has eight hemes. Therefore, in the present work: we chose myoglobin single crystal, studied paramagnetic susceptibility and discussed the anisotropy axis of the crystal.

MEASUREMENT

Myoglobin was prepared from whale muscle and washed with 1 % salin solution and hemolized by the addition of deionized water[9)] The single crystal was grown by dialyzing solution (1 % myoglobin in 1.1 M $(NH_4)_2SO_4$ solution, pH6.8) against 2.5 M $(NH_4)_2SO_4$ solution at pH6.8 with a small amount of EDTA. These inner and outer solutions were deoxygenated by adding dithionite. The completeness of deoxygenation was checked by measuring absorption spectra in the region from 400 to 650 nm.

The single crystals of about 2 mm^3 volume were used as samples. In the present study, we used the type A crystal of the deoxymyoglobin, since the nature of this form has been elucidated crystallographically by the X-ray diffraction studies. The character of this crystal is monoclinic with a space group of P_{21}, where the dyad screw axis corresponds to the b axis. Unit cell dimensions are given as a = 64.6 A, b = 33.1 A, c = 34.8 A and β = 105.5. There are two myoglobin molecules in the unit cell, as shown Fig. 1a.

The measurement of the paramagnetic anisotropy of the single crystal is carried out by measuring the torque exerted on the crystal, when the crystal is subjected to the homogeneous magnetic field. The sensitivity is about 10^{-4} dyn·cm.

THEORETICAL BACKGROUND OF TORQUE MEASUREMENT

Let us consider the general case where the three principal values of the magnetic susceptibility tensor are different from each other. These principal directions of the tensor are taken to be x, y and z and the corresponding principal values of the tensor denoted as χ_x, χ_y and χ_z. Since the sample crystal is of monoclinic type and the dyad screw axis corresponds to the b axis, the paramagnetic susceptibility tensor per unit cell may be expressed by the following form with the (a, b, c^*) coordinate system

$$\chi = \begin{pmatrix} \chi_{aa} & 0 & \chi_{ac^*} \\ 0 & \chi_{bb} & 0 \\ \chi_{ac^*} & 0 & \chi_{c^*c^*} \end{pmatrix} \qquad (1)$$

Elements of its tensor are written in terms of the three principal values (χ_x, χ_y, χ_z) as follows,

$$\left.\begin{aligned} \chi_{aa} &= U_{aa,x}\chi_x + U_{aa,y}\chi_y + U_{aa,z}\chi_z \\ \chi_{bb} &= U_{bb,x}\chi_x + U_{bb,y}\chi_y + U_{bb,z}\chi_z \\ \chi_{c^*c^*} &= U_{c^*c^*,x}\chi_x + U_{c^*c^*,y}\chi_y + U_{c^*c^*,z}\chi_z \\ \chi_{ac^*} &= U_{ac^*,x}\chi_x + U_{ac^*,y}\chi_y + U_{ac^*,z}\chi_z \end{aligned}\right\} \qquad (2)$$

Now we define the new coordinate system (X, Y and Z). The relation between the coordinate system (X,Y,Z) and (a, b, c^*) coordinate system is obtained by, just, rotating around the c^* axis by an angle of θ. Then the new coordinate system (a′, b′, c^*) is rotated around the b axis by an angle of φ. Finally the (a″, b′, $c^{*\prime}$) system is rotated around c^* axis by an angle of ψ. Thus the (X, Y, Z) coordinate system is obtained. If the Y axis is taken to be in the direction of the magnetic field, the rotated angle of the magnetic field corresponds to ψ . The torque value $L(\psi)$ can be obtained by differentiating the static magnetic energy with respect to ψ . Thus the torque value per unit cell is

$$L(\psi)=\frac{1}{2}H^2\{(\chi_z-\frac{\chi_x+\chi_y}{2})(k_1\sin 2\psi+k_2\cos 2\psi)+(\chi_x-\chi_y)(k_3\sin 2\psi+k_4\cos 2\psi)\} \quad (3)$$

The temperature dependence of torque value is attributed only to the anisotropy terms such as ($\chi_z-(\chi_x+\chi_y)/2$) and ($\chi_x-\chi_y$). The coefficients are functions of the orientation of each heme, and of the direction of the rotating magnetic field plane (θ,φ).

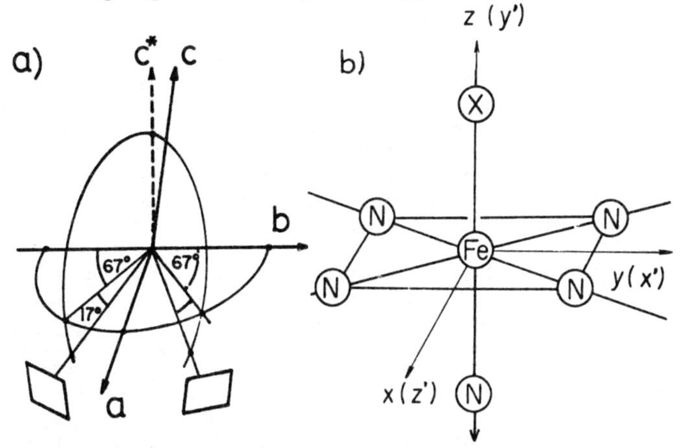

Fig. 1(a). The relation between two heme planes and the crystal coordinate system (a,b,c*). (b) The coordinate systems (x,y,z) and (x',y',z') for one heme.

RESULT

In the present measurements, torque curves were measured in the three different planes of the sample crystal. These planes are normal to the a axis (case A), to the b axis (case B) and to the c* axis (case C). Fig. 2 shows experimental data of torque curves (case B) which were measured at 4.2 and 2 K. Torque curves are of sinusoidal shape with one period for each half rotation of the magnetic field, and these amplitudes are proportional to the square of magnetic field strength H, which is consistent with Eq. 3. This liner relation means that the paramagnetic saturation effect does not occur in this temperature region, even under a

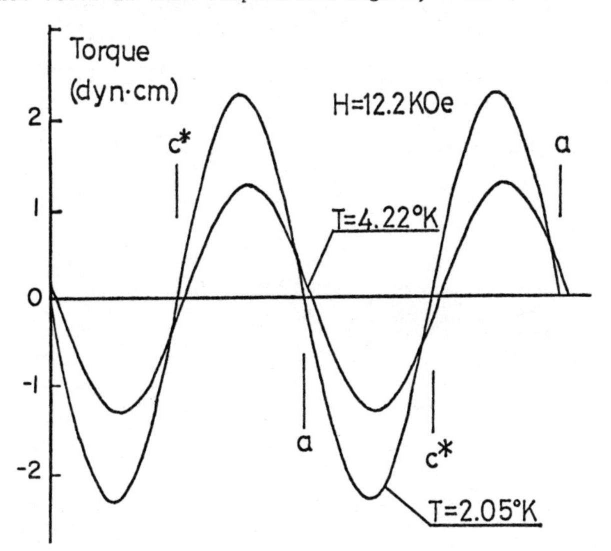

Fig. 2. The temperature dependence of the torque curves measured at 2.05 and 4.2 K (case B).

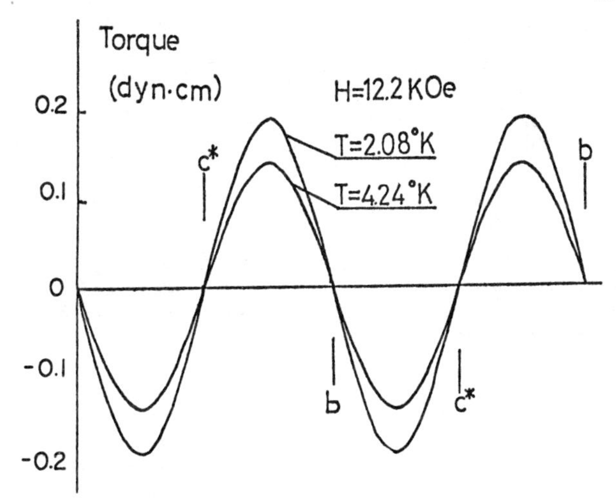

Fig. 3. The tenperature dependence of the torque curves measured at 2 and 4.2 K (case A).

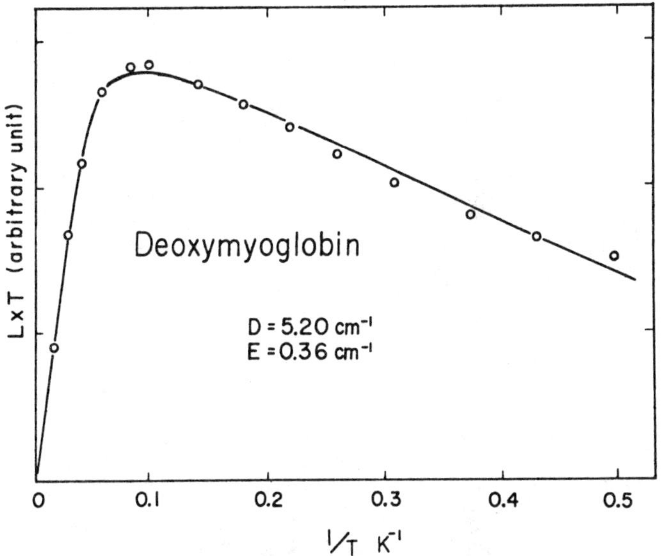

Fig. 4. The temperature dependence of the torque value (case B). The vertical axis L×T (Torque×Temperature) is proportional to $\Delta(n^2)$. Solid line represents the theoretical curve for the pair of D = 5.20 cm^{-1} and E = 0.36 cm^{-1}.

magnetic field of 12 kOe. The temperature dependence of torque curves is shown in Fig. 2 and Fig. 3. In the case B (Fig. 2), two torque curves are quite different in their amplitudes and moreover in their phases. As can be easily seen from Eq. 3, the difference in amplitude is due to the temperature dependence of each of three principal value of the magnetic susceptibility tensor, while the phase shift is due to the differences in temperature dependence of these three principal values, then the direction of the magnetic field is close to the c* axis or a axis, the torque values were nearly equal to zero: it means that $\chi_{aa}>\chi_{c^*c^*}\gg\chi_{ac^*}$. In the case of A (Fig. 3), the phase shift of the torque curves could not be observed. This tendency can be explained by Eq. 3, where the direction of the magnetic field is in the bc* plance, so that $\varphi=0$ and $\theta=\frac{\pi}{2}$. Thus the coefficients k_2 and k_4 become zero, and Eq. 3 is reduced to the formula

$$L(\psi)=\frac{1}{2}H^2\{k_1(\chi_z-\frac{\chi_x+\chi_y}{2})+k_3(\chi_x-\chi_y)\}\sin 2\psi$$

The torque values were zero, when the direction of the magnetic field was parallel to the b axis or c* axis

($\chi_{bb} > \chi_{cc^*}$). In the case C, the temperature dependence of phase shift was not observed as in the case A. ($\chi_{bb} > \chi_{aa}$). The temperature dependence of torque value is shown in Fig. 4, where the torque values are plotted in term of L x T (torque x temperature) for the case of B. According to Eq. 3, this quantity is proportional to $\Delta(n^2)$, which correspinds to

$$C_1\left(n_z^2 - \frac{n_x^2+n_y^2}{2}\right) + C_2(n_x^2 - n_y^2) \equiv \Delta(n^2) \qquad (4)$$

the effective Bohr magneton number n_α (α = x,y and z) being defined by $\chi_\alpha = \beta^2 n_\alpha^2/3kT$. In the other cases of A and C, the same tendencies of temperature dependence of the torque curves were observed.

ANALYSIS AND DISCUSSIONS

The general form of the spin Hamiltonian including the C_2 symmetry was described[3] as

$$\mathcal{H} = DS_z^2 + E(S_x^2 - S_y^2) + \beta\vec{S}\cdot\tilde{g}\cdot\vec{H} \qquad (5)$$

The coefficients (D, E and g) were represented as functions of spin orbit interaction and energy differences between sublevels of the d-level. In order to consider the E term of Eq. 5, we use the new parameters F and α instead of D and E[10], which are defined as D = $F\cos\alpha$ and E = $F\sin\alpha/\sqrt{3}$. Then the spin Hamiltonian Eq. 5 becomes

$$\mathcal{H} = F\cos\alpha S_z^2 + F\sin\alpha/\sqrt{3}\,(S_x^2 - S_y^2) + \beta\vec{S}\cdot\tilde{g}\cdot\vec{H} \qquad (6)$$

We can calculate the temperature dependence of the three principal values of the magnetic susceptibility on the basis of this equation. In the present experiment, the magnetic field is 12.2 kOe, so that the Zeeman term becomes less than 1.2cm^{-1}. From previous magnetic susceptibility measurements[3], it can be expected that the value of F falls to around 5 cm^{-1}. Therefore, this Zeeman term may be treated as a perturbation to an anisotropic term which can be diagonized using eigenfunction of S_z (S = 2). As the paramagnetic saturation effect was not observed, each energy level (λ_i; i =1-5) is calculated to the second order in the magnetic field strength. By averaging these moments with use of the Boltzmann factor, we obtained the magnetic susceptibility as follows.

$$\chi_\alpha = \frac{M_\alpha}{H_\alpha} = \frac{1}{H_\alpha}\cdot\frac{\Sigma m_{i\alpha}\exp(-\frac{\lambda_i}{kT})}{\Sigma \exp(-\frac{\lambda_i}{kT})} \qquad (7)$$

The explicit forms of χ_x, χ_y and χ_z are shown in the Appendix of reference 4. The effective Bohr magneton number is calculated from the above expressions of magnetic susceptibility. Generally speaking, each component of the effective Bohr magneton number is a function of temperature (T) and the coefficients (F, α, g_x, g_y and g_z) of the spin Hamiltonian Eq. 6. The values of g_x and g_y can be written using F and α . Thus, the curve of ($\Delta(n^2)$ vs $1/T$) is a function F, C_1, and C_2. Its temperature dependence is determined by the values of F and α , and its magnitude dependes on the values C_1 and C_2. Since intramolecular structure of the deoxymyoglobin has already been elucidated by X-ray diffraction studies, the orientation of the hemes in the unit cell are known from the molecular coodinate (Fig. 1 a). In the recent analysis, the coordinate system (x, y, z) for one heme is defined with the z axis perpendicular to heme plane (see Fig. 1 b). The values of F and α were ajusted so as to minimize the deviation between the experimental data and the calculated curve. These treatments were carried out by an electronic computer, and we obtained two sets of the F and α values such as F = 5.20 cm^{-1}, α = 0.12 rad. and F = 5.20 cm^{-1}, α = 2.2 rad.. The former set gives D = 5.16 cm^{-1} and E = 0.4 cm^{-1}, while the later gives D = -3.10 cm^{-1} and E = 2.38 cm^{-1}. However, by rotating the coordinate system such as ($x \to z'$, $y \to x'$ and $z \to y'$) (see Fig. 1 b), the latter values of D and E can be transformed to the former values of D and E. In other words, in the former case, the principal axis of the anisotropy is perpendicular to the heme plane and in the latter case the principal axis is in the heme plane. On the other hand, the principal values of the susceptibility tensor of the single crystal can be determined from the easy direction of the torque curve; ($\chi_{bb} > \chi_{aa} > \chi_{c^*c^*}$). Now, as seen from Fig. 1, the planes of the two hemes in an unit cell are almost parallel to each other, Thus these heme planes are nearly perpendicular to the a axis. This fact leads to the result that the pricipal axis of the anisotropy of one heme lies in the plane, as reported by the Mossbauer effect. However, our conclusion is only tentative, since quantitative analysis of the single crystals used was necessary for the determination of absolute values of the susceptibility tensor. Further quantitative experiment and the analysis are now going on, and the result will be reported shortly.

REFERENCES

1) A. Tasaki, J. Otsuka and M. Kotani, Biochim. Biophys Acta, 140 284 (1967)
2) N. Nakano, K. Nakano and A. Tasaki, ibid., 251 307 (1971)
3) N. Nakano, J. Otsuka and A. Tasaki, ibid., 236 222 (1971)
4) N. Nakano, J. Otsuka and A. Tasaki, ibid., 278 355 (1972)
5) T. Iizuka and M. Kotani, ibid., 181 275 (1969)
6) T. Iizuka and M. Kotani, ibid., 194 351 (1969)
7) G. Lang and W. Marshall, Proc. Phys. Soc. , 158 20 (1969)
8) B. H. Huynh et al., J. Chem. Phys 61 3750 (1974)
9) J. C. Kendrew and R. G. Purrich, Proc. Roy. Soc. London, Ser A 238 305 (1956)
10) J. Otsuka, J. Phys. Soc. Japan 24 886 (1968)

$NiI_2 \cdot 6H_2O$: A DISORDERED LINEAR CHAIN MAGNET*

C.R. Stirrat[+], P.R. Newman[++], and J.A. Cowen
Department of Physics
Michigan State University
East Lansing, Michigan 48824

ABSTRACT

Electron spin resonance, magnetic susceptibility and nuclear magnetic resonance measurements are reported on $NiI_2 \cdot 6H_2O$. Although there is only one molecule in the trigonal unit cell, the ESR results suggest that the Ni^{2+} ion resides in either of two almost equivalent sites along the trigonal axis: (0,0,0) or (0,0,½). The appearance of three different zero field splittings, and the broad and structured spectral lines are qualitatively interpreted as due to a distribution of crystalline electric fields arising from a relative z-directed displacement between adjacent chains of Ni^{2+} ions. Measurements of the electronic magnetic susceptibility are shown not to exhibit any evidence of disorder. The susceptibility is Curie like at high temperatures and goes through a peak at T_c = 0.2K indicating a transition to a magnetically ordered state with a net moment along the c-axis. The ^{127}I NMR in the paramagnetic state also shows no evidence of disorder. The electric field gradient at the ^{127}I nucleus is very small (ν_q = 5.7 MHz) and there is a significant transferred hyperfine interaction with the Ni^{2+} spin.

INTRODUCTION

Nickel iodide hexahydrate ($NiI_2 \cdot 6H_2O$) in which the nickel ions lie along chains parallel to the trigonal c-axis, appeared to be a transition element salt that might exhibit low-dimensional magnetic behavior.

The crystal structure has been determined by Gaudin-Louer, Weigel, et.al.[1,2] The unit cell contains one chemical formula unit. The cell is trigonal with a = 7.638Å and c = 4.876Å and has the space group P$\bar{3}$m1[3]. They propose that the Ni^{2+} ion occupies the "a" ($\bar{3}$m) position at (0,0,0) and the I^- ions occupy the "d" (3m) positions at (1/3, 2/3, z) and (2/3, 1/3, $\bar{z}$) with z = 0.2142. The six water molecules form a regular octahedron around the Ni^{2+} ion, with the three-fold axis of the octahedron being parallel to the [001] axis.

Measurements of electron spin resonance, magnetic susceptibility, and ^{127}I nuclear magnetic resonance on aligned single crystals are reported.

EXPERIMENTAL

Single crystals of $NiI_2 \cdot 6H_2O$ were grown from a saturated aqueous solution at room temperature. The resultant dark green hexagonal platelets cleave easily perpendicular to the c-axis and are very hygroscopic.

Measurements of the electron spin resonance (ESR) were made at x-band (9.2 GHz) and k-band (25 GHz) in applied fields to 20 kOe. The sample was immersed in liquid helium pumped to 1K.

The near zero applied field magnetic susceptibility was measured over the temperature range from 0.01K to 4K using conventional mutual inductance coils operated at 17 Hz.

The ^{127}I NMR was measured using a pulsed NMR spectrometer operated from 2 to 25 MHz. A boxcar integrator was then used to sample the free induction decay signal as a function of applied field. All measurements were corrected for shape-dependent demagnetizing effects.

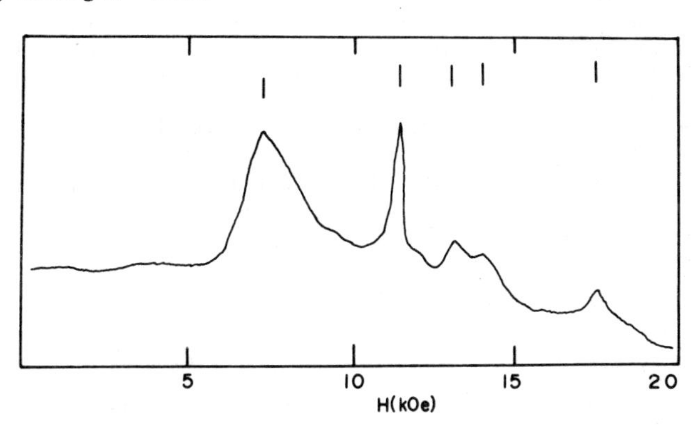

1. ESR pattern with $\vec{H}$ parallel to c at 1.1K and 9.23 GHz.

RESULTS - ESR

The ESR spectrum near 1K at all frequencies shows axial symmetry about the trigonal (c) axis. The on-axis spectrum is extremely complex and structured but may be characterized by perhaps five rather broad and structured resonances (Fig. 1). As one rotates the applied field away from the c axis, four of the lines associate into two groups of two (Fig. 2). Behavior is seen in both the x-band and k-band data. Although it is extremely difficult to accurately parametrize the spectra due to the width and structure of the resonances, the data can be fit to the following Hamiltonian for an S=1 system in an octahedral ligand field with a rhombic distortion[4]

$$H = -DS_z^2 - E(S_x^2 - S_y^2) + g\mu_B \vec{H} \cdot \vec{S} \qquad (1)$$

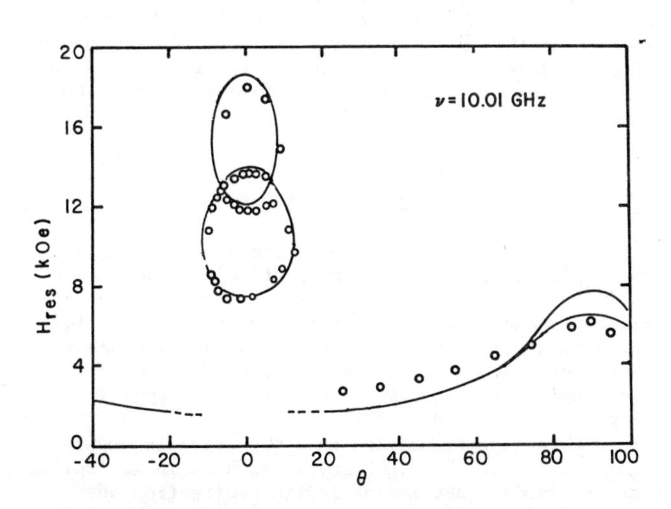

2. Angle dependence of the center of the ESR lines at 1.1K and 10.01 GHz. The solid lines are calculated using the parameters of Table I.

The parameters which give the best fit to the data are given in Table I. The ESR spectrum appears to result from a distribution of disorder in the local coordination of the Ni^{2+} ions. This distribution, which is reflected by the effective D values, is apparently not smooth but peaks at the three values given in Table I. The disorder may arise from the relative displacement along the c-axis of adjacent chains of Ni^{2+} ions. We believe that although there is only one Ni^{2+} ion per unit cell, its location need not be restricted to the "a" one-fold special position at (0,0,0). It is also possible to shift by z/2 so that the Ni^{2+} ion resides at (0,0,½). Let us consider the effect of disordering chains of Ni^{2+} ions with respect to the two special positions in the unit cell.

TABLE I

ESR	SUSCEPTIBILITY
$g_1 = 2.27 \pm 0.08$	$g_{ave} = 2.21 \pm 0.05$
$g_2 = 2.27 \pm 0.08$	
$D_1/k = 2.27 \pm 0.1K$	$D_{ave}/k = 2.15 \pm 0.05K$
$D_2/k = 1.60 \pm 0.1K$	
$D_3/k = 2.6 \pm 0.1K$	$E_{ave}/k = 0.0 \pm 0.05K$
$E_1/k = 0.00 \pm 0.05K$	
$E_2/k = 0.00 \pm 0.05K$	$4J_{ave}/k = 0.095 \pm .01K$
$E_3/k = 0.00 \pm 0.06K$	

Electron spin Hamiltonian parameters for $NiI_2 \cdot 6H_2O$ as obtained from ESR and susceptibility results

Each Ni^{2+} ion which is octahedrally coordinated with six waters of hydration is also surrounded by six I^- ions which lie outside the octahedron. In turn, each I^- has three Ni^{2+} ions as neighbors. Although the exact details of the mechanisms are not at present understood, it is possible that the number of chains of nearby Ni^{2+} ions which are dislocated (shifted by ½ a unit cell along c) relative to a given chain of Ni^{2+} ions will, through the connecting I^- ions disturb the ligands surrounding each Ni^{2+} ion. This effect will then lead to a distribution of crystalline electric fields at the Ni^{2+} ion. Also, as we indicate in the section on the ^{127}I NMR, there is a significant amount of delocalized spin density at the I^- sites, the disorder in the crystal may also produce a distribution of exchange values.

SUSCEPTIBILITY

The susceptibility results are shown in Fig. 3 and 4 plotted as χ and χ^{-1} respectively. The susceptibility like the ESR is completely isotropic in the plane perpendicular to c and is labeled χ_a. The inverse susceptibility exhibits the Curie-Weiss behavior characteristic of an S=1 system with positive zero field splitting ($m_s = \pm 1$ having lower energy) in the temperature range 0.5 to 4K.

There is a transition to a magnetically ordered state at $T_c = 0.2K$. The very large peak in χ_c implies that there is a net moment in the c direction but neither the temperature nor the angle dependence of χ well below 0.2K are easily understood in terms of a model of the ordered state.

If one had been able to interpret the ESR data so as to obtain the range of values of q, D_1, D_2, D_3 and J as well as the population of the sites, it would have been possible to form the density operator and calculate the susceptibility. In the absence of such a program, the best one can do is to fit the experimental data with a Hamiltonian consisting of a single (average) g, D and J as follows.[5]

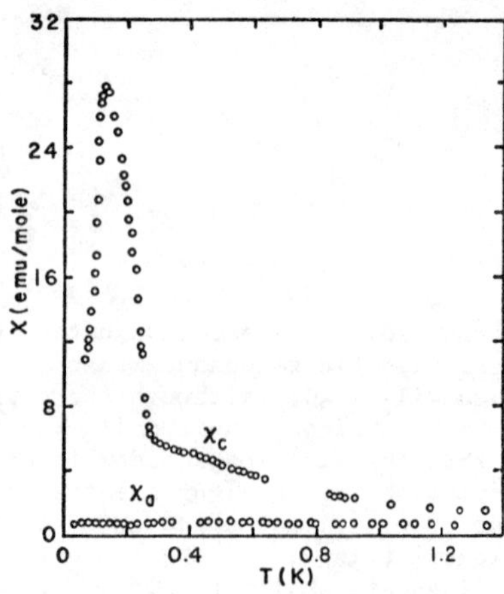

3. Magnetic susceptibility of $NiI_2 \cdot 6H_2O$ between 4 and 0.01K with the measuring fields parallel and perpendicular to the c axis.

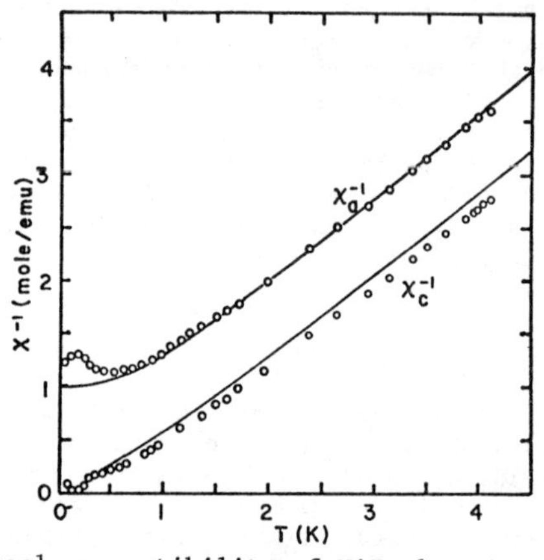

4. Reciprocal susceptibility of $NiI_2 \cdot 6H_2O$ between 4 and 0.01K with the measuring fields parallel and perpendicular to the c axes. The solid lines were calculated using the parameters in Table I.

$$H = -DS_z^2 + g\mu_B \vec{H} \cdot \vec{S} + 2zJ\langle \vec{S}_i \rangle \cdot \vec{S}_j \qquad (2)$$

then

$$\chi_i = \frac{2N_0 g^2 \mu_B^2 \delta_i}{1 - 2zJ\delta_i} \qquad (3)$$

where i = a or c, and

$$\delta_a = D^{-1}[\exp D/kT - 1]/[2 \exp D/kT + 1] \qquad (4)$$

$$\delta_c = (kT)^{-1}[\exp D/kT]/[2 \exp D/kT + 1] \qquad (5)$$

The best fit is obtained with

g = 2.21, D/k = 2.15K, 2zJ/k = 0.095 K

This is not very good agreement with the ESR parameters but it is satisfactory considering the complexity of the system.

NUCLEAR MAGNETIC RESONANCE

The ^{127}I NMR results are shown in Figs. 5 and 6 for H parallel and perpendicular to the c-axis respectively. In the space group $P\bar{3}m1$ the two iodine nuclei occupy the special positions 3m and are related by inversion so that with I = 5/2 and 100% abundance of

TABLE II

^{127}I Nuclear Spin Hamiltonian Parameters for $NiI_2 \cdot 6H_2O$ at T = 1K

$\nu_q = 5.70 \pm 0.0$ MHz,	$\eta = 0.00 \pm 0.03$
$A_\parallel = 2.30 \pm 0.2$ MHz,	$A_\perp = 0.55 \pm 0.2$ MHz

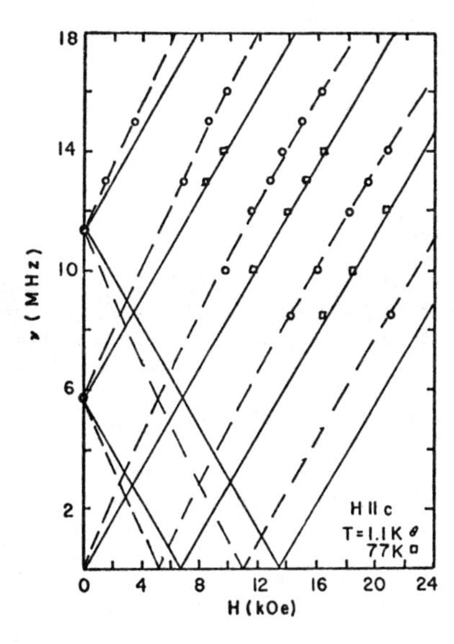

5. ^{127}I NMR with H applied parallel to the c axis. The solid line is the theoretical calculation without a hyperfine term, the dashed line includes hyperfine term, □ : T = 78K, O: T = 1.1K.

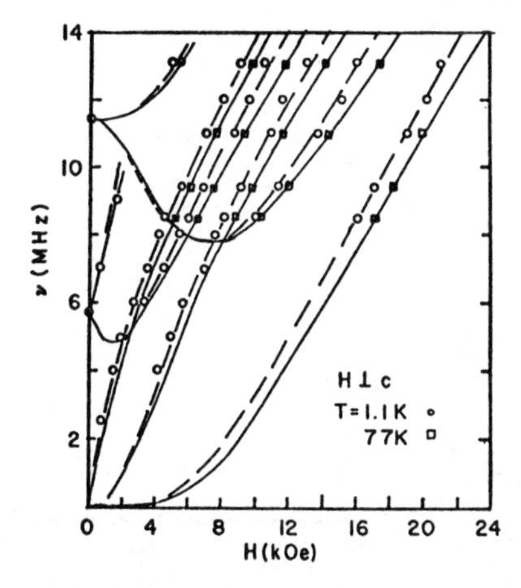

6. ^{127}I NMR with H applied perpendicular to the c axis. The solid line is the theoretical calculation without a hyperfine term, the dashed line includes hyperfine term, □ : T = 78K, O: T = 1.1K.

^{127}I, one expects, in zero applied field, two pure quadrupole resonance lines. In the frequency range 2 to 20 MHz there are just two zero field resonances - at 5.70 and 11.39 MHz. At high temperatures the nuclear spin Hamiltonian has the form:[6]

$$H_N = \frac{h\nu_q}{2}\left[I_z^2 + \frac{\eta}{3}(I_x^2 - I_y^2)\right] - \gamma_N h\vec{I}\cdot\vec{H} \qquad (6)$$

where

$$\nu_q = \frac{3e^2qQ}{2I(2I-1)h}$$

and η is a measure of the non-axial nature of the field gradient.

The solid lines in Figs. 5 and 6 are the theoretical curves for

$\nu_q = 5.70 \pm 0.01$ MHz and $\eta = 0.00 \pm 0.03$

The squares are data points which were obtained at 78K. The signals are very weak but they clearly agree well with the theoretical curves. The NMR, like the ESR and the susceptibility, shows axial symmetry (η=0). It is interesting to note that with a quadrupole moment of $-0.69 \times 10^{-24}cm^2$, the ^{127}I pure quadrupole resonance typically occurs in the range 100 to 400 MHz[7] so that the iodine here sees a very small electric field gradient consistent with the symmetry of the site: 3m.

At temperatures near 1K the NMR lines are shifted by a hyperfine interaction so that a term

$$H_N = \vec{I}\cdot\vec{\vec{A}}\cdot\vec{S} \qquad (7)$$

must be added to the Hamiltonian to calculate the resonance fields. $\vec{\vec{A}}$ is the hyperfine tensor which is assumed to have axial symmetry so that $A_{zz} = A_{\parallel}$ and $A_{xx} = A_{yy} = A_{\perp}$. The experimental values are given in Table II. In addition, at low temperatures the applied fields must be corrected for demagnetizing effects using

$$H_i = H_{oi} - (N_i - 4\pi/3)M_i \qquad (8)$$

where again i = a or c, H_{oi} is the applied field, N_i is the demagnetizing factor and M_i the magnetization which is given by $M_i = g_i\mu_B\langle S_i\rangle$.

At low fields and high temperatures one normally approximates this by $M_i = \chi_i H_i$ but in this system M_i saturates at relatively low fields so that one must compute $\langle S_i\rangle$ using a density operator derived from the

CONCLUSION

$NiI_2\cdot6H_2O$ appears to be a system in which the possibility of magnetic linear chain behavior is masked by a crystallographic disorder due to two very nearly equivalent ("a" and "b") sites in the unit cell where the Ni^{2+} may reside. It is possible to understand, at least qualitatively, the extremely complex ESR spectrum in terms of a relative displacement of the chains of Ni^{2+} ions along the z axis which interact with one another through the intervening I^- ions. The relative number of misaligned chains leads to a distribution of crystalline electric field D-values which in turn leads to a distribution of ESR lines.

The disorder does not appear to affect the electronic susceptibility. This measurement is an "averaging" probe of the magnetic behavior and thus might not be expected to indicate microscopic details as in the ESR data. The susceptibility does however behave as an S=1 system in an "average" octahedral ligand field. The peak in the susceptibility at low temperatures indicates a transition to a magnetically ordered state at $T_c \simeq 0.2K$.

In contrast to the complexity of the electronic magnetism, the nuclear interactions of the ^{127}I nucleus are relatively straightforward. Because they are related by a crystallographic symmetry operation, the two ^{127}I nuclei produce only 1 NMR spectrum. The electric field gradient at the ^{127}I site is rather small and therefore the pure quadrupole resonances are at easily observable frequencies.

* Supported in part by NSF Science Development Grant GU 2648.

+ Present Address: Johns Hopkins University, Applied Physics Lab., Laurel, Maryland.

++Present Address: Department of Physics, University of Pennsylvania, Philadelphia, Pennsylvania.

1. M. Gaudin-Louer, D. Weigel, C.R. Acad. Sc. Paris, 264B, 895, (1967).
2. M. Louer, D. Grandjean, and D. Weigel, Jour. Sol. St. Chem., 7, 222-228, (1973).
3. International Tables for X-ray Crystallography (N.F.M. Henry and K. Lonsdale, eds.) Vol. I, pp. 2 252 and 270, (Kynoch Press, Birmingham, England, 1952); ibid. Vol. III, p. 269 (1962).
4. A. Abragam and B. Bleaney, Electron Paramagnetic Resonance of Transition Ions (Clarendon, Oxford, 1970), Chap. 3.
5. ibid. Chap. 9.
6. A. Abragam, The Principles of Nuclear Magnetism (Clarendon, Oxford, 1961), Chap. 7.
7. S.L. Segel and R.G. Barnes, Catalog of Nuclear Quadrupole Interactions and Resonance Frequencies in Solids, Part I. (Report 1S-520 Ames Laboratory, Ames, Iowa, 1962).

SINGLE CRYSTAL SUSCEPTIBILITY MEASUREMENTS IN $CsNiF_3$ AND $CsFeCl_3$

P. A. Montano
Dept. of Physics, West Virginia University
Morgantown, WV 26506

ABSTRACT

The low field susceptibility in single crystals of the quasi-one-dimensional ferromagnets $CsNiF_3$ and $CsFeCl_3$ has been measured. The susceptibility at low temperatures in $CsNiF_3$ shows a marked field dependence, consequently the low field measurements were carried out between 30 and 100 Oe. A strong anisotropy in the susceptibility was observed at low temperatures below 30K. The experimental data were analyzed using a classical field model, given a value of D=1~2K and an intrachain exchange interaction of J=6.5±1.0K. Below 10K the interchain interaction becomes significant and corrections to the one dimensional susceptibility are necessary. The high temperature susceptibility for $CsNiF_3$ was analyzed using high temperature series expansions; a J=6.5±0.3K was obtained, in full agreement with the low temperature results. Susceptibility measurements were carried out in the isomorphous crystal $CsFeCl_3$ (S=2). A strong anisotropic susceptibility was observed below 100K. The high temperature data were analyzed using the molecular field approximation (MFA) and a positive J_{intra} was found. The susceptibility measurements are in qualitative agreement with recent Mössbauer measurements.

INTRODUCTION

It is the purpose of this work to report zero field susceptibility measurements in the quasi-one-dimensional ferromagnets $CsNiF_3$[1] and $CsFeCl_3$[2]. The experimental results were compared with different theoretical models. The susceptibility measurements were carried out on powder as well as single crystal samples. For the measurements the Curie-Faraday method was used. A temperature range between 2K to 300K was covered. The magnetic field was varied between 30 Oe and 16 kOe. Due to the strong magnetic field dependence of the susceptibility in $CsNiF_3$[3] below 30K, the single crystal and powder results reported here were carried out between 30 and 100 Oe. A linear relationship between the field and the magnetization was observed for $CsFeCl_3$ down to temperatures of 2K and up to 10 kOe fields.

EXPERIMENTAL RESULTS AND DISCUSSION

$CsNiF_3$: In this crystal the Ni^{2+} (S=1) ions are ferromagnetically coupled along the c-axis[1]. In addition, there is a single ion anisotropy which favors orienting of the spins perpendicular to the c-axis. A weak interchain coupling causes $CsNiF_3$ to undergo a phase transition at 2.61K to an ordered three-dimensional magnetic state[4]. The Hamiltonian describing the linear chain in the presence of a magnetic field is given by $H=-2J\sum_i S_i S_{i+1}+D\sum_i (S_i^z)^2 - g\mu_B\sum_i Ho^z Sz - g\mu_B\sum_i Ho\,(Sx+Sy)$. Here J is the exchange interaction, D is the single ion anisotropy and Ho is the external magnetic field. In the high temperature region the susceptibility was analyzed using the molecular field approximation (MFA):

$$\chi_\perp = \frac{2Ng^2\mu_B^2}{3k_B(T-\theta_\perp)} + \frac{8N\mu_B^2}{\Delta^1}$$

and

$$\chi_\parallel = \frac{2Ng^2\mu_B^2}{3k_B(T-\theta_\parallel)} + \frac{8N\mu_B^2}{\Delta_0}$$

where $\theta_\parallel=(8J-D)/3k_B$, $\theta_\perp=(8J+D/2)/3k_B$. The last terms in the righthand side of $\chi_\parallel$ and $\chi_\perp$ represent the temperature independent Van Vleck paramagnetism. The experimental results for χ_{powder} and $\chi_\perp$ were analyzed between 46K<T<300K. They gave an isotropic behavior with $\theta_\parallel=\theta_\perp$, θ=17K (Figure 1). The best value obtained for the exchange was J=6.4±4K with $g_\parallel$= 2.22-2.23, $g_\perp$= 2.24±0.02. The experimental results were also analyzed using the high temperature expansion[5] of the susceptibility for a pure Heisenberg Hamiltonian with S=1. The best fits for χ_p and χ , for temperatures between 50<T<295K, were J=6.5±0.3K, $g_\parallel$= 2.22±0.02 and $g_\perp$= 2.24±0.02. The above value of J is, within the experimental error, in good agreement with the value reported by Lebesque et al[6].

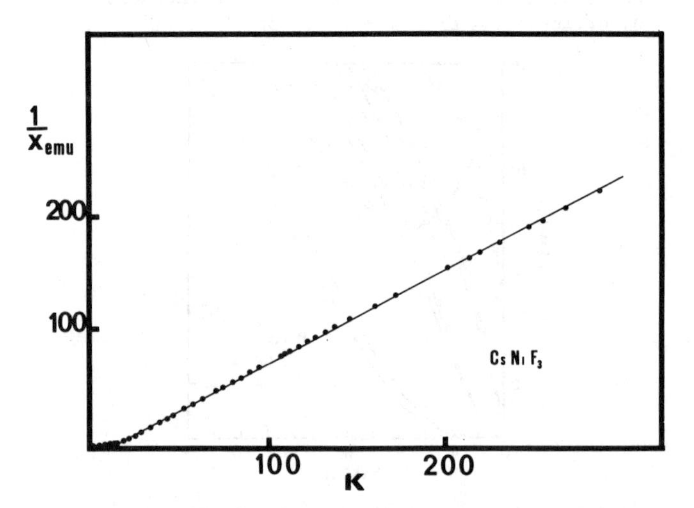

Fig. 1. Inverse susceptibility for $CsNiF_3$, powder sample.

Below 30K an anisotropy in the susceptibility becomes evident and the magnetization starts to be strong field dependent. The zero field susceptibility results were analyzed using the classical spin field model of Scalapino and McGurn. The best value for the single ion anisotropy was $D_{classical}$= 3K±1K and for J_{cl}=13 2K (D=1.5K, J=6.5±1.0K, J_{cl}= S(S+1)J). In Figure 2 the anisotropy $\chi_\perp-\chi_\parallel$ is plotted vs temperature. One expects the fit to the experimental data to be better at low temperatures, where the correlation between the spins extends over several lattice spacings. The discrepancies between the experimental results can be caused by the classical approximation, g-factor changes, and the interchain interaction. A simple MFA for the interchain interaction slightly improved the agreement at low temperature (2zJ'=-0.15, J' interchain interaction, z number of nearest neighbor chains). In conclusion, a good agreement between the values of the intrachain interaction was obtained from different models in two different temperature regions, J=6.5±0.5K. A value for the single ion anisotropy between 2 the 1K was obtained.

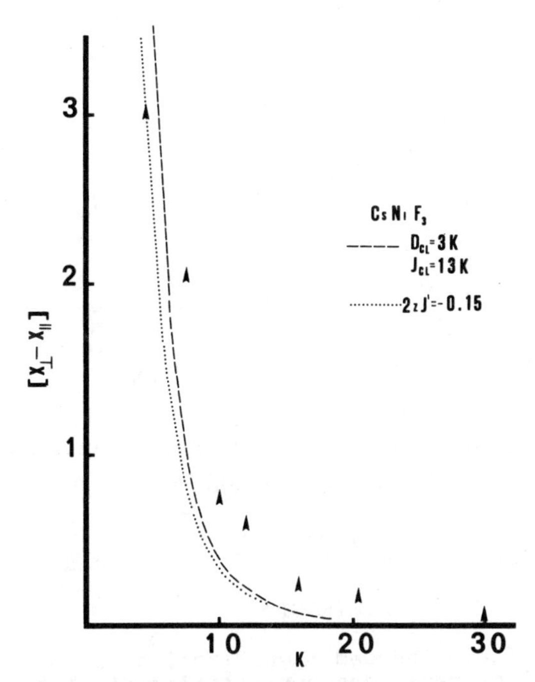

Fig. 2. Anisotropy of the susceptibility for $CsNiF_3$ vs T (X in emu/mole).

$CsFeCl_3$: The space group of hexagonal $CsFeCl_3$ is $2D_{6h}^4$. The Cs^+ and Cl^- ions form a hcp lattice, and the Fe^{2+} ion is subjected to the crystal field produced chiefly by the distorted octahedral of neighboring Cl^- ions. The ^5D term of Fe^{2+} free ion is then split by the cubic component of the crystalline field into an orbital ^5Eg doublet and a $^5T_{2g}$ orbital triplet, where the latter is the ground state. There is a further splitting caused by the spin-orbit interaction $-\lambda\vec{L}\cdot\vec{S}$ (L=2, S=2) and the residual trigonal component of the crystal field $-\Delta(L_z^2-2)$, where Δ is the crystal field strength and the sign of Δ depends upon the nature of the distortion. In the temperature range of this measurements only the $^5T_{2g}$ level will be populated. Single crystal Mössbauer measurements in $CsFeCl_3$ indicate that $\Delta<0$ and that the ground state is a singlet[2]. The Fe^{2+} ions form an infinite chain along the hexagonal axis, with a much stronger magnetic interaction between the ions along the c-axis than between those in the same layers perpendicular to this axis. In Figure 3 the zero field inverse susceptibility for the powder sample of $CsFeCl_3$ vs temperature is plotted. Single crystal measurements on $CsFeCl_3$ indicate the presence of a marked anisotropy below 100K. The high temperature data was analyzed using the molecular field approximation (for all 15 levels of the T_{2g} ground state). The best fit from powder and single crystal results was obtained for $-\Delta/\lambda=.9$ ($\lambda=78cm^{-1}$, a value expected for λ of Fe^{2+} with octahedral coordination to the chlorines[8]) and $J=3.5\pm1.5cm^{-1}$. The fit was in excellent agreement with the experimental data between 30K<T<300K. (Figure 3 and Figure 4) In Figure 4 we have plotted $\chi_\perp$ and $\chi_\parallel$ vs T; one observes that the MFA gives a poor fit to χ at low temperatures. In the figure a simple pair model (S=1) calculation is included (with $D=13cm^{-1}$, $g_\parallel=2.36$, $g_\perp=3.8$, $\alpha_\parallel^2=0.5\alpha_\perp^2$ and corrections for Van Vleck paramagnetism). The pair model gives a relatively better fit for $\chi_\perp$ but a very poor one for $\chi_\parallel$ between 4K and 30K. In conclusion, our susceptibility measurements agree with the Mössbauer results including the fact that the ground state is a singlet and that $-\Delta/\lambda\simeq0.9$; a positive value for the interchain exchange interaction of $3.5\pm1.5cm^{-1}$ was found. No evidence of a transition temperature could be detected from the magnetization measurements. $CsFeCl_3$, like $RbFeCl_3$ behaves at low temperatures as a quasi one dimensional ferromagnet.

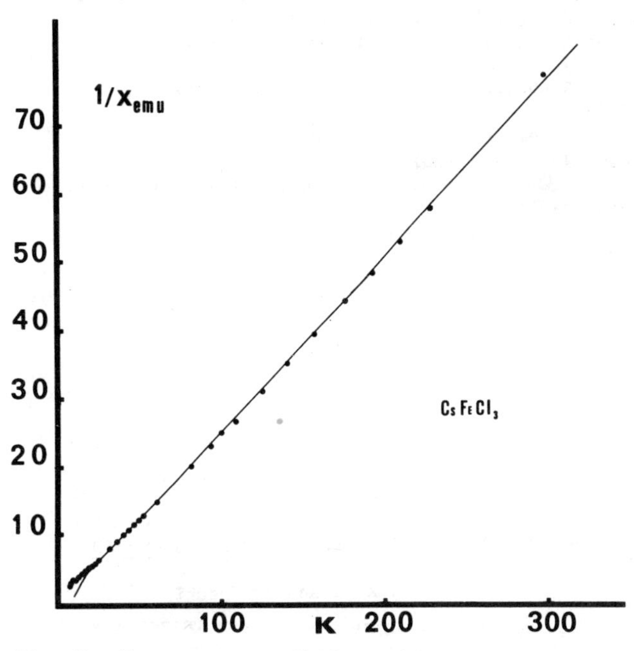

Fig. 3. Inverse susceptibility of a powder $CsFeCl_3$ sample vs T (X in emu per mole).

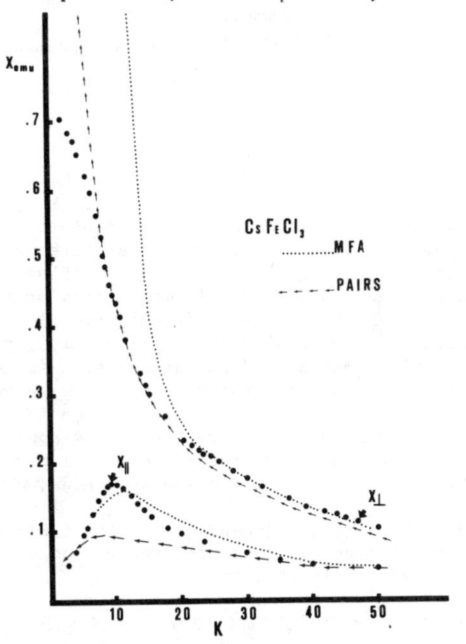

Fig. 4. Susceptibility of a single crystal sample of $CsFeCl_3$.

End of last page.

REFERENCES

[1]M. Steiner, W. Kruger, and D. Babel, Solid State

Commun. 9, 227 (1971).
[2]P. A. Montano, H. Schechter, E. Cohen, and J. Makovsky, Phys. Rev. B9, 1066 (1974).
[3]A. R. McGurn, P. A. Montano, and D. J. Scalapino, Solid State Commun., 15, 1463 (1974).
[4]M. Steiner, and B. Dorner, Solid State Commun., 12, 537 (1973).
[5]G. S. Rushbrook and P. J. Wood, Mol. Phys. 1, 257 (1958).
[6]J. V. Lebesque, J. Snel and J. J. Smit, Sol. St. Comm. 13, 371 (1973).
[7]A. R. McGurn, and D. J. Scalapino, Phys. Rev. B11, 2552 (1975).
[8]J. P. Suchet and F. Bailly, Ann. Chem. Phys. 10, 517 (1965).

MEASUREMENT OF $(ME)_E$ EFFECT OF Cr_2O_3 BY SQUID

E. Kita, G. Kaji and A. Tasaki
Faculty of Engineering Science, Osaka University,
Toyonaka, Osaka, Japan
and
K. Siratori
Faculty of Science, Osaka University, Toyonaka, Osaka, Japan

ABSTRACT

To study the origin and mechanism of the magnetoelectric effect, static measurement of the $(ME)_E$ effect were carried out in the low temperature region. For this purpose, we had designed a sensitive magnetometer by utilizing a SQUID. rf-biased SQUID was operated in 22MHz and its sensitivity was approximately 1×10^{-8}emu. Two platelets of different crystal orientations were cut from a single crystal of Cr_2O_3 prepared by Verneuil method and magnetoelectric susceptibilities along the normal of the platelets were measured. The induced magnetization due to the $(ME)_E$ effect was recorded against strength of the applied electric field at each temperature between 1.6 and 4.2 K. From the results on the two specimens, $\alpha_{/\!/}$ and $\alpha_\perp$ were deduced.

A large temperature dependence of the $(ME)_E$ susceptibility was found in this temperature range. The obtained susceptibilities were

$$\alpha_{/\!/}(1.6K) = 6.5\times10^{-6},\quad \alpha_{/\!/}(4.2K) = 3.9\times10^{-6}\text{cgs/g},$$

$$\alpha_\perp(1.6K) = -0.1\times10^{-6},\quad \alpha_\perp(4.2K) = -1.5\times10^{-6}\text{cgs/g}.$$

These results are much different from the previous reports and seem difficult to be explained within the framework of the present theories of the ME effect. Impurities or defects in the crystal may be important.

ANTIFERROMAGNETISM IN $NiNb_2O_6$

I. Yaeger and A. H. Morrish
Department of Physics, University of Manitoba,
Winnipeg, Canada
and
B. M. Wanklyn
Clarendon Laboratory, Oxford, England

ABSTRACT

The field induced magnetization and principal magnetic susceptibilities of flux grown $NiNb_2O_6$ single crystals were measured from ambient down to 1.4K. At $T_N = 6.0 \pm 0.3$K an antiferromagnetic ordering was observed. An AF-PM phase transition was induced by external magnetic fields parallel to the crystallographic c axis in the temperature range below T_N. The experimental results are interpreted in terms of anisotropic bilinear interactions between Ni^2 ions within the manifold of the ground term $^3A_2$. A two-sublattice uniaxial antiferromagnet model is used to describe the Ni^{2+} spins in the mean field approximation. The bilinear Ni^{2+}-Ni^{2+} interactions are essentially of the exchange type with a small dipolar contribution. Along all three orthorhombic directions the ferromagnetic intrasublattice interaction is found to be stronger than the intersublattice antiferromagnetic coupling. The Ni^{2+} g tensor, taken to be isotropic, is $g = 2.4 \pm 0.1$ and the uniaxial anisotropy term $-KS_z^2$ in the effective spin Hamiltonian with S=1 yields $K = 7.9 \pm 0.7$ K/spin. The parameters obtained in the analysis of the susceptibility data are used to calculate the second order AF-PM phase boundary which is compared to the experimental one.

A complete account of this work will be published elsewhere.

NEAR ONE-DIMENSIONAL ANTIFERROMAGNETISM IN OXOVANADIUM CARBOXYLATES $VO(OOCR)_2$ [†]

Vladimir T. Kalinnikov* and Yu. V. Rakitin
Kurnakov Institute of General and Inorganic Chemistry, Academy of Sciences, Moscow, USSR

and

Van H. Crawford[††] and William E. Hatfield

Department of Chemistry, University of North Carolina, Chapel Hill, North Carolina 27514

ABSTRACT

The spin levels, the dependencies of the effective magnetic moments μeff, and the spin specific heat C_m as functions of kT/J for open linear Heisenberg chains with individual ion spins S_i of 1/2 to 5/2 and length $N \leq 11$ have been calculated. A comparison of the results with other well known models has been made. The magnetic susceptibility and EPR spectra of oxovanadium carboxylates $VO(OOCR)_2$ ($R=CH_3$, CH_2Cl, $CHCl_2$, CCl_3, $C(CH_3)_3$, C_2H_5, C_3H_7, iso-C_3H_7, C_4H_9, and C_6H_{13}) have been studied, and the data have been interpreted in terms of the calculations. The results suggest that the chains contain approximately 50 monomeric subunits. The magnetic susceptibility data for all the compounds are consistent with the Heisenberg model for antiferromagnetic linear chains with a small antiferromagnetic interchain interaction and the presence of a small amount of monomeric impurities.

INTRODUCTION

Our knowledge of exchange in cluster compounds and low-dimensional systems such as infinite chains has improved considerably during recent years. This improvement is mainly due to application of theoretical models to the interpretation of magnetic and thermodynamic properties of real systems. The interpretation of the magnetic properties of oligomeric chain molecules relies on models derived from calculations on finite chains undergoing exchange. Here we shall confine ourselves to the consideration of isotropic exchange interactions since these are of the most importance practically. The corresponding effective spin-Hamiltonian can be written as follows:

$$\hat{H} = -2J\sum_i S_i S_{i+1} \qquad (1)$$

Hamiltonian (1) has been applied in several studies of magnetic and thermodynamic properties of finite exchange chains.[1-4] However, the use of cyclic boundary conditions ($N+1 \equiv 1$) makes the results inapplicable to real finite exchange chains which are mostly open systems.

Historically, a start on the problem was made by Earnshaw, Figgis, and Lewis[5] where magnetic properties of a wide range of exchange chains were calculated using the Van Vleck equation for spin energy levels:[6]

$$E(S) = -\frac{zJ}{N-1}\left[S(S+1) - NS_i(S_i+1)\right] \qquad (2)$$

This equation is valid for dimeric, trigonal trinuclear, and tetrahedral cluster compounds only, rather than for linear chains, whether closed or open, and this approach leads to a difference of about 50% between the calculated and correct values of magnetic moments in the middle temperature region ($kT/|J| \sim 1$).[7]

Thus, the calculation schemes reported so far cannot be used in interpretation of magnetic properties of finite open chains. This present work has been undertaken to determine spin levels and magnetic moment vs. temperature dependences for finite chains with isotropic (Heisenberg-type) exchange coupling having various values of S and N.

THEORY, MAGNETIC AND THERMODYNAMIC PROPERTIES

The Heisenberg spin-Hamiltonian for open chains has the form:

$$\hat{H} = -2J\sum_{i=1}^{N-i} S_i S_{i+1} \qquad (3)$$

The corresponding energy levels are characterized by total spin values, S_1, determined according to the vector coupling scheme:[1]

$$\prod_{i=1}^{N} D^{(S_i)} = \sum_i D^{(S_i)} \qquad (4)$$

As some of the S_i levels may occur in expansion (4) more than once, the energy levels having the same value of S_i are further divided according to quantum numbers ν_i. The wave functions corresponding to a given S_i value may be defined in terms of a representation which involves squares of intermediate-spin operators as quantized values. The form of intermediate-spin operators depends on the chosen scheme of composition of moments. In the present case, it is convenient to use the scheme where every successive term is derived from the preceding one by the addition of an ion spin following the order in which ions are arranged into chains. Then, intermediate-spin operators can be written as:

$$\hat{S}_{12} = \hat{S}_1 + \hat{S}_2,\ \hat{S}_{12,3} = \hat{S}_{12} + \hat{S}_3, \ldots,\ \hat{S} = \hat{S}_{\ldots n-2,\, n-1} + \hat{S}_n \qquad (5)$$

Generally, Hamiltonian (3) and the intermediate-spin operators will not commutate, and the former will have a non-diagonal form in the representation

$$|S_1S_2(S_{12})\ldots S_{n-1}(S_{\ldots n-2,n-1})S_nS\rangle.$$

The corresponding matrices contain blocks interrelating various sets $|S_1S_2(S_{12})\ldots S_{n-1}(S_{\ldots n-2,n-1})S_nS\rangle$ which have the same S_i value. The diagonalization procedure involves solution of secular determinants of the form:

$$|\langle\Psi|\hat{H}|\bar{\Psi}\rangle - E\sigma(S_{12},\bar{S}_{12})\ldots(S_{\ldots n-2,n-1},\bar{S}_{\ldots n-2,n-1}) \times (S,\bar{S})| = 0, \qquad (6)$$

where $\langle\Psi|\hat{H}|\Psi\rangle = \langle S_1S_2(S_{12})\ldots S|\hat{H}|S_1S_2(\bar{S}_{12})\ldots S\rangle$.

Calculations of the matrix elements

$$\langle\Psi|\hat{H}|\Psi\rangle$$

can be considerably simplified by using irreducible tensor operators.[8] Belinskii[9] has obtained the corresponding expressions in the form most suitable for our purposes.

The calculation procedure is relatively simple to develop into an algorithm. A program has been written in ALGOL for computation of the matrix elements. The matrices so obtained have been computer diagonalized to determine eigenvalues of Hamiltonian (3). The problem has been solved for chains with S of 1/2, 1, 3/2, 2, and 5/2 and N_{max} of 11, 7, 6, 5, and 5, respectively.

The calculated energy values have been used to determine the effective magnetic moment vs. temperature dependences according to the formula:

$$\mu^2_{eff}(kT/J)=\frac{g^2}{N}\ \frac{\Sigma_S S(S+1)(2S+1)\exp[-E'(S)/kT/J)]}{\Sigma_S\ (2S+1)\exp[-E'(S)/(kT/J)]} \qquad (7)$$

where the sums are taken over all permitted S values including those repeatedly occurring in expansion (4). The symbol E'(S) stands for $E(S)|_{J=1}$ and g is g-factor. For simplicity, we assumed g=2 in our calculations.

While extensive consideration of the results of the calculations cannot be given here, some representative results are particularly pertinent to the present discussion. Theoretical curves for closed-chain systems[3] have been compared with those for open-chain ones (this work). The difference is not great and, according to expectations, decreases as the chain length increases. The reason for that is of very general nature: closed-chain models differ from the corresponding open-chain ones in that the former involve correlations between the terminal atoms, which have a weaker effect the greater the chain length.

Jotham[10] has suggested a statistical model as an approximation to Heisenberg chains. A comparison of the results obtained by his approach and by the rigorous solution reveals that both lead to the same conclusions for the case of N=2, merely because Jotham's probabilities for parallel and antiparallel spin orientations are identical to those characteristic of dimeric species. Increase of N results in deterioration of the agreement. At high N values, qualitative inconsistency of the models becomes evident with the difference being greatest in the ferromagnetic region.

Several works have been published where the $\mu_{||}(T)$ curves obtained within the Ising model for external field H parallel to the z axis have been applied to interpretation of magnetic properties of linear chains with isotropic exchange. The theoretical predictions for the Ising model have been compared with the solutions obtained in this work for Heisenberg exchange chains of various length with S_i of 1/2.

The Ising curves, as a rule, lie well above the Heisenberg ones in the antiferromagnetic region. We have undertaken an attempt to reduce the difference by changing the exchange integral value of the Ising model, in order to obtain the same effective moment values for both models in the most characteristic region of $kT/|J|$ of about 1 for chains having N of 2 to 11. It is noteworthy that in all cases this can be done by dividing J by ca. 0.8. The improvement is quite appreciable for short chains. The agreement, however, becomes worse as N increases. We conclude that the Ising model can only be applied in the case of very short Heisenberg chains.

As for ferromagnetic region, the difference between the theoretical $\mu(kT/J)$ curves for the Ising model and for the Heisenberg one is extremely great and altogether impossible to reduce by adjusting the exchange integral value.

The spin specific heat vs. reduced temperature, kT/J, dependences can be determined from the calculated energy values according to the formula:

$$C'_M(X)=\frac{R}{NX^2}\cdot\frac{[\Sigma_i\exp(-E'_i/X)]\cdot[\Sigma_i\omega_i E_i'^2\exp(-E'_i/X)]-[\Sigma_i\omega_i E'_i\exp(-E'_i/X)]^2}{[\Sigma_i\omega_i\exp(-E'_i/X)]^2} \qquad (8)$$

where X stands for kT/J, R is universal gas constant, $E'_i = E_i|_{J=1}$ is reduced energy and ω_i is the degree of degeneracy of the ith spin state, respectively, and the summation is over all energy levels.

Since we have yet to study the heat capacities we will not go into great detail on the results of these calculations. Our principal conclusions follow: The temperature dependences of the C'_M values which represent magnetic contribution to specific heat remain nearly the same within a given S_i series. Hence the maxima positions can be used for the evaluation of the exchange integral values in the absence of dimeric species, even when the chain length is not known or the substance contains chains of various lengths. For the same reason, however, the specific heat data have little structural implications and provide no information on the number of monomer units per cluster molecule.

EXPERIMENTAL RESULTS FOR VANADYL CARBOXYLATES

Vanadyl carboxylates have been known for about 10 years.[11] Attempts at the preparation of single crystals of these compounds suitable for an X-ray study have been a failure. The structural information available is that drawn from the magnetochemical measurements,[12] infrared,[11,13] and electron diffuse reflectance[14] spectra. Nearly all the compounds studied exhibit a rapid decrease of effective magnetic moments at low temperatures. In principle, temperature dependence of magnetic moments may be due to spin-orbit coupling. However, this factor can hardly play an important role in vanadyl complexes because the axial component of the ligand field (the V=O bond) removes orbital degeneracy from the ground state, $^2T_2$.[15]

Antiferromagnetic exchange coupling is another factor that may cause the low-temperature decrease of magnetic moments. The di- and trimeric cluster models involving ions with quenched orbital moments and spin values of 1/2, however, fail to account for magnetic behavior of the compounds in question. It has been concluded from that that the anomalies observed for vanadyl carboxylates arise from exchange interactions of many paramagnetic V(IV) ions involved in a complex molecule with linear or reticular structure. This conclusion agrees with the deductions from the IR[11,13] and electron diffuse reflectance[13] spectra.

The magnetic behavior of vanadyl carboxylates depends only slightly on the nature of the radical R and can be described as follows: The magnetic susceptibility curve $\chi'_M(T)$ features a broad maximum at T of about 250°K, reaches its minimum at about 30 to 50°K, and then arises abruptly as temperature decreases. The increase continues down to the lowest temperature attained in our experiment, 1.8°K. On the other hand, the effective magnetic moment $\mu_{eff}(T)=\sqrt{8\chi'_M(T)\cdot T}$ decreases smoothly from ca. 1.2μ_B at 300°K to ca. 0.2μ_B at helium temperatures. The EPR spectra of all the compounds studied consist of a slightly asymmetric singlet with a peak width decreasing smoothly from ca. 100 Oe at 295°K to 50 Oe at 77°K. The effective g-factor values are in the range 1.96±0.02.

The exchange integral values have been determined using the best fit procedure which minimizes the error functional:

$$F = [\frac{1}{L}\Sigma^L_{i=1}(\chi_i-\chi^o_i)^2\chi_i^{-2}]^{1/2} \qquad (9)$$

where χ_i and χ^o_i are experimental and theoretical magnetic susceptibilities at the T_i temperature, respectively, and L is the number of temperature points. The minimization has been carried out by varying the exchange parameter J and concentration of monomer admixture, with the g-factor fixed at the value of 1.96 obtained from the EPR measurements. The best fit procedure has been applied to the experimental points corresponding to temperatures exceeding 80°K, to remove the low tem-

TABLE I

The Best Fit Parameters of Heisenberg Infinite Chain Model for Oxovanadium(IV) Carboxylates (T=80-300°K); g is Fixed at the EPR Value of 1.96.

R	$-J\ (cm^{-1})$	Monomer impurity, %	F
CH_3	120	0	1.3
CH_2Cl	130	0	2.3
$CHCl_2$	131	1.5	3.0
CCl_3	121	0.6	2.8
$C(CH_3)_3$	114	0	3.0
C_2H_5	118	0.2	1.8
C_3H_7	117	0.1	1.7
C_3H_7-iso	114	0	2.0
C_4H_9	108	0	1.8
C_6H_{13}	110.5	0.2	0.5

perature effects neglected at this stage of the analysis. The results appear in Table 1. The F values are quite small, being of about 2% for most compounds studied, which provides a convincing evidence that, as far as exchange interactions are concerned, the vanadyl chains may be regarded as sufficiently long ones.

A comparison of the experimental and theoretical curves shows that at the higher temperatures they fit quite well. At the lower temperatures, the experimental curve goes somewhat below the theoretical one and then intersects it and rises abruptly. The reason for that may be as follows: At the higher temperatures, the isolated chain model provides a satisfactory description of the magnetic properties of vanadyl acetate. Antiferromagnetic intermolecular (interchain) interactions gain in importance as temperature lowers, and at still lower temperatures, monomer admixtures (or terminal effects) play a predominant role in determining the shape of the curve.

The intermolecular exchange integral values, J', can be determined using the Oguchi theory[16] which relates T_N to the J'/J ratio for linear antiferromagnetic chains. The vanadyl carboxylates have T_N of about 100° K whence J'/J is of the order $8 \cdot 10^{-2}$. This value somewhat exceeds those characteristic of most one-dimensional structures studied where the J'/J values fall in the range 10^{-3} to 10^{-2}.[17,18] It should, however, be noted that our value may prove somewhat overestimated because of neglect of monomer admixtures which would cause an increase in the measured T_N value.

Further examination of the data in Table 1 reveals a discernible (despite a considerable variance) trend to a slow decrease of the exchange integral values as the length of the n-alkyl chain of the radical R increases. Bulky radicals such as cynnamate and benzoate have a similar effect. This trend may be due to some changes in the intramolecular exchange mechanism, although other explanations are possible. A decrease in the energy of interchain interactions ranks among the most probable ones. In fact, from steric considerations, the bulkier the radical R, the weaker the interchain interaction and hence the lower the observed effective value of J.

REFERENCES

†The research at the University of North Carolina was supported by the National Science Foundation through Grants No. GP42487X and MPS-74-11495-A01 and by the UNC Materials Research Center through NSF Grant No. GH-33632.

*Exchange Scientist at the University of North Carolina under agreement between the National Academy of Sciences of the USA and the Academy of Sciences of the USSR.

††Present address: Department of Chemistry, State University of New York, Oswego, New York.

1. J.C. Bonner and M.E. Fisher, Phys. Rev., A135, 640 (1964).
2. C. Weng, Ph.D. Dissertation, Carnegie-Mellon University (1968).
3. C.K. Majumdar, V. Mubayi, and C.S. Jain, Chem. Phys. Lett., 21, 175 (1973).
4. R.L. Orbach, Phys. Rev., 115, 1181 (1959).
5. A. Earnshaw, B.N. Figgis, and J. Lewis, J. Chem. Soc., A 1657 (1966).
6. J.H. Van Vleck, The Theory of Electric and Magnetic Susceptibilities, Oxford, 1932.
7. B.S. Tsukerblat, A.V. Ablov, V.M. Novotortsev, V.T. Kalinnikov, V.V. Kalmykov, and M.I. Belinskii, Dokl. Akad. Nauk SSSR, 210, 1144 (1973) (Russ.)
8. A.R. Edmonds, Angular Momentum in Quantum Mechanics, Princeton University Press, 1957.
9. M.I. Belinskii, Thesis, Kishinev, Institut Khimii Akademii Nauk Mold. SSR (The Institute of Chemistry at the Academy of Sciences of the Moldavian S.S.R.), (1974) (Russ.).
10. R.W. Jotham, Phys. Stat. Sol., (8), 55 K125 (1973).
11. V.T. Kalinnikov, V.V. Zelentsov, M.N. Volkov, Izv. vyssh.uchebn. zavd., Khimiya i Khim. tekhnologiya, 9, 729 (1966); Zh. strukt. Khim., 8, 63 (1967)(Russ).
12. a) V.V. Zelentsov, V.T. Kalinnikov, Dokl. Akad. Nauk SSSR, 155, 395 (1964)(Russ.).
    b) V.V. Zelentsov, V.T. Kalinnikov, M.N. Volkov, Zh. strukt. khim., 6, 647 (1965)(Russ.).
    c) V.V. Zelentsov, V.T. Kalinnikov, M.N. Volkov, Zh. neorg. khim., 10, 1506 (1965)(Russ.).
    d) V.T. Kalinnikov, T.G. Aminov, Dokl. Akad. Nauk SSSR, 177, 633 (1967)(Russ.).
    e) V.T. Kalinnikov, V.V. Zelentsov, O.D. Ubozhenko, T.G. Aminov, Dokl. Akad. Nauk SSSR, 187, 1089 (1969)(Russ.).
    f) V.T. Kalinnikov, V.V. Zelentsov, O.N. Kuz'micheva, T.G. Aminov, Zh. neorg. khim., 15, 661 (1970) (Russ.).
13. V.T. Kalinnikov, V.V. Zelentsov, M.N. Volkov, S.M. Shostakovskii, Dokl. Akad. Nauk SSSR, 159, 882 (1964) (Russ.).
14. V.T. Kalinnikov, V.V. Zelentsov, Zh. neorg. Khim., 12, 3404 (1967) (Russ).
15. C.J. Ballhausen, H.B. Gray, Inorg. Chem., 1, 111 (1962).
16. T. Oguchi, Phys. Rev., 133, A1098 (1964).
17. L.J. deJongh and A.R. Miedema, Adv. Phys., 23, No. 1 (1974).
18. H.T. Witteveen, Ph.D. Dissertation, Leiden (1973).

SPIN DYNAMICS OF A DIMERIZED HEISENBERG CHAIN†

M. Drawid* and J.W. Halley*
Becton Center, Yale University, 15 Prospect Street,
New Haven, Connecticut, 06520

ABSTRACT

We describe an approximate Green function decoupling theory for the dynamics of a dimerized Heisenberg chain. The theory retains the rotational invariance of the spin system in the approximations as Bulaevskii's approximate theory does not. We find that, within the approximations used, the excitation spectrum of the chain can be described as consisting of two branches: A band with a gap of k = 0 and a second, acoustic, band with no gap. The acoustic band does not appear in the Bulaevskii theory. We discuss the possible relevance of this result to experiments.

INTRODUCTION

The question of whether a gap exists in the spin wave spectrum of a dimerized linear antiferromagnetic Heisenberg chain remains open. It has attracted renewed interest as a consequence of recent experimental and theoretical work[1] on the system TTF $CuS_4C_4(CF_3)_4$ which apparently undergoes a spin Peierls transition at a temperature of ∿ 12°K.

The basic question is summarized in Figure 1, where a qualitative sketch of $\alpha = J_2/J_1$ versus gap Δ appears. The possibilities include curves of types a, b, c and d. The Bulaevskii theory used to interpret the recent experiments gives a result of type d, with the gap going to zero linearly with $1 - \alpha$. This theory can easily be shown to break spin rotational invariance in the approximation but gives a result qualitatively consistent with the exact one[5] when $\alpha = 1$. Perturbation theories[4] which are based on expansions in α give similar results. The theory described here gives the curve a. The possibilities b and c are not really excluded by existing calculations. Numerical calculations on finite chains are not conclusive,[2] but have been interpreted as supporting curves of type d.

While curve a may seem improbable, we offer the following qualitative arguments which may suggest the contrary. First, as $\alpha \to 0$ the system might be said to be going from 1 dimensional to zero dimensional behavior. If the transition follows the patterns established in going from n to n - 1 dimension behavior when n = 3 and 2, then, at zero temperature, one would expect the transition at $\alpha = 0$ and not $\alpha = 1$. Secondly, consider a single pair of spins coupled by an exchange integral $J > 0$ and calculate the correlation function $\tau = \langle\sigma_1^z \sigma_2^z\rangle$ in the ground state as a function of J. Obviously $\tau = -1$ for any finite J and $\tau = 0$ for $J = 0$ so that τ jumps from -1 to 0 discontinuously as $J \to 0$. It may be that this fundamental discontinuity might manifest itself in the $\alpha \to 0$ limit of the more complex system. Finally, if the acoustic mode[5] at $\alpha = 1$ is associated with spin rotation invariance,[6] then it is unclear why dimerization, which does not break this invariance, should affect its qualitative character.

THEORY

In pursuing this question we have constructed a theory of the chain along lines described earlier by one of the present authors and several other workers.[7] This kind of theory is somewhat crude as a description of the uniform chain, but as a decoupling approximation it is similar in spirit to the Hartree Fock approximation employed by Bulaevskii. The theory gives a ground state energy for the uniform spin 1/2 chain of $E_g/2J$ = .740 compared to the exact result of .887 and .870 from the Bulaevskii theory. At the same time the theory has the advantage, from the present point of view, of retaining spin rotational invariance throughout.

We describe a version of the procedure appropriate to spin 1/2 which improves on published versions in the treatment of terms of the type $\langle\langle\sigma_r\sigma_{r'}\sigma_{r''};\ldots\rangle\rangle$ where $r = r' \neq r''$ and gives, as a consequence, a qualitatively correct result for the excitation spectrum in the limit that one exchange interaction is zero. We write the Hamiltonian as

$$H = 4\sum_{r\in 1}\left(J_1\vec{\sigma}_r^{(1)}\cdot\vec{\sigma}_{r+a}^{(2)} + J_2\vec{\sigma}_r^{(1)}\cdot\vec{\sigma}_{r-b}^{(2)}\right) \tag{1}$$

where the $\vec{\sigma}_r^{(1,2)}$ are Pauli spin matrices and (1) and (2) refer to sites of types 1 and 2 on the chain. We retain the difference $a \neq b$ in the lengths. The approximation scheme is based on the second order equation of motion

$$E^2\langle\langle\sigma_r;\sigma_{r'}\rangle\rangle = \frac{1}{2\pi}\langle[[\sigma_r, H],\sigma_{r'}]\rangle + \langle\langle[[\sigma_r, H], H];\sigma_{r'}\rangle\rangle \tag{2}$$

where the Zubarev[8] notation is used and σ without an arrow refers to the z component. One computes the commutators and decouples according to the scheme

$$\langle\langle\sigma_r^\mu\sigma_{r'}^\nu\sigma_{r''}^\lambda;\ldots\rangle\rangle = \langle\sigma_r^\mu\sigma_{r'}^\nu\rangle\langle\langle\sigma_{r''}^\lambda;\ldots\rangle\rangle + \langle\sigma_r^\mu\sigma_{r''}^\lambda\rangle\langle\langle\sigma_{r'}^\nu;\ldots\rangle\rangle + \langle\sigma_{r'}^\mu\sigma_{r''}^\nu\rangle\langle\langle\sigma_r^\lambda;\ldots\rangle\rangle \tag{3}$$

Many of the terms vanish because $\langle\sigma_r^\mu\sigma_{r'}^\lambda\rangle \propto \delta_{\mu\lambda}$. In addition, for the spin 1/2 case the terms with any 2 of r, r', r" equal are treated exactly and (3) is not used. We treat spin operators on the 2 sublattices separately and Fourier transform according to the definitions

$$\sigma_k^{(1)} = \sqrt{\frac{2}{N}}\sum_{r\in 1} e^{ikr}\sigma_r^{(1)}$$
$$\sigma_k^{(2)} = \sqrt{\frac{2}{N}}\sum_{r\in 2} e^{ikr}\sigma_r^{(2)} \tag{4}$$

We find coupled equations for the two Green functions

$$G_{11}(E,k) = \langle\langle\sigma_k^{(1)};\sigma_{-k}^{(1)}\rangle\rangle$$
$$G_{21}(E,k) = \langle\langle\sigma_k^{(2)};\sigma_{-k}^{(1)}\rangle\rangle \tag{5}$$

which are

$$\begin{pmatrix} E^2 - \mathfrak{F}_{11} & \mathfrak{F}_{12} \\ \mathfrak{F}_{12}^* & E^2 - \mathfrak{F}_{11}^* \end{pmatrix}\begin{pmatrix} G_{11} \\ G_{21}\end{pmatrix} = \begin{pmatrix} N_{11} \\ N_{21}\end{pmatrix} \tag{6}$$

in which

$$N_{11} = \frac{-2}{\pi}(J_1\tau_a + J_2\tau_b) \tag{7a}$$

$$N_{21} = \frac{2}{\pi}(e^{ika}J_1\tau_a + e^{-ikb}J_2\tau_b) \tag{7b}$$

$$\mathcal{F}_{11} = 2(\mathcal{G} + J_1J_2(\tau_a e^{-ik(a+b)} + \tau_b e^{ik(a+b)}) \quad (7c)$$

$$\mathcal{F}_{21} = 2(\mathcal{G}_a e^{-ika} + \mathcal{G}_b e^{ikb}) \quad (7d)$$

and we define

$$\mathcal{G} = J_1^2 + J_2^2 + 2\tau_2 J_1 J_2 \quad (7e)$$

$$\mathcal{G}_a = J_1^2 + J_2J_1(\tau_2 + \tau_b) \quad (7f)$$

$$\mathcal{G}_b = J_2^2 + J_2J_1(\tau_2 + \tau_a) \quad (7g)$$

$$\tau_a = \langle\sigma_r^{(1)}\sigma_{r+a}^{(2)}\rangle \quad (7h)$$

$$\tau_b = \langle\sigma_r^{(1)}\sigma_{r-b}^{(2)}\rangle \quad (7i)$$

$$\tau_2 = \langle\sigma_r^{(1)}\sigma_{r+a+b}^{(1)}\rangle \quad (7j)$$

The solutions of equation (6) can be written out using these results. The poles of the Green functions corresponding to the excitations of the system are given by

$$E^2 = \mathrm{Re}\mathcal{F}_{11} \pm \sqrt{|\mathcal{F}_{12}|^2 - (\mathrm{Im}\,\mathcal{F}_{11})^2} \quad (8)$$

and it is easy to show using equations (7c) through (7g) that the branch with a minus sign in (8) has no gap as $k \to 0$. Within the approximations made, this result is independent of any values which the static correlation functions τ_a, τ_b and τ_2 may take. For illustrative purposes we show the energies given by (8) with the values[9] $\tau_a = -.64$, $\tau_b = -.354$, $\tau_2 = -.135$ in Fig. 2.

To determine the static correlation functions self consistently one determines them by using the results of solving (6), together with the relation[6]

$$\langle\sigma_{r'}\sigma_r\rangle = i\int \frac{(\langle\langle\sigma_r;\ \sigma_{r'}\rangle\rangle_{E+i\varepsilon} - \langle\langle\sigma_r;\ \sigma_{r'}\rangle\rangle_{E-i\varepsilon})}{(e^{E\beta} - 1)}\, dE \quad (9)$$

It does not seem to have been previously noted in published work on this type of approximation that the static correlation functions are in fact over determined by this procedure. Leaving this question aside for the present discussion, we can write down equations which determine τ_a, τ_b and τ_2 using (6) and (9). The first of these is

$$1 = \frac{2}{N}\sum_k \langle\sigma_k^{(1)}\sigma_{-k}^{(1)}\rangle \quad (10)$$

It is not hard to show that the solutions to (6) reduce to previous results (except for terms treated differently as noted after (3)) in the limit $J_2/J_1 \to 1$ $a = b$. In the limit that $J_2 \to 0$ we have $E_1 \to 0$, $E_2 \to 2J$. This indicates that the approximations recover a dispersionless exciton band with the correct gap energy in this limit. Inserting equations (5) - (9) in (10) we find

$$|\tau_a| = \frac{1 + e^{-\beta J}}{1 + 3e^{-\beta J}}$$

compared to the exact result of

$$|\tau_a| = \frac{1 - e^{-\beta J}}{1 + 3e^{-\beta J}}$$

These results in the limit $J_2/J \to 0$ are independent of the values which the correlation functions τ_2, τ_b take in that limit. On the other hand the susceptibility depends in this theory on the values which τ_2 and τ_b take as a function of J_2/J_1. As noted above, τ_2 and τ_b are expected to be discontinuous in the limit $J_2/J_1 \to 0$. We could evaluate the result in the limit $J_2/J_1 \to 0$ while holding $\tau_2 = \tau_b = 0$. This would be equivalent to evaluating the result in the temperature region $J_2 \ll k_BT \ll J_1$. Unfortunately, Green function theories are known to be qualitatively unreliable when the temperature is larger than the exchange couplings and the results would not be meaningful. On the other hand, we can evaluate χ' in the limit $J_2/J_1 \to 0$, $k_BT = 0$. The Green function theory is qualitatively reliable, but the result will not correspond to the completely dimerized limit because, as we have emphasized, τ_2 and τ_b are expected to be discontinuous at $J_1/J_2 = 0$.

CONCLUSIONS

We note that these results are not necessarily in conflict with experiments because the present theory is not valid for $k_BT \gtrsim J_2$. For $J_1 \gg k_BT \gg J_2$ we speculate that the system should behave essentially as if $J_2 = 0$ and treatments based[4] on perturbation theory from the limit $J_2/J_1 = 0$ may be qualitatively correct. In other words, the energy levels when $J_1 \gg J_2$, but $J_2 \neq 0$, may consist of an acoustic spin wave band followed at higher energies by a gap and then by a second exciton like band. For $J_1 > k_BT \gg J_2$ the susceptibility due to the acoustic spin wave band would then saturate and the temperature dependence of the susceptibility would be like that of a system with a gap. If this picutre is correct it would suggest (1) that χ' should rise from zero at very low temperatures and (2) that the estimate of J_2/J_1 given by fitting the Bulaevskii theory[3] to experiments would be too high at low temperatures.

In conclusion, we find it interesting and possibly suggestive that a decoupling theory, similar in spirit to that of Bulaevskii, but retaining spin rotational invariance, gives an acoustic branch.

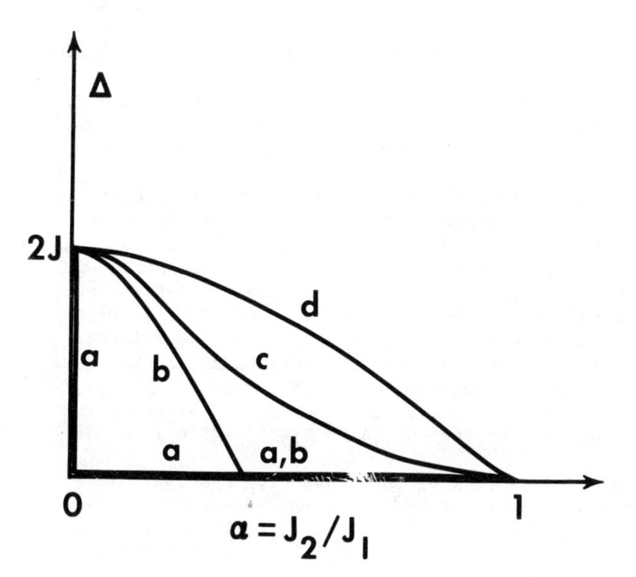

Figure 1: Qualitative sketch of four possible dependencies of the gap Δ on the parameter $\alpha = J_2/J_1$ in the dimerized chain.

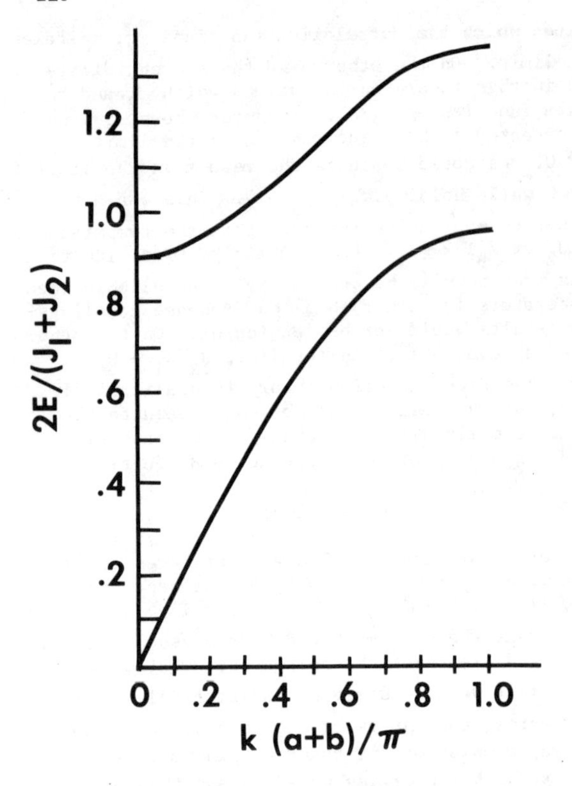

Figure 2: A plot of energies E_1 and E_2 given by equations (8) and (7), using values of τ_a, τ_b and τ_2 described in the text and footnote 8.

REFERENCES

†Research supported in part by NSF Grant 34890.

*Permanent address: Physics Dept., University of Minnesota, Minneapolis, Minn. 55455.

1. J.W. Bray, H.R. Hart, L.V. Interante, I.S. Jacobs, K.S. Kasper, G.D. Watkins, S.H. Wee and J. Bonner, Phys. Rev. Letters 35, 744 (1975); E. Pytte, Phys. Rev. B 10, 4637 (1974).
2. W. Duffy and K.P. Barr, Phys. Rev. 165, 647 (1968).
3. L.N. Bulaevskii, Zh. Eksp. Teor. Fiz. 44, 1008 (1963) [Sov. Phys. JETP 17, 684 (1963)].
4. R.M. Lynden-Bell and H.M. McConnell, J. Chem. Phys. 37, 794 (1962).
5. J. des Cloiseaux and J.J. Pearson, Phys. Rev. 128, 2131 (1962).
6. The Goldstone theorem does not apply directly here because the ground state is nondegenerate. We believe that the acoustic mode at $\alpha = 1$ may nevertheless be associated with the spin rotation invariance of the Hamiltonian in a more subtle way.
7. S. Scales and H.A. Garsch, Phys. Rev. Letters 28, 917 (1972); S.K. Lo and J.W. Halley in Magnetism and Magnetic Materials. 1971 AIP Conference Proceedings No. 5 (American Institute of Physics, N.Y., 1972).
8. Z.N. Zubarev, Sov. Physics, Usphekhi 3, 320 (1960).
9. These values are chosen by taking $|\tau_a| = \alpha|\tau_1| + 1 - \alpha$, $|\tau_b| = \alpha|\tau_1|$, $\alpha = .714$, and τ_1, τ_2 to satisfy the self consistency equations of reference 5 for the uniform chain, modified according to the remarks after equation (3). The results are illustrative only. The existence of the acoustic mode is independent of the values of τ_a, τ_b and τ_2.

CRYSTAL FIELD STUDY OF THERMAL AND MAGNETIC PROPERTIES OF Er^{3+} ION IN THIOINDATES*

S. P. Taneja, Vishwamittar and S. P. Puri
Department of Physics, Punjab Agricultural University, Ludhiana 141004 (Pb.) INDIA

With a view to check the validity of the crystal field parameters obtained through the approximation of actual low field by an octahedral field at the Er^{3+} site in $CdIn_2S_4$, $ZnIn_2S_4$ and $CdInGaS_4$: calculations have been carried out for Schottky specific heat, the g-values (assuming a small trigonal distortion in addition to an essentially cubic field) and paramagnetic susceptibility. The results are compared with the available experimental and free ion data. The crystal field (CF) hamiltonian for an Oh site with Z-axis along one of the cubic axes can be written as (neglecting the ion-ion interactions and taking J as good quantum number):

$$H_{CF} = \beta\ V_4^\circ\ (O_4^\circ + 5\ O_4^4) + V\ V_6^\circ\ (O_6^\circ - 21\ O_6^4).$$

The CF parameters V_n°, obtained from the analysis of optical spectra of Er^{3+} in three thioindates,[1,2] were used in the calculations. The calculated mean g-value $|\bar{g}_{cal}| = 6.74$ matches well with the experimental result $|\bar{g}_{exp}| = 6.63$ for Er^{3+} : $CdIn_2S_4$. The temperature dependence of gram-atomic paramagnetic susceptibility is calculated through the VanVleck formula and practically identical results are obtained for all the three host lattices. The susceptibility follows the Curie-Weiss law, with $K = 11.45/(T+4.5)$ down to 10°K. The computed magnitude of Curie constant, $C = 11.45$, is in nice accord with the experimental 11.98 for erbium metal.

1. M. R. Brown, K. G. Roots and W. A. Shand, J. Phys. C. 3, 1323 (1970).
2. W. A. Shand, J. Phys. C. 3, L 115 (1970).

*Work performed under the auspices of CSIR, New Delhi, India.

APPROXIMATIONS FOR THE MORI CONTINUED FRACTION IN PARAMAGNETS*

M. Drawid† and J.W. Halley†
Becton Center, Yale University, 15 Prospect Street,
New Haven, Connecticut, 06520

ABSTRACT

We point out that an approximation previously used[3] to describe the 1 dimensional Heisenberg paramagnet contains an inconsistency at low temperature. We show that using the form for $K_1(t)$ used in reference 3, one finds $K_3[s=\delta_1^{1/2}]/K_3[0]\alpha T^{+1/2}$ at low temperatures, so that the approximation $K_3[s] \approx K_3[0]$ fails in the region of the spin wave resonance away from k=0 at low temperatures. Here the Laplace transform R[s] of a response function related to the structure factor is given by $R[s] = R(0)/(s + \delta_1/(s + \delta_2/(s + K_3[s]))) = R(0)/(s + K_1[s])$ and $\delta_1 = <\omega^2>$; $\delta_2 = <\omega^4>/<\omega^2> - <\omega^2>$. We have constructed a fully selfconsistent version of the approximation and explored its consequences for the classical Heisenberg chain and the xy model. We conclude that there are serious consistency problems whenever $\delta_1 >> \delta_2$.

INTRODUCTION

A simple truncation of the Mori continued fraction,[1] introduced by Lovesey[2] in studies of classical liquids, has been applied by Lovesey and Meserve[3,4] to calculate the dynamical structure factor of paramagnetic systems in 1 and 3 dimensions. The results of this approximation are in quite impressive accord with experiments. In this note we explore the self consistency of the approximation procedure. We find that the Lovesey and Meserve approximations are in fact not fully self-consistent in their simplest form. We explore whether a more elaborate version of the same theory can be used self-consistently.

We review the formalism briefly, following the notation of references 3 and 4, except that our exchange constant J is 1/2 the J of those papers. The dynamical response function $S(k,\omega)$ which is directly proportional to the inelastic neutron cross-section for scattering from a magnetic insulator is

$$S(k,\omega) = \frac{1}{2\pi}\int_{-\infty}^{+\infty} dt\; e^{-i\omega t} <S^z_k(0)S^z_{-k}(t)> \tag{1}$$

It is related to a form factor $F(k,\omega)$ by the relation (7) below. Also

$$F(k,\omega) = \frac{1}{2\pi}\int_{-\infty}^{\infty} dt\; e^{-i\omega t}\; R_k(t)/R_k(0) \tag{2}$$

where the Laplace transform $R_k[s]$ of $R_k[t]$ was shown by Mori to be exactly equal to

$$R_k[s] = \frac{\chi_k}{s + \delta_1/(s + \delta_2/(s + \delta_3/\ \ldots} \tag{3}$$

where the δ_n are cumulant frequency moments of $F(k,\omega)$. Here χ_k is the static susceptibility. The first of the cumulant moments are related to the moments $<\omega^n> = \int_{-\infty}^{+\infty}\omega^n F_k(\omega)d\omega/\int_{-\infty}^{\infty} F_k(\omega)d\omega$ by the relations

$$\delta_1 = <\omega^2> \tag{4}$$

$$\delta_2 = \frac{<\omega^4>}{<\omega^2>} - <\omega^2> \tag{5}$$

δ_1 and δ_2 can be calculated as a function of temperature for many systems. It is not hard to show that

$$S(k,\omega) = \frac{\omega\chi_k F(k,\omega)}{1 - e^{-\omega\beta}} \tag{6}$$

and that

$$F(k,\omega) = \frac{\chi''(k,\omega)}{\pi\omega\chi_k} = \frac{1}{\chi_k}\ \mathrm{Re}(R_k[i\omega]) \tag{7}$$

where $\beta = 1/k_BT$ and $\chi''(k,\omega)$ is the imaginary part of the dynamic susceptibility.

The contribution of Lovesey and Meserve was to propose truncating the continued fraction in equation (4) by writing it as

$$R_k[s] = \frac{\chi_k}{s + K_1[s]} \tag{8a}$$

$$K_1[s] = \frac{\delta_1}{s + K_2[s]} \tag{8b}$$

$$K_2[s] = \frac{\delta_2}{s + K_3[s]} \tag{8c}$$

They make the three pole approximation, writing $K_3[s] \approx$ constant $= 1/\tau$. Lovesey and Meserve suggest that

$$K_3[s] \approx K_3[0] = \frac{1}{\tau} = \left(\frac{\pi\delta_2}{2}\right)^{1/2} \tag{9}$$

in the frequency range of interest and they estimate $K_3[0]$ as follows. It follows from (8b) - (8c) that

$$\frac{1}{\tau} = K_3[0] = \frac{\delta_2}{\delta_1}K_1[0] = \frac{\delta_2}{\delta_1}\int_0^{\infty} dt\; K_1(t) \tag{10}$$

One can show that the small t expansion of $K_1(t)$ is

$$K_1(t) = \delta_1(1 - \tfrac{1}{2}t^2\delta_2 + \ldots.) \tag{11}$$

Lovesey and Meserve now <u>assume</u> that

$$K_1(t) \simeq \delta_1\; f(t^2\delta_2) \tag{12}$$

even at long times in order to calculate $K_3[0]$ in (10). In particular, if $f(t^2\delta_2)$ is a Gaussian then one gets the value given in (9). This is the value used in all of the actual applications.

Our point concerning this argument is that if one assumes equation (12) and is interested in frequencies in the region of $\sqrt{\delta_1}$ and $\sqrt{\delta_2}$ then the resultant $K_3[s]$ will not necessarily satisfy the first equality in equation (9). The approximations may therefore be mutually inconsistent. We will find emperically below that this problem is particularly acute if $\sqrt{\delta_1} >> \sqrt{\delta_2}$ and is much less serious if $\sqrt{\delta_1} \approx \sqrt{\delta_2}$.

We illustrate the point by use of the Gaussian for $K_1(t)$. Eliminating $K_2[s]$ from (8b) and (8c) for finite s we have

$$K_3[s] = \frac{\delta_2 K_1[s]}{\delta_1 - sK_1[s]} - s \tag{13}$$

Using the Gaussian for $K_1(t)$ we have

$$K_1[s] = \delta_1 \int_0^\infty e^{-st} e^{1/2\ \delta_2 t^2}\, dt \tag{14}$$

$$= \delta_1 \sqrt{\frac{\pi}{2\delta_2}}\ (1 - \Phi(s/\sqrt{2\delta_2}))\ e^{s^2/2\delta_2}$$

where $\Phi(s/\sqrt{2\delta_2})$ is the error function. Inserting (14) in (13) we can compute $K_3[s]$ self consistently in this Gaussian approximation. By expanding the error function for large argument we find

$$K_3[s] \underset{s \to \infty}{\to} \frac{2\delta_2}{s} \tag{15}$$

in this Gaussian approximation. One sees from the equation (15) how problems might arise. Suppose $\delta_2 << \delta_1$ as is true for many systems at low temperature. Then if $s \approx \sqrt{\delta_1}$ (in the region of the collective mode or diffusion rate) then (15) is appropriate and $K_3[s]$ is not $\approx K_3[0]$ with s in this region.

We note in particular that, at low temperatures in a classical system $\delta_2 \propto T$ while δ_1 contains a T-independent term. As a consequence, for low enough T δ_1 can always be made much greater than δ_2. Then (15) can be used and

$$\frac{K_3[s = \sqrt{\delta_1}]}{K_3[s = 0]} = \sqrt{\frac{8\delta_2}{\pi\delta_1}} \quad \alpha\ T^{1/2}$$

It follows that (9) always fails in the region $s \approx \sqrt{\delta_1}$ for a classical system at low enough temperatures. This may be the reason why it is apparently difficult to account for the linear temperature dependence of the line widths in Heisenberg linear antiferromagnets[5] on the basis of the 3-pole approximation.

EXAMPLES

The preceding remarks apparently suggest that the difficulties we have pointed out may be ameliorated by doing the calculation self-consistently. Unfortunately, the following results appear to indicate that, at least when a Gaussian function is used for $K_1(t)$, the self-consistent results, while different, may not be of much use if one is in a regime where the 3-pole approximation itself is not nearly self-consistent.

The self-consistent calculation of $F(k,\omega)$ of equation (7) from the $K_1[s]$ of (14) (using (10)) is not completely trivial because one must make the analytical continuation indicated in (7). We do this by use of the convergent series[6]

$$\Phi(S/\sqrt{2\delta_2}) = \frac{2}{\sqrt{\pi}} \sum_{k=1}^{\infty} (-1)^{k+1} \frac{(s/\sqrt{2\delta_2})^{2k-1}}{(2k-1)(k-1)!} \tag{16}$$

into which the substitution $s \to i\omega$ can be meaningfully made. The resultant expression for $F(k,\omega)$ is

$$F(k,\omega) = \frac{e^{\omega^2/2\delta_2}\ \delta_1 \sqrt{\frac{\pi}{2\delta_2}}}{\frac{\delta_1^2\pi}{2\delta_2} + \omega^2 (e^{\omega^2/2\delta_2} - \sum_{k'=0}^{\infty} \frac{\delta_1(\omega^2/2\delta_2)^{k'}}{\delta_2(2k'+1)k'!})^2} \tag{17}$$

Equation (17) is the rigorous consequence of assuming that the function f in equation (12) is a Gaussian. We have evaluated (17) numerically using moments used by Lovesey and Meserve in some examples to which the three pole approximation has been applied.

Consider first the classical Heisenberg chain. The needed moments at low temperatures are given in reference 3. Taking S = 5/2, J = 7°K, T = 1°K we find $\sqrt{\delta_1/\delta_2} = 3.063$ when $ka = \pi/8$, $\sqrt{\delta_1/\delta_2} = 6.14$ when $ka = \pi/4$. The corresponding functions $F_k(\omega)$, from the 3 pole approximation and equation (17) are shown in Figures 1 and 2. For $\sqrt{\delta_1/\delta_2} = 3.063$ the two results are quite similar. For $\sqrt{\delta_1/\delta_2} = 6.14$ the shapes appear somewhat similar, but note that the scales for the 3 pole approximation and for the result of equation (17) are different. The tails of the self consistent F are very strongly suppressed (by two orders of magnitude here) by the exponential in the denominator when one gets away from the peak. Most of the weight is extremely close to the center of the line. These results indicate that caution must be used in interpreting apparently sucessful fits of experiment to the three pole approximation when the ratio of the peak position (roughly $\sqrt{\delta_1}$) to the peak width (roughly $\sqrt{\delta_2}$) is much more than about 4. The three pole curve in Figure 2 looks quite reasonable, but our calculation indicates that it is not even approximately self consistent. (This does not, of course, exclude the possibility that $K_3[s] \approx K_3[0]$ may be justifiable in some other way with τ chosen by use of a different function f in (12).)

The preceding example shows that the three pole approximation can sometimes be inconsistent when it looks reasonable. Is the self-consistent form (17) then better? Figure 2 is not encouraging. We can get a more definite idea about this by looking at the xy model where the exact answer is known. We show the exact answer,[7] the three pole approximation and equation (17) in Figure 3 for the xy model when $ka = \pi/2$ at, T = 0. The self consistent result has less similarity to the exact result than the three pole approximation. We have also made calculations for the xy model at infinite temperature both using equation (17) and self-consistently using the approximation $K_3(t) = \delta_3 \exp(-t^2\delta_4)$ suggested in reference 4. These results, which will be presented elsewhere, are in much better agreement with the exact correlation functions.

DISCUSSION

In most of the applications made to date, $\sqrt{\delta_1/\delta_2}$ is not large and our remarks serve mainly to indicate the limits of the possible usefulness of the approximation. At lower temperatures where $\sqrt{\delta_1/\delta_2}$ gets large, fluctuations will live longer and a different, longer range function f in (12) might be physically justified while removing the difficulties.

REFERENCES

*Research supported in part by NSF Grant 34890.

†Permanent address: Physics Dept., Univ. of Minnesotta, Mineapolis, Minn. 55455.

1. H. Mori, Prog. Theor. Phys. 33, 423 (1965) and 34, 399 (1965).
2. S.W. Lovesey, J. Phys. C: Prec. Phys. Soc., London 4, 3057 (1971).
3. S.W. Lovesey and R.A. Meserve, Phys. Rev. Letters 28, 614 (1972).
4. S.W. Lovesey and R.A. Meserve, J. Phys. C. 6, 79 (1973
5. F.B. McClean and M. Blume, Phys. Rev. B7, 1149 (1973).
6. I.S. Gradskteyn and J.M. Ryzhik, Table of Integrals Series and Products, Academic Press, N.Y. (1965), equation 8.253.1.
7. S. Katsura et al., Physica 46, 67 (1970).

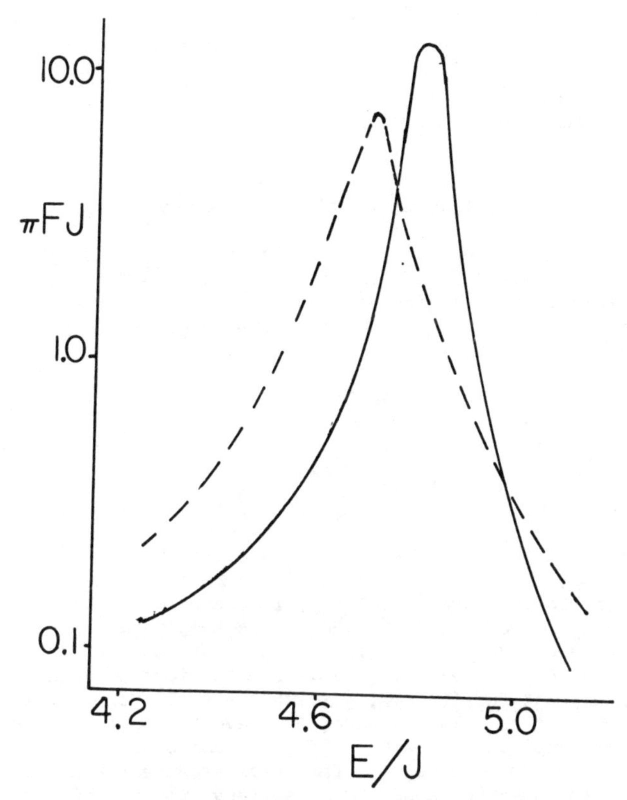

Figure 1: $\pi F_k J$ for $\sqrt{\delta_1/\delta_2} = 3.063$ in the 3-pole approximation and according to equation (17). Only the region near $E=\omega=\sqrt{\delta_1}$ is shown. Full curve: equation (17). Dashed curve: 3-pole approximation.

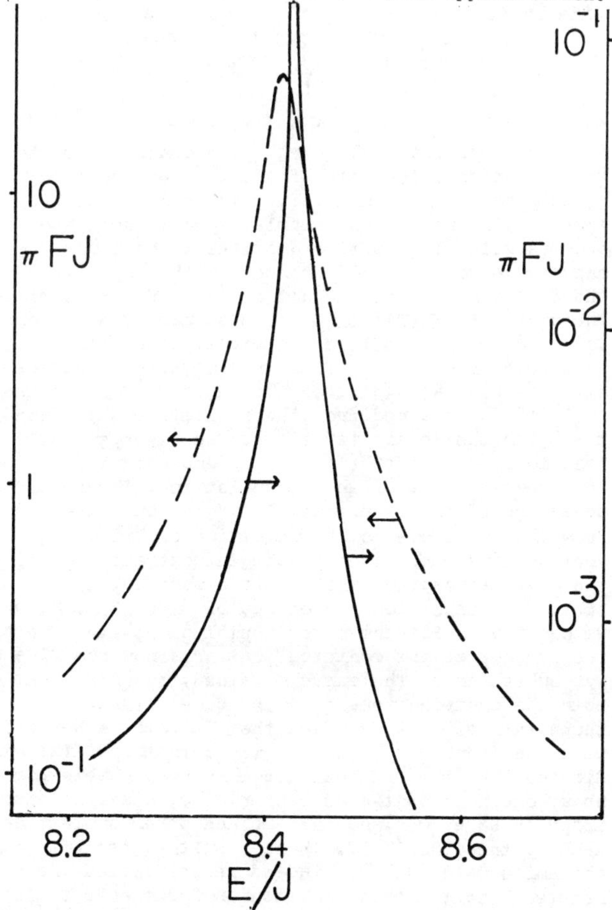

Figure 2: Same as Figure 1 with $\sqrt{\delta_1/\delta_2} = 6.14$. Note different scales.

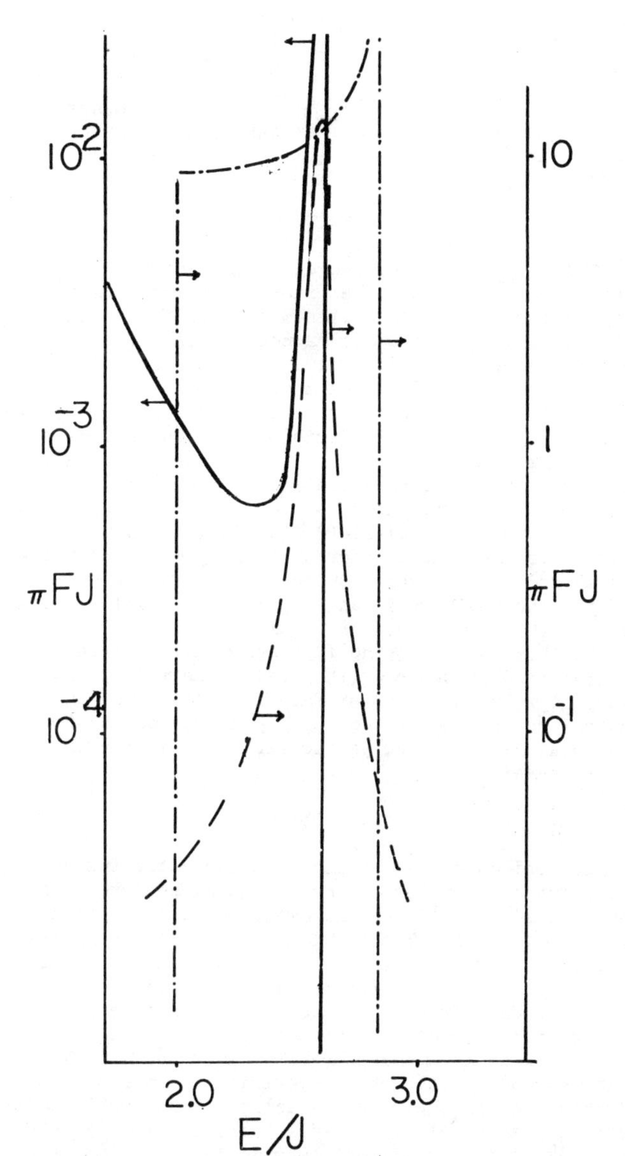

Figure 3: xy model (Longitudinal $F_k(\omega)$). $ka = \pi/2$, $T = 0$. Dashed and full curves as in Figures 1 and 2. Dash-dotted curve is exact result. Range of E/J is much larger relative to $\sqrt{\delta_2}$ than in Figures 1 and 2.

ON THE MAGNETIC ANISOTROPY IN HEXAGONAL FERRITES

G.Asti and S.Rinaldi
Laboratorio MASPEC del CNR, 43100 Parma, Italy

ABSTRACT

The temperature dependence of the anisotropy constants of some hexagonal ferrites is analyzed on the basis of the Callen and Callen model applied to the different sublattices whose magnetizations are taken from Mössbauer data. The analysis allows clarification of the role of the different mechanisms which cause the anisotropic properties of these materials. We consider also the peculiar effects of Co^{2+} ion in some hexagonal oxides (Co_2- Z, Co_xFe_{2-x}- W). Through a suitable coordinate transformation and considerations of symmetry, the contribution to the uniaxial anisotropy can be written $E_k = K^c[(1/3)-(2/3)\sin^2\theta+(7/12)\sin^4\theta]$, where K^c is the second order anisotropy constant relative to Co^{2+} in octahedral surroundings. This result explains the positive K_1 and negative K_2 usually associated with the presence of cobalt in hexagonal ferrites. On this basis, taking into account the different temperature dependence of the sublattice magnetizations, it is possible to explain the dependence on temperature and composition of the anisotropy of Co--hexagonal ferrites and particularly the spin reorientation transitions.

INTRODUCTION

The origin of the strong magnetocrystalline anisotropy characteristic of the hexagonal ferrites has been discussed in the earlier works of Casimir et al.[1] van Loef et al.[2] and Fuchikami[3]. The dipole-dipole interaction is relatively strong and in general favours the orientation of the moments perpendicular to the c-axis[1,2]. However, with the exception of some compounds having Y structure, this mechanism is not the dominant one. Most compounds exhibit a strong uniaxial anisotropy which is generally attributed to a strong contribution of single ion anisotropy from ferric ions located in trigonal bipiramidal lattice sites (5-fold coordinated)[1,4]. An attempt to calculate this contribution has been made by Fuchikami[3] who calculated the second order contribution of the spin orbit interaction of Fe^{3+} in a strong trigonal field. His results confirm the importance of this sublattice but also suggest that comparable contributions to the anisotropy may come from iron in other sublattices. Moreover he reports paramagnetic susceptibility data obtained by Tsushima and Kajiura on Fe^{3+} doped $BaAl_{12}O_{19}$ which confirm the above.

In the following we carry out an analysis of the temperature dependence of the anisotropy constants, using the Callen and Callen theory[5] for the single ion contribution to the anisotropy. Moreover we analyze in some detail the anisotropy properties of the Co -containing hexagonal ferrites which show peculiar behaviour.

RESULTS

a) M and W type hexaferrites

The Callen and Callen theory is valid for single ion anisotropy and connects the temperature behaviour of the anisotropy constants $\chi_{1,0}$ with that of the magnetization. $\chi_{1,0}$ are the coefficients for the expansion of the anisotropy energy in terms of normalized spherical harmonics and are related to the usual anisotropy constants relative to the expansion in powers of the sinus of the polar angle by the following relations[6]:

$$K_1 = -0.946175\,\chi_{2,0} - 4.231422\,\chi_{4,0} + \dots$$
$$K_2 = 3.702494\,\chi_{4,0} + \dots \qquad (1)$$

where we have neglected the terms of order higher than 4. The Callen and Callen theory predicts the following relations:

$$\chi_1^i(T) = \frac{\chi_{1,0}(T)}{\chi_{1,0}(0)} = \frac{I_{1+\frac{1}{2}}(X_i)}{I_{\frac{1}{2}}(X_i)}$$
$$m_i(T) = \frac{M_i(T)}{M_i(0)} = \frac{I_{3/2}(X_i)}{I_{\frac{1}{2}}(X_i)} \qquad (2)$$

between the magnetizations M_i and the single ion anisotropy of the i-th sublattice where the $I_1(X)$ are the hyperbolic Bessel functions. In order to carry out our analysis we must know exactly the temperature dependence of the magnetization of the various sublattices. These data are now available for the most important hexagonal ferrites using Mössbauer data. Moreover low temperature data for $BaFe_{12}O_{19}$ have been obtained by Streever[7] using N.M.R. The first compound we have considered is $BaFe_{12}O_{19}$ for which we have taken the experimental data on the anisotropy from Casimir et al.[1] and on the hyperfine fields from Albanese et al.[8] and from Streever[7]. Reliable data near 0°K cannot be found in the literature. Due to the slow variation of the anisotropy field H_A with temperature, we have considered it more convenient to extrapolate H_A and to deduce K_1 through reconstruction of the behaviour of M_s near 0°K from Streever's data[7]. In Fig. 1 the function $[\chi_2(T)-\chi_{2-dip}(T)]/[\chi_2(0)-\chi_{2-dip}(0)]$ is plotted against the reduced temperature T/T_c. This function represents the single ion contribution to the anisotropy. The data of the dipolar contribution χ_{2-dip} are theoretical values taken from van Loef's work.[2] As is known, the dipolar term is negative and becomes slightly positive just below the Curie temperature due to the rapid decrease with temperature of the magnetization of sublattice 12K . For comparison the curve of $\chi_2(T)/\chi_2(0)$ is also reported. In order to estimate the relative contributions of single ion anisotropy from ferric ions in the various lattice sites we have also plotted the functions $\chi_2^{2b}(T)$ and $\chi_2^{12K}(T)$ as obtained from the sublattice magnetizations for the 2b and 12K sites. The other sublattices lead to functions $\chi_2^i(T)$ with a temperature dependence very near to that of the 2b sublattice. The remarkable deviation of the experimental single ion anisotropy from the curve due to 2b sublattice contribution is worth noting. This seems to be in contradiction with the usual assumption that 2b is almost solely responsible for single ion anysotropy of $BaFe_{12}O_{19}$. However we can try to fit the experimental data with a weighted average of the contributions arising from 2b and 12k sublattices. The curve obtained, assuming that both the contributions have the same weight, is also shown in Fig. 1. We can see that it fits rather well with the experimental single ion anisotropy. This indicates that the 12K sublattice is responsible for about one half of the characteristic uniaxial anisotropy of $BaFe_{12}O_{19}$, and that 2b and other sublattices make up the other half. These results agree with Fuchikami's calculations; indeed he calculated the parameter D of the spin Hamiltonian for the 2b sublattice obtaining a value D=-.9 that is about one half of the value necessary to completely account for the anisotropy of $BaFe_{12}O_{19}$.

A similar analysis on $BaZn_2Fe_{16}O_{27}$ (Zn_2- W)[9], which closely resembles $BaFe_{12}O_{19}$, gave analogous results, as shown in Fig. 2.

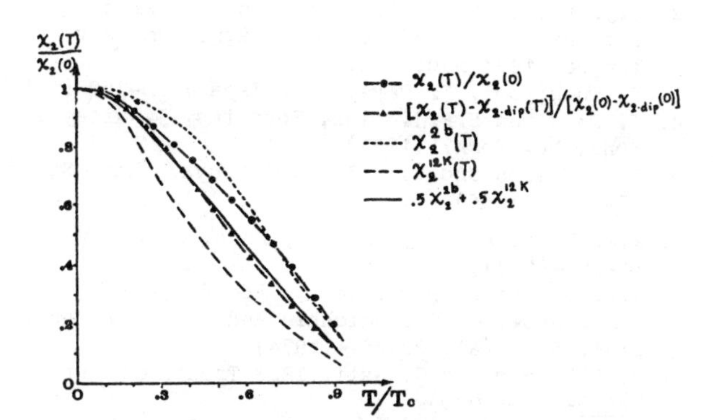

Fig. 1. Experimental and theoretical temperature dependence of the various contributions to the anisotropy constants for $BaFe_{12}O_{19}$.

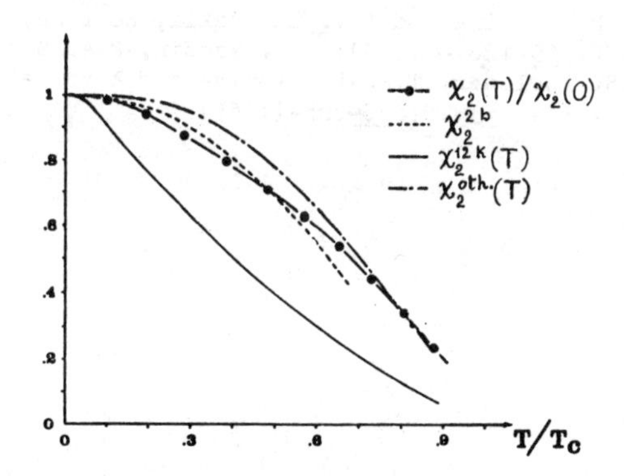

Fig. 2. Same as Fig. 1, for Zn_2-W.

b) Hexaferrites containing Co^{2+}

Among the large family of hexagonal ferrites containing a divalent metallic cation, the Co-ferrites form a peculiar class characterized by i) a stronger anisotropy than for analogous compounds containing other cations, ii) the presence of a strong term of fourth order in the expansion of the anisotropy energy and iii) a complex temperature dependence of the anisotropy field, leading to spin-reorientation phase transitions.

These characteristics are generally ascribed to the degeneracy, even in trigonal crystal field, of the ground state of the Co^{2+} ion, but their details are not yet completely understood. We are attempting here to explain at least qualitatively the characteristic properties of anisotropy of some of these compounds.

First of all we consider the fact that in all the hexagonal ferrites the Co^{2+} ions seem to enter into sublattices with octahedral coordination. The c-axis of the hexagonal structure has direction cosines $(1/\sqrt{3}, 1/\sqrt{3}, 1/\sqrt{3})$ in the orthogonal reference system (x,y,z) centered at the cation site and with axes containing the coordination anions. The splitting of the Co^{2+} levels in the cubic crystal field associated with this coordination octahedra, via spin orbit interaction, causes an anisotropy term of the type:

$$E_K = K^C(\alpha_1^2\alpha_2^2 + \alpha_1^2\alpha_3^2 + \alpha_2^2\alpha_3^2) \quad (3)$$

where the α are the direction cosines of the magnetization with respect to the (x,y,z) reference system.

Expressing (3) in terms of polar coordinates (θ, φ) referred to a polar axis parallel to the c-axis and to an azimuthal axis having direction cosines $(1/\sqrt{2}, 1/\sqrt{2}, 0)$ in the (x,y,z) reference system, one finds:

$$E_K = K_0 + K_1\sin^2\theta + K_2\sin^4\theta + \ldots = \quad (4)$$
$$= K^C(\tfrac{1}{3} - \tfrac{2}{3}\sin^2\theta + \tfrac{7}{12}\sin^4\theta + \tfrac{2}{3}\sin^3\theta\cos\theta\sin(3\varphi))$$

The last term has trigonal symmetry. However for hexagonal compounds of the W and Z type, we must also consider that for every octahedra there is the symmetric one rotated around the c-axis by 180°. Averaging over the two, the last term in eq. (4) vanishes.

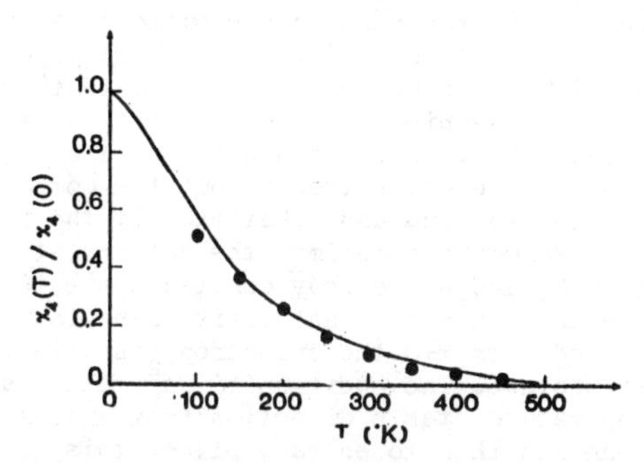

Fig. 3. Experimental and theoretical temperature dependence of χ_4 for Co_2-Z.

Equation (4) indicates that a positive single-ion cubic anisotropy contribution, due to the local environment, contributes in hexagonal symmetry to both a negative K_1 and to a positive K_2 term. It must be noted that if one calculates by means of eqs (1) the coefficient of the expansion in spherical harmonics, obviously $\chi_{2,0}$ goes to zero. However one must note that in the presence of trigonal distorsion of the octahedron, as one expects normally in these compounds, the ratio between K_1 and K_2 as derived in eq. (4) can be modified. Let us first consider the case of $Ba_3Co_2Fe_{24}O_{41}$ (Co_2- Z) for which we have complete Mössbauer and magnetic data from the work of Albanese et al.[10]. This compound exhibits a cone of easy magnetization at temperature below 230°K, an easy plane between 230 and 515°K and an easy axis above 515°K. In Fig. 3 we report the temperature dependence of $\chi_4(T)$ as obtained from experiments. For comparison the theoretical curve as obtained from (2), where M(T) is assumed to be proportional to the hyperfine field of sextet I of ref. 10, is also plotted. The agreement between the two curves indicates that the Co^{2+} ions contributing to χ_4 occupy sublattices whose magnetization decreases very rapidly with temperature.

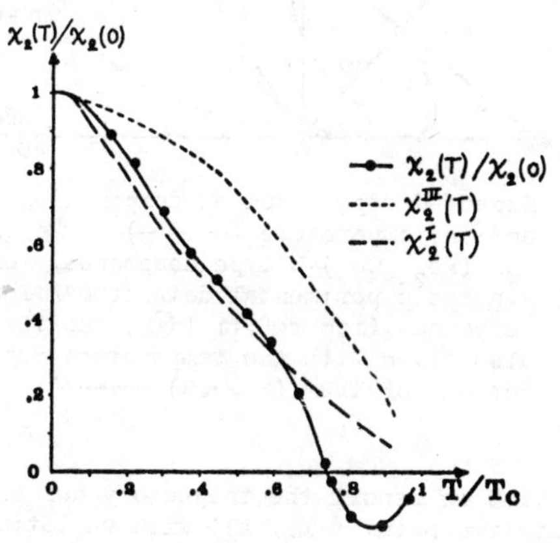

Fig. 4. Temperature dependence of χ_2 for Co_2-Z and of the expected contributions of two typical sublattices.

In Figure 4 we report the temperature dependence of the experimental $\chi_2(T)$ together with the $\chi_2^1(T)$ calculated from eq. 2 for two contributions: a planar term, $\chi_2^1(T)$, dominant at low temperature due to Co^{2+}; an axial term, $\chi_2^{III}(T)$, coming from sublattices such as those belonging to sextet III of ref. 10, whose magnetizations vary slowly with temperature. The lack of data on the behaviour of the dipolar energy does not allow a direct comparison between theory and experiment; however, even considering only the single-ion anisotropy, a change in sign of $\chi_2(T)$ at high temperature can be interpreted as the consequence of the faster temperature decrease of the Co^{2+} contribution with respect to the other contributions. Another interesting case is that of the series $Ba_2Co_xFe_{12-x}O_{27}$ (Co_xFe_{2-x} - W). Depending on the Co contents the compounds of this series undergo a variety of spin reorientations[11,12]. There is a great deal of disagreement among different authors about the composition assigned to the compounds that exhibit the same properties . Moreover sometimes the data relative to cone angles θ_0 and anisotropy constants are inconsistent. Nevertheless the qualitative behaviour is well established. For x=0 the anisotropy is uniaxial at any temperature. The addition of Co^{2+} changes the low temperature stable direction from an easy axis to a cone and then to an easy plane; this corresponds to a continuous increase of both χ_2 and χ_4 constants (see Fig. 5). The effect of the temperature on the various cases is that of inducing spin reorientation transitions of the type cone-axis or cone-plane--cone-axis or plane-axis. Let us consider now the typical compound $Co_{.9}Fe_{1.1}$-W; according to Bickford's data, at 78°K it is conical, then θ_0 increases with temperature up to 125°K where it becomes planar; at 305°K it begins again to be conical with θ_0 decreasing with T up to 370°K where it becomes uniaxial. The anisotropy constants, which are reported in ref. 11, have a behaviour similar to those of Co_2-Z. Again the change in sign at high temperature can be accounted for by the competition between i) the planar term due to Co which dominates at low temperature, ii) an axial term with a lower temperature dependence.

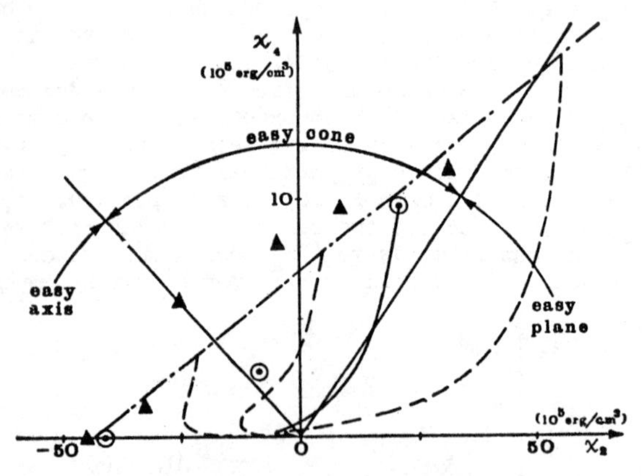

Fig. 5. Expected dependence on composition (·—·—·) and on temperature (— — —) of χ_2 and χ_4 for ($Fe_{2-x}Co_x$)-W type compounds. For comparison the experimental data at 0°K for various compounds from ref.11 (⊙) and 12 (▲) are also given with the temperature dependence for one of them (x = ·.9)(———).

In Fig. 5 we report the trajectory described by a representative point (χ_2, χ_4) with variation in temperature for $Co_{.9}Fe_{1.1}$- W, taken from experimental data, and for other typical compositions assuming analogous temperature dependence of the χ_i.

REFERENCES

1. H.B.G. Casimir, J. Smit, U. Enz, J.F. Fast, H.P.J. Wijn, E.W. Gorter, A.J.W. Duyvesteyn, J.D. Fast and J.J. de Jong, J. de Physique Rad., 20-360-(1959)
2. J.J. van Loef and A.B. van Groenou, Proc. International Conference on Magnetism, Nottingham 1964 pag. 646.
3. N. Fuchikami, J. Phys. Soc. Japan, 20-760-(1965)
4. J. Smit and H.P.J. Wijn, Ferrites, J. Wiley & Sons, New York, (1959)
5. H.B. Callen and E. Callen, J. Phys. Chem. Solids, 27-1271-(1966)
6. R.R. Birss and G.J. Keeler, Phys. Stat. Sol. (b), 64-357-(1974)
7. R.L. Streveer, Phys. Rev., 186-285-(1969)
8. G.Albanese, M. Carbucichhio and A. Deriu, Phys. Stat. Sol. (a), 23-351-(1974)
9. G. Albanese and G. Asti, IEEE Trans. on Mag., Mag 6-158-(1970)
G. Albanese, M. Carbucicchio and G. Asti, to be published in Appl. Phys. (vol.10 n°3 1976)
10. G. Albanese, A. Deriu and S. Rinaldi, J. Phys. C: Sol. State Phys., 9-1313-(1976)
11. L.R. Bickford Jr., J. Phys. Soc. Japan, 17-B1--272-(1962)
12. M. Perekalina and A.V. Zalesskii, Sov. Phys. JETP, 19-1337-(1964); I.I. Yamzin, R.A. Sizov, I.S. Zheluder, T.M. Perekalina and A.V. Zalesskii, Sov. Phys. JETP, 23-395-(1966)

DISPLACEMENT OF THE CURIE TEMPERATURE OF $FeBO_3$ UNDER PRESSURE

D. M. Wilson*
University of Oklahoma, Norman, OK., 73069

ABSTRACT

The spontaneous magnetization of $FeBO_3$ crystallites was determined in the critical region as a function of hydrostatic pressure to 3 kbar. At each pressure this magnetization could be described by the scaling law using T_C and a pressure independent critical exponent β as parameters. The Curie temperature, obtained from the fit of the data to the scaling law, varied linearly with pressure and at a rate of 0.53 ± 0.03 K/kbar. No evidence for any change in the canting angle with pressure was found.

INTRODUCTION

Iron borate, $FeBO_3$, is a magnetic insulator that can be simply described as an antiferromagnet with its sublattices canted at a small angle to each other. This canting gives rise to a weak ferromagnetic moment which provides the name for this class of magnetic materials. Dzyaloshinskii developed a theory for weak ferromagnetism based on the magnetic symmetry of the crystal and Landau's theory of a second order phase transition.[1] In addition to the normal superexchange interaction, proportional to T_C, which aligns the magnetic moments, another interaction, named after Dzyaloshinskii, is included to cause the canting. Pressure, which varies the inter-ionic spacing, is widely known to effect T_C and thus the superexchange interaction. In weak ferromagnets an effect on the Dzyaloshinskii interaction as evidenced by a change in the canting angle must also be considered. This paper is divided into two sections, experimental and results that describe how the magnetization is measured and analyzed.

EXPERIMENTAL

The sample, 10.19 g of iron borate, was tightly packed into a Be-Cu pressure vessel to keep the crystallites from rotating in the magnetic field. Because the pressure vessel contained a magnetic moment due to the presence of a small amount of cobalt, a correction had to be made for it, but this affected the precision only slightly below T_C. The bulk of the pressure vessel and associated temperature bath necessitated the use of a vibrating double coil magnetometer to measure the magnetization. The relative precision of a measurement was typically 0.1%. The applied field, limited to 5 kOe, was known to 3 Oe. The sample temperature was taken to be the temperature of the fluid circulating around the pressure vessel. A copper constantan thermocouple, calibrated to an accuracy of 0.2 K, measured this temperature to a precision of 0.03 K. However, during measurements of the magnetization as a function of pressure the sample temperature could be held constant to within 0.01 K with the aid of a sensitive thermistor located in one end of the Be-Cu vessel.

A couple of corrections had to be made before the magnetization could be determined as a function of pressure. As a hydraulic pump applied pressure, the sample became compacted and displaced towards the closed end of the sample holder. The amount of compaction was directly related to the countermovement necessary to recenter the sample and obtain a maximum signal. Because the calibration constant for the magnetometer was known as a function of position along the length of the sample space, a correction due to compaction could be calculated. The uncertainty in the pressure was 0.02 kbar over the available 3 kbar range. Additional details on the experiment can be found elsewhere.[2,3]

RESULTS

The spontaneous magnetization of a single crystal of $FeBO_3$ obeys a scaling law near T_C.[4] Iron borate has a basal plane in which the anisotropy field is very low.[5] For such a material it is predicted that a powdered sample also follows the same scaling law but with the spontaneous magnetization, $\langle m_S \rangle$, reduced by a factor.[2,3] For our purposes this scaling law can be simplified to

$$\langle m_S \rangle = k s_1 (D/B)\{(T_C(p) - T)/T_C(p)\}^{\beta} \quad (1)$$

where k is a constant, s_1 is the factor that accounts for the reduced spontaneous magnetization of a powdered sample, and B and D are coefficients in an expansion of the free energy near T_C. For randomly oriented crystallites, as our sample appears to be, s_1 is $\pi/4$.

By measuring the spontaneous magnetization at different pressures one can extrapolate it to zero and determine T_C as a function of pressure. Shown in Fig. 1 is the magnetization in the critical region at atmospheric pressure. Since the magnetization is small the demagnetization correction to the applied field is small. In this case it is always less than 5% and usually less than 1%. Below the Curie temperature the magnetization is not initially a linear function of the magnetic field due to domains, anisotropy in the basal plane of iron borate, etc. Since these effects decrease rapidly as the magnetic field increases, the spontaneous magnetization is obtained from a linear extrapolation from higher fields. However, close to the Curie temperature a correction to the linear relationship between the

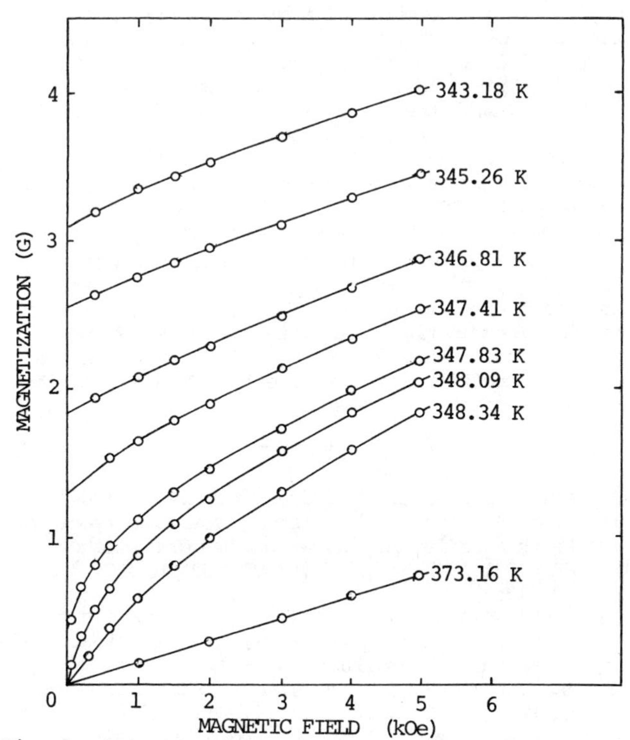

Fig. 1. Magnetization at various temperatures near T_C.

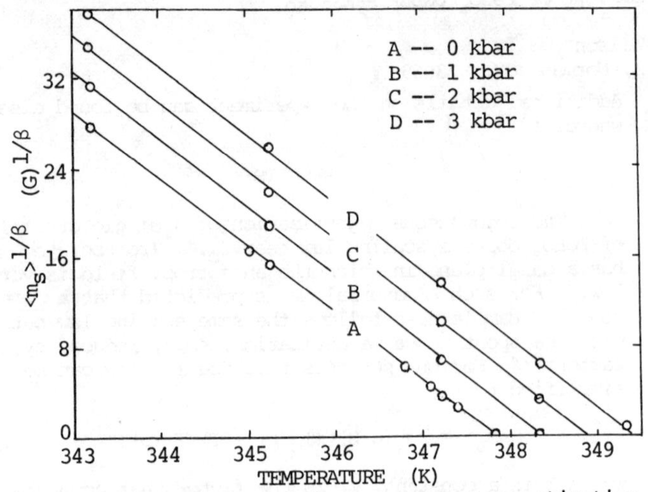

Fig. 2. Extrapolation of the spontaneous magnetization to find the Curie temperature at various pressures.

magnetization and the magnetic field is expected even for an ideal, single domain sample. This correction,[2] proportional to H^3, was experimentally found within three degrees of T_C. At moderate fields, 0.6 to 2 kOe, the effect of domains has become small and the H^3-correction is still negligible. From data more than three degrees below T_C an adjustment proportional to $<m_s>$ was found to estimate the effect of domains etc. on the magnetization. Applying this adjustment, one recovers a linear dependence on magnetic field at moderate values of the field and the spontaneous magnetization can be accurately obtained. The critical exponent β measured on a single crystal[4] using the Mossbauer effect, 0.354 ± 0.005, agrees within experimental error with that obtained using the above spontaneous magnetizations. This latter value of β is 0.350 ± 0.007.

When $<m_s>^{1/\beta}$ is plotted versus temperature as in Fig. 2, T_C can be obtained from a straight line intercept with the temperature axis. The dependence of T_C is linear on pressure to 3 kbar and is 0.53 ± 0.03 K/kbar. This pressure dependence can be contrasted with the published value of -1.7 ± 0.1 K/kbar for another weak ferromagnet, FeF_3.[6] The difference in the effect of pressure on the super-exchange interaction between $FeBO_3$ and FeF_3 may be due to differences in bond length, bond angles, or the substitution of fluorine for oxygen.

A log-log plot of Eq. (1) will have a slope of β. Using such a plot we find the systematic variation in β with pressure is less than 1%. The parallelism of the lines in Fig. 1 indicate that there is no pressure dependence in the ratio D/B. Thus the canting angle is uneffected by pressure to 3 kbar within the precision of the experiment.

REFERENCES

* Now at MPM Division, Pfizer, Easton, PA 18042

1. I. E. Dzyaloshinskii, Thermodynamic Theory of Weak Ferromagnetism in Antiferromagnetic Substances, JETP 32, 1547-62 (1957), Soviet Phys. JETP 5, 1259-72 (1957).
2. D. M. Wilson, The Magnetization of $FeBO_3$ in the Critical Region, (PhD Thesis, Univ. of Okla., 1975), Mich. Univ. Microfilm No. 75-21,839.
3. D. M. Wilson and S. Broersma, Critical Exponents of $FeBO_3$, AIP Conf. Proc. 24, 285-6 (1974).
4. M. Eibschutz, L. Pfeiffer, and J. W. Nielsen, Critical-Point Behavior of $FeBO_3$ Single Crystals by Mossbauer Effect, J. Appl. Phys. 41, 1276-7 (1970).
5. R. C. LeCraw, R. Wolfe, and J. W. Nielsen, Ferromagnetic Resonance in $FeBO_3$, a Green Room-Temperature Ferromagnet, Appl. Phys. Lett. 14, 352-4 (1969).
6. I. N. Nikolaev, L. S. Pavlyukov, and V. B. Marin, Displacement of the Neel Temperature in FeF_3 Under Pressure, Fiz. Tver. Tela 17, 3389-91 (1975).

MAGNETIC PROPERTIES OF THE HYDRIDES OF SELECTED RARE-EARTH INTERMETALLIC COMPOUNDS WITH TRANSITION METALS*

D. M. Gualtieri, K. S. V. L. Narasimhan and
W. E. Wallace
Department of Chemistry, University of Pittsburgh,
Pittsburgh, PA 15260

ABSTRACT

$ErFe_2$, $TmFe_2$ and $HoFe_2$ have been found to form stable hydrides of composition $ErFe_2H_{3.9}$, $TmFe_2H_{4.3}$, and $HoFe_2H_{4.47}$ when exposed to hydrogen at 3.6 x 10^6 Pa at room temperature. X-ray diffraction reveals an expansion of the $ErFe_2$ lattice parameter from 7.282 ± 0.001 to 7.828 ± 0.001 Å, the expansion of the $TmFe_2$ lattice parameter from 7.238 ± 0.001 to 7.839 ± 0.001 Å, and the expansion of the $HoFe_2$ lattice parameter from 7.284 ± 0.001 to 7.880 ± 0.001 Å with hydrogen absorption. Magnetization measurements as a function of temperature and field were made on these hydrides. The compensation temperature observed at 480 K for $ErFe_2$ and 233 K for $TmFe_2$ was shifted to 42 K and 8 K, respectively, upon hydriding. Magnetic moments measured at 4.2 K in an applied field of 120 kOe are 6.45 and 5.60 μ_B/FU for $TmFe_2$ and $HoFe_2$, respectively. These large moments arise due to the reduction of the iron moment. Results on hydrides of $NdCo_5$ and $DyCo_{5.2}$ are also presented.

INTRODUCTION

Rare earth compounds with transition metals such as iron, cobalt or nickel have been subjects of intense study for the past several years in this laboratory.[1] Rare earth compounds with cobalt were found to possess the highest energy product for permanent magnet applications and those with nickel, iron and cobalt were found to absorb and desorb large quantities of hydrogen at moderate pressures. Kuijpers[2] measured the magnetic properties of RCo_5 (R = rare earth) hydrides by encapsulating these in an epoxy to prevent the desorption of hydrogen from the samples. Recently[3] it was found in this laboratory that it is possible to 'poison' the surface of rare earth transition metal hydrides with SO_2. This process effectively prevents both the absorption and desorption of hydrogen. Utilizing this technique, an attempt has been made to characterize the magnetic properties of the hydrides of rare earth compounds with iron, cobalt and nickel. In this publication we report the magnetic properties of the hydrides of $HoFe_2$, $ErFe_2$ $TmFe_2$, $NdCo_5$ and $DyCo_{5.2}$

EXPERIMENTAL

All compounds were prepared by induction melting the elements in their proper proportion in a water cooled copper boat under a titanium gettered argon atmosphere. Required excess Tm was added to compensate for the loss of Tm during melting. All the ingots formed, except $DyCo_{5.2}$, were annealed at 950°C for one week in evacuated quartz tubes. $DyCo_{5.2}$ was annealed for 15 minutes at 1150°C in the induction boat. X-ray diffraction of the powders revealed no secondary phases, and a least-squares refinement of the 2Θ values was used to obtain the lattice constants shown in Table I.

TABLE I.

Magnetic moment per formula unit (μ_B/fu), compensation temperature (T_{comp}), Curie temperature (T_{Curie}) and lattice constants for the RFe_2 compounds and hydrides.

Compound	μ_B/fu	T_{comp} (K)	T_{Curie} (K)	a(Å)
$HoFe_2$	5.11	NONE	614	7.284
$HoFe_2H_{4.5}$	2.35*	60	287	7.880
$ErFe_2$	4.75	480	596	7.282
$ErFe_2H_{3.9}$	5.60+	42	280	7.828
$TmFe_2$	2.52	225	610	7.238
$TmFe_2H_{4.3}$	6.45+	18	270	7.839

* Measured at 21 kOe, 4.2 K.

\+ Measured at 120 kOe, 4.2 K.

Hydriding of the samples was carried out in an apparatus similar to that of Kuijpers, an additional provision being made to admit SO_2 to the sample chamber. The details of the apparatus are given elsewhere.[3] The samples were exposed to SO_2 after hydriding and used for magnetic and x-ray measurements. Lattice constants for the hydrides are also shown in Table I.

Magnetic measurements were carried out on the powders using the Faraday method in the temperature range 4.2 K to 1000 K in fields up to 21.2 kOe. The measurements above room temperature were made on hydride specimens wrapped in tantalum foil to allow desorption of hydrogen at elevated temperatures, where SO_2 is desorbed, without injury to the apparatus. Above a temperature of about 350 K, hydrogen desorption takes place, and thus the Curie temperature of several of the hydrides could not be obtained. Saturation magnetization measurements were obtained on $ErFe_2$ and $TmFe_2$ using a 120 kOe superconductive magnet in conjunction with a moving sample magnetometer. The powder samples were enclosed in a plastic container at the end of a long bakelite rod connected to a drive mechanism. The sample was moved through two pairs of precision wound coils connected in series opposition. The induced voltage was measured using an integrating digital voltmeter. An average of several travels of the sample through the coils was used for computing the magnetic moment. The apparatus was calibrated with high purity nickel and $MnCl_2 \cdot 4H_2O$ powders. The accuracy of such measurement is better than 1%.

RESULTS AND DISCUSSION

RFe_2 compounds (R = Tm, Er, Ho) absorb nearly four moles of hydrogen per mole of the intermetallic. The rare earth sublattice moments in the unhydrided materials[4] are coupled antiparallel to the iron sublattice moments, and as the temperature is increased the rare earth moments disorder much faster than the iron moments, often giving rise to a compensation temperature (T_{comp}). $TmFe_2$ and $ErFe_2$ both have a T_{comp}, whereas $HoFe_2$ does not show a T_{comp}. The absence of a compensation temperature is an indication of the dominance of the rare earth sublattice either because of a stronger intra-rare earth coupling and/or a strong polarization effect of the 3d moments on the rare earth through the conduction electrons. The Curie temperature (T_c) in these materials is controlled by the transition metal sublattice.

Magnetic properties of the RFe_2 hydrides reveal that the T_{comp} and T_c are drastically affected by the hydrogen absorption (see Figs. 1 and 2). $TmFe_2$ hydride has a T_{comp} of 8 K, as compared to the unhydrided sample T_{comp} of 225 K. The T_c of $TmFe_2$ is also decreased from 610 K to 270 K with hydriding. A similar decrease in T_c and T_{comp} was observed for $ErFe_2$. In the case of $HoFe_2$ T_{comp} was observed for the hydride whereas no compensation temperature appears for the unhydrided material. These results can be explained by the weakening of the R-R exchange due to the lattice expansion and decreasing Fe-Fe exchange due to the transfer of electrons from hydrogen to the 3d bands of iron. Kuijpers[2] has shown that the magnetic moment of cobalt in a series of RCo_5 compounds decreases with increasing hydrogen absorption. Thus it is not surprising that the Curie temperatures of the hydrides are drastically affected. Measured magnetic moments at 120 kOe and 4.2 K for $TmFe_2$ and $HoFe_2$ are much higher in the hydrided state than that expected for antiparallel coupling of the rare earth and iron moments. These compounds do not show a tendency to saturate even at the highest field applied (see Figs. 3 and 4). If we assume each hydrogen atom absorbed donates one electron to the lattice, the moment of the iron sublattice will be decreased, resulting in an increased moment for the compound. This would suggest that the compensation temperature be shifted to a higher temperature, but the weakening of the rare earth exchange (due to lattice expansion) results in a rapid disorder of the moments of the rare earth sublattice and T_{comp} occurs at a lower temperature.

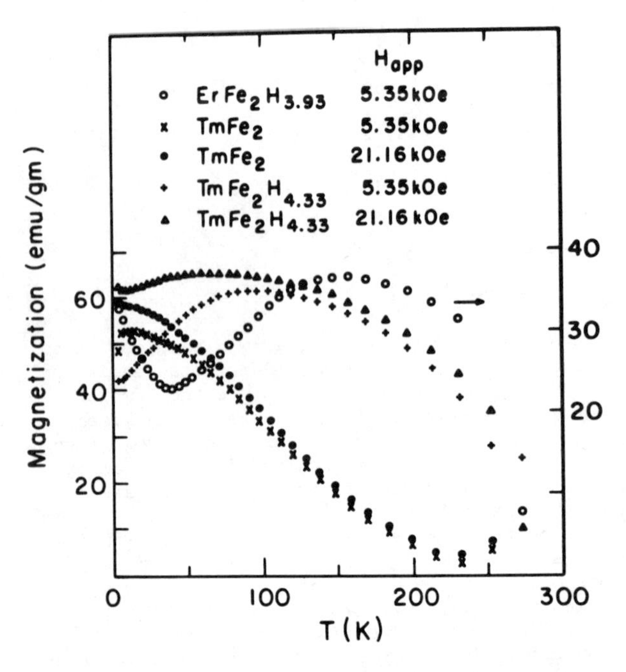

Fig. 1. Magnetization as a function of temperature for $TmFe_2$ and the hydrides of $ErFe_2$ and $TmFe_2$.

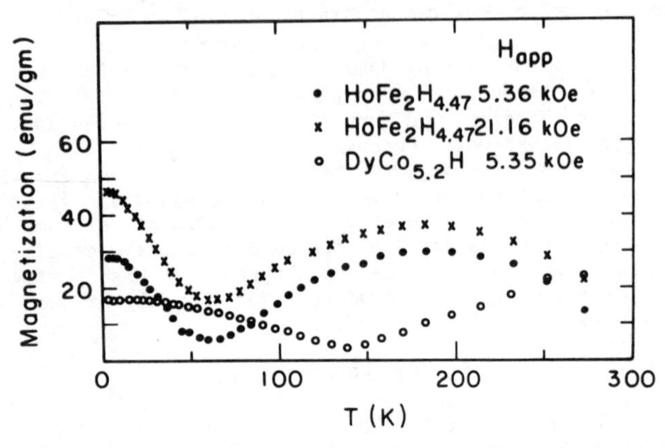

Fig. 2. Magnetization as a function of temperature for the hydrides of $HoFe_2$ and $DyCo_{5.2}$.

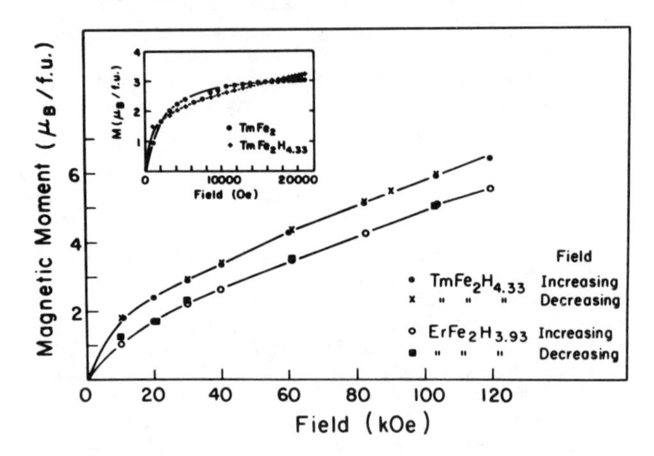

Fig. 3. Magnetization as a function of applied field for the hydrides of $TmFe_2$ and $ErFe_2$ at 4.2 K. The insert compares the magnetization of $TmFe_2$ and its hydride.

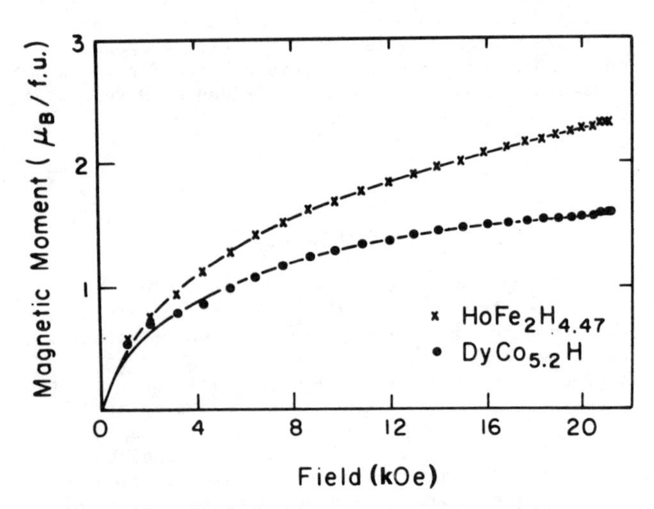

Fig. 4. Magnetization as a function of applied field for the hydrides of $HoFe_2$ and $DyCo_{5.2}$, measured at 4.2 K.

Magnetic measurements on hydrides of $NdCo_5$ and $DyCo_{5.2}$ were carried out to study the effect of hydriding on the spin reorientation temperature. The easy direction of magnetization in $NdCo_5$ changes from the c-axis to the basal plane at 66 K in the hydride, as compared with 270 K for the unhydrided compound.[1] For $DyCo_{5.2}$ this change occurs at 350 K for the unhydrided compound,[1] but the preference for basal plane alignment diminishes with modest hydrogen uptake as evidenced by x-ray diffraction of oriented powders at room temperature. In this case, complete hydriding was not possible because of the high equilibrium pressure of the compound, and only the low hydrogen concentration α-hydride could be formed. Calculations reveal that the change in the easy direction of magnetization occurs due to the effect of crystal fields on the rare earth ions[5] which becomes important at lower temperatures. The shifting of the easy direction change to lower temperatures in the case of the hydrides may be caused by the lattice expansion that changes the distance of nearest neighbor atoms and the weakening of the cobalt exchange due to filling of the 3d bands. Further work is in progress involving magnetic and Mössbauer measurements of RCo_5, R_2Fe_{17}, RCo_3 and RFe_3 compounds.

CONCLUSIONS

Magnetic measurements of $HoFe_2$, $ErFe_2$ and $TmFe_2$ reveal that the compensation and Curie temperatures are drastically affected by hydrogen absorption. The magnetic moments are higher for the hydrided compounds. The change in the easy direction of magnetization in $NdCo_5$ and $DyCo_{5.2}$ is shifted to lower temperatures by hydrogen absorption.

REFERENCES

* This work was assisted by a grant from the Army Research Office - Durham.

1. W. E. Wallace, Rare Earth Intermetallics, Academic Press, 1973.
2. F. A. Kuijpers, "RCo_5-H and Related Systems," Thesis Philip. Res. Reports Suppl., 1973, no. 2, p. 72.
3. D. M. Gualtieri, K. S. V. L. Narasimhan and T. Takeshita, "Control of the Hydrogen Absorption and Desorption of Rare Earth Intermetallic Compounds," J. Appl. Phys. (in press).
4. W. E. Wallace and E. A. Skrabeck, "Magnetic Moments and Iron Nuclear Hyperfine Fields in Laves Phases Containing Lanthanides Combined with Iron and Cobalt," Rare Earth Research II, edited by K. S. Vores, Gordon and Breach, New York, 1964, pp. 431-441.
5. J. E. Greedan and V. U. S. Rao, "An Analysis of the Rare Earth Contribution to the Anisotropy of RCo_5 and R_2Co_{17} Compounds," J. Sol. State Chem. 6, 387-95 (1973).

JAHN-TELLER TRANSITION IN $PrCu_2$

K. Andres
Bell Laboratories, Murray Hill, N.J. 07974
and
P.S. Wang, Y.H. Wong and B. Lüthi
Rutgers University, New Brunswick, N.J. 08903
and
H.R. Ott
Laboratorium für Festkörperphysik der ETH,
8093 Zürich, Switzerland

ABSTRACT

Previous specific heat measurements[1] on $PrCu_2$ indicated that below 7 K the two lowest lying singlet states of the $^3H_4$ manifold of Pr^{3+} spontaneously increase their splitting due to a Jahn-Teller effect. $PrCu_2$ has an orthorhombic structure which can be viewed as a distortion of the hexagonal AlB_2 structure. Further measurements on single crystals yield the following results: Large anisotropies are observed in the magnetic susceptibility, the electric resistivity and the elastic constants. At the transition the susceptibility shows relatively small anomalies, while marked anomalies are observed in the resistivity and thermal conductivity. The anisotropy of the resistivity remains unchanged through the transition, contrary to what one would expect from single ion anisotropic Coulomb charge scattering. Ultrasonic measurements above the transition temperature show softening of certain shear modes, and thermal expansion measurements down to 1,5 K show that the crystal distorts below 10 K.

INTRODUCTION

It has been shown previously[2] that among all (RE)Cu_2 compounds only $LaCu_2$ crystallizes in the hexagonal AlB_2 structure and that all the others (including $EuCu_2$ and $YbCu_2$, where the rare earth ion is divalent) crystallize in an orthorhombic structure[3]. The unit cell of this structure, which has the symmetry Imma, contains two inequivalent rare earth sites whose crystal field symmetry is lower than orthorhombic. The low symmetry causes all 2J + 1 = 9 crystal field levels of Pr^{3+} in $PrCu_2$ to be singlets. First indications of a Jahn-Teller effect, which causes an increase in separation of the two lowest singlet states, came from measurements of the specific heat in polycrystalline samples which increases sharply below 8 K, noncharacteristic of an ordinary Schottky anomaly. Recently, we have been successful in growing single crystals of $PrCu_2$, and this has enabled us to study the transition in more detail.

RESULTS

Susceptibility: The magnetic susceptibility in the three principal directions is shown in Fig. 1. It is quite anisotropic and shows relatively small but distinct features below 10 K. This is in qualitative agreement with the assumption that the two lowest singlet states are Jahn-Teller active and have a non zero matrix element of $\underline{J_z^2}$ (J_z = angular momentum operator) between them. The matrix element of $\underline{J_z}$ between those two states must then be zero and the Van Vleck susceptibility of the ground state arises from matrix elements of $\underline{J_z}$ to higher excited states. It is possible to explain qualitatively the observed anomalies with the three level model indicated in Fig. 1 and by assuming different susceptibility anisotropies of the ground state and the first excited state.

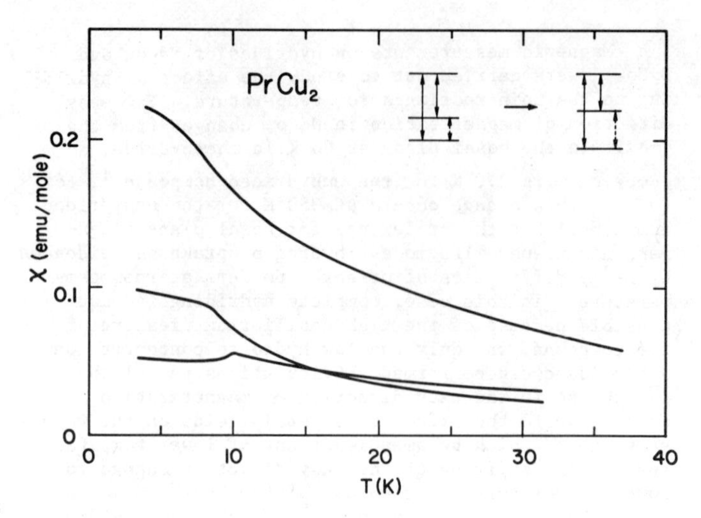

Fig. 1. Magnetic susceptibilities in the three principal directions measured in a field of 1 kOe. solid arrow = J_z-interaction, broken arrows = J_z^2-interaction.

Resistivity and thermal conductivity: The temperature dependent part of the resistivity is shown in Fig. 2. Although a strong anomaly is observed through the transition, the anisotropy remains independent of temperature. The anisotropy shown in fig. 2 is the same as the anisotropy of the room temperature resistivity as well as the residual resistivity. This suggests that it is entirely due to the anisotropic structure of the Fermi surface and not due to anisotropies in the scattering cross section between conduction electrons and f-electrons. The latter would be expected from Coulomb charge-scattering between conduction- and f-electrons[3]; its contribution must be unimportant in $PrCu_2$. The thermal conductivity

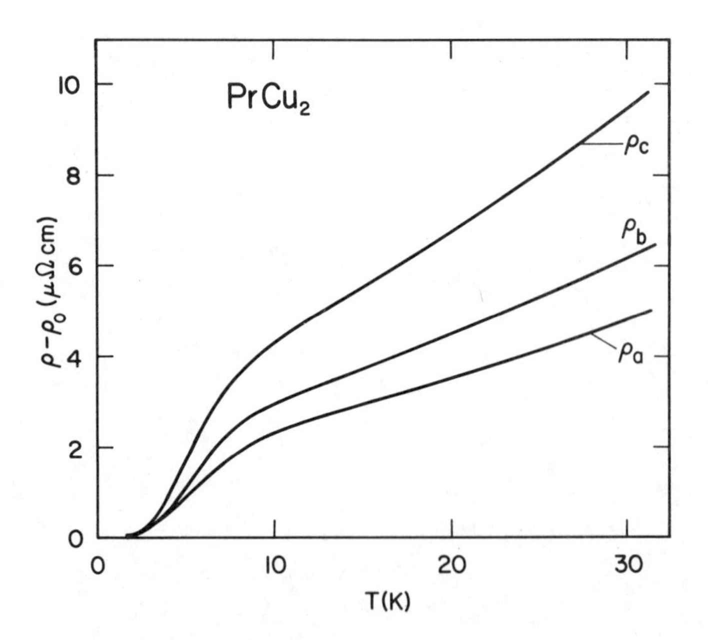

Fig. 2. Temperature dependence of the electrical resistivity in the three principal directions at low temperatures.

shows an anomalous decrease around 7 K and the Lorenz number has a minimum at 3 K where it is reduced to about 60% of its ideal valuc. This indicatcs the existence of phonon-like soft modes which can scatter conduction electrons by small angles, so that the effective electrical relaxation time is longer than the thermal relaxation time. Since we expect the elastic constant of at least one accoustic mode to go to zero at the Jahn-Teller transition (see below), the existence of relatively soft $k \neq 0$ phonons is to be expected.

Thermal expansion: Measurements in the three principal directions show an expansion of the a-axis ($\frac{\Delta a}{a} = 1.3 \times 10^{-4}$) and a contraction in the b- and c-axis

$$\left(\frac{\Delta b}{b} = -5 \times 10^{-4}, \frac{\Delta c}{c} = -1.7 \times 10^{-4}\right)$$

as the crystal cools through the transition, these changes starting at about 10 K and being completed at about 5 K (see fig. 3). This confirms that there is an actual Jahn-Teller distortion and that the specific heat anomaly cannot solely be due to 4f quadrupole-quadrupole interactions without ion-lattice couplings.

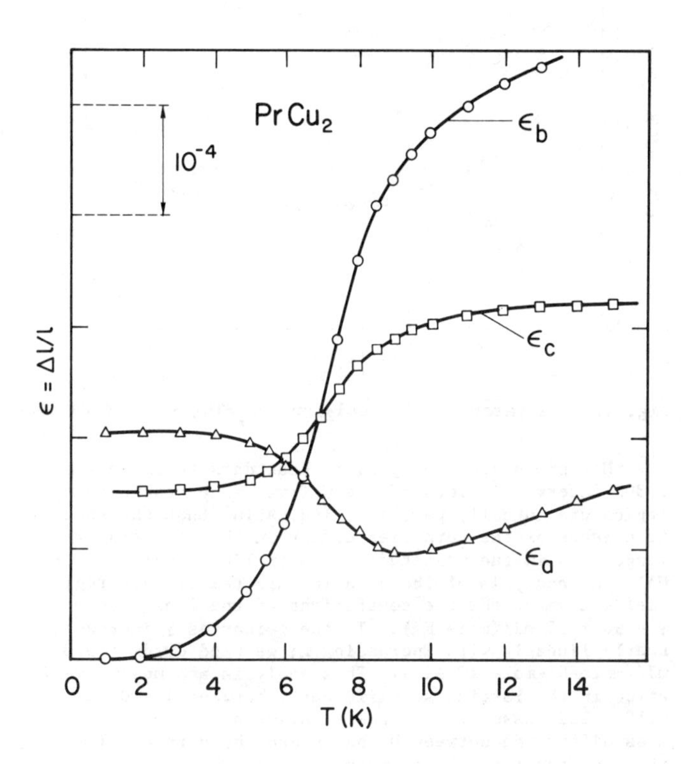

Fig. 3. Observed strains along the three principal axes below 15 K.

Sound velocity: We have propagated both longitudinal and shear sound waves along the a-axis of the crystal, and we observe a striking softening of the elastic constant for the shear wave with the shear polarisation vector along the b-axis (Fig. 4). Unfortunately, we loose the echo around 10 K due to increased attenuation, so that we cannot follow it all down to the transition. The elastic constant of this mode is c_{66}; if it goes to zero at the transition, we would expect the crystal to suffer a shear distortion in the a-b plane such that the angle between a and b would be

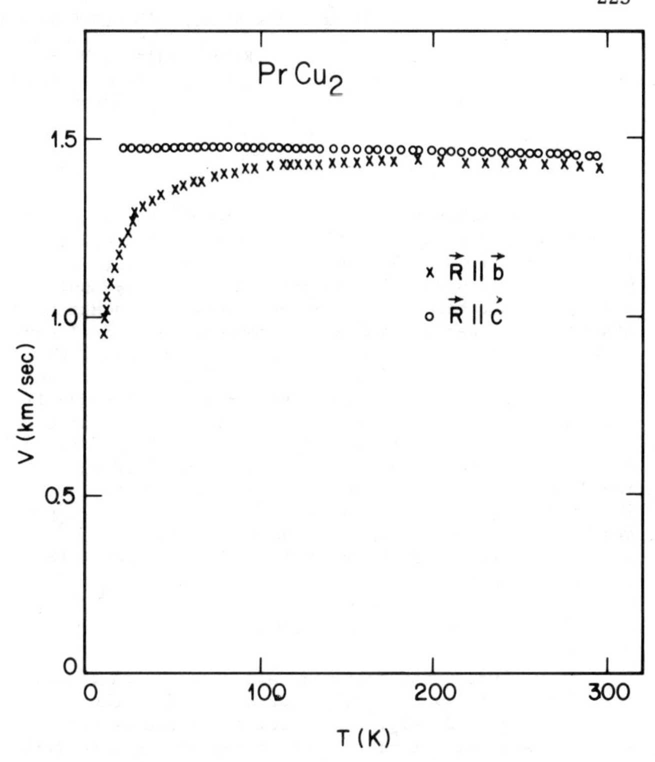

Fig. 4. Temperature dependence of shear sound waves propagating along the a-axis. $\vec{R}$ denotes the polarisation vector of the shear waves.

different from 90°, the magnitude of the a- and b-axis remaining unchanged to first order. This would be in contradiction to what we observe in the thermal expansion. However, it is still possible that it is yet another mode which actually goes to zero at the transition, such as a shear mode propagating along $\vec{a} + \vec{b}$ with the shear polarisation in the a-b plane. Its elastic constant is

$$\sqrt{(c_{11} + c_{22} - 2c_{12})/4\rho}$$

and it would lead to a Jahn-Teller distortion consisting of a lengthening of the a-axis and a shrinkage of the b-axis (or vice versa). Further experimental investigations are in progress.

CONCLUSIONS

All our observations on single crystals of $PrCu_2$ confirm that there is indeed a cooperative Jahn-Teller effect occuring in this material which is driven by the strain-coupling of the singlet crystal field states of the Pr^{3+}-ion. To our knowledge this is the first example of a Jahn-Teller effect in an intermetallic compound with a nondegenerate ground state.

REFERENCES

1. M. Wun and N.E. Phillips, Phys. Letters 50A 195 - 196 (1974)
2. A.R. Storm and K.E. Benson, Acta Cryst. 16 701 - 702 (1963)
3. A.C. Larson and D.T. Cromer, Acta Cryst. 14 73 - 74 (1961)
4. M.J. Sablik and P.M. Levy, preprint

SPECIFIC HEAT AND ELECTRICAL PROPERTIES OF THE SPIN FLUCTUATION SYSTEM $U_{1-x}Th_xAl_2$*

R. J. Trainor, M. B. Brodsky and H. V. Culbert†
Argonne National Laboratory
Argonne, Illinois 60439

ABSTRACT

The pseudobinary intermetallic compounds $U_{1-x}Th_xAl_2$ have been studied by specific heat (2-30K) and electrical resistivity (1.5-300K) measurements for x = 0.0, 0.02, 0.05, 0.10 and 0.95. The specific heat of UAl_2 contains a large linear term indicative of a very high density of states at the Fermi level and a pronounced $T^3 \log T$ contribution due to ferromagnetic spin fluctuations. With increasing x, the specific heats show both an increase in the magnitude of the linear term and a strong modification of the $T^3 \log T$ term. The T^2 resistivity in UAl_2 changes to a $T^{3/2}$ dependence by x = 0.10. These effects are discussed in terms of reduced overlap of 5*f* wavefunctions and increased impurity scattering with increasing x. The specific heat of the x = 0.95 sample exhibits a strongly field-dependent maximum near 2K associated with interactions between local moments formed on the U sites.

INTRODUCTION

Spin fluctuation phenomena associated with incipient ferromagnetism have recently been observed in the intermetallic compound UAl_2.[1,2] At low temperatures UAl_2 exhibits a prominent field-independent $T^3 \log T$ contribution to the specific heat,[1] a large resistivity[2] proportional to T^2, and a static susceptibility[1] varying as $(1 - aT^2)$. Each of these features is predicted by ferromagnetic spin fluctuation theory;[3] they are mutually consistent with a characteristic spin-fluctuation temperature T_{sf} of about 25K for UAl_2.[4] This very low T_{sf} is why UAl_2 is the only uniform metal for which all of the above features have been experimentally observed (e.g., for Pd $T_{sf} \sim 300K$) and is due to the direct overlap of 5*f* wavefunctions which produces extremely narrow bands at the Fermi energy.

In this paper we report a study of the specific heats (2.0-30K) and electrical resistivities (1.5-300K) of the pseudobinary intermetallic compounds $U_{1-x}Th_xAl_2$ for x = 0, 0.02, 0.05, and 0.10. The work is directed at observing both the consequences of varying the *f-f* wavefunction overlap and the effects of impurity scattering on the spin-fluctuation properties of UAl_2.

We have also measured the temperature dependences of the specific heat (1.0-4K, in zero and applied fields up to 20 kOe) and magnetization (2-300K) of $Th_{0.95}U_{0.05}Al_2$, a system sufficiently dilute that the U ions possess local 5*f* moments.[5] Special attention is directed at the previously reported low-temperature anomaly in the specific heat of this system.[6]

EXPERIMENTAL

UAl_2 and $ThAl_2$ have different crystal structures, the cubic Laves phase type C15 and hexagonal type C32 respectively. However, for small amounts of UAl_2 in $ThAl_2$ (up to at least 18 at.%)[6] and $ThAl_2$ in UAl_2 (up to ≈ 30 at.% as determined by our x-ray studies) the crystal structure of the host is maintained in the alloys. All samples were prepared by arc-melting, and subsequent x-ray patterns showed only the appropriate single phase. Lattice constants for the U-rich composition were found to increase slightly with increasing $ThAl_2$ concentration, with the overall variation between x = 0 and x = 0.10 being only ≈ 0.1%.

Specific heats were measured using pulse techniques. Magnetization measurements were made in fields between 3 and 14.5 kOe, using a Faraday method, and resistivities were measured by a conventional four-lead d.c. technique. These techniques have been described in detail elsewhere.[7]

EXPERIMENTAL RESULTS

The low temperature specific heats for x = 0 and 0.10 are shown in Fig. 1, plotted as C/T versus T^2 up to about 9K. The striking features of these data are the large magnitudes of C/T and the nonlinearity of the curves at low temperatures. The upturn in C/T for UAl_2 has been previously shown to be associated with a contribution dominated by a term varying as $T^2 \log T$.[1,4] With increasing $ThAl_2$ concentration, the magnitude of the upturn systematically decreases and by x = 0.10 is barely perceptible (see Fig. 1).

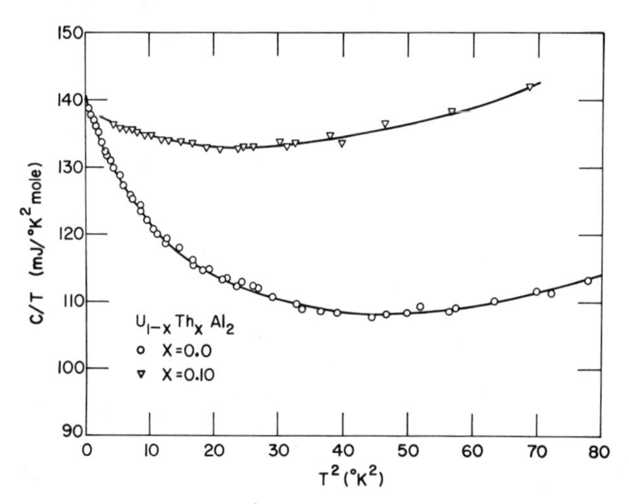

Fig. 1. C/T versus T^2 for UAl_2 and $U_{0.9}Th_{0.1}Al_2$ below **9K**.

Not shown are the specific heat data taken in the 9-30K interval. Plots of the C/T vs. T curves in this region are mutually parallel, indicating that the specific heats of the ternaries differ in this temperature range only in the magnitude of their linear terms. For UAl_2, an analysis of the data in this temperature region[4] yields a value for the coefficient of the linear term of $\gamma = 90 \pm 10$ mJ/(mole-K^2). In the ternaries γ increases nearly linearly with increasing x; we find $d\gamma/dx = 3.0$ mJ/(mole-K^2-mole %$ThAl_2$). This analysis assumes no variation in the lattice specific heats between x = 0 and 0.10. This assumption appears valid in view of the small mass difference between U and Th and the very small variation of the lattice parameters.

The low temperature electrical resistivities $\Delta\rho = \rho - \rho_0$, where ρ_0 is the residual resistivity, are shown in Fig. 2 for the compositions x = 0 and x = 0.10. For UAl_2 $\Delta\rho = AT^2$ below 3K, with A = 0.25 μΩ-cm/K^2, as reported previously.[2] With increasing x, $\Delta\rho$ changes systemically from a T^2 dependence to a $T^{3/2}$ dependence (attained by x = 0.10). At 300K, the incremental resistivities $\rho_{300} - \rho_0$ are 125, 120, 58 and 43 μΩ-cm for x = 0, 0.02, 0.05 and 0.10, respectively. The residual resistivities obey a Nordheim-rule dependence [i.e., $\rho_0 \propto x(1-x)$] only roughly with concentration with

ρ_0 = 20, 28, 69 and 94 μΩ-cm for x = 0, 0.02, 0.05 and 0.10 respectively. UAl_2 is a congruent-melting line compound, and samples with a high degree of atomic order are easily prepared. However, UAl_2 is an extremely brittle metal with a relatively poor thermal conductivity, and the rapid cooling following the arc-melting preparation produces microcracks in the sample. These microcracks are at least partially responsible for the relatively high ρ_0 in UAl_2 and the observed deviations from Nordheim's rule.

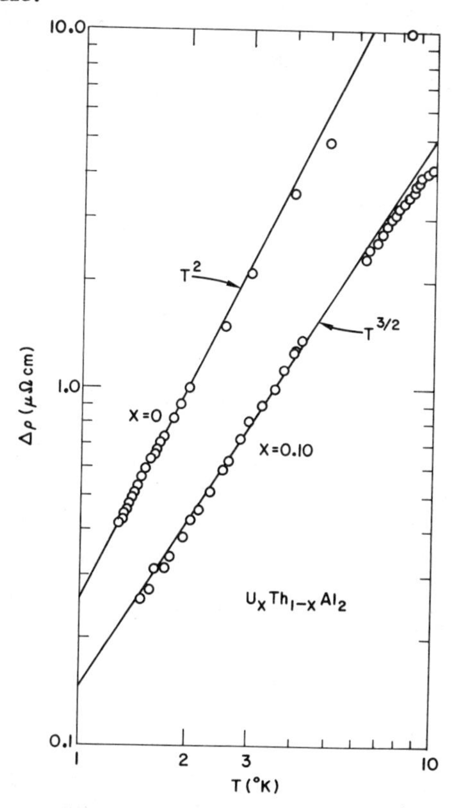

Fig. 2. Log Δρ versus log T for UAl_2 and $U_{0.9}Th_{0.1}Al_2$ at low temperatures.

The specific heats for the Th-rich end of the series have been reported previously (up to 18 at.% UAl_2 in $ThAl_2$) by Scott et al.[6] The magnetic susceptibilities have also been measured.[5] At high temperatures the susceptibilities were found to contain both a Curie-Weiss term and a large temperature-independent contribution. The Curie-Weiss fit gave effective moments of μ_{eff} = 2.38 μ_B, which were nearly independent of U concentration. The low temperature specific heats show two interesting features: an enhancement of γ roughly proportional to U concentration [$d\gamma/dx \simeq 2.0$ mJ/(mole-K^2-mole % UAl_2)] and upturns in C/T below ≈ 6-7K.[6] To study the nature of the upturns we have repeated the specific heat measurement on $Th_{.95}U_{.05}Al_2$ in applied fields of 0, 4 and 20 kOe. The data have been analyzed by assuming that the total specific heat C is composed of three terms

$$C = C_{el} + C_{lat} + C_M, \quad (1)$$

where $C_{el} = \gamma T$ and $C_{lat} = \beta T^3$ are the electronic and lattice terms for $ThAl_2$ (taken from the data of ref. 6), and C_M is a magnetic term due to the U spins which also contains the enhanced linear term. The temperature and field dependences of C_M are shown in Fig. 3 between 1 and 4K. The data show rounded maxima in C_M, which are strongly field dependent. If the enhancement of γ reported previously is subtracted from C_M, then the maximum in zero field occurs at 1.6K, instead of 2.0K as shown in Fig. 3. The peak in C_M for an applied field of 20 kOe lies above the temperature investigated on this sample (4K).

The magnetization vs. field curves for $Th_{0.95}U_{0.05}Al_2$ are linear for temperatures above 10K. Below ≈ 10K, however, these curves are nonlinear, and an Arrott plot (M^2 versus H/M) was constructed (not shown). The values of H/M at $M^2 = 0$, plotted against temperature, indicate an ordering temperature of 1.7K, close to the temperature of the zero-field specific heat maximum (see Fig. 3).

DISCUSSION

U-rich alloys: The effects of nonmagnetic impurities on the specific heat of a ferromagnetic spin-fluctuation system have been studied theoretically by Fulde and Luther[8] and more recently by Strauss and Ron.[9] It is found that when a mean free path ℓ due to scattering centers is introduced into the system, the $T^3 \log T$ term is strongly modified. Introducing a characteristic temperature $T_i \propto \ell^{-1}$, one finds $T^3 \log T \rightarrow T^3 \log (T + T_i)$ for finite ℓ. In the case of small ℓ (i.e., $T_i >> T$) the $T^3 \log T$ term goes over to a T^3 term, and the characteristic spin-fluctuation upturn in C/T disappears.

We have analyzed the data for $U_{1-x}Th_xAl_2$ assuming that the non-lattice contribution to the specific heat, $C - C_{lat}$, can be written (below 4K) as

$$C = aT + bT^3 \log[(T + T_i)/T_{sf}] \quad (2)$$

where a is found from the extrapolation of the C/T curve to $T^2 = 0$, b and T_{sf} are assumed not to vary appreciably between x = 0 and x = 0.10, and T_i is taken as zero for pure UAl_2. For x = 0.02, 0.05 and 0.10 we obtain values of T_i = 2.8, 4.0 and 6.0K.

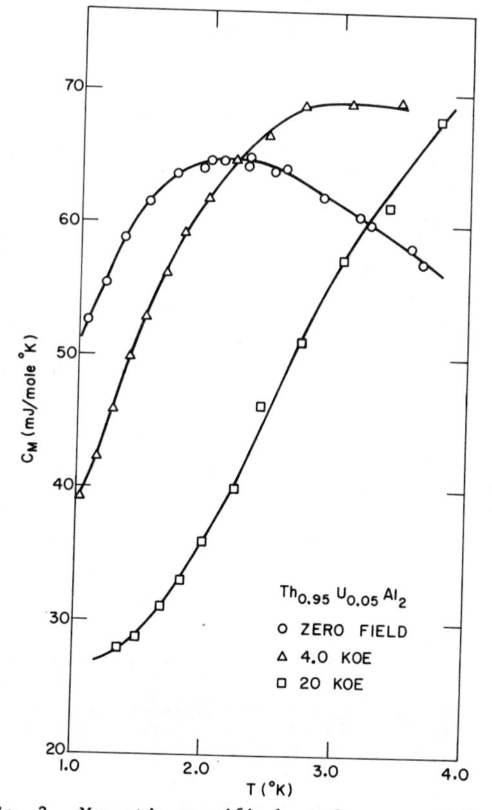

Fig. 3. Magnetic specific heat C_M versus T for $Th_{0.95}U_{0.05}Al_2$ in various applied fields.

Values of T_i can also be estimated from the residual resistivities. From ref. 8, we write

$$T_i = (E_F/S)\ (p_F \ell)^{-1}\ (h/k_B), \qquad (3)$$

where E_F is the Fermi energy, S is the Stoner exchange enhancement factor, and p_F is the Fermi momentum. Using a parabolic-band model, the mean free path is given by the Drude formula, $\ell = (m^* v_F / ne^2 \rho_0)$. Here m^* is the electronic effective mass, v_F is the Fermi velocity, and n is the number of current carriers. Combining this with (3) enables the determination of T_i from ρ_0, if m^* and n are known. We take S = 4, as is appropriate for UAl_2.[1] Assuming $(m^*/m) = (\gamma/\gamma_{free})$, where γ is taken from the specific heat data and γ_{free} is the γ-value calculated for free electrons, we obtain values of T_i = 0.9, 1.3, 3.1 and 4.2K for x = 0.0, 0.02, 0.05 and 0.10, respectively for n = 1. Considering the crudeness of this calculation and the questionable nature of the ρ_0 data (due to the microcracks), the agreement between the T_i values determined independently from C and ρ_0 is good, thus substantiating both the validity of the Fulde-Luther theory and the existence of ferromagnetic spin fluctuations in UAl_2.

The temperature dependence of the electrical resistivity of a spin fluctuation system containing nonmagnetic impurities has not been explored theoretically. However, in the case of magnetic impurities in exchange enhanced metals (e.g., Fe in Pd) transitions from $\Delta\rho \propto T^2$ to $\Delta\rho \propto T^{3/2}$ with increasing impurity concentration have been observed.[10] Theoretical treatments employing a two-current model, in which the resistivity is determined by a competition between electron-magnon scattering (which gives the T^2 term) and impurity scattering (which does not conserve momentum and gives a $T^{3/2}$ term), have been partially successful in understanding these systems.[11] It seems reasonable that a similar theory could be developed to understand the resistivity data presented here.

The absence of long range magnetic order in a number of light-actinide intermetallic compounds has been attributed to broadening of the 5*f* states by direct overlap of 5*f* wavefunctions as a consequence of relatively short interactinide distances.[12] This mechanism is clearly operable in systems with the cubic Laves phase structures.[13] When the degree of *f-f* overlap is reduced, either by dilation of the lattice or by substitution of non-5*f* ions for the actinide constituents, it is therefore expected that the system will exhibit a stronger tendency toward magnetism. This is clearly seen in the $U_{1-x}Th_xAl_2$ system, in which we observe an increase in γ with reduced U concentration. A similar effect has been observed in the susceptibilities of $U_{1-x}Y_xAl_2$ alloys.[14] Thus, even though neither $ThAl_2$ nor UAl_2 magnetically order, it appears possible that some solid solutions of the two could order, if the appropriate composition could be made as a single-phase material.

$\underline{Th_{0.95}U_{0.05}Al_2}$: The upturns originally observed in the C/T plots of ref. 6 bear some resemblance to the upturn observed in pure UAl_2.[1] However, the anomaly in UAl_2 has been shown to be nearly independent of applied magnetic fields up to 43 kOe, supporting the spin fluctuation prediction and precluding a local-moment interpretation.[1] In contrast, we find a strong field dependence for the specific heat of $Th_{0.95}U_{0.05}Al_2$. In fact, if we identify the temperature at which C_M is a maximum as T_M and take μ_{eff} from the reported susceptibility data,[5] we find the shift ΔT_M in T_M due to the 4 kOe field to be $k_B \Delta T_M \simeq 0.9\ \mu_{eff} \Delta H$, near the free spin value. Thus the localized moments responsible for the anomaly reside in very small effective fields.

The specific heat anomalies shown in Fig. 3 are similar to those found for spin glass systems, although lower temperature data would be desirable to observe the expected $C_M \propto T$ behavior as $T \to 0$. However, the total entropy associated with the specific heat anomaly in $Th_{0.95}U_{0.05}Al_2$ excluding the enhanced linear term is found to be only ≈ 100 mJ/mole-K, much less than (0.05)R log 2. This is a reasonable result, since much of the entropy of the 5*f* electrons is removed via the enhanced linear term (which is much too large to have its origin in the 6*d* bands), and indicates that if the peak in C_M is due to a spin-glass-like transition, then not all of the *f* electrons are localized. Thus, it appears that in this system there may be clusters (or U-rich regions) in which the U 5*f* electrons are delocalized by *f-f* overlap; hence the concentration-dependent enhancement of γ.

REFERENCES

*Work supported by the U. S. Energy Research and Development Administration.

†Present address: Michael Reese Hospital, Chicago, Illinois 60616.

1. R. J. Trainor, M. B. Brodsky, and H. V. Culbert, Phys. Rev. Lett. 34, 1019 (1975).

2. A. J. Arko, M. B. Brodsky, and W. J. Nellis, Phys. Rev. B5, 4564 (1972).

3. See E. P. Wohlfarth and P. F. deChatel, Comments Solid State Phys. 5, 133 (1973) and the references listed therein.

4. R. J. Trainor, M. B. Brodsky, and G. S. Knapp, Proc. 5th Int. Conf. on Plutonium and Other Actinides, to be published.

5. V. Jaccarino, J. H. Wernick, and H. J. Williams, Bull. Am. Phys. Soc. 7, 8, 556 (1962).

6. W. R. Scott, V. Jaccarino, J. H. Wernick, and J. P. Maita, J. Appl. Phys. 35, 1092 (1964).

7. R. J. Trainor, G. S. Knapp, M. B. Brodsky, G. J. Pokorny, and R. B. Snyder, Rev. Sci. Instr. 46, 1368 (1975); J. W. Ross and D. J. Lam, Phys. Rev. 165, 617 (1968); M. B. Brodsky, N. J. Griffin, and M. D. Odie, J. Appl. Phys. 40, 895 (1969).

8. P. Fulde and A. Luther, Phys. Rev. 170, 570 (1968).

9. M. Strauss and A. Ron, Phys. Stat. Sol. (b) 67, 405 (1975).

10. G. Williams and J. W. Loram, J. Phys. Chem. Solids 30, 1827 (1969),

11. D. L. Mills, A. Fert, and I. A. Campbell, Phys. Rev. B4, 196 (1971).

12. H. H. Hill, in Pu 1970 and Other Actinides, ed., W. N. Miner, (Am. Inst. of Min., Met., Pet. Eng., New York, 1970), pp. 2-19.

13. See for example A. T. Aldred, B. D. Dunlap, D. J. Lam, and I. Nowick, Phys. Rev. B10, 1011 (1974).

14. K. H. J. Buschow and H. J. van Daal, AIP Conf. Proc. 5, 1464 (1971).

DE HAAS-VAN ALPHEN EFFECT AND THE BAND STRUCTURE OF UGe_3*

A. J. Arko and D. D. Koelling
Argonne National Laboratory, Argonne, Illinois 60439

ABSTRACT

De Haas-van Alphen measurements and band structure calculations are reported for UGe_3, a compound characterized by spin-fluctuation phenomena. Only a partial construction of the Fermi surface is possible from the complex experimental data. Band structure calculations are at present plagued by numerical difficulties but, while not agreeing in detail with the data, nevertheless do show some of the features of the Fermi surface. Considerable structure in N(E) is predicted within 0.1 eV of E_F. Large mass enhancements in UGe_3 are implied by these preliminary results.

INTRODUCTION

Systems having itinerant 5f electrons at the Fermi level have been the source of much study.[1] Few band structure calculations were attempted[2] primarily because of the lack of hard experimental data such as those provided by the de Haas-van Alphen (dHvA) effect with which to compare the results. Recently we have succeeded in measuring the dHvA effect[3,4] in several systems having the cubic $L1_2$ (ordered $AuCu_3$) structure, and have obtained at least qualitative if not quantitative agreement with band structure calculations. Thus the reliability of band structure calculations for 5f electron systems was demonstrated. It was found in URh_3 and UIr_3 that the uranium 5f-electrons are strongly bonded to the transition metal d-electrons resulting in very broad bands at the Fermi level. Because of this bonding, these two materials are not magnetic in spite of the fact that they appear in the magnetic region of the Hill plots[5] (which correlate superconducting and magnetic properties with actinide separation because of the direct f-f interaction). This bonding is also the basis for understanding the transport properties and relatively high melting temperatures.

In this paper we report dHvA measurements and band structure calculations for UGe_3. This compound also has the $L1_2$ crystal structure with a lattice constant of 4.206 Å such that it falls in the magnetic region of the Hill plots. Furthermore there are no d-states on the Ge so that the f-d bond is not a relevant mechanism. It has been reported to display spin-fluctuation phenomena.[6] From the T^2 term in the resistivity the spin fluctuation temperature (T_{SF}) has been estimated[7] to be greater than 500°K. This is the first nearly magnetic actinide compound to be investigated via the dHvA effect. While the work is in its early stages, we believe that the comparison of the data to the theoretical band structure will yield valuable information about the underlying physics of the nearly-magnetic phenomena believed present.

EXPERIMENTAL

Single crystals of UGe_3 as large as 2 mm on a side were grown by precipitation from a Bi flux containing ≈ 5 at.% U-Ge_3 mixture. The temperature was lowered from 1000°C to 300°C at 4°C/hr. The resistance ratio of the crystals was not measured; however, the Dingle temperature at one angle was 0.5°K indicating a very pure specimen. Measurements[8] were made in the (100) and (110) planes at fields as high as 71 kG, and temperatures as low as 0.5°K. Using a rotating probe capable of two degrees of freedom only one mounting of the sample was needed to obtain the data. An on-line PDP-11 mini-computer was used to calculate both fast and slow Fourier transforms greatly facilitating the analysis of the very complex spectrum.

BAND CALCULATIONS

The calculations were performed using a relativistic variant of the APW method formulated using energy-derivative orbitals.[9] The potential used was constructed using the standard overlapping charge density model from uranium in the configuration $f^3d^2s^1$ and germanium with the configurations s^1p^3 (referred to below as A) and s^2p^2 (B). The former appeared to give better results and thus we will focus on it here. The uranium configuration was chosen on the basis of our experience with U metal and the other U $L1_2$ materials. Exchange was treated using the Kohn-Sham-Gaspár (i.e., $\rho^{1/3}$ with $\alpha = 2/3$) model. As the trial energy energy for the APW orbitals was fixed at the Fermi energy, our results are only valid within 1-2 eV of the Fermi energy. Calculations were performed at 18 points in the irreducible 1/48'th of the unit cell. Further calculations were performed using a Fourier series fit of 13 star functions. (Because the coefficients must be the same for all elements of a star, the plane waves of the star are summed into "star functions".) As UGe_3 exhibited several anticrossings (kinks) near the Fermi energy, these fits were very marginally acceptable with rms errors of 0.01 Ry. Nonetheless, they allow us to see the gross features of the bands and have been used pending a better interpolation (LCAO?).

RESULTS AND DISCUSSION

The frequencies (in 10^6 Gauss) vs angle are shown in Fig. 1 for the two planes investigated, while the measured effective masses are listed in Table I. A very complex set of frequencies is observed. Except for the branches μ_i, the effective masses are generally very large as one would expect in a system having 5f character at the Fermi level. No effective masses were measured on the branches labeled χ_i but are probably greater than unity as judged from the signal amplitude.

Frequency branches μ_i have the lowest mass and result from a lens-shaped surface (probably an electron surface) located either at M or at X (see Fig. 2 for symmetry labels). The low mass would seem to suggest an s-p type sheet associated with these frequencies. Because of the potential sensitivities, the band structure cannot help us determine the location of μ_i. Bands obtained using potential A (shown in Fig. 3) give no closed surfaces at X or M. A very slight change in potential, however, can produce a closed surface at either location. Potential B was successful in producing pockets at X and M but gave a poorer fit overall. In Fig. 2 we have placed the surfaces at X without sufficient justification, in order to show the relative magnitude.

The symmetry branches γ and ρ are consistent with surfaces centered either at Γ or at R. The band structure calculations indicate the existence of a large hole surface at R and a somewhat smaller hole surface at Γ. On the basis of this we have assigned ρ to the point R, and γ to the point Γ. The locations and the relative sizes of the surface associated with μ_i, γ and ρ are shown in Fig. 2.

Frequency branches χ_i are almost certainly a

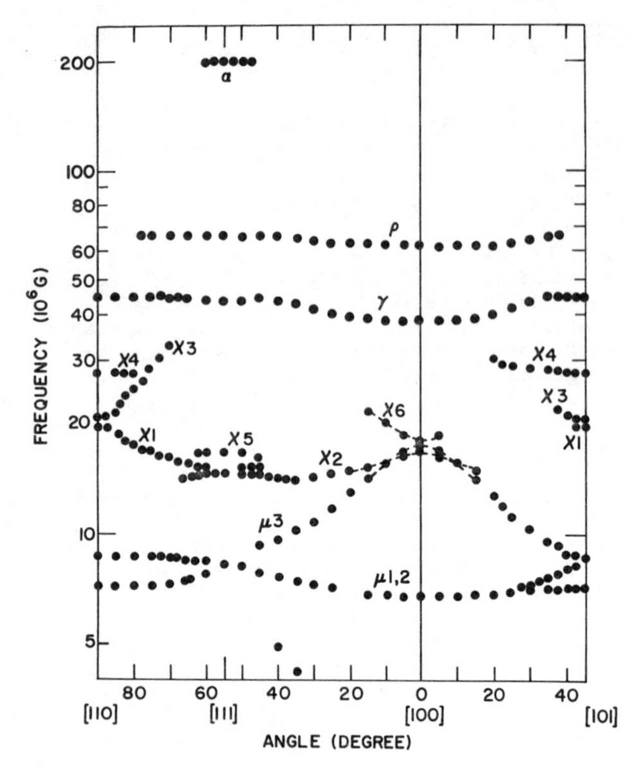

Fig. 1 de Haas-van Alphen spectrum for UGe_3 in 10^6 Gauss.

Table I. Effective masses in UGe_3 for several orbits. m^*_{EXP} is the measured mass while m^*_{th-A} is the mass obtained using potential A.

Frequency Branch	Orientation	m^*_{EXP}	m^*_{th-A}
α	[111]	2.3	
μ	[100]	0.41	
γ	[100]	3.7	1.45
ρ	[100]	4.1	

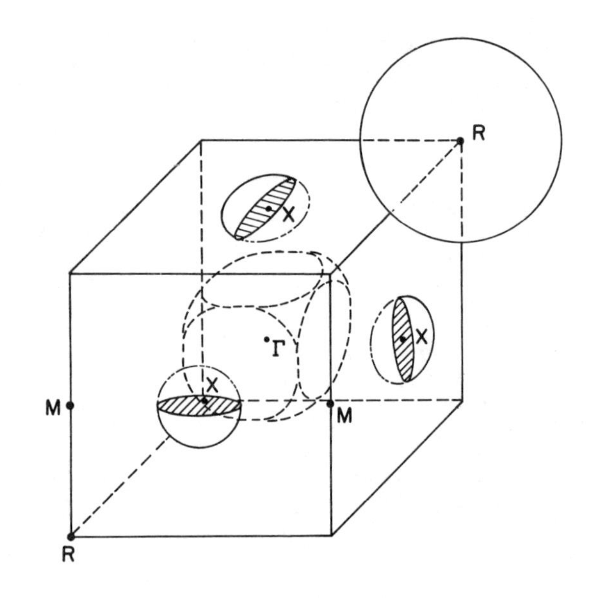

Fig. 2 Brillouin zone for UGe_3 with a partial construction of the Fermi surface from the data. Placement of various surfaces and exact shapes not uniquely determined (see text).

series of hole and neck orbits associated with a complex multiply-connected Fermi surface. While the magnitudes and cutoffs do not agree with the calculations, a multiply-connected, Γ-centered electron surface (which we will call χ), with necks along the Γ-M line, is presently indicated in the bands. Indeed, the surface which we have associated with γ is a hole surface entirely surrounded by χ. One can see from the bands in Fig. 3 that very small shifts in the bands or E_F will result in drastic changes for this surface. At present the band in question almost touches E_F at the point M.

A small hole surface at Γ is predicted by the bands in Fig. 3 which is not observed in the data. E_F would have to be shifted up by 0.015 Ry in order to eliminate this surface. Such a shift would improve the agreement with experiment for all pieces of the Fermi surface and is within the error of the calculation.

It would appear then that the bands shown in Fig. 3 correctly describe the qualitative features of the large segments of the Fermi surface, but do poorly on fine detail. Unlike URh_3 which had essentially only one band at the Fermi level and relatively minor sensitivity to small changes in potential, UGe_3 has several band crossing with large fitting errors and greater sensitivity to small changes in potential. Coupled to the fact that the bands are relatively flat (large dk_F/dE), this yields large variations in the predicted Fermi surface. Nevertheless the present bands can serve as a useful guide even with the limited agreement that is obtained.

An interesting feature in Fig. 3 is that very flat bands exist at or near E_F at the points M and Γ. These give rise to a sharp peak in the density of states very near E_F. Both potentials investigated gave rise to the same peak. This is the first actinide intermetallic compound studied by us which displays a peak in N(E) at E_F, and may be responsible for the weak many-body effects in UGe_3. The calculated value for the electronic specific heat term (the bare γ) varies from ≈ 7 to 10 mJ/mole depending on the exact location of E_F with respect to the peak. This is in contrast to the measured value[10] of 20.4 mJ/mole K^2. The calculated effective masses differ from the measured masses by the same factor of ≈2 - 2.5. We have sufficient confidence in the bands at this point to state that the adjustment of the fine details is not likely to increase the γ-value above 11 mJ/mole K^2. Thus we believe that there exist very large mass enhancements in UGe_3, with $\lambda = (\lambda_{Ph} + \lambda_{SF}) \gtrsim 1$, coming from both the electron-phonon interaction, λ_{Ph}, and electron-electron interactions, λ_{SF}. Such enhancements are expected in spin-fluctuation systems and give rise to the upturn in specific heat[11] near T = 0 provided T_{SF} is sufficiently low. The value obtained for λ in URh_3 is ≈ 0.5. Since no spin-fluctuation effects exist in this material we will assume that this is the value for the phonon mass enhancement, λ_{Ph}. Assuming a similar value of λ_{Ph} in UGe_3 (not totally justified in light of the very different electronic character)

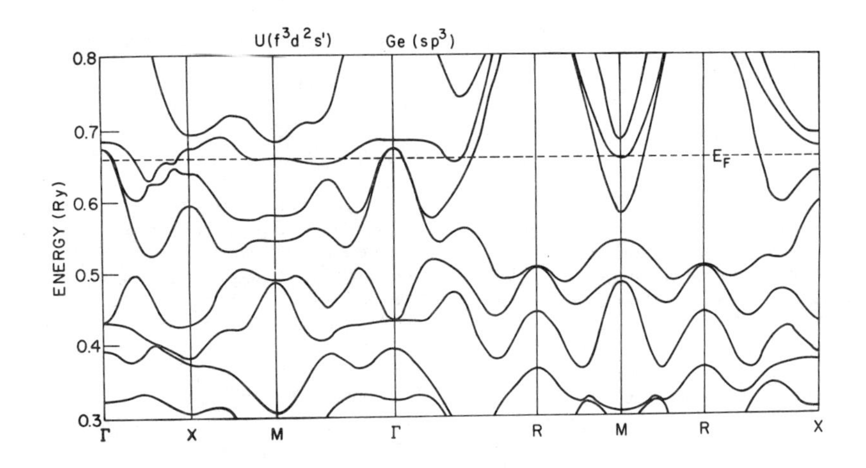

Fig. 3 Band structure for UGe_3 utilizing potential A.

we are left with $\lambda_{SF} \geq 0.5$. These estimates, crude as they are, are in general agreement with the values obtained in the spin-fluctuation system $UA\ell_2$.

We wish to thank Dr. R. J. Trainor and Dr. M. B. Brodsky for useful discussions, and J. Barhorst for technical assistance.

REFERENCES

*Work supported by the U. S. Energy Research and Development Administration.

1. For a review, see The Actinides: Electronic Structure and Related Properties, edited by A. J. Freeman and J. B. Darby (Academic Press, New York, 1974).
2. For details, see Ref. 1, Ch. 2, Vol. I and Ch. 1, Vol. II.
3. A. J. Arko, M. B. Brodsky, G. W. Crabtree, D. Karim, D. D. Koelling, L. R. Windmiller, and J. B. Ketterson, Phys. Rev. B12, 4102 (1975).
4. A. J. Arko, M. B. Brodsky, G. W. Crabtree, D. Karim, L. R. Windmiller, and J. B. Ketterson, Proceedings of the 5th International Conference on Plutonium and Other Actinides, Baden-Baden, Germany (1975).
5. H. H. Hill, Nuc. Metall. 17, 2 (1970).
6. K. H. J. Buschow and H. J. van Dall, AIP Conf. Proc. No. 5, Magnetism and Magnetic Materials-1971, C. D. Graham and J. J. Rhyne, eds. (AIP, New York, Pt. 2), pp. 1464-1477.
7. M. B. Brodsky, Phys. Rev. B9, 1381 (1974).
8. For measurement details, see R. W. Stark and L. R. Windmiller, Cryogenics 8, 272 (1968).
9. P. M. Marcus, Intern. J. Quant. Chem. 1S, 567 (1967); O. K. Andersen, Phys. Rev. B12, 3060 (1975); D. D. Koelling and G.O. Arbman, J. Phys. F 5, 2041 (1975).
10. M. H. van Maaren, H. J. van Dall, and K. H. J. Buschow, Solid State Commun. 14, 145 (1974).
11. R. J. Trainor, M. B. Brodsky, and H. V. Culbert, Phys. Rev. Letters 34, 1019 (1975).

CRYSTAL STRUCTURE AND MAGNETIC PROPERTIES OF RCu_4Ag (R = RARE EARTH) INTERMETALLIC COMPOUNDS*

T. Takeshita, S. K. Malik,[†] A. A. El-Atttar and
W. E. Wallace
Department of Chemistry, University of Pittsburgh
Pittsburgh, Pennsylvania 15260

ABSTRACT

The RCu_5 intermetallic compounds crystallize either in the cubic $AuBe_5$ type structure or in the hexagonal $CaCu_5$ structure depending upon the particular rare earth ion involved. The effect of substituting Ag for one Cu on the structure stability of RCu_5 compounds has been investigated. It is observed that the compounds RCu_4Ag (R = Nd to Tm) crystallize in the $AuBe_5$ type structure. The magnetization of these compounds with R = Nd to Ho has been studied in temperature range 4.2 to 300 K and in magnetic fields up to 21 kOe. The compounds with Nd, Sm, Gd and Tb are found to order antiferromagnetically with Néel temperature of 7, 7, 9 and 16.5 K, respectively. The Dy and Ho compounds exhibit ferromagnetic behavior in magnetic fields larger than about 1 kOe. In the paramagnetic region, the magnetic susceptibility of all the compounds except that of $SmCu_4Ag$ follows the Curie-Weiss law. The paramagnetic Curie temperatures of these compounds can be explained on the basis of RKKY theory with a suitable choice of the Fermi wave vector.

INTRODUCTION

Intermetallic compounds having the formula AB_5 generally crystallize either in the hexagonal $CaCu_5$ type structure or in the cubic $AuBe_5$ type structure. In the RCu_5 (R = rare earth) series, both these structures are observed, depending upon the rare earthion. The compounds with R = La to Sm possess the $CaCu_5$ type structure,[1] whereas those with R = Ho to Tm have the $AuBe_5$ type structure.[2] Compounds with R = Gd or Tb crystallize in either of the two structures, depending upon the heat treatment.[2] We have investigated the structural and magnetic properties of the RCu_5 compounds when Cu is replaced by Ag, Al, Ni, etc. In this paper we report the results on RCu_4Ag (R = Nd to Tm) compounds, all of which crystallize in $AuBe_5$ type structure. The results of magnetization measurements on some of these compounds in the temperature interval 4.2 to 300 K and in magnetic fields up to 21 kOe are presented.

EXPERIMENTAL

The RCu_4Ag (R = La to Tm) compounds were prepared by induction melting of stoichiometric amounts of the constituent elements in a water-cooled copper boat under a continuous flow of purified argon gas. The purity of the starting materials was as follows: rare earths 99.9%, Cu 99.999% and Ag 99.999%. The alloy ingots were turned and melted several times to ensure homogeneity. The weight loss during melting was negligible. Powder X-ray diffraction patterns of the as cast materials were obtained on a Picker X-ray diffractometer using CuK_α radiation. The magnetization measurements were performed using Faraday method in magnetic fields up to 21kOe and in the temperature interval 4.2 to 300 K.

RESULTS AND DISCUSSION

X-ray investigations of the as-cast samples revealed that the RCu_4Ag compounds with R = Nd to Tm were of single phase. The X-ray patterns could be indexed on the basis of cubic $AuBe_5$ (or $MgCu_4Sn$ type structure). The lattice constants obtained by an iterative least-square fit of the observed $\sin^2\theta$ values are listed in Table I. The X-ray patterns of the compounds with La, Ce and Pr showed lines in addition to those expected from $AuBe_5$ type structure. The additional phase(s) could not be eliminated by prolonged annealing. The substitution of Ag for Cu in RCu_5 has thus changed hexagonal $NdCu_5$ and $SmCu_5$ to the cubic $AuBe_5$ (or $MgCu_4Sn$) type and also stabilized the cubic morphology of $GdCu_5$ and $TbCu_5$. Since the atomic radius of Ag is larger than that of Cu (both monovalent), it appears that there is some conbribution of the atomic size to the stability of this type of crystal structure, which may overcome any electronic effects.

Results of magnetic measurements are shown in Figs. (1) to (3). Figure (1) shows the magnetization

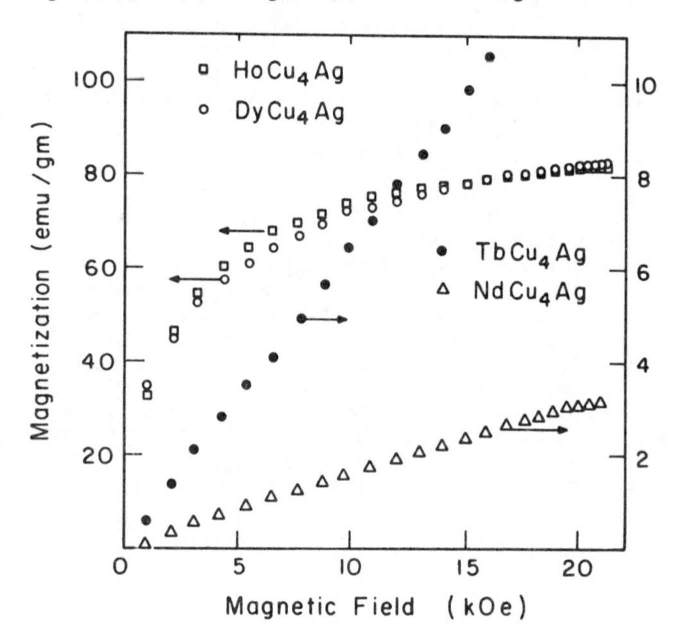

Figure 1. Magnetic versus applied field at 4.2 K for some RCu_4Ag (R = Nd, Tb, Dy, Ho) compounds.

σ versus applied field H curves at 4.2 K for some of the compounds. The σ-H curves for Dy and Ho compounds exhibit behavior characteristic of ferromagnetic systems. On the other hand, the magnetization of Nd and Tb compounds [Fig. (1)] and also those of Sm and Gd compounds (not shown) varies linearly with H. Figure (2) shows a plot of σ versus temperature at 21 kOe for the Gd and Tb systems. The pronounced maximum is associated with the onset of antiferromagnetism. Néel temperatures of 7, 7, 9 and 16.5 K are observed for the system with R = Nd, Sm, Gd and Tb, respectively. The Curie temperatures of Dy and Ho compounds [Fig. (3)] are found to be 25 and 22 K, respectively. From a plot of σ vs 1/H the saturation moment values (extrapolated to infinite field) have been obtained. In the paramagnetic state, the magnetic susceptibility (χ) of all the compounds except $SmCu_4Ag$ follows Curie-Weiss behavior. Experimental results on the ordering temperatures (T_C or T_N), saturation moments (μ_{sat}), effective magnetic moment (μ_{eff}) in paramagnetic state and paramagnetic Curie temperatures (θ_p) are collected in Table I. The μ_{sat}

Table I. Lattice constants, magnetic ordering temperatures, saturation magnetic moments, paramagnetic Curie temperatures and effective magnetic moments of RCu_4Ag compounds.

Compound	Lattice Constant (Å)	Magnetically Ordered State T_C (K)	T_N (K)	μ_{sat} (μ_B)	$g_J J$	Paramagnetic State θ_p (K)	μ_{eff} (μ_B)	$g_J \sqrt{J(J+1)}$
$NdCu_4Ag$	7.234		7	-		-9.0	3.39	3.62
$SmCu_4Ag$	7.186		7	-		χ vs. T nonlinear		0.84
$GdCu_4Ag$	7.163		9	-		-7.0	7.86	7.94
$TbCu_4Ag$	7.148		16.5	-		9.5	9.38	9.72
$DyCu_4Ag$	7.112	25		8.80	10.0	20.0	9.65	10.63
$HoCu_4Ag$	7.101	22		8.33	10.0	12.0	9.91	10.6

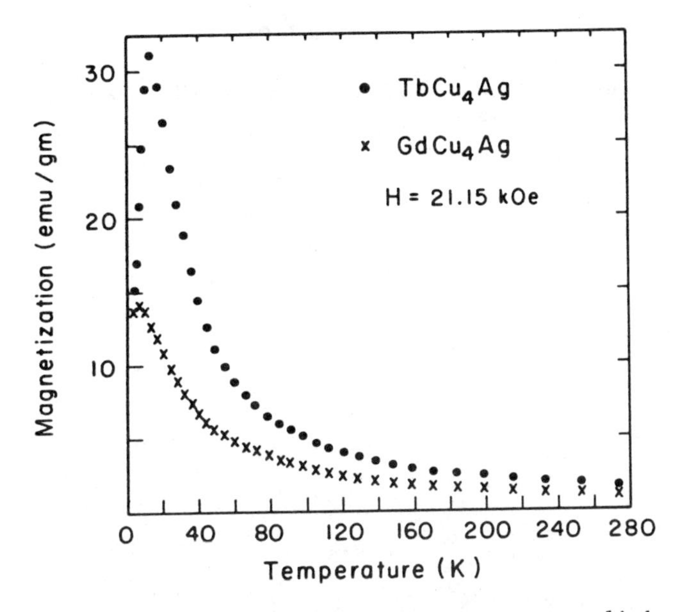

Figure 2. Magnetization versus temperature at applied field H = 21 kOe for $GdCu_4Ag$ and $TbCu_4Ag$.

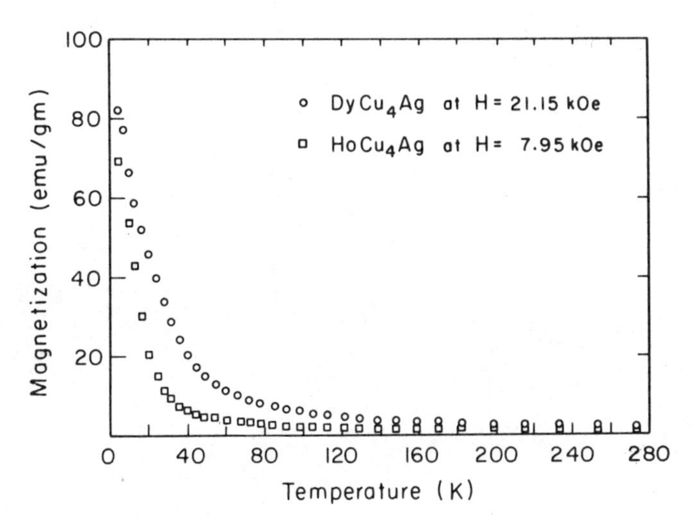

Figure 3. Magnetization versus temperature for $DyCu_4Ag$ (H = 21 kOe) and $HoCu_4Ag$ (H = 7.95 kOe).

and μ_{eff} values in RCu_4Ag compounds are somewhat smaller than the corresponding values for free trivalent rare earth ions. This may be because of crystal fields or conduction electron polarization effects or a combination of both.

It is interesting to intercompare the magnetic properties of the RCu_4Ag compounds with the cubic RCu_5 compounds.[3] $TbCu_5$ and $TbCu_4Ag$ are antiferromagnets; $HoCu_5$ and $HoCu_4Ag$ are ferromagnets. However, while $DyCu_5$ shows a metamagnetic transition, $DyCu_4Ag$ appears to be a ferromagnet. The magnetization of $DyCu_4Ag$ at 6 kOe does not reveal any maximum as a function of temperature. The possibility cannot be excluded that $DyCu_4Ag$ and $HoCu_4Ag$ are also antiferromagnets, as other RCu_4Ag compounds, with critical fields less than 1 kOe so that the antiferromagnetic state has not been detected.

The dominant interaction responsible for magnetic ordering in rare earth intermetallic compounds is thought to be the indirect coupling between rare earth spins operating through the conduction electron spins, the so-called RKKY[4] interaction [$\mathcal{H}_{ex} = -2J(0)\vec{S}\cdot\vec{G}$]. The paramagnetic Curie temperature is given by

$$k_B\theta_p = -\frac{3\pi Z^2}{E_F}(g_J-1)^2 J(J+1)J(0)^2 \sum_{n\neq m} F(2k_F R_{nm}) \quad (1)$$

where the symbols have the same meaning as in ref. (4). The function $F(x) = (x \cos x - \sin x)/x^4$ is the oscillatory RKKY function which depends on the number of electrons per atom in the conduction band, and is independent of the lattice constant 'a' of cubic compounds. The sign of θ_p depends only on the sign of $F(x)$; θ_p is positive when $F(x)$ is negative. Thus, in a series of rare earth compounds, if the number of electrons per atom is assumed to be constant, the θ_p values for all the compounds should have the same sign. From Table I we note that the θ_p values for RCu_4Ag compounds are small but both positive and negative.

The values of the Ruderman-Kittel sum $\sum F(2k_F R)$ are plotted in Fig. (4) as a function of k_F/k_F^o for RCu_4Ag compounds. k_F^o is the free electron Fermi wave vector if both Cu and Ag contribute one electron and rare earth contributes three electrons to the conduction band of the alloy. The θ_p values of RCu_4Ag compounds can be reconciled with RKKY theory only by allowing k_F to vary slightly from compound to compound, which in the sense of free electron model implies slightly different number of conduction electron per atom in each of these compounds. Further, we must be in that region of k_F/k_F^o where $\sum F(2k_F R)$ is changing sign. From Fig. (4), this means that we are close to either $k_F/k_F^o = 0.78$ or $k_F/k_F^o = 1.05$. Since both Cu and Ag are assumed to contribute the same number of electrons to the conductance band, the values of

$\sum F(2k_F R)$ are identical for RCu_4Ag and RCu_5. From interpretations of simultaneous measurements of the Knight shift and θ_p values in RCu_5 compounds and the θ_p value of $DyCu_4Zn$ it has been concluded that k_F values in the vicinity of $k_F/k_F^o = 0.78$ are most approrpiate to RCu_5 and hence to RCu_4Ag compounds. The value of J(0) in RCu_4Ag compounds which can be obtained from θ_p values and $\sum F(2k_F R)$ will depend on the exact choice of k_F but is typically of the same order of magnitude[3] as in RCu_5 compounds.

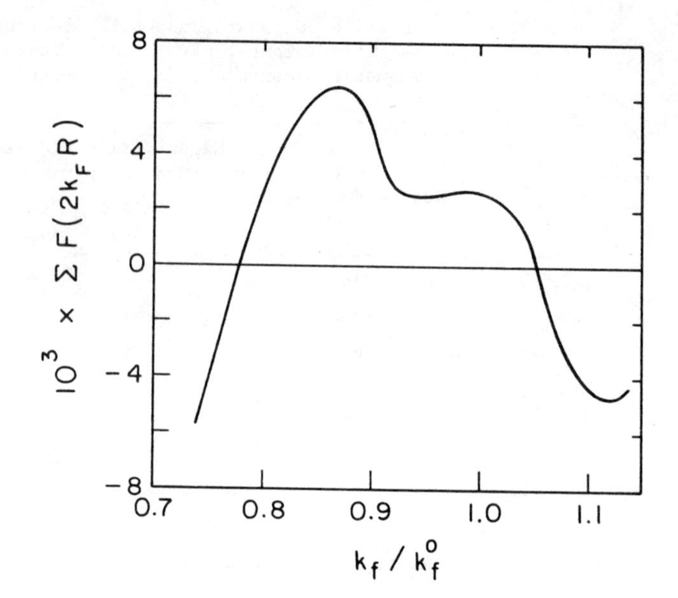

Figure 4. Plot of $\sum F(2k_F R)$ versus k_F/k_F^o for RCu_4Ag compounds.

REFERENCES

* This work was assisted by a contract with the U. S. Energy Research and Development Administration.

† On leave of absence from Tata Institute of Fundamental Research, Bombay 400 005, India.

1. K. H. J. Buschow, A. S. van der Goot, Acta Cryst. B27, 1085-1088 (1971).
2. K. H. J. Buschow, A. S. van der Goot and J. Birkhan, J. Less Comm. Metals 19, 433-436 (1969).
3. K. H. J. Buschow, A. M. van Diepen and H. W. de Wijn, J. Appl. Phys. 41, 4609-4612 (1970).
4. M. A. Ruderman and C. Kittel, Phys. Rev. 96, 99-102 (1954); T. Kasuya, Progr. Theoret. Phys. (Kyoto) 16, 45-57 (1956); K. Yosida, Phys. Rev. 106, 893-898 (1957).

MAGNETIC PROPERTIES OF $R(Fe_{1-x}Mn_x)_2$ COMPOUNDS (R = Gd, Tb, Dy, Ho or Er)*

M. Merches, K. S. V. L. Narasimhan and W. E. Wallace
Department of Chemistry, University of Pittsburgh
Pittsburgh, PA 15260
and A. Ilyushin
Department of Physics, Moscow State University
Moscow, U.S.S.R.

RFe_2 compounds crystallize in the cubic $MgCu_2$ type structure whereas RMn_2 compounds crystallize in the cubic structure when R = Gd, Tb, Dy and hexagonal $MgZn_2$ type of structure when R = Ho or Er. Solid solution between RFe_2 and RMn_2 was studied using x-ray, magnetic and Mössbauer investigations. When R = Tb and Dy complete homogeneity exists between the end members whereas when R = Gd, Ho or Er only a limited solubility is found. Saturation magnetization at 4.2 K reveals that the moment increases when x is increased whereas the Curie temperature decreases. Magnetization versus temperature on the substituted ternaries revealed the presence of a broad maximum at low temperatures with a very small moment at the lowest temperature measured. This maximum is not present when the specimen is cooled in a field.

INTRODUCTION

A great deal of interest has been shown in recent years in the magnetic properties of Laves phase RFe_2 compounds (R = rare earth) because of their varied magnetic properties in the crystalline[1,2] as well as amorphous state.[3] Recently Ilyushin and Wallace[4,5] reported the structural and Mössbauer studies on rare earth iron-manganese Laves phase ternaries and noted that the solid solubility of manganese in the Laves phase RFe_2 is limited when R = Gd, Ho or Er whereas a complete solubility exists when R = Tb or Dy. The present investigation deals with the magnetic properties of $R(Fe_{1-x}Mn_x)_2$ ternaries from 4.2 K to above their Curie temperatures.

EXPERIMENTAL

Stoichiometric amounts of rare earth (99.9% pure), Fe and Mn (99.99% pure) were induction melted together in a water cooled copper boat under a flowing atmosphere of purified argon gas. The ingots thus obtained were annealed at 750 C for two weeks. Annealed specimens were tested for single phase homogeneity using x-ray, metallographic and thermomagnetic analysis techniques. The last mentioned technique is useful for identifying any magnetic impurity phase by its Curie temperature. Lattice constants were calculated from the x-ray diffraction pattern obtained by using a CuK_α radiation and a Picker diffractometer.

Magnetic measurements were obtained on loose powders encapsulated in a pyrex or quartz tubes using the Faraday method. The temperature range covered was from 4.2 K to 1000 K in fields up to 21 kOe. Saturation magnetization was calculated from an extrapolation of magnetization (σ) versus inverse field plots and the Curie temperature was calculated from σ^2 vs temperature plots.

RESULTS AND DISCUSSION

Fig. 1 shows the range of stability of the Laves phase structure. It is interesting to note that although both $GdMn_2$ and $GdFe_2$ crystallize in the cubic structure, the ternaries $Gd(Mn_xFe_{1-x})_2$ consists of two phases when 0.2 < x < 0.8. Currently, further work is being done to elucidate the two phase region.

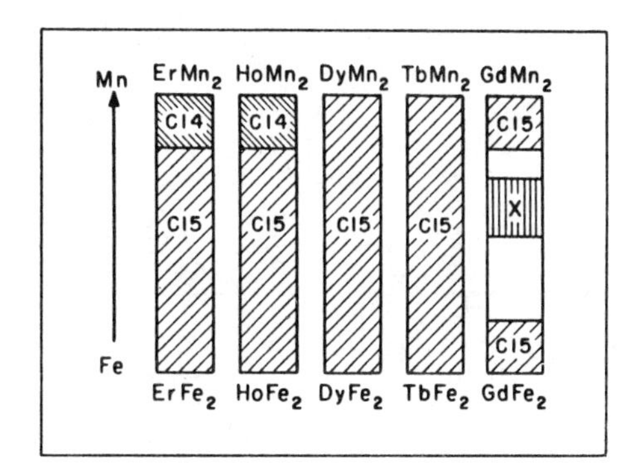

Fig. 1 Range of stability of the cubic C15 Laves phase and hexagonal C14 phases.

Table I shows the saturation magnetic moment for all the compositions investigated. The moment contribution of the transition metal sublattice can be evaluated from the total moment observed for the compound by assuming the moment of the rare earth sublattice to be the same in the ternaries as in the binary RFe_2 compounds and an antiferromagnetic coupling of the rare earth and transition metal moments. Table II shows the moment of the transition metal sublattice computed using these considerations. This table also shows the results reported for $Y(Fe_{1-x}Mn_x)_2$ ternaries by

Table I. Saturation moment (μ_B/F.U.) and Curie temperatures (K) for $R(Fe_{1-x}Mn_x)_2$ ternaries.

R		x = 0	0.2	0.4	0.6	0.8	1.0
Gd	μ	2.80[a]	3.86	4.82[b]	5.14[b]	5.23	-
	T_C	785[a]	670	540[b]	323[b]	208	-
Tb	μ	4.72[a]	6.18	6.01	6.17	6.58	-
	T_C	711[a]	-	395	261	146	-
Dy	μ	5.50[a]	5.64	6.50	7.03	7.76	-
	T_C	635[a]	488	318	230	110	-
Ho	μ	5.50[a]	6.67	7.36	7.70	9.11[c]	-
	T_C	612[a]	448[d]	298[d]	159	62[c]	-
Er	μ	4.80[a]	-	6.71	6.43	7.62[c]	7.86[e]
	T_C	590[a]	444[d]	225	136	49[c]	25[e]
Y	μ	2.91[a]	2.0[f]	1.2[f]	0.4[f]	-	-
	T_C	550[a]	390[f]	260[f]	110[f]	-	-

a) Reference 9
b) Two phase region
c) C_{14} type structure
d) References 4 and 5
e) Reference 10
f) Reference 6

Table II. Magnetic moment(μ_B) for transition metal sublattice in $R(Fe_{1-x}Mn_x)_2$

Gd = 7.0 μ_B [d)] Tb = 8.0 μ_B [a)]

Ho = 9.5 μ_B [b)] Er = 7.76 μ_B [c)]

Dy = 10.0 μ_B [d)]

R \ X	0	0.2	0.4	0.6	0.8
Gd	4.2	3.14	2.18[e)]	1.86[e)]	1.77
Tb	3.28	1.82	1.99	1.83	1.42
Dy	4.5	4.36	3.50	2.97	2.24
Ho	4.0	2.83	2.14	1.8	0.39[f)]
Er	2.96	-	1.05	1.33	0.14[f)]
Y	2.9	2.0	1.2	0.4	-

a) Reference 7
b) Reference 11
c) Reference 10
d) Free ion moment
e) Two phase region
f) C14 type structure

Kirchmayr.[6] The moments reported by Kirchmayr are from extrapolation to 0 K of moments measured at 80 K. The decrease in the transition metal sublattice moment as the content increases can be attributed to a simple dilution of iron sublattice by the Mn atoms or by the antiparallel coupling of the Mn and Fe moments. Mössbauer measurements would be helpful to extract the iron moment, but it was found that the spectrum was too complex to interpret. Neutron diffraction on these ternaries may be helpful in the understanding of the magnetic structure of these compounds.

Magnetization (σ) versus temperature in an applied field of 6 kOe for some of the compounds showed unusual behavior at low temperatures. Magnetization goes through a maximum with decreasing temperature (see Figs. 2 and 3). At higher fields (see Fig. 4)

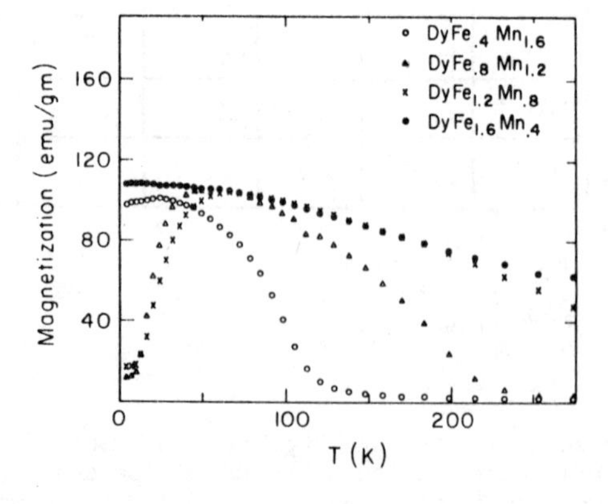

Fig. 2 Magnetization versus temperature in a field of 4400 Oe for various $Dy(Fe_{1-x}Mn_x)_2$ ternaries. Samples in the middle range of composition exhibit a broad maximum.

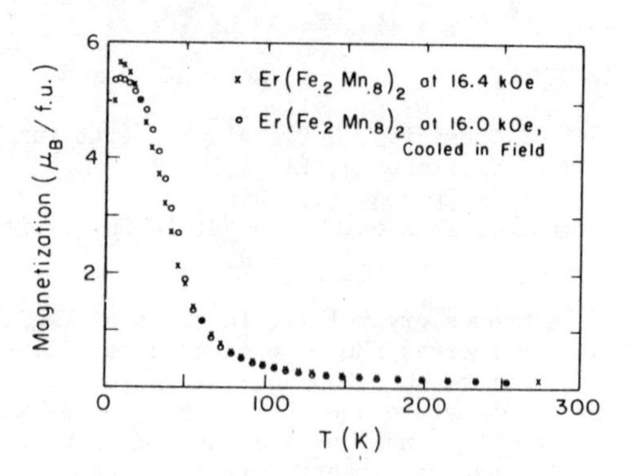

Fig. 3 Magnetization versus temperature for two $Er(Fe_{.2}Mn_{.8})_2$ samples cooled at 4.2 K with and without an applied field. The maximum present in absence of the field disappears when the field is present during the cooling period.

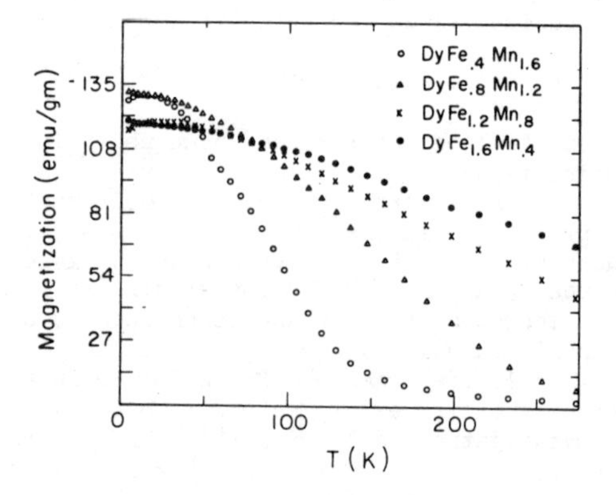

Fig. 4 Magnetization versus temperature in a field of 21,160 Oe for various $Dy(Fe_{1-x}Mn_x)_2$ ternaries. Increasing the field from 4,000 to 21,000 Oe, the broad maximum in the magnetization versus temperature curves disappears.

or samples cooled in a magnetic field (see Fig. 3) this anomaly was not observed. The typical variation of σ versus applied field is shown in Fig. 5 for

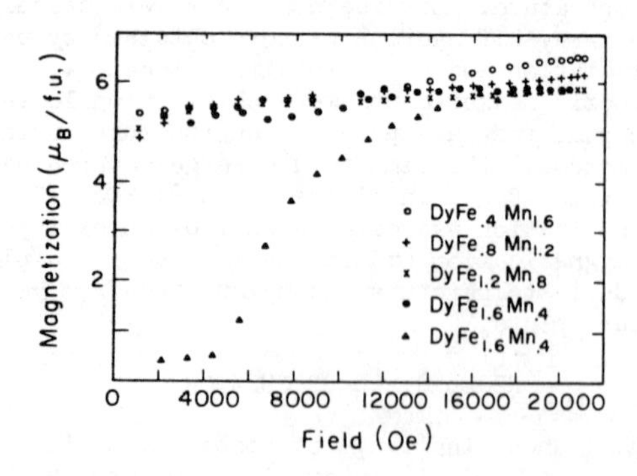

Fig. 5 Magnetization versus field for various $Dy(Fe_{1-x}Mn_x)_2$ ternaries. Hysteresis observed for $DyFe_{1.6}Mn_{.4}$ is also shown as (Δ) for increasing field and (o) for decreasing field.

$DyFe_{1.6}Mn_{.4}$. σ increases slowly until a critical field is reached beyond which magnetization increases rapidly and shows a tendency to saturate at the maximum field applied. Reversing the field the magnetization shows a simple ferromagnetic behavior. For the sake of clarity the results are shown only for $DyFe_{1.6}Mn_{.4}$. Nesbitt et al.[7] also observed a similar behavior for $TbMn_2$. These authors found a shifted hysteresis loop on a specimen cooled in a magnetic field. All the above results can be understood by a theory proposed by Kouvel[8] to explain the magnetic behavior of Cu-Mn and Ag-Mn alloys. According to this theory at low temperatures the magnetic unit consists of a mutually interacting assembly of ferromagnetic and antiferromagnetic domains giving a net moment of zero. This arises owing to the varying Mn-Mn distances in the lattice due to the disordered arrangements. Such a random distribution of Fe and Mn in the transition metal sublattice has been noted in the Mössbauer spectrum of $R(Fe_{1-x}Mn_x)_2$ ternaries. In a field cooled state the ferromagnetic domains align along the field direction and a strong anisotropy develops in the antiferromagnetic domain. This gives rise to a metastable state which prevents the magnetization from reverting to a zero value. A shifted hysteresis loop is a good indication of the mechanism postulated above. This was observed in the field cooled specimen of $TbMn_2$. Currently, hystersis loop measurements are being made on the ternary compounds.

CONCLUSIONS

Magnetic measurements on $R(Fe_{1-x}Mn_x)_2$ compounds were undertaken to study the effect of Mn substitution on the magnetic ordering of RFe_2 compounds. Results indicate that Mn carries a small moment in these alloys and a definite conclusion about the coupling between the Fe and Mn moments could not be obtained. Magnetization versus temperature on some of the ternaries shows a maximum in the magnetization at low temperatures but the maximum is not present when the sample is cooled in a magnetic field. These effects are attributed to the presence of ferromagnetic and antiferromagnetic domains in the sample.

REFERENCES

1. U. Atzmony, M. P. Dariel, E. R. Bauminger, D. Lebenbaum, I. Nowik and S. Ofer, Phys. Rev. Lett. 28, 244 (1972).
2. W. E. Wallace, Rare Earth Intermetallics, Academic Press, 1973, p. 180.
3. A. E. Clark, AIP Conf. Proceed., No. 18, Part 2, 1015 (1973).
4. A. S. Ilyushin and W. E. Wallace, J. Solid State Chem. 17, 131 (1976).
5. A. S. Ilyushin and W. E. Wallace, J. Solid State Chem. 17, 373 (1976).
6. H. Kirchmayr, J. App. Phys. 39, 1088 (1963).
7. E. A. Nesbitt, H. J. Williams, J. H. Wernick and R. C. Sherwood, J.A.P. 34, 1347 (1963).
8. J. S. Kouvel, J. Phys. Chem. Solids 24, 795 (1963).
9. K. N. R. Taylor, Adv. Phys. 20, 551 (1971).
10. G. P. Felcher, L. M. Corliss and J. M. Hastings, J. App. Phys. 36, 1001 (1965).
11. J. M. Moreau, C. Michel, M. Simmons, T. J. O'Keefe and W. J. James, J. Phys. C1, 670 (1971).

* This work was assisted by a grant from the Army Research Office.

MAGNETIC PROPERTIES OF SOME RARE EARTH SILICIDES OF THE TYPE RFe_2Si_2*

S. G. Sankar, S. K. Malik[†] and V. U. S. Rao
Department of Chemistry, University of Pittsburgh
Pittsburgh, Pennsylvania 15260

and

R. Obermyer
Physics Department, Pennsylvania State University
McKeesport, Pennsylvania 15132

ABSTRACT

RFe_2Si_2 compounds (where R = Rare Earth, Y or Th) crystallize in tetragonal $BaAl_4$-type structure. Magnetic properties of these compounds with R = Y, Tb, Ho and Er have been examined in the temperature range 4.2-950 K. $TbFe_2Si_2$ exhibits a Néel temperature of 5.5 K indicating that the rare earth sublattice orders antiferromagnetically. In $HoFe_2Si_2$ and $ErFe_2Si_2$, the rare earth sublattice does not order at temperatures down to 4.2 K. The iron sublattice orders magnetically at nearly 800 K in all the compounds investigated. Saturation magnetization studies on YFe_2Si_2 indicate partial ordering of iron moments confirming earlier work that a majority of iron atoms are diamagnetic. Crystal field interactions seem to be significant in $TbFe_2Si_2$.

INTRODUCTION

A number of compounds of the composition RM_2X_2 (where R = rare earth, Y or Th, M = transition metal and X = Si) have been reported to crystallize in tetragonal $BaAl_4$-type structure.[1] The body centered tetragonal cell has two formula units and the space group is I4/mmm. Detailed structural analyses indicate that the R, M and X atoms occupy preferentially the Ba, Al(1) and Al(2) sites, respectively. The origin of magnetism in these materials has been a subject of considerable importance. Recent work by Felner et al[2] in some RFe_2Si_2 compounds revealed that most of the iron is diamagnetic and a partial ordering of iron sublattice occurs. From a detailed neutron diffraction analysis of $NdFe_2Si_2$, the Nd sublattice was found to order antiferromagnetically at 15.6 K.[3] The present work has been undertaken to extend the investigations of Felner et al.[2] on RFe_2Si_2 to include Tb, Ho and Er compounds. Magnetic properties of YFe_2Si_2 have also been examined to ascertain the nature of transition metal interactions in the lattice.

EXPERIMENTAL

Samples were prepared by melting several times stoichiometric quantities of the elements in a water-cooled copper boat under purified argon atmosphere. The purities of the starting metals were: 3N (with respect to other rare earths) for R and better than 4N for Fe and Si. The resultant samples were annealed at ∿700°C for a few days. X-ray diffraction analyses were performed on these samples to establish their homogeneity. Saturation magnetization measurements were performed at 4.2 K in fields up to 120 kOe employing an extraction technique. Thermomagnetic analyses were carried out using Faraday method in fields up to 20 kOe and in the temperature interval 4.2 to 950 K.

RESULTS AND DISCUSSION

Results of magnetic measurements are shown in Figs. 1-8. The temperature dependence of magnetization for all these compounds was recorded on an X-Y recorder. For the sake of clarity, only a few data points are shown in the figures. Lattice constants and magnetic data obtained in the current study are given in Table 1. Variation of magnetization with the external field at 4.2 K for $TbFe_2Si_2$ (Fig. 1) indicates that the sample achieves a saturation moment of 6.9 μ_B. This value is much smaller than the gJ value expected for free Tb^{3+} ion suggesting the presence of a sizeable crystal field interaction in $TbFe_2Si_2$. Malik et al.[4] reported that $PrFe_2Ge_2$ and $TbFe_2Ge_2$ undergo a metamagnetic and a spin-flop transition respectively in the presence of applied fields of approximately 15 kOe at 4.2 K. Spin-flop transitions were also reported by Felner et al.[2] in $NdFe_2Si_2$ and $NdFe_2Ge_2$ at about 12 kOe. However, we do not observe a metamagnetic behavior in $TbFe_2Si_2$. The transition from antiferro-to ferromagnetic ordering under the influence of the external field appears to take place rather smoothly in this compound. Unpublished measurements by Malik in $GdFe_2Si_2$ also reveal such a smooth transition.

From the temperature dependence of magnetization (Fig. 2) we note that the Tb sublattice in $TbFe_2Si_2$ orders antiferromagnetically at nearly 5.5 K in an external field of 5 kOe. However, at higher field strength this transition is not observed. The magnetic susceptibility obtained in two applied field strengths as a function of temperature is shown in Fig. 3. The experimental values are compared with the susceptibility calculated for a tripositive Terbium free ion. These results reveal the strong

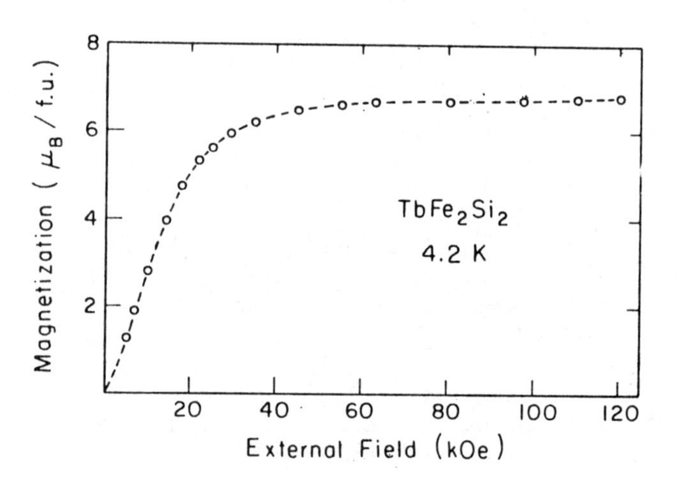

Fig. 1: Variation of magnetization with external field for $TbFe_2Si_2$ at 4.2 K.

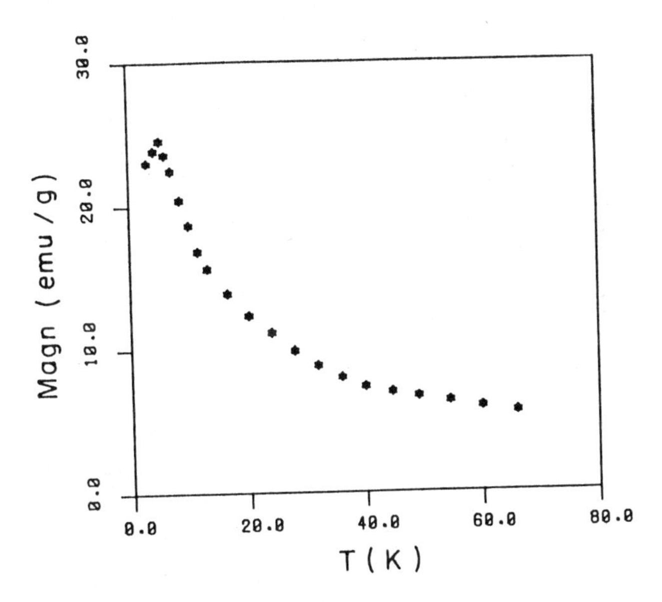

Fig. 2: Temperature dependence of magnetization of $TbFe_2Si_2$ at low temperatures in an external field of 5 kOe.

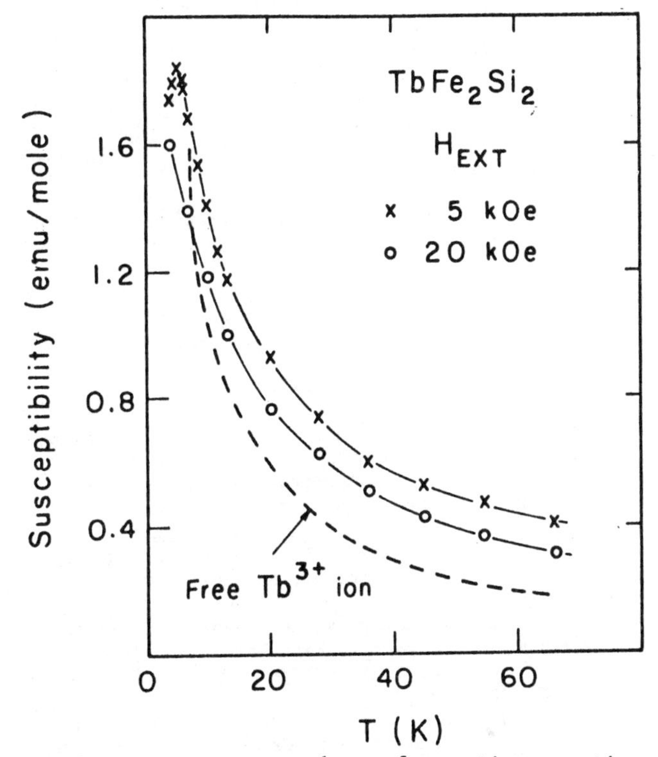

Fig. 3: Temperature dependence of magnetic susceptibility of $TbFe_2Si_2$ in two external fields.

dependence of magnetic susceptibility on the applied field indicating that iron atoms possess magnetic moment. Thermomagnetic analysis carried out to high temperatures (Fig. 4) indeed confirm the ferromagnetic ordering of the iron sublattice. The Curie temperature for the iron sublattice is found to be 820 K.

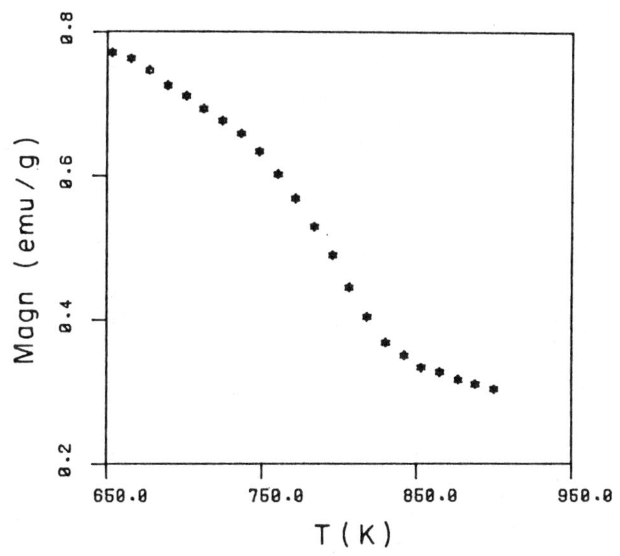

Fig. 4: Temperature dependence of magnetization of $TbFe_2Si_2$ at high temperatures.

Magnetization results of $HoFe_2Si_2$ and $ErFe_2Si_2$ reveal that the rare earth sublattice does not order in the temperature range investigated (Fig. 5). The iron sublattice orders magnetically at nearly 820 K as shown in Figures 6 and 7.

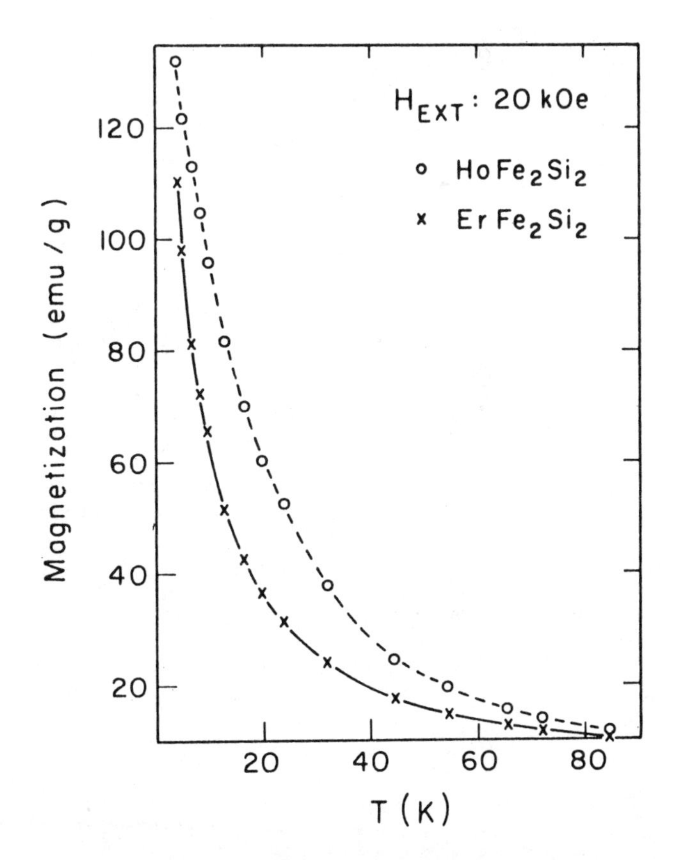

Fig. 5: Temperature dependence of magnetization of $HoFe_2Si_2$ and $ErFe_2Si_2$ at low temperature.

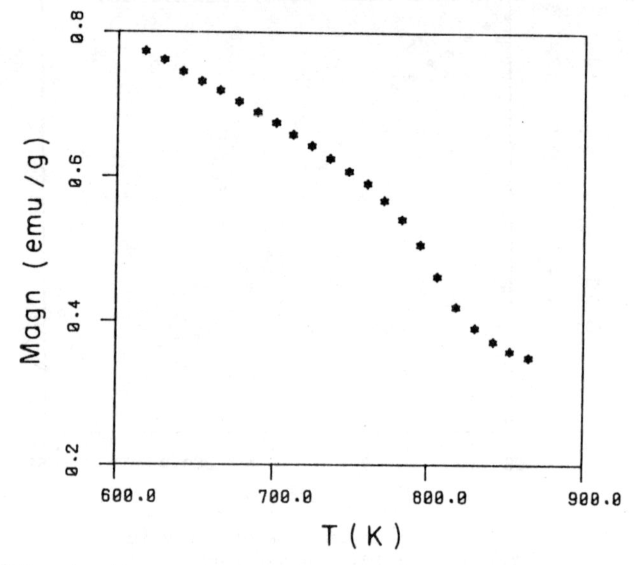

Fig. 6: Temperature dependence of magnetization of $HoFe_2Si_2$ at high temperature in an external field of 5 kOe.

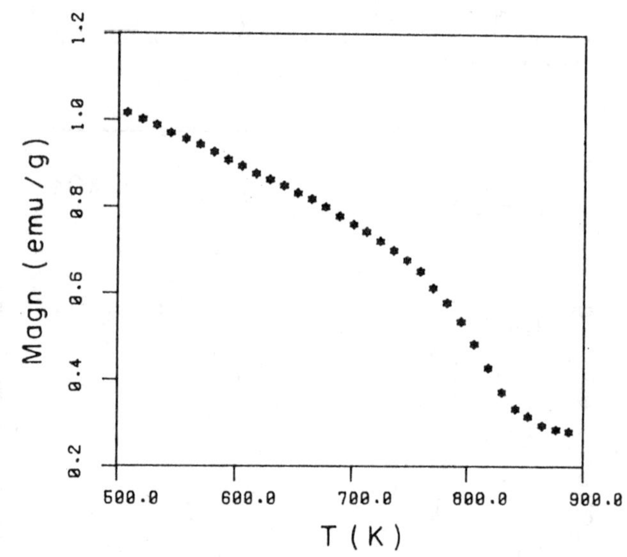

Fig. 7: Temperature dependence of magnetization of $ErFe_2Si_2$ at high temperatures in an external field of 5 kOe.

Magnetization data were obtained for YFe_2Si_2 with a view to examine the nature of transition metal interaction in these compounds. The results shown in Fig. 8 indicate that the iron sublattice orders ferromagnetically at nearly 790 K. The saturation magneti-

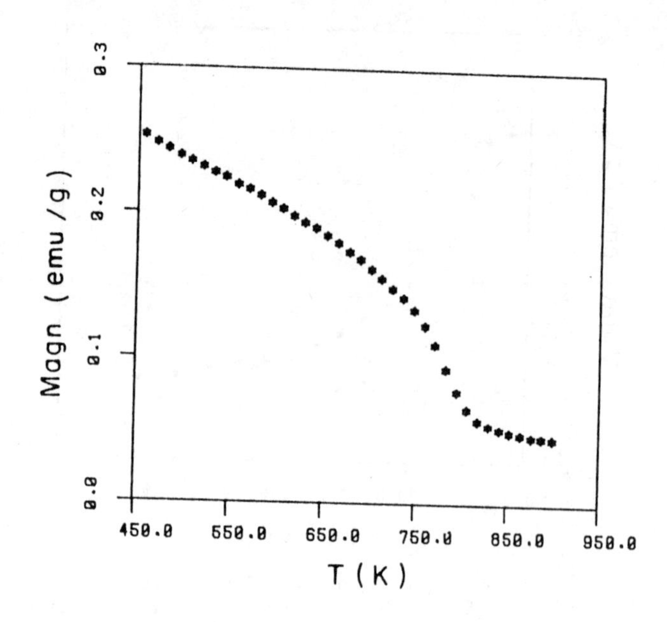

Fig. 8: Temperature dependence of magnetization of YFe_2Si_2 at high temperatures in an external field of 5 kOe.

zation data reveal that the transition metal possess a moment of 0.3 μ_B per iron atom. These results are in agreement with the earlier work by Felner *et al.*[2] and by Pinto and Shaked[3] who reported the existence of a weak ferromagnetic sublattice while most of the iron is diamagnetic. It may be pointed out that the ordering temperatures of the iron sublattice obtained in our study are much higher than those reported by Felner *et al.*[2] for the lighter lanthanide counterparts. The reason for this discrepancy is not clear. Futher work is in progress to examine this aspect.

REFERENCES

* Supported by U. S. Energy Research and Development Administration.

† On leave from Tata Institute of Fundamental Research, Bombay, India.

1. W. Rieger and E. Parthé, "Ternary Alkaline Earth and Rare Earth Metal Silicides and Germanides with $ThCr_2Si_2$ Structure", Mh. Chem. 100, 444-54 (1969).
2. I. Felner, I. Mayer, A. Grill and M. Schieber, "Magnetic Ordering in Rare Earth Iron Silicides and Germanides of the RFe_2X_2 Type", Solid State Comm. 16, 1005-9 (1975).
3. H. Pinto and H. Shaked, "Neutron Diffraction Study of $NdFe_2Si_2$", Acta Cryst. 7, 3261-6 (1973).
4. S. K. Malik, S. G. Sankar, V. U. S. Rao and R. Obermyer, "Magnetic Behavior of Some Rare Earth Germanides of the Type RFe_2Ge_2", A. I. P. Proceedings 29, 585 (1976).

TABLE I

STRUCTURAL AND MAGNETIC PROPERTIES OF RFe_2Si_2 COMPOUNDS

R	a (Å)	c (Å)	T_N K	$\mu^{4.2\,K}_{120\,kOe}$ μ_B	gJ	T_c K
Tb	3.921	9.953	5.5	6.9	9	820
Ho	3.901	9.921	-	8.5	10	820
Er	3.889	9.899	-	7.7	9	825
Y	3.907	9.927	-	-	-	790

SUBHARMONIC GENERATION IN IRON*

J.B. Holmes and G.C. Alexandrakis
Department of Physics, University of Miami
Coral Gables, Florida 33124

ABSTRACT

In this paper we report on the first observation of microwave subharmonic generation in ferromagnetic metals. It has been accomplished in iron at room temperature through a new transmission resonance experiment which will be described here. As in all transmission resonance experiments, the sample forms the common wall of two microwave cavities, one of which is the excitation and the other the detection cavity. The subharmonic has been found in the geometry in which both the static and the microwave fields in the excitation cavity are mutually parallel and also parallel to the sample surface. The detected subharmonic signal strength is about 10^{-17} watts for iron samples a few microns thick. Of course thinner samples will give stronger signals. The samples used were 99.9% pure. The nonlinear phenomena described here, as opposed to the linear transmission resonance phenomena, are sensitive to the exchange stiffness constant and could be used for its determination as a function of temperature.

INTRODUCTION

Let us assume that a ferromagnetic sample in foil form, several μm thick, lies on a xz-plane, and that there is a static magnetic field along the z-axis applied on the sample, see Fig. 1. Suppose further that we make this sample to form the common wall of two microwave cavities, one of which is going to be the input and the other the output cavity. We will call the geometry in which the static magnetic field is parallel to the microwave field, with both fields parallel to the sample surface at the input cavity, the $H_{\parallel,\parallel}$ geometry.

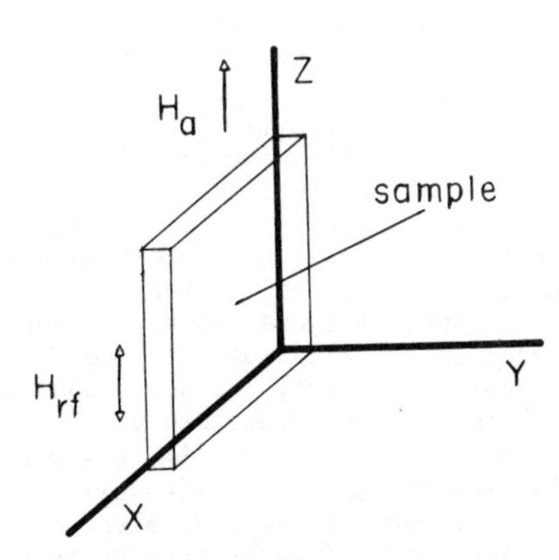

Fig. 1 Shows the $H_{\parallel,\parallel}$ geometry. Both the static and microwave fields are along the z-axis. The sample lies on the xz-plane.

It has been found earlier[1,2] in the $H_{\parallel,\parallel}$ geometry, when the sample is placed between two identical cavities where the output cavity is also oriented to have its microwave field parallel to the z-axis, that as the static field changes from zero to about 10 kG the transmitted amplitude of an X-band microwave frequency through the sample changes from a low value at zero applied static field to a value about 40db higher and essentially constant from about 5 kG to 10 kG. This effect has been observed in most ferromagnetic metals. The effect was taken to be broadly reminiscent of parallel pumping in insulators[3] but not necessarily involving a threshold value for the applied microwave field. Possible excitation of low k spin waves at half the applied microwave frequency (subharmonic generation) in the x and y-direction was speculated[1] to be connected with these observations.

It was thought, in other words, that at low applied static field values where the transmitted power is low, power is pumped from the applied r.f. field H_z into the $H_{x,y}$-components through excitation of transverse spin waves at half the applied frequency. This process was then thought to cease for some reason at high static field values where the transmitted power is observed to increase and remain approximately constant.

A nonlinear theory of the effect was subsequently put forward[2,4,5] which predicted at least qualitatively the transmitted amplitude behavior mentioned above. It furthermore predicted subharmonic generation. We therefore decided to search experimentally for the subharmonic.

THE EXPERIMENT

If spin waves are excited in the $H_{\parallel,\parallel}$ geometry and the sample lies in the xz-plane, with the microwave and static fields along the z-axis, the spin disturbance along the x-axis will propagate along the y-axis across the sample thickness. Since the spin waves are expected to have half the applied frequency we are going to try to detect them in the following way, see Fig. 2.

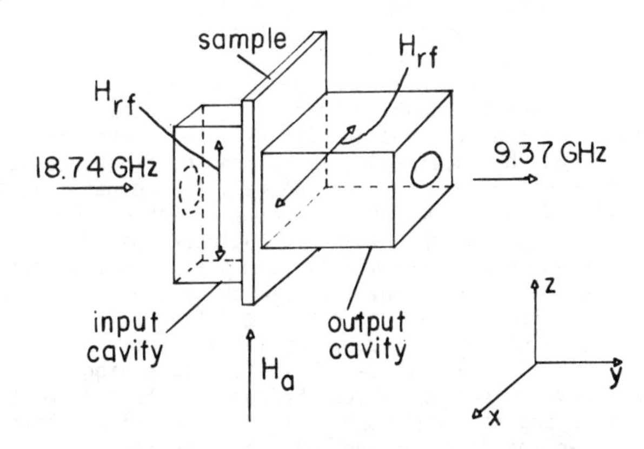

Fig. 2 Shows the cavity geometry used for the detection of the subharmonic. The input cavity is tuned at 18.74 GHz, the output at 9.37 GHz. Both cavities operate at the TE_{101} mode.

The sample is made to form again the common wall between two microwave cavities where the input cavity in the $H_{\parallel,\parallel}$ geometry is fed at 18.74 GHz. But now the output cavity is chosen to resonate at 9.37 GHz with its microwave field normal to the input H_{rf}. Thus if there indeed is a magnetization component excited along the x-axis at half the applied frequency and propagates along the y-axis, it will excite the receiving cavity.

Both cavities are chosen to resonate at the TE_{101} mode. The cavities are coupled by teflon plungers to the wave guide sections[6,7] fitting them and are tuned by teflon screws.[6,7]

Having settled the cavity geometry the question now is how will the system be fed so that the detection will be phase sensitive. To accomplish this we used a basic superheterodyne receiver spectrometer used in other transmission resonance experiments[6,7]. A small part of the power produced by a phase stabilized klystron operating at 9.37 GHz is used to feed the heterodyne receiver, to which the output cavity is connected. The rest of the klystron power, a few hundred mW, is fed into a frequency doubler using a HP 5082-0320 step recovery diode. The 18.74 GHz frequency produced at the doubler is then fed through Ku wave guides to the input cavity. Thus all microwave fields are coherent.

Two frequency doublers were built and tested using the HP 5082-0320 diodes. In the one doubler an E-type coupling was used and in the other, H-type. The E-type coupler was built by modifying an N-type coaxial to Ku wave guide coupling section. The H-type coupler was built by modifying an X-band tunable crystal mound. Both doublers were stable and produced several mW of power at 18.74 GHz. The output from the doubler was amplified to a few hundred mW by a Ku band microwave amplifier before it was applied to the excitation cavity tuned at 18.74 GHz.

The question of spectral purity of the power reaching the excitation cavity has to be now considered, as well as the response of the receiver to the different frequencies that are possibly reaching it. The receiver of about 10^{-18} watts sensitivity is built to only detect the 9.37 GHz frequency because it is fed by the 9.37 GHz plus the IF-frequency of 30 MHz and the first detection step is a phase sensitive one at 30 MHz. The IF-strip band width is about 1 MHz. The last detection step occurs at the lock-in amplifier by having amplitude modulated the signal originating at the detection cavity at the lock-in reference frequency, which is 85 Hz in our case.

The key problem then is to insure that no power at 9.37 GHz is reaching the cavity system other than the one which might be generated inside the sample as subharmonic of the 18.74 GHz input frequency. Of course, right after the doubler there is a large amount of power at the 9.37 GHz and it has to be prevented from reaching the excitation cavity and therefore possibly the detection cavity. Fortunately this problem is very easily solved. We found the attenuation of the 9.37 GHz in a 7.5 cm section of Ku wave guide to be 30 db. Our cavities were approximately 10 m away from the doubler, therefore the 9.37 GHz power reaching the detection cavity through this channel is totally negligible. The question of power at 9.37 GHz leaking into the cavity system from the outside is also met by having cavities and wave guides assembled throughout with indium "O rings" and shielded completely. The receiver channel between the detector and the receiving cavity is also isolated by about 150 db and has never shown an indication of absorption resonance at the receiving cavity.

Regarding the higher harmonics of the 9.37 GHz frequency present at the doubler, there are several reasons why they should not interfere with the detection of the subharmonic in our experiment. First the 18.74 GHz frequency was the strongest signal generated. Without the microwave amplifier, an absorption resonance corresponding to the 18.74 GHz frequency was the only resonance detected. Second, the microwave amplifier will selectively amplify only in the Ku-bandwidth. The 18.74 GHz frequency is the only harmonic in this range. Third, the input cavity will resonate only at the even harmonics of the fundamental 9.37 GHz frequency. Finally and most importantly, the receiver will detect only the 9.37 GHz (see previous discussion) which corresponds to the subharmonic of only the 18.74 GHz frequency.

In view of the above reasoning and evidence, we were confident that the detection system was free of any suprious effects of major consequences. We then started the search for the subharmonic.

Different iron samples 99.9% pure[8] were etched with acid solutions from a 18 μm stock and tried. When samples a few μm thick were used, we observed a relatively sharp signal, about 100 G wide, in the neighborhood of 2.2 kG of applied static field. The signal strength is about 10^{-17} watts. See Fig. 3, where four signal traces are shown corresponding to four phase differences introduced in the phase sensitive detection loop.

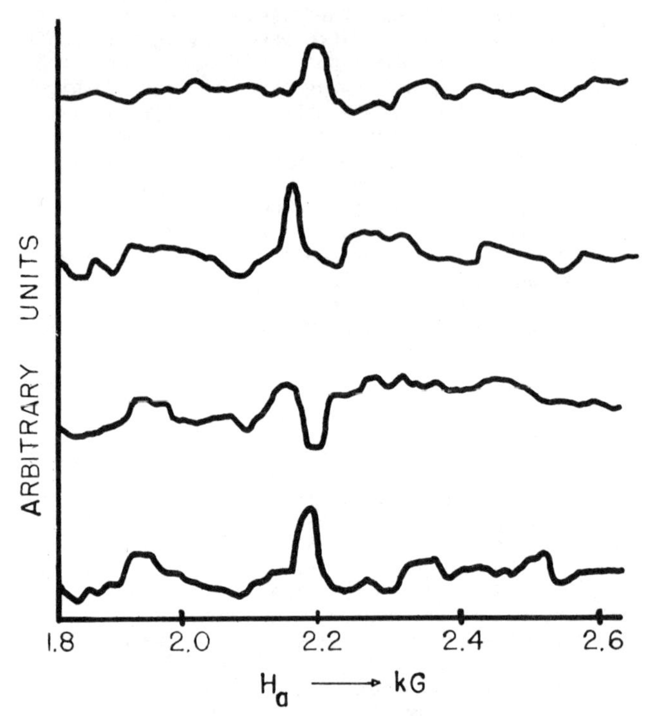

Fig. 3 Shows different traces of the subharmonic signal corresponding to different phase values introduced in the receiver. The signal strength is about 10^{-17} watts.

The applied static field value at which the resonance appears settles completely the question of it being a new signal and not a spurious effect. We calculated the expected position of absorption or transmission resonances in iron due to possible cavity misalignment and/or unwanted frequency participation in the effect. None of these resonances appear in the neighborhood of 2.2 kG value. Iron oxide resonances are also easily excluded. We also found pinholes in the sample to be of no consequence.

Finally when we rotated the receiving cavity by a $\pi/2$ angle from its standard orientation, that is of having its microwave field axis along the x-direction, so that its magnetic field direction came along the z-axis, there was no subharmonic signal observed.

In view of all these factors we feel confident that we have indeed observed for the first time subharmonic generation in ferromagnetic metals.

DISCUSSION

Part of the motivation behind this work has been the desirability, in the authors' point of view, of finding a flexible method of measuring the exchange stiffness constant in bulk and, ultimately, the exchange integral for ferromagnetic metals and alloys over a wide temperature range. It was thought that transmission resonance experiments might provide a good method for doing this.

Initially, transmission signals were observed for the geometry in which the applied static field was normal to both the sample surface and the microwave magnetic polarization $H_{\perp,\perp}$ geometry. A linear theory was devised which fits the data reasonably well, but the theory was very insensitive[9,10,11] to the exchange stiffness constant, A.

However, the nonlinear theory mentioned above that qualitatively describes the transmission of the incident frequency and predicts the subharmonic signal shows both signals to be very sensitive to A. Thus we hope that both signals in the $H_{\parallel,\parallel}$ geometry can be useful in determining A.

We have observed the transmitted signal at the incident frequency in practically all ferromagnetic materials. We have only observed the subharmonic so far in iron. We plan to continue our search in other ferromagnetic metals and alloys.

In regard to the comparison in the subharmonic case, the theory predicts satisfactorily[5] the position and strength using accepted parameters, namely, resistivity equals 10×10^{-6} ohm-cm, the transverse relaxation time equals 10^{-10} seconds, and $A = 2.75 \times 10^{-6}$ ergs/cm.[12] The sample thickness was taken to be 3 μm. A more exact comparison must wait for stronger subharmonic signals which we plan to obtain using thinner samples.

In regard to the comparison for the transmitted signal at the applied frequency, we have comparisons for iron and nickel. The agreement in nickel appears to be very good. In fact, we are now trying to fit the nickel experimental results with the theory. The agreement so far is very encouraging. For iron the agreement is still only qualitative. In both of these metals a resistivity several times greater than the static value is required by the theory in order to predict reasonable signal strengths and shapes compared to experiment.[5] All other parameters including A have values consistent with values given in the literature.

The fact that a reasonable shape and correct signal strength is predicted for the subharmonic case using static resistivities is consistent with experiments in the $H_{\perp,\perp}$ case where the microwave field in the metal is perpendicular to the static applied field. The requirement of higher resistivity values in the description of the transmission signal at the applied frequency could very well be due to the fact that the magnetic component of the microwaves inside the metal is now parallel to the applied static field.[13]

In conclusion we feel that the subharmonic effect described here appears interesting and after further work with thinner samples might be helpful in studying the dependence of A on temperature and spin relaxation processes in ferromagnetic metals and alloys.

REFERENCES

*Work partially supported by the National Science Foundation and Friends of Physics, Inc., Miami, Fla.

1. G.C. Alexandrakis, O. Horan, T.R. Carver, and C.N. Manikopoulos, Phys. Rev. Lett. 25, 1758 (1970).
2. O.L.S. Lieu and G.C. Alexandrakis, AIP Conference Proceedings, No. 24, 512 (1974), C.D. Graham, Jr., G.H. Lander, and J.J. Rhyne, Editors.
3. F.R. Morgenthaler, J. Appl. Phys. 31, 95S (1960); E. Schloman, J.J. Green, and U. Milano, J. Appl. Phys. 31, 386S (1960); C. Kittel, Quantum Theory of Solids (Wiley, New York, 1964); M. Sparks, Ferromagnetic Relaxation Theory (McGraw-Hill, New York, 1964) contains an extensive bibliography on this subject.
4. O.L.S. Lieu, G.C. Alexandrakis, and M.A. Huerta, to be published.
5. J.B. Holmes and G.C. Alexandrakis, Physics Letters A, in press.
6. R.B. Lewis and T.R. Carver, Phys. Rev. 155, 309 (1967).
7. G.C. Alexandrakis, T.R. Carver, and O. Horan, Phys. Rev. B, 5, 3472 (1972).
8. Iron foil supplier, A.D. Mackey, New York.
9. O. Horan and G.C. Alexandrakis, to be published.
10. Y.J. Liu and R.C. Barker, AIP Conference Proceedings, No. 24, 505 (1974), C.D. Graham, Jr., G.H. Lander, and J.J. Rhyne, Editors.
11. J.F. Cochran, B. Heinrich, and G. Dewar, AIP Conference Proceedings, No. 24, 505 (1974), C.D. Graham, Jr., G.H. Lander, and J.J. Rhyne, Eds.
12. B.E. Argyle, S. Charap, and E.W. Pugh, Phys. Rev. 132, 2051 (1963); F. Keffer, Handbuck der Physik XVIII/2, 179, 371 (Springer-Verlag, Berlin, 1966); C. Herring, Rado and Suhl, Magnetism IV, Academic Press, New York, 1966, p. 357.
13. R. Bozorth, Ferromagnetism (Van Nostrand, Princeton, N.J., 1951), 749.

STATIC AND DYNAMIC MAGNETOELASTIC COUPLING IN METALLIC FERROMAGNETS*

M. O'Donnell and H. A. Blackstead
Physics Department, University of Notre Dame, Notre Dame, Indiana 46556

ABSTRACT

A new magnetoelastic interaction, which is linear in both applied field and strain and originates in the crystal field, is proposed for ferromagnets. With the inclusion of this term, the results of spin transmission experiments on Ni, Fe and Gd can be easily explained and certain apparent anomalies in the magnetoelastic properties of Gd can be understood. In addition, with an rf magnetic field applied parallel to the magnetization, the generation of longitudinal phonons at the driving frequency is predicted for metallic ferromagnets. The results of pulse-echo experiments on Gd single crystals have demonstrated that rf field frequency phonons can be excited by an rf field applied parallel to the magnetization.

RESULTS

A number of experimental results on ferromagnetic Gd, Fe and Ni have raised serious questions relating to the completeness of the Hamiltonian currently used to describe ferromagnetic systems. G. C. Alexandrakis et al[1,2] have demonstrated that spin transmission resonance signals can be detected from Gd, Fe and Ni foils at the driving frequency with the application of a dc field parallel to the rf field in the plane of the foil. In addition, some recent experiments on Gd single crystals and films[3] and Ni films[4] have demonstrated that longitudinal phonons can be excited in these materials with the application of an rf field polarized parallel to the magnetization. And finally, the volume magnetostriction of Gd near and above the Curie temperature has been shown to be linear in applied field. In the ferromagnetic state the linearity in applied field is observed for fields greater than those required to saturate the magnetization.[5,6] To explain these apparent anomalies, we propose a magnetoelastic interaction which is first order in strain, first order in applied field, and time reversal invariant. This 'new' term has the general form:

$$\sum_{n}\sum_{i,j,k,l} F_{ijkl} H_i J_{nj} \epsilon_{kl}$$

where F_{ijkl} is a coupling tensor, H_i is the ith component of the applied field, J_{nj} is the jth component of the total angular momentum at the nth site, and ϵ_{kl} is a Cartesian strain. This term arises from the inclusion of the effects of spin-orbit coupling in the ferromagnetic state of a material.[7] We will now examine the implications for a ferromagnetic system resulting from the inclusion of this term.

It is immediately evident that this term provides a contribution to the magnetostriction which is proportional to the applied field. As a result, consideration of a term of the proposed form provides an alternative explanation for the origin of "forced magnetostriction", as well as the anomalous linearity in the magnetostriction of gadolinium. Although Gd is usually considered a pure S-state ion, experimental evidence has shown that it possesses basal plane magnetocrystalline anisotropy which is adequately described by single ion theory.[8] Since $\vec{L}\cdot\vec{S}$ vanishes, j j coupling appears to provide the only opportunity for the magnetic moment to be coupled to the lattice with a single ion description. Using third order perturbation theory[7], it is straightforward to show that j j coupling leads to both the single ion anisotropy terms and the proposed magnetoelastic term. Therefore, the linearity in applied field of the magnetostriction of Gd now can be understood.

If a ferromagnetic material is subjected to an applied rf magnetic field, the rf perturbation Hamiltonian contains two terms: first, the standard Zeeman term, and second, a term which is proportional to both the applied field and the strains of the lattice. Therefore, if an rf field is applied parallel to the magnetization, the only first order terms that will contribute to the absorption of rf power correspond to the direct excitation of phonons.

This conclusion is easily verified by writing out the explicit form of the perturbation Hamiltonian for the case in which H_{rf} is parallel to the static field applied in the z direction.

$$H_{pert} = \sum_n (g\mu_B H_{rf} J_{iz} + \sum_{j,k,l} F_{3jkl} H_{rf} J_{nj} \epsilon_{kl})$$

where

$$H_{rf} = H^{o}_{rf}\, e^{-i\omega t}\, e^{-y/\Delta} \cos(y/\Delta)$$

Here Δ is the familiar rf skin depth. Since the rf magnetic field has finite Fourier components only in the y direction, it is clear that momentum conservation (for finite momenta) requires that excitations resulting from the perturbation must propagate along the y axis.

In the appropriate small oscillation approximation, ϵ_{kl} may be written as:

$$\epsilon_{kl} = \bar{\epsilon}_{kl}(\vec{\alpha}) + \epsilon_{kl}(\vec{q},\omega)$$

Here the components of $\vec{\alpha}$ are the direction cosines of the magnetization and $\bar{\epsilon}_{kl}(\vec{\alpha})$ is a time independent equilibrium strain. The angular momentum operators and the frequency and wave vector dependent strain $\epsilon_{kl}(\vec{q},\omega)$ may be replaced by magnon and phonon operators $\alpha_{\vec{q}}$ and $\beta_{\vec{q},p}$, respectively. The index p is used to label the transverse and longitudinal phonon modes. Crystal symmetry requires that the tensor F_{ijkl} have only six independent, finite terms for hexagonal systems, and only three independent, finite terms for cubic systems.

The magnon expansion for J_z contains no terms linear in magnon operators, but does contain a constant plus terms of second order and higher. As a result, the Zeeman term does not give rise to an excitation at the rf field frequency, and will not be considered further.

Denoting the y component of the wave vector by q, and the Fourier transform of the rf field by f (q), we write the perturbation Hamiltonian as:

$$H_{pert} = H^{o}_{rf} \; e^{-i\omega t} \sum_{q} f(q) \, [\, C_1(\beta_{q,l} + \beta^{\dagger}_{q,l}) + C_2 \, a_q + C_2^{*} \, a^{\dagger}_q \,]$$

Here C_1 and C_2 are constants which have absorbed the various coupling constants, number densities, etc. The first term in H_{pert} gives rise to a non-resonant direct coupling of the rf field to longitudinal phonons. The second term gives rise to a "resonant" response of the phonon intensity as a consequence of the general features of coupled magnon - phonon excitations in a polycrystalline system. In particular, the magnon operator a_q must be expressed as a linear combination of all mixed mode operators, denoted by $\gamma_{q,i}$

$$a_q = \sum_j S_{1j}(q) \, \gamma_{q,j}$$

where i is an index used to label the four normal modes which are mixtures of the magnon and three phonon modes, and the matrix S diagonalizes the coupled Hamiltonian.

Varying an applied magnetic field will cause the uncoupled longitudinal phonon dispersion curve to cross the magnon dispersion curve. For a somewhat larger applied field, the transverse phonon dispersion curve will also cross the magnon dispersion curve. Inclusion of finite coupling terms lifts the degeneracy of the modes, and for appropriate q values, the various modes will be strongly mixed. As a consequence, magnon and phonon mode labels are both misleading and incorrect. The apparent phonon "resonance", illustrated in figure 1, is a consequence of the field dependent coupling terms which cause absorbed energy to appear in a "phonon like" mode, and the width of the "resonance" is closely related to the field range over which the branches (for fixed q) are strongly coupled. This "resonance" does not correspond to an enhanced photon absorption, as illustrated in figure 1.

This result provides an explanation for the experimental findings of G. C. Alexandrakis et al. The transmitted signal detected by Alexandrakis with parallel fields can result from the creation of phonons within a skin depth of the exposed surface of the foil, and the subsequent reconversion of these phonons into microwave photons at the other surface.

The results of experiments on Gd[3] single crystals and Gd[9] and Ni[4] polycrystalline films have demonstrated that, indeed, phonons can be generated at the driving frequency if the magnetization and applied rf field are parallel.

In order to appreciate the necessity of including the proposed term in the ferromagnetic Hamiltonian, it is essential to recognize that the standard magnetoelastic Hamiltonian cannot lead to phonon excitation at the rf field frequency with parallel applied fields.[3]

In conclusion, we argue that the inclusion of a magnetoelastic term which is linear in applied field and strain appears to be justified for ferro-

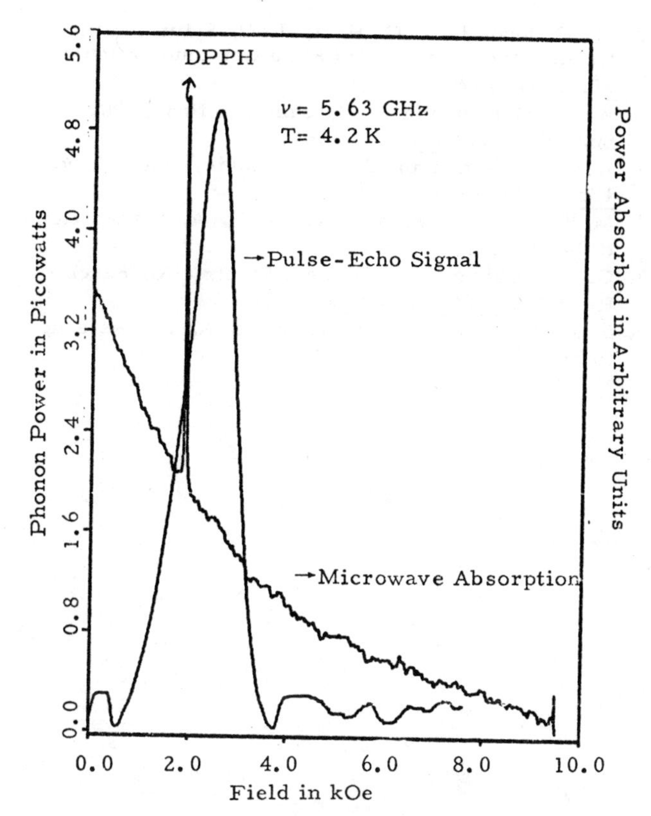

Fig. 1. Microwave power absorption and longitudinal phonon pulse-echo signals at 5.63 GHz with parallel rf and dc fields applied in the film plane. The nickel film (thickness = 2.0 kÅ) was evaporated onto a sapphire substrate heated to 250°C. The sharp peak in the absorption curve at 2.06 kOe is due to a dpph sample present in the microwave cavity, which is used as an aid in tuning the system. The pulse-echo data shows a strong peak at 2.54kOe, the field at which the phonon and magnon normal modes are strongly mixed. As discussed in the text, the absorption curve does not exhibit a resonant response. In the pulse-echo experiment, peak pulse power of approximately 1kW was incident on the cavity. In the absorption experiment, approximately .5W of power was incident on the cavity

magnetic Gd, Fe and Ni, and possibly for all ferromagnets. The additional contributions to the equilibrium strains from this term may be the primary origin of forced magnetostriction in all ferromagnetic systems. In addition, the inclusion of this term can give rise to larger transition rates from the phonon-like modes than the Zeeman perturbation.

REFERENCES

*Work supported in part by the National Science Foundation

1. G. C. Alexandrakis, T. R. Carver, O. Horan, Phys Rev. B5, 3472 (1972).
2. G. C. Alexandrakis, O. Horan, T. R. Carver, C. N. Manicopoulos, Phys. Rev. Letters 25, 1758 (1970).
3. M. O'Donnell, J. T. Wang, and H. A. Blackstead, Phys. Rev. Letters 36, 606 (1976).

4. M. O'Donnell, S. C. Hart, J. G. Sylvester and H. A. Blackstead, Solid State Communications 18, 1141 (1976).
5. W. D. Corner and F. Hutchinson, Proc. Phys. Soc. (London) 75, 781 (1960).
6. W. E. Coleman and A. S. Pavlovic, Phys. Rev. 135, a426 (1964).
7. M. O'Donnell, W. Wilcox, and H. A. Blackstead, to be published
8. C. D. Graham Jr., General Electric Research Report No. 66 - C - 218 (1966).
9. S. C. Hart, M. O'Donnell and H. A. Blackstead, to be published.

INELASTIC SCATTERING OF PHONONS BY MAGNONS IN FERROMAGNETS*

J.R.L.de Almeida and Cid B.de Araujo **
Departamento de Fisica,Universidade Federal
de Pernambuco,50000 Recife ,Brasil

ABSTRACTS

The Raman scattering of phonons by magnons is analyzed by quantum statistical methods. The cross-section for the scattering,the non-linear threshold and the stationary regime above the threshold,are determined by using the Green's function formalism and the representation of coherent states.The relaxation of the modes is described in a formal way by considering them interacting with a Markoffian heat-bath.New results for the cross-section and the coherent properties of the states are obtained and compared with previous treatments. The possibility of experimental investigation of this effect is discussed and estimates are made using the parameters for YIG.

INTRODUCTION

Non-linear interactions of phonons and magnons have been studied by several authors in studies of ferromagnetic resonance.[1-3] Altough several processes involving magnons are well known,the non-linear scattering of acoustic phonons due to the magneto-elastic interactions received little attention.[4,5] The literature on this effect is covered by ref.4 and 5 and only the spontaneous scattering of 2 phonons-1 magnon was investigated.In fact the possibility of stimulated scattering in ferromagnets is very large due to the strong magneto-elastic coupling and the very small damping of magnons and phonons in a number of materials.

In this paper we study the spontaneous and the stimulated scattering in cubic ferromagnets.

HAMILTONIAN AND CROSS-SECTION OF THE SCATTERING

The process which we consider involves one exciting phonon which interacts with the spin system giving origin to a scattered phonon and a magnon.In this section we deduce the Hamiltonian and the cross-section for the process in terms of Green's functions.The relaxation of the modes is taken into account by considering them interacting with a Markoffian heat-bath.6

The Hamiltonian is deduced from the local energy density function.[1]The relevant terms for cubic symmetries are

$$(\rho_0 U)' = 2[B_{ijkl13}\alpha_1\alpha_3 + B_{ijkl23}\alpha_2\alpha_3]\epsilon_{ij}\epsilon_{kl} , \quad (1)$$

where

$$\alpha_1 = \frac{m_x}{m_s} ; \quad \alpha_2 = \frac{m_y}{m_s} ; \alpha_3 \simeq 1 , \quad (2)$$

where m_s is the saturation magnetic moment per unit of mass and ϵ_{ij} are the strain tensor elements.The B-coefficients can be measured by ultrasonic techniques [6] and satisfies [1,6] : $B_{144}=B_{255}$; $B_{564}=B_{465}$; $B_{244}=B_{344}=B_{155}=B_{355}$;(we used here the abbreviated Voigt notation [1]). Equations (1)-(2) are valid in the small signal elastic and magnetic disturbances approximation.

The Hamiltonian is obtained by integrating eq.(1) over the sample volume, and assumes the form

$$H' = \sum_{kqq'\mu\nu} \left(\beta_{qq'\mu\nu} a^{+}_{q\mu} a_{q'\nu} c_k \delta(\vec{q}=\vec{q}'+k) + h.c.\right) , \quad (3)$$

where $a_{q\mu}$ ($a^{+}_{q\mu}$) are phonon operators and c_k (c^{+}_k) are magnon operators. The coupling is obtained in the usual way [1] by doing the expansion of the strain tensor elements in terms of phonon operators and the spins in terms of magnon operators. It contains all possible polarization selection rules for the scattering.From the analysis of $\beta_{qq'\mu\nu}$ one can see that taking the (xz)-plane as the scattering plane the following processes are allowed: (1)longitudinal phonon $\rightarrow$ transverse phonon + magnon ; (2) longitudinal phonon $\rightarrow$ longitudinal phonon + magnon; (3) transverse phonon $\rightarrow$ transverse phonon + magnon; (4) transverse phonon $\rightarrow$ longitudinal phonon + magnon.

For simplicity we will assume a collinear scattering. Also,since the pumped phonon mode will have a well defined wavevector $\vec{q}$ and polarization μ ,we reduce eq.(3) to the form

$$H' = \sum_{k\nu} \left(\beta_{q,q-k,\mu\nu} a^{+}_{q\mu} a_{q-k,\nu} c_k + h.c. \right) , \quad (4)$$

The total Hamiltonian of the system will be $H = H_m + H_{ph} + H' + H''$, where

$$H_m = \sum_k \omega_m(k) c^{+}_k c_k , \quad (5)$$

$$H_{ph} = \sum_{q\mu} \omega_{ph}(q\mu) a^{+}_{q\mu} a_{q\mu} , \quad (6)$$

$$H'' = \sum_{\alpha} \omega_\alpha R^{+}_\alpha R_\alpha + \sum_{\alpha k\mu} \left(g^{*}_{k\alpha} R^{+}_\alpha c_k + j^{*}_{k\alpha} R^{+}_\alpha a_{k\mu} + h.c. \right) (7)$$

where H_{ph} (H_m) is the phonon (magnon) Hamiltonian and H'' describes the interaction of phonons and magnons with the reservoir. R_α and ω_α are the operators of the heat-bath quasi-particles and their energies,respectively. $g_{k\alpha}$ ($j_{k\alpha}$) is the coupling between magnons (phonons) to the heat-bath.The hypotheses concerning the reservoir are [7]: (i) the reservoir has a large thermal capacity and a short memory; (ii) the pump and the interaction with the reservoir influence the system independently; (iii) the various modes are coupled to the reservoir each one independently of the other.

The cross-section for the process is obtained in the Born approximation.[8] In terms of Green's functions it is given by

$$\frac{d^2\sigma}{d\Omega\, d\omega_{k'}} = \frac{4\pi V S^2}{N_q V_q}|\alpha(k)|^2 \Big\{ D(\omega_{q-k})\, n_q (n_{q-k}+1)(n_{k'}+1) - D(\omega_q)(n_q+1)\, n_{q-k}\, n_{k'} \Big\} \mathrm{Im}\langle\langle S_i^+ ; S_i \rangle\rangle_{-\omega_{k'}+i\epsilon} \quad (8)$$

where V_q is the group velocity of the sound waves, $D(\omega_q)$ is the density of states in terms of the frequency, V is the volume of sample and N_q is the number of incident phonons in the volume V.

Eq.(8) was calculated assuming the following decoupling approximation : $\langle S_i (S_z)^2 S_i \rangle \simeq \langle S_z^2 \rangle \langle S_i S_i \rangle = S^2 \langle S_i S_i \rangle$. Calculation of $\langle\langle S_i^+ ; S_i \rangle\rangle$ can be done by using Holstein - Primakoff approximation for the spins.

Considering collinear scattering along x-direction and assuming phonons linearly polarized along z ,we obtain for the process: transverse phonon → transverse phonon + magnon with rotation of $\pi/2$ in the vibration plane

$$\frac{d^2\sigma}{d\Omega\, d\omega_{k'}} = \frac{V S^2}{\pi N_q V_q}|\beta_{q,q-k}|^2 \Big\{ D(\omega_q)(n_q+1)\, n_{q-k}\, n_{k'} - D(\omega_{q-k})\, n_q (n_{q-k}+1)(n_{k'}+1) \Big\} \frac{\eta_k}{[\omega_m(k') - \omega_m(k) - \beta^2(n_q - n_{q-k})]^2 + \eta_k^2} \quad (9)$$

Since losses were introduced in the Hamiltonian, eq. (9) gives not only the magnitude of the cross-section but also the line shape of the scattering.

In conclusion,we point out that eq.(9) leads to an integrated cross-section of the same order of magnitude as obtained in previous treatments.[4,5] Such cross-section is as sizeable as the three-phonon anharmonic interaction.[5]

COHERENCE PROPERTIES OF THE PARAMETRIC PHONON STATES

In the preceeding section we were interested in the cross-section for the spontaneous scattering.Since the Hamiltonian (4) describes a non-linear process we must consider the possibility of a stimulated scattering. So,it becomes important to calculate the threshold and study the statistical properties of the system below and above the critical point. This can be done by solving a system of equations involving phonon and magnon amplitudes and studying the evolution of the scattered phonon states as a function of the pumping. The system of equations mentioned above results from Heisenberg's equations of motion using the total Hamiltonian and treating operators as c-numbers ($a_q \to U_q e^{-i\omega_q t}$, $c_k = V_k e^{-i\omega_k t}$). Assuming exact energy and momentum matching conditions,and considering only three interacting modes,we obtain

$$\dot{U}_q = -i\beta U_Q V_k - \eta_q U_q + F_q e^{i\omega_q t} , \quad (10)$$

$$\dot{U}_Q = -i\beta^* U_q V_k^* - \eta_Q U_Q + F_Q e^{i\omega_Q t} , \quad (11)$$

$$\dot{V}_k = -i\beta^* U_q U_Q^* - \eta_k V_k + G_k e^{i\omega_k t} , \quad (12)$$

where q (Q) refers to the wavevector of pumping (scattered) phonon modes and η_q, η_k are the relaxation rates for phonons and magnons. F and G are "Langevin fluctuating forces".[7]

Under the assumption that in the beginning of the process the external excitation is much larger than any excitation in the crystal ($|U_q|^2 >> |V_k|^2$), we obtain from eq.(10)-(12) the critical number of incident phonons to reach the threshold for the stimulated scattering which is given by

$$|U_q^c|^2 = \frac{\eta_Q(\eta_k + \eta_q)}{\beta^2} . \quad (13)$$

For YIG (yttrium iron garnet) at room temperatures,eq.(13) leads to a critical pumping power of 10^5 watts/ cm^2 which is very large.However since the life time of phonons is very sensitive to the temperature,one may have threshold of about 100 mWatts / cm^2 for temperatures around 4.2 °K.

To study the coherence properties of the scattered phonon states let us investigate the behavior of the density matrix which takes into account the probability distribution for different values of U_Q , V_k . We can write a Fokker-Planck type equation which is stochastically equivalent to the system (10)-(12).[9,10] This Fokker-Planck equation describes the behavior of $P(U_Q)$,the diagonal elements of the density operator in the coherent states representation.[10] Around the threshold,where $|U_Q|^2 \simeq |V_k|^2$, the equation assumes the form

$$\frac{\partial P}{\partial y} + \frac{1}{x}\frac{\partial}{\partial x}[(D - x^2)x^2 P] = \frac{1}{x}\frac{\partial}{\partial x}\left(x\frac{\partial P}{\partial x}\right) + \frac{1}{x^2}\frac{\partial^2 P}{\partial \phi^2} , \quad (14)$$

where

$$x = \left(\frac{\beta^2}{\eta_k^2 \Gamma_k}\right)^{1/2} r \; ; \; y = \sqrt{\beta\Gamma_k \eta_k^{-1}}\, t \; ; D = \frac{\beta|U_q|^2 - \eta_Q\eta_k}{\eta_k\sqrt{\Gamma_k}} , \quad (15)$$

and (r , ϕ) are the polar coordinates of U_Q. Eq.(14) was obtained under the assumption that $\ddot{U}_Q < \eta_Q \dot{U}_Q$ which allows the description of steady state and slow transients(compared to $1/\eta_Q$) in the system. It shall be studied for the stationary regime($\partial P/\partial y = 0$).According to symmetry P is independent of ϕ [10] and so $\partial^2 P/\partial\phi^2 = 0$ .Integrating eq.(14) under the previous conditions we obtain

$$P(x) = N \exp\left(\frac{D}{2}x^2 - \frac{1}{4}x^4\right) , \quad (16)$$

where N is a constant of normalization.

Eq.(16) indicate that below the threshold $P(x)$ is a Gaussian function,and just above the threshold ($U_q \gtrsim 1.01\, U_q^c$) it behaves like a δ - function which means that the distribution of the parametric phonon states becomes coherent.[9,10]

An important consequence of the above calculations is that the possibility of observing the Raman scattering of phonons by magnons in a "direct way" (standard ultrasonic experiments) as suggested a few years ago in this Conference[5] may be further clarified. The existence of a threshold determines the occurence of two distinct regimes on the coherence of the scattered phonon states. The "direct way" of observation may be only expected to work above the threshold since the scattered phonon distribution is thermal below it . From the behavior of the ρ distribution one can expect to detect the scattered phonons for amplitudes $U_q \gtrsim 1.01\, U_q^c$ .This condition can be attained in the case of YIG at low temperatures.For pumping below the threshold it appears that light scattering techniques might be appropriate.

We wish to acknowledge useful discusions with J.R.Rios Leite.

REFERENCES

* Work partially supported by CNPq ,CA-PES and BNDE

** Present address:Gordon McKay Laboratory,Harvard University,Cambridge, Mass. 02138

1.B.A.Auld,in Adv in Microwaves,vol.3(Academic Press,New York,1968)
2.F.Keffer,in Handbuch der Physik,ed.by S.Flugge(Springer-Verlag,Berlin,1966) vol.XVIII/2
3.R.W.Damon,in Magnetism ,ed.by G.T.Rado and H.Suhl (Academic Press,New York , 1963)vol.I
4.I.A.Akhiezer,JETP Letters 5 ,160 (1968)
5.B.A.Huberman and E.Burstein ,AIP Conf. in Magnetism and Magnetic Materials (1971) p.1350
6.D.E.Eastman,Phys.Rev.148,530 (1966); J.Appl.Phys.37 ,996(1966);J.Appl.Phys. 37 ,2312(1966)
7.U.Balucani,F.Barocchi and Tognetti,Phys. Rev.A 5 ,442 (1972)
8.C.Kittel,Quantum Theory of Solids(John Wiley and Sons,New York,1963)
9.Cid B.de Araujo,Phys.Rev.B 10 ,3961(1974) Phys.Status Solidi(b)68 ,K117 (1975) ; Cid B.de Araujo and S.M.Rezende,Phys. Rev.B 9 ,3074 (1974)
10.R.J.Glauber,Phys.Rev.131 ,2766(1963) ; J.R.Klauder and E.C.G.Sudarshan,Fundamentals of Quantum Optics (Benjamin, New York,1968)

EPR STUDIES OF Dy^{3+} IN THE SINGLET-GROUND-STATE PARAMAGNETS, PrX AND TmX (X=P, As, Sb)

K. Sugawara
Case Western Reserve U*, Cleveland, OH 44106
and C. Y. Huang
Los Alamos Scientific Lab†, Los Alamos, NM 87545

ABSTRACT

In order to complement our previous study of the S-state Gd^{3+} EPR, we have investigated the EPR of Dy^{3+} diluted in singlet-ground-state paramagnets, PrX and TmX (X=P, As) from 5 K to 50 K. The g-values at 5 K have been found to be 7.1 ± 0.2 and 9 ± 1 for PrAs and TmAs, respectively, and they approach the theoretical g-value for Γ_6, 6.6, at high temperatures (T≈50 K). In the temperature range investigated, the observed g-shifts are proportional to the magnetic susceptibilities of the respective hosts. The position of the first-excited state of Dy^{3+} has been determined by comparing the temperature-dependent intensities of the Dy^{3+} EPR with those of Gd^{3+} impurities in the same host. We have also measured the g-values of Dy^{3+} in PrP, PrSb and TmP at 5 K. From the g-shifts at low temperatures we have computed the exchange coupling constants between Dy^{3+} and its surrounding rare-earth ions (Pr^{3+} and Tm^{3+}). The temperature variation of the linewidth of Dy^{3+} in PrAs is discussed.

In recent years, there has been a great deal of interest in the study of the interaction of magnetic impurities in some Van Vleck paramagnets by EPR. However, to date, only S-state impurities[1,2] have been considered. For this reason, in this paper we report our recent results of non-S-state Dy^{3+} ($^6H_{15/2}$) diluted in some metallic Van Vleck paramagnets, PrX and TmX (X=P, As, Sb).

When an impurity magnetic ion with Lande g-factor λ is in a magnetic host of the magnetic ions with Lande g factor λ', the g-shift of the magnetic impurity ion can be expressed as[1-3]

$$\Delta g = g_0\,(\lambda-1)(\lambda'-1)\,\chi\,(\sum_i J_i)/(\lambda\lambda'\,\mu_0^2), \qquad (1)$$

where g_0 is the g-value in the absence of the exchange interaction, χ the magnetic susceptibility of the host, J_i the exchange interaction of the impurity ion with the i-th host magnetic ion, and μ_0 the Bohr magneton. Eq. (1) indicates that Δg decreases with increasing temperature and the g-value approaches g_0 at high temperature.

We have investigated the EPR of 0.05 at. % Dy^{3+} in PrX and TmX. The powder samples used in this work were prepared by a method similar to that described in Ref. 2. The EPR experiments were performed from 5 K to 50 K employing a conventional EPR spectrometer operated at 9.2 GHz. The energy level scheme for Dy^{3+} in the octahedral crystal-field is shown in Fig. 1(a). Based on our previous investigation[4] of Dy^{3+} in nonmagnetic LaX and YX, the ground-state is assumed to be the Γ_6 doublet with g_0 = 6.6. Fig. 2(a) displays our measurements for the temperature variation of the g-value of Dy^{3+} in PrAs. The temperature dependence of the g-shift, g(T)-6.6, is in reasonable agreement with that of the magnetic susceptibility of PrAs measured by Tsuchida, et al[5] Since the observed g-value approaches 6.6 at high temperatures, the ground-state of Dy^{3+} in PrAs is thus identified as Γ_6. Fig. 2(b) depicts the temperature variation of the g-value of 0.1 at. % Dy^{3+} in TmAs. The g-value at 5 K is 9.0 ± 1.0, which is considerably shifted from the value in the absence of the exchange interaction, g_0 = 6.6. Again, within our experimental error, the temperature dependence of the g-shift is proportional

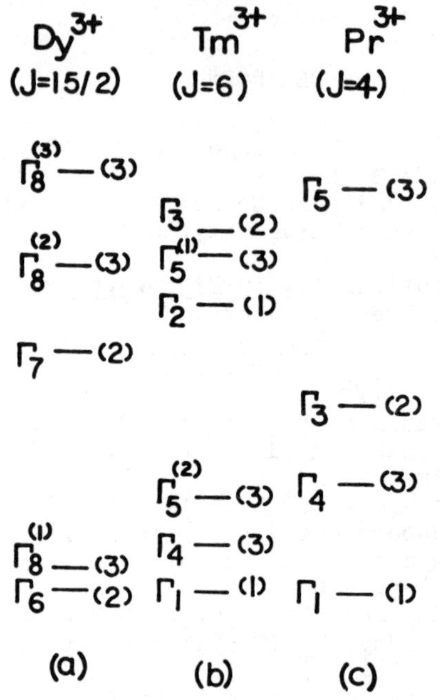

Fig. 1. The crystal-field energy level schemes of Dy^{3+}, Tm^{3+}, and Pr^{3+} in rare-earth-group VA compounds.

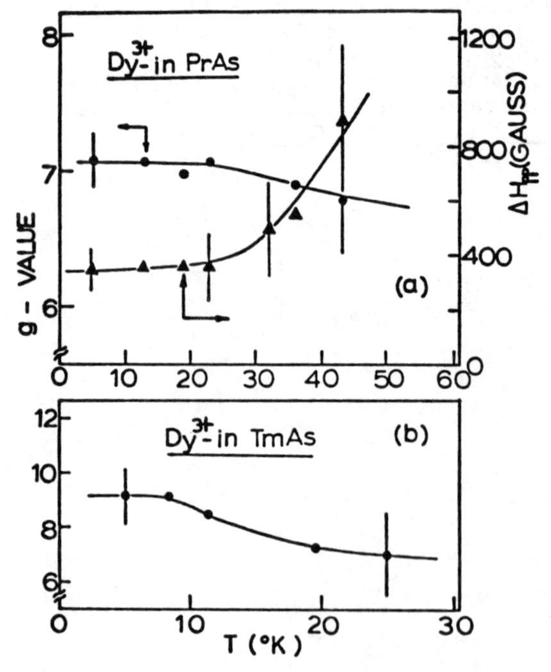

Fig. 2. (a) The temperature dependence of the g-value (solid circles) and the line-width (solid triangles) of Dy^{3+} in PrAs. (b) The temperature dependence of the g-value of Dy^{3+} in TmAs.

Table I

The g-values of 0.05 at. % Dy^{3+} in PrP, PrAs and PrSb, and 0.1 at. % Dy^{3+} in TmP and TmAs observed at 5 K, the exchange coupling constants, $\sum_i J_i(Dy^{3+})$, calculated by using Eq. (1), and $\sum_i J_i(Gd^{3+})$ obtained in Ref. 2.

Host	g	$\sum_i J_i(Dy^{3+})$	$\sum_i J_i(Gd^{3+})$
PrP	9.0 ± 1.0	-63 ± 26 K	-1.7 K
PrAs	7.1 ± 0.2	-12 ± 4 K	-3.7 K
PrSb	8.3 ± 0.5	-30 ± 9 K	-12 K
TmP	9.3 ± 0.9	+11 ± 3 K	+4.1 K
TmAs	9.0 ± 1.0	+ 9 ± 3 K	+1.0 K

to that of the susceptibility[6] of TmAs. We have also obtained the g-values of 0.05 at. % Dy^{3+} in PrP and PrSb, and 0.1 at. % Dy^{3+} in TmP. These results and the exchange coupling constants, $\sum_i J_i(Dy^{3+})$, computed by means of Eq. (1) are summarized in Table I. It is interesting to notice that the exchange interactions are negative for the praseodymium compounds and positive for the thulium compounds. For comparison, the exchange coupling constants of Gd^{3+} in the same hosts,[2] $\sum_i J_i(Gd^{3+})$, are also given in Table I. Even though $\sum_i J_i(Dy^{3+})$ and $\sum_i J_i(Gd^{3+})$ have the same sign in the same host, the former is considerably greater than the latter.

Since the zero-field splitting of Gd^{3+} is very small, the ratio of the EPR intensity of Gd^{3+} to that of Dy^{3+} in the same host is proportional to the partition function of Dy^{3+}, and hence the crystal-field splitting of Dy^{3+} can be obtained. This method of determining the crystal-field splitting has been successfully applied to some magnetic ions in nonmagnetic metallic hosts.[4] As a result, we have performed EPR on 0.05 at. % Gd^{3+} and 0.05 at. % Dy^{3+} in PrAs from 5 K to 50 K. As expected, the line of Gd^{3+} ($g \simeq 2$) was well separated from that of Dy^{3+} ($g \simeq 7$). From the ratio of the intensities of these two absorption lines we have found the first excited-state of Dy^{3+} in PrAs, $\Gamma_8^{(1)}$, is $\sim$ 35 K above the Γ_6 ground-state. This value is close to that[4] of Dy^{3+} in YAs (30 K).

Also shown in Fig. 2(a) is the temperature variation of the linewidth of 0.05 at. % Dy^{3+} in PrAs. The solid curve can be expressed by

$$\Delta H = 350 + aT + b/(\exp \Delta'/T - 1) \text{ gauss}, \qquad (2)$$

where $a \simeq 1$, $b \simeq 8 \times 10^4$ and $\Delta' \simeq 220$ K. The first term is the residual linewidth and the second one is the Koringa term. The last term describes the temperature dependence of the relaxation rate of the transitions to and from the excited state[7,8] at energy Δ'. The value of Δ' coincides with the splitting between the ground-state and the second excited-state, Γ_7. (It is not clear why the first excited-state does not contribute to relaxation.) Since the exchange interaction between Dy^{3+} and Pr^{3+} is strong, the angular-momentum fluctuations of Pr^{3+} should also contribute to the Dy^{3+} linewidth. Further investigation is needed in order to understand the relaxation mechanism.

REFERENCES

* Supported by the National Science Foundation.
† Work performed under the auspices of the U.S.E.R.D.A.

1. C. Rettori, D. Davidov, A. Grayevskey, and W. M. Walsh, Phys. Rev. B 11, 4450 (1975).
2. K. Sugawara, C. Y. Huang, and B. R. Cooper, Phys. Rev. B 11, 4455 (1975).
3. M. T. Hutchings, C. G. Windsor, and W. P. Wolf, Phys. Rev. 148, 444 (1966).
4. K. Sugawara and C. Y. Huang, to appear in J. Phys. Soc. Japan 41, No. 1 (1976).
5. T. Tsuchida and W. E. Wallace, J. Appl. Phys. 43, 2885 (1965).
6. G. Busch, A. Menth, O. Vogt, and F. Hullinger, Phys. Letters 19, 622 (1966); G. Busch, O. Vogt, O. Marince, and A. Menth, Phys. Letters 14, 262 (1965).
7. L. L. Hirst, Phys. Rev. 181, 597 (1969).
8. K. Sugawara and C. Y. Huang, J. Phys. Soc. Japan 40, 295 (1976).

TRANSMISSION RESONANCE EXPERIMENTS WITH NICKEL AND PERMALLOY PLATELETS

B. Heinrich and J.F. Cochran
Simon Fraser University
Burnaby, British Columbia
Canada V5A 1S6

J.H. Liaw and R.C. Barker
Yale University
New Haven, Connecticut 06520
U.S.A.

ABSTRACT

A series of transmission resonance experiments at 24GHz has been initiated using nickel and permalloy single crystal platelets mounted over 0.5-0.8 mm holes in copper diaphragms placed between two resonant cavities. Preliminary experiments have been done on samples in the thickness range of 0.5 to 5 microns and a $4\pi M$ range from 6.1×10^3 to 7.5×10^3 Oe. In the thin film limit the platelet is never opaque, but a transmitted power dip is found at the FMR field, but because of the transparency of the film, the antiresonance $4\pi M$ below the FMR field is not observed. In the thick film limit a broad FMAR transmission peak is observed, and the sample becomes opaque about the FMR field. In both limits agreement of the experimental transmitted power versus applied field with the theory at parallel resonance is excellent. In the experiments to date there is no evidence of the interference between the electromagnetic and spin wave modes expected if the surface spins are pinned.[1,2] This is consistent with FMR absorption experiments, where in the few cases where they have been observed, spin wave spectra have been very weak.[3]

INTRODUCTION

Nickel and permalloy platelets grown by means of vapor transport are nearly perfect metal single crystals having atomically smooth surfaces. These properties make them very suitable for the study of ferromagnetic resonance absorption[4,5,6]. Transmission studies on metal platelets at microwave frequencies are also of interest for at least two reasons: (1) platelets which are thick compared to the skin-depth $\delta = (c^2\rho/2\pi\omega)^{\frac{1}{2}}$, are expected to exhibit a transmission maximum at ferromagnetic antiresonance[7,8] (FMAR) from which it should be possible to deduce the intrinsic magnetic damping of the material, and (2) specimens comparable with the skin-depth may, if the spins are suitably pinned at the surfaces, exhibit an oscillatory component in the magnetic field dependence of the transmission amplitude due to interference between the propagating spin-wave and the propagating electromagnetic wave. This latter possibility was advanced by Phillips[9], and has been further discussed by Liu and Barker[1], and by Cochran, Heinrich, and Dewar[2]. This interference structure differs from structure due to standing spin-waves in that it occurs in specimens which are thick compared with the spin-wave damping length, therefore the standing spin-wave amplitude in the specimen is small compared with the spin-wave amplitude at the rear surface of the specimen. The structure in the transmission amplitude due to interference between the spin-wave and the electromagnetic wave has a spacing in applied magnetic field which is inversely proportional to specimen thickness and to the exchange stiffness constant, but unlike standing spin-waves this magnetic field spacing varies slowly with external field and is therefore quasi-periodic. Moreover, the period of the interference pattern is very insensitive to the strength of the surface pinning of the RF magnetization, although the amplitude of the structure is very sensitive to the surface pinning strength. These properties make this interference effect a potentially powerful tool for the investigation of surface-pinning and exchange stiffness in ferromagnetic metals.

EXPERIMENTAL DETAILS

We report the results of measurements on three platelets: 1) a Ni-Fe platelet grown at Yale, and approximately 3×10^{-4} cm thick (Liaw permalloy, abbreviated LPY), 2) a Ni-Fe platelet approximately 6×10^{-5} cm thick (PY1) obtained from General Electric Co.[12], and 3) a pure, nickel platelet approximately 8×10^{-5} cm thick (Ni1), also obtained from the General Electric Co.[10]. The thick specimen (LPY) was soldered over a 0.6 mm aperature in a 0.005" thick copper diaphragm using indium as a solder. It is extremely difficult to solder specimens thinner than 5×10^{-4} cm without damaging them, and, indeed, LPY became slightly deformed during the soldering process. We found it impossible to solder thinner specimens over an aperature in a copper diaphragm without either rupturing the platelets or producing unacceptably large deformations in them. Therefore, the thinner specimens were mounted on clean copper diaphragms using colloidal silver dag[11]. Best results were obtained when the copper diaphragms were cleaned by heating them to 300°C for one hour in an atmosphere of dry hydrogen gas.

The specimens, mounted on copper diaphragms, were mounted between identical transmitter and receiver cavities of a 24GHz system which uses homodyne detection to attain a small signal sensitivity of 10^{-17} watts. This system will be described in detail in a forthcoming publication[2]. The transmission amplitude ratios observed using the platelets mounted over 0.6 mm aperatures in the copper diaphragms were typically between 10^{-3} and 10^{-4} i.e. the transmitted power for an incident power of 1/3 watt was of order 10^{-6} - 10^{-8} watts. At these relatively large power levels the sensitivity of the apparatus was limited by Klystron instabilities to a signal to noise amplitude ratio of approximately 500 to 1.

RESULTS

The variation of transmission with magnetic field, with the applied field in the specimen plane, is shown for LPY in Fig. 1. Near the FMR field of 5.1 koe the absorptipn becomes so large that the transmission signal becomes effectively zero. At low fields the transmission amplitude becomes a maximum at 0.70 koe, the field which corresponds to ferromagnetic anti-resonance (FMAR). The solid curves shown in the figures are the magnetic field dependence of the transmission amplitudes calculated using an extension of the theory of Ament and Rado which is described by Cochran, Heinrich, and Dewar[2] This theory includes exchange, and magnetic damping is described by a term of the Landau-Lifshitz form i.e. $d\vec{M}/dt)_{damping} = -\lambda/M_s^2\,(\vec{M}\times\vec{M}\times\vec{H}_{eff})$.
Calculated curves were fitted to the data using ω/γ, $4\pi M_s$, d/δ, and λ as fitting parameters. For the specimen of Fig. 1, the values of ω/γ, and $4\pi M_s$ were determined primarily by the FMR and FMAR fields; the calculated curves were very insensitive to the valve chosen for λ, but the detailed shape of the curves, especially for fields larger than 6 koe, was very sensitive to the value chosen for d/δ. From the value for $4\pi M_s$ and the data reported by Bozorth[12] we concluded that LPY was an alloy of 5% iron in

nickel. The corresponding room temperature resistivity[12] should be $\rho = 10.0\times10^{-6}\Omega$ cm. This value of resistivity gives a skin-depth $\delta = \sqrt{c^2\rho/\pi\omega} =$ 1.03μm at 23.95 GHz, from which we deduce a specimen thickness of 2.85 μm, in good agreement with the value 2.875±0.5% μm deduced from X-ray absorption measurements.

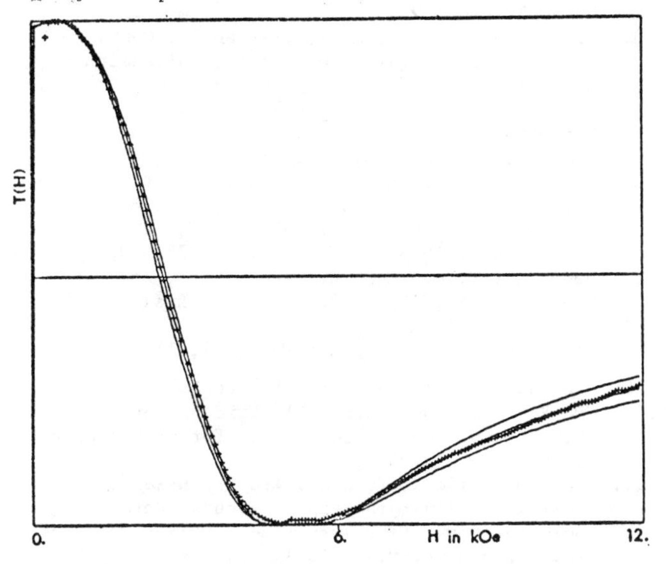

Fig. 1 The magnetic field dependence of the transmission amplitude in the sample LPY. The applied field is in the specimen plane. Calculated curves (solid lines) were fitted to the data using the parameters $4\pi M_s$ = 7.181 kG, ω/γ = 7.881 koe, $\lambda = 2.1\times10^8$ sec^{-1}. The solid curves drawn for the three different values of d/δ demonstrate the sensitivity of the fit to the particular choice of the parameter d/δ. (d/δ = 2.67, 2.76, 2.86)

For the two thinner specimens the absorption at FMR was much smaller than that for LPY, and the transmission was nearly field independent apart from the absorption dip centered at the FMR field. These specimens were mounted using Ag dag,[11] and the leakage signal around the specimens and through the colloidal silver paste was as large as 30% of the high field transmission signal amplitude. In order to remove this background amplitude (whose strength was a priori unknown) we subtracted the transmission amplitude at an external field far removed from FMR from both the experimental data and the calculated transmission signal. As a result, the absorption dips at FMR appear as peaks in Figs. 2, 3, and 4. The experimental data for PY1 are shown in Fig. 2 for the external field in the specimen plane, and in Fig.3 for the field perpendicular to the specimen plane. The calculated curves were fitted to the data by varying ω/γ, $4\pi M_s$, λ, and d/δ - the same values of these parameters were used to generate the curves shown in both Figs. 2 and 3. The resulting value of $4\pi M_s$ indicated[12] that PY1 was an alloy of 2.6% Fe in Ni. The corresponding resistivity[12] is $\rho = 9\times10^{-6}\Omega$cm, leading to an estimate of $d = 0.57\times10^{-4}$cm. This thickness is in good agreement with the value of 0.537±4%μm deduced from X-ray absorption measurements. The value $\lambda = 2.0\times10^8$ sec^{-1} required to fit the data is slightly less than the value $\lambda = 2.27\times10^8$ Hz which is obtained from a linear extrapolation between $\lambda = 2.45\times10^8$ Hz for pure Ni[4,13] and $\lambda = 0.75\times10^8$ Hz for 25% Fe in Ni[14]. We attribute the discrepency between the experimental data and the calculated curve in Fig. 3 to a small field-dependent component of the leakage signal near FMR.

The experimental data for the transmission through the pure Ni platelet with the transmission at 9.815 koe subtracted is shown in Fig. 4 for the field parallel to the specimen plane. Only the thickness of the specimen was used as an adjustable parameter to fit the calculated curve to the data. Values of ω/γ, $4\pi M_s$, and λ were obtained from transmission studies on thick polycrystalline Ni foils[13]. The calculated curve is in excellent agreement with the data except near FMR where the minor discrepency is very likely due to a small component of field dependent leakage signal - the fit between theory and experiment could not be improved by adjusting the other parameters in the theory. The quality of the fit obtained was very sensitive to small changes in thickness, and therefore we could conclude that d = 0.8 ± .05 μm.

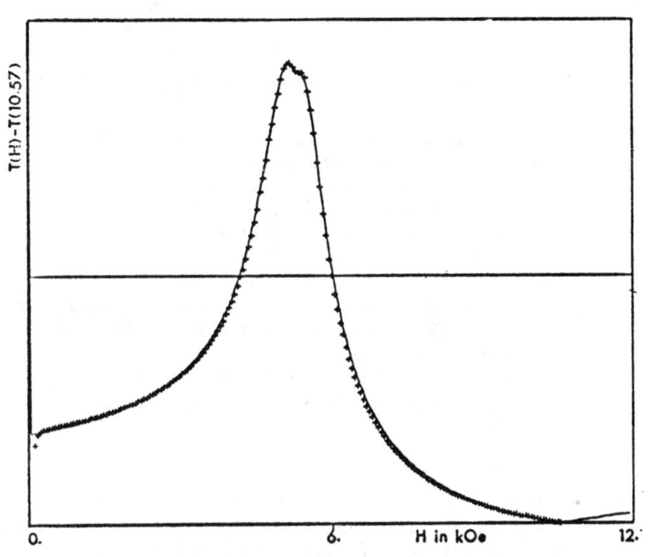

Fig. 2 The variation of transmission with magnetic field, with the applied field in the specimen plane, for PY1. The solid line was calculated using the values $4\pi M_s$ = 6.491 kG, ω/γ = 7.957 koe, $\lambda = 2\times10^8$ sec^{-1}, d/δ = 0.58 .

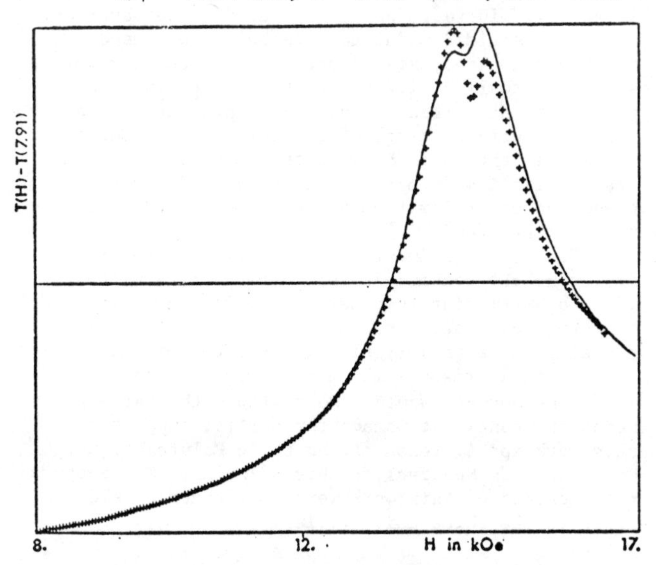

Fig. 3 The variation of transmission with magnetic field, with the applied field perpendicular to the specimen plane, for PY1. The solid line was calculated using the same parameters as were used for the calculated curve of Fig. 2

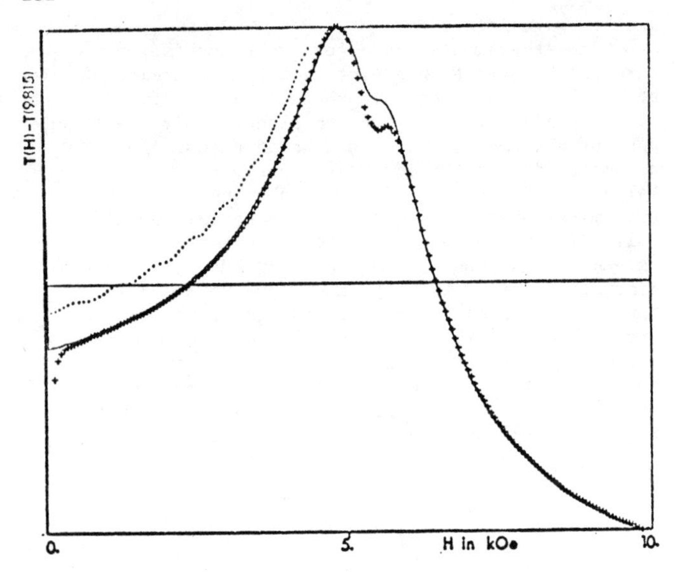

Fig. 4 The transmission through the pure Ni sample for the applied field parallel to the specimen plane. The solid line was calculated using values $4\pi M_s = 6.055$ kG, $\omega/\gamma = 7.847$ koe, $\lambda = 2.45\times10^8 sec^{-1}$, $\rho = 7.90\times10^{-6}$ ohm cm, d = 0.8±.5 μm. Values $4\pi M_s$, ω/δ, λ were obtained from transmission studies on thick pollycrystalline Ni[13]. The dotted line (which has been displaced upwards to make it more visible) represents the calculated transmission for infinite pinning on both surfaces of the sample.

Again, this value of d is in good agreement with 0.805±2%μm deduced from X-ray absorption measurements.

The thickness of PY1 and the nickel platelet were well-suited for the observation of spin-wave and electromagnetic wave interference. If the spins in these specimens had been completely pinned the interference structure should have been ~5% of the maximum signal variation as shown by the dotted curve in Fig. 4. In fact, no such structure was observed, even when a field modulation technique was used which would have revealed structure in the derivative signal as small as 1:2000 of the main derivative peak. If the surface pinning is represented by a uniaxial surface energy of the form $E_s = K_s \mathrm{Cos}^2\alpha$, where α is the angle between the magnetization and its equilibrium direction, then this null result places an upper limit on the pinning parameter of $K_s = 0.1$ ergs/cm^2.

Further experiments will be directed toward improving the sensitivity of the apparatus with a view to attempting to observe interference structure in permalloy platelets. We also plan to subject the platelets to a variety of surface treatments in an attempt to increase surface pinning[15,16,17,18].

The authors would like to thank the National Research Council of Canada for partial support of this work and A. Yelon of the Ecole Polytechnique, Université de Montreal for his stimulus. The cooperative aspects of this work were done under a NATO grant to the three universities.

REFERENCES

1. Liu, Y.J, Barker, R.C., presented at the 20th Annual Conf. on Magnetism and Magnetic Materials, San Francisco, 1974. AIP Conf. Proc., No. 24, p. 505, 1975.
2. Cochran, J.F., Heinrich, B., and Dewar, G., to be published.
3. Liaw, J.H., Barker, R.C., 21st Annual Conf. on Magnetism and Magnetic Materials, Philadelphia, Dec., 1975.
4. Rodbell, D.S., Physics 1, 279, 1965.
5. Vittoria, C., Barker, R.C., and Yelon, A. Phys. Rev. Letters 19, 792, 1967.
6. Phillips, T.G., Rupp, Jr., L.W., Barker, R.C. and Vittoria, C., J. de Physique, 32, C1-1162, 1971.
7. Heinrich, B. and Meshcheryakov, V.F, Zh ETF Pis. Red. 9, 618, 1969. [JETP Lett. 9, 378, 1969.]
8. Heinrich, B. and Meshcheryakov, V.F., Zh. Eksp. Teor. Fiz. 59, 424, 1970. [Soviet Physics JETP 32, 232, 1971.]
9. Phillips, T.G., J. App. Physics 41, 1109, 1970.
10. The authors are very greatful to Dr. R.W. De Blois of the General Electric Research Laboratory, Schenectady, N.Y., for providing us with these platelets.
11. Acheson Colloids (Canada) Ltd. Brantford, Ont. Colloidal Silver in Methyl Isobutyl Ketone.
12. Bozorth, R.M. Ferromagnetism. D. Van Nostrand Co., Princeton, N.J., 1951.
13. Dewar, G., Cochran, J.F. and Heinrich, B. To be published.
14. Bailey, G.C. and Vittoria, C. Phys. Rev. B8, 3247, 1973.
15. Massenet, O., Biragnet, F. Juretschke, H. Montmory, R. and Yelon, A. I.E.E.E. Trans. Magnetics 2, 553 (1966).
16. Hoffman, F., Stankoff, A. and Pascard, H. J. Appl. Phys. 41, 1022, 1970.
17. Monod, P., Hurdequenit, H., Janossy, A., Obert, J. and Chaumont, J. Phys. Rev. Lett. 29, 1327, 1972.
18. Janossy, A. and Monod, P. Proceedings ICM-73, V, 459. Moscow.

TWO-MAGNON-SCATTERING CONTRIBUTIONS TO FMR LINEBROADENING IN POLYCRYSTALLINE GARNETS

P. Röschmann
Philips GmbH Forschungslaboratorium Hamburg, Hamburg 54, West Germany

I ABSTRACT

Measurements of the FMR linewidth, ΔH, of different Walker modes and of the lineshift, δf, of the UPR mode have been made as a function of frequency and temperature on spheres of polycrystalline garnets of pure YIG and CaVIn substituted YIG with reduced anisotropy field H_a. The experimental results can be explained on the basis of anisotropy and porosity (p) induced two-magnon-scattering contributions, ΔH_a and ΔH_p. Excellent quantitative agreement is found with Schlömann's theory for ΔH_a and δf, apart from the region close to the top of the spinwave manifold (SWM). The deviations are related to the inhomogeneity ratio, $(4\pi M_s/H_a)$, and may be explained by the pertubation of the SWM, by approach to saturation effects, and by the density of magnetostatic modes (MSM) evaluated from numeric calculations. The frequency and field dependence of the separated porosity scattering contribution ΔH_p show an excellent quantitative agreement with ΔH_{eff} measurements reported by Vrehen, Broese van Groenou and Patton but contradict known theories. ΔH_p consists of a nearly field independent fraction arising from scattering to high-k states and a usually much larger fraction, showing a distinct dependence on $\cos\Theta_o$ of low-k spinwaves. Possible coupling models, based on these results, are presented. The experimental findings provide a guide for the evaluation of H_a and p of polycrystalline ferrites from ΔH-measurements.

II INTRODUCTION

The linewidth, ΔH, of the ferrimagnetic resonance (FMR) in polycrystalline ferrites is generally orders of magnitude larger than the linewidth, ΔH_s, of single crystals of the same chemical composition. The broadening of the FMR line results from inhomogeneous local magnetic fields, which arise in polycrystalline ferrites due to the random orientation of the crystal anisotropy energy axes of different grains, ΔH_a, and due to the demagnetizing fields around pores or around second phase material inclusions, ΔH_p. The FMR linewidth of polycrystalline ferrites may be regarded as a sum of the contributions:

$$\Delta H = \Delta H_s + \Delta H_a + \Delta H_p \tag{1}$$

However, the line broadening does not reach the magnitude of the magnetic field inhomogeneities. The long range dipolar coupling between the precessing spins leads to collective modes of spin oscillations which counteract the tendency of locally varying spinprecession resonance frequencies at the nonuniformities of the internal field. The inhomogeneous broadened FMR is "dipole narrowed"[1]. Thus the eigenmodes of the sample have to be considered in terms of the collective modes of motion of the rf. magnetization such as the spinwaves[2] and the magnetostatic modes[3]. In a perfectly homogeneous material these modes are orthogonal and no energy is exchanged between them. The width of the spectrum of these modes is approximately $\gamma\cdot 4\pi M_s/2$, (γ=2.8 MHz/Oe, the gyromagnetic ratio; $4\pi M_s$ the saturation magnetization). If an inhomogeneous effective magnetic field, $H_e \ll 4\pi M_s$, is present in the material and one of the collective modes is excited initially, its energy is scattered to other, mainly to degenerate, eigenmodes. This coupling process is referred to as two-magnon-scattering[4]. It is the main FMR-relaxation channel in polycrystalline ferrites and leads also to a frequency shift of the FMR line. The spinwave (SW) model ceases to be valid for inhomogeneous fields, $H_e \geq 4\pi M_s/2$, corresponding to the width of the spinwave manifold (SWM). In this case the SWM is strongly perturbed and the dipolar interactions between the magnetizations in neighbouring grains become negligible. Hence, the different crystallites resonate independently and the so-called "independent grain" (IG) model applies.

Both, the SW- and IG-model have been theoretically studied in detail by Schlömann[5,6] and Vrehen et al.[7] for inhomogeneities due to cubic anisotropy. The main features of the results may be summarized as follows: (i) for $H_a \ll 4\pi M_s$ the SW model predicts line broadening only within the limits of the SWM for low wavenumbers k, ΔH_a is proportional to the density ρ of degenerate longwavelength spinwave states and to the "dipole narrowed" inhomogeneity parameter $P_a = H_a^2/4\pi M_s$; (ii) for $H_a > 4\pi M_s$ the IG model predicts broadening of the FMR line approximately proportional to $0.8\cdot 4\pi M_s$ also at frequencies or fields outside of the SWM.

Experimentally these theories have been tested by measurements of ΔH at the FMR as a function of frequency and temperature[8,9,10] and by off-resonance measurements of the complex rf. susceptibility from which a loss parameter in terms of an effective linewidth, ΔH_{eff}, is deduced[7,11]. For $H_a \approx 4\pi M_s$, good agreement was found for ΔH_{eff} measurements with the IG-model[7,11]. Excellent quantitative agreement between theory and experiment was obtained for ΔH_{eff}[7] and ΔH at the FMR[10] for $H_a \ll 4\pi M_s$ with Schlömann's result[5] from the SW-model (rearranged):

$$\Delta H_a = \frac{38\pi}{105}\frac{H_a^{\,2}}{4\pi M_s}\rho\left[\frac{d\Omega_k}{d\Omega_H} + \frac{1}{19}\right]. \tag{2}$$

In eq. (2), $\Omega_H = H_i/4\pi M_s$, is the normalized internal static field, and $\Omega_k = f_k/\gamma\cdot 4\pi M_s$, is the normalized frequency of spinwaves given by:

$$\Omega_k^2 = (\Omega_H + D^* k^2)(\Omega_H + D^* k^2 + \sin^2\Theta_k), \tag{3}$$

where Θ_k is the angle between H_i and the wavevector k of spinwaves and $D^* = D/4\pi M_s$ is the normalized exchange constant.

The factor in brackets in eq. (2) is a coupling function of the initially excited $k \approx 0$ mode to the degenerate modes of the low-k part of the SWM. The low-k part of the SWM is defined by the condition: $D^* k^2 \ll 1$. For YIG, having $D^* = 2.8\cdot 10^{-12}$ cm^2, we assume $k < 5\cdot 10^4$ cm^{-1} ($\Theta_k \to \Theta_o$) and eq. (3) reduces to:

$$\Omega_k^2 \approx \Omega_H^{\,2} + \Omega_H \sin^2\Theta_o. \tag{4}$$

The derivative $d\Omega_k/d\Omega_H$ in eq. (2) calculated from eq. (4) is:

$$\frac{d\Omega_k}{d\Omega_H} = \frac{1}{2}\left(\frac{\Omega}{\Omega_H} + \frac{\Omega_H}{\Omega}\right). \tag{5}$$

with $\Omega = f/\gamma\cdot 4\pi M_s$, the normalized frequency of the initially excited mode. The spinwave density function, ρ, for $k \to 0$ in eq. (2) is[12]:

$$\rho = \frac{\Omega}{\Omega_H \cos\Theta_o}. \tag{6}$$

The theoretical treatment of porosity induced scattering is much more involved because the local fluctuations are not randomly distributed as is the case for anisotropy. It is necessary, first, to find an appropriate model for the form of the pores and inclusions. The assumption of a spherical pore seems to be adequate for porous polycrystalline ferrites. The inhomogeneous static magnetic fields H_p around a spherical pore lie between $-2/3\cdot 4\pi M_s$ and $+1/3\cdot 4\pi M_s$. The pattern of the local resulting field depends on the applied static field. Over a considerable volume fraction of the porous ferrite sample the effective fields lie in the transition region between the IG- and the SW-model, $1<|H_p/4\pi M_s|<0.1$. Thus neither of these models is strictly applicable. However, a large part of the sample endures perturbations, $H_p/4\pi M_s<0.1$, making the SW-model preferable. However, the predictions of the SW-model for porosity induced line broadening have to be modified due to the strong and locally varying pertubation fields at the vicinity of the pores. Moreover, the field inhomogeneities around pores have significant Fourier components also at high wave numbers, thus exchange dominated spin waves have to be considered, too. Due to the complex interdependence of the perturbed static and rf. fields around pores a satisfactory theoretical result for the problem of two-magnon-scattering at pores is not available yet. Calculations[4,13,14] and experiments[15] have shown that ΔH_p is in general proportional to the porosity inhomogeneity parameter $p \cdot 4\pi M_s$, where p is the ratio of the volume of pores and inclusions and the volume of the ferrite sample. However, the observed frequency dependence of ΔH_p measured at the FMR[10,16] and the field dependence of ΔH_{eff} with predominant porosity broadening[11,17] are almost inverse to the density of states function ρ, eq. (6), which indicates strong modifications in the dispersive region of the SWM.

In this paper, two-magnon-scattering processes are investigated. For this purpose extensive FMR linewidth and some FMR lineshift measurements on spheres of dense polycrystalline YIG and CaVIn substituted YIG's are presented. Experimental results are compared with theoretical predictions and with reported ΔH_{eff} data from off-resonance measurements. Some limits of the theory for ΔH_a are found and more experimental insight into the problem of porosity linebroadening is gained.

III EXPERIMENTAL AND RESULTS

The resonance linewidth of the different spheres has been measured at frequencies between 300 MHz and 10 GHz in a nonresonant microwave coupling circuit where the ferrimagnetic sphere is located close to the short-circuit of a modified 50Ω coaxial line. The FMR linewidth was determined from the Lorentzian shaped, reflected microwave signal using a microwave network analyzer. The evaluation procedure for strong and weak r.f. coupling to the FMR is described in ref. 10.

In order to extend the range that is accessible by changing the frequency of the UPR- or 110-mode above and within the SWM, two non-uniform low order magnetostatic modes, the 220- and 210-modes, have been included in the FMR linewidth investigation. For any given field, H_i, the frequencies of the 220- and 210-modes lie above, resp. below, that of the 110-mode. The normalized frequencies of these modes are given by:[3]

$$f_{nmo}/\gamma 4\pi M_s = \Omega_{nmo} = \Omega_H + F_{nmo} \quad (7)$$

where n,m are the mode indices according to Walker's notation and F_{nmo} for n - m = 0 or 1 is given by:

$$F_{nmo} = m/(2n + 1). \quad (8)$$

The field $\Omega_{H\ crit}$ at which these modes reach the top of the SWM: $k \to 0$, $\Theta_o = \pi/2$, is found from eq. (4) with $(\Omega_k - \Omega_H) = F$:

$$\Omega_{H\ crit} = \frac{F^2}{1-2F} \quad (9)$$

and the cosine of the spinwave angle Θ_o is:

$$\cos^2\Theta_o = 1 - 2F - \frac{F^2}{\Omega_H} \quad . \quad (10)$$

At $\Omega_{H\ crit} = \Omega_H$ we find $\cos\Theta_o = 0$ and the theoretical results eq. (2) and (6) become infinite, thus a strong maximum of the linewidth is predicted at this point.

The investigations were carried out on highly polished spheres of dense polycrystalline pure YIG and CaVIn-substituted YIG with diameters between 0.6 and 0.7 mm. The In-substitution on octahedral lattice sites in YIG reduces the anisotropy field to very low values. The Ca-substitution on dodecahedral lattice sites is necessary to prepare polycrystalline pure phase In-substituted YIG materials with practically theoretical density, $p \approx 0.05\%$. Vanadium is used for charge compensation. The preparation techniques and the material parameters of the investigated garnets are compiled in ref. 18.

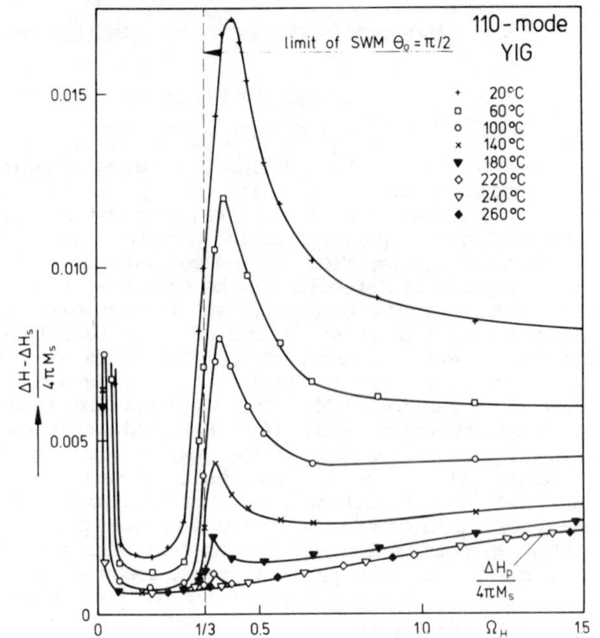

Fig. 1. Normalized FMR linewidth broadening of the 110-mode versus Ω_H for a polycrystalline YIG sphere at different temperatures.

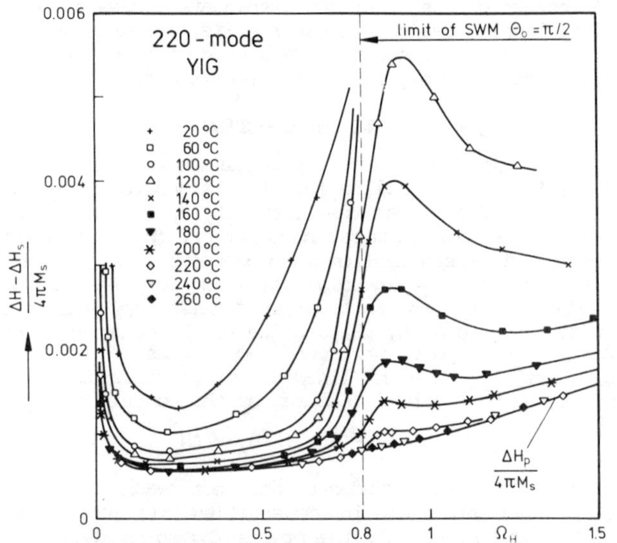

Fig. 2. Normalized FMR linewidth broadening of the 220-mode versus Ω_H for a polycrystalline YIG sphere at different temperatures.

In all samples the grain sizes range from 2 μm to 30 μm. Annealing at temperatures between 1000 °C and 1100 °C in oxygen atmosphere yielded in many cases a considerable reduction of ΔH, particularly in the low field range. The scatter of the ΔH data measured on different spheres made from the same material usually remained within a few tenth of an Oersted, the same holds for the different sphere diameters between 0.3 mm and 1.5 mm.

The investigations on pure YIG have been carried out at different temperatures between 20 °C and 260 °C. $\Delta H_s(f,T)$, $H_a(T)$ and $4\pi M_s(T)$ of single crystal YIG are used as reference data[10] to determine ΔH_a and ΔH_p. In the CaVIn-substituted YIG mainly H_a is varied by different amounts of indium substitution.

Fig. 1 presents linewidth data for the 110-mode, measured as a function of frequency at different temperatures on polycrystalline YIG; the results $\Delta H(f,T)$ have been normalized by subtracting the single crystal linewidth data $\Delta H_s(f,T)$ and by dividing by $4\pi M_s(T)$, and are plotted versus the normalized internal field Ω_H calculated with eq. (7) and (8). The linewidth data measured for the 220- and 210-mode have been treated in

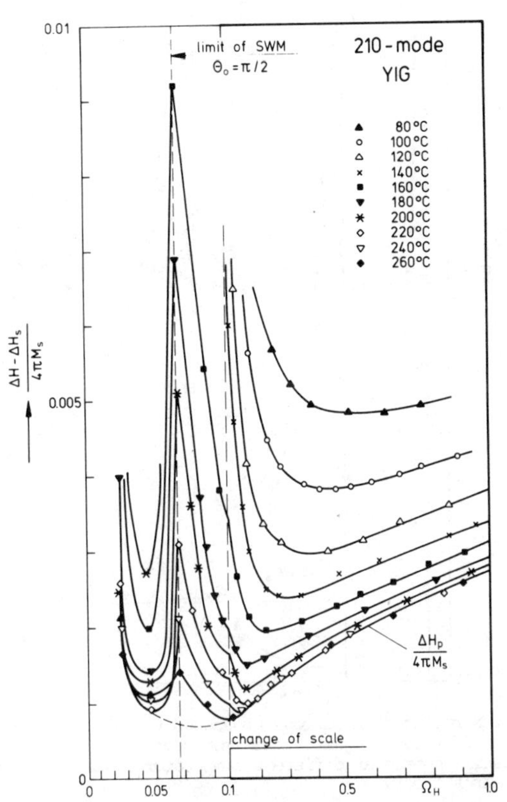

Fig. 3. Normalized FMR linewidth broadening of the 210-mode versus Ω_H for a polycrystalline YIG sphere at different temperatures.

the same way and are presented in fig. 2 and 3. The normalized linewidths in fig. 1-3 have a pronounced temperature dependent maximum occurring very close to the field $\Omega_{H\ crit}$, where each of the modes reaches the top of the SWM and the maximum of the SW density. At temperatures above 200 °C the anisotropy field is negligible and the linewidth peaks vanish. The data for T>220 °C coincide to one limiting curve which represents the remaining normalized porosity broadening contribution $\Delta H_p/4\pi M_s$. The differences between these limiting curves and the reduced linewidth curves for the various investigated temperatures result from anisotropy broadening and are discussed in the next section.

Fig. 4 shows $\Delta H(f)$ data for a CaVIn-substituted YIG sphere for the 210-, 110- and 220-mode plotted versus Ω_H. The resonance frequencies of the investigated modes and the fields ($\Omega_{H\ crit}$) at which they enter the SWM are indicated in the insert of fig. 4. The observed maxima of the linewidth lie close to $\Omega_{H\ crit}$. The unknown values of H_a and ΔH_a of this polycrystalline substituted YIG have been obtained with Schlömann's result, eq. (2), using trial values for H_a. The result of this operation is indicated by the hatched line for $(\Delta H-\Delta H_a)$. Measured ΔH_s data of $Y_3Fe_{5-x}Ga_xO_{12}$ single crystal spheres for $0<x<0.7$ with $4\pi M_s>600$ Gauss are included in fig. 4. The data fall on one curve when plotted versus Ω_H and may serve as a reference for intrinsic (and surface pit) losses[19]. Assuming $\Delta H_p \propto p\cdot 4\pi M_s$[13,14], the residual porosity, p=0.07%, was derived by comparing the evaluated ΔH_p at Ω_H=1.5 with the results for pure YIG (figs. 1,9, p=0.17%). Fig. 5 presents ΔH results of the 110-mode for a group of $CaV_{0.5}In_y$-substituted YIG's plotted versus Ω_H. With increasing amount of Indium substitution H_a and ΔH_a decrease. Hence, ΔH_a decreases just like ΔH in pure YIG where H_a is varied by changing the temperature.

Experimental lineshift data were obtained using the set-up for linewidth measurements. The phase angle of the complex reflection factor at the FMR, measured with the microwave network analyzer, was recorded versus the linearly varied applied magnetic field at different constant frequencies in steps of 100 MHz between 1.6 GHz

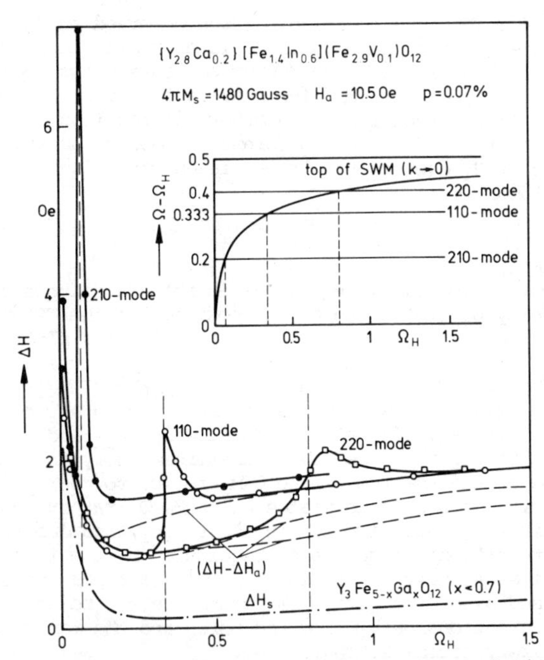

Fig. 4. Measured FMR linewidth ΔH of different magnetostatic modes versus Ω_H of a polycrystalline CaVIn-substituted YIG sphere. The insert demonstrates where the investigated modes reach the top of the SWM.

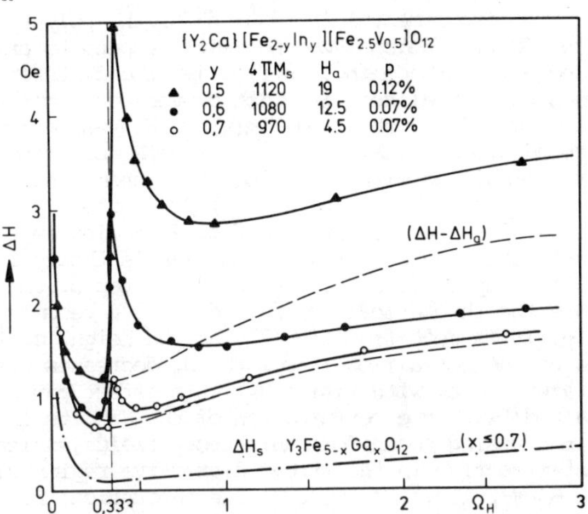

Fig. 5. Measured FMR linewidth ΔH of the 110-mode versus Ω_H of polycrystalline CaVIn-substituted YIG's. H_a and ΔH_a decrease with rising amount of indium substitution.

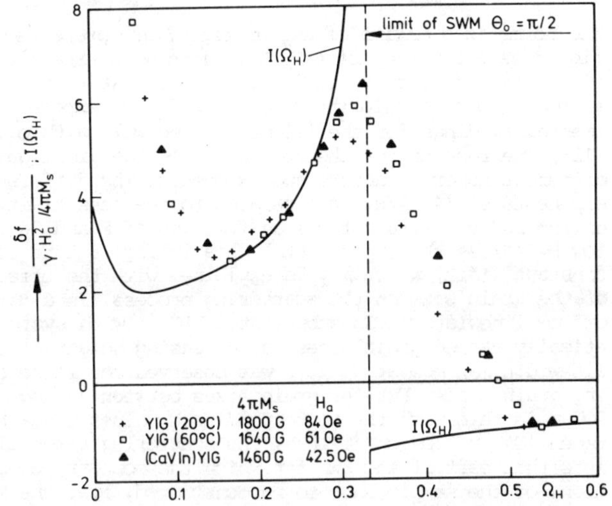

Fig. 6. Normalized experimental lineshift δf of the UPR-mode versus Ω_H for spheres of polycrystalline YIG and $Ca_{0.6}V_{0.3}In_{0.3}$ substituted YIG compared with Schlömann's theory.

and 4.2 GHz. The positions of the observed zeros (resonance) of the phase angle deviate from a linear pattern of 100 MHz marks. These deviations correspond to δf. The shift, δf, of the FMR line results from coupled resonance modes repelling each other. δf resulting from coupling of the UPR mode to spinwave modes at nonuniformities of H_i due to anisotropy fields was calculated by Schlömann[5]:

$$\delta f = \gamma \cdot (H_a^2/4\pi M_s) \cdot I(\Omega_H). \qquad (11)$$

The function $I(\Omega_H)$, eq. (50a) in ref. 5, depends on the density of states and on the position of the UPR mode with respect to the SWM. Normalized δf data plotted versus Ω_H are presented in fig. 6 for polycrystalline YIG at 20 °C and 60 °C and for a $Ca_{0.6}V_{0.3}In_{0.3}$-substituted YIG. The experimental data in fig. 6 agree quite well with eq. (11) at fields below $\Omega_{H\ crit}$ and well inside of the SWM. The deviations from theory in the region around the top of the SWM are due to the finite density of available states at $\cos\Theta_0 \approx 0$ (rather than the infinite density implied in the theory) and due to the perturbation[6] of spinwave frequencies by the inhomogeneities. At very low fields, $\Omega_H < 0.15$, a rapid increase of δf and also of ΔH (fig. 1) occurs. Here, the relative perturbations H_e/H_i from the pores and the random anisotropy fields strongly rise.

IV ANISOTROPY SCATTERING

The experimental ΔH results for the polycrystalline pure YIG will be analyzed now with respect to anisotropy induced linebroadening. In fig. 1-3 the differences between the normalized linewidth curves for temperatures T<220 °C and the observed limiting $\Delta H_p/4\pi M_s$ curves correspond to $\Delta H_a/4\pi M_s$. The evaluated ΔH_a contributions were normalized with the anisotropy inhomogeneity parameter P_a using $H_a(T)$ and $4\pi M_s(T)$ of single crystal YIG. The results are presented in fig. 7 showing an excellent quantitative agreement with Schlömann's theory for ΔH_a, eq. (2), apart from the region at the top of the SWM where the theoretical spinwave density ρ becomes infinite. It is seen from fig. 7 that the height of the maxima of the normalized linewidth ΔH_a decreases with temperature, i.e. with rising H_a. This effect can be explained[7] with the perturbation of the SWM itself caused by the rising nonuniform anisotropy fields, which is most effective in the strong dispersive region close to the $\Theta_0 = \pi/2$ limit of the SWM. The theoretical result for the maximum of ΔH_a at $\Omega_{H\ crit}$ of ref. 7 for the UPR-mode is:

$$(\Delta H_a)_{max} \ / \ (H_a^2/4\pi M_s) = 1.74\ (4\pi M_s/H_a)^{\frac{1}{2}}. \qquad (12)$$

The normalized maxima of ΔH_a of fig. 7 are presented in fig. 8 as a function of $4\pi M_s/H_a$, along with experimental ΔH_a results for the CaVIn-substituted YIG's from fig. 4 and 5 and from a forthcoming paper[19]. The experimental findings for the 110-mode agree well with eq. (12). The results for the 220- and 210-mode yield quite different numeric factors and exponents. The different exponents of $(4\pi M_s/H_a)$ are related to the sensitivity of the SWM to perturbations as function of F and Ω_H. For low H_a values the experimental data in fig. 8 tend to a constant limit, which may be explained with the effects of the grain size on the scattering process. In a study of low linewidth CaVSn-substituted YIG [20] with systematically varied grain sizes, a decreasing height of the linewidth maximum at $\Omega_{H\ crit}$ was observed for increasing grain sizes. For the grain sizes between 2µ and 30 µm in this work the corresponding k values range between 10^4 and 10^3 cm^{-1}. Thus, the scattering takes place into that part of the SWM for which the boundary conditions of the sample have to be considered, i.e. the MSM. A detailed study[21] showed that the maximum of the density of MSM, d_{max}, at $\Omega_H = \Omega_{H\ crit}$ depends on F:

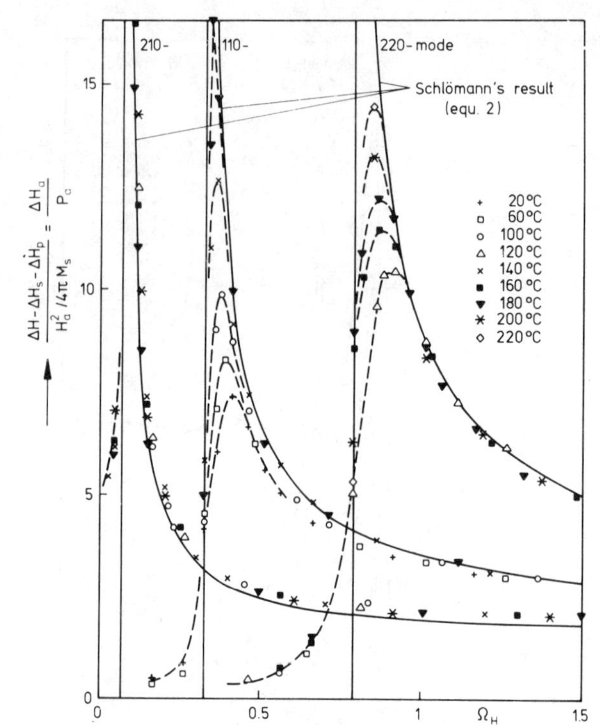

Fig. 7. Normalized anisotropy broadened FMR linewidth contributions of different magnetostatic modes versus Ω_H for a polycrystalline YIG sphere at different temperatures compared with Schlömann's theory.

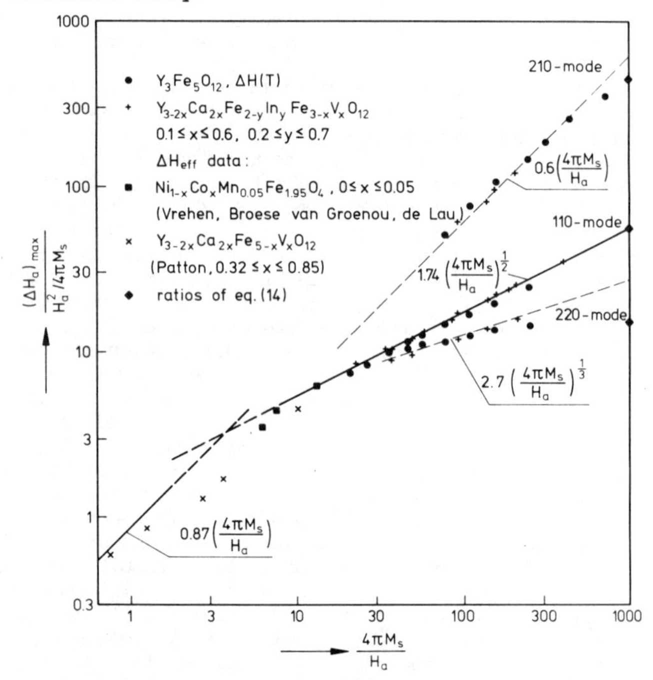

Fig. 8. Normalized maxima of the experimental ΔH_a results at the top of the SWM versus the inhomogeneity ratio $4\pi M_s/H_a$ for different magnetostatic modes and various polycrystalline ferrite compositions. The solid lines represent the theoretical results for $(\Delta H_a)_{max}$, eq. (12) (ref. 7) and eq. (15) (ref. 6). The ΔH_{eff} data for $H_a > 0.1 \cdot 4\pi M_s$ (of ref. 7 and 11) indicate strong perturbations of the SWM.

$$d_{max} \propto \frac{(1 - 2F)^2}{F(1 - F)}. \qquad (13)$$

Eq. (2) yields with eq. (9) and $\rho \to d_{max}$:

$$\frac{(\Delta H_a)_{max}}{H_a^2/4\pi M_s} \propto \left(\frac{1 - 2F}{F}\right)^2 + \left(\frac{1 - 2F}{1 - F}\right)^2. \qquad (14)$$

The ratios of the normalized $(\Delta H_a)_{max}$ from eq. (14) for the 220-, 110- and 210-modes are 0.29:1:7.6. These ratios are presented in fig. 8 at low H_a, fitted for the 110-mode. They explain qualitatively the different values of $(\Delta H_a)_{max}$ observed for the investigated modes (see also fig. 4).

Additionally fig. 8 presents $(\Delta H_{eff})_{max}$ data from ref. 7 and 11 for different ferrites with low porosities and $0.75<4\pi M_s/H_a<13.6$ in order to demonstrate the limits of validity of the IG- and SW-model. The perturbations of the SWM increase more strongly than predicted by eq. (12) if $H_a \gtrsim 0.1 \cdot 4\pi M_s$; a higher value, $H_a \gtrsim 4\pi M_s/3$ was observed by Van Hook and Euler[22] for the transition from the SW-model to the IG-model when the UPR frequency lies well inside the SWM ($\Omega_H>3$). These different results are consistent with the field dependent sensitivity of the SWM to perturbations. For $H_a \approx 4\pi M_s$ the ΔH_{eff} data agree well with Schlömann's result[6] for large anisotropy:

$$\Delta H_a = 0.87\ H_a \ . \qquad (15)$$

As can be seen from fig. 1-3 and 7, anisotropy line broadening is also observed in the region above the top of the SWM, i.e. $\Omega_H<\Omega_{H\ crit}$. This is the region of surface modes (spinwaves and MSM), their densities[12,21] must be regarded in the region closely below $\Omega_{H\ crit}$ for two-magnon-scattering. Scattering to exchange dominated states, $k>10^5$, seems to be negligible because of the close agreement of ΔH_a with the theory based on low-k states. However, at very low internal fields, $\Omega_H<0.2$, the observed increase of the linewidth contribution ΔH_a above the limiting $\Delta H_p/4\pi M_s$ curves in fig. 1-3 is likely caused by scattering to high-k states due to domain built up in the crystallites[23]. A rapid rise of ΔH occurs at $\Omega_H \approx H_a/4\pi M_s$, where the sample is no longer saturated.

V POROSITY SCATTERING

The porosity contributions ΔH_p of the 210-, 110- and 220-mode for the polycrystalline YIG as inferred from the results in fig. 1-3 are presented in fig. 9 in normalized units for an extended range of Ω_H. The value of p = 0.17% used for the normalization of the data in fig. 9 was acquired by a comparison of the ΔH_p data from resonance measurements with off-resonance ΔH_{eff} results of Vrehen and Broese van Groenou (ref. 24). Their ΔH_{eff} data measured at 9 GHz on YIG with p between 0.6% and 4% were corrected for anisotropy broadening and normalized with $p \cdot 4\pi M_s$. Fig. 10 presents their results as a function of Ω_H together with ΔH_p obtained from the 210-,

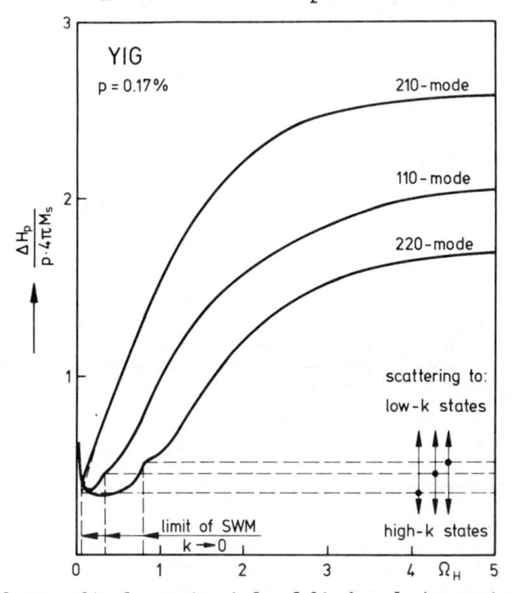

Fig. 9. Normalized porosity induced linebroadening contribution ΔH_p versus Ω_H of a polycrystalline YIG sphere for different magnetostatic modes.

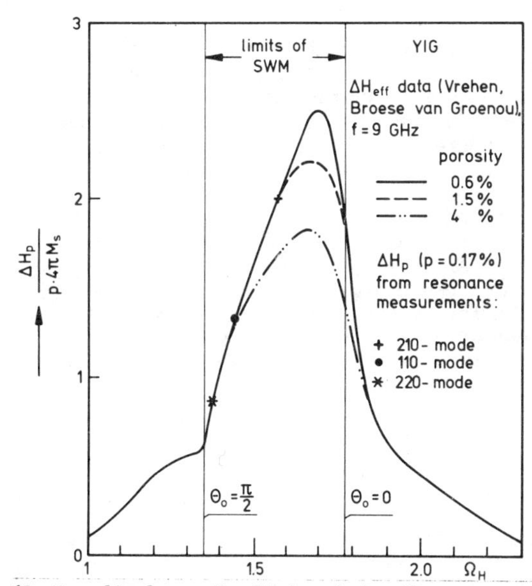

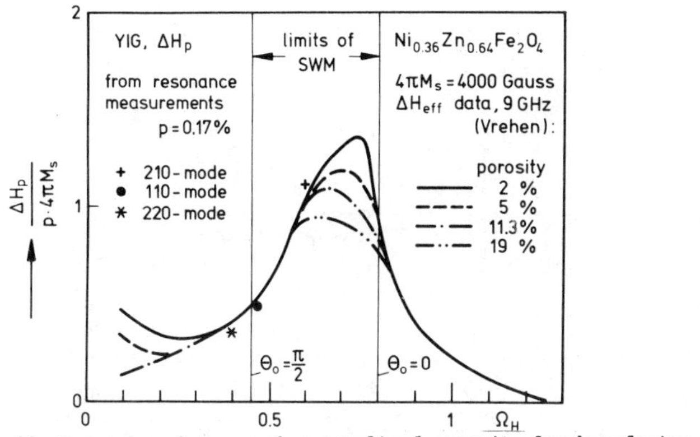

Fig. 10. Normalized porosity linebroadening ΔH_p versus Ω_H for polycrystalline YIG. Off-resonance ΔH_{eff} data for different porosities (ref. 24), corrected for anisotropy broadening, are compared with ΔH_p from resonance measurements of the 210-, 110- and 220-mode. Excellent agreement is found between the different experimental methods.

110- and 220-mode. Excellent agreement is found between these entirely different experimental methods if p=0.17% is used. The same value of p was obtained from comparison with Patton's[11] ΔH_{eff} data for YIG measured at 9.96 GHz and with Vrehen's[17] ΔH_{eff} data for NiZn ferrites. The comparison of ΔH_p of YIG with ΔH_{eff} of NiZn ferrites is given in fig. 11 and shows also a remarkably good agreement between the resonance and off-resonance measurement techniques at quite low Ω_H for materials with quite different $4\pi M_s$.

Fig. 10 and 11 demonstrate that porosity scattering is not neccessarily linearly related to p; towards the lower limit of the SWM, $\Theta_0 \to 0$, the normalized ΔH_p curves are not independent of p. The strong absorption just below the lower limit of the SWM, indicates strong perturbations of the eigenfrequencies of spinwaves and MSM, since in a homogeneous medium[25,26] there are no modes available in this region.

At frequencies above the top of the (k→0) SWM scattering takes place only to exchange dominated high-k spinwave states as can be seen from fig. 9 and 10,11. The different values of ΔH_p observed for the 210-, 110-, 220-mode in fig. 9 at the top of the SWM reflect the increase of k_{max}, obtained from eq. (3) with $\Theta_k = 0$:

$$k_{max} = (F/D^*)^{\frac{1}{2}} = 0.6 \cdot 10^6 \cdot F^{\frac{1}{2}}\ cm^{-1} . \qquad (16)$$

Fig. 11. Comparison between the normalized porosity linebroadening contributions ΔH_p derived from ΔH (FMR) of YIG and from off-resonance ΔH_{eff} of NiZn ferrites with different porosities (ref. 17). The data for materials with different $4\pi M_s$ agree well.

At frequencies within the SWM additional porosity induced scattering to low-k states occurs, which is related to the dispersion properties of spinwaves with $k<5\cdot10^4$ cm^{-1}, eq. (4). This allotment of ΔH_p into contributions due to independent scattering to high-k and to low-k states is indicated in fig. 9 and can be explained with some reasonable arguments on the structure of the effective inhomogeneous fields H_p around pores. Expanding the locally strong variations of H_p around a single pore into a Fourier integral of equivalent wavelengths yields a maximum of the equivalent spectrum at high-k values, which is the required scattering reservoir for two-magnon-scattering into high-k spinwaves. The scattering, to low-k spinwaves results from periodic perturbations of the internal static field due to an average distance a_p between neighbouring pores. Assuming an isotropic distribution of the pores with approximately equal diameter d_p gives:

$$a_p \approx d_p \cdot p^{-\frac{1}{3}}. \qquad (17)$$

For pore sizes in the μm range and porosities between 10^{-1} and 10^{-3} the main part of the equivalent Fourier spectrum of the periodic perturbations with recurrence distance a_p lies in the required range $k<5\cdot10^4$ cm^{-1}. This contribution, $\Delta H_p(k\to0)$, has been extracted from the results of fig. 9 and is presented in fig. 12 as a function of $\cos\Theta_o$. Results for a $Ca_{0.2}In_{0.7}V_{0.1}$ substituted YIG, $H_a\approx2$ Oe, $4\pi M_s$ = 1460 G, p = 0.28%, from ref. 19 have been included and agree well with findings on YIG. For $\Omega_H\to\infty$, $\Delta H_p(k\to0)$ is proportional to the respective maximum value of $\cos\Theta_o$.

For finite Ω_H, however, all modes behave differently. This may be explained by the different sensitivity[10] of $\cos\Theta_o$ (density of states) on local field variations and by the different penetration depth into the SWM of the modes shown in fig. 12. The frequency and field dependence of ΔH_p contradict the known theoretical predictions based on the SW-model and the density of states function[14,15]. $\Delta H_p(\Omega_H)$ of fig. 9 behaves almost contrarily to $\Delta H_a(\Omega_H)$ of fig. 7. However, if $H_a\approx4\pi M_s$, it is comparable with the effective fields H_p around pores. For this case Patton's "effective linewidth results for predominant anisotropy broadening exhibit similar characteristic features as the experimental findings for porosity broadening, i.e. the absence of a maximum of the linewidth at the top of the SWM and an increase of the linewidth in the range towards the bottom of the SWM. A satisfying theory for the field or frequency dependence of ΔH_p has not been found yet.

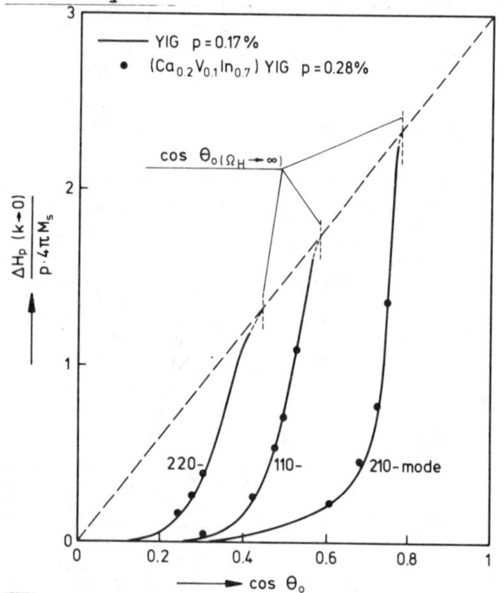

Fig. 12. Normalized porosity induced FMR linebroadening contribution ΔH_p (k→0) versus $\cos\Theta_o$ for the 220-, 110- and 210-mode in polycrystalline YIG and CaVIn-substituted YIG

VI CONCLUSIONS

The broadening of the FMR line in polycrystalline ferrites can be separated into non interacting anisotropy and porosity induced two-magnon-scattering processes. ΔH measured at the FMR agrees quantitatively with off-resonance ΔH_{eff} data. The good agreement between these entirely different experimental methods provides an expanded understanding of the influence of the in-inhomogeneity parameters involved in FMR losses.

For $H_a<0.1\cdot4\pi M_s$ the experimental ΔH_a results are proportional to the "dipole narrowed" inhomogeneity parameter $P_a = H_a^2/4\pi M_s$. Excellent quantitative agreement was found with Schlömann's theory based on the (k→0) SW-model apart from a narrow region close to the top of the SWM. With rising normalized inhomogeneity, $H_a/4\pi M_s$, the experimentally observed deviations from theory extend deeper into the SWM. This effect results from broadening of the spinwave frequencies by the inhomogeneity. Finally for $H_a>4\pi M_s$ the IG-model applies.

The frequency and field dependence of ΔH_p contradicts theoretical predictions based on the SW-model and behaves almost inverse to the density of states function ρ. ΔH_p is not always proportional to the porosity parameter p over the whole range of the SWM. Deviations from a linear relation to p arise at the bottom of the SWM, they extend towards the top of the SWM with rising p. Porosity induced scattering takes place to low-k and to high-k states which can be qualitatively explained with some reasonable arguments on the structure of pores. However, a satisfactory theory to describe the frequency and field dependence of ΔH_p has not been found yet. The experimental results provide a guide to determine the values of H_a and p of polycrystalline ferrites with quite good accuracy[19] from ΔH-measurements.

ACKNOWLEDGEMENTS

I wish to thank Dr. G. Winkler for providing the polycrystalline garnet spheres and Dr. W. Tolksdorf for the single crystal spheres. The outstanding properties of their garnet materials are the base of many experimental findings of this work. The help with the measurements by Mrs. H. Runge and J. Kanzenbach is gratefully acknowledged. The author thanks Prof. C. E. Patton, Colorado State University, and Dr. E. Schlömann, Raytheon Res. Div., and Dr. H. Dötsch for valuable discussions and helpful suggestions, Dr. W. Schilz and Dr. H. J. Schmitt for the critical reading of the manuscript and Mrs. U. Maurer and Mrs. U. Zacher for the careful preparation of the drawings and the text.

REFERENCES

1. S. Geschwind, A.M. Clogston, Phys. Rev. 108, 49(1957)
2. A.M. Clogston, H. Suhl, L.R. Walker, P.W. Anderson, J. Phys. Chem. Solids 1, 129(1956)
3. L.R. Walker, Phys. Rev., 105, 390(1957)
4. M. Sparks, Ferromagnetic Relaxation Theory, McGraw-Hill, New York 1964
5. E. Schlömann, J. Phys. Chem. Solids 6, 242 and 257 (1958)
6. E. Schlömann, Phys. Rev., 181(1969)
7. Q.H.F. Vrehen, A. Broese van Groenou, I.G.M. de Lau Phys. Rev. B, 1, 2332(1970)
8. C.R. Buffler, J. appl. Phys., 30S, 172 S (1959)
9. C.E. Patton, J. appl. Phys., 43, 2872(1972)
10. P. Röschmann, IEEE Trans. Magn., MAG-11, 1247(1975)
11. C.E. Patton, Phys. Rev., 179, 352(1969)
12. R.W. Damon, J.R. Eshbach, J. Phys. Ch. Solids, 19, 308(1961)
13. E. Schlömann, AIEE Spec. Publ., T-91, 600(1956)
14. M. Sparks, J. Appl. Phys. 36, 1570(1965)
15. Ch. Greinacher, Z. ang. Phys., 20, 381(1966)
16. H. Brand, H.W. Fieweger, IEEE Trans. MTT-13, 712(1965)
17. Q.H.F. Vrehen, J. appl. Phys., 40, 1849(1969)
18. G.Winkler,P.Hansen,P.Holst, Philips Res. Repts.,27, 151(1972)
19. P. Röschmann, G. Winkler, submitted to J. of Magn. Magn. Mat.
20. T.Inui,H.Takamizawa,N.Ogasawara,T.Fuse,AIP Conf.Proc.,24(1975)
21. P.Röschmann, H.Dötsch, to be published in Philips Res. Repts.
22. H.J. Van Hook, F. Euler, J. Appl. Phys., 40, 4001(1969)
23. A.Globus,P.Duplex,M.Guyot, IEEE Trans. Magn.,MAG-7, 617(1971)
24. Q.H.F.Vrehen,A.Broese van Groenou, J.de Phys.,32S,C1-156(1971)
25. C.E. Patton, IEEE-Trans. Magn., MAG-8, 433(1972)
26. K.Motizuki,M.Sparks,P.E.Seiden, Phys. Rev., 140, A972(1965)

MAGNETIC CONTROL OF SURFACE ELASTIC WAVES IN A NOVEL LAYERED STRUCTURE

A. K. GANGULY, K. L. DAVIS, D. C. WEBB and C. VITTORIA
Naval Research Laboratory, Washington, D. C. 20375

ABSTRACT

In this paper we describe theoretical and experimental results for magnetoelastic surface wave propagation in a new layered structure consisting of a ferromagnetic film deposited on a piezoelectric substrate between two interdigital transducers. The surface wave velocity in this structure can be varied continuously by adjusting an external d.c. magnetic field. The velocity change occurs predominantly in the bias field range where the static magnetization vectors are undergoing rotation. We consider three different orientations of the external field both tangential and normal to the plane of the film. Propagation characteristics are significantly different in the three orientations. The calculated velocity changes are in good agreement with changes measured on a Ni-$LiNbO_3$ delay line.

INTRODUCTION

Magnetoelastic coupling in ferromagnetic materials offers a means of changing the velocity of acoustic waves continuously by applying an external d.c. magnetic field. This mechanism has been studied in bulk wave,[1] Love wave[2] and Rayleigh wave[3] geometries. These devices are intrinsically dispersive and have narrow-band operation. In this paper we describe the main characteristics of surface wave propagation in a new layered structure in which acoustic velocity is changed by adjusting an applied magnetic field in a manner resulting in low dispersion and broad-band operation. The basic structure consists of a magnetostrictive thin film deposited on a surface acoustic wave (SAW) substrate between two interdigital transducers. The surface acoustic wave is launched in the conventional manner and propagates into the layered region. Due to magnetoelastic coupling in the film and the substrate, the surface wave velocity in the layered structure changes as an external magnetic field is applied. This change in velocity is caused mainly by changes in the elastic constants of the magnetostrictive material as the magnetization vectors rotate with applied field. We are able to operate at frequencies where dispersion is very small since it arises mainly from acoustic velocity mismatch between the film and the substrate.

We will consider three orientations of the external field H_0: (i) field in the plane of the film and parallel to the direction of surface wave propagation, (ii) field in the plane of the film but normal to the direction of propagation and (iii) field perpendicular to the plane of the film. We make the following simplifying assumptions: (a) the film is magnetically and elastically isotropic; (b) the exchange effects may be neglected since the wavelengths of interest to us are much greater than atomic spacing; (c) the electric and magnetic fields may be treated under quasi-static approximation because the wavelengths are smaller than free space electromagnetic wavelengths. The last approximation is valid only in insulating materials. However, we expect the approximation to hold for a thin metal film when its thickness is smaller than the skin depth.[5]

EQUATIONS OF MOTION AND DISPERSION RELATION

The geometrical configuration of the layered structure is shown in Fig. 1. The origin of the coordinate system is taken at the film-substrate interface. h is the thickness of the film. The film is parallel to the x_1x_2-plane. The substrate occupies the region $x_3 < 0$. x_1-axis is along the direction of propagation. The three orientations of the external d.c. magnetic field (H) to be considered are along the three coordinate axes.

In order to study coupled magnetoelastic waves we must solve the equations of motion for the magnetization vector ($\vec{M}$) and the particle displacement ($\vec{U}$) in conjunction with Maxwell's equations. The equations of motion may be routinely obtained from the free energy (W) expression. W may be written as the sum of the following terms[6,7]:

$$W_z = -\mu_o M_o \vec{\alpha} \cdot \vec{H} \quad \text{(Zeeman energy)}, \tag{1}$$

$$W_d = 1/2\, \mu_o M_o^2 \alpha_3^2 \quad \text{(demagnetizing energy)}, \tag{2}$$

$$W_e = 1/2\, \Sigma\, c_{ijk\ell} S_{k\ell} \quad \text{(elastic energy)}, \tag{3}$$

$$W_k = B \sum_{ij} \alpha_i \alpha_j S_{ij} \quad \text{(magnetoelastic energy)}, \tag{4}$$

$$W_\sigma = -\frac{3\lambda\sigma}{2} (\vec{\alpha} \cdot \vec{n})^2 \quad \text{(energy due to uniaxial stress } \sigma\text{)}. \tag{5}$$

In the above $\vec{M} = M_o \alpha$, M_o the saturation moment, $\vec{\alpha}$ the direction cosines of $\vec{M}$, μ_o vacuum permeability, H the magnetic field, $c_{ijk\ell}$ the element of the elastic tensor, S_{ij} element of strain tensor, B the magnetoelastic coupling constant of an isotropic medium. B is related to the magnetostrictive constant λ by $B = -3\lambda c_{2323}$. The term W_σ arises from an intrinsic stress σ which always exists in the plane of the film in deposited film structures.[8] This stress induces a preferential direction along which the magnetization vectors tend to align. Let $\vec{n}$ denote a unit vector along the "easy" axis. For $\lambda\sigma<0$, $\vec{n}$ is in a plane normal to the stress axis and for $\lambda\sigma>0$, $\vec{n}$ is parallel to the stress axis. Here we will consider the case $\lambda\sigma<0$. The Ni-$LiNbO_3$ experimental structure to be discussed later belongs to this class. The case $\lambda\sigma>0$ is discussed in a forthcoming paper.[9] In reality the intrinsic stress σ is radial rather than uniaxial. However for $\lambda\sigma<0$, the problem can be simplified by noting that from the symmetry of the stress, the magnetization vectors must initially point along $\pm x_3$ directions.

We will write all the field quantities, $\vec{M}$, $\vec{U}$, S, and $\vec{H}^{eff} = -(\mu_o M_o)^{-1} \nabla_\alpha W$, as the sum of d.c. and rf parts. Under linear approximation, we have for the d.c. and rf components of $\vec{H}^{eff}$:

$$\vec{H}_o^{eff} = \vec{H}_o - \hat{x}_3 \alpha_3 M_o + |\beta| M_o \vec{n}(\vec{\alpha}_o \cdot \vec{n}) - (2B/\mu_o M_o) S^o \cdot \vec{\alpha}, \tag{6}$$

$$\vec{h}^{eff} = \vec{h} + |\beta| \vec{n}\, (\vec{m} \cdot \vec{n}) - (2B/\mu_o M_o)\, (s \cdot \vec{\alpha}_o + S^o \cdot \vec{m}/M_o), \tag{7}$$

where $\beta = 3\lambda\sigma/\mu_o M_o^2$ and $\vec{x}_3$ is a unit vector along x_3-axis. At equilibrium $\vec{M}_o \| \vec{H}^{eff}$. In the expression for $\vec{h}^{eff}$ only the term $\vec{h}$ enters in Maxwell's equations and the electromagnetic boundary conditions.

The equilibrium values of S^o and $\vec{\alpha}_o$ may be obtained by minimizing the energy W with respect to α_{oi} ($i = 1,2,3$) under the restrictions $|\vec{M}|$ = constant and $\sum_{i=1}^{3} S_{ii}^o = 0$. We then have[7]

$$S_{ii}^{(o)} = 3\lambda(\alpha_{oi}^2 - 1/3)/2$$
$$S_{ij}^{(o)} = S_{ji}^{o} = 3\lambda\alpha_{oi}\alpha_{oj}/2. \quad (8)$$

The values of α_{oi} are shown below for the three orientations of H_o in the case $\lambda\sigma<0$.

$$\begin{array}{ll} H_o \| x_1 \text{ or } x_3 & H_o \| x_2 \\ \alpha_{01} = \sin\theta & \alpha_{01} = 0 \\ \alpha_{02} = 0 & \alpha_{02} = \sin\theta \\ \alpha_{03} = \cos\theta & \alpha_{03} = \cos\theta \end{array} \quad (9)$$

where θ is the angle between $\vec{M}_o$ and the x_3-axis. For H_o along x_1 or x_2-axis, the magnetization vectors have equal probability of pointing in the $\pm x_3$ direction and the resultant demagnetizing field is zero. Then

$$\sin\theta = H_o^{(e)}/|\beta|M_o \quad \text{for } H_o \le |\beta|M_o$$
$$= 1 \quad \text{for } H_o > |\beta|M_o. \quad (10)$$

From Eqs. (8-10) and (6), we get

$$H_o^{eff} = (|\beta|+2\beta_1/3)M_o \quad \text{for } H_o \le |\beta|M_o$$
$$= H_o^{(e)}+2\beta_1 M_o/3 \quad \text{for } H_o > |\beta|M_o, \quad (11)$$

where $\beta_1 = B^2/\mu_o M_o^2 c_{2323}$. For H_o along x_3-axis, the magnetization vectors in the $-x_3$ direction are initially reversed by the displacement of 180° walls.[7] This process continues until $H_o = M_o$, i.e., the external field is just sufficient to overcome the demagnetizing field. In this case

$$\theta = 0 \text{ for } H_o > M_o \quad (12)$$

and

$$H_o^{eff} = H_o+(|\beta|-1+2\beta_1)M_o. \quad (13)$$

The equation of motion for $\vec{M}$ is

$$\frac{d\vec{M}}{dt} = -\mu_o|\gamma|(\vec{M}\times\vec{H}^{eff}) \quad (14)$$

where γ is the gyromagnetic ratio.

We assume that the time dependence of the rf quantities are of the form $e^{-i\omega t}$. Under linear approximation, we get

$$i\omega\vec{m} = |\gamma|\mu_o\vec{M}_o \times \left(\vec{h}^{eff} - \frac{H_o^{eff}}{M_o}\vec{m}\right). \quad (15)$$

The components of $\vec{m}$ can be obtained after some algebraic manipulations by substituting Eqs. (7-13) in Eq. (15). Then we get the following equation for the rf part of the magnetic induction vector $\vec{b} = \mu_o(\vec{h}+\vec{m})$

$$b_i = \mu_o \sum_j \mu_{ij} h_j - \sum_{jk} p_{ijk} S_{jk}, \quad (16)$$

where μ_{ij} is the relative permeability tensor and P_{ikj} a third-rank tensor due to magnetoelastic coupling. The tensors μ and p are given below for the three orientations of the magnetic field:

(a) H_o along x_2 axis

$$\mu = \begin{pmatrix} 1+\eta\tilde{\omega}_o/\omega & -i\eta\cos\theta & i\eta\sin\theta \\ i\eta\cos\theta & 1+\eta(\omega_o/\omega)\cos\theta & -\eta(\omega_o/\omega)\sin\theta\cos\theta \\ -i\eta\sin\theta & -\eta\frac{\omega_o}{\omega}\sin\theta\cos\theta & 1+\eta(\omega_o/\omega)\sin^2\theta \end{pmatrix}, \quad (17)$$

$$p = \frac{\eta B}{M_o}\begin{pmatrix} 0 & -ia_1\frac{\omega}{\omega_o} & ia_1\frac{\omega}{\omega_o} & -ia_2\frac{\omega}{2\omega_o} & \frac{\tilde{\omega}_o}{\omega}c_1 & \frac{\tilde{\omega}_o}{\omega}c_2 \\ 0 & a_1c_1 & -a_1c_1 & a_2c_1 & ic_1^2 & ic_1c_2 \\ 0 & -a_1c_2 & a_1c_2 & -a_2c_2 & ic_1c_2 & -ic_2^2 \end{pmatrix} \quad (18)$$

where $a_1=(\omega_o/\omega)\sin 2\theta$, $a_2=(\omega_o/\omega)\cos 2\theta$, $c_1=\cos\theta$, $c_2=\sin\theta$ $\eta=\omega\omega_m/(\omega_o\tilde{\omega}_o-\omega^2)$, $\omega_m=\mu_o|\gamma|M_o$, $\omega_o=\mu_o|\gamma|(H_o^{eff}+\beta_1 M_o/3)$, $\tilde{\omega}_o=\omega_o-\beta_1\omega_m\sin^2\theta$. In Eq. (18) the elements of the tensor p have been shown in the compressed matrix notation, i.e., $p_{ijk} = p_{iq}$.

(b) H_o along x_1 axis. μ and p for this case can be obtained from Eqs. (17) and (18) by a 90° clockwise rotation of the coordinate system about the z-axis.

(c) H_o along x_3 axis. In this case μ and p may be obtained from case (b) by setting θ = o.

From Eqs. (1-5) and (16) the components of the stress tensor $\tau_{ij} = \partial W/\partial S_{ij}$ are given by

$$\tau_{ij} = \sum_{k\ell}\left\{C_{ijk\ell} - \frac{B}{\mu_o M_o}(\alpha_{oi}p_{jk\ell}+\alpha_{oj}p_{ik\ell})\right\}S_{k\ell} + \frac{B}{M_o}\sum_k\left\{\alpha_{oi}(\mu_{jk}-\delta_{jk})+\alpha_{oj}(\mu_{ik}-\delta_{ik})\right\}h_k, \quad (19)$$

Equations (16) and (19) give the constitutive equation for the magnetoelastic fields.

The propagation characteristics of the layer modes will be determined by solving the following equations in the three regions (i) $-\infty<x_3<0$ (piezoelectric substrate), (ii) $0<x_3<h$ (ferromagnetic film) and (iii) $x_3>h$ (free space):

$$\nabla\cdot\vec{b} = 0 \quad (20)$$

$$\nabla\cdot\vec{D} = 0 \quad (21)$$

$$\rho\frac{\partial^2\vec{u}}{\partial t^2} = \nabla\cdot\vec{\tau} \quad (22)$$

where ρ is the density of the material and $\vec{D}$ the rf electric displacement vector. Equation (22) applies in regions (i) and (ii) only. The constitutive equations in the three regions are as follows:

$x_3<0$	$0<x_3<h$	$x_3>h$
$\tau=c\cdot S-e\cdot\vec{E}$	Eqs. (16) and (19)	$\vec{b}=\mu_o\vec{h}$
$\vec{b}=\mu_o\vec{h}$	and	$D=\epsilon_o\vec{E}$
$D=\epsilon\cdot\vec{E}+e\cdot S$	$D=\epsilon\cdot\vec{E}$	

(23)

ε denotes dielectric tensor, e piezoelectric tensor and $\vec{E}$ rf electric field. Under quasistatic approximation, we write $\vec{E}$ and $\vec{h}$ in terms of electric and magnetic potentials ϕ and ψ respectively

$$\vec{E} = -\nabla\phi \text{ and } \vec{h} = -\nabla\psi. \qquad (24)$$

Equations (20-24) are solved with the boundary conditions: (a) all field variables vanish at $x_3 = -\infty$, (b) $\tau_{13}, \tau_{23}, \tau_{33}, D_3, b_3$, ϕ and ψ are continuous at $x_3 = 0$ and (c) $\tau_{13} = \tau_{23} = \tau_{33} = 0$ but D_3, b_3, ϕ and ψ are continuous at $x_3 = h$.

The system of Eqs. (20-24) are solved by utilizing the state space approach of Fahmy and Adler[10] which reduces the wave propagation problem to a first order matrix differential equation. Details of this solution are discussed in Ref. 9.

RESULTS

The measured and computed phase shifts (θ_{ph}) of the acoustic waves are shown in Fig. 1 as a function of the bias field for a .85 micron Ni-film on YZ-cut $LiNbO_3$ delay line. The operating frequency is 210 MHz. The phase shift through the device was measured using a vector voltmeter.[4] θ_{ph} is calculated from the relation $\theta_{ph} = -\omega\ell\{V(0)-v(H_o)\}/v(0)v(H_o)$ where ℓ, the length of the film in the direction of propagation, is 1.5 cm. Since acoustic propagation is faster in the substrate than in the layer, the dispersion curves contain many branches.[9] The phase shifts in Fig. 1 refer to the lowest branch which is found to be the dominant mode. The parameters for N_i used are: $\lambda = -3.75\times10^{-5}$, $\mu_o M_o = .5$ Weber/m^2, $B = 1.07\times10^7$ N/m^2 and $\sigma = 6.2\times10^8$ N/m^2.

As stated earlier, the magnetization vector in the Ni-$LiNbO_3$ structure initially points along the x_3 direction.[3] When H_o acts along the x_1-axis, the moments rotate toward this direction. The velocity of acoustic waves first decreases, attains a minimum at $\theta = 42°$ and increases again towards a saturation value as the applied field is increased. When H_o is applied along the x_3 direction, the velocity does not change very much for $H_o<M_o$. For $H_o>M_o$ the velocity increases monotonically toward a saturation value. In the case of H_o acting along the x_2 axis the moments rotate from a direction where they couple with acoustic waves to a direction where they do not. As the applied field is increased, the velocity drops by a very small amount and then increases to its saturation value. The velocity changes are the same for propagation along $\pm x_1$ directions.

We have demonstrated the possibility of continuously varying the acoustic velocity in a ferromagnetic film - piezoelectric substrate structure. This device is essentially non-dispersive and has broad-band operation. Although the velocity change in Ni-$LiNbO_3$ structure is only .02°/o, it will be possible to attain a change of a few percent with the class of materials[11,12] (e.g., $Tb_xDy_{1-x}Fe_2$ and $Tb_xHo_{1-x}Fe_2$) which possess giant magnetostrictive constants. This tuning scheme should then have many important device applications such as velocity compensation in adaptive array systems, or tunable SAW oscillators and resonators.

REFERENCES

1. W. Strauss, Proc. IEEE, 53, 1485-1495 (1965).
2. H. Van de Vaart, J. Appl. Phys., 42, 5305 (1973); H. Mathews and H. Van de Vaart, Appl. Phys. Lett. 15, 373 (1969).
3. J. P. Parekh, J. Appl. Phys. 45, 434 (1974); 45, 1860 (1974).
4. A. K. Ganguly, K. L. Davis, D. C. Webb, C. Vittoria and D. W. Forester, Elec. Letters, 11, 610 (1975).
5. C. Vittoria, M. Rubinstein and P. Lubitz, Phys. Rev., B12, 5150 (1975).
6. R. C. LeCraw and R. L. Comstock, Physical Acoustics, vol. III B, W. P. Mason, Ed. (Academic Press Inc., New York, 1965), pp. 127-199.
7. S. Chicazumi, "Physics of Magnetism" (John Wiley & Sons, Inc., New York, 1964), pp. 161-185.
8. K. L. Chopra, "Thin Film Phenomena" (McGraw Hill Book Company, New York, 1969) pp. 266-327.
9. A. K. Ganguly, K. L. Davis, D. C. Webb and C. Vittoria, J. Appl. Phys., May, 1976 (to be published).
10. A. H. Fahmy and E. L. Adler, Appl. Phys. Lett. 22, 495 (1973).
11. N. C. Koon, A. I. Schindler and F. L. Carter, Phys. Lett. 37B, 413 (1973).
12. A. E. Clark and H. T. Savage, IEEE Trans. on Sonics and Ultrasonics, SU-22, 50 (1975).

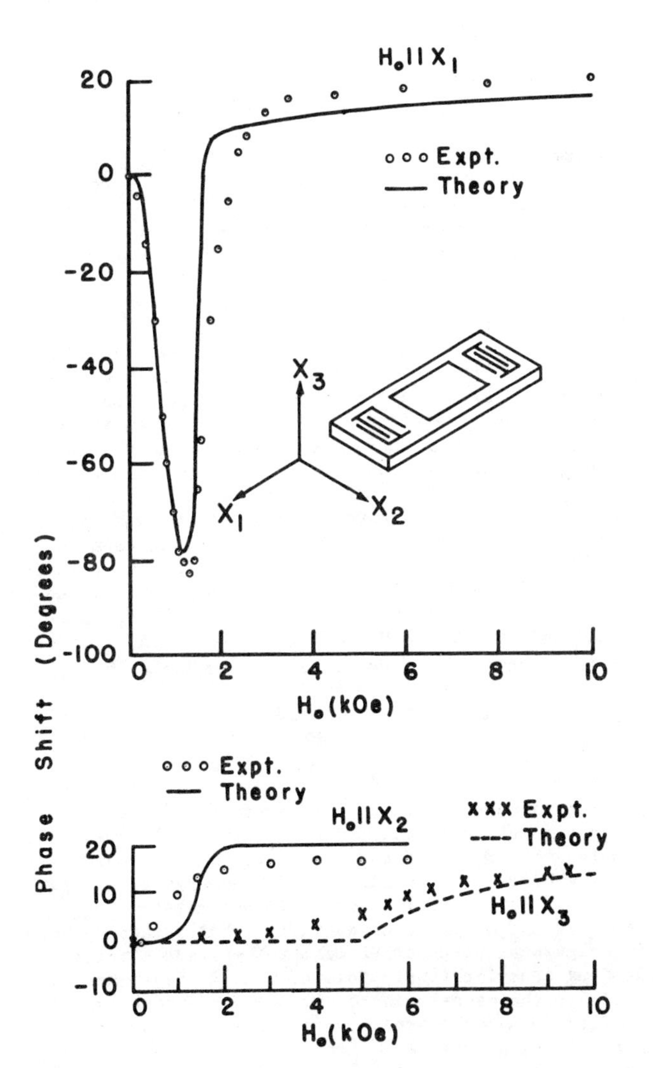

Fig. 1. Phase shift vs applied field for three orientations

MAGNETOELASTIC SURFACE WAVE ATTENUATION IN FERROMAGNETIC CONDUCTORS*

Y. J. Liu
The American University, Washington, D. C. 20016
and
H. T. Savage
Naval Surface Weapons Center, Silver Spring, Maryland 20910

ABSTRACT

We have studied the effect of conductivity and magnetic damping on magnetoelastic Rayleigh-type surface wave propagation along a tangential saturating field in the <100> direction of a cubic ferromagnetic medium. We are to investigate potential utilization of highly magnetostrictive rare earth-iron alloys in current surface wave technologies. Material parameters appropriate to Ga-YIG have been used, and conductivity and magnetic damping were treated as parameters in sample calculations. We point out two "loss bands" of frequency near ferromagnetic resonance in which propagation of surface waves is highly attenuating. The attenuation depends strongly on the conductivity and magnetic damping in the lower band, but is otherwise not a strong function of either parameter. To avoid complexities acoustic damping is set equal to zero in the present investigation.

INTRODUCTION

Recent work has shown the rare earth-iron (RFe) alloys to have a huge room temperature magnetostriction[1] and large magnetomechanical coupling factor[2]. $Tb_{.27}Dy_{.73}Fe_2$, for example, has a magnetostriction of 1500 ppm and a coupling factor of 0.6 at the room temperature. These characteristics have lead to development work on the RFe for use in sonic and ultrasonic transducers.

Surface acoustic wave (SAW) devices have desirable characteristics such as ease of excitation, tapping, and detection. Magnetoelastic SAW (M-SAW) have additional useful characteristics. Some of these are field-dependent velocities, nonreciprocity, and convenient nonlinearities. M-SAW propagation in low-loss insulator of YIG has been treated[3] and observed[4]. The delay times were observed to be a function of the bias field. An M-SAW isolator, by performing the marriage of M-SAW and magnetostatic surface waves, has also been demonstrated[5]. However, the insertion losses were found to be quite high (>60 db). Part of this is due to small magnetoelastic interaction in YIG which is less than 0.01 of that in the RFe. Therefore, we intend to investigate M-SAW in the RFe. Since the RFe are conductors, it is desirable to develop a theory which takes the conductivity into account.

THEORY

Parekh and Bertoni[3] have treated a similar but simpler magnetoelastic system, appropriate for YIG which has zero conductivity and sufficiently small damping. In RFe_2, however, the existence of conductivity enhances the exchange effect and causes elastic damping. We therefore want to incorporate in our analysis not only the conductivity, but exchange interactions and Landau-Lifshitz (L-L) and elastic damping contributions. This results in seven normal modes in the magnetoelastic system[6], of which two are of electromagnetic type, two of exchange type, and three of elastic type. Each mode is an admixture and has a complex propagation constant.

Inclusion of magnetocrystalline anisotropy is important but generally tedious. However, if we confine $\vec{M}_0$ to a symmetry direction, it is justifiable that it can be simply described by a magnetic field along $\vec{M}_0$. This "anisotropy field" has, up to the second order, a magnitude of $2K_1/M$ for $\vec{M}_0$ along <100> and $-4(K_1+K_2/3)/(3M)$ for $\vec{M}_0$ along <111>, where $M\equiv|\vec{M}_0|$.

We consider a cubic semi-infinite medium with a (100) interface. We orient the coordinate axes along cube edges, and launch the surface wave in the x_1-direction with no x_2-dependence as shown in Fig. 1.

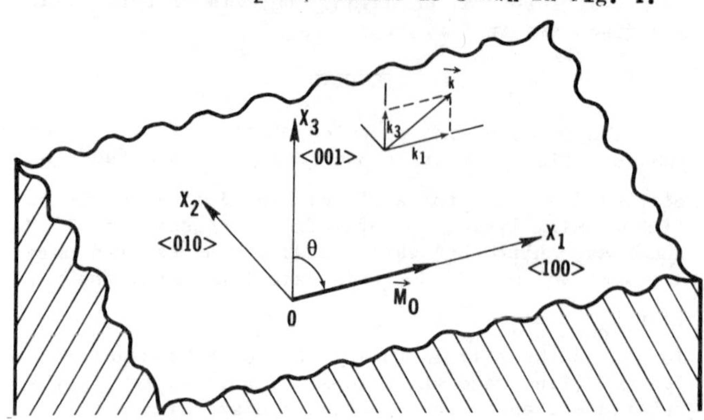

Fig. 1 Geometrical configuration used in the study.

We employ the L-L equation of motion:

$$\frac{1}{\gamma}\frac{\partial\vec{M}}{\partial t} = \vec{M}\times\{\vec{H}_{eff} + \frac{\eta}{M}\vec{M}\times\vec{H}_{eff}\} \tag{1}$$

where $\gamma(<0)$ is the gyromagnetic ratio and $\eta\equiv-\lambda/(\gamma M)$ the dimensionless L-L damping coefficient. $\vec{H}_{eff}$ is the sum of internal magnetic field, anisotropy field, exchange field, and magnetostrictive field, for which we denote respectively by $\vec{H}_i$, $\vec{H}_A$, $\vec{H}_e$, and $\vec{H}_L$. We have

$$\vec{H}_A = -\nabla_M F_A \tag{2}$$

$$\vec{H}_e = \frac{2A}{M^2}\nabla^2\vec{M} \tag{3}$$

$$\vec{H}_L = \bar{\bar{g}}\cdot\vec{e} \tag{4}$$

where F_A, A, $\bar{\bar{g}}$, $\vec{e}$ are respectively the magnetocrystalline anisotropy energy, the exchange constant, the magnetostriction constants (3×6 components), and the strain (6 components: e_{11}, e_{22}, e_{33}, e_{23}, e_{31}, e_{12}). $\vec{H}_i$ satisfies the Maxwell's equations:

$$\nabla\times\vec{E} = -\frac{1}{c}\frac{\partial}{\partial t}(\vec{H}_i + 4\pi\vec{M}) \tag{5}$$

$$\nabla\times\vec{H}_i = \frac{4\pi\sigma'}{c}\vec{E} \tag{6}$$

where $\sigma'\equiv\sigma-i\omega\varepsilon/(4\pi)$ is the effective complex conductivity at frequency ω. Note that $\sigma'\simeq\sigma$ for the RFe_2 at all frequencies of interest here.

For the elastic motion, we have

$$\rho\frac{\partial^2\vec{u}}{\partial t^2} = \nabla\cdot\bar{\bar{\tau}} + \vec{M}_0\cdot\nabla\vec{H}_i \tag{7}$$

where ρ is the mass density, $\vec{u}$ the lattice displacement vector, and $\bar{\bar{\tau}}$ the magnetoelastic stress tensor expressed by, with the summation convention used,

$$\tau_{ij} = c_{ijp}e_p + \varepsilon_{ijk}m_k + \xi_{ijp}\dot{e}_p \tag{8}$$

where $i,j,k=1\to3$, $p=1\to6$. Note that p is a double index used in the conventional way: 1≡11, 2≡22, 4≡23 or 32, etc. c_{ijp} denotes the cubic elastic stiffness, ε_{ijk} the magnetoelastic coupling, and ξ_{ijp} the elastic damping. The second term on the right-hand side of

Eq. (7) represents the magnetic force.

Under rf steady state, all the varying quantities ($\vec{m}$, $\vec{h}$, etc.) have an $\exp\{i(\vec{k}\cdot\vec{r}-\omega t)\}$ dependence. The vector $\vec{m}$ is in general not orthogonal to the static vector $\vec{M}_0$ due to lattice distortion:

$$\vec{m} = \vec{m}_\perp - \vec{M}_0 \nabla\cdot\vec{u} \tag{9}$$

However, we here consider systems with small volume magnetostriction, in which case $\vec{m}$ reduces to

$$\vec{m} = \vec{m}_\perp \equiv m_\theta \hat{e}_\theta + m_\phi \hat{e}_\phi \tag{10}$$

as it is in a purely ferromagnetic system, where the carets denote unit vectors.

In the special case of $\theta=\pi/2$, cf. Fig. 1, Eqs. (5) and (6) combine to give

$$\begin{pmatrix} h_1 \\ h_2 \\ h_3 \end{pmatrix} = -\frac{4\pi}{1+\Delta k^2} \begin{pmatrix} 0 & \Delta k_1 k_3 \\ 1 & 0 \\ 0 & 1+\Delta k_3^2 \end{pmatrix} \begin{pmatrix} m_2 \\ m_3 \end{pmatrix} \tag{11}$$

where $\Delta \equiv ic^2/(4\pi\sigma'\omega)$, and $\overline{\overline{g}}$ (3×6) and $\overline{\overline{\varepsilon}}$ (6×3) simplify to

$$g_{26} = g_{35} = -\frac{b_2}{M}, \text{ all other } g_{ip} = 0 \tag{12}$$

$$\varepsilon_{11} = \frac{2b_1}{M}, \ \varepsilon_{53} = \varepsilon_{62} = \frac{b_2}{M}, \text{ all other } \varepsilon_{pi} = 0 \tag{13}$$

where b_1, b_2 are the first two magnetoelastic constants in a cubic system. Then, by defining $c'_{pq} = c_{pq} - i\omega\xi_{pq}$, p,q=double indices, and going through some algebraic manipulations, we obtain the following matrix equation:

$$(a_{mn})(y_n) = 0 \qquad m,n=1\rightarrow 5 \tag{14}$$

where $y_1 \equiv u_1$, $y_2 \equiv u_2$, $y_3 \equiv u_3$, $y_4 \equiv m_3$, $y_5 \equiv m_2$

$$a_{11} = -\eta a_{21} = -i\eta b_2 k_3$$

$$a_{22} = \eta a_{12} = -a_{13} = \eta a_{23} = i\eta b_2 k_1$$

$$a_{24} = H + H_\sigma(1+\Delta k_3^2) \qquad a_{14} = -\frac{i\omega}{\gamma} - \eta a_{24}$$

$$a_{15} = H + H_\sigma \qquad a_{25} = \frac{i\omega}{\gamma} + \eta a_{15}$$

$$a_{31} = \rho\omega^2 - c'_{11}k_1^2 - c'_{44}k_3^2$$

$$a_{53} = \rho\omega^2 - c'_{44}k_1^2 - c'_{11}k_3^2$$

$$a_{33} = a_{51} = -(c'_{12} + c'_{44})k_1 k_3$$

$$a_{42} = \rho\omega^2 - c'_{44}k^2 \qquad a_{45} = i\left(\frac{b_2}{M} - H_\sigma\right)k_1$$

$$a_{34} = i\left(\frac{b_2}{M} - H_\sigma \Delta k_1^2\right)k_3$$

$$a_{54} = i\left\{\frac{b_2}{M} - H_\sigma(1 + \Delta k_3^2)\right\}k_1$$

$$a_{32} = a_{35} = a_{41} = a_{43} = a_{44} = a_{52} = a_{55} = 0$$

with the abbreviations

$$H_\sigma \equiv \frac{4\pi M}{1+\Delta k^2} \qquad H = H_0 + \frac{2A}{M}k^2 + \frac{2K_1}{M}$$

where $H_0 \equiv |\vec{H}_0|$, $\vec{H}_0$ being the static part of $\vec{H}_i$. For nontrivial excitations it is necessary that

$$\| a_{mn} \| = 0 \tag{15}$$

which expands into a seventh degree polynomial in k_1^2 or k_3^2.

BOUNDARY CONDITIONS

From Eq. (15) there exist seven bulk modes in either of the $\pm x_3$-directions corresponding to each propagation mode k_1. Only the seven modes whose amplitudes diminish in the $-x_3$-direction, namely $\mathrm{Im}(k_3)<0$, are physically acceptable, regardless of their directions of <u>phase</u> propagation. The total amplitude of an rf variable is the sum of those of the seven modes: $h_1 = \sum_{n=1}^{7} h_{1n}$, $m_2 = \sum_{n=1}^{7} m_{2n}$, etc.

However, the values of k_1 are not arbitrarily set. Only those $\pm k_1$ pairs such that the corresponding seven modes satisfy some boundary conditions simultaneously at the interface can exist and propagate.

There are seven boundary conditions consistent to the starting equations, two electromagnetic, two exchange, and three elastic. The consideration of electromagnetic boundary conditions amounts to the treatment of impedance mismatch at $x_3=0$. The air-sample interface requires $e_1=h_2$ and $e_2=-h_1$ in the employed system of units. Using Eqs. (6) and (11), these relations read

$$\sum_n (1 - Z_{3n}) \frac{m_{2n}}{Q_n} = 0 \tag{16}$$

$$\sum_n \{(1 - Z_{3n})\Delta k_1 k_{3n} + Z_1(1 + \Delta k_{3n}^2)\}\frac{m_{3n}}{Q_n} = 0 \tag{17}$$

where $Q_n = 1+\Delta(k_1^2+k_{3n}^2)$, $Z_1 = -ick_1/(4\pi\sigma')$, and $Z_{3n} = -ick_{3n}/(4\pi\sigma')$. By adopting a perpendicular surface anisotropy model[7], the exchange boundary equations can be written:

$$\sum_n k_{3n} m_{2n} = 0 \tag{18}$$

$$\sum_n \left(ik_{3n} - \frac{K_s}{A}\right) m_{3n} = 0 \tag{19}$$

where $-\infty < K_s < \infty$ is a phenomenological constant. And, with a stress-free interface, the elastic equation $\hat{e}_3\cdot\overline{\overline{T}}=0$ gives rise to

$$c'_{44} \sum_n i(k_{3n}u_{1n} + k_1 u_{3n}) + \frac{b_2}{M}\sum_n m_{3n} = 0 \tag{20}$$

$$c'_{44} \sum_n k_{3n}u_{2n} = 0 \tag{21}$$

$$c'_{12} \sum_n k_1 u_{1n} + c'_{11} \sum_n k_{3n}u_{3n} = 0 \tag{22}$$

Note that n runs from 1 to 7 in Eqs. (16) through (22).

DISPERSION RELATIONS

Four of the five rf variables u_1, u_2, u_3, m_3, and m_2 associated with each of the seven modes can be expressed in terms of the fifth by the use of Eq. (14). Taking m_2 to be the fifth variable for each mode, we turn the seven boundary equations (16) through (22) into a set of homogeneous linear equations:

$$(A_{mn})(m_{2n}) = 0 \qquad m,n=1\rightarrow 7 \tag{23}$$

which requires

$$\| A_{mn} \| \equiv f(\omega,k_1) = 0 \tag{24}$$

for nontrivial excitations.

Equation (24) contains implicitly the dispersion relations of M-SAW. Due to algebraic complexity, one needs to search numerically for the values of k_1 at a given ω. The solution is nonunique and of transcendental type[8], but only the root or roots situated most closely to the $\mathrm{Re}(k_1)$ axis and satisfying $\mathrm{Re}(k_1)\times\mathrm{Im}(k_1)>0$ will be of practical significance and adopted.

With the solution of Eq. (24), Eq. (23) can be used to determine all the seven constituent strengths in terms of one. This is then the only unknown remained to be determined by the external sources.

It is noted that since $f(\omega,k_1)$ is an even function of k_1, as indicated by the expansion of Eq. (15), the propagation of M-SAW in this configuration is reciprocal, as it is in the insulator case[3].

PROPAGATION LOSSES

The introduction of conductivity in a real physical system does not only complicate the electromagnetic propagation, but introduces a significant source of damping in the elastic lattice by interacting with the conduction electrons. The elastic damping, unfortunately, can by no means be fully described by the simple term appearing in Eq. (8). Generally speaking, however, it is not expected to be significant at frequencies below the microwave region. We set to zero the elastic damping at the present time in an attempt to investigate energy losses due to other mechanisms.

In a surface wave system, ferromagnetic resonance (FMR) is not strictly defined in the conventional sense. In the configuration with surface wave propagating parallel to the bias field, "FMR" is spread over a range of frequency near $\omega=\gamma(H_0+2K_1/M)$. This causes a broad "loss band" for the surface wave propagation. The energy loss is a strong function of σ and λ. The structure and location of this FMR band is complicated by the interaction of electromagnetic modes with elastic modes, resulting in a band shift to the low side.

Any waves with k_1 satisfying Eq. (24) are in principle excitable in the medium. Only waves that are composed of evanescent modes in the $-x_3$-direction, however, can propagate on the surface with the least attenuation. Similar to an optical total internal reflection problem, these waves must necessarily propagate with $|\mathrm{Re}(k_1)|$ greater than the magnitude of the dominant characteristic k or k's of the medium in which evanescent waves are excited. These surface waves thus propagate slower than any of the bulk elastic or electromagnetic modes, and very little energy is leaked into the medium.

It is not until one goes beyond the FMR frequency that energy is expected to scatter into the medium. The scattering takes place as the spin-wave mode sweeps into the region where $|\mathrm{Re}(k_3)|>|\mathrm{Im}(k_3)|$, and has a rapidly growing real part[9]. As the frequency rises further, this mode becomes more loosely coupled to the rest of the system due to the boundary conditions. Its amplitude diminishes and the leakage is rapidly reduced. This mechanism establishes a second loss band situated right above the FMR frequency. Its band width broadens as the exchange coupling increases. Note that this loss band has been previously described in exchange-free insulators[3,10].

NUMERICAL RESULTS

We employ the parameter values found in ref. 3 in our numerical procedure. In addition, we use $A=0.36\times 10^{-6}$ erg/cm, ε(dielectric constant)=10, $K_1=0$, and $K_s=0$. σ and λ are two changing parameters in the study. No numerical approximations are used in seeking the roots.

In Fig. 2 we plot the M-SAW propagation constant k_1 as a function of frequency at a number of values of σ and λ. The attenuating part, $\mathrm{Im}(k_1)$, characterizes the two loss bands described in the previous section. It is convincingly demonstrated that the behavior of

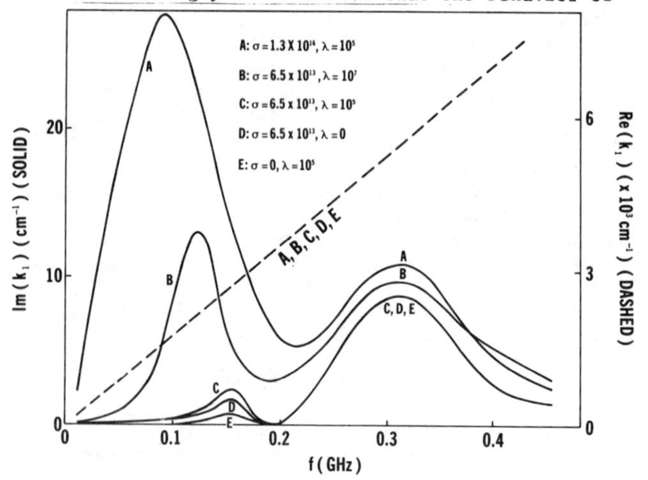

Fig. 2 Sample dispersion characteristics. H_0=50 Oe $4\pi M$=300 G

the lower band (of FMR nature) depends strongly on σ and λ, while that of the upper band does not.

In addition, we found two closely located low-loss modes by solving Eq. (24) in certain frequency ranges. This has helped us understand that the observation of multiple modes in YIG[11] may not be inconsistent with an appropriate theory. The double mode found in our case is represented in terms of k_1 in Fig. 3. It shows that one of the modes starts to disappear at around 60 MHz.

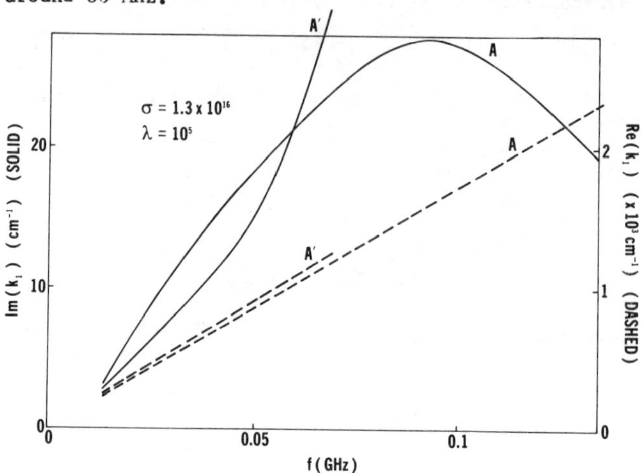

Fig. 3 Two close solutions to Eq. (24). H_0=50 Oe $4\pi M$=300 G

CONCLUSIONS

The purpose of this study is to gain more insight into the physics of M-SAW propagation, especially to understand the roles conductivity and magnetic damping play in a magnetoelastic system. The existence of an FMR loss band has been pointed out in addition to the spin-wave loss band previously reported in YIG. The loss due to conductivity has been shown to be small outside of the FMR band, which makes the device applications of the RFe_2 look promising.

ACKNOWLEDGEMENT

The authors wish to thank J.R. Cullen, E.R. Callen and A.E. Clark for helpful discussions. Communications with R. C. Barker in the course of the work have also proven to be fruitful.

REFERENCES

*Work supported by the Office of Naval Research, Naval Electronics Command, and NSWC IED Fund.

1. A.E. Clark and H.S. Belson, IEEE Trans. Mag., MAG-8, 477 (1972).
2. H.T. Savage, A.E. Clark, and J.M. Powers, IEEE Trans. Mag., MAG-11, 1355 (1975).
3. J.P. Parekh and H.L. Bertoni, J. Appl. Phys., 45, 434 (1974).
4. F.W. Voltmer, R.M. White, and C.W. Turner, Appl. Phys. Lett., 15, 153 (1969).
5. M.F. Lewis and E. Patterson, Appl. Phys. Lett., 20, 276 (1972).
6. C. Vittoria, J.N. Craig, and G.C. Bailey, Phys. Rev., B10, 3945 (1974).
7. C.H. Bajorek and C.H. Wilts, J. Appl. Phys., 42, 4324 (1971).
8. J.P. Klozerberg, B. McNamara, and P.C. Thonemann, J. Fluid Mech., 21, 545 (1965).
9. For a detailed discussion of the ferromagnetic system, see Y.J. Liu, Ph.D. thesis, Yale U. (1974).
10. M.R. Daniel, J. Appl. Phys., 44, 1404 (1973).
11. T.L. Tsai, G. Komoriya, G. Thomas, and J.P. Parekh, IEEE Ultrasonics Symposium Proceedings, 489 (1974).

IMPROVED MICROWAVE DEVICES USING NARROW LINEWIDTH GARNETS

J. Guidevaux
Thomson-CSF, Département Ferrites UHF, 93100 Montreuil France

and

A. Hermosin and R. Sroussi
Thomson-CSF, Laboratoire Central de Recherches, 91401 Orsay France

ABSTRACT

Two specially interesting materials have been developed in a new class of polycrystalline microwave garnets introduced by Winkler et al.[1] with linewidth < 10 oe. The following characteristics have been measured

composition	4π Ms gauss	T_c,°C	Δ Hoe
$Y_{2.9}Ca_{0.1}Fe_{4.2}Ge_{0.1}In_{0.7}O_{12}$	1520	130	7
$Y_{1.6}Ca_{1.4}Fe_{3.9}V_{0.7}In_{0.4}O_{12}$	900	190	7

We used these new materials in coaxial and waveguide circulators in different frequency bands between 3.7 GHz and 8.5 GHz. The figure of merit of these circulators is about 600, to be compared to 200 for circulators equiped with classical garnets. As expected, the insertion losses have been lowered, typically they are less than 0.1 dB. These circulators in a four-port configuration, are very attractive for parametric amplifiers applications. Finally the feasability of magnetically tunable pass-band filters employing these polycrystalline materials is demonstrated.

INTRODUCTION

Among the non reciprocal microwave ferrite devices, certainly the junction circulator is the one which has got the most important development during the last fifteen years. In ten percent bandwidth they have 20 dB isolation and 0.2 or 0.3 dB insertion loss. These values are sufficient for usual applications. However, in some particular applications circulators with lower insertion loss are required. For instance, in the microwave link, the power capability of the transmitter and the sensitivity of the receiver depend on the quality of many circulators used. This is the same thing for the parametric amplifier noise temperature which increases about 7°K for 0.1 decibel of loss. Therefore many efforts have been spent in order to decrease the losses of ferrite materials. These materials have now very good dielectric properties and we must pay attention to the magnetic losses. The magnetic resonance linewidth ΔH fixes the figure of merit of the devices[2] : the smaller the ΔH, the bigger the figure of merit, i.e. the lower the insertion losses. As far as we know, only Ito et al.[3] and B. Desormière, J. Guidevaux[4] have studied a low loss circulator equiped with a bulk single crystal and LPE epitaxial garnet, respectively. This paper is devoted to the results obtained with polycrystalline garnets in the diagrams CaGeIn, CaVIn, CaZr - YIG with very narrow linewidth.

Very low insertion loss is difficult to measure with a good accuracy, therefore we also compare the figure of merit obtained with circulators using narrow resonance linewidth garnet with the one obtained by circulators equiped with classical materials (Y Al I G and Mn Mg ferrite) measured in the same frequency band with the same apparatus.

At last we show that these materials are suitable for the realization of magnetically tunable band-pass filters analogous to YIG filters.

MATERIALS

One of the most significant improvement in microwave ferrite materials has been the introduction of a new class of ultra narrow linewidth polycrystalline garnets[1]. Commercially available polycrystalline garnets present typically a linewidth $\Delta H > 30$ oe to be compared to 0.3 oe for single crystal YIG. These new materials form an intermediate class with $\Delta H < 10$ oe, which associates very low magnetic losses with the flexibility of solid-solution ceramic processing.

The study of these new garnets has a twofold aspect : a) influence of the different substituted cations on the magnetization and specially on the magnetocrystalline anisotropy. b) technological improvement leading to porosity free ceramics. Finally, measurement of very narrow linewidth implies a careful polishing of the spherical samples.

Three main families of garnet materials have been studied corresponding to the following chemical formulae

(1) $Y_{3-x}\ Ca_x\ Fe_{5-x}\ Zr_x\ O_{12}$

(2) $Y_{3-2x}\ Ca_{2x}\ Fe_{5-x-y}\ V_x\ In_y\ O_{12}$

(3) $Y_{3-x}\ Ca_x\ Fe_{5-x-y}\ Ge_x\ In_y\ O_{12}$

These families of garnet materials offer a large choice of materials with $\Delta H < 10$ oe and with saturation magnetization between 300 and 2000 gauss, for microwave applications. Ca Zr - YIG is the best choice when high (1800 - 2000 gauss) saturation magnetization is needed. Ca V In - YIG is well adapted for the intermediate range and Ca Ge In - YIG is useful for applications where $4\pi M_s$ = 900 gauss is necessary. Most of the useful materials have $\Delta H < 5$ oe. Besides, the dielectric loss tangent for every material is about 10^{-4}.

Two materials have been developed for the circulators and filters studied in this work. Table 1 gives the main magnetic properties.

Table I

	$4\pi M_s$ gauss	T_c,°C	Δ H Oe
$Y_{2.9}\ Ca_{0.1}\ Fe_{4.2}\ Ge_{0.1}\ In_{0.7}\ O_{12}$	1520	130	7
$Y_{1.6}\ Ca_{1.4}\ Fe_{3.9}\ V_{0.7}\ In_{0.4}\ O_{12}$	900	190	7

CIRCULATORS

With these materials we realized different circulators in the telecommunication frequency bands. In each case the mechanical sizes have been kept identical to those of conventional circulators equiped with classical material (Y Al I G and Mn Mg ferrite). Only the garnet diameter have been optimized in order to take into account saturation magnetization and applied magnetic field discrepancy.

Figures 1 to 4 show the swept frequency insertion loss measurement of four-port and three-port coaxial circulators operating between, 3.7 and 4.2 GHz, 5.9 and 6.4 GHz, 6.4 and 7.1 GHz and a three-port waveguide circulator operating between 7.7 and 8.5 GHz. The insertion losses are typically less than 0.1 dB for all devices. Isolation and VSWR, not showed, are always better than 25 dB and 1.10 respectively. It is interesting to calculate in each case the highest figure of merit : i.e. the ratio of the isolation maximum to the insertion loss minimum and compare it with the figure of merit of circulators equiped with classical ferrites or garnets of similar saturation magnetization.

Table II shows the results and allows to appreciate advantages of narrow linewidth materials.

Table II

Frequency (GHz)	Figure of merit max classical material	narrow linewidth material
3.7 - 4.2	210	1150
5.9 - 6.4	135	430
6.4 - 7.1	175	620
7.7 - 8.5	300	640

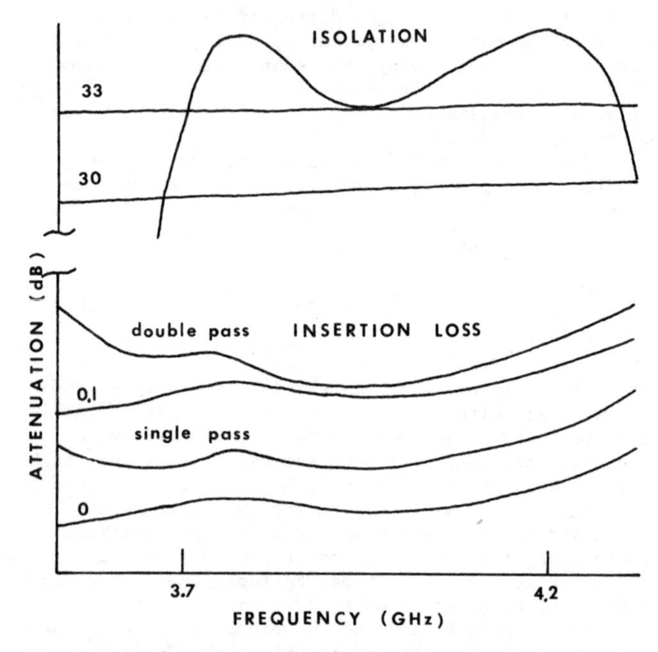

Fig. I. Performances of a 4 GHz four-port coaxial circulator for parametric amplifier.
CaVIn-YIG 900 gauss

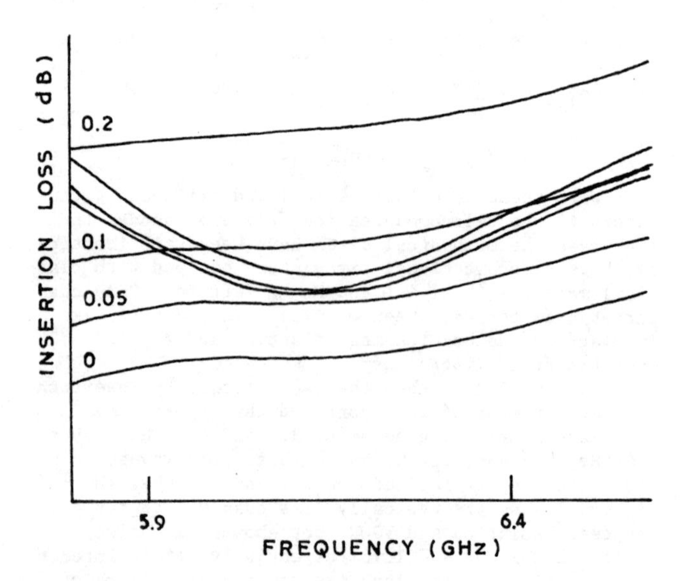

Fig. 2 Insertion loss of a 6 GHz three-port coaxial circulator. CaGeIn-YIG 1520 gauss.

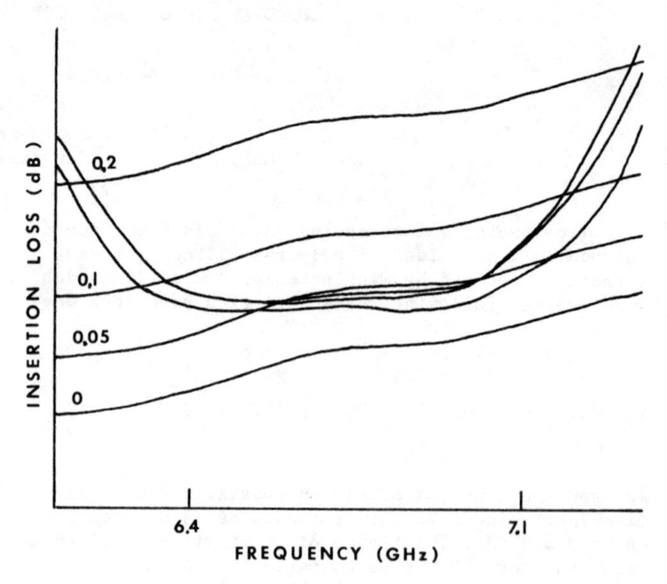

Fig. 3. Insertion loss of a 7 GHz three-port coaxial circulator. CaGeIn-YIG 1520 gauss.

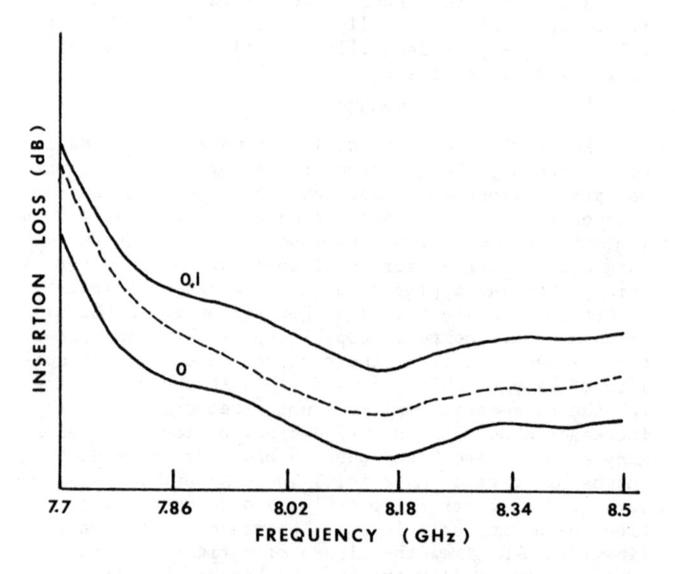

Fig. 4. Insertion loss of a 8 GHz three-port waveguide circulator. CaGeIn-YIG 1520 gauss.

These results can be also compared to the theoretical prediction pointed out by R.L. Comstock[5]. In many respects the junction circulator is similar to a symmetrical transmission cavity. The loaded Q of the circulator mode is

$$Q_L = \frac{0,71}{4\pi M_s} \frac{\omega}{\gamma}$$

It is typically in the range $1 < Q_L < 4$. The unloaded Q (for below resonance operation) is approximately

$$Q_o = \frac{2\omega^2}{4\pi M_s \Delta H \gamma^2}$$

ω is the operating angular frequency, γ the gyromagnetic ratio
$4\pi M_s$ the saturation magnetization and ΔH the resonance linewidth. Insertion loss is given by :

$$P_i = 20 \log_{10} \left(1 - \frac{Q_L}{Q_o}\right)$$

Table III lists the theoretical and experimental values for the central frequency. We observed a good agreement because dielectric and connector losses are neglected in the predicted values.

Table III

Classical materials		
Frequency (MHz)	Insertion loss (dB) Calculated	Measured
6150	0.08	0.15
6750	0.08	0.3
4350	0.13	0.2
Narrow Linewidth materials		
6150	0.014	0.05
6750	0.011	0.05
4350	0.021	0.05

No problems occur for isolation and reflected power measurement : usual substitution and reflectometry method were used, but insertion loss in the range < 0.1 dB is difficult to measure exactly. We have used a dual channel audio substitution method which makes swept frequency measurements independent of source amplitude changes. With this method, swept frequency measurements are possible in a reproducible way. Nevertheless, this is a substitution method and a jack-jack adapter is necessary for reference levels. The adapter loss, unknown exactly, must be added to the circulator loss but all coaxial microwave devices are generally tested this way. Therefore, the performances pointed out are comparable to each other. In order to get rid of this uncertainty a 7 millimeters sexless connectors circulator has been tested (no adapter is necessary for reference level). Its insertion loss between 3.7 and 4.2 GHz was less than 0.1 dB with a minimum close to 0.05 dB at the central frequency. The adapter loss can be estimated < 0.02 dB. In our case insertion loss measurements have been recorded with a sensitivity of 0.01 dB/cm so the results are reproducible only with fresh connectors engaged with a torque wrench.

In addition, measurements were performed from -20°C to + 70°C, a shift of transmission curves occurs, isolation remains > 20 dB and loss increases up to 0.2 dB at the frequency limits.

TUNABLE FILTERS

It is known that a ferrimagnetic material used with a spherical shape properly magnetized and set within a microwave transmission line behaves like a very small resonator. This property has been employed for a long time in YIG filters.

Generally these filters use highly polished spheres of yttrium iron garnet single crystals. So far this material was the only one whose intrinsic properties allowed to obtain sufficient coupling factor suitable for good filtering. The coupling factor of such a sphere, increases with the material volume and decreases with its resonance linewidth. These values are[6], for a good single crystal, respectively about 0.25 oe and 3.3 $10^{-2}mm^3$. The same coupling factor can be obtained with a less than 10 oersted resonance linewidth polycrystal with a volume of $1.32 mm^3$ i.e. about 1.35mm diameter suitable within most waveguide or coaxial structure. Isotropic spheres of such dimension are easy to handle and the filter tuning becomes very simple. The material is cheap and it is not necessary to polish the spheres which can be directly sized by batch production abrasion process. The materials used in this work have an extremely low anisotropy, so the filter frequency temperature drift, which depends on the magnetocrystalline anisotropy changes, will be very low.

Since the critical field depends on ΔH, the non linear power absortion threshold is higher. Figure 5 shows a transmission curve obtained with only one sphere 1.8 mm diameter in a classical coaxial structure. With this material at 2.400 MHz the filter is not saturated before 2.5 dBm.

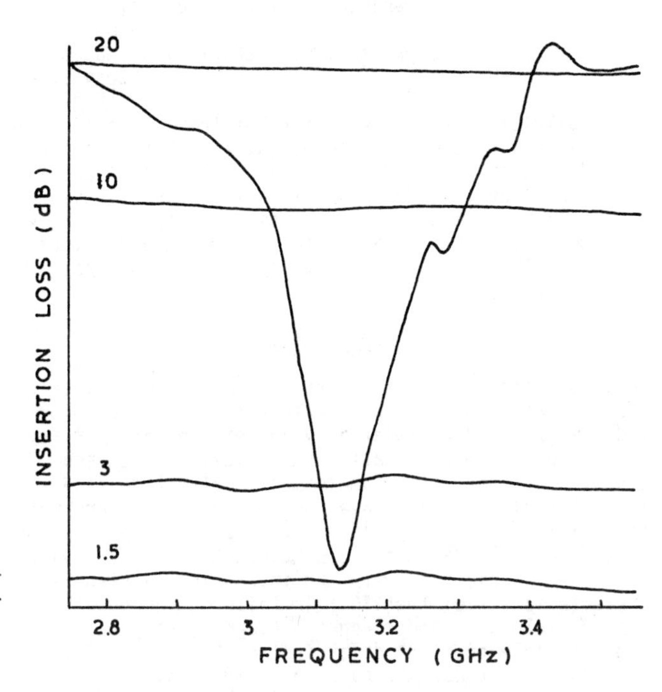

Fig. 5. Transmission curve of a single stage coaxial filter CaGeIn-YIG 1520 gauss 1.8 mm diameter

CONCLUSIONS

Experimental results pointed out in this paper show performances improvement due to polycrystalline narrow linewidth garnet. From now on the four-port circulator described has allowed a 10°K decreasing of a 4 GHz parametric amplifier noise temperature. Moreover a possible application of these materials in designing tunable filters is potentially very interesting.

REFERENCES

This work was supported by the Direction des Recherches et Moyens d'Essais, Paris, France, (Devices) and by the Délégation Générale à la Recherche Scientifique et Technique, Paris, France, (Materials).

1. G. Winkler, P. Hansen, P. Holot, Philips Res. Rep. 27 151 - 171, 1972.
2. B. Lax. K.J. Button, Microwave Ferrites and Ferrimagnetics McGraw-Hill 1962 Chap. 12 p. 540.
3. Y. Ito, Proceeding of the Institute of Electronics and communication Engineers Japan 1970.
4. B. Désormière, J. Guidevaux, Brevet Français TH-CSF N° 74 11 699.
5. R.L. Comstock, C.E. Fay, J.Appl. Phys., 36, 3, 1253 - 1258, (1965)
6. B. Désormière, Electronics Letters, 2, 7, 235 (1966)

DELAY LINES BASED ON MAGNETOSTATIC VOLUME WAVES IN EPITAXIAL YIG

Z.M. Bardai,[†] J.D. Adam,[†] J.H. Collins[†] and J.P. Parekh[††, †††]

[†]Department of Electrical Engineering, University of Edinburgh, Edinburgh EH9 3JL, Scotland
[††]Department of Engineering, State University of New York Maritime College, Fort Schuyler, Bronx, N. Y. 10465
[†††]Department of Electrical Sciences, State University of New York at Stony Brook, Stony Brook, N. Y. 11794

ABSTRACT: This paper describes the realisation and operation of a magnetostatic volume wave (MSVW) delay line utilizing an epitaxial YIG film which exhibits useful constant delay characteristics. The delay line comprises a normally magnetized YIG film which is separated from a ground plane by a dielectric layer. Experimental results have been obtained using a 4.5 μm thick LPE-grown YIG film and a 115 μm thick glass dielectric layer, with 30 μm wide and 3 mm long Al microstrip lines deposited on the latter to form input and output couplers. The device has exhibited, in agreement with theory, a constant delay of about 350 nsec over a 200 MHz bandwidth in the 1 GHz to 5 GHz frequency range.

I. INTRODUCTION

Magnetostatic wave delay lines formed from bulk Yttrium Iron Garnet (YIG) have used either the backward volume mode of propagation in axially magnetised rods[1] or forward volume wave propagation in normally magnetised discs.[2] In contrast, magnetostatic delay line studies in epitaxial YIG films have focused on the surface wave mode (MSSW)[3] with propagation transverse to an in-plane bias field. To date low loss tapped MSSW delay devices have been fabricated, with the intrinsically dispersive characteristics of the YIG film tailored through the use of dielectric and conducting layers, to give approximately constant delay in the 100 nsec to 600 nsec range over 200 MHz bandwidths in L and S-band. Linearly dispersive characteristics have also been demonstrated.[3]

MSSW devices show decreasing bandwidth with increasing operating frequency. Figure 1 shows the frequency dependence of the propagation limits of both MSSW and magnetostatic volume waves (MSVW). The maximum MSSW bandwidth is determined by the propagation limits

$$\omega_{sh} = \gamma(H + 2\pi M) \quad (1)$$

and

$$\omega_{s\ell} = \gamma[H(H + 4\pi M)]^{\frac{1}{2}} \quad (2)$$

where ω_{sh}, $\omega_{s\ell}$, γ, H and $4\pi M$ are the MSSW upper and lower frequency limits, the gyromagnetic ratio (2.8 MHz/Oe), internal dc magnetic field, and saturation magnetisation (1750 Oe for YIG), respectively. At low microwave frequencies, using pure YIG maximum bandwidths of 1.5 GHz are possible centred on 2 GHz. However, bandwidths of only 300 MHz are available at 10 GHz. The MSSW bandwidth may be extended by metallising the YIG surface[4] or through the use of suitably biased multiple YIG film structures.[5]

A simpler means of obtaining wide band operation at high microwave frequencies is through the use of MSVW. The potential bandwidth of MSVW is given by the propagation limits (see Fig. 1)

$$\omega_{vh} = \gamma[H(H + 4\pi M)]^{\frac{1}{2}} \quad (3)$$

and

$$\omega_{v\ell} = \gamma H \quad (4)$$

where ω_{vh} and $\omega_{v\ell}$ are the upper and lower frequency bounds of the MSVW spectrum, respectively. Thus, bandwidths of 1.5 GHz are available centred on 2 GHz increasing to 2.2 GHz centred on 10 GHz.

This paper describes initial studies of MSVW propagation in normally magnetised epitaxial YIG films in the 1 GHz to 5 GHz range. A theoretical study of a simple YIG-dielectric-ground plane configuration, Fig. 2, has shown that approximately constant delay characteristics in the range 100 nsec to 1 μsec may be obtained over bandwidths of 200 MHz. Test delay lines fabricated from YIG films, grown by liquid phase epitaxy (LPE) on gadolinium gallium garnet (GGG),[6] have verified the theoretical predictions.

II. THEORY

The magnetostatic boundary value problem to be solved for the YIG/dielectric/ground plane configuration under normal magnetisation is shown in Fig. 2(a). The MSVW correspond to sinusoidal variation of fields in the YIG region, hyperbolic variation in the dielectric layer, and exponential decay away from the YIG film in the dielectric region above it. Assuming a time-harmonic variation exp $(j\omega t)$, the MSVW dispersion relation is readily shown to be

$$\tan k\beta d\,[(1 - \beta^2) - \exp(-2kt_1)(1 + \beta^2)] + 2\beta = 0 \quad (5)$$

where $\beta = (-\mu)^{\frac{1}{2}}$, with $\mu = 1 + \Omega_H/(\Omega_H^2 - \Omega^2)$ and in the usual notation $\Omega = \omega/\omega_m$, $\Omega_H = \omega_0/\omega_m$, ω is the signal frequency, $\omega_m = \gamma 4\pi M$, $\omega_0 = \gamma H$, d is the thickness of the YIG film and t_1 is the thickness of the dielectric layer.

For the limiting values of $t_1 = 0$ (metallised YIG slab) and $t_1 \rightarrow \infty$ this relation reduces respectively to

$$\tan k\beta d = \beta^{-1} \quad (6)$$

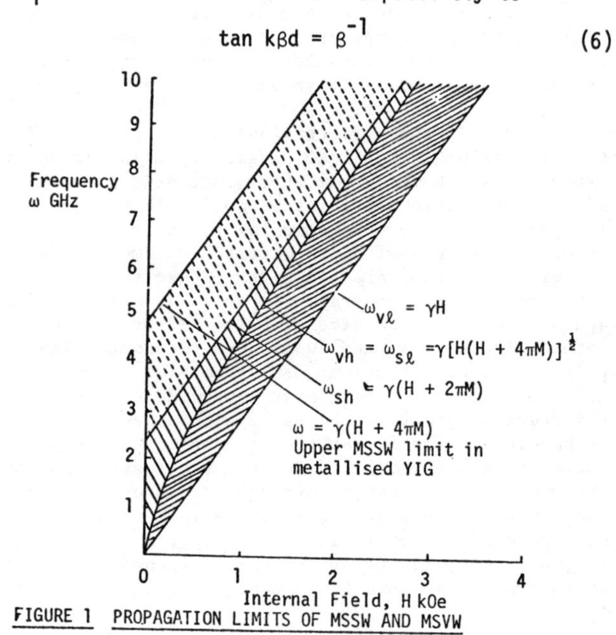

FIGURE 1 PROPAGATION LIMITS OF MSSW AND MSVW

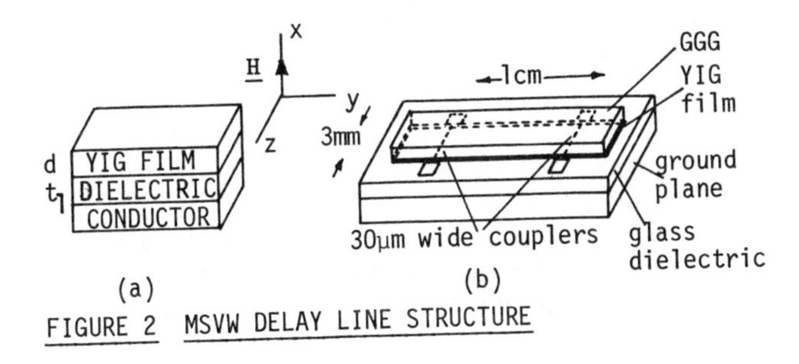

FIGURE 2 MSVW DELAY LINE STRUCTURE

and

$$\tan k\beta d/2 = \beta^{-1} \tag{7}$$

Equation 7 is the dispersion relation for MSVW propagation in a normally magnetised plate derived by Damon and van de Vaart.[2] As is evident from equation 5 the effect of the dielectric will be significant only for $kt_1 < 5$. The multiple roots of equation 5 give rise to a number of MSVW modes propagating at a given frequency with different group velocities. These higher order modes can lead to spurious echoes.

A real solution k to the dispersion equation (5) for a given frequency ω exists over the entire MSVW spectrum, i.e., $\omega_{v\ell} = 2.25$ GHz $< \omega < \omega_{vh} \sim 4$ GHz. However, in order to emphasize the constant delay characteristic achievable with the chosen structure, a calculated plot of group delay (μsec/cm) vs. frequency is presented in Fig. 3 for the frequency range 2.25 GHz to 3 GHz for a 4.5 μm YIG film biased to 2550 Oe. The constant delay characteristic for the lowest order mode is evident with appropriate choice of dielectric thickness. The optimum dielectric thickness for this film is of the order of 50 μm. For the second mode, which has a much higher delay, the presence of the dielectric causes a delay shift (upper curve) which is independent of the dielectric thickness.

III. EXPERIMENTAL RESULTS

Figure 2(b) shows a diagram of the MSVW delay line. The dielectric was a microscope glass slide, 115 μm thick. The slide was metallised on both surfaces with approximately 2 μm of aluminum. The lower side formed the ground plane. On the upper surface 3 mm long, 30 μm wide microstrip conductors terminated in 0.5 mm square bonding pads were defined photolithographically and acted as transducers for the MSVW. The YIG film, grown by LPE[6] on a GGG substrate was cut to 11 mm x 3 mm and placed in contact with the transducers. No electromagnetic matching was employed at the transducer inputs. The magnetic bias field was applied normal to the film plane. Microwave pulses of 50 nsec duration, within the frequency range 1-5 GHz, were applied to the input of the delay line under test and the output observed on a sampling oscilloscope.

Theoretical and experimental delay versus frequency characteristics of a 4.5 μm pure YIG film on a 115 μm glass dielectric are shown in Fig. 4(a) over the frequency range 1-5 GHz. The delayed pulse exhibits a constant delay of 350 nsec ±5% over a 200 MHz bandwidth in agreement with the computed characteristics. In the calculation, a saturation magnetization $4\pi M = 1750$ Oe was assumed, and an anisotropy field contribution of +100 Oe was taken to provide a fit between experiment and theory. The constant delay value is seen to be maintained over a constant bandwidth and to be almost invariant with the bias field. CW power saturation was observed at input levels above 5 dBm.

Figure 4(b) shows the insertion loss over the frequency range 1-5 GHz. The minimum insertion loss at 1 GHz is approximately 30 dB and occurs at the centre of the constant delay region. This insertion loss consists in part of the propagation loss, estimated as 60 dB/μsec[7] for the 0.8 Oe linewidth film used, and a 6 dB loss introduced through the bi-directionality of the microstrip couplers. An improvement in the total insertion loss should therefore be possible through recourse to better quality, lower linewidth films.[8] The insertion loss increases rapidly at longer delays (>400nsec), outside the constant delay region, since the MSVW half wavelength then exceeds the transducer width of 30 μm.

The occurrence of minimum insertion loss at a frequency in the centre of the constant delay region, together with the constancy of the delay with bias field, permits a tubable bandpass filter with constant delay to be realised.

As stated above, the dispersion relation has multiple roots at any given frequency, each one giving rise to a delayed signal output with increasingly longer delays. Second order modes were observed at long delays, much greater than three times the first mode delay, which decreased with increasing signal frequency, at a fixed bias field to a nearly constant value of 1.2 μsec. This is in agreement with the second mode characteristic shown in Fig. 3. The minimum attenuation of the second delay mode was 60 dB. In general, the miminum delay of the second mode was greater than three times the first mode delay. Figure 5 is an oscilloscope trace of the first and second delay modes with the input pulse as reference. The operating frequency is 2.5 GHz and the bias field 2.5 kOe. The main delay is at ~350 nsec and the second mode is at ~1.2 μsec.

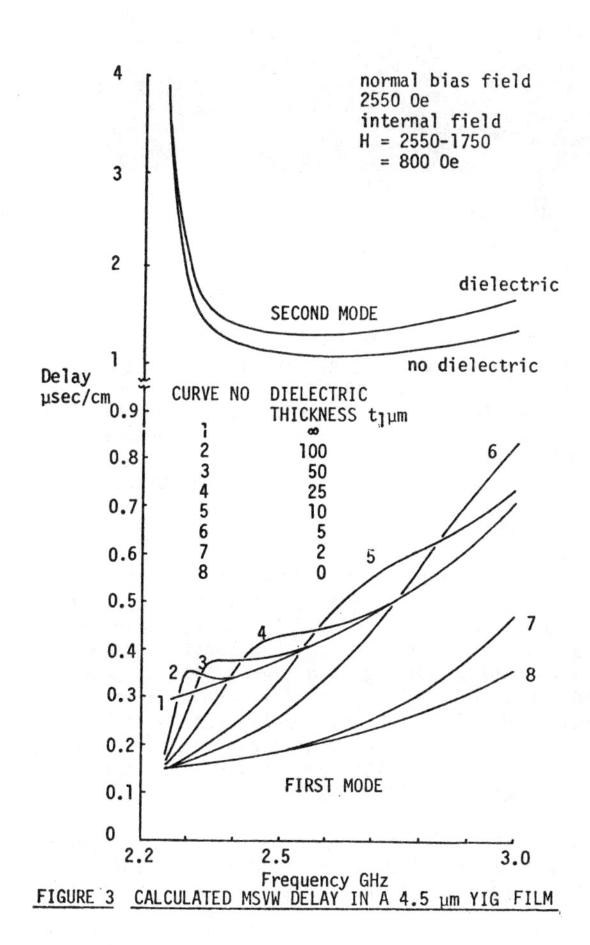

FIGURE 3 CALCULATED MSVW DELAY IN A 4.5 μm YIG FILM

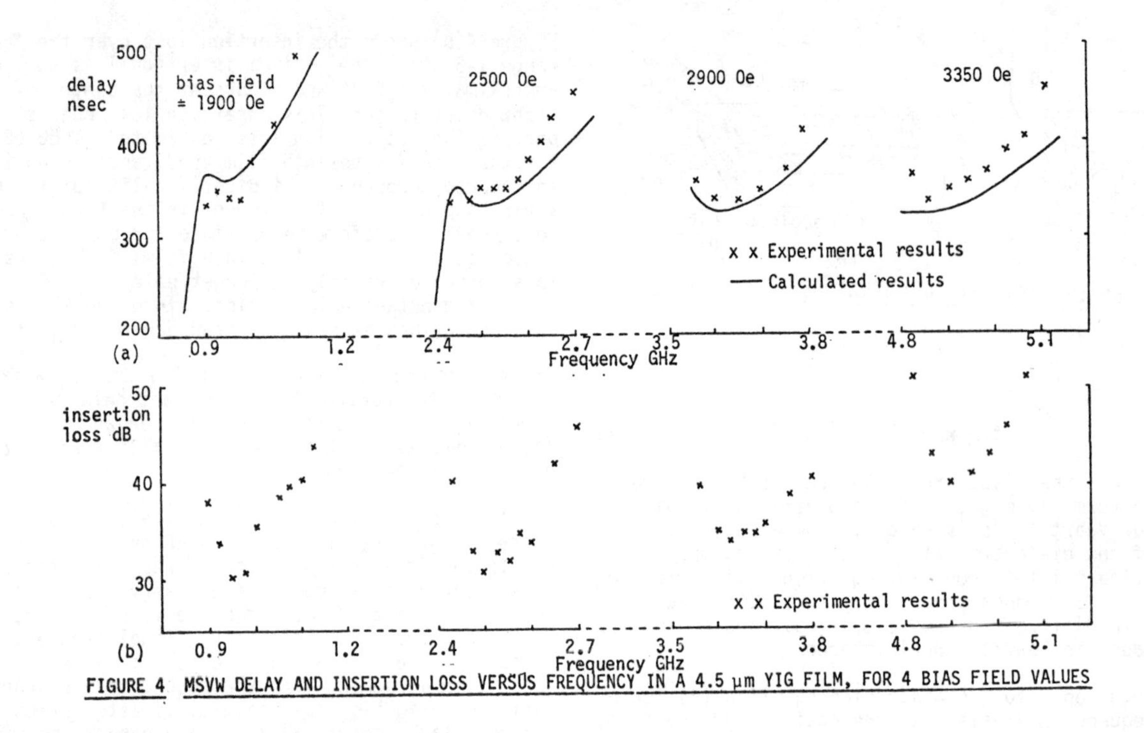

FIGURE 4 MSVW DELAY AND INSERTION LOSS VERSUS FREQUENCY IN A 4.5 μm YIG FILM, FOR 4 BIAS FIELD VALUES

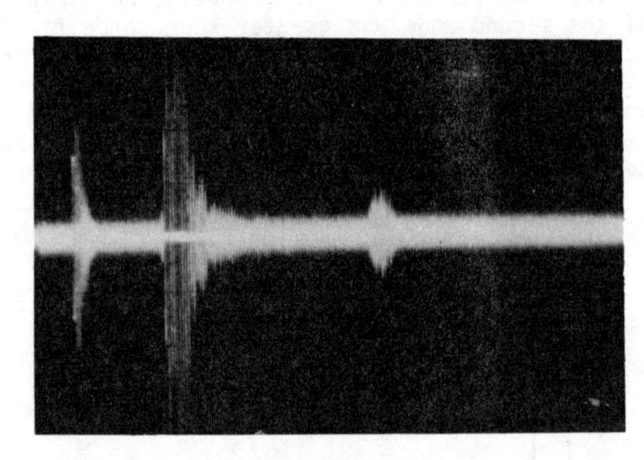

FIGURE 5 OSCILLOSCOPE TRACE SHOWING INPUT PULSE, 1st MODE DELAY (350 nsec) AND 2nd MODE DELAY (1.2 μsec)

IV. CONCLUSIONS

MSVW propagating in a normally magnetised epitaxial YIG-dielectric-ground plane structure have been shown theoretically and experimentally to yield constant delay characteristics over ~200 MHz bandwidths at frequencies up to 5 GHz. Similar constant delay characteristics have been obtained previously for MSSW propagation in a transversely magnetised structure of the same dimensions but with lower applied bias fields.

The centre frequency of the MSVW constant delay region tunes linearly, and the delay magnitude remains approximately unchanged with bias field variations. In contrast the MSSW constant delay characteristics are obtained over a limited bias field range for a given geometry. MSVW has the disadvantage of second mode excitations with delays which may be as short as three times the desired first mode delay. Multiple delays of this type have not been observed with MSSW.

In the future, MSVW devices based on similar layered structures will allow tapped tunable constant delay X-band devices to be fabricated, which, if made from narrow linewidth epitaxial YIG[8] ($\Delta H \sim 0.2$ Oe) will be competitive with presently available bulk acoustic wave devices at these frequencies. Samarium Cobalt magnets will be required to produce the 5 kOe bias fields required for MSVW at X-band. The unique feature of MSVW in a normally magnetised film compared with other magnetostatic modes, namely isotropic propagation in the film plane, will allow definition of long delay paths through use of etching techniques.

ACKNOWLEDGEMENTS

The authors wish to acknowledge contract support from the British Ministry of Defence Procurement Executive.

REFERENCES

1. R. W. Damon and H. van de Vaart, "Propagation of Magnetostatic Spin Waves at Microwave Frequencies, II Rods", J. Appl. Phys. 37, 2445, 1966.
2. R. W. Damon and H. van de Vaart, "Propagation of Magnetostatic Spin Waves in a Normally Magnetised Disc", J. Appl. Phys. 36, 3453, 1965.
3. J. D. Adam, J. H. Collins and J. M. Owens, "Microwave Device Applications of Epitaxial Magnetic Garnets", The Radio and Electronic Engineer, 45, 738, 1975.
4. S. R. Seshadri, "Surface Magnetostatic Modes of a Ferrite Slab", Proc. IEEE 58, 506, 1970.
5. A. K. Ganguly and C. Vittoria, "Magnetostatic Wave Propagation in Double Layers of Magnetically Anisotropic Slabs," J. Appl. Phys. 45, 4665 (1974)
6. J. D. Adam, J. M. Owens and J. H. Collins, "Ferromagnetic Resonance Linewidth of Thick Yttrium Iron Garnet Films Grown by Liquid Phase Epitaxy", Electronics Letters 9, 325, 1973.
7. C. Vittoria and N. D. Wilsey, "Magnetostatic Wave Propagation Losses in an Anisotropic Insulator," J. Appl. Phys. 45, 414 (1974)
8. M. T. Elliott, "Effects of Lead Incorporation on the Ferromagnetic Resonance Linewidths of Liquid Phase Epitaxial Grown Yttrium Iron Garnet", Paper 5E-2, 21st Conference on Magnetism and Magnetic Materials, Philadelphia, Pa., Dec 9-12, 1975.

FINITE WIDTH EFFECTS IN MAGNETOSTATIC SURFACE WAVE PROPAGATION

R. W. Patterson and T. W. O'Keeffe
Westinghouse Research Laboratories, Pittsburgh, PA 15235

ABSTRACT

A model is proposed for the propagation of magnetostatic surface waves in a layered structure consisting of a ground plane and a YIG film separated by a dielectric. The standard assumption of an infinite sample width is replaced with an assumption of a sinusoidal transverse distribution of rf fields. Instead of a single solution, a multiplicity of solutions is found, each with a distinct dispersion relation. Experimental observation of the first two finite-width modes is in excellent agreement with predicted results derived from the model.

INTRODUCTION

Magnetostatic surface wave (MSSW) propagation in a three-layer structure, consisting of a ground plane, a dielectric and a YIG film, was first investigated by Bongianni.[1] His analysis related MSSW dispersion characteristics to the values of the magnetic bias field and the dielectric and magnetic film thicknesses. We have investigated the additional effects produced by a finite sample width.

The geometry we shall be considering is sketched in Fig. 1. The finite width is in the z direction, which Bongianni assumed to be infinite. We introduced finite width effects by assuming that rf quantities vary as cos $k_z z$, where $k_z = n\pi/w$, w is the sample width, and n is an integer. Sparks[2] and Adam[3] have used such a distribution to describe MSSW in isolated thick samples, where demagnetizing fields are more significant than in our thin films. In our case, the cosine distribution is the simplest variation consistent with magnetization pinning at the sample edges ($z = \pm w/2$), and does not represent a complete solution of the boundary value problem.

Introducing $k_z = n\pi/w$ into Bongianni's analysis results in a dispersion relation given by

$$\exp(2Md) = \frac{\mu_1 - s\mu_2 - R}{\mu_1 + s\mu_2 + R} \cdot \frac{\mu_1 + s\mu_2 - R\tanh(Nt)}{\mu_1 - s\mu_2 + R\tanh(Nt)}, \quad (1)$$

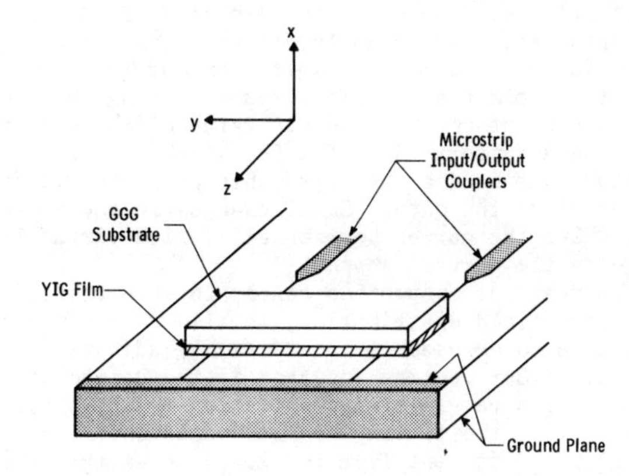

Figure 1: Device geometry showing shorted tapered microstrip coupling lines

where $M = (k_y^2 + k_z^2/\mu_1)^{1/2}$,

$N = (k_y^2 + k_z^2)^{1/2}$,

$R = N/M$,

μ_1 and μ_2 are the diagonal and off-diagonal elements of the permeability tensor, t is the thickness of the dielectric, d is the thickness of the YIG film and $s = \pm 1$ for propagation in the $\pm\hat{y}$ direction. The internal magnetic field is assumed to lie in the $+\hat{z}$ direction. Equation (1) reduces to Bongianni's result (Eq. (19) of ref. 1) when n is set equal to zero, which is equivalent to assuming an infinite width.

The derivative $\partial\omega/\partial k$, obtained from (1), is the group velocity of MSSW; the inverse of the group velocity represents the transit time per centimeter of a MSSW delay line. Figure 2 shows the delay per centimeter obtained in samples of different widths, with all other geometrical parameters kept equal. One curve shows the characteristics of an infinite width (n = 0) mode, while the other two show the characteristics of the n = 1 modes in samples 2 mm and 1 mm wide. The $w = \infty$ curve displays a pair of singularities between which is a narrow region of backward waves

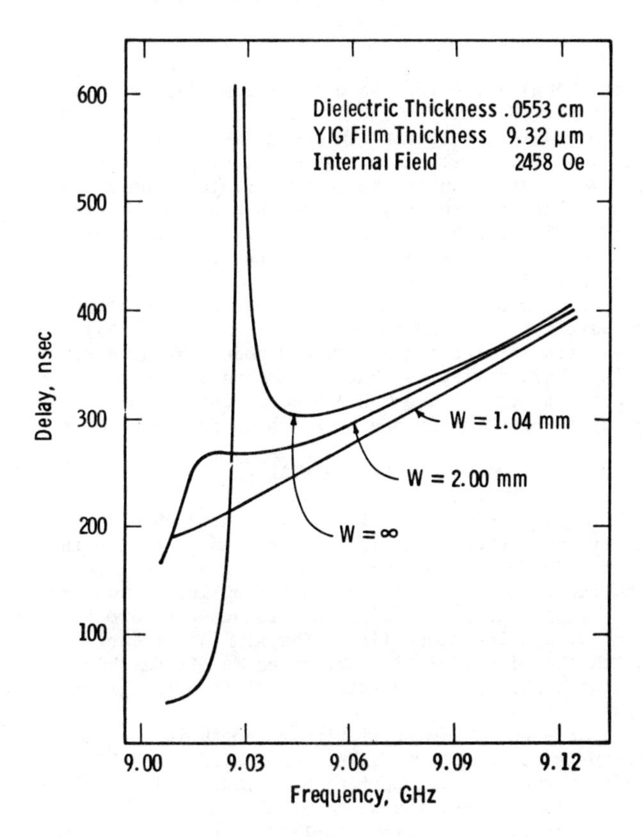

Figure 2: Delay vs frequency characteristics for three different sample widths

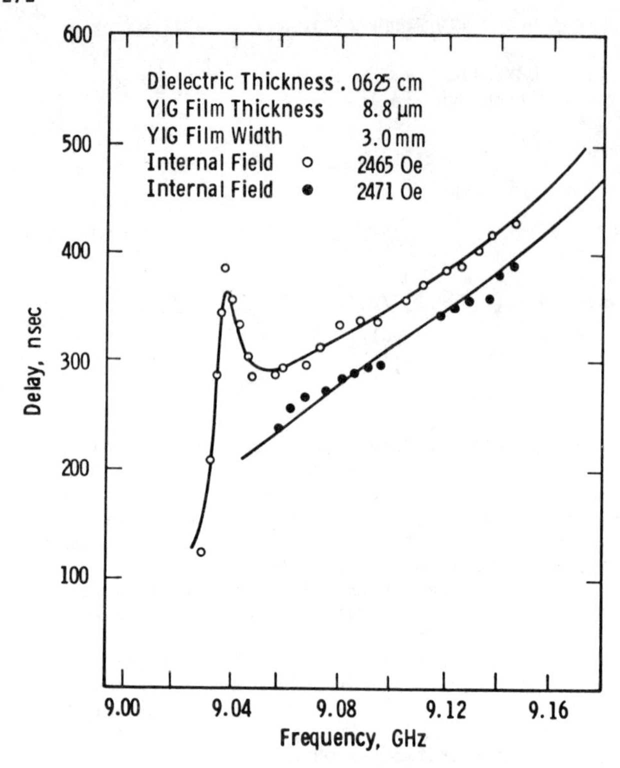

Figure 3: Delay vs frequency characteristics corresponding to the n=1 mode (open circles) and the n=2 mode (solid circles). The internal field magnitudes were used as adjustable parameters to fit the data.

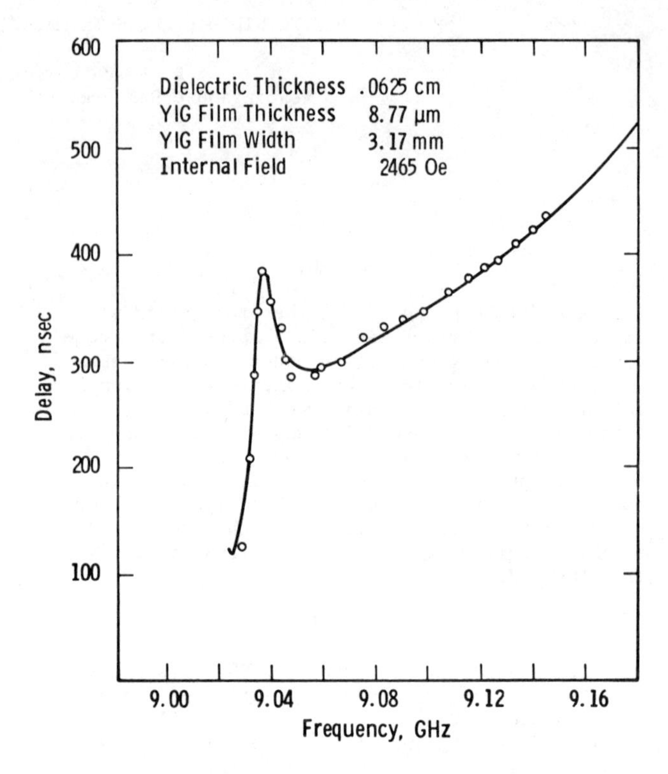

Figure 4: Best fit of the n=1 data of Fig. 3 to the theory, using film thickness, film width and internal field magnitude as adjustable parameters. Note slight change in "optimized" values from measured values used in Fig. 3.

(not shown) where the phase and group velocities are oppositely directed. The w = 2 mm curve displays a region 24 MHz wide in which the delay is 266 ± 3 nsec. The w = 1 mm curve has a region slightly greater than 100 MHz wide in which the delay deviates from a linear dependence on frequency by less than 2 nsec. Either bandwidth could be extended somewhat, at the expense of the other, by appropriate choice of the dielectric and magnetic thicknesses.

In our observations of MSSW propagation at 9 GHz, we have not observed any behavior similar to that predicted for the infinite width mode. We have seen delay characteristics that correspond to the behavior predicted for n = 1 and n = 2 modes in various samples of different widths, one of which is described below.

EXPERIMENTAL

Shorted microstrip lines, separated by one centimeter and sketched in Fig. 1, are used to excite the MSSW. Beneath the sample the lines are 50 μm wide, to achieve more efficient broadband coupling.[4] Away from the sample the microstrip line broadens to form a nominal 50Ω impedance line. The alumina dielectric on which the microstrip is fabricated is 625 μm thick, and also serves to separate the YIG from the ground plane.

At 9 GHz the electrical wavelength in the microstrip is comparable to sample dimensions ($\lambda/4 \simeq 3.5$ mm), so that the exciting current distribution is markedly non-uniform across the width of the sample (w = 3 mm). Consequently, the narrow couplers were made long enough to allow the centerline (z = 0) of the sample to be located as far as $\lambda/2$ from the short. This position provides an even symmetrical current distribution across the sample width which preferentially excites the n = 1 mode. Positioning the centerline of the sample $\lambda/4$ from the short provides an odd symmetrical distribution which preferentially excites the n = 2 mode. Because the sample width is about $\lambda/4$, placing the sample edge adjacent to the short provides a non-symmetrical current distribution which excites several modes, which in turn produce strong interference effects.

Delay vs frequency data for the n = 1 and n = 2 modes of a 3 mm wide sample, together with the curves obtained from (1) for those modes, are shown in Fig. 3. A network analyzer was used to observe the variation of the phase difference between the input and output of the delay line, from which the experimental delay data were obtained. The slightly different magnetic field values for the two modes result from our inability to reset the field exactly after repositioning the sample. The measurement of the absolute value of the magnetic field had an uncertainty of ±.5%, so the internal field magnitude was used as an adjustable parameter to fit the curves to the data. Small changes in the magnetic field shift the curves horizontally, with virtually no change in the curves' shapes.

To test our assumption concerning the cosine transverse field distribution, we allowed the width of the sample to be used as an additional adjustable parameter. Our feeling was that if the cosine distribution were a very poor approximation to the actual distribution, the fitting program would choose a width appreciably different from the measured width. Furthermore, since our interferometrically-determined thickness value was subject to a potential error of a few percent, we also allowed the sample thickness to be adjusted. The curve resulting from a "best fit" to the

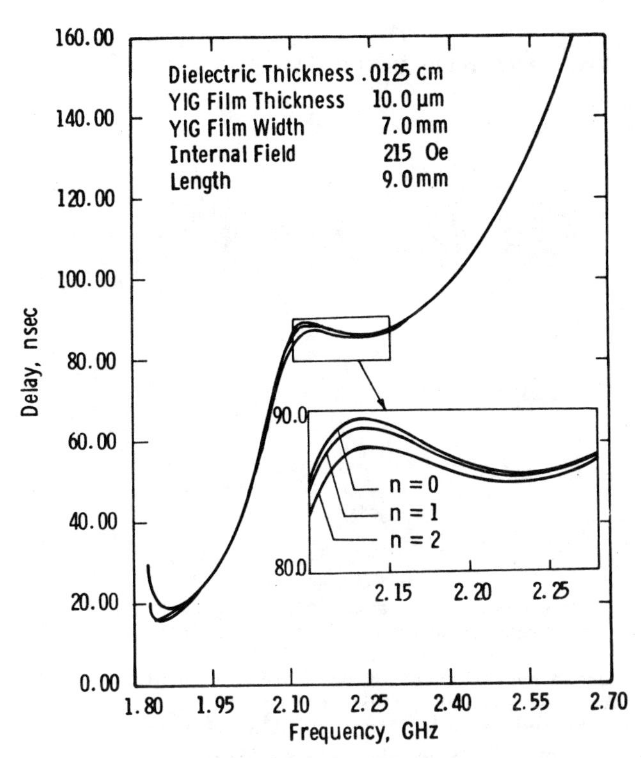

Figure 5: Delay vs frequency characteristics for three modes at 2 GHz. The differences are more apparent in the expanded scale insert.

n = 1 data of Fig. 3, using three adjustable parameters, is shown in Fig. 4. Note that the "optimized" value of the thickness differs by only 0.03 µm, and the width differs by only 0.2 mm from the measured values. This agreement suggests that the cosine distribution is at least an adequate approximation to the actual distribution.

DISCUSSION

Previous workers[1,5,6] investigating MSSW propagation characteristics in the 2-4 GHz range have not reported any disparity between Bongianni's n = 0 theory and experimental results. The principle reason is that at these lower frequencies, the n = 0 mode has delay characteristics quite similar to the n ≠ 0 modes, for commonly employed geometries. Figure 5 is a plot of the delay characteristics of the n = 0, 1 and 2 modes for a sample with the dimensions used by Bongianni.[1] The differences, which can be more readily seen in the insert of Fig. 5, are less than the experimental errors generally encountered.

The frequency at which finite width effects become significant depends on sample geometry. The extent to which the n ≠ 0 mode differs from the n = 0 mode appears to be indicated by the extent to which the factor R in (1) differs from unity. If k_z^2/μ_1 is negligible compared to k_y^2, then (1) reduces to the n = 0 case. For a given value of magnetic bias field there exist lower and upper frequency limits for MSSW propagation corresponding to $k_y = 0$ and $k_y = \infty$, respectively. Consequently, as the frequency approaches the upper frequency limit, all the n ≠ 0 modes tend to coalesce with the n = 0 mode, as can be seen in Fig. 2. Between the lower and upper frequency limits, μ_1 increases from zero to some maximum value, and this maximum value increases as the bias field is reduced for operation at lower frequencies.[1] In the midranges of the passbands, μ_1 is approximately an order of magnitude smaller at 9 GHz than it is at 2 GHz. Consequently, obtaining width effects at 2 GHz comparable to those obtained at 9 GHz would require the sample to be about 1/3 as wide, all other geometrical factors being equal.

ACKNOWLEDGMENTS

The authors would like to express their appreciation to G. W. Roland, who grew the garnet films, and to J. A. Kerestes, who fabricated them into usable geometries.

REFERENCES

1. W. L. Bongianni, "Magnetostatic Propagation in a Dielectric Layered Structure," J. Appl. Phys. 43, 2541-2548, 1972.

2. M. Sparks, "Magnetostatic Surface Modes of a YIG Slab," Electron. Lett. 5, 618-619, 1969.

3. J. D. Adam, "Delay of Magnetostatic Surface Waves in YIG," Electron. Lett. 6, 718-719, 1970.

4. A. K. Ganguly and D. C. Webb, "Microstrip Excitation of Magnetostatic Surface Waves: Theory and Experiment," IEEE Trans. Microwave Theory Tech. MIT-23, 998-1006, 1975.

5. A. K. Ganguly, C. Vittoria, and D. Webb, "Interaction of Surface Magnetic Waves in Anisotropic Slabs," AIP Conf. Proc. 24, 495-496, 1975.

6. J. D. Adam, Z. M. Bardai, J. H. Collins, and J. M. Owens, "Tapped Microwave Nondispersive Magnetostatic Delay Lines," AIP Conf. Proc. 24, 499-500, 1975.

MAGNETOELASTIC EFFECTS ASSOCIATED WITH ELASTIC SURFACE WAVE PROPAGATION IN EPITAXIAL GARNET FILMS+

G. Volluet, B. Désormière and B.A. Auld++
THOMSON-CSF - Research Center, Domaine de Corbeville
91401 Orsay, France

ABSTRACT

Surface wave delay lines have been fabricated on epitaxial garnet films, using a ZnO coating and interdigital transducers for elastic wave excitation. Amplitude and phase delay variations of the delayed signal have been measured as a function of an in-plane magnetic field, at frequencies of 210 MHz and 335 MHz. For pure YIG films, the strongest effects are observed when the films are not magnetically saturated, exhibiting stripe domain patterns. The observed absorptions are explained by the gyromagnetic resonances driven by the effective field associated with the elastic strains. This effective field was determined from the relevant terms of the magnetoelastic energy; the stripe domain resonances were computed only for a (1,0,0) oriented film. An "easy-plane" film of GdGa doped YIG was also used and good agreement was found between gyromagnetic resonances and acoustic absorptions. Also the motion of stripe domains induced by an elastic wave has been observed. The drift velocity has been measured as a function of incident power. A discussion of this new effect is given.

INTRODUCTION

In contrast to magnetoelastic bulk waves, magnetoelastic surface waves have not been extensively studied until recently[1,2,3], despite the growing success of elastic surface wave devices. The main reason for this lack of interest seems to lie in the difficulty of getting good conditions for the experiments; in particular, the elastic surface waves are more conveniently generated at comparatively low frequencies, where the magnetic substrates are rather lossy. However such a magnetoelastic effect can be used to design new surface wave devices, such as variable delay lines, acoustic isolators and circulators. This paper is mainly concerned with experimental results obtained for the amplitude and phase delay variations encountered by an elastic surface wave propagating in an epitaxial ferrimagnetic film. A film was preferred to a bulky magnetic substrate since its magnetic state can be known and controlled, as shown hereafter. Also are presented the first results obtained for the driving of magnetic stripe domains by elastic surface waves.

ELASTIC WAVE TRANSMISSION EXPERIMENTS

As in the previous work of Lewis et al[1], we first used (2,1,1)-Yttrium Iron Garnet (YIG) films grown on $Gd_3Ga_5O_{12}$ by the liquid phase epitaxy process. The elastic surface waves were launched and received by interdigital transducers engraved in piezoelectric ZnO films wich had been sputtered on the magnetic films. The amplitude variation of the 2 µs delayed pulse was recorded as a function of the biasing field, as shown in Figure 1. Here the film thickness was 8 µm, the operating frequency 210 MHz, the propagation and field directions were <0,1,$\bar{1}$> and <$\bar{1}$,1,1>, respectively. In contrast to Lewis'results, a serie of resonances are observed, three narrow and one broad ("dashed curve"). Reversing the biasing field direction results in some non-reciprocal effect, as shown by the solid line; in particular the third resonance vanishes. All these absorptions occur when the magnetic film is not saturated, in contrast to Lewis' assumptions : this is

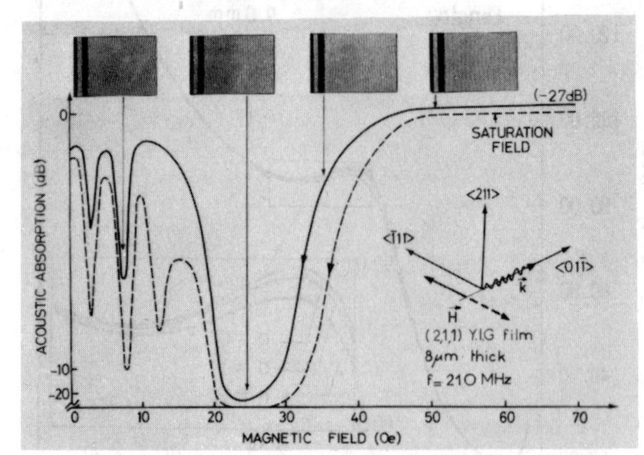

Fig. 1 - Acoustic absorption versus biasing field. Magnetic domains are shown in inset.

evidenced by the inset of magnetic domain pictures shown on this Figure, where the domain have been visualized by the Faraday effect. Similar results are obtained when the magnetic field is applied along the propagation direction <0,1,$\bar{1}$>. However the absorption spreads over a larger field range (up to about 100 Oe) in accordance with the magnetic anisotropy in the film plane : the saturation fields were found to be 56 Oe and 125 Oe for fields along <$\bar{1}$,1,1> and <0,1,$\bar{1}$> directions respectively. Another difference for this second experiment (field along <0,1,$\bar{1}$>) is that no change was observed when the field was reversed. From these results we may conclude that these magnetoelastic interactions are strongly dependent on the domain pattern. Also the resonant character of these absorption spectra suggests that gyromagnetic resonances can explain such a strong effect. To check these assumptions an experiment was designed in order to observe these gyromagnetic resonances with a magnetic excitation. The result obtained with a microstrip circuit is shown in Figure 2, where the acoustic absorption at 210 MHz was also plotted. The agreement is quite good, according to the experimental accuracy. It must be noted that all

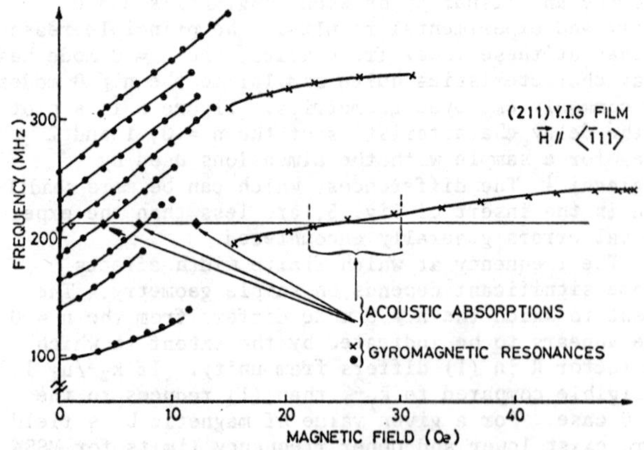

Fig. 2 - Gyromagnetic resonances in a YIG film.

the resonances vanish before magnetic saturation, suggesting that they are of the stripe domain resonance type[4]. The calculation of such stripe domain resonances was undertaken following the previous work of Smit and Beljers[4] about uniaxial materials, but it was not possible to solve the problem in the present case, due to its complexity : this results from the complicated film orientation ((2,1,1) plane) which is not a symetry plane for the magnetic anisotropy. Then a more tractable geometry was studied : a (1,0,0) film. The calculations were performed in this case and a typical result is shown in Figure 3, together with the acoustic absorption, measured as previously explained at 210 MHz

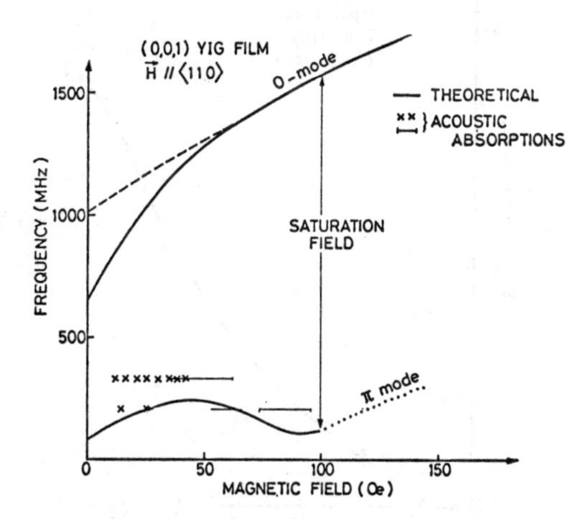

Fig. 3 - Stripe domain resonances versus magnetic field

and 335 MHz. Only the two extreme cases of in-plane precession (O-mode) and anti-phase precession (π-mode) in the adjacent domains were computed. With acoustic excitation, the phase variation is imposed by the acoustic wavelength, which does not coincide in general with the domain periodicity, so that a detailed comparison is presently meaningless. Also, open stripe domains were assumed for the computation, although for the rather thick film used (17 μm thick) closure domains are present which probably influence the gyromagnetic spectrum. However, the acoustic absorptions are situated as expected between the two computed curves.

Another approach to this study of magnetoelastic interactions was to use magnetic films with smaller in-plane saturation fields. Such films have been grown previously for integrated optics experiments[5]; the stress or growth anisotropy induced by the epitaxy process is used to force the magnetization to lie in the film plane. Using the same experimental structure as above, but now with an "easy-plane" film (of composition $Gd_{.45}Y_{2.55}Ga_{.9}Fe_{4.1}O_{12}$ and (1,1,1) orientation) we obtained the amplitude variation of the delayed signal shown in Figure 4, with the orientation of the biasing field as a parameter. It is seen that no fine structure occurs; also, no difference was observed in

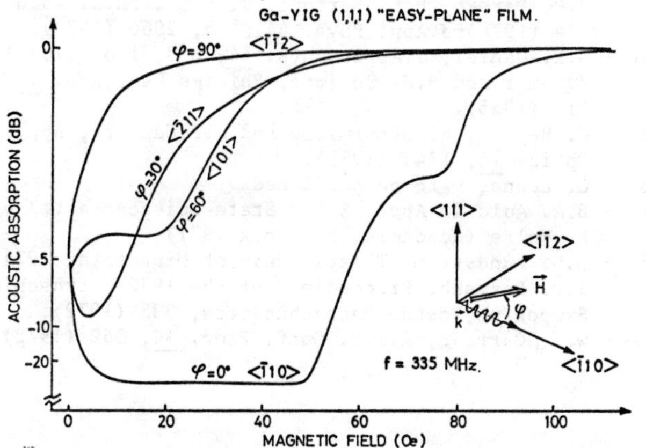

Fig. 4 - Acoustic absorption for various biasing field directions

these spectra for reverse biasing fields. These two effects, observed for pure YIG films, can be attributed to the presence of stripe domains. As before, the resonance spectra were computed, assuming a single domain film, but allowing the static magnetization to differ from the biasing field direction. The magnetization direction was found by minimizing the total energy[6], then the resonance frequency was computed with the general expression :

$$\frac{\omega}{\gamma} = \frac{1}{M\cos\theta}\left\{ f''_{\theta^2}\, f''_{\psi^2} - f''^{2}_{\theta\psi} \right\}^{1/2}$$

where f''_{θ^2}, f''_{ψ^2}, $f''_{\theta\psi}$ are the second derivatives of the energy density with respect to the polar angles of the magnetization (ψ , the azimuth angle and θ, the elevation angle from the film plane) and ω,γ and M are the angular frequency, gyromagnetic ratio and magnetization intensity, respectively. The result of this computation is shown in Figure 5, together with the experimental results. The material parameters have been

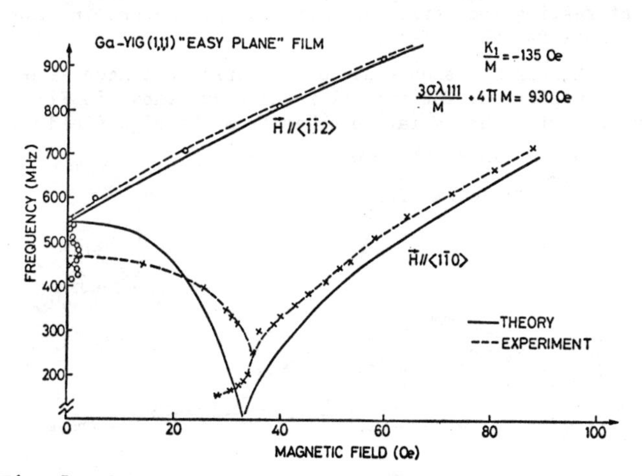

Fig. 5 - Computed and measured gyromagnetic resonances

determined through X-ray and Magnetooptical measurements[5] except for the cubic anisotropy field, which was determined by fitting the theoretical and experimental results obtained for the resonance frequency for H parallel to ⟨1̄,1̄,2⟩. It is then seen that a resonant interaction can occur at the operating frequency of 335 MHz in the saturation state only for a field parallel to the hard axes in the plane. The rather broad absorption is mainly caused by the comparatively large value of the linewidth (18 Oe at 300 MHz). However this result shows the possibility of having a strong magnetoelastic effect in a saturated film at reasonable frequencies. To interpret such experiments would require the set of coupled equations governing the elastic propagation and magnetization motion to be solved, which in the present case would involve quite intricate calculations. But one can avoid this long process and derive usefull conclusions from simple arguments : the magnetoelastic interaction can be viewed from the magnetic sub-system as an effective field[7] :

$$h_i^e = -\frac{\partial f_{mel}}{\partial \alpha_i}\cdot\frac{1}{M}$$

where f_{mel} is the magnetoelastic energy density and α_i the components of a unit vector parallel to the magnetization.

This effective field acts as a driving field for the magnetization precession, the amplitude of which depends, for a given driving field intensity, on the position of the elastic mode relative to the magnetic normal modes. For instance, in the above case of a (1,1,1) oriented film with propagation along <$\bar{1}$,1,0>, one finds :

$$\begin{cases} h_1^e = \frac{\sqrt{2}}{3}\left(\frac{B_1 - B_2}{M}\right) e_{23} \cos\phi \\ h_2^e = \frac{\sqrt{2}}{3}\left(\frac{B_1 - B_2}{M}\right) e_{23} \sin\phi - \frac{2}{3}\left(\frac{B_1 + 2B_2}{M}\right) e_{22} \cos\phi \\ h_3^e = -\frac{2}{3}\left(\frac{B_1 + 2B_2}{M}\right) e_{23} \cos\phi + \frac{\sqrt{2}}{3}\left(\frac{B_1 - B_2}{M}\right) e_{22} \sin\phi \end{cases}$$

where <$\bar{1}$,$\bar{1}$,2>, <$\bar{1}$,1,0> and <1,1,1> corresponds to the 1, 2, 3 axes, respectively B_1, B_2 are the magneto-elastic constants, e_{ij} the strain tensor components and ϕ the angle between the magnetization and the <$\bar{1}$,1,0> axis. From these expressions it can be shown for instance that the acoustic absorption is smaller for the magnetization along <$\bar{1}$,0,1> (ϕ=60°) than along <$\bar{1}$,1,0> (ϕ=0) although they are magnetically equivalent. It is then possible, with these simple arguments, and also with the resonance computations, to find the interesting geometries and material parameters in order to optimize the effect.

Measurements of phase delay variations have also been performed. The typical results are shown in Figure 6.with the variation taken from the high field

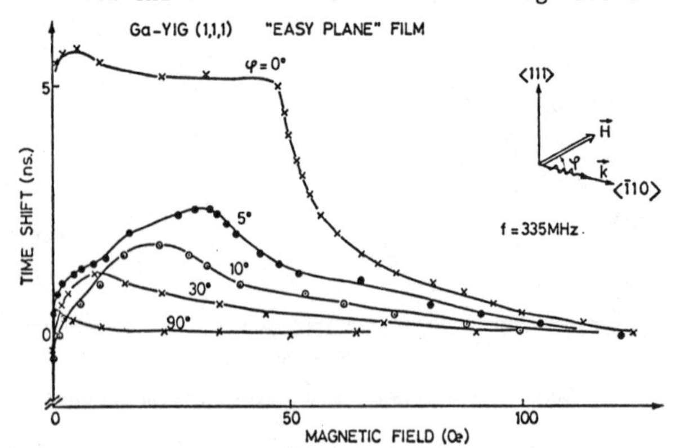

Fig. 6 - Time shift measurements of delayed signal as a function of in-plane magnetic field

value of delay. The largest variation (5.5nsec) was obtained for a field in the film plane and parallel to the propagation direction : it corresponds to a value of .32x10^{-2} for $\Delta k/k$,where k is the wavenumber.This has to be compared with a previous measurement reported by M.S.Lundstrom[8] who found 7x10^{-4}at 49 MHz with a pure YIG plate.Calculations of this Δk effect have also been performed by J.P.Parekh[9] who obtained a maximum value of 3x10^{-2} for $\Delta k/k$,when the field makes an approximate angle of 40°with the propagation in contrast to our measurement.But the film orientation and material parameters were different ((100) oriented Ga : YIG film), so that a detailed comparison is not possible.

STRIPE DOMAIN DRIFT EXPERIMENTS

The drift of magnetic stripe domains,subjected to an elastic surface wave,was observed for the first time to our knowledge.For this experiment a(0,0,1) YIG film was used in order to have a regular stripe domain pattern,with a width dependence on the applied field.The propagation direction for the elastic wave was the <1,1,0> axis and the biasing field was applied along the <$\bar{1}$,1,0> axis in order to set the stripes parallel to the wavefront.An optical set-up was designed to measure the drift velocity : a 632.8nm laser light was focused on the film ;the polarized light was analyzed with a polarizer crossed on the light transmitted by one kind of domains and then displayed on a diffusing screen.A photomultiplier was then set behind this screen,with a diaphragm so that only the light coming from one stripe was received.The signal so obtained was then fed to an oscilloscope and a frequency meter. From this frequency measurement and the determination of the stripe width,the drift velocity was inferred.This velocity drift has been plotted in Figure 7 versus the acoustic power.One can notice the small values of the acoustic power necessary to move the stripes. However a saturation level occurs also for a rather small value.

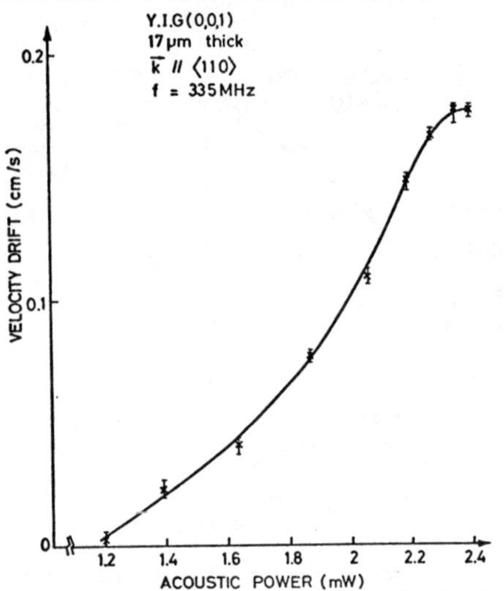

Fig. 7 - Measurements of the domain drift velocity versus acoustic power.

Evaluation of this effect has been attempted by considering the influence of the elastic wave on the wall energy,as previously explained by W.J.Carr Jr[10].This, however,did not permit an explanation of the experimental results,in particular the fourth power dependence of the initial variation of the drift velocity on the acoustic power. The simplifications introduced in this evaluation are probably excessive, in particular neglecting the wall mass and the interactions between domains. Work is in progress to find a more refined model capable of explaining the observed behavior.

ACKNOWLEDGMENT

The authors are grateful to Dr. Hepner for his help with the numerical calculations and to D.Cohen for his technical assistance.

REFERENCES

+ - Work supported by D.R.M.E.
++ - Present address : W.W. Hansen Laboratories, Stanford University, Stanford, CA 94305

1 - M.F. Lewis and E. Patterson, Appl. Phys. Letters 20, n° 8, 276 (1972).
2 - J.P. Parekh, H.L. Bertoni, J. Appl. Phys. 45,n°1, 434 (1974)-J.Appl.Phys. 44,n° 6, 2866 (1973).
3 - M.R. Daniel, J.Appl. Phys. 44, n°3, 1404 (1973).
4 - J. Smit and H.J. Beljers, Philips Res. Rep. 10, 113 (1955).
5 - G. Hepner, B. Désormière and J.P.Castéra, Appl. Optics 14, 1749 (1975).
6 - G. Hepner - to be published
7 - B.A. Auld in Appl. Solid State Science edited by R. Wolfe (Academic, New York 1971)
8 - M.S. Lundstrom, Thesis (Univ.of Minnesota, 1974).
9 - J.P. Pareskh, Proceedings of the 1972 Ultrasonics Symposium,Boston,Massachussetts, 333 (1972).
10 - W.J. Carr Jr, A.I.P. Conf. Proc. 10, 369 (1972).

NEW MAGNETOACOUSTIC MODES IN YIG SUBSTRATES*

Goh Komoriya and G. Thomas
State University of New York at Stony Brook, Stony Brook, New York 11794
and
J. P. Parekh
State University of New York Maritime College, Fort Schuyler, Bronx, New York 10465

ABSTRACT

Previously unreported magnetoacoustic modes have been detected on the surface of a $1.5 \times 0.1 cm^2$ YIG disk. The waves were excited by meander-line pairs along the [100] and [011] directions on a $(01\bar{1})$ surface. The meander line excited both a magnetoelastic Rayleigh wave and the new modes for both the [100] and [011] directions. A drop of water on the surface of the substrate was used to absorb the energy in the surface wave modes while the velocity and dispersion of the new mode were being examined. For the [100] direction the new modes were approximately 8dbm down from the surface wave, while for the case of the [011] direction nearly all of the energy was carried in the new modes. The velocity of the surface wave was $(v_s/1.07) \times (1 \pm 0.015)$ (i.e. within 1.5% of the theoretical value) while the velocity of the first of the new modes was $v_s(1 \pm 0.0$[illegible]$)$ for the [100] direction and is dispersive for the [011] direction. If the excitation frequency is held constant the maximum amplitude of the new wave is excited at approximately 20G above the magnetic field which produces the maximum excitation for the surface wave. For excitation along the [011] direction, the new wave splits into two modes for magnetic fields between 300G and 330G. The new mode could also be excited by a Rayleigh wave produced by IDT's on a ZnO film for the [100] direction.

INTRODUCTION

Magnetoelastic (M.E.) surface waves have received increasing attention in recent years.[1-11] Much of this interest has been stimulated by the theoretical studies which have predicted strong nonreciprocal and dispersive behavior of M.E. surface waves. Experimental observation at relatively low frequency magnetic fields have shown some nonreciprocal behavior[6,11] and have shown variability of the delay time with magnetic field.[8,11] It has been known for many years that magnetoacoustic bulk waves exhibit strong nonlinear effects. M.E. surface wave devices have been shown to have very high internal efficiency as convolvers[10,11] even though the overall insertion losses are high.

EXPERIMENTAL SET-UP AND RESULTS

Meander line pairs were deposited on the $(01\bar{1})$ surface of a $1.5 \times 0.1 cm^2$ YIG disk so that the direction of propagation could be along the [011] or [100] directions. The meander lines were formed by etching the lines into a 1000 to 1200 Å copper film or a 1500 to 2000 Å aluminum film using standard photolithographic techniques. Each of the meander lines consisted of 10 pairs of lines spaced to excite 70 MHz or 140 MHz Rayleigh waves. The meander lines used for excitation and detection were separated by approximately 1.18 cm, 1.0 cm, 0.8 cm, and 0.6 cm for both the 70 MHz and the 140 MHz lines.

The YIG disk was mounted inside an aluminum box into which BNC couplers were fastened. The box was attached to a jig so the sample could be rotated in either the plane of the disk or the plane perpendicular to the surface of the disk. The box, containing the YIG disk, was placed between the poles of a 7" electro-magnet. The field was measured and found to vary less than 1 gauss over an area at least as large as the sample when the box was removed. No attempt was made to match the impedance of the transducers to the rest of the system.

The exciting signal on rf oscillator was passed through a mixer which was pulsed with a rectangular pulse whose duration and repetition rate could be varied. Typically a 0.25μs pulse was used. The mixer produced reasonably good bursts of rf power. The signal detected at the other meander lines was passed through a second mixer which was used to down-convert the frequency so that it could be easily amplified and displayed on an oscilloscope. Delay times were measured directly from the display on the oscilloscope. The delayed pulse of rf power detected by one of the meander lines was smoothed considerably as can be seen in the oscilloscope traces, hence the delay time was measured from the beginning of the exciting pulse to the maximum of the delayed pulse.

The magnetic field was increased to values well above the saturation magnetization and then lowered until a delayed signal could be measured at the detector approximately 1.18cm away from the exciting transducer. Several distant modes could be observed.

For the [100] direction of propagation, two modes which propagate faster than the Rayleigh-wave mode could be observed. In Fig. (1) are the oscilloscope traces of the delayed pulse at 143.5 MHz and 360G. Fig. (1a) is the trace for an unloaded surface; Fig. (1b) is for exactly the same conditions except that a drop of water has been placed in the path connecting the two transducers. As can clearly be seen, the large amplitude Rayleigh wave has been absorbed nearly completely by the water droplet, while the other two modes remain unchanged. In Fig. (2), the insertion loss and delay time for these modes are shown as a function of magnetic field. In Fig. (3) the variation of the insertion loss and delay time is shown as a function of the angle of in plane rotation of the magnetic field away from alignment with the direction of propagation of the waves.

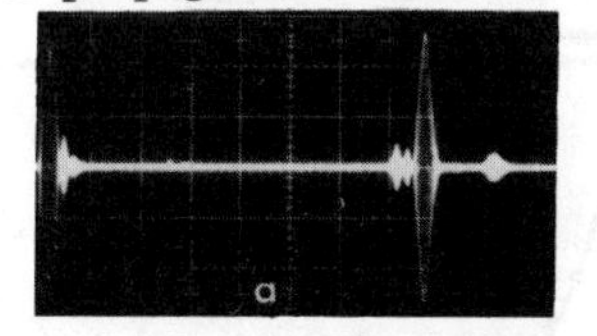

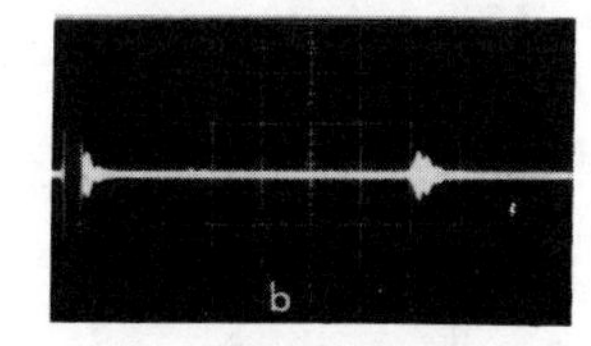

Fig. 1. Oscilloscope traces of the delayed magnetoacoustic waves propagating in the [100] direction. In both traces the magnetic field is 360G and the frequency is 143.5 MHz. The horizontal scale is 0.5μs/cm and the vertical is 0.1V/cm. The trace in (a) is for an unloaded surface while in (b) a drop of distilled water has been placed between the excitation and detection transducers.

In the case of propagation in the [011] direction these new modes become even more complicated. In Fig. (4) are the oscilloscope traces of the delayed pulse for 143.5 MHz and 295G. Fig. (4a) is the trace for the unloaded surface and Fig. (4b) the trace for the same conditions except that a drop of distilled water has been placed in the path between the two transducers. As can be seen the two traces remain almost unchanged. There is a small amount of energy carried by the Rayleigh wave which is absorbed by the water droplet (see arrows). There exist at least three modes other than the Rayleigh wave and perhaps as many as seven modes can be detected. (The direct-air transmission of the exciting pulse may be seen to the left of the delayed signal. As can be seen, it is a fairly clean pulse, with only a small amount of "ringing" after the main pulse.)

In Fig. (5) the variation of the insertion loss and delay time of the two strongest delayed signals of Fig. (4) is shown as a function of magnetic field. As can be seen, there is a considerable amount of dispersion as a function of magnetic field for this direction of propagation. In Fig. (6) the variation of the insertion loss and delay time as a function of the deviation of the direction of the magnetic field from alignment with the direction of propagation is shown. As can be seen, there is a change of nearly 20 db in the insertion loss for a change of less than 6° from alignment of the magnetic field.

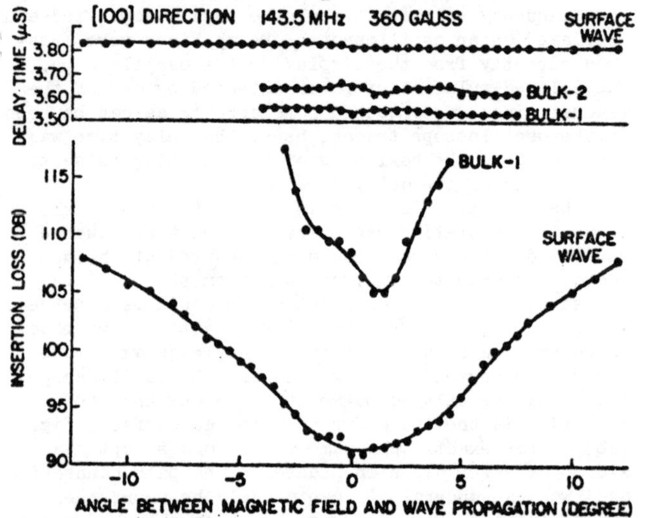

Fig. 2. The variation of the insertion loss and delay time as a function of magnetic field for the waves displayed in Fig. (1). The curve labeled Bulk wave corresponds to the new magnetoacoustic wave.

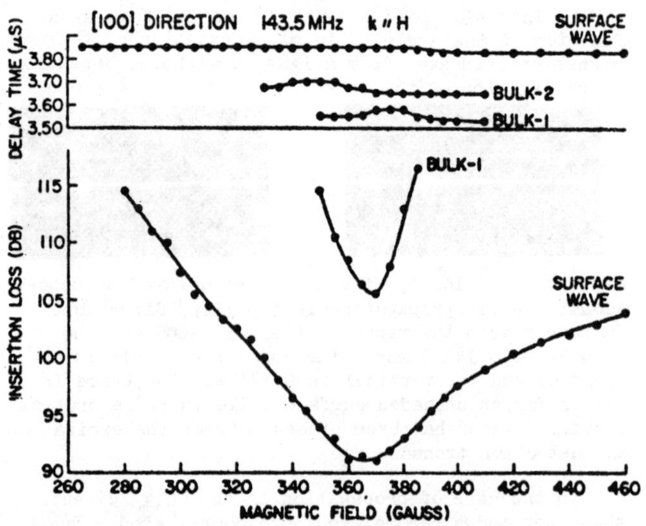

Fig. 3. The variation of the insertion loss and delay time as a function of the deviation of the direction of magnetic field from alignment with the direction of wave propagation for the waves displayed in Fig. (1).

Next we attempted to detect the delayed signal at a transducer placed midway between the two end transducers. Only the surface wave mode could be detected (with almost an identical amplitude as at the end transducer) for propagation along the [100] direction and neither surface nor the new mode could be detected for propagation along the [110] direction. (The absence of the surface mode in the case of the [110] propagation may only indicate that the surface wave mode is extremely weak, as can be seen in Fig. 4).

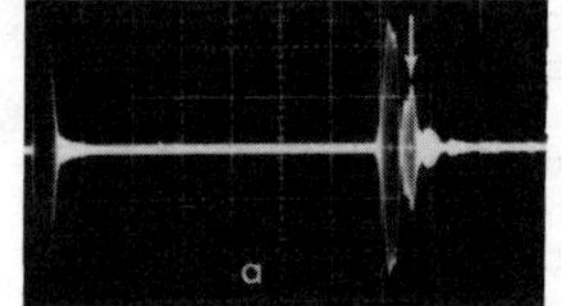

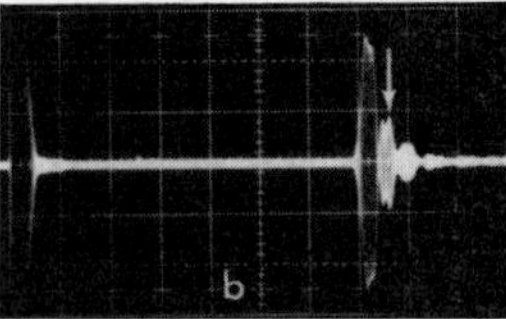

Fig. 4. Oscilloscope traces of the delayed magneto-acoustic waves propagating in the [001] direction. In both traces the magnetic field is 295G and the frequency is 143.5 MHz. The horizontal scale is 0.5μs/cm. The trace in (a) is for an unloaded surface (the arrow marks the presence of the M.E. surface wave). While in (b) a drop of distilled water has been placed between the excitation and detection transducers. (The arrow in (b) marks the position of the M.E. surface wave in trace (a).)

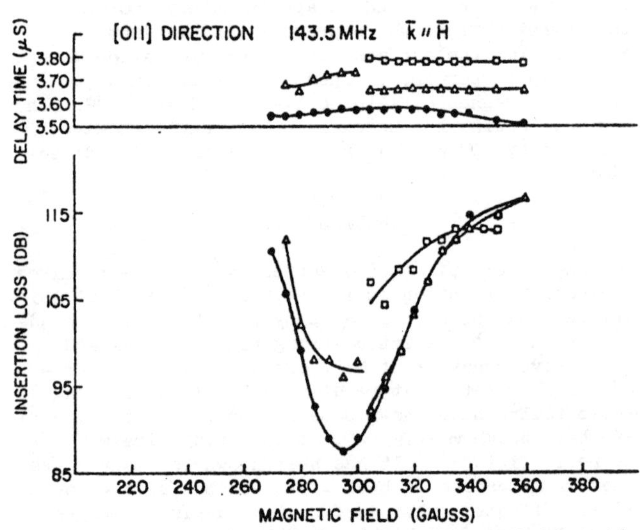

Fig. 5. The variation of insertion loss and delay time as a function of magnetic field for the waves displayed in Fig. (4).

As can be seen in Figs. (1) through (6), the new modes have a propagation velocity which is faster than the Rayleigh wave. If we assume that the peak of the exciting pulse of rf power occurs half way between the beginning and end points of the pulse we calculate the velocity of the Rayleigh wave to be $(v_s/1.07)$ (1 ± 0.015) and the velocity of the first (and strongest) of the new modes to be $v_s(1 \pm 0.02)$ where v_s is the shear velocity YIG.

For a fixed magnetic field (i.e. 360G) the frequency at which the maximum excitation of the strongest of the new modes occurred was at 154 MHz, that is, 10.5 MHz above the frequency at which the maximum excitation of the Rayleigh wave occurred. For a synchronous wavelength (i.e. twice spacing between the line pairs of the meander lines) the resonant frequency for this mode should be 7% higher. This is because the velocity of the first of the new modes is approximately 7% higher than that of the Rayleigh wave. Experimentally the resonant frequency is found to be 7.3% above that of the Rayleigh wave, which is well within experimental error.

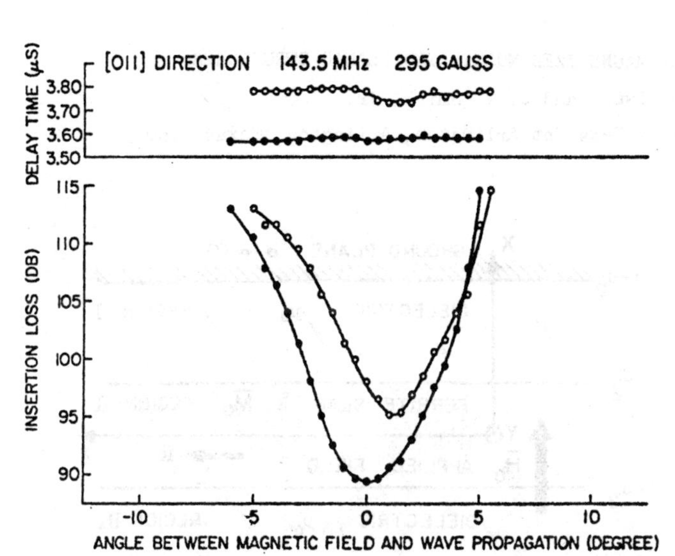

Fig. 6. The variation of the insertion loss and delay time as a function of the deviation of the direction of magnetic field from alignment with the direction of wave propagation for the waves in Fig. (4).

The results of this experiment cannot be attributed solely to the meander lines. Almost identical results to those obtained for the [100] direction were obtained when the Rayleigh wave was excited by IDT's. 1.7μ thick ZnO films were dc sputtered on gold pads which were evaporated onto the (01$\bar{1}$) surface of a 1.5x0.5x0.1cm^3 YIG plate, exciting waves in the [100] direction. The IDT transducers had a peak efficiency at 83 MHz. The surface waves excited by these transducers had a tuned insertion loss (~17db) at high magnetic fields. Even for fields which produced maximum interaction with the magnetic lattice of the YIG, the insertion loss was only about 40 db. When a drop of distilled water was placed between the excitation and detection transducers, however, the surface wave was absorbed and the presence of at least one of the new modes was observable. The insertion loss for the new modes was approximately 60db.

DISCUSSION

The presence of the new modes may explain some of the observations in previous experiments. For example, in Ref. 8 large variation in the delay time with magnetic field was reported. If we assume for the moment that two modes - a surface wave and one of the new guided modes - were present, then there would be an apparent shift in the delay time as a function of magnetic field, even without dispersion in either. As can be seen in Figs. (2) and (5) the relative amplitude of the two modes changes as a function of magnetic field. Hence, if the pulse length of the exciting pulse is long enough (or if the spacing between the exciting and the detecting transducers is short enough) the two modes will overlap and the maximum of the combined pulse will be a function of their relative amplitudes. Now as the magnetic field changes, so will the maximum of the combined pulse.

The new modes are some form of magnetoacoustic waves: They can be excited by means of purely magnetic coupling (i.e. meander lines are line sources which couple to the magnetic ions) or by means of purely elastic coupling (i.e. ZnO films and IDT's). On the other hand, they do not seem to have the same particle displacement throughout the sample, since the meander lines which were more than about 2mm away from the edge of the disk could not detect the signal. If the excitation efficiency is a strong function of internal magnetic field and the effective demagnetization factor changes rapidly with position it is possible that the conditions at the interior transducers are not right for the detection of the sample. However, the signal at the end transducer is observable for a change of at least 30% in the magnetic field; hence, it does not seem likely that the conditions at the transducer which is approximately 1cm from the exciting transducer cannot be brought into a range where detection is possible.

There are several types of magnetoacoustic waves which might be responsible for at least part of the observations of the new modes. The most likely are the SH and SV plate modes. The velocity of propagation of the SH and SV plate modes approach v_s for thick plates, and, since the first of the new modes has a propagation velocity nearly equal to v_s, they cannot be ruled out. However, if the new modes discussed in this paper are the result of SH and SV plate modes they cannot be the simple modes present in a purely elastic plate.[12] There are at least two reasons for this assertion: (1) Elastic plate modes[12] have substantial particle desplacement at the surface. Yet neither the double-sided tape used to fasten the sample to the aluminum box nor water placed on the surface absorbed measurable energy from the new modes; and (2) The Rayleigh wave could be detected by meander lines placed between the end transducers while the new modes could only be detected at the end transducers.

A more plausible explanation is that these new modes are in fact, a new type of guided mode which perhaps only exist in samples with non uniform magnetic field.

REFERENCES

* Work supported by the NSF under Grant ENG 73-03848.

1. J. P. Parekh and H. L. Bertoni, J. Appl. Phys. 45, 1860 (1974); J. Appl. Phys. 45, 434 (1974); J. Appl. Phys. 44, 2866 (1973) and Appl. Phys. Letters 20, 362 (1972).
2. H. van der Vaart and H. Mathews, Appl. Phys. Letters 16, 222 (1970).
3. H. Mathews and H. van der Vaart, Appl. Phys. Letters 15, 373 (1969).
4. F. W. Voltmer, R. M. White and C. W. Turner, Appl. Phys. Letters 15, 153 (1969).
5. M. F. Lewis and E. Patterson, Appl. Phys. Letters 20, 276 (1972).
6. M. R. Daniel, J. Appl. Phys. 44, 1404 (1973).
7. J. Parekh, 1972 Ultrasonics Symposium Proc. IEEE.
8. T. L. Tsai, G. Komoriya, G. Thomas and J. P. Parekh, 1974 Ultrasonics Symposium Proc., IEEE, p. 489 (1974).
9. S. Shen, J. P. Parekh, and G. Thomas, 1974 Ultrasoncis Symposium Proc., IEEE p. 201
10. M. S. Lundstrom and W. P. Robbins, 1974 Ultrasonics Symposium Proc., IEEE p. 348.
11. J. P. Parekh, S. Shen and G. Thomas, 1975 Ultrasonics Symposium Proc., IEEE p. 201.
12. B. A. Auld, Acoustic Fields and Waves in Solids, Vol. II, Wiley-Inter Sc., New York p. 66 (1973).

MAGNETOSTATIC PROPAGATION FOR UNIFORM NORMALLY-MAGNETIZED MULTILAYER PLANAR STRUCTURES

Ming-Chi Tsai, Hung-Jen Wu, J. M. Owens and C. V. Smith, Jr.

Department of Electrical Engineering, The University of Texas at Arlington, Arlington, Texas, 76019

ABSTRACT

Magnetostatic propagation in layered structures with magnetization in the plane and perpendicular to the direction of propagation has been studied theoretically and experimentally. Problems in device application of this geometry include low power saturation levels and limited delay times for reasonable film sizes. Both problems are minimized by use of normal magnetization giving volume modes having isotropic propagation in the film plane.

This work reports theoretical and experimental studies for a normally-magnetized ground plane-dielectric-YIG-dielectric-ground plane structure. The dispersion relation is derived for the magnetostatic limit and this geometry. This theory indicates that with at least one ground plane within a magnetostatic wavelength of the YIG, significant modification of the dispersion characteristic can be obtained. Nondispersive regions of approximately 10% bandwidths are predicted. Experimental studies with an LPE-YIG film thickness of 10μm and a dielectric thickness of 8 to 20 times the YIG thickness show nondispersive propagation over bandwidths > 100MHz at 1.6GHz. Agreement between theory and experiment is good. Saturation characteristics are better than those of similar magnetostatic surface wave devices.

INTRODUCTION

The first theory of magnetostatic waves in a ferrite slab was given by Damon and Eshbach[1]. This analysis treated, in particular, surface wave modes with a bias field perpendicular to the propagation direction. Damon and Van de Vaart[2] treated the case of normal magnetization of a slab with propagation in the plane of the slab. This mode has the advantage of isotropic propagation in the plane and more uniform energy density in the ferrite material than for surface modes. Bongianni[3] analyzed propagation in a dielectric layered thin film structure. In this structure Bongianni was able to show that nondispersive propagation was possible over limited bandwidths by proper choice of a YIG-dielectric-ground plane geometry. Studies of this mode by Adam, Bardai, Collins and Owens[4] have shown that output power saturation occurs at levels of less than -20dbm, limiting device applications (e.g., local oscillator delay lines).

This paper reports on theoretical and experimental studies of propagation in a normally magnetized ground plane-dielectric-YIG-dielectric-ground plane structure. Nondispersive propagation is observed for appropriate geometries with > 10% bandwidths. Saturation powers are greater than those for smilar surface wave structures.

THEORY

The model geometry considered for this problem is shown in Figure 1. An insulating ferrite slab is located parallel to and between perfectly conducting ground planes. The spaces between the ferrite slab and ground planes are filled with dielectric media. A static uniform magnetic induction of magnitude B_o is applied normal to the plane of the ferrite slab. It is assumed that the magnetic intensity internal to the ferrite slab has a magnitude H_o and produces a saturated magnetization of the ferrite slab with magnitude M_o parallel to the applied induction.

The development of the general magnetostatic equations for a ferrite medium has been considered in detail by others[1,2]. For a lossless ferrite and exp(jωt) time dependence, the perturbation complex

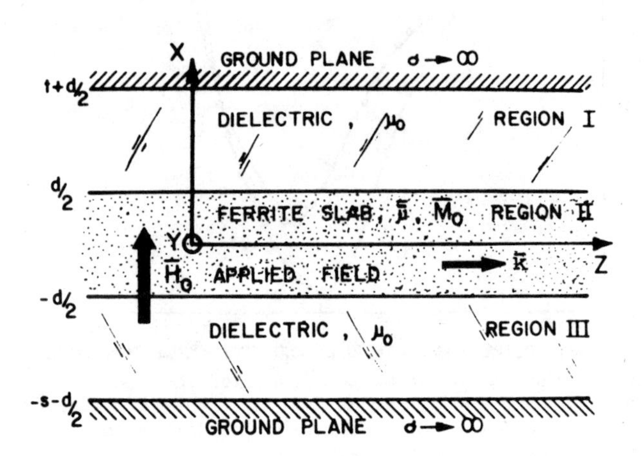

FIG. I Theoretical and experimental model geometry

amplitudes satisfy the magnetostatic equations

$$\text{curl } \bar{h} = 0 \tag{1}$$

$$\text{div } \bar{b} = 0 \tag{2}$$

and

$$\bar{b} = \bar{\bar{\mu}}\, \bar{h} \tag{3}$$

where $\bar{\bar{\mu}}$ is the Polder tensor[5]. This tensor, for normal magnetization, has the form

$$\bar{\bar{\mu}} = \mu_o \begin{pmatrix} 1 & 0 & 0 \\ 0 & 1+C & -jD \\ 0 & jD & 1+C \end{pmatrix} \tag{4}$$

with $C = \gamma^2 M_o H_o [(\gamma H_o)^2 - \omega^2]^{-1}$ (5)

and $D = \gamma\omega M_o [(\gamma H_o)^2 - \omega^2]^{-1}$. (6)

γ is the gyromagnetic ratio. These equations, as should be expected, reduce to those for the dielectric regions when the saturation magnetization M_o is set to zero. The boundary conditions for this system are continuity of the normal component of induction and tangential components of intensity at the dielectric-ferrite interface and vanishing of the normal component of induction at the ground planes.

A solution representing wave propagation along the z direction and independent of y can be obtained by assuming exp(-jkz) dependence and utilizing $h_z(x)$ as a potential for the x-variation; thus, from (1) and (3)

$$h_x = -(jk)^{-1}\, dh_z/dx \tag{7}$$

$$b_x = -\mu_o (jk)^{-1} dh_z/dx \tag{8}$$

$$b_y = -j\mu_o D\, h_z \tag{9}$$

and

$$b_z = \mu_o (1+C) h_z \tag{10}$$

Substitution of these results into (2) obtains

$$d^2h_z/dx^2 - k^2(1+C)h_z = 0 \text{ (ferrite)} \tag{11}$$

and $d^2h_z/dx^2 - k^2h_z = $ (dielectric) (12)

The solution satisfying the boundary conditions $b_x = 0$ at the ground planes has the form

$$h_z(x) = \begin{cases} h_I \cosh [k(t+d/2-x)] , & I \quad (13) \\ h_{II}^{(1)} \sin (pkx + h_{II}^{(2)} \cos (pkx), & II \\ h_{III} \cosh [k(s+d/2+x)] , & III \end{cases}$$

where $p = j(1+C)^{\frac{1}{2}}$ (14)

Matching the remaining four conditions at the dielectric-ferrite interfaces gives four linear homogeneous equations for the amplitudes h_I, $h_{II}^{(1)}$, $h_{II}^{(2)}$ and h_{III}; the necessary condition for a nontrivial solution yields the dispersion relation

$$\tan(pkd) \frac{p^2 - \tanh(kt)\tanh(ks)}{p[\tanh(kt) + \tanh(ks)]} = 1 \qquad (15)$$

Examination of this dispersion relation indicates even symmetry in both ω and k which implies propagation is isotropic in the plane of the ferrite slab. This result has been confirmed in experiments by reversing the direction of the applied static field. Taking advantage of symmetry a portion of a normalized dispersion relation is presented in Figure 2 for the particular case of $t/d \to \infty$, $s/d = 10$ and the

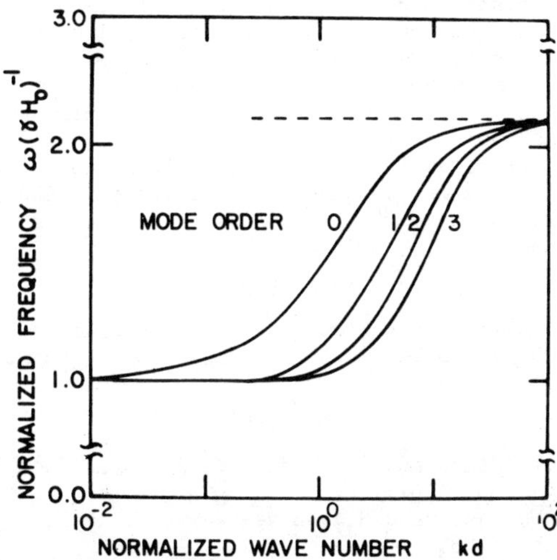

FIG. 2 Normalized dispersion relation for $M_o/H_o = 3.46$, $t/d \to \infty$ and $s/d = 10$.

four lowest order modes. The general shape for this normalized dispersion relation is the same for all cases with the invariant point $\omega/\gamma H_o = 1$, $kd = 0$ and the asymptote $\omega/\gamma H_o = \sqrt{(1+M_o/H_o)}$ for large values of kd. Characteristically, the variation in the dispersion relation occurs for kd values near 1 to 10. It should be noted that the upper limit for magnetostatic waves is $k \cong 10^7 \, m^{-1}$ for YIG.

For magnetostatic delay lines the parameter of significance is the group delay time

$$\tau = (d\omega/dk)^{-1} \text{ sec/m} \qquad (16)$$

Due to the analytic complexity of the dispersion relation a simple form for the group delay is not attainable; consequently, two methods were applied to calculate τ. The first method, following Bongianni[3], computes increments Δω and Δk during numerical solution of the dispersion relation and applies

$$\tau = \lim_{\Delta\omega \to 0} \Delta k/\Delta\omega \qquad (17)$$

The second method uses an explicit expression for τ obtained by differentiation of the dispersion relation and solving for $(d\omega/dk)^{-1}$. This expression must be evaluated during the numerical solution of the dispersion relation. Both methods yield the same results except at the end points where the increment method must fail. A plot of normalized group delay time as a function of normalized frequency is shown in Figure 3 for the particular case of $t/d \to \infty$, $s/d = 10$ and the four lowest order modes.

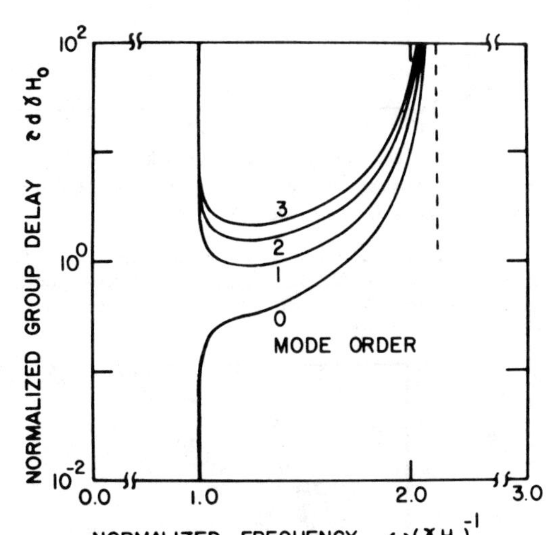

FIG. 3 Normalized group delay time for $M_o/H_o = 3.46$ $t/d \to \infty$ and $s/d = 10$.

In order to determine a suitable magnetostatic delay line structure, the dispersion relation was investigated for the lowest order mode as a function of s/d with $t/d \to \infty$ and $M_o/H_o = 3.46$. The results are presented in Figure 4; clearly, there is significant variation in the dispersion relation only in a limited range. On the expanded scale of time delay versus frequency in Figure 5, it is evident that regions of essentially constant group delay exist for a suitable choice of geometry and other parameters.

EXPERIMENTAL TECHNIQUE AND RESULTS

The propagation characteristics of the normally magnetized magnetostatic waves in a dielectric layered structure was measured by pulse delay techniques. Figure 6 shows the experimental setup, for this measurement, as well as the device structure. A CW microwave source was gated with an HP3312A PIN switch to provide an RF pulse of about 50 nsec duration with CW leakage greater than 50db below the peak pulse amplitude. The output signal was coupled directly to a Tektronix 564/S2 sampling oscilloscope. The delay line was fabricated from a 10μm thick LPE YIG film grown from a PbO/BaO Flux on a 1.50 cm diameter Gadolinium Gallium Garnet substrate 0.5mm thick. Input and output coupling was provided by 50μm diameter Au wires laid over the YIG surface and shorted to ground. Spacing between the couplers was 1 cm. The dielectrics used were mylar sheets with thicknesses of 80, 125, 160, 200 μm corresponding to s/d ratios of 8, 12, 16, 20. Dow Corning 704 silicon oil was used between layers to insure uniform dielectric continuity and the entire sandwich was compressed together using polyuathane sponge.

The time delay of the wave packet was measured for different s/d ratios with $t/d \to \infty$ and compared with

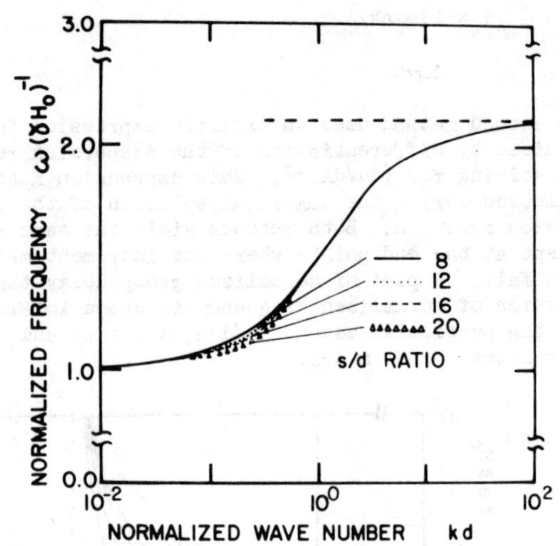

FIG. 4 Normalized dispersion relation for $M_0/H_0 = 3.46$, $t/d \rightarrow \infty$, Mode 0 and s/d = 8,12,16,20.

theory. Results are shown in Figure 5. The fit of experiment data to theoretical curves was good. Insertion loss was 45-50 db due to the fact the couplers were unmatched.

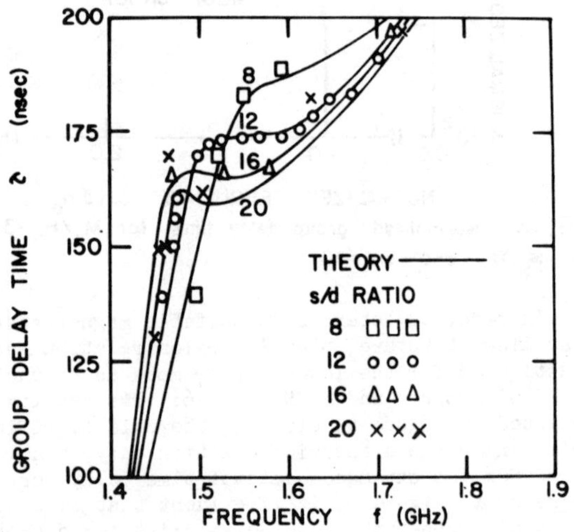

FIG. 5 Group delay time versus frequency for Mode 0, $M_0/H_0 = 3.46$, $t/d \rightarrow \infty$ and s/d = 8,12,16,20 with a 10 μm YIG slab on a GGG substrate.

Deviation from monotonic behavior can be clearly seen and "flat" delay regions with fractional bandwidth of >10% identified. As in the magnetostatic surface wave case, a decrease in dielectric thickness results in an increase in time delay, as well as a change in the shape of the nondispersive region. Reversing of the direction of the magnetic bias field resulted in no significant change in the propagation characteristics.

Network analyzer measurement of the couplers indicated an input impedance of ≅ 5Ω yielding an excess coupler loss due to mismatch of 14-18 db. Taking into account the 3db bi-directional loss at each coupler this indicates that the propagation loss is in the range of 8-10db which is consistent for material with $\Delta H \cong 40$ amps m^{-1}.[6] This is a typical linewidth for such material. Measurement of saturation power input and output indicate for a matched line, output saturation power should be in the range of -2 to -4 dbm for a 150nsec delay, 100nsec wide pulse at 1.5 GHz.

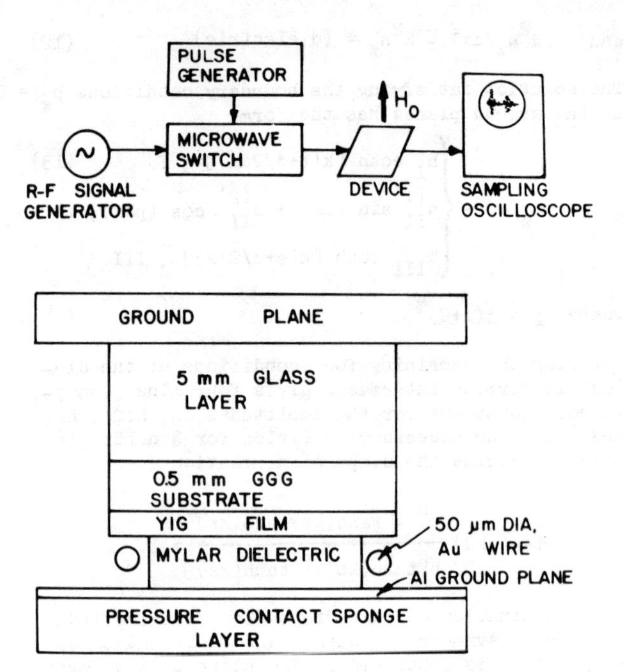

FIG. 6 Schematic of experimental measurement configuration

CONCLUSIONS

This work demonstrates nondispersive magnetostatic delay line structures utilizing normally magnetized ground plane-dielectric-YIG-dielectric-ground plane thin film structures. Agreement between theory and experiment for a representative structure is very good. The magnetostatic modes propagate as bulk volume waves. The advantages of such structures for practical devices include isotropic propagation in the plane of ferrite film, better use of film volume with approximately 10 db better saturation powers than surface wave devices, and possibility of meander structures and wave guides giving long delays in limited spaces.

ACKNOWLEDGMENT

The authors acknowledge the financial support of the Graduate School, The University of Texas at Arlington.

REFERENCES

1. R.W. Damon and J.R. Eshbach, J. Phys. Chem Solids 19, 308 (1961).
2. R.W. Damon and H. Van de Vaart, J. Appl. Phys. 36, 3453 (1965).
3. W.L. Bongianni, J. Appl. Phys. 43, 2541 (1972).
4. J.D. Adam, Z.M. Bardai, J.H. Collins and J. M. Owens, AIP Conf. Proc. 24, 499 (1975).
5. D. Polder, Phil. Mag. 40, 99 (1949).
6. J.C. Sethares and M.R. Stiglitz, IEEE Trans. Mag 10, 787 (1974).

S-BAND LITHIUM FERRITE PHASE SHIFTERS BY ARC PLASMA SPRAYING

W. Wade* and R. Babbitt*
U. S. Army Electronics Technology and Devices Laboratory
(ECOM) Fort Monmouth, New Jersey 07703

ABSTRACT

The arc plasma fabrication of ferrite phase shifters has been extended from C-band to S-band. Two S-band lithium ferrite compositions, Ampex 3-601 and Ampex 3-750, were sprayed around a dielectric for phase shifter applications in the 3 to 4 GHz range. The larger dielectric and greater ferrite wall thickness of the S-band phase shifter necessitated a moderation of the arc plasma parameters from those established for C-band. The remanence of arc plasma fabricated phase shifters was 250 gauss and 380 gauss for the 3-601 and 3-750 compositions respectively, which were significantly less than the stress free Br's for these compositions. These remanence values produced differential phase shifts of 55°/in. and 81°/in. which were in relative agreement with the 77°/in. differential phase shift produced by a commercial YIG phase shifter with a remanence of 355 gauss. Most significantly, the insertion loss of the arc plasma fabricated lithium ferrite phase shifters was the same as for the commercial YIG phase shifter, i.e. 0.4 dB for a 2.3 inch length. Also, the temperature dependence of Br for the arc plasma phase shifters was more stable than the commercial YIG.

INTRODUCTION

It has been demonstrated that the arc plasma spray (APS) process is capable of fabricating ferrite phase shifters.[1,2] Originally, C-band lithium ferrite phase shifters were sprayed and evaluated. More recently the APS process has been extended for S-band (3 to 4 GHz) phase shifters. The use of lithium ferrite for S-band offers a significant cost advantage over yttrium-iron garnet compositions. However, lithium ferrite has supposedly higher microwave losses at S-band frequencies; but work by C. F. Jefferson and R. West[3] indicates that low loss S-band lithium ferrites would be achieved with proper compositional substitutions, and more recent developments by P. Baba et.al.[4] have significantly improved the microwave characteristics of lithium ferrite. Recent discussions with G. Argentina of Ampex indicate that most S-band lithium ferrite phase shifters have insertion losses greater than 1 dB. Also, the substitution of cobalt to increase the power capacity of lithium ferrite, further increases microwave losses.

This effort, the arc plasma spraying of S-band lithium ferrite, was done with two lithium ferrite compositions, Ampex 3-750 and Ampex 3-601. The saturation magnetization ($4\pi Ms$) of the 3-750 and 3-601 compositions were 750 and 600 gauss respectively. Also, the 3-601 composition had a cobalt substitution which increased its ΔH_k from 2 to 6 oersteds (reported by Ampex). The phase shifter performances of these compositions, after arc plasma spraying, are compared to a conventionally sintered gadolinium substituted yttrium-iron garnet phase shifter.

ARC PLASMA SPRAY PROCEDURE

The same equipment and technique developed for spraying C-band phase shifters were used for S-band. The APS parameters, except for spray distances, were similar. A greater spraying distance was required to compensate for the large dielectric cross sectional dimensions, Figure 1, as compared to 0.15" x 0.12" for C-band. If the closer spray distance, as set for C-band, was used, enough heat would build up behind the S-band dielectric to maintain the deposited ferrite in, or close to, a molten state. This condition generally produces cracks and a second compositional phase which adversely affect magnetic properties. Two different combinations of APS parameters were found acceptable for spraying S-band compositions. These combinations are:

#1 Arc Gas-Argon/Helium (85/3)cfh
Arc Current - 320 Amps
Spray Distance - 2½ in.

#2 Arc Gas-Argon/Helium (55/5)cfh
Arc Current - 350 Amps
Spray Distance - 3 in.

The spray distances of conditions 1 and 2 may be compared to 1-7/8" and 2-1/2" respectively, for C-band. Spraying condition #1 produced deposits with densities from 91% to 99% of theoretical, and condition #2 from 82% to 98%. The variations of density are attributed to differences in powder characteristics, i.e. the smaller particle size powders deposit with higher densities for a fixed set of spray parameters. Longer anneal times were found to be more beneficial for S-band than for C-band phase shifters. This may be due to the thicker deposits required for S-band. Anneal cycles for 2 hours at 1000°C produced the best device performance (lower microwave losses).

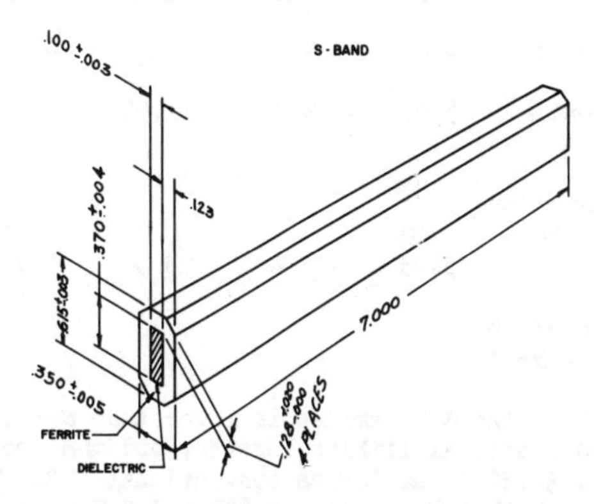

FIG. 1 S-Band Phase Shifter Dimensions

MAGNETIC AND DEVICE RESULTS

The $4\pi Ms$ of APS deposited 3-601 composition ranged from 520 to 580 gauss and for 3-750 composition from 710 to 770 gauss. The variations of $4\pi Ms$ are attributed to density.

Initially, the dielectric losses and hysteresis properties were determined from stress free samples, i.e. the lithium ferrite deposit was cut free of the dielectric before annealing. The annealing, then, would relieve any stresses generated by coefficient of thermal expansion mismatch between the ferrite and dielectric. Low dielectric losses, 0.0004, were readily obtained for any anneal cycle above 950°C. Stress free 3-601 samples had remanences (B_r) of

400 gauss and 3-750 samples had remanences of 520 gauss, which are comparable to values of conventional samples reported by Ampex, Table I. Similar to C-band results, coercive forces of arc plasma deposited samples were higher than for conventionally sintered samples.

TABLE I

Hysteresis Results of
Stress Free APS Lithium Ferrite

SAMPLE	H_c (Oe)	B_r (gauss)
Conventional 3-601	0.9	400
APS 3-601	1.7	410
Conventional 3-750	0.6	530
APS 3-750	1.9	526

Magnetic loss, μ'', and ΔH_k were measured on stress free arc plasma sprayed 3-601 and 3-750 compositions by Dr. J. Green of Raytheon. The results of these measurements are presented in Table II, and can be compared to measured results for a conventional gadolinium doped yttrium-iron garnet and a conventional cobalt substituted lithium ferrite.

TABLE II

Magnetic Loss and ΔH_k of APS Lithium Ferrite

Sample	ΔH_k (Oe)	μ''	ΔH_{eff} (Oe)	$\Delta H_k/\Delta H_{eff}$
APS 3-750	2.96	0.0031	9.49	0.312
APS 3-601	8.85	0.0061	23.32	0.379
*Conventional YIG	5.97	0.0033	11.55	0.519
**Conventional Li Ferrite	5.05	0.0084	25.50	0.198

*Trans-Tech G-600
**Ampex 3-755

The μ'' of the APS samples is lower than measured for conventional lithium ferrite, and the arc plasma 3-750 is as low as conventional YIG. The ΔH_k for both APS samples, 3-601 and 3-750, is 50% higher than that reported by Ampex of 6 Oe. and 2 Oe. respectively, for these compositions. Finally, the $\Delta H_k/\Delta H_{eff}$ ratios for the APS samples are significantly higher than for conventional lithium ferrite (3-755) and begin to approach that of the conventional garnet. The measurements were made on early low loss arc plasma samples; better APS samples are expected with optimization of spraying parameters.

Figure 2 shows a section of an arc plasma fabricated S-band phase shifter. Stress free B_r values were not realized for arc plasma deposited phase shifters since a dielectric with a close coefficient of thermal expansion match has not been established for the S-band compositions. The stresses generated by the thermal expansion mismatch between ferrite and dielectric significantly reduce remanences. Figure 3 shows the hysteresis loop of arc plasma deposited lithium ferrite with mismatch stresses (A) and stresses relieved (B). The match of thermal expansion between the S-band compositions and their dielectrics is not as good as was established for C-band compositions, hence reduced B_r's had to be tolerated.

FIG. 2 Arc Plasma S-Band Phase Shifter

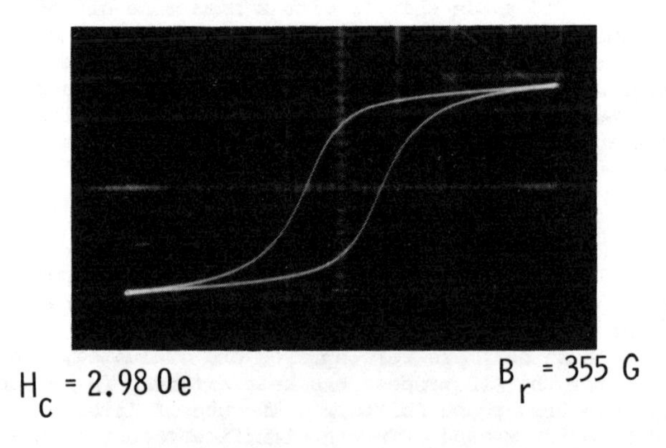

H_c = 2.98 Oe B_r = 355 G

A. As Deposited and Annealed

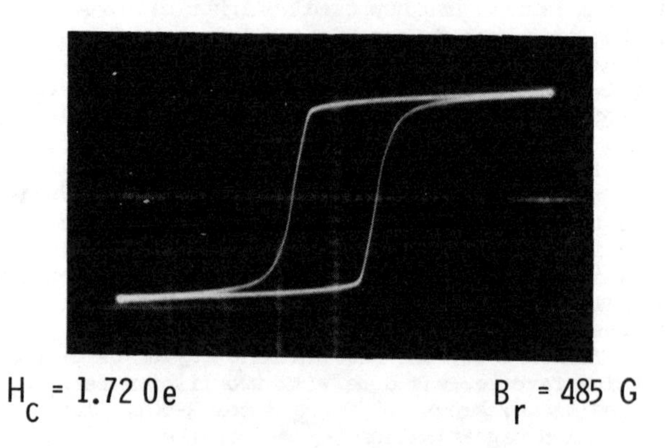

H_c = 1.72 Oe B_r = 485 G

B. Stress Relief Annealed at 1000°C

FIG. 3 Stress Effects on Arc Plasma Deposited 3-601

Table III gives the hysteresis properties achieved for the best ferrite-dielectric combinations available for S-band. The dielectric which resulted in the highest B_r's for the 3-750 composition was a lithium titanate with a coefficient of expansion (α) of 17.6 x 10^{-6}/°C, and this same dielectric and a

TABLE III
Hysteresis Properties of APS Lithium Ferrite

Sample	Dielectric	$(10^{-6}/^{o}C)$	K	H_c(Oe)	B_r(gauss)	$4\pi Ms$(gauss)
Conventional YIG	Ba Titanate	9.4	38.0	0.72	355	680
APS 3-750	Li Titanate	17.6	27.4	1.72	390	735
APS 3-601	Li Titanate	17.6	27.4	1.59	290	570
APS 3-601	Li Titanate	15.1	26.8	1.30	280	550

TABLE IV
S-band Phase Shifter Results at 15 Amp Drive

Sample	Dielectric	$(10^{-6}/^{o}C)$	K	$\Delta\phi$/in.	Insertion Loss (dB)
Conventional YIG	Ba Titanate	9.4	38.0	77	0.40
APS 3-750	Li Titanate	17.6	27.4	83	0.50
APS 3-601	Li Titanate	17.6	27.4	63	0.70
APS 3-601	Li Titanate	15.1	26.8	47	0.40

lithium titanate with a coefficient of $15.1 \times 10^{-6}/^{o}C$ produced equivalent Br's with the 3-601 composition. Included in Table III are the hysteresis properties of a 680 gauss conventionally fired gadolinium and aluminum doped yttrium-iron garnet phase shifter. Table IV gives the device results for these phase shifters.

The insertion loss is for 2.33" lengths for all samples; it is as low for arc plasma deposited lithium ferrite as for the conventional yttrium-iron garnet. A 7 inch 3-601 APS phase shifter, which had not been completely annealed, produced 360° phase shift and had an insertion loss of 1.0 dB. Based on experience with a more complete anneal a lower insertion loss is expected.

The temperature dependence of B_r was measured on phase shifters of the type in Table IV. The results of the temperature dependence measurements are plotted in Figure 4. These measurements show the arc plasma lithium ferrite phase shifters to have at least as good a temperature stability as the conventional YIG. The temperature variations of B_r for 3-601 and 3-750 are ±14% and ±7.4% respectively as compared to ±15% for YIG. Some of the temperature stability of the APS phase shifters is attributable to thermal stresses since these temperature stabilities are better than those determined for stress free samples.

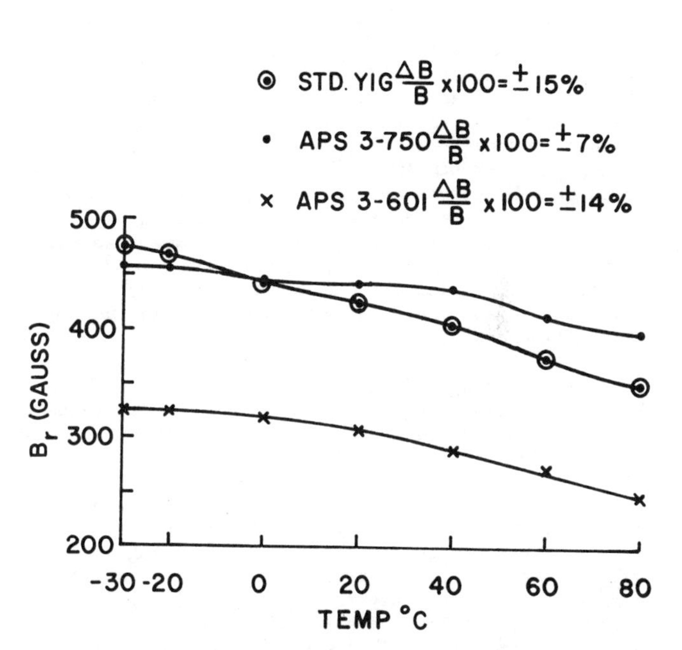

FIG. 4 Temperature Stability

CONCLUSION

Based on the $\Delta\phi$/in. of Table IV, to spray a phase shifter with 360° differential phase shift, a 7 inch length of 3-601 and a $4\frac{1}{2}$ inch length of 3-750 would be required. Based on the current maximum deposition rate of 80 mils/min/linear in., 17 minutes spraying time would be required for 3-601 and 11 minutes for 3-750. Both these spray times should prove economical for the fabrication of S-band phase shifters. In addition, the arc plasma lithium ferrite phase shifters would have low loss and good temperature stability.

ACKNOWLEDGMENTS

The authors wich to thank Messrs. T. Collins, J. Klimek, G. Sands, and Mrs. J. Mitchell for their assistance in this research project; and Dr. J. Green of Raytheon for measurements and helpful discussions.

REFERENCES

* This project has been funded as part of the US Army Manufacturing Methods and Technology Program

1. R. W. Babbitt, "Arc Plasma Sprayed C-band Lithium Ferrite Phase Shifters," IEEE Trans., Vol. Mag-11, No. 5, pp. 1253-1255, September 1975.
2. R. W. Babbitt, D. H. Harris, R. J. Janowecki, M. C. Willson and J. T. Cheng, "Arc Plasma Technology for Electronic Devices," Electronic Materials and Processing Symposium, 3 Oct 1973.
3. C. F. Jefferson and R. G. West, "Ferrite System for Application at Lower Microwave Frequencies," JAP, Supplement to Vol. 32, No. 3, pp. 390-S - 391-S, March 1961.
4. Paul D. Baba, et.al., "Fabrication and Properties of Microwave Lithium Ferrites," IEEE Trans., Vol., Mag-8, No. 1, pp. 83-93, Mar. 72.

MAGNETISM IN AMORPHOUS ALLOYS

T. Mizoguchi*
IBM Watson Research Center, Yorktown Heights, N. Y. 10598

ABSTRACT

The present state of knowledge and understanding of the basic magnetic properties of amorphous alloys is reviewed. It covers magnetic moments, exchange interaction, temperature dependence of the magnetization and anisotropy for two categories: metal-metalloid and rare earth-transition metal amorphous alloys. Some apparent discrepancies in our present knowledge are pointed out.

I. INTRODUCTION

"A theory of the physical properties of solids would be practically impossible if the most stable structure for most solids were not a regular crystal lattice," so Ziman started his book.[1] This is partially true also for experimental investigation, in which microscopic information can be obtained from the bulk properties of single crystals. So far major efforts have been confined to understanding the properties of crystalline solids. Now amorphous solids have become an important field of research. These solids represent a new state of matter and have served as a candidate for basic solid state physics study. They may also provide a bridge between the liquid and the crystalline state.

Amorphous solids have not been novel, for man has known glass since the Mesopotamian age. Quite recently some experimentalists began to study amorphous metallic alloys. One question is how does a random atomic arrangement affect the physical properties compared to those of a crystalline solid? However, it is not sufficient to only compare crystalline and amorphous states because the amorphous state can have advantages for basic solid state research on its own merits. We can vary the composition continuously in amorphous alloys without a restriction of crystal structure. In some cases we can get combinations of elements in amorphous alloys which cannot be realized in the crystalline phase.

Generally speaking the amorphous phase is not a stable ground state of a solid, which entails serious experimental limitations. Some properties of amorphous alloys may be attributed to the nature of non-equilibrium systems.

Amorphous metals and alloys so far obtained can be classified into three categories which are closely related to preparation methods.

1) Amorphous pure metals and alloys which can be obtained by evaporating on a cooled substrate. The amorphous state in this case is sustained by a substrate and stable only at low temperatures. There will be little reference in this article to this category.

2) Metal-metalloid alloys prepared by rapid quenching from the liquid state, or by electrodeposition or chemical deposition from aqueous solutions.

3) Metal-metal alloys prepared by sputtering, evaporation or rapid quenching. Typical combination is rare earth-transition metals. Zr-Cu system may also be included in this category.

The structure of the amorphous state[2] seems, in most cases examined so far, to be reasonably described as a random dense packing of hard spheres (RDPHS). The state is apparently stabilized by a mixture of atoms with appropriately different sizes, and also by a mixture of metallic bonding and some covalent bonding or electron transfer. Since transition metals and rare earth metals are good candidates for elements of amorphous alloys, the magnetism is inevitably important and interesting.

It would be interesting to compare the atomic arrangement with that of a spin system. As paramagnetism corresponds to the liquid state, the amorphous phase may be compared to the spin glass state in which the spins are frozen in random directions without any long range order. A spin glass state is not unique but one of many possible states, so that there may be transitions between many almost equally low lying states. The same is true for the structure of an amorphous phase. However, once it is heated up, it irreversibly changes into the most stable crystalline phases in contrast to the reversible change of the spin glass to the paramagnetic state. Why do they differ in this point? It is because the spin glass state is realized in a given random distribution of atomic positions and only directions of the spins are variables governed by the magnetic interaction. If the position of the moments were also allowed to be adjusted by magnetic interaction, the spin glass state would not occur. (Let us recall the bubble lattice.)

In the following, interesting aspects of the magnetic properties of amorphous alloys are reviewed in each section with some emphasis on apparent discrepancies in present knowledge.

II. MAGNETIC MOMENT

Many experimental studies of ferromagnetism in amorphous solids employ alloys of 3d transition metals with about 20-30 at.% of light metalloid elements of group IIIA, IVA and VA (B, C, Si and P). These small metalloid atoms are supposed to occupy relatively larger holes in the dense random packing structure of the metal atoms, transferring electrons to the unfilled d band of the transition metals. This accounts for the reduction of magnetization of these amorphous alloys as compared to the corresponding metallic crystalline alloys without metalloid elements.

In order to get a simple understanding, let us make the crude assumption that the glass former atom offers its s electrons to the conduction band and p electrons to the d holes of the transition metals. The magnetic moment of the transition metal atom, M, in $M_{1-x-y}G_xE_y$ can be expressed as follows:

$$\mu = \frac{m(1-x-y)-nx-gy}{1-x-y} \tag{2.1}$$

where G and E represent glass former elements, m is the original number of down spin holes of a 3d atom, n and g are the number of transferred electrons from the G and E atoms, respectively. Let us assume m to be about same as in the fcc phase (0.6, 1.6 and 2.6 for Ni and Co and Fe, respectively), in which the number of nearest neighbor metallic atoms is nearly same as that in the amorphous phase. The agreement between the observed moment in many amorphous alloys and the one calculated from Eq. (2.1) is fairly good in spite of the crude assumption.[3]

The average magnetic moments, $\bar{\mu}$, of 3d atoms were

studied in quasibinary amorphous $(T_{1-x}M_x)_{80}B_{10}P_{10}$ ferromagnetic alloys, where T represents Fe or Co, M represents V, Cr, Mn, Fe, Co or Ni.[4] They are plotted in Fig. 1 as a function of the average outer electron concentration, N, of metallic atoms, along with corresponding crystalline data (broken lines) on the so-called Slater Pauling curve which extends over both bcc and fcc phases with some anomalies in the Invar region.[5,6] Though the electron transfer from the glass former atoms reduces the magnetic moment of the quasibinary Fe-Ni and Fe-Co amorphous alloys compared to the Slater Pauling curve, $\bar{\mu}$ decreases with increasing N with the slope of -1, which does not conflict with a simple rigid band model. Transition metals such as Mn, Cr and V which have fewer 3d electrons than Fe seem to couple antiparallel to Fe moments.[4] It is interesting that, roughly speaking, the atomic moment of the 3d elements can be considered to increase linearly from $0\mu_B$ for Ni to $5\mu_B$ for V with decreasing atomic number in these amorphous alloys. Randomness of the potential may help localization of 3d electrons, which are fully polarized by sufficiently large intra-atomic exchange.

It is interesting to point out that the behavior of the magnetization of Fe and Co base amorphous alloys resembles that of Co and Ni base metallic crystalline alloys, respectively. This shift may be attributed to the forced insertion of electrons into the 3d transition metals in this amorphous system.

The pressure effects on the magnetization of these amorphous iron alloys were measured with a microbomb technique.[7] $(d\bar{\mu}/dP)\bar{\mu}$ of $(Fe_{1-x}M_x)_{80}B_{10}P_{10}$ is about $-2\times10^{-2}(\text{K bar})^{-1}$ for M=Fe, Co and Ni, while it is about order of magnitude greater for M=Cr and Mn. From the thermodynamic relation between the pressure dependence of magnetization and volume magnetostriction:

$$\rho \frac{dM}{dP} = \frac{1}{V}\frac{\partial V}{\partial H}, \tag{2.2}$$

where ρ is the density. The volume magnetostriction is estimated to be about 10^{-9} Oe^{-1} for M=Fe, Co and Ni, which is about the same order as that for the crystalline Fe-Ni alloy.

For heavy rare earth transition metal amorphous alloys, the magnetization is sensitive to a slight change of composition because of ferrimagnetic coupling between rare earth and transition metal moments. (For example, a 2% composition difference around $Gd_{25}Fe_{75}$ gives a change of the net magnetization by a factor of about 2.)

An extensive survey of R_xFe_{1-x} amorphous alloys[8] (where R=Gd, Tb, Dy, Ho, Tm, Yb and Lu) indicates a decrease of Fe moment with increasing R content, x, with the rate dependent on the effective rare earth spin (g-1)J, as

$$\mu_{Fe} = [2.2 - \frac{x}{1-x}(3.2-0.70(g-1)J)]\mu_B . \tag{2.3}$$

Considerable reduction of Fe moment in the R-Fe amorphous alloys (for example, $\mu_{Fe} \simeq 0.7$ for $Gd_{0.33}Fe_{0.67}$) can be attributed to the charge transfer from the R atom with the addition of a moment induced by R.

An intensive study has been done for amorphous $Gd_{1-x-y}Co_xMo_y$ in order to get the proper magnetic properties for practical application for bubble devices.[9] In this alloy the value of Co moment was found to be greatly reduced by Mo as follows:

$$\mu_{Co} = [1.65 - 5.0y/(x + y)]\mu_B , \tag{2.4}$$

$(\mu_{Co} \to 0 \text{ for } y/x \to 0.5)$.

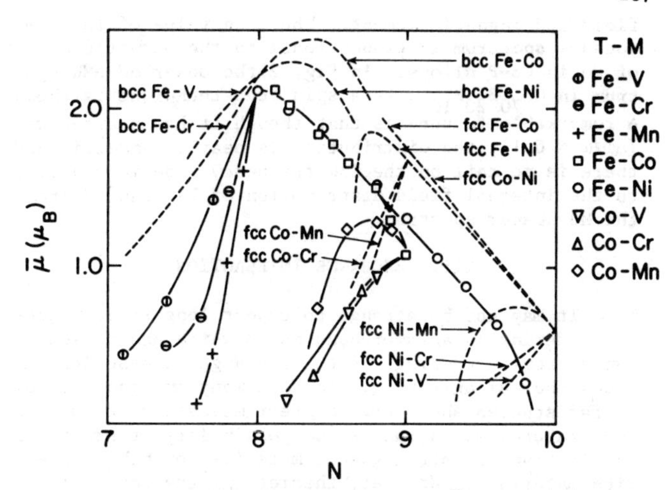

Fig. 1. The average magnetic moment, $\bar{\mu}$, of 3d atoms in quasibinary amorphous $(T_{1-x}M_x)_{80}B_{10}P_{10}$ ferromagnetic alloys as a function of the average outer electron concentration, N, of the metallic atoms, along with that for crystalline alloys (broken lines).

The magnetic moments of dilute 3d atoms in some amorphous alloys were obtained from the Curie-Weiss susceptibility. In an amorphous $Pd_{80}Si_{20}$, 3d atoms of Cr, Mn, Fe and Co have 3.6, 5.7, 5.9 and $4.4\mu_B$ effective moment, respectively.[10,11] In an amorphous $Cu_{57}Zr_{43}$ alloy, which shows temperature independent Pauli paramagnetism in itself, all 3d atoms (V, Cr, Fe, Co and Ni) except Mn were found to lose their localized moment.[12,13] Only Mn impurity gives the well behaved Curie Weiss susceptibility with an effective moment of about $1.6\mu_B$.

The local environment in an amorphous solid differs site by site because of the random arrangement of atoms, in contrast to the regular crystalline lattice. Therefore the magnetic moment of an atom is not expected to be identical on every site but has a distribution in a certain range, which may be observable through hyperfine interactions. The Mössbauer absorption spectrum of amorphous $Fe_{75}P_{15}C_{10}$ was fitted to an assumed combination of Lorentzian and Gaussian distributions of the hyperfine field. In this distribution a considerable portion of Fe nuclei seems to feel zero hyperfine field.[14]

Nuclear magnetic reasonance of Co was observed in amorphous Co-P[15,16] and $Co_{70}B_{20}P_{10}$,[17] which gave more direct information about the distribution of hyperfine

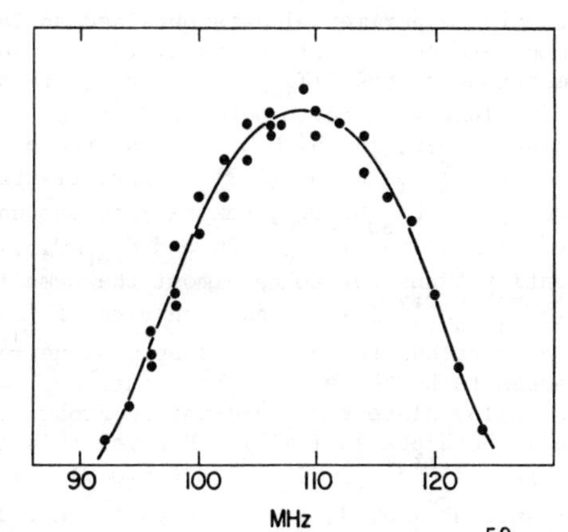

Fig. 2. The spin echo NMR spectrum of Co^{59} (0.99 KOe/MHz) in an amorphous $Co_{70}B_{20}P_{10}$ alloy at 4.2K.

field and magnetic moment. The mean value of the resonance spectrum is proportional to the magnetic moment of Co in Co-P alloys. In Fig. 2 the observed NMR spectrum in $Co_{70}B_{20}P_{10}$ with a spin echo technique is shown. A remarkable feature is that though it has a broad width ($\Delta\omega/\omega \simeq 0.1$), the distribution is nearly symmetric and there is no tail on the low frequency side in contrast to the internal field distribution of Fe deduced from the Mössbauer spectrum.

III. EXCHANGE INTERACTION

It may not be strange to expect long range magnetic ordering in an amorphous solid which has no long range atomic ordering, if the exchange interaction exceeds thermal agitation. In fact many amorphous alloys so far studied show ferro or ferrimagnetism at room temperature, which raises the possibility of practical application as soft magnetic materials or bubble device materials. However, theoretical treatment of spin ordering in amorphous solid is much more difficult than in regular crystalline lattices which have translational symmetry.

Let us define

$$J(\vec{q}) = \sum_j J_{ij}(|\vec{r}_j-\vec{r}_i|)\exp[i\vec{q}\cdot(\vec{r}_j-\vec{r}_i)] \ , \qquad (3.1)$$

where J_{ij} is the exchange interaction between atoms at the position $\vec{r}_i$ and $\vec{r}_j$. We define $\vec{Q}$ as the $\vec{q}$ which gives the maximum in $J(\vec{q})$. In the case of a crystalline Bravais lattice, the most stable spin configuration is, in general, a screw structure of the wave vector $\vec{Q}$. The ferro or antiferromagnetic state is realized for $\vec{Q}=0$ or $\vec{Q}=\vec{K}/2$, respectively, where $\vec{K}$ is a reciprocal lattice vector.

For an amorphous phase, the above discussion can not be applied straightforwardly. However it would be interesting to afford a physical insight into the stable spin configuration in amorphous alloys, especially in the case of $\vec{Q} \neq 0$. It should be noted that in an amorphous alloy $J(\vec{q})$ in Eq. (3.1) depends not only on $\vec{q}$ but also i, that is, the origin in the summation. Therefore a vector $\vec{Q}$ differs from site to site not only in magnitude but also quite randomly in direction. These situations very likely prevent a coherent ordering of spin systems except in the case of $\vec{Q}=0$.

In the molecular field approximation, the Curie temperature is expressed as

$$T_C = \frac{2S(S+1)}{3k} \sum_{<ij>} J_{ij} \ . \qquad (3.2)$$

Let us examine experimental data obtained so far. The Curie temperatures, T_C, of the quasibinary transition metal amorphous alloys, $(T_{1-x}M_x)_{80}B_{10}P_{10}$, are plotted in Fig. 3, along with corresponding T_C of metallic crystalline alloys, in which hcp Co has the highest T_C. Let us compare $\sum J_{ij}$ in Eq. (3.2) in pure crystalline Co and amorphous $Co_{80}B_{10}P_{10}$, taking into account the change of spin value. (S_{cry}=0.86 and S_{amr}=0.5.) Then this quantity turns out to be almost the same in both phases ($\sum J_{ij}^{amr}/\sum J_{ij}^{cry} \simeq 1.1$). Assuming a short range exchange interaction, the average J between nearest neighbor Co seems to be the same in both pure hcp Co and the amorphous alloy since the coordination number is almost the same in both phases (z=12). However this is not the case for Fe ($\sum J_{ij}^{amr}/\sum J_{ij}^{cry} \simeq 12J^{amr}/8J^{cry} \simeq 0.7$). The nearest neighbor interaction, J, is estimated to be roughly half in amorphous $Fe_{80}B_{10}P_{10}$ compared to that in pure bcc Fe, taking into account the difference of coordination number.

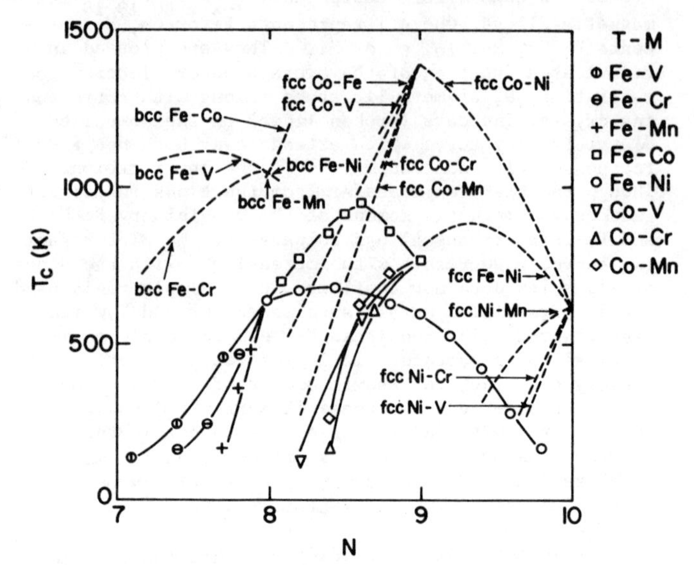

Fig. 3. The Curie temperature of quasibinary amorphous $(T_{1-x}Mn)_{80}B_{10}P_{10}$ alloys as a function of the metallic atoms, along with that for crystalline alloys (broken line).

There is a significant discrepancy between the T_C of crystalline bcc alloys and that of fcc alloys as seen in Fig. 3, suggesting the complexity of the effect of crystalline structure on T_C. On the other hand, in amorphous alloy systems, the T_C is a smooth function of the alloy composition in the whole range. It may be interpreted with a pair interaction model as the first approximation. The exchange interaction between Fe and Co seems to be most strong in this amorphous alloy system.

The pressure dependence of T_C is relatively large for M=Mn or Cr in this Fe base alloy ($dT_c/dp \simeq -2$K/Kbar) compared to that for M=Ni(-0.7K/Kbar).[7]

The series of amorphous rare earth-iron alloys RFe_2 (where R=Gd, Tb, Dy, Ho, Er, and Y) have been studied by J. J. Rhyne.[18] The observed T_C exhibits a smooth decrease from 500K for $GdFe_2$ to 0 for YFe_2 with decreasing De Gennes factor, $G=(g-1)^2J(J+1)$, of the rare earth atom. This behavior of T_C of the amorphous alloys should be compared to that of the corresponding crystalline Laves phase compounds, for which YFe_2 has T_C of 535K and which show a much weaker dependence on G.

A two sublattice molecular field calculation gives an adequate fit to the observed T_C in the crystalline Laves RFe_2 compound with the exchange constant of J_{Fe-Fe}=832K, J_{R-Fe}= -137K and J_{R-R}=98K. The fit for the amorphous alloys is less satisfactory and yields J_{Fe-Fe}=0, J_{R-Fe}= -129K and J_{R-R}=73K. It is quite puzzeling that J_{Fe-Fe} anomalously vanishes in amorphous R-Fe alloys. There are other works in which T_c of amorphous YFe_2 was reported as finite (∿120K).[8] It should be carefully reexamined which properties are unique in an amorphous alloy of a certain composition and others depend on the preparation condition, thermal treatment, etc.

It is interesting that for RFe_2 system T_C in the amorphous state is lower than that in the crystalline state while for RCo_2 system the situation is reversed. For example, T_C for crystalline compounds vary between 0 K for YCo_2 and 409 K for $GdCo_2$, whereas T_C is higher than 450 K for all amorphous $R_{0.33}Co_{0.67}$ alloys.[8]

For $(Gd_{1-x}Co_x)_{1-y}Mo_y$ amorphous alloys (with x=0.79 ∿0.88 and y=0.08∿0.16), Hasegawa et al. gave the following exchange interactions: J_{Co-Co}≃370±200K, J_{Co-Gd} ≃ -89±0.7K J_{Gd-Gd}≃1.4K.[9] In this case the Co-Co interaction dominates the others and the Gd-Gd interaction is very small. The paramagnetic Curie temperatures of amorphous Gd-Cu and Gd-Al suggest J_{Gd-Gd}≃0.5∿0.6 K.[19]

IV. TEMPERATURE DEPENDENCE OF MAGNETIZATION

The overall temperature dependence of the magnetization of an ordered magnetic system can often be treated with a molecular (mean) field approximation (MFA). However, at low temperature we know that a model involving spin wave excitations is better. Near T_C, that is, in the critical region, the MFA also fails because fluctuations become dominant.

Taking the Heisenberg model, the molecular field acting on spin S_i is expressed with the thermal average of surrounding spins S_j as follows:

$$H_m(i) = \langle H_{eff}(i)\rangle = -\frac{1}{g\mu_B}\sum_j 2J_{ij}\langle S_j\rangle \;, \tag{4.1}$$

where J_{ij} represents the exchange integral between S_i and S_j. Since in the simple case of a crystalline ferromagnet the thermal average of every spin should be identical, the molecular field can be unique. In amorphous alloys, in which J_{ij} has some distribution, the molecular field also differs from site to site. Then a second average over the site should be taken to make it simple.

The fluctuation of the exchange integral causes a faster decrease of the reduced magnetization with reduced temperature compared to the case without fluctuations. In Fig. 4 the temperature dependence of the magnetization for an amorphous $Co_{70}B_{20}P_{10}$ is shown with the expected curve from the MFA. A measure of the deviation of the exchange integral from the average can be estimated from the overall temperature dependence.[20] For an amorphous $Co_{70}B_{20}P_{10}$ $\delta\equiv(\langle\Delta J^2\rangle)^{1/2}/\langle J\rangle\simeq 0.3$.

For heavy rare earth-transition metal amorphous alloys, the temperature dependence of the net magnetization is peculiar because of the ferrimagnetic coupling between rare earth and transition metal moment. The direct observation of the temperature dependence of a sublattice magnetization has been done using polar Kerr rotation which is due mostly to the Co atoms in Gd-Co-Mo alloys.[21] The anomalously large Hall effect observed in amorphous rare earth-transition metal alloys changes its sign at the compensation point, reflecting the spin reversal in the external field.[22]

The spin wave excitations cause a low temperature magnetization of:

$$M(T)=M_o(1-BT^{3/2} - CT^{5/2} - \quad) \;. \tag{4.2}$$

This has a $T^{3/2}$ dependence in the lowest order, which comes from the first term, Dq^2, in the spin wave dispersion relation:

$$E(q)=h\omega_q=Dq^2+Eq^4+ \quad . \tag{4.3}$$

The next term, Eq^4, gives an additional $T^{5/2}$ term in the temperature dependence of the magnetization. The coefficient B of $T^{3/2}$ term can be related to the stiffness constant D in Eq. (4.3) by

$$B = \frac{2.612g\mu_B}{M_o}\left(\frac{k}{4\pi D}\right)^{3/2} \;, \tag{4.4}$$

where k is Boltzmann's constant.

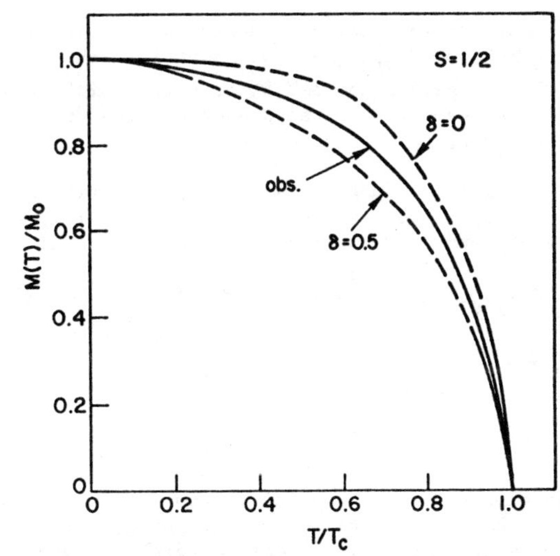

Fig. 4. The reduced magnetization M(T)/M(0), of an amorphous $Co_{70}B_{20}P_{10}$ versus the reduced temperature T/T_C, with the expected curves from the molecular field approximation.

In Fig. 5, $(M_o-M(T))/(M_oT^{3/2})\simeq B+CT$ for amorphous $Co_{70}B_{20}P_{10}$, and $(Fe_{93}Mo_7)_{80}B_{10}P_{10}$ are plotted versus T.

There were considerable discrepancies between D values obtained directly from inelastic neutron scattering and from low temperature magnetization measurements for amorphous ferromagnetic alloys.[23,24] (By contrast for crystalline ferromagnetics, there is good consistency.) A recent experiment on amorphous $(Fe_{1-x}Mo_x)_{80}B_{10}P_{10}$ is quite interesting.[25] As seen from Fig. 5 the temperature dependence of magnetization is adequately expressed with only the $T^{3/2}$ term for $T<T_C/3$. On the other hand, well defined spin wave excitation was observed with inelastic neutron scattering which gave an anomalous deviation of spin wave energy from the q^2 dependence. The D value estimated from the low temperature magnetization in only about 2/3 of that from neutron scattering.

Although well defined spin waves have been observed for long wave length ($q\lesssim 0.25Å^{-1}$), they probably become diffusive for higher q in amorphous ferromagnets. The magnetic excitation cannot be interpreted by only long wave length spin waves but also the low energy tail of diffusive modes must be taken into account.

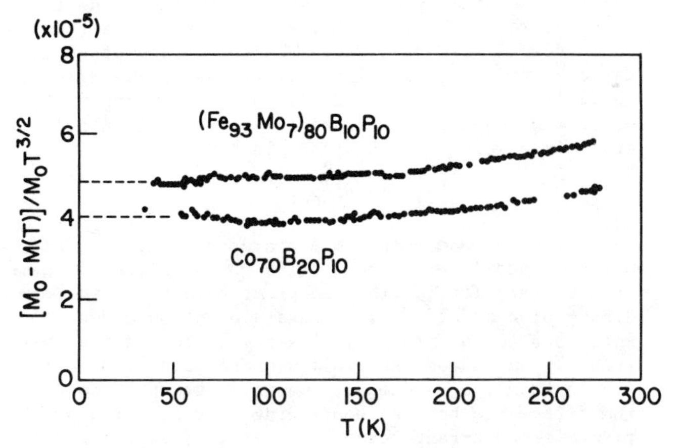

Fig. 5. $(M_o-M(T))/M_oT^{3/2}$ for amorphous $C_{70}B_{20}P_{10}$ and $(Fe_{93}Mo_7)_{80}B_{10}P_{10}$ versus the temperature T, where M_o and M(T) represent the magnetization at 0 K and T, respectively.

Spin wave dimensional resonance was observed in the same sample of $Co_{75}P_{25}$ as for magnetization measurement.[26] It gave D=138±meV $Å^2$ at 4.2K, which is in substantial agreement with the expected value from the coefficient of B.

The T_C of a ferromagnet is the typical critical point of a second order phase transition. Near T_C the following asymptotic relations have been proposed:

$$\lim_{H\to 0} M \equiv M_s \propto (T_C-T)^\beta , \qquad (T<T_C) \qquad (4.5)$$

$$\lim_{H\to 0}(H/M) \equiv \chi_o^{-1} \propto (T-T_C)^\gamma , \qquad (T>T_C) \qquad (4.6)$$

$$M \propto H^{1/\delta} , \qquad (T=T_C) \qquad (4.7)$$

Experimental studies for typical ferromagnetic crystals show well behaved critical phenomena. However if the system is heterogeneous it is not certain that well defined critical behavior will occur. (Consider, for example, superparamagnetism.) In amorphous alloys, microscopic circumstances may differ from site to site slightly and the exchange interaction has a distribution. How do amorphous ferromagnets behave near T_C?

A detailed experimental study for an amorphous $Co_{70}B_{20}P_{10}$ showed that there was a definite second order transition in this amorphous ferromagnet with the critical indices $\beta = 0.402\pm0.007$, $\gamma = 1.342\pm0.025$, $\delta = 4.39\pm0.05$.[27] A clear peak of the specific heat was also observed at T_C as shown in Fig. 6. The amorphous alloy can be an isotropic ideal ferromagnet in the critical region in which the fluctuation of the magnetization becomes long ranged, so that the microscopic randomness may be averaged out. In Fig. 7 $(d \ln M_s/dT)^{-1}$ and $(d \ln \chi_o^{-1}/dT)^{-1}$ are plotted versus T for an amorphous $Co_{70}B_{20}P_{10}$ and a disordered crystalline $Cu_{20}Ni_{80}$ alloy. For the latter the asymptotic relation of critical behavior, (Eq. (4.5) and Eq. (4.6)), hold only in a narrower temperature region around T_C, $(|T-T_C|=0.006)$ in which the critical fluctuation would melt out the static heterogeneity of spin clusters.

An amorphous $Gd_{37}Al_{63}$ alloy was found to show a clear transition to the spin glass state at about 16K, at which the susceptibility showed sharp cusp in a low DC field.[28] (The lowest field used in the measurement was about 0.1 Oe.) The susceptibility around the transition temperature decreases with increasing field and the cusp becomes broadened out. Below the transition temperature, remarkable thermal hysteresis and relaxation effects were observed. The spin glass state so far observed in crystalline alloys was realized in relatively dilute disordered system. The amorphous structure seems to be favorable for the spin glass state in a more concentrated spin system.

V. ANISOTROPY

A useful and accessible characterization of atomic arrangements in amorphous solids can be given in terms of a radial distribution function; however in the ordinary process of Fourier conversion of observed coherently scattered intensity of x-rays, electrons or neutrons by amorphous materials, we assume a priori an isotropic atomic arrangement. Cargill attempted to see the difference between scattering in reflection and transmission arrangements in order to detect anisotropy of atomic arrangement in an amorphous Gd-Co film.[29] Although direct experimental confirmation of anisotropic arrangement of atoms in amorphous alloys with diffraction techniques is still not convincing, there exists magnetic anisotropy other than the bulk shape

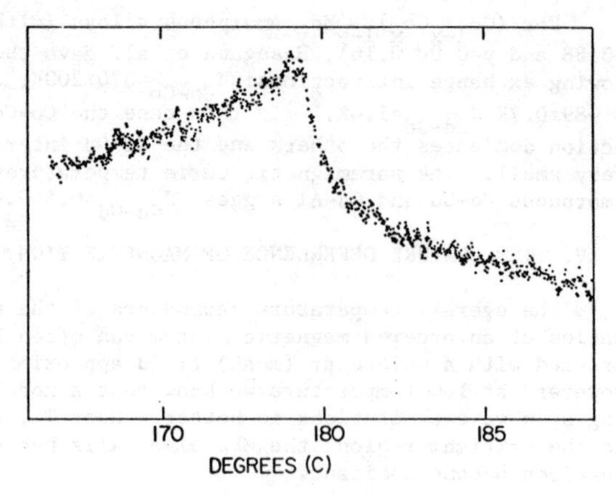

Fig. 6. The specific heat C_P of an amorphous $Co_{10}B_{20}P_{10}$ alloy around the Curie temperature obtained with AC calorimetry.

effect in some kinds of amorphous alloys. Since some magnetic properties are very structure sensitive, it can be used as a powerful tool for structural study.

Initially amorphous alloys were expected to have no anisotropy. In such a case, the magnetic moment would be expected to change its direction very gradually, as seen from the expression of domain wall width,

$$\delta = \pi(A/K)^{1/2} , \qquad (5.1)$$

where A is the exchange stiffness constant and K is the anisotropy energy.

In rapidly quenched amorphous alloys, the local internal stress plays an important role for anisotropy through the magnetostriction and makes interesting domain patterns. A zero magnetostrictive amorphous alloy was found to be remarkably soft magnetic material.[30]

One of the most striking discoveries in amorphous magnetism is the perpendicular anisotropy in sputtered Gd-Co films by P. Chaudhari, J. J. Cuomo and R. J. Gambino.[31] They expected stress induced anisotropy with a substrate, but they could observe stripe domains even after removing a substrate. The magnetic anisotropy in amorphous alloys is closely related to preparation conditions. For example, amorphous Gd-Co films prepared by sputtering can have sufficiently large per-

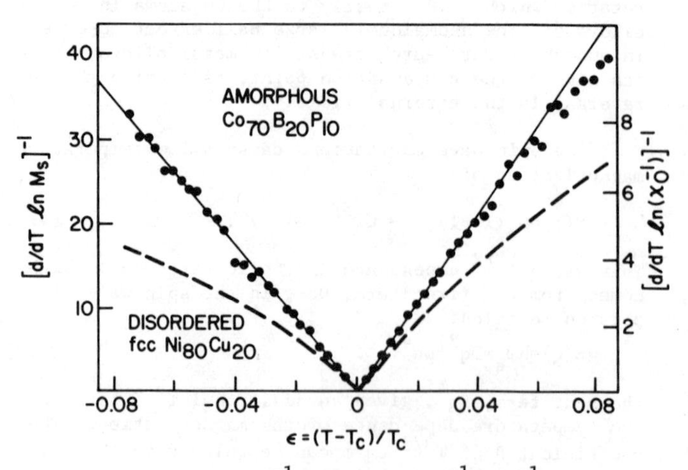

Fig. 7. $(d \ln Ms/dT)^{-1}$ and $(d \ln \chi_o^{-1}/dT)^{-1}$ versus T for an amorphous $Co_{70}B_{20}P_{10}$ and a disordered Crystalline $Cu_{20}Ni_{80}$ alloy, where M_s, χ_o and T represent the spontaneous magnetization, susceptibility and temperature, respectively.

pendicular anisotropy to sustain stripe or bubble domains; however, evaporated Gd-Co appears to have a hard axis perpendicular to the plane through most of the composition range.[32] It is interesting to point out that for Gd-Fe both methods afford easy perpendicular anisotropy.[33,34]

As the mechanism of magnetic anisotropy we can cite the following: 1) stress induced anisotropy through magnetostriction, 2) magnetic dipole interaction, 3) anisotropy via the spin orbit coupling.

The first mechanism is possible even in amorphous materials, and is in fact important in quenched metal-metalloid alloys, but not for all amorphous alloys.

The contribution of classical dipole interactions can be classified in three types. The first is a bulk shape effect for which surface poles are responsible. The second may be called as an internal shape effect and comes from anisotropic heterogeneity of the material. Small angle x-ray scattering for electrodeposited amorphous Co-P films have been interpreted in terms of oriented, ellipsoidal scattering regions of different composition, which gives anisotropy consistent with the observed in-plane saturation field.[35] The third contribution of dipole interaction is attributed to microscopic atomic pairs. If the angular distribution of pairs is not uniform, it makes a net anisotropy.

The spin orbit coupling makes spins couple to the lattice, giving single ion anisotropy and also an anisotropic exchange interaction. The latter was originally proposed by Van Vleck and has pseudo-dipolar type. Since the microscopic local symmetry around an ion is quite low in amorphous materials, considerable crystal field splitting may be expected.

The important characteristic of the amorphous state is the random orientation of the local anisotropy axis. Even if it is isotropically distributed, it is supposed to reduce the Curie temperature and the magnetization and also makes coercivity large.[36,37]

The temperature dependence of the uniaxial anisotropy, K_u, of $Gd_{1-x-y}Co_xMo_y$ films was obtained from a ferromagnetic resonance experiment, ($K_u \simeq 7\times10^4$ erg/cm^3 at 80 K) and fitted to a quadratic equation of the form

$$K_u = C_{11}M_1^2 + C_{12}M_1M_2 + C_{22}M_2^2 \quad , \qquad (5.2)$$

where M_1 and M_2 represent the sublattice magnetization of Co and Gd respectively.[38] The determination of C_{ij} shows that the Gd-Co coupling is dominant and that even the contribution of Gd is of considerable importance.

The effect of ion implantation into an amorphous Gd-Co film is interesting because it may give a possibility to control the anisotropy of amorphous alloys by changing microscopic atomic arrangements without heating up the film.[39] A relatively low level of radiation damage seems to both destroy anisotropy, and induce it when the sample is in a strong external field.

Although it is not easy to distinguish the actual mechanism of magnetic anisotropy in amorphous alloys, the net anisotropy must be due to slight deviation from the isotropic random distribution of the atoms in amorphous structure. The deviation can be attributed to the preparation condition which does not have a spherical symmetry. The progressive efforts to control the anisotropy in amorphous alloys will give a new aspect in solid state physics. Thus we might control microscopic structure of matter instead of a given crystalline lattice.

ACKNOWLEDGEMENTS

The author wishes to thank S. Maekawa, G. S. Cargill III, J. C. Slonczewski, P. Chaudhari, T. R. McGuire and A. P. Malozemoff for valuable discussions.

REFERENCES

1. J. M. Ziman, Principles of the Theory of Solids (Cambridge University Press, London 1964).
2. G. S. Cargill III, in Solid State Physics (F. Seitz, D. Turnbull and H. Ehrenreich, eds.), Academic Press New York, Vol. 30, 1975.
3. K. Yamauchi and T. Mizoguchi, J. Phys. Soc. Japan 39, 541 (1975).
4. T. Mizoguchi, K. Yamauchi and H. Miyajima, in Amorphous Magnetism (H. D. Hooper and A. M. deGraaf, eds.), Plenum Press, New York, 1973, p. 325. Proc. Int. Conf. on Magnetism (Moscow) 1973.
5. S. Chikazumi, T. Mizoguchi and T. Yamaguchi, J. Appl. Phys. 39, 939 (1968).
6. T. Mizoguchi, J. Phys. Soc. Japan 24, 1170 (1968).
7. K. Yamauchi, PhD. Thesis, Gakushuin University, 1975.
8. N. Heiman, K. Lee and R. I. Potter, AIP Conf. Proc. 29, 130 (1975).
9. R. Hasegawa, B. E. Argyle and L. J. Tao, AIP Conf. Proc. 24, 110 (1974).
10. R. Hasegawa, J. Appl. Phys. 41, 4096 (1970).
11. R. Hasegawa and C. C. Tsuei, Phys. Rev. B2, 1631 (1970).
12. T. Mizoguchi and T. Kudo, AIP Conf. Proc. 29, 167 (1976).
13. F. R. Szofran, G. R. Gruzalski, J. W. Weymouth, D. J. Sellmyer and B. C. Giessen, to be published.
14. T. E. Sharon and C. C. Tsuei, Phys. Rev. B5, 1047 (1972).
15. J. Durand, M. F. Lapierre and C. Robert, to be published in J. Phys. F. Metal Phys.
16. K. Raj, J. I. Budnick, R. Alben, G. C. Chi and G. S. Cargill III, presented at Int. Topical Conf. on Structure and Excitation of Amorphous Solids (1976).
17. T. Mizoguchi et al, to be published.
18. J. J. Rhyne, AIP Conf. Proc. 29, 182 (1976).
19. T. Mizoguchi, T. McGuire, R. Gambino and S. Kirkpatrick, to be published.
20. K. Handrich, Phys. Stat. Sol. 32, K55 (1969).
21. B. E. Argyle, R. J. Gambino and K. Y. Ahn, AIP Conf. Proc. 24, 564 (1974).
22. K. Okamoto, T. Shirakawa, S. Matsushita and T. Sakurai, IEEE Trans. Magnetics MAG-10, 799 (1974).
23. R. W. Cochrane and G. S. Cargill II, Phys. Rev. Lett. 32, 476 (1974).
24. H. A. Mook, D. Pan, J. D. Axe and L. Passell, AIP Conf. Proc. 24, 112 (1975); J. D. Axe et al. idid. 24, 119 (1975).
25. J. D. Axe, T. Mizoguchi, G. Shirane and K. Yamauchi, to be published.
26. J. R. McColl, D. Murphy, G. S. Cargill III and T. Mizoguchi, AIP Conf. Proc. 29, 172 (1976).
27. T. Mizoguchi and K. Yamauchi, J. de Physique 35, C4-287 (1974).
28. T. Mizoguchi, T. McGuire, E. S. Kirkpatrick and R. J. Gambino, to be published.
29. G. S. Cargill III, AIP Conf. Proc. 24, 138 (1975).
30. H. Fujimori, Y. Obi, T. Masumoto and H. Saito, Proc. 2nd Int. Conf. on Rapidly Quenched Metals, F-14 (1975).
31. P. Chaudhari, J. J. Cuomo and R. J. Gambino, IBM J. of Res. and Dev. 11, 66 (1973).
32. R. C. Taylor and A. Gangulee, to be published.
33. R. C. Taylor, J. Appl. Phys. 47, 1164 (1976).
34. N. Imamura, Y. Mimura and T. Kobayashi, IEEE Trans. Mag. MAG-12, 55 (1976).
35. G. C. Chi and G. S. Cargill III, 2nd Int. Conf. on Rapidly Quenched Metal (1975).
36. R. Harris, M. Plischke and M. J. Zuckermann, Phys. Rev. Lett. 31, 160 (1973).
37. R. W. Cochrane, R. Harris, M. Plischke, D. Zobin and M. J. Zuckermann, to be published.
38. P. Chaudhari and D. C. Cronemeyer, AIP Conf. Proc. 29, 113 (1976).
39. R. Hasegawa, R. J. Gambino, J. J. Cuomo and J. F. Ziegler, J. Appl. Phys. 45, 4036 (1974).

* Permanent address: Department of Physics, Gakushuin University, Mejiro, Tokyo, Japan.

MAGNETOELASTIC PHENOMENA IN AMORPHOUS ALLOYS

B. S. Berry and W. C. Pritchet
IBM Research Center, Yorktown Heights, New York 10598

ABSTRACT

Vibrating-reed measurements have been used to study the amorphous alloys $Fe_{75}P_{15}C_{10}$ and $Fe_{40}Ni_{40}P_{14}B_6$. The ΔE-effect is enhanced to large values by stress-relief anneals, and can be increased even further in a selected direction by magnetic annealing. The maximum ΔE-effect so far observed (0.8) is thought to be close to the theoretical limit. The temperature dependence of the ΔE-effect contrasts markedly with that seen in crystalline Ni, a result explained by the absence of magnetocrystalline anisotropy in the amorphous alloys. In addition to the ΔE-effect, vibrating reeds are shown to be subject to another type of magnetoelastic effect, termed the pole-effect, which can be used to determine the temperature dependence of the reduced magnetization. The kinetics of thermally activated atomic movement have been studied both by internal friction measurements and by magnetic annealing experiments. In both alloys the activation energy for magnetic annealing is found to be only about one-half of that for stress-induced ordering.

1. INTRODUCTION

This paper deals with the investigation of amorphous ferromagnetic alloys by dynamical measurements on vibrating reed samples. This has proved a most fruitful area of study, and has therefore successfully competed for a good share of our attention from the time some years ago when Professor Pol Duwez first tempted us with samples of his splat-quenched foils. It is self-evident that within this interesting group of new materials the ferromagnetic alloys are the most attractive. In this symposium paper we shall attempt to give a general overview of our work on these alloys, highlighting both the large magnetoelastic effects that are possible, and the ability of amorphous alloys to respond to magnetic annealing treatments.

From an experimental viewpoint, an obvious advantage of a specimen in the form of a reed or thin strip is the ease with which it is obtained from the foil or ribbon shapes in which most liquid-quenched amorphous alloys are produced. The vibrating-reed apparatus employed is basically an internal friction apparatus developed initially for thin film samples.[1] It enables mechanical damping and resonant frequency measurements to be made in the fundamental flexural mode of the reed and, importantly, at a series of higher overtones. The primary magnetoelastic phenomena sensed by such measurements are the ΔE- effect and the associated damping losses. However, the measurements also provide significant information in a number of other areas, some of which will also be described below.

2. PROCEDURE

The reed samples were typically of dimensions 2.5cm x 0.3cm x 0.005cm and possessed useable cantilever modes in the range 50Hz to 5000Hz. Mounted reeds were tested and treated in vacuum at temperatures from 80°K to about 700°K, with or without an applied longitudinal or transverse magnetic field of about 100 Oe maximum strength. So far, four different amorphous ferromagnetic alloys have been examined. Using their nominal atomic compositions, they are $Fe_{75}P_{15}C_{10}$, prepared by the piston and anvil technique at Caltech, electrodeposited $Co_{80}P_{20}$ from Harvard, and the Metglas alloys 2826 and 2605 from the Allied Chemical Corporation ($Fe_{40}Ni_{40}P_{14}B_6$ and $Fe_{80}B_{20}$, respectively). As already reported,[2] no ΔE-effect has been observed in the $Co_{80}P_{20}$ alloy, a result which is now thought to involve either a high internal stress or a strong growth anisotropy. In this paper we shall concentrate on the alloys $Fe_{75}P_{15}C_{10}$ and $Fe_{40}Ni_{40}P_{14}B_6$, which in many respects behave similarly.

3. RESULTS AND DISCUSSION

3.1 The ΔE-Effect and the Pole-Effect: Two Frequency-Related Phenomena.

The uniaxial elastic constant E (Young's modulus) of a ferromagnetic material may be smaller in a demagnetized or partially magnetized state than the value E_s at saturation. The difference $\Delta E = E_s - E$ arises from the magnetostrictive strain contributed by stress-induced domain movements. This strain adds to the ordinary elastic strain, thereby increasing the compliance of the material and reducing its modulus. Equivalent dimensionless measures of the ΔE-effect are $\Delta E/E = (E_s - E)/E$, or $(E^{-1} - E_s^{-1})/E_s^{-1}$. These quantities essentially represent the ratio of the in-phase magnetostrictive strain to the purely elastic strain. When measured at low frequencies, such that rate effects do not impede the full manifestation of the ΔE-effect, the magnitude ΔE/E assumes its fully-relaxed equilibrium value and then provides its most significant measure of the strength of the intrinsic magnetoelastic coupling.

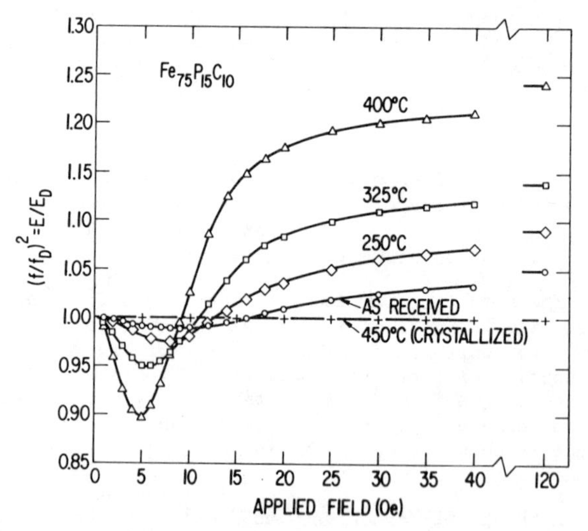

Fig. 1. Room temperature ΔE-behavior of amorphous $Fe_{75}P_{15}C_{10}$, as deduced from frequency measurements on a reed vibrating in the third tone, near 1200 Hz. Cumulative anneals of 2 hr. duration at temperatures up to 400°C produced a significant enhancement of the ΔE-effect, whereas a crystallization anneal at 450°C eliminated the ΔE-effect completely.

This low-frequency limit is realized very well in vibrating-reed experiments.[2] Figure 1 shows how E changes, relative to its initial demagnetized value, on application of a longitudinal magnetizing field to a reed of $Fe_{75}P_{15}C_{10}$. The change in E is inferred from the change in a resonant frequency of the sample, using the proportionality $f^2 \propto E$. The data illustrate a number of points, which may be summarized as follows. Starting from the as-received condition, the ΔE-effect can be enhanced significantly by annealing. This increase is dramatically different from the elimination of the ΔE-effect which occurs when the alloy crystallizes, and suggests, along with other evidence, that the material remains substantially amorphous up to 400°C. The tendency of specimens to curl during these anneals, and the magnitude of the ΔE-effect finally reached, are consistent with internal stress-relief as the explanation of the increase in the ΔE-effect. We have estimated[2] that annealing reduces the internal stress by about a factor of five, to a level of $1x10^8$ dynes/cm^2. Another feature of the results is the modulus minimum that occurs around fields of 5-10 Oe. This minimum, which corresponds to a maximum value of the ΔE-effect, occurs near a magnetization of 0.5 M_s. We have shown that the modulus minimum is a general feature of the relaxed behavior of soft magnetic materials. The minimum reflects the dominant effect of the "macroscopic" contribution to the ΔE-effect, i.e. the contribution that can be linked to a stress-induced change in the macro-magnetization of the sample.[2] This contribution disappears completely for both the demagnetized and saturated conditions, and thus necessarily exhibits a maximum at some intermediate magnetization. While Fig. 1 refers only to the effect of thermal annealing, reference to either Figs. 7 or 10 shows that magnetic annealing can also have a very substantial effect on the ΔE-effect. The domain models thought to explain this behavior have been described elsewhere.[3] The largest room-temperature ΔE-effect we have so far observed is 0.8, in magnetically annealed $Fe_{75}P_{15}C_{10}$. This value is probably close to the maximum value theoretically possible, and is thought to involve the combined effect of both stress-induced wall migration and domain-rotation mechanisms.

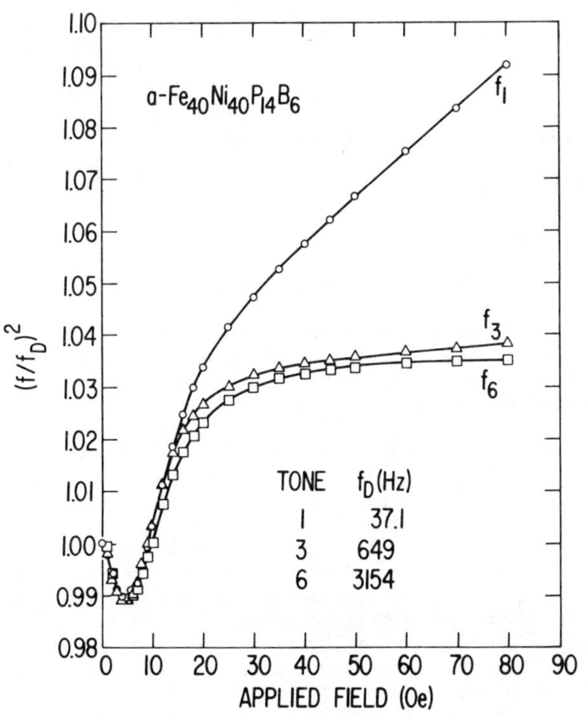

Fig. 2. Effect of a longitudinal magnetizing field on the resonant frequencies of a reed of amorphous $Fe_{40}Ni_{40}P_{14}B_6$, tested at room temperature in an as-received condition. The frequencies of each tone are normalized to the values f_D in the demagnetized condition. The pronounced increase in the curve for the fundamental mode at high fields is a manifestation of the pole-effect.

The ΔE-effect described above is an intrinsic material property. We have also observed another effect which can alter the frequency of a vibrating reed, and which in contrast to the ΔE-effect can be thought of as an extrinsic magnetoelastic effect. This new phenomenon, which we have termed the "pole-effect," does not affect the different tones of the reed with equal strength, and is usually only important in the fundamental mode and at high fields. Consequently, to avoid overlap with the pole-effect, it is usually most convenient to study the ΔE-effect via the use of overtones. To illustrate the pole-effect, Fig. 2 shows frequency data taken at various tones on a reed of $Fe_{40}Ni_{40}P_{14}B_6$. Whereas all the curves start off in the same way, exhibiting the modulus minimum caused by the ΔE-effect, the curve for the fundamental mode shows no tendency to saturate and departs steadily from the curves for the higher tones. The explanation of this difference resides in the interaction of the applied field with the magnetic poles at the ends of the reed. This interaction, which tends to keep the reed straight and aligned with the field, increases the overall flexural stiffness of the reed beyond that provided by its elasticity, and thereby raises the flexural vibration frequencies. An analysis of the pole-effect has been performed with the simplifying assumption that the poles are concentrated at the ends of the reed. For fields in the post-saturation range the theory yields the observed linear relationship between f^2 and the applied field H, and shows that the slope of this line is proportional to the saturation magnetization M_s. The theory also satisfactorily accounts for the much weaker pole-effect present at higher tones. Because of the assumptions made in the

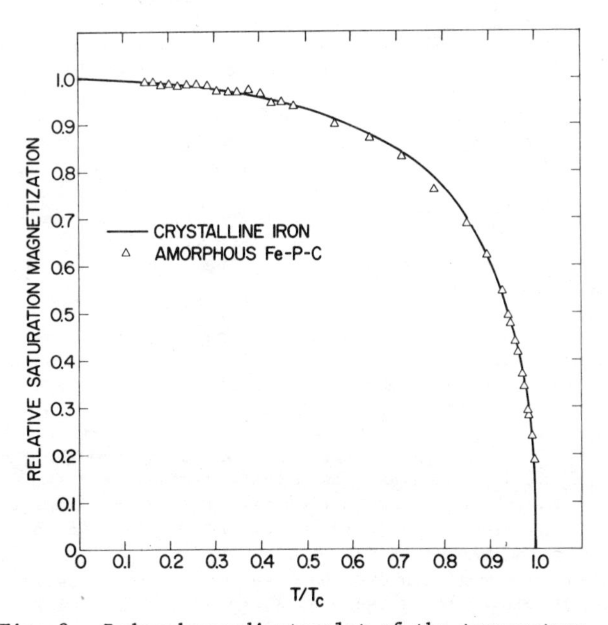

Fig. 3. Reduced coordinate plot of the temperature dependence of the saturation magnetization of amorphous $Fe_{75}P_{15}C_{10}$, as deduced from the use of the pole-effect. The sample had previously been annealed for 1 hr. at 350°C. The solid curve shows the behavior of crystalline iron.

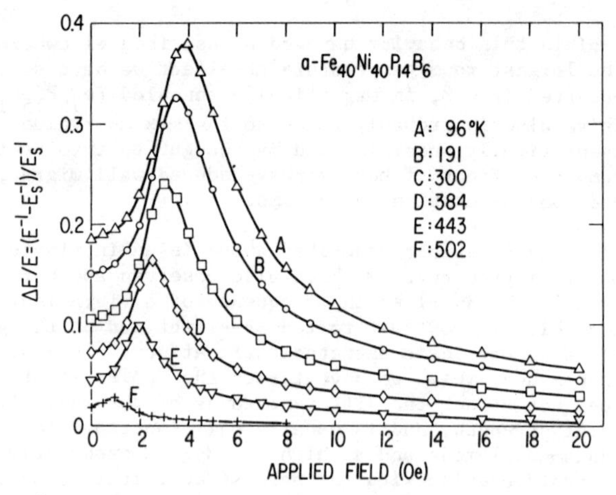

Fig. 4. Temperature dependence of the ΔE-effect in a reed of amorphous $Fe_{40}Ni_{40}P_{14}B_6$, after an anneal of 15 min. at 350°C. From measurements in the third tone at approximately 700 Hz.

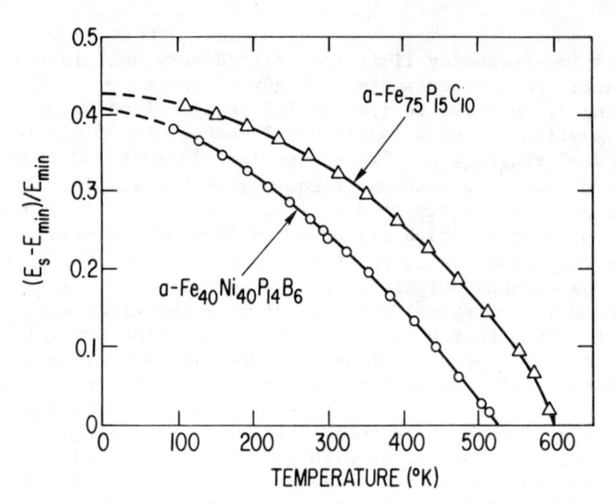

Fig. 5. Temperature dependence of the peak ΔE response exhibited by stress-relieved reeds of two amorphous alloys.

theory, the pole-effect cannot be used to obtain an accurate absolute measure of M_s. However, it can be expected to provide an accurate relative comparison of M_s at different temperatures, and hence to provide a novel mechanical method of obtaining the temperature dependence of the reduced magnetization. Figure 3 shows the results so obtained on an amorphous $Fe_{75}P_{15}C_{10}$ sample with a Curie temperature of 603°K. These data, which are in reasonable agreement with the results of other workers using direct techniques, show that when reduced coordinates are employed the behavior of the amorphous alloy is remarkably similar to that of crystalline iron.

3.2 Temperature Dependence of the ΔE-Effect

Several interesting insights are provided by measurements of the ΔE-effect as a function of temperature. Figure 4 shows illustrative results for stress-relieved $Fe_{40}Ni_{40}P_{14}B_6$. Here, frequency data for the third tone have been converted to the quantity ΔE/E, with the consequence that the minimum in the frequency data (c.f. Fig. 1 or 2) is now seen as a maximum in the ΔE-effect. The ΔE-effect increases smoothly with decreasing temperature, and continues to gain substantial strength on cooling below room temperature. All the curves of Fig. 4 are of the same general shape, although the maxima (which correspond to a magnetization level of $\sim 0.5\ M_s$) are obtained with smaller fields at higher temperatures. The temperature dependence of the peak ΔE-response is shown in Fig. 5 for two different amorphous alloys. These curves are obviously quite similar to each other, and clearly bring out the monotonic variation of ΔE/E with temperature referred to earlier. An interesting contrast exists between this behavior and the behavior of crystalline nickel, which has long been known to exhibit a pronounced maximum in its ΔE-effect near 200°C. We have been able to explain the behavior of nickel with a model in which the ΔE-effect is controlled by both the internal stress σ_i and the magnetocrystalline anisotropy energy, K_c. The initial rise in the ΔE-effect on cooling below T_c is explained by the increase in the saturation magnetostriction with decreasing temperature, in a regime where the ΔE-effect is dominated by the σ_i term. The subsequent sharp decline is due to a sharp increase of K_c, which at lower temperatures takes over as the dominant factor limiting the magnitude of the ΔE-effect. Since in the case of crystalline nickel the existence of the K_c term is at the root of the temperature maximum in the ΔE-effect, we may argue that the lack of such a maximum for the amorphous alloys is a natural consequence of their lack of magnetocrystalline anisotropy. We are also led to the prediction that the curves shown in Fig. 5 should match the temperature dependence of the linear magnetostriction. As yet, this comparison has not been made.

Another interesting aspect of the temperature dependence of the ΔE-effect is shown in Fig. 6. Here the ΔE-effect is manifest by the differences in the curves for the modulus minimum, and the demagnetized and saturated conditions. The decrease in f_s with increasing temperature reflects the normal temperature dependence of the saturated modulus, E_s. This trend is however reversed for the frequencies f_D and f_{min}, due to the counteracting influence of the decrease in the ΔE-effect with increasing temperature. These data show that it should not be difficult to obtain an amorphous oscillator with an accurately temperature-compensated frequency. Finally, it should be noted that convergence of the f_D and f_{min} curves to the f_s curve provides a simple means of determining the Curie point, which in the case of the material of Fig. 6 is found to be 525°K.

3.3 Internal Friction Contributions

The stress-induced domain movements responsible for the ΔE-effect cause a dissipation of mechanical energy and a magnetic contribution to the total internal friction exhibited by a vibrating reed. The high degree of correlation that can exist between the internal friction and the ΔE-effect is illustrated by Fig. 7. To obtain the widely different ΔE-behavior shown in the upper section of Fig. 7, the sample was magnetically annealed along different directions. The lower section shows the strikingly similar pair of curves obtained from complementary internal friction measurements. (The measure of internal friction used here is the logarithmic decrement of free decay, δ.) Figure 7 indicates a relatively high degree of reversibility in the stress-induced domain motion, since the out-of-phase strain responsible for the internal friction is only about 2% of the in-phase magnetostrictive strain producing the ΔE-effect. It is not necessary to dwell here on the separation of the magnetic loss into various components, since this was recently treated in another paper.[2] We shall simply remark that at low amplitudes in the audio frequency range the dominant loss does not fit the customary separation into eddy-current and hysteretic compo-

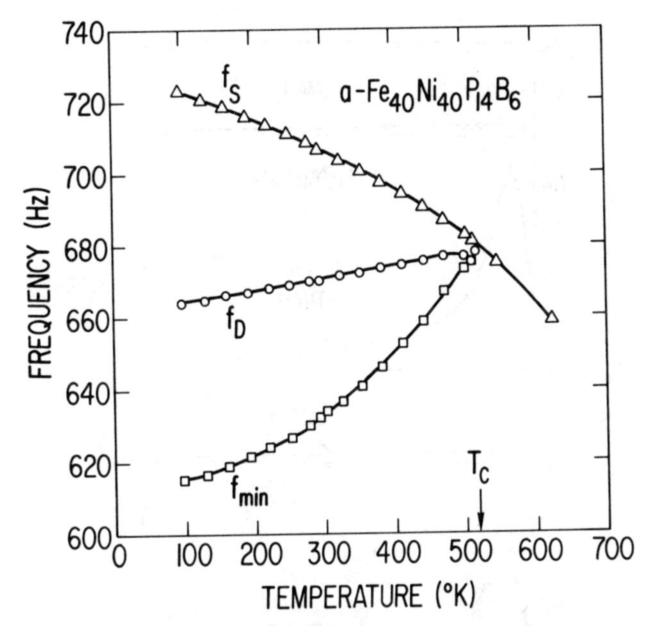

Fig. 6. Temperature dependence of the third-tone resonant frequency of an annealed (15 min. at 350°C) reed of amorphous $Fe_{40}Ni_{40}P_{14}B_6$. The frequencies at saturation and after demagnetization are denoted by f_S and f_D respectively; f_{min} is the frequency at which the modulus minimum occurs. The data indicate a Curie temperature of 525°K.

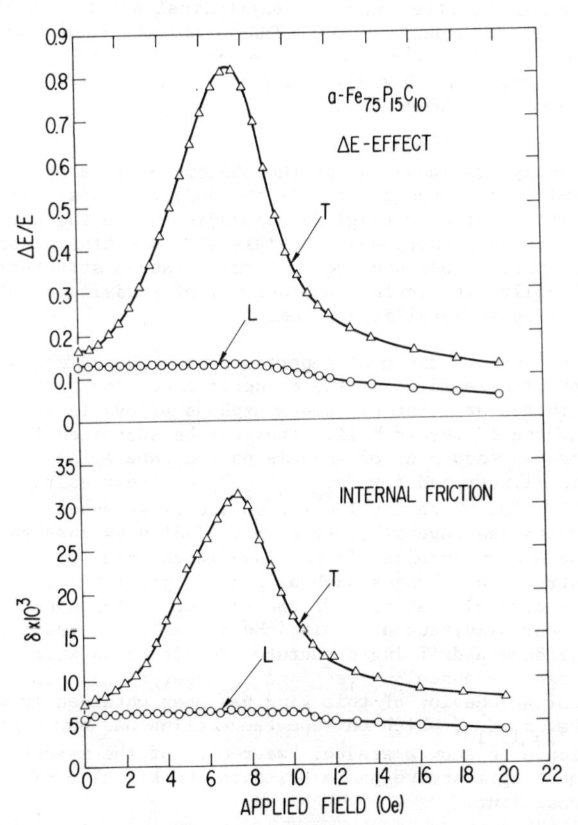

Fig. 7. Correlation of the internal friction and ΔE-effect exhibited by amorphous $Fe_{75}P_{15}C_{10}$. The curves labelled T and L were obtained respectively after transverse and longitudinal magnetic annealing treatments. Measurements were performed at room temperature on a reed vibrating in the third tone at about 1100 Hz.

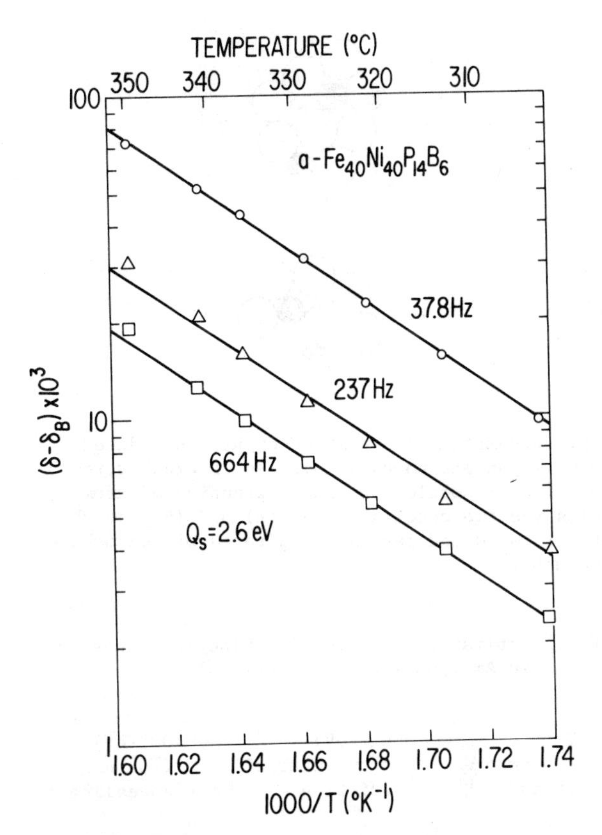

Fig. 8. Thermally activated internal friction in amorphous $Fe_{40}Ni_{40}P_{14}B_6$. Measurements were conducted up to the temperature of the prior stabilizing anneal. The background damping δ_B was obtained from extrapolation of data obtained below 300°C. The activation energy Q_s (2.6 eV) is obtained from the 1/T shifts between the curves for the different frequencies, which correspond to the first three reed frequencies.

nents.[4] While very large ΔE-effects (note the value of 0.8 reached in Fig. 7) show the potential for enormous eddy-current damping at higher frequencies, the "anomalous" audio loss is expected to remain dominant to about 1MHz.[2]

In addition to the internal friction produced by magnetic losses, two other internal friction contributions have been identified in vibrating-reed experiments. One of these, which is best seen in saturated samples at low or moderate temperatures, is the transverse thermal current damping produced by the thermoelastic effect.[5] This effect has particular utility for the determination of the thermal diffusivity of the sample.[6] The second non-magnetic contribution appears at elevated temperatures as an exponentially rising loss exhibiting a frequency-shift behavior characteristic of a thermally-activated process. As illustrated by Fig. 8 this contribution has been observed to start in $Fe_{40}Ni_{40}P_{14}B_6$ at about 300°C and to increase by an order of magnitude in a 50°C interval. This type of internal friction has been observed in all the amorphous alloys we have examined. Our most detailed study has been made on the $Pd_{80}Si_{20}$ alloy, where the internal friction measurements have been supplemented by creep and elastic after-effect measurements which show directly that the first stage of low-stress thermally activated deformation behavior is both linear in stress and recoverable with time. This is the type of viscoelastic response known as anelasticity, and is a characteristic manifestation of an internal relaxa-

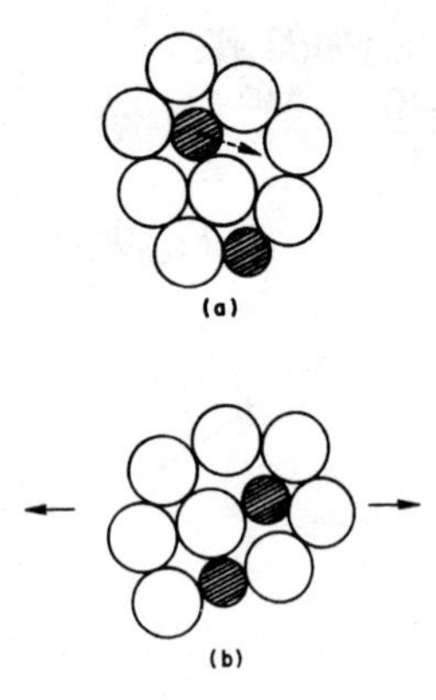

Fig. 9. Schematic illustration of stress-induced ordering in an amorphous alloy. An external stress changes the proportion of time a grouping of atoms occupies the two configurations (a) and (b), by changing the forward and reverse jump rates of the mobile shaded atom.

TABLE 1. Activation Energies for Kinetic Processes in Amorphous Ferromagnetic Alloys

ALLOY nominal atomic composition	Q_s (eV) from internal friction	Q_m (eV) from magnetic annealing
$Fe_{75}P_{15}C_{10}$	2.2	1.0
$Fe_{40}Ni_{40}P_{14}B_6$	2.6	1.35
$Fe_{80}B_{20}$	2.6	-

tion process. In a structurally disordered amorphous alloy, it seems this type of behavior can only be ascribed to a stress-induced ordering mechanism of the type indicated schematically in Fig. 9. The activation energies Q_s obtained from the internal friction data are therefore taken to represent activation energies for local atomic motion in the amorphous alloy. The values obtained for several ferromagnetic alloys are listed in Table 1. While little is so far known about diffusion in metallic glasses, the numbers appear to be quite reasonable, lying between the values for interstitial and substitutional diffusion in comparable crystalline alloys. However, as we shall soon see, there is a surprising difference between these energies and those involved in another process involving directional ordering, namely the phenomenon of magnetic annealing.

3.4 Magnetic Annealing of Amorphous Alloys

In addition to their use as indicators of magnetoelastic phenomena, the ΔE-effect and associated internal friction are also of interest as tools which can be employed to study the response of amorphous alloys to magnetic annealing treatments. Indeed, the ΔE-effect appears to have been the first property used to show that a uniaxial anisotropy can be induced in an amorphous alloy by magnetic annealing.[7] This discovery is of considerable significance. From a technological viewpoint, it offers the promise of improved properties. From a scientific viewpoint, it helps to confirm the directional short-range ordering model of magnetic annealing[8] in a new and unique way;

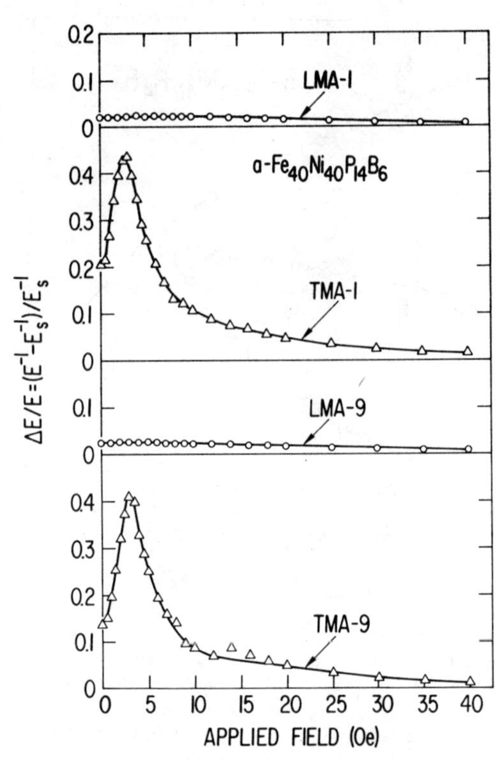

Fig. 10. Illustration of the sustained reversibility of the magnetic annealing response in amorphous $Fe_{40}Ni_{40}P_{14}B_6$. The curves show the ΔE-effect obtained after the first pair of longitudinal and transverse magnetic anneals (LMA-1, TMA-1) and the essentially identical response after the ninth consecutive pair of treatments. From vibrating reed measurements in the third tone near 700 Hz.

namely by demonstrating the phenomenon in a material in which short-range order is the only type possible. It also provides a tangible contradiction to the half-truth that amorphous materials should exhibit isotropic properties because they possess a random structure. Finally, it provides another way of studying atomic motion in metallic glasses.

One of the most important and perhaps most surprising features of the magnetic annealing response exhibited at least by some amorphous alloys is the high degree of reversibility that can be sustained in repeated sequences of anneals on the same sample. This is illustrated for $Fe_{40}Ni_{40}P_{14}B_6$ by the results of Fig. 10, which involve a sequence of 18 anneals. Such sustained reversibility clearly indicates that the mechanism involved is not associated with irreversible structural changes such as phase separation, clustering, or crystallization. Of course, such effects may also occur simultaneously with the directional ordering, and produce a drifting structural condition on which the magnetic annealing response is superposed. Evidence for a behavior of this kind has been obtained from $Fe_{75}P_{15}C_{10}$, which on repeated cycling has now been found to show a gradual "wear-out" of the magnetic annealing response and a drift in the kinetics of annealing.

Initially, it appeared to us that the activation energy for magnetic annealing, Q_m, was comparable to the value Q_s given by the internal friction experiments.[3] This conclusion was based on an estimate of Q_m that involved guessing the pre-exponential frequency factor. The value used seemed reasonable at the time, on the basis of numbers that applied to crystalline materials. However, it turned out to be several orders of magnitude larger than the result obtained by Luborsky,[9] who has performed the first detailed study of magnetic annealing kinetics on an amorphous material with the result Q_m=1.4eV for the alloy $Fe_{40}Ni_{40}P_{14}B_6$. Prompted by this work, we have investigated the internal friction response of the same alloy, and obtained the result Q_s=2.6eV, almost double the value reported for Q_m. To establish whether a similar discrepancy existed for $Fe_{75}P_{15}C_{10}$, a technique was needed that could be applied to short reeds, rather than toroids wound from long ribbons. To this end, we employed the technique of measuring the change in the remanent ΔE-effect caused by transverse magnetic annealing of an initially longitudinally annealed sample. As a check on the method, measurements were first made on $Fe_{40}Ni_{40}P_{14}B_6$, with the results shown in Fig. 11. As expected, the data show that the ΔE-effect passes through a maximum value as a function of hard-axis annealing time. This maximum can be expected at a time $t_{1/2}$ (roughly equal to 0.69 of the relaxation time, τ) at which the anisotropy vanishes in the longitudinal direction before starting to build in the transverse direction. As shown in Fig. 12, the method gives results in excellent agreement with Luborsky's data. Figure 12 also shows the results obtained for $Fe_{75}P_{15}C_{10}$. In this case we encountered a drift in the kinetics which necessitated a progressive correction to the results of successive measurements. However, the results leave no doubt that Q_m is approximately 1.0eV, showing that in this material also Q_m is only about $0.5Q_s$ (c.f. Table 1). We thus conclude that in both materials the directional ordering that proceeds under stress involves different atomic motions from those that occur during magnetic annealing. Hence, we also conclude that the directionally-ordered state produced by stress is different from that induced by magnetization. Because of the lower activation energies involved, it seems more likely that magnetic annealing involves movement of the metalloid atoms P, B or C. Finally, it should be noted that the data of Fig. 12 have been fitted with straight lines. Actually, a closer inspection of the data suggests the rate plots are actually somewhat curved. While linear rate plots are usually found for atomistic processes in crystalline materials, there is less reason to expect such linearity for amorphous materials. Until this question is resolved, the value of the pre-exponential factor derived from a linear fit must be regarded with caution.

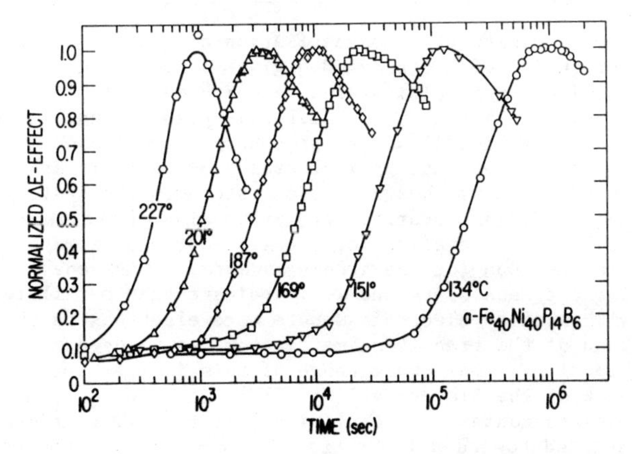

Fig. 11. Kinetics of the change in the ΔE-effect produced by magnetic annealing along a previously induced hard-axis direction. The ΔE-effect is measured for the remanent state obtained by brief interruptions of the hard-axis field. From vibrating reed data on amorphous $Fe_{40}Ni_{40}P_{14}B_6$, taken at the third tone near 700 Hz.

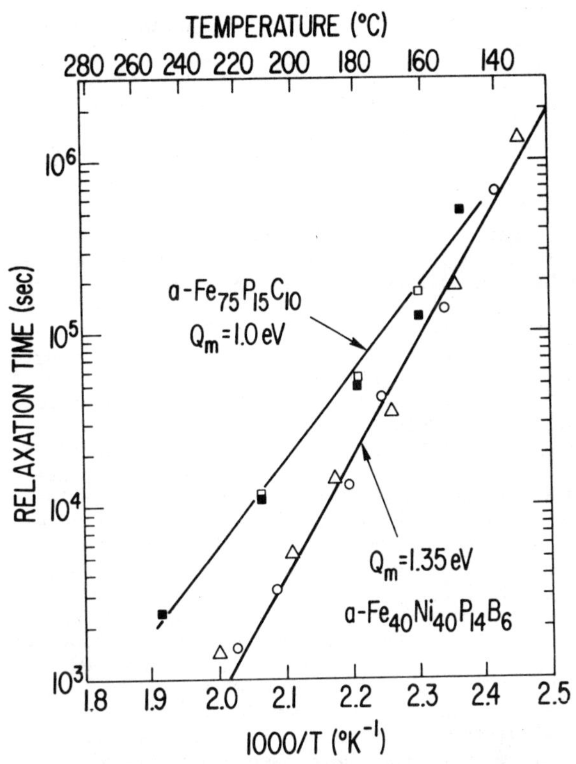

Fig. 12. Rate plot of magnetic annealing data in two amorphous magnetic alloys. The two sets of data points for $Fe_{40}Ni_{40}P_{14}B_6$ compare the present ΔE-measurements (triangles) with Luborsky's results (circles). For $Fe_{75}P_{15}C_{10}$ the filled black squares show the results obtained directly from the ΔE-measurements, whereas a nearby unfilled square indicates an estimated corrected value that allows for drift during the sequence of measurements.

REFERENCES

1. B. S. Berry and W. C. Pritchet, IBM J. Res. and Devlop. 19, 334 (1975).
2. B. S. Berry and W. C. Pritchet, J. Appl. Phys. (1976) to be published.
3. B. S. Berry and W. C. Pritchet, Phys. Rev. Lett. 34, 1022 (1975).
4. A. S. Nowick and B. S. Berry, Anelastic Relaxation in Crystalline Solids (Academic, New York, 1972) Chap. 18.
5. A. S. Nowick and B. S. Berry, Ref. 4. pp. 493-503.
6. B. S. Berry and W. C. Pritchet, J. Appl. Phys. 44, 3122 (1973).
7. B. S. Berry and W. C. Pritchet, U.S. Patent 3,820,040 June 25, 1974.
8. B. D. Cullity, Introduction to Magnetic Materials (Addison-Wesley, Reading, Mass., 1972), Chap. 10.
9. F. E. Luborsky, 21st Annual Conference on Magnetism and Magnetic Materials. AIP Conf. Proc. 1976 - to be published.

ADVANCES IN FERROMAGNETIC METALLIC GLASSES

R. Hasegawa, R. C. O'Handley and L. I. Mendelsohn
Materials Research Center, Allied Chemical Corporation, Morristown, NJ 07960

ABSTRACT

Recent developments in transition metal glasses have resulted in the synthesis of an iron-rich glass (nominal composition, $Fe_{80}B_{20}$) in continuous ribbon form. The as-quenched glass shows a saturation moment of 1.99 μ_B/Fe atom at 4.2 K, an induction of 16 kG at 300 K, a ferromagnetic Curie temperature of 650 K, a saturation magnetostriction of 30 x 10^{-6} and a coercivity H_c of about 100 mOe. Large domains, elongated parallel to the ribbon axis, are observed. Both Mössbauer and FMR data indicate that the magnetic anisotropy (K = 3 x 10^4 erg/cm^3) is in the ribbon plane. Evidence is presented that boron donates less electrons to the d-band of transition metal atoms compared with other glass-forming metalloids such as phosphorous. Some of the low-field properties include: the permeability at 20 G, $\mu(20)$ = 1,700, μ_{max} = 102,000 and core-loss, W ≃ 0.3 to 0.4 watts/kg at f=1 kHz and B_{max} = 1 kG. Field annealing results in improvements in these properties: A typical field-annealed toroid shows $H_c \sim 40$ mOe, $\mu(20)$ = 4,000, μ_{max} = 320,000 and W = 0.1 watts/kg at f = 1 kHz and B_{max} = 1 kG. These properties represent significant improvements over existing iron-base metallic glasses and their commercial crystalline alloy counterparts.

INTRODUCTION

Encouraged by the observation[1] of ferromagnetic ordering in a Fe-Pd-Si metallic glass obtained by rapid-quenching from the liquid and by a semiclassical prediction for magnetic ordering due to Gubanov,[2] an iron-rich ferromagnetic metallic glass was first realized in an alloy of composition $Fe_{80}P_{12.5}C_{7.5}$.[3] With recent advances in materials synthesis, metallic glasses are now available in continuous ribbon form. This latter development has generated considerable interest in both fundamental and applied studies of such materials.[4] For example, technologically promising magnetic properties have been found in several metallic glasses, e.g. $Fe_{80}P_{16}C_3B_1$[5] (METGLAS® #2615),[6] $Fe_{80}P_{13}C_7$,[7] $Fe_{40}Ni_{40}P_{14}B_6$[5] (METGLAS #2826) and cobalt-rich alloys with near zero magnetostriction having compositions $Co_{72}Fe_3P_{16}B_6Al_3$[8] and $Co_{70}Fe_5Si_{15}B_{10}$.[7] All these metallic glasses have a general composition $(TM)_{100-x}M_x$ with x ≃ 20-25, which lies in the vicinity of a deep eutectic in the transition metal (TM)-metalloid (M) phase diagram. Typically, for melt quenched metallic systems, TM and/or M represent a combination of elements. Save for alloys such as Pd-Si and Pt-Si,[9] binary combinations were unsynthesized until recently. Characterization of such binary metallic glasses is important to better understand the physical properties of more complicated systems. The purpose of this report is to discuss some selected physical properties of some new metallic glasses, especially $Fe_{80}B_{20}$ (METGLAS 2605),[10] containing boron as the only metalloid.

EXPERIMENTAL

Ribbons approximately 2 mm wide and 40 μm thick of composition $(TM)_{80}B_{20}$ were prepared by rapid-quenching from the melt and were checked by x-ray diffraction to confirm their glassiness. Thermomagnetization data were taken by a vibrating sample magnetometer in the temperature range 4.2-1050 K with magnetic fields up to 9 kOe. The crystallization behavior of $Fe_{80}B_{20}$ was studied by combining both magnetization and x-ray data. For this purpose, several strips about 2 cm long were crystallized at 800 K in vacuum.

Mössbauer spectroscopy in transmission geometry was used to obtain average hyperfine field and isomer shifts (with respect to pure iron) and observe rf induced sideband effects in the spectra.[11] From the Mössbauer line intensity ratio and the data taken by a ferromagnetic resonance technique (FMR),[12] the direction and the magnitude of the magnetic anisotropy at room temperature were determined. The samples for the Mössbauer study were strips about 2.5 cm long and the FMR measurements at 9.268 GHz were taken on a disk of diameter 1.5 mm and thickness 0.015 mm.

Magnetic domains were observed by scanning electron microscopy (SEM) using an accelerating voltage of 50 kV. The maximum contrast of the secondary electron image is obtained when the sample (1 cm long) tilt-axis is parallel to the direction of magnetization.[13] Samples for the SEM and the FMR studies were polished lightly mechanically and electropolished in a solution of 1/3 nitric acid and 2/3 methanol by volume.

A conventional B-H hysteresigraph was used to determine the coercivity, remanence and saturation induction of samples in both straight strip (about 30 cm long) and toroidal form (average circumference of 12 cm). AC coreloss for the toroids was determined from the B-H loops for low frequencies (f < 10^3 Hz) and from the complex input power in the exciting coil for high frequencies ($10^3 \leq f \leq 10^5$ Hz). Stress effects were studied on straight strips about 1.5 meters long in a 1 meter long solenoid. An ac bridge with a semiconductor strain-gauge and a lock-in-amplifier was used to determine the magnetostriction of the materials.

The effects of thermal and field annealing on the low field properties were studied on samples annealed at 600 K for 1 hour.[14] Supplementary information concerning annealing effects was provided by the temperature dependence of the electrical resistivity in the range 300-700 K.

RESULTS AND DISCUSSION

A. Magnetization and Electron Transfer

The saturation magnetization of 190.0 emu/g observed at 4.2 K for $Fe_{80}B_{20}$ gives a moment μ_s = 1.99 μ_B per iron atom. This value is compared in Table I with those of other iron-rich metallic glasses containing other metalloids such as phosphorous and carbon.[15,16] It is noticed that μ_s increases progressively as boron is replaced by phosphorous or carbon in the system $Fe_{80}(P,C,B)_{20}$. Furthermore, the value of the isomer shift (Table I) is smaller for the iron-boron base metallic glasses than for the iron-phosphorous base ones.[17,18] These trends of μ_s and I.S. indicate that metalloids with more sp electrons donate more electrons to the d-band of the transition metal atoms as pointed out previously.[18] One consequence of this is that the peak of the Slater-Pauling (S-P) curve for metallic glasses containing metalloids with more sp electrons is shifted toward a lower transition metal electron concentration. The results[19] of magnetic moments for the $(TM)_{80}B_{20}$ system are summarized in Fig. 1. The shifts of the S-P curve noted above are clearly demonstrated in this figure and suggests that boron and phosphorous donate about 1.6 and 2.4 electrons per atom to the common d-band of TM. Further evidence for the electron transfer discussed here has been recently established by NMR measurements on glassy $(Ni\text{-}Pt)_{75}P_{25}$ alloys.[20]

The temperature dependence of the magnetization σ at H = 9 kOe for $Fe_{80}B_{20}$ is shown in Fig. 2 for 300 ⩽ T ⩽ 1050 K. The Curie temperature, T_c, of the metallic glass is well defined as evidenced by a set of straight lines in the plot H/σ versus σ^2 in the vicinity of T_c[21] (see Fig. 3). The value of T_c (=647 K) thus determined is close to that (=649 K) obtained by a differential scanning calorimetry (DSC) on a similar sample. The validity of the Belov-Goriaga-Arrott method in determining T_c for the present metallic glass and others[17] indicates that the thermodynamic equation of state near T_c is of the same form for glassy ferromagnets as it is in crystalline ferromagnets. Additional evidence for this is the fact that the magnetization behavior of glassy rare-earth transition metal films can be described relatively accurately by Brillouin functions in the vicinity of T_c.[22]

As seen in Table I, the value of T_c for $Fe_{80}B_{20}$ is higher than any other metallic glasses containing 80 at.% Fe. This leads to a smaller temperature dependence of the magnetic properties such as the saturation induction at 300 K, an advantage for magnetic device applications.

Table I Saturation magnetization μ_s in μ_B per Fe atom, isomer shift I.S. (mm/sec), Curie temperature T_c (K) and crystallization temperature T_{cr} (K) of iron-rich glassy alloys.

Alloy	μ_s	I.S.	T_c	T_{cr}	Reference
$Fe_{80}B_{20}$*	1.99	0.07±.006	647	658	present work
$Fe_{80}P_{10}B_{10}$	2.05	--	637	--	15
$Fe_{80}P_{13}C_7$	2.10	0.19	587	690	16
$Fe_{80}P_{16}C_3B_1$**	2.13	0.14±.005	565	600***	17,18

*METGLAS #2605
**METGLAS #2615
***T_{cr} of as low as 568 K has been observed. See C.-L. Chien and R. Hasegawa (this conference).

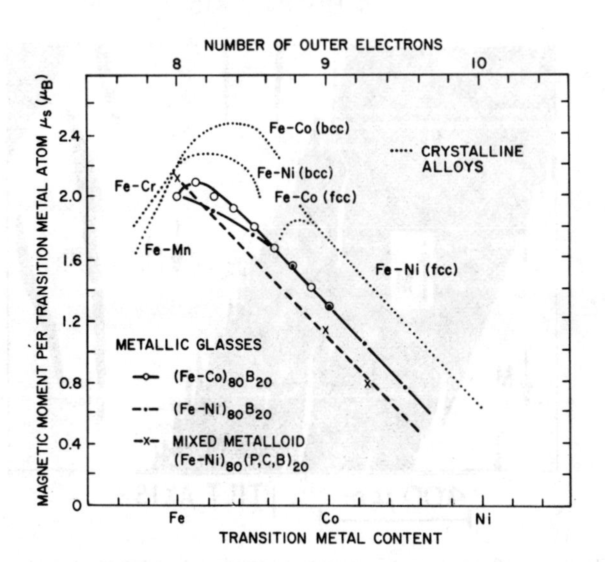

Fig. 1. Magnetic moment versus transition metal content.

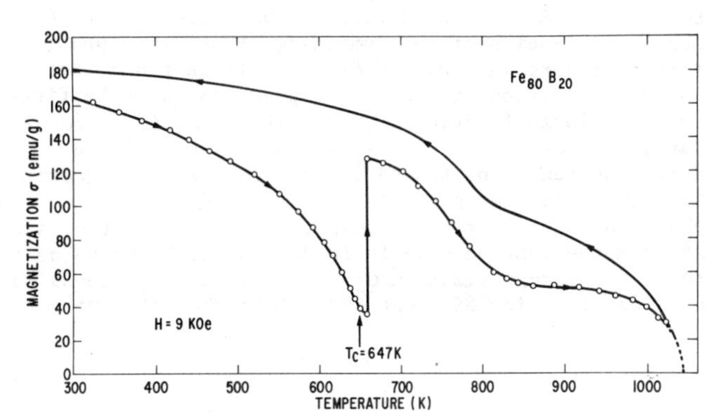

Fig. 2. Magnetization at 9 kOe as a function of temperature.

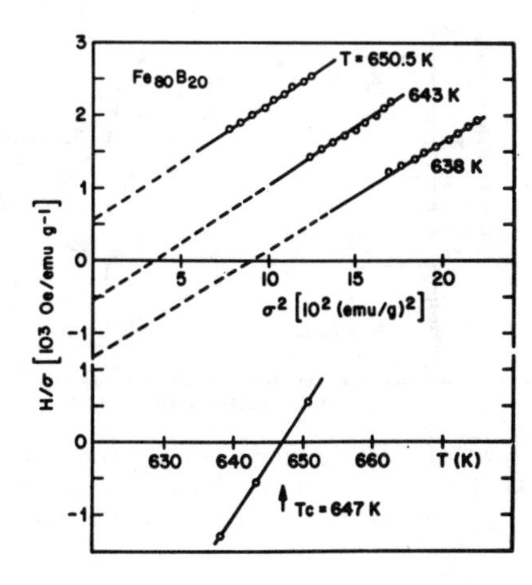

Fig. 3. Belov-Goriaga-Arrott plot. The intercept at the vertical axis, i.e. $(H/\sigma)_{\sigma^2=0}$ is plotted against T in the lower part of the figure, giving T_c = 647 K.

B. Crystallization Behavior

The transition from the glass-to-crystalline state for $Fe_{80}B_{20}$ takes place around 660 K as indicated by an increase of the magnetization at this temperature (see Fig. 2). The magnetization, σ, at 9 kOe increases steadily at 657 K from 35 emu/g to 125 emu/g within an hour beyond which the increase of σ with time is negligibly small. This does not necessarily mean that the crystallization is complete within an hour at T = 660 K. As noticed in Fig. 2, σ decreases, as expected, by heating the sample beyond 660 K, but remains relatively constant around 900 K, implying further crystallization taking place. This is confirmed in the magnetization versus temperature behavior upon cooling from 1050 K, i.e. the cooling curve lies above the heating curve (Fig. 2). As noted in Table I, the crystallization temperature, T_{cr}, for the present metallic glass is the highest among the iron-rich glasses. This is favorable in light of the thermal stability of the material.

As the observed magnetic transition temperatures, one (T_{c1}) near 800 K and the other (T_{c2}) at ∿1040 K, for the crystallized $Fe_{80}B_{20}$ are about the same on heating and cooling, the crystalline phases nucleated at ∿660 K retain their structures with their volume ratios depending on temperature and time. One of the crystalline phases has been identified as α-Fe both from x-ray diffraction and magnetic behavior (T_{c1} = 1043 K). The magnetization of the α-Fe phase varies as a Brillouin function of S = 1, which makes it possible to extract the magnetization of the remaining crystalline phase. This procedure is shown in Fig. 4. It is found that the magnetization of the second phase can also be fitted to a Brillouin function of S = 1 with a Curie temperature, T_{c2}, of 793 K. Extrapolations of the fitted Brillouin functions toward T = 0 give partial magnetizations σ_1 = 139 emu/g and σ_2 = 50 emu/g for the α-Fe phase and the other phase respectively. Since the saturation magnetization of α-Fe is 222 emu/g,[23] the value of the partial magnetization σ_1 given above corresponds to a weight fraction of about 63%. Then the value σ_2 suggests that the yet unknown phase has a saturation magnetization of about 135 emu/g. This value and T_{c2} of 793 K are not those of orthorhombic FeB (σ_s = 93.9 emu/g; T_c = 630 K) or tetragonal Fe_2B (σ_s = 174 emu/g; T_c = 1000 K),[24] which may be expected as equilibrium interstitial phases. X-ray diffraction data given in Table II, on the other hand, indicate that the lattice spacings, d, for the unidentified phase are close to those of the postulated orthorhombic Fe_3B.[25] It is likely that this phase is stabilized by such impurities as carbon.

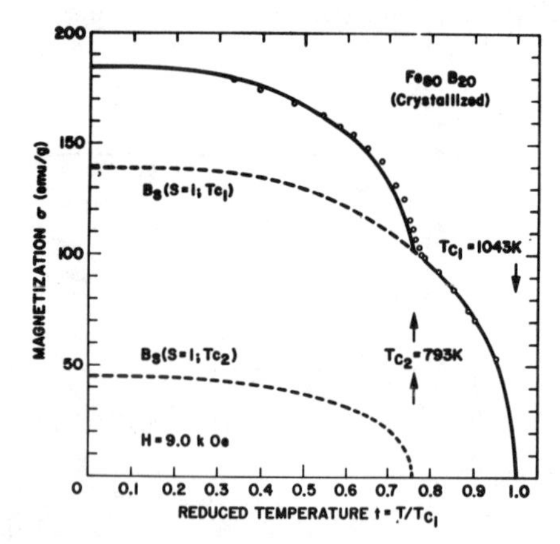

Fig. 4. Partial and total magnetizations of the crystallized $Fe_{80}B_{20}$ versus temperature reduced to T_{c1}. $B_s(S, T_c)$ is a Brillouin function.

Table II X-ray diffraction data taken with Mo*K*α radiation for the minor phase of crystallized $Fe_{80}B_{20}$ and the d-spacings calculated for different (hkℓ) reflections from the postulated orthorhombic Fe_3B structure.

unidentified phase		Fe_3B (orthorhombic)	
d(Å)	Intensity*	d(Å)	hkℓ
4.152	0.30	4.209	011
3.655	0.28	3.702	101
2.950	0.39	3.060	111
2.683	0.51	2.666	102
2.508	0.76	2.515	021
2.194	0.17	2.190	121
2.080	1.00	2.104	022
1.918	0.38	1.902	122
1.873	0.45	1.865	113
1.831	0.23	1.811	030
1.749	0.16	1.752	212
1.687	0.28	1.677	130
1.475	0.15	1.484	300
1.216	0.18	1.213	322

*Normalized to the strongest intensity for hkℓ = 022.

The crystallization behavior discussed above is important in the sense that it may shed light on the local short range order in the glassy state of the material as in the case of Mn-P-C.[26] It appears that the local atomic arrangement of $Fe_{80}B_{20}$ metallic glass has no similarity to that of any equilibrium phase based on Fe and B.

C. Magnetic Anisotropy and Domains

Room temperature Mössbauer spectra for $Fe_{80}B_{20}$ metallic glass consist of six well defined lines having an intensity ratio of 3:4:1:1:4:3. This suggests that the direction of the magnetic anisotropy is in the plane of the ribbon. The FMR data give the magnitude of the anisotropy, K = 2.8 x 10^4 erg/cm^3, directed on the average about 60° away from the ribbon axis. The off-ribbon axis direction of the anisotropy is consistent with the magnetic domain patterns shown in Fig. 5. The wide (∿600 μm in width) domains observed lie parallel to the ribbon axis with wall serrations oriented at ∿20 to 30° off the ribbon axis. Small, elongated domains exhibit the same orientation, which is the direction of magnetization. Due to the demagnetizing effect from the sample shape (1 cm x 35 μm), the magnetization lies closer to the ribbon axis than the expected anisotropy direction. SEM patterns for different sections of the ribbon sample are similar to the one described above and rarely show domains indicative of the magnetic anisotropy having components perpendicular to the ribbon plane.

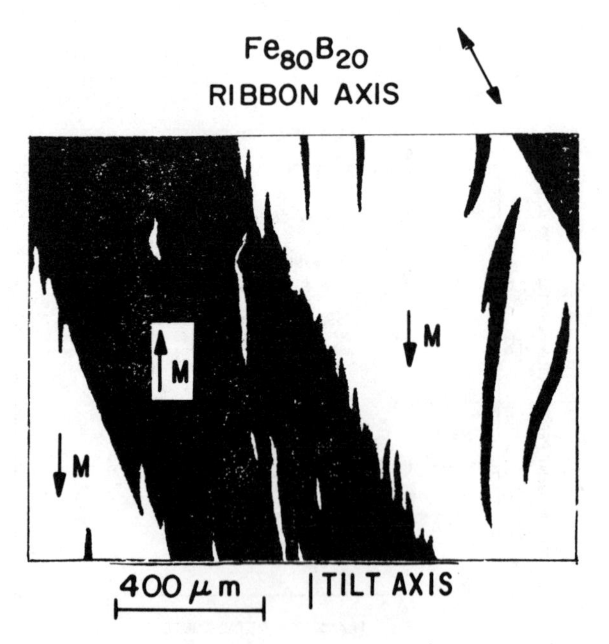

Fig. 5. Scanning electron micrograph (reproduced). The arrow shows the direction of magnetization.

D. Low Field DC Properties

Hysteresis loops taken at f = 0.01 Hz are shown in Fig. 6 for the toroidally wound as-quenched (6a) and heat-treated (6b) samples of $Fe_{80}B_{20}$. It is found that the as-quenched glass has a room temperature saturation induction B_s of 16 (±3%)kG, remanence B_r of 7-8 kG and coercivity H_c of 80-120 mOe. As in the case of many of the soft-magnetic materials, thermal and field annealing reduces H_c and increases B_r for the present metallic glass ribbon. Table III compares the room temperature values of B_s, B_r and H_c for $Fe_{80}B_{20}$ with those of other metallic glasses containing 80 at.% Fe and a commercial 50% Fe-Ni crystalline alloy.[27] It is noticed that $Fe_{80}B_{20}$ exhibits the highest B_s and B_r and the lowest H_c among the metallic glasses listed and is comparable magnetically with the commercial crystalline alloy.

A positive saturation magnetostriction of 30×10^{-6} has been observed for $Fe_{80}B_{20}$; this is also confirmed by a large rf induced sideband effect in the Mössbauer spectra.[11,17] The relatively large value of the magnetostriction is responsible for the initial stress sensitivity of $\sim$-6 mOe/kg mm^{-2} for H_c and $\sim$2kG/kg mm^{-2} for B_r (Fig. 7). A stress of about 10 kg/mm^2 is sufficient to achieve the dc properties of a field-annealed toroid of $Fe_{80}B_{20}$. However, the remanence to saturation ratio, B_r/B_s, reaches a value of about 0.8 under a stress of 25 kg/mm^2; B_r/B_s = 0.9 and 0.96 for METGLAS #2615 and #2826 respectively, have been reported for the same stress.[5] The somewhat unfavorable high stress behavior of $Fe_{80}B_{20}$ is probably due to the inclined magnetic anisotropy direction (60° off the ribbon axis). An attempt to control the magnetostriction of $(TM)_{80}B_{20}$ metallic glasses has been made recently, and a near zero magnetostriction alloy has been indeed found at the composition $Co_{74}Fe_6B_{20}$.[28]

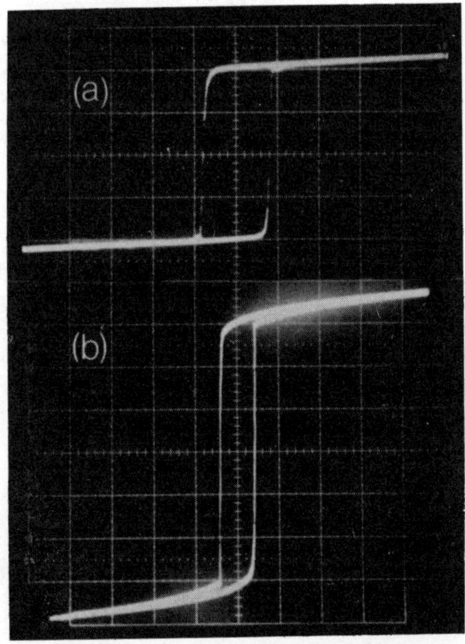

Fig. 6. Hysteresis loops for (a) as-quenched and (b) field annealed toroid of $Fe_{80}B_{20}$. The horizontal axis (applied field):100 mOe/div; the vertical axis (induction):4 kG/div.

Table III Saturation induction B_s(kG), remanence B_r(kG) and coercivity H_c(Oe)

Alloy	B_s	B_r	H_c	Ref.
$Fe_{80}B_{20}$ (METGLAS #2605)	16.0	12.3*	0.04*	present work
$Fe_{80}P_{16}C_3B_1$ (METGLAS #2615)	14.9	6.25*	0.05*	5, 17
$Fe_{80}P_{13}C_7$	14.0	3.5+	0.06+	7
Crystalline 50% Fe-Ni	16.0	14.6	0.1	27

*Field-annealed toroid; +straight strip.

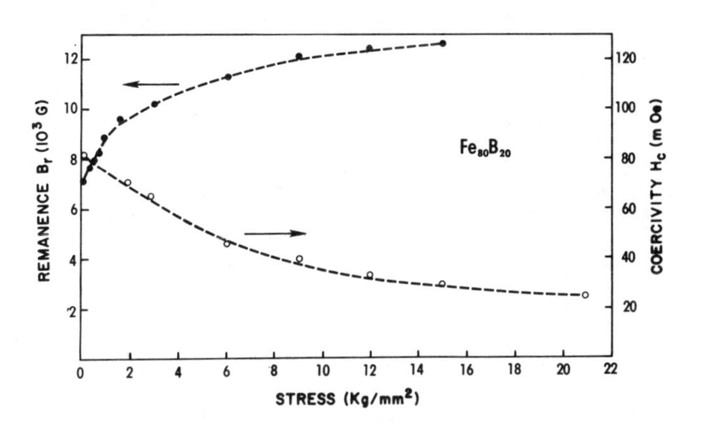

Fig. 7. Effect of stress on remanence and coercivity of a straight strip.

The temperature dependence of H_c for a toroidally wound sample is shown in the upper part of Fig. 8. It appears that H_c of as-quenched material decreases with temperature faster than expected above 460 K, remains relatively constant between 500 and 530 K, and decreases relatively fast between 530 and 600 K. This behavior is probably attributable to the stress-relief of the material since more direct measurements[29] of this effect show a similar temperature dependence and the internal stress introduced during quenching contributes to the total coercivity of the material. On the other hand, H_c versus T for the as-quenched strips does not show the two-step stress-relief which is not yet well understood, as seen in the lower part of Fig. 8. The coercivity H_c is relatively unchanged with temperature between 450 and 510 K, above which a gradual decrease of H_c is observed up to T $\sim$ 580 K. The rate of H_c decrease above 580 K depends on the samples: In one strip, a sharp drop of H_c from 60 to 20 mOe was observed between 580 and 590 K. In any case, the stress-relief seems to be complete in the vicinity of 600 K. Thus the field annealing would be expected to be most effective around this temperature, which is evidenced by the decrease of H_c and increase of B_r by approximately 50% in $Fe_{80}B_{20}$ annealed around 600 K.

The apparent difference in the H_c versus T behavior for the as-quenched toroids and straight strips (Fig. 8) is not well understood. One possible explanation might be as follows: The internal stress quenched-in during the material synthesis may start to redistribute itself around 450 K in straight strips in such a way that the stress induced anisotropy along the ribbon axis increases as T increases. This continues up to about 550-580 K, at which gradual and/or sudden stress-relief takes place. The stress redistribution behavior could be different in a toroid which is under tension on one side and under compression on the other side of the ribbon, resulting in the difference observed. Although the details of the stress-relief mechanism are not clear, the effect does not seem to be related to the embrittlement of the material. As shown in Ref. 29, metallic glasses containing phosphorous as the major metalloid, such as METGLAS #2826, become brittle at lower temperatures ($\sim$400

K) than does $Fe_{80}B_{20}$ while both materials have almost identical stress-relief behavior. The higher embrittlement temperature of about 500 K of $Fe_{80}B_{20}$ represents one of the significant improvements made recently in the development of metallic glasses.

In an attempt to better understand the thermal annealing effect, the temperature dependence of the resistivity ρ was studied (Fig. 9). This was done because the electron mean free path is, in general, short in metallic glasses[30] and ρ reflects certain aspects of their local short range atomic arrangements. On heating the sample, it is noticed (Fig. 9) that ρ tends to be concave to the temperature axis especially above 540 K. Although a slight decrease in ρ was observed when the sample was kept at 600 K for 1 hour, the major effect of the annealing is to increase the slope of ρ versus T, which is linear in T up to about 610 K. This finding suggests that thermal annealing in the vicinity of 600 K tends, on the average, to make the local atomic arrangements more ordered while retaining a glassy structure. The nature of the atomic short range order is not yet clarified, although some indication exists that transition metal atoms tend to cluster together during annealing.[17]

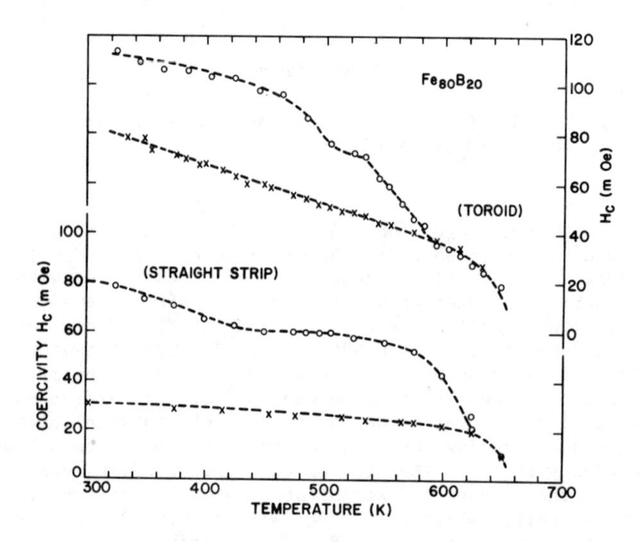

Fig. 8. Coercivity versus temperature for a toroid and a straight strip. Circles and crosses represent heating and cooling temperature cycles.

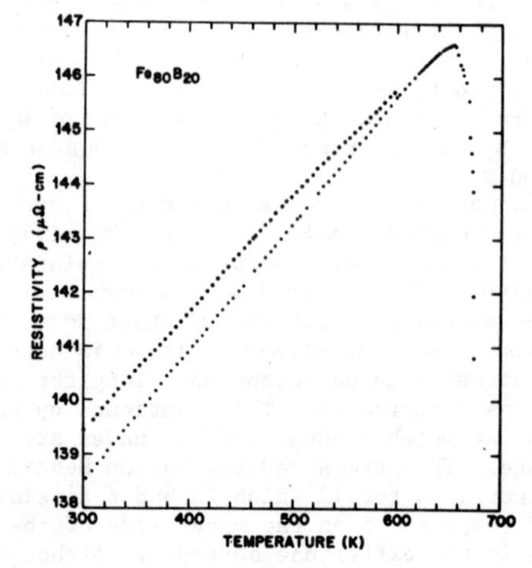

Fig. 9. Resistivity as a function of temperature. The upper curve corresponds to the initial heating. After the annealing, ρ versus T (lower curve) is reversible up close to the crystallization temperature (∿660 K).

E. Low Field AC Properties

Although permeabilities such as $\mu(B)$, defined as B/H, and μ_{max} ($=B_r/H_c$) are, strictly speaking, dc properties, they are often used as measured of the ac performance of soft magnetic materials and, therefore, are included in this section. Table IV compares the results of $\mu(20)$, $\mu(100)$ and μ_{max} of $Fe_{80}B_{20}$ with those of a commercial 50% Fe-Ni alloy,[27] having the same saturation induction (16 kG). Low induction permeabilities of both as-quenched and field-annealed $Fe_{80}B_{20}$ are higher than those of the crystalline alloy. The value of μ_{max} listed in the table is one of the highest ones obtained for the field-annealed $Fe_{80}B_{20}$ and its lower limit is 110,000 largely due to its coercivity reaching as high as 80 mOe, thus being comparable to the crystalline counterpart.

Since the resistivity of $Fe_{80}B_{20}$ is about three times larger than that of the crystalline Fe-Ni alloy, eddy current losses are expected to be lower in the metallic glass. However, the large and coarse domains observed in as-quenched $Fe_{80}B_{20}$ (see Fig. 5) tend to increase the actual eddy current induced during flux reversal. Thus the resultant ac core loss for the as-quenched material is somewhat higher than the crystalline alloy as shown in Fig. 10. When field-annealed,

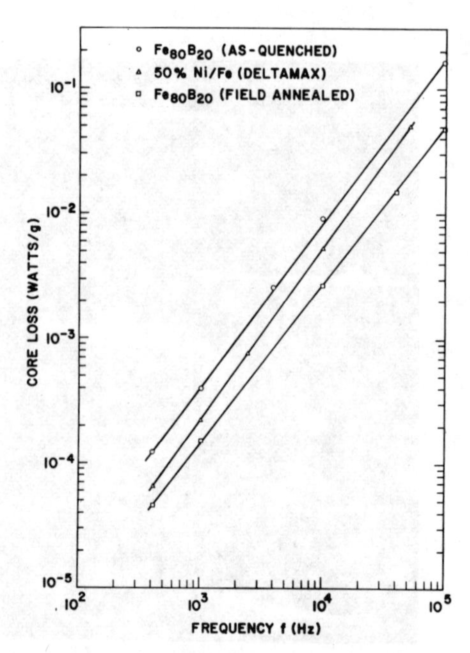

Fig. 10. Frequency dependence of the core-loss at the maximum induction level of 1 kG.

Table IV Permeabilities, resistivity ρ ($\mu\Omega$-cm) and 60Hz core loss (watts/kg) at room temperature

	$Fe_{80}B_{20}$ as-quenched	$Fe_{80}B_{20}$ field-annealed	50% Fe-Ni (crystalline)
$\mu(20)$	1,700	4,000	500
$\mu(100)$	4,600	11,000	2,000
μ_{max}	102,000	320,000	146,000
ρ	130-140		45
core-loss (at B=13 kG)	--	0.53	0.77

the metallic glass toroid exhibits core losses lower than the commercial one. The frequency dependence of the loss is given by a power law f^n with n = 1.3 and 1.4 for $Fe_{80}B_{20}$ and 50% Fe-Ni respectively. The smaller value of n for the metallic glass suggests a favorable ac performance at higher frequencies. In the low frequency region, the high saturation induction level and the low core-loss are advantageous in such applications as various transformers. As an example, core-losses at 60 Hz and 13 kG induction level are compared in Table IV between $Fe_{80}B_{20}$ and the crystalline Fe-Ni. It should be mentioned that the loss level of the metallic glass is about a factor of three lower than the value for 50 μm thick grain-oriented Fe-Si alloy sheets.

CONCLUSION

Synthesis of the new metallic glass $Fe_{80}B_{20}$ has led to a better understanding of the role of the metalloids such as boron, carbon and phosphorous in the magnetic properties of transition metal-metalloid glasses. The evidence noted above strongly indicate that the metalloid having more sp electrons donates more electrons to the transition metal atom. An analysis based on the magnetic moment studies on various metallic glasses suggests that approximately 1.6 and 2.4 electrons per boron and phosphorous atom respectively are transferred to the d-band of transition metal atoms.

The relatively high Curie point (∿650 K) and crystallization temperature (∿660 K) of the present metallic glass are advantageous from the standpoint of magnetic and thermal stability required in magnetic applications. Field annealed toroids of $Fe_{80}B_{20}$ exhibit the lowest coercivity (∿40 mOe), the highest remanence (∿12 kG) and saturation induction (∿16 kG) among the reported metallic glasses containing 80 at.% Fe. In addition, their ac core losses are lower by a factor of 1.5 and 3 than commercial crystalline 50% Fe-Ni and grain oriented Fe-Si alloys, respectively. The properties relevant to magnetic applications may be further improved by optimizing the field-annealing procedure. Nevertheless, the results reported above represent significant improvements of magnetic and also mechanical properties over existing iron-rich metallic glasses and some crystalline counterparts.

ACKNOWLEDGEMENTS

Drs. R. Ray and S. Kavesh led the effort to synthesize $Fe_{80}B_{20}$. Collaboration with Prof. C.-L. Chien in the Mössbauer study is greatly appreciated. X-ray, FMR, SEM and DSC data were supplied by Drs. P. Harget, H. Yue, L. Tanner and S. Takayama. The authors would like to thank Drs. L. A. Davis and J. J. Gilman for their support of the present work.

REFERENCES

1. C. C. Tsuei and Pol Duwez, J. Appl. Phys. 34, 435 (1966)
2. A. I. Gubanov, Fiz. Tverd. Tela 2, 502 (1960)
3. Pol Duwez and S. C. H. Lin, J. Appl. Phys. 38, 4096 (1967)
4. J. J. Gilman, Physics Today 28, 46 (1975)
5. T. Egami, P. J. Flanders and C. D. Graham, Jr., AIP Conf. Proc. #24, 697 (1975)
6. Registered trademark of Allied Chemical Corporation
7. M. Kikuchi, H. Fujimori, Y. Obi and T. Masumoto, Jpn. J. Appl. Phys. 14, 1077 (1975)
8. R. C. Sherwood, E. M. Gyorgy, H. S. Chen, S. D. Ferris, G. Norman and H. J. Leamy, AIP Conf. Proc. #24, 745 (1975)
9. S. Takayama, J. Mat. Sci. 11, 164 (1976)
10. R. Hasegawa, R. C. O'Handley, L. E. Tanner, R. Ray and S. Kavesh; and R. C. O'Handley, L. I. Mendelsohn, R. Hasegawa, R. Ray and S. Kavesh (both submitted to Appl. Phys. Lett.)
11. C.-L. Chien and R. Hasegawa, AIP Conf. Proc. #24, 127 (1975)
12. R. Hasegawa, AIP Conf. Proc. for the 21st Magnetism and Magnetic Materials Conference
13. H. J. Leamy, S. D. Ferris, G. Norman, D. C. Joy, R. C. Sherwood, E. M. Gyorgy and H. S. Chen, Appl. Phys. Lett. 26, 259 (1975)
14. The field annealing conditions are similar to those described by F. E. Luborsky, J. J. Becker and R. O. McCary, IEEE Trans. Magnetics, MAG-11, 1644 (1975).
15. T. Mizoguchi, K. Yamauchi and H. Miyajima in Amorphous Magnetism edited by H. O. Hooper and A. M. deGraaf (Plenum Publishing Corp. 1973) p. 325
16. C. C. Tsuei, G. Longworth and S. C. H. Lin, Phys. Rev. 170, 603 (1968)
17. R. Hasegawa and R. C. O'Handley, Proc. 2nd Int. Conf. on Rapidly Quenched Metals (Boston 1975)
18. R. Hasegawa and C.-L. Chien, Solid State Comm. 18, 913 (1976)
19. R. C. O'Handley, R. Hasegawa, R. Ray and C.-P. Chou (submitted to Solid State Comm.).
20. R. Hasegawa, W. A. Hines. L. T. Kabacoff and Pol Duwez (submitted to Physical Review Letters)
21. K. P. Belov and A. N. Goriaga, Fiz. Metal i Metalloved 2, 3 (1956); A. Arrott, Phys. Rev. 108, 1394 (1957)
22. R. Hasegawa, B. E. Argyle and L.J. Tao, AIP Conf. Proc. #24, 110 (1975)
23. R. M. Bozorth, Ferromagnetism (D. Van Nostrand Co., Princeton, NJ 1968)
24. M. C. Cadeville and A. J. P. Meyer, Comptes Rendus 255, 3391 (1962)
25. R. Fruchart and A. Michel, Mem. Soc. Chim., p. 422 (1959)
26. R. Hasegawa, Phys. Rev. B3, 1631 (1971)
27. 50% Fe-Ni Alloy (Deltamax): Tape Wound Cores, TC-101B (Arnold Engineering) p. 13-14
28. R. C. O'Handley, L. I. Mendelsohn and E. A. Nesbitt (this conference)
29. L. A. Davis, R. Ray, C.-P. Chou and R. C. O'Handley, Scripta Met. (June 1976)
30. R. Hasegawa, Phys. Rev. Lett. 28, 1376 (1972)

FORMATION AND STABILITY OF AMORPHOUS FERROMAGNETIC ALLOYS

D. Turnbull
Division of Engineering and Applied Physics, Harvard University, Cambridge, Mass. 02138

The formation of amorphous ferromagnetic alloys by rapid melt quenching and by condensation processes will be surveyed. The atomic transport and crystallization rates in these alloys will be discussed.

THERMODYNAMICS OF AMORPHOUS MAGNETIC METALS AND ALLOYS AT HIGH TEMPERATURE

M. Ausloos,
International Centre for Theoretical Solid State Physics-Belgium,
Sart Tilman, B-4000 Liège 1, Belgium

ABSTRACT

Thermodynamic properties of amorphous magnetic metals and alloys are derived along a perturbation scheme. The free energy is explicitly treated. The zero order approximation is considered to be that of a structurally ordered magnet, and a perturbation approximation is performed considering the topological disorder and the "random" exchange integrals. The crystalline structure concept is replaced by a radial distribution function $g(r)$ describing the amorphous structure. The coefficients of the high temperature expansion are related to moments of the perturbation potential defined however in terms of a weight function which varies with the system. A model exchange integral representative of the isotropic nature of magnetic systems is used to calculate the first few terms of the expansion. In a first approximation, they are given as polynomials in the density. Spin correlation functions appear as factors. The convergence of the series is discussed.

INTRODUCTION

There has been much theoretical work on substitutional magnetic alloys. These materials present the advantage of having a known lattice structure. Approximations developed for disordered (non-magnetic) alloys can thus be easily adapted to include magnetic interactions.[1,2] Localized spin models as well as "itinerant magnetism" have been considered.[3]

Another class of random magnetic materials has received less attention. Topologically disordered (single element or alloy) materials are more difficult to treat, because of the lack of lattice structure. However, some experimental evidence indicates that the distribution of spins is topologically disordered. Some of them are insulating liquids,[4] but others present semi-metallic conductivities.[5]

In such cases, the spins can still be considered as localized,[6] but free to move in some small volume in order to adapt to their environment. Spin-spin interaction will be assumed to be of the Heisenberg type with appropriate exchange integrals, which have *a priori* to depend on the distance between spins, but also on the short range order (number and *type* of nearest neighbors). The fluctuations in the average number of nearest neighbors is often connected with the change of the average distance between pairs of spins, hence, with the change of the average exchange interaction through its dependence on the spin-spin distance. Such correlations have been investigated by Slechta who separated the various effects in analyzing different models.[7] He limited his interests to small clusters, however.

Alternative approaches are either phenomenological or use Brout cumulant expansion (for the spin $\frac{1}{2}$ Ising model) followed by spatial configuration averaging to take into account the topological disorder. We will use a related method though we will attempt to put on an equal footing spatial configuration and spin averaging.

PERTURBATION THEORY

The present work considers that N spins interact through a Heisenberg Hamiltonian

$$H = \sum_{1\leq \ell<m\leq N} \vec{S}_{\ell'}.\vec{S}_{m'}\, J_{\ell'm'}(\vec{r}_{\ell'},\vec{r}_{m'}) \ . \quad (1)$$

The position $\vec{r}_{\ell'}$ of the spin $\vec{S}_{\ell'}$ is arbitrary. We compare this topologically disordered system to a three dimensional Heisenberg one

$$H_o = \sum_{1\leq \ell<m\leq N} J(\ell,m)\vec{S}_\ell.\vec{S}_m \ , \quad (2)$$

where the position $\vec{r}_\ell$ of the spin $\vec{S}_\ell$ corresponds to a given lattice structure. The latter system will be refered to as the "unperturbed" system. The perturbation is then taken as $H_1 \equiv H - H_o$.

In this work, we suppose that the topologically disordered system can be obtained by simple deformation of the ordered structure. Furthermore, we suppose that there is a one to one relation between spins, i.e. the number of nearest neighbors (more generally, the value of the short range order parameters) remains constant during the deformation.

In fact, the number of nearest neighbors can be quite different in liquid-like structures and in lattice structures. We have presented elsewhere the case where the number of nearest neighbors is arbitrary.[8] However, other approximations have then to be introduced in order to allow for an analytical calculation.

The influence of the number of nearest neighbors can however be taken into account in further work along the lines proposed here, since we will describe the spin distribution by a pair correlation function.

The most important improvement arises from the fact that the free energy itself is the calculated quantity. Usually the Boltzmann factor $\exp(-\beta H)$ is expanded with respect to the perturbation H_1 , i.e.

$$\exp(-\beta H) \simeq \exp(-\beta H_o)\ (1+\beta H_1+\ldots) \ . \quad (3)$$

Here, we develop the free energy

$$-\beta A = \langle \ell n \ \mathrm{Tr} \exp -\beta(H_o + H_1)\rangle \quad (4)$$

as a functional series of the perturbing potential H_1 , i.e.

$$-A\beta = A_o+A_1\beta+A_2\beta^2/2!+\ldots+A_i\beta^i/i!+\ldots \ . \quad (5)$$

Other thermodynamic functions can similarly be obtained.[10]

Obviously,

$$A_o = -\langle \ell n \ \mathrm{Tr} \exp -\beta H_o \rangle \quad (6)$$

is a quantity which has received much attention. A review of various methods to calculate A_o has been recently given by Rushbrooke et al.[9] It is therefore supposed to be known here.

The coefficients $A_1 \ldots A_i$ are given by relations similar to those relating cumulants to (density) moments in the theory of statistics, e.g.

$$A_1 = -M_1/M_o \quad , \tag{7a}$$

$$A_2 = (M_2M_o - M_1^2)/M_o^2 \quad . \tag{7b}$$

The general term A_s can receive an exact analytical form in terms of the perturbing potential if we define $\bar{M}_i = M_i/M_o$, then ,

$$A_s = -\sum_{\ell=1}^{s} \sum_{(k_i r_i)} (-)^{\sum r_i} \left(\sum_{i=1}^{s} r_i - 1\right)! \prod_{i=1}^{s} \cdot \left[\frac{\bar{M}_{k_i}^{r_i}}{r_i!(k_i!)^{r_i}}\right] \tag{8}$$

under the condition

$$\Sigma_i \; k_i r_i = s \quad . \tag{9}$$

Notice that A_s is not the usual (density) cumulant since the moment M_i , (or $\bar{M}_i$) differs from that of the density defined in terms of the total Boltzmann factor $\exp(-\beta H)$.[11] Indeed, the explicit calculation of the coefficients in expansion (5) indicates that the M_i's are defined by

$$M_i = \langle \int \ldots \int d\vec{r}_i \ldots d\vec{r}_N [H-H_o]^i \prod_{\ell m} \exp\{-\beta H_o(r_{\ell m})\} \rangle \; , \tag{10}$$

where the average $\langle \ldots \rangle$ indicates that the trace is taken with respect to the spatial configuration, but spin configuration averages have also to be included. An illustrating way of representing the various contributions would be to use a diagrammatic representation.

Here, we give the explicit analytical form of the most simple terms in (5) for a realistic model. In order to represent the liquid like structure of amorphous magnets,[12,13,14] the spins are supposed to be localized, and surrounded by an exclusion (hard) sphere. The exchange integrals in (1) and (2) have thus the form

$$J_{\ell' m'}(r_{\ell'}, r_{m'}) = \begin{cases} \infty & \text{if } |\vec{r}_{\ell'} - \vec{r}_{m'}| < d \\ -\lambda\phi(|\vec{r}_{\ell'} - \vec{r}_{m'}|) & \text{if } |\vec{r}_{\ell'} - \vec{r}_{m'}| > d, \end{cases} \tag{11}$$

where d is the diameter of an exclusion sphere. Only nearest neighbor interactions are considered. Furthermore, the isotropic nature of the liquid is taken into account by assuming that the exchange integral is spherically symmetric. In so doing, the discrete amorphous short range ordered structure is supposed to be equally described by a radial distribution function $g(r)$, but also by higher order correlations functions. This allows for a spin to be partially delocalized.

Such a feature is used in the next section in order to relax the approximation that the reference Hamiltonian was here one for localized spins sitting at given lattice sites, i.e. we allow for a broadening of the spin distribution functions per site. Hence, the product in (10) which was only a product of δ functions is considered below to be given e.g. by a product of cut-off functions, i.e.

$$\prod_{\ell m} \exp[-\beta H_o(|\vec{r}_{\ell m}|)] \rightarrow \prod_{\ell m} \theta(|\vec{r}_\ell - \vec{r}_m| - d), \tag{12}$$

where $\theta(x)$ is the step function.

APPLICATION TO THE MODEL INTERACTION (11)

Notice that the sign and form of the Heisenberg exchange integral for the ordered system does not appear anymore because of (12).

Furthermore, the spin distribution now appears in (10) only in $(H-H_o)^i$. Hence, the average $\langle \ldots \rangle$ can be decoupled, and the various spin correlation functions factorize out from the spatial integration.

It is furthermore possible to give simple expressions for the A_i , if one considers their density expansion, as in the case of atomic liquids.[15] Let us only write the leading contributions to the density (ρ = N/V) expansion

$$A_1 \simeq \{\langle S_1 S_2 \rangle\}(E_{11}\rho + E_{12}\rho^2 + \ldots) \quad , \tag{13}$$

$$A_2 \simeq \{\langle S_1 S_2 S_3 S_4 \rangle\}(E_{21}\rho + \ldots) \quad , \tag{14}$$

where $\{\langle S_1 S_2 \ldots S_k \rangle\}$ stands for all spin correlation functions up to the k-th order (with appropriate normalization), excluding the possibility of two spins sitting in the same sphere. On the other hand,

$$E_{11} = 4\pi\lambda \int_d^\infty dr \; r^2 \phi(r) \quad , \tag{15}$$

$$E_{12} = 4\pi\lambda(V_o/V) \int_d^{2d} dr \; r^2 v_L(r)\phi(r) \; , \tag{16}$$

$$E_{21} = 4\pi\lambda^2 \int_d^\infty dr \; r^2 \phi^2(r) \quad , \tag{17}$$

with

$$v_L(r) = (1-r/2d)^2 \; (1+r/4d) \quad . \tag{18}$$

Higher order terms require more lengthy writing and will be given elsewhere.[16] Comparison of (16) to (15) however indicates that E_{ij} is small when $j > i$. In (16), V_o is the hard sphere exclusion volume ($= 4\pi d^3/3$), while V_o/V is related to the packing fraction and is an adjustable parameter here. The above expansion can however be supplemented in numerical calculation by introducing known pair and higher order distribution functions in place of the product in (10), as in the case of atomic liquids.[10]

CONVERGENCE OF THE EXPANSIONS

The convergence criterion of (5) is

$$(s+1)|A_s| > \beta |A_{s+1}| \quad . \tag{19}$$

To evaluate (19) analytically is however quite difficult in general. It is however expected that the series is asymptotic, and only the first few terms are relevant.[15,17] This is further borne out by simulation work of the equation of state of close packed solids. If one truncates the series to the second order term (this was shown to be adequate for ordinary liquids,[18] but is likely to be an approximation here) the "convergence" criterion becomes

$$2\langle S_i S_j\rangle(E_{11}+E_{12}\rho+\dots) \geq \beta\{\langle S_i S_j S_k S_\ell\rangle\}E_{21} \quad , (20)$$

which is approximately given by

$$k\ \lambda\ \chi_T > S^2\ C_H \quad , \quad (21)$$

where C_H and χ_T are respectively the specific heat and the magnetic susceptibility.

REFERENCES

1. R.A. Tahir-Kheli, Phys. Rev. B6, 2808, 2826, 2838 (1972).
2. E.N. Foo and D. Wu, Phys. Rev. B5, 98 (1972).
3. H. Aoki and H. Kamimura, J. Phys. Soc. Japan 40, 6 (1976).
4. B.Z. Kaplan and D.M. Jacobson, Nature 259, 654 (1976).
5. T.E. Sharon and C.C. Tsuei, Phys. Rev. B5, 1047 (1972).
6. J.L. Finney, Proc. Roy. Soc. (London) A319, 479 (1970).
7. J. Slechta, Phys. Stat. Sol. (b) 67, 595 (1975).
8. M. Ausloos and P. Clippe, J. Phys. C: Solid State Phys. 9, (1976).
9. G.S. Rushbrooke, G.A. Baker Jr. and P.J. Wood, in Phase Transitions and Critical Phenomena 3, 245 (1974), C. Domb and M.S. Green eds. (Academic Press, London).
10. R. Zwanzig, J. Chem. Phys. 22, 1420 (1954).
11. F. Yonezawa and T. Matsubara, Prog. Theor. Phys. 35, 357 (1966).
12. J. Logan, Phys. Stat. Sol. (a) 32, 361 (1975).
13. L. Weiss, Wiss. Z. Techn. Univ. Dresden 23, 1026 (1974).
14. J.J. Rhyne, Phys. Rev. Letters 29, 1562 (1972).
15. P. Clippe and R. Evrard, J. Chem. Phys. 64, 3217 (1976).
16. P. Clippe and M. Ausloos, work in progress.
17. B.J. Alder, A.A. Young and M.A. Mark, J. Chem. Phys. 56, 3013 (1972).
18. P. Clippe and R. Evrard, Mol. Phys. 29, 645 (1975).
19. M. Ausloos, Phys. Letters 55A, 368 (1976).

Note added in proof : In the case of (non-magnetic) liquid metals, a similar theory can be worked out,[19] and leads to attractive results. It is therefore easy to study thermodynamic properties of liquid magnetic metals following such a procedure and that outlined here.

MAGNETIC PROPERTIES OF Fe AND Co BASED AMORPHOUS ALLOYS

N. Kazama,* M. Kameda and T. Masumoto
The Research Institute for Iron, Steel and Other Metals
Tohoku University, Sendai, Japan

* Present Address: IBM Research Laboratory,
San Jose, California 95193

ABSTRACT

Detailed magnetic measurements were carried out on amorphous alloys: Fe-Co-P-C, Fe-Cr-P-C and Co-Fe-Si-B. The saturation magnetizations (Ms) are obtained for these alloys using the formula $dM/dH = Ms(a/H^2) + X_o$. The spin-wave dispersion coefficients (D) are also obtained from the low temperature magnetization measurements.

The contribution of the structural disorder to the magnetic properties is found directly by comparing the results for the amorphous $Fe_{80}P_{13}C_7$ and $Co_{75}Si_{15}B_{10}$ with those for the corresponding single-phase crystalline bcc Fe-P-C and hcp Co-Si-B, which were obtained by careful thermal treatments. The results show that the structural disorder has influence on the magnetic excitations rather than the magnetic moment.

INTRODUCTION

Amorphous transition metal alloys usually contain a large amount of glass former atoms such as P, Si, C, B, etc., which make the amorphous phase stable up to high temperature, for example Fe-P-C and Co-B-Si alloys have their crystallization temperature at 670 and 770K respectively. These amorphous materials, if heated to a sufficiently high temperature, transform to stable equilibrium phases. We found the formation of a metastable single phase for Fe-P-C and Co-Si-B alloys by the prolonged aging at the incipient state of crystallization, which appears to cause short-range ordering of atoms prior to crystallization. At this stage, we can directly observe the effect of the structural disorder on the magnetic properties of these amorphous ferromagnets, thus separating out the effect of chemical disorder.

EXPERIMENTAL RESULTS

Amorphous ferromagnets: $(Fe_{1-x}Co_x)_{80}P_{13}C_7$, $(Fe_{1-x}Cr_x)_{80}\text{-}P_{13}C_7$ and $(Co_{1-x}Fe_x)_{75}Si_{15}B_{10}$ were prepared by means of the so-called "Roller method." Ribbon shape specimens produced with a high speed rotation of about 3000 rpm were all about 50μm thick and 3mm wide and were used for the present measurements. On the basis of the information (Fig. 1) obtained from previous thermal study[1], we were able to produce single phase crystalline alloys by prolonged aging of these amorphous materials for 1×10^4 min. at 320°C and 4×10^4 min. at 350°C for Fe-P-C and Co-Si-B alloy respectively. These phases were identified by X-ray diffraction (Mo-Kα radiation) to be bcc-$Fe_{80}P_{13}C_7$ with the lattice parameter $a_o = 2.86_1$ Å and hcp - $Co_{75}Si_{15}B_{10}$ with $a_o = 2.51_7$ Å and Co = 4.08_1 Å as shown in Fig. 2. We cannot obtain further information from the reflection intensities because of the preferred orientation,

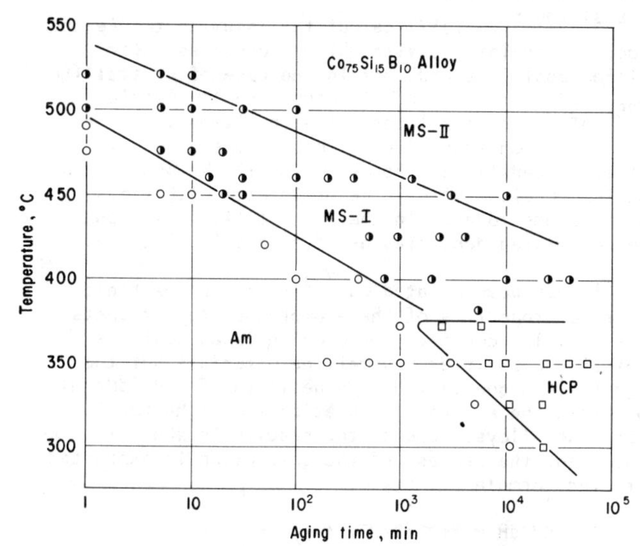

Fig. 1 Temperature-time transformation diagram of amorphous Co-Si-B system on aging. Am, Ms-I, MS-II and hcp correspond to the amorphous, metastable phase I, II and hcp-single phase respectively.

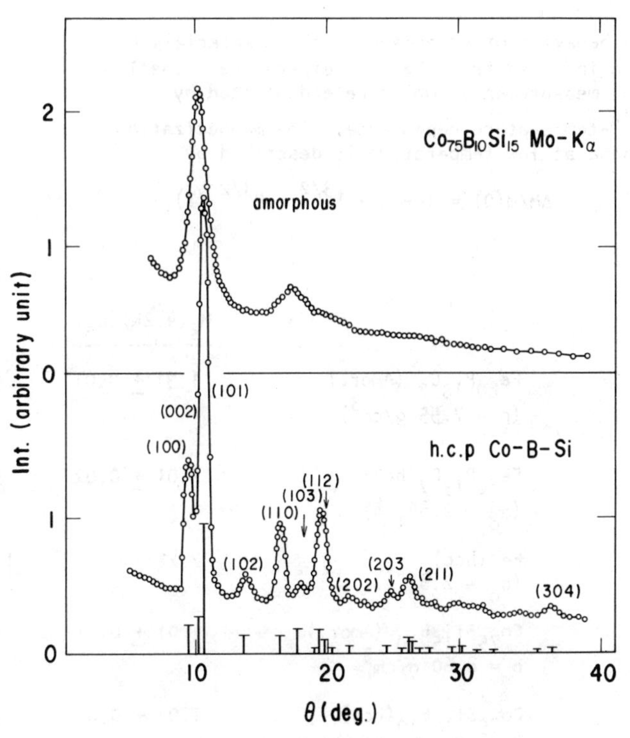

Fig. 2 X-ray scattering patterns for the amorphous state and the crystalline state of $Co_{75}Si_{15}B_{10}$ alloy.

however, by comparing the experimentally measured densities with those calculated from various models, we can arrive at the most likely atomic arrangements. Using the Archimidean method, we measured the densities of amorphous $Fe_{80}P_{13}C_7$ and $Co_{75}Si_{15}B_{10}$ and found them to be 7.55 and 8.46 g/cm^3 respectively. These results are in excellent agreement with densities 7.55 and 8.64 g/cm^3 that in the case of crystalline $Fe_{80}P_{13}C_7$, calculated assuming the Fe and P occupy bcc lattice site while C occupies interstitial positions and that in the case of crystalline $Co_{75}Si_{15}B_{10}$, the Co and Si occupy the hcp lattice site and B occupies interstitial positions. Other realisitic configuration of these atoms give quite different densities. For example, if the bcc or hcp lattice consists of only Fe or Co atom with glass former atoms located in the interstitial position, the calculated densities are 8.78 and 10.37 g/cm^3.

It has been pointed out[2,3] that the low-field magnetic properties of these amorphous ferromagnets have very low coercive force which is as small as that of the purified crystalline materials but the high-field magnetization change in the field (dM/dH) to obtain the accurate saturation moment Ms for amorphous alloys. One of the results is shown in Fig. 3 and the process of the saturation is expressed with the formula

$$dM/dH = Ms\ (a/H^2) + \chi_o$$

where a and χ_o are the parameters which have no temperature dependence below room temperature. We observed the maximum values of a = 167 Oe and χ_o = 5.6×10^{-4} emu/g, for X = 0.25 in $(Fe_{1-X}Co_X)_{80}P_{13}C_7$ alloys.

The behavior of spin-wave of these materials has been inferred from the low temperature magnetization measurements, which were dominated by a $T^{3/2}$-temperature dependence. The magnetization change at low temperature is described by

$$\Delta M/M(0) = 1 - \beta \cdot T^{3/2} \cdot F^{3/2}(T),$$

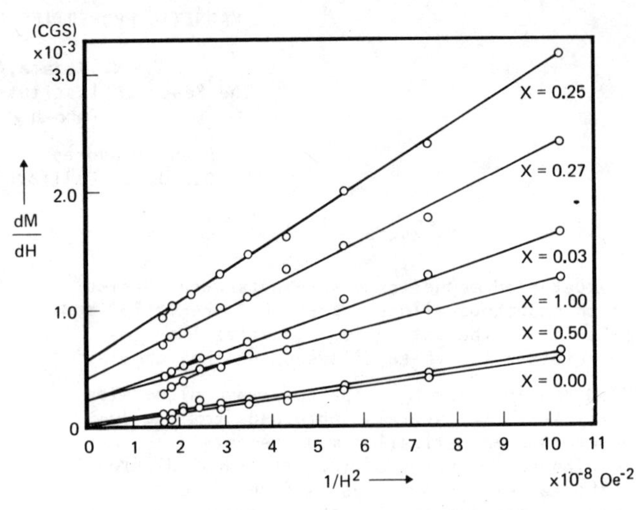

Fig. 3 The magnetization change (dM/dH) in the magnetic field (H) for $(Fe_{1-x}Co_x)_{80}P_{13}C_7$ alloy system.

where

$$\beta = (g\mu_B/8\pi^{3/2}\rho) \cdot (kB/D)^{3/2}$$

and

$$F^{3/2}(T) = \sum_{n=1}^{\infty} n^{3/2} e^{-nTg/T},$$

with the density; ρ and the energy gap; Tg, which depends on the applied magnetic and the anisotropy field. Since we cannot estimate the anisotropy, Tg = 0.4K is used which corresponds to the applied field. For the crystalline alloys, the same densities as that of amorphous state are used. It was found that the $T^{3/2}$ dependence is more applicable to the amorphous state than to the crystalline state. The results are shown in Fig. 4 and Table 1.

TABLE I

	M_s(4.2K) (μ_B)	T_c(K)	$\beta \times 10^{-7}$ ($K^{3/2}$)	D(meV Å^2)
$Fe_{80}P_{13}C_7$ (Amor.) (ρ = 7.55 g/cm^3)	1.91 ± 0.01	568 ± 0.4	219	98
$Fe_{80}P_{13}C_7$(bcc) (a_o = 2.86$_1$ Å)	2.01 ± 0.02	593 ± 3	167	113
Fe (bcc) (a_o = 2.87 Å)	2.22	1044	---	281
$Co_{75}Si_{15}B_{10}$ (Amor.) ρ = 8.60 g/cm^3	1.01 ± 0.01	660 ± 6	169	189
$Co_{75}Si_{15}B_{10}$(hcp) (a_o = 2.51$_7$) (Co = 4.08$_1$)	1.03 ± 0.02	675 ± 10	127	213
Co (hcp) (a_o = 2.507) (Co = 4.069)	1.70	1070	17.2	510

CONCLUSION

We can discuss the following from Table 1 and Fig. 4. The difference between the results for the amorphous and crystalline state shows the effect of structural disorder on the magnetic properties. Furthermore the difference between the crystalline alloys having glass former atoms and the pure metals show the effect of chemical disorder on these materials.

For the case of Co-based alloys the results point out that the magnetic moment is influenced almost entirely by charge transfer from the glass former atoms[4,5]. For Fe-based alloys, it appears that while charge transfer is responsible for most of the change in moment, structural disorder plays a role. The spin-wave coefficients (D) are also reduced to 1/3 the value of the corresponding binary alloys by the glass former atoms. However, we could still observe considerable effect of the structural disorder on them through the change of D-value. The concentration dependence of D is the same as the Curie temperature for the Fe-based alloys, but it is worth noting that the value of D increases in spite of decreasing T_c beyond 20 atomic percent Co.

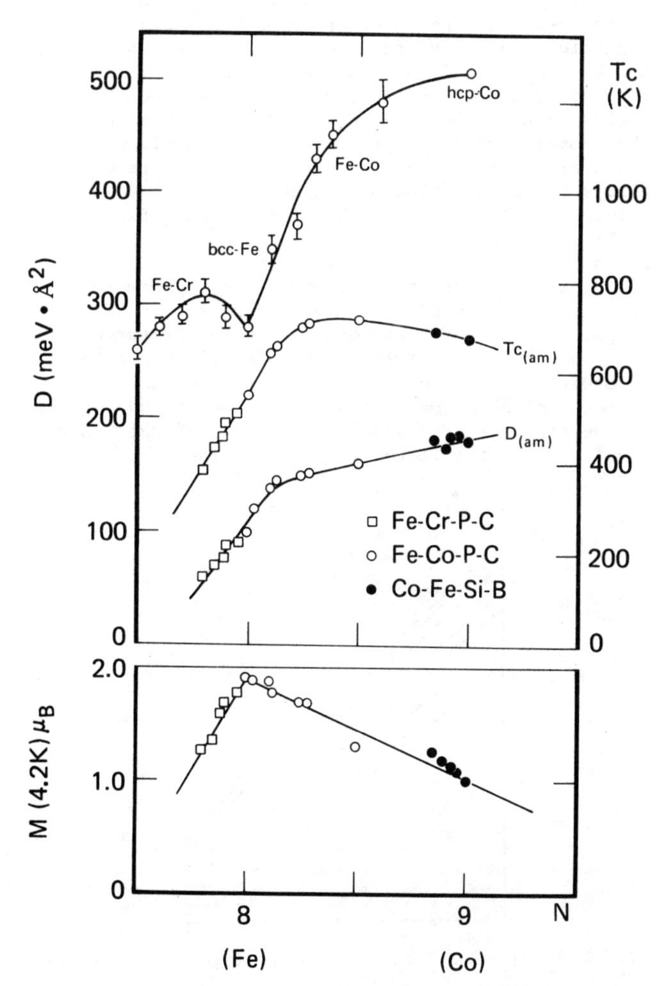

Fig. 4 Concentration dependence of spin-wave coefficient (D) for amorphous and crystalline binary alloys, magnetic moment (M) at 4.2K and Curie temperature T_c(K).

ACKNOWLEDGMENT

The authors wish to thank Dr. N. Heiman for his critical reading.

REFERENCES

1. T. Masumoto, Y. Waseda, H. Kimura and A. Inoue, Proc. of 2nd International Conf. on Rapidly Quenched Metals, Boston, Nov. (1975).
2. R. C. Sherwood, E. M. Gyorgy, H. C. Chen, S. D. Ferris, G. Norman and H. J. Leamy, AIP Conf. Proc., 24 (1975), 745.
3. T. Egami, P. J. Flanders and C. D. Graham, AIP Conf. Proc., 24 (1975), 697.
4. G. S. Cargill III and R. W. Cochrane, J. Phys. 35 (1974), 269.
5. K. Yamauchi and T. Mizoguchi, J. Phys. Soc. Hapan, 39 (1975), 541.

SPIN-WAVES IN AMORPHOUS Fe_xGe_{1-x} THIN FILMS

G. Suran[+], H. Daver[++], J. Sztern[+]

[+]Laboratoire de Magnétisme de Bellevue, C.N.R.S.

[++]Groupe de Transition de Phases de Grenoble, C.N.R.S.

ABSTRACT

Magnetic properties of Fe_xGe_{1-x} amorphous thin films were studied by F.M.R. between 4.2 and 300° K for $0.5 < x < 0.65$. From the resonance spectra, the saturation magnetization, g factor, the spin-wave stiffness constant D and their temperature dependences could be determined. For $x > 0.6$ $M_s(T)$ follows the Bloch law and D(T) a T^{α} law where $3/2 \leqslant \alpha \leqslant 2$. Near 0°K D(0) deduced from magnetization measurements and from S.S.W. spectra are in good agreement. For $x < 0.6$ and with decreasing Fe content the S.S.W. spectra progressively disappears and $M_s(T)$ deviates largely from the Bloch law.

INTRODUCTION

In reference (1) we reported some magnetic properties of amorphous Fe_xGe_{1-x} thin films. Measurements were made by F.M.R. and the composition range studied were $0.4 < x < 0.5$. The results for this concentration range were the following : 1) the critical concentration corresponding to the appearance of ferromagnetism is $x_c = 0.4$, 2) $M_s(T)$ follows a linear law as a function of T for a large range of temperature. We also observed an anomalous behaviour of both g factor and resonance linewidth ΔH for temperatures lower than 100°K : both values increase progressively as T is lowered to 4.2°K. The magnitude of ΔH is related to the Fe concentration in the film decreasing as the Fe content increases : for example at 100°K, ΔH = 600 Oe for x = 0.43 and ΔH = 70 Oe for x = 0.496. These results were related to the local structure of the film. Electron diffraction studies showed[2] that the short range order of amorphous Fe_xGe_{1-x} alloys change continuously as a function of concentration : for higher Ge concentration ($x < 0.2$) the short range order is close to polytetrahedral one, while for Fe rich one ($x > 0.6$) it can be described by the dense random packing hard sphere (D.R.P.H.S.) structure.

In order to find out if the peculiarities observed are related to the concentration range studied, we investigated films with higher Fe concentration ($0.5 < x < 0.65$) and the results are reported here. Another interest of this range of concentration is that ΔH of uniform resonance mode is lower than 100 Oe for a wide range of temperature and well resolved higher order modes could be detected which were identified as standing spin-wave (S.S.W.) modes. So for $x > 0.5$ the spin wave stiffness constant D(0) near 0°K and its temperature dependence D(T) could be studied on films with various Fe concentration.

EXPERIMENTAL

The Fe_xGe_{1-x} film were obtained by coevaporation under 10^{-7} Torr. The films were deposited onto glass substrates cooled down to 77°k [1,2]. In order to avoid compositional gradient along the films thickness[1], the deposited thickness ranged between 600 and 800 A, was measured with "Talystep 1" the precision being of ±30A. Magnetic properties were determined by ferromagnetic resonance (F.M.R.) in the Ku band at 17.3 GHz and in the range 4.2 to 300K. At any given temperature the spectra for applied field perpendicular ($H_{\perp}$) and parallel ($H_{//}$) to the film plane were recorded. The corresponding F.M.R. conditions are :

$$\frac{\omega}{\gamma} = H_{\perp} - 4\pi M_s + D\left(\frac{n\pi}{d}\right)^2 \quad (1)$$

$$\left(\frac{\omega}{\gamma}\right)^2 = H_{//}\left|H_{//} + 4\pi M_s + D\left(\frac{n\pi}{d}\right)^2\right| \quad (2)$$

where d is the thickness and D = 2A/M. $4\pi M_s$ and g factors were calculated from $H_{\perp}$ and $H_{//}$ corresponding to the uniform mode (n = 0). The modes n = 1,2......correspond to S.S.W. modes. Eq (1) with n integer can be used to identify the various S.S.W. modes if the effective surface anisotropy $(K_s)_{eff}$ is small and volume (chemical) inhomogeneities are negligible. ΔH is the as measured peak-to-peak derivative linewidth.

RESULTS

The Fe_xGe_{1-x} films for $0.4 < x < 0.65$ can be roughly divided into three groups according to their magnetic properties. The first group with $x < 0.5$ whose magnetic properties are given in (1) does not present a S.S.W. spectra and only the main resonance is observed. The second group for $0.5 < x < 0.6$ shows the ΔH of main resonance to be 60 ± 10 Oe at 100°K and independent of the composition. The spin wave spectra could be observed systematically but the dispersion spectra were not always found to follow a quadratic law. The low temperature anomalies for T < 100°K of ΔH and g factor are still observed though much less pronounced than for films of Group 1. Two samples with x = 0.53 and x = 0.58 -the properties of which are reported here- illustrate this group : one observes (see table I) a slight increase of g factors at low temperatures for both samples. For T < 100°K the S.S.W. spectra for film with x = 0.53 could not be observed due to the increase of ΔH of main line form 60 Oe at 100°K to 130 Oe at 4.2°K. For sample with x = 0.58 the S.S.W. spectra did not follow the quadratic law.

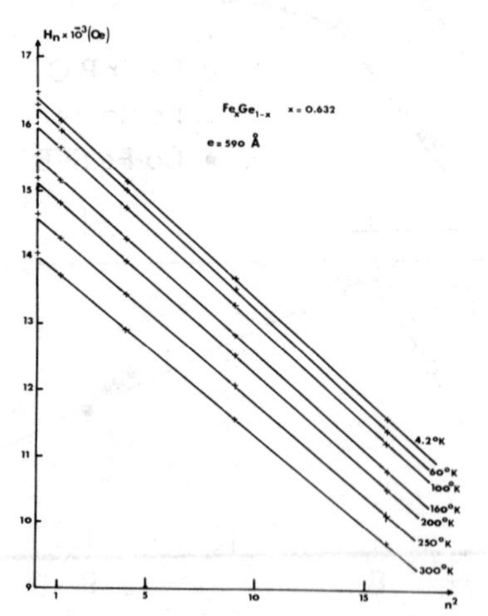

Fig. 1. Field for resonance of the S.W. modes plotted against n^2 for various temperatures.

The third group corresponds to x > 0.6. On these samples one observes a well resolved S.S.W. spectra for all compositions and at all temperatures. The g factor is g = 2.08 ± 0.01 (Table I) and independent of T and composition. These films are rather thin so only 4 to 5 higher order modes were present in the spectra. At all temperatures and on all samples the magnetic resonant fields follow fairly well the quadratic dispersion law (fig. 1), when the succesive modes are numbered as n = 0, 1, 2, 3, 4. As modes with both parity are excited the boundary conditions are assumed to be slightly asymmetric.

The effective surface anisotropy $(K_s)_{eff}$ is small as the intensities of various S.W. modes are weak with respect to the main line and I_n decrease rapidly with mode number. For sample of fig. 1 $I_1/I_0 \simeq 5/100$ and $I_2/I_0 \simeq 1/100$ (I_n intensity of the nth (mode), I_n is independent of T and $(K_s)_{eff} < 10^{-2}$ erg cm^2. This results show that $(K_s)_{eff}$ is temperature independent and does not perturb the temperature dependence of D(T).

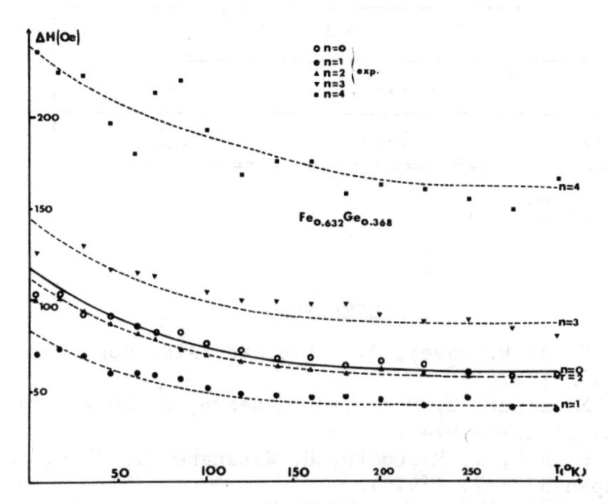

Fig. 2. Temperature dependence of the derivative linewidth of the various modes. n = 0 corresponds to the main resonance.

The linewidth of the S.W. modes as a function of T is reported in fig. 2. The higher order modes vary as a function of T in the same way as the main line and one observes a slight increase of ΔH for T < 100°K. At a given T ΔH_n increases slightly with mode number up to n = 3 and than abruptly for n = 4, these results being characteristic for all samples studied. It is difficult to conclude if the large increase of ΔH_4 is related to the amorphous structure or to the chemical inhomogeneities.

The temperature dependence of $4\pi M_s$ for composition x > 0.5 is shown in fig. 3. For samples with x > 0.6 the Bloch law $M_s = M_o(1-BT^{3/2})$ is perfectly obeyed up to 300°k which is the upper limit of our cryostat. For x < 0.6 one observes a deviation from Bloch law and the larger is the deviation the lower the Fe content. The inset of fig. 3 shows that the reduced magnetization ΔM/M follows the Bloch law up to 0.15 for x = 0.53 and up to 0.3 for x = 0.58.

The temperature variation of D(T) for 100°K < T < 300°K is $D(T) = Do(1-CD^{\alpha})$ with $3/2 \leqslant \alpha \leqslant 2$. All samples seemed to obey satisfactory a $T^{3/2}$ law (Fig. 4). However a T^2 law could be used with about the same accuracy for samples with x = 0.632 and x = 0.596. For T < 100°K D(T) has an anomalous behaviour : D(T) appears to be only weakly dependent on temperature and even decreases near 4,2°K for some samples. The uncertainties upon the value of α are due to the anomalous behaviour of D for T < 100°K and to the insufficient number of experimental points between 100°K and 300°K, so a least square fit is not sufficiently sensitif to determine the exact value of α.

In order to check the validity of D deduced from S.S.W spectra, we calculated D from the slope B of Bloch law. B is related to D by(6) :

$$B = 0.0587 \frac{g\mu_B}{M_o} \left(\frac{k_B}{D}\right)^{3/2} \quad (3)$$

where k_B is the Boltzmann's constant and μ_B is the Bohr magneton. Though eq.(3) holds good for a crystalline structure it is believed that it can be extended, in certain case, to amorphous compound also[6]. The value of D near 0°K as deduced from magnetization measurement D^M and from S.S.W. spectra D^{SSW} (Table I) agree very well and are within ± 20 %.

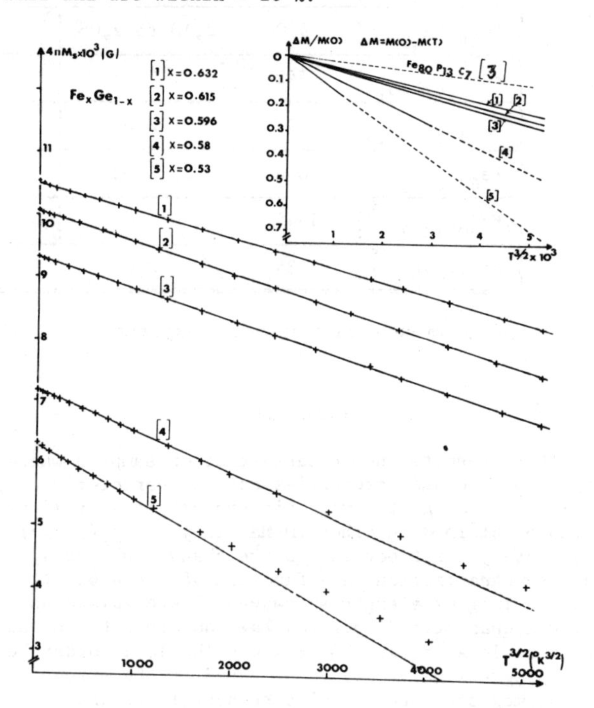

Fig. 3. M(T) versus $T^{3/2}$ for five amorphous FeGe film. The solid line corresponds to the $T^{3/2}$ law. The inset shows ΔM/M versus $T^{3/2}$.

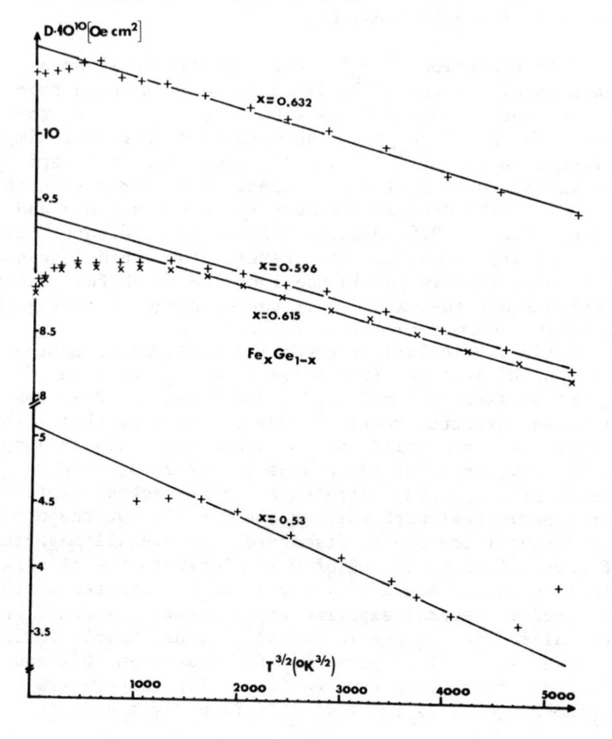

Fig. 4. The spin wave coupling constant D plotted against $T^{3/2}$

TABLE I

Composition, magnetization M(0), magnetization slope B, D(0) deduced from B (D^B), from spin-wave spectra (D^{SSW}) and exchange stiffness constant slope C for amorphous FeGe and Fe.

Composition	M_o (Gauss)	g_{eff}	$B \times 10^{+5}$ ($°K^{3/2}$)	D^B (meV $Å^2$)	D^{SSW} (meV $Å^2$)	$C \times 10^{+5}$ ($°K^{3/2}$)
$Fe_{53}Ge_{47}$	500	2,05 to 2,03	13,8	54,9	53,2(at 100°K)	6,62
$Fe_{58}Ge_{42}$	570	2,10 to 2,06	9,2	66,4	–	–
$Fe_{59,6}Ge_{40,4}$	742	2,08	5,36	80,3	101,8	2,27
$Fe_{61,5}Ge_{38,5}$	800	2,08	4,95	80,5	102,1	2,28
$Fe_{63,2}Ge_{36,8}$	836,5	2,08	4,45	84	120,4	2,11
$Fe_{80}P_{13}C_7$ (3)	1345	–	–	98 (3)	(∿115) (a)	–
Fe cryst. (5)	1735	2,06	–	∿320	350 ± 30	2,2

(a) From neutron scattering experiments at 113°K on $Fe_{75}P_{15}C_{10}$.[4]

DISCUSSION

These results show clearly that on samples where the microstructural properties can be characterised by a D.R.P.H.S. model the magnetic properties ar similar to those obtained on other TM-MA alloys[6] : $M_s(T)$ follows closely the Bloch law so the change in the spontaneous magnetization as a function of T can be attributed to long wavelength spinwaves. These spinwaves follow a quadratic dispersion law and are not too sensitive to local order. For $x < 0.6$ the local order becomes progressively more complicated as the structure is intermediate between polytetrahedral one and D.R.P.H.S. model. One can presume that available sites for Fe atoms, the dispersion of number on nearest neighbour and distance flluctuation between Fe-Fe paires increase so the microscopic random nature manifests itself in the spin dynamics.

The evolution of $M_s(0)$ and D(0) (Table I) as a function of concentration seems to confirm these hypothesis. For $x > 0.6$ M(0) is 2.05 ± 0,1 μ_B per Fe atom (including $Fe_{75}P_{15}C_{10}$) so the metalloid acts as a simple diluent, while for $x < 0.6$ μ_B/Fe atom decreases rapidly. The variation of D(0) = f(x) seems to be similar to that of M(0) : D(0) decreases slowly for $0.6 < x < 0.8$ and rapidly for $x < 0.6$. However further measurements would be necessary to confirm this result. The rapid decrease of D could explain partly the increase in ΔH for concentration lower than $x < 0.5$ as the exchange narrowing of linewidth becomes here less important[7].
Theoretical calculations predict a temperature dependence of D(T) of the form $D(T) = D(1-CT^{\alpha})$ where $\alpha=5/2$ in the Heisenberg localized[(8)] model and $\alpha = 2$ in the itinerant electron model[(9)]. One can presume that these calculations are valid for amorphous materials as long as the wavelength of spin waves is large compared to the short range order structure. In the actual state of our experimental work neither the $\alpha = 3/2$ nor the $\alpha = 2$ experimental law can be discarded. The overall magnitude of temperature variation of D. is closer to the itinerant electron model, but a $T^{3/2}$ law is to be compared to that obtained in several experiments by S.S.W. resonance on crystalline ferromagnetic films[(10)]. For example Philips observed also a $T^{3/2}$ power law for D between 80°K and 300°K on crystalline pure Fe films. The temperature dependence of D(T) for T < 100°K is not yet understood.

We acknowledge the technical assistance of M. TESSIER.

REFERENCES

1. G. Suran, H. Daver, J.C. Bruyère, AIP. Conf. Proc. p.162 (1976)
2. O. Massenet, H. Daver, J. Geneste, J. Phys., Paris, CA, 35 279 (1974).
3. N. Kazama, T. Masumoto, H. Watanabe, J. Phys. Soc. Japan 37 1171 (1974).
4. J.D. Axe, L. Passell, C.C. Tsuei, AIP Conf. Proc. 24 119 (1974).
5. T.G. Philips, Proc. Roy. Soc. A 222 224 (1966).
6. G.S. Cargill, AIP Conf. Proc. 24 138 (1974).
7. A.M. Clogston, H. Suhl, L.R. Walker, P.W. Anderson, J. Phys. Chem. Solids 1 129 (1956).
8. F.J. Dyson, Phys. Rev. 102 1230 (1956).
9. Isuyama, Kubo, J. Appl. Phys. 23 1074 (1964).
10. R. Weber, P. Tannenwald, J. Phys. Chem. Solids 24 1357 (1963),
T. Philips, H. Rosenberg, Phys. Rev. Letters 11 198 (1963).

NEUTRON SCATTERING STUDY OF SPIN WAVES IN THE AMORPHOUS FERROMAGNET $(Fe_{0.3}Ni_{0.7})_{0.75}P_{0.16}B_{0.06}Al_{0.03}$

J. W. Lynn and G. Shirane
Brookhaven National Laboratory*
Upton, New York 11973

R. J. Birgeneau
Massachusetts Institute of Technology†
Cambridge, Massachusetts 02139

and

H. S. Chen
Bell Laboratories
Murray Hill, New Jersey 07974

ABSTRACT

The neutron inelastic scattering technique has been used to measure the collective magnetic excitation spectrum in the ferromagnetic ($T_C \sim 255K$) metallic glass $(Fe_{0.3}Ni_{0.7})_{0.75}P_{0.16}B_{0.06}Al_{0.03}$. In this system only the iron atoms appear to be magnetic, and the observation of relatively well-defined spin waves is striking in comparison with other amorphous systems where the spin waves rapidly broaden with decreasing iron concentration. The spin waves obey the conventional quadratic dispersion relation $E_{sw}=DQ^2$, with $D=35meV-\AA^2$. Contributions from fourth and higher order terms in the dispersion relation are not in evidence for wave vectors in the region investigated ($Q \lesssim 0.25$ $\AA^{-1}$).

INTRODUCTION

Metallic glasses form a new class of materials which have interesting and technologically important magnetic and metallurgical properties. A variety of experimental techniques has recently been employed to investigate the magnetism of these systems, but it has proven difficult to achieve a consistent picture of the underlying microscopic magnetic properties.[1] We have therefore undertaken a systematic investigation of the amorphous ferromagnetic systems $(Fe\text{-}Co\text{-}Ni)_{0.75}P_{0.16}B_{0.06}Al_{0.03}$. The structure of these materials can be characterized as a dense random packing of spheres. They also have well-defined ferromagnetic transition temperatures.[2] Moreover, in the glasses containing Fe and Ni, the Ni atoms appear not to possess magnetic moments. Thus the magnetic properties of these alloys may be more easily understood since there is only one magnetic species present. In this note we wish to report our initial neutron scattering measurements for $(Fe_{0.3}Ni_{0.7})_{0.75}P_{0.16}B_{0.06}Al_{0.03}$. This relatively low magnetic concentration was chosen so that the ferromagnetic transition would be below room temperature, which offers several advantages. Firstly, the critical magnetic region is in an experimentally convenient temperature range which is well below the temperatures at which repacking and recrystallization occur. Secondly, the low transition temperature implies that the exchange energy is relatively small, so that the spin-wave energies are in a more accessible energy range for neutron scattering measurements.

SAMPLES

The samples were made by the centrifugal spinning technique[3] in the form of ribbons 1 mm wide, 0.02 mm thick and 5 m long. The ribbons were then cut into strips 2 cm long and stacked together in the shape of a plate 2 x 2 cm^2. The total weight of the sample was 10 gm. A special isotope of boron (^{11}B) was used in the starting materials to avoid the high absorption cross section for neutrons of naturally occurring boron.

A diffraction pattern taken on the sample showed a broad peak in the scattering function S(Q) centered at 3.12 $\AA^{-1}$, which corresponds closely to the position expected on the basis of dense random packing of spheres. No sharp peaks characteristic of a crystalline phase were observed, and no preferred directional or structural anisotropies were detected. To determine the magnetic transition temperature, the intensity of the elastic scattering was measured at Q = 0.06 $\AA^{-1}$ as a function of temperature. The critical scattering was found to peak at $T \sim 255$ K, which we took to be the nominal value of the ferromagnetic transition temperature. This value is in good agreement with the value expected on the basis of bulk magnetization measurements[2] for this alloy concentration and is also consistent with magnetization measurements taken on this sample.

RESULTS AND DISCUSSION

The neutron scattering experiments were performed on a triple-axis spectrometer at the Brookhaven High Flux Beam Reactor. Pyrolytic graphite crystals were used as both monochromator and analyzer. A fixed incident neutron energy of 14.8 meV was employed for the measurements with $Q \lesssim 0.20$ $\AA^{-1}$, with horizontal collimation before and after the monochromator and analyzer respectively of 20 minutes full width at half maximum (denoted 20'-20'-20'-20') or a combination of 20 and 10 minutes, in this case 20'-10'-10'-20'. For Q in the range 0.20 to 0.25 $\AA^{-1}$ we used incident neutron energies of 40.5 meV and 10'-10'-10'-10' collimation. A pyrolytic graphite filter was placed before the monochromator to suppress higher order wavelengths in the incident beam.

The measurements of the inelastic scattering spectra were taken by fixing the magnitude of the momentum transfer Q and varying the energy transfer E. Some typical measurements are shown in Fig. 1. The peak centered at zero energy transfer, whose width is instrumental in origin, has a magnetic contribution but is primarily due to scattering from the cryostat and elastic nuclear scattering from the sample. The peaks on either side are due to the creation ($E > 0$) and annihilation ($E < 0$) of collective magnetic excitations in the system. These relatively well-defined spin waves are striking in comparison with other amorphous systems studied where the spin waves rapidly broaden with decreasing iron concentration[4,5] and become overdamped at magnetic concentrations considerably higher

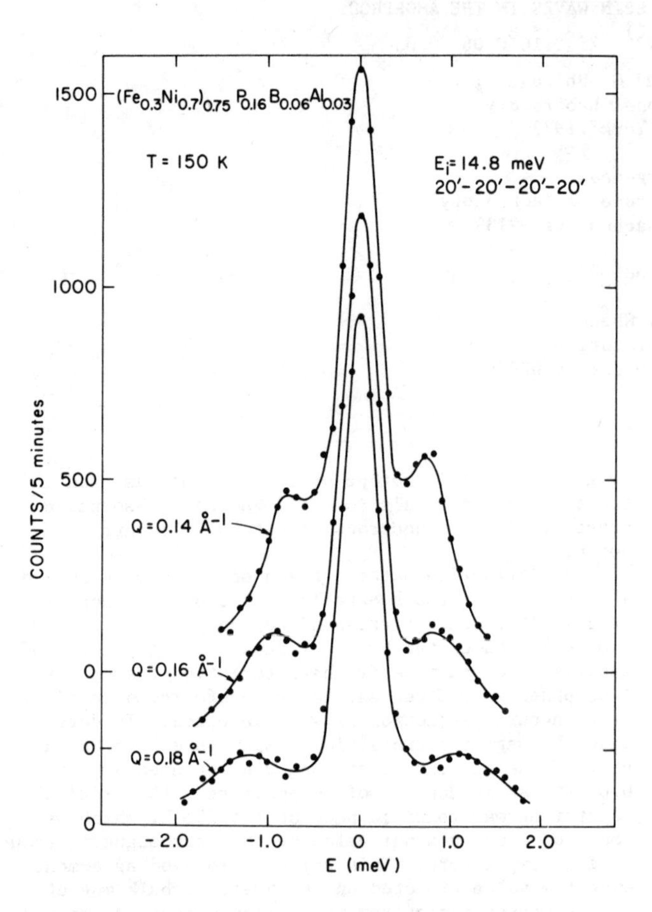

Fig. 1. Constant-Q scans at several values of Q below the ferromagnetic transition temperature. The peak centered at E = 0, which is limited in width by the instrumental resolution, has a magnetic contribution but is primarily due to nuclear elastic scattering and scattering from the cryostat. The peaks on each side are the spin-wave excitations. The solid curves are the least-squares fits as explained in the text.

than in the present system.

With increasing Q the excitations shift to higher energies and broaden. The observed peak intensities of these excitations are also seen to decrease rapidly with increasing Q. Firstly, this is due to the decrease in the thermal population of the spin-wave states with increasing energy. In addition, the line widths are rapidly increasing, and since for this energy and temperature range the integrated intensity in a constant Q scan is

$$\int I(E)dE \propto T/E \qquad (1)$$

the peak intensity decreases as the line widths increase. Thus with increasing Q the signal-to-noise ratios become rapidly less favorable.

The spin-wave energies, corrected for instrumental resolution, are shown in Fig. 2. Over this wave vector range the data are seen to obey

$$E_{sw} = DQ^2 \qquad (2)$$

very closely. The slope of the line yields a value for the spin-wave stiffness constant D of (35±3)meVÅ^2. The error bars in the figure represent one standard deviation.

To obtain some information about the line widths of the spin-wave excitations, we performed least-squares fits of the inelastic portion of the spectra to a dispersion relation of the form of Eq. (2), with the spectral weight function[6] taken to be of the Lorentzian form

$$F(Q,E) = \frac{1}{\pi}\left\{\frac{\Gamma(Q)}{\Gamma(Q)^2 + (E - E_{sw}(Q))^2}\right\}. \qquad (3)$$

The solid curves in Figs. 1 and 3 are the results of these least-squares fits. For the data in Fig. 1 we obtained values for the line widths Γ (HWHM) of 0.06, 0.23, and 0.45 meV at Q = 0.14, 0.16, and 0.18Å^{-1} respectively. The ratios of the line widths to the excitation energies for these values of Q are then Γ/E_{sw} = 0.09, 0.26, and 0.40. Thus as Q increases the spin waves become rapidly broader relative to their energies.

Fig. 3 shows data at 100 K and 220 K. These measurements were taken with somewhat better resolution than those in Fig. 1. At the lower temperature the intrinsic widths of the spin waves at Q = 0.12 Å^{-1} are much smaller than the instrumental resolution, whereas for Q = 0.16 Å^{-1} we find Γ = 0.23 meV. At 220 K the spin waves have obviously broadened and renormalized to lower energies. From the least-squares fits we obtain D = 22 meV-Å^2 for the spin-wave stiffness parameter, and Γ = 0.05 meV and Γ = 0.31 meV for the spin-wave line widths (HWHM) at Q = 0.12 and 0.16 Å^{-1} respectively. Note also that the scattering at E = 0 increases in intensity with increasing temperature, and

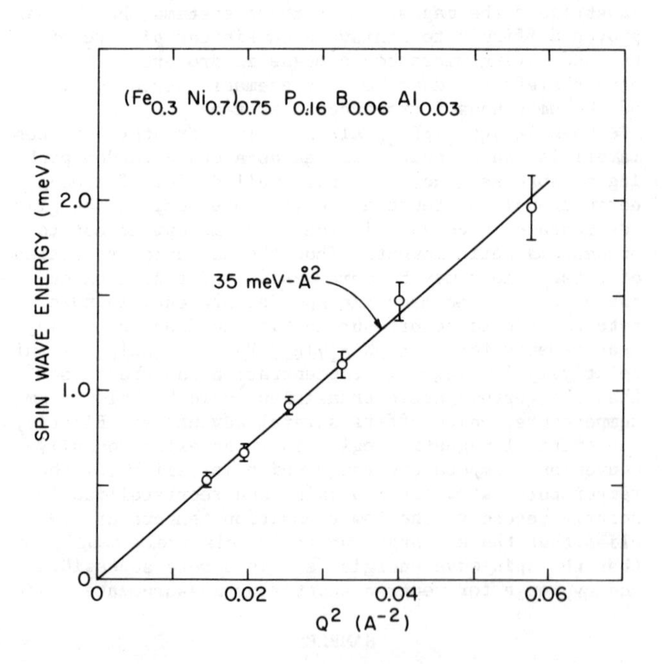

Fig. 2. The spin-wave energy at low temperatures plotted against Q^2, showing that the spin waves obey a quadratic dispersion relation over this wave vector range.

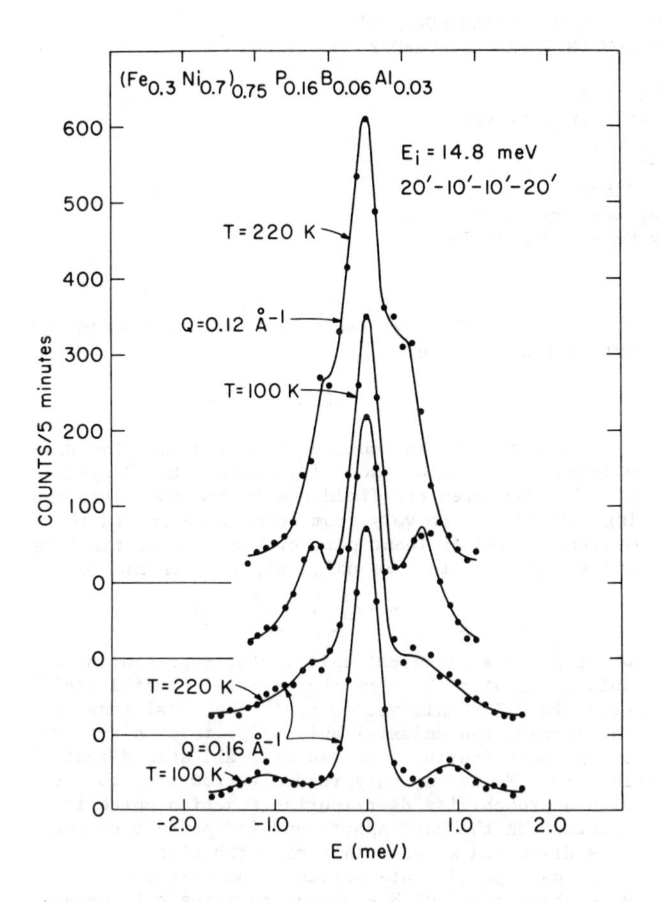

Fig. 3. Constant-Q scans at Q = 0.12 $Å^{-1}$ and Q = 0.16 $Å^{-1}$ at 100 and 220 K. The spin waves have broadened and renormalized to lower energies at the higher temperature. The solid curves are the least-squares fits to the data as explained in the text.

this increase cannot be attributed simply to an overlapping of the elastic and inelastic contributions to the scattering. Thus the description of the magnetic states may be more complicated than simply well-defined spin-wave states with a quadratic dispersion curve as given by Eq. (2). Clearly, further work will be needed in order to properly characterize the nature of this scattering. Further measurements are in progress.

ACKNOWLEDGMENTS

We wish to acknowledge stimulating discussions with J. D. Axe, L. Passell, and J. J. Rhyne, and would like to thank M. Iizumi and C. J. Glinka for providing some of the data analysis programs.

REFERENCES

* Work at Brookhaven performed under the auspices of the U. S. Energy Research and Development Administration.

† Work at M. I. T. supported by the National Science Foundation MRL Grant No. DMR12-03027-ADS.

1. See, for example, Amorphous Magnetism, Ed.H.O. Hooper and A.M. deGraaf, Plenum Press, New York (1973)
2. R.C.Sherwood, E.M.Gyorgy, H.S.Chen, S.D.Ferris, G.Norman and H.J.Leamy, AIP Conf.Proc. no.24, 745-46 (1974).
3. H. S. Chen and C. E. Miller, Mat.Res.Bull. 11, 49-54 (1976).
4. J. D. Axe, L. Passell and C. C. Tsuei, AIP Conf. Proc. no. 24, 119-120 (1974).
5. J. D. Axe, 21st Annual Conference on Magnetism and Magnetic Materials (to be published); and private communication.
6. W. Marshall and S. W. Lovesey, Theory of Thermal Neutron Scattering, Oxford University, London (1971), Ch. 13.

Note added in proof: Recently, R. C. Sherwood and E. M. Gyorgy have performed magnetization measurements on material taken from the sample used in these neutron scattering experiments. Their experiments were performed in a field of 15 kOe. The results are qualitatively similar to those obtained by Leamy et al (AIP Conf. Proc. 29, 211, 1976) in amorphous $(Co_{0.5}Ni_{0.5})_{0.75}P_{0.16}B_{0.06}Al_{0.03}$. Somewhat surprisingly, the D obtained from the 15 kOe measurements appears to be significantly larger (E. M. Gyorgy, private communication) than the value obtained directly from the spin-wave dispersion relation. This, in turn, implies that the field-dependent effects in these amorphous alloys cannot be interpreted using conventional spin-wave theory. Indeed, this is already evident in the measurements of Leamy et al in the Co-Ni alloy. Clearly, very low field magnetization measurements in $(Fe_{0.3}Ni_{0.7})_{0.75}P_{0.16}B_{0.06}Al_{0.03}$ are essential. Similarly, neutron scattering studies of the spin dynamics in a field would be most interesting.

HYSTERESIS CURVES FOR A DENSE RANDOM PACKING MODEL OF AN AMORPHOUS MAGNET WITH RANDOM UNIAXIAL ANISOTROPY

Mely Chen Chi
Physics Department, Wesleyan University
Middletown, CT. 06457
and
Richard Alben
Department of Engineering and Applied Science
Yale University, New Haven, CT. 06520

ABSTRACT

We consider the magnetic field behavior of a simple model for an amorphous ferromagnet with large random single ion anisotropy. The structure is represented by an interior portion of the dense random packing model of Finney. The spins are taken as classical unit vectors and are relaxed by a numerical procedure to direction which minimize applied field, anisotropy and near neighbor exchange energies. The hysteresis curves for the model resemble those found for $TbFe_2$ at low temperature, provided the anisotropy is about the same size as the exchange or larger. The coercivity increases abruptly with increasing anisotropy and then remains almost flat up to the "Ising" limit of infinite anisotropy. We have examined the energy of locally stable states and find that states of partially reversed moment, or states with spin glass character, are of markedly higher energy than aligned type states.

INTRODUCTION

It has been observed[1,2] that the magnetic behavior of amorphous $TbFe_2$ and of other amorphous heavy rare-earth transition metal materials is indicative of large but random single ion anisotropy. In particular the effective rare earth moment is significantly lower than for the crystalline Laves phase compounds and very large coercive forces are observed at low temperature.[1,2] Harris et al[3] have proposed a model to describe these materials. It is a Heisenberg model in which each spin has the same exchange interactions, but is subjected to a local uniaxial anisotropy field with random direction. Both classical and quantum mechanical spins have been considered and the properties of the model, including the effect of temperature, have been treated within a molecular field approximation. Recently[4] a treatment based on the presence of random local fields in addition to the average molecular field has been presented. This generalized model shows a discontinuous change in magnetization at a critical field which was identified as the coercive field. The low field state of the model was identified as being spin glass like. The local fields in this approximation are not related in any way to the specific orientation of neighboring spin. This seems to be a serious drawback since the issue of local correlations among spin directions is crucial to the existence of a possible spin glass like state.[4]

In this paper, we treat the simplest case of the random anisotropy model: classical spin and T = 0°K. Instead of using average or random molecular fields, we consider directly the interaction among neighbors. Relaxation is allowed to take place only by single spin processes. Our results show hysteresis loops, magnetization curves and variation of coercivity with anisotropy, some aspects of which appear to be applicable to real alloys at low temperature. We find that the demagnetized states of our model systems are of considerably higher energy than the approximately aligned states. This indicates that, although anisotropy may make it more difficult to bring differently oriented domains into alignment, the demagnetized state is probably best described as a polydomain and not as a spin glass.

MODEL

We adopt the assumption of Ref. 3 that the disordered structure of the alloys causes the "crystal field" — the electric field due to the ions surrounding each atom — to vary from point to point in the material. The dominant part of the crystal field in the amorphous alloys is uniaxial, i.e. of the form

$$H_c = - D\left((\vec{n}_i \cdot \vec{s}_i)^2 - \tfrac{1}{3}\right) \qquad (1)$$

where $\vec{s}_i$ is a classical unit vector representing the spin of an atom and D is the uniaxial crystal field strength. The unit vector $\vec{n}_i$ is the local easy axis. The form of the uniaxial anisotropy is such that the energy contribution averaged over all spin directions is zero. We let $\vec{n}_i$ vary randomly from atom to atom with a probability distribution function which is constant on the unit sphere and independent of the easy direction at any other magnetic site.

The magnetic interaction is described by a Heisenberg model with exchange coupling only between the near neighbor spins. The total Hamiltonian is then of the following form:

$$H = - D\sum_i\left((\vec{n}_i\cdot\vec{s}_i)^2 - \tfrac{1}{3}\right) - g\beta\vec{H}^{ext}\cdot\sum_i \vec{s}_i - J\sum_{<i,j>} \vec{s}_i\cdot\vec{s}_j \qquad (2)$$

where $\vec{H}^{ext}$ is the external magnetic field, J is the exchange and the sum in the last term is over neighbor pairs. We take an interior portion of the dense random packing model of Finney[5] to give the positions of the different sites, identifying as neighbors all atoms within 1.25 diameter of each other. The average number of neighbors for interior atoms is about 11. The boundary conditions are "quasiperiodic", i.e. spins at the surface are paired with other surface spins until all have found 11 neighbors.

COMPUTATIONAL METHOD

Our goal is to find a state which is an energy minimum, at least for small perturbations of the spins. Our approach to this problem is to successively minimize the total energy of the systems with respect to the direction of each spin. After all the spins have been adjusted, we compute the energy of the model and repeat the minimization process until the relative change of energy of the system is less than 5×10^{-6}. This relaxed spin state is used as the starting configuration for the next value of the applied field. In this way we calculate hysteresis loops. Initial magnetization curves are computed by starting with random directions for the spins.

This computational procedure is quite efficient and except for unstable portions of the hysteresis curve, converges quite well. There are some drawbacks however. Firstly only states reachable by a series of energy lowering single spin adjustments can be reached. Paths requiring complex adjustments of groups of spins are excluded. It is thus possible that some of our "minimum" energy states are unstable against certain small multispin perturbations. From past experience with these methods, we believe that this does not seriously affect our results, but a proof of this is difficult to obtain. Another drawback is that in real materials, there is always some probability of traversing paths where a local energy increase occurs. This fact means that our results can only be valid at relatively low temperature.

RESULTS

Our calculations were done on clusters of 100 and 996 spins. The cluster with 100 spins give essentially the same results as obtained with the cluster of 996 spins for the same physical situation. Figure 1 shows the variation of the shape of the hysteresis loops with the ratio of the anisotropy parameter D to the exchange J. For reference we also show a low temperature hysteresis loop from Refs. 7, 2.

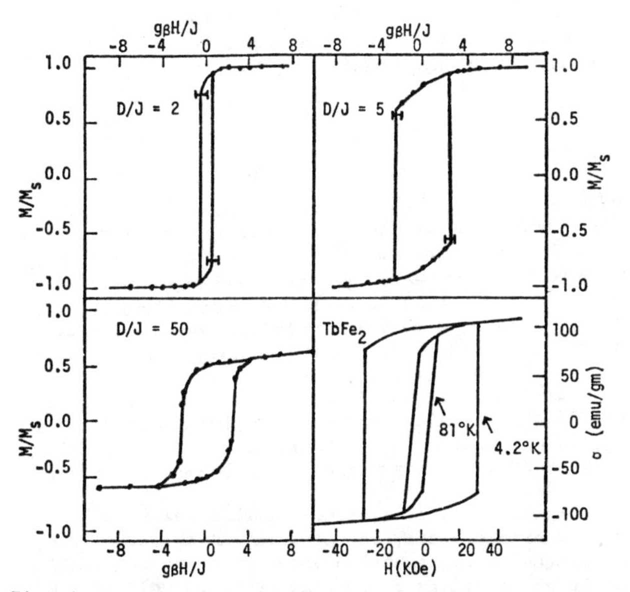

Fig. 1 Hysteresis loops for different values of the anisotropy parameter D compared with and experimental curves for $TbFe_2$ at 4.2°K and 81°K (Refs. 7,2). The calculated loops are for a cluster of 100 spins.

We note that the remanence decreases as D/J increases. This is shown quantitatively in Fig. 2. In the Ising limit, infinite anisotropy, the remanence cannot be greater than 0.5.

The coercivity also changes with D/J. This is shown in Fig. 3. The behavior is rather striking. For $D \lesssim 4J$ the coercivity is quite low despite the large single site anisotropy. For $D > 5J$, the coercivity jumps to a value which is about 20% of the maximum local exchange field [$H_{exchange} \approx (J/g\beta)\cdot 11$] and remains at this value as D increases to the Ising limit. This threshold behavior can be explained as follows. A system with a sufficiently small value of D/J is expected to form domains of size L determined by exchange energy and statistics of anisotropy energies. Following the arguments of ref. 8, the energy of the system is minimized for domain size $L \sim (\frac{J}{D})^2$. The average anisotropy energy D^{AVE} within such domains will be reduced with increasing domain size L, i.e. with averaging over increasing numbers of atoms

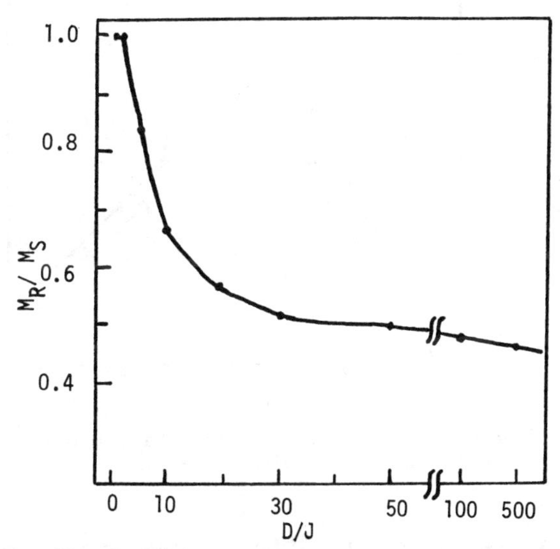

Fig. 2 Variation of the remanent magnetization with anisotropy parameter D divided by the exchange J.

N per domain: $D^{AVE} \sim \frac{D}{\sqrt{N}} \sim \frac{D}{L^{3/2}}$. From these equations it follows that $(\frac{D^{AVE}}{J}) \sim (\frac{D}{J})^4$ for sufficiently small values of $\frac{D}{J}$. Because the coercivity should be less than D^{AVE} as well as less than the local exchange field, $H_{exchange}$, the dependence of the coercivity on $\frac{D}{J}$ shows a threshold behavior with increasing $\frac{D}{J}$. The behavior of the energy over the hysteresis loop is illustrated in Fig. 4. We note the presence of two locally stable states for H less than H_c.

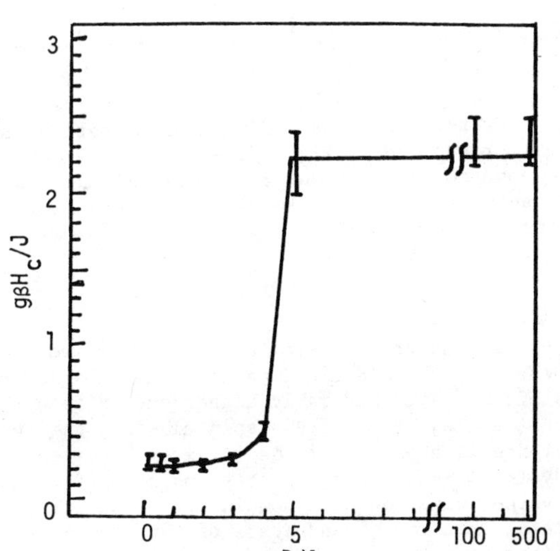

Fig. 3 Variation of the coercivity H_c with anisotropy parameter D divided by the exchange J.

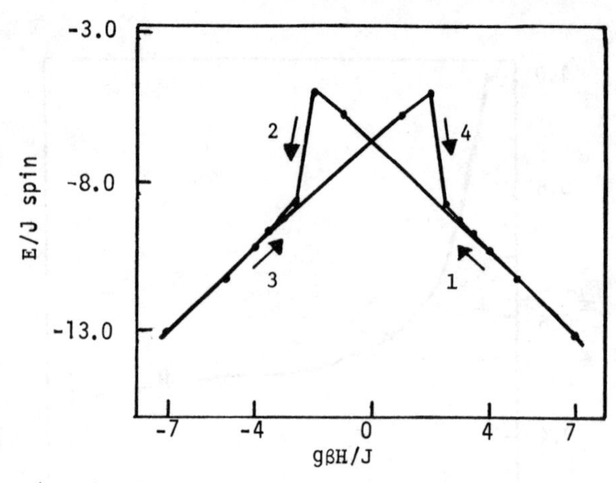

Fig. 4 Behavior of the energy around the hysteresis loop for the 100 spin model with D/J = 5.

In Fig. 5 we show results indicating initial magnetization curves as well as hysteresis loops for a case of very large anisotropy. For the large model two initial magnetization curves corresponding to different random starting conditions for the spins and different directions for the local anisotropy axes are shown. (Only the initial magnetization curves can be distinguished on this graph.) We note that for the large model we can form a locally stable initial state of low moment. In the small model, the system finds a state where the moment is quite considerable.

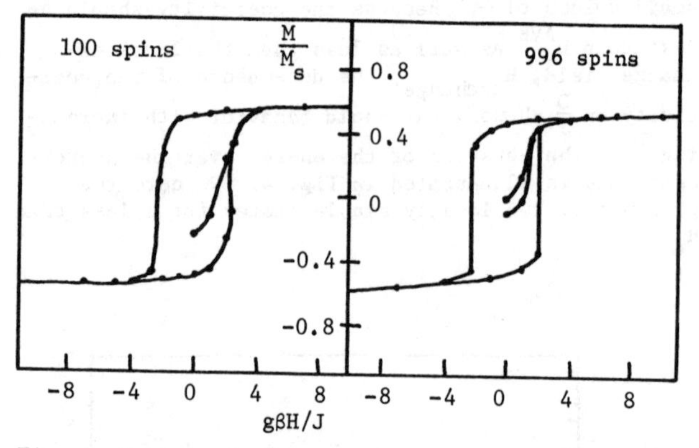

Fig. 5 Hysteresis loops and initial magnetization curves for two models with D/J = 100. Two results corresponding to different random directions for local anisotropy and the initial spins are shown for the 996 atom model.

In any case, as can be seen in Fig. 6 where we show the behavior of the energy, we note that the low moment state has a considerably higher energy than the remanent state. For the 996 spin cluster, the energy difference is about 3% of the total exchange energy of the system. Such a large energy could be accounted for by the formation of a domain like disorder and indicates that the ground state of this system is not described as a spin glass.

SUMMARY

We have presented results on hysteresis loops and magnetization curves for a simple model system with random uniaxial anisotropy. In particular we have

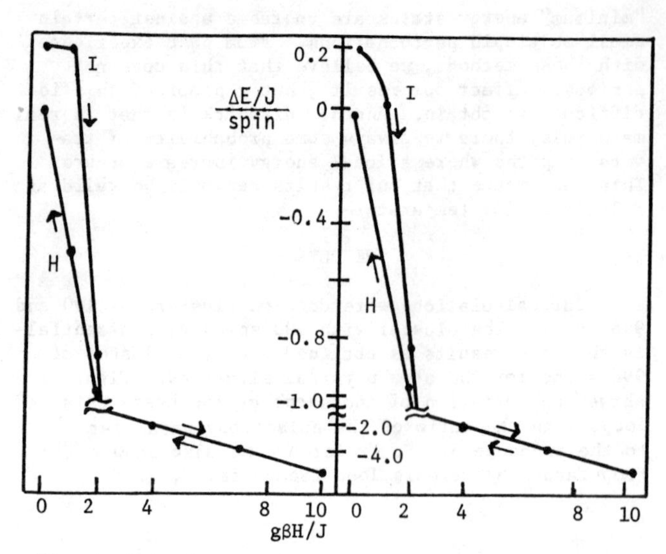

Fig. 6 Behavior of the energy for the initial magnetization curves and the remanent state for the 996 atom model hysteresis loops of Fig. 5.

found a threshold behavior for the development of a large coercivity as the anisotropy parameter D is increased. This coercivity takes a value about 20% of the maximum local exchange field and stays essentially constant for very large D. We have found that the remanent state, with relatively high moment in zero field, is considerably lower in energy than locally stable initial states with low moment.

Although this model is a highly idealized representation of a multicomponent amorphous metallic ferromagnet, we believe that some features of our results may be useful in understanding the behaviors of real materials.

Nonetheless, considerable work remains to be done on this simple model, particularly in characterizing the processes of magnetization change and examining the local stability of configurations. Only then would an extension to more complex models of real systems be justified.

We thank E. M. Gyorgy, B. I. Halperin, G. C. Chi and G. S. Cargill III for helpful discussions.

REFERENCES

* Partially supported by the N.S.F.

1. J. J. Rhyne, J. H. Schelleng and N. C. Koon, Phys. Rev. B 10, 4672 (1974).
2. H. T. Savage, A. E. Clark, S. J. Pickart, J. J. Rhyne and H. A. Alperin, IEEE Trans. Mag-10, 807 (1974).
3. R. Harris, M. Plischke, and M. J. Zuckermann, Phys. Rev. Lett. 31, 160 (1973).
4. R. Harris and D. Zobin, in Magnetism and Magnetic Materials-1975, edited by J. J. Becker, G. H. Landers and J. J. Rhyne (AIP, New York, 1976),p 156.
5. J. L. Finney, Proc. Roy. Soc. Ser. A319, 479 (1970).
6. G. S. Cargill III, Solid State Phys. 30 (1975), (New York: Academic Press).
7. A. E. Clark, Appl. Phys. Letters 23, 642 (1973).
8. Y. Imry and S. K. Ma, Phys. Rev. Lett. 35, 1399 (1975).

MAGNETIC PROPERTIES OF AMORPHOUS RARE EARTH-TRANSITION METAL ALLOYS CONTAINING A SINGLE MAGNETIC SPECIES

Neil Heiman and Kenneth Lee
IBM Research Laboratory, San Jose, CA 95139

ABSTRACT

The analysis of magnetic data on most amorphous rare earth-transition metal binary alloys studied to date is complicated by the fact that both constituents are magnetic. To avoid this problem we have studied amorphous films of $Gd_{1-x}Cu_x$, with x=0.76, 0.82, 0.86 and Lu_{1-x},Fe_x, $La_{1-x}Fe_x$ and $Y_{1-x}Fe_x$ with 0.67<x<0.85. The GdCu alloys have T_c=53K, 40K, 35K implying a Gd-Gd exchange about 1/2 that of Gd metal. Significant differences between the magnetic properties of LaFe, LuFe, and YFe alloys were observed. LaFe has a well defined T_c near 275K whereas LuFe and YFe fail to exhibit well defined magnetic transitions. The bulk magnetization and the internal field are considerably higher for LaFe than for LuFe or YFe. These results suggest that the magnetic properties of the Fe depend strongly upon the size of the rare earth atom. This interpretation is supported by results for amorphous Nd-Fe alloys which also have high T_c and large H_{INT} since Nd also has a large atomic size.

INTRODUCTION

The analysis of the magnetic properties of amorphous rare earth (RE)-transition metal (TM) alloys is complicated by the fact that in most cases both the RE and TM are contributing to the net magnetism. A further complication is that these alloys are generally ferrimagnetic so the bulk magnetization is the small difference between two large sublattice magnetizations. It is thus difficult to isolate the magnetic interactions of the individual constituents. We have therefore prepared and studied amorphous RE-TM alloy films containing only a single magnetic constituent. To isolate the RE-RE interaction, we have investigated sputter deposited amorphous films of $Gd_{1-x}Cu_x$; and in order to isolate the Fe-Fe interaction, we have prepared amorphous films of Y_{1-x}, $Lu_{1-x}Fe_x$, and $La_{1-x}Fe_x$ by coevaporation of the elements. In all cases 0.65≤0.85. We also prepared amorphous Y-Fe alloys by sputter deposition to determine whether preparation methods affect the magnetic properties of these alloy films. Sputter deposited films were on the order of 2.5μm thick while evaporated films were generally thinner (≈0.5μm). Films were determined to be amorphous by x-ray diffraction, and compositions were determined by electron microprobe analysis. All films were measured with a vibrating sample magnetometer (VSM) from 4.2K to 300K in fields up to 18kOe. In addition the Fe alloys were examined by Mossbauer effect at 4.2K and 300K.

AMORPHOUS GdCu ALLOYS

Three amorphous $Gd_{1-x}Cu_x$ compositions with x=0.76, 0.81, 0.86 exhibited ferromagnetic-like behavior. One difficulty encountered was that the 4.2K VSM signal for these samples was not completely saturated at 18kOe. This makes determination of the Gd moments uncertain. M vs. T curves for these alloys (see Fig. 1) indicate Curie temperatures (T_c) for these alloys of approximately 53K, 40K, and 35K. Measurements on a more Gd rich sample of amorphous GdCu to be reported by Mizoguchi et al.[1] show results in general agreement with ours.

Crystalline Gd-Cu alloys in this same composition range are antiferromagnetic and have different crystal

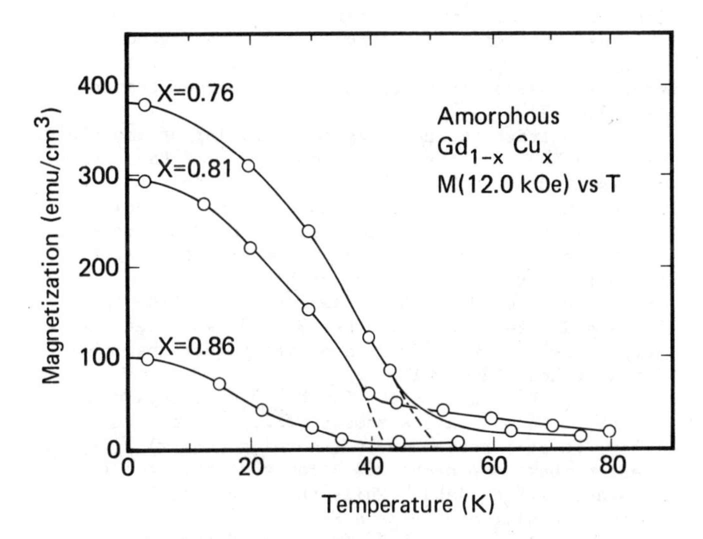

Fig. 1 Magnetization (at 10kOe) vs. temperature for three amorphous Gd-Cu alloys.

structures[2] from the other Gd-TM alloys. The different crystal structures are most likely responsible for the antiferromagnetic ordering in crystalline GdCu since in all other Gd-TM alloys the Gd-Gd exchange is ferromagnetic. In the absence of long range crystal ordering both amorphous Gd-Ni[3] and Gd-Cu alloys are ferromagnetic with similar T_c. The present Gd-Cu results coupled with earlier amorphous Gd-Ni and Ho-Ni results[3] suggest that the RE-RE exchange coupling is ferromagnetic with an exchange constant $J_{RE-RE} \approx 1.5\times10^{-16}$ ergs. This is about 1/2 the magnitude of J_{Gd-Gd} in Gd metal or in most crystalline Gd alloys.

AMORPHOUS ALLOYS OF Fe WITH Y, Lu, La

For studying amorphous RE-Fe alloys with a non-magnetic RE, we investigated not only Y-Fe alloys but also Lu-Fe and La-Fe alloys because: 1) The reported properties of Y-Fe[3,4] and Lu-Fe[5] are unusual and 2) the systematic variation of amorphous RE-Fe magnetic properties reported either by Rhyne[6] or Heiman et al.,[5] while attributed to the variation of the RE de Gennes factor (DGF), could also be influenced by atomic size since both parameters change monotonically through the RE series from Gd to Lu.

Evaporated amorphous alloy films of $Y_{0.33}Fe_{0.67}$, $Y_{0.25}Fe_{0.75}$, and $Y_{0.17}Fe_{0.83}$ were measured by VSM to have a moment of 0.2, 0.8, and 1.9μ_B/Fe atom, respectively. Because these results differed from those of Rhyne et al.[4] and because we encountered difficulties in evaporating Y, we also prepared amorphous Y-Fe alloy films by sputter deposition. The VSM results for sputter deposited $Y_{0.31}Fe_{0.69}$ films were nearly identical to those of Rhyne et al.[4] for $Y_{0.33}Fe_{0.67}$. The M vs. H curves failed to saturate in fields as high as 18kOe; however, the high field susceptibility and the value of M was less than in Rhyne's results. Similarly samples of $Y_{0.25}Fe_{0.75}$ and $Y_{0.15}Fe_{0.85}$ also failed to saturate. For all samples M decreased gradually with temperature over the entire range of 4.2K to 300K with no indication of a transition temperature.

Mossbauer effect measurements show internal magnetic fields (H_{INT}) for the sputter deposited Y-Fe alloys which deviate slightly from the trends in H_{INT} reported previously[5,6] but are still in general agreement. On the other hand, our H_{INT} measurement for amorphous $Y_{0.31}Fe_{0.69}$ is considerably smaller than reported by Pala and Forester.[8]

Mossbauer effect measurements on evaporated amorphous $Lu_{1-x}Fe_x$ alloy samples (x=0.67, 0.75) also show H_{INT} values in agreement with previous trends (see Fig. 3).

In contrast to the irregular behavior of amorphous Y-Fe and Lu-Fe alloys, amorphous alloys of $La_{1-x}Fe_x$ alloys (x=0.69, 0.76) behave as well-defined ferromagnets. The magnetizations of these samples easily saturate in a field of a few hundred Oersteds, and M. vs. T curves show a normal sharp ferromagnetic transition at $T_c \approx 275K$. The Mossbauer effect measurements of H_{INT} also show departures from the Lu-Fe and Y-Fe results in that $H_{INT} \approx 300kOe$ (see Fig. 2.) Surprisingly these numbers are even higher than those for amorphous Gd-Fe alloys.

The 4.2K magnetization for $La_{1-x}Fe_x$ alloys with x=0.69, 0.76 indicate a moment of $1.36\mu_B$/Fe atom and $1.53\mu_B$/Fe atom respectively (assuming $\rho=(1-x)\rho_{La}+x\rho_{Fe}$.) These numbers are about 30% below that expected from Mossbauer effect data. While this might imply incomplete alignment of the Fe moments, such an interpretation is unlikely in light of the ease of saturation.

In spite of the number of experimental uncertainties and possible disagreements among researchers on some points, it is clear that the magnetic behavior of amorphous Y-Fe and Lu-Fe alloys is complex and not indicative of a well-defined ferromagnetic material while amorphous La-Fe alloys are clearly ferromagnetic with a well-defined and fairly high T_c.

SYSTEMATICS OF MAGNETIC PROPERTIES OF AMORPHOUS RE-Fe ALLOYS

It has previously been established[5,6] that T_c for amorphous Re-Fe alloys at fixed Fe concentration decreases monotonically through the Re series from Gd through Tb, Dy, Ho, Er, Tm, Yb, to Lu (or Y). At the same time, the magnetic transition becomes progressively less well defined as one proceeds through the series until at Lu (or Y) there no longer appears to be a well defined ferromagnetic state. Since the DGF also decreases monotonically through the series, it was suggested that the T_c variation correlates directly with the variation of the DGF. Unfortunately, due to the lanthanide contraction, the size of the RE atom also decreases monotonically through the RE series. That amorphous alloys of Fe with La, which has the same DGF as Lu and Y but a much larger atomic size, are clearly ferromagnetic with well defined and reasonably high T_c establishes the fact that the size of the RE element plays an important role in determining the magnetic properties of these alloys. To further test the size dependence, we prepared amorphous alloys of $Nd_{1-x}Fe_x$ ($0.65 \le x \le 0.85$). We found that while Nd has a DGF whose values lies between that of Er and Tm, T_c for the Nd-Fe alloys were considerably higher ($\approx$375K vs. $\approx$175K). A plot of T_c vs. DGF for amorphous $RE_{0.33}$-$Fe_{0.67}$ alloys (Fig. 3) clearly shows two branches - one for heavy RE where atomic size increases with increasing DGF and one for light RE where atomic size decreases with increasing DGF. It is worth noting that H_{INT} from Mossbauer effect results (Fig. 2) shows a stronger dependence on RE size than on DGF. The dependence of the magnetic properties on RE atomic size can be seen directly in Fig. 4 where T_c values for amorphous RE-Fe_2 alloys are plotted as a function of RE atomic radius. A line showing the

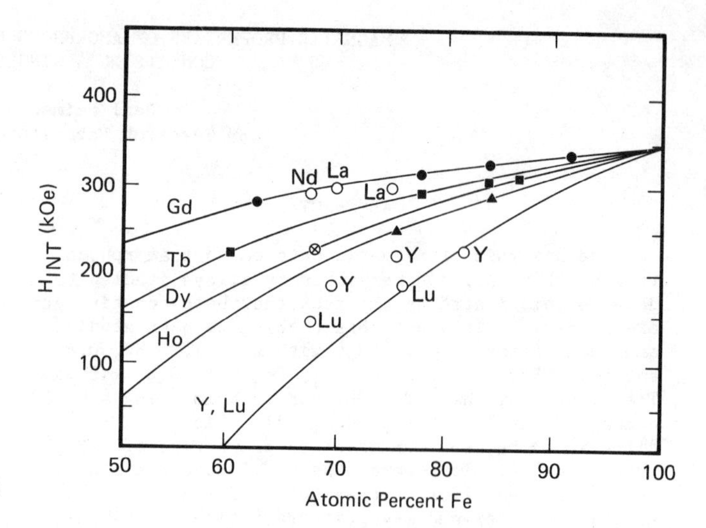

Fig. 2 H_{INT} as determined by Mossbauer effect measurements for amorphous RE-Fe alloys as a function of the species and concentration of RE.

variation of H_{INT} is also shown. The ordinates are scaled so that the end points of each line are coincident. In doing so we wish to indicate that in a rough qualitative way, the difference between the two lines is a measure of the RE-RE and RE-TM contributions to T_c.

It is also worth noting that when the values of the Fe spin are determined from the Mossbauer data from amorphous La-Fe and Nd-Fe alloys and used with the molecular field model proposed by Heiman et al.,[5,9] the correct values of T_c are obtained. All these results suggest that the Fe-Fe exchange interaction depends upon the size of the RE element and that the magnetic behavior of these alloys follows from that; however, two difficulties arise with such an

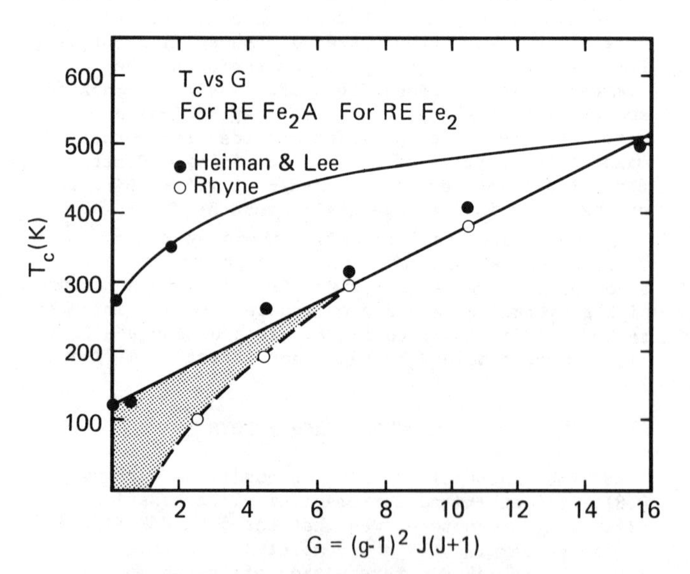

Fig. 3 T_c of amorphous $RE_{0.33}FE_{0.67}$ alloys vs. the deGennes factor, $(g-1)^2$ J(J+1), showing the differences between large size RE (La, Nd) and smaller size RE (Lu, Yb, Er, Ho). The discrepancy between our data and that of Rhyne (dashed line) is not due to different intrinsic behavior but rather to the different interpretation of T_c for those alloys with a poorly defined transition.

interpretation. First, Y has a metallic radius near that of Gd and generally forms crystalline alloys with lattice spacings similar to Gd alloys. Yet, the magnetic properties of amorphous Y-Fe alloys are nearly the same as those for Lu-Fe. On the other hand, Y is a 4d metal whereas the RE are 5d metals so that a direct comparison may not be valid. Second, it would appear from the Mossbauer effect data and a strict interpretation of the mean field model that the Fe spin rather than the exchange coupling constant is a function of RE size. Although the total Fe-Fe exchange interaction may depend on RE size due to Bethe interaction curve arguments, there is no ready physical explanation for a size dependence to the Fe moment. This latter difficulty can be perhaps removed by noting that an alloy is more properly described by a band model so that dividing the total exchange interaction into a spin and an exchange coupling constant is artificial and that in reality one observes the effects due to the total band splitting.

In conclusion, it appears that (with the possible exception of Y-Fe) the properties (T_c, H_{INT}, M vs. T) of amorphous RE-Fe alloys can be explained by assuming that the Fe-Fe exchange varies systematically (via the interatomic distance dependence of the exchange interaction) with the size and concentration of the RE element.

SUMMARY

(1) Amorphous $Gd_{1-x}Cu_x$ (x=0.67, 0.75, 0.83) alloys are ferromagnetic with T_c=53K, 40K, 35K, respectively. These results together with previous amorphous Gd-Ni and Ho-Ni data establish that in amorphous RE-TM alloys the RE-RE coupling is ferromagnetic with about 1/2 the coupling strength of the crystalline RE metals.

(2) For amorphous Y-Fe and Lu-Fe alloys, the field and temperature dependence of the magnetization is strange and H_{INT} follows previous trends. These results imply strong fluctuations in the Fe-Fe exchange coupling resulting in the absence of a well defined ferromagnetic state.

(3) In amorphous La-Fe alloys, the presence of normal ferromagnetic behavior with a high T_c ($\approx$275K) and large H_{INT} demonstrates that the size of the RE plays a significant role in determining the magnetic properties of amorphous RE-Fe alloys. This interpretation is further supported by the high T_c ($\approx$375K) and large H_{INT} of amorphous Nd-Fe alloys.

(4) These facts combine to suggest that many of the magnetic properties of the amorphous RE-Fe alloys can be explained by assuming that the Fe-Fe exchange is a function of the size and concentration of the RE element due to distance dependent exchange interactions.

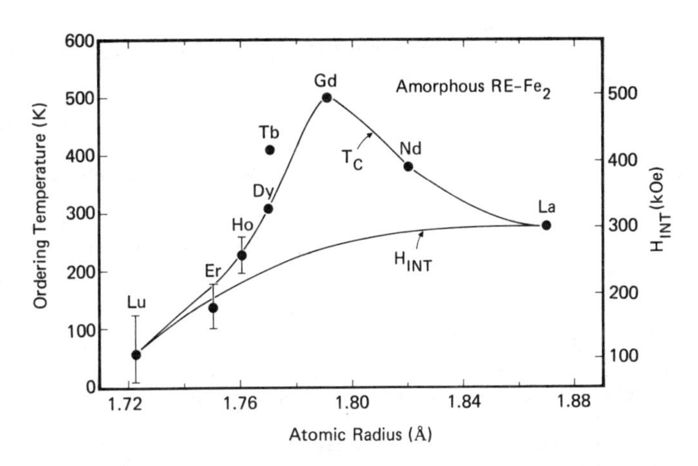

Fig. 4 T_c for amorphous $RE_{0.33}Fe_{0.67}$ alloys vs. RE atomic size. A line showing the variation of H_{INT} for these alloys is also included. The ordinates are scaled so that the end points of T_c and H_{INT} coincide.

ACKNOWLEDGMENT

We wish to thank D. F. Kyser for his characterization of the composition and W. Parrish for x-ray diffraction analysis of our samples.

REFERENCES

1. T. Mizoguchi, T. McGuire, and R. Gambino, International Conference of Magnetism 1976 (to be published).
2. K. N. R. Taylor, Advances in Physics 20, 551 (1971) - and references cited therein.
3. K. Lee and N. Heiman, AIP Conf. Proc. 24, 108 (1975).
4. J. J. Rhyne, J. H. Schelleng, and N. C. Koon, Phys. Rev. B10, 4672 (1974).
5. N. Heiman, K. Lee, and R. I. Potter, 21st Annual Conference on Magnetism and Magnetic Materials - 1975 (to be published).
6. J. J. Rhyne, 21st Conference on Magnetism and Magnetic Materials - 1975 (to be published).
7. N. Heiman and K. Lee, Physics Letters, 55, 297 (1975).
8. W. P. Pala and D. W. Forester, This Conference.
9. N. Heiman, K. Lee, R. I. Potter, and S. Kirkpatrick, J. Appl. Phys. (to be published).

MAGNETIC ORDER, CLUSTER PHENOMENA AND HYPERFINE STRUCTURE IN AMORPHOUS YFe_2

W. P. Pala*
American University, Washington, D.C. 20016
D. W. Forester
Naval Research Laboratory, Washington, D.C. 20375
and
R. Segnan*
American University, Washington, D.C. 20016

ABSTRACT

Mössbauer spectra of ^{57}Fe measured in bulk-sputtered YFe_2 (a-YFe_2) display temperature dependent behavior characteristic of long-range magnetic order for $T<T_c = 55 \pm 5K$. The spectral features are quite similar to those of other magnetically ordered amorphous rare-earth-Fe_2 (R-Fe_2) alloys. Spectral resolution is maintained up to $T\simeq 0.8\ T_c$ with no evidence for superparamagnetic cluster phenomena. The average magnetic hyperfine field, $\bar{H}_{eff}(T)$, closely follows an S=1 Weiss molecular field dependence with $\bar{H}_{eff}(5K) = 233$ kOe. This field is unexpectedly large and comparable with that measured in the heavy R-Fe_2 alloys. It does not follow the general trend of decreasing $\bar{H}_{eff}(0)$ with decreasing rare-earth moment. Our interpretation of the ordering scheme below T_c is similar to that in other R-Fe_2 alloys, which is not in agreement with earlier interpretations.

INTRODUCTION

A bulk sample of a-YFe_2 prepared by direct-current rapid sputtering techniques at Battelle Northwest Laboratories was furnished to us by Dr. A. E. Clark. Samples prepared in this way have been shown to be crystallographically amorphous by neutron diffraction[1] and X-ray scattering.[2] Their structure has been shown to be dense random-packing[3] with the separation distances following the Goldschmidt radii, which are 1.80Å for yttrium and 1.27Å for iron.[4] Small angle neutron scattering on this sample observed an intense magnetic component at low temperatures.[5] A broad peak at about 40K for $q=.04Å^{-1}$ suggests the possibility of short range order or anisotropy effects. The correlation length at 175K was determined to be 6Å. Magnetization studies[3] on this sample at temperatures below 50K have shown that a small hysteresis and remanent magnetization develops which increases to 1.8 kOe and 9.4 emu/g at 4.2K. These results and susceptibility data[3] were interpreted as indicative of weakly interactive small magnetic clusters, with paramagnetic order above 50K. Spin resonance measurements[6] at 300K indicate one resonance which is very similar to an ordinary ferromagnetic resonance from a system with a sizable $4\pi M_s$ of 5400 Oe. In contrast with this the magnetization data[3] gives an effective $4\pi M_s$ of 156 Oe at 1.853 kOe, the applied field used in the resonance experiment.

We present here the first of our experimental results using the Mössbauer effect on the iron site in this sample.

EXPERIMENTAL

Preparation of the sample from the bulk state to the powdered absorber and sealing of it into an air tight holder was done in an argon dry-box to prevent oxidation of the sample. 82 mg of powdered sample were combined with Methocel as an organic binder and pressed into a half inch diameter pellet which was sealed in a plexiglass holder for mounting in the dewar.

The Mössbauer spectrometer was used in the constant acceleration mode and also in the zero velocity single channel mode for a thermal scan to help in determining the Curie temperature of the system. Mössbauer spectra were taken in the range 4.2K to 300K with emphasis on the region under 80K. The higher temperature work was done using a Co^{57} in chromium source and the lower liquid helium work was done using a Co^{57} in rhodium source controlled at 78K.

RESULTS

Nuclear hyperfine magnetic (HFS) and quadrupolar interaction data were obtained from the spectra. Fig. 1 shows representative spectra from the paramagnetic region showing the quadrupolar splitting (QS). The QS decreases monotonically from $0.495 \pm .006$ mm/sec at 80K to $0.423 \pm .006$ mm/sec at 300K, which is lower than the $DyFe_2$ value of 0.51 mm/sec at 300K.

Fig. 2 shows the onset of order and the growth of a hyperfine field as a function of temperature. These spectra display the same overall features as magnetically ordered amorphous rare-earth-Fe_2 alloys[7,8,9] and are quite similar to amorphous $Fe_{44}Pd_{36}P_{20}$.[10] The hyperfine structure is seen to collapse smoothly, with no sharp transition indicative of superparamagnetism, into the growing quadrupolar splitting.

The average magnetic hyperfine field, $\bar{H}_{eff}(T)$, is displayed in Fig. 3 against an S=1 Weiss Molecular Field Model. $\bar{H}_{eff}(0)$ was determined to be 233 ± 6 kOe from this plot. The correlation between the S=1 dependence and the data is fairly good indicating an

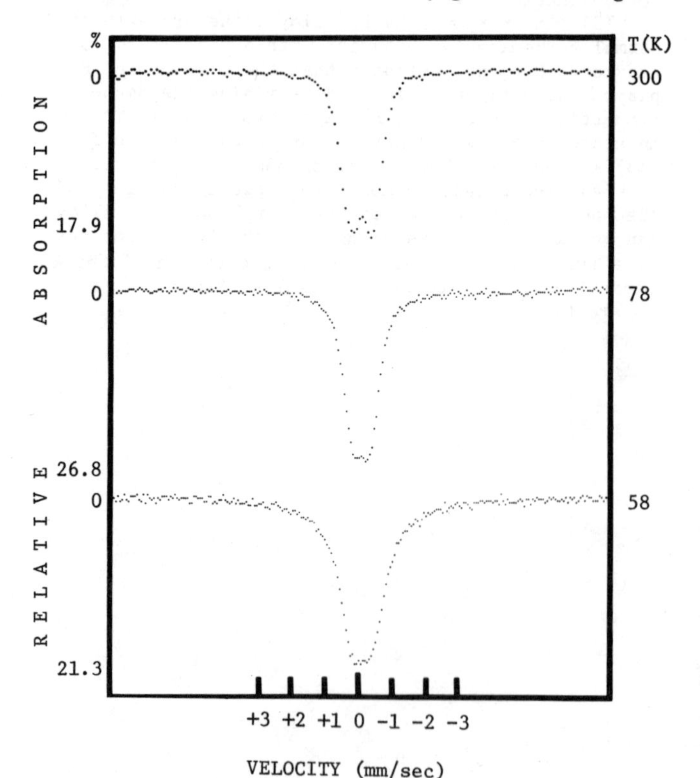

Fig. 1. ^{57}Fe Mössbauer spectra in amorphous-YFe_2.

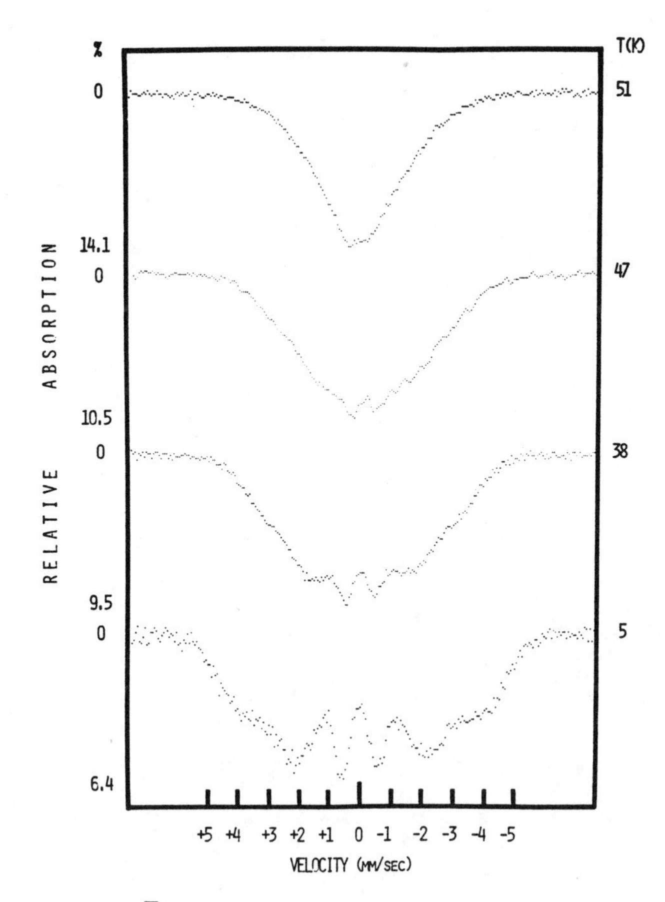

FIG. 2. ^{57}Fe MÖSSBAUER SPECTRA IN AMORPHOUS-YFe_2.

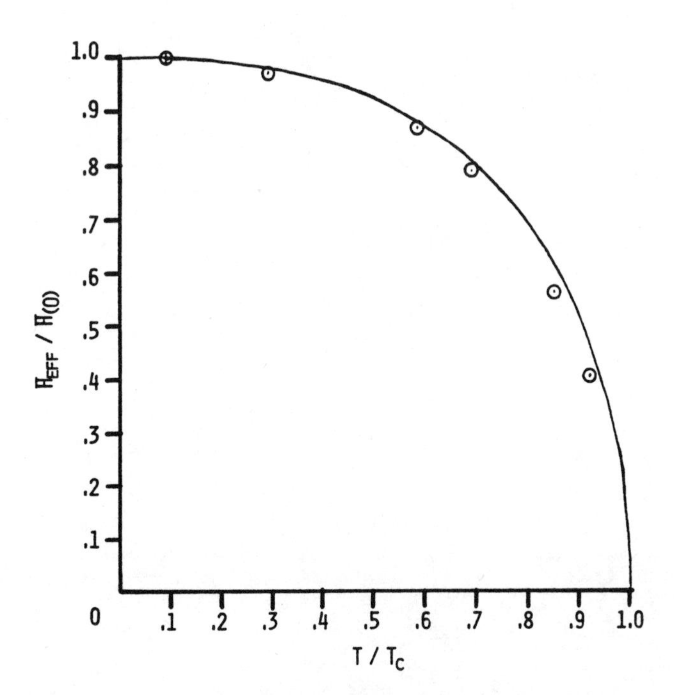

FIG. 3. REDUCED HYPERFINE FIELD V.S. REDUCED TEMPERATURE OF YFe_2 WITH T_C = 55K AND H(0) = 233 kOe. THE TEMPERATURE DEPENDENCE OF THE $H_{EFF}(T)$ CLOSELY FOLLOWS A WEISS MOLECULAR FIELD MODEL FOR SPIN S= 1 (SOLID LINE).

iron moment in YFe_2 of similar magnitude to that in the other amorphous R-Fe_2 alloys. This $\overline{H}_{eff}(0)$ is slightly less than the $GdFe_2$ value of 248 kOe which has a deGennes factor (G) of 15.75, and very close to the $DyFe_2$ value of 225 kOe with a G= 7.08.[7] Previous studies on a-RFe_2 alloys[8,11] have shown that the iron moment is dependent on the value of the deGennes factor,

$$G = (g-1)^2 J(J+1), \tag{1}$$

of the rare-earth. Since the G for YFe_2 is zero this does not follow the general trend of decreasing $H_{eff}(0)$ with decreasing rare-earth moment. If we scale with the atomic radii of the rare-earth YFe_2 falls closer to the series.[7]

The Curie temperature was determined to be T_c= 55K $\pm$ 5K from the HFS(T) data and a thermal scan of the γ-ray transmission at zero relative velocity. The thermal scan showed a sharp transition region of about 10K in breadth and a slower tailing-off region up to 78K, in agreement with the HFS(T) data.

DISCUSSION

One possible model to explain the data presented can be developed using short-range order below T_c and the spin-cluster model of these materials after Rhyne and Pickart, et al.[1,3,5] At 78K we have a sharply defined quadrupolar splitting with Mössbauer field dependence data (H/T) showing evedence of short lifetime clusters of approximately 7Å size.[12] As the temperature is lowered the size of the clusters increases and their lifetimes become longer until near 55K the intra-cluster order becomes static, as the magnetic energy overtakes the thermal energy, resulting in the observed HFS in the transition region.

The extension of order to longer distances is arrested and the random distribution of the now larger clusters is frozen in as the temperature is lowered below T_c. These clusters would then appear in the Mössbauer data to be associated with long-range magnetic order, but due to the absence of strong inter-cluster coupling they would not show long-range order by magnetization or neutron scattering.

The spin resonance data should indicate these microscopic clusters, partially explaining the large M_s observed.

ACKNOWLEDGEMENT

The authors are indebted to Dr. J. J. Rhyne for many helpful discussions.

REFERENCES

* Work supported in part by Office of Naval Research Contract No. N00014-75-C-0736.

1. J. J. Rhyne, S. J. Pickart, and A. Alperin, Phys. Rev. Lett. 29, 1562(1972).
2. G. S. Cargill, AIP Conf. Proc. 18, 631(1974).
3. J. J. Rhyne, J. H. Schelleng, and N. C. Koon, Phys. Rev. B10, 4672(1974).
4. J. J. Rhyne, data from neutron scattering (private communication).
5. S. J. Pickart, J. J. Rhyne, and H. A. Alperin, Phys. Rev. Lett. 33, 424(1974).
6. S. M. Bhagat and D. K. Paul, AIP Conf. Proc. 29, (to be published 1976).
7. D. W. Forester, Int. Conf. on Structure and Excitations of Amorphous Solids (to be published in AIP Conf. Proc., 1976).
8. N. Heiman, K. Lee, and R. I. Potter, 1975 Conf. on Mag. and Mag. Mats. (to be published in AIP Conf. Proc., 1976).
9. D. W. Forester, R. Abbundi, R. Segnan, and D. Sweger, AIP Conf. Proc. 24, 115(1975).

10. T. E. Sharon and C. C. Tsuei, Phys. Rev. B5, 1047 (1972).
11. N. Heiman and K. Lee, Phys. Lett. A55, 297(1975).
12. D. W. Forester, W. P. Pala, and R. Segnan (paper in preparation).

EXCHANGE ANISOTROPY IN METALLIC GLASSES

R. C. Sherwood, E. M. Gyorgy, H. J. Leamy and H. S. Chen
Bell Laboratories, Murray Hill, New Jersey 07974

ABSTRACT

Magnetic properties of centrifugal spin quenched ribbons of metallic glasses in the system Fe-Mn-P-B-Al were studied from room temperature to 1.5°K, with applied fields up to 15.3 kOe. The magnetization vs. temperature curve for $(Fe_{.5}Mn_{.5})_{.75}P_{.16}B_{.06}Al_{.03}$ has a broad maximum near 20°K, observed after cooling in zero applied field. Cooling from higher temperatures in the presence of 15.3 kOe applied field, essentially eliminated the broad maximum and established an asymmetrical displaced hysteresis loop at 1.5°K. The intrinsic coercive force was nearly zero on one side, and 7200 oersteds on the other, while the remanence was zero on one side and 3.5 e.m.u./g on the other. Magnetically cycling the field cooled sample several times with ±15.3 kOe at 1.5°K changed the magnetic state of the specimen, resulting in a nearly symmetrical hysteresis loop, with lower coercivity and remanence. Cooling in zero field established a symmetrical hysteresis loop with lower coercivity of ±1600 Oe. Magnetic saturation was not achieved by any treatment, and the maximum value of magnetization obtained for the field cooled sample was only 7.5 e.m.u./g. These magnetic properties are interpreted as evidence for ferromagnetic-antiferromagnetic "exchange anisotropy" in this alloy.[1]

INTRODUCTION

The magnetic interactions in amorphous alloys containing various combinations of Fe, Co, and Ni with the glass forming elements P, B, Si and C have been found to be ferromagnetic, as discussed by Cargill[2] in a recent review. In contrast, Fe-Mn-P-C amorphous alloys appear to have a mixed magnetic state, with properties which indicate ferromagnetic and antiferromagnetic interactions. Measurements by Sinha[3] and Mizoguchi et al.[4] have shown that the magnetic moment of the Fe alloy is reduced by approximately three Bohr magnetons per added Mn atom. Sinha[3] has also shown that some of the Mn rich alloys exhibit complex magnetic properties, with maxima in the magnetization vs. temperature curves which indicate ferrimagnetic behaviour. In this investigation we have studied mixed alloys in the system Fe-Mn-P-B-Al, to gain further understanding of the ferrimagnetic properties of these amorphous materials.

EXPERIMENTAL

Centrifugal spin quenched specimens were prepared in ribbon form, typically 0.5 mm wide, 20 μm thick and 100 m long, as described in detail by Chen and Miller.[5] The amorphous nature of the ribbons was verified by X-ray or electron diffraction. As quenched Fe-Mn alloys were all quite brittle compared with corresponding Fe-Ni or Fe-Co alloys, and have higher crystallization temperatures, as determined by differential thermal analysis. The temperature dependence of the magnetization, σ-T, and hysteresis, σ-H, was measured on pieces of the ribbons with a pendulum magnetometer.

RESULTS

The, σ-T, curve for $(Fe_{.5}Mn_{.5})_{.75}P_{.16}B_{.06}Al_{.03}$, which is representative of compositions which exhibit ferrimagnetic behaviour, is shown in Fig. 1.

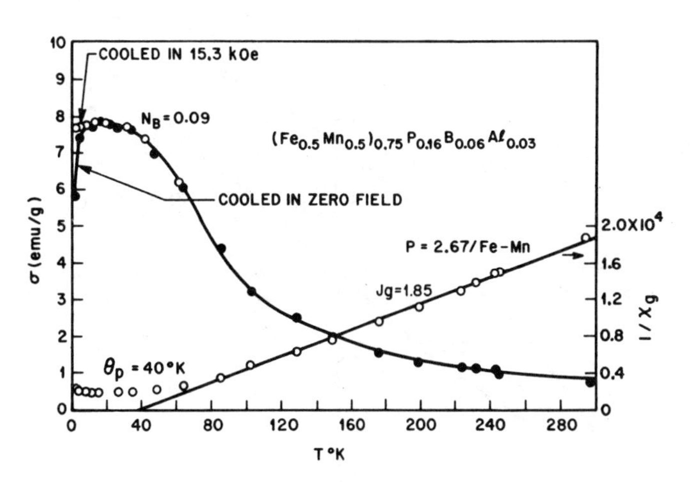

Fig. 1. Magnetization vs. temperature for $(Fe_{.5}Mn_{.5})_{.75}P_{.16}B_{.06}Al_{.03}$. H = 15.3 kOe

The broad maximum near 20°K in the σ-T curve, observed after the specimen was cooled in zero field, was nearly eliminated by cooling from temperatures above the maximum, in a field of 15.3 kOe. The average effective moment, $P = g\{J(J+1)\}^{1/2}$, derived from the slope of the linear portion of the 1/χ curve, is 2.67 μ_B/Fe-Mn atom. If we assume g = 2, that is, no orbital contribution to the moment, we can derive, Jg = 1.85, from the effective moment, which can be compared with the maximum moment of 0.09 μ_B/Fe-Mn atom measured at low temperatures. We will assume that this difference is mainly caused by the presence of antiferromagnetic interactions, however from susceptibility data alone, it is not possible to assign unambiguously, either definite moment values, or interaction signs, to the Fe or Mn atoms. In this alloy system, it is possible that the Fe or Mn atoms do not have moments of constant magnitude. The magnetic moments may be affected locally, by structural disorder, interatomic distance, the presence of itinerant magnetism, or by the kinds of elements which surround a given atom as near neighbors; but the extent of these effects in these alloys is not clearly understood at present.

Hysteresis loops measured at 1.5°K after cooling with and without an applied field are shown in Fig. 2. The total magnetization, σ, may be described by a ferromagnetic component, σ_o, and a field dependent susceptibility, χH, so that $\sigma = \sigma_o + \chi H$. It is evident that field cooling increases σ_o, and induces a magnetic state which is characterized by a hysteresis loop which is displaced on both the σ and H axis, with a displaced intrinsic coercive force, ${}_iH_c$, of 7200 Oe. The field cooled loop does not close at the end of the first field cycle, which indicates that the magnetic state which was frozen in by field cooling is metastable, and has been changed by the application of a reverse field.

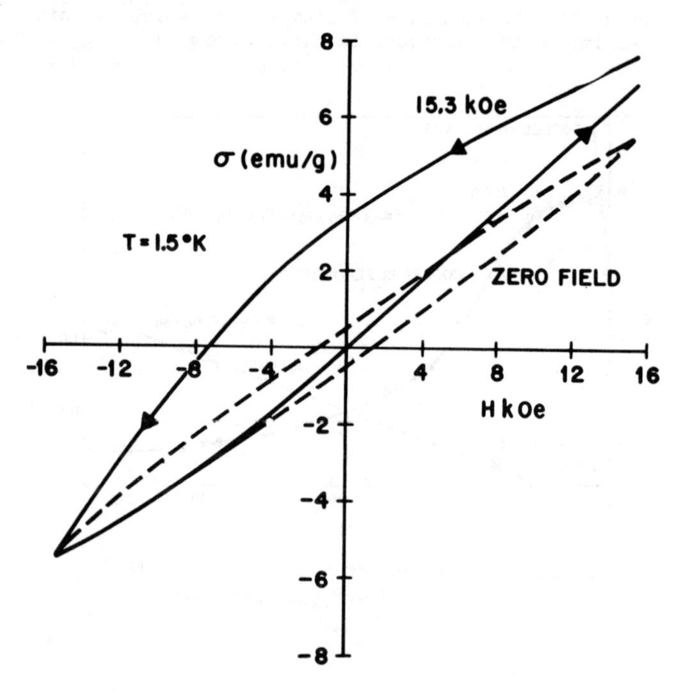

Fig. 2. Hysteresis loops at 1.5°K after cooling in zero field, and in an applied field of 15.3 kOe.

Subsequent field cycling with ± 15.3 kOe drastically changed the magnetic state of the specimen, and after 20 cycles a nearly symmetrical closed loop was obtained. The rates and directions of the changes in coercive force and remanence occasioned by field cycling were nearly identical, as is shown in Fig. 3.

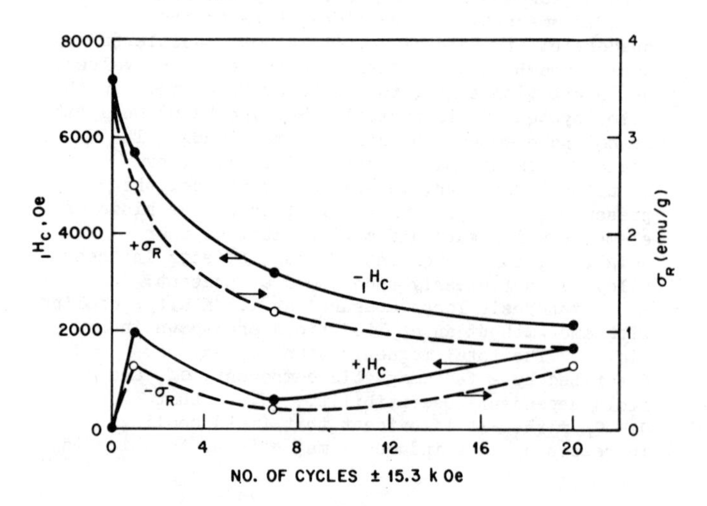

Fig. 3. Changes in coercive force and remanence of the field cooled specimen, after magnetic cycling with ± 15.3 kOe at 1.5°K.

The unusual nature of the field induced anisotropy and magnetization is suggested by the hysteresis loops shown in Fig. 4. The specimen was cooled in a field and σ-H was measured in the perpendicular direction, which might be expected to be magnetically hard, with low values of magnetization. Instead a symmetrical loop was obtained as expected, but with magnetization higher than that obtained by cooling the specimen in zero field. It appears that field cooling establishes a ferromagnetic component which can be magnetized to comparatively higher values in any direction.

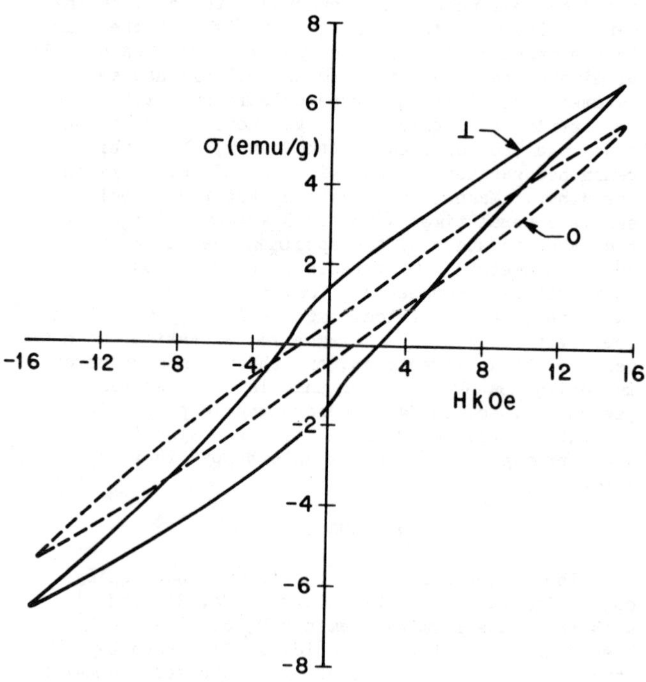

Fig. 4. Hysteresis loops at 1.5°K after cooling in zero field, and after cooling in a 15.3 kOe field, applied perpendicular to the measurement direction.

DISCUSSION

The magnetic properties of this alloy appear to be closely related to the properties of crystalline alloys which exhibit "exchange anisotropy" behaviour as described by Kouvel and co-authors[1], and to the "mictomagnetic" systems recently reviewed by Mydosh.[6] It may be appropriate to apply some of the features of Kouvel's "ensemble of domains" model to this alloy. A postulate of the model is the existence of Fe rich and Mn rich regions, which are expected to occur naturally as a consequence of the random constitution of the alloy. The magnetically ordered system consists of a microscopic mixture of antiferromagnetic and ferromagnetic (possibly Fe rich) domains. The domains mutually interact, and form units or ensembles of domains, which in turn are coupled to other ensembles. Cooling in zero field freezes a random arrangement of the domain structure, with essentially zero net magnetization. Cooling in the presence of a magnetic field produces a different spin structure. The magnetic field alignes the magnetization of the ferromagnetic domains along the field direction, and the surrounding coupled antiferrimagnetic spin structure is frozen with spin directions which are consistent with the interactions and spin directions at the boundaries of the ferromagnetic domains. This is a metastable state, and we may conjecture that field cooling increases the size or extent of the ferromagnetically aligned regions, compared with their size or extent when cooled in zero field. Once formed, it is possible that their size may be decreased by the application of a reverse field, which rearranges the spin structure and returns some of the spins to antiferromagnetic alignment. Behaviour similar to that shown in Fig. 4, has been observed in other mictomagnetic systems, and it has been suggested that after field cooling some of the ferromagnetic domains are coupled ferromagnetically by interdomain exchange interactions, even though they are separated by considerable distances.[1,7]

To account for the asymmetric remanence and coercive force, Kouvel[1] assumed the development of a strong antiferromagnetic anisotropy, which stabilizes the directions of the antiferromagnetic domain moments. The directionally stabilized antiferromagnetic spin system is magnetically coupled to the ferromagnetic domains, thus establishing an energy barrier against demagnetization. The exact nature or origin of the anisotropy is not known, and further work is needed to clarify this.[8] In any case, the present results would seem to provide an example of "exchange anisotropy" behaviour in an amorphous transition metal alloy, which cannot have long range magnetocrystalline anisotropy.

ACKNOWLEDGEMENT

The authors wish to thank E. Coleman for X-ray diffraction results.

REFERENCES

1. J. S. Kouvel, J. Phys. Chem. Sol. 24, 795 (1963).
2. G. S. Cargill III, AIP Conf. Proc. 24, 138 (1975).
3. A. K. Sinha, J. Appl. Phys., 42, 338 (1971).
4. T. Mizoguchi, K. Yamauchi, and T. Miyajima, Amorphous Magnetism, 325, ed. Hooper and deGraff, Plenum Press, New York (1973).
5. H. S. Chen, and C. E. Miller, Mat. Res. Bull., 11, 49 (1976).
6. J. A. Mydosh, AIP Conf. Proc. 24, 131 (1975).
7. P. A. Beck, Met. Trans., 2, 2015 (1971).
8. D. A. Smith, J. Phys. F: Metal Phys., 5, 2148 (1975).

MAGNETIC ANISOTROPY IN THERMALLY EVAPORATED Gd-Co

A. Onton and Kenneth Lee
IBM Research Laboratory, San Jose, California 95193

ABSTRACT

The x-ray diffraction spectrum of thermally evaporated a-GdCo is reported and found to be similar to that of bias sputter deposited a-GdCo which possesses an easy axis of magnetization normal to the film plane. The thermally evaporated a-GdCo films are shown to possess magnetic anisotropy with $K_u \sim -5 \times 10^5$ erg/cm^3 (hard axis normal to the film plane). Off axis atomic incidence experiments show that the hard axis K_u can be tilted toward the direction of incidence. No significant easy axis of magnetic anisotropy in the columnar growth direction was found. The results are consistent with the origin of the magnetic anisotropy being magneto-structural (cf., magneto-crystalline) in thermally evaporated a-GdCo.

INTRODUCTION

The rare earth (RE) 3d-transition metal (TM) form amorphous alloys with no long range structural order. Thin films of these alloys h have interesting magnetic properties, one of which is the presence of positive uniaxial anisotropy, K_u, oriented normal to the film plane.[1] With increasing data available on films prepared under different conditions, however, it has become apparent that anisotropy is very preparation condition dependent.

To review briefly, it was demonstrated by Chaudhari et al[1] that bias sputter deposition of a-GdCo produces films possessing intrinsic (i.e., not substrate strain related, for example) magnetization anisotropy with easy axis perpendicular to the film plane while sputter deposition with no substrate bias produces films with magnetization oriented in the plane of the film. Onton et al[2] showed that the magnetization anisotropy of sputter deposited films correlates with structural properties of the amorphous films observable in x-ray diffraction. Heiman et al[3] reported that perpendicular magnetization anisotropy with $K_u > 2\pi M_s^2$ was a general feature of thermally evaporated a-RE-TM films with the exception of a-GdCo, a fact subsequently investigated in detail by Taylor[4] for the a-GdFeCo system.

For a study leading toward a unified understanding of magnetic anisotropy in a-RE-TM films, a-GdCo is an ideal test case since its anisotropy is critically dependent on preparation conditions. The objective of this study has been to compare the structure of thermally evaporated a-GdCo to that of bias and 0-bias sputter deposited a-GdCo and to investigate in some detail the magnetization of thermally evaporated a-GdCo. In addition, non-perpendicular incidence of atoms during growth was studied in order to evaluate the effect of columnar growth on the anisotropy of thermally evaporated a-RE-TM films.

SAMPLE PREPARATION

Nominally 5000 Å thick films of a-GdCo were thermally evaporated from independent elemental sources on SiO_2, and Si (100) or [510] oriented substrates held at 270K during the deposition. The growth rate was approximately 3-4 Å/sec. For angular deposition, the substrate holder was tilted. The base vacuum before deposition was $\sim 5 \times 10^{-8}$ Torr. The composition of the resulting film was checked by electron microprobe analysis.

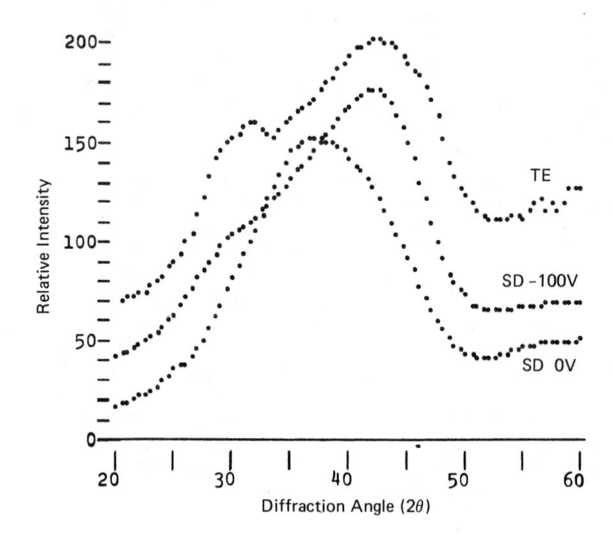

Fig. 1 X-ray diffraction spectra of amorphous $GdCo_3$ prepared by thermal evaporation, and sputter deposition with and without substrate bias. All the curves have been normalized to the same maximum intensity and are shifted vertically successively by 25 units. The sputter deposited films contain in addition some of the sputtering gas, Ar; $\lesssim 1$ at % for 0V, > 7 at % for -100V bias.

FILM STRUCTURE

The structure of the films was monitored by $Cu_{K\alpha}$ x-ray diffraction in the diffraction angle (2θ) range 15-60°. This region contains the strong lines of the diffraction patterns of the crystalline phases of GdCo and the first broad maximum of the undulatory diffraction pattern of a-GdCo. The data on a film of nominally a-$GdCo_3$ grown on a [510] Si substrate are shown in Fig. 1 together with similar data obtained with sputter deposited (with 0 and -100V substrate bias) samples of the same composition. It has been reported[2] for sputter deposited a-GdCo that the differences in x-ray diffraction shown in Fig. 1 correlate with the degree of magnetic anisotropy. The broad symmetric peak centered at $2\theta \sim 37$-38° correlates with the magnetic isotropy of the zero substrate bias case while the split peak with maximum shifted to $2\theta \sim 42$-43° and a shoulder near $2\theta \sim 30°$ correlates with the presence of perpendicular magnetic anisotropy ($K_u > 2\pi M_s^2$) of the bias sputter deposited films. Clearly the thermally evaporated a-GdCo exhibits the structural (Fig. 1) features which are associated with the magnetic anisotropy of bias sputter deposited films. Thus the structural data appear initially to be 'nconsistent with the previous report[3] that thermally evaporated a-GdCo had in-plane magnetization ($K_u < 2\pi M_s^2$).

MAGNETIC PROPERTIES

With the applied field, H, perpendicular to the film plane, the magnetization curve of the thermally evaporated a-$GdCo_3$, whose x-ray diffraction pattern is shown in Fig. 1, had a linear section for low H and a well defined saturation field H_t. At 295K $4\pi M_s$ was $\sim$ 3.8 kGauss, and $H_t \sim 6.8$ kOe. This implies a hard axis anisotropy perpendicular to the film plane with $Q' \equiv (H_k'/4\pi M_s)$ ~ -0.8 and $K_u' \equiv (1/2)\, H_k' M_s \sim -5 \times 10^5$ erg/cm^3, where $H_k' = 4\pi M_s - H_t$. The primes (') denote that no correction for domain structure magnetostatics has been made in deriving these values (maximum possible correction, $\Delta Q = -1$).

Another set of films was prepared with measured compositions: $Gd_{0.10}Co_{0.90}$. In this case the substrates were placed at various angles to the incident atomic beam to obtain atomic beam incidence angles (ϕ) between 0° and 60°, $\pm$ 10° during evaporation. Fig. 2 defines the angle of the incident atomic beam during evaporation (ϕ) in relationship to the direction of the applied magnetic field (α, ξ) for the magnetization measurement. The normal incidence sample had $4\pi M_s$ = 4.5 kGauss, $Q' \sim -0.7$, and $K_u' \sim -5 \times 10^5$ erg/cm^3; again a hard axis of magnetization perpendicular to the film plane. The in-plane magnetization was measured for the normal incidence sample as a function of in-plane applied field direction at 30° intervals. Within experimental error, the in-plane magnetization was isotropic.

The magnetization of a non-normal angle of incidence sample with ϕ = 60° on a Si (100) substrate is shown in Fig. 3. The data were taken with the direction of the applied magnetic field varied both in the plane of incidence ($-90 \leq \alpha \leq 90$, $\xi = 0$) and in-plane ($\alpha = 90$, $0 \leq \xi \leq 180$).

By comparing the α = - 30°, 0°, and 30° data

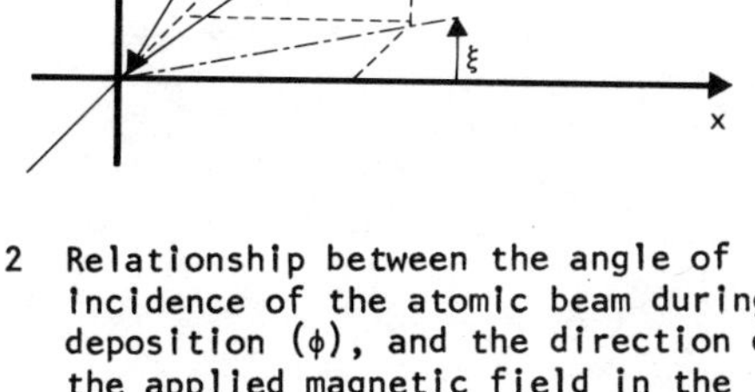

Fig. 2 Relationship between the angle of incidence of the atomic beam during deposition (ϕ), and the direction of the applied magnetic field in the magnetization measurements (α, ξ). n is the normal to the substrate plane, and the x-z plane is the plane of incidence in which the angle ϕ is measured.

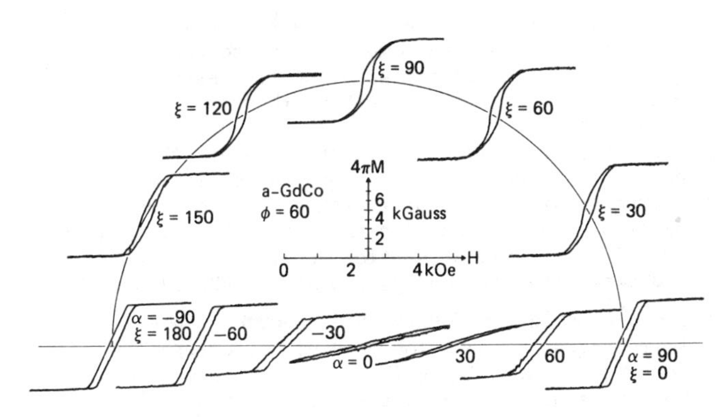

Fig. 3 Magnetization curves of a a-$Gd_{0.1}Co_{0.9}$ thermally evaporated film with atomic beam angle of incidence ϕ = 60°, measured from the normal to the substrate. The measurements shown are with the applied field in the plane of incidence ($\xi = 0$, $\alpha - 90 \leq \alpha \leq 90$ and in plane ($0 < \xi < 180$, $\alpha = 90$). At angles where the apparent saturation is at M less than other directions it takes up to 10 kOe to reach full M_s.

it is apparent that the hard axis, K_u, has been tilted toward the + α direction in the plane of incidence (i.e., in the direction of the incident atomic beam). From this data it is estimated that the angle of the hard K_u axis with respect to the normal is 9.5°. This was confirmed to $\pm$ 5° by direct measurements at higher applied fields near α = 10°. Similar data on samples prepared at ϕ = 30° and 45° are consistent with those at ϕ = 60° (Fig. 3).

The measurement of in-plane magnetization at ($\alpha = 90$, $\xi = 90$) is with the applied field perpendicular to the hard K_u axis. Thus, if the latter anisotropy were the only one, the loop should be square. However, only about (1/3) of the loop actually has vertical sides. That kind of loop is also characteristic of the in-plane measurements on normal incidence films. The "S" shaped portions of these "easy" loops are consistent with randomness in the distribution of anisotropy axis directions; presumably varying spatially within the film on a microscopic scale.

In summary, our thermally evaporated a-GdCo films are characterized by hard axis magnetization anisotropy normal to the film plane. They are magnetically isotropic in-plane, but appear to have a randomly directed anisotropy component.

DISCUSSION AND CONCLUSIONS

The conclusions that emerge from this study are that structurally, insofar as is evident from x-ray diffraction data such as in Fig. 1, the thermally evaporated a-GdCo is similar to bias sputter deposited a-GdCo. In a previous study of a-GdCo the distinctive x-ray signature of bias sputter deposited a-GdCo was shown to correlate with magnetic anisotropy. This correlation of magnetic anisotropy and the distinctive x-ray structural signature is found to extend to thermally evaporated a-GdCo. A major difference however does exist in the nature of the magnetic anisotropy of bias sputter deposited and thermally evaporated GdCo. Whereas bias sputter deposited a-GdCo has an easy axis of magnetization normal to the film plane, thermally evaporated a-GdCo (at perpendicular incidence) has a hard uniaxial magnetization axis normal to the film plane. Our conclusion on hard axis magnetic anisotropy is consistent with recent ferromagnetic resonance measurements of Frait et al[6] on thermally evaporated a-GdCo.

These results continue the findings that magnetic anisotropy in a-GdCo correlates with a prevalent structural unit within the amorphous matrix.[3] Consistent with this, radial distribution function analysis of sputter deposited films[5] indicates that the modulation of the RDF beyond r = 6 Å is larger for the bias sputter deposited films, pointing to a longer range structural coherence in the magnetically anisotropic films. What remains unclear is the reason for the difference in the nature of the magnetic anisotropy in sputter deposited and thermally evaporated films.

The data in Fig. 3 exclude the possibility that shape anisotropy, due to a tendency for columnar growth caused by self-shadowing, is a major effect in the anisotropy of thermally evaporated a-GdCo. For example, no significant assymetry in the value of H_t about $\alpha = 10$ is found (cf., Fig. 3 and Fig. 4). If there were any significant (relative to $2\pi M_s^2$) positive K_u directed along $\alpha = +60$ (the incident beam direction) it would show up as a low value of H_t there compared to H_t at $\alpha \sim -40$.

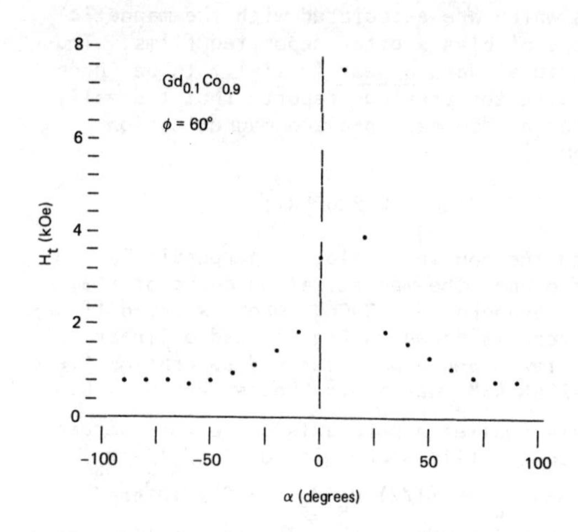

Fig. 4 Saturation field, H_t, as a function of applied field direction, α, measured from the normal to the film in the plane of incidence of the atomic beam. The angle of incidence, ϕ, is 60° for these data (cf. partial data in Fig. 3).

ACKNOWLEDGEMENTS

We would like to thank D. F. Kyser for electron microprobe analysis of film compositions, W. Parrish for making available an x-ray diffractometer, S. Lawrence for sample preparation, J. A. Sandate for measurements of magnetization, and C. J. Breen for x-ray diffractometer measurements.

REFERENCES

1. P. Chaudhari, J. J. Cumomo, and R. J. Gambino, IBM J. Res. Develop. 17, 66 (1973).
2. A. Onton, N. Heiman, J. C. Suits, and W. Parrish, IBM J. Res. Develop., to be published (July 1976).
3. N. Heiman, A. Onton, D. F. Kyser, K. Lee and C. R. Guarnieri, Magnetism and Magnetic Materials 1974, AIP Conf. Proceedings No. 24 (American Institute of Physics, N.Y., 1975), p. 573.
4. R. C. Taylor, Magnetism and Magnetic Materials 1975, AIP Conf. Proceedings (American Institute of Physics, N.Y.), to be published.
5. C. N. J. Wagner, N. Heiman, T. C. Huang, A. Onton and W. Parrish, Magnetism and Magnetic Materials 1975, AIP Conf. Proceedings (American Institute of Physics, N.Y.), to be published.
6. Z. Frait, I. Nagy, and T. Tarnoczi, Physics Letters 55A, 429 (1976).

GROWTH INDUCED ANISOTROPY IN SPUTTERED GdCo FILMS

Sotaro Esho and Shozo Fujiwara
Central Research Laboratories, Nippon Electric Co., Ltd., Nakahara-ku
Kawasaki, 211 Japan

ABSTRACT

Substrate bias effects on growth induced anisotropy in amorphous GdCo films have been investigated. Perpendicular anisotropy is induced with negative bias voltages higher than -30V. The induced anisotropy constant has close relations with Ar content and film density. With increasing bias voltages, both anisotropy constant and Ar content increase. Film density changes from 8.7 to 6.9 gm/cm^3 for bias voltages of zero and -170V, respectively. The change in density may result from Ar inclusion and resultant void formation. An electric field distortion near the substrate surface gives a weak in-plane anisotropy in addition to the perpendicular anisotropy. There is a critical thickness required to induce the perpendicular anisotropy at a given bias voltage. The critical thickness decreases with increasing bias voltages. The growth induced anisotropy of bias sputter deposited GdCo films could be explained by assuming shaped void formation.

INTRODUCTION

The growth induced anisotropy in rare earth-transition metal amorphous films has been intensively studied. Perpendicular anisotropy, K_u, was first observed in GdCo films prepared by bias sputter deposition[1]. Gambino et al[2] proposed pair ordering as the main source of K_u in bias sputter deposited GdCo films. After that, Heiman et al[3] revealed that K_u in RE-TM amorphous films varied according to film constituents and preparation methods. Their results suggested that, for K_u in amorphous RE-TM films, local anisotropy was the more common mechanism than pair ordering and shape anisotropy.

The growth induced anisotropy in GdCo films depends on preparation methods. It is found that bias sputter deposition induces a K_u sufficient to cause magnetization normal to the film plane[1,4], but evaporated films show no perpendicular magnetization[3]. Thus, for GdCo films, the most important process parameter to induce K_u is considered to be substrate bias. The purpose of this study is to reveal substrate bias effects on K_u in GdCo films.

EXPERIMENTAL

GdCo films were prepared by diode RF sputtering on Si substrates. A 9 cm diameter Co disc was used for a target. Small gadolinium pellets were placed on the disc to vary the composition of the target[4,6]. The substrates were fixed to a water-cooled anode which was located 4.5 cm above the target. The target was presputtered with a shutter for one hour to clean the target surface at an Ar pressure of 3×10^{-2} Torr. The RF power supplied to the target was adjusted to give a target self-bias of -1.5 kV DC. This self-bias gave a sputtering rate of about 350 Å/min. A negative DC bias was applied to the films via the anode during deposition. Bias voltage, V_b, ranged from 0 (grounded) to -170V DC.

Thickness and weight of the films were measured with a Taylor-Hobson talysurf and a microbalance, respectively. The films used were in a 4.0 to 5.5 μm thickness range. The saturation magnetization, $4\pi M_s$, and the perpendicular anisotropy, K_u, of the films were determined from torque measurements[5]. The Kerr rotation, θ_k, and hysteresis loop were measured by means of the polar Kerr effect[6]. The film composition ratio (Gd/Co) was determined by electron microprobe analysis within an accuracy of ±3%. The composition ratio (Gd/Co) of films studied was in a 0.24 to 0.29 range.

RESULTS AND DISCUSSION

Perpendicular anisotropy constant, K_u, clearly depends on the bias voltage, as shown in Fig. 1. Since the Gd/Co ratio or saturation magnetization of the films is changed by applying the bias voltage[1,4], the target composition was adjusted for each bias voltage in such a manner that the saturation magnetization was within a range of 200 to 600 G. The sign of the anisotropy constant changes at the bias voltage of about -30V. As the bias voltage increases, the perpendicular anisotropy is linearly induced. The increase in K_u has close relations with an Ar content included in films. The Ar content was estimated from characteristic X-ray intensity of Ar ($K\alpha$) detected by means of microprobe. Figure 2 shows the dependence of the X-ray intensity on the bias voltage. The X-ray intensity or Ar content shows an abrupt increase at bias voltages higher than -50V. However, the X-ray intensity tends to saturate at the highest bias voltage used here. It should be remarked that the threshold bias voltage for Ar content is nearly equal to the value at which the perpendicular anisotropy can first be observed. The tailing of the curve at lower bias voltages indicates a V_b^2 dependence of Ar content[7].

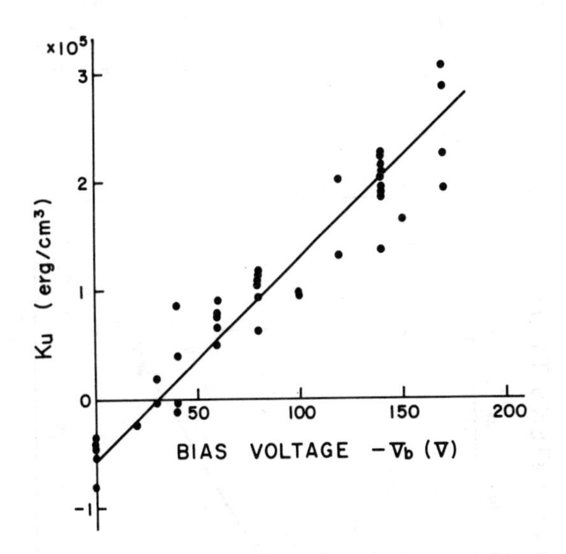

Fig. 1. Bias voltage dependence of perpendicular anisotropy constant K_u.

Figure 3 shows the bias voltage dependence of film density. Film densities obtained at bias voltages less than -50V are in agreement with an ideal density obtained from the linear interpolation between 12-fold coordinated pure Co (8.9 gm/cm^3) and pure Gd (7.95 gm/cm^3). This is consistent with the result obtained by Tao et al[8]. But, at higher bias voltage, a marked deviation from the ideal density can be observed. The decrease in density is considered to arise from the Ar inclusion. However, if the decrease only attributes to the increase in Ar content, the high bias films should contain approximately 25 at % Ar. This value seems to be too high to be expected. Therefore, a part of the decrease in density is considered to be due to void formation. The existence of voids could also be expected from the film chemical stability. The Kerr rotation θ_k of the films was measured as a function of time after removing the films from the sputtering chamber. The time required for θ_k to become zero changes from a half day to several weeks, depending on the bias voltages, as shown in Fig. 4. Since the decrease in θ_k is considered to result from oxidation of constituents, the deterioration behavior suggests that the void concentration increases with increasing bias voltages.

From these experimental results, the growth induced anisotropy of the films studied here could be explained in terms of a shape anisotropy due to voids. If voids are elongated along the film thickness direction, the shaped voids could produce a perpendicular anisotropy. However, the shape anisotropy ($2\pi M_s^2$) only provides K_u values of an order of magnitude less than the observed levels of K_u. The discrepancy between $2\pi M_s^2$ and K_u may be overcome by assuming that a shell surrounding the void has a high Co-rich composition or large saturation magnetization. The composition difference between the shell and the environment could be assumed from the experimental fact that Gd atoms are preferentially resputtered by applying negative bias voltages[9]. To perform a shape anisotropy order estimate for the shell, it is further assumed that the void volume is negligibly small. In this case, the shape anisotropy energy per unit volume may be given by[10]

$$E_a = -\frac{1}{2}(N_{/\!/} - N_\perp)\,\Delta M_s^2 \beta\,(1-\beta)\ ,$$

where $N_{/\!/}$ and $N_\perp$ are the demagnetizing factors of the shell parallel and perpendicular to the long shell axis, ΔM_s is the magnetization difference between the shell and the environment, and β is the partial volume of the shells. If ΔM_s is assumed to be higher than 400 emu/cm^3, E_a gives a shape anisotropy energy comparable to the observed levels of K_u with reasonable assumptions for $N_{/\!/}$, $N_\perp$ and β. Such a large magnetization difference could be expected in the GdCo system, since the magnetization shows a sharp dependence on composition near the compensation composition[1]. Two mechanisms, pair ordering[2] and local anisotropic coordination[3], have been proposed to explain the anisotropy of the films. Although these mechanisms could not be ruled out, the anisotropy is more likely due to a shaped void formation than to pair ordering or local coordination, from the results obtained here.

In order to investigate effects of electric field configurations on the perpendicular anisotropy, an extra electrode was placed near the anode surface, as the illustration inserted in Fig. 5. The electrode consists of a 1 cm high, 2 cm wide stainless steel block. Film properties can be widely varied by applying various DC voltages to the electrode. Figure 5 shows a result obtained by applying a negative voltage of -140V to both the anode and the extra electrode. Film properties were measured as a function of the distance d from a film edge where the electrode was placed. As the distance decreases, the Ar content decreases while the Gd concentration increases. The decrease in the Ar content suggests that the bias electric field direction is disturbed by the extra electrode potential and then part of Ar ions are captured by the electrode. The electric field distortion gives no anisotropy canting, but induces a

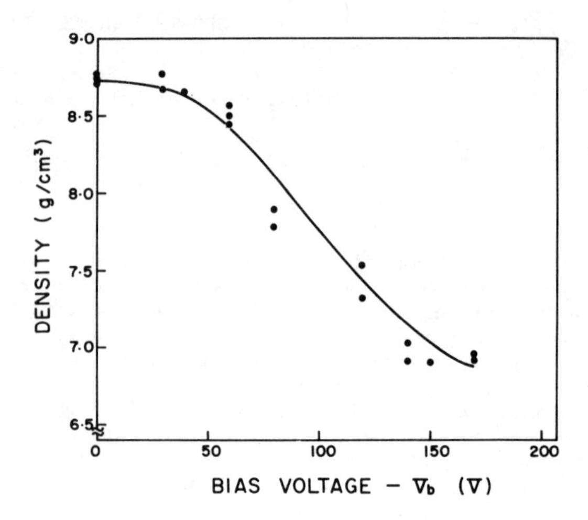

Fig. 3. Bias voltage dependence of film density.

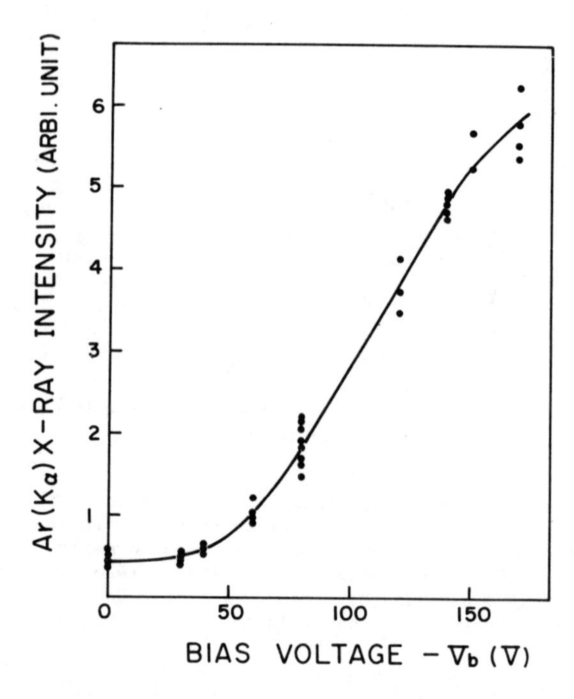

Fig.2. Bias voltage dependence of Ar($K\alpha$) X-ray intensity of films detected by electron microprobe analyzer.

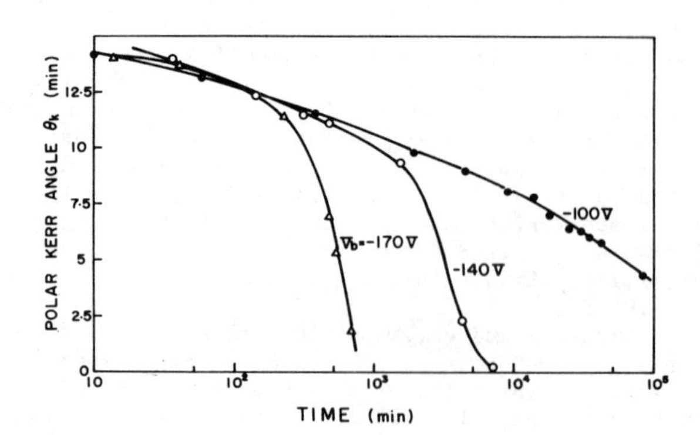

Fig. 4. Change in polar Kerr angle as a function of time after removing from sputtering chamber.

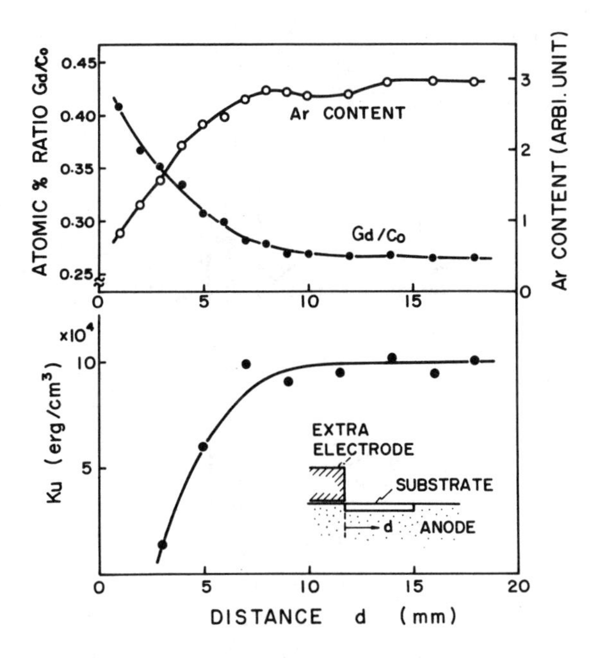

Fig.5. Effect of extra electrode on Ar content, atomic % ratio Gd/Co and K_u.

weak in-plane uniaxial anisotropy in addition to the perpendicular anisotropy. The formation of the in-plane anisotropy suggests that the voids are deformed in the film plane by oblique incidence of Ar ions.

Chaudhari et al[1] found out a fact that there was a critical thickness required to induce the perpendicular anisotropy. It is important in actual practice to investigate the bias voltage dependence of the critical thickness. Magnetic properties and composition of films were measured as a function of film

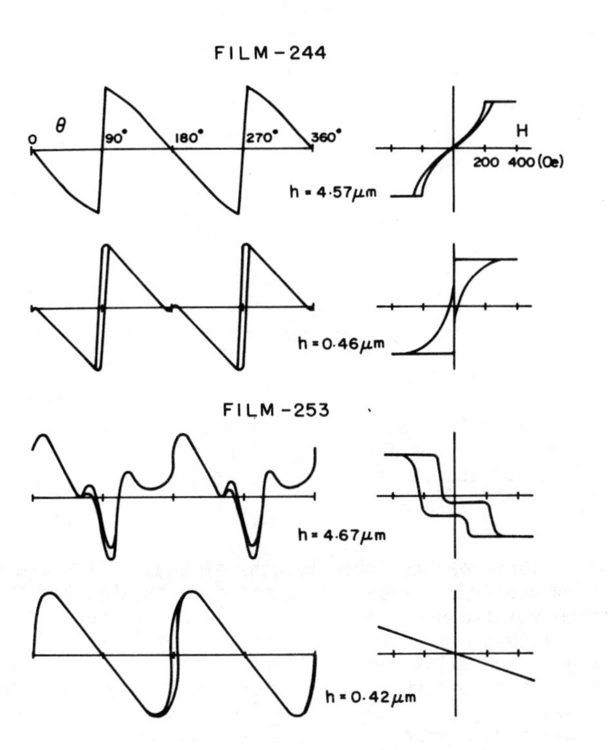

Fig. 6. Change in torque curves and hysteresis loops obtained before and after etching films. θ : angle between applied field and direction normal to film plane.

thickness by eliminating a surface layer of about 4000Å thickness for each ion milling step. Figure 6 illustrates the changes in torque curves and hysteresis loops of two typical films, before and after etching the films. Film 253 (V_b = -30V) shows an anomalous hysteresis loop and torque curve at the initial or not etched thickness of 4.67 μm. The anomaly is continuously observed to the thickness of 1.81 μm. At a thickness less than 1.12 μm, the film possesses a negative K_u. On the contrary, film 244 (V_b = -140V) shows no marked change in torque curve and hysteresis loop in a wide thickness range of from 0.46 to 4.57 μm (initial thickness). The anomaly could be explained by a layering structure of magnetization vectors. Since Gd/Co ratio and Ar content fluctuations in the films along film thickness could not be detected within the resolution limit of the techniques, the layering structure seems to be due to an anisotropy constant deviation along film thickness. This means that a critical thickness is required to induce the perpendicular anisotropy. The critical thickness, in general, decreases with increasing bias voltage. The origin of the critical thickness is not clear at this time, but relatively high bias voltages are found to be desired to induce a uniform perpendicular anisotropy.

CONCLUSION

The perpendicular anisotropy in GdCo films depends on Ar content included in the films. The Ar content is increased by applying a negative bias voltage to substrates during sputtering. The increase in the Ar content leads to a decrease in film density and a rapid film deterioration. An electric field distortion near the substrate surface gives an in-plane uniaxial anisotropy. The in-plane anisotropy might be due to oblique incidence of Ar ions. These results suggest that the growth induced anisotropy arises mainly from shaped voids caused by Ar inclusion.

ACKNOWLEDGEMENT

The authors wish to thank T. Furuoya for his encouragement and H. Nakamura for her assistance in preparing the films.

REFERENCES

1. P. Chaudhari, J.J. Cuomo and R.J. Gambino, IBM J. Res. Develop. 17, 66 (1973).
2. R.J. Gambino, J. Ziegler and J.J. Cuomo, Appl. Phys. Letters, 24, 99 (1974).
3. N. Heiman, A. Onton, D.F. Kyser, K. Lee and C.R. Guarnieri, AIP Conf. Proc. 24, 573 (1974).
4. H.C.Bourne, Jr., R.B. Goldfarb, W.L. Wilson, Jr., and R. Zwingman, IEEE Trans. on Mag. MAG-11, 1332 (1975).
5. S. Chikazumi, J. Appl. Phys. 32, 81s (1961).
6. S. Esho, Japan. J. Appl. Phys. 15, 93s (1976).
7. J.J. Cuomo and R.J. Gambino, J. Vac. Sci. Technol. 12, 79 (1975).
8. L.J. Tao, R.J. Gambino, S. Kirkpatrick, J.J. Cuomo and H. Lilienthal, AIP Conf. Proc. 18, 641 (1973).
9. J.J. Cuomo, P. Chaudhari, R.J. Gambino, J. Electron Mater. 3, 517 (1974).
10. S. Chikazumi, Physics of Magnetism (Wihly, New York, 1964), P. 372.

MAGNETIC PROPERTIES OF BIAS-SPUTTERED $Gd_{1-x}Fe_x$ AMORPHOUS FILMS†

R. Zwingman,* W. L. Wilson, Jr., and H. C. Bourne, Jr.
Dept. of Elec. Engr., Rice University, Houston, Texas 77001

ABSTRACT

The room temperature magnetic properties of amorphous $Gd_{1-x}Fe_x$ films prepared by dc biased rf sputtering are reported. For low magnetization films made at a fixed bias, the wall energies are found to be nearly constant, regardless of the film magnetization. Also, no trend in H_k or K_u versus $4\pi M_s$ and little trend in K_u with bias voltage can be discerned. The uniaxial perpendicular anisotropy appears to be induced by compressive stresses from the glass substrates. Such stresses are attributed to the deposition process and not to thermal contraction effects between the film and the substrate. The exchange constant is estimated to be 1.8×10^{-7} erg/cm; the anisotropy energies are on the order of 4×10^4 erg/cc.

EXPERIMENTAL TECHNIQUES

The films were prepared by dc biased rf sputtering from a gadolinium-iron target onto glass substrates. A solid iron plate with small gadolinium disks distributed on it was used as the target. Typical film thicknesses were .75μm from a four hour deposition. The dc target voltage was typically set at -580vdc which delivered about 40 watts to the 2.875 inch diameter target. The dc substrate bias source was a regulated voltage supply as previously reported by Bourne, et. al. [1].

The argon flow was adjusted to give a typical sputtering pressure of 10μm.. Argon presence in the films was confirmed by microprobe fluorescence. The argon content in the films was estimated from Fe and Gd x-ray fluorescence counts and the interferometrically determined film thickness. .07 - .14 atomic fraction argon was estimated for films with $x \approx .77$. Argon atomic fractions estimated by this method can be in error as much as 50%.

The anisotropy was determined by the longitudinal susceptibility technique of Shumate, et. al. [2] except that the Hall effect was used to detect the ac magnetization component, $m_\perp$, normal to the film. The anisotropy measurement system is shown in fig. 1. A film was first saturated in-plane by H_{dc} then a normal ac magnetic field was applied. It produced $m_\perp$ which generated a Hall voltage, V_h, proportional to $m_\perp$. V_h versus H_{dc} was plotted directly; $H_k - 4\pi M_s$ was obtained from the inverse of the V_h versus H_{dc} graph. A typical susceptibility trace and its inverse are shown in fig. 2. For this film $H_k - 4\pi M_s = 200 \pm 70$oe and $4\pi M_s = 1170$G.

The saturation magnetization was determined by using the stripe period, P_o, the collapse field, H_{col}, the initial susceptibility, χ_i, and the field needed to drive the film to one-half of its saturation value, $H_{.5}$ [3,4]. H_{col}, χ_i and $H_{.5}$ were obtained from the normal hysteresis traces taken by Hall effect measurements. The values of $4\pi M_s$ for each film calculated by using H_{col}, χ_i and $H_{.5}$ agreed to within 10% of each other.

MAGNETIC PROPERTIES

Stripe domains are observed by Kerr contrast microscopy in films with about 76 to 78 atomic % iron. Stripe period data can be obtained in the range bounded by high coercivity domain structures and unobservably small stripe domains. The former structure is characteristic of films with low magnetizations, $4\pi M_s < 400$G, and extremely high H_k fields, while the latter is characterized by high magnetizations, $4\pi M_s > 1100$G, and $H_k \sim 4\pi M_s$.

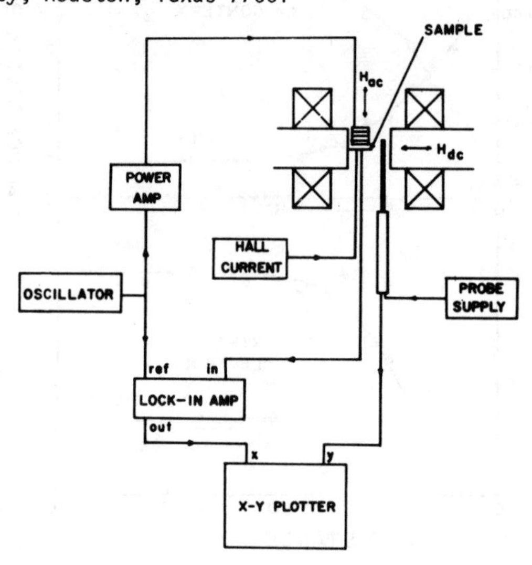

Fig. 1: Susceptibility Measurement System

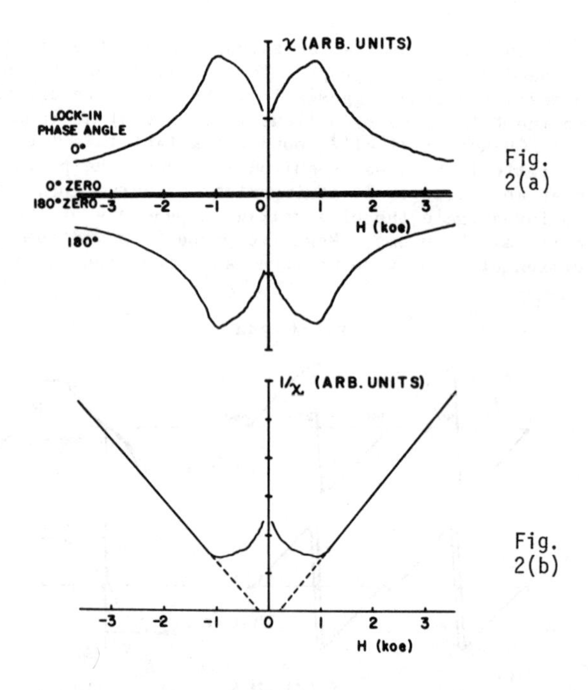

Fig. 2(a): Susceptibility from Hall voltage as a function of in-plane field, H_{dc}. Fig. 2(b) Inverse of Fig. 2(a). Extrapolation to field axis from $H_{dc} >> H_k$ gives $H_k - 4\pi M_s = 200 \pm 70$oe.

The characteristic lengths of these films are determined from stripe period and film thickness data from relations by Shaw, et. al. [3]. The characteristic length dependence on magnetization is shown in fig. 3 for films made at -70vdc and -100vdc substrate bias. Fitted curves to the characteristic length data gives slopes of -1.96 and -1.82 for the -70vdc and -100vdc data respectively. A -2 slope would indicate a material with a wall energy independent of magnetization.

The wall energies are shown in fig. 4 for films made at -70vdc and -100vdc. The averages of the wall

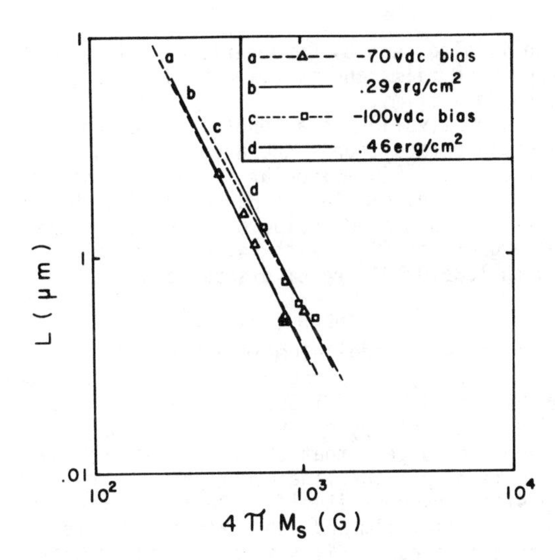

Fig. 3: Characteristic length dependence on $4\pi M_s$: (a) -70vdc, (b) wall energy is .29erg/cm^2, (c) -100vdc, (d) wall energy is .46erg/cm^2.

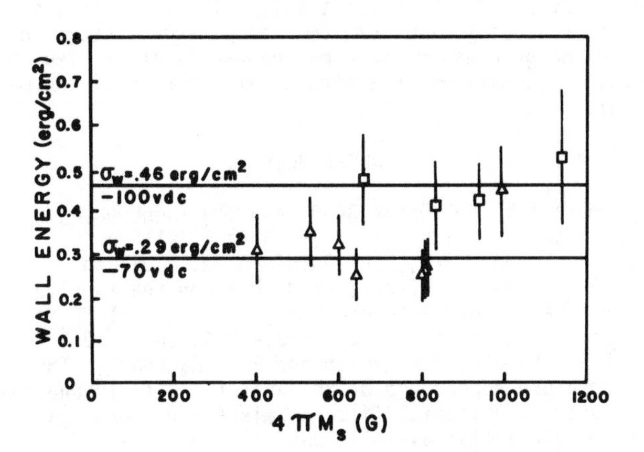

Fig. 4: Wall energy dependence on $4\pi M_s$ for different biases.

energies are .29erg/cm^2 and .46erg/cm^2 for films made at -70vdc and -100vdc respectively. The averages are 1) within the respective estimated error margins for all but one of the films and 2) higher for films made at -100vdc than at -70vdc. This evidence indicates that films made at a fixed bias have nearly the same wall energies but films made at higher biases have higher wall energies.

The dependence of the anisotropy field on magnetization is shown in fig. 5 along with a line representing the data of Imamura, et. al. [5]. There appears to be no trend in H_k or K_u with $4\pi M_s$ in these films. The anisotropy energies are about one-tenth of the anisotropy energies of Imamura, et. al. They vary from 2.3 to 5.6x10^{4}erg/cc with an average of 3.4x10^4 erg/cc for films made at -70vdc and from 1.7 to 8.8x10^{4}erg/cc with an average of 4.5x10^{4}erg/cc for films made at -100vdc. The average K_u of films made at -100vdc is higher than the average K_u of films made at -70vdc. However, there is a wide range of K_u values for the films made at -100vdc; this fact makes the interpretation of the anisotropy dependence on bias difficult. Anisotropy dependence on argon content in the films can not be discerned due to the considerable scatter in the argon percentage estimates.

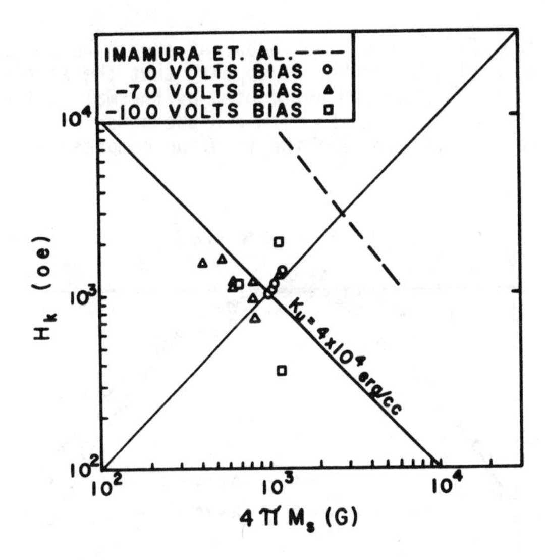

Fig. 5: Anisotropy field dependence on magnetization.

Low coercivity stripe domains can be seen in the films with K_u between 2.3 and 8.8x10^{4}erg/cc. The range of Q valves is about .9 to 4 for low coercivity stripe domains. Q≃.9 is associated with $4\pi M_s$≃1100G and stripe periods, P_o, of about 1μm while Q≃4 is associated with $4\pi M_s$=400G and P_o=4μm.

Fig. 6(a) is a photomicrograph of a GdFe film for which part of the film is detached from the substrate. The detached part of the film forms a bulge indicating that the attached part of the film is in a state of compressive stress. No stripe domains can be seen in

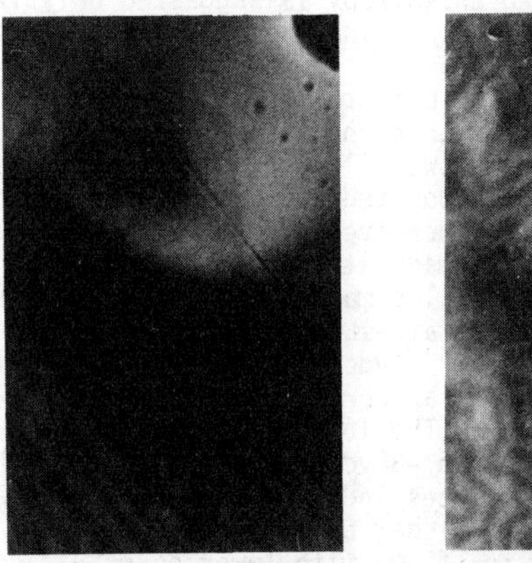

Fig. 6(a): GdFe

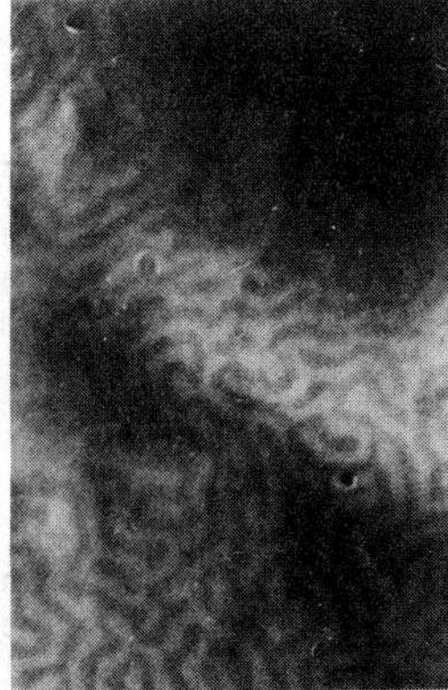

Fig. 6(b): GdCo

Fig. 6: Stripes and absence of stripes on attached and detached (upper portions of each fig.) parts of films.

the detached part of fig. 6(a) indicating that relieving the stress in the detached part has eliminated the uniaxial anisotropy. Such effects are consistently seen in several GdFe films made by dc bias sputtering. Stripes persist in a detached part of a GdCo film deposited under similar conditions, fig. 6(b). The results of these tests suggest that 1) the normal anisotropy is stress induced in these GdFe films and 2) stress has little effect in GdCo films.

Thermal contraction effects between the film and substrate are investigated by measuring the stress both at room temperature and at a temperature close to the deposition temperature. The film temperature during deposition is about 100°C [6]. The film stress is estimated by measuring the substrate curvature with and without the film, a technique described by Maissel and Glang [7]. There is no apparent stress relieved by heating a film to 100°C. It appears that the stress is deposition induced and not caused by thermal contraction effects between the film and the substrate. Further tests indicate that the in-plane compressive

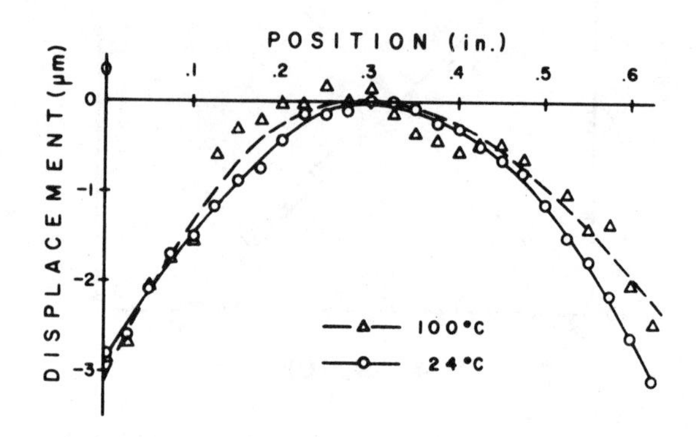

Fig. 7: Substrate curvature at 24°C and 100°C due to film.

stress is isotropic. Compressive stress levels of 5-10x10^{9}dyne/cm^2 are observed for films with V_b between -30vdc and -100vdc. The magnetostrictive coefficient is estimated to be about 4x10^{-6} by using K_u=4x10^{4}erg/cc and a compressive stress of 7.5x10^9 dyne/cm^2.

Considerable scatter is seen in the K_u with stress data. Stress induced anisotropy is suggested only from the disappearance of stripes in detached films, as discussed above.

The target which is used to make films at -70vdc is richer in gadolinium than the target used to make films at 0vdc. This allows films of similar $4\pi M_s$, H_k and composition to be deposited at -70vdc and 0vdc and allows a comparison of coercive properties to be made. The coercivity of films made at 0vdc was 39oe while it was less than 3 oe for films made at -70vdc. Similar magnetization films made at -100 vdc have coercivities similar to films made at -70vdc. The film resistivity dependence on bias shows an abrupt drop from ∿1000μΩ-cm to ∿250μΩ-cm at -30vdc. The lower resistivities for biases more negative than -30vdc are attributed to the higher backsputtering rates induced by the higher biases. Consequently, the higher biases produce cleaner films and give rise to films with lower coercivities and resistivities. Films made at 0vdc are not used in wall energy or anisotropy studies because of the unknown effect of impurities on these magnetic properties.

The exchange constant for bias sputtered GdFe films with observable and low coercivity stripes can be determined from the wall energy and anisotropy energy data of each film. However, there is considerable error in an exchange constant evaluated by this method [8]. The exchange constant is found to be 1.8±.9x10^{-7} erg/cm. An error of 40% could be attributed to the 10% uncertainty in $4\pi M_s$ measurements. Since there is 50% scatter in the exchange constant data, which is more than can be attributed to measurement uncertainty, there may actually be fluctuations in the exchange constant of bias sputtered GdFe films. High coercivities in sputtered GdCo films are attributed by Hasegawa to inhomogeneities in the exchange constant [9]. Such inhomogeneities may be attributed to variations in the ordering length of the material. Exchange constant fluctuations may be the cause of the scatter in the room temperature magnetization versus composition data. Scatter in the magnetization dependence on composition is also seen by Chaudhari, et. al. [10] for sputtered GdCo films and by Imamura, et. al. [5] for sputtered GdFe films.

The magnetization as a function of composition over a broad composition range agrees well with Imamura, et. al. [5]. Molecular field calculations of the magnetization, similar to Hasegawa [11], based on this data give a minimum value of exchange constant of 2.5x10^{-7}erg/cm for films with x≃.76. This value compares with 1.8±.9x10^{-7}erg/cm discussed above.

CONCLUSIONS

Properties of gadolinium-iron films prepared by dc biased rf sputtering are reported. Films with magnetizations between 400G and 1100G can have sufficient normal uniaxial anisotropy to produce stripe domains. Evidence seems to suggest that the normal anisotropy is due to an isotropic in-plane compressive stress due to the deposition process itself. Thermal contraction effects between the film and substrate contribute very little to the stress. Films made at a fixed bias have nearly the same wall energies but a higher bias appears to produce a higher wall energy. Wall energy is not found to depend upon magnetization. No trend in anisotropy with magnetization can be discerned and little trend with bias can be seen. Low coercivity films can be made if the bias and magnetization are greater than -30vdc and 400G respectively. The exchange constant is about 1.8x10^{-7}erg/cm. Such large scatter in the exchange constant data may be partly attributed to actual fluctuations in a structural ordering of the material.

REFERENCES

†Supported by NSF Grant 34584 and NASA Grant 44-006-001
*Now with Burroughs Corp., San Diego, California.

1) H.C. Bourne, Jr., R.B. Goldfarb, W.L. Wilson, Jr. and R. Zwingman, "Effects of dc bias on the fabrication of amorphous GdCo rf sputtered films," IEEE Mag. Trans., Vol. 11, No. 5, pp. 1332-1334, Sept, 1975.
2) P.W. Shumate, D.H. Smith and F.B. Hagedorn," The temperature dependence of the anisotropy field and coercivity in epitaxial films of mixed rare-earth garnets," JAP 44(1): 449-454, Jan. 1973.
3) R.W. Shaw, D.E. Hill, R.M. Sandfort and J.W. Moody, "Determination of magnetic bubble film parameters from strip period domain measurements," JAP 44(5): 2346-2349, May 1973.
4) J.A. Cape and G.W. Lehmann, "Magnetic domain structures in thin uniaxial plates with perpendicular easy axis," JAP 42(13): 5732-5756, Dec. 1971.
5) N. Imamura, Y. Mimura and T. Kobayashi, "Amorphous Gd-Fe alloy films prepared by rf cosputtering technique," IEEE Mag. Trans., Vol. 12, No. 2, pp. 55-61, March 1976.
6) C.T. Chen, private communication.
7) L.I. Maissel and R. Glang, editors, Handbook of Thin Film Technology, McGraw-Hill Book Company, 1970, Chapter 12.
8) J.T. Carlo, D.C. Bullock and F.G. West, "A ferrimagnetic model for the exchange constant on magnetic bubble garnets," IEEE Mag. Trans., Vol. 10, No. 3, pp. 626-629, Sept. 1974.
9) R. Hasegawa, R.J. Gambino, J.J. Cuomo and J.F. Ziegler, "Effect of thermal annealing and ion radiation on the coercivity of amorphous Gd-Co films," JAP 45(9):4036, Sept. 1974.
10)P. Chaudhari, J.J. Cuomo and R.J. Gambino, "Amorphous films for magnetic bubble and magneto-optic applications," 19th Conf. on Magnetism and Magnetic Materials, Boston, Mass., Nov. 1973.
11)R. Hasegawa, "Static bubble domain properties of amorphous Gd-Co films," JAP 45(7): 3109-3112, July 1974.

EFFECT OF SUBSTRATE BIAS ON THE ANISOTROPY IN LOW VOLTAGE MULTITARGET SPUTTERED Gd-Co AND Gd-Fe THIN FILMS

M. L. Covault, S. R. Doctor, C. S. Comstock, Electrical Engineering Dept., Iowa State University, Ames, Iowa 50011, and D. M. Bailey, Ames Laboratory-ERDA, Ames, Iowa 50011

ABSTRACT

By use of a low voltage multitarget sputtering system we were able to study binary films having the same composition, independent of the dc substrate bias, by applying separate voltages to each target. Using this technique we have been able to obtain Gd-Co films possessing a perpendicular anisotropy sufficient to cause the magnetization to be normal to the film plane under grounded substrate conditions as well as when negatively biased. However, when the substrates were allowed to electrically float only in-plane magnetization was observed. In contrast, Gd-Fe films possessing perpendicular anisotropy could be obtained with or without an applied dc bias. The preparation conditions and x-ray (FeK_α radiation) diffraction analysis of these films are reported and their significance to the origin of the net uniaxial anisotropy is discussed.

INTRODUCTION

The origin of the perpendicular uniaxial anisotropy in amorphous rare-earth (RE) transition-metal (TM) thin films is a very interesting phenomena that is not well understood. It is important to keep in mind that in the current scientific literature amorphous is most often used to describe the state of a solid in which there is no long-range order. It would be improbable for a total structural disorder to exist in a metastable amorphous solid since the atoms making up the material tend to arrange themselves at least in such a way as to satisfy a high number of chemical bonds. The resultant short-range order, which appears to be between 5 and 10 average atomic diameters[1], can cause the amorphous state to have some similarity with the corresponding crystalline state.

The model originally outlined by Gambino et al.[2,3] to describe the origin of this anisotropy is based on the pair ordering of the Co atoms in Gd-Co films. They estimated that the number of excess nearest-neighbor pairs in the film plane required to account for the observed anisotropy is approximately 10^{20} to 10^{21} pairs/cm^3. This corresponds to 0.1% to 1.0% more Co-Co pairs in-plane than out-of-plane. This excess of in-plane pairs is attributed to the sputter deposition process. More recently, Heiman et al.[4,5] have observed a perpendicular anisotropy in several other RE-TM amorphous thin films prepared by thermal evaporation. However, they reported that for Gd-Co this magnetic anisotropy could only be obtained by substrate bias sputter deposition. Subsequently, Onton et al.[6] found that the x-ray diffraction pattern from Gd-Co films prepared with a zero bias voltage possessed one broad peak while a Gd-Co film deposited with a -100 volt bias had a broad peak with a shoulder attached. As they increased the bias voltage in a negative sense, the center of the peak shifted to higher 2θ angles. They logically concluded that the change in structure in films possessing perpendicular anisotropy from those that do not was due to the change in short-range ordering and not to pair ordering.

We agree with this conclusion and hope to shed some light on this subject by reporting on the x-ray diffraction dependence on substrate bias for the Gd-Co and Gd-Fe systems. It is interesting to note that, contrary to what has been previously reported, we have been able to obtain a uniaxial perpendicular anistropy field greater than the magnetization in zero biased Gd-Co films. SI units are used in this paper.

FILM PREPARATION

A schematic drawing depicting the geometry of the multitarget sputtering system is shown in Figure 1. The targets are 0.64 cm diameter rods that are 20 cm long. They are spaced 7.62 cm apart and 10.16 cm from the substrate surface. To prevent material sputtered from one target from reaching the other, grounded stainless steel shields are located between them and outside the targets' Crookes dark space. The substrate holder is a water cooled copper block capable of holding three 2.54 cm by 7.62 cm substrates. This particular arrangement results in a composition difference of as much as 25% between the two ends of a single binary alloy film. The compositional control of the film is obtained by the application of independent dc voltages to the targets. By varying the voltage to each target, a wide range of composition profiles can be obtained. A typical gradient for a Gd-Co film determined by an electron

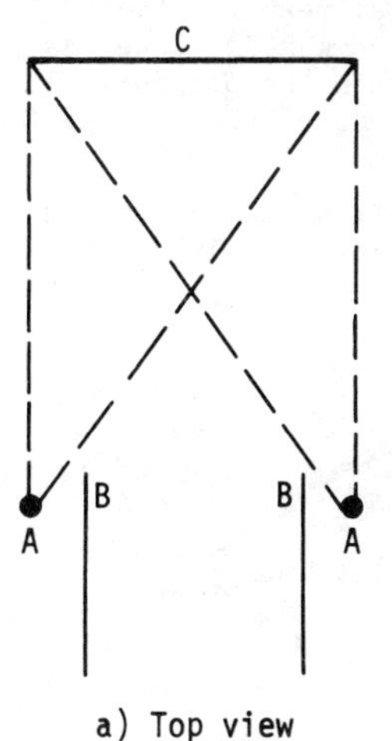

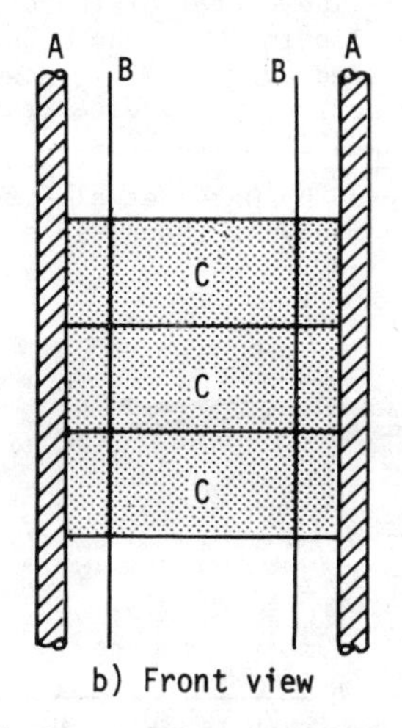

Fig. 1. Schematic drawing of the sputtering geometry showing only the targets (A), their shields (B), and the substrates (C).

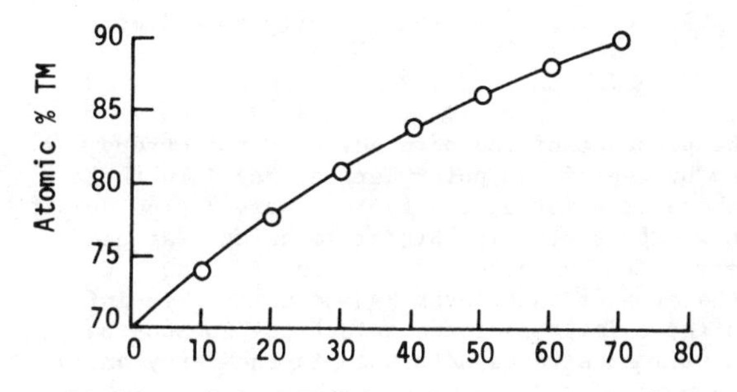

Fig. 2. Film composition along the length of the substrate.

microprobe is shown in Figure 2. The microprobe analyses of the films were roughly 5% accurate. A negative potential of 100 to 200 V was usually applied to the Gd target while 300 to 400 V was applied to the TM target. The resultant target current ranged from 75 to 150 mA depending on the plasma density near the targets. The purity of the metals used was better than 99.95%.

The cylindrical glass vacuum chamber which rests on a stainless steel feedthrough assembly is pumped by an oil (DC705) diffusion pump with a freon cooled chevron baffle. A mechanical backing pump is used. The chamber is evacuated to 2×10^{-5} Pa prior to sputtering. The depositions are performed in an argon atmosphere at a pressure of 0.1 Pa, as measured by an ionization gauge. This low pressure minimizes argon entrapment in the film and provides for a high gas throughput which leads to a low background pressure of impurities. The continuous flow of argon passes through a titanium gas purifier before entering the sputtering chamber. The plasma is excited by a 27MHz current loop located in the sputtering chamber. This loop is surrounded by glass to prevent sputtering of the loop material. The plasma is thus maintained independently of the potentials applied to the targets and substrate holder. Two magnetic coils in a Helmholtz configuration located outside the vacuum chamber are positioned symmetrically around the sputtering region to control the plasma density in the vicinity of the target. The resultant magnetic field is parallel to the axes of the target rods and is in the plane of the deposited film. The field intensity can be varied from 0 to 5×10^3 A/m, a value of 2.2×10^3 A/m was typically used.

Silver conductive paint was used to provide electrical contact between the substrates and the holder to which the dc bias voltage was applied. In the absence of an applied bias the substrates floated at the potential acquired from the plasma. All potentials reported were measured with respect to ground. Both metallic coated and uncoated Corning 0211 glass and AlSiMag glazed alumina substrates were used. The metal coating was a 1KÅ thermally evaporated Chromium film. Vacuum grease was applied to the substrate backs to provide thermal contact to the holder. During the twenty minute target pre-sputter time the substrates were covered with stainless steel shutters. An Inficon XTM rate and thickness monitor equipped with an rf/dc film sensor was located next to the substrate holder. The sensor's cross-field electron stripper prevented the sensing crystal from being heated by electron bombardment thus allowing control of the film growth during deposition.
Typical deposition rates for the RE-TM films ranged from 30 to 80 Å/min with a substrate temperature of 75° C. Films of thickness 1.5 to 2 μm were usually deposited. The film thickness variation along the length of the substrate, as determined by a Sloan Dektak stylus surface profilometer, was found to be small (<9%) for the composition profiles studied.

EXPERIMENTAL RESULTS AND DISCUSSION

The presence of the perpendicular anisotropy was observed by use of the polar Kerr effect. In those cases where it existed, the typical maze stripe domain pattern was present. The Bitter technique was used to determine the presence of in-plane domains. We found that the Cr underlayer had no noticeable influence on the existence of the anisotropy so non-coated Corning 0211 substrates were used in the x-ray analysis.

When the substrate is allowed to float in a "triode" sputtering system it assumes a potential, V_f, (typically in our system $V_f = +15$ to $+30$ V) which is more negative than the plasma potential, V_p. Because of this potential difference at least some acceleration of the argon ions into the film being deposited will occur even when the substrate is floating. The effective substrate bias is therefore $V_b = V_a - V_p$ where V_a is the applied bias. As V_a is increased in a positive sense it was observed in our laboratory that the magnetization of the Gd-Co films became oriented in the plane when the potential difference ($V_f - V_a$) reached the range 0 to 2 V. The floating potential, V_f, was used as a relative measure of V_p because it could easily be measured before V_a was applied. This suggests that the energy of the ions bombarding the film must be great enough to break up weakly bonded atom pairs or clusters to give rise to the proper short-range ordering needed for $H_k > M_s$. However, in the case of Gd-Fe this ion bombardment was not required to induce a perpendicular anisotropy field greater than the magnetization. Even with the substrate potential greater than the plasma potential, $H_k > M_s$. Both alloy systems were studied in the bias range of $V_a = -10$ to ($V_f + 15$) V. At these potentials and at the argon pressure used, little argon is expected to be incorporated into the films. By observing the composition range over which $H_k > M_s$ it is apparent that the perpendicular anisotropy field in the Gd-Co films decreases as the energy of the bombarding ions decreases, finally becoming zero near a substrate potential of V_f. The composition dependence of the magnetization can be approximated by

$$M_s = |2.93 \times 10^6 - 3.68 \times 10^6 \mathbf{X}|$$

where **X** is the fractional Co concentration[7].

To observe the structural differences and similarities of these films it is informative to look at their x-ray diffraction patterns. Because of the geometry of the Norelco diffractometer used, the resultant diffracted intensity of the films was due only to ordering perpendicular to the film surface. Filtered FeK_α radiation and a scintillation counter were used. The scattering from the film has been corrected for contributions from the substrate. The area analyzed was just to the right of the room temperature compensation point (79 at. % Co, 74 at. % Fe) of figure 2. This area varies in composition by 3 at. % and in thickness by less than 1%. There was little noticeable change in the diffraction patterns for the composition range (75 to 84 at. % Co for Gd-Co films and 69 to 78 at. % Fe for Gd-Fe films) for which $H_k > M_s$. This may be due to an averaging over the compositional gradient of the film area analyzed.

The center of the broad x-ray peak for the Gd-Co samples in which the magnetization was in-plane was located at $2\theta = 45°$. However, when the perpendicular anisotropy was greater than the magnetization, the peak shifted to $2\theta = 50°$ as shown in Figure 3. As argued by Onton et al.[6] this shift of intensities and accompanying change in shape is more likely due to

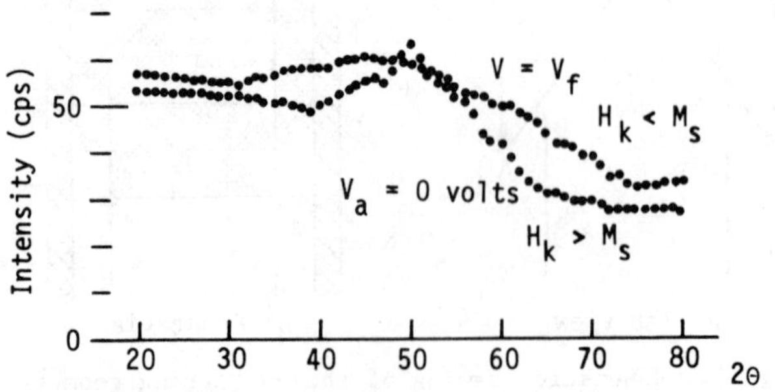

Fig. 3. X-ray pattern of sputtered amorphous Gd - Co films.

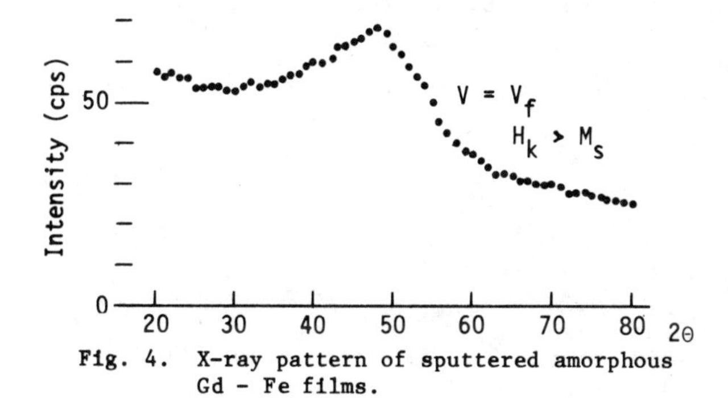

Fig. 4. X-ray pattern of sputtered amorphous Gd - Fe films.

short-range ordering rather than to pair ordering. In the Gd-Fe films, the peak intensity occurred at approximately $2\theta = 48^{\circ}$ in all cases as shown in Figure 4. This fact as well as the similarity in the shapes of the patterns suggest that the short-range order giving rise to the anisotropy is similar in the two systems. To determine if this is really the case the radial distribution functions need to be calculated for the amorphous Gd-Co and the Gd-Fe films of the same composition. Cargill[1] has determined the RDF's for $Gd_{36}Fe_{64}$ and $Gd_{18}Co_{82}$ and found that the short-range order is significantly different from that of their respective crystalline Laves phase structures.

By observing the domain structure before and after removal from their substrates we have ruled out stress as the origin of the anisotropy in these particular films. Another possible mechanism contributing to the observed anisotropy in our films may be their compositional gradient. Although it may enhance the magnetic anisotropy, it does not seem to support one model over the other. Incidently, neutron irradiation ($E > 0.1$ MeV, $\Phi = 8 \times 10^{17}$ n/cm^2, $T = 80$ °K) did not destroy the perpendicular anisotropy in the Gd-Co films deposited at $V_a = 0$ volts[8].

CONCLUSIONS

We have found that in order to induce a perpendicular anisotropy field which is greater than the magnetization in Gd-Co films, ion bombardment of at least a minimum energy is required during deposition. No inducement appears to be needed in the Gd-Fe system. The x-ray diffraction intensities indicate that the ordering is similar in Gd-Fe and Gd-Co amorphous films possessing this anisotropy. The change of position and shape of the x-ray pattern for the Gd-Co films where the magnetization becomes in-plane tend to support the short-range ordering model and not the pair-ordering model.

ACKNOWLEDGMENTS

This research was supported by funds made available through the Engineering Research Institute, Iowa State University, Ames, Iowa 50011. We would like to thank Ames Laboratory-ERDA for providing the Fe and Co targets and their Rare-Earth Center for preparing the Gd target.

REFERENCES

1. G. S. Cargill, Solid State Physics, Vol. 30, Academic Press, New York, 1975, pp. 227-320, "Structure of Metallic Alloy."
2. R. J. Gambino, J. Ziegler, and J. J. Cuomo, "Effects of Ion Radiation Damage on the Magnetic Domain Structure of Amorphous Gd-Co Alloys," Appl. Phys. Lett., 24, 99 (1974).
3. R. J. Gambino, P. Chaudhari, and J. J. Coumo, "Amorphous Magnetic Materials," AIP Conf. Proc., 18, 578 (1974).
4. N. Heiman, and K. Lee, "Magnetic Properties of Ho-Co and Ho-Fe Amorphous Films," Phys. Rev. Lett., 33, 778 (1974).
5. N. Heiman, A. Onton, D. F. Kyser, K. Lee, and C. R. Guarnieri, "Uniaxial Anisotropy in Rare Earth (Gd, Ho, Tb) - Transition Metal (Fe, Co) Amorphous Films," AIP Conf. Proc., 24, 573 (1975).
6. A. Onton, N. Heiman, J. C. Suits, and W. Parrish, "Structure and Magnetic Anisotropy of Amorphous Gd-Co Films," Proc. of the International Conference on the Electronic and Magnetic Properties of Liquid Metals, Mexico City, 1975 (to be published).
7. D. C. Cronemeyer, "Perpendicular Anisotropy in $Gd_{1-x}Co_x$ Amorphous Films Prepared by RF Sputtering," AIP Conf. Proc. 18, 85 (1974).
8. C. W. Chen (unpublished work).

THE MULTIPLE-TARGET METHOD FOR SPUTTERING AMORPHOUS FILMS FOR BUBBLE-DOMAIN DEVICES

C. T. Burilla, W. R. Bekebrede and A. B. Smith
Sperry Research Center, Sudbury, Massachusetts 01776

ABSTRACT

Previously, sputtered amorphous metal alloys for bubble applications have ordinarily been prepared by standard sputtering techniques using a single target electrode. We report for the first time the deposition of these alloys using a multiple target rf technique in which we use a separate target for each element contained in the alloy. One of the main advantages of this multiple-target approach is that the film composition can be easily changed by simply varying the voltages applied to the elemental targets. In our apparatus, the centers of the targets are positioned on a 15 cm-radius circle. The platform holding the film substrate is on a 15 cm-long arm which can rotate about the center, thus bringing the sample successively under each target. The platform rotation rate is adjustable from 0 to 190 rpm. That this latter speed is sufficient to homogenize the alloys produced is demonstrated by measurements we have made of the uniaxial anisotropy constant in $Gd_{0.12}Co_{0.59}Cu_{0.29}$ films. The anisotropy is 6.0 x 10^5 ergs/cm^3 and independent of rotation rate above ~25 rpm, but it drops rapidly for slower rotation rates, reaching 1.8 x 10^5 ergs/cm^3 for 7 rpm. The film quality is equal to that of films made by conventional methods. We have observed coercivities of a few oersteds in samples with stripe widths of 1 - 2 μm and magnetizations of 800 - 2800 G.

INTRODUCTION

The development of materials suitable for use in cylindrical domain device structures has concentrated primarily on garnet single crystal films. However, an alternative approach which has generated increasing interest recently is the use of an atomically disordered (amorphous) thin film of some appropriate magnetic alloy. Although many alloy compositions and preparative procedures have been investigated, the major attention has been given to rf sputtering of amorphous magnetic Gd-Co films. Chaudhari et al[1] have described their preparation, and Hasegawa[2] has studied their static bubble domain properties. The potential advantages of amorphous materials over the garnets are lower material cost, fewer problems associated with the preparation of large numbers of films and a simpler chemical system. As Chaudhari et al[1] have pointed out, there is a wide latitude in the choice of suitable substrates, ranging from amorphous materials (glass or fused quartz) to single crystals (NaCl, Si, etc.). The use of silicon substrates is attractive from the standpoint of being able to design monolithic integrated circuits incorporating bubble memories. As regards the scaling up of the film preparation, large quantity sputtering systems have been built for other applications, but the LPE process currently used for garnet film growth would be difficult and/or expensive to scale up because of the high temperatures required and the toxic nature of the commonly employed solvent system. Finally, it would be easier to achieve reproducibility of composition in a metal alloy system consisting of a limited number of elements, compared with the currently used garnet compositions where there may be as many as six ingredients, each with a different segregation coefficient.

In the Gd-Co binary system, the necessary low saturation magnetization required for operation in a bubble domain memory device can be obtained by adjusting the film composition to a value near the compensation point. Unfortunately, in this region there are undesirably large variations in magnetic properties with temperature. In an effort to achieve less temperature-sensitive behavior, a number of ternary alloys has been investigated in which a third element is added to the Gd-Co binary. These elements include Mo, Au, Cr, Ni and Cu[3-7].

The work described here involves primarily the Gd-Co-Cu ternary system and is partly a study of the magnetic properties of this ternary alloy and is partly an evaluation of a novel sputtering system applied for the first time to the preparation of amorphous magnetic thin films. All the sputtered bubble films described in the literature have been prepared in single-target systems utilizing a composite or alloy target in which the only control over film composition was the limited amount obtained by varying the rf bias on the substrate. Furthermore, the bias control was primarily effective in adjusting the Co/Gd ratio; its control over the third element depends upon the resputtering yield of that element relative to Co and Gd. For example, in the deposition of Gd-Co-Cr films[5], it was found that the Co/Cr ratio remained constant within the bias voltage range sufficient to adjust the Co/Gd ratio. It is clear, therefore, that in the initial stages of the development of a new ternary system, where it is necessary to vary the individual component concentrations over a broad range, it would be necessary to prepare many alloy targets, an expensive and time-consuming process. The multiple target sputtering system described in this paper was designed to avoid this problem by providing three separate elemental targets, which are individually powered. By changing the relative power to these targets, any desired film composition in the ternary system under study can be prepared. Although this concept has been used before[8,9], this is its first reported application to the preparation of amorphous bubble films.

APPARATUS

The rf sputtering apparatus is a Cooke Vacuum Products, Inc. Model C70-6-4B, modified to provide rotation of the substrate holder, as described below. The vacuum system employs a conventional diffusion pump and liquid nitrogen trap. The argon pressure during sputtering is maintained at a constant value by a Granville-Phillips automatic pressure controller. The vacuum chamber is a 60 cm-diameter stainless steel cylinder containing three 15 cm-diameter, water-cooled target electrodes arranged 90° apart on a 15 cm-radius circle. The 15 cm-diameter, water-cooled substrate carrier is mounted in a "J" head configuration and supported on a Ferrofluidics Corporation rotary shaft seal so that the "J" can be rotated up to a maximum of 190 rpm, thereby sequentially exposing the substrate to the three target electrodes. The relative arrangement of targets and substrate holder is shown schematically in Fig. 1.

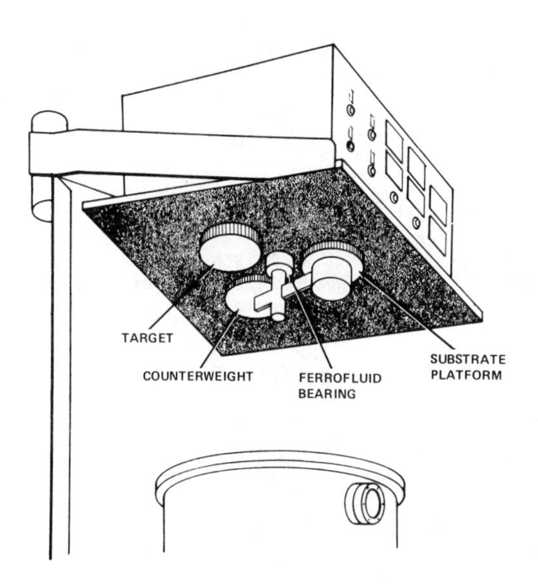

Fig. 1 Sketch showing the basic elements of the multiple-target sputtering system. (For clarity, the shutters and two of the targets are omitted from this drawing.)

All targets are powered from a single 13.56 MHz rf generator using the matching network and metering circuit shown in Fig. 2. This is a standard type of matching circuit, with appropriate modification to permit connecting any number of targets and the substrate platform. Meters are provided which can be

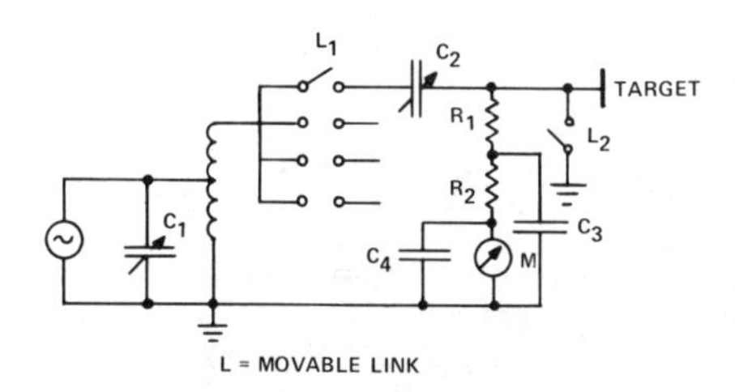

Fig. 2 Matching network used with multiple-target sputtering system. (Only one target is shown; the circuit consisting of C_2, C_3, C_4, R_1, R_2 and M is duplicated for each target and for the substrate platform. Unused targets are grounded using link L_2. Component values are: C_1 = 900 pF, C_2 = 300 pF, R_1 = 30MΩ, R_2 = 10 kΩ, M = 0-100μA dc.)

used as a guide in adjusting the power division among the targets or for resetting a previously established set of conditions. Instead of measuring the rf target voltage directly, the meter in Fig. 2 responds to the dc component produced by the rectifying action of a target in a plasma[10]. We find this method of metering gives a satisfactory indication while avoiding problems of pickup in the meter leads which can seriously affect measurements of rf voltages. Adjustment of the power distribution to the targets is accomplished by varying the relative settings of the C_2 capacitors, while C_1 is adjusted to provide the best match to the generator. The interaction between the capacitor adjustments is sufficiently small that any desired combination of meter readings can be conveniently achieved with less than 2% reflected power.

FILM PREPARATION

Amorphous alloy films were deposited onto 25 mm x 25 mm Corning 7059 glass plates using the rf sputtering system described above. Prior to sputtering, the vacuum chamber was evacuated to a pressure $< 7 \times 10^{-7}$ torr before back filling with ultra-high purity argon. During deposition, the argon pressure was held constant at 1.0×10^{-2} torr ± 2% by the automatic pressure control system. The target-to-substrate distance was 38 mm as measured when the substrate platform was directly under any given target. Substrate platform rotation rates between 7 and 190 rpm produced amorphous films which showed uniaxial anisotropy perpendicular to the film plane. All targets were sputter cleaned for 30 minutes and the substrates were sputter etched for the same length of time before film deposition was initiated. For all runs, the platform was rf biased, using voltages sufficient to produce a 100- to 200-volt dc component. Total power input to the matching network was of the order of 1.2 kW. This power level gave a deposition rate of about 160 Å per minute. A typical sputtering run of two hours duration yielded films about 2 μm thick.

By varying the voltage applied to each target, and keeping the bias voltage on the substrates constant, alloy films of $(Gd_{1-x}Co_x)_{1-y}Cu_y$ in the composition range of $0.75 \leq x \leq 0.90$ and $0 \leq y \leq 0.40$ were obtained. Ternary alloys utilizing molybdenum as the non-magnetic component have also been produced in a similar manner.

Average film compositions were determined by x-ray fluorescence and atomic absorption spectrophotometry. Film thickness was measured by interference microscopy.

RESULTS

An important question about this method of depositing films is what effects the platform rotation rate has on film properties. Therefore, a number of films was deposited with different rotation rates but having the composition $Gd_{0.12}Co_{0.59}Cu_{0.29}$. The uniaxial anisotropy constant K_u of these films was then measured, the assumption being that K_u would be more sensitive to rotation rate than any of the other magnetic parameters. Although the exact mechanism that determines K_u in amorphous films is not understood, this anisotropy must be caused by some sort of short-range order which presumably would be modified at low rotation rates. In our measurements, K_u was determined using the Kurtzig-Hagedorn technique[11], together with the corrections of Druyvesteyn et al[12] which were determined from $4\pi M$ and ℓ measurements obtained by the Fowlis-Copeland technique[13]. The results for K_u as a function of rotation rate are shown in Fig. 3. All of these samples had $4\pi M$ within a fairly narrow range of 2030 to 2770 G; additional measurements of K_u as a function of $4\pi M$ at fixed rotation rate verified that over this range $4\pi M$ variations do not have any significant effect on K_u. The figure shows that rotation rates greater than ~ 25 rpm are sufficient to homogenize the films, while the anisotropy drops rapidly for slower rotation. A rate of 25 rpm corresponds to a deposition rate of ~ 6 Å per revolution.

To further evaluate GdCoCu films prepared by the multiple target method, samples with a wide range of magnetic properties have been prepared using rotation rates greater than 50 rpm. In these films, the magnetization ($4\pi M$) ranges from 800 to 2800 G and the stripe width from ~ 1 to ~ 2 μm. Measurements have also been made of the coercivity of representative films using a vibrating sample magnetometer. In samples having the approximate composition

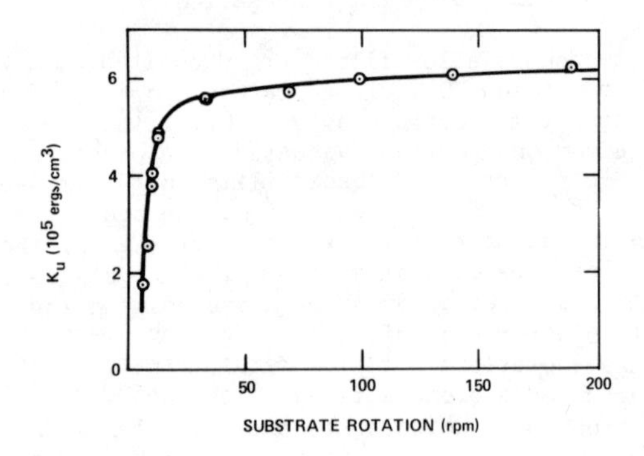

Fig. 3 Uniaxial anisotropy constant K_u as a function of substrate platform rotation rate for films of $Gd_{0.12}Co_{0.59}Cu_{0.29}$. (The magnetization of all these films lies between 2030 and 2760 G. The stripe widths of the films range from 0.9 to 1.3 μm.)

$Gd_{0.13}Co_{0.73}Cu_{0.14}$, the Gd/Co ratio was varied slightly between sputtering runs to yield the following results for magnetization and coercivity, H_c:

$4\pi M$(G)	H_c(Oe)
800	4.7
1130	2.6
1600	2.0

These results indicate that the coercivity follows approximately an inverse proportional relationship as would be expected[14].

We have also measured the temperature dependence of collapse field and stripe width for films made with this technique. For the composition $Gd_{0.14}Co_{0.71}Cu_{0.15}$ we have obtained the results shown in Fig. 4 by direct microscope measurements employing a hot stage. In addition to these results, the approximate compensation temperature was also determined by cooling the sample further and noting the temperature at which the stripe width becomes very large. Using this technique, we estimate that compensation occurs at $\sim -120^{\circ}$C. Of course, the sample of Fig. 4 is merely illustrative of what can be obtained by this sputtering method; for

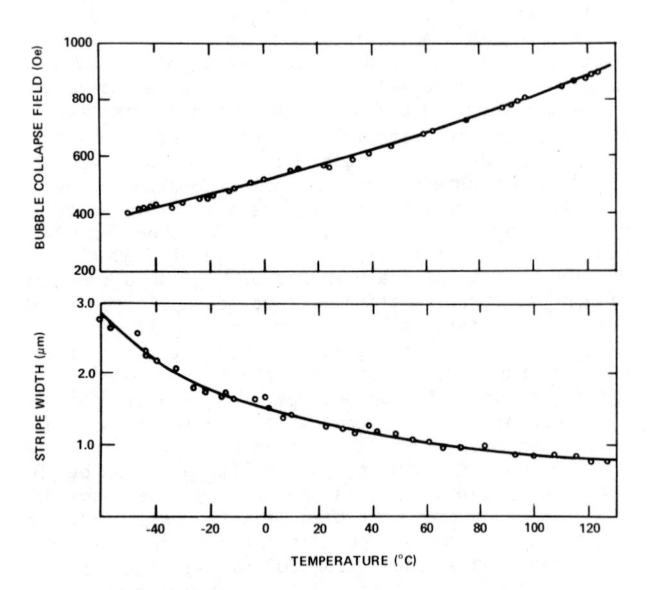

Fig. 4 Stripe width and collapse field vs. temperature for a $Gd_{0.14}Co_{0.71}Cu_{0.15}$ film.

bubble device use, the relative concentrations of Gd, Co, and Cu would have been chosen differently so as to minimize the temperature dependence of stripe width.

CONCLUSION

GdCoCu-alloy films have been prepared using a multiple-target sputtering method, and the static bubble properties of these films have been measured. The results demonstrate that this new technique is well suited for preparing bubble domain materials and has a number of practical advantages over previous methods of preparation.

ACKNOWLEDGMENTS

The authors would like to thank M. Kestigian for useful technical discussions, A. J. Doppler and W. T. Maloney for advice in the use of their vibrating sample magnetometer, and W. W. Goller for expert technical assistance in making the bubble-domain measurements.

REFERENCES

1. P. Chaudhari, J. J. Cuomo, and R. J. Gambino, IBM J. Res. Dev. 17, 66 (1973).
2. R. Hasegawa, J. Appl. Phys. 45, 3109 (1974).
3. R. Hasegawa, J. Appl. Phys. 46, 5263 (1975).
4. R. Hasegawa, R. J. Gambino, and R. Ruf, Appl. Phys. Lett. 27, 512 (1975).
5. J. Schneider, IBM J. Res. Dev. 19, 587 (1975).
6. R. Hasegawa and R. C. Taylor, J. Appl. Phys. 46, 3606 (1975).
7. V. J. Minkiewicz, P. A. Albert, R. I. Potter, and C. R. Guarnieri, 21st Annual Conference on Magnetism and Magnetic Materials. Paper No. 8A-2, Philadelphia, Pa., December 1975.
8. C. Corsi, I. Alfieri, and G. Petrocco, Infrared Physics 12, 271 (1972).
9. C. Corsi, J. Appl. Phys. 45, 3467 (1974).
10. J. L. Vossen and J. J. O'Neill, Jr., RCA Rev. 29, 149 (1968).
11. A. J. Kurtzig and F. B. Hagedorn, IEEE Trans. MAG-7, 473 (1971).
12. W. F. Druyvesteyn, J. W. F. Dorleijn, and P. J. Rijnierse, J. Appl. Phys. 44, 2397 (1973).
13. D. C. Fowlis and J. A. Copeland, AIP Conf. Proc. 5, 240 (1971).
14. R. J. Gambino, P. Chaudhari and J. J. Cuomo, AIP Conf. Proc. 18, 578 (1973).

THE STRUCTURE OF SPUTTERED VERY THIN FILMS OF Gd-Co AND Gd-Co-Mo AMORPHOUS FILMS

J. F. Graczyk

IBM Watson Research Center, Yorktown Heights, N. Y. 10598

ABSTRACT

We have investigated the structure of very thin films of Gd-Co alloys sputtered onto substrates of NaCl held at room temperature. The composition of these films ranged from .18 to .24 atom fraction of Gd and thickness from 80Å to a maximum of 600Å. We find that the structural intensity modulations, as a function of the momentum transfer vector $|\bar{s}| = 4\pi\sin\theta/\lambda(\text{Å})^{-1}$, are very similar for comparable film thicknesses and only weakly dependent on composition. The scattered electron intensity of the 80-200Å thick films show considerable difference to the intensity scattered from thicker films and to the x-ray data of Cargill and Wagner et al. These differences are in the ratio of the intensities of the first two amorphous haloes and the amplitude and shape of the modulations in the range of s from 3.7 to 7.8. For the thin films the ratio of I(2.05)/I(2.95)=.97 and for the thicker films this ratio reduces to 0.73. Similar qualitative results for the scattered intensity are also observed for the Gd-Co-Mo amorphous alloys. The radial distribution functions of the thin films have strong radial correlation peaks at r_1=2.50Å a weak peak at 3.17Å and a strong peak at 3.62Å. The remaining correlation peaks for higher r values can be qualitatively explained as distances arising from higher order pairs of the dense random packings of atoms. The radial distribution function of the thicker films shows an increase in the peak height at 3.17Å and a decrease of the peak at 3.62Å. We conclude from these results that the initial few layers of these amorphous films have a modified structure as compared to the thicker films. This is due to the high reactivity of Gd with O_2 which produces a large number of Gd-O pairs (r=3.62Å), which as we show here decreases with increasing film thickness.

I. INTRODUCTION

The short range order in amorphous Gd-Co films, since their discovery by Chaudhari, Cuomo and Gambino,[1] prepared by sputter deposition with and/or without bias applied to the substrate has been investigated by Cargill[2] and Wagner, Heiman, Huang, Onton and Parrish[3] by x-ray diffraction. In their studies they utilized very thick samples with thicknesses greater than several microns. The structure of very thin films with thicknesses on the order of 80 to 600Å have thus far not been reported. In this paper we report some results of our studies of the structure of these very thin films. We show that these show a structural change for increasing film thickness. The nature of the differences in the short range order structure is shown in this report. We then compare our experimental data to models for the structure. We find that only data for homogeneous films thicker than 350Å agrees qualitatively with recent models of dense random packings of hard spheres with two sizes of sphere diameters. These structures have been recently reported by Cargill[2] and Cargill and Kirkpatrick.[4] We finally present a tentative model for the structure of films of thickness less than 120Å.

II. FILM PREPARATION

Films of Gd-Co and Gd-Co-Mo were sputtered in an Argon atmosphere with variable bias onto freshly cleaved substrates of NaCl. The thickness of the films ranged from 80 to a maximum of 600Å. The composition of the films was determined by microprobe measurements with a precision of ±.03 and α-backscattering measurements of concentration profiles, with a depth resolution of no better than 240Å. For these very thin films the results of these two measurements showed: 1) The composition of these very thin films was not homogeneous through the film thickness, showing for most films enrichment of Co at the outer surface of the film. 2) The oxygen concentration ranged from .1 to .24 atom fraction which was found to increase with time of storage of the film. 3) The Argon concentration varied between 0.06 and 0.10 atom fraction.

III. ELECTRON DIFFRACTION-EXPERIMENTAL RESULTS

A. Experimental Procedure

The scattered intensity from these very thin films was obtained with 80 KeV electrons. The angular divergence of the incoming beam was approximately equal to $8*10^{-4}$ radians. The energy resolution of the filter was approximately 1.6eV and the scattered electron intensity was recorded over the range of $|\vec{s}|=4\pi\sin\theta/\lambda$ in $(\text{Å})^{-1}$ from 0.1 to 20. The angular resolution of the detection system was set approximately at $3*10^{-3}$ radians. A more detailed description of the apparatus as well as data treatment procedures has already been described elsewhere.[5]

B. Electron Diffraction

The elastic single scattering of electrons by a binary amorphous alloy for angles not very close to zero can be expressed as:[6]

$$F(s) = s\left(\frac{I-\langle f^2\rangle}{\langle f\rangle^2}\right) = \int_0^\infty 4\pi r(\rho^R(r)-\rho_o)\sin sr\,dr = \int_0^\infty G(r)\sin sr\,dr$$

with $|\vec{s}|=4\pi\sin\theta/\lambda$ in $(\text{Å})^{-1}$ and, F(s) is the reduced radial distribution function with ρ_o the average atomic density of the alloy and $\rho^R(r)$ is the experimentally derived total atomic distribution function. For a binary alloy $\rho^R(r)$ will consist of three contributions from the partial distribution functions for each atomic specie such as $\rho_{AA}(r)$, $\rho_{BB}(r)$ and $\rho_{AB}(r)$. Each of these partial distribution functions will be modified by the scattering factor f_A and more specifically the s dependence of the scattering factors.[6] To a first approximation we may neglect the s dependence of the scattering factors and weight the $\rho^R(r)$ or the radial distribution function $J(r)=4\pi r^2\rho(r)dr$ by a simplified correction function:

$$\rho(r)=X_A(f_A^2/\langle f\rangle^2)\rho_{AA}(r)+2X_AX_B(f_Af_B/\langle f\rangle^2)\rho_{AB}(r)+X_B(f_B^2/\langle f\rangle^2)\rho_{BB}(r)$$

with

$$\langle f\rangle^2=(X_Af_A+X_Bf_B)^2 \text{ and } \langle f^2\rangle=X_Af_A^2+X_Bf_B^2$$

where X_A is the atom fraction of specie A and X_B is the atom fraction of specie B. The values of the scattering factors f_A and f_B are taken for s=0.

C. Experimental Results

In the course of our experiments we have observed changes in the scattered intensity from these films, for increasing film thickness. We show in Figure 1 the scattered intensity from a film of $Gd_{.18}Co_{.82}$ of 120Å and 350Å in thickness. For films of nominal thickness less than 350Å, the high intensity first halo at s=2.05 is also found for other compositions of Gd ranging from .24 to .18. The effect of composition on the scattered intensity is to shift the structural intensity modulations towards lower s values with increasing Gd concentration. For thicknesses greater than 350Å the first halo decreases drastically in intensity with the second halo shifting toward lower s values and increases in amplitude. Also the well defined third halo for the thin films centered at s=5.3 is not observed for thicknesses greater than 350Å. For these we observe a leveling off of the intensity in the region of s from 3.8 to 7.5. For all films the intensity modulations for higher values of s were very weak in comparison to other amorphous binary alloys. The derived reduced radial density functions for selected alloys are shown in Fig. 2. We show the G(r) of a) films of 120Å in thickness with X_{Gd}=.18 and X_{GD}=.24, b) a film 350Å thick with X_{Gd}=.18 and c) the G(r) derived by Cargill from x-ray data for a very thick film of $Gd_{.18}Co_{.82}$. The very thin films were examined by dark field electron microscopy which showed some presence of crystallites of the order of $\sim$ 25Å in size, at films asperities. The scattered intensity from these regions of the film was not analyzed. The volume fraction of these "anomalous" regions was very small for the thicker films. The presence of these crystallites we believe is a result of local severe oxidation resulting in the precipitation of Co. We shall not elaborate on this point any further in this paper. We show in Table I the values of the experimental correlation distances and coordination numbers as well as interatomic distances for models utilized for the determination of the short range order.

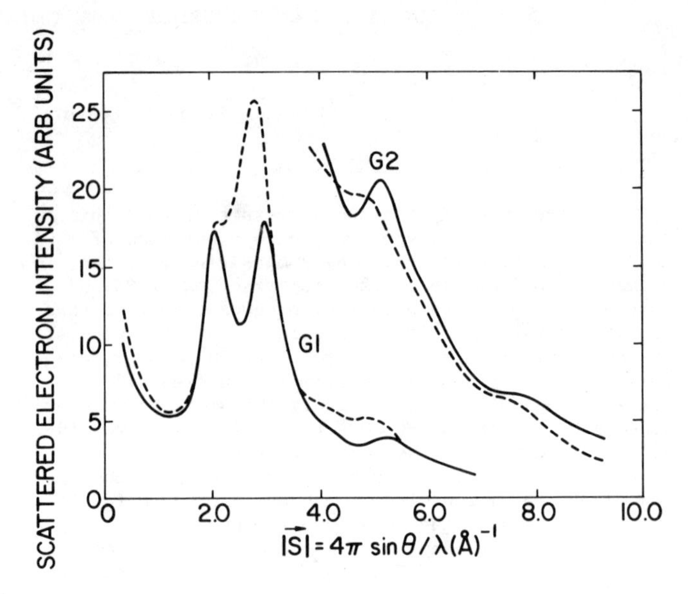

Figure 1. Elastically scattered electron intensity versus scattering vector S, with intensity gain expansion for two films: a) (——)$Gd_{.18}Co_{.82}$ 120Å thick and b) (---) 350Å thick.

TABLE I

Experimental Correlation Distances and Simply Corrected Coordination Numbers

Thin Films Results and R_{ij} of Models (Å)	Interatomic Distances (Å)					Coordin. Numbers		
	r_1 $\Delta r_1=\pm0.02$	r_2	r_3	r_4	r_5	C.N. (r_1)	C.N. (r_2)	C.N. (r_3)
$Gd_{.24}Co_{.76}$ ∿ 120Å	2.50	3.17	3.62	4.27	4.92	7.8	3.4	5.0
$Gd_{.18}Co_{.82}$ ∿ 120Å	2.50	3.14	3.59	4.20	4.93	8.2	3.0	4.1
$Gd_{.18}Co_{.82}$ ∿ 350Å	2.50	2.97	3.63	4.27	4.92	8.1	5.6	2.3
$Gd_{.18}Co_{.82}$ From Ref. 2.	2.50	2.95	3.45	4.20	4.79			
Gd(h.c.p) a=3.62 c=5.75	3.55	5.025	5.80	6.15	6.80			
Co(h.c.p) a=2.51 c=4.07	2.50	3.54	4.09	4.34	4.79			
Ar(f.c.c) a=5.43	3.83	5.41	6.61	7.66	8.55			
Gd_2O_3(hex) a=3.76 c=5.89	1.88	2.66	3.26	3.76	4.20			
$GdCo_5$(hex) a=4.97 c=3.97	2.488	3.505	4.30	4.978	5.505			
Gd H.S. Diam=3.55 3.60								
Co H.S. Diam=2.50								
Gd-O-Co Diam=3.60								

IV. DISCUSSIONS AND CONCLUSIONS

Comparing the reduced radial density functions G(r) for the various films we observe that the initial 120Å layer of the film shows different short range order from that of the thick films. All alloys investigated show similar trend for the initial 120Å layer of deposition and therefore very thick samples will have an anomalous layer. This finding is significant as it may have implications on the interpretation of other physical measurements on thick films. The G(r) for the very thin films shows a near neighbor distance at r_1=2.50Å which is the Goldschmidt diameter for Co and therefore this distance can be associated with Co-Co pairs as in the case for the thick films. The height of this peak for the thin films is larger than the peak in the data of Cargill and is associated with the enrichment of Co at the surface of the film. The second correlation peak occurs at r_{12}=3.17Å for the thin films as compared to the thick films for which R_{Co-Gd}=2.95Å. We see from the G(r) that r_{12} shifts toward lower r values for the 350Å thick film. We have associated this distance as a gaussian weighted sum of R_{Co-Gd}=2.95 and a substantial fraction of Gd-O pairs at 3.26Å giving the centroid at 3.17Å. The oxygen therefore breaks some of the Gd-Co pairs and produces Gd-O pairs. This is also consistent with the peak at 3.62Å (not observed for the thick films). This peak results

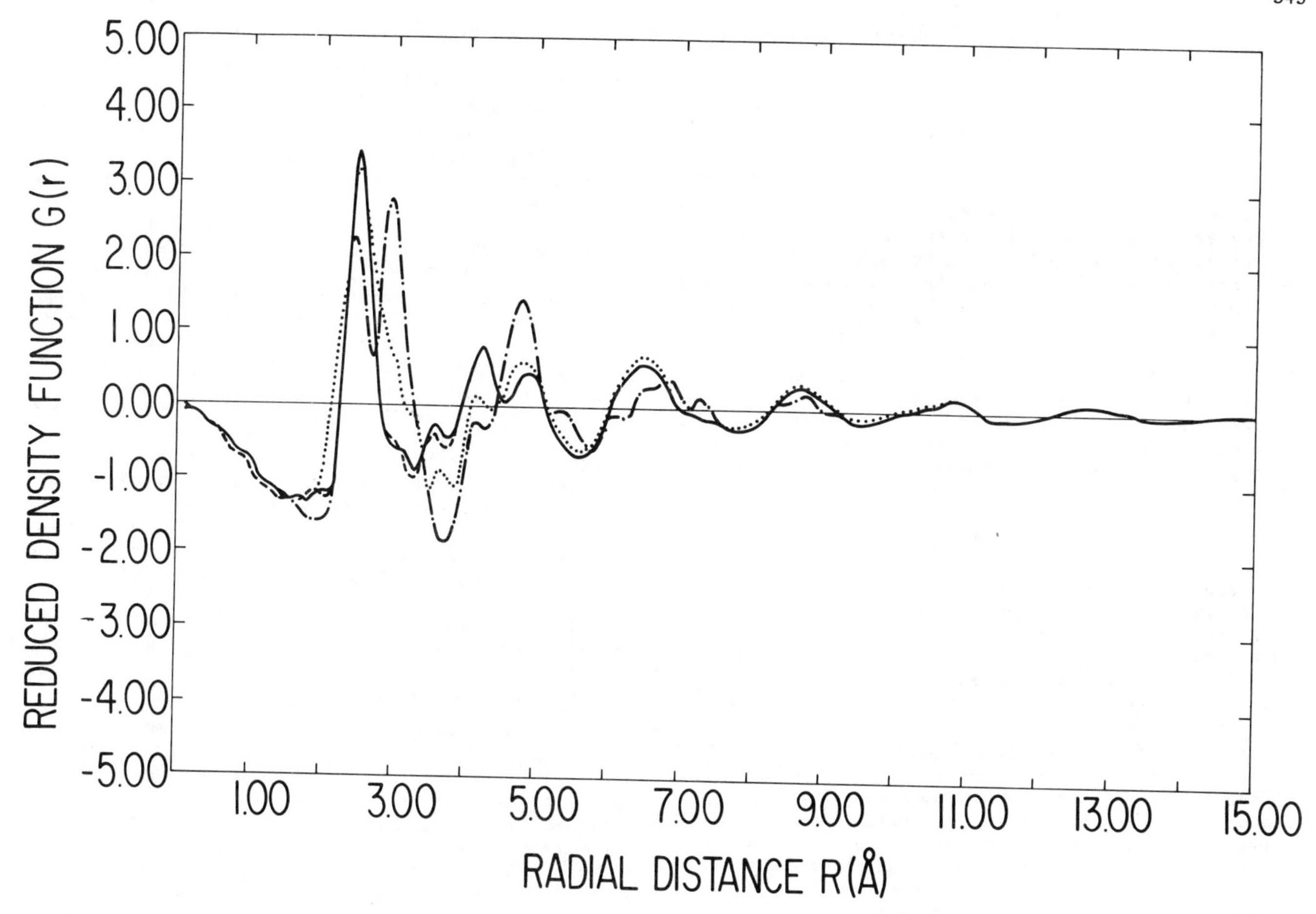

Figure 2. Reduced radial density function versus r for the following films: a) (---) 120Å thick with X_{Gd}=.18 and for (——) X_{Gd}=.24, b) (•••••) X_{Gd}=.18 350Å thick and c) (-•-•-) the G(r) from x-ray data of Ref. 2.

similarly, as a gaussian sum of R_{Gd-Gd} at 3.55Å and Gd-O-Gd at 3.76Å giving the centroid at 3.62Å. The height of this correlation peak is also consistent with this model, it decreases as the film thickness increases. Films with thicknesses greater than 600Å show features in their G(r) which can qualitatively be compared with good agreement to the x-ray results of Cargill[2] and Wagner et al.[3] Comparison of the short range order of these thin films to a variety of other models including phase mixtures as shown in Table I shows no better agreement with the data than the modified structure described above. We have observed in some films regions for which the oxygen content was extremely high resulting in phase separation and precipitation of Gd_2O_3 and Co. We have also performed tilting experiments with no evidence for layering, if films asperities discussed earlier were eliminated by selecting a proper area of the film or utilizing small selected area diffraction volumes.

In conclusion therefore we have shown in this paper that the structure of the initial layers of these amorphous alloys is different from the structure of the thick films. This modified structure is a result of the initial transients in the sputtering process and the high reactivity of Gd with O_2. As a result of this reaction the first ∿ 200Å of the film contains a smaller number of Gd-Co pairs relative to the thick films, and an increase of Gd-O pairs.

ACKNOWLEDGEMENTS

Thanks are due to J. D. Kuptsis and his colleagues for the microprobe measurements, J. D. Baglin for the α-backscattering experiments, P. Silano for the film preparations and Dr. P. Chaudhari and R. J. Gambino for helpful discussions.

REFERENCES

1. P. Chaudhari, J. J. Cuomo and R. J. Gambino, IBM J. Res. Dev. 17, 66 (1973).
2. G. S. Cargill III, in Solid State Physics, F. Seitz, D. Turnbull and H. Ehrenreich (eds.). (Academic Press, New York, 1975) Vol. 30, p. 227.
3. C. N. J. Wagner, N. Heiman, T. C. Huang, A. Onton and W. Parrish, AIP Conf. Proc. to be published. (Presented at Conf. on Magnetism and Magnetic Materials, Philadelphia, Dec. 1975).
4. G. S. Cargill III and E. S. Kirkpatrick, AIP Conf. Proc. to be published. (Presented at the Int. Topical Conf. on Structure and Excitations in Amorphous Solids, Williamsburgh, Va., May 1976).
5. J. F. Graczyk and P. Chaudhari, Phys. Stat. Sol.(b) 58, 163 (1973).
6. B. E. Warren, X-Ray Diffraction, Addison-Wesley, Reading, Mass., 1969.

Magnetic and Galvanomagnetic Properties of Amorphous Thin Film Gd-Fe-Ni Alloys

T. R. McGuire, R. C. Taylor and R. J. Gambino
IBM Research Center, Yorktown Heights, N. Y. 10598

ABSTRACT

Amorphous films with composition $Gd_{18}Fe_{82-x}Ni_x$ where x varies from 12 to 68 have been prepared by evaporation. By comparing Gd-Fe-Ni with binary Gd-Fe, Gd-Au and Y-Fe amorphous alloys using both magnetization and spontaneous Hall data we analyze Gd-Fe-Ni first as a two sublattice system of Gd and Fe-Ni followed by separate consideration of the Fe and Ni. Based on a small decrease associated with the Fe magnetization it is calculated that the magnetization of nickel is strongly dependent on the composition, showing a decrease in moment with increasing nickel content, diminishing to almost zero at 66% Ni. For all compositions the Hall coefficient of Ni is found to be negative. Gd also has a negative Hall coefficient while Fe is positive.

INTRODUCTION

The addition of nickel to binary amorphous alloys of Gd with Co causes the magnetization temperature dependence to be significantly altered.[1] We have now investigated Ni added to Gd-Fe to gain additional understanding of the behavior of Ni. Both magnetization and Hall effect data were used to analyze this behavior.

Binary amorphous alloys of Gd-Co studied by Gambino, et al.[2] and of Gd-Fe by Taylor[3] are found to be ferrimagnetic. There is an unbalanced antiparallel alignment of the Gd and 3d metal which results in a net magnetic moment. Detailed analysis of magnetization data for these alloys also shows that there is a large decrease of the Co moment as the Co concentration is decreased. For Fe the decrease in moment is much smaller.[4] The change in moment with concentration is attributed to charge transfer from the Gd to the 3d metal.[5]

The Hall effect in amorphous magnetic materials is known to be large, amounting to a change in Hall resistivity of several percent of the sample resistivity. Lin[6] found a 5.8% effect for amorphous Fe-C-P and Okamoto, et al.[7] a 2% change for Gd-Co. Gambino and McGuire[8] have reported results for Gd-Co-Mo indicating that the Hall effect must be described in terms of both the Gd and Co sublattices. In this paper we employ a similar approach using sublattice Hall data to sort out the behavior of the various magnetic ions in the alloy.

Pure Ni in the amorphous state has been reported by Tamara and Endo[9] to be ferromagnetic with a smaller moment than crystalline nickel. Binary Gd-Ni amorphous alloys have not been reported, but as previously noted[1], Gd-Co-Ni has been investigated by Hasegawa and Taylor[1]. They find that the temperature dependence of magnetization can be satisfactorily described by a two sublattice model in which the Co and Ni form one sublattice which interacts antiferromagnetically with the Gd sublattice.* These data were analyzed on the assumption that the moment of nickel did not change appreciably from the value of pure nickel (0.6 μ_B) while that of cobalt was reduced, similar to the decrease found for Co in the binary Gd-Co alloys. Meyer et al[10] fit their data for the Gd-Co-Ni system using the moment of pure Co and a slightly reduced moment for Ni.

EXPERIMENTAL

The amorphous Gd-Fe-Ni alloy films were prepared by co-evaporation of the elements from three electron gun sources. Film compositions were regulated by individual control of the evaporation rate of each element using quartz crystal oscillator automatic rate controllers. Deposition on ambient temperature glass substrates at the rate 10 Å/sec was made to a thickness of about 5000 Å. Electron microprobe analysis was employed to determine film composition.

The saturation magnetization was measured for each sample from 4.2 to 295°K using a force balance magnetometer. Since the film thickness was known and the sample area could be accurately determined the magnetization data are given in volume units of $4\pi M$ (gauss).

Hall effect measurements were made using the van der Pauw[11] method. This method is exceptionally convenient for films since the shape of the sample is not important as long as the thickness is uniform. In this case four spring held contacts were symetrically located on the outer periphery of a square cut film. Both the resistivity and Hall resistivity were measured. The resistivity was measured using the averaging technique described by van der Pauw and will be referred to as the average resistivity (ρ_{av}). A saturating magnetic field must, of course, be applied normal to the plane of the film to obtain the saturation value of the Hall effect.

The Hall resistivity (ρ_H) in ferromagnetic materials is composed of two parts as given in the following relation,

$$\rho_H = R_oB + R_s\,4\pi M,$$

where R_o and R_s are respectively the ordinary and spontaneous Hall coefficients. The ordinary term comes from the Lorenz force on the conduction electrons caused by the internal field B. The second term is the extraordinary or spontaneous Hall effect which comes about from preferential scattering[12,13] of the conduction electrons from the individual magnetic ions which for a ferromagnetic material is proportional to $4\pi M$. The Hall coefficients have the units Ω cm/G. In amorphous materials[6,7,8] R_s is ~100 times larger than R_o, and in this paper we have neglected the R_o term. The large value for R_s can be seen to arise from excess scattering caused by the atomic disorder of the amorphous material. This raises both the average resistivity (ρ_{av}) and the Hall resistivity of the specimen.

Because the Gd-3d metal alloys are ferrimagnetic and composed of two or more sublattices, the magnetization of each sublattice and its contributing Hall resistivity must be considered individually. We have reported previously[8] that for the system Gd-Co-Mo the Hall resistivities of the Gd and Co sublattices are additive even though the magnetic moment vectors of the sublattices point in opposite directions. This is because the sign of the Hall effect is opposite; Gd is negative and Co positive. Using this same approach we interpret the Gd-Fe-Ni data by determing values for the magnetization and Hall resistivity for the three different sublattices of Gd, Fe and Ni and show that a consistant magnetic structure emerges.

RESULTS AND DATA ANALYSIS

The magnetic measurements for the Gd-Fe-Ni system are shown in Fig. 1. Initially we will treat the data as fitting a pseudo-two sublattice system composed of Gd and Fe-Ni. For the two compositions with the greatest amount of Fe, (Fe_{70} and Fe_{66}) the Fe-Ni magnetization dominates over the measured temperature range (4.2 to 295°K). However, the two samples with lower Fe show compensation points (T_{comp}), i.e. temperatures at which the two sublattices exactly balance each other. Below T_{comp} the Gd sublattice has the dominant moment.

The magnetization - temperature curves shown in Fig. 1 are projected from high field (6 to 18 kG) measurements to zero field. When the magnetic moment at these fields is nearly saturated there is no problem

* The term "sublattice" has no crystallographic meaning but refers only to the magnetic substructure or spin system of the alloy.

associated with using the H = 0 values fo $4\pi M$. However, near compensation the moment-field plots show some slope resulting in an increased moment at the higher field. This slope is due in part to overcoming anisotropy and in part to the applied field disturbing the balanced moment of the two sublattices. For compositions 1 and 2 which are near compensation we do not use the H = 0 value of $4\pi M$ but instead the values measured at 18 kG (Table II) because the Hall resistivity is measured at high fields.

As shown in Fig. 2 the Hall resistivity is positive when the Fe-Ni sublattice is dominant. When Gd has the greater magnetization and is pointed in the direction of the applied field the Hall resistivity is negative. This behavior follows a pattern in which Fe has a positive Hall coefficient[6] and Gd a negative Hall

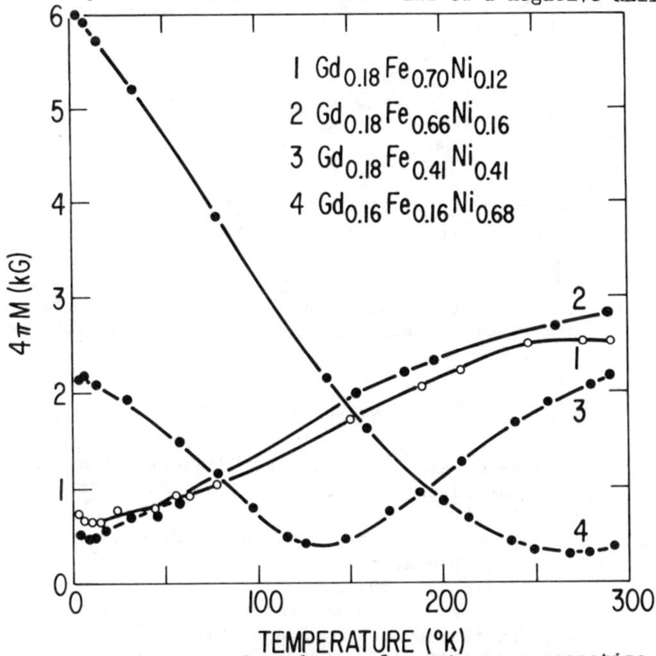

Fig. 1 Temperature dependence of spontaneous magnetization of Gd-Fe-Ni samples. Compositions 3 and 4 show compensation temperatures (T_{comp}).

TABLE I

Sample at%	ρ_H μΩcm	ρ_{av} μΩcm	ρ_H/ρ_{av} %	ρ_H/ρ_{av}* %
$Gd_{26.6}Fe_{73.3}$	-11.7	211	-5.5	-
$Gd_{21}Fe_{79}$	11.2	195	5.8	-
$Gd_{72}Au_{28}$	- 3.0	105	-2.9	-3.7
$Gd_{48}Au_{52}$	- .5	102	- .27	-4.5
YFe_2	4.2	226	1.87	6.0

* Estimated for the pure element

TABLE II

Sample at%	$4\pi M$ G	ρ_H/ρ_{av} %	$4\pi M$(Fe-Ni) G	ρ_H/ρ_{av}(Fe-Ni) %
$Gd_{18}Fe_{70}Ni_{12}$	900	4.70	16600	4.85
$Gd_{18}Fe_{66}Ni_{16}$	1300	3.25	17300	3.09
$Gd_{18}Fe_{41}Ni_{41}$	2200	-1.12	11600	.49
$Gd_{16}Fe_{16}Ni_{68}$	6000	- .60	3630	- .05

coefficient[8]. We will now attempt to assign quantitative values to Fe and Gd thus enabling us to derive the Ni sublattice behavior. In doing this we will only use 4.2°K data so that samples are under saturated conditions. The Hall resistivity, since it comes from electrons scattered by a magnetic atom, will be normalized by using the ratio ρ_H/ρ_{av} where ρ_{av} is given in Fig. 3 for these materials.

The values of ρ_H/ρ_{av} for Gd and Fe are given in Table I. For Fe we use +6% determined from both the YFe_2 projected to pure iron and from the two Gd-Fe alloys. For Gd we use -4% which comes from the Gd-Au samples and also from the Gd-Fe alloys.

It is also necessary to have magnetization data for the Gd and Fe. For Gd we choose $4\pi M$ = 24,700G corresponding to 7 μ_B per Gd. For Fe we use the value $4\pi M$ = 20,000 G, adjusting for the decrease[4,14] found for Fe as its concentration changes in relation to Gd.

Table II lists the results of analyzing the magnetic and Hall data. Columns 1 and 2 give the measured values of $4\pi M$ and ρ_H/ρ_{av} as taken from Figs. 1, 2, and 3. Columns 3 and 4 show the values calculated for the Fe-Ni sublattice as follows:

Using the weight percent (wt%) of each element we determine $4\pi M$(Fe-Ni) by

$$4\pi M = \pm(\text{Gd-wt\%})(24700) \mp (\text{Fe-wt\%})(4\pi M(\text{Fe-Ni}))$$

Using atomic percent (at%) of each element we determine ρ_H/ρ_{av}(Ni) from

$$\rho_H/\rho_{av} = \pm(\text{Gd-at\%})(-4) \mp (\text{Fe-at\%})(\rho_H/\rho_{av}(\text{Fe-Ni}))$$

The equations for $4\pi M$ and ρ_H/ρ_{av} are written with (±) signs preceding the Gd contribution where (+) corresponds to the Gd sublattice pointed parallel to the applied field and where (-) corresponds to when it is antiparallel. Similar (+ or -) correspondence also holds for the Fe-Ni sublattice with respect to the applied field.

We now consider the problem of treating Gd-Fe-Ni as a three sublattice system. This is done by using the values $4\pi M$ = 20,000 G (pure Fe) and ρ_H/ρ_{av} = +6% as discussed previously and letting them vary linearly with x for Fe_xNi_{1-x}. The results of this procedure are shown graphically in Figs. 4 and 5.

Figure 4 shows the linear plot of $4\pi M$ for Fe and the values of Fe-Ni as listed in Table II. In all cases the nickel contributes in a positive sense to the Fe-Ni sub-system, i.e., the Ni and Fe moments are parallel. In the sample with the highest nickel concentration (sample 4) the Ni moment is small.

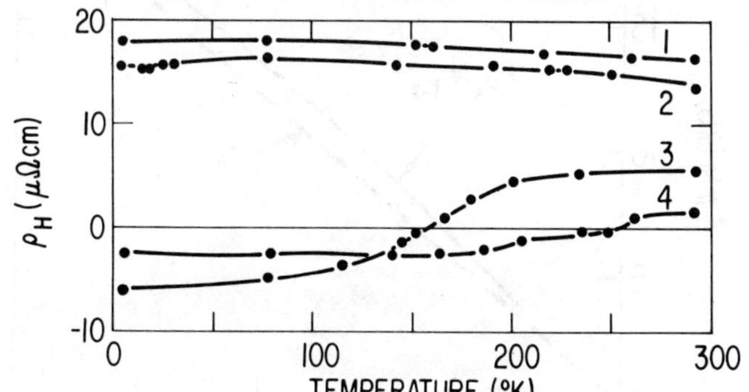

Fig. 2 Temperature dependence of the spontaneous Hall resistivity (ρ_H) of Gd-Fe-Ni samples whose compositions are shown on Fig. 1 and in Table II. A change in sign for ρ_H occurs at approximately T_{comp}. The values of ρ_H shown on the graph represent changing the saturation magnetization from + $4\pi M$ to - $4\pi M$, and it is necessary to divide by 2 to obtain values used in Table II.

Figure 5 shows the Hall resistivity ratios for Fe-Ni as taken from Table II. Also illustrated is the assumed Fe contribution which is linear with x with ρ_H/ρ_{av} = 6% as estimated for pure Fe. Subtracting the Fe contribution from the ρ_H/ρ_{av} (Fe-Ni) gives the Hall resistivity ratios for Ni which are found to be negative. The Ni has a maximum value for $|\rho_H/\rho_{av}|$ at x $\approx$ .5. We interpret this maximum as the expected increase of $|\rho_H/\rho_{av}|$ with increasing Ni concentration, followed by a decrease in $|\rho_H/\rho_{av}|$ associated with the reduction of the nickel moment at the highest Ni concentrations.

SUMMARY

We find that the sign of the Hall resistivity for Ni is negative for the four Gd-Fe-Ni compositions and over a limited range Ni has about the same magnitude of $|\rho_H/\rho_{av}|$ as Fe. We also observe that the Ni and Fe moments are parallel in the Fe-Ni sublattice. Both the magnetization and Hall data indicate, however, that the Ni moment goes to zero as the Fe concentration goes to zero. A similar result was observed by Wallace et al[15] in crystalline $DyNi_{5-x}Co_x$ alloys. The spontaneous Hall resistivities for amorphous Ni and Gd have the same negative sign as they do in the crystalline form. Similarly Fe is positive in both the amorphous and crystalline states.

We are grateful to A. P. Malozemoff for a critical discussion of the paper. We also wish to thank H. R. Lilienthal, L. J. Buszko and R. R. Ruf for their assistance in measurements and sample preparation.

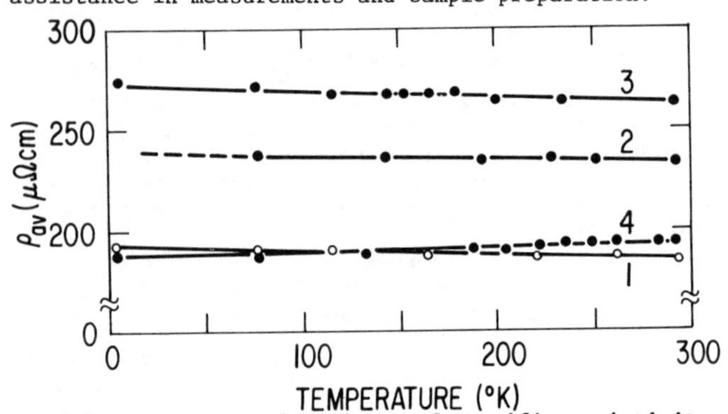

Fig. 3 Temperature dependence of specific resistivity for Gd-Fe-Ni samples.

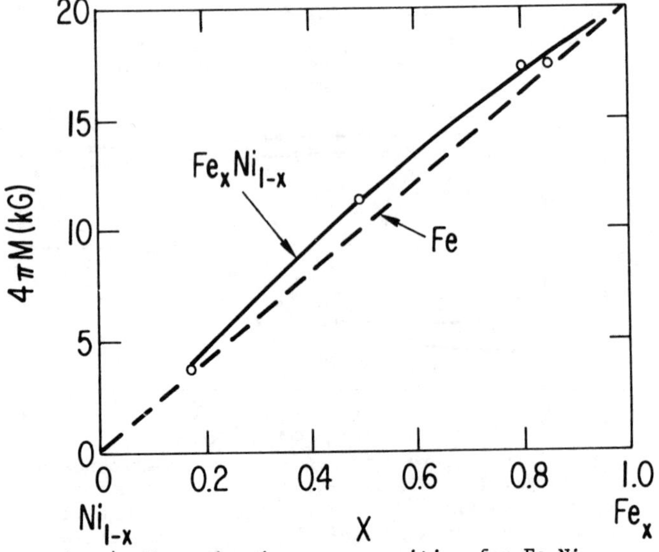

Fig. 4 Magnetization vs composition for Fe_xNi_{1-x} sublattice at 4.2°K.

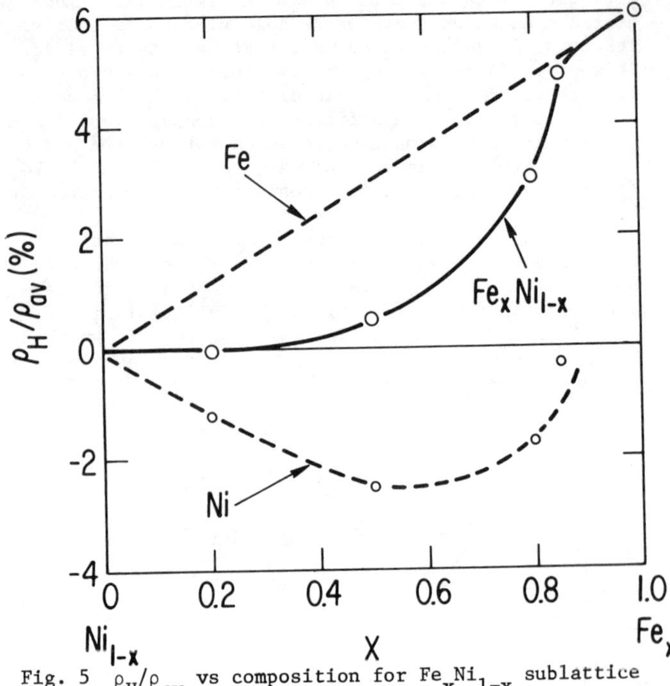

Fig. 5 ρ_H/ρ_{av} vs composition for Fe_xNi_{1-x} sublattice at 4.2°K showing breakdown into Fe and Ni contributions.

REFERENCES

1. R. Hasegawa and R. C. Taylor, "Magnetization of Amorphous Gd-Co-Ni Films", J. Appl. Phys. 46, 3606-3608 (1975).
2. R. J. Gambino, P. Chaudhari and J. J. Cuomo, "Amorphous Magnetic Materials", AIP Conf. Proc. No. 18, 578-592 (1974).
3. R. C. Taylor, "Magnetic Properties of Amorphous Gd-Fe Films Prepared by Evaporation", J. Appl. Phys. 47, 1164-1167 (1976).
4. N. Heiman, K. Lee and R. I. Potter, "Exchange Coupling in Amorphous Rare Earth-Iron Alloys", AIP Conf. Proc.
5. L-J. Tao, S. Kirkpatrick, R. J. Gambino and J. J. Cuomo, Solid State Commun. 13, 1491 (1973).
6. S. C. H. Lin, "Hall Effect in an Amorphous Ferromagnetic Alloy", J. Appl. Phys. 40, 2175-2177 (1969).
7. K. Okamoto, T. Shirahawa, S. Matsushita, and Y. Sakurai, "Hall Effect in Gd-Co Sputtered Films", IEEE Trans. MAG 10, 799-802 (1974).
8. R. J. Gambino and T. R. McGuire, "Hall Effect in Amorphous Thin Film Magnetic Alloys", 7th Int. Col. on Magnetic Films, Regensburg (1975); T. R. McGuire R. J. Gambino and R. C. Taylor, To be published.
9. K. Tamara and H. Endo, "Ferromagnetic Properties of Amorphous Nickel", Phys. Letters 29A, 52 (1969).
10. R. Meyer, H. Jouve and J. P. Rebouillat, "Effect of Ni Substitution in Gd-Co Amorphous Thin Films", IEEE Trans. on Mag., Mag-11, 1335-1337 (1975).
11. L. J. van der Pauw, "A Method of Measuring Specific Resistivity and Hall Effect of Discs of Arbitrary Shape", Philips Res. Repts. 13, 1-9 (1958).
12. J. Smit, "The Spontaneous Hall Effect in Ferromagnetics", Physica 21, 877-887 (1955); Physica 24, 39-51 (1958).
13. L. Berger, "Side Jump Mechanism for the Hall Effect in Ferromagnets", Phys. Rev. B2, 4559-4566, (1970); B5, 1862-1869, (1972).
14. A. Gangulee and R. C. Taylor, To be published.
15. "Rare Earth Intermetallics", W. E. Wallace, Academic Press, New York. 1973. page 129.

THE KERR AND THE HALL EFFECTS IN AMORPHOUS MAGNETIC FILMS

T. Shirakawa, Y. Nakajima, K. Okamoto, S. Matsushita, and Y. Sakurai
Dept. of Control Eng., Faculty of Eng. Science, Osaka Univ., Toyonaka Osaka 560 JAPAN

ABSTRACT

The temperature dependences of the extraordinary Hall effect (EHE) and the polar Kerr effect (PKE) hysteresis loops are measured in amorphous rare-earth (RE) transition metal (TM) films. The polarity of each effect corresponds to that of its TM component. The temperature dependence of Hall voltage is similar to that of the Kerr rotation angle, and, in Gd-Co-Mo films, it is similar to that of Co sublattice moment calculated by Hasegawa's two sublattice model. From these results, both EHE and PKE in RE-TM films are considered to mainly associate with the TM sublattice moment.

INTRODUCTION

The rare-earth (RE) transition metal (TM) amorphous films have attracted special interest recently not only for their potential ability for magnetic bubble and other applications,[1] but also for their anomalous behaviour of the extraordinary Hall effect (EHE) and the polar Kerr magneto-optic effect (PKE).[2~6] EHE and PKE of these films abruptly change their polarity at the magnetic compensation temperature (Tcomp). This fact is considered that the EHE and PKE mainly depend on either RE or TM sublattice moment which ferri-magnetically couples with each other.

In order to confirm this consideration and decide which sublattice is dominant for each effect, we investigate on following points in RE-TM amorphous films.

(1) Polarity change of EHE and PKE at Tcomp when the RE (or TM) element is substituted by another RE (or TM) element.

(2) Comparison between the temperature dependences of the Hall voltage, the Kerr rotation, and each sublattice moment.

EXPERIMENTAL TECHNIQUES

RE (Gd, Ho, Tb) - TM (Co, Fe, Ni) films were prepared both by rf-sputtering technique described previously[2] and by thermal evaporation using RE-TM alloy source and tungsten boat.

The Hall voltage is measured by using the sample with the electrodes for EHE and electrical resistivity measurements (See Fig.1). The electrodes are made on 18 × 8 × 0.2 mm glass substrate by successive deposi-

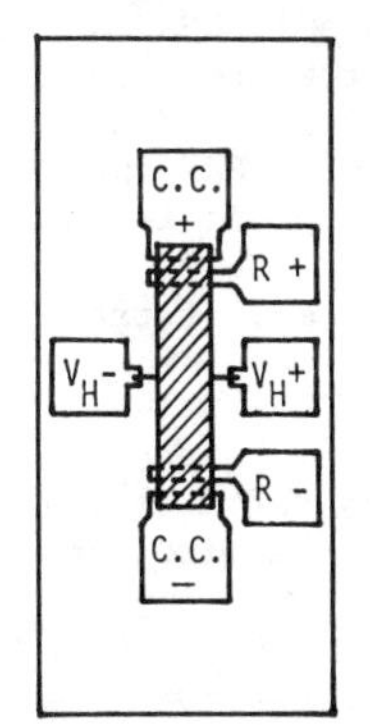

⊗ Positive direction of the applied magnetic field

C.C.± ; Control Current electrode

R ± ; Resistivity Measurement electrode

V_H ± ; Hall Voltage Measurement electrode

Fig.1. A sample with electrodes.

tion of Cr, Cu, and Au. The sample is formed by photo etching from the amorphous film deposited over the electrodes. The polarity of EHE voltage (V_H) is defined as shown in Fig.1, as the case of semiconductors.

PKE is measured in a system equipped with a polarized He-Ne laser, a beam splitter, and differentially connected two photodetectors with polarizers followed by X-Y recorder for plotting output voltage (V_K) proportional to the Kerr rotation. The polarity of V_K is so defined that the PKE coefficient of Ni film might be (-).

RESULTS AND DISCUSSIONS

(1)

EHE and PKE are measured in amorphous magnetic films with various combinations of RE and TM.

The temperature dependences of EHE and PKE hysteresis loop in a Gd-Co sputtered film are presented in Fig.2. A right-ascendent hysteresis loop indicates that the sign of the coefficient of the effect is (+). The sign of EHE coefficient changes from (-) to (+) as the temperature increases, and that of PKE changes from (+) to (-). Similar results are obtained in other RE-

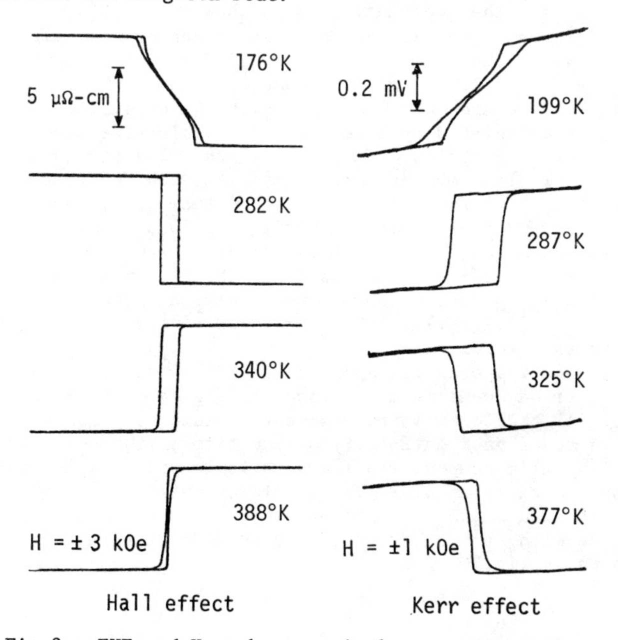

Fig.2. EHE and Kerr hysteresis loops patterns in a Gd-Co sputtered film.

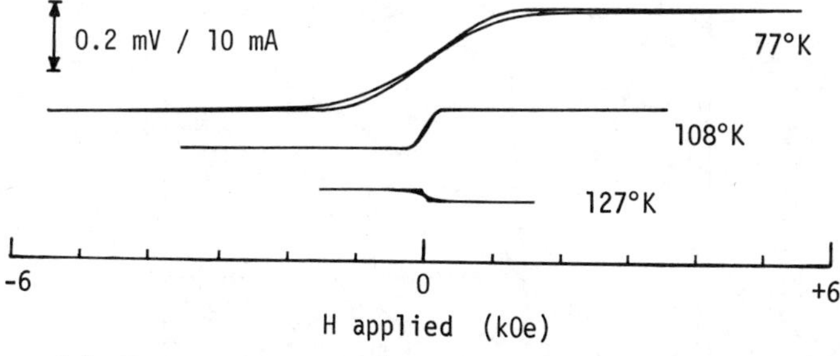

(a) Reversal of the EHE hysteresis loop in a Gd-Ni sputtered film with temperature.

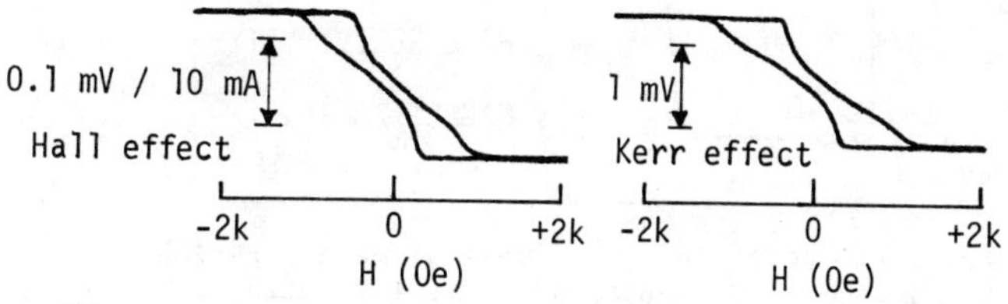

(b) Comparison of EHE and PKE hysteresis loops in a Gd-Ni evaporated film of Ni-rich composition at 291°K.

Fig.3. Hysteresis loops in Gd-Ni films.

Co and RE-Fe films. However, the EHE hysteresis loop of Gd-Ni sputtered film, as shown in Fig.3a, changes its sign from (+) to (-) with the temperature. These results in EHE are just opposite to those of RE-Co and RE-Fe films. The PKE properties is studied in a Gd-Ni evaporated film of Ni-rich composition which is equivalent to the condition above Tcomp. The sign of PKE in this sample is (-), as shown in Fig.3b, so it is considered to change from (+) to (-) with the temperature.

The fact that the polarity change in EHE is caused by the replacement of TM element, not by that of RE element, suggests the contribution of TM to EHE. And it is also interesting that the signs of EHE and PKE above Tcomp, where the TM sublattice moment is dominant and so directed along with the applied field, correspond to those of EHE and PKE coefficient of transition metals which are listed in Table 2.

Table.1. Sign variations of EHE and PKE in RE-TM films with the temperature.

COMBINATIONS	SIGN VARIATIONS	
	EHE	PKE
Gd - Co Tb - Co Ho - Co Dy - Co*	- → +	+ → -
Gd - Fe Tb - Fe		
Gd - Ni Tb - Ni**	+ → -	

Table.2. Signs of EHE and PKE coefficients of transition metals.

ELEMENTS	COEFFICIENTS	
	EHE	PKE
Co	+	-
Fe		
Ni	-	

* By Ref.7. No data of PKE.
** No data of PKE.

(2)

The temperature dependences of the absolute values of V_H and V_K are observed in some RE-TM amorphous films. The temperature dependences of the saturation EHE resistivity (ρ_H) and the saturation magnetization (M_s) in a Gd-Co sputtered film are shown in Fig.4. The magnetization changes gradually with temperature and takes a minimum value at 109°K. But, $|\rho_H|$ is nearly constant except for the neighborhood of Tcomp.

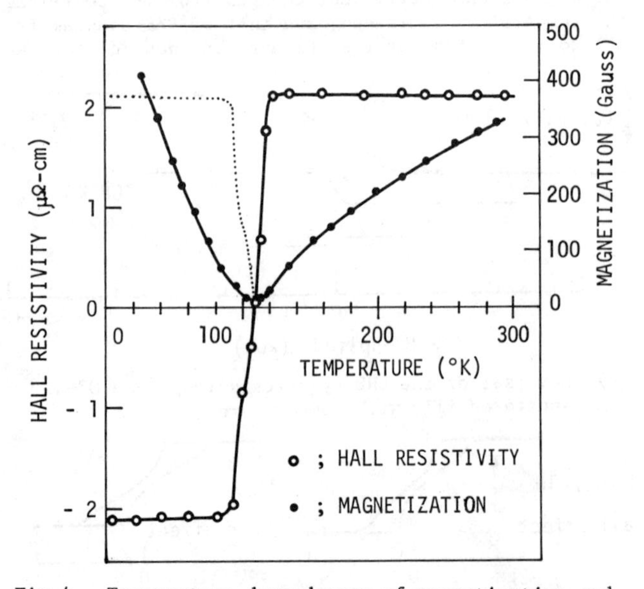

Fig.4. Temperature dependences of magnetization and ρ_H of a Gd-Co sputtered film. (17 at.% Gd)

In the case of crystalline magnetic metals such as Gd, ρ_H decreases in the low temperature. This phenomenon has relation to the reduction of electrical resistivity (ρ) in the low temperature, because ρ_H and ρ originate from the scattering of electron by disordered magnetic spins. However, in the case of amorphous films, the magnetic atoms are, also the spins are, disordered even at 0°K. So, it is expected that ρ of the Gd-Co amorphous film is constant as well as $|\rho_H|$ over a wide range of temperature.

The temperature dependences of ρ, ρ_H, and V_K in a Gd-Co film are indicated in Fig.5, and ρ is constant as

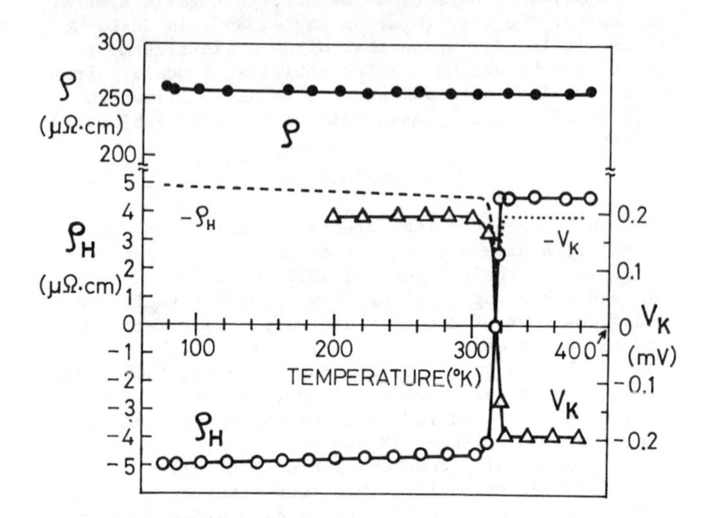

Fig.5. Temperature dependences of ρ, ρ_H and V_K in a Gd-Co sputtered film. (21.5 at.% Gd)

was expected. This constancy supports the assumption that the electron scattering by disordered magnetic spins is constant, i.e. ρ_H/M_H, M_H is the sublattice moment considered to contribute to EHE, is constant. Therefore, the constancy of $|\rho_H|$ is supposed because EHE is contributed by the sublattice moment which is nearly constant. As TM sublattice moment is considered to be nearly constant in the temperature range sufficiently lower than the Curie temperature, EHE is considered to associate with TM sublattice moment, as well as PKE for the constancy of $|V_K|$ shown in Fig.5.

In order to get the temperature dependence over the whole temperature range from 77°K to Curie temperature, the Hall voltage is measured in Gd-Co-Mo films. The results are shown in Fig.6 with the Co sublattice moment calculated by Hasegawa's two sublattice model.[8] The Hall voltage is normalized by its value at Tcomp. The sublattice moments are calculated with the fractions X, Y. The fractions X, Y in $[Gd_{1-X}(Co_{1-Y}Mo_Y)_X]$ are decided from Tcomp and Curie temperature which are obtained by Hall voltage measurement. 2.0×10^{-16} erg and -2.3×10^{-15} erg are used for the exchange constants J_{Gd-Gd} and J_{Gd-Co} respectively, and the equation $J_{Co-Co} = [5.1(1-Y)X/(1-XY)-2.8] \times 10^{-14}$ erg is directly used to determine J_{Co-Co}.

The temperature dependences of $|V_H|$ and $|V_K|$ in another Gd-Co-Mo film are shown in Fig.7a, and they do not fit the Co sublattice moment. This is probably because of poor uniformity of the film or the deviation of the film composition from available range. According to Argyle et al.[5] $|V_K|$ should agree with the Co sublattice moment. As shown in Fig.7a, $|V_H|$ and $|V_K|$ agree with each other, and also in a Tb-Fe film as shown in Fig.7b.

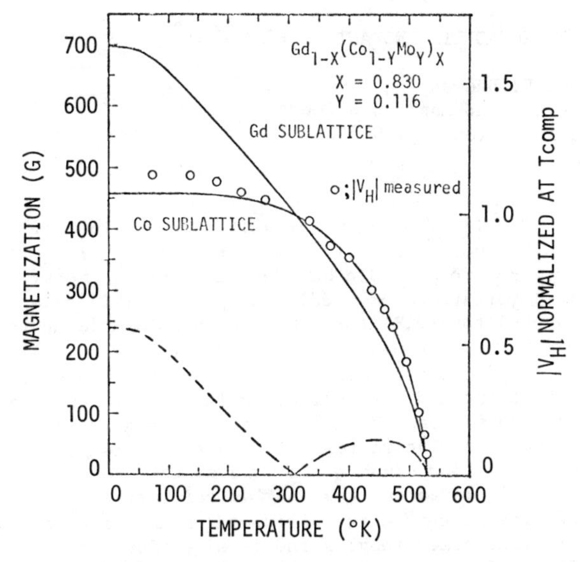

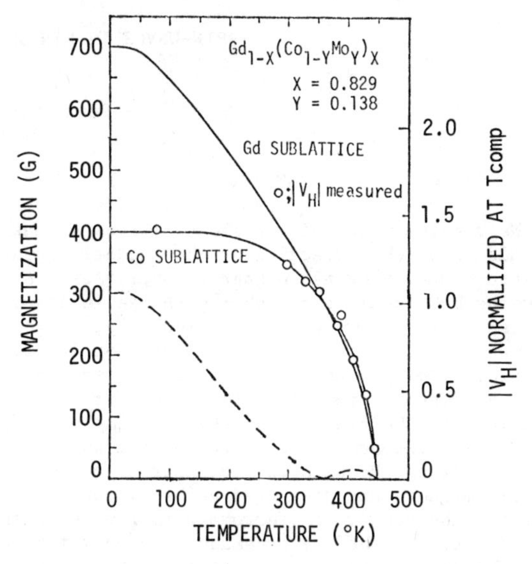

Fig.6. Comparison between V_H and calculated sublattice moments in Gd-Co-Mo sputtered films.

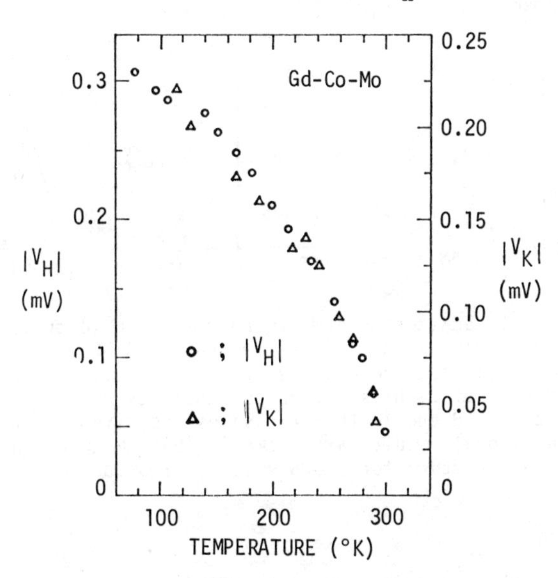

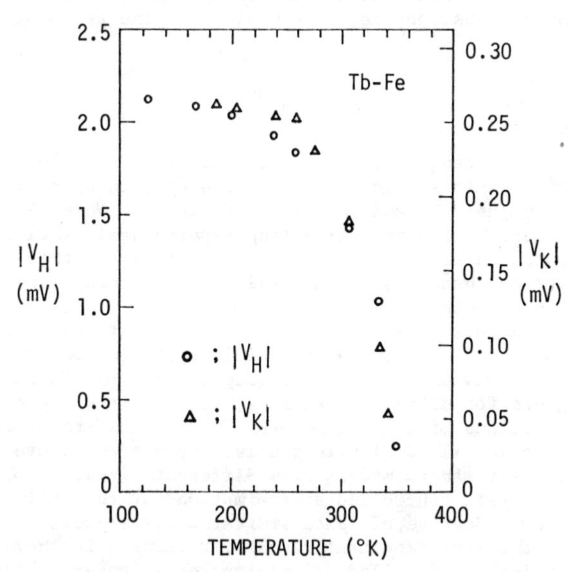

Fig.7. Comparison between V_K and V_H in a Gd-Co-Mo and a Tb-Fe films.

CONCLUSION

The temperature dependences of EHE and PKE hysteresis loops are measured in RE-TM films. It is observed that the polarity of EHE in RE-Co and RE-Fe films are different from that in RE-Ni film, and that above Tcomp the polarities of EHE and PKE in RE-TM films correspond to those in the films of their TM component. The temperature dependences of $|\rho_H|$ and $|V_K|$ in Gd-Co and Gd-Co-Mo films correspond to those of TM sublattice moment. Though $|\rho_H|$ increases with Gd fraction, both EHE and PKE are considered to associate with TM sublattice moment from the characteristics of polarity and temperature dependence.

REFERENCES

1) See, for examples ; P. Chaudhari, J. J. Cuomo, and R. J. Gambino, IBM J. Rec. Dev., 17, 66, 1973 ; P. Chaudhari, J. J. Cuomo, and R. J. Gambino, Appl. Phys. Lett., 22, 337, 1973 ; S. Matsushita, K. Sunago, and Y. Sakurai, Japan J. Appl. Phys., 15, 717, 1976.
2) T. Shirakawa, K. Okamoto, K. Onishi, S. Matsushita, and Y. Sakurai, IEEE Trans. on Mag., MAG-10, No.3, 795, 1974.
3) K. Okamoto, T. Shirakawa, S. Matsushita, and Y. Sakurai, IEEE Trans. on Mag., MAG-10, No.3, 799, 1974.
4) K. Okamoto, T. Shirakawa, S. Matsushita, and Y. Sakurai, AIP. Conf. Proc., 24, 113, 1974.
5) B. E. Argyle, R. J. Gambino, and K. Y. Ahn, AIP. Conf. Proc., 24, 564, 1974.
6) A. Ogawa, T. Katayama, M. Hirano, and T. Tsushima, AIP. Conf. Proc., 24, 575, 1974.
7) A. Ogawa, T. Katayama, M. Hirano, and T. Tsushima, Proc. 7th Conf. Solid State Devices, Tokyo, 1975 (Japan J. Appl. Phys.) 15 suppl., 87, 1976.
8) R. Hasegawa, B. E. Argyle, and L-J. Tao, AIP Conf. Proc., 24, 110, 1974.

SPIN-WAVES IN RANDOMLY DILUTED FERROMAGNETS AND ANTIFERROMAGNETS*

R. Alben and D. Beeman†
Department of Engineering and Applied Science
Yale University, New Haven, CT. 06520

ABSTRACT

We consider the spin-wave excitations of magnetic crystals in which a fraction of the magnetic spins are replaced randomly by non magnetic impurities. A simple Heisenberg nearest neighbor exchange is assumed. The treatment follows the usual linear spin wave approximation and so is only appropriate for low temperatures. However the disorder is treated exactly by use of a computer simulation technique for finite, but large models. We present results for simple, body centered and face centered cubic as well as the simple square lattice. Results for both ferromagnets and antiferromagnets are presented and show contrasting behaviors. The spectra of ferromagnets remain essentially continuous before breaking into many fine peaks below the percolation concentration. The spectra of the antiferromagnets develop large peaks, with some complex substructure, even well above the percolation limit.

In recent years there has been considerable theoretical interest in the excitation spectra of diluted ferromagnets[1-3] and antiferromagnets[4-6]. Very recently there have been stimulating experimental results on diluted antiferromagnets[7-9] and hopefully further results on these systems as well as on ferromagnets will be forthcoming.

In this paper we present a variety of computer simulation [5,6] results which illustrate the behavior of spin waves in dilute ferromagnets and antiferromagnets for different lattices, wave vectors and concentrations of magnetic spins. The results are exact for large (∿10,000) site models. By averaging over different random models with different sizes, we obtain an accurate controlled approximation for the infinite system. Because of space limitations, we present here just the most important results, reserving further examples and detailed discussion for a longer publication.

We consider the Heisenberg hamiltonian for a dilute system:

$$H = \pm |J| \sum_{i,j} p_i p_j S_i S_J , \tag{1}$$

where J is the constant exchange interaction, the sum is over pairs of neighboring sites i and j and p_i is 0 or 1 depending on whether the site i is occupied by atom of spin S or a non magnetic atom. We make the usual linear spin wave approximations for ferromagnets[1] (-) and antiferromagnets[5,6] (+). The results are thus expected to apply best for large S and for very low temperature, in which case even the excitations of isolated clusters should be well described.

The densities of states and scattering law S(Q,E) for finite random models are computed by use of the equation of motion method.[6,10] This yields the exact spectra convoluted with gaussian broadening function whose width can be chosen. Except for Fig. 10, where a narrower resolution was used, we use a resolution of 0.04 units of "normalized" energy, E/JSz, with z the number of neighbors.

In Fig. 1 we show a result illustrating the way we distinguish true structure from finite size artifacts due to periodic boundary conditions and statistical variations of model properties. This case corresponds to S(Q,E) for a half diluted simple cubic ferromagnet for a zone edge Q-vector. For the high energy part of the spectrum, there are few contributing states and statistical variations are considerable. More cases than the three shown would have to be averaged to discern real structure in this part of the spectrum. For E/6JS ≲ 1.1 however the density of states is large and the statistical variation is less. The fine structure is reproduced in the independent different sized samples and is a genuine, if subtle, effect of cluster statistics valid for the infinite system.

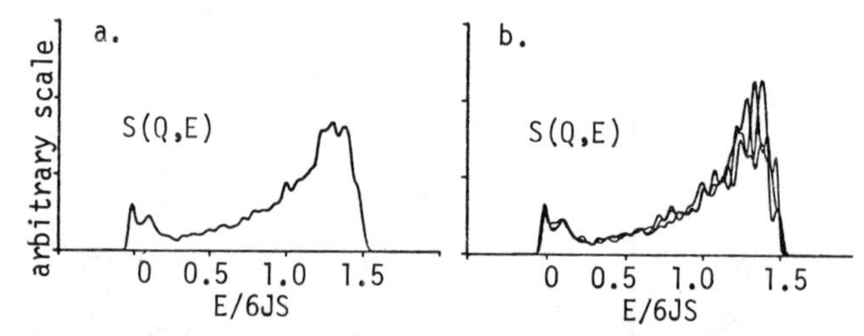

Fig. 1 Scattering law S(Q,E) for a simple cubic ferromagnet with a concentration C = 0.50 of magnetic spins. The wave vector Q = (1,1,1)π/a. The average of results for three models of different sizes (6,8 and 10 thousand sites) is given in (a). Individual results are shown in (b). Much of the fine structure for E/6JS < 1.1 is reproducible.

In Fig. 2 we show the S(Q,E) result of Fig. 1 compared with three analytical approximations. Since the spectrum is rather smooth, the exact moment result[2] is bound to fit well. The agreement for the effective medium result of Ref. 3 also is quite good and indicates that this theory is better for this case than that of Ref. 1.

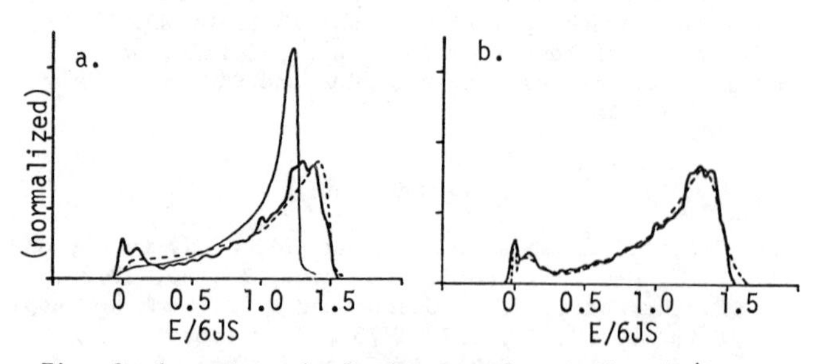

Fig. 2 Computer simulation result of fig. 1 (heavy line) compared to analytical approximations. (a) light line is CPA result of Harris et al[1], broken line is the effective medium theory of Theumann and Tahir-Kheli.[3] (b) broken line is the theory of Nickel[2] based on exact moments.

In Fig. 3 we show the density of states for simple cubic and body centered cubic ferromagnets. In both cases there is a striking change in the character of the spectra above, and below the percolation thresholds (about 0.3 for these lattices).

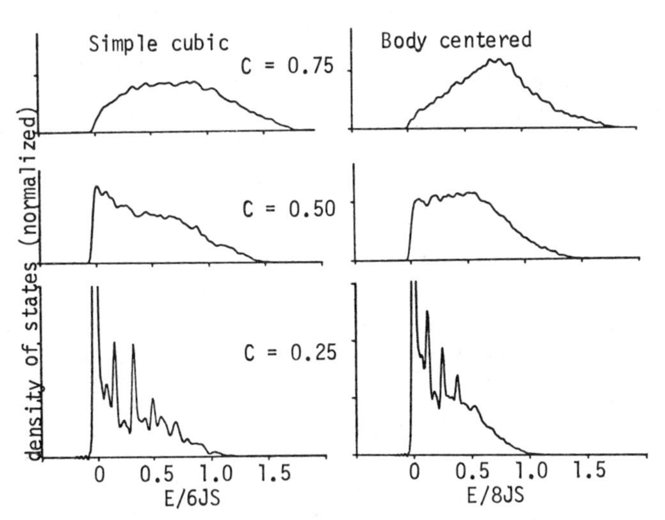

Fig. 3 Density of states for simple and body centered cubic ferromagnets with varying concentrations C of magnetic spins.

In Fig. 4 we show the density of states for the pure face centered cubic ferromagnet compared with

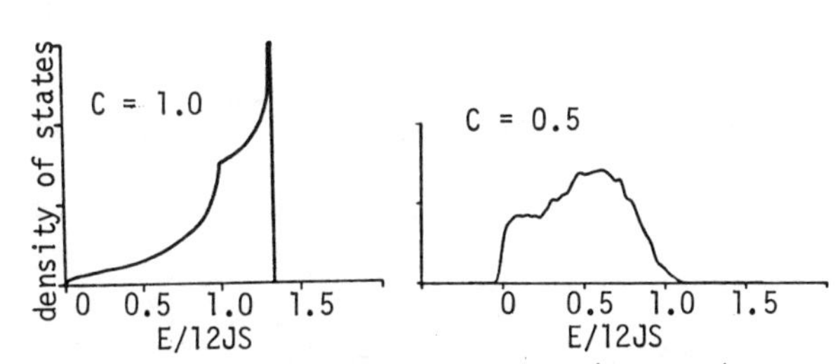

Fig. 4 Density of states for pure (C = 1.0) and half diluted (C = 0.5) face centered cubic ferromagnet.

the result for C = 0.5. In Fig. 5 we show the behavior of S(Q,E) with varying Q for the simple cubic lattice with C = 0.25, a concentration below the percolation threshold. Despite the fact that the spectrum is broken up into peaks associated with isolated clusters, there is a "hump" of intensity which moves toward higher energy as Q approaches the zone boundary and which might be interpreted as a "spin wave dispersion" although such an interpretation would not be very useful.

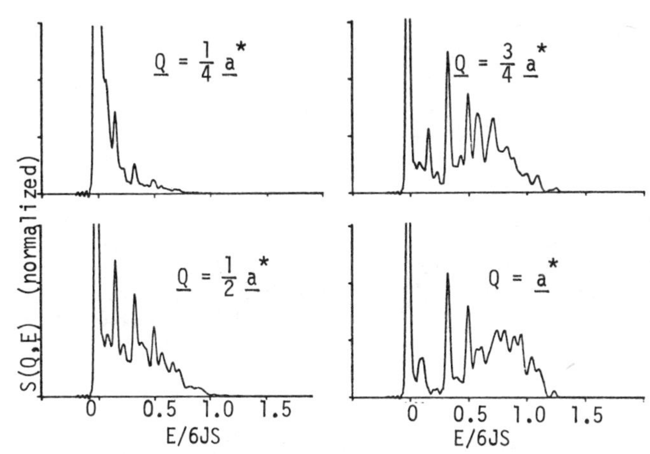

Fig. 5 S(Q,E) for a simple cubic ferromagnet with C = 0.25 for Q varying along the [111] direction ($\underline{a}^* = [1,1,1]\pi/a$).

We have also examined the two dimensional square ferromagnet[5] which should describe the layers in the system $K_2Cu_xZn_{1-x}F_4$. A result for the density of states is shown in Fig. 6. The behavior here is quite similar to that for the three dimensional lattices. Here the percolation threshold is C = 0.6. The cluster structure which appears for C = 0.5 is spaced more widely than for the higher coordinated lattices described in Fig. 3.

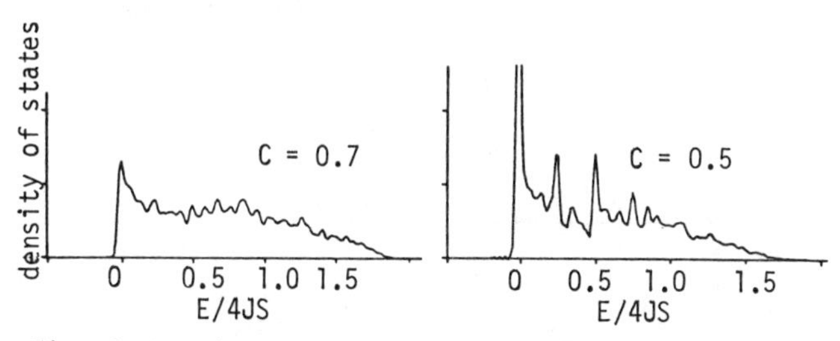

Fig. 6 Density of states for the simple square 2-dimensional ferromagnet for concentrations C above and below the percolation concentration of 0.6.

Most experimentally accessible insulating magnetic alloys are antiferromagnets[7] and some of these systems have already been studied with computer simulation techniques.[5,6] We will briefly present results which describe the basic lattices.

In Fig. 7 we show a result for S(Q,E) for a body centered cubic antiferromagnet with C = 0.5 compared with the CPA theory of Buyers et al.[4] We note, as in Ref. 5, the multipeaked structure which is qualitatively reproduced by this CPA approach. The CPA theory of Ref. 4 is based on accurately treating the diagonal disorder associated with the dilution problem. The off diagonal disorder is less satisfactorily treated. For the antiferromagnet it appears that the structure associated with the diagonal disorder is indeed the dominant effect. Note however that the number of cluster related peaks at low frequency is not given by the simple "Ising" cluster argument.

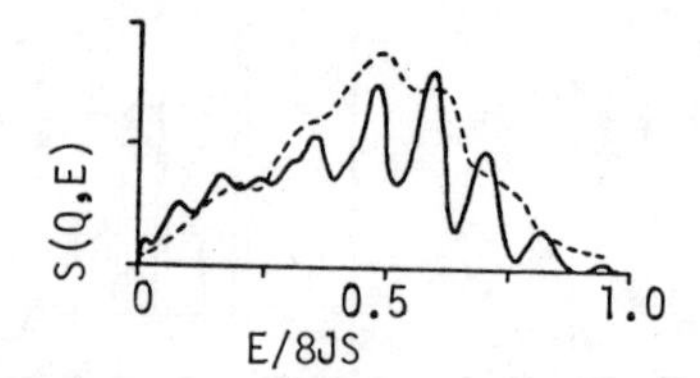

Fig. 7 S(Q,E) for Q = [1,0,0]π/a for body centered cubic antiferromagnet with C = 0.5. Solid line gives computer simulation results. All visible structure is reproducible. Broken line is the CPA result of Buyers et al[4] for the analogous rutile structure system.

In Fig. 8 we show the behavior of the simple cubic and body centered cubic[5] antiferromagnets for varying concentration. The behavior is very different from that of the ferromagnet in that cluster related peaks appear at even moderate dilution. The interpretation of this lies in the nature of the spin dynamics of nearest neighbor antiferromagnets. Except for E = 0, the propagation of modes is hindered since the precessional motion "natural" to one sublattice is not "natural" for the other. Thus a relatively small amount of disorder is sufficient to break the spectrum up into rather strongly localized modes.

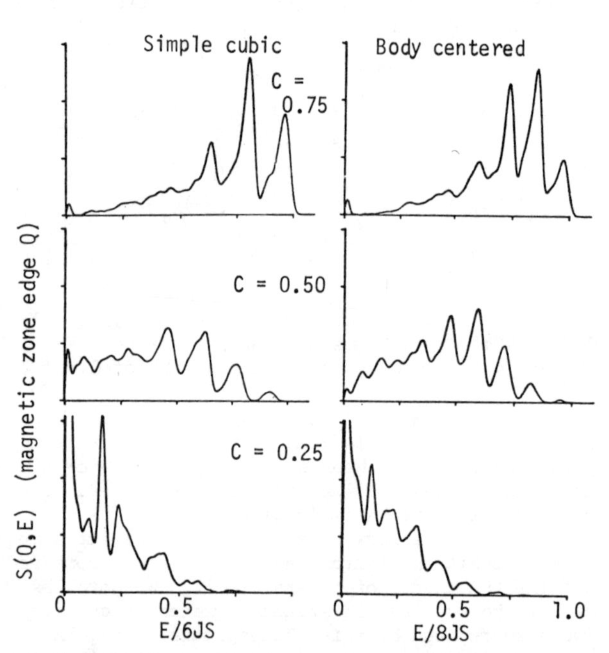

Fig. 8 S(Q,E) for simple and body centered cubic antiferromagnetics with varying concentration C of magneticspins. Q corresponds to [1,1,1]π/a for simple cubic, [100]π/a for body centered cubic.

In Fig. 9 we show S(Q,E) for two different values of Q. This illustrates that the peaks reflect the underlying density of states. They do not move but rather are weighted differently with changing Q. In the last figure (Fig. 10) we show a result for the square antiferromagnet. The resolution here is 0.02 units. This is sufficient to reveal the presence of a fine structure on the basis "Ising" cluster peaks. Indeed it is apparent that there are endless levels of peaks on subpeaks, which are an integral part of the behavior of all these model disordered magnets.

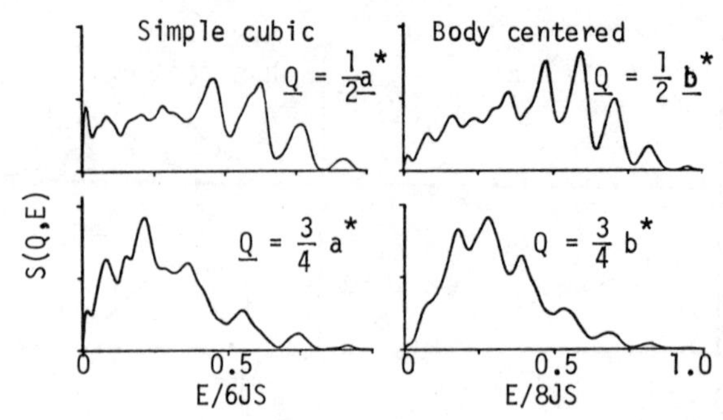

Fig. 9 S(Q,E) for different Q's for half diluted (C = 0.5) antiferromagnets. Note that the peaks do not shift appreciably with Q; rather the overall spectrum is weighted toward lower E as Q approaches a magnetic Bragg point.

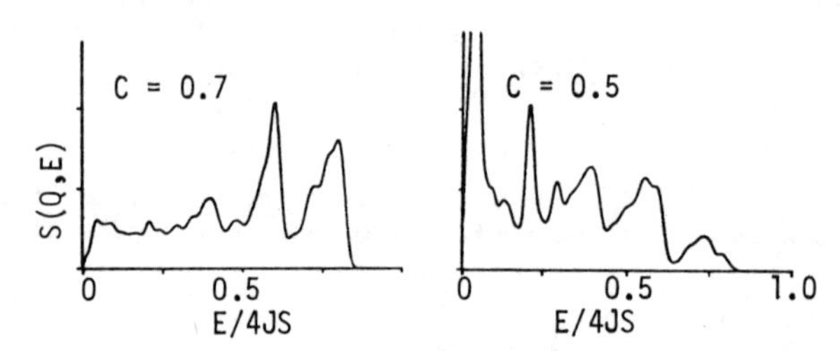

Fig. 10 S(Q,E) for magnetic zone edge Q for two dimensional square antiferromagnet.

REFERENCES

* Work supported by the NSF.

† Permanent address: Physics Department, Harvey Mudd College, Claremont, California.

1. A. B. Harris, P. L. Leath, B. G. Nickel and R. J. Elliott, J. Phys. C7, 1693 (1974).
2. B. G. Nickel, J. Phys. C7, 1719 (1974).
3. A. Theumann and R. A. Tahir Kheli, Phys. Rev. B12, 1796 (1975).
4. W. J. L. Buyers, D. E. Pepper and R. J. Elliott, J. Phys. C6, 1933 (1973).
5. W. K. Holcomb and A. B. Harris, AIP Conf. Proc. 24, 102 (1975); D. L. Huber, Solid State Commun. 14, 1153 (1974); Phys. Rev. B10, 4621 (1974); S. Kirkpatrick and A. B. Harris, Phys. Rev. B12, 4980 (1975).
6. R. Alben and M. F. Thorpe, J. Phys. C8, L275 (1975).
7. R. A. Cowley and W. J. L. Buyers, Rev. Mod. Phys. 44, 406 (1972).
8. O. W. Dietrich, G. Meyer, R. A. Cowley and G. Shirane, Phys. Rev. Lett. 35, 1735 (1975).
9. R. A. Cowley, R. J. Birgeneau, H. J. Guggenheim, and G. Shirane (unpublished).
10. R. Alben, M. Blume, H. Krakauer and L. Schwartz, Phys. Rev. B12, 4090 (1975).

EVIDENCE FOR SPLIT BANDS FROM ELECTRONIC SPECIFIC HEAT AND ELECTRICAL TRANSPORT DATA IN Fe-V AND IN OTHER IRON AND NICKEL ALLOYS†

L. Berger
Carnegie-Mellon University, Pittsburgh, Pennsylvania 15213

ABSTRACT

In concentrated alloys where the nuclear charges or exchange fields of the components differ sufficiently, the coherent potential approximation predicts that each alloy component has its own 3d band, having minimal overlap with the 3d bands of other components. As we pointed out earlier, this "split-band model" applies well to the spin-down bands of f.c.c. Ni-Fe, Ni-Fe-Cu, etc., and explains many of their properties. For example, several physical quantities exhibit anomalies when the Fermi level ε_F crosses the boundary T between nickel and iron bands. A systematic search for b.c.c. iron-base alloy series where the spin-up band might be split, and where ε_F might cross a boundary T between bands, yields Fe-V. As expected, the electronic specific heat coefficient γ of Fe-V has a minimum when ε_F is at T, around 15 at % V. The electrical resistivity ρ, the slope $\rho^{-1}\, d\rho/dH$, and the anomalous Hall coefficient R_s have sharp anomalies close by. The magnetoresistance $\Delta\rho/\rho_o$ is unusually large in Fe-V. A scattering resonance in the solvent band close to T leads to large $\alpha = \rho_\uparrow/\rho_\downarrow$, hence to these large $\Delta\rho/\rho_o$. The observed decrease of $\Delta\rho/\rho_o$ when ε_F crosses T into the vanadium band reflects a decrease of value of the matrix elements of spin-orbit interaction. Similar series such as Fe-Cr, Fe-Mo, Fe-W, Fe-Ti, Fe-Nb, are also discussed.

INTRODUCTION

As we showed a few years ago[1], there are many experimental data pointing to the validity of a very simple model for the band structure of f.c.c. Ni-Fe, Ni-Fe-Cu, Ni-Fe-Cr, Ni-Fe-Mo, Ni-Fe-W, Ni-Fe-V, Ni-Fe-Cu-Mo, etc. In this "split-band model", each alloy component has its own 3d band, distinct on the energy scale from the 3d bands of other components. By counting 3d states and counting electrons, it is easy to find the alloy compositions at which the Fermi level ε_F is located at the boundary (called T) between the nickel band and the iron band. When ε_F crosses the point T, it is found experimentally that the electronic specific heat coefficient γ goes through a minimum, and that the linear magnetostriction λ_s and the anomalous Hall coefficient R_s change sign. Also, the ferromagnetic anisotropy of resistance $\Delta\rho/\rho_o$ (often called magnetoresistance) has a large maximum close by, in the case of Ni-Fe and Ni-Fe-Cu.

There is also a good theoretical basis for the validity of this model. In the "split-band limit" where the nuclear electrostatic energies ε_A, ε_B of electrons sitting on atoms of two components A, B differ by more than one band width W, the coherent potential approximation predicts[2] the bands of the alloy to split indeed in the manner indicated above. Moreover, in the magnetic alloys mentioned above, elastic neutron diffraction data show the local magnetic moment on Fe atoms to be much larger than the moment on Ni. This causes a considerable difference of intratomic (Hund) exchange field, of the right sign to help in splitting the spin-down band. Therefore, the general condition for the validity of our split-band model is:

$$\left|(\varepsilon_A-\varepsilon_B) \pm J_{dd}(\mu_A-\mu_B)/4\mu_b\right| \geq W \tag{1}$$

where J_{dd} is the Slater intratomic exchange integral, μ_A and μ_B are the local moments, and μ_b is the Bohr magneton. The ± sign should be + for the splitting of the spin-down band, and - in the case of the spin-up band.

Note that the validity of our split-band model is now confirmed by detailed coherent potential approximation calculations[3] for Ni-Fe, which show this splitting very clearly in the spin-down band!

Recently, we have carried out[4] numerical calculations of γ, of the g factor (or orbital angular momentum), and of the anomalous Hall effect of Ni-Fe, on the basis of the split-band model. We obtained good agreement with existing data. For example, it is found both theoretically and experimentally that the total orbital angular momentum $\sum L_z$ per atom at 0 K has a minimum at the composition where ε_F crosses T.

The purpose of the present paper is to show that split-band effects are also observable in certain concentrated b.c.c. iron-rich alloys.

THE SPLIT-BAND MODEL FOR Fe-V and Fe-Cr

Many features of the band structure of iron and of concentrated b.c.c. iron-rich Fe-Co, Fe-Cr, Fe-V, Fe-Ni are already understood.[5,6] For electron concentrations between 6.0 and 8.3 e.a., the Fermi level of spin-down electrons remains fixed in a region of small density of states in the middle of the 3d band. The spin-up Fermi level moves across the upper part of the 3d band (Fig. 1a). The behavior of the saturation magnetization M_s as well as some features of the

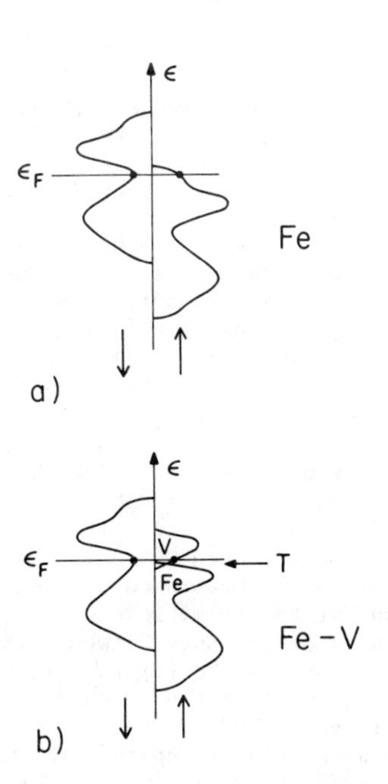

Fig. 1. a) Band structure of b.c.c. iron. b) Split-band model for b.c.c. Fe-V alloys.

behavior of γ are compatible[6] with the rigid band model in Fe-Co and Fe-Cr.

However, when going to Fe-Cr, and then to Fe-V, the difference of nuclear charges increases and reaches Z = -3, so that $\varepsilon_V-\varepsilon_{Fe}$ is positive and large. Moreover, neutron diffraction data indicate[7] $\mu_V \simeq -0.4\mu_b$, $\mu_{Fe} \simeq +2.2\mu_b$, so that $-J_{dd}(\mu_V-\mu_{Fe})$ is also positive and large. By Eq.(1), this may lead to a splitting of a vanadium band off the top of the spin-up iron band (Fig. 1b). Calculations by Gomes and Campbell[8], though restricted to dilute alloys, reached a similar conclusion. Moreover, a detailed calculation by the coherent potential approximation has been performed[9] for concentrated Fe-Cr; although the authors mention only a "deformation" of the spin-up band, our examination of their Fig. 15 suggests that a considerable degree of splitting into Cr and Fe bands is already taking place in this Z = -2 case.

Existing data[10] for M_s in Fe-V and Fe-Cr follow the Slater-Pauling curve. This shows that the V band (or Cr band) is magnetized as completely as possible (Fig. 1b).

In the split-band model, the number of 3d states in the spin-up vanadium band is 5 c_V, where c_V is the atomic concentration. On the other hand, the number of holes in the whole spin-up 3d band of the alloy is[6] $0.3-Z\ c_V$, where Z = -3. Therefore ε_F will be at the boundary T between Fe and V bands if

$$5\ c_V = 0.3 - Z\ c_V \qquad (2)$$

This gives $c_V = 0.15$.

Among binary iron-rich alloys, only for Fe-V, Fe-Cr, Fe-Ti, Fe-Mo, Fe-W and Fe-Nb series does it seem possible to find a simultaneous solution of Eqs. (1) and (2). For Fe-Cr (Z = -2), Eq.(2) would give $c_{Cr} = 0.10$; also, neutron diffraction data[7] indicate $\mu_{Cr} \simeq -0.7\mu_b$ and $\mu_{Fe} \simeq +2.2\mu_b$, favorable values for the existence of a solution of Eq.(1). The bands are probably not so completely split as in Fe-V, since Z = -2 only.

OBSERVATION OF SPLIT-BAND EFFECTS IN Fe-V AND Fe-Cr

In iron-rich alloys, neither the spin-up nor the spin-down band are completely full or completely empty. Under these conditions, a change of sign of the anomalous Hall effect or of the magnetostriction is not expected when ε_F crosses T. The total orbital angular momentum is not expected to exhibit any simple behavior there either.

The electronic specific heat coefficient γ is still expected to have a minimum when ε_F crosses T, in the quasi-gap between vanadium and iron bands. The existence of a minimum in that region is confirmed by existing data[11] for γ (Fig. 2). But not enough data points exist to fix its exact location.

Relatively sharp anomalies of the electrical resistivity[12] ρ, of the anomalous Hall coefficient[12] R_s, and of the field slope[13] $\rho^{-1}d\rho/dH$ above saturation are observed at 11 at % of vanadium (Fig. 2). These are probably also associated with the passage of ε_F through T at that concentration. Transport properties play the role of a spectroscope, seeing selectively what happens at the Fermi level.

The small discrepancy between 11% and the predicted 15% shows that Eq.(2) is able to predict the position of point T in the 3d band of the alloy with an error of only ≈0.05 eV.

It is also interesting that the Curie point T_c has a maximum[12] at 10 or 15 at % of vanadium, but the interpretation of that fact is not simple.

Physical properties of the Fe-Cr alloys are qualitatively similar to those of Fe-V, even though the band splitting is not so complete. A similar statement can be made[1] when we compare f.c.c. Ni-Co to Ni-Fe.

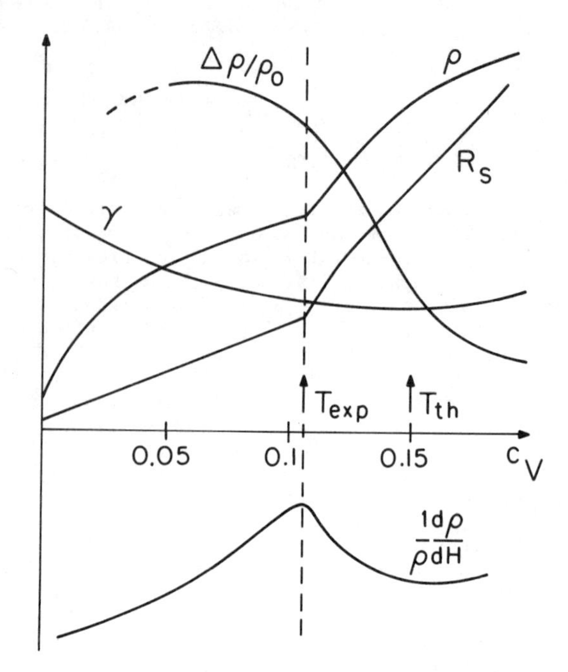

Fig. 2. Observation of anomalies when ε_F crosses the boundary T between iron and vanadium bands. T_{th} indicates the predicted location of the anomaly, and T_{exp} the experimentally observed location. The scale on the vertical axis is arbitrary. The temperature is 4 K for γ and $\Delta\rho/\rho_o$, and 300 K for other quantities.

EXPLANATION OF LARGE MAGNETORESISTANCE IN Fe-V AND Ni-Fe

In f.c.c. Ni-Fe and Ni-Fe-Cu, the ferromagnetic anisotropy of resistivity $\Delta\rho/\rho_o$ reaches large values, of about +0.2 at 4 K. This maximum seems to happen when ε_F is in the nickel band close[1] to T. As soon as ε_F moves into the iron band, $\Delta\rho/\rho_o$ drops to much smaller values.

More generally, among f.c.c. nickel-rich alloys, $\Delta\rho/\rho_o$ seems to be large in series where ε_F comes close to T(Ni-Fe, Ni-Fe-Cu, Ni-Co), but remains small in series where ε_F is far away from any such boundary between split bands (Ni-Cu, Ni-Sn).

It is interesting that the values $\Delta\rho/\rho_o \leq +0.1$ observed[13] in the split-band Fe-V system at low temperatures are considerably higher than values found so far in those b.c.c. alloys (Fe-Co, b.c.c. Fe-Ni, Fe-Si) where ε_F is far away from any boundary T. Moreover, as in Ni-Fe, $\Delta\rho/\rho_o$ seems to drop to much smaller values as soon as ε_F moves from the solvent band (band of majority component, i.e. here iron) to the solute band (band of vanadium). This happens for $c_V>0.15$, as shown on Fig. 2.

Campbell, Fert and Jaoul have shown[14] that large $\Delta\rho/\rho_o$ values in dilute f.c.c. nickel alloys are associated with large values of $\alpha=\rho_\downarrow/\rho_\uparrow$, where $\rho_\downarrow$ and $\rho_\uparrow$ are the resistivities for spin-down and spin-up electrical conduction respectively; for b.c.c. iron alloys, where it is the spin-up electrons which have the highest mobility[15], this should be replaced by $\alpha=\rho_\uparrow/\rho_\downarrow$. And existing data[16,15] of deviations from Mathiessen's rule suggest[15] indeed that α is larger in Fe-V, Fe-Mn, Fe-Mo, Fe-Re, and Fe-Cr than in Fe-Co and in other iron-based series.

Since the spin-down Fermi level is fixed[6] in a gap of the 3d band, $\rho_\downarrow$ is probably rather constant through a series of alloys. Then a large α would require a large value of $\rho_\uparrow$.

Gomes [8] has developed a tight-binding formalism for scattering in dilute iron alloys. His calculated α values for Fe-Mn, Fe-Cr, Fe-V are much larger than for Fe-Co, in agreement with the data mentioned above. Note that his value $\alpha = 4$ for Fe-V seems too small to explain the observed $\Delta\rho/\rho_0 \simeq 0.1$, however.

From his Eq.(13b), one can show that the resistivity may be written in the form:

$$\rho = \frac{2\pi m c_V}{e^2 \hbar N_a} \frac{5|V_{sd}|^2}{V_{dd}} \operatorname{Im}(1 - V_{dd} G_d/5)^{-1} \quad (3)$$

where we have kept only the contribution of s-d scattering in the numerator, and neglected s-band effects in the denominator. Also, V_{sd} and V_{dd} are matrix elements of the vanadium impurity scattering potential, and G_d is the iron d-band Green's function. And N_a is the number of atoms per unit volume. Applying Eq.(3) to spin-up electrons, we calculate the function G_d for a simple shape of band. We assume V_{dd} to be just strong enough to split a vanadium bound state off the top of the spin-up 3d band of iron (Fig. 3). The dependence of $\rho_\uparrow$ on ϵ_F according to Eq.(3) is shown on Fig. 3. The scattering resonance responsible for the large α and $\Delta\rho/\rho_0$ values observed at $c_V < 0.15$ is inside the solvent band (iron band). Fig. 3 shows that it is located close to the band top (point T).[17]

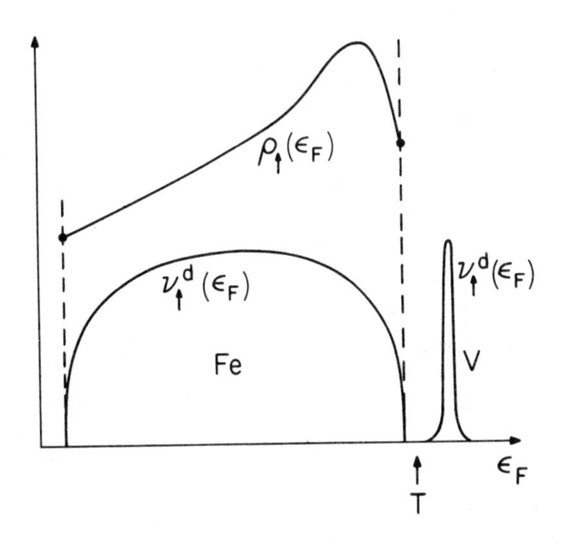

Fig. 3. $\nu_\uparrow^d(\epsilon_F)$ is the spin-up 3d density of states of a dilute Fe-V alloy, with a vanadium bound state. $\rho_\uparrow(\epsilon_F)$ is the spin-up resistivity according to the formalism of Gomes. It exhibits a scattering resonance.

When ϵ_F goes into the solute (vanadium) band for $c_V > 0.15$, α is probably also large. However, the proportionality between α and $\Delta\rho/\rho_0$ fails, and $\Delta\rho/\rho_0$ becomes smaller. This is because the spin-up solute-band states are much more localized on vanadium atoms than most spin-down states. This means a smaller degree of overlap between spin-up and spin-down states, which leads to a decrease in the value of matrix elements of the spin-orbit interaction $L_xS_x + L_yS_y$ between these states, and consequently to a decrease of $\Delta\rho/\rho_0$. In the extreme case of completely localized vanadium states, the new value of $\Delta\rho/\rho_0$ obtained after ϵ_F crosses into the vanadium band is predicted to be in a ratio of $c_V/c_{Fe} \simeq 0.18$ to the old value; we use the Smit theory of $\Delta\rho/\rho_0$ as in Ref. 14. This explains the experimental variation of $\Delta\rho/\rho_0$ in Fe-V (Fig. 2), and similarly in Ni-Fe.

We are grateful to T. R. McGuire for pointing out the existence of magnetoresistance data in Fe-V.

REFERENCES

†Work supported by the U.S. National Science Foundation.

1. H. Ashworth, D. Sengupta, G. Schnakenberg, L. Shapiro, and L. Berger, Phys. Rev. 185, 792 (1969). The idea of orbital degeneracy, also introduced in this paper, is now considered unnecessary by us.
2. B. Velicky, S. Kirkpatrick, and H. Ehrenreich, Phys. Rev. 175, 747 (1968).
3. H. Hasegawa and J. Kanamori, J. Phys. Soc. Japan 31, 382 (1971); 33, 1599 (1972).
4. L. Berger, to be published.
5. N. F. Mott, Adv. Phys. 13, 325 (1964).
6. L. Berger, Phys. Rev. 137, A220 (1965).
7. M. F. Collins and G. G. Low, Proc. Phys. Soc. (London) 86, 535 (1965).
8. A. A. Gomes, J. Phys. Chem. Solids 27, 451 (1966); A. A. Gomes and I. A. Campbell, J. Phys. (London) C1, 253 (1968).
9. H. Hasegawa and J. Kanamori, J. Phys. Soc. Japan 33, 1607 (1972).
10. M. V. Nevitt and A. T. Aldred, J. Appl. Phys. 34, 463 (1963).
11. C. H. Chang, C. T. Wei, and P. A. Beck, Phys. Rev. 120, 426 (1960).
12. N. Sueda, Y. Fujiwara and H. Fujiwara, J. Sci. Hiroshima Univ. A33, 267 (1969); A. V. Cheremushkina and M. I. Koroleva, Fiz. Tverd. Tela 5, 455 (1963) [Transl.: Sov. Phys.-Solid State 5, 330 (1963)].
13. N. Sueda and H. Fujiwara, J. Sci. Hiroshima Univ., A35, 59 (1971).
14. I. A. Campbell, A. Fert, and O. Jaoul, J. Phys. C3, S95 (1970).
15. I. A. Campbell, A. Fert and A. R. Pomeroy, Phil. Mag. 15, 977 (1967).
16. S. Arajs, F. C. Schwerer, and R. M. Fisher, Phys. Stat. Solidi 33, 731 (1969). See their Table I. The quantity of interest is Δ(300 K)/Δρ, and not Δ(300 K).
17. A calculation of conductivity in concentrated alloys by F. Brouers and A. V. Vedyayev, Phys. Rev. B5, 348 (1972), also shows a s-d scattering resonance there. See curve sd on their Fig. 3.

MAGNETIC SUSCEPTIBILITY OF $(Cu_3Pt)_{1-c}Mn_c$: EFFECT OF ATOMIC ORDER AND DISORDER*

C. L. Foiles
Michigan State University, East Lansing, Mich. 48824

ABSTRACT

The magnetic susceptibilities of Mn impurities in AgMn and Cu_3PtMn are dependent upon the atomic states of the alloys. In the former system, crystalline and amorphous alloys produce systematic differences in behavior which are readily explained from a free electron perspective. Data are presented for the latter system in atomically ordered and disordered states, and numerous parallels in behavior for the two systems are noted. However, a striking difference which contradicts free electron predictions for Cu_3PtMn also occurs. X-ray measurements of the long range order in Cu_3PtMn are presented to verify that the contradiction is real rather than an artifact due to preferential occupation of particular sites by the Mn.

INTRODUCTION

The presence of dilute amounts of Mn impurities in diamagnetic hosts leads to a fascinating diversity in alloy properties. It is generally accepted that magnetic interactions between these impurities play a major, and perhaps dominant, role in determining the properties. Thus, experiments which simultaneously control the interactions and permit a straight-forward interpretation of resultant behavior have particular value.

The alteration of the atomic nature of the alloy with attendant changes in the magnetic susceptibility of the impurity appears to be such an experiment. Korn achieved this alteration by using CuMn and AgMn alloys in crystalline and amorphous states.[1,2] At higher temperatures the impurity susceptibility obeyed a Curie-Weiss form with the Curie-Weiss temperature (θ) providing a measure of impurity-impurity interactions. Kok and Anderson[3] showed that when an RKKY interaction between magnetic impurities was combined with the appropriate atomic sampling of this interaction, the predicted and observed behaviors for θ agreed. Kok extended this model to low temperatures and explained the observed deviations from Curie-Weiss behavior.[4]

In an earlier paper we reported initial efforts to obtain analogous changes in atomic nature and impurity susceptibility by using alloys of the form $(A_3B)_{1-c}Mn_c$ where the binary host could be atomically ordered or disordered.[5] Many similarities in magnetic behavior occurred but a crucial difference also occurred; the signs for θ differed. In this paper we present additional susceptibility data and initial X-ray data for $(Cu_3Pt)_{1-c}Mn_c$. The crucial difference occurs once again and the X-ray data appear to eliminate one explanation for this difference.

RESULTS AND INTERPRETATION FOR AgMn

Since comparisons with the behavior of AgMn will be frequent, a brief summary of the properties of this system is essential.[2] The magnetic susceptibility of the Mn has the form

$$\chi(T) = \frac{c\,p^2}{3k_B(T-\theta)} \qquad (1)$$

at higher temperatures. In this equation c is the Mn concentration, p is the effective moment and k_B is the Boltzmann constant. Higher temperatures will be defined more precisely in the next paragraph. In both alloy states p is within 10% of 5 Bohr magnetons and the parameter of interest is θ. Lines B and C in figure 1 summarize the behavior of θ for AgMn. When the alloys are in crystalline states, line B, θ is positive (i.e.--indicative of an apparent ferromagnetic interaction) and linear in c. This behavior is valid over a range of lower concentrations but fails at higher concentrations due to deviations from the Curie-Weiss form. For the amorphous alloys, line C, θ is approximately zero and independent of c at lower concentrations but does become negative at higher concentrations. The division between lower and higher concentration regimes is approximately the same for both alloy states and we hereafter denote the lower regime as the "sub-c_1" regime.

Kok and Anderson[3] show that the preceding behavior is consistent with predictions based upon a

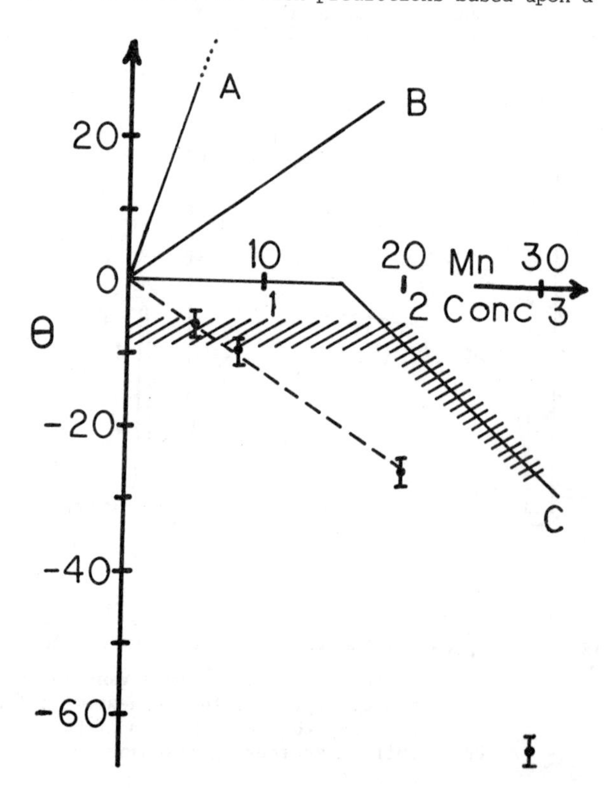

Figure 1. Behavior of θ. The variations of θ in crystalline CuMn (A), crystalline AgMn (B) and their amorphous states (C) are depicted as lines. The horizontal scale for these data ranges from 0 to 30 at.%. The shaded region and the data points are the variations of θ for Cu_3PtMn in atomically disordered and ordered states respectively. The horizontal scale for these data ranges from 0 to 3 at.%. Deviation from Curie Weiss behavior also occurs for the ordered Cu_3PtMn (3 at.%) sample.

simple free electron model of RKKY coupling and a random distribution of Mn impurities. In a crystalline environment the placement of impurities on lattice sites leads to a systematic sampling of the RKKY interaction. At higher temperatures, defined as temperatures for which the thermal energy is much greater than any exchange energy coupling the impurities, this systematic sampling produces θ. The sign of θ is determined by a lattice sum appropriate to the crystalline state and the c dependence is associated with the random distribution of impurities. Sampling of the RKKY coupling in an amorphous system produces a θ which is essentially zero.

Korn studied a single, amorphous CuMn sample containing 5 at.% Mn.[1] θ was zero. This result combined with the documented similarities in magnetic behavior for CuMn and AgMn[6] make it highly probable that the behavior of θ in crystalline CuMn parallels that in AgMn. Line A in figure 1 depicts θ in crystalline CuMn and the line becomes dotted at 5 at.% to denote the limit of comparable data for amorphous CuMn.

Cu_3PtMn RESULTS AND DISCUSSION

Measurements of magnetic susceptibility between 77.3°K and room temperature yield impurity susceptibilities which obey eqn. (1). p is constant within experimental error (5.4 ± 0.4 Bohr magnetons) and the interesting variations once again occur in the behavior of θ. Figure 1 summarizes the behavior of θ as the atomic order is altered and as c is varied. The shaded region depicts the variation of θ, with associated error limits, for four atomically disordered (DOS) samples. Once again, θ is independent of c at lower concentrations but does change at higher concentrations. The data points depict the behavior for the same four samples following thermal anneals to produce atomically ordered (OS) samples. The dashed line represents a linear fit to these data.

The similarities in θ behavior for AgMn and Cu_3PtMn (and by inference CuMn) alloys are easily discerned in figure 1. However, clear differences in behavior also occur. First, θ is not zero for the DOS alloys; it has a negative value. Second, the sub-c_1 regime is much smaller in the present study. For AgMn this region extends to nearly 15 at.% while it ends at essentially 2 at.% for Cu_3PtMn. Third, the signs of θ for the crystalline and OS alloys are different. In crystalline CuMn and AgMn, $\Delta\theta/\Delta c$ is approximately 5.5 and 1.4°K/at.% respectively while in OS Cu_3PtMn $\Delta\theta/\Delta c \simeq -13$°K/at.%. This last result corrects an earlier conclusion.[7]

Of the preceding differences, only the change in sign for θ poses a fundamental problem. Compare crystalline AgMn or CuMn with Cu_3PtMn in the OS. The former can be viewed as OS forms of Ag_3AgMn or Cu_3CuMn. Using the model of Kok and Anderson[3] would involve the same lattice sum to determine the sign of θ. This lattice sum is a function of the number of conduction electrons/atom (e/a) and has the same sign for $1/4 < e/a < 3/2$.[8] Since e/a for AgMn and CuMn is 1 and since e/a for Cu_3PtMn is probably near 1 and certainly within the above range, the sign discrepancy is serious.

Two extenuating circumstances prevent an immediate dismissal of the simple RKKY model in the present OS-DOS system; (1) if Mn impurities preferentially occupy certain sites in OS Cu_3Pt, the lattice sum becomes that of another structure and (2) if the extra zone gaps in Cu_3Pt (relative to Cu viewed as Cu_3Cu) cause an anisotropic coupling, the form of the lattice sum changes. The consequences of (1) are easily evaluated. Placing the Mn impurities only on Pt sites causes the appropriate lattice sum to be that of a simple cubic structure with a basis of 4 atoms/unit cell. Using reasonable e/a values, this sum predicts a negative θ and thus agrees with observations. Random placement of Mn impurities yields a face centered cubic sum while placement of Mn impurities only on the Cu sites yields a body centered tetragonal sum. Both of these latter sums predict a positive θ. The consequences of (2) are beyond the scope of this paper.

Fortunately, X-ray studies which determine the long range order parameter, S, provide a test for situation (1). Cu and Mn occupy nearby positions in the periodic table and thus have very similar X-ray form factors. Pt has a very different form factor. Thus, if the Mn in $(Cu_3Pt)_{1-c}Mn_c$ goes onto Cu sites or goes onto all sites randomly, the long range order perceived by X-rays decreases slowly as a function of c. However, if the Mn occupies only Pt sites, then both the Mn atoms and the Pt atoms displaced to Cu sites contribute to disorder and long range order decreases rapidly as a function of c.

The long range order parameter was determined using standard X-ray procedures. A GE XRD-6 diffractometer with a proportional counter was used to obtain fundamental lines and superlattice lines. The integrated intensities of the former lines are proportional to $(3f_{Cu} + f_{Pt})^2$ while those of the latter lines are proportional to $(f_{cu} - f_{Pt})^2S^2$ where f is the X-ray form factor. The presence of anti-phase domains has no influence on these integrated intensities[9]. The ratio of these intensities provides a measure of S^2. Our results for S are given in figure 2.

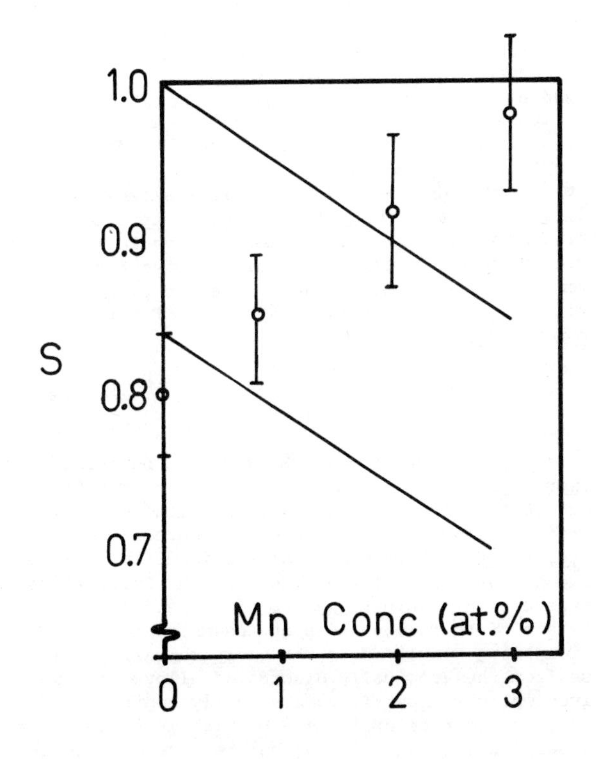

Figure 2. Long Range Order in Cu_3PtMn. The observed variation of S with Mn concentration is compared with two ideal cases. The line beginning at S=1.0 assumes perfect order is possible and places all Mn impurities at Pt sites. The line beginning at S=0.84 assumes a constant number of misplaced Cu atoms and places all Mn impurities at Pt sites.

Before discussing the significance of these data some comments about the determination of the error bars is necessary. There were sufficient counts in each peak so that random statistical error was 1%. The actual integrated intensities were measured using chart recordings and a planimeter. The largest instrumental error was in estimating the background and obtaining reproducible areas for the peaks. A number of trials indicated this error could be as large as 4% but was typically less. Thus, the ratio of intensities in the worst case could be in error 10% and hence the value of S would have a 5% error. This value was used to determine the given error bars. Independent trials using the same samples gave results lying within these limits.

The reliability of the absolute values was more difficult to assess. The given values were determined from ratios of the {110} and {220} lines taken at room temperature. Dispersion was included but no corrections for a Debye-Waller factor were included. Since no information is available for this factor and since work on the isomorphic system Cu_3Au indicates different values for the same system[10], an estimate of the influence of this factor was difficult. Nevertheless, we attempted to obtain a crude estimate by using the fundamental line Debye-Waller factor for the Cu_3Au system as determined by Gehlen and Cohen[10]. All the values of S were lowered by approximately 0.06 but no changes in the general patterns occurred.

The conflicts between predicted and observed long range order are clearly displayed in figure 2. Assuming the Mn impurities only occupy the Pt sites (for the situation of perfect order at c=0 and for the situation of a constant background disorder leading to S=0.84 at c=0[11]) requires S to decrease with increasing c. The experimental data do not exhibit such behavior. In fact, the data suggest an increase in S as c increases. Although questions about the reliability of the results for the c=0 sample[11] indicate assignment of a value for this increase may be premature, it is clear that no decrease in S with increasing c occurs. Moreover, if the value of S for the 3% Mn sample is corrected for the thermal factor as estimated above and the full lower error extreme is used, the resultant value barely verges upon being consistent with the maximum order achievable in a 3% sample where all Mn impurities are on Pt sites. Supplemental data for the sample indicate that this extreme limit for the experimental value of S is very improbable.[12]

CONCLUSIONS

Strong similarities in the magnetic behavior of Mn impurities for a series of crystalline-amorphous alloys (AgMn) and for a series of atomically ordered-disordered alloys (Cu_3PtMn) are documented. For each series, the effective moment of the impurity is insensitive to the state of the alloy but the Curie-Weiss temperature, θ, exhibits systematic dependences upon the state of the alloy and upon the Mn concentration, c. Comparing results from the amorphous alloys with those from the atomically disordered alloys (a comparison which we justify upon a purely empirical basis), each series exhibits a θ which is independent of c below a concentration c_1 but becomes more negative above c_1. Comparing results from the crystalline alloys with those from the atomically ordered alloys (a comparison which has both a theoretical and an empirical basis), each series gives a θ which is linearly dependent upon c below c_1. We suggest these detailed similarities indicate that related effects in the Mn-Mn interactions are occurring.

The differences in c_1 values and in $|\Delta\theta/\Delta c|$ values for the two alloy series cause no fundamental problems for a free electron interpretation of behavior: however, the difference in signs for $\Delta\theta/\Delta c$ does raise fundamental questions. X-ray measurements of the long range order parameter are given for the Cu_3PtMn alloys. The dependence of this parameter upon c and the absolute value of this parameter for the c=0.03 sample lead us to conclude that the sign discrepancy is real and not the artifact of long range order among the Mn impurities.

ACKNOWLEDGEMENTS

A portion of the magnetic susceptibility measurements were done by Dr. T.W. McDaniel. The permission to use these data plus many helpful discussions with Dr. McDaniel are gratefully acknowledged.

REFERENCES

* Work supported by National Science Foundation.

1. Dietrich Korn, Z. Physik 187, 463 (1965).
2. Dietrich Korn, Z. Physik 214, 136 (1968).
3. W.C. Kok and P.W. Anderson, Phil. Mag. 24, 1141 (1971).
4. W.C. Kok, Phil. Mag. 30, 351 (1974).
5. T.W. McDaniel and C.L. Foiles, Solid State Comm. 14, 835 (1974).
6. D.P. Morris and J. Williams, Proc. Phys. Soc. (London) 73, 422 (1959).
7. In reference 5 it was erroneously concluded that θ displays only a weak dependence upon c in $(Cu_3Pt)_{1-c}Mn_c$. That conclusion was based upon data for samples with c < 1 at.% and their associated error bars.
8. D. Mattis, The Theory of Magnetism (Harper Row, New York, 1965).
9. B.E. Warren, X-ray Diffraction (Addison-Wesley, Reading, Pa., 1969).
10. P.C. Gehlen and J.B. Cohen, J. Appl. Phys. 40, 5193 (1969).
11. The intensity ratio for the c=0 sample gave S=0.8 but the reliability of this result is unclear. Back-reflection X-ray photographs of this sample yield discrete spots superimposed on faint rings rather than simple rings. This indicates the grains are larger than desired (the approximate guidelines for the appearance of powder patterns suggests many grains exceed 10^{-3} cm in size) and raises questions about the proper Lorentz polarization factor to use in analyzing the intensity data. The fact that the {110} and {220} line intensities do not scale properly relative to the {111} intensity is additional support for these questions.
12. For the 3% Mn sample back reflection X-ray photographs give no indications of grain size problems. Integrated intensities for the {100}, {110}, {200} and {220} lines scale properly, within experimental error, relative to the {111} line intensity. The intensity ratios for the {100}-{200} and the {110}-{220} produce S values which differ by only 0.02. Finally, although the intensity data are too limited to warrant claiming a specific Debye-Waller factor, they are consistent with the procedure for estimating this error as discussed in the text and they indicate that the stated error is an upper limit.

MAGNETIC INTERACTIONS IN DISORDERED Ni-Mn ALLOYS NEAR THE 25% Mn COMPOSITION*

T. Satoh,** C. E. Patton, and R. B. Goldfarb

Department of Physics, Colorado State University, Ft. Collins, Colorado 80523

ABSTRACT

The magnetization of $Ni_{1-x}Mn_x$ alloys ($0.22 < x < 0.32$), disordered by quenching from 1000°C, has been studied as a function of field (0 - 100 kOe), temperature (4 - 300K), and field cooling history. For zero-field cooling, the temperature dependence of the initial susceptibility indicates an extrapolated Curie temperature which is negative for $x > 0.25$, near zero for $x \sim 0.25$ and positive for $x < 0.25$. Magnetization data (4 - 300K) for field cooling (8 kOe) and zero-field cooling indicate that the net moment induced by field cooling peaks at the Ni_3Mn composition. Further high field data (0 - 100 kOe) indicate that measuring fields in excess of 50 kOe at 4K are sufficient to eliminate the difference in magnetization for field cooling and zero-field cooling. The data can be qualitatively explained by a simple three phase model, in which Mn-rich, stoichiometric Ni_3Mn, and Mn-deficient units with varying properties couple to yield the observed behavior.

INTRODUCTION

Disordered Ni-Mn alloys near 25 at.% Mn show a decrease in magnetization at low temperature, and displaced hysteresis loops after field cooling.[1] These properties were qualitatively explained by an exchange anisotropy concept, related to competing interactions between antiferromagnetic Mn-Mn pairs and ferromagnetic Mn-Ni and Ni-Ni pairs[2] and local fluctuations in composition.[1] This exchange coupling was believed to be quite stable, since pulsed fields of 140 kOe did not destroy the displaced-loop behavior.[3]

Recent neutron diffraction data on alloys 5 - 20 at.% Mn show that the Ni and Mn moments both decrease in magnitude with increasing Mn content.[4] If the moment data were extrapolated to 25 at.% Mn, both atomic moments μ_{Ni} and μ_{Mn}, would be near zero. These results appear to be in conflict with the exchange anisotropy model of relatively large, strongly coupled atomic moments responsible for the observed magnetic properties.

In order to obtain more detailed information about the relative strength of the exchange interaction, the origin of the smaller induction at low temperature for zero-field cooling, and the effect of field cooling, a systematic study of the magnetic behavior of disordered Ni-Mn alloys in the 22 - 32 at.% Mn composition range has been made. Data for magnetization, field (0 - 8 kOe and 0 - 100 kOe) and temperature (4 - 300K) were obtained for five compositions: 22.1, 24.6, 26.2, 29.0, and 32.1 at.% Mn.

The data provide significant new insight into the properties of Ni_3Mn: (1) the strength of the exchange coupling is less than 50 kOe, (2) the magnetization behavior is consistent with a three-phase model, in which Mn-rich, stoichiometric Ni_3Mn, and Mn-deficient units, with varying properties, couple to yield the observed results.

EXPERIMENT

The alloy samples were cast into ingots in an argon atmosphere after rf-induction melting. The ingots were homogenized by sequential cold-working. Compositions were determined by atomic absorption methods. Samples for measurement were spheres, 2.5 - 3 mm in diameter. The alloys were put into a disordered state by sealing them in vacuo in quartz tubes (3 mm i.d. and 0.5 mm wall thickness), holding at 1000°C for 3 hours, and quenching in ice-water. Thermocouple measurements indicated that the time spent in the ordering temperature range of 600 - 350°C during the quenching process was less than 3 sec. Neutron diffraction did not detect the presence of any kind of atomic order for samples quenched in this manner.[5]

Magnetization was measured by the vibrating sample technique as a function of temperature (4 - 300K) and magnetic field (0 - 8 kOe or 0 - 100 kOe). During any one measurement sequence, the temperature was changed continuously in one direction, i.e., increasing or decreasing. Samples were cooled in the presence of a cooling field H_{cool} (referred to as field cooling) or in zero field (zero-field cooling), from room temperature to 4K.

RESULTS

Figure 1 shows the initial susceptibility as a function of temperature, as determined from the high field extrapolations on Arrott plots (M^2 vs. H/M), for different alloy compositions. The internal field was obtained by compensating for the sample demagnetizing field. The increases in $1/\chi$ for the 24.6, 26.2, 29.0, and 32.1 at.% Mn alloys are related to the decrease in magnetization at low temperature, originally reported by Kouvel et. al.[6] The 22.1 at.% Mn sample shows ferromagnetic ordering below 250K. The data show that the asymptotic Curie temperature Θ_C (temperature intercept of the $1/\chi$ extrapolation to $1/\chi = 0$) is close to zero for 25 at.% Mn, is negative for Mn in excess of 25 at.%, and is positive for less Mn. Every set of data, with the exception of 22.1 at.% Mn, suggests the presence of ferromagnetic and antiferromagnetic interactions, in addition to one or more transition temperatures. For 24.6 at.% Mn, the nearly zero Θ_C implies a rather weak exchange coupling. The stoichiometric composition represents a crossover point from positive to negative Θ_C.

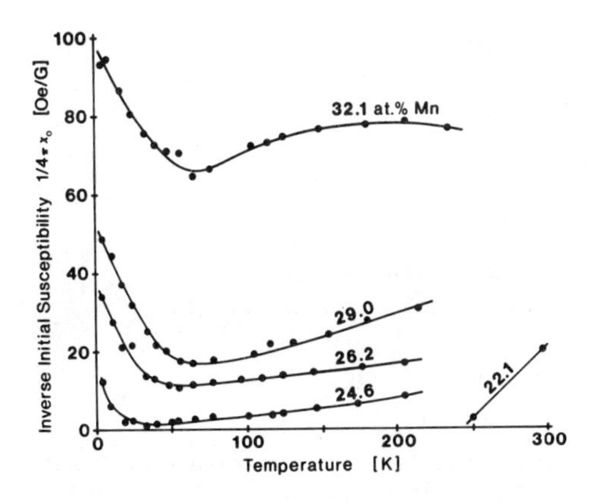

Fig. 1. Inverse initial susceptibility vs. temperature for five Ni-Mn alloy compositions, obtained from Arrott plots (M^2 vs. H/M) of magnetization data.

Further study of the magnetic behavior upon field cooling revealed some additional features. Figure 2 shows a representative curve of magnetization vs. temperature after field cooling in 8 kOe [curve (a) and (c)] and after zero-field cooling [curve (b) and (c)], for a 24.6 at.% Mn alloy. The magnetization simply decreases smoothly from a peak value at 4K for field cooling. For zero-field cooling, the magnetization starts at a much lower value at 4K, increases sharply with increasing temperature and merges with the field-cooling curve at about 50 - 60K.

Figure 3 summarizes the results of a detailed study of the field-cooling effect shown in Fig. 2 as a function of composition. The triangles indicate the induction $4\pi M$ measured at a temperature of 4K and a field of 8 kOe, after cooling to 4K in H_{cool} = 8 kOe. The squares are similar data obtained after cooling in zero field. Both curves show a steep increase with decreasing Mn content below 25 at.% Mn and a shallow, more gradual fall-off above 25 at.% Mn. The difference, $4\pi M_{cl}$ (field cooling) - $4\pi M_o$ (zero-field cooling), is rather interesting in that it shows a peak near stoichiometry. This peak is apparently a result of the competition between two tendencies: one for the net moment (with or without field cooling) to be large for low Mn content, and one for the relative contribution of field cooling to be large for high Mn content. The figure shows, for example, that the induction is large and the field-cooling effect is small at 22.1 at.% Mn, whereas the induction is small but the field-cooling effect accounts for almost all of it at 32.1 at.% Mn.

Measurements were made in fields up to 100 kOe in an attempt to estimate the exchange anisotropy field associated with the field-cooling effect. The results are summarized in Fig. 4. Curves of the induction $4\pi M$ at various measuring fields for cooling in 100 kOe or in zero field are plotted as a function of temperature. The characteristic fall-off in $4\pi M$ at low temperature in 15 kOe and 30 kOe measuring fields for zero-field cooling, and the absence of a fall-off for field cooling, follow the results of Fig. 2. For measurement fields of 50 and 100 kOe, however, the inductions for field cooling and zero-field cooling are identical. (In these cases, only one curve is shown.)

DISCUSSION

The above experimental results point to several properties of Ni-Mn disordered alloys: (1) the average magnetic interaction shifts from antiferromagnetic to ferromagnetic at the Ni_3Mn composition, with rather weak magnetic interactions ($\Theta_C \sim 0$) at Ni_3Mn; (2) the net field-cooling-induced moment peaks at stoichiometry; (3) the exchange field connected with field-cooling is weak, on the order of 50 kOe or less.

Based on the tendency expressed in Fig. 1 and property (1) above, the fact that the magnetic interactions are relatively short range allows the molecular field to be expressed as an average exchange interaction J between nearest neighbors, with $J > 0$ for $x < 0.25$, $J \sim 0$ for $x \sim 0.25$, and $J < 0$ for $x > 0.25$. Here x is the fraction of Mn atoms. It should also be possible to use the data to estimate the values of the nearest-neighbor exchange parameters J_{11}, J_{12}, and J_{22} for Mn-Mn, Mn-Ni, and Ni-Ni pairs, respectively. Consider the following equation relating J and the average spin S to the actual nearest-neighbor exchange interactions and individual spins.

$$JS(S+1) =$$
$$2[x^2J_{11}\langle\hat{S}_1\cdot\hat{S}_1\rangle + 2x(1-x)J_{12}\langle\hat{S}_1\cdot\hat{S}_2\rangle + (1-x)^2J_{22}\langle\hat{S}_2\cdot\hat{S}_2\rangle]$$

The J and S parameters can be obtained from the susceptibility data in Fig. 1. The $\langle\hat{S}_i\cdot\hat{S}_j\rangle$ can be estimated if the average Ni moment and Mn moment were known. If such data were available over the range of Mn composition in this study, the J_{11}, J_{12}, and J_{22} exchange parameters could be obtained. Such determinations would be quite useful in further analysis of the Ni-Mn system. Unfortunately μ_{Mn} and μ_{Ni} have been determined only for $x \leq 0.20$.[4]

The experimental results summarized above can, however, be qualitatively explained by a conceptually simple three phase model. Each phase consists of small, atomic-sized regions made up of Mn moments with different nearest-neighbor environments. Each phase (A) region consists of a Mn atom with more than three nearest-neighbor Mn atoms. These small units are mostly antiferromagnetic with a high Néel temperature T_N near 50K. Each phase (B) region consists of a Mn atom with three nearest-neighbor Mn atoms. These units are exposed to a weak molecular field and are superparamagnetic with an ordering temperature less than 10K. Their moments are large, possibly 0.1 - 0.2 μ_B at 4K and 10 kOe. Each phase (C) region consists

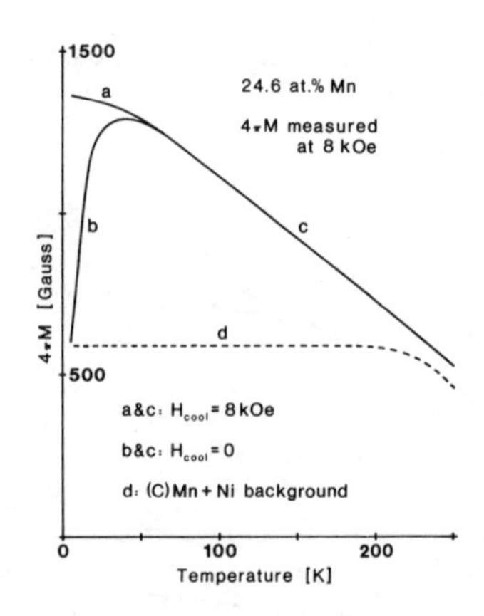

Fig. 2. Induction $4\pi M$ vs. temperature at 8 kOe measuring field for field cooling [curve (a) and (c)] and zero-field cooling [curve (b) and (c)]. Measurements were made after cooling to 4K. The data are for a slightly ordered 24.6 at.% Mn alloy and a cooling field of 8 kOe [curve (a) and (c)].

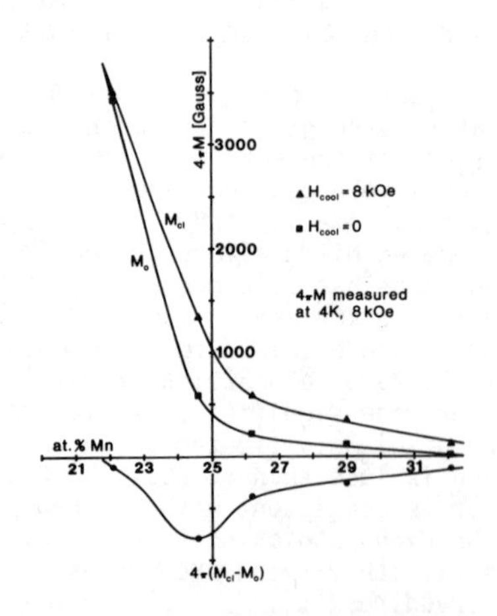

Fig. 3. Induction $4\pi M$ vs. Mn content for field cooling and zero-field cooling. $4\pi M_{cl}$ (triangles) denotes the induction measured at 8 kOe after cooling in 8 kOe. $4\pi M_o$ (squares) denotes the induction measured at 8 kOe after zero-field cooling. All data are at 4K.

of a Mn atom with less than three nearest-neighbor Mn atoms. These units are ferromagnetic with a high T_C. The Ni atoms are ferromagnetic and contribute to the magnetization as does phase (C). The separation into three phases is rather arbitrary, and is done primarily to illustrate the basic mechanisms. The Mn atoms have, of course, a statistical distribution of nearest-neighbor environments. It is assumed that each of the Mn nearest neighbors is of the phase (B) variety.

Based on this model, consider now the interactions and ordering during field cooling. Phase (C) is ferromagnetic and exhibits the usual Curie-Weiss behavior. The magnetic moments of phase (B) tend to be aligned with the cooling field. The spins of the phase (A) regions are coupled to the phase (B) regions. This coupling is a variation of the original exchange anisotropy concept developed by Meiklejohn and Bean[7] and adapted to Ni-Mn by Kouvel and co-workers.[1,3] As the temperature drops below T_N(A), the phase (A) regions freeze in, with antiferromagnetic alignment axes parallel to the moments of the phase (B) regions. Below T_N(A), the phase (B) regions are tightly coupled to the phase (A) regions and tend to be parallel to the field direction, corresponding to curve (a) in Fig. 2.

Now consider the case of zero-field cooling where the phase (B) moments are randomly oriented as the temperature falls below T_N(A). As before, the phase (A) regions freeze in with axes parallel to the individual phase (B) moments to which they are coupled. Below T_N(A), the phase (B) moments remain locked in their random orientations due to their tight coupling to the phase (A) regions, leading to the behavior of curve (b) in Fig. 2.

The key points are: (1) for field cooling the phase (B) moments tend to be field aligned; for zero-field cooling they are random. (2) As temperature falls below T_N(A) the (A) regions freeze-in with axes defined by their companion (B) regions. (3) Below T_N(A), the (B) regions remain tightly coupled to these frozen-in (A) regions.

The increase in $4\pi M$ down to 50K [curve (c) in Fig. 2] is due to paramagnetic (A) and the superparamagnetic (B) in combination. The drop in $4\pi M$ below 50K [curve (b)] for zero-field cooling is due to the antiferromagnetic ordering of (A) with random axes and the (B) regions tightly coupled to these (A) regions. Curve (d) in Fig. 2 represents the background contribution from the ferromagnetic phase (C) and Ni.

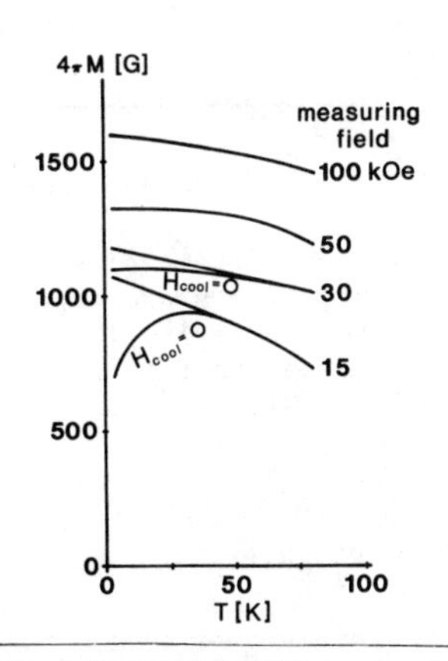

Fig. 4. High-field magnetization vs. temperature for the disordered 24.6 at.% Mn sample for different measuring fields as indicated. Where branched curves are shown, the lower branch is for zero-field cooling and the upper branch is for cooling in 100 kOe. Where a single curve is shown, field cooling and zero-field cooling gave identical data.

The above qualitative model can account for the peak in the field-induced moment at 25 at.% Mn in Fig. 3. For alloys with $x < 0.25$, the amount of phase (A) clusters is reduced which leads to a smaller field cooling effect. The concurrent increase in the number of ferromagnetic phase (C) interactions leads to the rapid increase in induction for both field cooling and zero-field cooling. Mn in excess of 25 at.% reduces the amount of phase (C) units, and increases the amount of phase (A) units. More of the available average moment is coupled to the phase (A) units and is affected by field cooling. Thus, while the net induction is reduced, a greater proportion of it is affected by field cooling.

The results in Fig. 4 can also be understood on the basis of the model. After zero-field cooling, a measuring field of 50 kOe is sufficient to rotate the phase (B) moments into the field direction, overcoming their strong coupling to the random phase (A) regions. Thus, 50 kOe may be interpreted as a crude measure of the exchange anisotropy coupling strength between (A) and (B).

According to the model, one would also expect an increase in the amount of phase (A) units to increase the exchange field. Data similar to that in Fig. 4 for the 29.0 at.% Mn alloy show that this is true. The exchange anisotropy field increases from 50 kOe to 75 kOe. This tendency was also checked by annealing a 24.6 at.% Mn sample at 500°C for 3 minutes to induce partial short range order and decrease the amount of phase (A). This treatment caused the exchange field to decrease by a factor of two, in addition to doubling the magnetization.

In conclusion, the present data on disordered Ni-Mn alloys in the 22 - 32 at.% Mn composition range can be qualitatively explained by considering the statistical variation in the local Mn environments. The three phase model represents a qualitatively useful way to make some explicit correlations with the data.

ACKNOWLEDGEMENTS

The authors are grateful to Professor C. D. Graham, Jr., University of Pennsylvania, for making available his high-field facility and collaborating on the high-field measurements. Dr. J. J. Rhyne, National Bureau of Standards, Washington D. C., is gratefully acknowledged for neutron diffraction measurements on the alloy samples.

REFERENCES

* Supported by the National Science Foundation, Grant DMR73-02665.
** On leave from the Electrotechnical Laboratory, Tokyo, Japan.

1. J. S. Kouvel and C. D. Graham, "Exchange Anisotropy in Disordered Nickel-Manganese Alloys," J. Phys. Chem. Solids 11, 220 (1959).
2. W. J. Carr, "Intrinsic Magnetization in Alloys," Phys. Rev. 85, 590 (1952).
3. J. S. Kouvel, C. D. Graham, and I. S. Jacobs, "Ferromagnetism and Antiferromagnetism in Disordered Ni-Mn Alloys," J. Phys. Radium 20, 198 (1959).
4. J. W. Cable and H. R. Child, "Magnetic Moment Distribution in NiMn Alloys," Phys. Rev. B 10, 4607 (1974).
5. J. J. Rhyne, private communication.
6. J. S. Kouvel, C. D. Graham, and J. J. Becker, "Unusual Magnetic Behavior of Disordered Ni_3Mn," J. Appl. Phys. 29, 518 (1958).
7. W. H. Meiklejohn and C. P. Bean, Phys. Rev. 102, 1413 (1956); Phys. Rev. 105, 904 (1957).

DOES $CoO\cdot Al_2O_3\cdot SiO_2$ GLASS ORDER MAGNETICALLY? A MÖSSBAUER STUDY

L.H. Bieman, P.F. Kenealy, and A.M. de Graaf,* Wayne State University, Detroit, MI 48202

ABSTRACT

The susceptibility of concentrated cobalt aluminosilicate glass exhibits superparamagnetic type behavior below ∿50°K, which terminates in a relatively sharp peak at low temperatures, seeming to indicate the onset of a long-range antiferromagnetic ordering.[1] The low temperature specific heat[2] and sound velocity[3] do not show cooperative type cusps or discontinuities around these temperatures. In order to shed light on this discrepancy a Co^{57} Mössbauer study of a 25 at.% Co glass (peak temperature ∿7°K) was made. Below ∿9°K, a dramatic spectral broadening was observed indicating magnetic hyperfine fields as large as 50 Tesla. It will be shown that all four experimental results can be qualitatively understood by assuming that the glass consists of small magnetically ordered regions (domains) whose net moments relax in a superparamagnetic fashion. The Mössbauer spectra above 4.2°K can be fitted by using a superparamagnetic relaxation time of the form $\tau = (2 \times 10^{-9}\ \text{sec}) \exp(28°K/T)$. The Mössbauer spectrum at 4.2°K has been used to obtain the distribution of hyperfine fields in the glass sample.

INTRODUCTION

The observation of a relatively sharp peak in the low-temperature magnetic susceptibility of concentrated cobalt- and manganese aluminosilicate glasses[1] seems to indicate the onset of some type of long range magnetic ordering in these systems. However, the sound velocity[3] shows a broad dip at these temperatures, rather than the sharp anomaly characteristic of a magnetic phase transition. Also, the specific heat[2] does not exhibit a cooperative-type peak, but a broad maximum superimposed on a nearly linear background. It has been suggested[1-3] that these observations may be accounted for by assuming that the glass below ∿50°K consists of small magnetically ordered regions (domains) whose net magnetic moments, μ, relax in a superparamagnetic fashion with a relaxation time

$$\tau = \tau_0 \exp(KV/kT) . \qquad (1)$$

Here, K is the anisotropy energy density of the domains, and V their volume. At temperatures sufficiently below KV/k the domain moments freeze in random anisotropy directions, resulting in a sharply reduced susceptibility.

The purpose of the present paper is to describe the results of a Co^{57} Mössbauer study of a cobalt aluminosilicate glass sample with a susceptibility peak temperature of ∿7°K. It will be shown that the Mössbauer data are consistent with the superparamagnetic domain model.[4] By fitting the spectra above 4.2°K, a value of KV/k = 28°K was obtained, which compares favorably with the values of this parameter deduced from susceptibility and sound velocity data.[1,3] The value of τ_0 is 2×10^{-9} sec, which falls in the range of values quoted in the literature.[5] From the Mössbauer spectrum at 4.2°K the distribution of hyperfine fields in the glass sample was obtained.

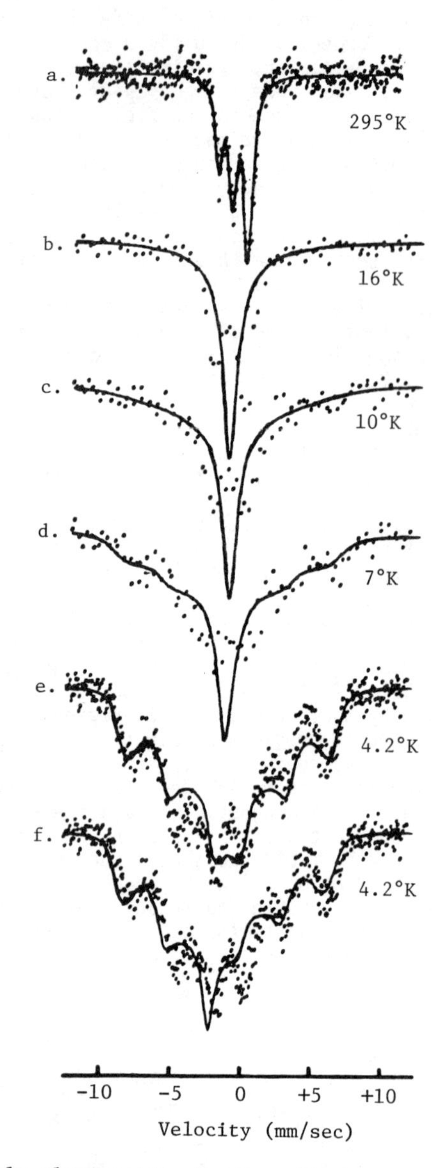

Fig. 1. Cobalt aluminosilicate glass Mössbauer spectra at various temperatures (dots are experimental points, solid lines are fitted curves). The two spectra at 4.2°K illustrate alternative fits to the spectrum (e: without EFG, f: with EFG).

EXPERIMENTAL

The Mössbauer source was prepared by firing a mixture of cobalt carbonate,[6] laced with 5 millicuries of Co^{57}, aluminum oxide, and pure silica sand on a quartz dish in an arc-image furnace. The sample contained 67.6 mol.% CoO, 19.4 mol.% Al_2O_3, and 13.0 mol.% SiO_2, which corresponds to 25 at.% Co. The sample had a mass of 74 mg, and an area of approximately 0.5 cm^2. Similarly prepared glasses showed no crystallinity when examined by powder X-ray diffraction and electron microscopy. An Fe^{57}, 90% enriched $Na_4Fe(CN)_6 \cdot 10\ H_2O$ absorber[6] of thickness 0.25 mg Fe^{57}/cm^2 was used for the measurements. Data was taken on a standard Mössbauer spectrometer with the source placed in a variable temperature cryostat, while the absorber was kept at room temperature. A laser interferometer was used to calibrate the velocity of the spectrometer.

RESULTS AND DISCUSSION

Experimental Mössbauer spectra between room temperature and 4.2°K are shown in Fig. 1. A three peak spectrum is observed at room temperature indicating the presence of electric field gradients at the Co sites in the glass. This spectrum is tentatively interpreted as a Fe^{2+} doublet superimposed on Fe^{3+} doublet. The effect of the internal magnetic field at the Co sites becomes noticeable below about 20°K. At 4.2°K the spectrum shows a six line Fe^{57} magnetic hyperfine pattern. It should be noticed that the central part of this spectrum has greater intensity than the wings, in contrast to the characteristic Fe^{57} magnetic hyperfine spectrum. In addition this spectrum is asymmetric. In order to fit the spectrum at 4.2°K the distribution of internal magnetic fields shown in Fig. 2 was required. The flat part of this distribution gives rise to the bell shape of the spectrum, while the peak around 48.5 Tesla produces the six line structure. The fit shown in Fig. 1e was obtained with this distribution, but neglecting effects due to the electric field gradients (EFG). To correctly include the effects of the EFG requires detailed knowledge of the structure of the glass, which is lacking. The effects of the EFG were included approximately by requiring that the six line hyperfine spectrum reduces to a doublet[7] in the absence of an internal magnetic field. The merit of this procedure is that it reproduces the asymmetry of the spectrum at 4.2°K as seen in Fig. 1f.

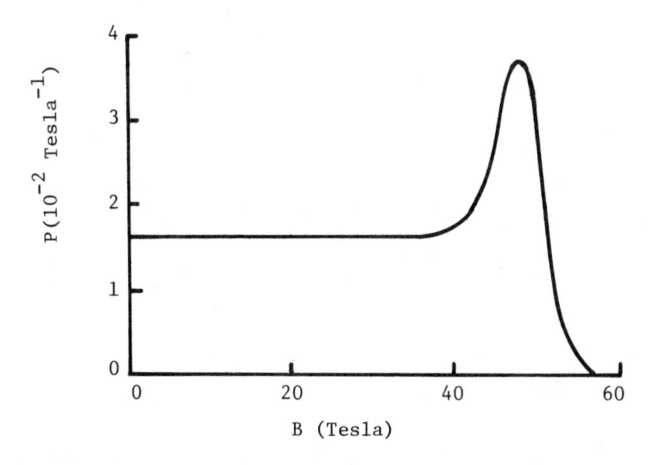

Fig. 2. Probability distribution of magnetic hyperfine fields in a 25 at.% Co aluminosilicate glass as obtained from fitting the Mössbauer spectrum at 4.2°K.

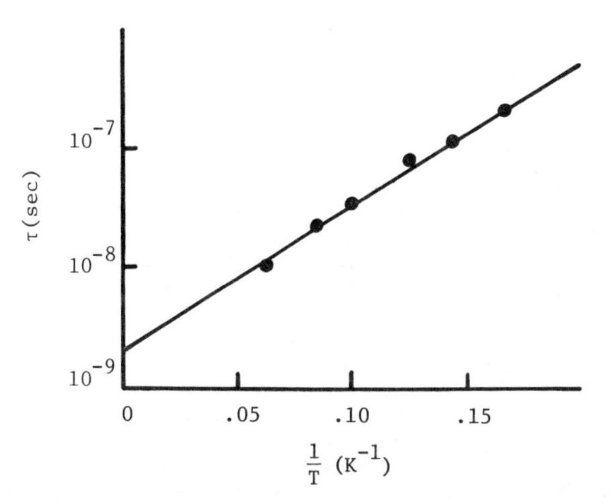

Fig. 3. Relaxation time vs. 1/T of the superparamagnetic domains in a 25 at.% Co aluminosilicate glass. The data (dots) were obtained from fitting the Mössbauer spectra above 4.2°K. The solid line represents the linear least square fit of the data.

As the temperature is raised above 4.2°K, the magnetic hyperfine splitting disappears, as shown in Fig. 1. In the framework of the superparamagnetic domain model this is caused by the decrease of the domain relaxation time given by Eq. (1). When this relaxation time becomes much smaller than the Fe^{57} nuclear lifetime, the magnetic hyperfine splitting averages to zero. The spectra above 4.2°K were fitted by making use of the magnetic field distribution obtained from the 4.2°K spectrum, and including the superparamagnetic relaxation phenomenon following a procedure outlined by Abragam.[8] Effects due to the EFG were neglected. In an attempt to compensate for this omission, data points in the central parts of the spectra (where the EFG have the strongest effect) were not included in the fitting routine. Representative fits are shown in Figs. 1b, 1c, and 1d. The relaxation times obtained from this fitting procedure are plotted vs 1/T in Fig. 3. The agreement with the temperature dependence predicted by Eq. (1) is excellent. From the slope of $\ln\tau$ vs. 1/T a value of KV/k = 28°K is obtained. This value compares favorably with the value 20.5°K obtained from susceptibility data[1] and with the values 40.9°K and 35.6°K obtained from sound velocity[3] data taken on a 25.8 at.% cobalt aluminosilicate glass, also using the superparamagnetic domain model. Thus far no specific heat measurements have been performed on a cobalt aluminosilicate glass. However, such measurements on two manganese aluminosilicate glasses have yielded similar values.[2]

ACKNOWLEDGMENTS

We would like to thank Dr. G. B. Beard for his assistance and constant interest, and Dr. R. B. Hahn for his help with the preparation of the source.

REFERENCES

*Supported in part by the National Science Foundation.
[1]R.A. Verhelst, et al., Phys. Rev. B 11, 4427 (1975).
[2]R.W. Kline, et al., 21st Annual Conference on Magnetism and Magnetic Materials, Philadelphia (1975).
[3]T.J. Moran, et al., Phys. Rev. B 11, 4436 (1975).

[4]The susceptibility and Mössbauer results are suggestive of a spin-glass phase below about 9°K. Recent spin-glass theories [for a recent review see J. Mydosh, AIP Conf. Proc. 24, 131 (1974)] would explain these results. However, these theories also predict cooperative type cusps in the specific heat and sound velocity, which are not observed.
[5]I.S. Jacobs and C.P. Bean, Magnetism, Editors G.T. Rado and H. Suhl (Academic Press, 1963), p. 275.
[6]The radioactive $CoCO_3$ and the absorber were purchased from New England Nuclear.
[7]The width of this doublet was made equal to the separation of the two outer lines of the room temperature spectrum. The justification for this choice is that at ∿20°K, the intensity of the middle line has become very small as compared with that of the two outer lines.
[8]A. Abragam, Nuclear Magnetism (Oxford, 1962), pages 447-451.

FERROMAGNETIC TO MICTOMAGNETIC TRANSITION REGION IN ORDERED Fe-Al ALLOYS

G.P. Huffman and G.R. Dunmyre
U.S. Steel Corp., Research Laboratory
Monroeville, Pa. 15146

ABSTRACT

Some highlights of a Mossbauer study of ordered Fe-Al alloys containing from 23 to 32 at.% Al are presented. From the temperature dependence of the hyperfine fields, it is concluded that the transition from ferromagnetism to mictomagnetism occurs between 27 and 32 at.% Al.

DISCUSSION

Previously, we have reported Mossbauer results for the temperature and concentration dependence of the Fe^{57} hyperfine fields in Fe-Al alloys of CsCl-type order containing from 35 to 50 at.% Al,[1] and for the anomalous temperature dependence of the hyperfine fields in both CsCl and DO_3 ordered alloys containing approximately 30 at.% Al.[2] The CsCl alloys may be described as two interpenetrating simple cubic lattices, one of pure Fe, the other having composition $Al_{100-C}Fe_C$, where

$$C = 100 - 2\,P_{Al} \qquad (1)$$

P_{Al} being the total atomic percent of Al in the alloy. From 35 to 50% Al, most Fe atoms on the Fe sublattice (Fe_F) do not have enough Fe nearest neighbors (nearest neighbor is henceforth abbreviated as nn) to have a significant magnetic moment, while all Fe atoms on the $Al_{100-C}Fe_C$ sublattice (Fe_A) have 8 Fe nn and a moment of $2.18\mu_B$.[3] It was shown in reference 1 that the essentially "antiferromagnetic" behavior of these alloys arises principally from negative RKKY exchange interactions between second nn Fe_A spins. Both the "Néel" temperature* and the hyperfine fields extrapolated to 0°K depend linearly on concentration, and the hyperfine fields decrease in a linear fashion with the square of the temperature.[1]

Near P_{Al} = 30 at.%, the hyperfine fields observed for alloys ordered in both the CsCl and DO_3 structures decrease rapidly with increasing temperature, pass through a sharp minimum near 120 to 130°K, and exhibit a broad maximum between 250 and 300°K.[2] This behavior was qualitatively explained by a model in which giant cluster moments interact with one another via antiferromagnetic second nn indirect exchange interactions. At $P_{Al} \approx 30$ at.%, the number of Fe_F atoms having significant moments (those with four or more Fe_A nn) is approximately equal to the number of Fe_A spins and all nn Fe_F and Fe_A spins are aligned parallel by a strong d-electron exchange interaction, producing large cluster moments. As the temperature is lowered from approximately 250 to 60°K, second nn Fe_A spins in the regions between the clusters become antiferromagnetically aligned and exert exchange anisotropy fields on the cluster moments, causing them to fluctuate rapidly. This model, described in more detail in reference 2, qualitatively explains the hyperfine field minima and numerous other interesting magnetic properties observed near P_{Al} = 30 at.%.[5,6,7]

Since the unusual magnetic properties of Fe-Al alloys near P_{Al} = 30 at.% clearly arise from a competition between ferromagnetic and antiferromagnetic exchange interactions, they represent an excellent example of mictomagnetism, as defined by Beck.[3] Alloys containing <25 at.%, on the other hand, exhibit fairly normal ferromagnetic behavior.[5] In the present work, Mossbauer studies were performed on a series of ordered Fe-Al alloys containing 23, 24.5, 25.8, 26.8, 28.5, 29.8, 30.8, and 31.8 at.% Al; several of these alloys were investigated in both the CsCl and DO_3 types of structure. More detailed results for all of these samples will be given in a future paper. In the current brief summary, we present only a few of the more interesting observations.

*There is no true long range antiferromagnetic order in these mictomagnetic alloys,[4] and the observed transition temperatures are perhaps better viewed as spin freezing temperatures.

RESULTS

The spectra obtained from alloys containing 23 to 26.8 at.% Al could be readily resolved into five or six magnetic hyperfine components, one for each significant configuration of Fe and Al nn. The dependence of the hyperfine fields and isomer shifts on the number of Fe and Al nn is shown in Fig. 1 for the DO_3 alloy containing 25.8 at.% Al. The dependence of the

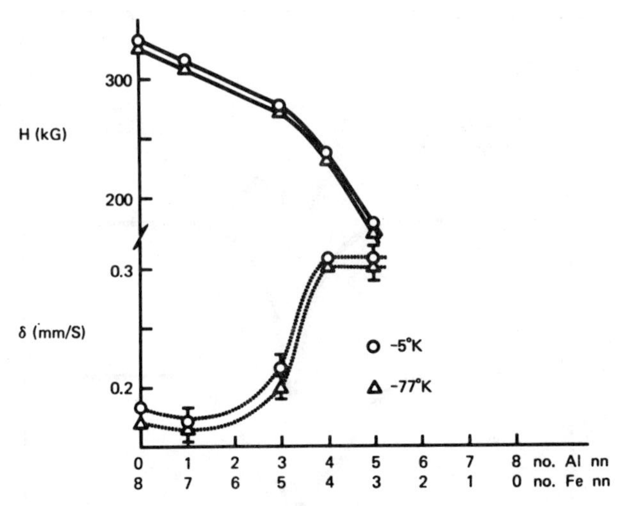

Fig. 1. Hyperfine fields (H) and isomer shifts (δ; measured with respect to metallic Fe at room temperature) as a function of nn configuration for a DO_3-ordered 25.8 at.% Al alloy.

hyperfine fields on nn configuration mirrors that of the magnetic moment,[3] while the variation of the isomer shift with nn number appears to be in accord with a recent band theoretical calculation for Fe_3Si and Fe_3Al.[8] Similar variations with nn number were observed for all of the samples studied. In the current work, normal ferromagnetic behavior is indicated by a linear variation of the hyperfine fields with $T^{3/2}$, and the results for the DO_3-ordered 25.8 at.% Al alloy are shown in Fig. 2.

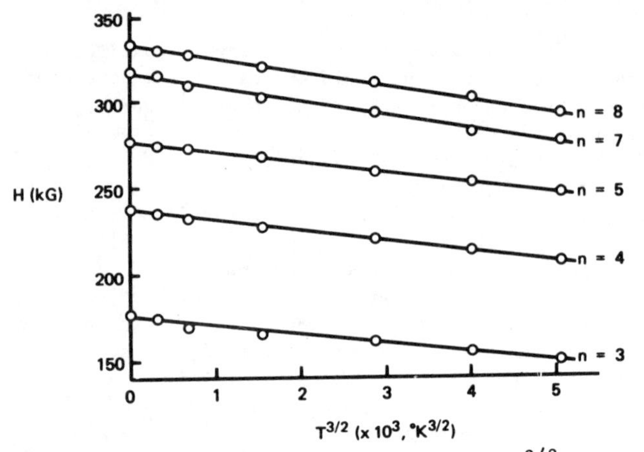

Fig. 2. Hyperfine fields as a function of $T^{3/2}$ for a DO_3-ordered 25.8 at.% Al alloy; n denotes the number of Fe nn.

A $T^{3/2}$ dependence of the hyperfine fields was also observed for DO_3-ordered alloys containing 23, 24.5 and 26.8 at.% Al and for a CsCl-ordered 25.8 at.% Al alloy.

Significant deviations from the $T^{3/2}$ dependence of the hyperfine fields are first observed for the 26.8 at.% Al alloy ordered in the CsCl structure, and become particularly striking near P_{Al} = 30 at.%, where the hyperfine fields exhibit sharp minima as a function of temperature. Some typical spectra for a CsCl ordered 29.8 at.% Al alloy are shown in Fig. 3.

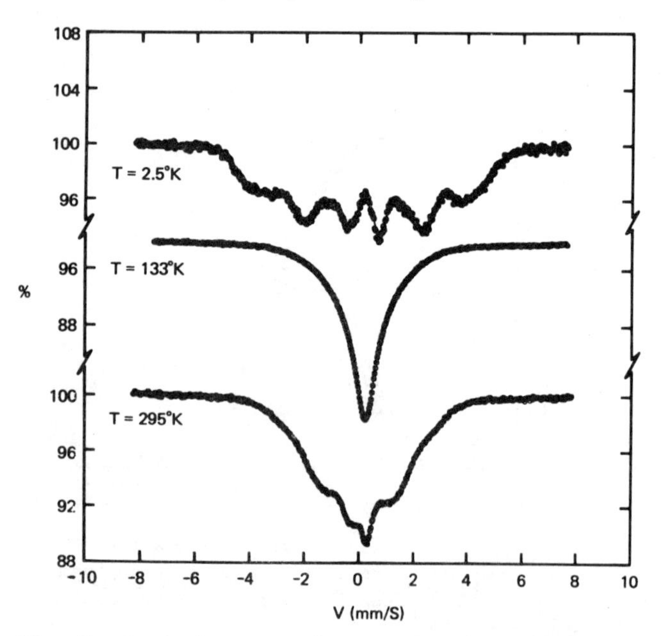

Fig. 3. Typical spectra for a CsCl-ordered 29.8 at.% Al alloy.

Reasonable least squares fits (solid curves) to such a spectra were obtained assuming five or six magnetic hyperfine components. As discussed in reference 2, each component of the spectra obtained near liquid He temperature can be tentatively assigned to a particular nn configuration, but this identification loses its meaning for spectra such as that shown in Fig. 3 for T=133°K. Nevertheless, the average hyperfine field, determined by weighting the component fields according to relative intensity serves as a meaningful indicator of the rapid cluster moment relaxation process taking place. The temperature dependence of the average hyperfine fields for six samples in the ferromagnetic to mictomagnetic transition region are shown in Fig. 4.

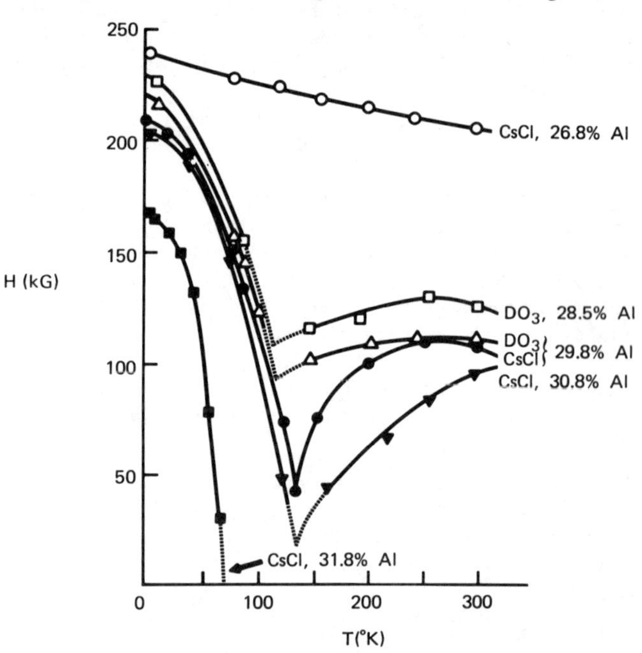

Fig. 4. Average hyperfine fields as a function of temperature for five alloys in the ferromagnetic to mictomagnetic transition region:

⊙ 26.8 at.% Al,CsCl	⊡ 28.5 at.% Al,DO_3
Δ 29.8 at.% Al,DO_3	● 29.8 at.% Al,CsCl
▼ 30.8 at.% Al,CsCl	■ 31.8 at.% Al,CsCl

The CsCl-ordered 26.8 and 31.8 at.% Al alloys appear to be quite close to the boundaries of the transition region, with the 26.8 at.% alloy exhibiting only a slight inflection in its hyperfine field-temperature curve, and the 31.8 at.% alloy showing no minimum and having a hyperfine field temperature dependence similar to that of the 35 to 50 at.% Al alloys.[1] For $P_{Al} \geq$ 31.8 at.%, the clusters are presumably small enough that their blocking temperatures are below the "Néel" points of the antiferromagnetically aligned regions separating them.

Recently, Shull and Beck have made A-C susceptibility measurements on ordered Fe-Al alloys,[7] and, between 27 and 31 at.% Al, they observe χ vs. T curves which decrease from their room temperature levels to rather flat minima at temperatures somewhat above the characteristic mictomagnetic or spin-glass cusp. These minima occur at approximately the same temperatures as the hyperfine field minima shown in Fig. 4.

Finally, since the collapse of the Mossbauer spectra at intermediate temperatures illustrated in Figs. 3 and 4 is clearly due to the rapid relaxation of superparamagnetic spin clusters, it should be possible to analyze the spectra to determine the cluster moment relaxation times. Results of this type have been obtained using the model for superparamagnetic Mossbauer spectra derived by Wickman.[9] Although application of this model to the current alloys is not entirely straightforward, the preliminary results seem quite reasonable. For example, the average cluster spin relaxation time determined for the CsCl-ordered 29.8 at.% Al alloy at 133°K is 2.2×10^{-9} sec. Further results obtained by this method of analysis will be presented elsewhere.

ACKNOWLEDGMENT

The authors are grateful to Professor P.A. Beck and Dr. R.D. Shull for a discussion of their A-C susceptibility results prior to publication.

REFERENCES

1. G.P. Huffman, "Mossbauer Study and Molecular Field Theory of the Magnetic Properties of Fe-Al Alloys," J. App. Phys. 42, pp. 1606-1607 (1971).
2. G.P. Huffman, Amorphous Magnetism, pp.283-290, (ed. by H.O. Hooper and A.M. deGraaf; Plenum, 1973).
3. P.A. Beck, "Some Recent Results on Magnetism in Alloys," Met. Trans. 2, pp. 2015-2024 (1971).
4. R. Nathans and S. Pickart, Magnetism, Vol. III, p. 235 (Academic, 1963).
5. A. Arrott and H. Sato, "Transitions from Ferromagnetism to Antiferromagnetism in Iron-Aluminum Alloys," Phys. Rev. 114, pp. 1420-1426 (1956).
6. J.S. Kouvel, "Exchange Anisotropy in an Iron-Aluminum Alloy," J. Appl. Phys. 30, 313S (1959).
7. R.D. Shull and P.A. Beck, "Magnetic Properties of Fe-Al Alloys," 21st Conf. on Magnetism and Magnetic Materials, paper 6A-11 (1975).
8. A.C. Switendick, "A Theoretical Model for Site Preference of Transition Metal Solutes in Fe_3Si," to be published in Solid St. Comm.
9. H.H. Wickman, "Mossbauer Paramagnetic Hyperfine Structure," Mossbauer Effect Methodology, Vol. 2, pp. 39-66 (ed. by I.J. Gruevrman, Plenum, 1966).

MONTE CARLO STUDIES OF CLASSICAL THREE DIMENSIONAL HEISENBERG SPIN GLASSES*

W. Y. Ching and D. L. Huber
Dept. of Physics, Univ. of Wisconsin, Madison, Wis. 53706

ABSTRACT

We report the results of Monte Carlo studies of classical three dimensional Heisenberg spin glasses. We calculate the internal energy, susceptibility, specific heat, effective order parameter, and spin-spin autocorrelation function for 4x4x4 arrays. The data, averaged over twelve runs, qualitatively support the characterization of the spin glass phase given by the theory of Edwards and Anderson.

INTRODUCTION

In order to account for the magnetic and thermodynamic properties of a variety of so-called spin glass alloys[1] Edwards and Anderson (EA)[2] proposed a classical model involving a Heisenberg interaction $J_{ij}\vec{S}_i\cdot\vec{S}_j$ between spins on sites i and j. The novel feature of their model is the assumption that the exchange integral J_{ij} is a random function having a Gaussian distribution centered about zero, with no correlation between different bonds. Using what is basically a mean field approach they showed that the system underwent a phase transition which gave rise to discontinuities in the slope of the susceptibility $\chi(T)$ and the specific heat $C_H(T)$ at the freezing temperature T_f. The ordered state below T_f was characterized by a finite value for the parameter $q = [\langle\vec{S}_i\rangle^2]_{AV}$, where < > denotes a thermal average for a fixed configuration of interactions, and $[\ ]_{AV}$ refers to an average over the distribution of exchange integrals. Subsequently, the analysis of EA has been extended to quantum mechanical spins by Fischer,[3] and to an analogous model involving Ising interactions by Sherrington and Kirkpatrick.[4] Recently, Harris et al.[5] have investigated the critical properties of the spin glass Hamiltonian using renormalization group techniques.

In assessing the significance of these studies it is important to bear in mind that there are two fundamental problems which are being addressed. The first is the development of realistic yet tractable models for alloy systems, the second is the detailed analysis of the features of the various models. The work we are reporting falls in the second category. We have undertaken a study of the thermodynamic properties of small clusters of exchange coupled classical spins using Monte Carlo techniques.[6]

Following EA the exchange interaction was taken to be of the form $J_{ij}\vec{S}_i\vec{S}_j$ where the $\vec{S}_i$ are spins of unit magnitude and i and j refer to nearest neighbor sites on a simple cubic lattice. The exchange integrals were postulated to have a Gaussian distribution with mean zero and rms width ΔJ. We have calculated the internal energy, susceptibility, specific heat, and the effective EA order parameter for 4x4x4 arrays with periodic boundary conditions. In addition, we have obtained data on the site-averaged spin-spin autocorrelation function. We partially compensate for the relatively small size of the array by averaging our data over twelve computer runs each with different configurations of exchange integrals and different sequences of random numbers. The finite size of the array of course precludes any quantitative analysis of the critical behavior of the model. Nevertheless within the limitations of our approach our results are qualitatively consistent with the picture of the spin glass provided by the EA analysis. Our findings are also consistent with the spin glass behavior reported by Binder and Schröder in their Monte Carlo studies of two dimensional Ising spin glasses.[7]

Since we made use of standard Monte Carlo techniques[6] we will not discuss the calculations in detail except to mention that the computer runs involved 1200 Monte Carlo steps per spin. The thermodynamic functions we obtained by averaging the data beginning with the 200th step, while the time-dependent functions were calculated with the origin in time being the 200th step. The runs were at consecutively lower temperatures with the final configuration of one run being the initial configuration of the next.

RESULTS

Our results for U(T) (the internal energy), $\chi(T)$, and $C_H(T)$ over the interval $0.1 \leq kT/\Delta J \leq 2.0$ are displayed in Figs. 1, 2, and 3 respectively. We have plotted the data obtained by averaging over twelve runs. The error bars represent the root mean square (rms) deviations. We also give the predictions of the EA theory, which are shown as solid curves. Not surprisingly the data at high temperatures compare favorably with the theory which is asymptotically exact as $T \to \infty$. In the case of U(T) the rms deviations remain relatively constant down to $kT/\ J = 0.1$, and there is reasonably quantitative agreement between experiment and theory over the entire temperature interval.

In the case of both $\chi(T)$ and $C_H(T)$ there is a significant increase in the rms deviations below $kT/\Delta J = 0.7$. This increase probably reflects the presence of metastable states,[7] which are beginning to make a significant contribution to the averages. Particularly noticeable is the pronounced peak in $\chi(T)$ which occurs for $kT/\Delta J \approx 0.4$. There is also a suggestion of a peak in $C_H(T)$ at the same temperature. In order to show this more clearly we have numerically differentiated U(T). The result is the broken curve in Fig. 3. It is worth noting that the peaks in $\chi(T)$ and $C_H(T)$ occur at roughly one half the freezing temperature in the EA theory, $.8165\ \Delta J/k$.

In Fig. 4 we have plotted an effective order parameter, $q_{eff}(T)$, which is defined by

$$q_{eff}(T) = N^{-1}\ \Sigma_i\ [\langle\vec{S}_i\rangle^2] \qquad (1)$$

where the sum is over the N spins in the array. The angular brackets < > represent a thermal average in the Monte Carlo sense, while the square brackets signify an average over the 12 computer runs. As before we have indicated the predictions of EA by the solid line. In spite of the small size of the array and the finite length of the computer runs the behavior of $q_{eff}(T)$ is qualitatively similar to the behavior of the order parameter in the EA theory in that it falls from approximately one to approximately zero in the interval $0 \leq kT/\Delta J \lesssim 0.7$.

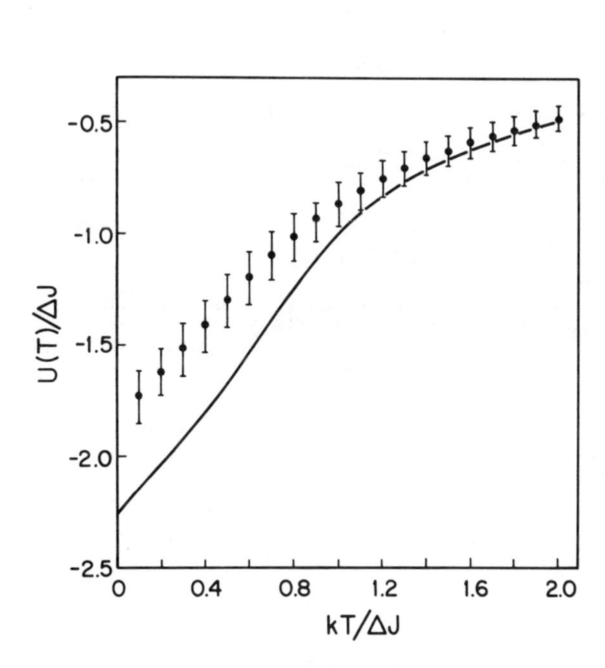

Fig. 1. $U(T)/\Delta J$ vs $kT/\Delta J$. Here and in Figs. 2, 3, and 4 the predictions of the mean field treatment of the EA model are shown as solid curves. The latter are obtained from Ref. 3 in the limit $S = \infty$ with $kT_f/\Delta J = \sqrt{2/3}$.

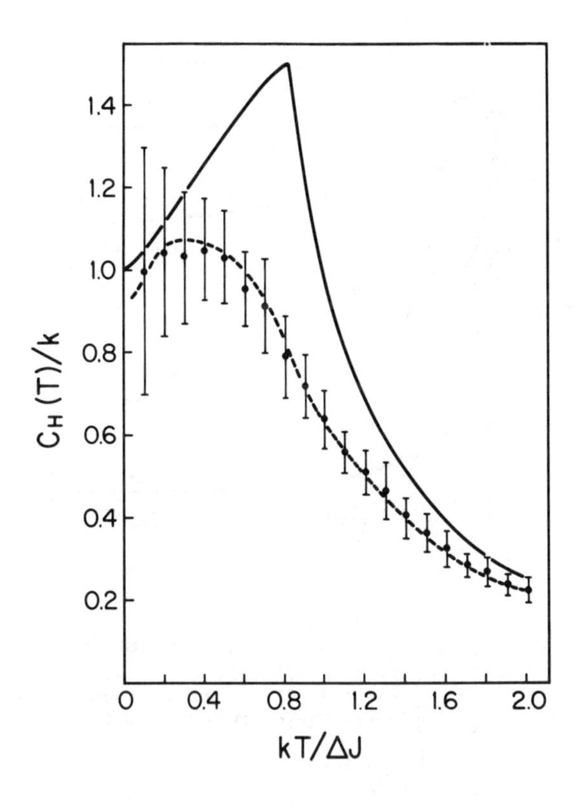

Fig. 3. $C_H(T)/k$ vs $kT/\Delta J$. The broken curve is obtained by numerically differentiating the data for $U(T)$ shown in Fig. 1.

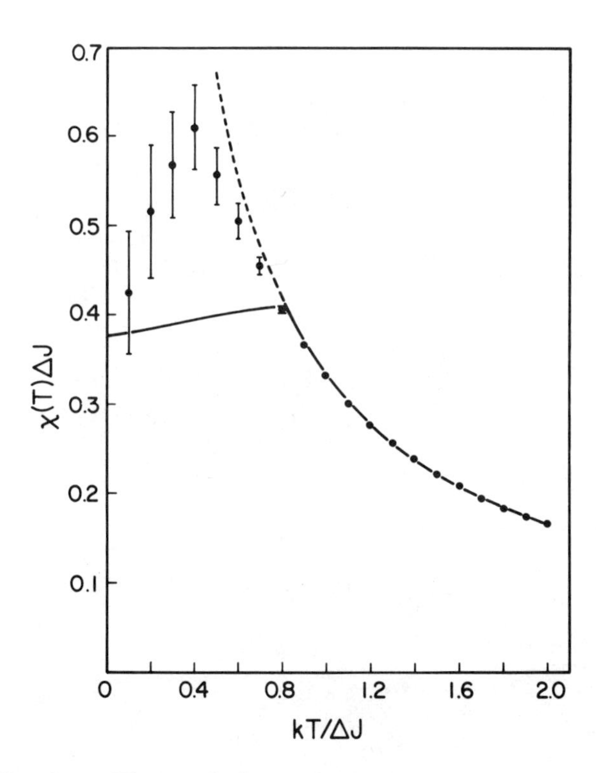

Fig. 2. $\chi(T)\Delta J$ vs $kT/\Delta J$. The broken curve is $\Delta J/3kT$.

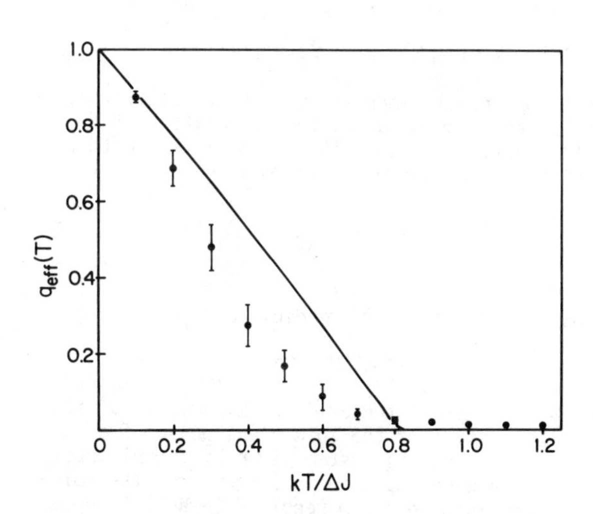

Fig. 4. $q_{eff}(T)$ vs $kT/\Delta J$.

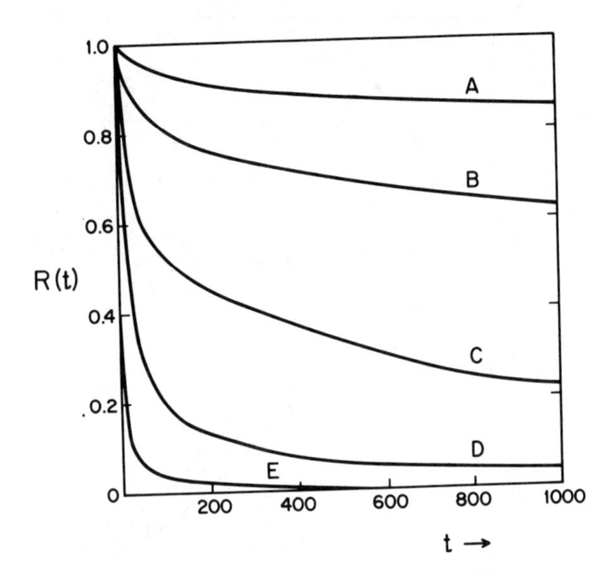

Fig. 5. Spin-spin autocorrelation function, R(t), vs t for kT/ΔJ = 0.1 (A), 0.2 (B), 0.4 (C), 0.7 (D), and 1.3 (E). Time is measured in units of Monte Carlo steps per spin.

In Fig. 5 we have plotted the effective spin-spin autocorrelation function, R(t), defined by

$$R(t) = N^{-1} \Sigma_i \, [\, \vec{S}_i \cdot \vec{S}_i(t)] \tag{2}$$

for a variety of temperatures. Particularly noticeable is the slow decay of the correlations for kT/ΔJ < 0.7. Such behavior is consistent with Fig.4 in that for the infinite system we have

$$\lim_{t\to\infty} [\langle \vec{S}_i \cdot \vec{S}_i(t) \rangle]_{AV} = [\langle \vec{S}_i \rangle^2]_{AV} = q(T), \tag{3}$$

assuming the system is ergodic.

DISCUSSION

We have presented evidence in support of the characterization of the spin glass phase of a classical Heisenberg system given by the EA theory. Our results along with the analogous results for the equivalent Ising problem reported in Ref. 7 suggest that the mean field theories of model spin glass systems give a qualitatively correct characterization of the thermodynamic behavior. In view of the uncertainties about mapping real alloys onto model Hamiltonians Monte Carlo calculations may very well provide a more stringent test of theories for the model systems than is gotten by comparison with data obtained from experimental studies of the conventional type.

REFERENCES

* Research supported by the National Science Foundation.

1. J. A. Mydosh, AIP Conf. Proc. 24, 131 (1975).
2. S. F. Edwards and P. W. Anderson, J. Phys. F: Metal Phys. 5, 965 (1975).
3. K. H. Fischer, Phys. Rev. Lett. 34, 1438 (1975).
4. D. Sherrington and S. Kirkpatrick, Phys. Rev. Lett. 35, 1792 (1975).
5. A. B. Harris, T. C. Lubensky, and J.-H. Chen, Phys. Rev. Lett. 36, 415 (1976).
6. D. P. Landau, AIP Conf. Proc. 18, 819 (1974).
7. K. Binder and K. Schröder (preprint).

NEW DIRECTIONS IN CRITICAL PHENOMENA

D. P. Landau*
University of Georgia, Athens, Georgia 30602

ABSTRACT

The present state of knowledge in critical phenomena is reviewed and new directions for research in the field are suggested. This discussion was inspired by the conference workshop on critical phenomena. Areas of interest include experiments and calculations related to scaling theory and the renormalization group, multicritical phenomena, random system behavior, characteristics of lower dimensional models and surfaces, and critical and multicritical dynamics.

I. INTRODUCTION

The past several decades have seen both intense study and success in the area of critical phenomena. The field is clearly an established, mature one in which many of the basic problems have been "solved" (i.e. at least to a high degree of numerical accuracy if not exactly), and it now seems appropriate to consider the future research topics. Are there important, interesting challenges remaining or should most of us allow the field to settle in a quiescent, old age while we search in other areas of physics for more exciting problems? In this paper we hope to consider (and in part to answer) this question. The discussion combines the author's personal (and necessarily subjective views) and those expressed during the Panel Workshop held during this conference. The multifaceted nature of many of the questions makes any grouping somewhat arbitrary and the reader should note that in some cases different aspects of the same problem appear in different sections.

II. "ORDINARY" CRITICAL POINTS

In this section, we shall consider those kinds of simple critical points which were studied in so much detail in the 50's and 60's. The asymptotic critical behavior of numerous simple lattice models has been studied in detail by series expansions[1] and more recently by the renormalization group method.[2] Developments in renormalization group techniques are particularly important since they appear to lay out the conceptual tools needed to really understand critical phenomena. The dependence of critical exponents on only a few of a system's characteristics, such as spin dimensionality n and lattice dimensionality d, is theoretically well accepted. The experimental picture, however, has been relatively disappointing as the critical exponents extracted from data on real systems show a range of values some of which seem inconsistent with theory. We now believe that many of the inconsistencies arise from the neglect of correction terms (confluent singularities) to the dominant asymptotic critical behavior in the analysis. Rather convincing evidence has been presented to show that this was the source of the apparent spin dependence of Ising model exponents.[3] Initial attempts to include such corrections for other models and in data analysis have proven quite promising.[4-7] We wish to emphasize that the choice of correction terms is non-trivial but it would seem that theory is now sufficiently advanced that the form for corrections could be guided by present theoretical estimates and scaling theory constraints. Both new experiments as well as reanalyses of old experimental data[8] seem desirable. Of course one lingering difficulty in experimental analysis is the problem of "rounding" near T_c. A method for "correcting" or at least interpreting the rounded region so that these data could be included in the analysis would be highly desirable.

It has long been speculated that fluid critical behavior should be the same as that of the Ising model. Much experimental work[9] suggested that fluid exponents were nearly universal but slightly different from the Ising values. In a very beautiful set of experiments Hocken and Moldover[10] showed that very close to T_c ($-1.5\times10^{-5}<(1-T/T_c)<5\times10^{-5}$!) Xe, SF_6, and CO_2 all have exponents and amplitude ratios which agree with the Ising model values. Corrections to scaling are obviously important in fluids but with the wide range of reduced temperature over which data are available it should be possible to study these corrections in detail.

Although many of the exponent scaling laws have been well verified, both theoretically and experimentally, there are still nagging questions concerning (hyperscaling) those relations which include the lattice dimensionality d:

$$d\nu=2-\alpha \tag{1a}$$

$$d\frac{(\delta-1)}{(\delta+1)} = 2-\eta \tag{1b}$$

The most recent series expansion results tend to indicate that hyperscaling (Eqns. 1a, 1b) are violated by a small amount in 3 dim. It is therefore particularly important that the effects of correction terms are considered in careful reanalyses for α and ν and δ. In addition, very beautiful recent work by Tracy and McCoy[11] has shown that the large y behavior (where $y=k\xi$) defines the exponent η and that both previous theoretical and experimental results did not satisfy the conditions necessary for the unambiguous determination of η. The renormalization group theory has hyperscaling "built in" and thus provides no independent information on this question. The strongest support for the validity of hyperscaling comes from data on He4 where the relation $d\nu=2-\alpha$ is tested. The results yield[12] 2.025±.003 and 2.026±.003 for the left hand and right hand sides respectively. Whether this result is peculiar to He4 or due simply to the nature of the system which allows high precision measurements to be made much closer to T_c than in real magnetic systems is simply not presently clear.

The expected critical behavior in real systems has been complicated (or perhaps clarified) by the realization that many "different" types of critical behavior may occur.[13-15] The presence of long-range dipolar interactions may give rise to new critical exponents. For example, a specific heat anomaly $\sim(\ln t)^{1/3}$ is expected for a uniaxial system and this behavior will be difficult to separate experimentally from a small value of α. In addition it has recently been shown[15] that many cubic systems with spatially degenerate antiferromagnetic structures may in fact behave as if they had effective spin-dimensionality $n\geq4$! Many real magnetic systems fall into this category, e.g. MnS_2, Eu, Cr, Dy, and "unusual" zero field critical behavior may therefore be expected.

One of the greatest theoretical challenges existing at the moment is the further development of real space renormalization group techniques.[16,17] The importance of this problem certainly extends to other areas (such as multicritical phenomena) but the most immediate application involves the comparison of critical temperatures and amplitudes (as well as exponents) associated with ordinary critical points.

III. RANDOM SYSTEMS

A wide range of models with various types of randomness built in offer a rich source of investigation. The application to real systems which may have random impurities, strains, and lattice distortions is obvious. The effect of random annealed (or mobile) impurities is known[18,19] to simply renormalize the usual critical exponents. The problem of fixed or quenched impurities is more complicated. This situation has been studied[20] in two- and three-dimensions by series expansions,[21,22] renormalization group,[23] and Monte Carlo.[24-27] The picture which seems to be emerging is that for $\alpha \lesssim 0$ the critical exponents are unaffected by quenched impurities but for $\alpha > 0$ the exponents assume new values. Present evidence suggests, however, that the crossover between pure and impure critical behavior occurs very close to T_c for small impurity concentrations and the impure critical behavior might be very difficult to observe. Because of this constraint investigations of (impure) exotic systems with unusually large values of α appear to offer the best hope for success. The elementary excitations of both mixed and impure magnetic systems have been examined using neutron scattering but studies of critical behavior are sparse. The mixed layer antiferromagnet $Rb_2Mn_{0.5}Ni_{0.5}F_4$ yields[28] d=2 Ising exponents within the experimentally accessible region, and a neutron study[29] of the impure system $Rb_2Mn_xMg_{1-x}F_4$ is underway. Clearly more experiments (including specific heat measurements) are called for.

One special region of the impurity concentration-temperature space is near the percolation limit[30] where the transition may be driven both thermally and via concentration. The nature of the critical exponents at the percolation limit remains to be determined. Crossover to finite temperature impure behavior as well as a breakdown of dynamic scaling have been proposed[30] and awaits testing by both experiment and other theoretical means. A rather different type of change occurs for Pt in Ni. Here $T_c \to 0$ at 50-50% concentration[31] because the Ni loses its moment. The connection with other kinds of "dilute" critical behavior is not transparent. Additional theoretical and experimental studies of impure itinerant systems are certainly needed.

One non-traditional problem attracting current interest is the spin-glass "freezing" which occurs for systems having relatively low magnetic site concentration. In this case experiment[32] is more advanced than theory and the spin-glass state has been observed in numerous real systems. From a theoretical point of view the problem is quite complicated. The relatively simple Edwards-Anderson model[33] has been recently studied by the Monte Carlo method.[34-36] The results indicate that the Edwards-Anderson model does qualitatively describe the spin-glass transition, but more sophisticated theories need to be developed.

One last type of "random" behavior is the gelation of polymer solutions. Here the state of knowledge is still relatively primitive and experiment and theory seem to disagree.[37] For solutions of different concentration a range of models including self-avoiding walks, and Cayley trees seem to be appropriate. One interesting question which arises concerns the possibility of two characteristic lengths becoming important. One length ξ_L measures the actual length along the polymer between sites and the second length ξ_c is the straight line distance between those sites. In the self-avoiding walk problem the two lengths are related but in the general case the significance of the two lengths remains to be determined. Predictions[38] have been made for the ferromagnetic susceptibility and neutron scattering function $S(\vec{\kappa},\omega)$ of magnetic polymers. Recent advances in small angle neutron scattering offer particular promise obtaining microscopic information regarding both magnetic and nonmagnetic polymers.[39]

IV. MULTICRITICAL PHENOMENA

The recent developments in critical phenomena of multicomponent systems have been truly striking. Since Griffiths[40] ushered in the "modern era" of multicritical phenomena, a broad range of experimental and theoretical techniques have been used to study multicritical points and the phase diagrams which occur in systems possessing them. Simple tricritical points in spatial dimension $d \gtrsim 3$ are now fairly well understood.[41-47] For the Gaussian model and a wide range of competing interaction Ising models the tricritical exponents are mean-field-like for d>3 and mean-field-like with logarithmic corrections for d=3. Mean-field theory predicts[48] that for certain ratios of competing interactions the tricritical point separates into a critical endpoint and a double critical point (bicritical endpoint). This behavior may be an artifact of mean-field theory but a search for such behavior is being carried out using the Monte Carlo method[49] and further studies by series expansions and renormalization group methods would be desirable. Experimental data on real magnetic systems is still rather sparse. Optical rotation[50] and neutron[51] scattering results on $FeCl_2$ are not totally consistent and one of the prime candidates for study, DAG, has been plagued with unwanted "staggered fields" produced by the unusual magnetic symmetry.[52] The only existing high resolution measurements are for DAG with the field along the [110] direction. Using a novel approach based on hysteresis effects, Giordano and Wolf[53] showed that the tricritical behavior of the magnetization and non-ordering susceptibility was consistent with theory. They also demonstrated that tricritical scaling[54] was valid although a pronounced asymmetry in the size of the scaling regions in the ordered and disordered phases was observed. A great many highly anisotropic magnetic systems exist[55] and tricritical points should be plentiful. Clearly more high resolution experimental results are needed to determine exponents, amplitudes and to test predictions of tricritical scaling functions.[56]

Tricritical behavior in 2-dim is less well understood although Monte Carlo calculations[57] and a renormalization group study[58] have given preliminary results. Experimental studies of real pseudo-2-dim magnetic systems are in progress[59] and the comparison between experiment and theory should be interesting.

The experimental data[60] for the He^3 molar concentration and concentration susceptibility have been analyzed[61] in terms of a scaling field theory. Values of the tricritical exponents extracted are in agreement with Riedel-Wegner predictions and scaling holds but the range of apparent tricritical scaling is much larger in the normal fluid than in the superfluid phase.

Tricritical points also exist in ternary and quaternary liquid mixtures.[62] These are examples of "non-symmetric" tricritical points since none of the three phases simultaneously becoming critical are related by symmetry. A pertinent modification of the BEG model has been studied[63] using mean-field theory but detailed experimental results are limited.[64] These experiments are quite difficult because temperature, pressure and composition must all be properly adjusted for the tricritical point to appear.

One particularly intriguing possibility[65] involves tricritical points in ferroelectrics where both the ordering and disordering fields (pressure and electric field) are experimentally accessible!

Recent developments concerning other types of multicritical points suggest a rich variety of problems. Bicritical points occur in a wide variety of anisotropic antiferromagnets.[66-74] Predictions have been made[66] for bicritical exponents, proper scaling axes, and the crossover behavior. Some experimental results[70-74] are available for n=2 and n=3 bicritical points but the agreement between theory and experiment is not complete on all counts. Further work is needed to determine if the low order theory needs substantial correction or if the experiments are wanting (e.g. because of a very small ssymptotic bicritical region). Bicritical points for d=2, n=3 are fascinating since lowest order renormalization group calculations[68,69] indicate T_b=0. It may be expected, however, that a more realistic Hamiltonian would include higher order symmetry breaking terms which reflect the underlying lattice structure (e.g. cubic interactions) which should raise T_b above 0. Monte Carlo studies[74] on an anisotropic square lattice show an extremely pronounced umbilicus with T_b appearing to be at finite T. Groundwork on bicritical equations of state and crossover scaling functions has been laid[75] and continued efforts appear promising. Clearly much remains to be done from both an experimental as well as theoretical point of view.

Many other systems and other types of multicritical points (tetracritical, polycritical, etc.) exist[76-82] and these areas are virtually untouched. The proliferation of multicritical points raises questions concerning proper classification schemes[83,84] and no unassailable catalogue has yet been devised.

V. TWO-DIMENSIONAL SYSTEMS

Systems with spatial dimensionality d=2 occupy a somewhat special position in the catalogue of models showing critical behavior.[85] For many Hamiltonians the d=2 case is sort of a borderline case in which long range order is barely stable and further decrease in lattice dimensionality may suffice to destroy the ordered state. One well known, and not completely understood, example involves n=2 and n=3 2-dim lattices which cannot show long range order. Instead a special type of phase transition showing only a divergent susceptibility was proposed by Stanley and Kaplan.[86] Extensive series expansions[87] and Monte Carlo studies[88] have given hints one way or the other but have not been able to exclude the alternative possibility. It now appears[69] that the Stanley-Kaplan transition does exist in the 2-dim xy model but probably not in the 2-dim Heisenberg model.

A substantial number of 2-dim models have been solved exactly yielding rather unusual results. The Baxter model[89] for example shows a continuous variation of critical exponents. The triangular Ising model with 3-spin interactions is expected to show a 1st order phase transition because of the presence of cubic terms in the Landau free energy whereas the exact solution[90] shows a continuous transition. The 3-component Potts model transition appears[91] to be 2nd order for d=2 and 1st order for d=4. What happens for $2<d<4$? How stable is the d=2 Potts model to perturbation? What is the effect of changing the number of components in the Potts model? Would the answer provide any useful physical insight? These questions are typical of those which may be raised about numerous 2-dim models.

Very nice experimental data[92] are available for a number of real (pseudo-) d=2 systems but further results would certainly be useful particularly for studying crossover[93] between d=2 and d=3 critical behavior. Perhaps the greatest hope for studying 2-dim phase transitions lies in adsorbed gas layers on solid substrates. True monolayers may be readily produced and lattice gas type solid-liquid-gas phase transitions as well as solid-solid transitions have been observed.[94] For the lighter gases, e.g. He^3, He^4, quantum effects should be readily observable. It should be noted that a suitable variable space for these studies is concentration-temperature which corresponds to magnetization-temperature (rather than magnetic field-temperature) in magnetic language. Monte Carlo studies[95] have shown that triple points and/or tricritical points may be expected as a simple consequence of competing interactions between the adsorbed gas atoms.

Surfaces of binary alloys[95] and of magnetic systems[95,96] are also challenging targets for further study.

VI. CRITICAL DYNAMICS

We have deliberately separated out all discussion of dynamics although clearly dynamic behavior near critical and multicritical points in impure as well as pure systems and in lower dimensional as well as d=3 systems are of interest. Qualitatively most theoretical methods are sufficient. Many more Universality Classes exist[97] for dynamic as opposed to static behavior and there are also more sources of confluent corrections. Simple gas-liquid phase transitions are well understood in terms of mode-mode coupling.[98] The dynamic scaling hypothesis[99] has been verified in a few cases in magnetic systems.[100] Measurements of the thermal conductivity of He^4 as a function of pressure[101] seem to violate universality and light scattering results[102] near T_λ do not agree completely with mode-mode coupling. Are these problems due to corrections to scaling?

Relatively little has been done about dynamic multicritical behavior. Mode-mode coupling appears to work near T_t in He^3/He^4 mixtures,[103] and a Monte Carlo study of a kinetic Ising model[104] yielded relaxation time behavior which was consistent with conventional theory. Dynamic bicritical behavior has been studied theoretically[105] but there are no experimental results available for comparison.

Experimental studies[106] of the dynamic behavior of displacive transitions have yielded unusual results. Recent molecular dynamics studies[107] suggest that the observed central peak may arise from cluster dynamics. This is clearly an interesting problem which is only partially solved.

Dynamic critical behavior is a complex problem but we have only begun to scratch the surface.

SUMMARY AND CONCLUSIONS

The discussion in the previous sections clearly shows the vitality remaining in the field. To be sure, the emphasis has shifted away from what we termed "ordinary" critical phenomena (i.e. Sec. II) in recent years but even this well studied area has not yielded all of its secrets. New techniques, including more sophisticated methods of generation and analysis of series expansions, computer simulation methods, and in particular various renormalization group methods, offer a variety of powerful tools for the theoretical study of new problems in critical phenomena. We feel that advances in theory will be an invaluable guide for the choice of systems to be studied experimentally as well as for the interpretation of experimental data.

REFERENCES

* Supported in part by the National Science Foundation

1. An excellent set of reviews including extensive references can be found in "Phase Transitions and Critical Phenomena" Vol. 3, ed. C. Domb and M. S. Green (Academic Press, N.Y. 1974).
2. See M. E. Fisher, Rev. Mod. Phys. 46, 597 (1974) and references therein.
3. W. J. Camp, D. M. Saul, J. P. Van Dyke, and

M. Wortis, Phys. Rev. (in press).
4. D. M. Saul, M. Wortis, and D. Jasnow, Phys. Rev. B 11, 2571 (1975).
5. W. J. Camp and J. P. Van Dyke, J. Phys. A9, 731 (1976).
6. G. Ahlers and A. Kornblit, Phys. Rev. B12, 1938 (1975).
7. M. Barmatz, P. C. Hohenberg, and A. Kornblit, Phys. Rev. B12, 1947 (1975).
8. High quality data on "nice" systems are certainly very difficult to come by. A reconsideration of some of the detailed data of magnetic behavior would seem most promising. See, e.g.: J. T. Ho and J. D. Litster, J. Appl. Phys. 40, 1270 (1969); D. D. Berkner and J. D. Litster, AIP Conf. Proc. 10, 894 (1973); P. Heller, Rep. Prog. Phys. 30, 731 (1967).
9. J. M. H. Levelt Sengers and J. V. Sengers, Phys. Rev. A12, 2622 (1975).
10. R. Hocken and M. R. Moldover, Phys. Rev. Lett. 37, 29 (1976).
11. C. A. Tracy and B. M. McCoy, Phys. Rev. B12, 368 (1975).
12. G. Ahlers (private communication).
13. A. I. Larkin and D. E. Khmel'nitskii, Soviet Physics JETP 29, 1123 (1969).
14. G. Ahlers, A. Kornblit, and H. J. Guggenheim, Phys. Rev. Lett. 34, 1227 (1975).
15. P. Bak and D. Mukamel, Phys. Rev. B13, 5086 (1976); D. Mukamel and S. Krinsky, Phys. Rev. B13, 5065 (1976).
16. Th. Niemeijer and J. M. J. Van Leeuwen, Phys. Rev. Lett. 31 1411 (1973); Physica 71, 17 (1974).
17. L. P. Kadanoff and A. Houghton, Phys. Rev. B11, 377 (1975).
18. I. Syozi, Prog. Theor. Phys. 34, 189, (1965).
19. H. Garelick and J. W. Essam, Proc. Phys. Soc. 92, 136 (1967).
20. M. E. Fisher and H. Au-Yang, J. Phys. C8, L418 (1975).
21. G. S. Rushbrooke, R. A. Muse, R. L. Stephenson and K. Pirnie, J. Phys. C5, 3371 (1972).
22. A. B. Harris, J. Phys. C7, 1671 (1974).
23. G. Grinstein and A. Luther, Phys. Rev. B13, 1329 (1976).
24. W. Y. Ching and D. L. Huber, Phys. Rev. B13, 2962 (1976).
25. E. Stoll and T. Schneider, AIP Conf. Proc. 29, 490 (1976).
26. R. Fisch and A. B. Harris, AIP Conf. Proc. 29, 488 (1976).
27. D. P. Landau, Proc. of 1976 Int. Mag. Conf. Physica (in press).
28. J. Als-Nielsen, R. J. Birgeneau, H. J. Guggenheim and G. Shirane, Phys. Rev. B12, 4963 (1975).
29. R. J. Birgeneau, R. Cowley, and G. Shirane (unpublished).
30. D. Stauffer, Z. Physik B22, 161 (1975). For reviews see A. R. Bishop, Prog. Theor. Phys. 53, 50 (1975); S. Kirkpatrick, Rev. Mod. Phys. 45, 574 (1973).
31. H. L. Alberts, J. Beille, D. Bloch, and E. P. Wohlfarth, Phys. Rev. B9, 2233 (1974).
32. V. Canella, J. A. Mydosh, and J. I. Budnick, J. Appl. Phys. 42, 1689 (1971); V. Canella and J. A. Mydosh, AIP Conf. Proc. 18, 651 (1974); J. A. Mydosh, AIP Conf. Proc. 24 131 (1975).
33. S. F. Edwards and P. W. Anderson, J. Phys. F5, 965 (1975).
34. K. Binder and K. Schroder, Phys. Rev. B (in press).
35. K. Binder and D. Stauffer, (to be published).
36. W. Y. Ching and D. L. Huber, (see the proceedings of this conference).
37. H. E. Stanley (private communication).
38. M. F. Thorpe, Phys. Rev. B13, 2186 (1976).
39. See the Proceedings of the Conf. on Neutron Scattering (Gatlinburg, June 6-10, 1976).
40. R. B. Griffiths, Phys. Rev. Lett. 24, 715 (1970).
41. M. Blume, V. J. Emergy, and R. B. Griffiths, Phys. Rev. A4, 1971 (1971).
42. E. K. Riedel and F. J. Wegner, Phys. Rev. Lett. 29, 349 (1972).
43. F. J. Wegner and E. K. Riedel, Phys. Rev. B7, 248 (1973); M. J. Stephen, E. Abrahams, and J. P. Straley, Phys. Rev. B12, 256 (1976).
44. D. M. Saul, M. Wortis, and D. Stauffer, Phys. Rev. B9, 4964 (1974); A. K. Jain and D. P. Landau, Bull. Am. Phys. Soc. 21, 209 (1976) and to be published.
45. M. Wortis, F. Harbus, and H. E. Stanley, Phys. Rev. B11, 2689 (1975).
46. M. E. Fisher and D. R. Nelson, Phys. Rev. B12, 263 (1975).
47. D. P. Landau, Phys. Rev. B (in press).
48. J. M. Kincaid and E. G. D. Cohen, Phys. Lett. 50A, 317 (1974).
49. D. P. Landau (unpublished).
50. J. A. Griffin and S. E. Schnatterly, AIP Conf. Proc. 24, 195 (1975).
51. R. J. Birgeneau, AIP Conf. Proc. 24, 258 (1975); R. J. Birgeneau, G. Shirane, M. Blume and W. C. Koehler, Phys. Rev. Lett. 33, 1098 (1974).
52. W. P. Wolf, AIP Conf. Proc. 24, 255 (1975).
53. N. Giordano and W. P. Wolf, AIP Conf. Proc. 29, 459 (1976); N. Giordano and W. P. Wolf, Phys. Rev. Lett. 35, 799 (1975).
54. E. K. Riedel, Phys. Rev. Lett. 28, 675 (1972); A. Hankey, H. E. Stanley, and T. S. Chang, Phys. Rev. Lett. 29, 278 (1972).
55. W. P. Wolf, J. de Physique (Suppl.) 32, Cl (1971).
56. D. R. Nelson and J. Rudnick, Phys. Rev. Lett. 35, 178 (1975); D. R. Nelson, AIP Conf. Proc. 29, 450 (1976).
57. D. P. Landau, Phys. Rev. Lett. 28, 449 (1972); B. L. Arora and D. P. Landau, AIP Conf. Proc. 10, 870 (1973), A. K. Jain and D. P. Landau (to be published).
58. T. W. Burkhardt, (to be published).
59. F. Rys (private communication).
60. G. Goellner, R. Behringer, and H. Meyer, J. Low Temp. Phys. 13, 113 (1973).
61. E. K. Riedel, H. Meyer, and R. P. Behringer, J. Low Temp. Phys. (in press).
62. R. B. Griffiths and B. Widom, Phys. Rev. A8, 2173 (1973).
63. D. Mukamel and M. Blume, Phys. Rev. A10, 610 (1973).
64. J. C. Lang and B. Widom (to be published).
65. V. H. Schmidt (presented at the Conference on Critical Phenomena in Multicomponent Systems, Athens, Ga. April 15-17, 1974).
66. M. E. Fisher, AIP Conf. Proc. 24, 273 (1975).
67. J. M. Kosterlitz, D. R. Nelson, and M. E. Fisher, Phys. Rev. B13, 412 (1976).
68. R. A. Pelcovits and D. R. Nelson, Physics Letters 57A, 23 (1976).
69. A. M. Polyakov, Phys. Lett. 59B, 79 (1975); A. A. Migdal, ZhETP 69, 1457 (1975); E. Brézin and J. Zinn-Justin, Phys. Rev. Lett. 36, 691 (1976).
70. H. Rohrer, AIP Conf. Proc. 24, 268 (1975); H. Rohrer, Phys. Rev. Letters 34, 1638 (1975).
71. N. F. Oliveira Jr., A. Paduan Filho, and S. R. Salinas, Physics Letters 55A, 293 (1975).
72. A. R. King and H. Rohrer, AIP Conf. Proc. 29, 420 (1976).
73. D. P. Landau and K. Binder, AIP Conf. Proc. 29, 461 (1976).
74. D. P. Landau and K. Binder (unpublished).
75. D. R. Nelson and E. Domany, Phys. Rev. B13, 236 (1976).
76. K. A. Muller and W. Berlinger, Phys. Rev. Lett.

35, 1547 (1975); W. B. Yelon, D. E. Cox, P. J. Kortman, and W. B. Daniels, Phys. Rev. B9, 4843 (1974); C. W. Garland, D. E. Bruins, and T. J. Greytak, Phys. Rev. B12, 2759 (1975).
77. P. H. Keyes, H. T. Weston, and W. B. Daniels, Phys. Rev. Lett. 31, 628 (1973).
78. M. R. H. Khajehpour, Y. L. Wang, and R. A. Kromhout, Phys. Rev. B12, 1849 (1975).
79. A. Aharony and A. D. Bruce, Phys. Rev. Lett. 33, 427 (1974).
80. A. Aharony (to be published).
81. M. E. Fisher and D. R. Nelson, Phys. Rev. Lett. 32, 1350 (1974).
82. D. Mukamel (to be published).
83. T. S. Chang, A. Hankey, and H. E. Stanley, Phys. Rev. B8, 346 (1973).
84. R. B. Griffiths, Phys. Rev. B12, 345 (1975).
85. An excellent review of 2-dim models (and other critical behavior as well) may be found in J. F. Nagle and J. C. Bonner, Annual Reviews of Physical Chemistry 27 (in press).
86. H. E. Stanley and T. A. Kaplan, Phys. Rev. Lett. 17, 913 (1966).
87. H. E. Stanley, Phys. Rev. Lett, 20, 589 (1968); D. S. Ritchie and M. E. Fisher, Phys. Rev. A7, 480 (1973); W. J. Camp and J. P. Van Dyke, J. Phys. C8, 336 (1974).
88. K. Binder nad D. P. Landau, Phys. Rev. B13, 1140 (1976).
89. R. J. Baxter, Phys. Rev. Lett. 26, 832 (1971).
90. R. J. Baxter and F. Y. Wu, Phys. Rev. Lett. 31 1294 (1976).
91. M. J. Stephen and L. Mittag, J. Phys. A7, L109 (1974); J. P. Straley and M. E. Fisher, J. Phys. A6, 1310 (1973); R. J. Baxter, J. Phys., C6, L445 (1973); I. G. Enting and C. Domb, J. Phys. A8, 1228 (1975); G. R. Golner, Phys. Rev. B8, 3419 (1973).
92. See for example, L. J. de Jongh and A. R. Miedema, Adv. Phys. 23, 1 (1974) and references therein; L. Bevaart, E. Frikkee, and L. J. de Jongh (to be published).
93. L. J. de Jongh and H. E. Stanley, Phys. Rev. Lett. 36, 817 (1976); F. Harbus and H. E. Stanley, Phys. Rev. B8, 2268 (1973).
94. M. Bretz, J. G. Dash, D. C. Hickernell, E. O. McLean, and O. E. Vilches, Phys. Rev. A8, 1589 (1973); D. M. Butler, G. B. Huff, R. W. Toth and G. A. Stewart, Phys. Rev. Letters 35, 1718 (1975); J. K. Kjems, L. Passell, H. Taub, and J. G. Dash, Phys. Rev. Letters 32, 724 (1974). J. K. Kjems (unpublished).
95. K. Binder and D. P. Landau (to be published).
96. K. Binder and P. C. Hohenberg, Phys. Rev. B9, 2194 (1974).
97. B. I. Halperin, P. C. Hohenberg, and S. Ma, Phys. Rev. B10, 139 (1974).
98. K. Kawasaki, in Phase Transitions and Critical Phenomena, C. Domb, M. S. Green (eds.) (Academic, N. Y. 1972); K. Kawasaki and J. Gunton, Phys. Rev. Lett. 29, 1661 (1972); M. Grover and J. Swift, J. Low Temp. Phys. 11, 751 (1973); E. D. Siggia, B. I. Halperin, and P. C. Hohenberg, Phys. Rev. B13, 2110 (1976).
99. B. I. Halperin and P. C. Hohenberg, Phys. Rev. 117, 952 (1969).
100. P. Heller, R. Nathans, and A. Linz, Phys. Rev. B1, 2304 (1970); O. W. Dietrich, J. Als-Nielsen, and L. Passell (to be published).
101. G. Ahlers, Proc. 12th Int. Conf on Low Temp. Physics (Keigaku, Tokyo, 1971), and private communication.
102. G. Winterling, F. S. Holmes, and T. J. Greytak, Phys. Rev. Lett. 30, 427 (1973).
103. P. Leiderer, D. R. Nelson, D. R. Watts, and W. W. Webb, Phys. Rev. Lett. 34, 1080 (1975).
104. H. Müller - Krumbhaar and D. P. Landau, Phys. Rev. B. (in press).
105. D. L. Huber, Physics Letters 49A, 345 (1974); D. L. Huber, Phys. Rev. B10, 3992 (1974).
106. T. Riste, E. J. Samuelson, K. Otnes, and J. Feder, Solid State Commun. 9, 1455 (1971); S. M. Shapiro, J. D. Axe, G. Shirane, and T. Riste, Phys. Rev. B6, 4332 (1972).
107. T. Schneider and E. Stoll, Phys. Rev. Lett. 31, 1254 (1976).

** The participants in the workshop were:

G. Ahlers	A. B. Harris
M. Blume	D. P. Landau
S. Friedberg	D. R. Nelson
B. I. Halperin	H. E. Stanley

THE MAGNETIC MEMORY FUNCTION OF IRON WHISKERS AT THE CRITICAL TEMPERATURE

B. Heinrich and A.S. Arrott
Department of Physics, Simon Fraser University
Burnaby, British Columbia, Canada V5A 1S6

ABSTRACT

The frequency dependence of the magnetic response of an iron whisker at the critical temperature is fitted using a magnetic viscosity field generated from a memory function of novel form. The characteristic relaxation time of the memory function peaks at T_c reaching the range of milliseconds. It is thought that the critical slowing down of magnetic fluctuations is seen through coupling of divergence free modes to the mode driven by the external field.

INTRODUCTION

The response of the iron whisker to an a.c. magnetic field is analysed to extract evidence for critical slowing down of magnetic fluctuations. If a small field is applied along the axis of the whisker and then suddenly removed, the magnetization will decrease rapidly even though the sample is right at the Curie temperature. This is because the magnetization creates a demagnetizing field that acts as a restoring force. To observe critical slowing down one must study fluctuations that do not produce demagnetizing fields. Fluctuations for which div $\vec{M} = 0$ in the bulk and $\vec{M}\cdot\hat{n} = 0$ at the surface should exhibit critical slowing down at the critical point. Unfortunately such modes do not couple to a detector coil. Yet such modes do couple to the modes which are excited and detected when the iron whisker is used as the core of a concentric coil transformer. Thus it is through mode-mode coupling that one can follow the process of critical slowing down.

Eddy currents are the basic loss mechanism in the a.c. response of an iron whisker. Fluctuations generate eddy currents even though div $\vec{M} = 0$. The energy is transferred readily from the driven mode to modes which have relaxation times comparable to the period of the driven field. The more modes that fall in that frequency range the more the energy is transferred and the greater the losses.

MODEL

The zero frequency response is determined by the magnetic stiffness coefficient α which relates the net magnetic moment $\overline{M}$ V of the whisker as seen by the detector coil to the field H_o produced by the primary current. The stiffness depends upon the intrinsic susceptibility and the demagnetizing factor D so that

$$\alpha\overline{\overline{M}} = \left(\frac{1}{\chi} + 4\pi D\right)\overline{\overline{M}} = H_o \tag{1}$$

The eddy currents appear as a viscosity field H_v which depends not only on $d\overline{M}/dt$ at the given instant but on $d\overline{M}/dt$ at all previous times with a weight given by the magnetic memory function $\phi(t)$. One writes phenomenologically

$$H_v(t) = \int_{-\infty}^{t} \phi(t-t')\,\frac{dM}{dt'}\,dt'. \tag{2}$$

It appears from our analysis that the form of $\phi(t)$ is not a simple exponential, nor a Gaussian nor a Lorentian function. Though those are mathematically convenient functions, they do not lead us to the observed response of the iron whiskers. Instead we propose that (for notation see ref. 1)

$$\phi(t) = \frac{\beta_o}{\pi\tau}\,\{(si(t/\tau))^2 + (ci(t/\tau))^2\}. \tag{3}$$

If we let $\overline{\overline{M}} = m\,e^{i\omega t}$, the viscosity field, written as

$$H_v(t) = i\omega m\,e^{i\omega t}\int_0^{\infty} \phi(t)\,e^{-i\omega t}\,dt, \tag{4}$$

will have contributions in phase with the magnetization as well as out of phase with it (except in the limit that $\phi(t)$ is a delta function). The magnetic viscosity coefficient β is defined as

$$\beta = \int_0^{\infty} \phi(t)\cos\omega t\,dt, \tag{5}$$

which for $\phi(t)$ given by Eq. 3 yields[1]

$$\omega\beta = \frac{\beta_o}{\tau}\,\ell n\,(1 + \omega\,\tau). \tag{6}$$

The magnetic response including the viscosity field is written as

$$(i\omega\beta + \alpha)m = h, \tag{7}$$

where h is the amplitude of the field, $H_o = h\,e^{i\omega t}$. The inphase contribution from H_v adds to the magnetic stiffness coefficient α. We do not have an explicit form for the integral

$$\alpha_\tau = \int_0^{\infty} \phi(t)\sin\omega t\,dt. \tag{8}$$

we have yet to evaluate it numerically for the full range of ω's. We have concentrated on analysing the magnetic measurements to compare with the functional form of β. Completeness would require us to analyse the magnetic stiffness as well.

EXPERIMENTS

We have shown recently[2] that the phase of the response of an iron whisker (210μm x 230μm x 8.7mm) has the functional form of Eq. 6. The curvature of the function is observed for temperatures within 20 millidegrees of the 1040 K Curie temperature when the externally applied fields are less than 0.1 oersteds. The frequency range of the measurements is 0.5 to 14 kHz. Analysis in terms of $\omega\beta$ rather than phase results in even better fits than shown previously for the phase.

The sample reported on here is the same whisker used previously but it has been etched to round the edges. The new dimensions are 8.7mm length by 185μm dia. We had evidence from experiments carried out several years ago that it makes a difference in the temperature dependence of the magnetic response whether a whisker has a rectangular cross section or a round one[3]. The results for $\omega\beta$ are shown in fig. 1. The data has been corrected for the various things that make it difficult to obtain accurate phase. For graphical clarity we have not shown the fits for higher temperatures. The frequency dependence of $\alpha = \alpha_\tau + 1/\chi + 4\pi D$ are shown in fig. 2 for several of the same temperatures. The temperature dependence of the parameters τ, β_o, and β_o/τ are shown in fig. 3. We include results from the experiment on the unetched whisker also.

CONCLUSIONS

Without claiming anything beyond phenomenology for our somewhat novel memory function, we wish to point out:

1. Our memory function does describe the loss component in the critical region. Qualitatively it looks like it also will fit the inphase component.
2. The characteristic time of the memory function τ increases rapidly and peaks at T_c. These times are many orders of magnitude longer than are accessible by quasi-elastic neutron scattering.
3. The measurements may be limited by temperature gradients in the whisker, but it seems unlikely that effects are smeared out by more than a few millidegrees, if that much.
4. The basic form of the memory function is not changed by a rather significant alteration of the size and shape of the whisker.

5. The memory function has a smooth fall off with time. It has an integrable singularity at the origin of time and a Lorentian-like tail at long times, the memory fades as $1/t^2$. The function differs quite noticeably from a Lorentian which would give an exponential fall off for $\omega\beta$. Near T_c, $\omega\tau$ varies from 1 to 28 in a single isotherm. This range provides a sensitive test of the functional forms.
6. The fluctuations for which div $\vec{M} = 0$ and $\vec{M}\cdot\hat{n}$ is specified at the surface can be indexed by two mode numbers[4]. The demagnetizing fields constrain the fluctuations in such a way that they have a two dimensional phase space. Desai[5] has given an argument which provides a connection between a two dimensional phase space and the form of our memory function.

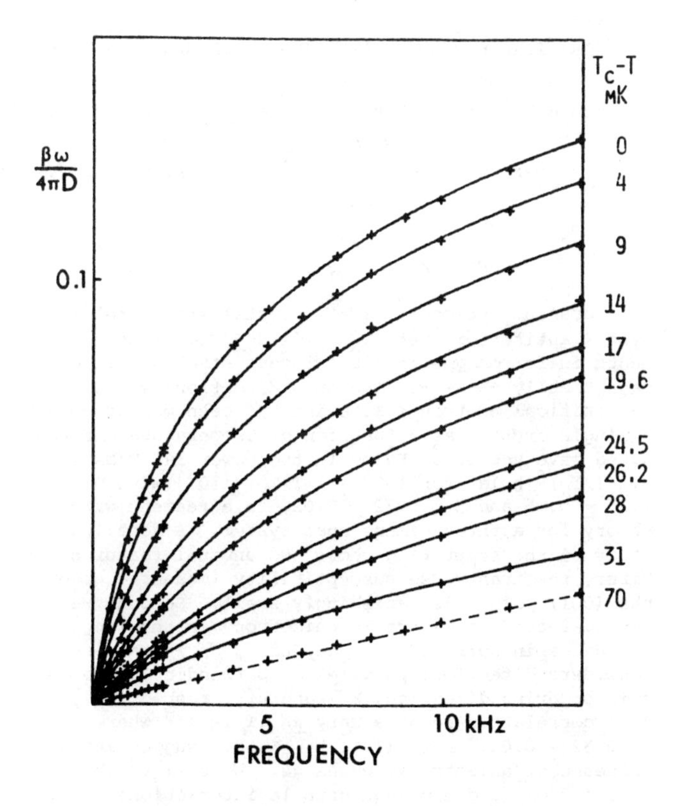

Fig. 1

The frequency dependence of $\omega\beta$, the magnetic viscosity, at several temperatures near T_c. The solid lines are fits to Eq. 6. The data are normalized to the inphase signal at frequencies where $\alpha = \alpha_o \equiv 4\pi D$, that is $\omega\beta/4\pi D = \phi \cdot \alpha/\alpha_o$. Thus only ω and the phase ϕ are measured absolutely.

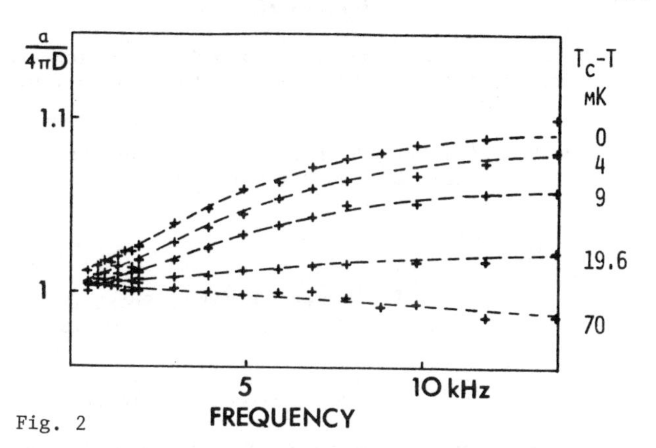

Fig. 2

The frequency dependence of the magnetic stiffness coefficient $\alpha = \alpha_\tau + 1/\chi + 4\pi D$ for some of the same temperatures shown in fig. 1. The data are normalized to $\alpha_o \equiv 4\pi D$, note the suppression of the zero for the ordinate.

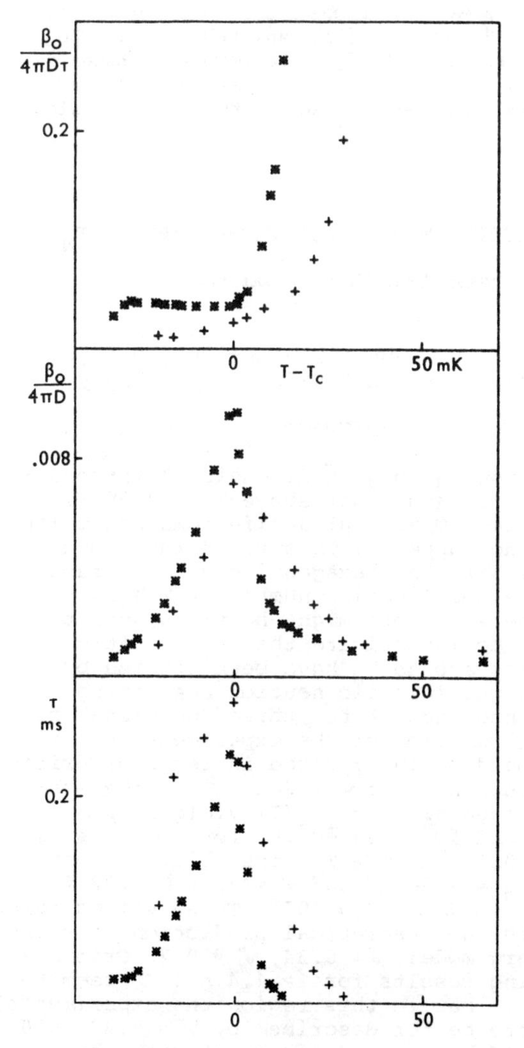

Fig. 3

The temperature dependence of the parameters used in fitting the magnetic viscosity: * are for the etched whiskers; + are for the as grown whisker. a) The coefficient of the memory function $\beta_o/\tau \times 1/4\pi D$; b) The initial viscosity coefficient $\beta_o \times 1/4\pi D$; c) The characteristic time τ.

REFERENCES

1. I.S. Gradshteyn and I.M. Ryzhik, Table of Integrals, Series, and Products, Academic Press (1965), p. 645 (6.256).
2. B. Heinrich and A.S. Arrott, AIP Conference Proc., Magnetism and Magnetic Materials, 1975, (in Press).
3. B. Heinrich and A.S. Arrott, AIP Conference Proc., No. 24, Magnetism and Magnetic Materials, 1974, P. 287.
4. A.S. Arrott, B. Heinrich, M.J. Press and T.L. Templeton, Conference Proc., AIP, this volume.
5. Private communication with Dr. R.C. Desai.

SPIN-ORIENTATION PHASE TRANSITIONS IN CUBIC FERRIMAGNETIC GdIG: MAGNETOOPTIC AND VISUAL INVESTIGATION

V. V. Eremenko, N. F. Kharchenko, and S. L. Gnatchenko
Physico-Technical Institute of Low Temperatures, Kharkov, USSR

The sharp and smooth magnetic transitions due to the magnetic field in the canting process of the magnetic sublattices of GdIG are investigated by measuring the Faraday rotation in the small section of the sample and by the visualization of the magnetic structure in the polarized light. The investigations were made near the magnetic compensation temperature at the orientation H along the [111] and [100] axes. The Faraday rotation of the different magnetic phases was measured in the vicinity of the phase transitions between the collinear and canted structures and also between different canted ones. The visual observations were used to construct the phase diagrams and the magnetic state coexistence regions. We have payed particular attention to the critical point (the case $H \| [100]$). Above the definite field the transition between the low- and high-temperature noncollinear states occurs smoothly. The experimental results are compared with the calculations carried out in the molecular field approximation making allowance for the three-sublattice structure of GdIG.

QUASIELASTIC NEUTRON SCATTERING AROUND T_N IN $CsNiF_3$: CROSSOVER IN THE EXPONENT ν.

M. Steiner, Hahn-Meitner-Institut, D-1000 Berlin 39, Postfach 390128, West-Germany

ABSTRACT

$CsNiF_3$ shows strong onedimensional ferromagnetic short rang order above $T_N = 2.66$ K, where threedimensional antiferromagnetic long range order appears. This short range order together with the hexagonal crystal structure makes the coordination number at $T \geqslant T_N$ very high. Therefore MFA might be valid near to T_N. In order to investigate that, the critical properties around T_N have been studied by means of quasielastic neutron scattering. Taking into account the threedimensional resolution function of the experiment, the results could be interpreted in terms of critical exponents β and ν. For $T \leq T_N$ the sublattice magnetization $M_s(T)$ yielded $T_N = 2.66_5$ K and $\beta = 0.34 \pm 0.03$ for $3 \cdot 10^{-3} \leqslant \epsilon \leqslant 3.55 \cdot 10^{-2}$. For $T \geqslant T_N$ the critical scattering $T_N = 2.66_3$ K and $\nu = 0.68 \pm 0.02$ for $1.5 \cdot 10^{-3} \leq \epsilon \leq 1.4 \cdot 10^{-2}$. These values agree well with the theoretical predictions for the Heisenberg model: $\beta = 0.34$, $\nu = 0.7$. Critical scattering results for $\epsilon > 1.4 \cdot 10^{-2}$ seem to indicate, that in this region the experimental points are better described by $\nu = 0.47 \pm 0.04$, what would be in good agreement with the MFA result $\nu = 0.5$.

CRITICAL NEUTRON SCATTERING EXPERIMENTS ON USb*

G. H. Lander
Argonne National Laboratory, Argonne, Illinois 60439
D. M. Sparlin
University of Missouri, Rolla, Missouri 65401
and
O. Vogt
ETH, Zurich, Switzerland

Uranium antimonide (NaCl crystal structure) is a type I antiferromagnet ($T_N = 241.2 \pm 0.1$ K) with moments in ferromagnetic (001) layers stacked antiferromagnetically + - + -. The spin direction is [001]. The critical neutron scattering has been measured from a single crystal as a function of temperature and reduced wave vector $\vec{q}$. Below T_N the power law $M/M_O = D(1-T/T_N)^{\beta}$ holds for $10^{-3} \leq 1-T/T_N \leq 10^{-1}$ with $D = 1.35 \pm 0.05$ and $\beta = 0.32 \pm 0.02$, in agreement with theory for a three dimensional system. A detailed study of the scattering shows two unusual features. First, the transverse susceptibility [measured around the (001) point] is vanishingly small. Second, we have detected considerable anisotropy in the longitudinal spin correlations, depending on whether one considers directions parallel or perpendicular to the spin ordering direction, k_{11}^c and k_{11}^a, respectively. Both correlation lengths vary as $(T/T_N-1)^{\nu}$ where $\nu \sim 0.32 \pm 0.01$. For all T, $k_{11}^c \sim 6k_{11}^a$ suggesting the presence of anisotropic exchange. We believe this arises from quadrupole-quadrupole interactions within the Γ_8 ground state of this $5f^3$ electron system.

*Work supported by the U. S. Energy Research and Development Administration.

THEORY OF BICRITICAL POINTS IN 2 + ε DIMENSIONS

Robert Pelcovits and David R. Nelson*
Department of Physics
Harvard University
Cambridge, Massachusetts 02138

ABSTRACT

Polyakov's theory of fixed length classical spins near two dimensions is applied to symmetry-breaking perturbations about the original isotropic spin system. Inclusion of a quadratic symmetry-breaking term allows an analysis of bicritical phenomena in 2 + ε dimensions. The crossover exponent ϕ is calculated to leading order in ε, and is found to diverge in d = 2. Crossover scaling theory is then applied to determine the shape of the bicritical phase boundaries. However, we find that higher order symmetry-breaking interactions (which occur in real magnetic crystals with cubic or hexagonal symmetries) are strongly relevant variables, and can drastically alter the bicritical behavior found when only quadratic symmetry-breaking terms were allowed.

We have employed a renormalization group technique due to Polyakov[1] to study the critical behavior of an anisotropic spin Hamiltonian with nearest neighbor spin coupling, namely

$$H = -J \sum_{\langle\vec{R},\vec{R}'\rangle} \vec{S}(\vec{R}) \cdot \vec{S}(\vec{R}') + \Delta \sum_{\langle\vec{R},\vec{R}'\rangle} S_n(\vec{R}) S_n(\vec{R}'). \quad (1)$$

The sums run over a d-dimensional lattice, $S_n(\vec{R})$ is the nth component of the spin $\vec{S}(\vec{R})$, and $\sum_{i=1}^{n} S_i^2(\vec{R}) = 1$. In the continuum limit we obtain a reduced Hamiltonian $\overline{H}$ of the form

$$\overline{H} = -H/k_BT = \int d\vec{R}\{\tfrac{1}{f}[\partial_\mu \vec{S}(\vec{R})]^2 + \tfrac{g_1}{f}[\partial_\mu S_n(\vec{R})]^2 + \tfrac{g_2}{f} S_n^2(\vec{R})\}, \quad (2)$$

with f proportional to k_BT/J and g_1, g_2 proportional to Δ/J. When (2) is transformed into momentum space, we imagine that the integrations are cut off by a Brillouin zone of radius Λ. The analysis reported here is restricted to $n \geq 3$.

Following Polyakov,[1] we have constructed recursion relations for f, g_1, and g_2 to first order in ε = d - 2. A new momentum cutoff $\tilde{\Lambda} < \Lambda$ is introduced, and the short wavelength components of $\vec{S}(\vec{R})$ (with wavelengths between $\tilde{\Lambda}$ and Λ) are integrated out. Linearizing about the nontrivial isotropic fixed point we find that g_1 is an irrelevant variable, which we will subsequently set to zero. However, g_2 is strongly relevant.

The recursion relations can be used to derive homogeneity relations for various thermodynamic quantities as a function of f and g_2 in 2 + ε dimensions. The resulting crossover scaling prediction for the susceptibility χ near the isotropic bicritical point[2] is

$$\chi(f,g_2) \approx t^{-\gamma}\Phi(g_2/t^{\phi}), \quad (3)$$

where γ is the isotropic susceptibility exponent and $t \propto k_B|T - T_c|/J$. We have calculated the anisotropic spin crossover exponent ϕ to $O(\varepsilon^2)$ with the aid of isotropic critical exponents determined to $O(\varepsilon^2)$ by Brézin and Zinn-Justin.[3] Our result is

$$1/\phi = \tfrac{1}{2}\varepsilon[1 + 2\varepsilon/(n-2)]. \quad (4)$$

In the interesting case ε = 0 (d = 2) and n = 3, the bicritical point occurs at T = 0 and we predict

$$\chi(f,g_2) = e^{4\pi/f}\Phi(g_2 e^{4\pi/f}). \quad (5)$$

The shapes of the Ising and XY bicritical lines which terminate at the bicritical point $(f^*,g_2^*) = (0,0)$ are given by relations of the form[4]

$$g_2 e^{4\pi/f} = \text{const.} \quad (6)$$

This result appears to be consistent with recent Monte Carlo simulations of anisotropic Heisenberg systems in two dimensions by Binder and Landau.[5]

To describe real quasi-two-dimensional crystals we must, of course, include higher order symmetry-breaking interactions. It is straightforward to extend the analysis to allow for, say, a cubic perturbation of the form

$$\overline{H}_v = \frac{v}{f}\int \sum_{i=1}^{n} S_i^4(\vec{R})d\vec{R} \quad (7)$$

In contrast to the situation for n = 3 near four dimensions, v is a strongly relevant perturbation, and we predict a crossover into a new type of cubic critical (and bicritical) behavior.[6] Thus, we expect that the asymptotic properties of layered spin-flopping magnetic crystals will not be controlled by the T = 0 bicritical fixed point described above.

A more realistic symmetry-breaking perturbation than (7) for d = 2, n = 3 would be

$$\overline{H}_v = \frac{v}{f}\int [S_x^4(\vec{R}) + S_y^4(\vec{R}) + \alpha S_z^4(\vec{R})]d\vec{R}, \quad (8)$$

where $\hat{z}$ is perpendicular to the plane of the lattice and $\alpha \neq 1$. However, (8) again contains strongly relevant perturbations, now leading, possibly, to an n = 2 cubic fixed point.

A more detailed exposition of the results summarized here will be presented elsewhere.[7]

REFERENCES

* Junior Fellow, Harvard Society of Fellows

1. A. M. Polyakov, Phys. Lett. 59B 79-81 (1975).
2. D. R. Nelson, J. M. Kosterlitz, and M. E. Fisher, Phys. Rev. Lett. 33, 813-817 (1974).
3. E. Brézin and J. Zinn-Justin, Phys. Rev. Lett. 36, 691-694 (1976). See also, A. A. Migdal, Zh. E. T. F. 69, 1457-1465 (1975).
4. See, e.g., E. K. Reidel and F. Wegner, Z. Phys. 225, 195-215 (1969), and P. Pfeuty, D. Jasnow, and M. E. Fisher, Phys. Rev. B10, 2088-2112 (1974).
5. K. Binder and D. P. Landau, Phys. Rev. B13, 1140-1155 (1976).
6. See, e.g., A. Aharony, Phys. Rev. B8, 4270-4273 (1973).
7. R. Pelcovits and D. R. Nelson, (to be published).

ONSET OF HELICAL ORDER

J.F. Nicoll, G.F. Tuthill, T.S. Chang and H.E. Stanley

Physics Department
Massachusetts Institute of Technology
Cambridge, Massachusetts 02139

Renormalization group techniques are used to treat the onset of helical order at higher order critical points for a large class of helical states First order perturbation results are given for all the critical point exponents for critical points of order θ and arbitrary anisotropic propagators. The critical point exponent η is calculated to second order for arbitrary isotropic propagators.

Recently several authors have used renormalization group techniques to discuss the onset of helical order in magnetic systems.[1-3] In particular, the existence of new types of critical behavior has been postulated for the "Lifshitz" point where the transition from a uniformly ordered to helically ordered state occurs[1].

Here we consider a class of systems which can exhibit phases of very general helical character, and report the renormalization group calculation of critical exponents for such systems. These critical points will be termed "generalized Lifshitz points". These systems are characterized by critical or fixed point propagators which differ from the usual k^2 dependence.[4]

By "helical" phases we mean magnetically ordered states with a periodic spatial structure whose periodicity need not be related to that of the lattice. Included thereby are spiral structures of various types, as well as states in which the magnetic moments are aligned uniformly in direction, but with sinusoidally varying magnitudes. The presence of these (and even more complicated phases) is well established, largely through neutron diffraction studies. Typical examples include the screw spiral structure of MnO_2, cone-spiral order in spinel-type compounds such as $MnCr_2O_4$, and sinusoidal phases in certain rare earth metals (Er, Nd and others).

We use continuum spin Wilson models with explicit wave-vector dependent terms in the Hamiltonian density which are coupled to the thermodynamic fields[6]. For motivational purposes we consider a free energy functional F of an isotropic single component magnetization M(x):

$$F=\int d_x^d \{a_1(\nabla M(\vec{x}))^2 + a_2(\nabla^2 M(\vec{x}))^2 + \ldots + A_1 M^2(\vec{x}) + A_2 M^4(\vec{x}) + A_3 M^6(\vec{x}) + \ldots\} . \quad (1)$$

Here the coefficients a_i and A_i are functions of the thermodynamic fields (temperature, pressure, magnetic field,...). At an ordinary critical point we have $A_1=0$, with $A_2>0$ and $a_1>0$; variations in the field variables which preserve these conditions sweep out a surface of second order ($\theta=2$) critical points. Similarly, $A_1=A_2=0$, $A_3>0$, $a_1>0$ characterizes a points of three phase criticality ($\theta=3$). For $A_1=A_2=\ldots=A_{\theta-1}=0$, $A_\theta>0$, we have a critical point of order θ.

For $a_1>0$, all of the competing phases are spatially uniform. However, if $a_1<0$, the free energy F will be minimized for some particular non-uniform phase. This is most easily seen by considering the Fourier transform representation of (1), where we include derivative terms up to order 2L and magnetization terms up to order 2θ:

$$F(\{M(\vec{k})\})=\int d^d k\left(\sum_{j=1}^{L} a_j k^{2j}\right) M^2(\vec{k}) + \sum_{i=1}^{\theta} A_i \left(\prod_{m=1}^{2i} \int d^d k_m M(\vec{k}_m)\right)\delta\left(\sum_{=1}^{2i} \vec{k}\right). \quad (2)$$

The minimum of F is obtained for a sinusoidally varying magnetization $M(\vec{k})$, where $\vec{k}$ is determined by minimizing the $\vec{k}$-dependent parts of (2). For L=2 and $a_1>0$, the system displays ordinary criticality between spatially uniform phases; for $a_1<0$ criticality is achieved between helical states of equal and opposite $\vec{k}$. As the Lifshitz point $(a_1=0)$[1] is approached, $\vec{k}$ approaches zero as $k\sim(-a_1)^{\beta_k}$ ($\beta_k=1/2$ in mean field theory). If both a_1 and a_2 vanish, it is necessary to include in (2) the k^6 term. In mean field theory, the conditions $A_1=\ldots=A_{\theta-1}=0$, $A_\theta>0$, and $a_1=a_2=\ldots=a_{L-1}=0$ specify a critical point with order θ and "Lifshitz character L". In the vicinity of such a point, there are L different values of the helicity wave-vector, each of which is associated with θ different values of $M(\vec{k})$. At the generalized Lifshitz point, all of these phases are simultaneously critical.

We will use as a model Hamiltonian for our renormalization group calculations $H=F(\{s(\vec{k})\})$, where $s(\vec{k})$ is the spin fluctuation variable. The critical propagator (the term in H proportional to $s^2(k)$) has leading dependence k^{2L}. It is also possible to have non-integral propagator exponents. By the introduction of a long range force[7] with interaction strength decaying like $r^{-(d+\sigma)}$, we can add a term proportional to $|k|^\sigma s^2(\vec{k})$ to H. If both terms are present in the critical region, then only $\tilde{\sigma} \equiv \min(\sigma,2L)$ is important; the critical propagator is proportional to $|k|^{\tilde{\sigma}}$.

The first order results in a perturbation expansion can be obtained (both in the isotropic propagator considered here and for the anisotropic case considered below) by utilizing the techniques developed for the simple L=1 case[8-11]. For a critical point of order θ of an isotropically interacting n-component spin system, the borderline dimension (above which mean field behavior holds) is given by $d_b=\tilde{\sigma}\,\theta/(\theta-1)$. Below this dimension, we obtain a correction to the pth eigenvalue (corresponding to s^{2p}) in terms of the unperturbed or Gaussian eigenvalues $\{\lambda_j\}$:

$$\lambda_p' = \lambda_p - 2\lambda_\theta \langle\theta,p;p\rangle_n / \langle\theta,\theta;\theta\rangle_n \quad (3a)$$

where[11]

$$\langle\theta;p;p\rangle_n \equiv \sum_{j=0}^{\theta/2} \binom{p}{j}\binom{p+\frac{n}{2}-1}{j}\binom{2p-2j}{\theta-2j} \qquad (3b)$$

For isotropic propagators, $\lambda_p = d+p(\tilde{\sigma}-d)$. The expansion parameter $\varepsilon_\theta(\tilde{\sigma}) \equiv \lambda_\theta$ is thus

$$\varepsilon_\theta(\tilde{\sigma}) = d+\theta(\tilde{\sigma}-d) = (\theta-1)(d_b-d). \qquad (4)$$

Thus, for the Ising model(n=1), we have

$$\lambda_p' = d+p(\tilde{\sigma}-d)-2\varepsilon_\theta(\tilde{\sigma})\binom{2p}{\theta}\Big/\binom{2\theta}{\theta} \qquad (5a)$$

while for $n=\infty$

$$\lambda_p' = d+p(\tilde{\sigma}-d)-\frac{2\varepsilon_\theta(\tilde{\sigma})\binom{p}{[1/2(\theta+1)]}}{\binom{\theta}{[1/2(\theta+1)]}} \qquad (5b)$$

This defines all the eigenvalues to first order. We now define the critical point exponent η by the shift in the exponent of the critical two-point correlation function

$$\Gamma_2(\vec{k}) \propto |\vec{k}|^{\tilde{\sigma}-\eta} \qquad (6)$$

For η, we find that if $\tilde{\sigma}\neq 2L$, there is no shift,i.e. that $\eta=0$ to $O(\varepsilon_\theta(\tilde{\sigma})^2)$. For $\tilde{\sigma}=2L$ we find:[9]

$$\eta_\theta(2L) = \frac{(-1)^{L+1}}{L\binom{2\theta}{\theta}^3}\, 4\varepsilon_\theta^2(2L)\, \frac{\theta\,\Gamma^2\left(\frac{d_b}{2}\right) C_n}{\Gamma\left(\frac{1}{2}(d_b-2L)\right)\Gamma\left(\frac{1}{2}(d_b+2L)\right)} \qquad (7a)$$

$$\text{with } C_n \equiv \left[\frac{\langle\theta,\theta;\theta\rangle_n}{\langle\theta,\theta;\theta\rangle_{n=1}}\right]^2 \prod_{j=1}^{\theta-1} \frac{2j+n}{2j+1} \qquad (7b)$$

Equation (7) agrees with the result of Ref. 1 for the special case $\theta=L=2$ and has also been verified for $\theta=2$ and all L by a differential renormalization group method[10]. Note that the combinatorial factor $C_n=1$ for n=1, and that for large n

$$C_n \to n^{-1}\binom{2\theta}{\theta}^3\Big/\binom{\theta}{\theta/2}^3 + O(n^{-2}),\ \theta\ \text{even}, \qquad (8a)$$

$$C_n \to \binom{2\theta}{\theta}^3\Big/\left\{(\theta+1)\binom{\theta}{1/2[\theta+1]}\right\}^3 + O(n^{-1}),\ \theta\ \text{odd}. \qquad (8b)$$

Other wave-vector associated exponents also have second order corrections; e.g., $\beta_k = 1/(2(L-1)) + O(\varepsilon^2(2L))$.[10]

Anisotropic Propagators

In the above we used a propagator isotropic in $\vec{k}$-space. However, the lattice structure of a real material can induce a preferred direction for the periodic behavior. Moreover, since catastrophic infrared divergences[10-11] set in at dimensions less than $d_{min}=2L$, the theory as developed above is unlikely to produce realistic predictions at d=3 if $L \geq 2$. These infrared divergences are, of course, intimately related to the appearance of infinitely many relevant Gaussian eigenvalues below d_{min}. For these reasons, we now consider anisotropic propagators.

We write the wave vector $\vec{k}$ as

$$\vec{k} = \vec{k}_1 + \vec{k}_2 + \ldots + \vec{k}_J \qquad (9)$$

where each $\vec{k}_i$ is a d_i-dimensional vector, so that $\sum d_i = d$. We consider a critical propagator G^{-1} of the form

$$G^{-1} = \sum_{i=1}^{J} |\vec{k}_i|^{\sigma_i} \qquad (10)$$

with $\sigma_1 \leq \sigma_2 \leq \ldots \leq \sigma_J \equiv \sigma_>$. For such systems, d_b for a critical point of order is determined by

$$\sum_{i=1}^{J} d_i/\sigma_i = \theta/(\theta-1) \qquad (11a)$$

and d_{min} by the condition

$$\sum_{i=1}^{J} d_i/\sigma_i = 1. \qquad (11b)$$

The introduction of anisotropy in the propagator lowers both d_b and d_{min}. For example, if only one component of $\vec{k}$ enters G^{-1} as k^{2L} and the remaining components have k^2 dependence, then (11) gives $d_b=(3\theta-1)/(\theta-1)-1/L$ and $d_{min}=3-1/L$. Thus, we have $d_b \geq 3 > d_{min}$ for all $\theta < 2L+1$.

For anisotropic systems, the critical point exponents $\{\eta_i\}$ are defined by examing the behavior of the two-point function for a wave-vector lying entirely in one of the d_i dimensional subspaces:

$$\Gamma_2(\vec{k}_i) \propto |\vec{k}_i|^{\sigma_i-\eta_i} \qquad (12)$$

There will also be difference values of the correlation length exponent ν_i in each of the subspaces. The following relationships between the exponents hold generally:

$$2-\alpha = \sum_{i=1}^{J} d_i\,\nu_i\ ;\ \gamma=(\sigma_i-\eta_i)\nu_i\ ;$$

$$\delta = \left\{\sum_{i=1}^{J} d_i/(\sigma_i-\eta_i) + 1\right\}\Big/\left\{\sum_{i=1}^{J} d_i/(\sigma_i-\eta_i)-1\right\}. \qquad (13)$$

The results (13) can be derived by constructing an exact differential[10-12] renormalization group equation suitable for the propagator (10). A similar approximate generator for a Hamiltonian $H(\vec{s},\ell)$ is[13]

$$\partial H/\partial\ell = \sum_{i=1}^{J} d_i\,\sigma_>/\sigma_i\} H + \tfrac{1}{2}\left(\sigma_> - \sum_{i=1}^{J} d_i\,\sigma_>/\sigma_i\right) \vec{s}\cdot\vec{\nabla}H + \nabla^2 H - \vec{\nabla}H\cdot\vec{\nabla}H \qquad (14)$$

where ℓ is the renormalization parameter[11]. The first order eigenvalue corrections are again given by (3); only the value of the unperturbed eigenvalues are changed. From (14), we see that the Gaussian eigenvalues are now:

$$\lambda_p = \sum_{i=1}^{J} d_i\sigma_>/\sigma_i + p\left(\sigma_> - \sum_{i=1}^{J} d_i\sigma_>/\sigma_i\right) \qquad (15)$$

The expansion parameter for $d<d_b$ is again $\varepsilon_\theta(d_i,\sigma_i)\equiv\lambda_\theta$. Thus, the corrected eigenvalues for the general anisotropic case are

$$\lambda'_p = \sum_{i=1}^{J} d_i\,\sigma_> /\sigma_i + p\Big(\sigma_> - \sum_{i=1}^{J} d_i\sigma_>/\sigma_i\Big) - 2\varepsilon_\theta(d_i,\sigma_i)\,\langle\theta',p;p\rangle_n/\langle\theta,\theta;\theta\rangle_n \quad . \tag{16}$$

References:

1. R.M. Hornreich, M. Luban and S. Shtrikman, Phys. Rev. Lett. 35, 1678 (1975); Phys. Lett. 55A, 269 (1975).
2. P. Bak, S. Krinsky, and D. Mukamel, Phys. Rev. Lett. 36, 52 (1976).
3. M. Droz and M.D. Continho-Filho, AIP Conf. Proc. 29, 465 (1975)
4. K.G. Wilson, Rev. Mod. Phys. 47, 773 (1975).
5. D.E. Cox, IEEE Trans. Magnetics 8, 161 (1972); and references contained therein.
6. This is in contrast to the approach used to describe antiferromagnetism in which a pair of competing spin fields (one for each sublattice) is considered from the outset, thus building in the lattice periodicity. Cf. D.R. Nelson and M.E. Fisher Phys. Rev. B11, 1030 (1975).
7. M.E. Fisher, S.K. Ma and B.G. Nickel, Phys. Rev. Lett. 29, 917 (1972).
8. J.F. Nicoll, T.S. Chang and H.E. Stanley, Phys. Rev. Lett., 33, 540; E33, 1525 (1974).
9. G.F. Tuthill, J.F. Nicoll, H.E. Stanley, Phys. Rev. B11, 4579 (1975).
10. J.F. Nicoll, G.F. Tuthill, T.S. Chang and H.E. Stanley (unpublished).
11. J.F. Nicoll, T.S. Chang, and H.E. Stanley Phys. Rev. A 13, 1251 (1976).
12. K.G. Wilson and J. Kogut, Phys. Rep. 12, 73 (1974).
13. The approximate generator is derived as in Ref. 11 from an exact generator similar to the Wilson incomplete integration generator (Ref. 12), except that we renormalize in such a way as to hold each of the propagator exponents separately fixed. Wave-vector dependent terms enter the calculation only at second order in perturbation expansions; by neglecting these terms we obtain (14). Note that to this order the wave-vector dependent portions of the Hamiltonian are invariant and therefore we need not write them explicitly. A similar exact generator can be obtained in Wegner-Houghton form, cf. Ref. 10-11.

ORDER-DISORDER AND α-γ TRANSITIONS IN FeCo*

P. Silinsky and M. S. Seehra+,
Physics Department, West Virginia University,
Morgantown, WV 26506.

Electrical resistivity (ρ) measurements in an annealed FeCo alloy (with 46.29 at.% Co) in the temperature range of 500-1350^0 K are reported. Special attention is given to the temperature regions of 900-1100^0 K and 1160-1300^0 K. A change in the slope of ρ vs T curve, associated with the order-disorder (OD) transition, yields a λ-type anomaly in the computed $\partial\rho/\partial T$ with maximum at $T_c \simeq 1006^0$ K. To our knowledge this is the first observation of the OD transition in FeCo using temperature-dependent electrical resistivity measurements. Similar to the observations in β brass [1], $\partial\rho/\partial T$ in the critical region is found to be proportional to the specific heat associated with the OD transition [2]. Near 1235^0 K, a discontinuous change in ρ of about 20% with hysteresis of about 12^0 K, is observed. This first order transition is associated with the α-γ (bcc-fcc) transition. However, the nature of the anomaly (a jump in ρ with increasing T) in FeCo is quite opposite to that observed in Fe. A detailed account of this work will be published elsewhere.

*Supported in part by the National Science Foundation.

+A. P. Sloan Research Fellow.

1. D. Simons and M. Salamon, Phys. Rev. Letters, 26, 750, (1971).
2. J. Orehotsky and K. Schröder, J. Phys. F: Metal Phys. 4, 196 (1974).

THE d=3, fcc, CONTINUOUS SPIN ISING MODEL

UNIVERSALITY AND CONFLUENT CORRECTIONS TO SCALING

M. Ferer* and R. Macy
Department of Physics
West Virginia University
Morgantown, West Virginia 26506

ABSTRACT

We have derived tenth-order, high-temperature series for the zeroth and second correlation function moments, μ_0 and μ_2 respectively, of several continuous spin (Landau-Wilson), nearest neighbor Ising models for the fcc lattice.

We have performed confluent singularity analysis, which allows for the contribution of sub-critical, confluent singularities, on the series for μ_0 and μ_2/μ_0. This analysis provides convincing evidence favoring the following universal values of the critical indices, $\gamma=1.250\pm0.012$ and $2\nu=1.275\pm 0.015$, in good agreement with known values for the spin ½ Ising model. Our analysis is consistent with the following value of the subcritical correction to scaling index, $\Delta=0.60\pm0.20$. Our results for γ and Δ are in good agreement with the results of previous workers.

Many of the most convincing renormalization group calculations applicable to magnetic systems do not use physical spins but, rather, replace sums over quantized spin projections with integrals over continuous spin variables.[1] In justifying the applicability of these calculations the universality hypothesis has been invoked in asserting that the continuous spin models have the same critical behavior as the physical spin models. However plausible this assertion may be; it should be tested. Camp and Van Dyke have derived tenth-order series for the susceptibility, μ_0, of a class of continuous spin Ising models in three dimensions, d=3; their analysis, which allowed for confluent contributions to the leading singularity, indicated that the value of the index γ, characterizing the scaling singularity of μ_0 for this class of models, was identical to the value found for the s=½, Ising model.[2] Because of the disagreement, over the validity of hyperscaling, between series results for the d=3, s=½ Ising model which violate the hyperscaling relation $d\nu=2-\alpha$[3] and renormalization group results which are consistent with hyperscaling,[1] we have investigated the second moment, μ_2, of the spin-spin correlation function for the continuous spin Ising models to determine if the value of the index ν, characterizing the correlation length of these models, is identical to its value for the s=½ model.

Using the Englert linked-cluster expansion,[4] we have derived tenth-order, high-temperature series for the spin-spin correlation function for a system of continuous spins, S_i, at the sites, i, of an fcc lattice with the Hamiltonian

$$-\beta H=\beta J\sum_{<ij>} S_iS_j - \lambda\sum_i (S_i^4-2S_i^2) \qquad (1)$$

for several values of the peaking parameter, λ = 0.25, 0.50, 1.00, 1.50, 2.00 . Numerical methods were used to determine the necessary bare semi-invariants. Using the correlation function series, we derived tenth-order series for the zeroth and second correlation function moments

$$\mu_n(\lambda)=M_2^0(\lambda)\sum_{j=o}^{\infty} a_{n,j}(\lambda)K^j,\ K=\beta JM_2^0(\lambda) \qquad (2)$$

where the normalization factor, M_2^0, is the second bare semi-invariant. In order that the reader might judge the validity of our interpretation of the analysis in a typical case, we present the series for $\mu_0(\lambda=\frac{1}{4})$ and for $\mu_2(\lambda=\frac{1}{4})$ in Table II; in subsequent tables the reader will find the analysis of these series upon which we based our conclusions for the $\lambda=\frac{1}{4}$ model.

In analyzing these series we used standard ratio methods[5] as well as confluent singularity methods.[6,7] Ratio methods neglect possible corrections to the scaling singularity assuming, for example, that $\mu_2/\mu_0 = A_2t^{-2\nu}$ where the reduced temperature $t=(1-K/K_c)$. These methods provide several sequences for each of the three parameters, e.g. for K_c, 2ν, and A_2 characterizing the singularity in the above expression; these sequences may be extrapolated using Neville Tables. Saul-Wortis methods allow for the contribution of one confluent correction to scaling assuming for example, that $\mu_2/\mu_0 = A_2t^{-2\nu}(1+B_2t^{\Delta_2})$; these methods provide several sequences for each of the five parameters, K_c, 2ν, Δ_2, A_2 and B_2. It should be noted that a confluent singularity, neglected in the ratio method, will add a correction $O(1/n^{1+\Delta_2})$ to the nth term of the ratio sequence for K_c, and a correction $O(1/n^{\Delta_2})$ for the index; of these two, the sequences for the index will have the larger and more slowly decreasing correction. Baker-Hunter methods allow for the contribution of any number of confluent corrections; and, given a value of K_c, they provide a Pade table for the index characterizing each singularity. The conclusions of our analysis are shown in Table I; the uncertainties quoted there for K_c^{-1}, γ, 2ν, Δ_1, and Δ_2 are intended to represent absolute confidence limits.

In analyzing the series to determine $K_c(\lambda)$, we rely most heavily on those ratio and Saul-Wortis sequences which are unbiased, in the sense of not requiring the value of any other parameter, e.g. a value of an index, in their determination.[5-7] These unbiased sequences provide the clearest estimate of the absolute uncertainty. Table IIIa shows the Neville Table extrapolation of the log-derivative sequence for $K_c^{-1}(\frac{1}{4})$ from the series for $\mu_0(\frac{1}{4})$ and for $\mu_2(\frac{1}{4})/\mu_0(\frac{1}{4})$. On the basis of these Neville Tables alone, we assert that $K_c^{-1}(\frac{1}{4}) = 10.615\pm0.002$. Table IIIb shows the unbiased Saul-Wortis (five-fit) sequence for $K_c^{-1}(\frac{1}{4})$ using the same two series. These sequences are more irregular and cannot be extrapolated with confidence; however, they do seem to be rapidly settling down to values near, but somewhat below, the ratio method estimate. On the basis of these sequences, we would assert that $10.611<K_c^{-1}(\frac{1}{4})<10.615$. Since the universality of the Ising value of the susceptibility index, $\gamma=1.25$, has been demonstrated elsewhere,[2] our primary concern is a determination of the value of ν consistent with $\gamma=1.25$. Thus we use γ-biased Saul-Wortis (four-fit) sequences, shown in Table IIIc for $\mu_0(\frac{1}{4})$, to determine the "best value" of K_c^{-1}. On the basis of the sequence in Table IIIc we assert that this "best value" for $\lambda=\frac{1}{4}$ is $K_c^{-1}(\frac{1}{4}) = 10.6131$. We must emphasize that, although the "best values" of $K_c^{-1}(\lambda)$ quoted in Table I are from γ-biased confluent singularity methods, these "best values" are consistent with the results of unbiased methods, and the quoted confidence limits reflect the variations observed in using the unbiased methods.

TABLE I

Values of the parameters characterizing the two leading singularities in $\mu_0 = A_1 t^{-\gamma_x}(1 + B_1 t^{\Delta_1})$ and in $\mu_2/\mu_0 = A_2 t^{-2\nu}(1 + B_2 t^{\Delta_2})$. The uncertainties in K_c^{-1} and in the indices reflect our absolute confidence limits. The values of the amplitudes, A_1, B_1, A_2, and B_2, are consistent with the "best value" of K_c, since they were determined from K_c- biased Saul-Wortis sequences.

λ	K_c^{-1}	γ	A_1	Δ_1	B_1	2ν	$2A_2$	Δ_2	B_2
¼	10.6131±0.0020	1.250±0.010	0.75	0.56±0.10	0.37	1.275±0.010	1.56	0.53±0.08	0.48
½	10.4793±0.0030	1.250±0.015	0.79	0.58±0.15	0.30	1.275±0.015	1.70	0.57±0.10	0.39
1	10.3031±0.0030	1.250±0.013	0.84	0.62±0.25	0.21	1.274±0.014	1.86	0.62±0.13	0.28
1½	10.1864±0.0030	1.249±0.012	0.88	0.65±0.20	0.15	1.275±0.010	1.98	0.68±0.20	0.22
2	10.1036±0.0030	1.249±0.012	0.90	0.56±0.30	0.12	1.274±0.010	2.06	0.74±0.30	0.19

TABLE II

Tenth-order, high temperature series for the zeroth and second correlation function moments for the λ=¼ model.

μ_0/M_2^0		μ_2/M_2^0	
1.0			
12.0	K	12.0	K
137.28749114590	K^2	288.0	K^2
1542.5529618138	K^3	5025.4029972302	K^3
17135.673789463	K^4	76650.184083781	K^4
188920.33943072	K^5	1083762.7873023	K^5
2071656.9604176	K^6	14590804.524471	K^6
22625810.479453	K^7	189795677.21250	K^7
246335743.41139	K^8	2406595753.9351	K^8
2675223109.3311	K^9	29919045469.184	K^9
28993440305.916	K^{10}	366151627090.80	K^{10}

Having determined $K_c(\lambda)$, we use K_c-biased ratio, Saul-Wortis, and Baker-Hunter methods to determine the indices. The "best value" of $K_c(\lambda)$, i.e. the one consistent with γ=1.25, provides the "best value" of the indices; our confidence limits on $K_c(\lambda)$ help establish realistic confidence limits for the indices. We, then, check these results for consistency with unbiased Saul-Wortis sequences. Table IVa shows the Neville tables of the log-derivative sequences for γ and 2ν from the series for μ_0(¼) and μ_2(¼)/μ_0(¼) using the "best value" of K_c(¼). From these tables and tables incorporating the uncertainty in K_c(¼), we assert $\gamma=1.240 \pm 0.007$ and $2\nu = 1.260 \pm 0.007$. These ratio method values exhibit non-universal λ-dependence which, as we shall see, is a fiction arising from the neglect of the confluent corrections. The ratio method values, already below the s=½ values $\gamma=1.250 \pm 0.003$ and $2\nu=1.276^{+0.004}_{-0.002}$,[3,8] would have been even lower if we had used the higher value of K_c^{-1}(¼) indicated by the ratio method. The unbiased Saul-Wortis sequences shown in Table IVb for λ=¼, although irregular, do not indicate any λ-dependence of the indices giving values in reasonable agreement with the s=½ values. The K_c-biased Saul-Wortis sequences for the indices, shown in Table IVc for the λ=¼ model, are much smoother. We use these latter sequences to determine the "best values" of the indices γ, 2ν, and the corresponding confluent correction indices Δ_1 and Δ_2; we assure that our confidence limits for these indices allow for our uncertainty in K_c as well as the variations observed in the five-fit sequences. These are the values in Table I.[9]

TABLE IIIa

Neville Tables of the log-derivative sequence for K_c^{-1} from the series for μ_0 and μ_2/μ_0 for the λ=¼ model.

μ_0			μ_2/μ_0		
10.881			12.000		
10.824	10.795		10.916	10.374	
10.717	10.610	10.548	10.847	10.777	10.912
10.681	10.627	10.639	10.732	10.561	10.417
10.661	10.621	10.616	10.695	10.619	10.677
10.649	10.618	10.613	10.671	10.612	10.603
10.641	10.617	10.615	10.657	10.613	10.615
10.635	10.616	10.615	10.647	10.615	10.617
10.631	10.616	10.615	10.641	10.615	10.615

TABLE IIIb / TABLE IIIc

IIIb μ_0	IIIb μ_2/μ_0	IIIc μ_0
10.91391	10.96454	10.76237
10.63566	10.64843	10.60558
10.60919	10.58565	10.61653
10.60538	10.60870	10.61493
10.61260	10.61401	10.61333
10.61190	10.61199	10.61322
10.61317		10.61305
		10.61306

IIIb Unbiased Saul-Wortis sequences for K_c^{-1}.

IIIc γ-biased Saul-Wortis sequences for K_c^{-1}.

TABLE IVa

Neville Tables of K_c-biased (K_c^{-1}=10.6131) log-derivative sequences for γ and 2ν from the series for μ_0 and μ_2/μ_0 respectively for the λ=¼ model.

γ			2ν		
1.1307			1.0000		
1.1593	1.1878		1.1307	1.2614	
1.1823	1.2283	1.2485	1.1630	1.2275	1.2106
1.1938	1.2284	1.2286	1.1885	1.2653	1.3031
1.2014	1.2319	1.2371	1.2019	1.2553	1.2404
1.2068	1.2340	1.2382	1.2111	1.2572	1.2610
1.2109	1.2352	1.2382	1.2177	1.2573	1.2576
1.2141	1.2362	1.2391	1.2227	1.2576	1.2583
1.2166	1.2369	1.2396	1.2266	1.2580	1.2597
1.2187	1.2376	1.2402	1.2298	1.2585	1.2604

TABLE IVb

Unbiased Saul-Wortis sequences for the scaling indices γ and 2ν as well as the corresponding confluent correction indices.

γ	Δ_1	2ν	Δ_2
5.6037	4.503	4.9330	3.837
1.2084	0.734	1.2100	0.760
1.2718	0.467	1.4085	0.436
1.2886	0.438	1.2950	0.481
1.2530	0.535	1.2709	0.543
1.2565	0.519	1.2811	0.507
1.2493	0.558		

TABLE IVc

K_c-biased (K_c^{-1}=10.6131) Saul-Wortis sequences for γ and 2ν as well as the corresponding confluent correction indices.

γ	Δ_1	2ν	Δ_2
1.7851	0.698	1.7670	0.687
1.2383	0.609	1.2576	0.588
1.2597	0.490	1.2982	0.450
1.2565	0.506	1.2780	0.513
1.2510	0.544	1.2747	0.530
1.2506	0.547	1.2754	0.526
1.2497	0.556	1.2745	0.532
1.2498	0.555		

We feel this study shows unambiguously[10] that, to within the quoted uncertainties, the value of the index 2ν is universal for the continuous spin Ising models considered, i.e. λ-independent and equal to its value for the s=½ model. The so-called "scaling" value 2ν = 1.250 is well outside our confidence limits. Unfortunately, this provides no new insight into the apparent violation of hyperscaling. Furthermore, although Δ = 0.60±0.20 is consistent with all values of Δ_1 and Δ_2 in Table I as well as with all values of the correction exponents for the spin-s Ising model,[11,12] this crude agreement does not exclude the seemingly unlikely possibilities that Δ_1 and Δ_2 have small but real non-universal variations and that Δ_1 and Δ_2 are unequal.

One of the authors (M. F.) wishes to thank M. E. Fisher for suggesting this project and wishes to gratefully acknowledge helpful discussions with M. Wortis and the hospitality of the University of Illinois where part of this work was completed.

REFERENCES

* This work was supported in part by a West Virginia University Faculty Senate Grant.
1. M. E. Fisher, Rev. Mod. Phys. 46, 597 (1974).
2. W. J. Camp and J. P. Van Dyke, AIP Conf. Proc. 24, 322 (1974).
3. M. A. Moore, D. Jasnow, and M. Wortis, Phys. Rev. Letters 22, 940 (1969); M. F. Sykes, D. L. Hunter, D. S. McKenzie, and B. R. Heap, J. Phys. A 5, 667 (1972).
4. F. Englert, Phys. Rev. 129, 567 (1963); M. Wortis, in Phase Transitions and Critical Phenomena, V. 3, Eds. C. Domb and M. S. Green (Academic, New York, 1974).
5. G. A. Baker, Jr. and D. L. Hunter, Phys. Rev. B 7, 3346 (1973); M. Ferer, M. A. Moore, and M. Wortis, Phys. Rev. B 4, 3954 (1971).
6. D. Saul, Ph. D. thesis, unpublished (University of Illinois, 1974).
7. G. A. Baker, Jr. and D. L. Hunter, Phys. Rev. B 7, 3377 (1973).
8. M. F. Sykes, D. S. Gaunt, P. D. Roberts, and J. A. Wyles, J. Phys. A 5, 640 (1972).
9. The results of a Baker-Hunter confluent singularity analysis (see ref. 7) are consistent with results presented in Table I.
10. In our Saul-Wortis analysis of μ_2/μ_0, we found no evidence of the worrisome, monotonic trends which were observed in the analysis of μ_2 for the spin-s Ising model and which are suggestive of non-asymptotic behavior.
11. D. Saul, M. Wortis and D. Jasnow, Phys. Rev. B 11, 2571 (1975); W. J. Camp and J. P. Van Dyke, Phys. Rev. B 11, 2579 (1975).
12. M. Wortis, Private communication; W. J. Camp, D. Saul, M. Wortis, and J. P. Van Dyke, to be published.

OPTICAL BIREFRINGENCE IN $Dy_3Al_5O_{12}$(DAG) IN FIELDS ALONG [111]

J. F. Dillon, Jr., Lynne D. Talley, and E. Yi Chen
Bell Laboratories, Murray Hill, New Jersey 07974 U.S.A.

ABSTRACT

In fields along [111] DAG approximates a very special Ising antiferromagnet--one in which applied fields couple to the antiferromagnetic order parameter (H_a-M_s coupling). We have measured the linear magnetic birefringence of DAG in H||[111] with light traveling along [1$\bar{1}$0] (perpendicular to the field). The H_a-M_s coupling allows us to produce antiferromagnetic DAG in either of the two time reversed states; A^+ which is absolutely stable in positive fields, and A^- which is only metastable. For these two states $\Delta n(H_a)$ is remarkably different (at 1.37 K and 3 kOe $\Delta n(A^+) - \Delta n(A^-) = 10^{-5}$). The metamagnetic transition, A^+ - P shows as a change in Δn of about $.75\times10^{-5}$. These are the first measurements of linear magnetic birefringence across the A^+ - A^- and A - P transitions. Both contrasts are substantial and make possible microscopy of the mixed phase.

INTRODUCTION

We have studied the linear magnetic birefringence (LMB) of $Dy_3Al_5O_{12}$(DAG) in magnetic fields along the [111] axis with light whose propagation direction $\vec{S}$ is perpendicular to that axis. This paper is a preliminary report of those measurements describing the general behavior of the LMB in the interesting portion of the magnetic phase diagram. LMB studies may be expected to give us new data with which to test and expand our current understanding of DAG. Further, we wish to pursue the question of the microscopic origin of the LMB in this basically cubic crystal which has no transition metal ions and does have an adjustable order parameter. Finally, we wish to lay a basis for the use of the birefringence for microscope studies of the two first order phase transitions in DAG.

It seems best to discuss DAG in terms of the magnetic phase diagram in "H_s","H_i",T space, Fig. 1, of an Ising antiferromagnet as given by Griffiths.[1] Here "H_s" is the staggered field conjugate to the antiferromagnetic order parameter $M_s = M_a - M_b$, and "H_i" the internal field conjugate to the ferromagnetic order parameter $M = M_a + M_b$. M_a and M_b are the magnetizations of the two sublattices. Our definitions of "H_s" and "H_i", and indeed our notation, follow the insight of Blume et al.[2] who realized that the symmetry properties of the garnet and many other structures allowed a coupling between an applied field along [111] and M_s. This we term H_a - M_s coupling, and note that it implies that both "H_s" and "H_i" may be considered "fictitious", since neither can be applied by itself. In the phase diagram the two antiferromagnetic states A^+ and A^- which are time reverse transforms of each other, may coexist on the shaded area in the

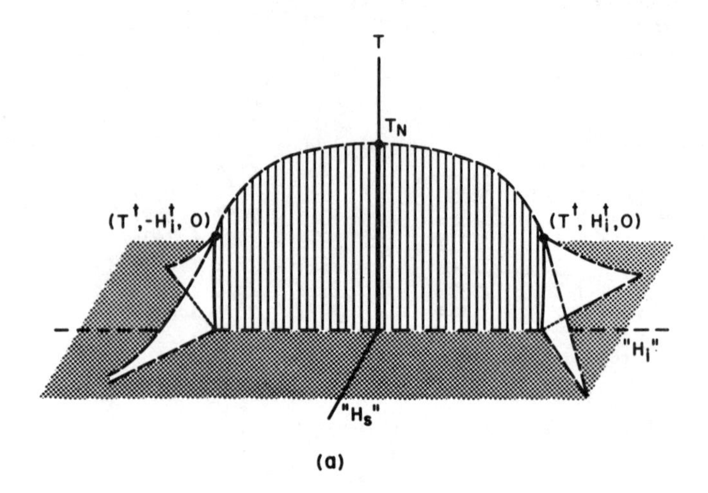

(a)

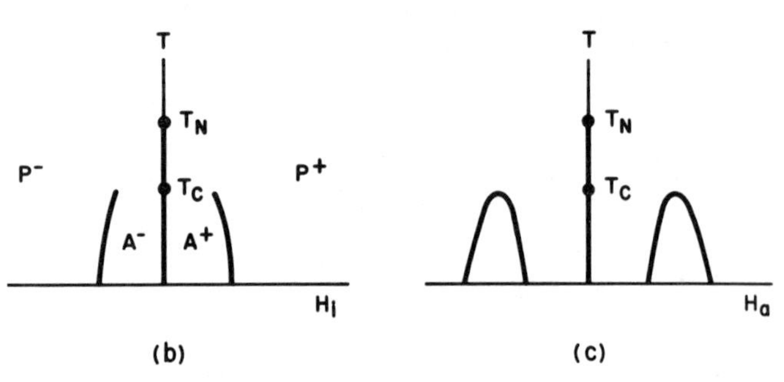

(b) (c)

Fig. 1 (a) Phase diagram of an Ising antiferromagnet plotted in "H_s","H_i",T space (see text). (b) For an antiferromagnet with coupling between applied field and staggered magnetization the (schematic) phase diagram in the demagnetized field H_i - T plane. (c) The corresponding phase diagram (schematic) in the applied field H_a - T plane. This is the diagram which applies to DAG in fields along [111].

"H_s" = 0 plane. This surface reaches up to the Néel point at ("H_s" = 0,"H_i" = 2.54 K)[3] a critical point in a line of critical points. Below 1.66 K the first order coexistence surface bifurcates and wings extend out into the "H_s" dimension. These are surfaces on which one of the antiferromagnetic states and the paramagnetic state can coexist. They terminate along the high temperature edge in a line of critical points.

Figure 1b is a plot of the phase diagram of DAG against internal field, H_i[111] = H_a[111] - NM and temperature.[4] As usual, NM is the product of the magnetization times the demagnetizing factor. Because of the H_a - M_s coupling, the H_i - T sheet furls through the diagram of Fig. 1a intersecting the A^+ - A^- coexistence surface along the temperature axis. The A^+ - P^+ and A^- - P^- coexistence lines correspond to the intersection of the H_i - T surface with the "wings". In Fig. 1c the phase diagram is plotted in the applied field-temperature plane where demagnetizing effects broaden the coexistence lines into areas in which both phases can coexist for a range of applied fields. In the present experimental geometry, the demagnetizing factor leads to a range of several hundred oersteds for the coexistence of the two phases, A and P.

Optical experiments can give very direct information on the various transitions and on some of the thermodynamic variables in this phase diagram. In earlier work utilizing a "parallel" geometry $(\vec{H}_a \| [111]) \| \vec{S}$ magnetooptical rotation measurements were used to follow magnetization as field and temperature were varied.[4-6] In addition, the rotation contrast of the two phases at the first order transitions made possible optical microscopy of the mixed phases. The structure of the mixed phase, the nucleation processes, and hysteretic processes could all be seen directly.

The experimental geometry used in the present work is "perpendicular", i.e. $(\vec{H}_a \| [111]) \perp \vec{S}$. We measure LMB at various temperatures as H_a is varied and find large effects representing the structure of the phase diagram. Elsewhere we will report the success of birefringence microscopy of the A-P and A^+ - A^- mixed phases.[7]

EXPERIMENTAL DETAILS

Though the birefringence has been observed in several samples, both melt and flux grown, the set of data reported here were taken on a flux grown DAG crystal. The polished $(1\bar{1}0)$ section used had major dimensions of about 5 mm by 4 mm and was 200μm thick. For the experimental observations this specimen was affixed to an opaque card with rubber cement. By glueing it in several ways it was ascertained that the strains introduced by the rubber cement did not impinge significantly on the measurements.

The sign of the birefringence of a uniaxial crystal is conventionally given by the sign of $n_e - n_o$, where the extraordinary ray is polarized along the unique axis, the ordinary ray perpendicular to it. In this case the axis is determined by the field which lies along [111], so we write $\Delta n = n_{\|} - n_{\perp}$. Retardation varies very little with wavelength (.4μm - .7μm) but for shorter wavelengths a given retardation is a greater fraction of a wavelength, and therefore easier to detect. We have chosen to use .546μm light because the subsequent microscopy would be done at this wavelength using the intense light from a filtered high pressure mercury arc.

The measurements of linear magnetic birefringence were made by a standard method in which a Babinet-Soleil compensator is used to cancel the retardation introduced by the sample (see for instance Lynch et al.[8]). The apparatus is depicted in Fig. 2. Light from a High Intensity Monochromator was polarized at -45° to the magnetic field. The crystal was mounted in a Dewar with flat glass windows so that [111] was parallel to the field and the major $(1\bar{1}0)$ faces were normal to the light beam. The Dewar tip was suspended between the pole faces of an electromagnet with a maximum field of 7 kOe. The Babinet-Soleil compensator axis was mounted parallel and an analyzer at +45° to the field. The retardation was modulated by a fused quartz plate[9] clamped between $BaTiO_3$ transducers so that the distortion axis was at 90° to the field. The 16.9 kHz frequency was determined by the resonant frequency of the whole structure. This phase modulation was converted to amplitude modulation by the analyzer, and the signal then detected by a photomultiplier and a PAR lock-in amplifier. The output of the PAR was used to determine the correct manual adjustment of the compensator.

The sample was immersed in superfluid helium whose temperature was set by controlling pressure. Note that each determination required several minutes, so the sample had sufficient time to reach the temperature of the bath, and in many cases time to reach magnetic equilibrium.

LINEAR MAGNETIC BIREFRINGENCE MEASUREMENTS

For temperatures below 2.54 K, we see the linear magnetic birefringence Δn behave as in Fig. 3. These data are for 1.375 K. Note that the birefringence depends on the past history of the sample, similar to effects first reported in Ref. 5. We concentrate first on the solid curve which represents the birefringence as a function of positive field for a crystal previously exposed to high positive fields. In the terms of Ref. 4, it is in the antiferromagnetic state which is stable in positive fields, A^+. The birefringence increases with a finite slope from zero field to a field of about 3.55 kOe where it abruptly breaks to decrease rapidly. The break point is labeled K and is believed to correspond to the first appearance of paramagnetic phase. From K to L the fraction of paramagnetic phase increases from 0 to 1. This A-P mixed phase region extends over only a few hundred gauss because the demagnetizing factor of the crystal along the field is so small. For a very thin (111) sheet with the field

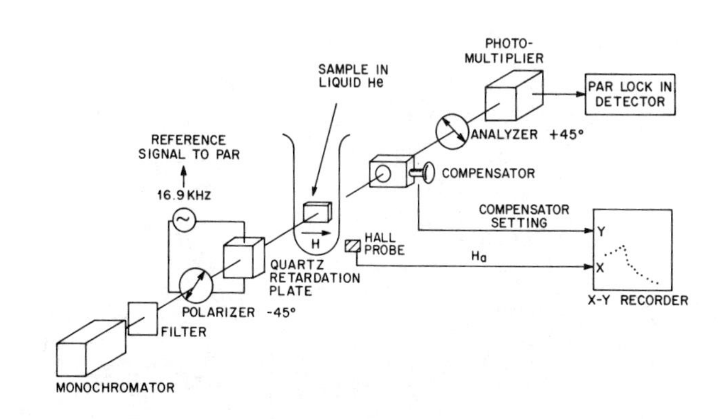

Fig. 2 Schematic of the experimental arrangement used to measure linear magnetic birefringence in DAG.

normal the A-P region would spread over about 6 kOe at 1.3 K. From L onward the spin system is all in the paramagnetic phase, and birefringence decreases rather steeply with field. Note the sign change. In the stable antiferromagnetic state increasing field makes birefringence go positive; in the paramagnetic state at the same temperature increasing field makes birefringence more negative.

The dotted curve of Fig. 3 is also the birefringence as a function of positive field, but for a crystal previously exposed to large negative fields. This represents a crystal in the metastable antiferromagnetic state A^-, at least it starts out as A^-. In this case the birefringence decreases monotonically with field up to an abrupt discontinuity at M. This is taken to represent the onset of a mixed phase which extends to the point L again. Over part of the range at least there is probably some mixture of A^-, A^+ and P. Since the relative amounts are not known and probably depend on the time intervals associated with the data-taking, it is now difficult to interpret the curve between M and L. Note that the first discontinuity M occurs at a substantially lower field for the A^- curve than for the A^+ curve. This we attribute to the different magnetizations of the two antiferromagnetic states which result in a higher internal field for A^- than for A^+ at the same applied field.

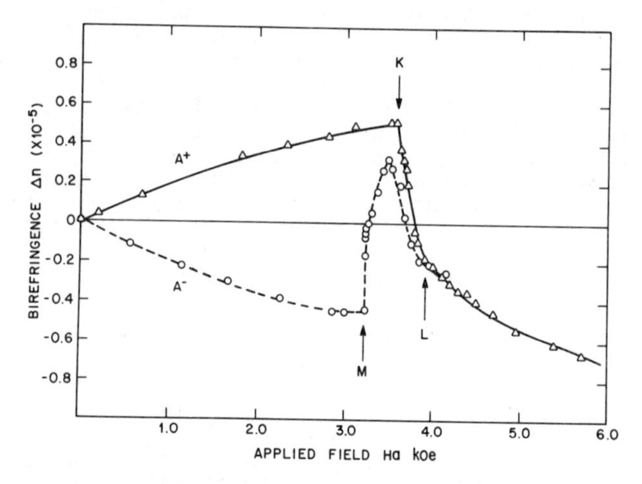

Fig. 3 Linear magnetic birefringence of DAG in fields along [111] at 1.375K. The solid line pertains to a crystal previously exposed to high positive fields and thus in the state A^+ up to the point K. The dashed curve pertains to a crystal previously exposed to high negative fields and thus in the state A^- up to the point marked M. Above L the crystal is in the paramagnetic state. Between K and L and between M and L the spin system is in a mixed state.

Salient features of our data include:

1. The $A^+ - A^-$ and $A^+ - P$ LMB contrasts are large enough to make practical optical polarization microscopy of the mixed phases. Note particularly that the $A^+ - A^-$ contrast develops approximately linearly with H_a and is appreciable at relatively low fields. The corresponding rotation contrast in the parallel geometry is only detectable as H_a approaches the transition.
2. At all temperatures up to T_N the $\Delta n(H_a)$ curves look very much the same. There is no special change in character at 1.66 K. The field values for the discontinuity decrease with increasing temperature, and the curves become somewhat more rounded.
3. There is a dramatic change in slopes. For A^+ $d(\Delta n)/dH_a$ starts positive. Above the $A^+ - P$ transition it is negative.

The burden of this paper is to report the general behavior of the birefringence and its size. We will defer a discussion of the detailed behavior with temperature of the sign of the birefringence until a later paper in which more complete data are presented. In particular, it seems important to have data on the magnetostrictive distortion of the crystal as field is varied.

ACKNOWLEDGMENTS

The authors greatly appreciate the preparation of the flux-grown crystals by S. Mroczkowski of Yale University. Valuable discussions with W. P. Wolf and N. Giordano are gratefully acknowledged.

REFERENCES

1. R. B. Griffiths, Phys. Rev. Lett. 24, 715 (1970).
2. M. Blume, L. M. Corliss, J. M. Hastings and E. Schiller, Phys. Rev. Lett. 32, 544 (1974).
3. W. P. Wolf, B. Schneider, D. P. Landau, B. E. Keen, Phys. Rev. B 5, 4472 (1972).
4. J. F. Dillon, Jr., E. Yi Chen, N. Giordano, W. P. Wolf, Phys. Rev. Lett. 33, 98 (1974).
5. J. F. Dillon, Jr., E. Yi Chen, W. P. Wolf, Proceedings of International Conference of Magnetism, Moscow 22-28, August 1973, VI, p. 38-48.
6. J. F. Dillon, Jr., E. Yi Chen, H. J. Guggenheim, AIP Conference Proceedings 24, 200 (1974).
7. J. F. Dillon, Jr., L. D. Talley and E. Yi Chen (in preparation).
8. R. T. Lynch, J. F. Dillon, Jr., L. G. Van Uitert, J. Appl. Phys. 44, 225 (1973).
9. L. F. Mollenauer, D. Downie, H. Engstrom, W. B. Grant, Appl. Optics 8, 661 (1969).

TEMPERATURE AND EXCHANGE FIELD DEPENDENCES OF THE MAGNETIC AND MAGNETOOPTICAL PROPERTIES OF $Y_3Fe_5O_{12}$ UNDER LOW AND INTENSE MAGNETIC FIELD

Maurice Guillot* and Henri Le Gall**

* C.N.R.S. Laboratoire de Magnétisme, BP 166 38042 Grenoble, France.

** C.N.R.S. Laboratoire de Magnétisme et d'Optique des Solides, 92190 Meudon-Bellevue, France

ABSTRACT

Faraday rotation (ϕ_F) measurement are reported in YIG at 1.15 and 0.6328 microns wavelengths under low and very intense magnetic fields (up to 4 megaOersteds). These results are discussed from the applied field and temperature dependences of the magnetooptical (M.O.) coefficients induced by the light beam in both the octahedral (a) and tetrahedral (d) sites. It is shown that under very intense magnetic field and in the visible range the M.O. coefficients A_m and D_m induced by the magnetic dipole transitions are independent on the applied field and ϕ^m follows the usual evolution of the magnetic structure which shows a first-order transition (ferri → non colinear structure) near H_a = 1.7 mega-Oersteds. On the other hand in the near infrared A_m and D_m have a strong magnetic field dependence in the mega-Oersteds range with increasing the d.c. field. It is shown that the magnetic and electric parts of ϕ_F are increasing and decreasing respectively when increasing the d.c. field and the experimental data are in agreement with a molecular field model where the magnetic intra-sublattices interactions J_{aa} and J_{dd} are vanishing.

The rare-earth iron garnets (RIG) have been subjected to extensive magnetooptical investigation in the past. Matthews et al [1] observed that the circular magnetic birefringence (CMB) of the Fe^{3+} ions in the near infrared range is higher in the octahedral site (or a site) than in the tetrahedral site (or d site). By assuming temperature independent magnetooptical (MO) coefficients, Cooper et al [2] have deduced from experiment a specific Faraday rotation approximatively twice higher in the a sites than in the d sites at 1.15 micron wavelength. By using temperature dependent MO coefficients we have shown more recently that the Faraday rotation is higher, both in the a and d sites, than the values published in preceeding papers, and decreases toward zero near the Néel temperature T_N = 560°K [3]. It is well-known that the Faraday rotation is induced by both the electric and magnetic dipole transition and is a linear function of the magnetization M_a and M_d of the Fe^{3+} ions in the a and d sites such as [3] :

$$\phi_F = \phi^e + \phi^m = -(A_e + A_m)\ |M_a| + (D_e + D_m)\ |M_d| \quad (1)$$

A_e, A_m and D_e, D_m are the MO coefficients induced by the electric and magnetic dipole transition in the octa- and tetrahedral sites respectively. In the equation 1 M_a and M_d are given in Bohr magneton per two unit formula $Y_3Fe_5O_{12}$. At 1.15 micron wavelength the electric and magnetic coefficients have opposite signs in a given site. As discussed in a previous study [3], the electric coefficients A_e and D_e have a temperature dependence which has been attributed to a corresponding influence of the exchange field on the optical absorption spectrum of the Fe^{3+} ions.

In the present study the temperature and exchange fields dependences of the Faraday rotation ϕ^m induced at 1.15 micron wavelength in YIG by the magnetic dipole transitions is investigated under low and very high applied magnetic fields H_a. It will be seen in the equation 1, that ϕ^m depends on H_a from the associated evolution of both the magnetic moments M_a and M_d and the MO coefficients A_m and D_m. These quantities are described, from the molecular field model, by using the Brillouin function $B_{5/2}$ such as :

$$M_i = M_{i,o}\ B_{5/2}\ (\mu_{i,o} \cdot H_{i,eff}/k_B T) \quad (2)$$

where the effective field is given by :

$$H_{i,eff} = \sum_j n_{ij}\ M_j \pm H_a \quad (3)$$

The plus and minus signs correspond to the sublatices d and a respectively. The molecular field coefficients n_{ij} are proportional to the exchange integrals J_{ij} which describe the superexchange interactions Fe^{3+} - O^{2-} - Fe^{3+} between the magnetic ions such as [4] : $J_{aa} = 4.668 \times 10^{-5}\ n_{aa}$, $J_{dd} = 14.005 \times 10^{-5}\ n_{dd}$, $J_{ad} = 9.333 \times 10^{-5}\ n_{ad}$ where J_{ij} and n_{ij} are given in cm^{-1} and $Oe.\ _B^{-1}$ units respectively. As shown by the equation 1 to 3 the magnetic part of the Faraday rotation is determined by the magnetic state of the Fe^{3+} ions which depends on the d.c. field and the correct choice of the molecular field coefficients. Most of magnetization values M_a, M_d and M (YIG) obtained from different experimental methods are similar. On the other hand the values of the exchange parameters J_{ij} deduced by fitting these magnetization data with a theoretical model, which uses, either the molecular field or the spin-waves, are different as shown on the Table I. By using the molecular field model Anderson has deduced two sets for J_{ij} by assuming that the intra-sublattice interactions J_{aa} and J_{dd} are either vanishing or not vanishing [5].. Both the two cases lead by calculation to a correct determination for the Néel temperature T_N, but only the former one (J_{ad} = 13.6 cm^{-1}, $J_{aa} = J_{dd}$ = 0) gives a macroscopic exchange constant D = 21.3 cm^{-1} in agreement with the specific heat and ferrimagnetic resonance data [10,11]. In addition we have shown [12] that a good agreement between experiments and a molecular field calculation for the temperature dependence of the sublattice and total magnetizations of the YIG is obtained when the simplified hypothesis of Anderson, which neglects the intrasublattice interactions, is used. In this case we have :

$$|M^{calc.}_{a,d} - M^{exp}_{a,d}|\ /\ |M^{exp}_{a,d}| \leq 0.025$$

Faraday rotation measurements, we have performed under very intense magnetic fields (up to

4 megaOersteds, T = 300°K, λ = 6 328 Å), confirms our choice of the exchange parameter $J_{ad} = 14\ cm^{-1}$, $J_{aa} = J_{dd} = 0$. Indeed we have observed a strong variation of the circular magnetic birefringence near 1.8 megaOersted associated with a transition from the ferrimagnetic to a non-colinear structure as predicted by the molecular field model when the d.c. field exceeds the value :

$$H_c = n_{ad}\ (M_d - M_a) \qquad (4)$$

When vanishing or non-vanishing intrasublattice exchange interactions are considered, the critical field given by the equation 4 is 1.7 or 3.5 megaOersted. The first value is in good agreement with the d.c. field evolution of our Faraday rotation measurements. We have calculated the evolution at room temperature of the sublattices and total magnetizations of YIG submitted to intense magnetic fields up to 2.5 megaOersted as shown on the figure 1. A first order transition is observed at H_a>1.7 megaOersted only when the intra-sublattice exchange interactions are vanishing. Such a transition is associated only with an orientation change of the sublattice moments and not to a change of their amplitude. We have extended, on the fig. 2, the calculation of the total magnetization of YIG $|\vec{M}_a+\vec{M}_d|$ under very high magnetic field to low temperatures. The slow increase of the magnetization with the d.c. field up to the critical field is due to the change of the moments amplitude corresponding to a paramagnetic process. The critical field increases by decreasing the temperature because of the corresponding increase of the total magnetization as indicated by the equation 4. It is to be noted that the d.c. field change of the magnetization cannot be neglected for magnetooptical investigation in YIG submitted to low and medium fields produced by usual electromagnets and superconductive coils.

In what follows we determine the influence of very high magnetic fields on the magneto-optical coefficients A_m and D_m associated with the magnetic dipole transitions. The magnetic part of the Faraday rotation is proportionnal to the off-diagonal component of the susceptibility tensor derived by Wangness for a two-sublattice system from the equation of notion of the magnetization |13|:

$$\phi^m = 2\pi\bar{n}\ \chi_{xy} \qquad (5)$$

$\bar{n}$ and c are the refractive index and the light velocity in vacuum. By extending the validity of the Wangsness analysis to ferrimagnetic samples submitted not to microwave but to optical excitations and to very high magnetic fields, we have deduced the following expression of the magnetic field dependence of the Faraday rotation in YIG |12| :

$$\phi^m(H_a) = -2\pi\bar{n}\ \gamma c^{-1}\ (M_d - M_a)\ \left(1 - \frac{\gamma^2}{\omega^2} P^2\ (H_a)\right) \qquad (6)$$

with $$P(H_a) = H_a + n_{ad}\ (M_d - M_a) \qquad (7)$$

γ and ω are the gyromagnetic factor and the optical frequency. For low magnetic fields the correcting term γ^2P^2/ω^2 in equation 6 is very small in the visible range and ϕ^m can be restricted to its well-known expression. On the other hand, under high magnetic fields, ϕ^m has the usual indirect dependence on the field, through the associated evolution of M_a and M_d as discussed above, and has a direct dependence by the correcting term which is proportional to H_a (eq. 7). The magnetic field dependence of the first-order MO coefficients $A_m(H_a)$ and $D_m(H_a)$ associated with the magnetic dipole transition are therefore given by :

$$-A_m = D_m = -\ 2\pi\bar{n}\gamma c^{-1}\ \left[1 - \frac{\gamma^2 P^2}{\omega^2}\right] \qquad (8)$$

At 1.15 micron wavelength the relative change of ϕ^m is small under magnetic fields up to 1 megaOersted and the correcting term may be neglected. From that we can conclude that the change of the "magnetic" Faraday rotation under very high magnetic fields at 1.15 micron wavelength will describe directly the evolution of the sublattices magnetization in YIG. On the other hand the correcting term has a strong increase by decreasing the frequency following a $1/\omega^2$ law and at 10 microns wavelength, for instance, a decrease by about 10 percents of the MO coefficients must be observed ($\omega^2P^2/\omega^2 = 0.1$). From that theoretical investigation we can compare on the figure 3 the magnetic field evolution of the total ($\phi^e+\phi^m$) and the magnetic (ϕ^m) Faraday rotations, as observed from our experiments up to 1.2 megaOersteds. From these curves it is concluded that the electric and magnetic contributions to the Faraday rotation at 1.15 and 1.08 micron wavelength have opposite magnetic field evolutions. The magnetic field dependence of the MO coefficients induced by the electric dipole transitions will be discussed in a forthcoming paper.

Acknowledgement : This work has been supported in part under a contract with the D.R.M.E. (Paris). The authors are pleased to acknowledge C. Gilly (Nuclear Center of Grenoble) who prepared the numerical calculations.

REFERENCES

|1| M. Matthews, S. Singh and R.C. Le Craw, Appl. Phys. lett., 7, 165, (1965)

|2| R.W. Cooper, W.A. Crossley, J.L. Page and R.F. Pearson, J. Appl. Phys., 39, 565, (1968) ; and W.A. Crossley, R.W. Cooper, J.L. Page and R.P. Van Stapele, Phys. Rev. 181, 896, (1969).

|3| G. Abulafya and H. Le Gall, Sol. Stat. Com., 11, 629, (1972).

|4| J.S. Smart in Magnetism ed. by G.T. Rado and H. Suhl (Academic Press, New York), Vol. III, 103, (1963).

|5| E.E. Anderson, Phys. Rev., 134, 1581 (1964).

|6| M. Guillot, Proc. Int. Conf. Magnetism, Moscow, (1973).

|7| R. Aleonard, Phys. Chem. Sol., 15, 167, (1960).

|8| P.J. Wojtowicz, J. Appl. Phys., 33, 1257, (1962).

|9| R.L. Gonano, Phys. Rev., 156, 521, (1967).

|10| H. Meyer and A.B. Harris, J. Appl. Phys., 31, 495, (1960).

|11| W.G. Nielsen, R.L. Comstock and R.L. R.L. Walker, Phys. Rev., 139, 472, (1965).

|12| M. Guillot and H.Le Gall, Phys. Stat. Sol. (to be published)

|13| R.K. Wangsness, Phys. Rev., 91, 1085 (1953)

Model	Exp Technique	J_{ad} cm^{-1}	J_{dd} cm^{-1}	J_{aa} cm^{-1}	D cm^{-1}	Ref
Molecular field	M(T), χ(T)	13.6 25.36	0 11.86	0 8.45	21.3 7.4	5-6
	χ(T)	25.2	10.3	6.1	14.5	7
Spin-wave	χ(T)	24.3	0	0	36.2	8
	NMR	22.4	2.3	<0.3	31.7	9

Table I

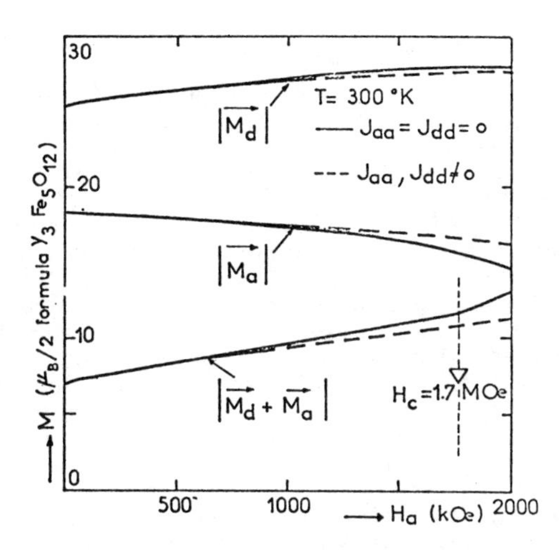

Fig. 1 Magnetic field dependence of the sublattice and total magnetization of YIG up to 2 MOe at room temperature

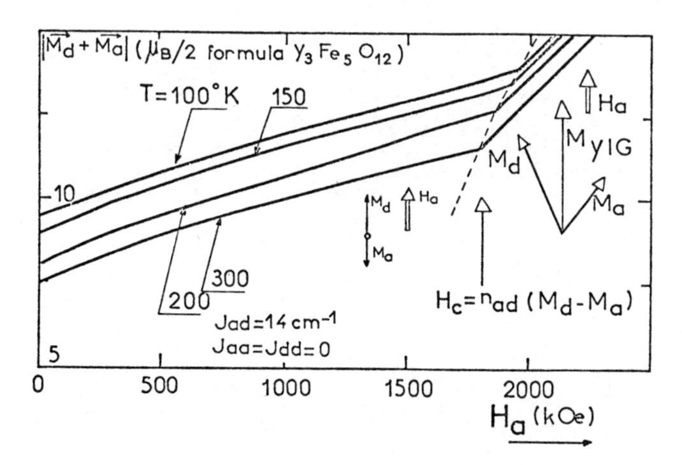

Fig. 2 Magnetic field dependence of the total magnetization of YIG up to 2 MOe at low temperature.

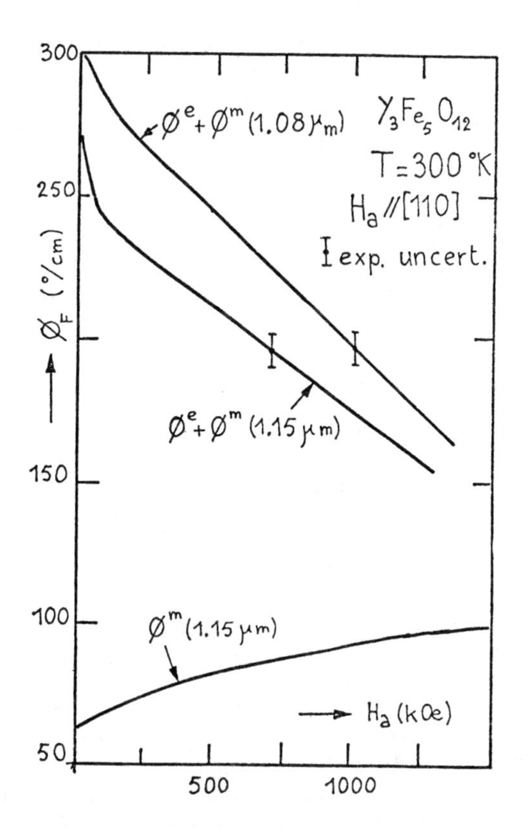

Fig. 3 Magnetic field dependence of the magnetic and total Faraday rotation in YIG up to 1.2 MOe.

SYNCHROTRON RADIATION - THE LIGHT FANTASTIC

P. M. Platzman
Bell Laboratories, 600 Mountain Ave.,
Murray Hill, New Jersey 07974

Synchrotron radiation, what it is and how it will contribute to our knowledge of magnetic systems will be discussed.

APPLICATIONS OF NEUTRON SCATTERING TO THE STUDY OF MAGNETIC MATERIALS

W. C. Koehler
Solid State Division, Oak Ridge National Laboratory*,
Oak Ridge, TN 37830

In this talk we shall review the types of interactions that neutrons undergo with condensed matter, and discuss those properties of neutrons that make them an ideal probe for the study of magnetism on a microscopic scale. Following a very brief survey of experimental methods, a few illustrative examples of specific investigations will be described in sufficient detail to illustrate the power of the techniques. Some personal views as to the future directions that may be taken by neutron scattering will be presented.

*Operated by Union Carbide Corporation for USERDA.

THE MAGNETIC MONOPOLE

P. B. Price
Department of Physics, University of California,
Berkeley, CA 94720

In a complex stack of balloon-borne track detectors, one fast particle left tracks consistent with those expected for a magnetic monopole of strength 137 e, traveling with about half the speed of light. The evidence has been criticized, and it has been argued that the tracks could have been left by a doubly fragmenting platinum nucleus traveling with 0.7 times the speed of light. Extensive tests and calibrations of the various detectors are still in progress. They appear, however, to rule out the doubly fragmenting platinum nucleus and to set extremely low confidence levels for any known particle to have produced the track. It can never be "proved" that this event was caused by a monopole, but the evidence is now strong that the event is unique and warrants further searches.

RECOVERY OF NONMAGNETIC METALS FROM WASTE

Ernst Schloemann
Raytheon Research Division, Waltham, MA 02154

ABSTRACT

Approximately 150 million tons of municipal solid waste are collected each year in the United States. This refuse contains many valuable materials that can be used again provided they are separated and purified sufficiently. The present paper reviews the sorting techniques that apply to this problem with special emphasis on the recovery of nonmagnetic metals.[1] Only approximately 0.5 to one percent of the municipal solid waste consist of such metals (mostly Aluminum), but because of their relatively high scrap value (up to $300/ton) they constitute a significant portion of the reclaimable value. Two new techniques for the recovery of nonmagnetic metals from municipal solid waste are described. The first separator consists of an inclined ramp on which the nonmagnetic metals are deflected to one side as they slide down. Deflection results from eddy currents induced by permanent magnets embedded in the ramp surface. A detailed theory has been developed that allows prediction of the deflection of metallic particles from the straight-down path as a function of all the relevant physical parameters. The theoretical predictions have been tested by comparison with experimental results.[2] Tests have been conducted on the heavy-fraction remaining after air classification of municipal solid waste.[3] In the second separator the permanent magnets are attached to the inner wall of a rotary drum. It splits the mixed feed material into three fractions consisting of 1) magnetic materials, 2) nonmagnetic metals and 3) nonmagnetic nonmetals.

REFERENCES

1. E. Schloemann, "Recovery of Nonmagnetic Metals from Waste," AIP Conf. Proc. "Materials Resources and Materials Science" (to be published).
2. E. Schloemann, "Separation of Nonmagnetic Metals from Solid Waste by Permanent Magnets I Theory, II Experiments on Circular Discs," J. Appl. Phys. 46, 5012-5021, 5022-5029 (1975).
3. D. B. Spencer and E. Schloemann, "Recovery of Non-Ferrous Metals by Means of Permanent Magnets," Resource Recovery and Conservation 1, 151-165 (1975).

Author Index to the Proceedings of the MMM-Intermag Conference

Numbers preceded by M refer to Magnetism and Magnetic Materials - 1976

Numbers preceded by T refer to IEEE Transactions on Magnetics, volume MAG-12, number 6, November 1976

AIP Conference Proceedings

		L. C. Number	ISBN
No. 1	Feedback and Dynamic Control of Plasmas (Princeton) 1970	70-141596	0-88318-100-2
No. 2	Particles and Fields - 1971 (Rochester)	71-184662	0-88318-101-0
No. 3	Thermal Expansion - 1971 (Corning)	72-76970	0-88318-102-9
No. 4	Superconductivity in d- and f-Band Metals (Rochester 1971)	74-18879	0-88318-103-7
No. 5	Magnetism and Magnetic Materials - 1971 (2 parts) (Chicago)	59-2468	0-88318-104-5
No. 6	Particle Physics (Irvine 1971)	72-81239	0-88318-105-3
No. 7	Exploring the History of Nuclear Physics (Brookline, 1967, 1969)	72-81883	0-88318-106-1
No. 8	Experimental Meson Spectroscopy - 1972 (Philadelphia)	72-88226	0-88318-107-X
No. 9	Cyclotrons - 1972 (Vancouver)	72-92798	0-88318-108-8
No. 10	Magnetism and Magnetic Materials - 1972 (2 parts) (Denver)	72-623469	0-88318-109-6
No. 11	Transport Phenomena - 1973 (Brown University Conference)	73-80682	0-88318-110-X
No. 12	Experiments on High Energy Particle Collisions - 1973 (Vanderbilt Conference)	73-81705	0-88318-111-8
No. 13	π-π Scattering - 1973 (Tallahassee Conference)	73-81704	0-88318-112-6
No. 14	Particles and Fields - 1973 (APS/DPF Berkeley)	73-91923	0-88318-113-4
No. 15	High Energy Collisions - 1973 (Stony Brook)	73-92324	0-88318-114-2
No. 16	Causality and Physical Theories (Wayne State University, 1973)	73-93420	0-88318-115-0
No. 17	Thermal Expansion - 1973 (Lake of the Ozarks)	73-94415	0-88318-116-9
No. 18	Magnetism and Magnetic Materials - 1973 (2 parts) (Boston)	59-2468	0-88318-117-7
No. 19	Physics and the Energy Problem - 1974 (APS Chicago)	73-94416	0-88318-118-5
No. 20	Tetrahedrally Bonded Amorphous Semiconductors (Yorktown Heights, 1974)	74-80145	0-88318-119-3
No. 21	Experimental Meson Spectroscopy - 1974 (Boston)	74-82628	0-88318-120-7
No. 22	Neutrinos - 1974 (Philadelphia)	74-82413	0-88318-121-5
No. 23	Particles and Fields - 1974 (APS/DPF Williamsburg)	74-27575	0-88318-122-3
No. 24	Magnetism and Magnetic Materials - 1974 (20th Annual Conference San Francisco)	75-2647	0-88318-123-1
No. 25	Efficient Use of Energy (The APS Studies on the Technical Aspects of the More Efficient Use of Energy)	75-18227	0-88318-124-X
No. 26	High-Energy Physics and Nuclear Structure - 1975 (Santa Fe and Los Alamos)	75-26411	0-88318-125-8
No. 27	Topics in Statistical Mechanics and Biophysics: A Memorial to Julius L. Jackson (Wayne State University-1975)	75-36309	0-88318-126-6
No. 28	Physics and Our World: A Symposium in Honor of Victor F. Weisskopf (M.I.T. 1974)	76-7207	0-88318-127-4
No. 29	Magnetism and Magnetic Materials - 1975 (21st Annual Conference, Philadelphia)	76-10931	0-88318-128-2
No. 30	Particle Searches and Discoveries - 1976 (Vanderbilt Conference)	76-19949	0-88318-129-0
No. 31	Structure and Excitations of Amorphous Solids (Williamsburg, Va., 1976)	76-22279	0-88318-130-4
No. 32	Materials Technology - 1975 (APS New York Meeting)	76-27967	0-88318-131-2
No. 33	Meson-Nuclear Physics - 1976 (Carnegie-Mellon Conference)	76-26811	0-88318-132-0
No. 34	Magnetism and Magnetic Materials - 1976 (Joint MMM-Intermag Conference, Pittsburgh)	76-47106	0-88318-133-9

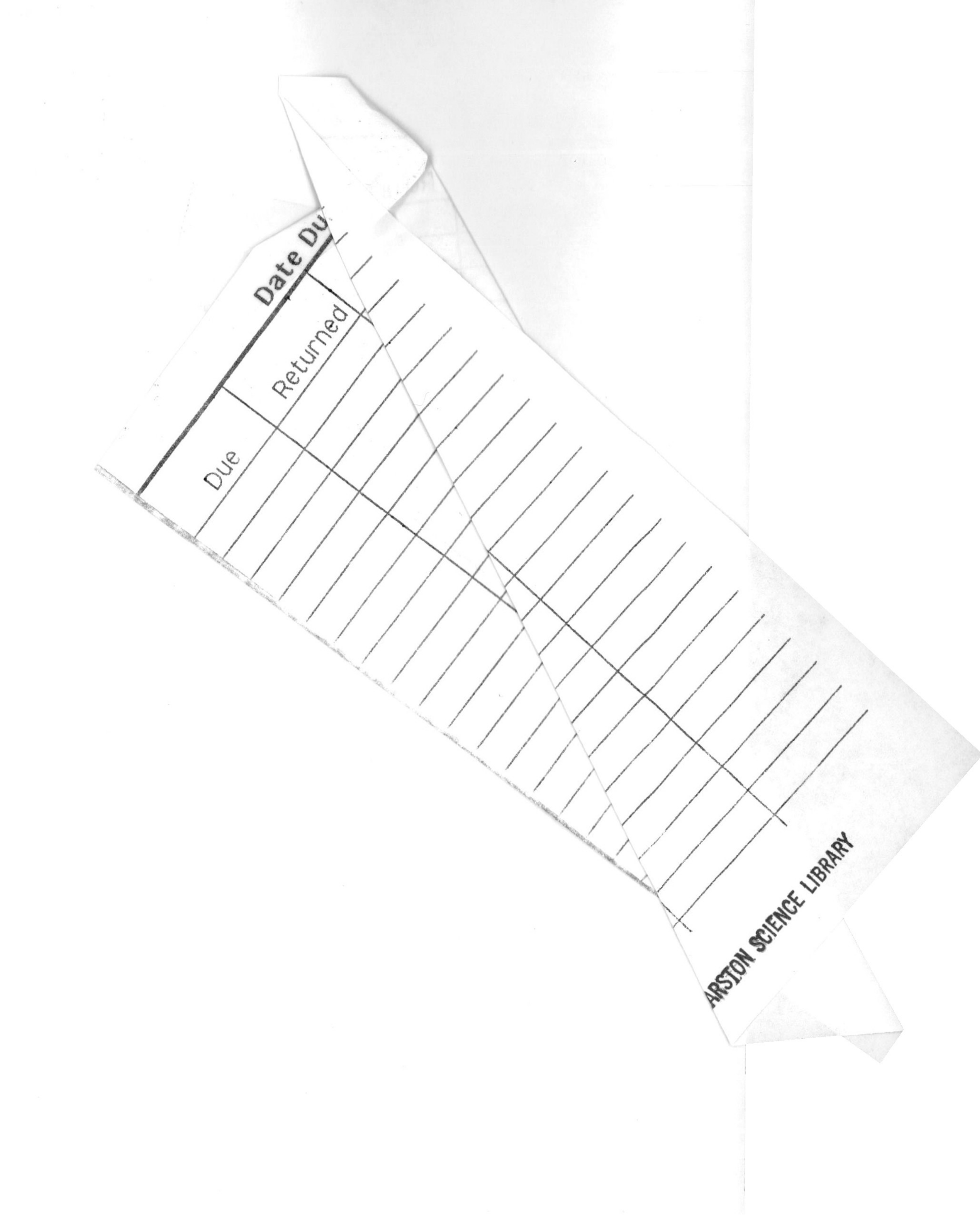
Date Du
Due
Returned
ARSTON SCIENCE LIBRARY